Selected Works of EMIL WOLF

With Commentary

World Scientific Series in 20th Century Physics

Published

Vol. 1 Gauge Theories — Past and Future
edited by R. Akhoury, B. de Wit, and P. van Nieuwenhuizen

Vol. 2 Scientific Highlights in Memory of Léon Van Hove
edited by F. Nicodemi

Vol. 3 Selected Papers, with Commentary, of T. H. R. Skyrme
edited by G. E. Brown

Vol. 4 Salamfestschrift
edited by A. Ali, J. Ellis and S. Randjbar-Daemi

Vol. 5 Selected Papers of Abdus Salam (with Commentary)
edited by A. Ali, C. Isham, T. Kibble and Riazuddin

Vol. 6 Research on Particle Imaging Detectors
edited by G. Charpak

Vol. 7 A Career in Theoretical Physics
edited by P. W. Anderson

Vol. 8 Lepton Physics at CERN and Frascati
edited by N. Cabibbo

Vol. 9 Quantum Mechanics, High Energy Physics and Accelerators — Selected Papers of J. S. Bell (with Commentary)
edited by M. Bell, K. Gottfried and M. Veltman

Vol. 10 How We Learn; How We Remember: Toward an Understanding of Brain and Neural Systems — Selected Papers of Leon N. Cooper
edited by L. N. Cooper

Vol. 11 30 Years of the Landau Institute — Selected Papers
edited by I. M. Khalatnikov and V. P. Mineev

Vol. 12 Sir Nevill Mott — 65 Years in Physics
edited by N. Mott and A. S. Alexandrov

Vol. 13 Broken Symmetry — Selected Papers of Y. Nambu
edited by T. Eguchi and K. Nishijima

Vol. 14 Reflections on Experimental Science
edited by M. L. Perl

Vol. 15 Encounters in Magnetic Resonances — Selected Papers of Nicolaas Bloembergen (with Commentary)
edited by N. Bloembergen

Vol. 16 Encounters in Nonlinear Optics — Selected Papers of Nicolaas Bloembergen (with Commentary)
edited by N. Bloembergen

Vol. 17 The Collected Works of Lars Onsager (with Commentary)
edited by P. C. Hemmer, H. Holden and S. K. Ratkje

Vol. 18 Selected Works of Hans A. Bethe (with Commentary)
edited by Hans A. Bethe

Vol. 19 Selected Scientific Papers of Sir Rudolf Peierls (with Commentary)
edited by R. H. Dalitz and R. Peierls

Vol. 20 The Origin of the Third Family
edited by O. Barnabei, L. Maiani, R. A. Ricci and F. R. Monaco

World Scientific Series in 20th Century Physics – Vol. 29

Selected Works of Emil Wolf

With Commentary

Emil Wolf
University of Rochester

World Scientific
Singapore • New Jersey • London • Hong Kong

Published by

World Scientific Publishing Co. Pte. Ltd.

P O Box 128, Farrer Road, Singapore 912805

USA office: Suite 1B, 1060 Main Street, River Edge, NJ 07661

UK office: 57 Shelton Street, Covent Garden, London WC2H 9HE

Library of Congress Cataloging-in-Publication Data
Wolf, Emil.
[Selections. 2001]
Selected works of Emil Wolf with commentary / Emil Wolf.
p. cm. -- (World Scientific series in 20th century physics ; vol. 29)
Includes bibliographical references.
ISBN 9810242042 (hardcover) -- ISBN 9810242050 (softcover)
1. Physics. 2. Optics. I. Title II. Series.

QC3 .W642 2001
535'.2--dc21 2001023741

British Library Cataloguing-in-Publication Data
A catalogue record for this book is available from the British Library.

The author and the publisher would like to thank the following organisations and publishers for their assistance and their permission to reproduce the articles found in this volume:

American Association of Physics Teachers (*Amer. J. Phys.*), American Geophysical Union (*Radio Science*), American Institute of Physics (*J. Math. Phys.*, *Phys. Teach.*), American Physical Society (*Phys. Rev.*, *Phys. Rev. Lett.*), Elsevier Science Publishers B. V. (*Opt. Commun.*, *Phys. Lett.*), National Physical Laboratory (India), Nature Publishing Group (*Nature*), Optical Society of America (*J. Opt. Soc. Amer.*, *Opt. Lett.*), Plenum Press, Royal Society (*Proc. Roy. Soc.*), Società Italiana di Fisica (*Nuovo Cimento*), Taylor & Francis Group (*J. Mod. Opt.*), University of Chicago Press (*Astrophys. J.*)

Printed in Singapore.

CONTENTS

Preface

Writing a preface for a volume containing a selection of one's own publications with commentary is like writing one's own scientific obituary. Until relatively recently, collected volumes of scientific writings followed the authors' death by many years, sometimes even centuries. However, modern technology, with its cellular phones, fax machines, e-mail and computers, has made it possible – whether for better or for worse – to produce publications at great speed, not imaginable even a few years ago.

I am flattered, of course, that World Scientific Publishing is honoring me by bringing out this volume. It consists of reprints of 69 of my papers, many written in collaboration with my students and colleagues, selected from more than 300 papers which I authored or co-authored over a period of about 45 years. The volume also includes five articles, not previously published, based on some of my lectures.

The reprints (some of which have been corrected for misprints and other small errors) are grouped into seven sections, dealing with various topics mainly in the field of optics, in which I have been and still am scientifically active, namely: diffraction, radiation theory, coherence and statistical optics, scattering and foundations of radiometry. Reprints of three semi-popular articles which deal with the history of physics are also included.

The papers in each section are reprinted in the order in which they appear in my publication list (included as Section 8 in this volume), which is arranged in the approximate order in which their manuscripts were submitted for publication. In the commentary which precedes each section, every paper included in this volume is briefly discussed, but not always in the order in which the paper is reprinted in this volume; it seemed more appropriate to discuss papers on closely related subjects together rather than in chronological order. Some of the papers could naturally be included in several different sections. This is particularly true of papers concerned with scattering of partially coherent light. Such papers could obviously be legitimately included either in the section on coherence and statistical optics or in the section on scattering. Similar remarks apply to papers dealing with radiometry with partially coherent light. A reader who is interested in such papers may have to search for them in several sections of this volume.

I doubt that many books have been written without a good deal of help and encouragement by others. This book is no exception. My Secretary, Ellen Calkins, typed and retyped numerous drafts of the lectures and of the commentaries and one of my students, Greg Gbur, expertly formatted them into a beautiful computer product. My friend, Professor Taco Visser, known for his "eagle eyes", carefully scrutinized the manuscript, corrected many errors and suggested improvements. Dr. Scott Carney and Mr. Sergey Ponomarenko also read some portions of the manuscript and have made helpful suggestions. My colleague, Professor Carlos Stroud, succeeded in improving the quality of some of the rather old photographs which are included in Section 7.2.

I am obliged to World Scientific Publishing not only for inviting me to prepare this volume, but also for having complied with all my wishes regarding the production. I would like specifically to acknowledge Dr. Wei Chen, who initiated this project, and Mr. H.T. Leong, who saw it to fruition. I am also obliged to the editors and publishers of the various journals and books for permission to reproduce the papers included in this volume.

As in connection with several other time-consuming projects with which I was involved, my wife, Marlies, cheerfully accepted the long periods of silence devoted to writing rather than to her company. To her particularly, and to all other persons mentioned in this preface, I wish to express my appreciation and offer my thanks.

Emil Wolf

Department of Physics and Astronomy
University of Rochester
Rochester, NY 14627, USA

December 2000

Section 1 – Diffraction

The majority of the papers in this Section (1.1 – 1.4, 1.8, 1.9) are concerned with the structure of focused fields. The first significant contribution to this subject was made by E. Lommel, in a classic paper published in 1885. The reader may wonder why, so much later, the subject is still attracting attention. The answer is that the focal region is an important but rather complicated domain in many optical fields, and that its understanding helps to utilize its rather remarkable properties, some of which have been finding new and useful applications until the present time.

The first paper in this Section (1.1), published almost a half-century ago, reveals a rather strange behavior of the surfaces of constant phase of a focused field in the neighborhood of the zeros of its intensity. This was one of the first examples of phase singularities, which are currently being investigated by many scientists. The investigations have given rise to a new field of optics, sometimes called *singular optics.*[*]

Most of the past researches on focusing have dealt with scalar wave fields. Papers 1.2 – 1.4 deal with the structure of focused electromagnetic fields. Papers 1.2 and 1.3 reveal the existence of a component of the electric field along the beam axis near focus, a phenomenon which in more recent times has attracted a good deal of attention because it offers the possibility of accelerating charged particles.[†] Paper 1.4 shows a fascinating behavior of the Poynting vector close to phase singularities.

It was assumed for a long time that the intensity of a focused field is maximum at the geometrical focus. However, soon after the first lasers were developed it was discovered that this is not always so and that in many cases the intensity maximum is closer – sometimes much closer – to the diffracting aperture. This fact had probably not been discovered earlier because the usual analysis of focused fields in pre-laser optics was entirely adequate for focusing by imaging systems then in use. Howver, the theoretical analysis hid a certain implicit assumption relating to the Fresnel number of the focusing geometry. Fresnel numbers of conventional imaging systems are very large compared to unity and under these circumstances the so-called Debye representation of the focused field is a very good approximation. Paper 1.8 examines the range of validity of the Debye approximation and shows that when the Fresnel number of the focusing geometry is of the order of unity or smaller the Debye representation is not appropriate to describe the field. The changes that the focused field undergoes as the Fresnel number decreases from large to small values are examined in paper 1.9.

The analysis presented in paper 1.10, published only a few months before this volume went to press, is an example of how relevant the classical theory of the focused field is to some modern technological problems. It demonstrates the possibility of improving the performance of compact disks by taking into account in their design the focusing properties of the scanning and read-out beams.

Three other papers in this Section deal with quite different subjects. Papers 1.5 and 1.6 are

---

[*] J.F. Nye, *Natural Focusing and Fine Structure of Light* (Institute of Physics Publishing, Bristol and Philadelphia, 1999); M.S. Soskin and M.V. Vasnetov, "Singular Optics", in *Progress in Optics*, ed. E. Wolf (Elsevier, Amsterdam), vol. 42 (2001), in press.

[†] R.H. Pantell, *Nucl. Instr. & Meth. in Phys. Res.*, A **393**, 1–5 (1997); M. Xie, *Proc. 1997 Particle Accelerator Conference*, Vancouver, B.C. Canada (IEEE, 1998), 660; M.A. Scully, Appl. Phys. B51, 238 (1990).

concerned with the theory of the so-called boundary diffraction wave. This subject was largely developed by A. Rubinowicz, who put on a more rigorous basis the so-called Young theory of diffraction at an aperture. The Young theory is different from the better-known theory of aperture diffraction based on the Huygens Principle. Papers 1.5 and 1.6 provide clarification of some aspects of the theory of the boundary diffraction wave and develops it further.

There is a widely held opinion that the well-known Kirchhoff diffraction theory, when applied to diffraction at an aperture in an opaque screen, is internally inconsistent and that one should therefore avoid its use whenever possible. In paper 1.7 it is shown that the theory can be interpreted in a completely consistent way and that, in some cases, its predictions provide better agreement with experiments than the "manifestly consistent" theories that have been proposed in more recent times.

Section 1 – Diffraction

REPRINTED FROM THE
PROCEEDINGS OF THE PHYSICAL SOCIETY, B, VOL. LXIX, p. 823, 1956

Phase Distribution near Focus in an Aberration-free Diffraction Image

BY E. H. LINFOOT† AND E. WOLF‡

†The Observatories, University of Cambridge

‡The Physical Laboratories University of Manchester

MS. received 12*th October* 1955, *and in revised form* 24*th February* 1956

Abstract. A satisfactory picture of the three-dimensional intensity distribution near focus in the aberration-free diffraction image of a monochromatic point object was first given by Zernike and Nijboer in 1949. No corresponding general picture of the phase relations between different parts of the image seems to have been worked out, although Gouy's discovery in 1890 of the so-called phase anomaly at focus aroused an interest in this topic which, as a succession of publications shows, has not decreased in the intervening 66 years.

In the present paper a sharpened version of Lommel's classical analysis is applied to obtain a general picture of the phase distribution near focus and to examine in detail, in the special case of an F/3·5 pencil, its peculiarities near the geometrical focal point, the Airy dark rings and the axial nodes (points of zero intensity) of the diffraction image. The 'phase anomaly' near focus and the singular behaviour of the phase along the axis of the pencil become more readily intelligible when they are considered against the background of this general picture.

§ 1. INTRODUCTION

A CENTRAL problem in the theory of image formation in optical instruments is the determination of the three-dimensional light distribution which represents the aberration-free image of a point source by an axially symmetric optical system. Analytical formulae for the intensity distribution in such an image were first given by von Lommel (1885) and almost simultaneously by Struve (1886). The first satisfactory diagram of the intensity distribution was published by Zernike and Nijboer (1949) more than sixty years later,† and similar diagrams for images affected by selected amounts of spherical aberration have been worked out by them (1949) and by A. Maréchal and his co-workers (1948).

Less progress seems to have been made in the complementary part of the problem, namely the study of the phase distribution in the three-dimensional diffraction image, even though modern developments in microwave optics have greatly enhanced its practical interest. Theoretical interest was never lacking; as early as 1890 Gouy discovered the so-called phase anomaly near focus, and the subject has since been treated by many authors‡, notably by Joubin (1892), Fabry (1893), Julius (1895), Zeeman (1897, 1900, 1901), Sagnac (1903, 1904 a, b), Debye (1909), Reiche (1909 a, b), Ignatowsky (1919), Fokker (1923, 1924), Picht (1930), Rubinowicz (1938), Breuninger (1938, 1939), Bouwkamp (1940) and Toraldo di Francia (1942).

† An inaccurate diagram, widely reproduced in optical textbooks, was given by Berek (1926). The isophotes in figure 2 were constructed from tables given by von Lommel (1885).

‡ For a review of the early papers, see Reiche (1909 a).

All these investigations were concerned with the aberration-free image and their joint outcome was, broadly speaking, to confirm Gouy's results, to determine the detailed behaviour of the phase distribution along particular rays through the geometrical focus, and to establish that the properties of the phase distribution along the axial ray are qualitatively different from those elsewhere. The last result seems to have been first established by Picht (1930) However, a satisfactory overall picture of the three-dimensional phase distribution, as distinct from the phase distribution along isolated rays, did not emerge.

Such a picture seems nowadays hardly less essential to a proper understanding of the diffraction image than that obtained by Zernike and Nijboer for the three-dimensional intensity distribution. In the present paper we go some way towards obtaining one by an application of Lommel's classical formulae which, for pencils of not too large numerical aperture, allow the calculation of the phases at points not too far from the geometrical focus. From the computed values of the phases at a sufficiently dense set of sample points, the phase distribution near the geometrical focus has been obtained by graphical methods.

It turns out that in the geometrical pencil of rays near focus the co-phasal surfaces are very nearly plane; the corresponding phenomenon for sound waves can be seen in schlieren photographs given by Pohl (1948). The singular behaviour of the phase along the axis becomes readily intelligible when its connection with the existence of points of zero intensity is made visible to the eye (see figures 5 and 6). The singular behaviour of the phase at those points of the geometrical focal plane which corresponds to the Airy dark rings becomes intelligible in the same way (figures 3 and 4).

§ 2. Phase near Focus

The notation and approximations are similar to those of our paper (1952) on telescopic star images. In figure 1, ABA′B′ represents a circular aperture

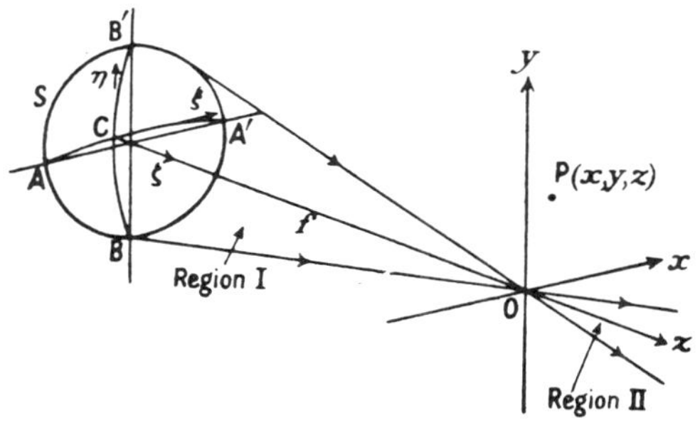

Figure 1.

through which issues a train of converging spherical waves of wavelength λ. The diameter AA′ of this aperture is of length $2a$; C is the pole of the wave surface S which momentarily fills it. O is the centre of curvature of S, CO $=f$ its radius of curvature. It is assumed throughout that $a^2 \ll f^2$; for example, $a^2/f^2 = 1/49$ in an F/3·5 pencil. P (x, y, z) is an arbitrary point in the space near O.

We define the variables u, v by the equations

$$u = ka^2 z/f^2, \qquad v = kar/f, \qquad \ldots\ldots(2.1)$$

where $k = 2\pi/\lambda$ and r stands for $\sqrt{(x^2+y^2)}$. In physical terms, $u/4\pi$ is the number of fringes of defocusing and v/π the number of fringes of lateral displacement of P relative to O. We note that $|v/u|\lesssim 1$ according as P lies in the geometrical cone of rays or in the geometrical shadow.

The complex displacement† at $P(x, y, z)$ in the space near O which results from waves of unit amplitude in S is given by the approximate formula

$$D_\lambda(\mathrm{P}) = \frac{2\pi i}{\lambda} a^2 \exp[ik(f-R')] \int_0^1 \exp(\tfrac{1}{2}iu\rho^2) \mathrm{J}_0(v\rho)\rho\, d\rho, \qquad \ldots\ldots(2.2)$$

where R' denotes the distance CP and J_0 is the Bessel function of order zero.

The equation (2.2) is valid, broadly speaking, at points P for which the number of fringes of defocusing and of lateral displacement are both O(1), that is to say do not exceed about 5 or 10. To define its domain of validity more precisely, we should need to specify the amount of inaccuracy which is regarded as tolerable and to examine the approximation errors in the same way as was done, for an F/15 pencil, in our paper on the intensity distribution already referred to.

The integral on the right of (2.2) can be evaluated in terms of the functions

$$U_n(u, v) = \sum_{m=0}^{\infty} (-1)^m \left(\frac{u}{v}\right)^{n+2m} \mathrm{J}_{n+2m}(v) \qquad \ldots\ldots(2.3)$$

introduced by von Lommel for the purpose; in fact‡

$$\int_0^1 \mathrm{J}_0(v\rho) \exp(\tfrac{1}{2}iu\rho^2)\rho\, d\rho = -iu^{-1} \exp(\tfrac{1}{2}iu)(U_2 + iU_1) \qquad \ldots\ldots(2.4)$$

and (2.2) therefore gives

$$D_\lambda(\mathrm{P}) = \frac{2\pi a^2}{\lambda f u} \exp\{i[k(f-R') + \omega + \tfrac{1}{2}u]\} \surd(U_1^2 + U_2^2), \qquad \ldots\ldots(2.5)$$

where

$$\cos\omega = \frac{U_2}{\surd(U_1^2+U_2^2)}, \quad \sin\omega = \frac{U_1}{\surd(U_1^2+U_2^2)} \qquad \ldots\ldots(2.6)$$

and $\surd$ denotes the positive square root here and elsewhere.

In equation (2.5)

$$\begin{aligned} R' = \mathrm{CP} &= \surd[x^2 + y^2 + (z+f)^2] \\ &= (z+f)\left[1 + \frac{r^2}{(z+f)^2}\right]^{1/2} \\ &= (z+f) + \frac{1}{2}\frac{r^2}{z+f} - \frac{1}{8}\frac{r^4}{(z+f)^3} - \ldots \end{aligned} \qquad \ldots\ldots(2.7)$$

Hence, by (2.1),

$$\begin{aligned} k(f-R') &= -\left(\frac{f}{a}\right)^2 u - \mathrm{O}\left(\frac{kr^2}{2f}\right) \\ &= -\left(\frac{f}{a}\right)^2 u - \frac{\lambda}{4\pi f}\left(\frac{f}{a}\right)^2 \mathrm{O}(v^2). \end{aligned} \qquad \ldots\ldots(2.8)$$

In the region of space where

$$u = \mathrm{O}(1), \qquad v = \mathrm{O}(1), \qquad \ldots\ldots(2.9)$$

the error term on the right of (2.8) is negligible in comparison with unity in a practical case; for example, if $\lambda = 2\times 10^{-5}$ inch the value of $\lambda f/4\pi a^2$ is

† Following Rayleigh, we give this name to the time-independent part $D_\lambda(\mathrm{P})$ of the wave displacement D (P) exp ($ikct$).

‡ See Watson 1922, p. 541.

approximately $F^2/157\,000$, where F is the focal ratio and f is the distance CO measured in inches. Thus we may safely use the approximation

$$k(f-R') = -\left(\frac{f}{a}\right)^2 u. \qquad \ldots\ldots(2.10)$$

It follows from (2.5) that, in the region where (2.9) holds, the amplitude $M(u, v)$ and the phase $\phi(u, v)$ of the complex displacement $D_\lambda(\mathrm{P})$ are given by the equations

$$M(u, v) = \frac{2\pi a^2}{\lambda f\,|u|}\sqrt{(U_1^2 + U_2^2)}, \qquad \ldots\ldots(2.11)$$

$$\phi(u, v) = \omega + \tfrac{1}{2}u - (f/a)^2 u \pmod{2\pi} \qquad \ldots\ldots(2.12)$$

respectively.†

From (2.12) and (2.6) it appears that $\phi(u, v)$, unlike $M(u, v)$, cannot be expressed in terms of the parameters u, v alone but has a structure which changes as the focal ratio varies. The multivalued function $\phi(u, v)$ has a branch point at each zero of the intensity $M^2(u, v)$ and is continuous elsewhere. At the focal point $u = v = 0$ (which is not a branch point) one of its values is $\pi/2$.

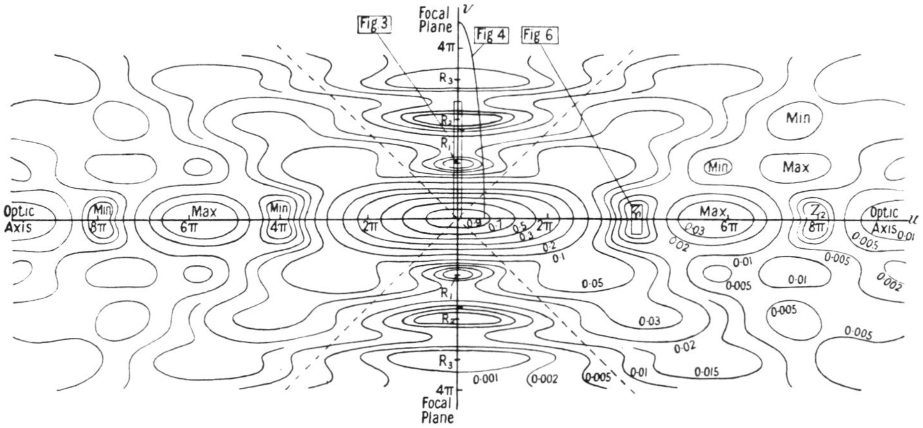

Figure 2. Key diagram showing the relation of figures 3, 4 and 6 to the intensity distribution in the diffraction image. The dotted lines represent the boundary of the geometrical cone of rays and the principal ray lies along the horizontal axis (u-axis). The contour lines are isophotes (lines of equal intensity) and the minima on the v-axis generate the Airy dark rings when the figure is rotated about the u-axis. The region covered by figure 5 is too large to be included in the diagram. The boundary of the quadrant labelled 'Fig. 4' appears elliptical because of the scale-distortion.

The co-phasal surfaces $\phi(u, v) = \text{constant}$ are surfaces of revolution about the axis $v = 0$. By (2.12), (2.6), (2.3) they possess the further symmetry expressed by the equation

$$\phi(-u, v) + \phi(u, v) = \pi \pmod{2\pi}. \qquad \ldots\ldots(2.13)$$

The form of the equiphase surfaces can be computed from (2.12) and (2.6). We first consider the region, very close to the geometrical focal plane, represented in figure 2 by the rectangle labelled 'Fig. 3'. This region is a thin, disc-shaped volume extending a little beyond the second Airy dark ring R_2. Close to the

† The notation (mod 2π) means that the two sides of the equation differ either by zero or by a multiple of 2π. ϕ itself is indeterminate to the extent of an additive multiple of 2π.

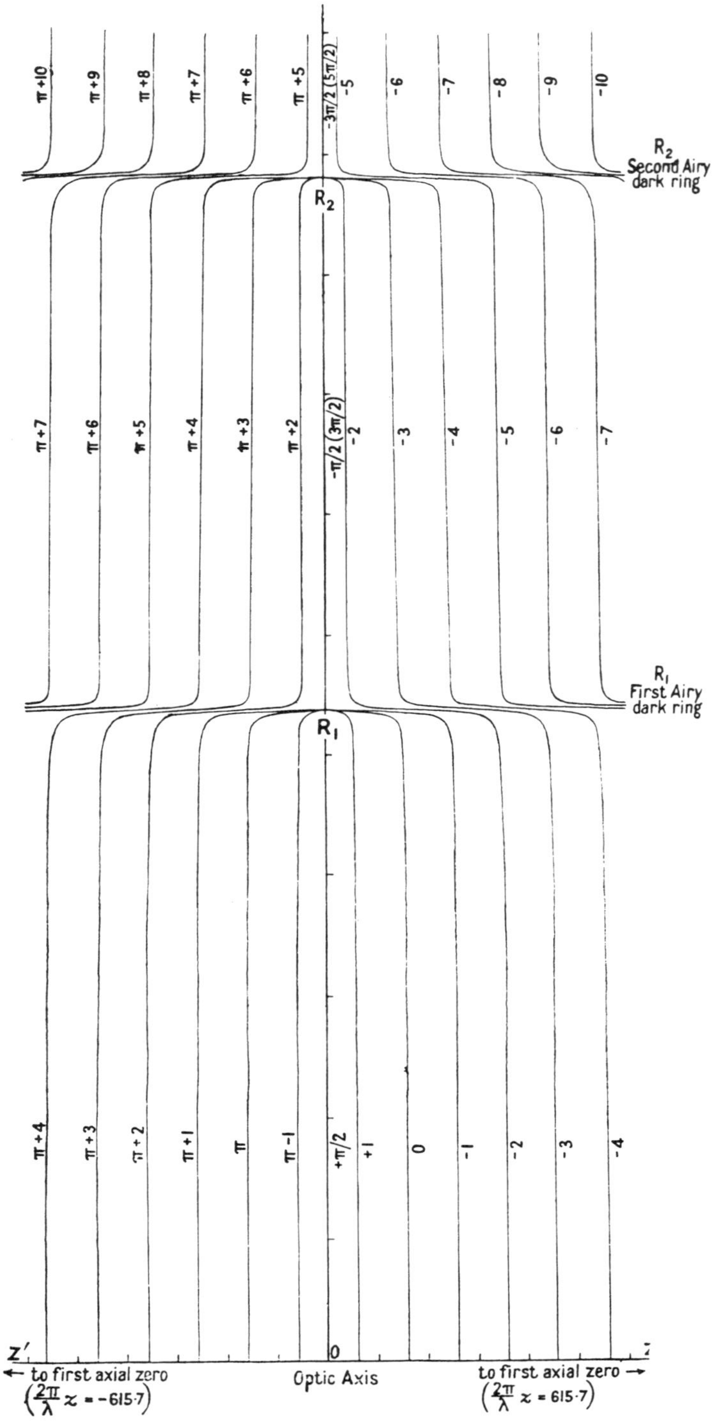

Figure 3. Co-phasal surfaces in the neighbourhood of the geometrical focal planeofan F/3·5 pencil. O is the geometrical focus, ZZ′ the axis of the pencil, OR_1 and OR_2 the radii of the first and second Airy dark rings.

geometrical focus O, the equiphase surfaces are found to be substantially plane except (see figure 3) at points whose distances from the optic axis are nearly equal to the radii OR_1, OR_2, ... of the successive Airy dark rings. However, in the annular regions where they are nearly plane the equiphase surfaces are spaced closer together, by a factor $1-a^2/4f^2$, than would be those in a parallel beam of light of the same frequency; moreover (see figure 2) the intensity is far from uniform over each equiphase surface, being greatest on the optic axis.

The co-phasal surfaces which near O represent a phase-range $-\pi/2$ to $+3\pi/2$ are almost exactly plane until they begin to draw level with the first Airy dark ring R_1; at this distance from the optic axis ZZ′ (see figure 3) they make a sudden swerve inward to unite at the points of zero intensity which constitute the first Airy dark ring. The co-phasal surfaces on either side of this set all make a similar inward swerve opposite R_1 which brings them closer to the focal plane by a distance (a little less than $\frac{1}{2}\lambda$) which corresponds to a phase change of π. At a distance from the axis equal to the radius of the second Airy dark ring R_2, the co-phasal surfaces corresponding to the phase-ranges $-3\pi/2$ to $-\pi/2$ and $3\pi/2$ to $5\pi/2$ of total length 2π suddenly swerve inward to meet in R_2, while the remainder again swerve towards the focal plane through a distance corresponding to a phase change π. This swerving of the equiphase surfaces where they face the Airy dark rings becomes gradually less and less sudden as their distance from the focal plane increases.

From (2.12), in which ω is a function of u, v alone, we see that as the focal ratio varies the form of the equiphase surfaces undergoes a transformation which is more complicated than a simple scale distortion. To display the phase distribu-

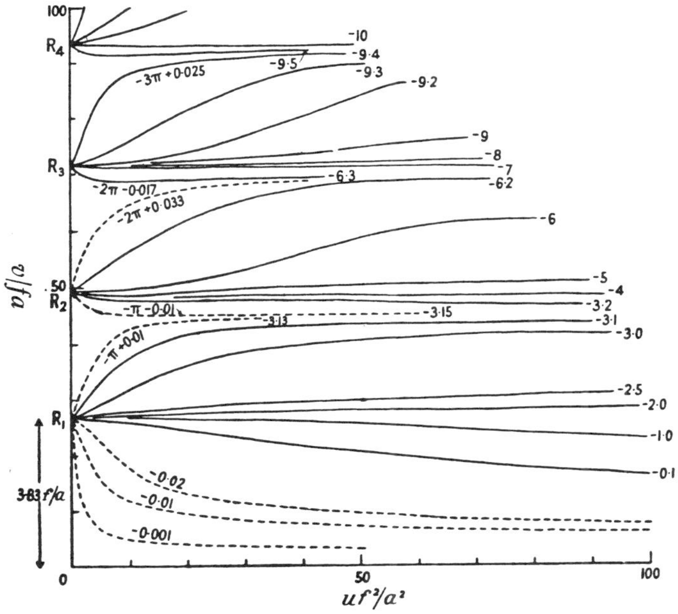

Figure 4. Difference $\phi-\tilde{\phi}$ between actual phase ϕ and phase $\tilde{\phi}=\frac{1}{4}u-(f^2/a^2)u$ of the fiducial plane wave. The numerical scales refer to an F/3·5 pencil; in this case, unit distance on the graph represents an actual distance $\lambda/2\pi$, the radius OR_1 of the first Airy dark ring is $3{\cdot}83f/a$ units $=0{\cdot}61\lambda f/a$, and the distance OZ_1 to the first axial node Z_1 (not shown) is $4\pi f^2/a^2=606$ units, nearly.

tion in the neighbourhood of the origin we therefore introduce a fiducial plane wave,† of wavelength $\lambda(1-a^2/4f^2)$, whose phase $\tilde{\phi}=\frac{1}{2}\pi+\frac{1}{4}u-(f/a)^2u$ agrees (see (2.12) and (3.8) below) with that of the actual wave along the part of the u-axis near the origin. Figure 4 shows the difference $\phi-\tilde{\phi}=\omega+\frac{1}{4}u-\frac{1}{2}\pi$ over the interior of a region which, for an F/3·5 pencil, is a sphere of radius 16λ centred on the geometrical focus. As will be seen from figure 2, this sphere is about 3·5 times as wide, and one-sixth as long, as the central bright nucleus of the diffraction image. A change in the value of the focal ratio $f/2a$ leaves the curves in figure 4 unaltered, but changes the numerical scales along the two axes in accordance with equations (2.1).

§ 3. The Phase Anomaly

In the approximation of geometrical optics, the aberration-free image of a point object is formed by a pencil of concurrent rays issuing from the exit pupil of an optical system. The orthogonal surfaces of this pencil, usually called the geometrical wave fronts, are spherical with the focal point O as common centre of curvature (see figure 1). In the region I, light is converging towards the focus O; in the region II the rays have passed through O and are diverging again. In this simplified picture, the geometrical wave fronts are regarded as surfaces of constant phase, and the phase difference between any two wave fronts is taken as $2\pi/\lambda$ times the optical distance between them, measured along the geometrical rays.

If the wave amplitude is taken as 1 over the suface of the geometrical wave front filling the exit pupil, the disturbance at the point $P(x, y, z)$ in the image space is represented in the simplified geometrical picture by the complex displacement function

$$\left.\begin{aligned} D_\lambda^*(P)&=f\frac{e^{ikR}}{R} \text{ in region I} \\ &=f\frac{e^{-ikR}}{R} \text{ in region II} \\ &=0 \text{ elsewhere in the image space} \end{aligned}\right\}, \qquad \ldots\ldots(3.1)$$

where $f=OC$ and $k=2\pi/\lambda$ as before, while

$$R=\sqrt{(x^2+y^2+z^2)}=\frac{f}{ka}\sqrt{\left\{\left(\frac{f}{a}\right)^2u^2+v^2\right\}}$$

is the distance of P from the focus O. In regions I and II, the phase attributed to the displacement at P is thus represented by the quantity

$$\phi^*(u, v)=-kR\operatorname{sgn} u, \qquad \ldots\ldots(3.2)$$

where $\operatorname{sgn} u$ is defined as ± 1 according as $u \gtrless 0$. In the remainder of the image space, there is complete darkness according to the present approximation, and the phase $\phi^*(u, v)$ is not defined.

The difference

$$\delta(u, v)=\phi(u, v)-\phi^*(u, v) \qquad \ldots\ldots(3.3)$$

between the phases predicted by (2.5) and (3.1) respectively is called the *phase anomaly*. It is defined only in the regions I and II, where by (3.2)

$$\delta(u, v)=\phi(u, v)+kR\operatorname{sgn} u \qquad \ldots\ldots(3.4)$$

† This expedient was suggested by a referee.

and by (2.13) and (3.4)

$$\delta(-u, v)+\delta(u, v)=\pi \ (\bmod 2\pi). \qquad ......(3.5)$$

When P tends to O, the geometrical phase $\phi^*(0,0)$ tends to zero, while the 'principal value' of $\phi(0,0)$ is $\frac{1}{2}\pi$ by (2.2); (3.4) then gives the well-known result

$$\delta(0,0)=\tfrac{1}{2}\pi. \qquad ......(3.6)$$

In the parts of the regions I and II which lie within 6λ of the geometrical focus, each surface of phase ϕ is substantially coincident with the 'best' plane approximation to the corresponding geometrical wave cap of phase $\phi-\frac{1}{2}\pi$, namely the plane which deviates from the geometrical wave cap by equal and opposite amounts at centre and edge. This has the consequence, already noted above, that consecutive surfaces of the same phase (mod 2π) in this region are spaced a distance $(1-a^2/4f^2)\lambda$ apart, in agreement with (2.12) and (3.9), instead of a distance λ apart.

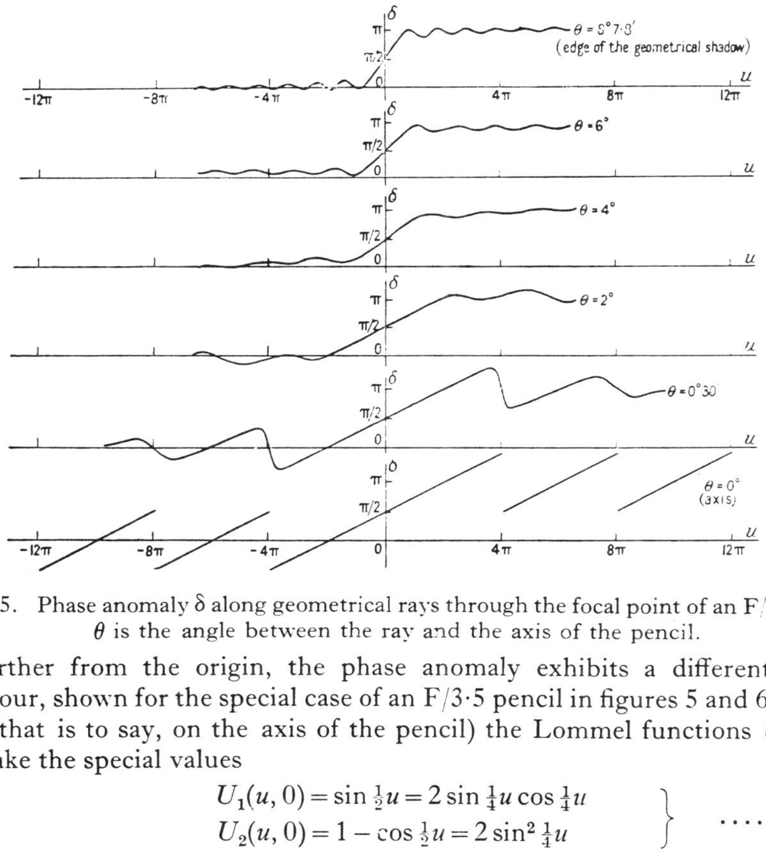

Figure 5. Phase anomaly δ along geometrical rays through the focal point of an F/3·5 pencil. θ is the angle between the ray and the axis of the pencil.

Further from the origin, the phase anomaly exhibits a different type of behaviour, shown for the special case of an F/3·5 pencil in figures 5 and 6. When $v=0$ (that is to say, on the axis of the pencil) the Lommel functions U_1, U_2 in (2.6) take the special values

$$\left.\begin{aligned} U_1(u,0)&=\sin\tfrac{1}{2}u=2\sin\tfrac{1}{4}u\cos\tfrac{1}{4}u\\ U_2(u,0)&=1-\cos\tfrac{1}{2}u=2\sin^2\tfrac{1}{4}u\end{aligned}\right\} \qquad ......(3.7)$$

and we obtain from (2.6)

$$\left.\begin{aligned}\omega&=\tfrac{1}{2}\pi-\tfrac{1}{4}u\,(\bmod 2\pi) \quad \text{if} \quad 2s\pi<\tfrac{1}{4}u<(2s+1)\pi\\ &=\tfrac{3}{2}\pi-\tfrac{1}{4}u\,(\bmod 2\pi) \quad \text{if} \quad (2s+1)\pi<\tfrac{1}{4}u<2s\pi\end{aligned}\right\} \qquad ......(3.8)$$

where $s=0, \pm 1, \pm 2, \ldots.$ Hence by (2.12), (3.1), (3.3)

$$\begin{aligned}\delta(u,0)&=\tfrac{1}{2}\pi+\tfrac{1}{4}u\,(\bmod 2\pi) \quad \text{if} \quad 8s\pi\leqslant|u|<8(s+\tfrac{1}{2})\pi\\ &=-\tfrac{1}{2}\pi+\tfrac{1}{4}u\,(\bmod 2\pi) \quad \text{if} \quad 8(s+\tfrac{1}{2})\pi\leqslant|u|<8(s+1)\pi.\end{aligned}$$

$$......(3.9)$$

There is a phase jump of π at each of the axial 'nodes' (points of zero intensity) $u=4s\pi$; $s=\pm 1, \pm 2, \ldots$. Between two consecutive nodes the phase anomaly $\delta(u, 0)$ varies linearly, as shown in the bottom curve of figure 5, until $|u|$ becomes so large that the approximation (2.2) begins to break down.

The computation of the phase distribution on the edge $v=|u|$ of the geometrical shadow is simplified by observing that the Lommel functions U_1, U_2 take for $v=u$ the simple forms†

$$U_1(u, u)=\tfrac{1}{2}\sin u$$
$$U_2(u, u)=\tfrac{1}{2}[J_0(u)-\cos u]. \qquad \ldots\ldots(3.10)$$

The resulting phase anomaly is shown, for the case of an F/3·5 pencil, in the top curve of figure 5.

Figure 5 illustrates the difference between the behaviour of the phase anomaly along the axial ray and along the non-axial rays of the pencil. Along each non-axial ray, the phase anomaly approaches π as the distance from the focus increases (Debye 1909). The different behaviour, described by (3.9), of the phase anomaly along the axial ray is by no means surprising, since this ray passes through the axial nodes. In figure 6 is shown the behaviour of the co-phasal surfaces near the first axial node.

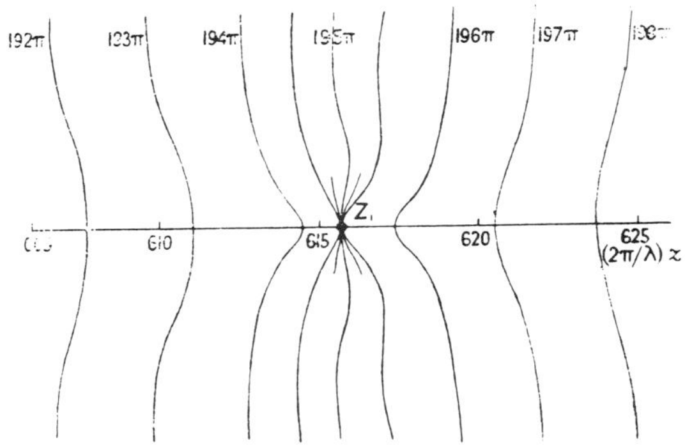

Figure 6. Co-phasal surfaces in the neighbourhood of the first axial node (point of zero intensity) Z_1 in an F/3·5 pencil.

ACKNOWLEDGMENTS

We should like to thank Mrs. I. Berrer for the care and skill with which she has carried out the heavy computations. One of us (E.W.) wishes to acknowledge gratefully financial assistance received from the Carnegie Trust for the Universities of Scotland towards computing expenses. He is also indebted to the Universities of Edinburgh and Manchester for awards of I.C.I. Research Fellowships.

REFERENCES

BEREK, M., 1926, *Z. Phys.*, **40**, 421.
BOUWKAMP, C. J., 1940, *Physica*, **7**, 485.
BREUNINGER, H. W., 1938, *Thesis*, University of Jena; 1939, *Ann. Phys., Lpz.*, **35**, 238.
DEBYE, P., 1909, *Ann. Phys., Lpz.*, **30**, 755.
FABRY, C., 1893, *J. Phys. Theor. Appl.* (3), **2**, 22.

† These follow from the Jacobi identities (Watson 1922, p. 22).

FOKKER, A. D., 1923, *Physica*, **3,** 334; 1924, *Ibid.*, **4,** 166.
GOUY, L. G., 1890, *C. R. Acad. Sci., Paris*, **110,** 1251; 1891, *Ann. Chim. (Phys.)*, **24,** 145.
IGNATOWSKY, B. C., 1919, *Trans. State Optical Institute, Petrograd*, **1,** III.
JOUBIN, P., 1892, *C.R. Acad. Sci., Paris*, **115,** 932.
JULIUS, W. H., 1895, *Archives Néerl. Sci.*, **28,** 221, 226.
LINFOOT, E. H., and WOLF, E., 1952, *Mon. Not. Roy. Astr. Soc.*, **112,** 452.
VON LOMMEL, E., 1885, *Abh. Bayer Akad. Wiss.*, **53,** 233.
MARÉCHAL, A., 1948, *Rev. Opt. (Theor. Instrum.)*, **27,** 73.
PICHT, J., 1930, *Z. Phys.*, **65,** 14.
POHL, R. W., 1948, *Optik*, 7th and 8th Ed. (Berlin : Springer), p. 353.
REICHE, F., 1909 a, *Ann. Phys., Lpz.*, **29,** 65; 1909 b, *Ibid.*, **29,** 401.
RUBINOWICZ, A., 1938, *Phys. Rev.*, **54,** 931.
SAGNAC, G., 1903, *J. Phys. (Théor. Instrum.)*, **2,** 721; 1904 a, *C.R. Acad. Sci., Paris*, **138,** 479, 619, 678; 1904 b, *Boltzmann Festschrift*, 528.
STRUVE, H., 1886, *Mém. de l'Acad. de St. Petersb.* (7), **34,** No. 5, 1.
TORALDO DI FRANCIA, G., 1942, *Ottica*, **7,** N.2.
WATSON, G. N., 1922, *Bessel Functions* (Cambridge : University Press).
ZEEMAN, P., 1897/98, *Versl. Gewone Vergad. Akad. Amst.*, **6,** 11; 1900, *Phys. Z.*, **1,** 542; 1901, *Archives Néerl. Sci.*, **4,** 314, 318.
ZERNIKE, F., and NIJBOER, B. R. N., 1949, Contribution to *Théorie des Images Optiques* (Paris : Éditions de la Revue d'Optique).

Reprinted without change of pagination from the
Proceedings of the Royal Society, A, *volume* 253, pp. 358–379, 1959

Electromagnetic diffraction in optical systems
II. Structure of the image field in an aplanatic system

BY B. RICHARDS†
The Computing Machine Laboratory, University of Manchester
AND E. WOLF‡
Department of Theoretical Physics, University of Manchester

(*Communicated by D. Gabor, F.R.S.—Received* 19 *February* 1959)

An investigation is made of the structure of the electromagnetic field near the focus of an aplanatic system which images a point source. First the case of a linearly polarized incident field is examined and expressions are derived for the electric and magnetic vectors in the image space. Some general consequences of the formulae are then discussed. In particular the symmetry properties of the field with respect to the focal plane are noted and the state of polarization of the image region is investigated. The distribution of the time-averaged electric and magnetic energy densities and of the energy flow (Poynting vector) in the focal plane is studied in detail, and the results are illustrated by diagrams and in a tabulated form based on data obtained by extensive calculations on an electronic computor. The case of an unpolarized field is also investigated.

The solution is not restricted to systems of low aperture, and the computational results cover, in fact, selected values of the angular semi-aperture α on the image side, in the whole range $0 \leqslant \alpha \leqslant 90°$. The limiting case $\alpha \to 0$ is examined in detail and it is shown that the field is then completely characterized by a single, generally complex, scalar function, which turns out to be identical with that of the classical scalar theory of Airy, Lommel and Struve.

The results have an immediate bearing on the resolving power of image forming systems; they also help our understanding of the significance of the scalar diffraction theory, which is customarily employed, without a proper justification, in the analysis of images in low-aperture systems.

1. INTRODUCTION

A knowledge of the structure of an electromagnetic field in the region of focus is of considerable theoretical as well as practical interest. As already indicated in part I of this investigation (Wolf 1959), information about the structure of this complex region is particularly desirable in connexion with the design and the analysis of performance of optical systems of wide angular aperture, both in the field of visible and microwave optics. The construction of large paraboloids for use in radio astronomy has made this problem also of topical interest.

Apart from the usual scalar treatments, the limitations of which were mentioned in part I, the only earlier treatments of this problem appear to be those of Ignatowsky (1919, 1920) and Hopkins (1943, 1945).§ The investigations of Ignatowsky have gone a considerable way towards the solution of the problem. He obtained

† Now at the Computer Section, I.C.I., Ltd, Wilton Works, Middlesbrough, Yorkshire.

‡ Now at the Institute of Optics, University of Rochester, Rochester, N.Y., U.S.A.

§ Since the present investigation was carried out, two related papers have appeared. Burtin (1956) determined the distribution of the electric energy density in the focal plane of a system with angular semi-aperture $\alpha = 45°$ on the image side, when the wave entering the system is linearly polarized. Focke (1957) considered an unpolarized wave and studied the energy density and the energy flow at points in the focal plane in systems of selected angular apertures.

formulae for the electric and magnetic field vectors in the image region of an aplanatic system† and also of a paraboloid of any angular aperture. Unfortunately his deductions from these formulae were chiefly confined to the study of the energy flow across the central bright nucleus of the image; the electric energy density (which is presumably what many detectors, e.g. a photographic plate, record) is not discussed. The researches of Hopkins were mainly concerned with the modification which the Airy pattern undergoes as the angular aperture of the system is increased. While his analysis is not based on the full Maxwell's equations, it does take into account the vectorial nature of the problem and his results are of practical interest in connexion with optical systems the angular semi-apertures of which do not exceed about 40°.

In the present paper a thorough investigation is made of the structure of the electromagnetic field near the focus of an aplanatic system which images a point source at infinity. Formulae are derived for the electric and magnetic vectors in the image space and a number of general properties of the electromagnetic field are deduced, both for polarized and unpolarized incident waves. The formulae are evaluated for a large number of the basic parameters of the problem and the results are presented in the form of diagrams and in tabulated form. Systems of selected angular semi-aperture α on the image side are considered, up to the limiting case $\alpha \to 90°$. The other limiting case, $\alpha \to 0$, is examined in detail and it is found that, in this case, the vector solution is completely characterized by one (generally complex) scalar function, which is found to be identical with that of the classical analyses of Airy (1835), Lommel (1885) and Struve (1886).

The results have an immediate bearing on the resolving power of systems with high angular aperture; they also help our understanding of the significance of the usual scalar methods of optical diffraction theory.

Finally, it may be mentioned that our basic formulae (equations (2·30) and (2·31)) are in agreement with those of Ignatowsky (1919); our deductions, however, go considerably further.

Some preliminary results of this investigation were reported in two previous notes (Richards 1956*a*; Richards & Wolf 1956).

2. Expressions for the field vectors in the image space of an aplanatic system

Consider an optical system of revolution, which images a point source. The imaging will be assumed to be aplanatic, i.e. axially stigmatic and obeying the sine condition. The source will be assumed to be at infinity in the direction of the axis, and to begin with it will be assumed that it gives rise to a linearly polarized monochromatic wave in the entrance pupil of the system. The case of an unpolarized wave will be considered later (§5) by averaging over all possible states of polarization. It is assumed that the linear dimensions of the exit pupil are large compared with the wavelength.

† By aplanatic system we mean one which, for a specified axial position of the object point is stigmatic and obeys the Abbe sine condition.

Expressions for the field in the image region of the system may be derived by an application of formulae (2·18) and (2·19) of part I. These formulae express the time-independent parts, **e** and **h** of the electric and magnetic vectors.

$$\mathbf{E}(x,y,z,t) = \mathscr{R}\{\mathbf{e}(x,y,z)\,\mathrm{e}^{-\mathrm{i}\omega t}\}, \quad \mathbf{H}(x,y,z,t) = \mathscr{R}\{\mathbf{h}(x,y,z)\,\mathrm{e}^{-\mathrm{i}\omega t}\}, \tag{2·1}$$

at any point $P(x,y,z)$ in the image space, which is not too close to the exit pupil, in the form

$$\mathbf{e}(x,y,z) = -\frac{ik}{2\pi}\iint_\Omega \frac{\mathbf{a}(s_x,s_y)}{s_z}\,\mathrm{e}^{\mathrm{i}k[\Phi(s_x,s_y)+s_x x+s_y y+s_z z]}\,\mathrm{d}s_x\mathrm{d}s_y, \tag{2·2}$$

$$\mathbf{h}(x,y,z) = -\frac{ik}{2\pi}\iint_\Omega \frac{\mathbf{b}(s_x,s_y)}{s_z}\,\mathrm{e}^{\mathrm{i}k[\Phi(s_x,s_y)+s_x x+s_y y+s_z z]}\,\mathrm{d}s_x\mathrm{d}s_y, \tag{2·3}$$

where $\mathscr{R}$ denotes the real part. Here **a** and **b** are the 'strength factors' of the unperturbed electromagnetic field $\mathbf{E}^{(i)} = \mathscr{R}\{\mathbf{e}^{(i)}\mathrm{e}^{-\mathrm{i}\omega t}\}$, $\mathbf{H}^{(i)} = \mathscr{R}\{\mathbf{h}^{(i)}\mathrm{e}^{-\mathrm{i}\omega t}\}$ which is incident on the exit pupil, i.e. **a** and **b** are defined by the relations

$$\mathbf{e}^{(i)} = \frac{\mathbf{a}}{\sqrt{(R_1R_2)}}\,\mathrm{e}^{\mathrm{i}k\mathscr{S}}, \quad \mathbf{h}^{(i)} = \frac{\mathbf{b}}{\sqrt{(R_1R_2)}}\,\mathrm{e}^{\mathrm{i}k\mathscr{S}}, \tag{2·4}$$

where R_1 and R_2 are the principal radii of curvature of the associated wave-front and $\mathscr{S}$ is the eikonal function. Further

$$k = \frac{\omega}{c} = \frac{2\pi}{\lambda} \tag{2·5}$$

is the vacuum wave number, and λ the vacuum wavelength (denoted in part I by k_0 and λ_0 respectively), it being assumed here that the refractive index of the image space is unity. Further, Φ is the aberration function of the system, s_x, s_y, s_z are the components of the unit vector **s** (with its positive direction in the direction of propagation of the light) along a typical ray in the image space, and Ω is the solid angle formed by all the geometrical rays which pass through the exit pupil of the system. According to (2·20) of part I, the strength factors are related by the formula (assuming that in the image space $\epsilon \simeq \mu \simeq 1$)

$$\mathbf{b} = \mathbf{s}\wedge\mathbf{a}. \tag{2·6}$$

In the problem under consideration, the imaging is aplanatic so that the wave-fronts in the image space are spherical, with a common centre at the Gaussian image point. Hence, for all vectors **s** in the solid angle Ω,

$$\Phi(s_x, s_y) = 0. \tag{2·7}$$

The strength factors **a** and **b** may be determined, with an accuracy sufficient for our purposes, by tracing rays through the system up to the region of the exit pupil, and by making use of the laws relating to the variation of the field vectors along each ray. Let AQ_0 be a typical ray entering the system at a height h from the axis and let θ be the angle which the corresponding ray QO in the image space makes with the axis (figure 1). Since the field in the object space is linearly polarized, the field obtained on refraction at the first surface σ_1 is also linearly polarized (the direction of polarization will, of course, be in general different for different rays); and if the angle of the incidence at σ_1 is small, the angle which the direction of vibrations of the electric (and also of the magnetic) vector makes with the meridional plane of the ray (the plane containing the ray and the axis of the system) will be effectively unchanged by refraction.† Moreover (see Born & Wolf 1959, § 3·1·3), in a homogeneous medium the direction of vibration remains unchanged along each ray. By successive repetition of these arguments from surface to surface it follows that the field in the region of the exit pupil is also linearly polarized and that, provided the angle of incidence at each surface of the system is small, the angles which the **e** and **h** vectors in the exit pupil make with the meridional plane of the ray are equal to the corresponding angle in the object space. Further, the magnitude of the field vectors in the region of the exit pupil may be determined from the geometrical law of conservation of energy, taking into account the fact that the system obeys the sine condition.

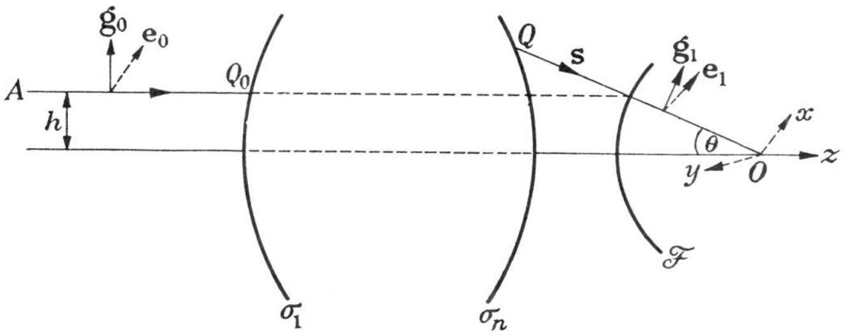

FIGURE 1. The meridional plane of a ray. The axis Ox is in the direction of the electric vector $\mathbf{e}_0$ in the object space.

Let $\mathscr{F}$ be the 'focal sphere', i.e. the sphere with centre at O and with radius f equal to the focal length of the system. Then, according to the sine condition

$$h = f \sin\theta. \tag{2·8}$$

This relation implies that the emergent ray meets the focal sphere at the same height at which the corresponding ray in the object space entered the system (see figure 1).

Consider now all the rays which enter the annulus bounded by circles of radius h and $h+\delta h$. Let δS_0 be the area of the annulus and δS_1 the corresponding area of the focal sphere, and let $\mathbf{e}_0$ and $\mathbf{e}_1$ be the electric amplitude vectors on δS_0 and δS_1 respectively. Then we may write

$$\mathbf{e}_0 = l_0 \hat{\mathbf{e}}_0 \, e^{ik\mathscr{S}_0}, \quad \mathbf{e}_1 = l_1 \hat{\mathbf{e}}_1 \, e^{ik\mathscr{S}_1}, \tag{2·9}$$

where $\hat{\mathbf{e}}_0$ and $\hat{\mathbf{e}}_1$ are unit vectors, in the direction of $\mathbf{e}_0$ and $\mathbf{e}_1$ respectively, l_0 and l_1 are amplitude factors and $\mathscr{S}_0$ and $\mathscr{S}_1$ are the corresponding values of the eikonal. Since the field is linearly polarized, l_0, l_1, $\hat{\mathbf{e}}_0$ and $\hat{\mathbf{e}}_1$ may be taken to be real, provided

† The truth of these statements follows from Fresnel formulae for refraction of a plane wave on a plane interface between two homogeneous media of different refractive indices. Within the accuracy of the present approximation (geometric optics limit, i.e. $k \to \infty$), these formulae also apply to the present case when neither the wave-fronts nor the refracting boundaries are necessarily plane.

that the origin of the phase is suitably chosen. According to the geometrical optics intensity law,

$$l_0^2\,\delta S_0 = l_1^2\,\delta S_1, \tag{2·10}$$

where we assumed that the refractive index of the object space, like that of the image space, is unity, and that the losses of energy due to reflexion and absorption within the system are negligible. Now from the figure,

$$\delta S_0 = \delta S_1 \cos\theta, \tag{2·11}$$

so that according to (2·10)

$$l_1 = l_0 \cos^{\frac{1}{2}}\theta. \tag{2·12}$$

Now $\mathbf{e}_1$ may be identified with the vector $\mathbf{e}^{(i)}$ of (2·4), with $R_1 = R_2 = f$ and $\mathscr{S} = \mathscr{S}_1$. Hence

$$\mathbf{a} = f l_0 \cos^{\frac{1}{2}}\theta\,\hat{\mathbf{e}}_1. \tag{2·13}$$

To determine $\mathbf{e}_1$ it is convenient to introduce two unit vectors $\mathbf{g}_0$ and $\mathbf{g}_1$ in the meridional plane of the ray, such that $\mathbf{g}_0$ is perpendicular to the ray in the object space and $\mathbf{g}_1$ is perpendicular to the ray in the image space, and both are directed away from the axis (see figure 1). Let Ox, Oy, Oz be Cartesian rectangular co-ordinate axes, with origin at the Gaussian focus, with Ox in the direction of $\mathbf{e}_0$ and with Oz along the axis of the system, pointing away from the plane of the exit pupil into the image space. Finally, let $\mathbf{i}$, $\mathbf{j}$ and $\mathbf{k}$ be unit vectors in the direction of the co-ordinate axes.

The electric and the magnetic vectors are orthogonal to the ray (see Born & Wolf 1959, § 3·1). Hence $\mathbf{e}_1$ lies in the plane of $\mathbf{g}_1$ and $\mathbf{g}_1 \wedge \mathbf{s}$, i.e.

$$\hat{\mathbf{e}}_1 = \alpha\mathbf{g}_1 + \beta(\mathbf{g}_1 \wedge \mathbf{s}), \tag{2·14}$$

where α and β are some constants. To determine α and β we make use of a result mentioned earlier, namely that as the light traverses the system, the angle between the electric (and also the magnetic) vector and the meridional plane of the ray remains constant. Moreover, each of these vectors remains on the same side of the meridional plane. These results imply that

$$\left.\begin{aligned} \mathbf{g}_1 . \hat{\mathbf{e}}_1 &= \mathbf{g}_0 . \mathbf{i}, \\ (\mathbf{g}_1 \wedge \mathbf{s}_1) . \hat{\mathbf{e}}_1 &= (\mathbf{g}_0 \wedge \mathbf{k}) . \mathbf{i}. \end{aligned}\right\} \tag{2·15}$$

On substituting from (2·14) into (2·15) we find that

$$\alpha = \mathbf{g}_0 . \mathbf{i}, \quad \beta = (\mathbf{g}_0 \wedge \mathbf{k}) . \mathbf{i} = \mathbf{g}_0 . (\mathbf{k} \wedge \mathbf{i}) = \mathbf{g}_0 . \mathbf{j}, \tag{2·16}$$

and from (2·12), (2·13) and (2·15) it follows that

$$\mathbf{a} = f l_0 \cos^{\frac{1}{2}}\theta[(\mathbf{g}_0 . \mathbf{i})\,\mathbf{g}_1 + (\mathbf{g}_0 . \mathbf{j})\,(\mathbf{g}_1 \wedge \mathbf{s})]. \tag{2·17}$$

It will now be convenient to introduce spherical polar co-ordinates r, θ, ϕ ($r > 0$, $0 \leqslant \theta < \pi$, $0 \leqslant \phi < 2\pi$), with the polar axis $\theta = 0$ in the z-direction, and with the azimuth $\phi = 0$ containing the electric vector in the object space. The components s_x, s_y, s_z of the unit vector $\mathbf{s}$ along a ray in the image space and the co-ordinates (x, y, z) of a point P in the image region may be expressed in the form

$$s_x = \sin\theta\cos\phi, \quad s_y = \sin\theta\sin\phi, \quad s_z = \cos\theta, \tag{2·18a}$$

$$x = r_P \sin\theta_P \cos\phi_P, \quad y = r_P \sin\theta_P \sin\phi_P, \quad z = r_P \cos\theta_P, \tag{2·18b}$$

so that the term in the exponent of the integrals (2·1) and (2·3) becomes

$$s_x x + s_y y + s_z z = r_P \cos\epsilon, \tag{2·19}$$

where

$$\cos\epsilon = \cos\theta\cos\theta_P + \sin\theta\sin\theta_P\cos(\phi-\phi_P). \tag{2·20}$$

Let (θ_0, ϕ_0) and (θ_1, ϕ_1) be the polar angles of $\mathbf{g}_0$ and $\mathbf{g}_1$ respectively. Then evidently (see figure 1)

$$\phi_0 = \phi_1 = \phi - \pi, \quad \theta_0 = \tfrac{1}{2}\pi, \quad \theta_1 = \tfrac{1}{2}\pi - \theta, \tag{2·21}$$

so that

$$\left.\begin{aligned} \mathbf{g}_0 &= -\cos\phi\mathbf{i} - \sin\phi\mathbf{j}, \\ \mathbf{g}_1 &= -\cos\theta\cos\phi\mathbf{i} - \cos\theta\sin\phi\mathbf{j} + \sin\theta\mathbf{k}. \end{aligned}\right\} \tag{2·22}$$

It follows on substitution from (2·22) and (2·18*a*) into (2·17) and (2·6) that the Cartesian components of the strength vectors **a** and **b** are

$$\left.\begin{aligned} a_x &= fl_0\cos^{\frac{1}{2}}\theta[\cos\theta + \sin^2\phi(1-\cos\theta)], \\ a_y &= fl_0\cos^{\frac{1}{2}}\theta[(\cos\theta - 1)\cos\phi\sin\phi], \\ a_z &= -fl_0\cos^{\frac{1}{2}}\theta\sin\theta\cos\phi, \end{aligned}\right\} \tag{2·23}$$

$$\left.\begin{aligned} b_x &= fl_0\cos^{\frac{1}{2}}\theta[(\cos\theta - 1)\cos\phi\sin\phi], \\ b_y &= fl_0\cos^{\frac{1}{2}}\theta[1 - \sin^2\phi(1-\cos\theta)], \\ b_z &= -fl_0\cos^{\frac{1}{2}}\theta\sin\theta\sin\phi. \end{aligned}\right\} \tag{2·24}$$

Finally, we also need an expression, in terms of θ and ϕ, for the quantity $\mathrm{d}s_x\mathrm{d}s_y/s_z$, which enters our basic diffraction integrals (2·2) and (2·3). This quantity represents the element $\mathrm{d}\Omega$ of the solid angle and is given by

$$\mathrm{d}s_x\mathrm{d}s_y/s_z = \mathrm{d}\Omega = \sin\theta\,\mathrm{d}\theta\,\mathrm{d}\phi. \tag{2·25}$$

On substituting from (2·23), (2·24), (2·25), (2·11) and (2·7) into (2·2) and (2·3) we obtain the following expressions for the Cartesian components **e** and **h**:

$$\left.\begin{aligned} e_x &= -\frac{\mathrm{i}A}{\pi}\int_0^\alpha\int_0^{2\pi}\cos^{\frac{1}{2}}\theta\sin\theta\{\cos\theta + (1-\cos\theta)\sin^2\phi\}\,\mathrm{e}^{\mathrm{i}kr_P\cos\epsilon}\,\mathrm{d}\theta\,\mathrm{d}\phi, \\ e_y &= \frac{\mathrm{i}A}{\pi}\int_0^\alpha\int_0^{2\pi}\cos^{\frac{1}{2}}\theta\sin\theta(1-\cos\theta)\cos\phi\sin\phi\,\mathrm{e}^{\mathrm{i}kr_P\cos\epsilon}\,\mathrm{d}\theta\,\mathrm{d}\phi, \\ e_z &= \frac{\mathrm{i}A}{\pi}\int_0^\alpha\int_0^{2\pi}\cos^{\frac{1}{2}}\theta\sin^2\theta\cos\phi\,\mathrm{e}^{\mathrm{i}kr_P\cos\epsilon}\,\mathrm{d}\theta\,\mathrm{d}\phi; \end{aligned}\right\} \tag{2·26}$$

$$\left.\begin{aligned} h_x &= \frac{\mathrm{i}A}{\pi}\int_0^\alpha\int_0^{2\pi}\cos^{\frac{1}{2}}\theta\sin\theta(1-\cos\theta)\cos\phi\sin\phi\,\mathrm{e}^{\mathrm{i}kr_P\cos\epsilon}\,\mathrm{d}\theta\,\mathrm{d}\phi, \\ h_y &= -\frac{\mathrm{i}A}{\pi}\int_0^\alpha\int_0^{2\pi}\cos^{\frac{1}{2}}\theta\sin\theta\{1 - (1-\cos\theta)\sin^2\phi\}\,\mathrm{e}^{\mathrm{i}kr_P\cos\epsilon}\,\mathrm{d}\theta\,\mathrm{d}\phi, \\ h_z &= \frac{\mathrm{i}A}{\pi}\int_0^\alpha\int_0^{2\pi}\cos^{\frac{1}{2}}\theta\sin^2\theta\sin\phi\,\mathrm{e}^{\mathrm{i}kr_P\cos\epsilon}\,\mathrm{d}\theta\,\mathrm{d}\phi. \end{aligned}\right\} \tag{2·27}$$

Here $\cos\epsilon$ is given by (2·19), α is the angular semi-aperture on the image side, i.e. 2α is the angle which the diameter of the exit pupil subtends at the geometrical focus and A is the constant

$$A = \frac{kfl_0}{2} = \frac{\pi fl_0}{\lambda}. \tag{2·28}$$

The integration with respect to ϕ can immediately be carried out with the help of the following formulae† which are valid for any integral value of n:

$$\left.\begin{aligned}\int_0^{2\pi} \cos n\phi\, e^{i\rho\cos(\phi-\gamma)}\, d\phi &= 2\pi i^n J_n(\rho)\cos n\gamma,\\ \int_0^{2\pi} \sin n\phi\, e^{i\rho\cos(\phi-\gamma)}\, d\phi &= 2\pi i^n J_n(\rho)\sin n\gamma.\end{aligned}\right\} \quad (2\cdot29)$$

Here $J_n(\rho)$ is the Bessel function of the first kind and order n. If in (2·26) and (2·27) we use the identities $\cos\phi\sin\phi = \frac{1}{2}\sin 2\phi$, $\sin^2\phi = \frac{1}{2}(1-\cos 2\phi)$ and apply (2·29), we finally obtain the following expressions for the components of the field vectors at a point P in the image region:

$$\left.\begin{aligned} e_x(P) &= -iA(I_0 + I_2\cos 2\phi_P),\\ e_y(P) &= -iAI_2\sin 2\phi_P,\\ e_z(P) &= -2AI_1\cos\phi_P,\end{aligned}\right\} \quad (2\cdot30)$$

$$\left.\begin{aligned} h_x(P) &= -iAI_2\sin 2\phi_P,\\ h_y(P) &= -iA(I_0 - I_2\cos 2\phi_P),\\ h_z(P) &= -2AI_1\sin\phi_P,\end{aligned}\right\} \quad (2\cdot31)$$

where

$$\left.\begin{aligned} I_0 = I_0(kr_P,\theta_P,\alpha) &= \int_0^\alpha \cos^{\frac{1}{2}}\theta\sin\theta(1+\cos\theta)\,J_0(kr_P\sin\theta\sin\theta_P)\,e^{ikr_P\cos\theta\cos\theta_P}\,d\theta,\\ I_1 = I_1(kr_P,\theta_P,\alpha) &= \int_0^\alpha \cos^{\frac{1}{2}}\theta\,\sin^2\theta J_1(kr_P\sin\theta\,\sin\theta_P)\,e^{ikr_P\cos\theta\cos\theta_P}\,d\theta,\\ I_2 = I_2(kr_P,\theta_P,\alpha) &= \int_0^\alpha \cos^{\frac{1}{2}}\theta\,\sin\theta(1-\cos\theta)\,J_2(kr_P\sin\theta\sin\theta_P)\,e^{ikr_P\cos\theta\cos\theta_P}\,d\theta.\end{aligned}\right\} \quad (2\cdot32)$$

Formulae (2·30) and (2·31) represent the analytic solution of our problem. They express the field at any point P (spherical polar co-ordinates r_P, θ_P, ϕ_P) in terms of the three integrals I_0, I_1 and I_2. We shall now study some consequences of these formulae.

3. The image field

It is convenient at this stage to introduce certain 'optical co-ordinates', which are a natural generalization of the co-ordinates (defined by (3·1*b*) below) employed

† These formulae may be derived as follows. We start from the integral representation of J_n:

$$\int_0^{2\pi} e^{i(n\phi+\rho\cos\phi)}\,d\phi = 2\pi i^n J_n(\rho)$$

(cf. Watson 1952, p. 20, (5)). We change ϕ into $\phi-\gamma$, multiply both sides of (2·30) by $e^{in\gamma}$ and express the resulting formula as follows:

$$\int_0^{2\pi}\cos n\phi\,e^{i\rho\cos(\phi-\gamma)}\,d\phi + i\int_0^{2\pi}\sin n\phi\,e^{i\rho\cos(\phi-\gamma)}\,d\phi = 2\pi i^n J_n(\rho)\,[\cos n\gamma + i\sin n\gamma].$$

Each side consists of two terms, one of which is an even function of γ and the other an odd function of γ. This is only possible if the even terms are equal to each other and so are the odd terms, and this implies (2·29).

frequently in connexion with diffraction in systems with low angular aperture. We define these optical co-ordinates by the formulae†

$$\left.\begin{aligned} u &= kr_P\cos\theta_P\sin^2\alpha = kz\sin^2\alpha,\\ v &= kr_P\sin\theta_P\sin\alpha = k\surd(x^2+y^2)\sin\alpha. \end{aligned}\right\} \qquad (3\cdot 1a)$$

From now on we shall omit the subscript P in the symbol ϕ_P for the azimuthal angle, and specify the point P of observation by the three parameters u, v and ϕ ($u \gtreqless 0$, $v \geqslant 0$, $0 \leqslant \phi < 2\pi$). The geometrical focal plane is given by $u = 0$, the axis by $v = 0$ and the edge of the geometrical shadow ($\surd(x^2+y^2) = \pm z\tan\alpha$) by $v = \pm u\sec\alpha$.

The integrals (2·32) are now regarded as functions of u and v,

$$\left.\begin{aligned} I_0(u,v) &= \int_0^\alpha \cos^{\frac{1}{2}}\theta\sin\theta(1+\cos\theta)J_0\left(\frac{v\sin\theta}{\sin\alpha}\right)e^{iu\cos\theta/\sin^2\alpha}\,d\theta,\\ I_1(u,v) &= \int_0^\alpha \cos^{\frac{1}{2}}\theta\sin^2\theta J_1\left(\frac{v\sin\theta}{\sin\alpha}\right)e^{iu\cos\theta/\sin^2\alpha}\,d\theta,\\ I_2(u,v) &= \int_0^\alpha \cos^{\frac{1}{2}}\theta\sin\theta(1-\cos\theta)J_2\left(\frac{v\sin\theta}{\sin\alpha}\right)e^{iu\cos\theta/\sin^2\alpha}\,d\theta. \end{aligned}\right\} \qquad (3\cdot 2)$$

We note that

$$I_n(-u,v) = I_n^*(u,v) \qquad (n = 0,1,2), \qquad (3\cdot 3)$$

where the asterisk denotes the complex conjugate. From (2·30), (2·31) and (3·3) we note the following relations which exist between the components of the field vectors at any two points $P_1(u,v,\phi)$ and $P_2(-u,v,\phi)$, which are symmetrically situated with respect to the focal plane:

$$\left.\begin{aligned} e_x(-u,v,\phi) &= -e_x^*(u,v,\phi), & h_x(-u,v,\phi) &= -h_x^*(u,v,\phi),\\ e_y(-u,v,\phi) &= -e_y^*(u,v,\phi), & h_y(-u,v,\phi) &= -h_y^*(u,v,\phi),\\ e_z(-u,v,\phi) &= e_z^*(u,v,\phi), & h_z(-u,v,\phi) &= h_z^*(u,v,\phi). \end{aligned}\right\} \qquad (3\cdot 4)$$

If $|e_x|$ denotes the amplitude and ψ_x the phase of e_x, the first relation in (3·4) implies that

$$|e_x(-u,v,\phi)| = |e_x(u,v,\phi)|, \qquad (3\cdot 5a)$$

and‡

$$\psi_x(-u,v,\phi) = -\psi_x(u,v,\phi)+\pi \quad (\mathrm{mod}\ 2\pi). \qquad (3\cdot 5b)$$

Relation (3·5*a*) shows that for any two points which are symmetrically situated with respect to the focal plane, the amplitudes of the e_x's are the same, and (3·5*b*) shows that there exists also a simple relation between the phases. The appearance of the additive factor π on the right of (3·5*b*) is connected with the well-known *phase anomaly* at focus (cf. Born & Wolf 1959, § 8·8, (43)). From (3·4) we see that relations of the form (3·5*a*) holds also for all the other components. However, relation of the form (3·5*b*) holds for the phases of e_x, e_y, h_x and h_y only. For the phase of the remaining two components we have, in place of (3·5*b*), relations of the following form:

$$\psi_z(-u,v,\phi) = -\psi_z(u,v,\phi) \quad (\mathrm{mod}\ 2\pi). \qquad (3\cdot 5c)$$

† When the angular aperture is small ($\alpha \ll 1$), formulae (3·1) reduce to

$$\left.\begin{aligned} u &\sim k(a/R)^2 z,\\ v &\sim k(a/R)\surd(x^2+y^2), \end{aligned}\right\} \qquad (3\cdot 1b)$$

where $\sin\alpha \sim \tan\alpha = a/R$, a being the radius of the exit pupil and R the distance between the exit pupil and the focal plane.

‡ The quantity mod 2π on the right of an equation denotes that the two sides of the equation are indeterminate to the extent of an additive constant $2m\pi$ where m is any integer.

Thus *the components of* **e** *and* **h** *in the direction of the axis of revolution of the system have no phase anomaly.*

From (2·30) and (2·31) it is also seen that

$$\left.\begin{aligned} h_x(u,v,\phi) &= -e_y(u,v,\phi-\tfrac{1}{2}\pi),\\ h_y(u,v,\phi) &= e_x(u,v,\phi-\tfrac{1}{2}\pi),\\ h_z(u,v,\phi) &= e_z(u,v,\phi-\tfrac{1}{2}\pi). \end{aligned}\right\} \tag{3·6}$$

Hence in any fixed plane of observation (u = constant), the **e** and **h** fields are the same but are rotated with respect to each other by 90° around the z-axis; this, of course, might have been expected, since the **e** and **h** fields in the object space have this relationship, and the laws relating to the transmission of these fields through the system are the same.

3·1. *Polarization of the image field*

To examine the state of polarization of the image field, we separate the real and imaginary parts of the integrals (3·2) and write

$$I_n(u,v) = I_n^{(r)}(u,v) + \mathrm{i}I_n^{(i)}(u,v) \qquad (n = 0,1,2), \tag{3·7}$$

where $I_n^{(r)}$ and $I_n^{(i)}$ are real. We also write

$$\mathbf{e}(u,v,\phi) = \mathbf{p}(u,v,\phi) + \mathrm{i}\mathbf{q}(u,v,\phi), \tag{3·8}$$

where **p** and **q** are real vectors; they are a pair of conjugate semi-diameters of the polarization ellipse of the electric vector. According to (2·30), (3·7) and (3·8)

$$\left.\begin{aligned} p_x(u,v,\phi) &= A(I_0^{(i)} + I_2^{(i)}\cos 2\phi), & q_x(u,v,\phi) &= -A(I_0^{(r)} + I_2^{(r)}\cos 2\phi),\\ p_y(u,v,\phi) &= AI_2^{(i)}\sin 2\phi, & q_y(u,v,\phi) &= -AI_2^{(r)}\sin 2\phi,\\ p_z(u,v,\phi) &= -2AI_1^{(r)}\cos\phi, & q_z(u,v,\phi) &= -2AI_1^{(i)}\cos\phi. \end{aligned}\right\} \tag{3·9}$$

Since, according to (3·4) the integrals I_0, I_1 and I_2 are all real when $u = 0$ (focal plane), it follows that

$$\left.\begin{aligned} p_x(0,v,\phi) &= 0, & q_x(0,v,\phi) &= -A[I_0(0,v) + I_2(0,v)\cos 2\phi],\\ p_y(0,v,\phi) &= 0, & q_y(0,v,\phi) &= -AI_2(0,v)\sin 2\phi,\\ p_z(0,v,\phi) &= -2AI_1(0,v)\cos\phi, & q_z(0,v,\phi) &= 0. \end{aligned}\right\} \tag{3·10}$$

From (3·10) we see that in the focal plane $\mathbf{p}\,.\,\mathbf{q} = 0$, i.e. the conjugate semi-diameters are at right angles to each other; hence in the focal plane, **p** and **q** are the semi-axes of the polarization ellipse of the electric vector. Moreover, the **p**-axis is perpendicular to the focal plane and the **q**-axis lies in the focal plane. Thus, *the polarization ellipse of the electric vector at any point in the focal plane is at right angles to the focal plane.* The angle $\chi(v,\phi)$ $(-\tfrac{1}{2}\pi < \chi \leqslant \tfrac{1}{2}\pi)$ between the plane of the polarization ellipse and the plane $\phi = 0$ (the xz-plane) is given by

$$\tan\chi(v,0) = \frac{q_y(0,v,\phi)}{q_x(0,v,\phi)} = \frac{I_2(0,v)\sin 2\phi}{I_0(0,v) + I_2(0,v)\cos 2\phi}, \tag{3·11}$$

and the two axes of the polarization ellipse are in the ratio

$$\rho(v,\phi) = \frac{|p_z(0,v,\phi)|}{+\sqrt{\{q_x^2(0,v,\phi)+q_y^2(0,v,\phi)\}}}$$

$$= \frac{2\,|I_1(0,v)\cos\phi|}{+\sqrt{\{I_0^2(0,v)+I_2^2(0,v)+2I_0(0,v)\,I_2(0,v)\cos 2\phi\}}}. \tag{3·12}$$

Along the y-axis ($\phi = \frac{1}{2}\pi$ or $\frac{3}{2}\pi$) only one of the components in (3·10), namely q_x, is different from zero. Hence *at each point of the y-axis* and, in particular, at the focus itself, *the electric field is linearly polarized in the x-direction*, i.e. *in the direction*

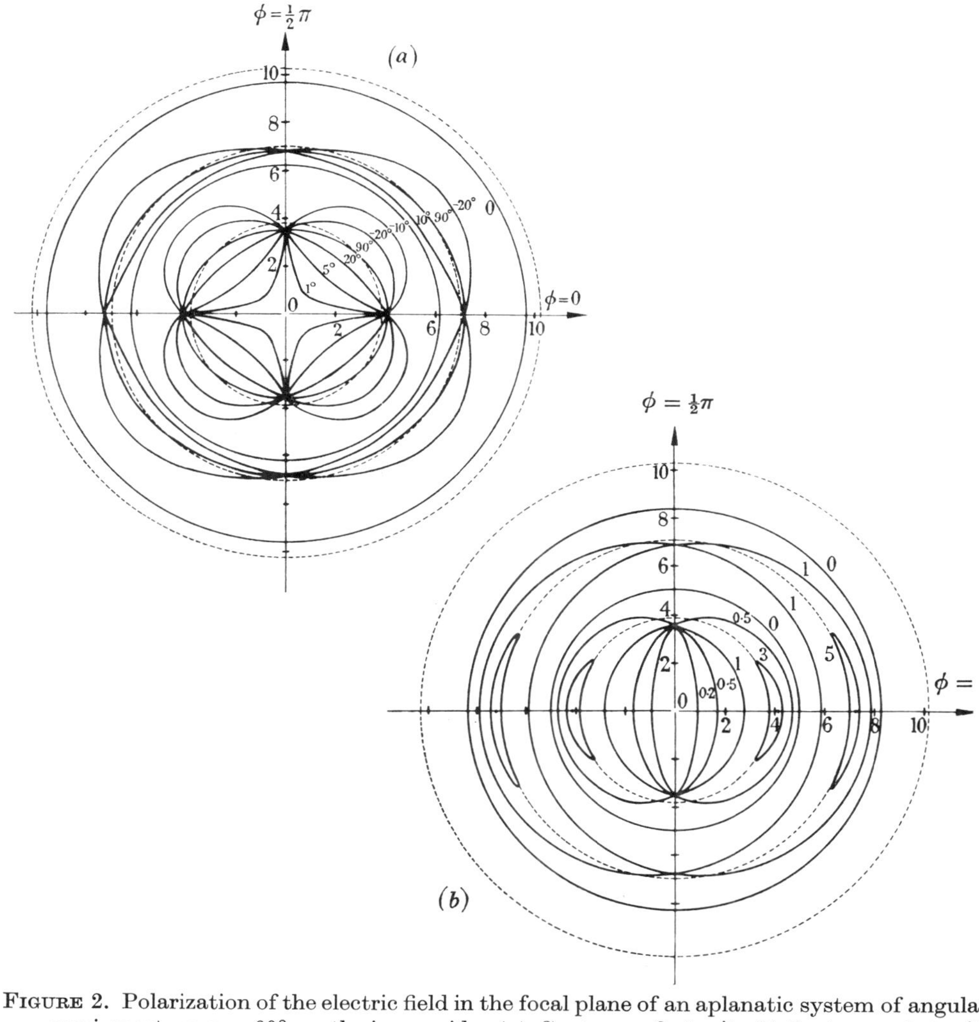

FIGURE 2. Polarization of the electric field in the focal plane of an aplanatic system of angular semi-aperture $\alpha = 60°$ on the image side: (a) Contours of $\chi(v,\phi)$; (b) Contours of $\rho(v,\phi)$. The field in the object space is linearly polarized with its electric vector in the azimuth $\phi = 0$. The field in the focal plane is, in general, elliptically polarized, with one axis (q) in the focal plane, and the other axis (p) perpendicular to the focal plane. $\chi(v,\phi)$ is the angle between the plane of the polarization ellipse of the electric vector and the meridional plane $\phi = 0$, and $\rho(v,\phi)$ is the ratio of the p and q axes. Radial distances are measured in v-units. The dashed lines indicate the dark rings of the Airy pattern represented by (4·12).

of vibrations of the electric field in the object space. Since **p** also vanishes when $I_1(0, v) = 0$, the electric field is also linearly polarized along circles centred on the focus, whose radii (measured in v-units) are given by the roots of this equation; however, along these circles, the direction of vibrations are not in general in the x-direction.

The behaviour of χ and ρ in the focal plane of an aplanatic system of angular semi-aperture $\alpha = 60°$ is shown in figure 2. The figure was computed from the formulae (3·11), (3·12) and (3·2) by means of the Manchester University Electronic Computer mark I. The details of these and other computations relating to this paper are given in a thesis by one of the authors (Richards 1956*b*).

Figure 2(*a*) shows that in the central nucleus of the image $\chi \sim 0°$, i.e. in this region the electric vector vibrates in planes nearly parallel to the plane which contains the direction of the electric vibrations in the object space and the axis of the system. According to figure 2(*b*) the axial component (e_z) is not negligible in all parts of the central nucleus.

The state of polarization of the field at points on the axis of the system ($v = 0$) is also of interest. Since $I_1(u, 0) = I_2(u, 0) = 0$ for all values of u, it follows from (2·30) that on the axis $e_y = e_z = 0$. Hence *the electric vector at each point on the axis of revolution in the image space is linearly polarized, and its direction is the same as the direction of the electric vector in the object space.*

3·2. *The energy density*

Let

$$\left.\begin{aligned} \langle w_e(u, v, \phi)\rangle &= \frac{1}{8\pi}\langle \mathbf{E}^2\rangle = \frac{1}{16\pi}(\mathbf{e}.\mathbf{e}^*), \\ \langle w_m(u, v, \phi)\rangle &= \frac{1}{8\pi}\langle \mathbf{H}^2\rangle = \frac{1}{16\pi}(\mathbf{h}.\mathbf{h}^*), \\ \langle w(u, v, \phi)\rangle &= \langle w_e(u, v, \phi)\rangle + \langle w_m(u, v, \phi)\rangle = \frac{1}{16\pi}(\mathbf{e}.\mathbf{e}^* + \mathbf{h}.\mathbf{h}^*), \end{aligned}\right\} \qquad (3·13)$$

denote the electric energy density, the magnetic energy density and the total energy density, each averaged over a time interval which is large compared to the basic period $T = 2\pi/\omega$. From (2·30) and (3·13), we have

$$\left.\begin{aligned} \langle w_e(u, v, \phi)\rangle &= \frac{A^2}{16\pi}\{|I_0|^2 + 4\,|I_1|^2\cos^2\phi + |I_2|^2 + 2\cos 2\phi\,\mathscr{R}(I_0 I_2^*)\}, \\ \langle w_m(u, v, \phi)\rangle &= \frac{A^2}{16\pi}\{|I_0|^2 + 4\,|I_1|^2\sin^2\phi + |I_2|^2 - 2\cos 2\phi\,\mathscr{R}(I_0 I_2^*)\}, \\ \langle w(u, v, \phi)\rangle &= \frac{A^2}{8\pi}\{|I_0|^2 + 2\,|I_1|^2 + |I_2|^2\}, \end{aligned}\right\} \qquad (3·14)$$

where $\mathscr{R}$ denotes the real part.

We note that because of (3·3) (as, of course, is also evident from (3·4))

$$\langle w_e(-u, v, \phi)\rangle = \langle w_e(u, v, \phi)\rangle, \qquad (3·15)$$

with similar relations involving $\langle w_m\rangle$ and $\langle w\rangle$. Hence *the distributions of the time-averaged electric energy density, magnetic energy density and total energy density are symmetrical about the focal plane* $u = 0$.

Further, we see that

$$\langle w_m(u,v,\phi)\rangle = \langle w_e(u,v,\phi-\tfrac{1}{2}\pi)\rangle. \tag{3·16}$$

Thus *the distribution of the time-averaged magnetic energy density is identical with the distribution of the time-averaged electric energy density, but the distributions are rotated with respect to each other by 90° about the axis of the system.* We also note that $\langle w\rangle$ is independent of ϕ, so that *the loci of constant time-averaged total energy density are surfaces of revolution about the axis of revolution of the system.*

Since, for all u, $I_1(u,0) = I_2(u,0) = 0$, we see from (3·14) that along *the axis of revolution* in the image space†

$$\langle w_e\rangle = \langle w_m\rangle = \tfrac{1}{2}\langle w\rangle = \frac{A^2}{16\pi}\,|I_0(u,0)|^2. \tag{3·17}$$

When $u = v = 0$, we have from (3·2)

$$\begin{aligned} I_0(0,0) &= \int_0^\alpha \cos^{\frac{1}{2}}\theta\,\sin\theta\,(1+\cos\theta)\,\mathrm{d}\theta \\ &= \tfrac{16}{15}\{1-\tfrac{5}{8}(\cos^{\frac{3}{2}}\alpha)(1+\tfrac{3}{5}\cos\alpha)\}, \end{aligned} \tag{3·18}$$

so that *at the focus* itself

$$\langle w_e\rangle = \langle w_m\rangle = \tfrac{1}{2}\langle w\rangle = \left(\frac{A}{15}\right)^2 \frac{16}{\pi}\{1-\tfrac{5}{8}(\cos^{\frac{3}{2}}\alpha)(1+\tfrac{3}{5}\cos\alpha)\}. \tag{3·19}$$

In figure 3, contours, computed from (3·14), of the time-averaged electric energy density in the focal planes of aplanatic systems of selected angular semi-apertures α are shown. The contours of the time-averaged magnetic energy density are, according to (3·16), identical with those for the time-averaged electric energy, but are rotated by 90° about the normal to the plane of the figure.

The first diagram in figure 3 represents the limiting case $\alpha \to 0$. We shall see later (§ 4) that in this case our solution reduces to that obtained on the basis of the usual scalar diffraction theory. Thus the first diagram is the ordinary *Airy diffraction pattern*: the contours are circles, the first zero contour being given by $v = 1·22\pi = 3·83$. As α increases, the pattern is seen to lose its rotational symmetry; the contours in the neighbourhood of the focus are then approximately elliptical, with their major axes in the direction of the electric vector in the object space ($\phi = 0$). Further away from the centre the contours are of a more complicated form and closed loops appear around certain points in the azimuths $\phi = 0, \tfrac{1}{2}\pi, \pi, \tfrac{3}{2}\pi$. For large values of α, the ellipticity of the contours becomes quite pronounced and the first minimum in the azimuth $\phi = 0$ is well outside the first zero of the Airy pattern, while in the azimuth $\phi = \tfrac{1}{2}\pi$ it is well inside. Hence, *if the wave in the object space is linearly polarized and detectors of electric energy* (e.g. *a photographic plate*) *are used, our solution predicts an increase in the resolving power in wide aperture systems for measurements in the azimuth at right angles to that of the electric vector of the incident wave.* This conclusion is in agreement with a prediction of Hopkins (1943, 1945) and appears to be supported by early experiments recorded by Carpenter (1901).

† Curves which illustrate the behaviour of these quantities along the axis of revolution are given in Richards (1956*a*, p. 358). This paper also contains curves which give the ratio (as function of α) between the values at focus computed from (3·19) and those computed by the application of the Huygens–Kirchhoff scalar diffraction theory.

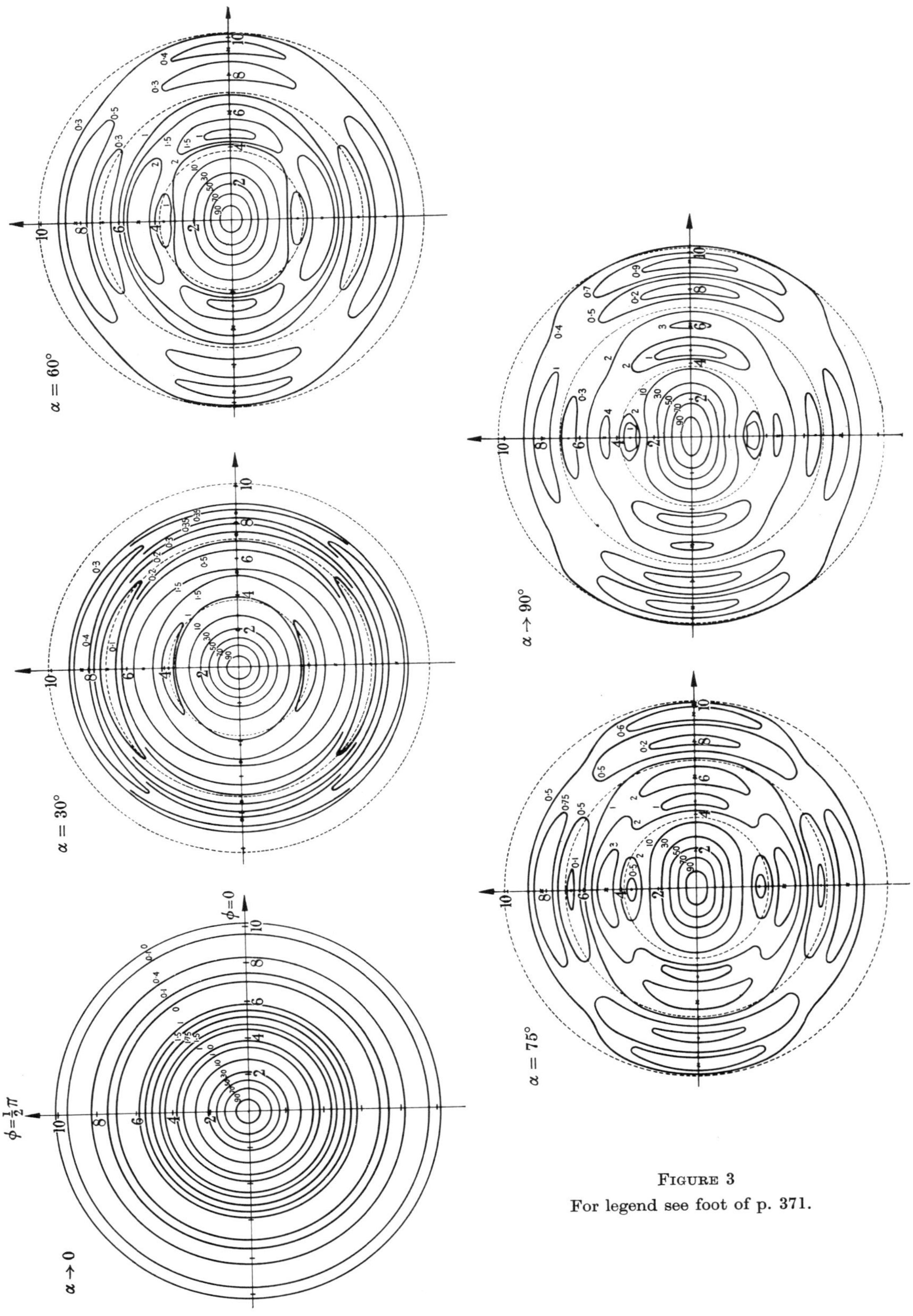

FIGURE 3
For legend see foot of p. 371.

The behaviour of the time-averaged electric energy density along the azimuths $\phi = 0$ and $\phi = \frac{1}{2}\pi$ in the focal plane is shown in detail in figure 4. It is seen that along the azimuth $\phi = 0$, the minima are not true zeros; the minima are, however, zeros along the azimuth $\phi = \frac{1}{2}\pi$. The behaviour of the time averaged total energy density, which as already noted is independent of the azimuth, is shown in figure 5*a*.

3·3. *The energy flow* (*Poynting vector*)

The time-averaged Poynting vector is given by

$$\langle \mathbf{S} \rangle = \frac{c}{4\pi} \langle \mathbf{E} \wedge \mathbf{H} \rangle = \frac{c}{8\pi} \mathscr{R}(\mathbf{e} \wedge \mathbf{h}^*). \tag{3·20}$$

On substituting from (2·30) and (2·31) into (3·20) we obtain the following expressions for the components of $\langle \mathbf{S} \rangle$:

$$\left.\begin{aligned} \langle S_x \rangle &= \frac{cA^2}{4\pi} \cos\phi \mathscr{I}\{I_1(I_2^* - I_0^*)\}, \\ \langle S_y \rangle &= \frac{cA^2}{4\pi} \sin\phi \mathscr{I}\{I_1(I_2^* - I_0^*)\}, \\ \langle S_z \rangle &= \frac{cA^2}{8\pi} \{|I_0|^2 - |I_2|^2\}, \end{aligned}\right\} \tag{3·21}$$

where $\mathscr{I}$ denotes the imaginary part.

We see that the magnitude of the time-averaged Poynting vector is given by

$$|\langle \mathbf{S} \rangle| = \frac{cA^2}{8\pi} \{(|I_0|^2 - |I_2|^2) + 4(\mathscr{I}[I_1(I_2^* - I_0^*)])^2\}^{\frac{1}{2}}. \tag{3·22}$$

Since this expression is independent of the azimuth ϕ, *the loci of constant* $|\langle \mathbf{S} \rangle|$ *are surfaces of revolution about the axis of the system.*

Along the axis of revolution in the image space ($v = 0$) (3·22) reduces to

$$|\langle \mathbf{S} \rangle| = \frac{cA^2}{8\pi} |I_0(u, 0)|^2; \tag{3·23}$$

on comparison with (3·17) we see that the relation $|\langle \mathbf{S} \rangle| = c\langle w \rangle$ holds at all points on the axis.

From (3·21) we readily deduce that *at each point in the image space the Poynting vector lies in the meridional plane which contains that point,* and that it makes an angle γ with the positive z-axis, given by

$$\left.\begin{aligned} \sin\gamma &= \frac{2|\mathscr{I}\{I_1(I_2^* - I_0^*)\}|}{[(|I_0|^2 - |I_2|^2)^2 + 4\{\mathscr{I}[I_1(I_2^* - I_0^*)]\}^2]^{\frac{1}{2}}}, \\ \cos\gamma &= \frac{|I_0|^2 - |I_2|^2}{[(|I_0|^2 - |I_2|^2)^2 + 4\{\mathscr{I}[I_1(I_2^* - I_0^*)]\}^2]^{\frac{1}{2}}}. \end{aligned}\right\} \tag{3·24}$$

FIGURE 3. Contours of the time-averaged electric energy density $\langle w_e \rangle$ in the focal plane of an aplanatic system of angular semi-aperture α on the image side.

The field in the object space is linearly polarized with its electric vector in the azimuth $\phi = 0$. The crosses, dots and circles indicate maxima, minima and saddle points respectively. Radial distances are measured in v-units. The values are normalized to 100 at the focus. The first figure, which represents the limiting case $\alpha \to 0$, is identical with the classical intensity pattern of Airy, characterized by (4·12). The dashed circles in the other figures represent the dark rings in the Airy pattern.

Using the relations (3·3) we see from (3·22) that

$$|\langle \mathbf{S}(-u, v, \phi)\rangle| = |\langle \mathbf{S}(u, v, \phi)\rangle|, \tag{3·25}$$

i.e. *the magnitude of the time-averaged Poynting vector is symmetric with respect to the focal plane.* Further, from (3·24), we have

$$\sin\{\gamma(-u, v, \phi)\} = -\sin\{\gamma(u, v, \phi)\}, \quad \cos\{\gamma(-u, v, \phi)\} = \cos\{\gamma(u, v, \phi)\},$$

so that, for $u \neq 0$,

$$\gamma(-u, v, \phi) = 2\pi - \gamma(u, v, \phi). \tag{3·26}$$

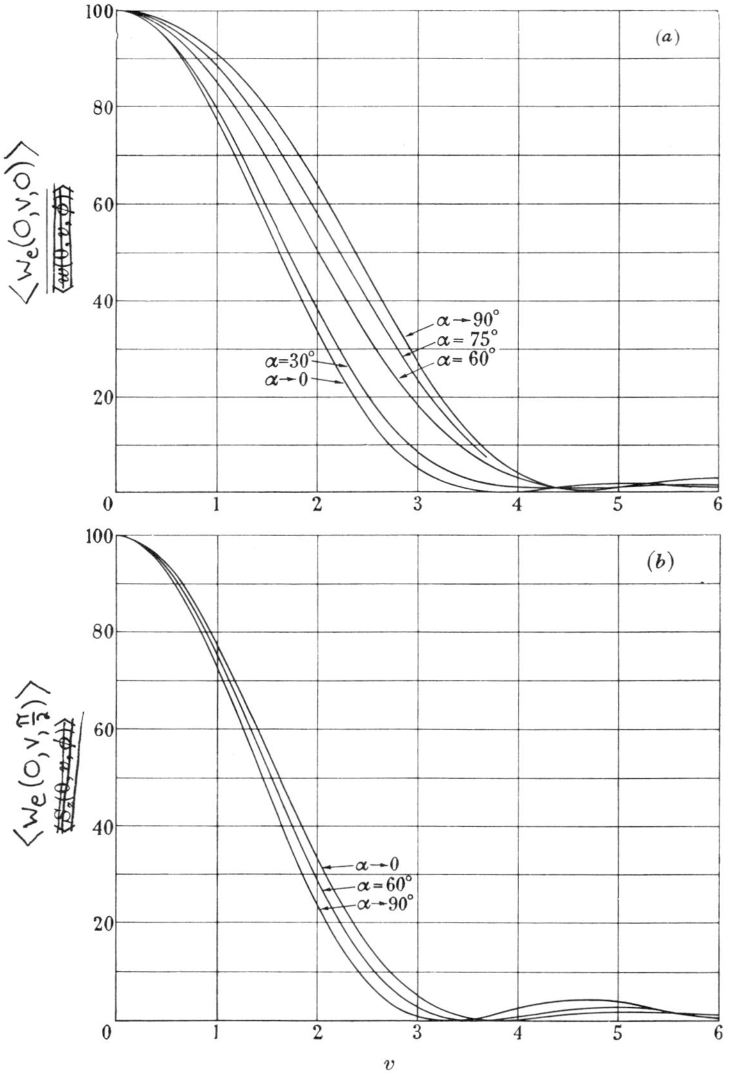

FIGURE 4. The variation of the time-averaged electric energy $\langle w_e \rangle$ along the two principal meridional sections of the focal plane in an aplanatic system of selected angular semi-aperture α in the image space: (*a*) Meridional section $\phi = 0$; (*b*) meridional section $\phi = \frac{1}{2}\pi$.

The field in the object space is linearly polarized with its electric vector in the azimuth $\phi = 0$. The values are normalized to 100 at the focus. The curves representing the limiting case $\alpha \to 0$ are identical with the classical Airy intensity curve given by (4·12).

When $u = 0$, we have from (3·21), since the integrals $I_n(0, v)$ $(n = 0, 1, 2)$ are all real,

$$\langle S_x \rangle = \langle S_y \rangle = 0, \tag{3·27}$$

i.e. *at any point in the focal plane, the time-averaged energy flow*, as represented by the Poynting vector, *is perpendicular to the focal plane.* Moreover,

$$\left.\begin{aligned} \langle S_z \rangle > 0 \quad &\text{if} \quad |I_0(0, v)| > |I_2(0, v)|, \\ \text{and} \qquad \langle S_z \rangle < 0 \quad &\text{if} \quad |I_0(0, v)| < |I_2(0, v)|. \end{aligned}\right\} \tag{3·28}$$

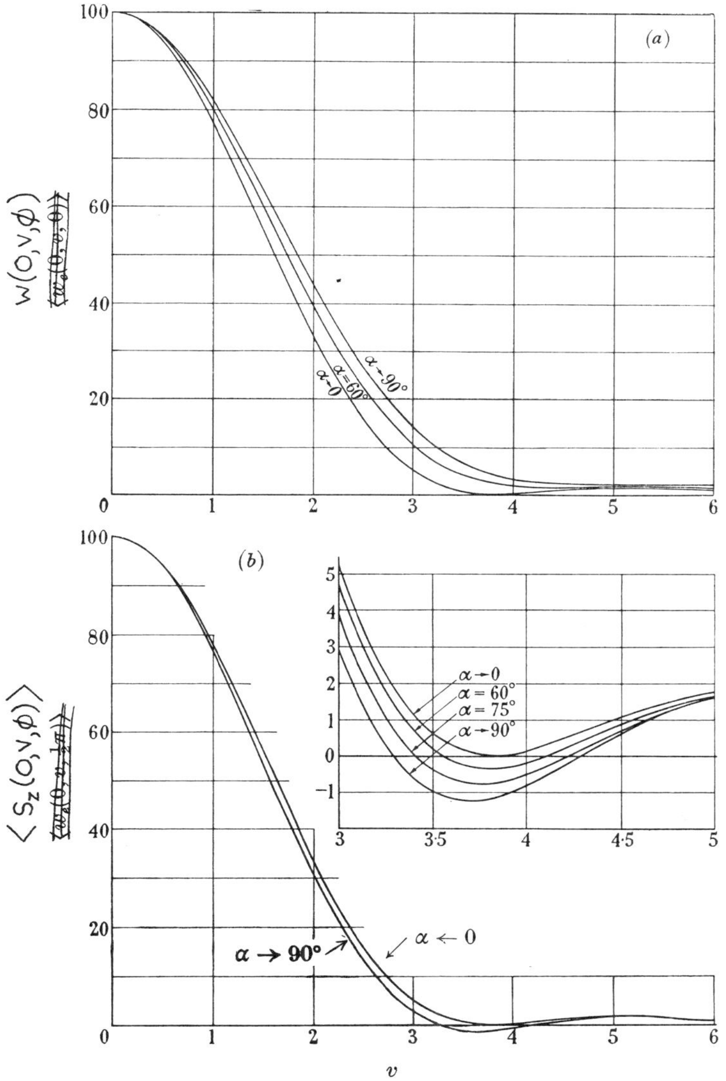

FIGURE 5. The variation of the time-averaged total energy density (a) and of the only non-vanishing component of the time-averaged energy flow (b) along any meridional section ϕ = const. of the focal plane of an aplanatic system of angular semi-aperture α on the image side. The curves represent the case of polarized as well as unpolarized wave. The curves in (a) also represent the variation of the time-averaged electric and magnetic energy densities of an unpolarized wave. The values are normalized to 100 at focus.

The second line implies that at points in the focal plane where $|I_2(0, v)| > |I_0(0, v)|$, the energy flow is directed back towards the object space.†

The computed curves, which show the behaviour of $\langle S_z \rangle$ in the focal plane of aplanatic systems of selected angular semi-apertures α are shown in figure 5. It is of interest to note that these (normalized) curves do not substantially depend on α. In particular, in the region of the Rayleigh limit, where the curves do not drop below 0·8 of their respective maxima, the differences (for any particular v-value in this region) are less than 1 %, in the full range ($0 \leqslant \alpha \leqslant 90°$) of the angular semi-aperture.

4. Approximate form of the solution for systems with a small angular aperture

The structure of images in systems with small angular aperture is usually investigated on the basis of a scalar diffraction theory. The intensity distribution in the focal plane of an aberration-free system of revolution which images a point source was first determined in this way by Airy (1835). His analysis was extended by Lommel (1885) and Struve (1886), who determined the distribution in the whole three-dimensional neighbourhood of the focus. It is of interest to examine the form which our solution takes when the angular semi-aperture α is small, and to compare it with these early solutions.

When α is small enough and u and v are not large compared with unity, we may replace the trigonometric factors in the amplitudes of the integrands in I_0, I_1 and I_2 in (3·2) by the leading term in their series expansions. In the exponents of the integrands we retain both the first and the second term in the expansion of $\cos\theta$ since, when u is not small compared to unity, the contribution of the second term is evidently also significant. The integrals then become

$$\left.\begin{aligned} I_0(u,v) &= 2\,\mathrm{e}^{\mathrm{i}u/\alpha^2}\int_0^\alpha \theta J_0\left(\frac{v\theta}{\alpha}\right)\mathrm{e}^{-\mathrm{i}u\theta^2/2\alpha^2}\,\mathrm{d}\theta,\\ I_1(u,v) &= \mathrm{e}^{\mathrm{i}u/\alpha^2}\int_0^\alpha \theta^2 J_1\left(\frac{v\theta}{\alpha}\right)\mathrm{e}^{-\mathrm{i}u\theta^2/2\alpha^2}\,\mathrm{d}\theta,\\ I_2(u,v) &= \tfrac{1}{2}\mathrm{e}^{\mathrm{i}u/\alpha^2}\int_0^\alpha \theta^3 J_2\left(\frac{v\theta}{\alpha}\right)\mathrm{e}^{-\mathrm{i}u\theta^2/2\alpha^2}\,\mathrm{d}\theta. \end{aligned}\right\} \tag{4·1}$$

For small x, $J_n(x) \sim x^n$ and we see that I_1 and I_2 are of lower order in α than I_0, so that these integrals may be neglected in comparison with I_0. Hence (2·30) and (2·31) reduce to

$$e_x = h_y = -\mathrm{i}AI_0, \tag{4·2}$$

$$e_y = e_z = h_x = h_z = 0. \tag{4·3}$$

Thus *in a system with a small angular aperture, the field in the image region is effectively linearly polarized; and the directions of the two field vectors in the image region are the same as their directions in the object space.* We see that in this case the image field is completely specified by one (complex) scalar wave (e_x or h_y), which is represented by the integral I_0 alone.

† This interesting fact was already deduced by Ignatowsky (1919).

To evaluate I_0 we introduce a new variable $\rho = \theta/\alpha$ and obtain

$$I_0(u, v) = 2\alpha^2 \, \mathrm{e}^{\mathrm{i}u/\alpha^2} \int_0^1 \rho J_0(v\rho) \, \mathrm{e}^{-\frac{1}{2}\mathrm{i}u\rho^2} \, \mathrm{d}\rho. \tag{4·4}$$

The integral on the right may be evaluated in terms of two of the Lommel functions

$$U_n(u, v) = \sum_{s=0}^{\infty} (-1)^s \left(\frac{u}{v}\right)^{n+2s} J_{n+2s}(v), \tag{4·5}$$

introduced by Lommel in the analysis already referred to. In fact†

$$\int_0^1 \rho J_0(v\rho) \, \mathrm{e}^{\frac{1}{2}\mathrm{i}u\rho^2} \, \mathrm{d}\rho = \frac{1}{u} \, \mathrm{e}^{-\frac{1}{2}\mathrm{i}u} \, [U_1(u, v) + \mathrm{i}U_2(u, v)]. \tag{4·6}$$

From (4·2), (4·4) and (4·6) we obtain the following expressions for the only two non-vanishing field components in the image region:

$$e_x = h_y = -2\mathrm{i}A\alpha^2 \frac{\exp\left\{\mathrm{i}u\left(\frac{1}{\alpha^2} - \frac{1}{2}\right)\right\}}{u} [U_1(u, v) + \mathrm{i}U_2(u, v)]. \tag{4·7}$$

The expressions (3·14) for the energy densities now become

$$\langle w_e \rangle = \langle w_m \rangle = \tfrac{1}{2}\langle w \rangle = \frac{A^2}{16\pi} |I_0|^2$$
$$= \frac{A^2\alpha^4}{4\pi} \frac{1}{u^2} [U_1^2(u, v) + U_2^2(u, v)]. \tag{4·8}$$

Formulae (3·21) for the components of the Poynting vector become

$$\langle S_x \rangle = \langle S_y \rangle = 0,$$
$$\langle S_z \rangle = \frac{cA^2}{8\pi} |I_0|^2 = \frac{cA^2\alpha^4}{2\pi} \frac{1}{u^2} [U_1^2(u, v) + U_2^2(u, v)]. \tag{4·9}$$

Thus, *when α is small, the energy flow*, as represented by the time-averaged Poynting vector, *is in the direction of the positive z-axis.* On comparison of (4·8) and (4·9) we also see that the relation

$$|\langle \mathbf{S} \rangle| = c\langle w \rangle \tag{4·10}$$

then holds everywhere in the image region.

The expressions on the right of (4·8) and (4·9) are proportional to the classical solution of Lommel and Struve for the 'intensity' in the image region of an aberration-free system.

† See Lommel (1885) or Born & Wolf (1959, § 8·8).

The U_n-series converge for all u and v values, but are convenient for computations only when $|u/v| < 1$. When $|u/v| > 1$, equation (4·4) may be evaluated in terms of two of the Lommel's V_n-functions defined by

$$V_n(u, v) = \sum_{s=0}^{\infty} (-1)^s (v/u)^{n+2s} J_{n+2s}(v).$$

The U and the V functions are related by the formula [cf. Watson 1952, p. 537]

$$U_{n+1}(u, v) - (-1)^{n-1} V_{n-1}(u, v) = \sin\{\tfrac{1}{2}(u + v^2/u - n\pi)\}.$$

When $u = 0$ (focal plane) we have from (4·7), since

$$\lim_{u\to 0}\frac{U_1(u,v)}{u} = \frac{J_1(v)}{v}, \qquad \lim_{u\to 0}\frac{U_2(u,v)}{u} = 0,$$

$$e_x = h_y = -\mathrm{i}A\alpha^2\left(\frac{2J_1(v)}{v}\right), \tag{4·11}$$

and (4·8) and (4·9) become

$$\langle w_e\rangle = \langle w_m\rangle = \tfrac{1}{2}\langle w\rangle = \frac{1}{2c}|\langle \mathbf{S}\rangle| = \frac{A^2\alpha^4}{16\pi}\left(\frac{2J_1(v)}{v}\right)^2. \tag{4·12}$$

The expression on the right of (4·12) is proportional to the classical solution of Airy for the 'intensity' in the image plane of an aberration-free system.

5. Images formed by an unpolarized wave

So far we have assumed that the wave entering the system is linearly polarized. We shall now briefly consider the case when the wave is unpolarized.

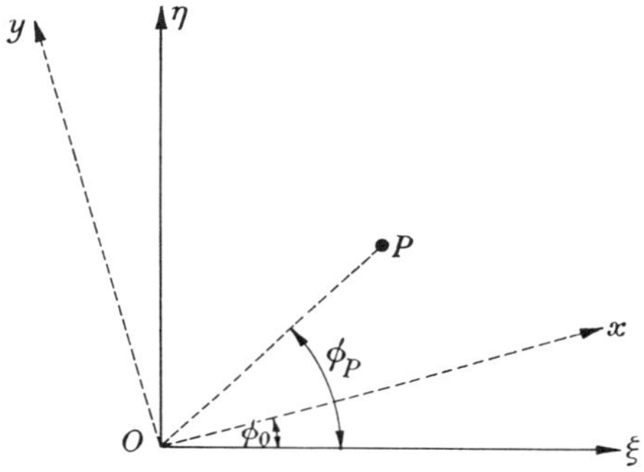

Figure 6. Change of reference system. Section at right angles to the axis of revolution.

An unpolarized incident wave may become partially polarized on refraction (or on reflexion) at the successive surfaces of the system. However, if the angles of incidence at each surface are small, as is assumed, the degree of polarization introduced in this way will also be small and we may, therefore, consider the wave emerging from the system to be effectively unpolarized. If further we assume that the wave is quasi-monochromatic, with mean angular frequency ω, the expressions for the time-averaged energy densities and energy flow may be obtained by averaging the corresponding expressions relating to the polarized wave over all possible states of polarization.

To carry out this averaging it is convenient to choose a new set of co-ordinate axes $O\xi$, $O\eta$ at right angles to each other and to the axis of revolution. Let ϕ_0 be the angle between $O\xi$ and the direction Ox of the electric vibrations of the incident wave, and let ϕ_P be the angle between $O\xi$ and the azimuth which contains the point P of observation (see figure 6). Then the first formula in (3·14) for the time-averaged electric energy density at a point P of observation becomes, since $\phi = \phi_P - \phi_0$,

$$\langle w_e\rangle = (A^2/16\pi)\{|I_0|^2+4\,|I_1|^2\cos^2(\phi_P-\phi_0)+|I_2|^2+2\mathscr{R}(I_0I_2^*)\cos[2(\phi_P-\phi_0)]\}. \tag{5·1}$$

If we denote by vertical bars an average over all possible values of ϕ_0 ($0 \leqslant \phi_0 < 2\pi$) it follows that when the incident wave is unpolarized the time-averaged electric energy density at P is†

$$\overline{\langle w_e \rangle} = \frac{1}{2\pi}\int_0^{2\pi} \langle w_e \rangle \, d\phi = \frac{A^2}{16\pi}\{|I_0|^2 + 2\,|I_1|^2 + |I_2|^2\}. \tag{5·2}$$

We note that $\overline{\langle w_e \rangle}$ is independent of ϕ_P so that the loci of the constant averaged electric energy are surfaces of revolution about the axis of the system, as was to be expected from symmetry. In a strictly similar manner we obtain expressions for $\overline{\langle w_m \rangle}$ and $\overline{\langle w \rangle}$, and we have in all

$$\overline{\langle w_e \rangle} = \overline{\langle w_m \rangle} = \tfrac{1}{2}\overline{\langle w \rangle} = (A^2/16\pi)\{|I_0|^2 + 2\,|I_1|^2 + |I_2|^2\}. \tag{5·3}$$

Comparison with the last expression in (3·14) shows that each term in (5·3) is also equal to one-half of the total energy density of the polarized wave. The distribution of this quantity in the focal plane has already been given in figure 5.

Next consider the energy flow. It is again convenient to transform first the expressions for the polarized case so that the co-ordinates which appear in them are referred to the fixed axes $O\xi$, $O\eta$, independent of the direction of polarization of the incident wave. Let $\langle S_\xi \rangle$ and $\langle S_\eta \rangle$ be the components of the time-averaged Poynting vector in the direction $O\xi$, $O\eta$. Then

$$\left.\begin{aligned} \langle S_\xi \rangle &= \langle S_x \rangle \cos\phi_0 - \langle S_y \rangle \sin\phi_0, \\ \langle S_\eta \rangle &= \langle S_x \rangle \sin\phi_0 + \langle S_y \rangle \cos\phi_0. \end{aligned}\right\} \tag{5·4}$$

On substituting for $\langle S_x \rangle$ and $\langle S_y \rangle$ from (3·21), where again ϕ is replaced by $\phi_P - \phi_0$ we find that $\langle S_\xi \rangle$ and $\langle S_\eta \rangle$ are given by the same formulae as $\langle S_x \rangle$ and $\langle S_y \rangle$ but with ϕ_P written in place of ϕ. Since these formulae are independent of ϕ_0, they apply not only to the polarized wave but also to the unpolarized one, i.e.

$$\overline{\langle S_\xi \rangle} = \langle S_\xi \rangle = (cA^2/4\pi)\,\mathscr{I}\{I_1(I_2^* - I_0^*)\}\cos\phi_P, \tag{5·5a}$$

$$\overline{\langle S_\eta \rangle} = \langle S_\eta \rangle = (cA^2/4\pi)\,\mathscr{I}\{I_1(I_2^* - I_0^*)\}\sin\phi_P. \tag{5·5b}$$

The remaining component, in the direction of the axis of revolution is, as before, given by the last formulae in (3·21),

$$\overline{\langle S_z \rangle} = \langle S_z \rangle = (cA^2/8\pi)\{|I_0|^2 - |I_2|^2\}. \tag{5·5c}$$

It immediately follows that the results expressed by (3·22), (3·23) and (3·24) hold also for the unpolarized wave.

When the angular semi-aperture α is sufficiently small ($\alpha \to 0$) one may, as in § 4, neglect I_1 and I_2 in comparison with I_0. The above formulae then become, when (4·4) and (4·6) are used:

$$\overline{\langle w_e \rangle} = \overline{\langle w_m \rangle} = \tfrac{1}{2}\overline{\langle w \rangle} = \frac{A^2\alpha^4}{4\pi}\frac{1}{u^2}[U_1^2(u,v) + U_2^2(u,v)], \tag{5·6}$$

$$\overline{\langle S_\xi \rangle} = \overline{\langle S_\eta \rangle} = 0, \tag{5·7a}$$

$$\overline{\langle S_z \rangle} = \frac{cA^2\alpha^4}{2\pi}\frac{1}{u^2}[U_1^2(u,v) + U_2^2(u,v)]. \tag{5·7b}$$

† The factor l_0 entering the constant $A = \pi f l_0/\lambda$ may now be related to the averaged electric energy density $\overline{\langle w_e \rangle}_i$ of the incident wave in the object space by the formula

$$\overline{\langle w_e \rangle}_i = (1/16\pi)(\mathbf{e}_0 \cdot \mathbf{e}_0^*) = (1/16\pi)\,l_0^2.$$

Here U_1 and U_2 are again two of the functions defined by (4·5). For points of observation in the focal plane ($u = 0$) these formulae reduce to

$$\overline{\langle w_e \rangle} = \overline{\langle w_m \rangle} = \tfrac{1}{2}\overline{\langle w \rangle} = \frac{1}{2c}\,|\langle \mathbf{S} \rangle| = \frac{A^2\alpha^4}{16\pi}\left(\frac{2J_1(v)}{v}\right)^2. \tag{5·8}$$

As already noted in connexion with the corresponding formulae (4·8), (4·9) and (4·12) the quantities on the right-hand sides of (5·6) and (5·7 *b*) are proportional to the classical solution of Lommel and Struve for the 'intensity' in the image region

TABLE 1. COMPARISON OF DATA RELATING TO THE STRUCTURE OF THE ELECTROMAGNETIC FIELD IN THE FOCAL PLANE ($u = 0$) OF APLANATIC SYSTEMS FOR SELECTED VALUES OF ANGULAR SEMI-APERTURE α ON THE IMAGE SIDE

$\langle w_e(u,v,\phi)\rangle$ and $\langle w_m(u,v,\phi)\rangle$ represent the time-averaged electric energy density and magnetic energy density respectively. $\langle S_z(u,v,\phi)\rangle$ represents the only non-vanishing component (in the direction of the axis of the system) of the time-averaged energy flow. All the quantities are normalized to 100 at the focus $u = v = 0$.

Quantities with a vertical bar refer to an unpolarized wave, the others to a linearly polarized wave in the object space, with its electric vector in the azimuth $\phi = 0$. The third row in each division of the table also represents one-half of the time-averaged total energy density $\langle w\rangle$, i.e.

$$\overline{\langle w_e \rangle} = \overline{\langle w_m \rangle} = \tfrac{1}{2}\overline{\langle w \rangle} = \tfrac{1}{2}\langle w \rangle.$$

The entries in the column $\alpha \to 0$ are identical with values given by the intensity formulae (4·12) of Airy.

value	quantity	v				
		$\alpha \to 0$	$\alpha = 30°$	$\alpha = 60°$	$\alpha = 75°$	$\alpha \to 90°$
80	$\langle w_e(0,v,0)\rangle = \langle w_m(0,v,\frac{1}{2}\pi)\rangle$	0·93	0·98	1·16	1·33	1·48
	$\langle w_e(0,v,\frac{1}{2}\pi)\rangle = \langle w_m(0,v,0)\rangle$	0·93	0·93	0·89	0·86	0·84
	$\overline{\langle w_e(0,v,\phi)\rangle} = \overline{\langle w_m(0,v,\phi)\rangle}$	0·93	0·95	1·00	1·03	1·05
	$\langle S_z(0,v,\phi)\rangle = \overline{\langle S_z(0,v,\phi)\rangle}$	0·93	0·93	0·93	0·93	0·91
60	$\langle w_e(0,v,0)\rangle = \langle w_m(0,v,\frac{1}{2}\pi)\rangle$	1·40	1·47	1·73	1·95	2·10
	$\langle w_e(0,v,\frac{1}{2}\pi)\rangle = \langle w_m(0,v,0)\rangle$	1·40	1·38	1·32	1·30	1·26
	$\overline{\langle w_e(0,v,\phi)\rangle} = \overline{\langle w_m(0,v,\phi)\rangle}$	1·40	1·42	1·50	1·55	1·58
	$\langle S_z(0,v,\phi)\rangle = \overline{\langle S_z(0,v,\phi)\rangle}$	1·40	1·40	1·39	1·38	1·36
40	$\langle w_e(0,v,0)\rangle = \langle w_m(0,v,\frac{1}{2}\pi)\rangle$	1·83	1·94	2·27	2·50	2·63
	$\langle w_e(0,v,\frac{1}{2}\pi)\rangle = \langle w_m(0,v,0)\rangle$	1·83	1·82	1·76	1·70	1·64
	$\overline{\langle w_e(0,v,\phi)\rangle} = \overline{\langle w_m(0,v,\phi)\rangle}$	1·83	1·87	1·99	2·05	2·10
	$\langle S_z(0,v,\phi)\rangle = \overline{\langle S_z(0,v,\phi)\rangle}$	1·83	1·83	1·82	1·82	1·79
20	$\langle w_e(0,v,0)\rangle = \langle w_m(0,v,\frac{1}{2}\pi)\rangle$	2·36	2·50	2·93	3·14	3·23
	$\langle w_e(0,v,\frac{1}{2}\pi)\rangle = \langle w_m(0,v,0)\rangle$	2·36	2·33	2·22	2·16	2·10
	$\overline{\langle w_e(0,v,\phi)\rangle} = \overline{\langle w_m(0,v,\phi)\rangle}$	2·36	2·41	2·56	2·66	2·73
	$\langle S_z(0,v,\phi)\rangle = \overline{\langle S_z(0,v,\phi)\rangle}$	2·36	2·36	2·34	2·33	2·27
1st min.	$\langle w_e(0,v,0)\rangle = \langle w_m(0,v,\frac{1}{2}\pi)\rangle$	3·83	4·15	4·75	4·75	4·70
	$\langle w_e(0,v,\frac{1}{2}\pi)\rangle = \langle w_m(0,v,0)\rangle$	3·83	3·77	3·56	3·40	3·29
	$\overline{\langle w_e(0,v,\phi)\rangle} = \overline{\langle w_m(0,v,\phi)\rangle}$	3·83	3·95	—	—	—
	$\langle S_z(0,v,\phi)\rangle = \overline{\langle S_z(0,v,\phi)\rangle}$	3·83	3·80	3·80	3·76	3·70
2nd min.	$\langle w_e(0,v,0)\rangle = \langle w_m(0,v,\frac{1}{2}\pi)\rangle$	7·02	7·03	8·10	8·10	8·00
	$\langle w_e(0,v,\frac{1}{2}\pi)\rangle = \langle w_m(0,v,0)\rangle$	7·02	6·95	6·85	6·70	6·55
	$\overline{\langle w_e(0,v,\phi)\rangle} = \overline{\langle w_m(0,v,\phi)\rangle}$	7·02	—	—	—	—
	$\langle S_z(0,v,\phi)\rangle = \overline{\langle S_z(0,v,\phi)\rangle}$	7·02	7·02	7·02	7·02	6·95

of an aberration-free system, and the expression on the right-hand side of (5·8) is proportional to the classical solution of Airy for the 'intensity' in the image plane.

Let us denote by suffix zero expressions (such as (5·6)) which refer to a low aperture system ($\alpha \to 0$). Then from (4·8), (4·9), (5·3), (5·6) and (5·7) the following relations are seen to hold:

$$\langle S_z\rangle_0 = \overline{\langle S_z\rangle}_0 = c\overline{\langle w\rangle}_0 \leqslant c\overline{\langle w\rangle} = c\langle w\rangle, \tag{5·9}$$

$$\overline{\langle w_e\rangle}_0 = \overline{\langle w_m\rangle}_0 = \tfrac{1}{2}\overline{\langle w\rangle}_0 \leqslant \tfrac{1}{2}\overline{\langle w\rangle} = \overline{\langle w_m\rangle} = \overline{\langle w_e\rangle}. \tag{5·10}$$

In practice, detectors of electric energy are usually used. The relation $\langle w_e\rangle_0 \leqslant \overline{\langle w_e\rangle}$ implies that the pattern recorded in any particular receiving plane (w = const.) will then be broader in a system with a wide angular aperture than in one with a low angular aperture. For patterns in the focal plane, this effect is seen in figure 5*a* which also shows that $\overline{\langle w_e\rangle}$ has no exact zeros, nor pronounced subsidiary maxima.

Finally, some of the main data which relate to the structure of the image in the focal plane of an aplanatic system are summarized in table 1. Data relating to both polarized and unpolarized incident waves are given.

The very extensive calculations on which the diagrams and the table in this paper are based were carried out on the Manchester University Electronic Computer mark I. We are indebted to Mr R. A. Brooker for helpful advice on computational techniques. We are also obliged to Miss B. Wood for help with construction of the contour diagrams shown in figures 2 and 3.

One of us (B. R.) wishes to acknowledge the award of a grant from the National Research and Development Corporation; the other (E. W.) is indebted to Manchester University for the award of an Imperial Chemistry Industries Research Fellowship during the tenure of which the main part of this work was carried out. The investigation was completed when he was a guest at the Institute of Mathematical Sciences (Division of Electromagnetic Research), New York University and was partially supported by the U.S. Air Force Cambridge Research Center, Air Research and Development Command, under contract No. AF 19 (604) 5238.

References

Airy, G. B. 1835 *Trans. Camb. Phil. Soc.* **5**, 283.
Born, M. & Wolf, E. 1959 *Principles of optics.* London: Pergamon Press.
Burtin, R. 1956 *Optica Acta*, **3**, 104.
Carpenter, W. B. 1901 *The microscope and its revelations*, 8th ed., p. 381. London: Churchill.
Focke, J. 1957 *Optica Acta*, **4**, 124.
Hopkins, H. H. 1943 *Proc. Phys. Soc.* **55**, 116.
Hopkins, H. H. 1945 *Nature, Lond.* **155**, 275.
Ignatowsky, V. S. 1919 *Trans. Opt. Inst. Petrograd*, vol. I, paper IV.
Ignatowsky, V. S. 1920 *Trans. Opt. Inst. Petrograd*, vol. I, paper V.
Lommel, E. 1885 *Abh. bayer. Akad. Wiss.* **15**, 233.
Richards, B. 1956*a* Contribution to *Symposium on astronomical optics and related subjects* (ed. Z. Kopal), p. 352. Amsterdam: North Holland Publishing Company.
Richards, B. 1956*b* Ph.D. Thesis, University of Manchester.
Richards, B. & Wolf, E. 1956 *Proc. Phys. Soc.* B, **69**, 854.
Struve, H. 1886 *Mém. Acad. St. Petersbourgh* (7), **34**, 1.
Watson, G. N. 1952 *Treatise on the theory of Bessel functions*, 2nd ed. Cambridge Univ. Press.
Wolf, E. 1959 *Proc. Roy. Soc.* A, **253**, 349.

PRINTED IN GREAT BRITAIN AT THE UNIVERSITY PRESS, CAMBRIDGE
(BROOKE CRUTCHLEY, UNIVERSITY PRINTER)

Reprinted from THE PHYSICAL REVIEW, Vol. 138, No. 6B, B1561–B1565, 21 June 1965
Printed in U. S. A.

Electromagnetic Field in the Neighborhood of the Focus of a Coherent Beam*

A. BOIVIN
Department of Physics, Laval University, Quebec, Canada

AND

E. WOLF
Department of Physics and Astronomy, University of Rochester, Rochester, New York

(Received 23 November 1964; revised manuscript received 17 February 1965)

An integral representation for the electromagnetic field in the region of focus of a coherent light beam that emerges from an aplanatic optical system has been derived by Ignatowsky (1919) and by Richards and Wolf (1959). In the present paper this representation is used to analyze the structure of the focal region in a typical case. Contours of the time-averaged electric energy density in the focal plane, in one defocused plane and in two meridional sections of the focal region of a system with angular semi-aperture 45° are presented. The meridional diagrams refer to axial sections through a cylindrical region around the axis near focus, of length 16λ and cross-sectional diameter 10λ, where λ is the wavelength of the light. It is found that the field has a strong longitudinal component at certain points of the focal plane and that longitudinal electric field strengths of the order of 10^5 V/cm could now be attained with focused laser beams. A diagram illustrating the complete behavior of the longitudinal component in the focal plane is also given.

1. INTRODUCTION

THE problem of determining the structure of the electromagnetic field in the region of focus of a coherent light beam has become of considerable interest, because of the possibility of producing very strong fields (of the order of several hundred thousands V/cm) by focusing the output of an optical maser. Such strong fields are being utilized, for example, for the generation of optical harmonics and for the study of high-energy scattering. It is evident that in order to utilize fully the potentialities promised by these developments, it is desirable to have some knowledge of the structure of the electromagnetic field in the focal region.

Early research on the structure of the focal region was based on the scalar-diffraction theory.[1] This work has proved very valuable in applications to instrumental optics. However, the results are not applicable to wide-aperture systems and, in any case, they do not exhibit vectorial features of the field, such as the state of polarization and the behavior of the Poynting vector. Investigations which take into account vectorial properties of the field are due to Ignatowsky,[2] Hopkins,[3] Burtin,[4] Focke[5] and Richards and Wolf.[6,7]

In this paper we present results of extensive computations relating to the field in the focal region of a monochromatic beam emerging from an aplanatic optical system. The computations are based on the vectorial integral representation of the field derived in Refs. 2 and 7. Some related results, both analytic and computational can also be found in these two references.

2. BASIC FORMULAS

Figure 1 illustrates the notation. The lens L represents an aplanatic system that converts the incident plane monochromatic wave into a converging spherical one, which is diffracted at the lens aperture. 2α is the angular aperture on the image side, i.e., the angle which the diameter of the lens subtends at the focus O, at distance f from the lens. OX, OY, OZ are Cartesian rectangular axes, with origin at the focus, with OX in the direction of vibration of the incident electric field, assumed to be linearly polarized, and with OZ along the axis of revolution, pointing away from the lens. (r,θ,ϕ) are spherical polar coordinates, with the polar axis along the axis of revolution of the system and with the azimuth $\phi=0$ in the OX direction.

Let

$$\mathbf{E}(P,t)=\mathrm{Re}\{\mathbf{e}(P)e^{-i\omega t}\},\quad \mathbf{H}(P,t)=\mathrm{Re}\{\mathbf{h}(P)e^{-i\omega t}\}, \quad (1)$$

represent the electric and magnetic fields, respectively,

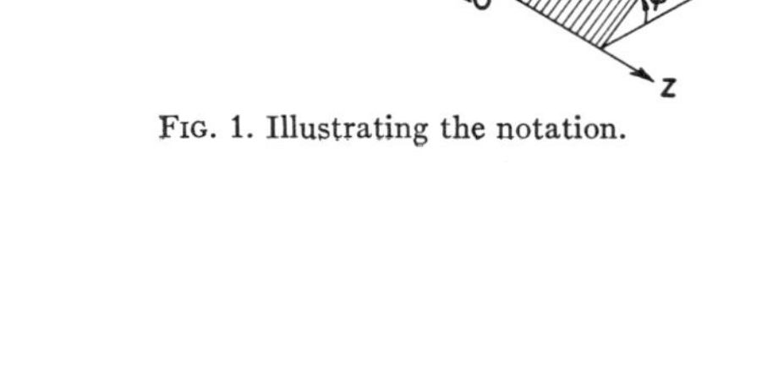

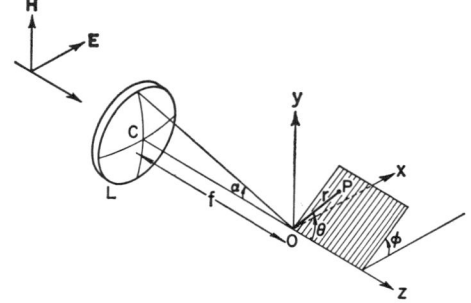

FIG. 1. Illustrating the notation.

* Research supported by the U. S. Army Research Office (Durham).

[1] For an account of this work see Sec. 8.8 in M. Born and E. Wolf, *Principles of Optics* (Pergamon Press, Inc., New York, 2nd ed., 1964).

[2] W. Ignatowsky, Trans. Opt. Inst. Petrograd (Leningrad) **1**, Paper IV (1919); **1**, Paper V (1920). See also V. A. Fock, *ibid.* **3**, 24 (1924).

[3] H. H. Hopkins, (a) Proc. Phys. Soc. **55**, 116 (1943); (b) Nature, **155**, 275 (1945).

[4] R. Burtin, Opt. Acta **3**, 104 (1956).

[5] J. Focke, Opt. Acta **4**, 124 (1957).

[6] B. Richards and E. Wolf, Proc. Phys. Soc. (London) **B69**, 854 (1956).

[7] B. Richards and E. Wolf, Proc. Roy. Soc. (London), **A253**, 358 (1959).

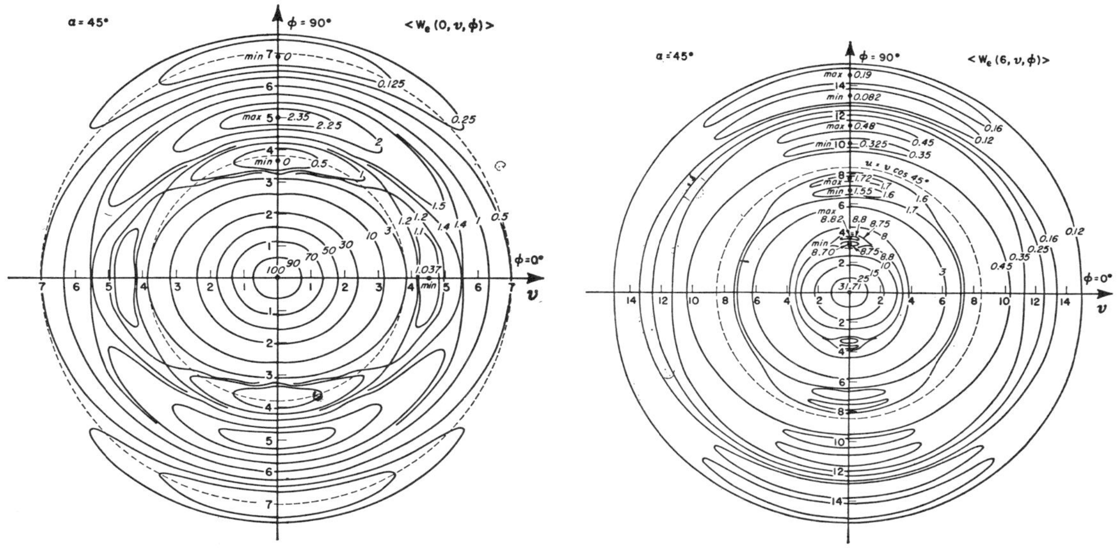

(a) (b)

FIG. 2. Contours of the time-averaged electric energy density $\langle w_e\rangle$ in the focal plane $u=0$ (a) and in the plane $u=6$ (b), in a system with semi-angular aperture $\alpha=45°$ on the image side.

at a typical point P in the focal region, at time t, with Re denoting the real part. Then it has been shown in Ref. 7 that the Cartesian components of the space-dependent parts $\mathbf{e}$ and $\mathbf{h}$ of the electric and magnetic fields are given by[8] (in Gaussian units and with suitable choice of the origin of the phase)

$$\left.\begin{aligned} e_x(P)&=-iA\,(I_0+I_2\cos 2\phi)\,,\\ e_y(P)&=-iAI_2\sin 2\phi\,,\\ e_z(P)&=-2AI_1\cos\phi\,,\end{aligned}\right\}\qquad(2)$$

$$\left.\begin{aligned} h_x(P)&=-iAI_2\sin 2\phi\,,\\ h_y(P)&=-iA\,(I_0-I_2\cos 2\phi)\,,\\ h_z(P)&=-2AI_1\sin\phi\,,\end{aligned}\right\}\qquad(3)$$

where

$$I_0\equiv I_0(kr,\theta;\alpha)=\int_0^\alpha \cos^{1/2}\theta'\,\sin\theta'(1+\cos\theta')\times J_0(kr\sin\theta'\sin\theta)e^{ikr\cos\theta'\cos\theta}d\theta'\,,$$

$$I_1\equiv I_1(kr,\theta;\alpha)=\int_0^\alpha \cos^{1/2}\theta'\,\sin^2\theta'\times J_1(kr\sin\theta'\sin\theta)e^{ikr\cos\theta'\cos\theta}d\theta'\,,\qquad(4)$$

$$I_2\equiv I_2(kr,\theta;\alpha)=\int_0^\alpha \cos^{1/2}\theta'\,\sin\theta'(1-\cos\theta')\times J_2(kr\sin\theta'\sin\theta)e^{ikr\cos\theta'\cos\theta}d\theta'\,,$$

and J_0, J_1, J_2 are Bessel functions of the first kind. The proportionality constant A in Eqs. (2) and (3) is given by

$$A=\pi fl/\lambda\,,\qquad(5)$$

where l represents the amplitude of the incident electric field in the object space, f is the focal length of the lens, $\lambda=2\pi c/\omega$ is the wavelength, and c is the vacuum velocity of light.

It is of interest to note that according to Eq. (2) and (3) the electric and magnetic fields have in general nonvanishing "longitudinal" components (z-components) in the focal region.

It is convenient to introduce the following "longitudinal" and "transversal" coordinates of the typical point P in the region of focus (see Fig. 1):

$$\left.\begin{aligned} u&=kr\cos\theta\sin^2\alpha=kz\sin^2\alpha\,,\\ v&=kr\sin\theta\sin\alpha=k(x^2+y^2)^{1/2}\sin\alpha\,,\end{aligned}\right\}\qquad(6)$$

where $k=2\pi/\lambda=\omega/c$. The position of P is then specified by the variables u, v, and ϕ. The three integrals defined by (4) now become functions of u, v (and of the parameter α which we will not show explicitly from now

[8] There are several misprints in Ref. 7. The last equation of (3.14) should read

$$\langle w(u,v,\phi)\rangle=(A^2/8\pi)\{|I_0|^2+4|I_1|^2+|I_2|^2\}.$$

Equation (3.19) should read

$$\langle w_e\rangle=\langle w_m\rangle=\tfrac{1}{2}\langle w\rangle=\left(\frac{A}{15}\right)^2\frac{16}{\pi}[1-\tfrac{5}{8}(\cos^{3/2}\alpha)(1+\tfrac{3}{5}\cos\alpha)]^2.$$

Formula (3.22) should read

$$|\langle\mathbf{S}\rangle|=\frac{cA^2}{8\pi}\{(|I_0|^2-|I_2|^2)^2+4(\mathrm{Im}[I_1(I_2^*-I_0^*)])^2\}^{1/2}.$$

The designations of the vertical axes in Fig. 4 and 5 of Ref. 7 have been interchanged. The vertical axes in Figs. 4(a) and (4b) represent $\langle w_e(0,v,0)\rangle$ and $\langle w_e(0,v,\frac{1}{2}\pi)\rangle$, respectively; the vertical axes in figures 5(a) and 5(b) represent $\langle w(0,v,\phi)\rangle$ and $\langle S_z(0,v,\phi)\rangle$, respectively.

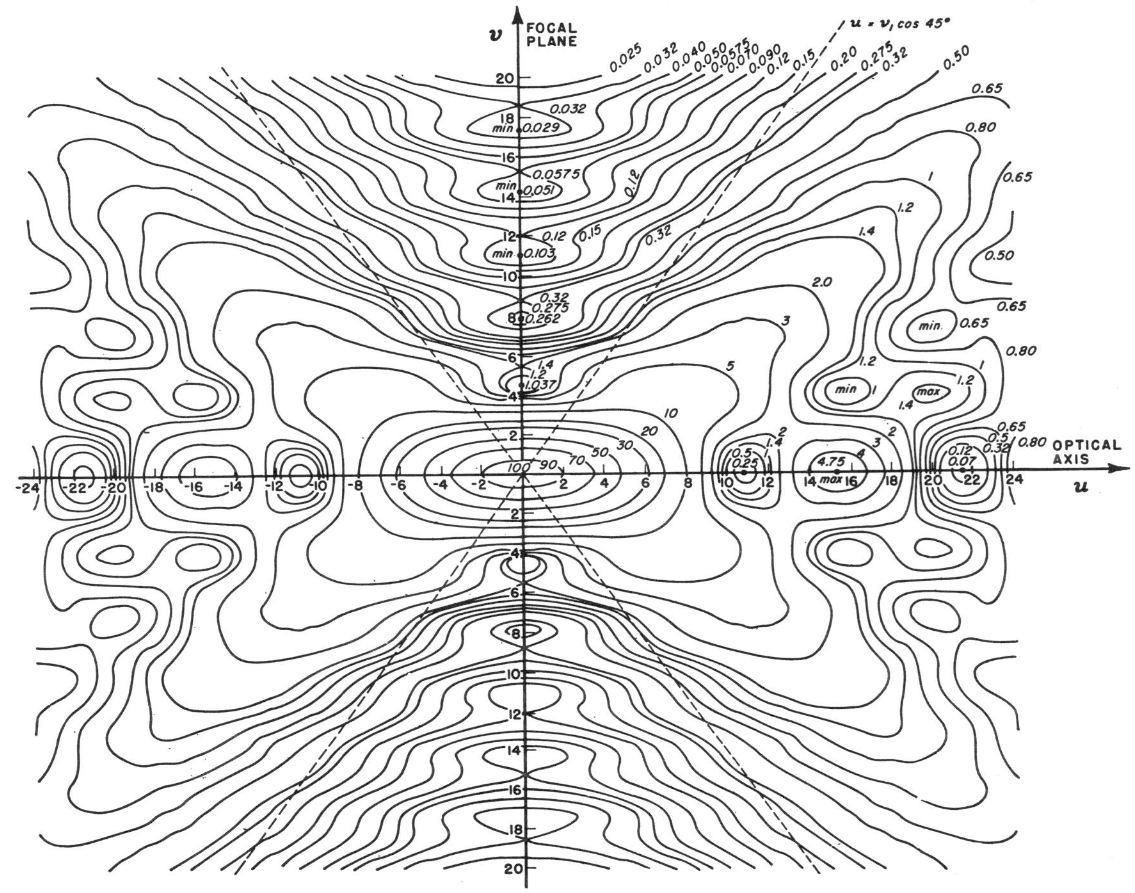

FIG. 3. Contours of the time-averaged electric energy density $\langle w_e \rangle$ near focus in the meridional plane $\phi=0$ ($\alpha=45°$).

on):

$$I_0(u,v)=\int_0^\alpha \cos^{1/2}\theta' \sin\theta' (1+\cos\theta') \times J_0\left(\frac{v \sin\theta'}{\sin\alpha}\right) e^{iu\cos\theta'/\sin^2\alpha} d\theta',$$

$$I_1(u,v)=\int_0^\alpha \cos^{1/2}\theta' \sin^2\theta' J_1\left(\frac{v \sin\theta'}{\sin\alpha}\right) e^{iu\cos\theta'/\sin^2\alpha} d\theta', \quad (7)$$

$$I_2(u,v)=\int_0^\alpha \cos^{1/2}\theta' \sin\theta' (1-\cos\theta') \times J_2\left(\frac{v \sin\theta'}{\sin\alpha}\right) e^{iu\cos\theta'/\sin^2\alpha} d\theta'.$$

3. RESULTS RELATING TO A TYPICAL CASE

The theoretical solution which was summarized in Sec. 2 was used to obtain detailed information about the structure of the focal region in an aplanatic system with angular semi-aperture $\alpha=45°$. First the values of the three integrals I_0, I_1, and I_2 were computed on an 7070 IBM electronic computer. The range covered was $0\leqslant u\leqslant 24$, $0\leqslant v\leqslant 20$ (in steps $\Delta u=0.25$, $\Delta v=0.5$) and corresponds, according to Eq. (6), to defocusing of amount $|z|\lesssim 8\lambda$ and to off-axis distance $r\lesssim 5\lambda$. With the help of these results and using graphical interpolation techniques, diagrams exhibiting the behavior of various quantities of physical interest were then constructed. The results are shown in Figs. 2–5. Each of the figures is normalized in such a way that at the focus, the time-averaged electric-energy density[9] $\langle w_e\rangle=100$, with

[9] To convert to actual values in physical units we use the results [cf. Ref. 7, Eqs. (2.28) and (3.17)] that at the focus

$$\langle w_e\rangle=\frac{1}{16\pi}\langle E^2\rangle=(A_G^2/16\pi)\,|I_0(0,0,\alpha)|^2 \quad \text{(Gaussian units)},$$

$$\langle w_e\rangle=\tfrac{1}{4}\langle E^2\rangle=\tfrac{1}{4}A_{\text{mks}}^2|I_0(0,0,\alpha)|^2 \quad \text{(mks units)},$$

where $A=\pi lf/\lambda$ and l is the strength of the linearly polarized electric field incident on the imaging system. In the present case ($\alpha=45°$), $I_0(0,0,\alpha)=0.5021$ and hence our normalization implies that we have taken

$$A_G=141.20 \text{ statV/cm},$$
$$A_{\text{mks}}=\ 39.83 \text{ V/m}.$$

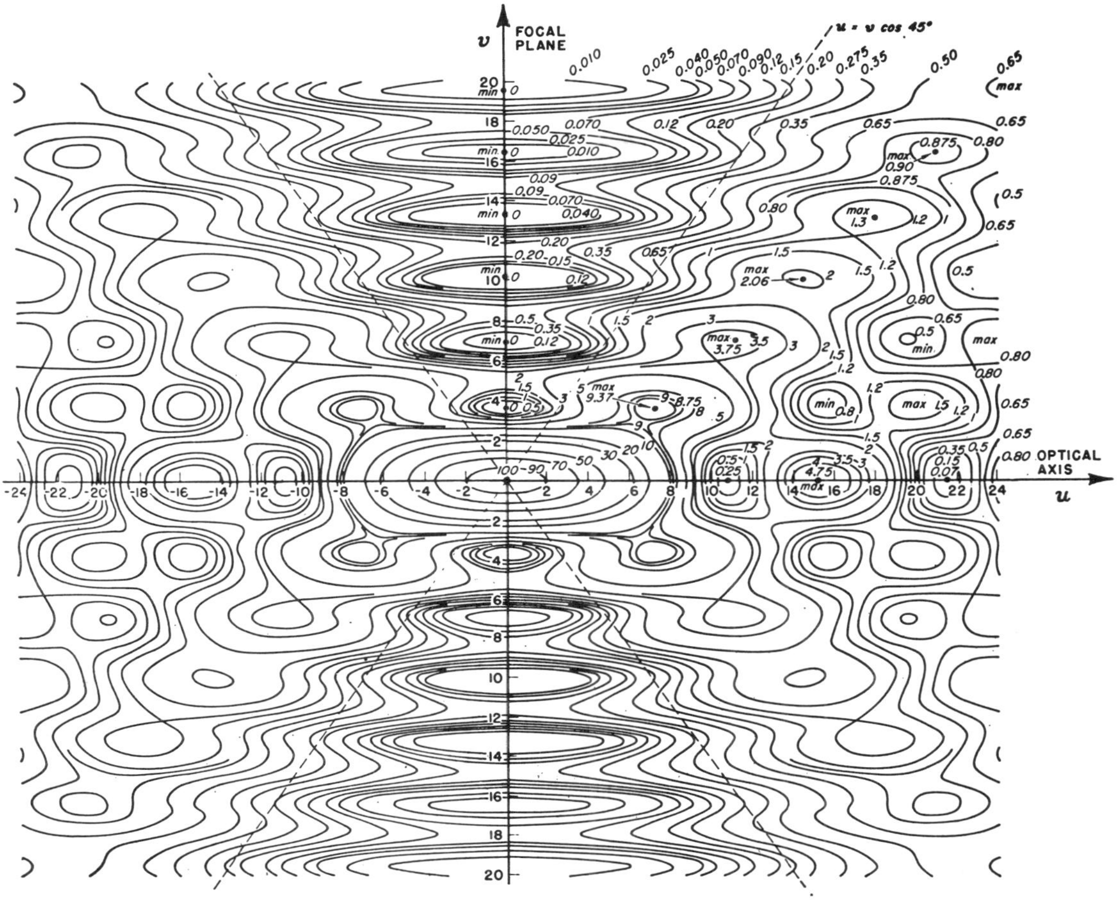

FIG. 4. Contours of the time-averaged electric energy density $\langle w_e \rangle$ near focus in the meridional plane $\phi = \frac{1}{2}\pi$ ($\alpha = 45°$).

the definition (appropriate to Gaussian units) $\langle w_e \rangle = (1/8\pi)\langle E^2 \rangle = (1/16\pi)\mathbf{e}\cdot\mathbf{e}^*$ being used in labeling the contours in Fig. 5.

It is apparent from Figs. 2 that the contours of the electric-energy density $\langle w_e(u,v,\phi)\rangle$ in the observing planes $u=0$ and $u=6$ are not rotational symmetrical as the elementary scalar theory predicts (cf. Ref. 1, Sec. 8.5.2 and Sec. 8.8). However, they exhibit symmetry with respect to the two meridional planes $\phi=0$ and $\phi=\frac{1}{2}\pi$. In the focal plane ($u=0$) the contours, inside the central core, are approximately elliptical, with their major axes in the direction of the electric vector of the incident wave ($\phi=0$). Along the meridional line $\phi=\frac{1}{2}\pi$ in the focal plane there are points at which the electric-energy density has zero value. On the other hand, the minima along the meridional line $\phi=0$ are not zero. The contours of the time-averaged magnetic-energy density are obtained by simply rotating the contours of Figs. 2 through 90° about the normals to the plane of the figures, while keeping the u and v axes fixed.

Figures 3 and 4 show the contours of the time-averaged electric-energy density near the focus in the meridional planes $\phi=0$ (containing the electric vector of the incident field) and $\phi=\frac{1}{2}\pi$ (at right angles to the electric vector of the incident field).[10] In each figure the two dashed lines $u=v\cos 45°$ indicate the boundaries of the geometrical shadow. Along the optical axis, $v=0$, the minima are not true zeros and the distribution is compressed toward the focus in comparison with the corresponding distribution in systems with small angular aperture 2α. In both figures the alternate rows of minima and maxima are seen to be aligned along mutually parallel straight lines. In Fig. 4 the very complicated behavior in the "wings" of the pattern should be noted. These two diagrams show that linear polarization of the incident field leads to a distribution which in the plane $\phi=0$ is similar to that associated

[10] Figures 3 and 4 also represent the contours of the time-averaged magnetic energy density $\langle w_m \rangle$ near focus m, the meridional planes $\phi=\frac{1}{2}\pi$ and $\phi=0$, respectively ($\alpha=45°$).

with spherical aberration and in the plane $\phi=\frac{1}{2}\pi$ is similar to one associated with amplitude filtering.

Figure 5 shows the behavior of the longitudinal component (z-component) of the electric field in the focal plane $u=0$ or more precisely, the contours of e_z, the longitudinal component being given by $E_z=\mathrm{Re}(e_z e^{-i\omega t})$. The striking resemblance of this diagram to the well known instantaneous picture of the lines of force due to an oscillating electric dipole[11] should be noted.

While e_z is seen to be zero at the focus itself, and on the azimuthal line $\phi=\frac{1}{2}\pi$, it has maxima of alternative signs along the azimuthal line $\phi=0$, the largest one being attained at the point $v=2.25$ (about half a wavelength from the axis) in the azimuth $\phi=0$. This maximum is about 27.9% of the maximum value which the transverse component attains (at the focus itself). At the points $v=3.98$, 7.10, 10.24, 13.45, $\cdots$ on the line $\phi=0$ the transverse electric field is strictly zero, so that at these points the electric field is ***purely longitudinal***. The values of the e_z component at these points are: $e_z=-9.70$, $+4.70$, -2.80, $+1.85$, $\cdots$. With the help of these results one can readily see that by focusing a beam emerging from an optical maser, longitudinal field strengths of the order of 10^5 V/cm can now be attained. It seems plausible that such strong longitudinal fields could be used for accelerating charged particles. However, because of the considerable complexity of the field in the focal region, the practical feasibility of such a proposal must await a more detailed study. In any case it should be born in mind that the present calculations refer to a wave whose amplitude distribution on emergence from the lens is that appropriate to aplanatism [see Ref. 7, Eq. (2.13)], whereas the amplitude distribution of a laser beam will be of a more complicated structure. One could, of course, employ a suitable "amplitude filter" to convert the laser output distribution into an aplanatic one.

[11] See, for example, G. Joos, *Theoretical Physics* (Blackie & Son Ltd., London, 1947), p. 327.

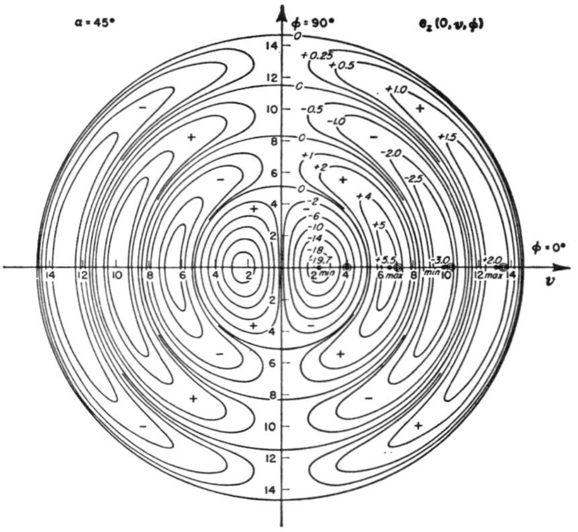

FIG. 5. Contours of the time-independent part e_z associated with the longitudinal component of the complex electric field $E_z=\mathrm{Re}\{e_z e^{-i\omega t}\}$ in the focal plane $u=0$ ($\alpha=45°$).

An extensive computational program is at present in progress and it is expected that it will provide detailed information about the longitudinal and transverse components of the field and about the Poynting vector in the focal region.

ACKNOWLEDGMENTS

We are indebted to Mrs. C. Reynolds for much advice and help with programming and to Mrs. G. Eyer for considerable assistance with the preparation of the figures. We also wish to acknowledge some help received from J. Dow in connection with the computations. Further, one of us (A.B.) is indebted to the John Simon Guggenheim Memorial Foundation for the award of a Fellowship, during the tenure of which the largest part of this work was carried out.

Reprinted from JOURNAL OF THE OPTICAL SOCIETY OF AMERICA, Vol. 57, No. 10, 1171–1175, October 1967
Printed in U. S. A.

Energy Flow in the Neighborhood of the Focus of a Coherent Beam*

A. BOIVIN
Department of Physics, Laval University, Quebec 10, Canada

AND

J. DOW† AND E. WOLF‡
Department of Physics and Astronomy, University of Rochester, Rochester, New York 14627

(Received 10 June 1967)

In a previous paper [Phys. Rev. **138**, B1561 (1965)] some new results were presented relating to the structure of the electromagnetic field near the focus of a coherent beam emerging from an aplanatic optical system. The present paper supplements the previous one by providing detailed analysis of the behavior of the Poynting vector in the focal region of such a beam. In particular, diagrams showing the flow lines and the contours of constant amplitude of the time-averaged Poynting vector are given. The energy flow is found to have vortices near certain points of the focal plane. Two diagrams showing the behavior of the flow near a typical vortex are also included.

INDEX HEADINGS: Diffraction; Image formation.

IN two previous papers,[1,2] the structure of the electromagnetic field in the focal region of a coherent beam was studied. In particular, some general properties of the field were established and contour diagrams were given, showing mainly the distribution of the energy density arising from the diffraction of a plane linearly polarized electromagnetic wave by an aplanatic optical system. Since then one of the predictions made in Ref. 2, namely the existence of a strong longitudinal component in some parts of the focal plane, has been verified by experiments with microwaves.[3] Recently, the theoretical analysis has been extended[4] to focused gaussian beams (which are of particular interest in connection with the output of lasers). Another related recent investigation was concerned with the structure of the focal region of a wide-angle spherical mirror.[5] The results were applied to provide estimates of the performance of the giant radar and radio telescope at Arecibo in Puerto Rico.

In the present paper, we extend the analysis carried out in Refs. 1 and 2 by discussing the detailed behavior of the Poynting vector in the focal region and obtain diagrams showing the flow lines and the magnitude of the Poynting vector in a typical case for an aplanatic system of angular semi-aperture $\alpha=45°$.

1. NOTATION AND BASIC FORMULAE

The present discussion is a direct continuation of that given in Refs. 1 and 2. We retain the notation of these papers.

In Fig. 1, the lens L represents an aplanatic system that converts a plane linearly polarized monochromatic wave (with time dependence $e^{-i\omega t}$ in the usual complex notation) into a converging one, which is diffracted at the lens aperture. 2α denotes the angular aperture on the image side, i.e., the angle which the diameter of the lens subtends at the focus 0, at distance f from the lens. $0X$, $0Y$, $0Z$ are cartesian rectangular axes, with origin

* Research supported by the U. S. Army Research Office (Durham).

† Present address: Department of Physics, Princeton University, Princeton, New Jersey.

‡ During the academic year 1966/1967, Guggenheim Fellow and Visiting Professor at the Department of Physics, University of California, Berkeley, California.

[1] B. Richards and E. Wolf, Proc. Roy. Soc. (London) **A253**, 358 (1959).

[2] A. Boivin and E. Wolf, Phys. Rev. **138**, B1561 (1965).

[3] A. I. Carswell, Phys. Rev. Letters **15**, 647 (1965).

[4] (a) A. L. Bloom: Paper 9A-6 presented at the International Quantum Electronics Conference, Phoenix, Ariz., April, 1966. [Abstract published in IEEE **QE-2**, 87 lxiii (1966)]. (b) D. J. Innes and A. L. Bloom, Spectra-Physics Laser Technical Bulletin, #5 (1966) (Published by Spectra-Physics, Inc., Mountain View, Calif.).

[5] A. Boivin and M. Gravel, J. Opt. Soc. Am. **56**, 1438A (1966).

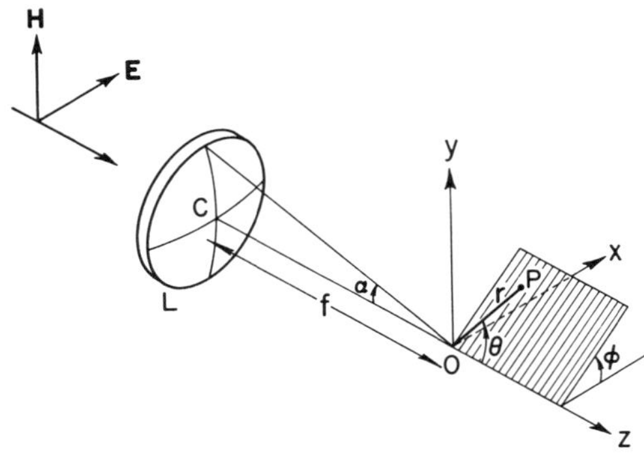

FIG. 1. Illustrating the notation.

at the focus, with $0X$ in the direction of vibration of the incident electric field, and with $0Z$ along the axis of revolution, pointing away from the lens. (r,θ,ϕ) are spherical polar coordinates, with the polar axis along the axis of revolution of the system and with the azimuth $\phi=0$ in the $0X$ direction.

As in the previous papers, it is convenient to specify each point in the focal region by the azimuthal angle and by the "longitudinal" and "transversal" coordinates

$$\begin{aligned} u &= kr\cos\theta\sin^2\alpha = kz\sin^2\alpha, \\ v &= kr\sin\theta\sin\alpha = k(x^2+y^2)^{\frac{1}{2}}\sin\alpha, \end{aligned} \tag{1}$$

where $k=\omega/c=2\pi/\lambda$, λ being the wavelength.

It has been shown in §3.3 of Ref. 1 that the cartesian components of the time-averaged Poynting vector are given by the following formulae:

$$\begin{aligned} \langle S_x\rangle &= (cA^2/4\pi)\cos\phi\,\mathrm{Im}\{I_1(I_2^*-I_0^*)\}, \\ \langle S_y\rangle &= (cA^2/4\pi)\sin\phi\,\mathrm{Im}\{I_1(I_2^*-I_0^*)\}, \\ \langle S_z\rangle &= (cA^2/8\pi)\{|I_0|^2-|I_2|^2\}, \end{aligned} \tag{2}$$

where Im denotes the imaginary part and I_0, I_1, I_2 are the integrals

$$\begin{aligned} I_0(u,v) &= \int_0^\alpha \cos^{\frac{1}{2}}\theta'\sin\theta'(1+\cos\theta')J_0(v\sin\theta'/\sin\alpha)e^{iu\cos\theta'/\sin^2\alpha}d\theta', \\ I_1(u,v) &= \int_0^\alpha \cos^{\frac{1}{2}}\theta'\sin^2\theta' J_1(v\sin\theta'/\sin\alpha)e^{iu\cos\theta'/\sin^2\alpha}d\theta', \\ I_2(u,v) &= \int_0^\alpha \cos^{\frac{1}{2}}\theta'\sin\theta'(1-\cos\theta')J_2(v\sin\theta'/\sin\alpha)e^{iu\cos\theta'/\sin^2\alpha}d\theta'. \end{aligned} \tag{3}$$

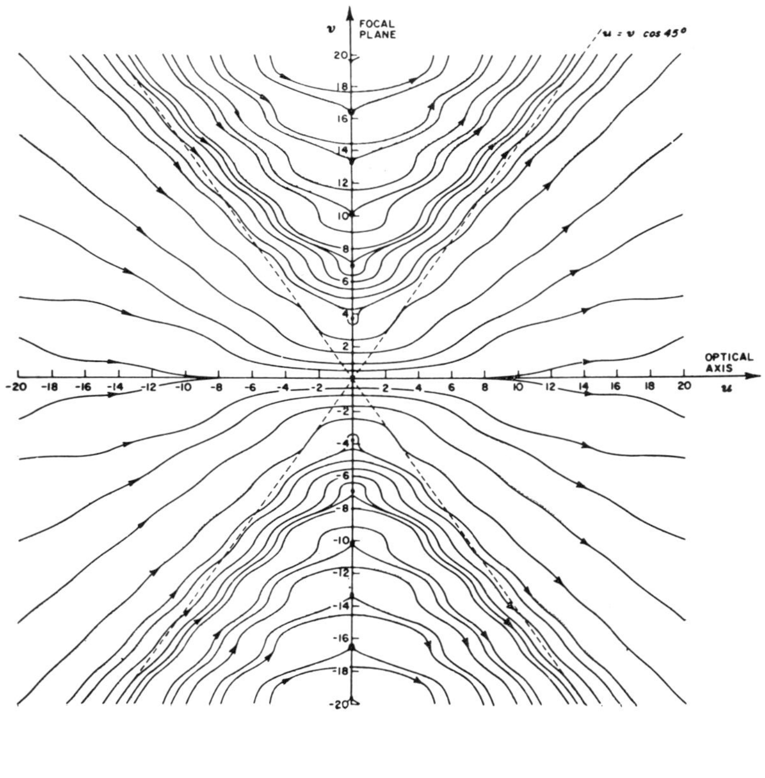

FIG. 2. Flow lines of the time-averaged Poynting vector in the neighborhood of the focus of an aplanatic system with angular semi-aperture $\alpha=45°$ on the image side. u and v denote the normalized longitudinal and transversal coordinates defined by Eq. (1).

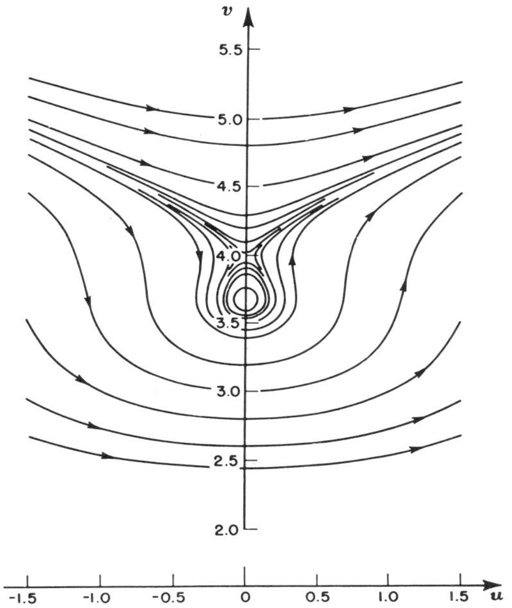

FIG. 3. Flow lines of the time-averaged Poynting vector, showing vortex behavior, in the neighborhood of the two singularities of the flow closest to the focus ($u=0$, $v=3.67$ and $u=0$, $v=3.99$). (Aplanatic system, $\alpha=45°$.)

Further,

$$A=\pi f l/\lambda, \tag{4}$$

where l represents the amplitude of the incident electric field in the object space.

2. BEHAVIOR OF THE POYNTING VECTOR

From Eq. (2) we see that the magnitude of the time-averaged Poynting vector is given by[6]

$$|\langle \mathbf{S}\rangle| = (cA^2/8\pi)((|I_0|^2-|I_2|^2)^2 +4\{\mathrm{Im}[I_1(I_2^*-I_0^*)]\}^2)^{\frac{1}{2}}. \tag{5}$$

Since this expression is independent of the azimuthal angle ϕ, the loci of constant $|\langle \mathbf{S}\rangle|$ are surfaces of revolution about the axis of the system (z axis). It is also readily seen from Eq. (2), that *at each point of the focal region the Poynting vector lies in the meridional plane which contains that point* and it makes an angle γ with the positive z axis, given by

$$\sin\gamma=\frac{2|\mathrm{Im}[I_1(I_2^*-I_0^*)]|}{[(|I_0|^2-|I_2|^2)^2+4\{\mathrm{Im}[I_1(I_2^*-I_0^*)]\}^2]^{\frac{1}{2}}},$$
$$\cos\gamma=\frac{|I_0|^2-|I_2|^2}{[(|I_0|^2-|I_2|^2)^2+4\{\mathrm{Im}[I_1(I_2^*-I_0^*)]\}^2]^{\frac{1}{2}}}. \tag{6}$$

[6] A misprint in the formula (3.22) for $|\langle \mathbf{S}\rangle|$ given in Ref. 1 is corrected here.

We note two symmetry relations with respect to the focal plane $u=0$:

$$|\langle \mathbf{S}(-u,v,\phi)\rangle| = |\langle \mathbf{S}(u,v,\phi)\rangle|, \tag{8a}$$

$$\gamma(-u,v,\phi)=2\pi-\gamma(u,v,\phi), \quad (u\neq 0). \tag{8b}$$

The first of these follows from Eqs. (5) and (3). The second follows from Eq. (6).

When $u=0$ (focal plane) the three integrals I_0, I_1, and I_2 defined by Eq. (3) are real and it then follows from Eq. (2) that $\langle S_x\rangle=\langle S_y\rangle=0$, i.e., at any point in the focal plane the Poynting vector is perpendicular to the focal plane.

The preceding results are already contained in Ref. 1 but are included here for the sake of completeness and also because some of the formulae were used in the computations of Figs. 3, 4, 5, and 6 of the present paper.

We see from Eq. (1), that

$$\frac{dv}{du}=\frac{1}{\sin\alpha}\frac{d}{dz}[(x^2+y^2)^{\frac{1}{2}}]. \tag{9}$$

Hence from Eqs. (9) and (2), the differential equation of the flow lines (lines such that at each point the Poynting vector is along the tangent to the line passing through that point), in a meridional plane is

$$\sin\alpha\frac{dv}{du}=\frac{[\langle S_x\rangle^2+\langle S_y\rangle^2]^{\frac{1}{2}}}{\langle S_z\rangle}=\frac{2\,\mathrm{Im}[I_1(I_2^*-I_0^*)]}{|I_0|^2-|I_2|^2}. \tag{10}$$

In view of Eq. (8b), the flow lines on the two sides of the focal plane near focus are reflections of each other in that plane. Moreover, rotation of a flow line through any angle about the z axis gives an equivalent flow line in some other meridional plane.

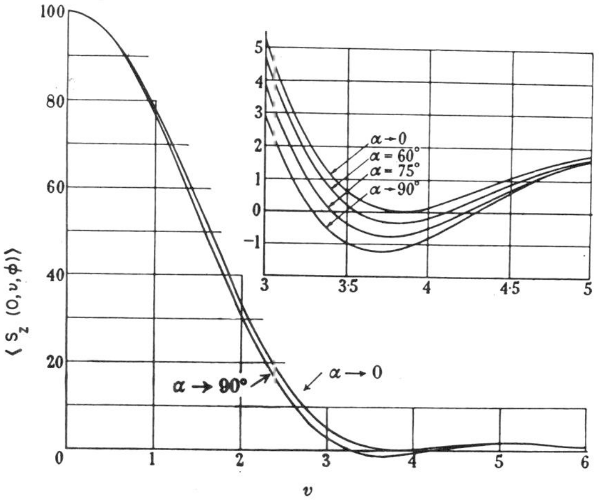

FIG. 4. The variation of the only nonvanishing component of the time-averaged energy flow $|\langle S_z\rangle|$ along any meridional section $\phi=$const of the focal plane in an aplanatic system of angular semi-aperture α on the image side. The values are normalized to 100 at the focus. [After B. Richards and E. Wolf, Proc. Roy. Soc. (London) A253, 358 (1959).]

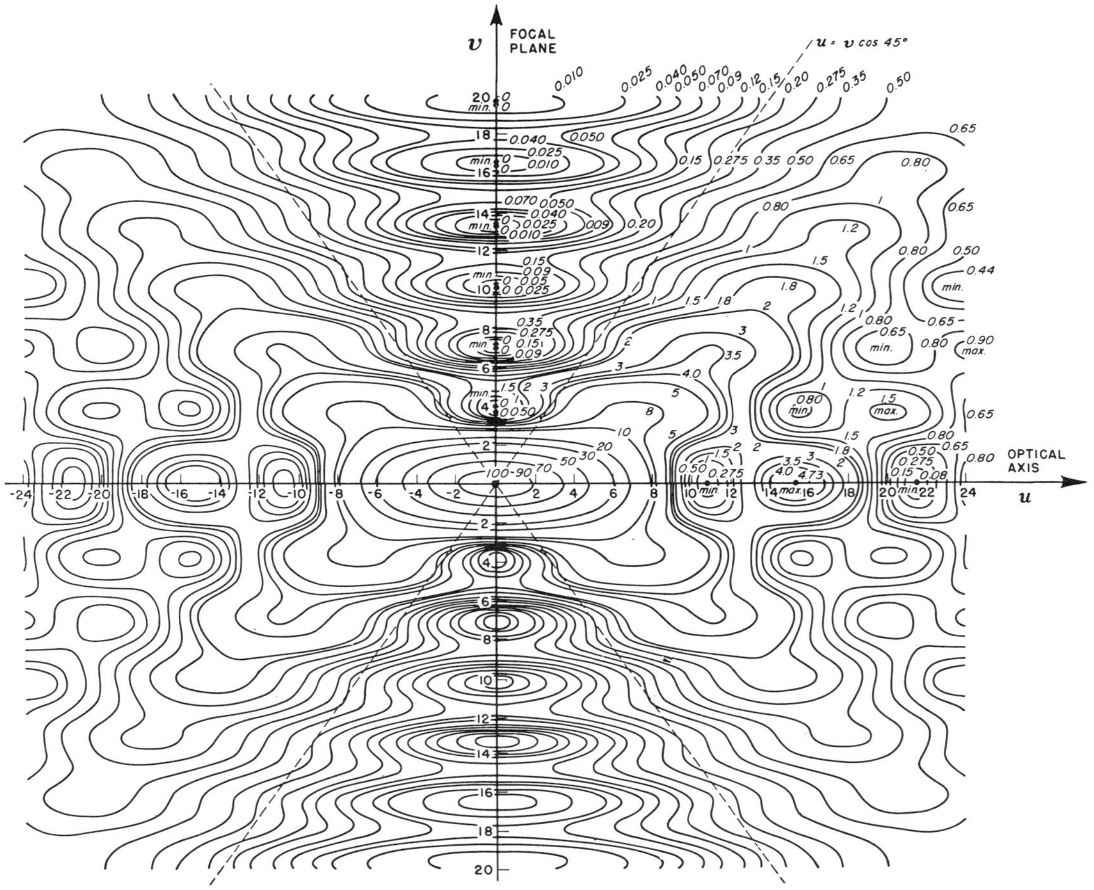

FIG. 5. Contours of constant magnitude of the time-averaged Poynting vector (constant $|\langle \mathbf{S} \rangle|$) in the focal region of an aplanatic system with angular semi-aperture $\alpha = 45°$. The values are normalized to 100 at the focus.

The differential Eq. (10) may be integrated numerically to give v as a function of u, subject to the specification of a starting point (u_0, v_0). Each starting point leads to a flow line $v = (u; u_0, v_0)$.

With the help of the Runge–Kutta method[7] we have performed the numerical integration for an aplanatic system with semiangular aperture $\alpha = 45°$ on the image side. In certain regions of the uv-plane, where $|I_0|$ is approximately equal to $|I_2|$, the slopes of the flow lines are very great and the numerical solutions for $v(u)$ are difficult to obtain. In such regions, we used the inverse of Eq. (10) and obtained the corresponding differential equation for $u = u(v)$, which could easily be integrated. Only at isolated points where the numerator and the denominator of Eq. (10) are both zero is the integration impossible, since dv/du is then not defined.

The flow lines were determined with the aid of an IBM 7074 digital computer, using increment $\Delta u = 0.1$ for the curves in Fig. 2 and $\Delta u = 0.01$ for the curves in

[7] See, for example, R. P. Agnew, *Differential Equations* (McGraw-Hill Book Company, New York, 1960), p. 306.

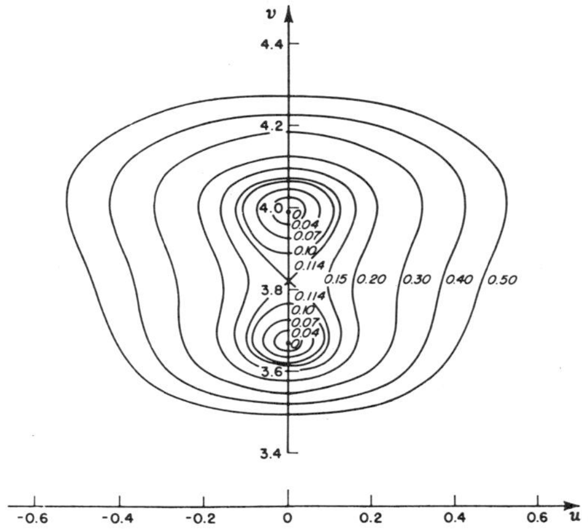

FIG. 6. Contours of constant magnitude of the time-averaged Poynting vector near the two singularities closest to the focus ($u=0$, $v=3.67$ and $u=0$, $v=3.99$). The values correspond to normalization 100 at the focus. (Aplanatic system, $\alpha = 45°$.)

Fig. 3, except in the regions of great slopes, where $(dv/du)\sin\alpha$ exceeds unity. In these regions the computer integrated the expression for du/dv, taking increments in v rather than u, with the corresponding spacings $\Delta v=0.1$ and $\Delta v=0.01$. The computed flow lines in the neighborhood of focus[8] are shown in Fig. 2.

The energy flow has singularities at points where the average Poynting vector vanishes. Such points are situated in the focal plane $u=0$ and are given by the v-roots of the equations [cf. Eq. (5)] $I_0-I_2=0$ or $I_0+I_2=0$. Numerical solutions of these equations give

$$I_0-I_2=0 \quad \text{when} \quad v=3.67,\ 6.92,\ 10.11,\ 13.27,\ 16.42,\ 19.57,\ \cdots$$

$$I_0+I_2=0 \quad \text{when} \quad v=3.99,\ 7.13,\ 10.26,\ 13.39,\ 16.52,\ 19.66,\ \cdots.$$

The flow lines in the neighborhood of the two zeros closest to the origin are shown in Fig. 3. We see that the zero $v=3.67$ is the center of a vortex and that at points in the focal plane situated between the two zeros the Poynting vector points into the half space from which the wave is incident ($\langle S_z\rangle<0$). The existence of regions where $\langle S_z\rangle$ becomes negative appears to have been first noted by Ignatowsky[9] and this fact was studied in detail by Richards and Wolf,[1] who computed $\langle S_z\rangle$ in the focal plane as a function of the distance from the focus, for aplanatic systems of three different angular apertures. We reproduce the relevant curves[10] in Fig. 4.

[8] The Poynting-vector flow lines, away from the singularities, resemble closely the contour lines for the fraction of total illumination within circles centered on axis in receiving planes $u=$const [E. Wolf, Proc. Roy. Soc. (London) A**204**, 533 (1951), Fig. 3a on p. 542]. The qualitative resemblance of these contour lines to flow lines appears to be a consequence of the principle of conservation of energy, applied to the region generated by the rotation of a portion of any particular contour line about the optical axis.

[9] V. S. Ignatowsky, Trans. Opt. Inst. Petrograd, Vol. 1, paper IV (1919).

[10] In Ref. 1, the designations of the vertical axis in Figs. 4 and 5 have been interchanged. The designation is corrected in our reproduction of Fig. 5(b) (shown here as Fig. 4).

In Fig. 5, the contours of constant $|\langle\mathbf{S}\rangle|$, normalized to 100 at the focus[11] are shown; their behavior in the neighborhood of the vortex closest to the origin is shown in Fig. 6.

Figure 5 closely resembles Figs. 3 and 4 of Ref. 2, which show the contours of constant time averaged electric energy density in the neighborhood of the focus. The similarity of these diagrams arises from the structure of the particular field under consideration and cannot be expected to hold generally.

ACKNOWLEDGMENTS

We wish to express our gratitude to Miss Beverly Johns, who did much programming on the basic integrals I_0, I_1, I_2, and to Miss Raymonde Verreault who constructed most of the contours of Fig. 5. We are also indebted to Mrs. G. Eyer and Mrs. G. Walker for drawing the figures shown in the present paper.

[11] This implies that the numbers on the contour in Figs. 5 and 6 represent the quantity $100|\langle\mathbf{S}(u,v)\rangle|/|\langle\mathbf{S}(0,0)\rangle|$, for an aplanatic system with $\alpha=45°$. The conversion of these dimensionless quantities to gaussian units or to the rationalized mks units may be effected by the relations [cf. Eq. (3.23) of Ref. 1]:

$$|\mathbf{S}(0,0)|=(cA^2/8\pi)|I_0(0,0)|^2 \text{ (gaussian units)},$$
$$|\mathbf{S}(0,0)|=(A^2/2Z_0)|I_0(0,0)|^2 \text{ (mks units)}.$$

Here A is given by Eq. (4), $I_0(0,0)=0.5021$ (with $\alpha=45°$) and $Z_0=c\mu_0=376.727$ ohms, μ_0 being the magnetic permeability of free space.

There are errors in the corresponding equations in footnote 9 of Ref. 2, relating to the values of the time-averaged electric energy density. The correct equations are

$$\langle w_e\rangle=(1/8\pi)\langle E^2\rangle=(A_G{}^2/16\pi)|I_0(0,0)|^2 \text{ (gaussian units)},$$
$$\langle w_e\rangle=\tfrac{1}{2}\epsilon_0\langle E^2\rangle=\tfrac{1}{4}\epsilon_0 A_{\text{mks}}{}^2|I_0(0,0)|^2 \text{ (mks units)},$$
$$A_G=141.20 \text{ statV/cm},$$
$$A_{\text{mks}}=1.339\times10^7 \text{ V/m},$$

where ϵ_0 is the permittivity of free space.

The caption to Fig. 5 of Ref. 2 does not specify the normalization of e_z. The figure displays (for an aplanatic system with $\alpha=45°$) the quantity $70.898e_z(0,v,\phi)/|\mathbf{e}(0,0,0)|$ which is identical with $e_z(0,v,\phi)$ when A is taken to be equal to $A_G=141.20$ statV/cm.

Reprinted from JOURNAL OF THE OPTICAL SOCIETY OF AMERICA, Vol. 52, No. 6, 615–625, June, 1962
Printed in U. S. A.

Generalization of the Maggi-Rubinowicz Theory of the Boundary Diffraction Wave—Part I†

KENRO MIYAMOTO*
Institute of Optics, University of Rochester, Rochester, New York

AND

EMIL WOLF
Department of Physics and Astronomy, University of Rochester, Rochester, New York

(Received March 20, 1961)

As a first step towards a generalization of the Maggi-Rubinowicz theory of the boundary diffraction wave, a new vector potential $\mathbf{W}(Q,P)$ is associated with any monochromatic scalar wavefield $U(P)$. This potential has the property that the normal component of its curl, taken with respect to the coordinates of any point Q on a closed surface S surrounding a field point P, is equal to the integrand of the Helmholtz-Kirchhoff integral; that is,

$$\mathrm{curl}_Q\mathbf{W}(Q,P)\cdot\mathbf{n}=\frac{1}{4\pi}\left\{U(Q)\frac{\partial}{\partial n}\left(\frac{\exp(iks)}{s}\right)-\frac{\exp(iks)}{s}\frac{\partial}{\partial n}U(Q)\right\},$$

where s is the distance QP and $\partial/\partial n$ denotes the differentiation along the inward unit normal $\mathbf{n}$ to S.

Further it is shown that the vector potential always has singularities at some points Q on S and that the field at P may be rigorously expressed as the sum of disturbances propagated from these points alone.

A closed expression for the vector potential associated with any given monochromatic wavefield that obeys the Sommerfeld radiation condition at infinity is derived and it is shown that in the special case when U is a spherical wave, this expression reduces to that found by G. A. Maggi and A. Rubinowicz in their researches on the boundary diffraction wave.

1. INTRODUCTION

THEORETICAL studies of diffraction of light by an obstacle whose linear dimensions are large compared to the wavelength are almost exclusively based on the classical principle associated with the names of Huygens and Fresnel. According to this principle, each point of the unobstructed part of a primary wave is assumed to be a center of a secondary disturbance and the diffracted field is considered to arise from the superposition of these secondary disturbances.

A different physical model for diffraction was suggested by Young[1] in 1802 prior to the appearance of Fresnel's celebrated memoir[2] on this subject. Young believed that the incident light undergoes a kind of reflection at the boundary of the diffracting body and he considered diffraction to arise from the interference between the direct light and the light propagated from each point of the boundary. But because of the early success of Fresnel's theory and also because the ideas of Young were only formulated in a rough qualitative manner, Fresnel's theory soon dominated the field and Young's explanation of diffraction has by and large been forgotten.

However, in the long period of time that has elapsed since then, there has accumulated a good deal of evidence, both experimental and theoretical, which suggests that Young's theory represents at least a nucleus of a sound physical model which may possibly

† This research was supported in part by the United States Air Force under contract, monitored by the Air Force Office of Scientific Research of the Air Research and Development Command.

Preliminary accounts of this work were presented at meetings of the Optical Society of America, held at Boston (1960) and at Pittsburgh (1961). Abstracts of these papers were published in J. Opt. Soc. Am. **50**, 1131 (1960), Abstract TA13 and *ibid.* **51**, 478 (1961), Abstract FB16.

* On leave of absence from Nippon Kokagu K.K., (Japan Optical Industry Company, Ltd.), Tokyo, Japan.

[1] T. Young, Phil. Trans. Roy. Soc. **20**, 26 (1802). Also *Miscellaneous Works of the late Thomas Young* (John Murray, London, 1855), Vol. 1, p. 151.

[2] *Oeuvrés complètes d'Augustin Fresnel* (Imprimerie Impériale, Paris, 1866), Vol. I, pp. 89, 129.

be simpler than that of Huygens and Fresnel. It is not possible to review here adequately all the supporting evidence, and in any case two excellent accounts of researches on this subject are already available[3,4] (see also reference 5, p. 448). It will suffice to say that Maggi[6] and Rubinowicz[7] showed independently that in the case of diffraction by an aperture in an opaque screen the Kirchhoff diffraction integral (which may be regarded as a mathematical refinement of the ideas of Huygens and Fresnel) may be decomposed in certain cases into the sum of two terms: One term represents a wave originating in every point of the boundary of the aperture (called the *boundary diffraction wave*[8]) and the other represents a wave propagated through the aperture in accordance with the laws of geometrical optics (called the *geometrical wave*).[9]

The researches of Maggi and Rubinowicz showed conclusively the basic correctness of Young's ideas. However, their analyses were restricted to cases when the wave incident upon the aperture is plane or spherical. Attempts to generalize these results to more general fields have so far not been very successful[10] and doubts have in fact been expressed[11] about the possibility of such a generalization.

As a physical model for diffraction, the Young-Maggi-Rubinowicz theory is intrinsically simple and physically appealing. It relates diffraction directly to the true cause of its origin, namely, the presence of the boundary of a diffracting body. It seems hard to believe that no proper generalization to more complicated fields exists. The possibility of such a generalization is not of academic interest alone as the following remarks will indicate. It is well known that the image of a small source formed by an optical system has as a rule a complicated structure. In consequence, calculations of light distribution in such an image are very laborious. The intensity has to be determined at many points (sometimes several hundreds) and the computations, based on the Huygens-Fresnel Principle involve the evaluation of very many double integrals, each extended over the exit pupil of the system. Since, as a rule, the phase of the integrand changes very rapidly from point to point over the entire region of the exit pupil, such calculations often represent a very formidable task even with the help of electronic computers. Now if a suitable generalization of the Young-Maggi-Rubinowicz theory were found there is hope that all these diffraction double integrals could be replaced by single integrals. In any case such a generalization would give a new insight into the physical process of image formation.

The present investigation is concerned with developing a general theory of the boundary diffraction wave. As a first step towards formulating this theory it will be shown that there exists a new potential $\mathbf{W}(Q,P)$, associated with any monochromatic wave field $U(P)$ which has the following properties:

(1) The normal component of its curl, taken with respect to the coordinates of any point Q on a closed surface S surrounding a field point P is equal to the integrand of the Helmholtz-Kirchhoff integral, i.e.,

$$\mathrm{curl}_Q \mathbf{W}(Q,P)\cdot\mathbf{n} = \frac{1}{4\pi}\left\{U(Q)\frac{\partial}{\partial n}\left(\frac{\exp(iks)}{s}\right) - \frac{\exp(iks)}{s}\frac{\partial}{\partial n}U(Q)\right\},$$

where s is the distance QP and $\partial/\partial n$ denotes differentiation along the unit inward normal $\mathbf{n}$ to S.

(2) $\mathbf{W}(Q,P)$ has always singularities at some points Q on S and the field at P may be rigorously expressed as the sum of disturbances propagated from these points alone.

Further a closed expression is derived for the vector potential $\mathbf{W}$ associated with any given monochromatic wave field U which obeys the Sommerfeld radiation condition at infinity. It is shown that in the special cases when U represents a plane or a spherical wave this expression reduces to those found by Rubinowicz[7,12,13] in his researches on the boundary diffraction wave.

Applications of these results to the general problem of the boundary diffraction wave within the framework of Kirchhoff's theory will be discussed in part II of this investigation.

2. NEW VECTOR POTENTIAL ASSOCIATED WITH THE HELMHOLTZ-KIRCHHOFF INTEGRAL

Consider a monochromatic scalar wave field of angular frequency ω, in free space:

$$V(x,y,z,t) = U(x,y,z)\exp(-i\omega t). \tag{2.1}$$

Here (x,y,z) are the Cartesian rectangular coordinates

[3] A. Rubinowicz, Nature **180**, 160 (1957).

[4] A. Rubinowicz, *Die Beugungswelle in der Kirchhoffschen Theorie der Beugung* (Polska Akademia Nauk, Warsaw, 1957).

[5] M. Born and E. Wolf, *Principles of Optics* (Pergamon Press, London and New York, 1959).

[6] G. A. Maggi, Ann. di Mat. IIa, **16**, 21 (1888).

[7] A. Rubinowicz, Ann. Physik **53**, 257 (1917).

[8] Since more complicated boundaries are in general encountered in the study of diffraction phenomena, the term "edge diffraction wave" would seem to be more appropriate but we conform here to the usual terminology.

[9] The equivalence of the formally different solutions of Maggi and Rubinowicz was first noted by F. Kottler [Ann. Physik **70**, 405 (1923); **71**, 457 (1923)] in his important papers dealing with the intricate concept of a black screen, the saltus interpretation of Kirchhoff theory and the generalization of the scalar Huygens-Fresnel Principle to electromagnetic waves. In this work the boundary wave also played an important role. Accounts of the researches of Maggi and Kottler are given in B. B. Baker and E. T. Copson, *The Mathematical Theory of Huygens' Principle* (Oxford University Press, Oxford, 1950), 2nd ed.

[10] R. S. Ingarden [Acta Phys. Polonica **14**, 77 (1955)] considered an extension of the Young-Maggi-Rubinowicz theory to waves which are of more complicated form, but which are restricted in that they have to obey the laws of geometrical optics. Even with this restriction no simple generalization was found by him.

[11] N. G. van Kampen, Physica **14**, 575 (1949).

[12] A. Rubinowicz, Phys. Rev. **54**, 931 (1938).

[13] A. Rubinowicz, Acta Phys. Polonica **12**, 225 (1953).

at a typical point P in the wave field and t denotes the time. The space-dependent part U satisfies the Helmholtz equation.

$$(\nabla^2+k^2)U=0, \tag{2.2}$$

where $k=\omega/c$. Let S be any closed surface bounding a volume $\mathcal{V}$, throughout and on the boundary of which U has continuous first- and second-order partial derivatives. According to the Helmholtz-Kirchhoff integral [reference 5, p. 376] the disturbance at any point P within $\mathcal{V}$ may then be expressed in the form

$$U(P)=\iint_S \mathbf{V}(Q,P)\cdot\mathbf{n}dS, \tag{2.3}$$

where

$$\mathbf{V}(Q,P)=\frac{1}{4\pi}\left\{U(Q)\,\mathrm{grad}_Q\frac{\exp(iks)}{s}-\frac{\exp(iks)}{s}\,\mathrm{grad}_Q U(Q)\right\}, \tag{2.4}$$

$\mathbf{n}$ is the unit *inward* normal to S, and s is the distance between P and a typical point Q on S (see Fig. 1). Further, grad_Q denotes the operator "gradient" taken with respect to the coordinates of Q.

It will now be shown that the vector $\mathbf{V}(Q,P)$ may always be expressed as the curl of a suitably chosen vector potential $\mathbf{W}(Q,P)$. For this purpose let us take the divergence of (2.4) and apply the identities $\mathrm{div}(u\mathbf{v})=u\,\mathrm{div}\mathbf{v}+\mathbf{v}\cdot\mathrm{grad}u$ and $\mathrm{div\,grad}=\nabla^2$. It then follows that

$$\mathrm{div}_Q\mathbf{V}(Q,P)=\frac{1}{4\pi}\left\{U(Q)\nabla_Q{}^2\frac{\exp(iks)}{s}-\frac{\exp(iks)}{s}\nabla_Q{}^2U(Q)\right\}. \tag{2.5}$$

Now both the functions $\exp(iks)/s$ and $U(Q)$ satisfy the Helmholtz equation (2.2), so that

$$\left.\begin{aligned}\nabla_Q{}^2\frac{\exp(iks)}{s}&=-k^2\frac{\exp(iks)}{s},\\ \nabla_Q{}^2U(Q)&=-k^2U(Q).\end{aligned}\right\} \tag{2.6}$$

On substituting from (2.6) in (2.5), it follows that

$$\mathrm{div}_Q\mathbf{V}(Q,P)=0. \tag{2.7}$$

This relation implies that whatever the nature of U is, $\mathbf{V}$ can always be expressed in terms of some vector potential $\mathbf{W}$ in the form

$$\mathbf{V}(Q,P)=\mathrm{curl}_Q\mathbf{W}(Q,P). \tag{2.8}$$

In terms of $\mathbf{W}$, the Helmholtz-Kirchhoff formula (2.3) becomes

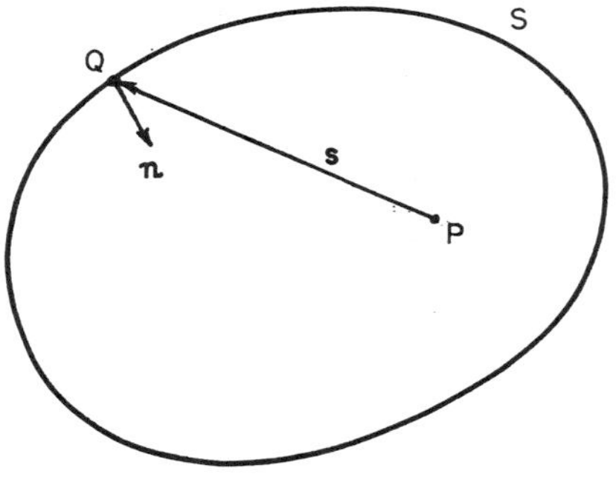

FIG. 1. Illustrating the Helmholtz-Kirchhoff integral theorem (2.3).

$$U(P)=\iint_S \mathrm{curl}_Q\mathbf{W}(Q,P)\cdot\mathbf{n}dS. \tag{2.9}$$

The vector potential $\mathbf{W}$ is, of course, not unique. For if $\mathbf{W}_1$ is an admissible potential, then, (because of the vector identity $\mathrm{curl\,grad}\equiv0$), the potential $\mathbf{W}_2=\mathbf{W}_1+\mathbf{grad}\phi$, where ϕ is an arbitrary scalar function is also admissible.

Considered as a function of Q (keeping P fixed for the moment), the vector potential $\mathbf{W}$ must have singularities on the surface S. For if $\mathbf{W}$ had no singularities on this surface, then Stokes' theorem, applied to (2.9) would imply that $U(P)=0$ at every point P inside the volume bounded by S. It will be assumed that all the singularities of $\mathbf{W}$ on S occur at discrete points[14] $Q_1, Q_2, \cdots, Q_N$. Let us surround all these points by small circles of radii $\sigma_1, \sigma_2, \cdots, \sigma_N$ and let $\Gamma_1, \Gamma_2, \cdots, \Gamma_n$ denote the boundaries of these circles. If S_- denotes the region of S which excludes the small circles, we have, by Stokes' theorem

$$\iint_{S-}\mathrm{curl}_Q\mathbf{W}\cdot\mathbf{n}dS=\sum_j\int_{\Gamma_j}\mathbf{W}\cdot\mathbf{l}dl, \tag{2.10}$$

where $\mathbf{l}$ is the unit vector along the tangent to Γ_j and dl is an element of Γ_j. The integral along Γ_j is described in the anticlockwise sense when the surface is viewed from the outside (see Fig. 2). From (2.9) and (2.10) it follows if we proceed to the limits $\sigma_j\to0$ ($j=1, \cdots N$), that the disturbance at P may be expressed in the form

$$U(P)=\textstyle\sum_j F_j(P), \tag{2.11}$$

where

$$F_j(P)=\lim_{\sigma_j\to0}\int_{\Gamma_j}\mathbf{W}\cdot\mathbf{l}dl. \tag{2.12}$$

[14] In some cases the singular points may not be discrete. In particular, in certain symmetrical situations one may expect that the singular points will form a continuous line. Such cases would need more refined analysis and will not be considered here.

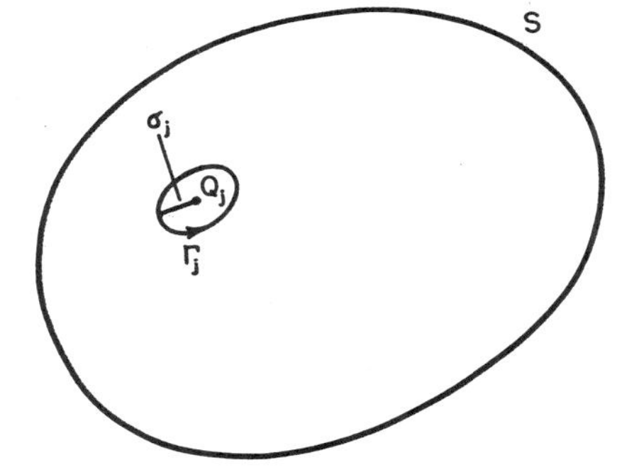

FIG. 2. Illustrating the notation relating the singular point of the vector potential.

It is of interest to note the successive reduction in the dimensionality of the representation of $U(P)$. The Helmholtz-Kirchhoff formula (2.3) is derived by transforming a volume integral into a surface integral (by means of Gauss' theorem). In the present section it was shown with the help of Stokes' theorem that the surface integral may be reduced to a set of line integrals which themselves degenerate into contributions from discrete points.

The main conclusion of the present section is the following: *If U represents a monochromatic wave disturbance whose first- and second-order partial derivatives are continuous within and on a closed surface S, then the value $U(P)$ at any point P inside S may be regarded as the sum of disturbances originating in a number of points (not necessarily discrete) $Q_1, Q_2, \cdots, Q_N$, which lie on S. These points are the singularities on S of the associated potential function* **W**, *and their location and also their number depends on the location of P inside S.*

In the next section the result just established will be illustrated for the special case when U is a plane wave. An explicit form of **W**, corresponding to any given field U will be derived in Sec. 4.

For the purpose of a later application, we note a trivial extension of the above result to the case when U has a discrete number of singularities $O_1, O_2, \cdots O_k$ inside the volume bounded by S but is regular at all other points within and on S. The effect of these singularities can be taken into account in the usual way: Each singularity O_j is surrounded by a small sphere S_j of radius ρ_j and the integration which led to (2.3) is taken throughout the volume inside S, excluding these spheres. Proceeding to the limit $\rho_j \rightarrow 0$ (all j), one obtains in place of (2.11) the more general formula

$$U(P)=\sum_j F_j(P)+\sum_j G_j(P), \qquad (2.13)$$

where the $F_j(P)$ are as before given by (2.12) and

$$G_j(P)=\lim_{\rho_j\rightarrow 0}\iint_{S_j}\mathbf{V}(Q,P)\cdot\mathbf{n}dS, \qquad (2.14)$$

Q being a typical point on the small sphere S_j and $\mathbf{n}$ is the unit inward normal to S_j.

3. VECTOR POTENTIAL FOR A PLANE WAVE

3.1 Homogeneous Plane Wave

Consider first a homogeneous plane wave, propagated in the direction specified by the unit $\mathbf{p}$:

$$U(P)=A\exp(ik\mathbf{p}\cdot\mathbf{r}). \qquad (3.1)$$

Here $\mathbf{r}$ is the position vector of P and A is a constant. If $\mathbf{r}'$ is the position vector of Q, the vector $\mathbf{V}$, defined by (2.4), which enters the integrand of the Helmholtz-Kirchhoff integral is easily found to be

$$\mathbf{V}(Q,P)=\frac{A}{4\pi}\exp(ik\mathbf{p}\cdot\mathbf{r}') \times\frac{\exp(iks)}{s}\left\{\left(ik-\frac{1}{s}\right)\hat{s}-ik\mathbf{p}\right\}, \qquad (3.2)$$

$\hat{s}$ denoting the unit vector in the direction PQ, i.e.,

$$\hat{s}=\mathbf{s}/s, \quad \mathbf{s}=\mathbf{r}'-\mathbf{r}. \qquad (3.3)$$

The formula (3.2) may be derived by means of (2.8) from the vector potential,

$$\mathbf{W}(Q,P)=A\exp(ik\mathbf{p}\cdot\mathbf{r}')\frac{\exp(iks)}{s}\frac{1}{4\pi}\frac{\hat{s}\times\mathbf{p}}{1+\hat{s}\cdot\mathbf{p}}. \qquad (3.4)$$

This expression is well known from previous investigations on the boundary diffraction wave.[15]

It is seen from (3.4) that **W** has only one singular point Q, on any convex surface S surrounding P, namely, the point on Q_1 for which

$$1+\hat{s}\cdot\mathbf{p}=0. \qquad (3.5)$$

This condition implies that Q_1 is the first intersection point with S of the line through the point P parallel to the direction $\mathbf{p}$ of the incident wave (see Fig. 3). According to the general result expressed by Eq. (2.11), the contribution $F_1(P)$ from Q_1 must be precisely equal

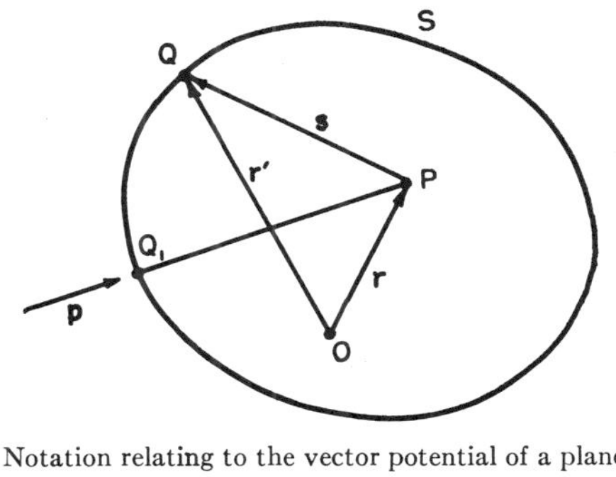

FIG. 3. Notation relating to the vector potential of a plane wave.

[15] B. B. Baker and E. T. Copson, *The Mathematical Theory of Huygens' Principle* (Oxford University Press, 1950), 2nd ed., p. 79.

to the value $U(P)$ of the field at P. We shall verify the correctness of this result by direct calculation.

Let Q_1 be surrounded by a small circle of radius σ_1 and with boundary Γ_1, situated on the surface S. Then, by (3.4),

$$\int_{\Gamma_1}\mathbf{W}\cdot\mathbf{l}dl=\frac{A}{4\pi}\int_{\Gamma_1}\exp(ik\mathbf{p}\cdot\mathbf{r}')\times\frac{\exp(iks)}{s}\frac{(\hat{s}\times\mathbf{p})\cdot\mathbf{l}}{1+\hat{s}\cdot\mathbf{p}}dl. \quad (3.6)$$

Let θ denote the angle between the direction $\mathbf{p}$ and QP, Q now being a typical point on the small circle Γ_1 (Fig. 4). Then

$$\frac{\hat{s}\times\mathbf{p}}{1+\hat{s}\cdot\mathbf{p}}=\frac{\sin\theta}{1-\cos\theta}\mathbf{k}=\frac{1}{\tan(\theta/2)}\mathbf{k}, \quad (3.7)$$

$\mathbf{k}$ being the unit vector

$$\mathbf{k}=\hat{s}\times\mathbf{p}/|\hat{s}\times\mathbf{p}|. \quad (3.8)$$

Also (see Fig. 4) $dl=\sigma_1 d\varphi/(\mathbf{k}\cdot\mathbf{l})$ where φ is the azimuthal angle of the typical point Q on Γ_1. Hence (3.6) becomes

$$\int_{\Gamma_1}\mathbf{W}\cdot\mathbf{l}dl\simeq A\exp(ik\mathbf{p}\cdot\mathbf{r})\times\frac{1}{4\pi}\int_0^{2\pi}\frac{\exp[ik(\mathbf{p}\cdot\mathbf{s}+s)]}{s}\frac{\sigma_1}{\tan\theta/2}d\varphi. \quad (3.9)$$

Now for σ_1 small enough, one has, with negligible errors,

$$\theta=\sigma_1/s. \quad (3.10)$$

Hence (3.9) gives, in the limit as $\sigma_1\rightarrow 0$

$$F_1(P)=\lim_{\sigma_1\to 0}\int_{\Gamma_1}\mathbf{W}\cdot\mathbf{l}dl =A\exp(ik\mathbf{p}\cdot\mathbf{r})\frac{1}{4\pi}\int_0^{2\pi}\frac{2s}{s}d\varphi \quad (3.11)$$
$$=A\exp(ik\mathbf{p}\cdot\mathbf{r}),$$

since $\mathbf{p}\cdot\mathbf{s}+s=0$ at Q_1. Thus $F_1(P)$ is precisely the value of the field at P,

$$F_1(P)=U(P), \quad (3.12)$$

in agreement with our general considerations.

3.2 Inhomogeneous Plane Wave

So far it has been assumed that the unit vector $\mathbf{p}$ is real. For the purpose of later analysis we must also consider the case where one of the Cartesian components of $\mathbf{p}$ is imaginary.

Let p_x, p_y, and p_z be the components of the unit vector $\mathbf{p}$ $(p_x^2+p_y^2+p_z^2=1)$ with respect to a set of Cartesian rectangular axes. Of special interest for the present analysis is the case when p_x and p_y are real and

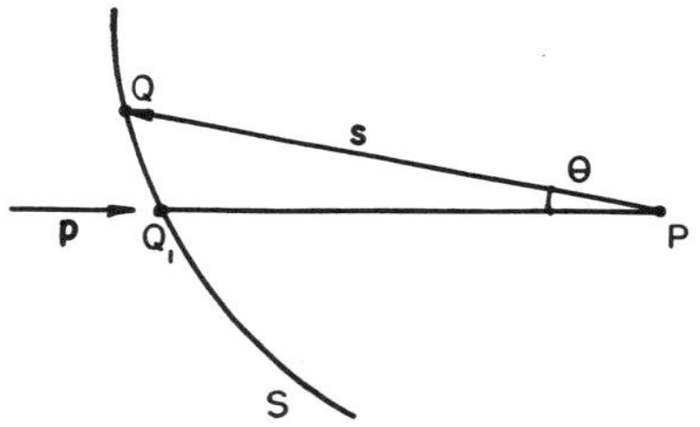

FIG. 4. Notation relating to the evaluation of the contribution from the singular point Q_1 for a *homogeneous* plane wave.

p_z is pure imaginary, i.e.,

$$p_z=\pm i|p_z|. \quad (3.13)$$

Then

$$p_x^2+p_y^2-|p_z|^2=1. \quad (3.14)$$

Both cases with the positive or negative signs in (3.13) will be of interest and we therefore treat the two cases together. If (x,y,z) are the coordinates of P, we have

$$U(P)=A\exp(ik\mathbf{p}\cdot\mathbf{r})=A\exp\{\mp k|p_z|z\}\times\exp[ik(p_xx+p_yy)]. \quad (3.15)$$

This represents an inhomogeneous plane wave, whose amplitude decreases or increases exponentially with z, according to whether the upper or lower sign is taken.

It is readily verified that (3.2) and (3.4) apply in the present case also. However, there is a difference in the location of the singular points. The singular points Q_j are still derived from (3.5), but because of (3.13) the corresponding $\hat{s}$ directions are given by

$$p_z\hat{s}_z=0, \quad (3.16a)$$

$$p_x\hat{s}_x+p_y\hat{s}_y+1=0. \quad (3.16b)$$

Since p_z is assumed to differ from zero, $\hat{s}_z=0$, i.e., the singular points now lie on the curve C of intersection between the surface S and the plane $z=z_p$, z_p being the z coordinate of P. If we set

$$p_x=(p_x^2+p_y^2)^{\frac{1}{2}}\cos\varphi,\quad p_y=(p_x^2+p_y^2)^{\frac{1}{2}}\sin\varphi, \quad (3.17a)$$

$$\hat{s}_x=\cos\varphi',\quad \hat{s}_y=\sin\varphi', \quad (3.17b)$$

the relation (3.16b) gives $(p_x^2+p_y^2)^{\frac{1}{2}}\cos(\varphi'-\varphi)=-1$. If we set

$$\chi=\varphi'-\varphi, \quad (3.18)$$

and make use of (3.14) it follows that

$$\cos\chi=-1/(1+|p_z|^2)^{\frac{1}{2}}. \quad (3.19)$$

Since there will be two values of χ, χ_1 and χ_2, say, in the range $0\leqslant\chi<2\pi$ which satisfy (3.19), there are now two singular points, Q_1 and Q_2 (see Fig. 5).

4. GENERAL EXPRESSION FOR THE VECTOR POTENTIAL

We shall now derive a general expression for the vector potential $\mathbf{W}(\mathbf{r}',\mathbf{r})$ associated with any given wave field $U(\mathbf{r}')$.

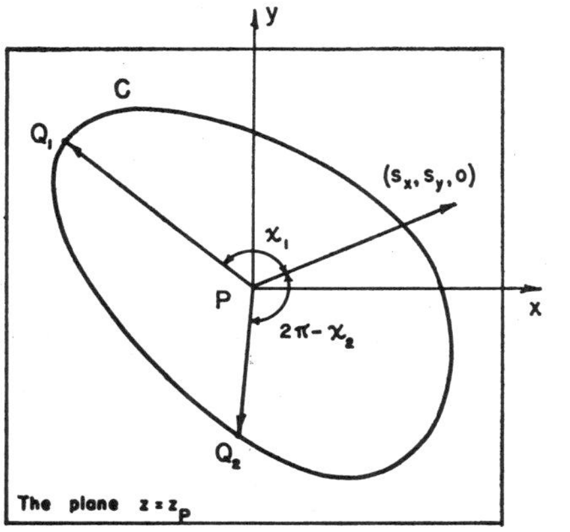

FIG. 5. Illustrating the locations of the singular points Q_1 and Q_2 for an *inhomogeneous* plane wave. (C is the curve in which the plane $z=z_p$ intersects the surface S.)

In general, the wave field may be expressed in the form of an *angular spectrum of plane waves*, i.e., in the form[16]

$$U(\mathbf{r}')=U_0^+(\mathbf{r}')+U_0^-(\mathbf{r}')+U_i^+(\mathbf{r}')+U_i^-(\mathbf{r}'), \quad (4.1)$$

where[17]

$$U_0^\pm(\mathbf{r}')=\iint_{p_x^2+p_y^2\leq 1} A_0^\pm(p_x,p_y) \times\exp(ik\mathbf{p}_0^\pm\cdot\mathbf{r}')dp_xdp_y, \quad (4.2a)$$

$$U_i^\pm(\mathbf{r}')=\iint_{p_x^2+p_y^2>1} A_i^\pm(p_x,p_y) \times\exp(ik\mathbf{p}_i^\pm\cdot\mathbf{r}')dp_xdp_y, \quad (4.2b)$$

where $A_0^\pm$, $A_i^\pm$ are (generally complex) amplitude factors and

$$p_0^\pm=[p_x,\ p_y,\ \pm(1-p_x^2-p_y^2)^{\frac{1}{2}}], \quad (4.3a)$$

$$p_i^\pm=[p_x,\ p_y,\ \pm i(p_x^2+p_y^2-1)^{\frac{1}{2}}]. \quad (4.3b)$$

For the sake of brevity, (4.1) will be written symbolically in the more compact form

$$U(\mathbf{r}')=\iint A(\mathbf{p})\exp(ik\mathbf{p}\cdot\mathbf{r}')dp_xdp_y. \quad (4.4)$$

Equation (4.4) expresses the field as a superposition of plane waves, generally of different amplitudes and directions of propagations, including complex ones. The waves for which the propagation vector $\mathbf{p}$ is given by (4.3a) are ordinary (homogeneous) waves (subscript 0); those for which $\mathbf{p}$ is given by (4.3b) are inhomogeneous waves (subscript i).

Assuming that the order of the operation curl_Q and integration over the (p_x,p_y) domain may be interchanged, the vector potential $\mathbf{W}$ associated with the wave field U, may be obtained by linear superposition of the potentials associated with all the plane waves forming the angular spectrum. Hence, using (3.4) and remembering that (3.4) applies to a homogeneous as well as to an inhomogeneous wave, it follows that the vector potential associated with U may be expressed in the form

$$\mathbf{W}(\mathbf{r}',\mathbf{r})=\frac{1}{4\pi}\frac{\exp(iks)}{s}\hat{s}\times\mathbf{w}(\mathbf{r}',\hat{s}), \quad (4.5)$$

where

$$\mathbf{w}(\mathbf{r}',\hat{s})=\iint A(\mathbf{p})\exp(ik\mathbf{p}\cdot\mathbf{r}')\frac{\mathbf{p}}{1+\hat{s}\cdot\mathbf{p}}dp_xdp_y. \quad (4.6)$$

Here $\mathbf{r}'$ and $\mathbf{r}$ are again the position vectors of Q and P, respectively, $\hat{s}=\mathbf{s}/s$ and $\mathbf{s}=\mathbf{r}'-\mathbf{r}$.

It is, of course, desirable to express $\mathbf{W}$ directly in terms of the wave field U rather than in terms of the spectral amplitudes A. For this purpose, a method based on operational techniques will be used.

Let us take the gradient of (4.4), with respect to the argument $\mathbf{r}'$:

$$\mathrm{grad}'U=ik\iint \mathbf{p}A(\mathbf{p})\exp(ik\mathbf{p}\cdot\mathbf{r}')dp_xdp_y.$$

It is seen that the operation of taking the gradient is equivalent to a multiplication by $ik\mathbf{p}$ under the integral signs. Hence (4.5) may be written symbolically (in the sense of Heaviside's operational calculus[18]) as

$$\mathbf{w}(\mathbf{r}',\hat{s})=\frac{(ik)^{-1}\,\mathrm{grad}'}{1+(ik)^{-1}\hat{s}\cdot\mathrm{grad}'}U(\mathbf{r}'). \quad (4.7)$$

$$\hat{s}\cdot\mathrm{grad}'=\partial/\partial\tau. \quad (4.8)$$

This operator represents differentiation in the direction of $\hat{s}$. Equation (4.7) then becomes

$$\mathbf{w}(\mathbf{r}',\hat{s})=[1/(ik+\partial/\partial\tau)]\,\mathrm{grad}'U(\mathbf{r}'). \quad (4.9)$$

Hence, as is clear from (4.6), $\mathbf{w}$ satisfies the differential equation

$$(\partial/\partial\tau)\mathbf{w}(\mathbf{R},\hat{s})+ik\mathbf{w}(\mathbf{R},\hat{s})=\mathrm{grad}U(\mathbf{R}), \quad (4.10)$$

where

$$\mathbf{R}=\mathbf{r}+\tau\hat{s}, \quad (4.11)$$

is the position vector of a typical point A on the line PQ (see Fig. 6).

[16] E. Wolf, Proc. Phys. Soc. (London) 74, 280 (1959).

[17] If there are singularities of the field in the region under consideration, $A(p_x,p_y)$ may change on crossing a plane containing each singularity [cf., expression (6.8) below for the spherical wave].

[18] R. Courant and D. Hilbert, *Methoden der Mathematischen Physik* (Springer-Verlag, Berlin, 1937), Vol. II, p. 187.

To solve (4.10) we multiply both sides by $\exp(ik\tau)$ and obtain

$$\frac{\partial}{\partial\tau}[\mathbf{w}\exp(ik\tau)]=\exp(ik\tau)\cdot\mathrm{grad}U. \quad (4.12)$$

On integrating (4.12) between the limits $\tau=s_0$ and $\tau=s$, it follows that

$$\mathbf{w}(\mathbf{r}+s\hat{s},\hat{s})\exp(iks)-\mathbf{w}(\mathbf{r}+s_0\hat{s},\hat{s})\exp(iks_0)$$

$$=\int_{s_0}^{s}\exp(ik\tau)\,\mathrm{grad}U(\mathbf{r}+\tau\hat{s})d\tau, \quad (4.13)$$

or, using (4.5),

$$\mathbf{W}(\mathbf{r}',\mathbf{r})=\frac{\hat{s}}{4\pi s}\times\left[\int_{s_0}^{s}\exp(ik\tau)\,\mathrm{grad}U(\mathbf{r}+\tau\hat{s})d\tau+\mathbf{w}(\mathbf{r}+s_0\hat{s},\hat{s})\exp(iks_0)\right]. \quad (4.14)$$

In the integral on the right of (4.14) let the variable of integration be changed from τ to $\mu=\tau-s$. Remembering that $\mathbf{r}+\tau\hat{s}=\mathbf{r}'+\mu\hat{s}$, $\mathbf{r}'+\mathbf{r}+s\hat{s}$ and formally proceeding to the limit $s_0\rightarrow\infty$, one finally obtains the following expression for the vector potential:

$$\mathbf{W}(\mathbf{r}',\mathbf{r})=\frac{\exp(iks)}{4\pi s}\hat{s}\times\int_{\infty}^{0}\exp(ik\mu)\,\mathrm{grad}U(\mathbf{r}'+\mu\hat{s})d\mu+\mathbf{W}_\infty, \quad (4.15)$$

where

$$\mathbf{W}_\infty=\frac{\hat{s}}{4\pi s}\times\lim_{s_0\rightarrow\infty}[\mathbf{w}(\mathbf{r}+s_0\hat{s},\hat{s})\exp(iks_0)]. \quad (4.16)$$

In (4.15) the operator "grad" operates on the argument $\mathbf{r}'+\mu\hat{s}$ of U.

The formula (4.15) shows that, apart from an additive contribution $\mathbf{W}_\infty$ which will be discussed below, the vector potential $\mathbf{W}(\mathbf{r}',\mathbf{r})$ has the form of a spherical wave, the vector amplitude of which is at right angles to the direction joining the two points $Q(\mathbf{r}')$ and $P(\mathbf{r})$. The vector amplitude may be considered to be the sum of contributions from each point A on the half line $A_\infty Q$ (see Fig. 6), the extension of which passes through P. Each contribution is proportional to the vector product of the unit vector $\hat{s}$ is in the direction PQ, with the gradient of the wave field at that point and contains an effective retardation term $\exp(ik\mu)=\exp(i\omega\mu/c)$, where μ/c is precisely the time needed for the light disturbance to propagate from A to Q. In general there is also a contribution from infinity, represented by the term $\mathbf{W}_\infty$.

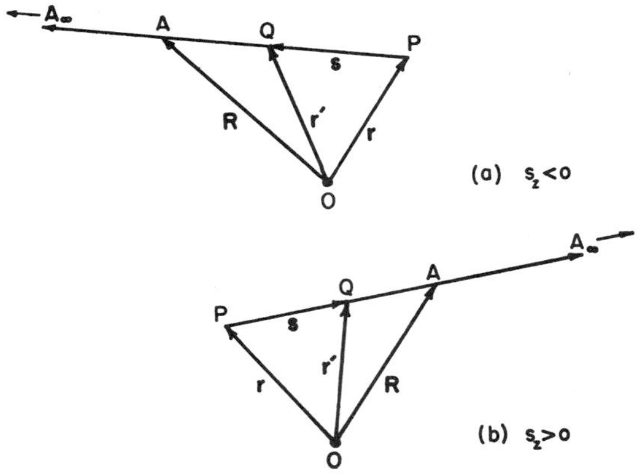

FIG. 6. Illustrating the meaning of the position vectors $\mathbf{r}$, $\mathbf{r}'$ and $\mathbf{R}$: $\mathbf{r}'=\mathbf{r}+s\,\hat{s}$, $\mathbf{R}=\mathbf{r}+\tau\hat{s}=\mathbf{r}'+\mu\hat{s}$, $\mu=\tau-s$.

It appears difficult to obtain a general closed expression for the residual contribution $\mathbf{W}_\infty$, though in any particular case this contribution can, in principle, be determined from (4.16) and (4.6). However, it is not difficult to show that if the field obeys the Sommerfeld radiation condition[15] in half of the space (in $z>0$ when $s_z>0$ and in $z<0$ when $s_z<0$) then $\mathbf{W}_\infty=0$. To show this consider the asymptotic behavior of the integral (4.6) as the point Q moves to infinity in the direction specified by a unit vector $\mathbf{u}$. It is readily shown with the help of the principle of stationary phase (see the Appendix) that for sufficiently large values of $kR=k|\mathbf{R}|$, $\mathbf{w}_0^+$ and $\mathbf{w}_0^-$ (the parts of $\mathbf{w}$ given by (4.6), involving the amplitudes A_0^+ and A_0^-, respectively) have the following behavior for large R (if formally R, $\mathbf{u}$ and $\hat{s}$ are for the moment treated as independent):

$$\mathbf{w}_0^+(R\mathbf{u},\hat{s})\sim-\frac{2\pi iu_z}{k}\left(\frac{\mathbf{u}}{1+\hat{s}\cdot\mathbf{u}}\right)A_0^+(u_x,u_y)\frac{\exp(ikR)}{R},\quad \text{when } u_z>0, \quad (4.17\text{a})$$

$$\sim-\frac{2\pi iu_z}{k}\left(\frac{-\mathbf{u}}{1-\hat{s}\cdot\mathbf{u}}\right)A_0^+(-u_x,-u_y)\frac{\exp(-ikR)}{R},\quad \text{when } u_z<0, \quad (4.17\text{b})$$

$$\mathbf{w}_0^-(R\mathbf{u},\hat{s})\sim\frac{2\pi iu_z}{k}\left(\frac{-\mathbf{u}}{1-\hat{s}\cdot\mathbf{u}}\right)A_0^-(-u_x,-u_y)\frac{\exp(-ikR)}{R},\quad \text{when } u_z>0, \quad (4.18\text{a})$$

$$\sim\frac{2\pi iu_z}{k}\left(\frac{\mathbf{u}}{1+\hat{s}\cdot\mathbf{u}}\right)A_0^-(u_x,u_y)\frac{\exp(ikR)}{R},\quad \text{when } u_z<0. \quad (4.18\text{b})$$

Next consider the behavior of $\mathbf{w}_i^+$ and $\mathbf{w}_i^-$ (the part of $\mathbf{w}$ involving the amplitudes A_i^+ and A_i^-). It is evident that with increasing R $\mathbf{w}_i^+$ increases exponentially to ∞ or decreases exponentially to zero according as $u_z<0$ or $u_z>0$. The opposite is the case for each plane wave contributing to $\mathbf{w}_i^-$, i.e., $\mathbf{w}_i^-$ decreases or increases exponentially according as $u_z<0$ or $u_z>0$. Obviously the waves whose amplitudes diverge as $R\rightarrow\infty$ cannot be admitted on physical grounds. Hence, if the representation (4.1) is to be valid for $z\rightarrow-\infty$ then $U_i^+=0$ and if it is to be valid for $z\rightarrow+\infty$ then $U_i^-=0$.

Returning to (4.16) let us decompose $\mathbf{W}_\infty$ into two parts,[19]

$$\mathbf{W}_\infty=\mathbf{W}_\infty^+ + \mathbf{W}_\infty^-, \tag{4.19}$$

where the first part on the right arises from contributions associated with the spectral amplitudes A_0^+ and the other from those associated with the spectral amplitudes A_0^-. Then using (4.17a) or (4.18b) according as $s_z<0$ $(u_z<0)$ or $s_z>0$ $(u_z>0)$ one immediately obtains (noting that $1+\hat{\mathbf{s}}\cdot\mathbf{u}\rightarrow 2$ as $R\rightarrow\infty$)

$$\left.\begin{aligned}\mathbf{W}_\infty^+&=0 \quad \text{when} \quad s_z>0,\\ \mathbf{W}_\infty^-&=0 \quad \text{when} \quad s_z<0.\end{aligned}\right\} \tag{4.20}$$

Moreover, if $s_z>0$ and the field obeys the Sommerfeld radiation condition as $R\rightarrow\infty$ in the half-space $z>0$, then $A_0^-=0$ and hence, using (4.18a),

$$\mathbf{W}_\infty^-=0 \quad (s_z>0). \tag{4.21}$$

Similarly, if $s_z<0$ and the field obeys the Sommerfeld radiation condition as $R\rightarrow\infty$ in the half-space $z<0$, then $A_0^+=0$ and (4.17b) gives

$$\mathbf{W}_\infty^+=0 \quad (s_z<0). \tag{4.22}$$

From (4.19)–(4.22) it may be concluded that *if* $s_z>0$ *(or* $s_z<0$*) and the incident field obeys the Sommerfeld radiation condition as* $R\rightarrow\infty$ *in the half-space* $z>0$ *(or* $z<0$*), then the residual term*

$$\mathbf{W}_\infty=0. \tag{4.23}$$

If the Sommerfeld radiation condition is not satisfied in the appropriate half-space, $\mathbf{W}_\infty$ will in general be nonzero. In such cases the value of $\mathbf{W}_\infty$ cannot be calculated with the help of (4.17b) or (4.18a); for in this case the factor $1-\mathbf{s}\cdot\mathbf{u}\rightarrow 0$ as $R\rightarrow\infty$ and (4.17b) and (4.18) now no longer describe correctly the asymptotic behavior of $\mathbf{W}_0^+$ and $\mathbf{W}_0^-$. In such cases $\mathbf{W}_\infty$ may in principle be determined from the defining equations (4.16) and (4.6). For a converging spherical wave this will be done in Sec. 6.

The general formula (4.15) for the vector potential $\mathbf{W}$ will now be applied to determine the potential for spherical waves (diverging and converging); it will be seen to lead to the expressions found by Rubinowicz in his researches concerned with the boundary diffraction wave. The full connection between the present theory and the theory of the boundary wave will be discussed in Part II of the present investigation.

[19] There are no contributions to $\mathbf{W}_\infty$ arising from A_i^+ or A_i^-, since one of these is neglected for physical reasons already explained, and the other obviously tends to zero as $s_0\rightarrow\infty$.

5. DIVERGING SPHERICAL WAVE

Consider first the case of a *diverging* spherical wave

$$U(\mathbf{r})=\exp(ikr)/r. \tag{5.1}$$

Then

$$\operatorname{grad}U(\mathbf{R})=\mathbf{R}\left(\frac{ik}{R}-\frac{1}{R^2}\right)\frac{\exp(ikR)}{R}, \tag{5.2}$$

with

$$\mathbf{R}=\mathbf{r}'+\mu\hat{s}, \quad R=|\mathbf{R}|. \tag{5.3}$$

Substitution into (4.15) gives

$$\mathbf{W}(\mathbf{r}',\mathbf{r})=\frac{\exp(iks)}{4\pi s}\hat{s}\times\mathbf{r}'\times\int_\infty^0\left(ik-\frac{1}{R}\right)\frac{\exp[ik(R+\mu)]}{R^2}d\mu+\mathbf{W}_\infty, \tag{5.4}$$

where $\mathbf{W}_\infty$ is given by expression of the form (4.24). To evaluate the integral in (5.4) we note that

$$R^2=r'^2+2\mu(\mathbf{r}'\cdot\hat{s})+\mu^2, \tag{5.5}$$

$$dR/d\mu=(\mathbf{r}'\cdot\hat{s}+\mu)/R. \tag{5.6}$$

Using (5.5) and (5.6) one can readily verify that the integrand of the integral in (5.4) is equal to the differential

$$\frac{d}{d\mu}\left\{\frac{1}{R+\mathbf{r}'\cdot\hat{s}+\mu}\frac{\exp[ik(R+\mu)]}{R}\right\}. \tag{5.7}$$

Hence (5.4) becomes, since $R=r'$ when $\mu=0$ and $R\rightarrow\infty$ as $\mu\rightarrow\infty$,

$$\mathbf{W}(\mathbf{r}',\mathbf{r})=\frac{\exp(iks)}{4\pi s}\hat{s}\times\left[\frac{\mathbf{r}'}{r'+\mathbf{r}'\cdot\hat{s}}\frac{\exp(ikr')}{r'}\right]+\mathbf{W}_\infty. \tag{5.8}$$

Now a divergent spherical wave obeys the Sommerfeld radiation condition both in $z>0$ and $z<0$ as $R\rightarrow\infty$. Hence according to (4.21) and (4.22)

$$\mathbf{W}_\infty=0. \tag{5.9}$$

Thus if we also substitute $\mathbf{s}/s$ for $\hat{s}$, we obtain from (5.8)

$$\mathbf{W}(\mathbf{r}',\mathbf{r})=\frac{1}{4\pi}\frac{\exp(ikr')}{r'}\frac{\exp(iks)}{s}\frac{\mathbf{s}\times\mathbf{r}'}{sr'+\mathbf{s}\cdot\mathbf{r}'}. \tag{5.10}$$

This is the required expression for the potential associated with a diverging spherical wave. It is identical with the formula found by Rubinowicz[13] for this particular case.

Consider now the singularities of $\mathbf{W}$ on a closed convex surface surrounding a typical field point $P(\mathbf{r})$. According to (5.10) the singularities are given by the condition

$$sr'+\mathbf{s}\cdot\mathbf{r}'=0. \tag{5.11}$$

If θ denotes the angle between $\mathbf{s}$ and $\mathbf{r}'$, (5.11) implies that $\cos\theta=-1$, i.e., $\theta=\pi$. Two cases must be distinguished:

(a) *The surface S does NOT enclose the singularity O of the spherical wave* (the singularity of U). Then evidently $\mathbf{W}$ has just *one* singular point on S, namely the point Q_1, in which the line OP intersects S the first time (Fig. 7). Since $\mathbf{W}$ has now only one singularity on S and the field is regular within the volume bounded by S, it follows from the general formula (2.11) that the disturbance $F_1(P)$ associated with Q_1 must be equal to $U(P)$:

$$F_1(P)=U(P)=\exp(ikr)/r, \tag{5.12}$$

where r is the distance OP. This result may also be verified by direct evaluation of the expression (2.12) for F_1 by means of calculations similar to those carried out in Sec. 3.1 in connection with the homogeneous plane wave.

(b) *The surface S encloses the singularity O of the spherical wave.* In this case no point Q_1 on S obeys the condition (5.11), so that $\mathbf{W}$ has now no singularity on the surface. However, the field U has one singularity (at O) within the volume bounded by S. According to (2.14) the disturbance $G_1(P)$ associated with O must be equal to $U(P)$, i.e.,

$$G_1(P)=U(P)=\exp(ikr)/r, \tag{5.13}$$

where as before r represents the distance OP. This result may again be verified by direct calculation, using the formula (2.14).

6. CONVERGING SPHERICAL WAVE

Let us now consider the case of a *converging* spherical wave

$$U(\mathbf{r})=\frac{\exp(-ikr)}{r}. \tag{6.1}$$

The general formula (4.15) now gives

$$\mathbf{W}(\mathbf{r}',\mathbf{r})=\frac{\exp(iks)}{4\pi s}\hat{s}\times\mathbf{r}\int_{\infty}^{0}\left(-ik-\frac{1}{R}\right)\times\frac{\exp[ik(-R+\mu)]}{R^2}d\mu+\mathbf{W}_\infty. \tag{6.2}$$

Using (5.5) and (5.6) it may be verified that the integrand of the integral in (6.2) is equal to the differential

$$\frac{d}{d\mu}\left\{\frac{1}{-R+\mathbf{r}\cdot\hat{s}+\mu}\,\frac{\exp[ik(-R+\mu)]}{R}\right\}. \tag{6.3}$$

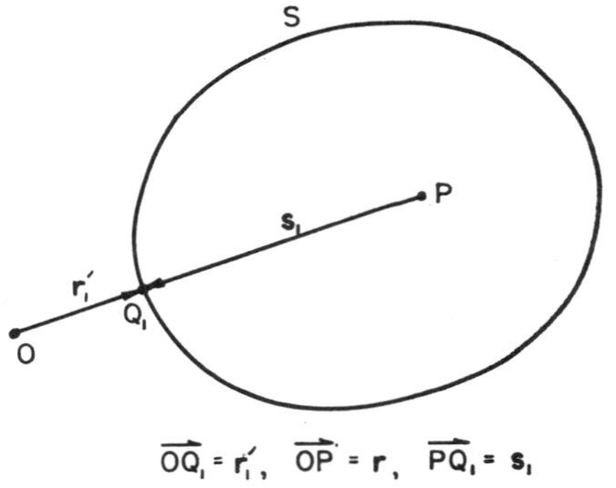

FIG. 7. Illustrating the location of the singular point Q_1 of $\mathbf{W}$ for a *diverging* spherical wave, when the surface S does not enclose the singularity O of the wave.

Hence

$$\int_{\mu_0}^{0}\left(-ik-\frac{1}{R}\right)\frac{\exp[ik(-R+\mu)]}{R^2}d\mu =\frac{1}{-r'+\mathbf{r}'\cdot\hat{s}}\,\frac{\exp(-ikr')}{r'} -\frac{1}{-R_0+\mathbf{r}'\cdot\hat{s}+\mu_0}\,\frac{\exp(-ikR_0)}{R_0}, \tag{6.4}$$

where R_0 is the value of R corresponding to $\mu=\mu_0$. Unlike the case of a divergent spherical wave, the contribution from the lower limit μ_0 of the integral does not vanish in the present case as $\mu_0\rightarrow\infty$. To evaluate the contribution we make appeal to Fig. 6 and find that

$$\begin{aligned}\mu_0&=[R_0^2-(\mathbf{r}\times\hat{s})^2]^{\frac{1}{2}}-(\mathbf{r}'\cdot\hat{s})\\&=R_0\left[1-\left(\frac{\mathbf{r}\times\hat{s}}{R_0}\right)^2\right]^{\frac{1}{2}}-(\mathbf{r}'\cdot\hat{s})\\&=R_0-\frac{1}{2}\frac{(\mathbf{r}\times\hat{s})^2}{R_0}+0\left(\frac{1}{R_0^3}\right)-(\mathbf{r}'\cdot\hat{s}).\end{aligned}$$

Hence as $\mu_0\rightarrow\infty$, one has (since then, also $R_0\rightarrow\infty$),

$$\frac{1}{-R_0+\mathbf{r}'\cdot\hat{s}+\mu_0}\,\frac{\exp[ik(-R_0+\mu)]}{R_0}\rightarrow-\frac{2}{(\mathbf{r}\times\hat{s})^2}\times\exp(-ik\mathbf{r}'\cdot\hat{s}) \tag{6.5}$$

and (6.2) reduces to

$$\mathbf{W}(\mathbf{r}',\mathbf{r})=\frac{\exp(iks)}{4\pi s}\hat{s}\times\mathbf{r}\left[\frac{1}{-r'+\mathbf{r}'\cdot\mathbf{s}}\,\frac{\exp(-ikr')}{r'}\right]+\frac{\hat{s}\times\mathbf{r}'}{2\pi s}\,\frac{\exp[-ik(\hat{s}\cdot\mathbf{r}')+iks]}{(\hat{s}\times\mathbf{r})^2}+\mathbf{W}_\infty. \tag{6.6}$$

Now the second term in (6.6) may be written in a somewhat different form. According to Fig. 6, $s-\mathbf{r}'\cdot\hat{s}=-\hat{s}\cdot\mathbf{r}$ and also, $\mathbf{s}\times\mathbf{r}'=\mathbf{s}\times\mathbf{r}$, since $\mathbf{r}'=\mathbf{r}+s\hat{s}$. Hence the second term in (6.6) becomes

$$\frac{1}{2\pi s}\frac{\hat{s}\times\mathbf{r}}{(\hat{s}\times\mathbf{r})^2}\exp(-ik\hat{s}\cdot\mathbf{r}). \quad (6.7)$$

Next we must evaluate the contribution $\mathbf{W}_\infty$. For this purpose we represent the converging spherical wave in the form of an angular spectrum of plane waves. This representation [analogous to one given by H. Weyl, Ann. Physik **60**, 481 (1919) for a divergent spherical wave] is well known. In fact

$$U(\mathbf{r})=\frac{\exp(-ikr)}{r}=\frac{ik}{2\pi}\int_{-\infty}^{\infty}\int\frac{\exp[ik(p_xx+p_yy+p_z^-|z|)]}{p_z^-}dp_xdp_y, \quad (6.8)$$

where

$$\begin{aligned}p_z^-&=-(1-p_x^2-p_y^2)^{\frac{1}{2}}, &&\text{when } p_x^2+p_y^2\leqslant 1,\\ &=i(p_x^2+p_y^2-1)^{\frac{1}{2}}, &&\text{when } p_x^2+p_y^2>1.\end{aligned} \quad (6.9)$$

Comparison of (6.8) and (4.4) immediately gives the spectral amplitudes $A(\mathbf{p})$. Then using (4.6) and (4.16) one obtains

$$\mathbf{W}_\infty=\frac{1}{4\pi s}\lim_{s_0\to\infty}\left[\hat{s}\times\int_{-\infty}^{\infty}\int\frac{\mathbf{p}}{1+\hat{s}\cdot\mathbf{p}}\frac{ik}{2\pi}\frac{1}{p_z^-}\times\exp\{ik[\mathbf{p}\cdot\mathbf{r}+(\hat{s}\cdot\mathbf{p}+1)s_0]\}dp_xdp_y\right]. \quad (6.10)$$

To evaluate this expression we note that

$$\begin{aligned}\frac{\partial}{\partial s_0}&\left[\hat{s}\times\int_{-\infty}^{\infty}\int\frac{\mathbf{p}}{1+\hat{s}\cdot\mathbf{p}}\frac{ik}{2\pi}\frac{1}{p_z^-}\times\exp\{ik[\mathbf{p}\cdot\mathbf{r}+(\hat{s}\cdot\mathbf{p}+1)s_0]\}dp_xdp_y\right]\\ &=\hat{s}\times\int_{-\infty}^{\infty}\int ik\mathbf{p}\frac{ik}{2\pi}\frac{1}{p_z^-}\times\exp\{ik[\mathbf{p}\cdot\mathbf{r}+(\hat{s}\cdot\mathbf{p}+1)s_0]\}dp_xdp_y\\ &=\hat{s}\times\exp(iks_0)\,\text{grad}\frac{\exp(-ikR)}{R}\\ &=\frac{\partial}{\partial s_0}\left[\frac{\hat{s}\times\mathbf{r}}{-R+\hat{s}\cdot\mathbf{r}+s_0}\frac{\exp[ik(-R+s_0)]}{R}\right],\end{aligned} \quad (6.11)$$

with $\mathbf{R}=\mathbf{r}+s_0\hat{s}$, where, on going from the second to the third expression, the representation (6.8) was again used and on going from the third to the fourth line the identity employed already in connection with (6.2) and (6.3) was used.

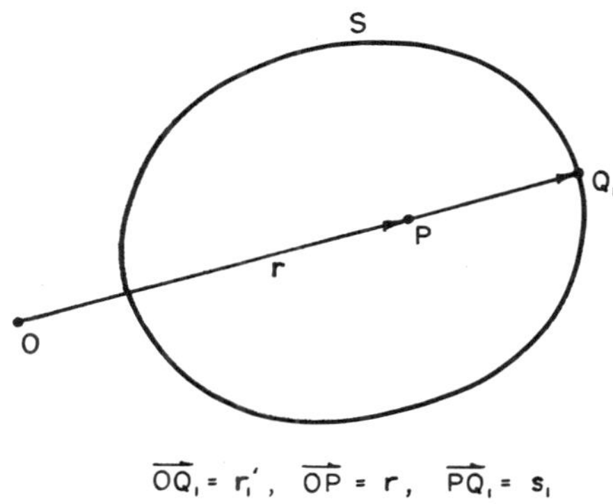

FIG. 8. Illustrating the location of the singular point Q_1' of $\mathbf{W}$ for a *converging* spherical wave, when the surface S does not enclose the singularity O of the wave.

The terms which follow the differentiation operator $\partial/\partial s_0$ in the first and the last line of (6.11) can evidently only differ by an additive term independent of s_0 and by considering the limiting form of the expressions as $s_0\to-\infty$ this term is readily seen to be zero. Hence (6.10) becomes

$$\mathbf{W}_\infty=\frac{\hat{s}\times\mathbf{r}}{4\pi s}\lim_{s_0\to\infty}\left[\frac{1}{-R+\hat{s}\cdot\mathbf{r}+s_0}\frac{\exp[ik(-R+s_0)]}{R}\right],$$

or, using (6.5),

$$\mathbf{W}_\infty=-\frac{1}{2\pi s}\frac{\hat{s}\times\mathbf{r}}{(\hat{s}\times\mathbf{r})^2}\exp(-ik\hat{s}\cdot\mathbf{r}). \quad (6.12)$$

On substituting from (6.12) into (6.6) it is seen, in view of (6.7), that the term $\mathbf{W}_\infty$ precisely cancels out the second term in (6.6), and if $\hat{s}$ is replaced by $\mathbf{s}/s$ in the first term, one finally obtains the following expression for the vector potential of a converging spherical wave:

$$\mathbf{W}(\mathbf{r}',\mathbf{r})=-\frac{1}{4\pi}\frac{\exp(-ikr')}{r'}\frac{\exp(iks)}{s}\frac{\mathbf{s}\times\mathbf{r}'}{sr'-\mathbf{s}\cdot\mathbf{r}'}. \quad (6.13)$$

It will be seen later (Part II) that (6.13) leads to the same expression for the boundary diffraction wave as found by Rubinowicz[12] in his study of the anomalous propagation of phase near focus of a converging spherical wave diffracted at an aperture.

Next let us consider the singularities of $\mathbf{W}$ on a closed convex surface S. In place of (5.11), one now has, according to (6.13), the following expression for the position of the singularities:

$$sr'-\mathbf{s}\cdot\mathbf{r}'=0. \quad (6.14)$$

If as before θ denotes the angle between $\mathbf{s}$ and $\mathbf{r}$ then (6.14) demands that $\cos\theta=1$, i.e., $\theta=0$. Two cases must again be distinguished:

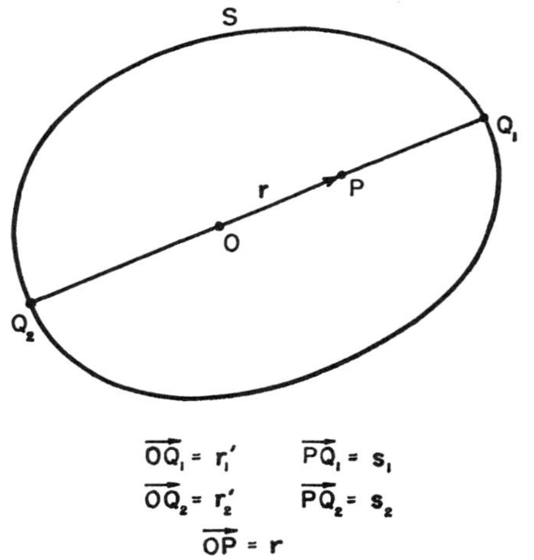

FIG. 9. Illustrating the locations of the two singular points Q_1 and Q_2 on S, for a *converging* spherical wave, when the surface S encloses the singularity O of the wave.

(*a*) *The surface does NOT enclose the singularity of the wave.* Then evidently there is one singular point Q_1 on S, namely the point in which the line OP (prolonged) intersects the surface for the second time (see Fig. 8). In this case one finds from (2.12) or by direct calculation that the contribution from Q_1 is

$$F_1(P)=U(P)=\exp(-ikr)/r, \tag{6.15}$$

where r is the distance OP.

(*b*) *The surface S encloses the singularity O of the spherical wave.* Evidently two points Q_1 and Q_2 on S now obey the condition (6.14), namely, the two points in which the line OP intersects S (see Fig. 9).

The contributions $F_1(P)$ and $F_2(P)$ from Q_1 and Q_2 are readily evaluated using (2.12). The calculations are somewhat similar to those carried out in Sec. 3.1 in connection with a plane wave and yield (with $r=OP$)

$$\left.\begin{aligned} F_1(P)&=\exp(-ikr)/r,\\ F_2(P)&=-\exp(ikr)/r.\end{aligned}\right\} \tag{6.16}$$

The contribution from the singularity O of the wave can be determined from (2.14) and is

$$G_1(P)=\exp(ikr)/r. \tag{6.17}$$

It is seen that

$$F_1(P)+F_2(P)+G_1(P)=\exp(-ikr)/r=U(P), \tag{6.18}$$

in agreement with the general result (2.13).

APPENDIX

Asymptotic Behavior of an Angular Spectrum of Plane Waves

We derive here the asymptotic approximation for large value of kR to the integral

$$\mathbf{I}^{\pm}(\mathbf{R})=\iint_{p_x^2+p_y^2\leq 1} \mathbf{B}^{\pm}(\mathbf{p}_x\cdot\mathbf{p}_y)\exp(ikR\mathbf{p}^{\pm}\cdot\mathbf{u})dp_xdp_y, \tag{A1}$$

where

$$\left.\begin{aligned}\mathbf{p}^{\pm}&=[p_x, p_y, \pm(1-p_x^2-p_y^2)^{\frac{1}{2}}],\\ \mathbf{u}&=\mathbf{R}/R.\end{aligned}\right\} \tag{A2}$$

The approximation may readily be obtained by the application of the principle of stationary phase (reference 5, p. 750). According to this principle, the asymptotic approximation to (A1) for large kR is

$$\mathbf{I}^{\pm}(\mathbf{R})\sim\frac{2\pi i\epsilon}{|\Delta^{\frac{1}{2}}|}\mathbf{B}^{\pm}(p_x',p_y')\frac{\exp(ikR\mathbf{p}^{\pm\prime}\cdot\mathbf{u})}{kR}. \tag{A3}$$

Here $\mathbf{p}'$ is the vector $\mathbf{p}$ which makes the phase factor

$$\phi=\mathbf{p}^{\pm}\cdot\mathbf{u}=p_xu_x+p_yu_y\pm(1-p_x^2-p_y^2)^{\frac{1}{2}}u_z, \tag{A4}$$

stationary within the domain of integration; i.e., p_x' and p_y' are roots of the equation

$$\partial\phi/\partial p_x=\partial\phi/\partial p_y=0. \tag{A5}$$

Further,

$$\Delta=\left[\frac{\partial^2\phi}{\partial p_x^2}\frac{\partial^2\phi}{\partial p_y^2}-\left\{\frac{\partial^2\phi}{\partial p_x\partial p_y}\right\}^2\right]', \tag{A6}$$

$$\epsilon=\begin{cases}+1 & \text{when}\quad \Delta>0,\quad (\partial^2\phi/\partial p_x^2)'>0,\\ -1 & \text{when}\quad \Delta>0,\quad (\partial^2\phi/\partial p_y^2)'<0,\\ i & \text{when}\quad \Delta<0,\end{cases} \tag{A7}$$

and the prime in (A6) and (A7) denotes values at the stationary points.

Let us consider first the integral $\mathbf{I}^+$. In this case one may readily verify that ϕ is stationary when

$$\left.\begin{aligned}\mathbf{p}^+&=\mathbf{u} &&\text{if}\quad u_z>0,\\ \mathbf{p}^+&=-\mathbf{u} &&\text{if}\quad u_z<0.\end{aligned}\right\} \tag{A8}$$

The factors Δ and ϵ become

$$\Delta=1/u_z^2, \tag{A9}$$

$$\left.\begin{aligned}\epsilon&=-1\quad\text{when}\quad u_z=>0\\ \epsilon&=+1\quad\text{when}\quad u_z=<0.\end{aligned}\right\} \tag{A10}$$

Substitution into (A3) then gives

$$\mathbf{I}^+(R)\sim-\frac{2\pi i}{k}u_z\mathbf{B}^+(\pm u_x, \pm u_y)\frac{\exp(\pm ikR)}{R} \tag{A11}$$

according as $u_z>0$ or $u_z<0$.

In a strictly similar manner one obtains the following asymptotic approximation for $\mathbf{I}^-(\mathbf{R})$:

$$\mathbf{I}^-(\mathbf{R})\sim\frac{2\pi i}{k}u_z\mathbf{B}^-(\mp u_x, \mp u_y)\frac{\exp(\mp ikR)}{R} \tag{A12}$$

according as $u_z>0$ or $u_z<0$.

Reprinted from JOURNAL OF THE OPTICAL SOCIETY OF AMERICA, Vol. 52, No. 6, 626–637, June, 1962
Printed in U. S. A.

Generalization of the Maggi-Rubinowicz Theory of the Boundary Diffraction Wave—Part II†

KENRO MIYAMOTO*
Institute of Optics, University of Rochester, Rochester, New York

AND

EMIL WOLF
Department of Physics and Astronomy, University of Rochester, Rochester, New York

(Received March 20, 1961)

With the help of the results derived in Part I of this investigation, a new representation is obtained for the field arising from diffraction of a monochromatic wave by an aperture in an opaque screen. It is shown that within the accuracy of the Kirchhoff diffraction theory, the diffracted field $U_K(P)$ may be expressed under very general conditions in the form

$$U_K(P)=U^{(B)}(P)+\Sigma_j F_j(P). \quad (1)$$

Here $U^{(B)}$ represents a disturbance originating at each point of the boundary of the aperture and $\Sigma_j F_j(P)$ represents the total effect of disturbances propagated from certain special points Q_j in the aperture. Expressions for $U^{(B)}$ and $F_j(P)$ are given, in terms of the new vector potential which was introduced in the previous paper.

In the special case when the wave incident upon the aperture is plane or spherical, the last term in (1) is found to represent a wave disturbance $U^{(G)}(P)$ which obeys the laws of geometrical optics. The formula (1) then becomes equivalent to a classical result of Maggi and Rubinowicz, which may be regarded as a mathematical refinement of early ideas of Young about the nature of diffraction. It is shown further, that even when the incident wave is not plane or spherical, the last term in (1) is, at least approximately, equal to a "geometrical wave."

The results suggest a new approach to the solution of certain types of diffraction problems and provide a new insight into the process of diffraction.

1. INTRODUCTION

IN Part I of this investigation[1] (to be referred to as I), it was shown that a vector potential $\mathbf{W}(Q,P)$ of a new kind may be associated with any given monochromatic wavefield $U(P)$. With the help of this potential, the Helmholtz-Kirchhoff integral, which represents the disturbance at a point P inside a volume bounded by a closed surface S, was then expressed in a new form, namely as the sum of line integrals taken along vanishingly small circles surrounding certain special points Q_j on S. These special points (which depend on the location of P) are the singularities of the potential on S. Expressed in more physical terms, this representation implies that the wave disturbance at P may be expressed as the sum of disturbances originating in certain special point sources situated on the surface S.

In the present paper the analysis will be extended to the case when the Helmholtz-Kirchhoff integral is taken over an *open* surface. As is well known, this case is encountered, for example, in the application of the Kirchhoff theory to diffraction at an aperture in an opaque screen. It will be shown that under certain very general conditions the "Kirchhoff field" $U_K(P)$ may be expressed in the form

$$U_K(P)=U^{(B)}(P)+\textstyle\sum_j F_j(P), \quad (1.1)$$

where $U^{(B)}(P)$ represents a disturbance originating in the boundary of the aperture and the $F_j(P)$ are contributions from those special points Q_j that are situated in the aperture.

When the field incident upon the aperture is a plane or a spherical wave, Eq. (1.1) will be shown to reduce to

$$U_K(P)=U^{(B)}(P)+U^{(G)}(P), \quad (1.2)$$

where $U^{(G)}(P)$ represents a wave propagated in accordance with the laws of geometrical optics. The representation (1.2) is identical with a well-known result derived for this special case by Maggi[2] and Rubinowicz[3] and represents a mathematical refinement of early ideas of Young[4] about the nature of diffraction.

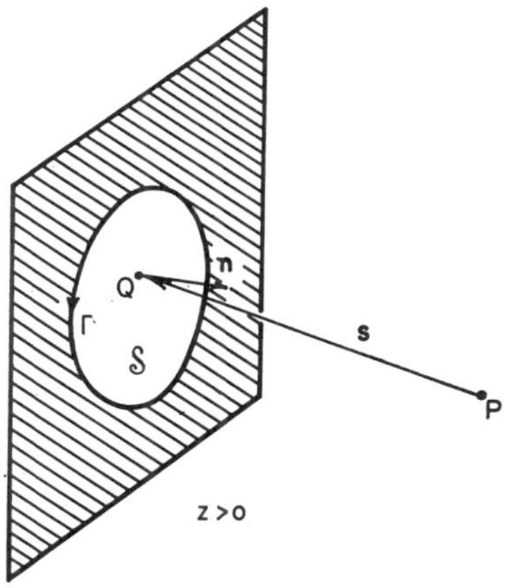

FIG. 1. Illustrating notation relating to Kirchhoff's integral (2.1).

† This research was supported in part by the United States Air Force under contract, monitored by the Air Force Office of Scientific Research of the Air Research and Development Command.

* On leave of absence from Nippon Kokagu K.K., (Japan Optical Industry Company, Ltd.), Tokyo, Japan.

[1] K. Miyamoto and E. Wolf, J. Opt. Soc. Am. **52**, 615 (1962).

[2] G. A. Maggi, Ann. di Mat., IIa, **16**, 21 (1888).

[3] A. Rubinowicz, Ann. Physik **53**, 257 (1917).

[4] T. Young, Phil. Trans. Roy. Soc. **20**, 26 (1802). For an account of Young's ideas see A. Rubinowicz, Nature **180**, 160 (1957).

It will also be shown that, whether or not the wavefield incident upon the aperture is plane or spherical, the special points Q_j are, at least to a good approximation, precisely those points in the aperture which make the phase of the integrand of the Kirchhoff diffraction integral stationary. Moreover, in this approximation, the total contribution from the points Q_j [represented by the second term on the right of (1.1)] will be shown to be equal to the geometrical optics field.

The present theory provides a new approach for treatment of a wide class of diffraction problems and gives a new insight into the behavior of light waves in media containing material obstacles whose linear dimensions are large compared to the wavelength.

2. TRANSFORMATION OF THE BASIC INTEGRAL OF KIRCHHOFF'S DIFFRACTION THEORY. THE BOUNDARY WAVE

Consider the diffraction of a monochromatic scalar wave $V(P,t)=U(P)\exp(-i\omega t)$ at an aperture $\mathcal{S}$ in a plane opaque screen. Within the accuracy of the Kirchhoff diffraction theory, the space-dependent part of the wave which represents the diffracted field at a point P in the half-space $z>0$ into which the wave is propagated (see Fig. 1) is given by

$$U_K(P)=\iint_{\mathcal{S}}\mathbf{V}(Q,P)\cdot\mathbf{n}dS, \tag{2.1}$$

where

$$\mathbf{V}(Q,P)=\frac{1}{4\pi}\left\{U(Q)\,\mathrm{grad}_Q\left(\frac{\exp(iks)}{s}\right)-\left(\frac{\exp(iks)}{s}\right)\mathrm{grad}_Q U(Q)\right\}. \tag{2.2}$$

Here Q denotes a typical point in the aperture $\mathcal{S}$, s denotes the distance QP, $\mathbf{n}$ is the unit vector normal to the plane of the screen and pointing into the half-space $z>0$, and $k=\omega/c$, c being the velocity of light.

The conditions under which the Kirchhoff integral (2.1) gives a good approximation to the true field (in spite of the physically incorrect and mathematically inconsistent boundary conditions) are well known[5] and will not be discussed here. Our purpose is to derive some new properties of the Kirchhoff solution, with the help of the results derived in Part I of this investigation.

It was shown in Sec. 2 of Part I that $\mathbf{V}(Q,P)$ may always be expressed in the form

$$\mathbf{V}(Q,P)=\mathrm{curl}_Q\mathbf{W}(Q,P), \tag{2.3}$$

where $\mathbf{W}(Q,P)$ is a certain vector potential. Moreover it was shown that the vector potential associated with a given field $U(P)$ may be expressed in the form

$$\mathbf{W}(Q,P)=\frac{\exp(iks)}{4\pi s}\hat{s}\times\int_{\infty}^{0}\exp(ik\mu)\,\mathrm{grad}U(\mathbf{r}'+\mu\hat{s})d\mu+\mathbf{W}_\infty. \tag{2.4}$$

Here s is the distance PQ, $\hat{s}$ is the unit vector pointing in the direction from P to Q, $\mathbf{r}'$ is the position vector of Q, and $\mathbf{W}_\infty$ is a certain residual contribution from infinity. This residual contribution vanishes if the incident field obeys the Sommerfeld radiation condition in the half space $z<0$ or $z>0$ according as $S_z<0$ or $S_z>0$.

It is evident from the discussion of Sec. 2 of Part I that even when U is regular at each point Q in the aperture plane $\mathcal{S}$ as will now be assumed, the vector potential $\mathbf{W}(Q,P)$ considered as function of Q may have singularities at some points $Q_1, Q_2, \cdots, Q_M$ in $\mathcal{S}$. Let us surround each point Q_j by a small circle of radius $\sigma_j(j=1, 2, \cdots, M)$ and apply Stokes theorem to the integral (2.1) after substituting from (2.3) and excluding these circles from the domain of integration. Then on proceeding to the limit $\sigma_j\rightarrow 0$ ($j=1, 2, \cdots, M$) one obtains

$$U_K(P)=U^{(B)}(P)+\textstyle\sum_{\mathcal{S}}F_j(P), \tag{2.5}$$

where

$$U^{(B)}(P)=\int_\Gamma \mathbf{W}\cdot\mathbf{l}dl, \tag{2.6}$$

and

$$F_j(P)=\lim_{\sigma_j\rightarrow 0}\int_{\Gamma_j}\mathbf{W}\cdot\mathbf{l}dl. \tag{2.7}$$

Here Γ denotes the boundary of the aperture, Γ_j denotes the boundary of σ_j, and $\mathbf{l}$ denotes the unit vector tangential to these boundary curves. The integral along Γ is described anticlockwise, that along Γ_j clockwise when viewed from P. The summation sign $\sum_{\mathcal{S}}$ implies summation over all the singularities Q_j of $\mathbf{W}(Q,P)$ in $\mathcal{S}$, P being kept constant.

The formula (2.6) shows that $U^{(B)}(P)$ is the sum of contributions from each element of the boundary and accordingly will be called the *boundary wave*. It is a generalization of the boundary diffraction wave of the earlier more restricted theories. It should be noted that in the Kirchhoff formula (2.1) only values of the incident *field* and its normal derivatives at points in the aperture enter. On the other hand, the expression (2.6) for the boundary wave requires the knowledge of the *potential* $\mathbf{W}(Q,P)$ at each point Q of the boundary of the aperture; this in turn seems to require the knowledge of the field at each point of the body of the truncated cone whose vertex is at P and whose generators are lines, passing through each point of the boundary of the aperture, on the side of the screen opposite to P. However, we shall see later (Sec. 4) that

[5] See, for example, M. Born and E. Wolf, *Principles of Optics* (Pergamon Press, New York, 1959) §8.3.2.

to a good approximation the boundary wave depends only on the field at the boundary of the aperture.[6]

The representation (2.5) may be expressed in a simpler form if either all, or none of the singularities of $\mathbf{W}$ in the plane $z=0$ are inside the aperture $\mathcal{S}$. To show this we recall the identity derived in Part I Eq. (2.13) according to which the incident field $U(P)$ may be expressed in the form,

$$U(P)=\sum_{\mathcal{S}} F_j(P)+\sum_{\mathfrak{v}} G_j(P). \tag{2.8}$$

Here F_j are the contributions from all the singularities of $\mathbf{W}$ on any *closed* convex surface S surrounding P and $G_j(P)$ are contributions from all the singularities of U inside the volume $\mathfrak{v}$ bounded by S. Each contribution F_j is given by a formula such as (2.7) and each contribution G_j is given by,

$$G_j(P)=\lim_{\rho_j\to 0}\iint_{S_j} \mathbf{V}(Q,P)\cdot\mathbf{n}dS, \tag{2.9}$$

where ρ_j is the radius of a sphere S_j centered on the singularity of U, and $\mathbf{n}$ is the unit inward normal to this sphere.

If we choose as S the usual surface used to derive the Kirchhoff formula (2.1) viz. a surface consisting of: (1) the aperture $\mathcal{S}$, (2) a portion $\mathcal{B}$ of the screen, and (3) a large hemisphere $\mathfrak{D}$ of radius R centered on some point in $\mathcal{S}$, containing P in its interior (see Fig. 2), then (2.9) becomes

$$U(P)=\sum_{\mathcal{S}} F_j(P)+\sum_{\mathcal{B}} F_j(P)+\sum_{\mathfrak{D}} F_j(P)+\sum_{\mathfrak{v}} G_j(P). \tag{2.10}$$

Let us now proceed to the limit $R\to\infty$. Then under the same assumptions as are implied in the derivation of (2.1), namely the absence of singularities in the half-space $z>0$, and the satisfaction by the incident field of Sommerfeld's radiation condition on the hemisphere $\mathfrak{D}$, the last two terms in (2.10) will be absent and one has

$$U(P)=\sum_{\mathcal{S}} F_j(P)+\sum_{\mathcal{B}} F_j(P). \tag{2.11}$$

One of the following situations must arise:

(I) *All of the singularities of* $\mathbf{W}$ *in the aperture plane are inside* $\mathcal{S}$. Then the last term in (2.11) will be absent and (2.5) gives

$$U_K(P)=U^{(B)}(P)+U(P). \tag{2.12}$$

(II) *No singularities of* $\mathbf{W}$ *are inside* $\mathcal{S}$. Then (2.5) reduce to

$$U_K(P)=U^{(B)}(P). \tag{2.13}$$

(III) *Some but not all of the singularities of* $\mathbf{W}$ *in the aperture plane are inside* $\mathcal{S}$. In this case (2.5) simplifies no further, but using (2.11) it may also be expressed in the form

$$U_K(P)=U^{(B)}(P)+U(P)-\sum_{\mathcal{B}} F_j(P). \tag{2.14}$$

These results will now be applied to cases of diffraction of a plane wave and a spherical wave at an aperture. It will be shown that our formulas correctly reduce to the classical results of Maggi and Rubinowicz.

3. THE MAGGI-RUBINOWICZ REPRESENTATION

3.1 Diffraction of a Plane Wave

Consider the diffraction of a plane wave

$$U(P)=A\exp(ik\mathbf{p}\cdot\mathbf{r}) \tag{3.1}$$

at an aperture $\mathcal{S}$ in an opaque screen. According to Eq. (3.4) of I, the vector potential $\mathbf{W}$ is now given by (see Fig. 3)

$$\mathbf{W}(Q,P)=\frac{1}{4\pi}A\exp(ik\mathbf{p}\cdot\mathbf{r}')\frac{\exp(iks)}{s}\frac{\hat{s}\times\mathbf{p}}{1+\hat{s}\cdot\mathbf{p}}, \tag{3.2}$$

where as before $\mathbf{r}'$ is the position vector of Q.

The vector potential (3.2) has now only one singularity Q_1 in the plane of the aperture, at the point given by $1+\hat{s}\cdot\mathbf{p}=0$. This is the point in which the line through P in the direction of propagation of the plane wave intersects the plane of the aperture. It follows that Q_1 lies inside the aperture or outside it according as P lies in the direct beam of light or in the geometrical shadow. It then follows from (2.12) and (2.13) that the diffracted field may be represented in the form

$$U_K(P)=U^{(B)}(P)+U^{(G)}(P), \tag{3.3}$$

where

$$\left.\begin{aligned}U^{(G)}(P)&=U(P)\text{ when }P\text{ is in the direct beam,}\\&=0\text{ when }P\text{ is in geometrical shadow.}\end{aligned}\right\} \tag{3.4}$$

According to (2.6) and (3.2), the boundary wave $U^{(B)}$ is now given by the integral

$$U^{(B)}(P)=\frac{A}{4\pi}\int_\Gamma \exp(ik\mathbf{p}\cdot\mathbf{r}')\frac{\exp(iks)}{s}\frac{(\hat{s}\times\mathbf{p})\cdot\mathbf{l}}{1+\hat{s}\cdot\mathbf{p}}dl, \tag{3.5}$$

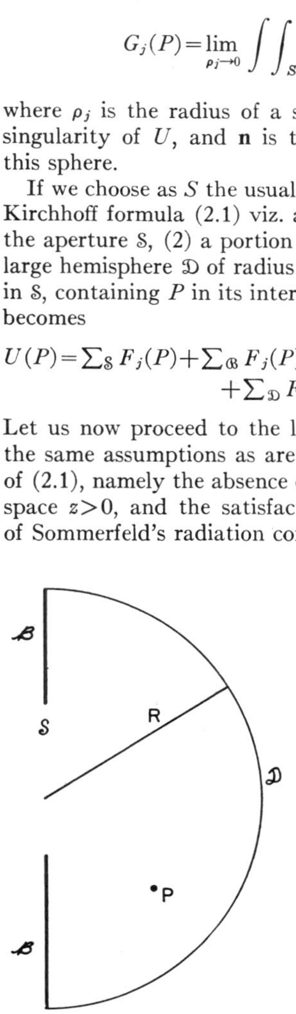

FIG. 2. Choice of the surface S.

[6] In the special cases when the wave incident on the aperture is plane or spherical, this is true without any approximation.

$\mathbf{r}'$ being the position vector of a typical point Q on the boundary Γ of the aperture.

The formula (3.3) is nothing but the *Maggi-Rubinowicz representation*[7] of the Kirchhoff diffraction integral. It expresses the diffracted field as a superposition of the boundary wave $U^{(B)}$ with the "geometrical wave" $U^{(G)}(P)$. Thus, in the direct beam, the field may be regarded as arising from the interference of the unperturbed incident field $U(P)$ with the boundary wave $U^{(B)}(P)$ given by (3.5) and in the shadow region as arising from the boundary wave alone. As is well known, these results embody the essential ideas of the early attempts of Young[4] to explain diffraction.

Similar results also hold (as in fact was shown in the investigations of Maggi and Rubinowicz) in the case when the wave incident upon the aperture is a spherical wave, diverging or converging. It will now be verified that these results also readily follow from our general solution.

3.2 Diffraction of a Divergent Spherical Wave

Consider Kirchhoff's diffraction of a divergent spherical wave,

$$U(P)=A\frac{\exp(ikr)}{r}, \tag{3.6}$$

with its singularity $r=0$ on the opposite side of the aperture plane to where P is situated.

In place of (3.2) one now has according to Eq. (5.10) of I,

$$\mathbf{W}(Q,P)=\frac{1}{4\pi}A\frac{\exp(ikr')}{r'}\,\frac{\exp(iks)}{s}\,\frac{\mathbf{s}\times\mathbf{r}'}{sr'+\mathbf{s}\cdot\mathbf{r}'}, \tag{3.7}$$

where $\mathbf{r}'$ is the position vector OQ. Again $\mathbf{W}(Q,P)$ has only one singularity Q_1 in the plane of the aperture, this singularity being the point of intersection with the plane of the aperture of the line joining the origin $O(r=0)$ to the point P of observation. It follows that

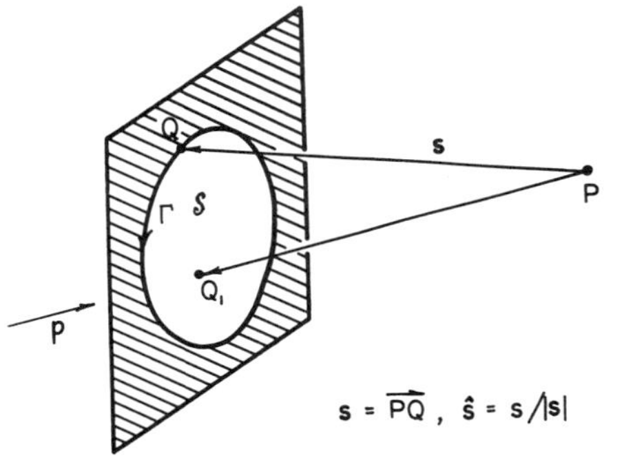

FIG. 3. Diffraction of a plane wave at an aperture in an opaque screen.

[7] See, for example, B. B. Baker and E. T. Copson, *The Mathematical Theory of Huygens' Principle* (Oxford University Press, Oxford, 1950), 2nd. ed., §2.1; or reference 5, §8.9.

(3.3) again applies, with $U(P)$ now given by (3.6) and

$$U^{(B)}(P)=\frac{A}{4\pi}\int_{\Gamma}\frac{\exp(ikr')}{r'}\,\frac{\exp(iks)}{s}\,\frac{(\mathbf{s}\times\mathbf{r}')\cdot\mathbf{l}}{sr'+\mathbf{s}\cdot\mathbf{r}'}dl, \tag{3.8}$$

in agreement with the results of Maggi[2] and Rubinowicz[3,8]

3.3 Diffraction of a Convergent Spherical Wave

Consider now Kirchhoff's diffraction of a convergent spherical wave

$$U(P)=A\frac{\exp(-ikr)}{r}, \tag{3.9}$$

with its singularity $(r=0)$ on the same side of the aperture plane as the point P of observation.

Since the assumption of the absence of singularities of U in the half space $z>0$ is not valid now, the diffracted field must now be determined from the formula (2.5), viz.,

$$U_K(P)=U^{(B)}(P)+\textstyle\sum_S F_j(P), \tag{3.10}$$

rather than from the formulas given at the end of Sec. 2.

According to Eq. (6.13) of I, the vector potential $\mathbf{W}$ associated with (3.9) is

$$\mathbf{W}(Q,P)=-\frac{1}{4\pi}A\frac{\exp(-ikr')}{r'}\,\frac{\exp(iks)}{s}\,\frac{\mathbf{s}\times\mathbf{r}'}{sr'-\mathbf{s}\cdot\mathbf{r}'}, \tag{3.11}$$

where $\mathbf{r}'$ is the position vector OQ, O being the singularity $(r=0)$ of the spherical wave. It is seen from (3.11) that the vector potential has again only one singularity Q_1 in the plane of the aperture, this singularity being the point of intersection with the plane of the aperture of the line OP. According to Eq. (6.16) of I, the contribution from this singular point Q_1 will be

$$F_j(P)=\begin{cases}A\exp(-ikr)/r & \text{if } P \text{ lies in region I}\\ -A\exp(ikr)/r & \text{if } P \text{ lies in region II,}\end{cases} \tag{3.12}$$

(see Fig. 4) where $r=OP$.

From (3.10) and (3.12) it follows that the diffracted field is given by

$$U_K(P)=U^{(B)}(P)+U^{(G)}(P), \tag{3.13}$$

where the geometrical wave $U^{(G)}(P)$ is given by

$$\left.\begin{aligned}U^{(G)}(P)&=A\exp(-ikr)/r && \text{in region I,}\\ &=-A\exp(ikr)/r && \text{in region II,}\\ &=0 && \text{in region III}\end{aligned}\right\} \tag{3.14}$$

and the boundary wave $U^{(B)}(P)$ is given by the line

[8] The form of the solution given here is the same as that of Rubinowicz. Maggi expressed the solution in a somewhat different form. The equivalence of the results of Maggi and Rubinowicz was first noted by F. Kottler [Ann. Physik **70**, 405 (1923)]. See also A. Rubinowicz, *Die Beugungswelle in der Kirchhoffschen Theorei der Beugung*, (Polska Akademia Nauk, Warsaw, Poland, 1957).

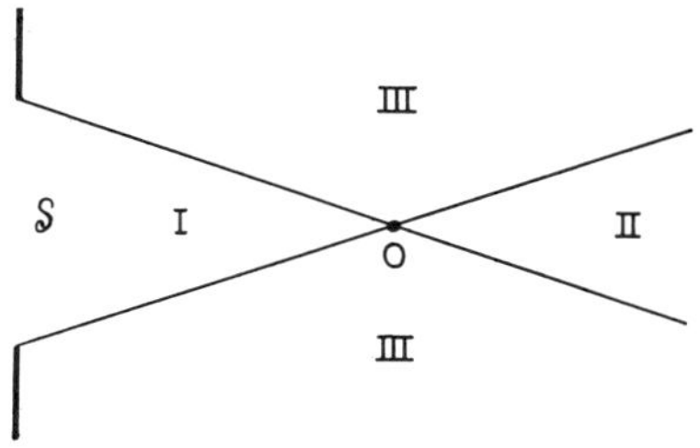

FIG. 4. Kirchhoff's diffraction of a convergent spherical wave. Regions in which the formulas (3.12) and (3.14) apply. The boundaries of the regions are formed by the edge of the geometrical shadow.

integral of (3.11) around the edge Γ of the aperture, viz.

$$U^{(B)}(P)=-\frac{A}{4\pi}\int_\Gamma \frac{\exp(-ikr')}{r'}\frac{\exp(iks)}{s}\times\frac{(\mathbf{s}\times\mathbf{r}')\cdot\mathbf{l}}{sr'-\mathbf{s}\cdot\mathbf{r}'}dl. \quad (3.15)$$

The formulas (3.13)–(3.15) are in agreement with the results of Rubinowicz[9] obtained in his analysis of the anomalous phase propagation near focus.

4. AN APPROXIMATE EXPRESSION FOR THE VECTOR POTENTIAL

It will now be shown that there is a simple approximate expression for the vector potential associated with any field which is adequately represented by geometrical optics. In Sec. 5 this expression will be used to derive an approximate generalization of the Maggi-Rubinowicz representation.

Let $A(\mathbf{r}')$ denote the amplitude and $k\phi(\mathbf{r}')$ the phase of the unperturbed incident field:

$$U(\mathbf{r}')=A(\mathbf{r}')\exp[ik\phi(\mathbf{r}')], \quad (4.1)$$

(A,ϕ real). According to Eqs. (4.5) and (4.7) of I, the associated vector potential **W** may be symbolically expressed in the following form:

$$\mathbf{W}(\mathbf{r}',\mathbf{r})=\frac{\exp(iks)}{4\pi s}\hat{s}\times\frac{(ik)^{-1}\,\mathrm{grad}'}{[1+(ik)^{-1}\hat{s}\cdot\mathrm{grad}']}U(\mathbf{r}'). \quad (4.2)$$

We have, from (4.1),

$$(ik)^{-1}\,\mathrm{grad}'U(\mathbf{r}')=\left(\mathrm{grad}'\phi+\frac{1}{ik}\frac{\mathrm{grad}'A}{A}\right)U(\mathbf{r}'). \quad (4.3)$$

Let us assume that the second term on the right is very much smaller than the first term. This will certainly be the case if[10]

$$|\mathrm{grad}'\phi(\mathbf{r}')|=1, \quad (4.4a)$$

$$|\mathrm{grad}'A(\mathbf{r}')|/|A(\mathbf{r}')|\ll k=2\pi/\lambda, \quad (4.4b)$$

λ being the wavelength. These two conditions will be satisfied in regions where geometrical optics is a valid approximation for the incident field. Under these assumptions the second term on the right-hand side of (4.3) may be neglected and it is seen that the effect of operating by grad′ on $U(\mathbf{r}')$ is equivalent to multiplying $U(\mathbf{r}')$ by (ik) $\mathrm{grad}'\phi(r')$. Further if the denomination in (4.2) is expanded in a series involving inverse powers of (ik) and one retains the leading term only, the following approximate expression for the vector potential is then obtained[11]:

$$\mathbf{W}(\mathbf{r}',\mathbf{r})=U(\mathbf{r}')\frac{\exp(iks)}{s}\frac{1}{4\pi}\frac{\hat{s}\times\mathrm{grad}'\phi(\mathbf{r}')}{1+\hat{s}\cdot\mathrm{grad}'\phi(\mathbf{r}')}. \quad (4.5)$$

This formula also readily follows from a series expansion, derived in Appendix A, for the exact potential. Moreover the analysis carried out in the Appendix suggests that (4.5) is the asymptotic approximation (for large values of k) to the exact vector potential of the incident field.[12]

The vector potential (4.5) is seen to be pointing in the direction at right angles to the line (specified in direction by the unit vector $\hat{s}$) which joins the points $P(\mathbf{r})$ and $Q(\mathbf{r}')$ and to the normal [specified by $\mathrm{grad}'\phi(r')$] to the surfaces of constant phase of U. It consists of three parts which may be interpreted as follows: (1) the incident wave $U(\mathbf{r}')$, (2) a secondary spherical wave $\exp(iks)/s$, (3) a vectorial inclination factor

$$\frac{1}{4\pi}\frac{\hat{s}\times\mathrm{grad}'\phi(\mathbf{r}')}{1+\hat{s}\cdot\mathrm{grad}'\phi(\mathbf{r}')}.$$

Evidently the singularities of the vector potential **W** are given by

$$1+\hat{s}\cdot\mathrm{grad}'\phi(\mathbf{r}')=0. \quad (4.6)$$

Now in the geometrical optics approximation $\mathrm{grad}'\phi$ represents the unit vector perpendicular to the surfaces of constant phase of U, i.e., the unit vector in the direction of a ray. Hence *the singularities in the aperture of the vector potential* $\mathbf{W}(Q,P)$ *given by Eq.* (4.5) *are precisely those points* Q_j *which lie on the rays* (*if any*) *that pass through* P (see Fig. 5).

5. AN APPROXIMATE GENERALIZATION OF THE MAGGI-RUBINOWICZ REPRESENTATION

As already mentioned, the Maggi-Rubinowicz representation, which expresses the Kirchhoff field as the sum of a boundary wave $U^{(B)}$ and a geometrical wave

---

[9] A. Rubinowicz, Phys. Rev. **54**, 931 (1938).

[10] The equation $|\mathrm{grad}'\phi|=1$ is the eikonal equation of geometrical optics for a medium of refractive index unity (cf. reference 5, Chap. III).

[11] The analogy between (4.5) and the formulas (3.2), (3.7), and (3.11) for the potential of a plane and a spherical wave should be noted. In fact (3.2), (3.7), and (3.11) follow from (4.5) exactly, on substituting for ϕ the appropriate expressions ($\phi=\mathbf{p}\cdot\mathbf{r}'$ for a plane wave and $\phi=\pm r'$ for a spherical wave).

[12] According to a private communication, the formula (4.5) was also derived by Dr. G. D. Wassermann (Newcastle on Tyne), using a different procedure.

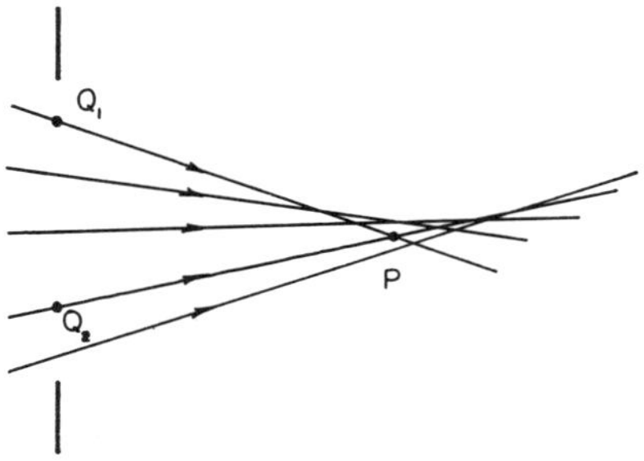

FIG. 5. Illustrating the physical significance of the singularities in the aperture of the vector potential $\mathbf{W}(Q,P)$ given by (4.5). The singularities are precisely those aperture points Q_j which lie on the geometrical rays that pass through P.

$U^{(G)}(P)$, viz.,

$$U_K(P)=U^{(B)}(P)+U^{(G)}(P), \tag{5.1}$$

has so far only been justified for cases when the wave incident upon the aperture is plane or spherical. In the present treatment a representation of the form

$$U_K(P)=U^{(B)}(P)+\sum_{\mathcal{S}} F_j(P), \tag{5.2}$$

was found, valid for any incident field. In (5.2) $F_j(P)$ are the contributions from certain special points Q_j in the aperture. It was also shown that in the special cases when the incident field is plane or spherical, one has rigorously

$$\sum_{\mathcal{S}} F_j(P)=U^{(G)}(P), \tag{5.3}$$

so that in these cases (5.1) and (5.2) are strictly equivalent.

The question arises whether (5.3) always holds. It will now be shown that if the approximate vector potential (4.5), instead of the exact one, is used for determining the contributions from the special points Q_j the answer is in the affirmative. Whether the same is true when the exact vector potential is used is a much more difficult question which will not be discussed here.[13]

Let the field incident upon the aperture be again represented in the form (4.1). Then, if the approximate vector potential (4.5) is used and we set

$$\text{grad}'\phi=\mathbf{p}, \tag{5.4}$$

the Kirchhoff field may be expressed in the form (5.2) with

$$U^{(B)}(P)=\int_\Gamma U(Q)\frac{\exp(iks)}{s}\frac{1}{4\pi}\frac{(\hat{s}\times\mathbf{p})\cdot\mathbf{l}}{1+\hat{s}\cdot\mathbf{p}}dl, \tag{5.5}$$

and

$$F_j(P)=\lim_{\sigma_j\to 0}\int_{\Gamma_j} U(Q)\frac{\exp(iks)}{s}\frac{1}{4\pi}\frac{(\hat{s}\times\mathbf{p})\cdot\mathbf{l}}{1+\hat{s}\cdot\mathbf{p}}dl, \tag{5.6}$$

where the integral in (5.6) is taken around a small circle σ_j centered on the singularity Q_j in the aperture.

[13] An attempt to extend the early work on the boundary wave to more general fields was also made by R. S. Ingarden, Acta Phys. Polon. **14**, 77 (1955).

Now it has been seen at the end of the preceding section that within the accuracy of the present approximation the singularities Q_j are precisely those points in the aperture which lie on the rays (if any) that pass through P (Fig. 5). The contribution (5.6) from these points is calculated in Appendix B and it is found that

$$F_j(P)=U(Q_j)\exp(iks_j)\Lambda_j, \tag{5.7}$$

where s_j is the distance Q_jP,

$$\Lambda_j=\left(\frac{r_jr_j'}{R_jR_j'}\right)^{\frac{1}{2}}\epsilon_j, \tag{5.8}$$

r_j,r_j' and R_j,R_j' are the principal radii of curvature of the wave front of the incident field at Q_j and P, respectively, and

$$\epsilon_j=\left.\begin{matrix}+1 & \text{if} & R_jR_j'>0, & R_j>0,\\ -1 & \text{if} & R_jR_j'>0, & R_j<0,\\ -i & \text{if} & R_jR_j'<0. & \end{matrix}\right\}. \tag{5.9}$$

Now (5.7) gives *precisely* the contribution to the field at P, as given by geometrical optics. The factor $\exp(iks_j)$ accounts for the change in phase associated with the passage of light from Q_j to P, the factor $(r_jr_j'/R_jR_j')^{\frac{1}{2}}$ expresses the corresponding change in the amplitude of the light in accordance with the geometrical intensity law and the factor ϵ_j accounts for the phase changes at foci (the phase anomaly). It follows that the second term on the right of (5.2) represents the field at P as given by geometrical optics, so that (5.2) reduces to (5.1). Thus we have obtained *an approximate generalization of the Maggi-Rubinowicz representation for cases when the incident field is not plane or spherical.*

Let us now examine the structure of the boundary wave. According to (5.5) the boundary wave arises from the superposition of contributions from each element dl of the boundary of the aperture. The contribution from element dl at a point $Q(\mathbf{r})$ of Γ is proportional to the field at this point and may be considered to be propagated in the form of a spherical wavelet $[\exp(iks)/s$; s is the distance $QP]$. The directional behavior of the secondary wavelet is represented by the inclination factor,

$$\frac{1}{4\pi}\frac{(\hat{s}\times\mathbf{p})\cdot\mathbf{l}}{1+\hat{s}\cdot\mathbf{p}}=\frac{1}{4\pi}\frac{\sin\theta}{1+\cos\theta}\cos\varphi. \tag{5.10}$$

Here θ is the angle between the direction of the vector PQ and the direction of the ray through Q, and φ is the angle between the tangent to Γ at Q and the normal to the plane containing PQ and the ray.

Next let us consider the asymptotic behavior, for large values of k, of the expression (5.5) for the boundary wave $U^{(B)}$. According to the principle of stationary phase, the leading term in the asymptotic expansion of $U^{(B)}$ comes from those points Q_j on the boundary Γ which make the phase of the integrand, i.e., the phase

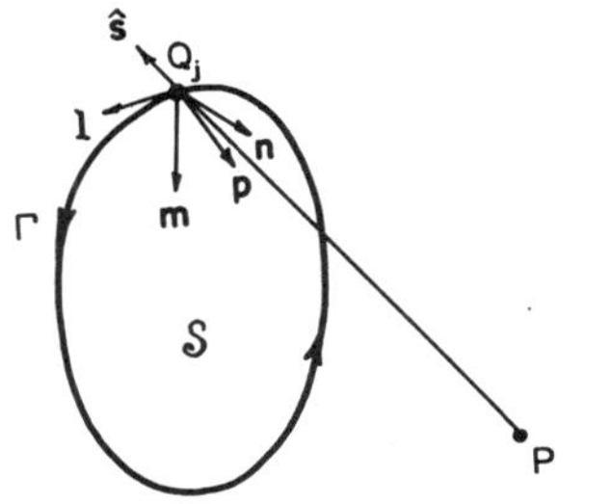

FIG. 6. A boundary stationary point Q_j (critical point of the second kind).

of $U(Q)\exp(iks)$, stationary, i.e., from points on Γ where

$$(\partial/\partial l)(s+\phi)=0, \tag{5.11}$$

l being the arc length of the boundary measured from a fixed point on it. The asymptotic approximation is calculated in Appendix C and is found to be

$$U_{\mathrm{I}}{}^{(B)}=\frac{1}{k^{\frac{1}{2}}}\sum_j U(Q_j)\frac{\exp(iks_j)}{4\pi s_j}\left[\frac{2\pi}{|\partial^2(s+\phi)/\partial l^2|}\right]_j^{\frac{1}{2}} \times\alpha_j\left[\frac{(\hat{s}\times\mathbf{p})\cdot\mathbf{l}}{1+\hat{s}\cdot\mathbf{p}}\right], \tag{5.12}$$

where $\alpha_j=\exp(\pm i\pi/4)$ according as $[\partial^2(s+\phi)/\partial l^2]_j \gtrless 0$ and the significance of the other symbols is shown in Fig. 6.

If the boundary Γ has discontinuities in its tangential derivative (corners), these will give rise to the next term in the asymptotic expansion of $U^{(B)}$. At each corner point Q_j the tangential derivative of the phase will change discontinuously. If superscripts + and − denote limiting values at the two sides of Q_j, the leading order contribution of such points to the asymptotic expansion of $U^{(B)}$ is shown in Appendix C to be

$$U_{\mathrm{II}}{}^{(B)}=\frac{1}{ik}\sum_j U(Q_j)\frac{\exp(iks_j)}{4\pi s_j} \times\left\{\frac{\sin\varphi(\mathbf{p}-\hat{s})\cdot\mathbf{n}}{[(\hat{s}+\mathbf{p})\cdot\mathbf{l}^+][(\hat{s}+\mathbf{p})\cdot\mathbf{l}^-]}\right\}_j, \tag{5.13}$$

where $\mathbf{n}_j$ denotes the unit normal to the aperture at Q_j pointing into the half-space containing P, and φ_j is the angle between the two tangents at Q_j (see Fig. 7).

Finally it is of interest to compare the foregoing results with the asymptotic behavior of the Kirchhoff integral. When the incident field is given by (4.1), the

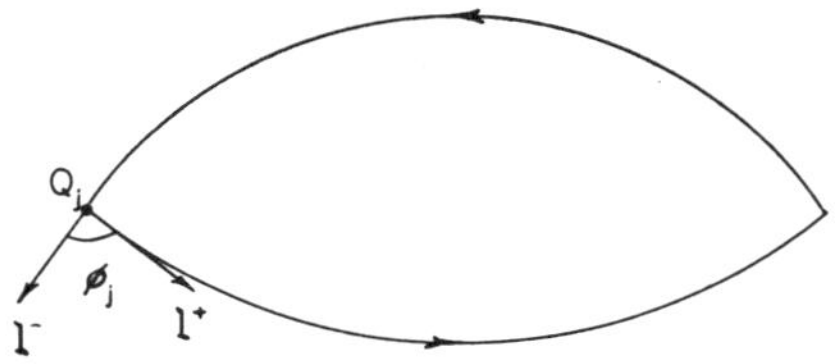

FIG. 7. A corner of the aperture Q_j (critical point of the third kind).

Kirchhoff integral (2.1) is readily expressed in the form

$$U_K(P)=\iint_{\mathcal{S}} g(Q,P)\exp[ikf(Q,P)]dS, \tag{5.14}$$

where

$$g(Q,P)=\left[\left(ik-\frac{1}{s}\right)(\hat{s}\cdot\mathbf{n})-\left(ik\mathbf{p}+\frac{\mathrm{grad}'A}{A}\right)\cdot\mathbf{n}\right]A(Q)$$
$$f(Q,P)=s+\phi(Q), \tag{5.15}$$

Q being a typical point in the aperture. Now the asymptotic behavior of integrals of the form (5.14) may be readily determined by the application of the two-dimensional form of the principle of stationary phase.[14] One then finds that the asymptotic expansion of U_K is completely governed by the behavior of the integrand in (5.14) at certain special points. These points are essentially of three types: (1) points inside the aperture at which the phase f of the integrand is stationary, (2) points on the boundary Γ of the aperture at which f is stationary with respect to a small displacement along Γ, (3) points (if any) where the boundary Γ has a discontinuously changing tangent (corners of Γ).

The three types of points are known as *critical points* of the first, second, and third kind, respectively. The contribution from each such point can itself be expanded in an asymptotic series. Let $U_K{}^{(\mathrm{I})}$ denote the leading term in the expansion arising from all critical points of the first kind and let $U_K{}^{(\mathrm{II})}$ and $U_K{}^{(\mathrm{III})}$ have a similar meaning with regard to the critical points of the second and third kind, respectively. The values of $U_K{}^{(\mathrm{I})}$, $U_K{}^{(\mathrm{II})}$ and $U_K{}^{(\mathrm{III})}$ are calculated in Appendix D and comparison with results established in the present section shows that:

(a) The singular points Q_j in the aperture of the vector potential **W** given by (4.5), are precisely the critical points of the first kind associated with the Kirchhoff integral, and moreover

$$\sum\nolimits_{\mathcal{S}} F_j=U^{(G)}=U_k{}^{(\mathrm{I})}. \tag{5.16}$$

(b) The tangential stationary points of the phase of the integrand of $U^{(B)}$, which as we saw give rise to the asymptotic approximation to $U^{(B)}$, are precisely the critical points of the second kind associated with the Kirchhoff integral, and their contributions are equal to each other,

$$U_{\mathrm{I}}{}^{(B)}=U_K{}^{(\mathrm{II})}. \tag{5.17}$$

(c) The leading order contribution to $U^{(B)}$ arising from corners (if any) of the aperture is equal to the leading order contribution to the Kirchhoff integral arising from critical points of the third kind and these contributions are again equal to each other,

$$U_{\mathrm{II}}{}^{(B)}=U_K{}^{(\mathrm{III})}. \tag{5.18}$$

[14] N. G. van Kampen, Physica **14**, 575 (1949). Brief explanation of the principle and references to other papers will be found in reference 5, p. 750.

Thus it may be concluded that our approximate decomposition of the Kirchhoff diffraction integral into the sum of a boundary wave and a geometrical wave leads to correct expressions for the asymptotic behavior of the Kirchhoff field.

APPENDIX A. AN EXPANSION FOR THE VECTOR POTENTIAL

Let $A(\mathbf{r}')$ be the amplitude and $k\phi(\mathbf{r}')$ the phase of the wave function which specifies the unperturbed field, i.e.,

$$U(\mathbf{r}')=A(\mathbf{r}')\exp[ik\phi(\mathbf{r}')]. \qquad \text{(A1)}$$

Then according to (A1) and Eq. (4.16) of I, the associated vector potential is given by

$$\begin{aligned}\mathbf{W}(\mathbf{r}',\mathbf{r})&=\frac{\exp(iks)}{4\pi s}\hat{s}\times\int_\infty^0\exp(ik\mu)\\&\qquad\times\operatorname{grad}U(\mathbf{r}'+\mu\hat{s})d\mu+\mathbf{W}_\infty\\&=\frac{\exp(iks)}{4\pi s}\hat{s}\times\int_\infty^0(ikA\operatorname{grad}\phi+\operatorname{grad}A)\\&\qquad\times\exp[ik(\mu+\phi)]d\mu+\mathbf{W}_\infty. \qquad \text{(A2)}\end{aligned}$$

The argument of A and ϕ on the right-hand side of (A2) is understood to be $\mathbf{r}'+\mu\hat{s}$.

The integral in (A2) may be developed into a series by successive integrations by parts. We have, for any differentiable function f,

$$(\partial/\partial\mu)f(\mathbf{r}'+\mu\hat{s})=\hat{s}\cdot\operatorname{grad}f(\mathbf{r}'+\mu\hat{s}). \qquad \text{(A3)}$$

In particular,

$$\begin{aligned}&(\partial/\partial\mu)\exp[ik(\mu+\phi)]\\&\qquad=ik(1+\hat{s}\cdot\operatorname{grad}\phi)\exp[ik(\mu+\phi)]. \qquad \text{(A4)}\end{aligned}$$

Using (A4) and (A3), the integral in (A2) may be developed by successive integrations by parts into a series. The first few steps give:

$$\begin{aligned}&\int_\infty^0(ikA\operatorname{grad}\phi+\operatorname{grad}A)\exp[ik(\mu+\phi)]d\mu\\&=\int_\infty^0\mathbf{H}(\mu)\frac{\partial}{\partial\mu}\exp[ik(\mu+\phi)]d\mu\\&=\mathbf{H}(\mu)\exp[ik(\mu+\phi)]\Big|_\infty^0\\&\quad+\int_\infty^0\left\{\frac{-\hat{s}\cdot\operatorname{grad}}{ik(1+\hat{s}\cdot\operatorname{grad}\phi)}\right\}\mathbf{H}\frac{\partial}{\partial\mu}\exp ik(\mu+\phi)d\mu\\&=\mathcal{T}^0\mathbf{H}\cdot\exp[ik(\mu+\phi)]\Big|_\infty^0+(1/ik)\mathcal{T}\mathbf{H}\cdot\exp ik(\mu+\phi)\Big|_\infty^0\\&\quad+\left(\frac{1}{ik}\right)^2\int_\infty^0\mathcal{T}^2\mathbf{H}\cdot\frac{\partial}{\partial\mu}\exp[ik(\mu+\phi)]d\mu, \qquad \text{(A5)}\end{aligned}$$

where

$$\mathbf{H}=\frac{A\operatorname{grad}\phi+(ik)^{-1}\operatorname{grad}A}{1+\hat{s}\cdot\operatorname{grad}\phi}, \qquad \text{(A6)}$$

and $\mathcal{T}$ is the operator

$$\mathcal{T}=\frac{-1}{1+\hat{s}\cdot\operatorname{grad}\phi}\hat{s}\cdot\operatorname{grad}. \qquad \text{(A7)}$$

This procedure may be continued indefinitely. It will be assumed that the contributions from the lower limit will, on vector multiplication by the factor $[\exp(iks)/4\pi s]\cdot\hat{s}$ cancel out the term $\mathbf{W}_\infty$ in (A2) and one then finally obtains

$$\mathbf{W}(\mathbf{r}',\mathbf{r})=\frac{\exp\{ik(s+\phi(\mathbf{r}'))\}}{4\pi s}\hat{s}\sum_{n=0}^\infty\frac{1}{(ik)^n}\left\{\frac{-\hat{s}\cdot\operatorname{grad}'}{1+\hat{s}\cdot\operatorname{grad}'\phi(\mathbf{r}')}\right\}^n\left\{\frac{A(\mathbf{r}')\operatorname{grad}'\phi(\mathbf{r}')+(ik)^{-1}\operatorname{grad}'A(\mathbf{r}')}{1+\hat{s}\cdot\operatorname{grad}'\phi(\mathbf{r}')}\right\}. \qquad \text{(A8)}$$

Leaving aside questions of convergence, the series in (A8) may be rearranged in order of ascending powers of $(1/ik)$, and one obtains an expansion of the following form for $\mathbf{W}$:

$$\mathbf{W}(\mathbf{r}',\mathbf{r})=\frac{\exp(iks)}{4\pi s}U(\mathbf{r}')\hat{s}\times\sum_{n=0}^\infty\frac{1}{(ik)^n}\mathbf{B}_n(\mathbf{r}',\hat{s}). \qquad \text{(A9)}$$

The functions $\mathbf{B}_n$ corresponding to the two lowest values of n are:

$$\left.\begin{aligned}\mathbf{B}_0&=\frac{\mathbf{p}}{1+\hat{s}\cdot\mathbf{p}},\\\mathbf{B}_1&=\frac{1}{A(\mathbf{r}')}\frac{\operatorname{grad}'A(\mathbf{r}')}{1+\hat{s}\cdot\mathbf{p}}\\&\qquad+\frac{1}{A(\mathbf{r}')}\left(\frac{-\hat{s}\cdot\operatorname{grad}'}{1+\hat{s}\cdot\mathbf{p}}\right)\frac{A(\mathbf{r}')\mathbf{p}}{1+\hat{s}\cdot\mathbf{p}},\end{aligned}\right\} \qquad \text{(A10)}$$

where

$$\mathbf{p}=\operatorname{grad}'\phi(\mathbf{r}'). \qquad \text{(A11)}$$

The series (A9) may be expected to be an asymptotic expansion of the vector potential $\mathbf{W}(\mathbf{r}',\mathbf{r})$, for large values of the parameter k. It is seen that the leading term in this expansion is precisely the expression (4.5).

APPENDIX B. CONTRIBUTION F_j FROM A SINGULARITY OF THE VECTOR POTENTIAL (4.5)

The approximate formula (4.5) for the vector potential $\mathbf{W}(Q,P)$ may be written in the form

$$\mathbf{W}(Q,P)=\frac{1}{4\pi}U(Q)\frac{\exp(iks)}{s}\frac{\hat{s}\times\mathbf{p}}{1+\hat{s}\cdot\mathbf{p}}, \qquad \text{(B1)}$$

where

$$\mathbf{p}=\operatorname{grad}\phi(Q). \qquad \text{(B2)}$$

Let Q_j be a singularity of $\mathbf{W}$ in the aperture (i.e., a point such that $1+\hat{s}\cdot\mathbf{p}=0$). According to (B1) and (2.7) the contribution F_j from this singularity to the Kirchhoff field is given by

$$F_j=U(Q_j)\exp(iks_j)\cdot\Lambda_j, \tag{B3}$$

where

$$\Lambda_j=\frac{1}{4\pi s_j}\lim_{\sigma_j\to 0}\int_{\Gamma_j}\frac{(\hat{s}\times\mathbf{p})\cdot\mathbf{l}}{1+\hat{s}\cdot\mathbf{p}}dl. \tag{B4}$$

Here s_j is the distance PQ_j and the integration is taken along the boundary Γ_j of a small circle of radius σ_j centered on Q_j and situated in the plane of the aperture.

Let θ_j denote the angle between $\mathbf{p}$ and $\hat{s}$ and let φ be the azimuthal angle of a typical point Q on Γ_j. Then one finds in a manner similar to that in Part I [cf. Eqs. (3.7)–(3.9)] that

$$\Lambda_j=\frac{1}{2\pi s_j}\lim_{\sigma_j\to 0}\int_0^{2\pi}\frac{\sigma_j}{\theta_j(\varphi)}d\varphi. \tag{B5}$$

An expression for $\theta_j(\varphi)$ may readily be obtained. Let r_j,r_j' be the principal radii of curvature at Q_j and $R_j=r_j-s_j$, $R_j'=r_j'-s_j$ the principal radii at P of the wave fronts of the incident field (see Fig. 8). Then $\theta_j(\varphi)$ is given by

$$\theta_j=\sigma_j\left[\frac{1}{s_j}-\frac{1}{\rho_j(\varphi)}\right], \tag{B6}$$

where $\rho_j(\varphi)$ is the radius of curvature of the azimuthal section φ of the wave front at Q. By elementary geometry

$$\frac{1}{\rho_j(\varphi)}=\frac{\cos^2\varphi}{r_j}+\frac{\sin^2\varphi}{r_j'}. \tag{B7}$$

From (B5), (B6), and (B7) it follows that

$$\begin{aligned}\Lambda_j&=\frac{1}{2\pi s_j}\int_0^{2\pi}\left[\frac{1}{s_j}-\frac{1}{\rho_j(\varphi)}\right]^{-1}d\varphi\\&=\frac{1}{2\pi s_j}\int_0^{2\pi}\left[\left(\frac{1}{s_j}-\frac{1}{r_j}\right)\cos^2\varphi\right.\\&\qquad\left.+\left(\frac{1}{s_j}-\frac{1}{r_j'}\right)\sin^2\varphi\right]^{-1}d\varphi.\end{aligned} \tag{B8}$$

The integral occurring in (B8) can be evaluated by elementary methods. In fact,[15]

$$\int[A\cos^2\varphi+B\sin^2\varphi]^{-1}d\varphi=\frac{1}{(AB)^{1/2}}\tan^{-1}\left[\frac{(AB)^{1/2}}{A}\tan\varphi\right], \tag{B9}$$

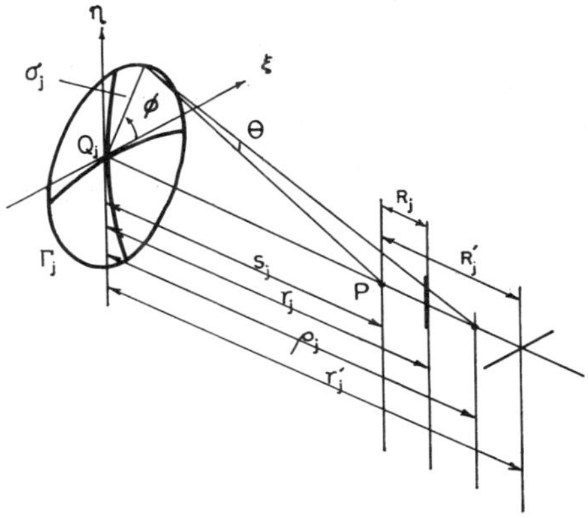

FIG. 8. Illustrating the notation relating to the evaluation of Λ_j.

and (B8) becomes

$$\begin{aligned}\Lambda_j&=\frac{1}{s_j}\left|\frac{1}{s_j}-\frac{1}{r_j}\right|^{-\frac{1}{2}}\left|\frac{1}{s_j}-\frac{1}{r_j'}\right|^{-\frac{1}{2}}\epsilon_j\\&=\left|\frac{r_jr_j'}{R_jR_j'}\right|^{\frac{1}{2}}\epsilon_j,\end{aligned} \tag{B10}$$

where

$$\left.\begin{aligned}\epsilon_j&=+1 &&\text{when}\quad R_jR_j'>0,\quad R_j>0,\\&=-1 &&\text{when}\quad R_jR_j'>0,\quad R_j<0,\\&=-i &&\text{when}\quad R_jR_j'<0.\end{aligned}\right\} \tag{B11}$$

Hence (B3) and (B10) show that the contribution to the Kirchhoff field $U_K(P)$ from the singularity at Q_j is

$$F_j=U(Q_j)\exp(iks_j)\left|\frac{r_jr_j'}{R_jR_j'}\right|^{\frac{1}{2}}\epsilon_j. \tag{B12}$$

APPENDIX C. ASYMPTOTIC BEHAVIOR OF THE EXPRESSION (5.5) FOR THE BOUNDARY WAVE

In this Appendix, the asymptotic behavior (for large values of k) of the expression (5.5) for the boundary wave, viz.,

$$U^{(B)}=\frac{1}{4\pi}\int_\Gamma U(Q)\frac{\exp(iks)}{s}\frac{(\hat{s}\times\mathbf{p})\cdot\mathbf{l}}{1+\hat{s}\cdot\mathbf{p}}dl \tag{C1}$$

will be examined.

For this purpose consider the integral

$$J(k)=\int_a^b g(l)\exp[ikf(l)]dl, \tag{C2}$$

where g and f are real functions of l, k is a real param-

[15] See, for example, R. Courant, *Differential and Integral Calculus* (Blackie and Son, London, 1942), Vol. I, 2nd ed., p. 214. When $AB<0$ the integral is indefinite. However, if the formula (B9) is applied formally one obtains

$$\int_0^{\pi/2}[A\cos^2\varphi+B\sin^2\varphi]^{-1}d\varphi=\frac{-\pi i}{2|AB|^{\frac{1}{2}}}$$

and this expression is used in determining the value of Λ_j when $R_jR_j'<0$.

eter, and a and b are real constants. It is well known[16,17] that the leading terms in the asymptotic expansion of J with respect to k arise from stationary points of f within the range of integration (i.e., from points $l=l_j$ such that $f'(l_j)=0$) and from the end points of the range of integration ($l=a$ and $l=b$). The leading term is

$$J_{\mathrm{I}}(k)=\left(\frac{2\pi}{k}\right)^{\frac{1}{2}}\sum_j \frac{\alpha_j}{|f''(l_j)|^{\frac{1}{2}}}g(l_j)\exp[ikf(l_j)], \quad \text{(C3)}$$

where

$$\left.\begin{aligned}\alpha_j&=\exp(i\pi/4) \quad \text{if} \quad f''(l_j)>0,\\ &=\exp(-i\pi/4) \quad \text{if} \quad f''(l_j)<0.\end{aligned}\right\} \quad \text{(C4)}$$

The second term in the asymptotic expansion is

$$J_{\mathrm{II}}(k)=\frac{1}{ik}\left\{\frac{g(b)}{f'(b)}\exp[ikf(b)] -\frac{g(a)}{f'(a)}\exp[ikf(a)]\right\}. \quad \text{(C5)}$$

Let us now apply these formulas to the expression (C1) for the boundary diffraction wave, bearing in mind that U is assumed to be of the form

$$U=Ae^{ik\phi}. \quad \text{(C6)}$$

The phase of the integrand is $k(s+\phi)$, so that the stationary points in the range of integration are given by

$$\partial(s+\phi)/\partial l=(\hat{s}+\mathbf{p})\cdot\mathbf{l}=0. \quad \text{(C7)}$$

The formula (C3) applied to (C1) now gives the following expression for the leading term in the asymptotic expression for the boundary diffraction wave (C1):

$$U_{\mathrm{I}}^{(B)}(P)=\frac{1}{k^{\frac{1}{2}}}\sum_j U(Q_j)\frac{\exp(iks_j)}{4\pi s_j} \times\left\{\frac{2\pi}{|\partial^2(s+\phi)/\partial l^2|}\right\}^{\frac{1}{2}}\alpha_j\left[\frac{(\hat{s}\times\mathbf{p})\cdot\mathbf{l}}{1+\hat{s}\cdot\mathbf{p}}\right]_j, \quad \text{(C8)}$$

where s_j is the distance Q_jP.

If the boundary Γ of the aperture has corner points, these will contribute the next term $U_{\mathrm{II}}^{(B)}$ to the asymptotic expansion. The contribution from these points is readily obtained by dividing the range of integration Γ into segments, each terminated by successive corners and applying the formula (C5). If Q_j is a corner point and Q_j^+ and Q_j^- refer to the limiting values as this point is being approached from the positive and negative sides of Q_j, respectively, one obtains

$$U_{\mathrm{II}}^{(B)}(P)=\frac{1}{ik}\sum_j U(Q_j)\frac{\exp(iks_j)}{4\pi s_j} \times\left[\frac{\hat{s}\times\mathbf{p}}{1+\hat{s}\cdot\mathbf{p}}\right]_j\left[\frac{1}{\partial(s+\phi)/\partial l}\right]_{Q_j^+}^{Q_j^-}. \quad \text{(C9)}$$

Now the inclination factors in (C9), i.e., the products of the third and fourth terms under the summation sign may, in view of (C7), be expressed in the form

$$\begin{aligned}&\frac{[\hat{s}\times\mathbf{p}]}{1+\hat{s}\cdot\mathbf{p}}\cdot\frac{\{[(\hat{s}+\mathbf{p})\cdot\mathbf{l}^+]\mathbf{l}^- -[(\hat{s}+\mathbf{p})\cdot\mathbf{l}^-]\mathbf{l}^+\}}{[(\hat{s}+\mathbf{p})\cdot\mathbf{l}^+][(\hat{s}+\mathbf{p})\cdot\mathbf{l}^-]}\\ &=-\frac{\hat{s}\times\mathbf{p}}{1+\hat{s}\cdot\mathbf{p}}\cdot\frac{(\hat{s}+\mathbf{p})\times(\mathbf{l}^+\times\mathbf{l}^-)}{[(\hat{s}+\mathbf{p})\cdot\mathbf{l}^+][(\hat{s}+\mathbf{p})\cdot\mathbf{l}^-]}\\ &=\frac{\hat{s}\times\mathbf{p}}{1+\hat{s}\cdot\mathbf{p}}\cdot\frac{(\hat{s}+\mathbf{p})\times\mathbf{n}\sin\varphi}{[(\hat{s}+\mathbf{p})\cdot\mathbf{l}^+][(\hat{s}+\mathbf{p})\cdot\mathbf{l}^-]}.\end{aligned} \quad \text{(C10)}$$

where $\mathbf{n}$ is the unit normal to the aperture at Q_j pointing towards the half-space containing P, and φ is the angle between the vectors $\mathbf{l}^+$ and $\mathbf{l}^-$ (see Fig. 7). Now using standard vector identities one can readily simplify the numerator of (C10):

$$\begin{aligned}(\hat{s}\times\mathbf{p})\cdot[(\hat{s}+\mathbf{p})\times\mathbf{n}]&=\{\mathbf{p}\times[(\hat{s}+\mathbf{p})\times\mathbf{n}]\}\cdot\hat{s}\\ &=\{(\hat{s}+\mathbf{p})(\mathbf{p}\cdot\mathbf{n})-\mathbf{n}(\mathbf{p}\cdot\hat{s}+1)\}\cdot\hat{s}\\ &=(1+\hat{s}\cdot\mathbf{p})(\mathbf{p}-\hat{s})\cdot\mathbf{n}.\end{aligned} \quad \text{(C11)}$$

Using (C10) and (C11) in (C9), one finally obtains the following expression for $U_{\mathrm{II}}^{(B)}$:

$$U_{\mathrm{II}}^{(B)}(P)=\frac{1}{ik}\sum_j U(Q_j)\frac{\exp(iks_j)}{4\pi s_j} \times\left\{\frac{(\mathbf{p}-\hat{s})\cdot\mathbf{n}\sin\varphi}{[(\hat{s}+\mathbf{p})\cdot\mathbf{l}^+][(\hat{s}+\mathbf{p})\cdot\mathbf{l}^-]}\right\}_j. \quad \text{(C12)}$$

APPENDIX D. ASYMPTOTIC BEHAVIOR OF THE KIRCHHOFF DIFFRACTION INTEGRAL

In this appendix the asymptotic behavior for large value of k of the Kirchhoff integral will be examined. The incident field will be assumed to be of the form given by the geometrical optics approximation (4.1), viz., $U(\mathbf{r}')=A(\mathbf{r}')\exp[ik\phi(\mathbf{r}')]$, where $\phi(\mathbf{r}')$ is assumed to obey the eikonal equation $(\mathrm{grad}\phi)^2=1$. The Kirchhoff integral (2.1) then takes the form

$$U_K(P)=\frac{1}{4\pi}\int\int_{\mathcal{S}}\left\{\left(ik-\frac{1}{s}\right)(\hat{s}\cdot\mathbf{n})-(ik\mathbf{p}+\mathrm{grad}\log A)\cdot\mathbf{n}\right\} \times A\frac{\exp[ik(s+\phi)]}{s}d\xi d\eta, \quad \text{(D1)}$$

(ξ,η) being the Cartesian coordinates of a typical point in the aperture $\mathcal{S}$, $\mathbf{p}=\mathrm{grad}\phi$ and the other symbols have the same meaning as before. As we are concerned only with the leading terms of the asymptotic approximation, the terms in the amplitude of the integral which are independent of k may be neglected and (D1)

---

[16] H. and B. S. Jeffreys, *Methods of Mathematical Physics* (Cambridge University Press, 1946), p. 474.

[17] J. Focke, Ber. sächs. Ges. Akad. Wiss **101** (1954), Heft 3.

reduces to

$$U_K(P)=\frac{ik}{4\pi}\int\!\!\int_{\mathcal{S}}(\hat{s}-\mathbf{p})\cdot\mathbf{n}\frac{A}{s}\exp[ik(s+\phi)]d\xi d\eta. \quad \text{(D2)}$$

The main contribution to the asymptotic approximation arises in general from points inside the aperture at which the phase of the integrand of (D2) is stationary, i.e., where

$$\frac{\partial(s+\phi)}{\partial\xi}=\frac{\partial(s+\phi)}{\partial\eta}=0, \quad \text{i.e.,} \quad \hat{s}+\mathbf{p}=0. \quad \text{(D3)}$$

The contribution is (cf. van Kampen,[14] p. 758; Focke,[17] p. 36; Born and Wolf,[5] p. 751)

$$U_K^{(\mathrm{I})}(P)=\frac{2\pi i}{k}\sum_j\frac{\epsilon_j}{(|\Delta_j|)^{\frac{1}{2}}}\times\left\{\frac{ik(\hat{s}-\mathbf{p})\cdot\mathbf{n}}{4\pi s}A\exp[ik(s+\phi)]\right\}_j, \quad \text{(D4)}$$

where

$$\Delta_j=\left\{\frac{\partial^2(s+\phi)}{\partial\xi^2}\frac{\partial^2(s+\phi)}{\partial\eta^2}-\left[\frac{\partial^2(s+\phi)}{\partial\xi\partial\eta}\right]^2\right\}_j, \quad \text{(D5)}$$

$$\left.\begin{aligned}\epsilon_j&=+1 \text{ if } \Delta_j>0, \quad \partial^2(s+\phi)/\partial\xi^2>0,\\ &=-1 \text{ if } \Delta_j>0, \quad \partial^2(s+\phi)/\partial\xi^2<0,\\ &=-i \text{ if } \Delta_j<0\end{aligned}\right\} \quad \text{(D6)}$$

and the subscript j refers to value at the critical point Q_j, the summation being performed over all such critical points. Using (D3), (D4) may be simplified. Recalling that $U=A\exp(ik\phi)$, one then obtains

$$U_K^{(\mathrm{I})}(P)=\sum_j U(Q_j)\exp(iks_j)\frac{(\mathbf{p}\cdot\mathbf{n})_j}{s_j(|\Delta_j|)^{\frac{1}{2}}}\epsilon_j. \quad \text{(D7)}$$

Now it may readily be shown (cf. Miyamoto,[18] Sec. 3) that

$$\frac{\mathbf{p}\cdot\mathbf{n}}{s_j(|\Delta_j|)^{\frac{1}{2}}}=\left(\frac{d\sigma_Q}{d\sigma_P}\right)^{\frac{1}{2}}, \quad \text{(D8)}$$

where $d\sigma_Q$ is the element of a wave front of the incident field at Q and $d\sigma_P$ is the corresponding element at P. By elementary differential geometry

$$d\sigma_Q/d\sigma_P=|r_jr_j'/R_jR_j'|$$

where r_j,r_j' and R_j,R_j' are the principal radii of curvature of the wave front at Q and at P, respectively. Thus (D8) may be expressed in the form

$$U_K^{(\mathrm{I})}(P)=\sum_j U(Q_j)\exp(iks_j)\left|\frac{r_jr_j'}{R_jR_j'}\right|^{\frac{1}{2}}\epsilon_j \quad \text{(D9)}$$

and comparison with (5.7) shows that $U_K^{(\mathrm{I})}(P)=F_j(P)$.

[18] K. Miyamoto, J. Opt. Soc. Am. **48**, 57 (1958), Sec. 3.

Next consider the main contribution from the critical point of the second kind, i.e., from points Q_j on the boundary of the aperture at which the phase of the integrand is stationary with respect to a small variation dl along the boundary:

$$\partial(s+\phi)/\partial l=0, \quad \text{i.e.,} \quad (\hat{s}+\mathbf{p})\cdot\mathbf{l}=0. \quad \text{(D10)}$$

The contribution is (cf. van Kampen,[14] p. 580; Focke,[17] p. 40; Keller, Lewis, and Seckler,[19]

$$\begin{aligned}U_K^{(\mathrm{II})}(P)&=\sum_j\frac{i}{k}\left[\frac{2\pi}{k|\partial^2(s+\phi)/\partial l^2|}\right]_j^{\frac{1}{2}}\left\{\frac{ik}{4\pi}(\hat{s}-\mathbf{p})\cdot\mathbf{n}\times\frac{A}{s}\exp[ik(s+\phi)]\right\}_j\left\{\frac{\alpha_j}{\partial(s+\phi)/\partial m}\right\}_j\\ &=\frac{1}{k^{\frac{1}{2}}}\sum_j U(Q_j)\frac{\exp(iks_j)}{4\pi s_j}\left[\frac{2\pi}{|\partial^2(s+\phi)/\partial l^2|}\right]^{\frac{1}{2}}\times\left[\frac{\mathbf{n}\cdot(\mathbf{p}-\hat{s})}{\partial(s+\phi)/\partial m}\right]_j\alpha_j, \end{aligned}\quad \text{(D11)}$$

where

$$\alpha=\begin{cases}\exp(+i\pi/4) & \text{when } [\partial^2(s+\phi)/\partial l^2]_j>0,\\ \exp(-i\pi/4) & \text{when } [\partial^2(s+\phi)/\partial l^2]_j<0,\end{cases} \quad \text{(D12)}$$

and $\partial/\partial m$ denotes differentiation in the direction specified by the unit vector

$$\mathbf{m}=\mathbf{n}\times\mathbf{l}, \quad \text{(D13)}$$

$\mathbf{n}$ being the unit normal to the aperture. Now by a straightforward but rather long calculation[20] one may verify that the term in brackets preceding the factor α_j in (D11) is equal to the inclination factor in (C8), i.e., that

$$\left[\frac{\mathbf{n}\cdot(\mathbf{p}-\hat{s})}{\partial(s+\phi)/\partial m}\right]_j=\left[\frac{(\hat{s}\times\mathbf{p})\cdot\mathbf{l}}{1+\hat{s}\cdot\mathbf{p}}\right]_j. \quad \text{(D14)}$$

Comparison of (D11) with (5.12) then shows that $U_K^{(\mathrm{II})}=U_{\mathrm{I}}^{(B)}$.

Finally let us consider the main contribution from the critical points of the third kind. This contribution is given by[21] (cf. van Kampen,[14] p. 541; Focke,[17] p. 421;

[19] J. B. Keller, R. M. Lewis, and B. D. Seckler, J. Appl. Phys. **28**, 570 (1957).

[20] To carry out this calculation use first the relation

$$\partial(s+\phi)/\partial m=(\hat{s}+\mathbf{p})\cdot\mathbf{m}$$

on the left of (D14). Next choose $\mathbf{n},\mathbf{l}$ and $\mathbf{m}=\mathbf{n}\times\mathbf{l}$ as base vectors and write

$$\hat{s}=\alpha\mathbf{n}+\beta\mathbf{l}+\gamma\mathbf{m},$$
$$\mathbf{p}=a\mathbf{n}+b\mathbf{l}+c\mathbf{m},$$

where $\alpha^2+\beta^2+\gamma^2=a^2+b^2+c^2=1$. The relation (D10) implies that $b=-\beta$. Straightforward vector algebra then shows that each term in (D14) is equal to $(a-\alpha)/(c-\gamma)$.

[21] This contribution is of the same order as the contribution of the second term in the series arising from a critical point of the first kind.

Keller, Seckler, and Lewis[19])

$$U_K^{(\mathrm{III})}(P)=\frac{1}{k}\sum_j\left\{\frac{\sin\varphi}{k[\partial(s+\phi)/\partial l]_+[\partial(s+\phi)/\partial l]_-}\right\}_j$$

$$\times\left\{\frac{ik}{4\pi}(\hat{s}-\mathbf{p})\cdot\mathbf{n}\frac{A}{s}\exp[ik(s+\phi)]\right\}_j, \quad \text{(D15)}$$

where the subscripts $+$ and $-$ denote limiting values at the two sides of the corner point Q_j. Since $\partial(s+\phi)/\partial l = (\hat{s}+\mathbf{p})\cdot\mathbf{l}$ and $A\exp(ik\phi)=U(Q)$, (D15) may be written as

$$U_K^{(\mathrm{III})}(P)=\frac{1}{ik}\sum_j U(Q_j)\frac{\exp(iks_j)}{4\pi s_j}$$

$$\times\left\{\frac{\sin\varphi\cdot(\mathbf{p}-\hat{s})\cdot\mathbf{n}}{[(\hat{s}+\mathbf{p})\cdot\mathbf{l}^+][(\hat{s}+\mathbf{p})\cdot\mathbf{l}^-]}\right\}_j. \quad \text{(D16)}$$

Comparison with (5.13) shows that $U_K^{(\mathrm{II})}=U_{(\mathrm{II})}^{(B)}$.

Reprinted from JOURNAL OF THE OPTICAL SOCIETY OF AMERICA, Vol. 56, No. 12, 1712–1722, December 1966
Printed in U. S. A.

Consistent Formulation of Kirchhoff's Diffraction Theory*

E. W. MARCHAND
Research Laboratories, Eastman Kodak Company, Rochester, New York 14650

AND

E. WOLF†
Department of Physics and Astronomy, University of Rochester, Rochester, New York 14627

(Received 15 July 1966)

Kirchhoff's diffraction theory, which is often criticized because of an apparent internal inconsistency, is shown to be a rigorous solution to a certain boundary-value problem that has a clear physical meaning. This new interpretation of Kirchhoff's theory is a direct consequence of the Rubinowicz theory of the boundary diffraction wave.

We argue that these true boundary conditions of Kirchhoff's theory are physically reasonable for diffraction at an aperture in a *black* screen whose linear dimensions are large compared with the wavelength. The boundary values which Kirchhoff's solution takes in the plane of the aperture and in the near zone on the axis for the case of a normally incident plane wave diffracted at a circular aperture are compared with previously published results of experiments with microwaves, and reasonable agreement is found. (Strict agreement cannot be expected since the screens used in the microwave experiments were not black.) Moreover, Kirchhoff's solution is found to be in closer agreement with the experimental results than the "manifestly consistent" Rayleigh–Sommerfeld theory.

We also suggest an extension of the Kirchhoff theory, which might provide a physically reasonable approximation to the solution of the problems of diffraction at an aperture in a black screen whose linear dimensions are of the order of magnitude of or smaller than the wavelength of the light.

INDEX HEADING: Diffraction.

1. INTRODUCTION

It is useful to begin with the basic formula of Kirchhoff's diffraction theory for the field obtained on diffraction of a spherical or plane monochromatic wave $U^{(i)}(Q)e^{-i\omega t}$ at an aperture $\mathfrak{A}$ in a plane black screen, (Ref. 1, Sec. 8.3.2) viz.

$$U_K(P)=\frac{1}{4\pi}\iint_{\mathfrak{A}}\left\{U^{(i)}(Q)\frac{\partial}{\partial n}\left(\frac{e^{iks}}{s}\right)-\frac{e^{iks}}{s}\frac{\partial U^{(i)}(Q)}{\partial n}\right\}d\mathfrak{A}. \tag{1.1}$$

Here s denotes the distance from a typical point Q in the aperture to P and $\partial/\partial n$ denotes differentiation with respect to the normal to $\mathfrak{A}$, pointing into the half-space containing the point P.

It is well known that, under conditions frequently encountered in optics, the Kirchhoff theory gives results in excellent agreement with experiment. On the other hand, there also exists a good deal of literature which stresses a well-known mathematical inconsistency implicit in the derivation of the basic formula (1.1) of the Kirchhoff theory. The inconsistency arises because, although (1.1) is usually derived from the exact identity of Helmholtz,[1,2] the boundary conditions assumed in the derivation are overspecified: both the field U and its normal derivative $\partial U/\partial n$ are prescribed in the plane of the aperture. Since the time-independent wave equation for U is elliptic, the specification of U or its normal derivative $\partial U/\partial n$ in the plane of the aperture (together with the specification of the asymptotic behavior of the solution at infinity in the appropriate half-space) is sufficient to specify the solution uniquely. The inconsistency in Kirchhoff's theory arises from the specification of both U and $\partial U/\partial n$ and has the consequence that, as the point of observation P approaches the plane of the aperture, the Kirchhoff solution $U_K(P)$ does not recover the assumed boundary conditions. This fact appears to have been first pointed out by Poincaré[3] (see also Refs. 2 and 4).

In view of the practical success of Kirchhoff's theory, many attempts have been made to justify the Kirchhoff solution as a consistent approximation. Thus Born,[5] for example, suggested that the Kirchhoff solution might perhaps be regarded as a first approximation in a sequence of iterative solutions which converges to an exact solution; but the nature of the exact solution has never been specified, and it appears from publications

* Some of the results contained in this paper were presented 9 October 1964, at a meeting of the Optical Society of America, in New York City [cf. J. Opt. Soc. Am. 54, 1405A (1964)] and were reported briefly in Acta Phys. Polon. 27, 147 (1965).

† During the academic year 1966/1967 Guggenheim Fellow at the Department of Physics, University of California, Berkeley, Calif. 94720.

[1] M. Born and E. Wolf, *Principles of Optics* (Pergamon Press, New York, 1965), 3rd ed.

[2] B. B. Baker and E. T. Copson, *The Mathematical Theory of Huygens' Principle* (Clarendon Press, Oxford, England, 1950), 2nd ed.

[3] H. Poincaré, *Théorie Mathématique de la Lumière* (Georges Carré, Paris, 1892), Pt. II, pp. 187–188.

[4] G. Toraldo di Francia, Atti Fond. Giorgio Ronchi **11**, 503 (1956).

[5] M. Boron, *Optik* (Julius Springer-Verlag, Berlin, 1933; reprinted 1965), p. 152.

by Franz[6] and Schelkunoff[7] (see also Bouwkamp[8]) that an obvious iteration scheme of which Kirchhoff's solution could be regarded as the first approximation does not, in fact, converge. Modified versions of the Kirchhoff theory have been suggested,[9] but these modifications have met with a rather limited success.

A completely different viewpoint concerning the validity of Kirchhoff's theory was taken by Kottler.[10] He showed that Kirchhoff's solution is an exact solution to a certain *saltus* problem [the problem of solving the time-independent wave equation subject to prescribed jumps (discontinuities)]. He made the suggestion that the solution to such a saltus problem correctly describes diffraction at a black screen. Kottler's interpretation has not been generally accepted, and his contention—that Kirchhoff's theory, when applied to diffraction at a *black* infinite half-plane, gives an accurate description of the diffracted field—does not appear to have been tested experimentally to this day.

In the present paper we show that the inconsistency in Kirchhoff's diffraction theory is only apparent, and that Kirchhoff's diffraction integral (1.1) represents, in fact, a rigorous solution to a certain boundary-value problem, which has a clear physical meaning. This new interpretation is a direct consequence of the Rubinowicz theory[11] of the boundary diffraction wave.

Apart from the question of a consistent mathematical formulation of Kirchhoff's theory, which we now put forward, there is still the question as to what extent the boundary conditions, which we show that the Kirchhoff solution obeys, are, in fact, physically reasonable. We give a heuristic argument to show that, for diffraction at an aperture in a black screen whose linear dimensions are large compared to the wavelength of the light, these boundary conditions are physically reasonable, apart possibly from the form of a certain inclination factor, the exact choice of which could be determined from experiment. If this conclusion is correct, the Kirchhoff theory (applied to diffraction at apertures of linear dimensions which are large compared with the wavelength) would describe the diffracted field very well, not only in the far zone at moderate angles of diffraction, but in fact everywhere in the plane of and beyond the aperture. We quote in Sec. 4 some tentative experimental evidence[12] which indicates that this is likely to be so. In any case, results of experiments indicate that, at least in the near zone, the Kirchhoff theory provides a far better approximation than the "manifestly consistent" Rayleigh–Sommerfeld formula of the first kind.

Finally, we also suggest a possible extension of the Kirchhoff theory which might provide a physically reasonable approximation to the solution of the problem of diffraction at an aperture in a black screen whose linear dimensions are of the order of magnitude of or smaller than the wavelength of the light.

2. THE BOUNDARY-VALUE PROBLEM WHICH KIRCHHOFF'S FORMULA (1.1) SOLVES EXACTLY

Consider a diverging spherical (or plane) scalar wave incident upon an aperture in a plane black screen (Fig. 1). Let x, y, z be the coordinates of a typical point in space, referred to cartesian coordinate axes, with origin 0 at some point in the plane of the aperture and with $0z$ perpendicular to the plane of the aperture and pointing into the space into which the light is propagated. Further let

$$U^{(i)}(x,y,z)=e^{ikr}/r \qquad (2.1)$$

represent the incident field at a point $P(x,y,z)$, i.e., the field that would be generated at P in the absence of the screen. Here $r=SP$ is the distance from the source S of the spherical wave to P. A time periodic factor $e^{-i\omega t}(\omega=kc,\ c=\text{vacuum velocity of light})$ is suppressed, as is customary.

In the presence of the diffracting screen, the field is modified. Let $\mathcal{A}$ denote the aperture and $\mathcal{B}$ the "dark side" of the screen. We may always express the field at a typical point $P'(x,y,0)$ in the plane of the aperture

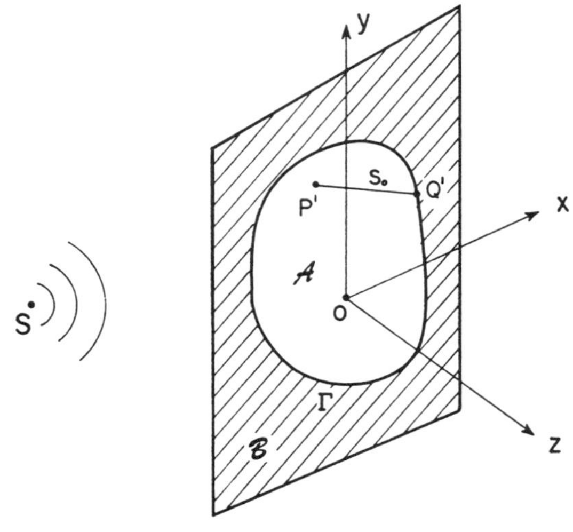

FIG. 1. Illustrating the notation.

[6] W. Franz, Z. Physik **125**, 563 (1949).
[7] S. A. Schelkunoff, Commun. Pure Appl. Math. **4**, 43 (1951).
[8] C. J. Bouwkamp, Rept. Progr. Phys. **17**, 35 (1954).
[9] Some of the modified theories are reviewed in Ref. 8, Sec. 4.
[10] F. Kottler, (a) Ann. Physik (Leipzig) **70**, 405 (1923). (b) in *Progress in Optics IV*, E. Wolf, Ed. (North-Holland Publ. Co. Amsterdam and John Wiley & Sons Inc., New York, 1965), p. 281.
[11] A. Rubinowicz, Ann. Physik (Leipzig) **53**, 257 (1917). For other references and an account of the theory, see Ref. 1, Sec. 8.9.
[12] The experimental evidence is tentative in the sense that it uses data obtained from diffraction experiments with screens that are not black.

in the form $U(x,y,0)=f_0(x,y)$, where

$$f_0(x,y)=U^{(i)}(x,y,0)+U^{(s)}(x,y,0) \text{ when } P'\in\mathfrak{A}$$
$$=U^{(s)}(x,y,0) \text{ when } P'\in\mathfrak{B}, \quad (2.2)$$

provided that the function $U^{(s)}(x,y,0)$ is chosen suitably. We now give a heuristic argument concerning the form of $U^{(s)}$ and show later that it leads to an expression which underlies a consistent formulation of Kirchhoff's theory.

Since the screen was assumed to be black, it absorbs all the radiation incident upon it. Thus the field $U^{(s)}(x,y)$ may be assumed to arise entirely from scattering of the incident radiation at the edge Γ of the aperture. Moreover, it seems reasonable to assume that, if the linear dimensions of the aperture are large compared with the wavelength, the scattered field $U^{(s)}$ consists of contributions from all the elements $\mathbf{dl}$ of Γ, each such contribution being proportional to the incident field $U^{(i)}$ at $\mathbf{dl}$ and being propagated as a spherical wave. This suggests that $U^{(s)}(x,y,0)$ can be expressed in the form

$$U^{(s)}(x,y,0)=\int_\Gamma U^{(i)}(\xi',\eta',0)\frac{e^{iks_0}}{s_0}\mathbf{K}(\xi',\eta';x,y)\cdot\mathbf{dl}, \quad (2.3)$$

where $(\xi',\eta',0)$ are the coordinates of the point Q' at which the element $\mathbf{dl}$ of the edge Γ is situated and s_0 is the distance $Q'P'$ (Fig. 1). Further, $\mathbf{K}$ is some vectorial inclination factor which may be expected to depend on the directions of incidence (SQ') and diffraction ($Q'P'$).

On substituting from (2.3) into (2.2) we obtain the following expression for the total field in the plane of the aperture:

$$f_0(x,y)=\epsilon_0(x,y)U^{(i)}(x,y,0)+\int_\Gamma U^{(i)}(\xi',\eta',0)\frac{e^{iks_0}}{s_0}\mathbf{K}(\xi',\eta';x,y)\cdot\mathbf{dl}, \quad (2.4)$$

where

$$\epsilon_0(x,y)=1 \text{ if } (x,y,0)\in\mathfrak{A}$$
$$=0 \text{ if } (x,y,0)\in\mathfrak{B}. \quad (2.5)$$

We demonstrate in the next section that Kirchhoff's integral is, in fact, an exact solution to a boundary-value problem, with the boundary condition in the plane of the aperture given by (2.4) and with the vectorial inclination factor[13]

$$\mathbf{K}(\xi',\eta';x,y)=\frac{1}{4\pi}\frac{SQ'\times Q'P'}{|SQ'||Q'P'|-SQ'\cdot Q'P'}. \quad (2.6)$$

More precisely we establish the following theorem:

The Kirchhoff integral (1.1) is a solution of the Helmholtz equation

$$\nabla^2U_K+k^2U_K=0, \quad (2.7)$$

which

(1) satisfies the boundary condition

$$U_K(x,y,z)\rightarrow f_0(x,y) \quad (2.8)$$

as $z\rightarrow +0$, where $f_0(x,y)$ is given by (2.4), with the inclination factor $\mathbf{K}$ given by (2.6), and

(2) obeys the Sommerfeld radiation condition in the half-space $z>0$ as $R=(x^2+y^2+z^2)^{\frac{1}{2}}\rightarrow\infty$.

3. PROOF OF THE NEW FORMULATION

Let us introduce the complex potential[14,15]

$$\mathbf{W}^{(i)}(Q,P)\equiv\mathbf{W}^{(i)}(\xi,\eta,\zeta;x,y,z),$$

where

$$\mathbf{W}^{(i)}(\xi,\eta,\zeta;x,y,z)=U^{(i)}(\xi,\eta,\zeta)\frac{e^{iks}}{s}\left[\frac{1}{4\pi}\frac{SQ\times QP}{|SQ||QP|-SQ\cdot QP}\right], \quad (3.1)$$

$$P=(x,y,z),\quad Q=(\xi,\eta,\zeta),$$
$$s=QP=[(x-\xi)^2+(y-\eta)^2+(z-\zeta)^2]^{\frac{1}{2}}.$$

By means of a rather complicated transformation[11,1] the Kirchhoff formula (1.1) may be shown to be expressible in the following form, valid everywhere in the half-space $z>0$, except at the shadow boundary:

$$U_K(x,y,z)=\epsilon(x,y,z)U^{(i)}(x,y,z)+\int_\Gamma \mathbf{W}^{(i)}(\xi',\eta',0;x,y,z)\cdot\mathbf{dl}. \quad (3.2)$$

Here $(\xi',\eta',0)$ are the coordinates of a typical point Q' at the edge Γ of the aperture, $\mathbf{dl}$ is the vector element of the edge at Q' and the integration is taken along Γ. Further,

$$\left.\begin{aligned}\epsilon(x,y,z)&=1, \text{ if } (x,y,z)\in A,\\ &=0, \text{ if } (x,y,z)\in B,\end{aligned}\right\} \quad (3.3)$$

where A denotes the "lit" region of the half-space $z>0$, i.e., that part of the half-space into which light is propagated in accordance with geometrical optics, and B denotes the shadow region (shown shaded in Fig. 2).

We note that the first term on the right of (3.2) is a continuous function of (x,y,z) at every point in the half-space $z\geq 0$, except on the shadow boundary and at the

[13] When the incident wave is a plane wave, propagated in a direction specified by the unit vector $\mathbf{p}$, (2.6) must be replaced by

$$\mathbf{K}(\xi,\eta;x,y)=\frac{1}{4\pi}\frac{\mathbf{p}\times\hat{s}}{1-\mathbf{p}\cdot\hat{s}},$$

where $\hat{s}=Q'P/|Q'P|$. A similar change must then be made in the expression for $\mathbf{W}^{(i)}$, given by Eq. (3.1).

In (2.6), and in several other places, an adjacent pair of capital letters indicates a vector. Also, $\hat{s}$ represents a vector, even though not printed in boldface type.

[14] A. Rubinowicz, Acta Phys. Polon. **12**, 225 (1953).

[15] K. Miyamoto and E. Wolf, (a) J. Opt. Soc. Am. **52**, 615 (1962); (b) **52**, 626 (1962).

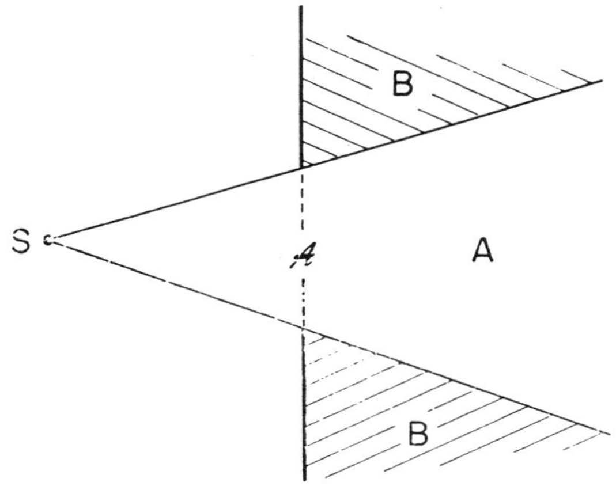

FIG. 2. Regions A and B relating to the definition (3.3) of $\epsilon(x,y,z)$. Region A is represented by the unshaded area lying on the right of the plane $z=0$.

edge Γ. Further, as is evident from (3.1), the integrand in the integral (3.2) is a continuous function of both sets of variables $(\xi',\eta',0)$ and (x,y,z), except when the vectors SQ' and $Q'P$ are parallel, i.e., when the point $P(x,y,z)$ lies on the shadow boundary, or when $Q'P=0$, i.e., when $x=\xi'$, $y=\eta'$, $z=0$, in which case the point P lies on the boundary curve Γ. It follows, therefore, that the integral in (3.2) is a continuous function of (x,y,z) in the half-space $z\geq 0$, except possibly when (x,y,z) is situated on the shadow boundary or on the boundary curve Γ. These results imply that as $z\rightarrow +0$, the limit $U_K(x,y,z)$ exists, except possibly when the point $(x,y,0)$ is situated on Γ, and moreover that (with this possible exception)

$$\lim_{z\rightarrow +0} U_K(x,y,z)=\epsilon_0(x,y)U^{(i)}(x,y,0) + \int_\Gamma \mathbf{W}^{(i)}(\xi',\eta',0;\, x,y,0)\cdot d\mathbf{l}. \quad (3.4)$$

But $\mathbf{W}^{(i)}(\xi',\eta',0;\, x,y,0)\cdot d\mathbf{l}$ is simply the integral in (2.4) with the vectorial inclination factor $\mathbf{K}$ given by (2.6). Hence (3.4) implies that

$$\lim_{z\rightarrow +0} U_K(x,y,z)=f_0(x,y), \quad (3.5)$$

where f_0 is given by (2.4), with $\mathbf{K}$ given by (2.6).

Moreover, since according to formula (1.1), the Kirchhoff solution $U_K(x,y,z)$ is expressible as a superposition of spherical waves e^{iks}/s and their derivatives, it follows that in the half-space $z>0$, U_K obeys the Helmholtz equation

$$\nabla^2 U_K+k^2U_K=0. \quad (3.6)$$

Finally, as is well known [cf. Ref. (15b) Appendix D, or Ref. (16)], the Kirchhoff solution U_K obeys the Sommerfeld radiation condition at infinity, in the half-space $z>0$.

We have now established the theorem enunciated at the end of Sec. 2.

It is natural to enquire whether the solution to the boundary-value problem just established is in some sense unique. We have not found an answer to this question for the general case, but in Appendices A and B a uniqueness theorem is formulated and demonstrated for the special case of diffraction of a plane wave incident normally on a circular aperture.

4. COMPARISON WITH EXPERIMENT

Many measurements have been reported in the literature for the field obtained by diffraction of a plane or a spherical wave at an aperture in an opaque screen. Most of these measurements correspond, unfortunately, only to situations where the screen is conducting, whereas our theory was formulated for a black screen. Hence comparison between our theory and available experimental data cannot be expected to show conclusively whether our considerations are entirely correct physically. Nevertheless, it seems worthwhile to make such a comparison, in order to obtain at least a rough indication about the correctness of our viewpoint.

In Fig. 3(a) curves are reproduced from a paper by Ehrlich, Silver, and Held,[17] which show, in a few typical cases, the behavior of the field across the aperture in a conducting screen, as obtained from experiments with microwaves of wavelength $\lambda=3.2$ cm. In Fig. 3(b) the corresponding curves obtained from formula (3.4) of the present paper are shown for comparison. These theoretical curves for the aperture field, calculated as they are from Kirchhoff's theory, ought, according to the usual criticism of Kirchhoff's theory, to be rather different from the experimental curves. Yet, as is clearly seen on comparison with Fig. 3(a), they exhibit the oscillatory nature of the measured field, which, in accordance with our predictions, may be regarded as arising from interference between the incident field and the scattered field, the latter being represented by the boundary diffraction wave. The theoretical curve is seen to exhibit, in each case, the correct number of maxima and minima of the measured field. The only appreciable difference between the theoretical and the experimental curves is in the exact values of the maxima and minima, and these differences may be expected to be due to the fact that the experiments refer to conducting screens, whereas our theoretical curves refer to a black screen.

[16] J. B. Keller, R. M. Lewis, and B. D. Seckler, J. Appl. Phys. 28, 570 (1957).

[17] M. J. Ehrlich, S. Silver, and G. Held, J. Appl. Phys. 26, 336 (1955). Similar experimental curves were published by J. Buchsbaum, A. R. Milne, D. C. Hogg, G. Bekefi, and G. A. Woonton, J. Appl. Phys. 26, 706 (1955); C. L. Andrews, Phys. Rev. 71, 777 (1947); J. Appl. Phys. 21, 761 (1950).

In the last-quoted paper, Andrews pointed out that his experimental results are consistent with Thomas Young's ideas about the origin of diffraction. Since the Rubinowicz theory of the boundary diffraction wave may be regarded as a mathematical refinement of Young's ideas, Andrews's views are essentially in qualitative agreement with those of the present paper.

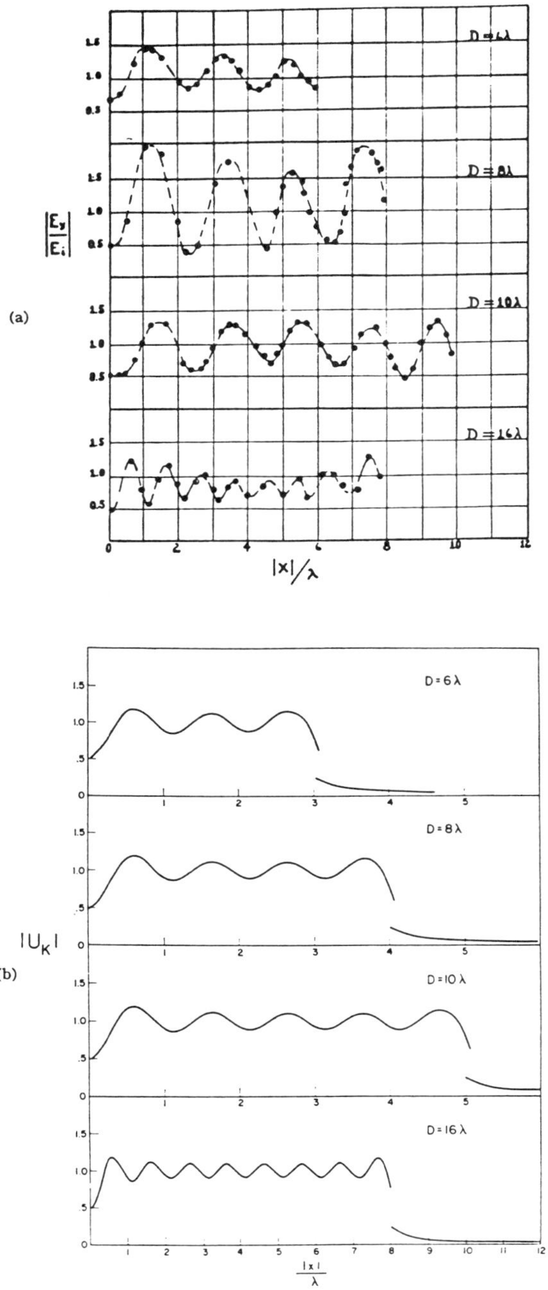

FIG. 3. (a) Measured principal component of the electric field in the plane of a circular aperture of diameter D in a conducting screen, as a function of the radial distance x (measured in wavelengths). The incident field is a plane wave of unit amplitude falling normally onto the plane of the aperture. [From M. J. Ehrlich, S. Silver, and G. Held, J. Appl. Phys. **26**, 342 (1955)]. (b) The field amplitude $|U_K|$, calculated from Eq. (3.4), in the plane of a circular aperture, of diameter D, as a function of the radial distance x (measured in wavelengths). The incident field is a plane wave falling normally onto the plane of the aperture.

In Fig. 4(a), curves are reproduced for the field measured along the axis near the aperture. In Fig. 4(b) the corresponding curves, computed on the basis of Kirchhoff's theory, are shown. Again it is seen that the Kirchhoff theory is in reasonable agreement with the experimental results for the electric field. It is this field rather than the magnetic one that is effective in measurements at optical frequencies.

We may, therefore, conclude that in the plane of the aperture and on the axis close enough to this plane, Kirchhoff's theory, for diffraction at an aperture of linear dimensions that are large compared with the wavelength, predicts results that are in substantial agreement with the experiment.

Finally, it is of interest to examine how the Rayleigh–Sommerfeld theory[18] compares with results of experiment. In the notation of the preceding sections of this paper, the Rayleigh–Sommerfeld formulae of the first and second kind, respectively, are:

$$U_{\mathrm{I}}(P)=\frac{1}{2\pi}\iint_{\mathfrak{A}} U^{(i)}(Q)\frac{\partial}{\partial n}\left(\frac{e^{iks}}{s}\right)d\mathfrak{A}, \tag{4.1}$$

$$U_{\mathrm{II}}(P)=-\frac{1}{2\pi}\iint_{\mathfrak{A}} \frac{\partial U^{(i)}(Q)}{\partial n}\frac{e^{iks}}{s}d\mathfrak{A}. \tag{4.2}$$

The first formula does not predict any interference effects in the plane of the aperture, since it recovers the assumed boundary conditions[19]:

$$\lim_{z\to+0} U_{\mathrm{I}}(x,y,z)=\epsilon_0(x,y)U^{(i)}(x,y,0), \tag{4.3}$$

where again $\epsilon_0(x,y)=1$ or 0, according as the point $(x,y,0)$ is inside the aperture or on the screen.

In the case of the second formula (4.2), we use the identity

$$U_K(P)=\tfrac{1}{2}[U_{\mathrm{I}}(P)+U_{\mathrm{II}}(P)], \tag{4.4}$$

from which it follows that

$$\lim_{z\to+0} U_{\mathrm{II}}(x,y,z)=\lim_{z\to+0} 2U_K(x,y,0)-\lim_{z\to+0} U_{\mathrm{I}}(x,y,0),$$

or, using (3.4) and (4.3),

$$\lim_{z\to+0} U_{\mathrm{II}}(x,y,z)=\epsilon_0 U^{(i)}(x,y,0)$$
$$+2\int_\Gamma \mathbf{W}^{(i)}(\xi',\eta',0;\,x,y,0)\cdot d\mathbf{l}. \tag{4.5}$$

[18] Some aspects of the Rayleigh–Sommerfeld theory are discussed in the papers by E. W. Marchand and E. Wolf, J. Opt. Soc. Am. **52**, 761 (1962); **54**, 587 (1964). In the first of these two papers, the theory was associated with the names of Rayleigh and Kirchhoff, but it seems more appropriate to associate it with the names of Rayleigh and Sommerfeld.

[19] N. Mukunda, J. Opt. Soc. Am. **52**, 336 (1962).

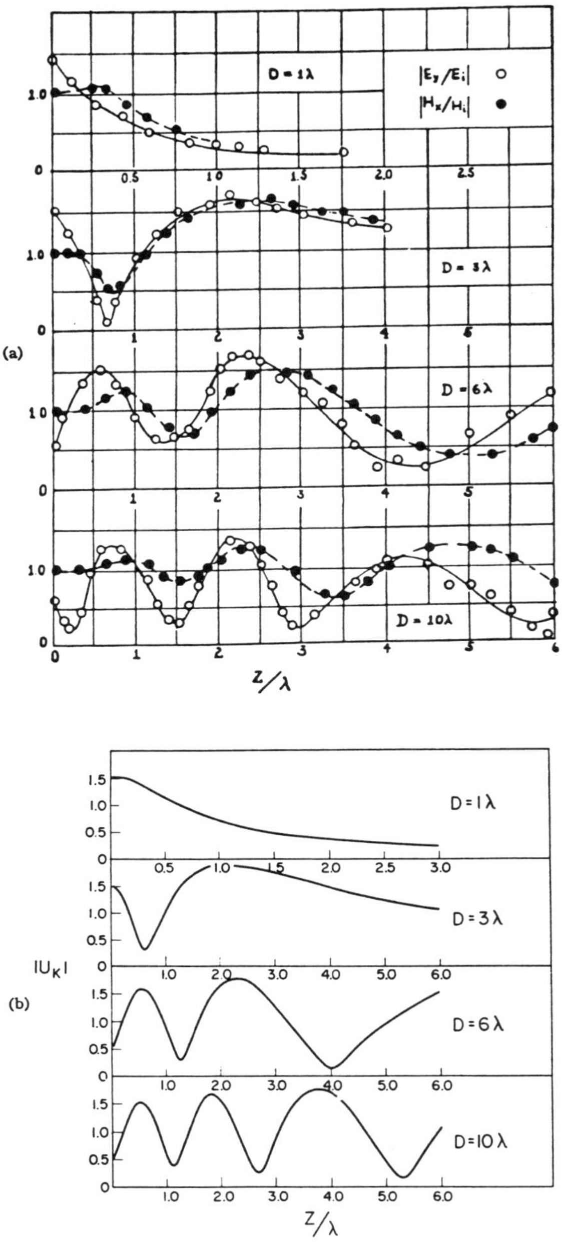

FIG. 4. (a) Measured axial distribution of the principal tangential component of the electric and magnetic fields **E** and **H**. for circular aperture of diameter D in a conducting screen, as a function of the distance z (measured in wavelengths) from the aperture. The incident field is a plane wave of unit amplitude falling normally on the plane of the aperture. [From M. J. Ehrlich, S. Silver, and G. Held, J. Appl. Phys. **26**, 345 (1955)]. (b) The axial distribution of the field amplitude U_K, calculated from (3.2), for circular aperture of diameter D, as a function of the distance z (measured in wavelengths) from the aperture. The incident field is a plane wave of unit amplitude, falling normally on the plane of the aperture.

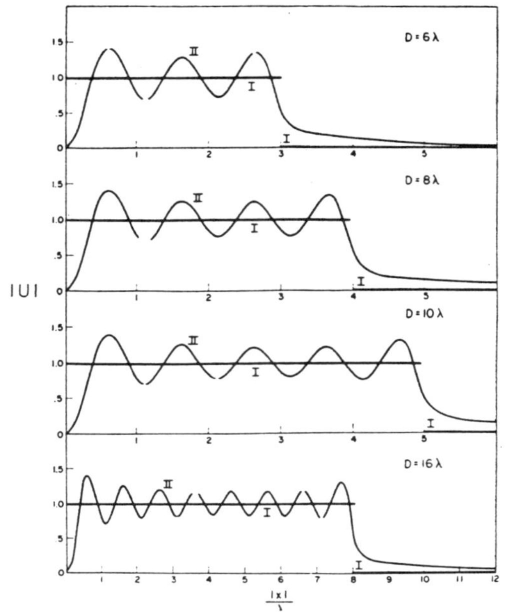

FIG. 5. The field amplitudes $|U_{\text{I}}|$ (denoted by I) and $|U_{\text{II}}|$ (denoted by II) given by the Rayleigh–Sommerfeld formulae of the first and the second kind, respectively, for the field in the plane of a circular aperture of diameter D, as a function of the radial distance x (measured in wavelengths). The incident field is a plane wave falling normally onto the plane of the aperture.

The behavior of U_{I} and U_{II} in the plane of the aperture, for the case of diffraction of a plane wave incident normally on a circular aperture, is shown in Fig. 5. The curves in this figure were computed from formulae (4.3) and (4.5). These curves should be compared with those given in Fig. 3(a) and (b). The comparison shows clearly that in the plane of the aperture, the "manifestly consistent" Rayleigh–Sommerfeld formula of the first kind gives results which do not agree at all with experiment and which are inferior to those obtained from Kirchhoff's theory.[20] The Rayleigh–Sommerfeld formula of the second kind is seen to lead to a much better agreement with results of experiments, but it leads to incorrect values in and near the center of the aperture. We believe that the comparison would be even more favorable to Kirchhoff's theory if the results, shown in Fig. 3(a), were obtained from experiments with a black screen.

5. EXTENSION TO SMALL APERTURES. MULTIPLY DIFFRACTED BOUNDARY WAVES

We have already stressed that, although our present formulation of the Kirchhoff theory is mathematically

[20] We have shown elsewhere [J. Opt. Soc. Am. **54**, 587 (1964)] that, if the linear dimensions of the aperture are large compared with the wavelength, the difference in the behavior of the Rayleigh–Sommerfeld solutions and of the Kirchhoff solution in the plane of the aperture does not lead to significantly different results for the predicted field in the far zone, in the neighborhood of the forward direction.

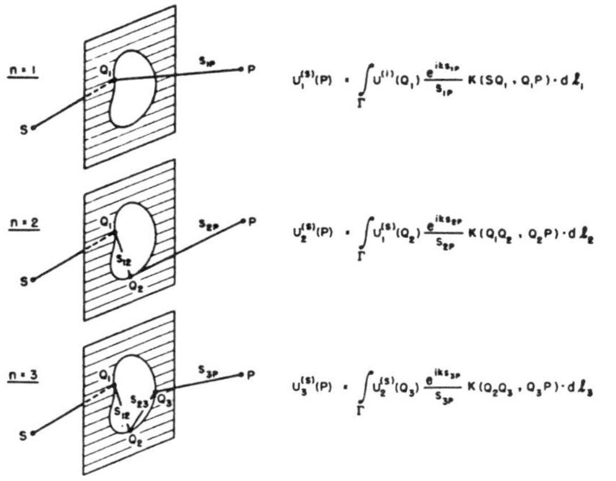

FIG. 6. Illustrating the notation relating to the singly ($U_1^{(s)}$), doubly ($U_2^{(s)}$), and triply ($U_3^{(s)}$) scattered boundary waves.

consistent, the question of whether or not the new boundary conditions, which we showed that it obeys, are physically reasonable, need separate consideration. In Sec. 2 we gave reasons to support these boundary conditions on physical grounds, for apertures whose linear dimensions are large compared with the wavelength, and in Sec. 4 we quoted experimental evidence to support this view.

It seems natural to enquire whether the Kirchhoff theory may be extended to problems of diffraction at an aperture in a black screen, whose linear dimensions are of the order of a wavelength or less. In such cases some of the radiation which is scattered by the edge Γ of the aperture may be expected to reach other elements of Γ in appreciable strength, then to be rescattered, and so to contribute to the total field.[21] Moreover the "rescattered radiation" is again rescattered, and this process continues. In order to describe this situation in a mathematically compact form we first rewrite the formula (3.2) in the form

$$U_K(P)=\epsilon(P)U^{(i)}(P)+U_1^{(s)}(P), \tag{5.1}$$

where

$$U_1^{(s)}(P)=\int_\Gamma U^{(i)}(Q_1)\frac{e^{iks_{1P}}}{s_{1P}}\mathbf{K}(SQ_1,Q_1P)\cdot\mathbf{dl}_1. \tag{5.2}$$

The meanings of some of the symbols which appear in this formula are indicated in the top diagram ($n=1$) of Fig. 6.

Assuming that the law of scattering from each element of the edge is given by the integrand of (5.2), we may expect that the following formulae, which take into account also the doubly, triply, . . . diffracted boundary waves, would describe the total field more accurately than the formulae (5.1):

$$U(P)=\epsilon(P)U^{(i)}(P)+\sum_{n=1}^{\infty}U_n^{(s)}(P), \tag{5.3}$$

where

$$\begin{aligned}
U_2^{(s)}(P)&=\int_\Gamma U_1^{(s)}(Q_2)\frac{e^{iks_{2P}}}{s_{2P}}\mathbf{K}(Q_1Q_2,Q_2P)\cdot\mathbf{dl}_2\\
&=\int_\Gamma\int_\Gamma U^{(i)}(Q_1)\frac{e^{ik(s_{12}+s_{2P})}}{s_{12}\cdot s_{2P}}\mathbf{K}(SQ_1,Q_1Q_2)\cdot\mathbf{dl}_1\\
&\qquad\times\mathbf{K}(Q_1Q_2,Q_2P)\cdot\mathbf{dl}_2,\\
&\vdots\\
U_n^{(s)}(P)&=\int_\Gamma U_{n-1}^{(s)}(Q_n)\frac{e^{iks_{nP}}}{s_{nP}}\mathbf{K}(Q_{n-1}Q_n,Q_nP)\cdot\mathbf{dl}_n\\
&=\int_\Gamma\cdots\int_\Gamma U^{(i)}(Q_1)\frac{e^{ik(s_{12}+s_{23}+\cdots s_{nP})}}{s_{12}s_{23}\cdots s_{nP}}\\
&\qquad\times\mathbf{K}(SQ_1,Q_1Q_2)\cdot\mathbf{dl}_1\mathbf{K}(Q_1Q_2,Q_2Q_3)\cdot\mathbf{dl}_2\\
&\qquad\times\cdots\mathbf{K}(Q_{n-1}Q_n,Q_nP)\cdot\mathbf{dl}_n.
\end{aligned} \tag{5.4}$$

Here $U_2^{(s)}$, $U_3^{(s)}$, represent the total contributions obtained from double, triple, . . . scattering from the edge Γ. In the integrals (5.4), $Q_1, Q_2, \ldots Q_n, \ldots$ denote typical points on the edge Γ of the aperture (Fig. 6) and the distances s_{ij} are defined as follows:

$$s_{12}=Q_1Q_2,\quad s_{23}=Q_2Q_3,\quad\cdots\quad s_{n-1,n}=Q_{n-1}Q_n,\qquad s_{nP}=Q_nP, \tag{5.5}$$

and $\mathbf{dl}_j$ is a vector element of the edge Γ at the point Q_j.

Although (5.3) contains formally a contribution from an infinite number of terms under the summation sign, only the first two terms are, in general, nonvanishing, if the inclination factor $\mathbf{K}$ is assumed to be given by Eq. (2.6). For in this case, for any three coplanar points Q_j, Q_{j+1}, Q_{j+2}, we have

$$\mathbf{K}(Q_jQ_{j+1},Q_{j+1}Q_{j+2})=0. \tag{5.6}$$

Since the factor $\mathbf{K}(Q_1Q_2;Q_2Q_3)$ appears in the expression for $U_n^{(s)}(P)$ whenever $n\geq 3$, and since Q_1, Q_2, Q_3 are coplanar,

$$U_3^{(s)}(P)=U_4^{(s)}(P)=\cdots U_n^{(s)}(P)=\cdots=0. \tag{5.7}$$

The expression (5.3) for the total field then reduces to

$$U(P)=U_K(P)+U_2^{(s)}(P), \tag{5.8}$$

where $U_K(P)$ is again the Kirchhoff solution given by (5.1).

APPENDIX A

Some Remarks on Uniqueness

The solution to a boundary-value problem of the type formulated in the present paper may also be de-

[21] See also R. W. Dunham, J. Opt. Soc. Am. 54, 1102 (1964).

rived with the help of the well-known Rayleigh diffraction integrals[8] of the first kind. We briefly consider this alternative approach, as it has a bearing on the problem of uniqueness that is not completely settled in the present paper.

The Rayleigh diffraction integral of the first kind expresses the solution $U(x,y,z)$ of the Helmholtz equation

$$\nabla^2 U + k^2 U = 0 \tag{A1}$$

for the half-space $z>0$, which attains prescribed boundary values

$$\operatorname*{Lim}_{z\to+0} U(x,y,z) = f(x,y) \tag{A2}$$

and satisfies appropriate boundary conditions at infinity. This diffraction integral has the form

$$U(x,y,z) = \frac{1}{2\pi}\int_{-\infty}^{\infty}\int_{-\infty}^{\infty} f(x',y') \times \left\{\frac{\partial}{\partial z'}\left(\frac{e^{iks}}{s}\right)\right\}_{z'=0} dx'dy', \tag{A3}$$

where

$$s = [(x'-x)^2+(y'-y)^2+(z'-z)^2]^{\frac{1}{2}}. \tag{A4}$$

Luneburg,[22] p. 311–319, has formulated a certain uniqueness theorem for the Rayleigh solution. He showed that if $f(x,y)$

(a) is continuous and has continuous partial derivatives outside some circle of radius R_0,

(b) is sectionally continuous in the xy plane,

(c) obeys the conditions

$$|f(x,y)| < \frac{B}{R}, \quad \left|\frac{\partial f(x,y)}{\partial x}\right| < \frac{B}{R}, \quad \left|\frac{\partial f(x,y)}{\partial y}\right| < \frac{B}{R}, \tag{A5}$$

for $R > R_0$, where $R = (x^2+y^2)^{\frac{1}{2}}$ and B is a constant, then the function $U(x,y,z)$ given by (A3) is the *only* solution of the boundary-value problem stated above which has the following properties:

(1) It is regular for $z>0$,

(2) When $R' = (x^2+y^2+z^2)^{\frac{1}{2}} > R_0$, $z>0$,

$$|U(x,y,z)| < \frac{C}{R'}, \quad \left|\frac{\partial U(x,y,z)}{\partial R'}\right| < \frac{C}{R'}, \tag{A6}$$

where C is a constant.

(3) If θ denotes the angle which any half-line through the origin makes with the positive semi-axis $0z$, then, associated with any three-dimensional sector $0<\theta<\pi/2-\delta(\delta<\pi/2)$ of the domain $R'>R_0$, $z>0$, there is a constant $D(\delta)$ such that for all points (x,y,z) of the three-dimensional sector,

$$|\partial U/\partial R' - ikU| < D/R'^2. \tag{A7}$$

The condition (A7) is essentially Sommerfeld's radiation condition, and (A6) are the associated conditions of finiteness.[2]

Were it possible to show that the function $f(x,y) = f_0(x,y)$, where $f_0(x,y)$, given by (2.4) and (2.6), satisfies the conditions (a), (b), and (c) of Luneburg's uniqueness theorem, it would follow from this uniqueness theorem and from the results of the previous section, that the Kirchhoff solution is the only solution of our boundary-value problem. Unfortunately, it is hard to determine, in general, whether f_0 satisfies the three conditions, since it seems very difficult to see whether, in general, f_0 has a finite discontinuity at the edge Γ of the aperture. However, in the special case when the aperture is circular and when $U^{(i)}$ is a plane wave incident normally upon the aperture all the conditions of Luneburg's theorem may be shown to be satisfied; this is done in Appendix B. In the more general case we should bear in mind that there is a possibility that Luneburg's theorem is valid even under somewhat more relaxed conditions on the boundary function f.

APPENDIX B

Application of Luneburg's Uniqueness Theorem to the Present Problem

Let us first consider the behavior of the integral (representing the scattered field) which occurs in (2.4) with **K** given by (2.6), i.e.,

$$f_0^{(s)}(P') = \int_\Gamma U^{(i)}(Q')\frac{e^{iks_0}}{s_0} \times \left[\frac{1}{4\pi}\frac{SQ'\times Q'P'}{|SQ'||Q'P'| - SQ'\cdot Q'P'}\right]\cdot \mathbf{dl}. \tag{B1}$$

Here $Q'(\xi,\eta,0)$ is a point on the edge Γ of the aperture, P' is a typical point $(x',y',0)$ in the plane of the aperture, and $s_0 = |Q'P'|$, where $S(x_0,y_0,z_0)$ denotes, as before, the source point.

We represent the coordinates (ξ,η) in parametric form:

$$\xi = \xi(l), \quad \eta = \eta(l), \tag{B2}$$

where l denotes the distance from a fixed point on the edge Γ to the point $(\xi,\eta,0)$, measured along the edge. The edge is assumed to be sufficiently smooth, so that the first derivatives $\xi'(l)$ and $\eta'(l)$, taken with respect to the parameter l, exist and are continuous. Then the vectorial edge element **dl** has the components

$$\xi'(l)dl, \ \eta'(l)dl, \ 0.$$

[22] R. K. Luneburg, *Mathematical Theory of Optics* (University of California Press, Berkeley and Los Angeles, Calif., 1964).

We also have

$$SQ'=(\xi-x_0,\ \eta-y_0,\ -z_0),$$
$$Q'P'=(x'-\xi,\ y'-\eta,\ 0).$$

Hence

$$(SQ'\times Q'P')\cdot \mathbf{dl}=\begin{vmatrix} \xi-x_0 & \eta-y_0 & -z_0 \\ x'-\xi & y'-\eta & 0 \\ \xi'(l) & \eta'(l) & 0 \end{vmatrix} dl. \quad \text{(B3)}$$

Also

$$|SQ'||Q'P'|-SQ'\cdot Q'P'$$
$$=[(\xi-x_0)^2+(\eta-y_0)^2+z_0^2]^{\frac{1}{2}}[(x'-\xi)^2+(y'-\eta)^2]^{\frac{1}{2}}$$
$$-[(\xi-x_0)(x'-\xi)+(\eta-y_0)(y'-\eta)]. \quad \text{(B4)}$$

It is evident that both the expressions (B3) and (B4) are continuous functions of x', y' and l and, moreover, that (B4) does not vanish[23] unless P' is on Γ. Also, the factor e^{iks_0}/s_0 is a continuous function of (ξ,η), except when P' is on Γ. Thus if P' is not on Γ, the integrand of (B1) and hence the integral (B1) itself is a continuous function of P'.

Calculation of the partial derivatives of the integral (B1) by differentiation under the integral sign shows, by a similar reasoning, that the first partial derivatives of the integral (B1) with respect to the coordinates x' and y', respectively, are also continuous functions of (x',y'), provided that P' is not on Γ.

Finally, in (2.4) we use these results and also the fact that $U^{(i)}(x',y',0)$ and its partial derivatives are continuous functions of (x',y') and that ϵ_0 is given by (2.5). It then follows that

$$f_0(x',y'),\quad \frac{\partial f_0(x',y')}{\partial x'},\quad \frac{\partial f_0(x',y')}{\partial y'}$$

are each continuous at all points of the plane of the aperture, except possibly on the edge Γ. Hence, in particular, they are continuous outside a circle of radius R_0, which contains the aperture in its interior. Thus f_0 satisfies the condition (a) of Luneburg's uniqueness theorem, as stated in Appendix A.

To show that f_0 satisfies the condition (b), viz., that it is sectionally continuous in the xy-plane, we need only to prove, in view of the results just established, that f_0 is either continuous everywhere on Γ or that it has a finite discontinuity there. Analysis of the general case (spherical incident wave and aperture of arbitrary form) appears to be rather difficult. We, therefore, restrict ourselves to the case when $U^{(i)}$ is a plane wave incident normally on the plane of the aperture, the aperture being assumed to be circular. In place of (B1) we then have [see footnote (13)]

$$f_0^{(e)}(P')=\frac{1}{4\pi}\int_\Gamma \frac{e^{iks_0}}{s_0}\frac{\mathbf{p}\times\mathbf{\hat{s}}}{1-\mathbf{p}\cdot\mathbf{\hat{s}}}\cdot\mathbf{dl}, \quad \text{(B5)}$$

where $\mathbf{p}$ is the unit vector in the direction of the positive z axis. The integral (B5) may be transformed into the form [see Eq. (3.12) in the reference quoted in footnote 20]

$$f_0^{(e)}(P')=-\frac{1}{4\pi}\int_0^{2\pi}\frac{\exp(ika\sigma)}{\sigma^2}[1-\rho\cos\psi]d\psi, \quad \text{(B6)}$$

where

$$\sigma=[1-2\rho\cos\psi+\rho^2]^{\frac{1}{2}},\qquad \rho=d/a. \quad \text{(B7)}$$

Here a denotes the radius of the aperture and d is the distance from the center of the aperture to P'. In order to determine the behavior of $f_0^{(e)}(P')$ as P' crosses the edge Γ of the aperture, we rewrite (B6) in the form

$$f_0^{(e)}(P')=-\frac{1}{4\pi}\left[\int_0^{\pi}\frac{(1-\rho^2)\exp(ika\sigma)}{1-2\sigma\cos\psi+\rho^2}d\psi+\int_0^{\pi}\exp(ika\sigma)d\psi\right]. \quad \text{(B8)}$$

Now the second integral is a continuous function of ρ when $\rho=1$ (edge of aperture), and it is readily shown that

$$\lim_{\rho\to1}\int_0^\pi \exp(ika\sigma)d\psi=\int_0^\pi \exp[2ika\sin\psi']d\psi'. \quad \text{(B9)}$$

When $\rho=1$ the integrand of the first integral in (B8), viz.,

$$A(\rho)=\int_0^\pi\frac{(1-\rho^2)\exp(ika\sigma)}{1-2\rho\cos\psi+\rho^2}d\psi, \quad \text{(B10)}$$

is indeterminate at one point in the range of integration, namely, at $\psi=0$. Hence it is convenient to split the range of integration into two parts,

$$A(\rho)=A_1(\rho)+A_2(\rho), \quad \text{(B11)}$$

where

$$A_1(\rho)=\int_\delta^\pi\frac{(1-\rho^2)\exp(ika\sigma)}{1-2\rho\cos\psi+\rho^2}d\psi, \quad \text{(B12)}$$

$$A_2(\rho)=\int_0^\delta\frac{(1-\rho^2)\exp(ika\sigma)}{1-2\rho\cos\psi+\rho^2}d\psi, \quad \text{(B13)}$$

and $0<\delta<\pi$. $A_1(\rho)$ is evidently continuous when $\rho=1$, since the integrand itself is continuous in the range $\delta\leq\psi\leq\pi$. Obviously

$$\lim_{\rho\to1}A_1(\rho)=0, \quad \text{(B14)}$$

[23] It is clear from an examination of the left-hand side of (B4) that, if $Q'P'\neq0$, the expression could only vanish if $\cos(SQ',Q'P')=1$, i.e., if the directions SQ' and $Q'P'$ were parallel. This is impossible, since S is situated off the plane of the aperture, whereas both Q' and P' are in this plane.

independently of δ. Next consider $A_2(\rho)$. Suppose that we choose δ so small that the term $\exp(ika\sigma)$ is essentally constant throughout the range of integration. Then in (B7) we may take $\psi \sim 0$ so that

$$\sigma \sim (1-2\rho+\rho^2)^{\frac{1}{2}} = |1-\rho|,$$

and hence

$$\exp(ika\sigma) \sim \exp(ik|1-\rho|).$$

Equation (B13) then gives

$$A_2(\rho) \sim (1-\rho^2)\exp(ika|1-\rho|)\int_0^\delta \frac{d\psi}{1+\rho^2-2\rho\cos\psi}$$
$$= 2\,\mathrm{signum}(1-\rho^2)\exp(ika|1-\rho|) \times \tan^{-1}\left|\frac{1+\rho}{1-\rho}\tan\frac{\delta}{2}\right|. \quad \text{(B15)}$$

Hence

$$\lim_{\rho\to 1+} A_2(\rho) = -\pi, \qquad \lim_{\rho\to 1-} A_2(\rho) = +\pi. \quad \text{(B16)}$$

From (B8), if (B9)–(B16) are used, it follows that

$$\operatorname*{Lim}_{\rho\to 1+} f_0^{(s)}(P') = \tfrac{1}{4} - \frac{1}{4\pi}\int_0^\pi \exp[2ika\sin\psi']d\psi',$$
$$\operatorname*{Lim}_{\rho\to 1-} f_0^{(s)}(P') = -\tfrac{1}{4} - \frac{1}{4\pi}\int_0^\pi \exp[2ika\sin\psi']d\psi'. \quad \text{(B17)}$$

We recall that the expressions (B17) represent the limiting behavior of the line integral in (2.4) (scattered field) as the field point approaches the edge of the aperture [under the assumption that the incident field $U^{(i)}(x,y,z)$ is a plane wave, e^{ikz}, incident normally on the plane of the aperture, the aperture being circular]. Hence it follows from (2.4) and (B17) that

$$\lim_{\rho\to 1+} f_0(x,y) = \tfrac{1}{4} - \frac{1}{4\pi}\int_0^\pi \exp[2ika\sin\psi']d\psi',$$
$$\lim_{\rho\to 1-} f_0(x,y) = \tfrac{3}{4} - \frac{1}{4\pi}\int_0^\pi \exp[2ika\sin\psi']d\psi'. \quad \text{(B18)}$$

Hence we have

$$\Delta f_0 = \lim_{\rho\to 1+} f_0(x,y) - \lim_{\rho\to 1-} f_0(x,y) = -\tfrac{1}{2}, \quad \text{(B19)}$$

independently of the radius a of the aperture. Thus (at least in the special case considered here) f_0 has a finite discontinuity at the edge Γ but is continuous everywhere else in the plane of the aperture, i.e., $f_0(x,y)$ is sectionally continuous in the xy plane, and hence satisfies the condition (b) of Luneburg's uniqueness theorem.

The behavior of the field at the edge Γ of the aperture [discontinuity of amount $-\frac{1}{2}$ in the boundary function $f_0(x,y)$ on Γ, which is predicted by Eq. (B19)] is seen clearly in Fig. 3(b). This figure shows the behavior of the boundary function in the whole xy plane, appropriate to selected values of the aperture radius.

Finally, we show that, whatever the nature of the incident field (spherical or plane wave) and whatever the shape of the aperture, $f_0(x,y)$ also obeys the conditions (c), (Eq. A5), of Luneburg's uniqueness theorem. From (2.4) and (2.6) we see that, for any point $(x',y',0)$ that lies outside Γ,

$$f_0(x',y') = \frac{1}{4\pi}\int_\Gamma U^{(i)}(\xi,\eta,0)\frac{e^{iks_0}}{s_0}\frac{(\hat{r}\times\hat{s})\cdot\mathbf{l}}{1-\hat{r}\cdot\hat{s}}dl, \quad \text{(B20)}$$

where $\hat{r}$ and $\hat{s}$ are the unit vectors

$$\hat{r} = \frac{SQ'}{|SQ'|}, \quad \hat{s} = \frac{Q'P'}{|Q'P'|} \quad \text{(B21)}$$

and $\mathbf{l}$ is the unit tangential vector to Γ. Hence by a well-known inequality for integrals,

$$|f_0(x',y')| \leq \frac{1}{4\pi}\mathrm{Max}\left\{|U^{(i)}(\xi,\eta,0)|\left|\frac{(\hat{r}\times\hat{s})\cdot\mathbf{l}}{1-\hat{r}\cdot\hat{s}}\right|\right\}\int_\Gamma \frac{1}{s}dl, \quad \text{(B22)}$$

where the symbol Max refers to the maximum value of the expression in the curly bracket, with respect to the position of the point $Q'(\xi,\eta,0)$ on Γ. The term $|U^{(i)}(\xi,\eta,0)|$ is uniformly bounded on Γ. Moreover, the numerator of the other factor in the curly bracket is less than or equal to unity, whereas the denominator of this factor is never zero, since the unit vectors $\hat{r}$ and $\hat{s}$ cannot be parallel. In fact, the factor $1/(1-\hat{r}\cdot\hat{s}\cdot)$ is uniformly bounded for all points Q' on Γ. Hence (B22) implies that

$$|f_0(x',y')| \leq M\int_\Gamma \frac{1}{s}dl, \quad \text{(B23)}$$

where M is some constant.

Now the integral on the right of (B23) may be interpreted as the Newtonian potential at the point $P'(x',y')$ of a uniform mass distribution on a curve coincident with Γ. Hence by a theorem in potential theory[24] if the point P' is sufficiently far away from Γ in the plane of the aperture, a constant B exists such that

$$|Rf_0| < B, \quad \left|R^2\frac{\partial f_0}{\partial x}\right| < B, \quad \left|R^2\frac{\partial f_0}{\partial y}\right| < B, \quad \text{(B24)}$$

[24] O. D. Kellogg, *Foundations of Potential Theory* (Dover Publications, Inc., New York, 1953), p. 144.

where $R=OP'$ is the distance of P' from the origin. If we choose R so large that $1/R^2<1/R$, we have

$$|f_0|<\frac{B}{R},\quad \left|\frac{\partial f_0}{\partial x}\right|<\frac{B}{R},\quad \left|\frac{\partial f_0}{\partial y}\right|<\frac{B}{R},\tag{B25}$$

and these are the required conditions (c) (Eq. (A5) of Appendix A on the function f_0.

In conclusion, we stress that in the present Appendix we have demonstrated that the Kirchhoff solution, as formulated in the present paper, obeys the conditions (a) and (c) of Luneburg's uniqueness theorem (cf. Appendix A) generally (spherical or plane incident wave, aperture of arbitrary form). As for condition (b), we have demonstrated it only for the case of a plane wave, incident normally on a circular aperture.

Volume 39, number 4 OPTICS COMMUNICATIONS 15 October 1981

CONDITIONS FOR THE VALIDITY OF THE DEBYE INTEGRAL REPRESENTATION OF FOCUSED FIELDS ★

Emil WOLF ☆ and Yajun LI
Department of Physics and Astronomy, University of Rochester, Rochester, N.Y. 14627, USA

Received 23 June 1981

A very simple sufficiency condition is obtained, under which the Debye diffraction integral may be expected to give a good approximation to the solution of a boundary value problem that is generally taken to represent a field in the region of focus. When the angular aperture of the focusing system is sufficiently small, the condition reduces to the requirement that the Fresnel number of the diffracting aperture, when viewed from the geometrical focus, is large compared to unity.

1. Introduction

Several theoretical methods are available for analyzing the structure of the field near the focus of a converging spherical wave diffracted at an aperture, for example techniques based on the Huygens-Fresnel principle [1] or on the Debye integral representation [2]. It has been generally assumed that all these methods give essentially the same results. However, evidence has been accumulating in recent years which indicates that under the somewhat extreme conditions that have seldom been encountered in conventional optics in the past, but that are becoming increasingly common with the use of laser beams, the predictions of the different theories may differ from each other. For example, it appears from numerous past investigations, based on different methods [1,3,4], that the intensity distribution in the focal region of a converging uniform spherical wave emerging from a conventional optical system with a circular exit pupil is symmetric about the focal plane; and more general symmetry properties of the field in the focal region were established not long ago by Collett and Wolf [5], by the use of the Debye integral representation. On the other hand, Holmes, Korka and Avizonis [6] showed from calculations based on the Fresnel diffraction integral that the intensity distribution in certain very large focal length infrared systems does not possess such symmetries. Recently Stamnes [7] and Stamnes and Spjelkavik [8] found discrepancies in some cases between the predictions of the Kirchhoff theory and of the Debye integral representation.

In order to clarify this rather confused situation, it is obviously necessary to obtain more precise conditions than are currently available for the range of validity of the different theories. In the present note we derive a simple condition under which the Debye integral representation (somewhat generalized so as to apply also to the diffraction of non-uniform converging waves at a circular aperture) may be expected to give a good approximation to the structure of a focused field.

2. The Debye integral representation of a focused field

Consider a converging monochromatic scalar wave, emerging from a circular aperture of radius a, in an opaque screen. Let O be the geometrical focus of the wave (i.e. the point towards which the wave would converge in the absence of the diffracting aperture), assumed to be situated on the normal to the aperture, through its centre C, at a distance f from it. We take a cartesian coordinate system, with the origin at the geometrical focus O and with the positive z-direction pointing from C to O (see fig. 1).

By starting from the Helmholtz-Kirchhoff integral formula with Kirchhoff-type boundary conditions [ref. [4], sec. 8.3] and then formally proceeding to the limit

★ Research supported by the U.S. Army Research Office.
☆ Also at the Institute of Optics.

Volume 39, number 4 OPTICS COMMUNICATIONS 15 October 1981

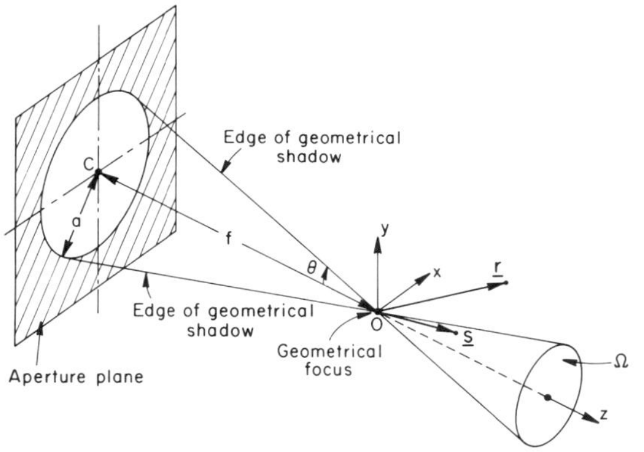

Fig. 1. Notation relating to the Debye integral (2.3).

as $f \to \infty$, Debye [2] obtained an expression for the field in the region of the geometrical focus that may be written in the form (a time-periodic factor $e^{-i\omega t}$ being suppressed ‡1)

$$U(x,y,z) = -\frac{ik}{2\pi} \iint_\Omega \frac{a(s_x, s_y)}{s_z} \times \exp[ik(s_x x + s_y y + s_z z)] \, ds_x \, ds_y . \tag{2.1}$$

Here

$$k = \omega/c \tag{2.2}$$

(c being the speed of light in vacuo), is the wave number and ($s_x, s_y, s_z > 0$) are the cartesian components of a unit vector $\boldsymbol{s}$. The integration extends over the solid angle Ω formed by all the unit $\boldsymbol{s}$-vectors which point along the geometrical rays that may be thought of as proceeding from each point in the aperture through the geometrical focus O and beyond. The amplitude factor $a(s_x, s_y)$, whose significance will become apparent shortly, is assumed to be real, except possibly for a constant phase factor. Actually Debye considered only the case when this amplitude factor is constant, appropriate to the case when the wave incident on the aperture is uniform. The generalization to an arbitrary but real function $a(s_x, s_y)$ makes it possible to study more general types of focused fields, e.g. those arising from focusing of gaussian laser beams. Other generalizations, that will not be considered in this note, were discussed by Picht [9], Luneburg [10] and Wolf [11].

We will re-write the Debye integral (2.1) in the more compact form

$$U(\boldsymbol{r}) = -\frac{ik}{2\pi} \iint_\Omega a(\boldsymbol{s}_\perp) \exp(ik\boldsymbol{s}\cdot\boldsymbol{r}) \, d\Omega, \tag{2.3}$$

where

$$d\Omega = ds_x \, ds_y / s_z , \tag{2.4}$$

is an element of the solid angle formed by the $\boldsymbol{s}$-directions and $\boldsymbol{s}_\perp \equiv (s_x, s_y, 0)$.

Since (2.3) expresses the field $U(\boldsymbol{r})$ as a superposition of plane monochromatic waves of the same wave number k associated with the fixed temporal frequency ω, it is clear that $U(\boldsymbol{r})$ satisfies, throughout the whole space, the Helmholtz equation

$$\nabla^2 U(\boldsymbol{r}) + k^2 U(\boldsymbol{r}) = 0 . \tag{2.5}$$

It is also known that it has the following asymptotic behavior [5] as $kr \equiv k(x^2 + y^2 + z^2)^{1/2} \to \infty$ in any fixed direction ‡2 specified by a unit vector $\hat{\boldsymbol{r}}(\hat{r}_x, \hat{r}_y, \hat{r}_z)$:

$$U(r\hat{\boldsymbol{r}}) \sim a(-\hat{\boldsymbol{r}}_\perp) \frac{e^{-ikr}}{r} \quad \text{when } \hat{r}_z < 0, \; -\hat{\boldsymbol{r}} \in \Omega, \tag{2.6a}$$

$$\sim 0 \quad \text{when } \hat{r}_z < 0, \; -\hat{\boldsymbol{r}} \notin \Omega, \tag{2.6b}$$

‡1 Debye employed a time-periodic factor $\exp(+i\omega t)$ so that our expression for the field is the complex conjugate of his. Moreover, Debye regarded the scalar [that we denote by $U(x, y, z)$] as a cartesian component of a Hertz vector, so that his original analysis could take into account the polarization properties of a focused electromagnetic field, that we ignore in the present note.

‡2 The asymptotic behavior indicated by eqs. (2.6), (2.7) and (3.14) arises from critical points of the first kind (interior stationary points) – cf. [4] Appendix III, sec. 3. Under certain circumstances, e.g., when the amplitude function $a(\boldsymbol{s}_\perp)$ exhibits special symmetry properties, contributions of other types of critical points may become dominant for some values of the argument $\boldsymbol{s}_\perp$. In particular, when $a(\boldsymbol{s}_\perp)$ is constant, i.e., when the incident spherical wave is uniform, critical points of the second kind (located at the edge of the aperture) will dominate the asymptotic behavior for $\boldsymbol{s}_\perp = 0$ and also for a sufficiently small neighborhood of $\boldsymbol{s}_\perp = 0$. This will result in a somewhat anomalous behavior of the Debye integral along certain directions in the half-space $z > -f$. Such anomalous behavior along the z-axis, and in a narrow cone of directions around this axis, in the case of an incident uniform spherical wave was noted already by Debye ([2], p. 771–772). It is analogous to the well-known Stokes phenomenon in the theory of asymptotic expansions of functions of a complex variable.

Volume 39, number 4 OPTICS COMMUNICATIONS 15 October 1981

and

$$U(r\hat{r}) \sim -a(\hat{r}_\perp)\frac{e^{ikr}}{r} \quad \text{when } \hat{r}_z > 0, \quad \hat{r} \in \Omega, \tag{2.7a}$$

$$\sim 0 \quad \text{when } \hat{r}_z > 0, \quad \hat{r} \notin \Omega, \tag{2.7b}$$

where $\hat{r}_\perp = (r_x, r_y, 0)$.

In practice the distance between the aperture and the geometrical focus is, of course, always finite and the following question thus arises: How large must it be in order that the Debye integral provides a reasonable approximation to the field in the focal region? We will now examine this question.

3. A sufficiency condition for the applicability of the Debye integral representation

The asymptotic behavior, expressed by eqs. (2.6) and (2.7) of the Debye integral (2.3) as $kr \to \infty$ in any fixed direction specified by a unit vector $\hat{r}$ may readily be obtained by the use of the principle of stationary phase [ref. [4], appendix III]. To bring into evidence a condition under which the principle applies to the present problem we first re-write (2.3) in the form

$$U(r\hat{r}) = -\frac{ik}{2\pi}\iint_\Omega a(\mathbf{s})\exp[ikr(\hat{r}\cdot\mathbf{s})]\,d\Omega. \tag{3.1}$$

Let us consider the case when $\hat{r}_z < 0$. i.e. when the point represented by the position vector $\mathbf{r} = r\hat{r}$ is situaated in the half-space $z < 0$ that contains the aperture. We then have

$$(-\mathbf{s})\cdot\hat{r} = \cos\psi, \tag{3.2}$$

where ψ $(0 \leqslant \psi \leqslant \pi/2)$ is the angle between the unit vectors $-\mathbf{s}$ and $\hat{r}$ (see figs. 2). Now the principle of stationary phase can be applied to obtain the asymptotic approximation [given by eqs. (2.6)] provided that the exponential term in the integrand of (3.1) oscillates rapidly as $\mathbf{s}$ explores the domain of integration (the solid angle Ω), i.e. provided that

$$kr\,\Delta(\cos\psi) \gg 2\pi, \tag{3.3}$$

where $\Delta(\cos\psi)$ is the maximum change (taken to be positive) that $\cos\psi$ undergoes in the process of integration.

Suppose now that $r\hat{r}$ is the position vector of a point in the aperture plane. Then $r \geqslant f$ and (3.3) will certainly be satisfied if

$$kf\,\Delta(\cos\psi) \gg 2\pi. \tag{3.4}$$

To determine $\Delta(\cos\psi)$ we must distinguish between two cases, namely when the point Q, represented by the position vector $\mathbf{r} = r\hat{r}$, is situated in the aperture and when it is situated on the screen. If α denotes the angle that the vector $\hat{r}$ makes with the negative z-axis, these two cases are distinguished by the inequalities $\alpha < \theta$ and $\alpha > \theta$, where θ is the angular semi-aperture of the focusing geometry.

3.1. $\alpha < \theta$ *(the point Q is within the aperture)*

Let Q′ be any point in the half-space $z > 0$ on the extension of the line QO (see fig. 2a). Consider now a circular cone K, with OQ′ as axis and with the generators formed by those $\mathbf{s}$-directions within the solid angle Ω, that make a fixed angle $\psi = \cos^{-1}[\mathbf{s}\cdot(-\hat{r})]$ with OQ′. Clearly only those $\mathbf{s}$-directions among them will contribute to the Debye integral (3.1) that are located on the portion of the body of this cone that is situated within the geometrical optics cone of rays, of solid angle Ω. If now we consider all the cones K, with the same axis OQ′ but with different vertex angles ψ, it is clear that the maximum relevant value of ψ is $\alpha+\theta$ and the minimum value is zero, i.e.

$$0 \leqslant \psi \leqslant \alpha + \theta. \tag{3.5}$$

Hence $1 \geqslant \cos\psi \geqslant \cos(\alpha+\theta)$ so that

$$\Delta(\cos\psi) = 1 - \cos(\alpha+\theta). \tag{3.6}$$

Now $\alpha+\theta \geqslant \theta$ and hence $\cos(\alpha+\theta) \leqslant \cos\theta$ and consequently

$$\Delta(\cos\psi) \geqslant 1 - \cos\theta \equiv 2\sin^2(\theta/2). \tag{3.7}$$

In view of (3.7) it follows that the condition (3.4) will necessarily be satisfied if

$$kf \gg \frac{\pi}{\sin^2(\theta/2)}. \tag{3.8}$$

3.2. $\alpha > \theta$ *(the point Q is on the screen)*

The analysis for this case is similar. The only difference arises from the fact that the line OQ′ is now outside the geometrical optics cone of rays (see fig. 2b). Consequently the minimum value of ψ is not zero now but has the value $\alpha - \theta$. We have, therefore, in place of (3.5), the inequalities

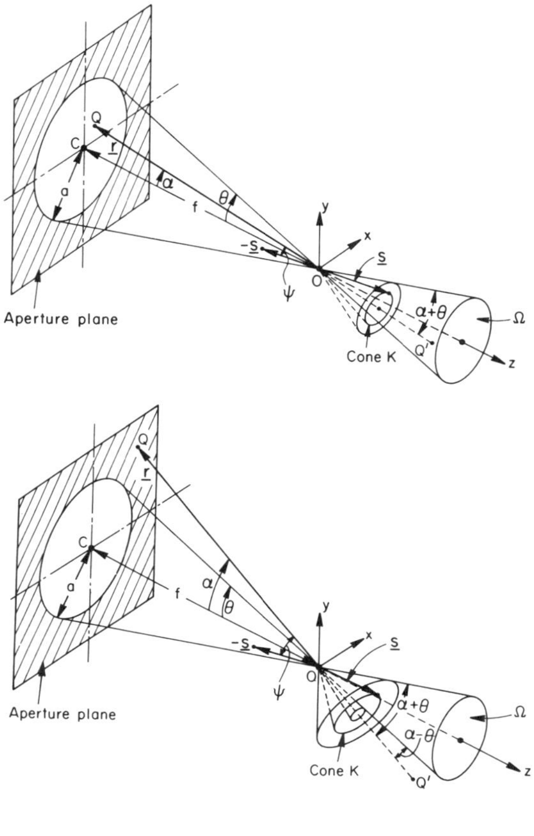

Fig. 2. Geometrical construction relating to determination of $\Delta(\cos\psi)$. (a) The case $\alpha < \theta$ (Q in the aperture). (b) The case $\alpha > \theta$ (Q on the screen).

$$\alpha - \theta \leqslant \psi \leqslant \alpha + \theta. \tag{3.9}$$

Hence $\cos(\alpha-\theta) > \cos\psi > \cos(\alpha+\theta)$ and we have, in place of (3.6),

$$\Delta(\cos\psi) = \cos(\alpha-\theta) - \cos(\alpha+\theta) \equiv 2\sin\alpha\sin\theta. \tag{3.10}$$

Since $\theta \leqslant \alpha < \pi/2$, $\sin\alpha \geqslant \sin\theta$ and hence (3.10) implies that

$$\Delta(\cos\psi) \geqslant 2\sin^2\theta. \tag{3.11}$$

It follows that the condition (3.4) will necessarily be satisfied if

$$kf \gg \pi/\sin^2\theta. \tag{3.12}$$

We note that (3.12) differs from (3.8) only in that $\sin^2\theta$ appears in place of $\sin^2(\theta/2)$. Since $\theta > \theta/2$ it is clear that when the requirement (3.8) is satisfied, so is (3.12). Moreover, it follows from eqs. (2.7) that in the half-space $z > 0$, sufficiently far from the geometrical focus, the Debye integral behaves as an outgoing spherical wave. Hence, in view of a well-known uniqueness theorem for the solution of the Helmholtz equation in a half-space [ref. [10], sec. 45], we have established the following

Theorem: When

$$kf \gg \frac{\pi}{\sin^2(\theta/2)} \tag{3.13}$$

the Debye integral (2.1) represents, to a high degree of accuracy, the solution of the Helmholtz equation (2.5), valid throughout the half-space $z > -f$, that satisfies, in general [‡2], *on the plane of the aperture the boundary conditions*

$$U(\boldsymbol{r}) = a(-\hat{\boldsymbol{r}}_\perp)\,\frac{e^{-ikr}}{r} \quad \text{when Q is in the aperture} \tag{3.14a}$$

$$= 0 \quad \text{when Q is on the screen} \tag{3.14b}$$

and that behaves as an outgoing spherical wave at infinity in the half-space $z > 0$. In (3.14) $\boldsymbol{r} = r\hat{\boldsymbol{r}}$ ($|\hat{\boldsymbol{r}}| = 1$) is the position vector of the point Q, refered to the geometrical focus O, and $\boldsymbol{r}_\perp = (\hat{r}_x, \hat{r}_y, 0)$, where $\hat{r}_x, \hat{r}_y$ are the first two cartesian components of the unit vector $\hat{\boldsymbol{r}}$.

4. The sufficiency condition for low-aperture systems

If the angular semi-aperture $\theta = \tan^{-1}(a/f)$ is sufficiently small, the condition (3.13) acquires a simple physical significance. To see this we use the trigonometric identity $\sin^2(\theta/2) \equiv \frac{1}{2}(1-\cos\theta)$ and the relation $\cos\theta = f/(a^2+f^2)^{1/2}$ (see fig. 1) and obtain at once the formula

$$\sin^2(\theta/2) = \tfrac{1}{2}[1 - f/(a^2+f^2)^{1/2}]. \tag{4.1}$$

If we expand the second term on the right in a binomial series we readily find that

$$\sin^2(\theta/2) = \tfrac{1}{4}(a/f)^2\,[1 - \tfrac{3}{4}(a/f)^2 + \ldots], \tag{4.2}$$

where the terms in the brackets on the right-hand side that are not shown explicitly are of fourth and higher order in (a/f). Suppose now that

Volume 39, number 4 OPTICS COMMUNICATIONS 15 October 1981

$$(a/f)^2 \ll 1, \tag{4.3a}$$

i.e. that the angular semi-aperture

$$\theta \equiv \tan^{-1}(a/f) \ll 45°. \tag{4.3b}$$

According to (4.2) we may then make the approximation

$$\sin^2(\theta/2) \approx \tfrac{1}{4}(a/f)^2 \tag{4.4}$$

and, if we also use the fact that $k = 2\pi/\lambda$, where λ is the wavelength, the condition (3.13) reduces to (since $2 > 1$)

$$a^2/\lambda f \gg 1. \tag{4.5}$$

The expression on the left of (4.5) is the *Fresnel number [12]* of the focusing geometry. Thus *when the angular semi-aperture θ is much less than 45° the condition (3.13) may be replaced by the requirement that the number of Fresnel zones [12] in the aperture, when viewed from the geometrical focus, be large compared with unity.*

5. Discussion

We have shown that when the condition (3.13) is satisfied, the Debye integral (3.1) represents, to a good approximation, a solution to a well-posed boundary value problem. It should, however, not be assumed that the Debye integral represents rigorously the field in the focal region, even in extreme cases of very large values of the asymptotic parameter kf; for the boundary conditions (3.14) are independent of the physical properties of the screen and so evidently can only be approximations to the true, but generally unknown, values that the diffracted field takes in the plane of the screen. Moreover polarization properties of the field have not been taken into account ‡3. However, when the condition (3.13) is satisfied and when, in addition, the radius a of the aperture is large compared with the wavelength, and the angular semi-aperture is not too large‡4 ($\theta \lesssim 30°$), the Debye integral may be expected to describe the focal region with about the same high degree of accuracy as the Huygens-Fresnel principle‡5. It is under these circumstances that the recently noted symmetry properties of the focal region [5] are likely to be valid. Conversely, one might expect that when not all the above conditions are satisfied the symmetry properties will no longer hold. In fact, the asymmetry of the intensity distribution along the axis near focus in some very large focal length infrared systems predicted by Holmes, Korka and Avizonis [6], and some of the discrepancies between calculations based on the Debye integral and on the Kirchhoff diffraction theory noted by Stamnes [7] and by Stamnes and Spjelkavik [8] may readily be shown to pertain to focusing under circumstances where the condition (3.13) is not satisfied.

‡3 See, however, remarks in footnote ‡1.

‡4 It is known that when $\theta < 30°$ polarization effects are not very significant in the vicinity of the focus, cf. ref. [13].

Acknowledgement

We are indebted to Dr. George Sherman for some helpful discussions relating to the matter discussed in footnote ‡2.

‡5 We thus evidently do not subscribe to the view expressed by Sommerfeld on p. 319 of his book cited in ref. [2], that the Debye integral "... exactly describes not only the phase jump but also the diffraction pattern in the vicinity of the focal point (or line)". We also consider his subsequent remark, namely "Debye's method is not limited to Kirchhoff's approximation but is based on the fundamental of wave optics" as rather misleading. We are of the opinion that Kirchhoff's theory, when applied to diffraction on apertures in opaque screens, is at least as accurate as Debye's theory and, in addition, has appreciably wider range of validity (cf. ref. [14]).

References

[1] See, for example, E.H. Linfoot and E. Wolf, Mon. Not. Roy. Astr. Soc. 112 (1952) 452.
[2] P. Debye, Ann. d. Phys. 30 (1909) 755. An account of the Debye representation in the English language is given in A. Sommerfeld, Optics (Academic Press, New York, N.Y., 1954) sec. 45.
[3] F. Zernike and B.R.A. Nijboer, contribution in: Théorie des images optique, eds. P. Fleury, A. Maréchal and C. Anglade (Revue d'Optique, Paris, 1949) p. 227.
[4] M. Born and E. Wolf, Principles of optics (Pergamon Press, Oxford and New York, 6th ed., 1980) sec. 8.8.
[5] E. Collett and E. Wolf, Optics Lett. 5 (1980) 264.
[6] D.A. Holmes, J.E. Korka and P.V. Avizonis, Appl. Optics 11 (1972) 565.

[7] J.J. Stamnes, J. Opt. Soc. Amer. 71 (1981) 15.
[8] J.J. Stamnes and B. Spjelkavik, submitted to Optics Comm.
[9] J. Picht, Ann. d. Phys. 77 (1925) 685. See also his Optische Abbildung (Vieweg, Braunschweig, 1931) ch. 8.
[10] R.K. Luneburg, Mathematical theory of optics (University of California Press, Berkeley and Los Angeles, 1964) sec. 46–48.
[11] E. Wolf, Proc. Roy. Soc. (London) A253 (1959) 349.
[12] A.E. Siegman, An introduction to lasers and masers (McGraw-Hill, New York) p. 337–338.
[13] B. Richards and E. Wolf, Proc. Roy. Soc. (London) A253 (1959) 358.
[14] E.W. Marchand and E. Wolf, J. Opt. Soc. Am. 56 (1966) 1712.

Y. Li and E. Wolf Vol. 1, No. 8/August 1984/J. Opt. Soc. Am. A 801

Three-dimensional intensity distribution near the focus in systems of different Fresnel numbers

Yajun Li

Institute of Electro-Optical Engineering, National Chiao Tung University, 1001 Ta Hsueh Road, Hsinchu, Taiwan 300, China

Emil Wolf*

Department of Physics and Astronomy, University of Rochester, Rochester, New York 14627

It was recently shown that, when a converging spherical wave is focused in a diffraction-limited system of sufficiently low Fresnel numbers, the point of maximum intensity does not coincide with the geometrical focus but is located closer to the exit pupil. In the present paper both qualitative and quantitative arguments are presented that elucidate the modifications that the whole three-dimensional structure of the diffracted field undergoes as the Fresnel number is gradually decreased. Contours of equal intensity in the focal region are presented for systems of selected Fresnel numbers, which focus uniform waves.

1. INTRODUCTION

In a number of publications that appeared in recent years[1–5] it was demonstrated that the classic theory regarding the structure of the focal region does not predict correct results under all circumstances. More specifically, it was found that, when a uniform, converging, monochromatic spherical wave is diffracted at a circular aperture, the classic theory is adequate for calculating the intensity distribution along the axis near the geometrical focus only if the Fresnel number of the focusing geometry [defined by Eq. (2.2) below] is large compared to unity.[6] Since before the invention of the laser the Fresnel number of a focusing system was generally much larger than unity, no discrepancies with predictions of the classic theory were then found. However, in recent years systems have been developed, especially for use with laser beams, whose Fresnel numbers are of the order of unity or even smaller. Hence it is desirable to extend the classic theory of focusing to systems of this kind.

In the present paper we present isophotes (contours of equal intensity) in the focal region in systems of different Fresnel numbers. We begin with some general qualitative observations, which indicate why one must expect changes in the structure of the focal region as the Fresnel number N of the system changes gradually from large values ($N \gg 1$) to very small values ($N \ll 1$). The qualitative arguments also indicate that the classic theory of focusing reaches its limit of validity for systems whose Fresnel numbers are of the order of unity and that it breaks down completely when the Fresnel number is much smaller than unity. The qualitative considerations, which also indicate the general trend of the changes in the structure of the focal region as the Fresnel number decreases, are confirmed by results of detailed computations, presented in Section 4.

Only focusing of uniform waves is considered in this paper. However, the results may be expected to be similar for focused laser beams in systems in which the size of the diffracting aperture is comparable with or smaller than the effective cross section of the laser beam that is incident upon it.

2. SOME QUALITATIVE CONSIDERATIONS

Let us consider a uniform, monochromatic, spherical wave converging to a focus F and diffracted at a circular aperture, of radius a, in an opaque screen. We denote by O the center of the aperture and by f the distance OF (Figs. 1 and 2). We assume that

$$a \gg \lambda, \qquad \left(\frac{a}{f}\right)^2 \ll 1, \tag{2.1}$$

where λ is the wavelength. Let

$$N = a^2/\lambda f \tag{2.2}$$

be the Fresnel number of the focusing geometry.

It follows from elementary calculations that the maximum, attained at the edge of the aperture, of the distance Δ between the spherical wave front through O and the plane of the aperture is, to a good approximation, given by

$$\Delta_{\max} = a^2/2f. \tag{2.3}$$

It is seen from Eqs. (2.2) and (2.3) that

$$N = \Delta_{\max}/(\lambda/2), \tag{2.4}$$

i.e., the Fresnel number is equal to the value of $\Delta_{\max}$ expressed in units of half wavelength. Expressed differently, N represents the number of Fresnel zones (Ref. 7, Sec. 8.2) that fill the aperture when the aperture is viewed from the geometrical focus F. To obtain some indication about the order of magnitude of the Fresnel number in some typical cases, let us consider two simple examples. For a conventional focusing lens we might have $a = 1$ cm, $f = 5$ cm, $\lambda = 0.5$ μm, and, according to Eq. (2.2), we then have $N = 4000$. On the other hand, for a long-focal-length infrared laser system we may

0740-3232/84/080801-08$02.00

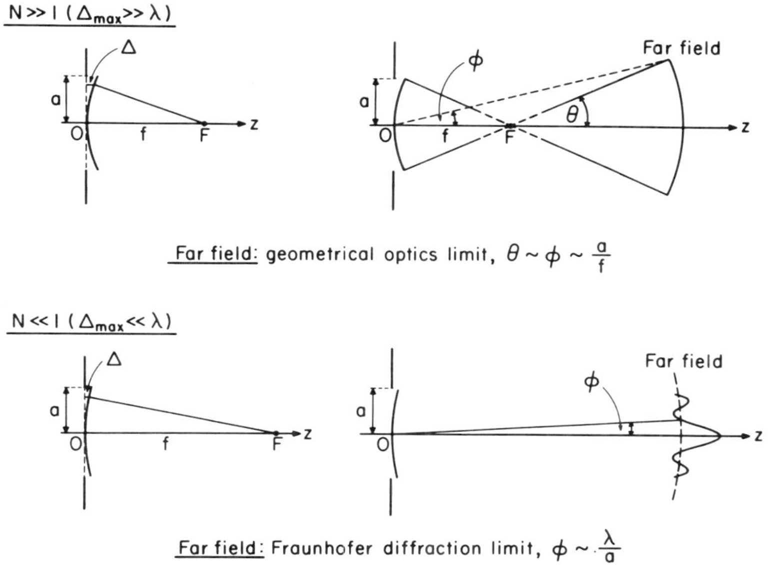

Fig. 1. Illustrating some qualitative differences in the structure of the far field generated by the diffraction of a uniform, converging monochromatic spherical wave at an aperture in an opaque screen in systems of large Fresnel numbers (upper figures) and low Fresnel numbers (lower figures).

have $a = 1$ cm, $f = 3$ m, $\lambda = 10\ \mu$m, and, according to Eq. (2.2), $N = 3.3$ in this case.

It is well known that for conventional focusing systems (systems with $N \gg 1$) the maximum intensity in the diffracted field is at the geometrical focus F. Moreover, except close to the axis and in the vicinity of the edge of the geometrical shadow, the far field has the appearance of a cutoff portion of a uniform spherical wave, diverging from the geometrical focus[8,9] F. This "spherical cap" subtends, in each cross section through the axis OF, the angle $2\theta = 2a/f$ at F; and, because it is located in the far zone, it subtends (within the accuracy of the asymptotic approximation of large distances from F) the angle $2\phi \sim 2\theta$ at the center O of the aperture:

$$\phi \sim \theta = a/f \tag{2.5}$$

(see the top right-hand diagram in Fig. 1). This structure of the far field is, of course, in agreement with predictions of geometrical optics.

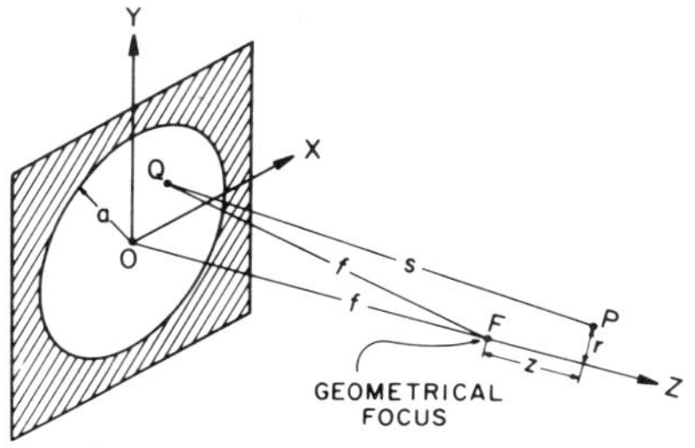

Fig. 2. Illustrating the notation used in determining the structure of the focal region. The point Q is located on a spherical wave front W of radius f, centered on F and passing through the center O of the aperture.

Let us now consider the other extreme case, $N \ll 1$. For simplicity we will assume that the radius a of the aperture and the wavelength λ have the same values as before and that the small value of N is achieved by making the distance f between the plane of the aperture and the geometrical focus be sufficiently large. According to Eq. (2.4) we now have $\Delta_{\max} \ll \lambda/2$, and hence we are essentially dealing with the diffraction of a *plane* wave at the circular aperture. In such a case the far field has quite a different structure from that in systems that we just discussed (for which $N \gg 1$). It can no longer be described by geometrical optics but rather must be described by the theory of Fraunhofer diffraction. According to a well-known formula pertaining to Fraunhofer diffraction at a circular aperture, namely, the Airy diffraction formula (Ref. 7, Sec. 8.5.2), the far field will now have a sharp maximum in the forward (axial) direction and will rapidly decrease to zero in an oscillatory manner on each side of it. The effective angular spread of the radiation is given by the order-of-magnitude relation

$$\phi \approx \lambda/a. \tag{2.6}$$

Since, when $N \ll 1$, we are effectively dealing with the diffraction of a plane wave at a circular aperture, it seems intuitively clear that the intensity maximum will now be located in the vicinity of the center of the aperture[10] and that the diffracted field, at distances sufficiently far away from the aperture, will behave as if it had been generated by a radiating source located in the region of the aperture.

Comparison of the two extreme cases that we have just briefly discussed indicates that, as we proceed from focusing systems of large Fresnel numbers N to focusing systems of small Fresnel numbers, the intensity maximum moves from its location at the geometrical focus F to some point in the vicinity of the center O of the aperture. It is clear that this transition is accomplished in a continuous manner. If we assume, as seems reasonable, that the shift of the intensity maximum from its location at the geometrical focus is a monotonic function[11] of N (with a and λ being kept fixed), we may conclude that with decreasing values of the Fresnel number the point of maximum intensity will move toward the center of the aperture.[12] This is indeed what was found from explicit calculations[3] and was readily confirmed by experiment.[13]

In Fig. 1 the two extreme situations of systems with large and small Fresnel numbers N are illustrated by the upper and lower diagrams, respectively. They may be regarded as representing, in a sense, two modes of behavior of the diffracted field. One of the modes dominates when $N \gg 1$, the other when $N \ll 1$. We may expect, roughly speaking, that in general there will be a mixture of these two modes of behavior and that both will significantly influence the structure of the field when the angles ϕ, given by Eqs. (2.5) and (2.6), are of the same order of magnitude, i.e., when

$$a/f \approx \lambda/a. \tag{2.7}$$

If we recall the definition (2.2) of the Fresnel number N we see that Eq. (2.7) is equivalent to the condition

$$N \approx 1. \tag{2.8}$$

Hence our rough qualitative arguments indicate that, *as the Fresnel number N of a focusing system is gradually decreased from large values ($N \gg 1$) to small values ($N \ll 1$), we may expect to find discrepancies with predictions of the classic theory of focusing when N attains values of the order of unity and that the classic theory will not be applicable when N has values much smaller than unity.*

We will see shortly that these rough qualitative predictions are in agreement with results of detailed calculations.

3. EXPRESSIONS FOR THE FIELD IN THE FOCAL REGION

We again consider diffraction of a uniform, monochromatic, converging spherical wave at a circular aperture of radius a in an opaque screen (Fig. 2). Let

$$V^{(i)}(Q, t) = A\frac{e^{-ikf}}{f}e^{-i\omega t} \tag{3.1}$$

be the field distribution of the incident wave at a typical point Q on the wave front through the center O of the aperture. In Eq. (3.1) A is a constant and t denotes the time. With the assumptions expressed by Eqs. (2.1), the diffracted field at a point P is, according to the Huygens–Fresnel principle, given by [with the time-periodic factor $\exp(-i\omega t)$ omitted]

$$U_N(P) = -\frac{i}{\lambda}\frac{Ae^{-ikf}}{f}\iint_W \frac{e^{iks}}{s}\mathrm{d}S, \tag{3.2}$$

where s denotes the distance QP and the integration extends over the wave front W filling the aperture. We attached the suffix N to the symbol $U(P)$ for the diffracted field to stress its dependence on the Fresnel number N, defined by Eq. (2.2).

Let (z, r, ψ) be the cylindrical coordinates of the field point P, referred to axes with origin at the geometrical focus F and with the z direction along OF (see Fig. 2). It is convenient to introduce the dimensionless variables[14] u_N and v_N, which, together with ψ, specify the position of the field point P:

$$u_N = 2\pi N\frac{z/f}{1 + z/f}, \tag{3.3a}$$

$$v_N = 2\pi N\frac{r/a}{1 + z/f}. \tag{3.3b}$$

Because the field is rotationally symmetric with respect to the z axis, the diffracted field will not depend on the azimuthal angle ψ.

It was recently shown[1,2,15] that, throughout a neighborhood of the geometrical focus, the diffracted field $U_N(P)$ may be expressed in the form

$$U_N(P) = B_N(u_N)\exp[i\Phi_N(u_N, v_N)] \times \int_0^1 J_0(v_N\rho)\exp(-iU_N\rho^2/2)\rho\mathrm{d}\rho, \tag{3.4}$$

where J_0 is the Bessel function of the first kind and zero order,

$$B_N(u_N) = -\frac{2\pi i}{\lambda}\left(\frac{a}{f}\right)^2\left(1 - \frac{u_N}{2\pi N}\right)A, \tag{3.5a}$$

and

$$\Phi_N(u_N, v_N) = \frac{1}{1 - u_N/2\pi N}\left[\left(\frac{f}{a}\right)^2 u_N + \frac{1}{4\pi N}{v_N}^2\right]. \tag{3.5b}$$

The range of validity of expression (3.4) is discussed in Ref. 15.

Before proceeding further we note that the expression on the right-hand side of Eq. (3.4) is of the same mathematical form as the expression obtained by Lommel in his classic paper on the structure of the focal region,[16] viz.,

$$U(P) = B\exp[i\Phi(u)]\int_0^1 J_0(v\rho)\exp(-iu\rho^2/2)\rho\mathrm{d}\rho, \tag{3.6}$$

where

$$u = \frac{2\pi}{\lambda}\left(\frac{a}{f}\right)^2 z, \tag{3.7a}$$

$$v = \frac{2\pi}{\lambda}\left(\frac{a}{f}\right) r, \tag{3.7b}$$

$$B = -\frac{2\pi i}{\lambda}\left(\frac{a}{f}\right)^2 A, \tag{3.8a}$$

$$\Phi(u) = \left(\frac{f}{a}\right)^2 u. \tag{3.8b}$$

Now the following relations between the quantities u_N, v_N, B_N, and Φ_N of the present theory and the quantities u, v, B, and Φ of Lommel's theory can readily be deduced:

$$u_N = \frac{u}{1 + u/2\pi N}, \tag{3.9a}$$

$$v_N = \frac{v}{1 + u/2\pi N}, \tag{3.9b}$$

$$B_N(u_N) = \left(1 - \frac{u_N}{2\pi N}\right)B, \tag{3.10a}$$

$$\Phi_N(u_N, v_N) = \frac{1}{1 - u_N/2\pi N}\left[\Phi(u_N) + \frac{1}{4\pi N}{v_N}^2\right]. \tag{3.10b}$$

It follows at once from Eqs. (3.9) and (3.10) that, as $N \to \infty$, with u and v being kept fixed,[17]

$$u_N \sim u, \qquad v_N \sim v, \tag{3.11}$$

$$B_N(u_N) \sim B, \qquad \Phi_N(u_N, v_N) \sim \Phi(u), \tag{3.12}$$

and the expression on the right-hand side of Eq. (3.4) then reduces to Lommel's expression (3.6), i.e., as $N \to \infty$,

$$U_N(P) \sim U(P), \tag{3.13}$$

where the symbol $\sim$ indicates limit in the asymptotic sense. Thus we conclude that Lommel's solution represents the asymptotic limit of our solution for focusing systems of large Fresnel numbers.

Since the mathematical structure of the integral in expression (3.4) and in Lommel's expression (3.6) is the same, we may develop the integral on the right-hand side of Eq. (3.4) in the same type of series as that first employed by Lommel in analyzing consequences of his solution.[16] We then obtain for the field $U_N(P)$ in the focal region the following expression:

$$U_N(P) = (1/2)B_N(u_N)\{\exp[i\Phi_N(u_N, v_N)]\} \times [C(u_N, v_N) - iS(u_N, v_N)], \tag{3.14}$$

where

$$C_N(u_N, v_N) = \frac{\cos(u_N/2)}{(u_N/2)} U_1(u_N, v_N) + \frac{\sin(u_N/2)}{(u_N/2)} U_2(u_N, v_N), \tag{3.15a}$$

$$S_N(u_N, v_N) = \frac{\sin(u_N/2)}{(u_N/2)} U_1(u_N, v_N) - \frac{\cos(u_N/2)}{(u_N/2)} U_2(u_N, v_N), \tag{3.15b}$$

and U_1 and U_2 are two of the Lommel functions:

$$U_n(u_N, v_N) = \sum_{s=0}^{\infty} (-1)^s \left(\frac{u_N}{v_N}\right)^{n+2s} J_{n+2s}(v_N), \tag{3.16}$$

$J_m(v_N)$ being a Bessel function of the first kind and of order m.

Although the formulas are valid in the neighborhood of the geometrical focus F, they are convenient for computations only when $|u_N/v_N| < 1$, i.e., when $|z|/r < f/a$; this is the region of the geometrical shadow. When $|u_N/v_N| > 1$, i.e., when $|z|/r > f/a$, the field point P is in the geometrically illuminated region, and it is then more convenient to use the following alternative expressions for the functions $C(u_N,v_N)$ and $S(u_N,v_N)$ that appear in Eq. (3.14):

$$C(u_N, v_N) = \frac{2}{u_N} \sin\frac{v_N^2}{2u_N} + \frac{\sin(u_N/2)}{(u_N/2)} V_0(u_N, v_N) - \frac{\cos(u_N/2)}{(u_N/2)} V_1(u_N, v_N), \tag{3.17a}$$

$$S(u_N, v_N) = \frac{2}{u_N} \cos\frac{v_N^2}{2u_N} - \frac{\cos(u_N/2)}{(u_N/2)} V_0(u_N, v_N) - \frac{\sin(u_N/2)}{(u_N/2)} V_1(u_N, v_N), \tag{3.17b}$$

where V_0 and V_1 are two of the Lommel functions[18]:

$$V_n(u_N, v_N) = \sum_{s=0}^{\infty} (-1)^s \left(\frac{v_N}{u_N}\right)^{n+2s} J_{n+2s}(v_N). \tag{3.18}$$

4. THE STRUCTURE OF THE FOCAL REGION

The close formal similarity of expression (3.4) of the present theory and expression (3.6) of Lommel's theory makes it possible to draw at once some interesting conclusions.

We note that the variables u and v of Lommel's theory depend linearly on the two cylindrical coordinates (z, r) of the field point [Eqs. (3.7)]. On the other hand, the variables u_N and v_N of the present theory are nonlinear functions of z and r [Eqs. (3.3)]. In the linear regime, the effect of changing any of the parameters λ, a, or f is equivalent to *scaling*; i.e., if one of the parameters is changed, the field distribution $U(P)$ is modified only in the simple manner of becoming spatially magnified (with magnification $\gtrless 1$) in the longitudinal (z) direction or the transverse (r) direction or both. This is, however, not so with the field distribution $U_N(P)$. The nonlinear relationship (3.3) between the cylindrical coordinates (z, r) and the dimensionless variables (u_N, v_N) implies a more drastic change. This is illustrated in Fig. 3. In the top left-hand diagram of Fig. 3 a rectangular mesh formed by the z- and r-coordinate lines has been superimposed upon the well-known isophote diagram (diagram formed by contours of equal intensity $I = |U|^2$) in the region of focus, calculated from Lommel's theory. Because of the linearity, the mesh also represents lines of constant values of u (vertical lines) and of constant values of v (horizontal lines). The other three diagrams in Fig. 3 show the corresponding lines u_N = constant, v_N = constant in focusing systems of Fresnel numbers N = 100, N = 10, and N = 1, respectively, plotted in the z, r plane.

In Fig. 4 the isophotes $I_N \equiv |U_N|^2$ = constant, normalized to unity at the geometrical focus, are shown for focusing systems of selected Fresnel numbers. We see that with decreasing N, the point of maximum intensity moves from its coincidence with the geometrical focus toward the center of the aperture and that the isophote diagrams approach more and more those of a complicated radiation field generated by a source in the vicinity of the center of the aperture. These quantitative results are in agreement with the conclusions that we reached by rough qualitative arguments in Section 2.

Finally we note that, when the point $P(z, r, \psi)$ is located in the geometrical focal plane $z = 0$, one has, according to Eqs. (3.3), (3.7), and (2.2),

$$u_N = u = 0, \qquad v_N = v, \tag{4.1}$$

$$B_N(0) = B, \qquad \Phi_N(0, v_N) = \Phi(0) + \frac{1}{4\pi N} v^2. \tag{4.2}$$

On substituting from Eqs. (4.1) and (4.2) into Eq. (3.4) and on comparing the resulting expression with Eq. (3.6) we see that

$$U_N(P)|_{z=0} = U(P)|_{z=0} \exp\left(\frac{iv^2}{4\pi N}\right). \tag{4.3}$$

This formula implies that, irrespective of the value of the Fresnel number N, the amplitude of the field at any point in the geometrical focal plane has the same value as given by Lommel's theory and that the corresponding phase factor differs from the phase factor of Lommel's theory by the amount $v^2/4\pi N$. It is an immediate consequence of the first of these two results that, *irrespective of the Fresnel number N of the system, the intensity distribution in the focal plane*

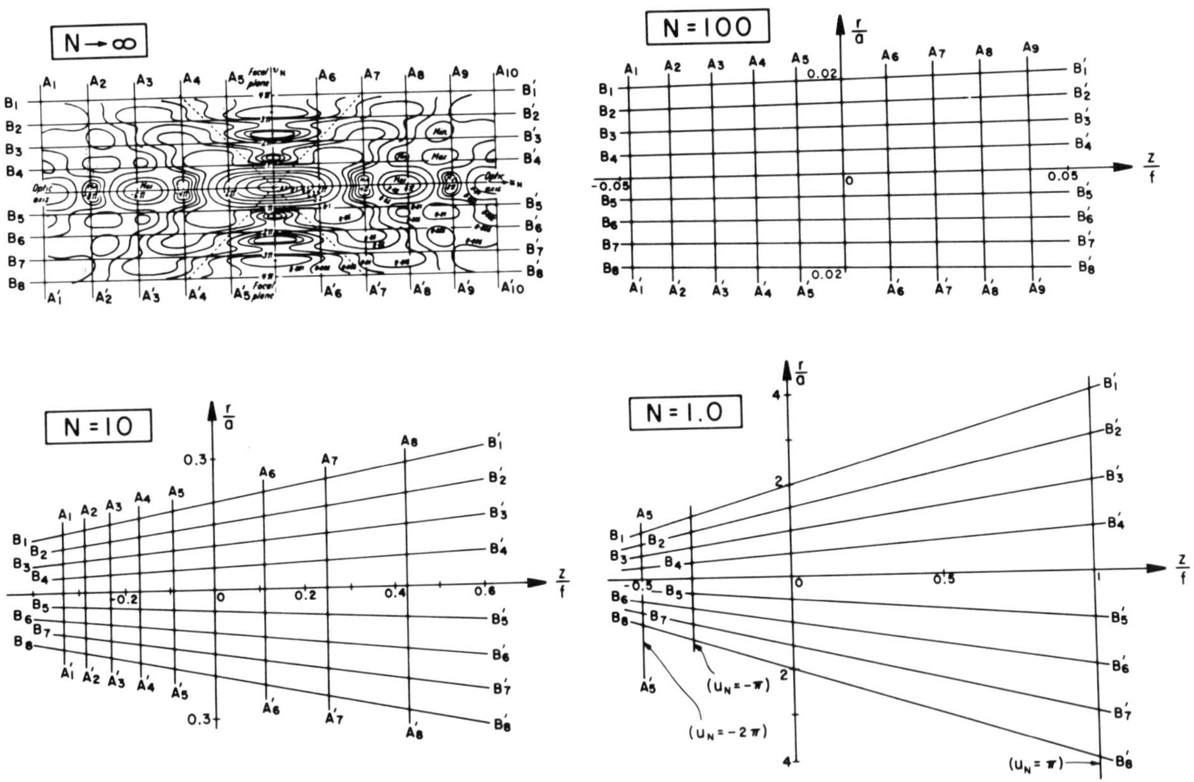

Fig. 3. Illustrating consequences of the nonlinear transformation [Eqs. (3.3)]. Some of the coordinate lines u_N = constant (vertical) and v_N = constant (horizontal or inclined to the z axis), plotted in the z, r plane, for systems of different Fresnel numbers. Effects of the transformation on the isophote diagrams that are included in the top left-hand figure ($N \rightarrow \infty$) are shown in Fig. 4.

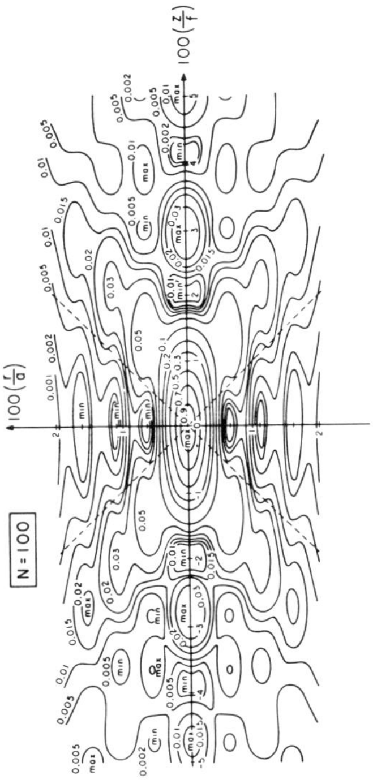
N = 100

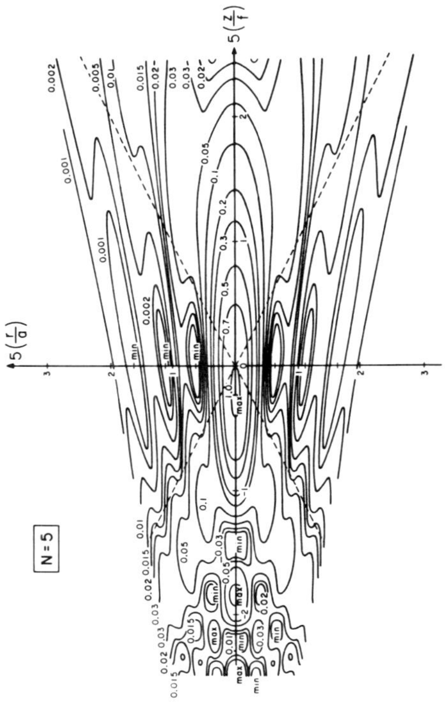
N = 5

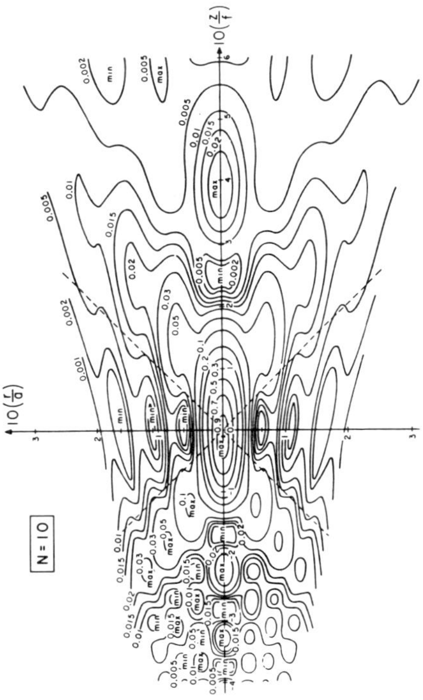
N = 10

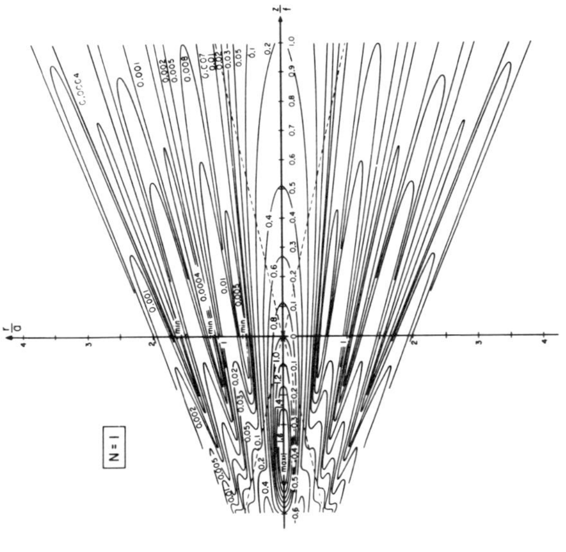

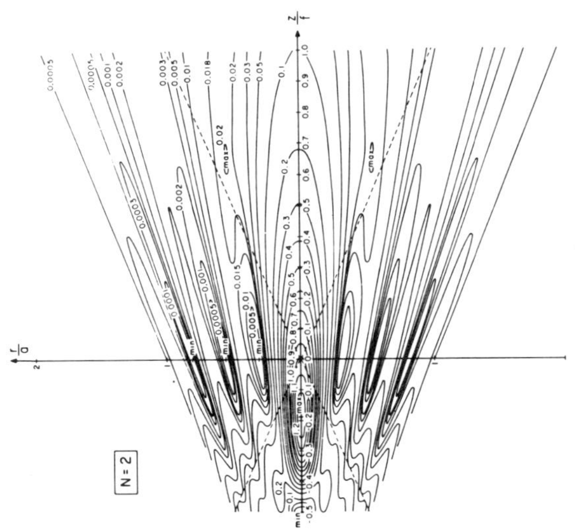

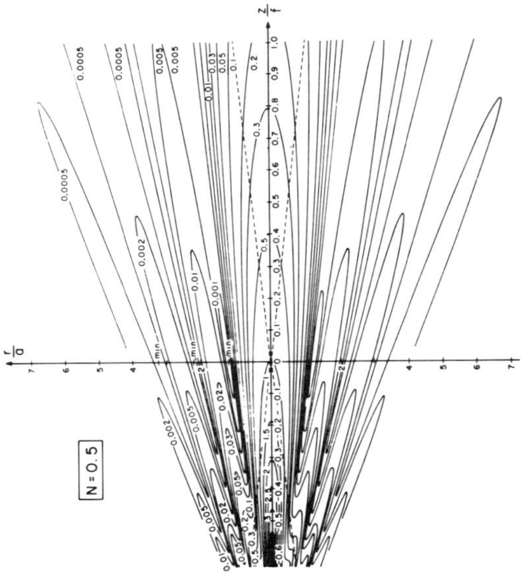

Fig. 4. Isophotes (contours of the intensity) in a meridional plane in the neighborhood of the geometrical focus of a uniform, converging, monochromatic spherical wave diffracted at a circular aperture in an opaque screen, in systems of different Fresnel number N. The intensity is normalized to unity at the geometrical focus. The dotted lines represent the boundary of the geometrical shadow. The first isophote diagram (for $N = 100$) is essentially the same as that obtained from Lommel's classic theory (Fig. 2 of Ref. 19 and Fig. 8.41 of Ref. 7).

is given by the Airy-pattern formula [Ref. 7, p. 441, Eqs. (25) and (22)]

$$I_N(P)|_{z=0} \equiv |U_N(P)|^2_{z=0} = \left[\frac{2J_1(v)}{v}\right]^2 I_0, \tag{4.4}$$

where

$$I_0 = \left(\frac{\pi a^2 |A|}{\lambda f^2}\right)^2 \tag{4.5}$$

is the intensity at the geometrical focus.

ACKNOWLEDGMENTS

The main part of this research was carried out while E. Wolf was on leave of absence at Schlumberger-Doll Research in Ridgefield, Connecticut. Preliminary results were presented on October 22, 1982, at the Annual Meeting of the Optical Society of America in Tucson, Arizona [J. Opt. Soc. Am. **72,** 1818 (A) (1982)].

* E. Wolf is also with the Institute of Optics, University of Rochester.

REFERENCES

1. J. H. Erkkila and M. E. Rogers, "Diffracted fields in the focal volume of a converging wave," J. Opt. Soc. Am. **71,** 904–905 (1981).
2. J. J. Stamnes and B. Spjelkavik, "Focusing at small angular apertures in the Debye and Kirchhoff approximations," Opt. Commun. **40,** 81–85 (1981).
3. Y. Li and E. Wolf, "Focal shifts in diffracted converging spherical waves," Opt. Commun. **39,** 211–215 (1981).
4. Y. Li, "Dependence of the focal shift on Fresnel number and f number," J. Opt. Soc. Am. **72,** 770–774 (1982).
5. M. P. Givens, "Focal shifts in diffracted converging spherical waves," Opt. Commun. **41,** 145–148 (1982).
6. Similar results were also found in connection with focused Gaussian beams. For a discussion of this subject and for pertinent references, see Y. Li and E. Wolf, "Focal shift in focused truncated Gaussian beams," Opt. Commun. **42,** 151–156 (1982).
7. M. Born and E. Wolf, *Principles of Optics*, 6th ed. (Pergamon, Oxford, 1980).
8. E. Wolf and Y. Li, "Conditions for the validity of the Debye integral representation of focused fields," Opt. Commun. **39,** 205–210 (1981).
9. G. C. Sherman and W. C. Chew, "Aperture and far-field distributions expressed by the Debye integral representation of focused fields," J. Opt. Soc. Am. **72,** 1076–1083 (1982).
10. In this connection it seems worthwhile to mention that none of the currently available theories of diffraction at an aperture in an *opaque* screen appears to be adequate to predict the structure of the diffracted field in the near zone of the aperture; nor are any experimental results available to elucidate this question.
11. For the same reason as indicated in Ref. 10, this assumption may not be valid when the maximum reaches the immediate vicinity of the aperture.
12. A somewhat paradoxical aspect of this situation should be noted. As the Fresnel number N becomes smaller, with a and λ being kept fixed, the distance f increases, i.e., the geometrical focus F moves farther away from the aperture; however, the point of maximum intensity moves in the opposite direction, i.e., closer to the aperture.
13. Y. Li and H. Platzer, "An experimental investigation of diffraction patterns in low-Fresnel-number focusing systems," Opt. Acta **30,** 1621–1643 (1983).
14. The choice of these variables is suggested by detailed calculations given, for example, in Ref. 15. As will be seen shortly, they reduce to the usual dimensionless variables of the classic theory of Lommel in the limit of large Fresnel numbers.
15. Y. Li, "Encircled energy for systems of different Fresnel numbers," Optik **64,** 207–218 (1983).
16. E. Lommel, "Die Beugungserscheinungen einer kreisrunden Oeffnung und eines kreisrunden Schirmchens," Abh. Bayer. Akad. Math. Naturwiss. Kl. **15,** 233–328, (1885). The main part of Lommel's analysis is presented (in English) in Sec. 8.8 of Ref. 7.
17. This limiting procedure is equivalent to keeping the field point $P(z, r, \psi)$ as well as the wavelength λ and the angular semiaperture a/f fixed and letting $f \to \infty$. That the resulting expression is to be interpreted as an asymptotic rather than an ordinary approximation is strongly suggested by recent discussions of the Debye representation of focused fields,[8,9] which is known to be equivalent to Lommel's representation under the usual circumstances.
18. When the suffix n is a negative integer, Lommel's original definition of the function V_n differs from that given by Eq. (3.18) by a factor $(-1)^n$.
19. E. H. Linfoot and E. Wolf, "Phase distribution near focus in an aberration-free diffraction image," Proc. Phys. Soc. London Sect. B **69,** 823–832 (1956).

15 August 2000

ELSEVIER

Optics Communications 182 (2000) 265–272

OPTICS COMMUNICATIONS

www.elsevier.com/locate/optcom

Interference of converging spherical waves with application to the design of compact disks

Charles M.J. Mecca [a], Yajun Li [b], Emil Wolf [c,*]

[a] WEA Manufacturing, Inc., 1400 East Lackawanna Avenue, Olyphant, PA 18448, USA
[b] 527 Race Place, Oakdale, NY 11769, USA
[c] Department of Physics and Astronomy and Rochester Theory Center for Optical Science and Engineering, University of Rochester, Rochester, NY 14627, USA

Received 8 May 2000; accepted 26 May 2000

Abstract

Two-point-source model is proposed to simulate the reflection of light from the data surface of a compact disk (CD). The returning field is modeled as arising from the interference of two converging spherical waves. The resulting field in the focal region exhibits maximum extinction (destructive interference) of light when the pit depth is optimized. For systems of small transverse magnification, the optimum pit depth is found to be one-half wavelength rather than a quarter wavelength as is customary assumed by the usual plane wave model.

Keywords: Interference of converging waves; Compact disk; Data surface; Information pit; Transverse magnification; Order of interference; Focused fields

1. Introduction

The wave incident on a compact disk (CD) is focused on its data surface, on which numerous binary data are recorded in pits that are impressed along the CD surface and are covered with a very thin metal layer to ensure high reflection. A laser beam, focused by an objective lens, is used to read the data. It was claimed in several publications [1–3], based on elementary properties of interference between the incident and a reflected waves, that the maximum extinction of the returned light is obtained when the light reflected by a pit is in antiphase with the light reflected by the surrounding land, namely, when the pit depth is a quarter-wavelength. This criterion has been applied to CD manufacturing for many years. However, recent findings reveal that the quarter wavelength criterion may not predict optimum result under all circumstances [4].

To simulate the waves reflected from the CD data surface, we propose a two-point-source model. More specifically, two spherical (not plane) waves return from the disk and travel to the objective lens. One spherical wave is due to the reflection of the read-out beam from the pit, the other due to reflection from the surrounding land. The returning field then results from the superposition of two spherical waves, which first diverge from the disk and are then focused by the objective lens onto the photodetector. Conse-

* Corresponding author. Tel.: +1-716-275-4397; fax: +1-716-473-0687; e-mail: ewlupus@pas.rochester.edu

0030-4018/00/$ - see front matter
PII: S0030-4018(00)00772-0

quently, one is dealing with interference of two converging spherical waves with slightly different foci, slightly different focal lengths and slightly different cone angles rather than with interference of two plane waves. The focal regions of the two converging spherical waves are overlapping. In the common region, interference of the two focused spherical waves takes place.

In the present paper, we provide a model describing the interaction of two converging spherical waves in the focal region. This model is next applied to investigate how the binary data recorded on the CD

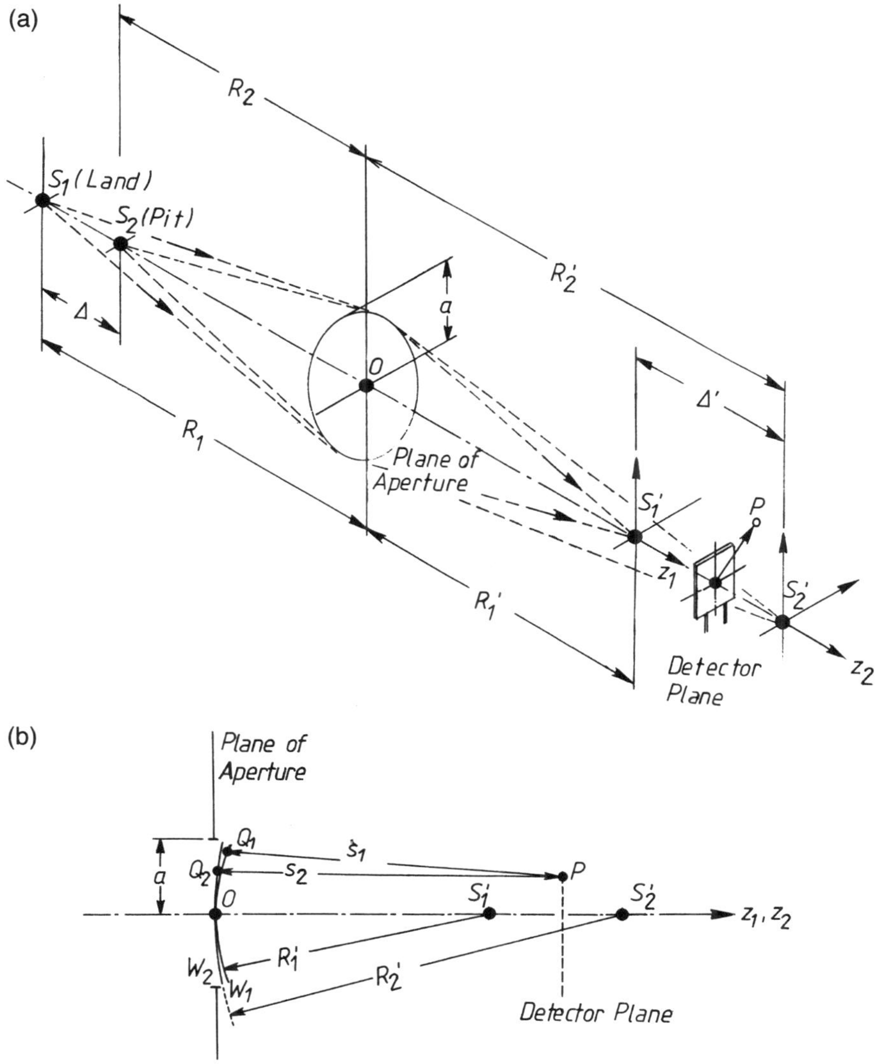

Fig. 1. The two-point-source model of light interaction with a CD: two monochromatic point sources S_1 and S_2, generating light of the same wavelength λ, are located on the axis of the objective lens. The point source S_1 is associated with reflection of the light from the land. The point source S_2 is associated with the reflection of the light from the pit. Δ denotes the pit depth. (a) Notation; (b) two wavefronts W_1 and W_2 of radii R_1' and R_2' are centered on the points S_1' and S_2' respectively and pass through the center O of a circular aperture of radius a.

surface are transferred into a series of light pulses. We then calculate the optimum pit depth for which destructive interference leads to a maximum extinction of the light in the focal region.

2. Imaging of two point sources

Consider two point sources S_1 and S_2 of light of the same wavelength λ are placed close to each other on the axis of a thin lens of focal length f which fills an aperture of radius a as shown in Fig. 1. Let

$$\Delta = R_1 - R_2 \tag{1}$$

be the separation of points S_1 and S_2, with R_1 and R_2 denoting the radii of curvature of the two spherical wavefronts immediately behind the thin lens. On the other side of the lens, two spherical wavefronts emerge, converging to points S_1' and S_2'. We denote their radii of curvature by R_1' and R_2', respectively. For the point source S_1, the radii of curvature R_1 and R_1' satisfy the lens relation

$$\frac{1}{R_1} + \frac{1}{R_1'} = \frac{1}{f}, \tag{2}$$

and for the point source S_2, the radii of curvature R_2 and R_2' satisfy a similar relation

$$\frac{1}{R_2} + \frac{1}{R_2'} = \frac{1}{f}. \tag{3}$$

The separation of the image points S_1' and S_2' is given by (see Fig. 1(a))

$$\Delta' = R_1' - R_2' = -M_T^2 \times \Delta, \tag{4}$$

where M_T is the transverse magnification of the system. Here we have assumed that the two spherical wave systems have the same transverse magnification.

3. Diffraction integrals

As we have indicated, we consider two spherical waves, say $V_1^{(i)}$ and $V_2^{(i)}$, generated by the two point sources S_1 and S_2, emerging from the aperture. At typical points Q_1 and Q_2 on the wavefronts which pass through the center O of the aperture, the field distributions can be expressed in the form

$$V_1^{(i)}(Q_1,t) = A\frac{e^{-ikR_1'}}{R_1'}e^{-i\omega t}, \tag{5a}$$

$$V_2^{(i)}(Q_2,t) = A\frac{e^{-i(kR_2'+\theta_0)}}{R_2'}e^{-i\omega t}, \tag{5b}$$

where A is a constant amplitude, t denotes the time and

$$\theta_0 = k\Delta \tag{6}$$

is the phase shift introduced by the spatial separation Δ of the two point sources S_1 and S_2.

According to the Huygens–Fresnel principle (Ref. [5], Section 8.2), the diffracted fields at a point P of the detector plane (see Fig. 1(b)) are given by the expressions (with time-periodic factor $\exp(-i\omega t)$ omitted)

$$U_1(P) = -\frac{i}{\lambda}\frac{Ae^{-ikR_1'}}{R_1'}\iint_{W_1}\frac{e^{iks_1}}{s_1}dS, \tag{7a}$$

and

$$U_2(P) = -\frac{i}{\lambda}\frac{Ae^{-i(kR_2'+\theta_0)}}{R_2'}\iint_{W_2}\frac{e^{iks_2}}{s_2}dS, \tag{7b}$$

where s_1 and s_2 denote the distances Q_1P and Q_2P and the integrals extend over the wavefronts W_1 and W_2 filling the aperture.

Let (z_1,r_1,ψ_1) and (z_2,r_2,ψ_2) be the two sets of cylindrical coordinates of the point P in the focal regions of the two converging spherical waves originating from the point sources S_1 and S_2. The origins of the two coordinate systems are at S_1' and S_2', i.e., at the image points of the point sources S_1 and S_2. The z-coordinates of the two focusing systems have a separation Δ',

$$z_1 = z_2 + \Delta', \tag{8}$$

along the common direction $OS_1'$and OS_2' (see Fig. 1(a)). The radial distances from the z-axes are

$$r_1 = r_2 = r \tag{9a}$$

and the azimuthal angles are also equal, i.e.,

$$\psi_1 = \psi_2 = \psi. \tag{9b}$$

It is convenient to introduce the Lommel parameters (u_1,v_1) and (u_2,v_2) which, together with the angle ψ, specify the position of the field point P:

$$u_1=\left(\frac{a}{R'_1}\right)^2 kz_1, \tag{10a}$$

$$v_1=\left(\frac{a}{R'_1}\right)kr_1; \tag{10b}$$

$$u_2=\left(\frac{a}{R'_2}\right)^2 kz_2, \tag{11a}$$

$$v_2=\left(\frac{a}{R'_2}\right)kr_2. \tag{11b}$$

Because the fields are rotationally symmetric about the z-axis, the diffracted fields are independent of the azimuthal angle ψ.

The photodetector which changes the intensity variations of the light into an electrical signal, is assumed to be placed at a point $z_1 = z_0$ in the focal regions of the two converging spherical waves. We then obtained from Eqs. (10) and (11) the following expressions for the Lommel parameters in the detector plane:

$$u_1=\left(\frac{a}{R'_1}\right)^2 kz_0, \tag{12a}$$

$$v_1=\left(\frac{a}{R'_1}\right)kr_1; \tag{12b}$$

$$u_2=\left(\frac{a}{R'_2}\right)^2 k(\Delta'+z_0), \tag{13a}$$

$$v_2=\left(\frac{a}{R'_2}\right)kr_2=\left(\frac{R'_1}{R'_2}\right)v_1. \tag{13b}$$

Assuming, as is usually the case, that the focusing system of focal length f has a high numerical aperture, the field in the region of the geometrical focus may be expressed in the form (Ref. [5], Section 8.8)

$$U(P)=-\frac{2\pi i a^2 A}{\lambda f^2}e^{i\left(\frac{f}{a}\right)^2 u}\int_0^1 J_0(v\rho)e^{-\frac{1}{2}iu\rho^2}\rho d\rho, \tag{14}$$

We now apply Eq. (14) to the two converging spherical waves discussed in Section 2.

On substituting $u=u_1$, $v=v_1$ and $f=R'_1$ into Eq. (13), we obtain for the diffracted field of the first converging spherical wave the expression

$$U_1(P)=-\frac{2\pi i a^2 A}{\lambda (R'_1)^2}e^{i\left(\frac{R'_1}{a}\right)^2 u_1}\times\int_0^1 J_0(v_1\rho)e^{-\frac{1}{2}iu_1\rho^2}\rho d\rho. \tag{15}$$

Similarly, on substituting $u=u_2$, $v=v_2$ and $f=R'_2$ into Eq. (13), we obtain the following expression of the diffracted field for the second converging spherical wave:

$$U_2(P)=-\frac{2\pi i a^2 A}{\lambda (R'_2)^2}e^{i\left[\left(\frac{R'_2}{a}\right)^2 u_2-\theta_0\right]}\times\int_0^1 J_0(v_2\rho)e^{-\frac{1}{2}iu_2\rho^2}\rho d\rho. \tag{16}$$

The subscripts 1 and 2 affixed to the symbol $U(P)$ for the diffracted fields indicate that they originated from the point sources S_1 and S_2, respectively.

The field distribution in the focal region arising from the superposition of the two fields given by Eqs. (15) and (16) can be expressed as

$$\begin{aligned}U(P)&=U_1(P)+U_2(P)\\&=-\frac{2\pi i a^2}{\lambda (R'_1)^2}\left\{e^{i\left(\frac{R'_1}{a}\right)^2 u_1}\times\int_0^1 J_0(v_1\rho)e^{-\frac{1}{2}iu_1\rho^2}\rho d\rho\right.\\&\quad+e^{i\left[\left(\frac{R'_2}{a}\right)^2 u_2-\theta_0\right]}\left(\frac{R'_1}{R'_2}\right)^2\\&\quad\left.\times\int_0^1 J_0(v_2\rho)e^{-\frac{1}{2}iu_2\rho^2}\rho d\rho\right\}.\end{aligned} \tag{17}$$

According to Eq. (17), the intensity distribution $I(P)=|U(P)|^2$ in the focal region is, therefore,

$$\begin{aligned}I(P)=\Bigg|&2\left\{\sqrt{I_{02}}\,e^{i\left(\frac{R'_2}{a}\right)^2 u_1}\int_0^1 J_0(v_1\rho)e^{-\frac{1}{2}iu_1\rho^2}\rho d\rho\right.\\&+\sqrt{I_{02}}\,e^{i\left[\left(\frac{R'_2}{a}\right)^2 u_2-\theta_0\right]}\\&\left.\times\int_0^1 J_0(v_2\rho)e^{-\frac{1}{2}iu_2\rho^2}\rho d\rho\right\}\Bigg|^2,\end{aligned} \tag{18}$$

where

$$I_{01} = \left(\frac{\pi a^2 |A|}{\lambda (R'_1)^2}\right)^2 \text{ and } I_{02} = \left(\frac{\pi a^2 |A|}{\lambda (R'_2)^2}\right)^2 \quad (19)$$

are constants.

4. The validity of Eq. (17) in the overlapping focal region

Eq. (17) is valid under the condition that the focal regions of two converging spherical waves overlap. Let us examine this situation.

The three-dimensional intensity distribution near the focus can be represented by isophotes, namely by contour lines of equal intensity (see Fig. 2 or Fig. 8.41 of Ref. [5]). Important for our consideration is the tubular structure in the bright central portion of the figure, which indicates the focal depth. The focal depths of the focused fields originated from the point sources S_1 and S_2 are of the order of (Ref. [5], see Eq. (27) in Section 8.8)

$$|\Delta z_1| = \frac{\lambda}{2}\left(\frac{R'_1}{a}\right)^2 \quad (20a)$$

and

$$|\Delta z_2| = \frac{\lambda}{2}\left(\frac{R'_2}{a}\right)^2 \quad (20b)$$

Hence Eq. (17) will be a good approximation provided that

$$|\Delta'| \lesssim |\Delta z_1| \quad (21a)$$

and

$$|\Delta'| \lesssim |\Delta z_2| \quad (21b)$$

On substituting from Eqs. (4), (20a) and (20b) into Eqs. (21a) and (21b), we obtain the following estimates for the range of validity of our theory:

$$M_T^2 \times \Delta \lesssim \frac{\lambda}{2}\left(\frac{R'_1}{a}\right)^2 \quad (22a)$$

and

$$M_T^2 \times \Delta \lesssim \frac{\lambda}{2}\left(\frac{R'_2}{a}\right)^2. \quad (22b)$$

With the choice $\Delta = \lambda/2$ and $M_T \approx (R'_1/R_1) \approx (R'_2/R_2)$, Eqs. (22a) and (22b) become

$$a \lesssim R_1 \quad (23a)$$

and

$$a \lesssim R_2 \quad (23b)$$

The numerical aperture of commonly used objective lens lies in the range between about 0.43 to 0.5, i.e., R_1 and R_2 are in the range between $1.73a$ and $2.10a$. Hence, the inequalities in Eqs. (23a) and (23b) are satisfied. We may, therefore, conclude that the focal regions of two converging spherical waves overlap and Eq. (17) correctly describes the combined effects of the two waves in the overlapping focal regions.

5. Interference effects in the focal region

To obtain a better insight into the structure of the region of superposition, we rewrite Eq. (18) in the form

$$I(P) = I_1(P) + I_2(P) + 2\sqrt{I_1(P) I_2(P)} \times \cos(\phi_1 - \phi_2), \quad (24)$$

where

$$I_1(P) = I_{01}\left|2\int_0^1 J_0(v_1 \rho) e^{-\frac{1}{2} i u_1 \rho^2} \rho \, d\rho\right|^2, \quad (25a)$$

and

$$I_2(P) = I_{02}\left|2\int_0^1 J_0(v_2 \rho) e^{-\frac{1}{2} i u_2 \rho^2} \rho \, d\rho\right|^2, \quad (25b)$$

are the intensity distributions in focal regions of the two converging spherical waves that represent the waves returning from the pit and from the surrounding land respectively on the data surface of a CD. After a long calculation, the phase factors ϕ_1 and ϕ_2 can be expressed in the form

$$\phi_1 = \left(\frac{R'_1}{a}\right)^2 u_1, \quad (26a)$$

$$\phi_2 = \left(\frac{R'_2}{a}\right)^2 u_2 - \theta_0. \quad (26b)$$

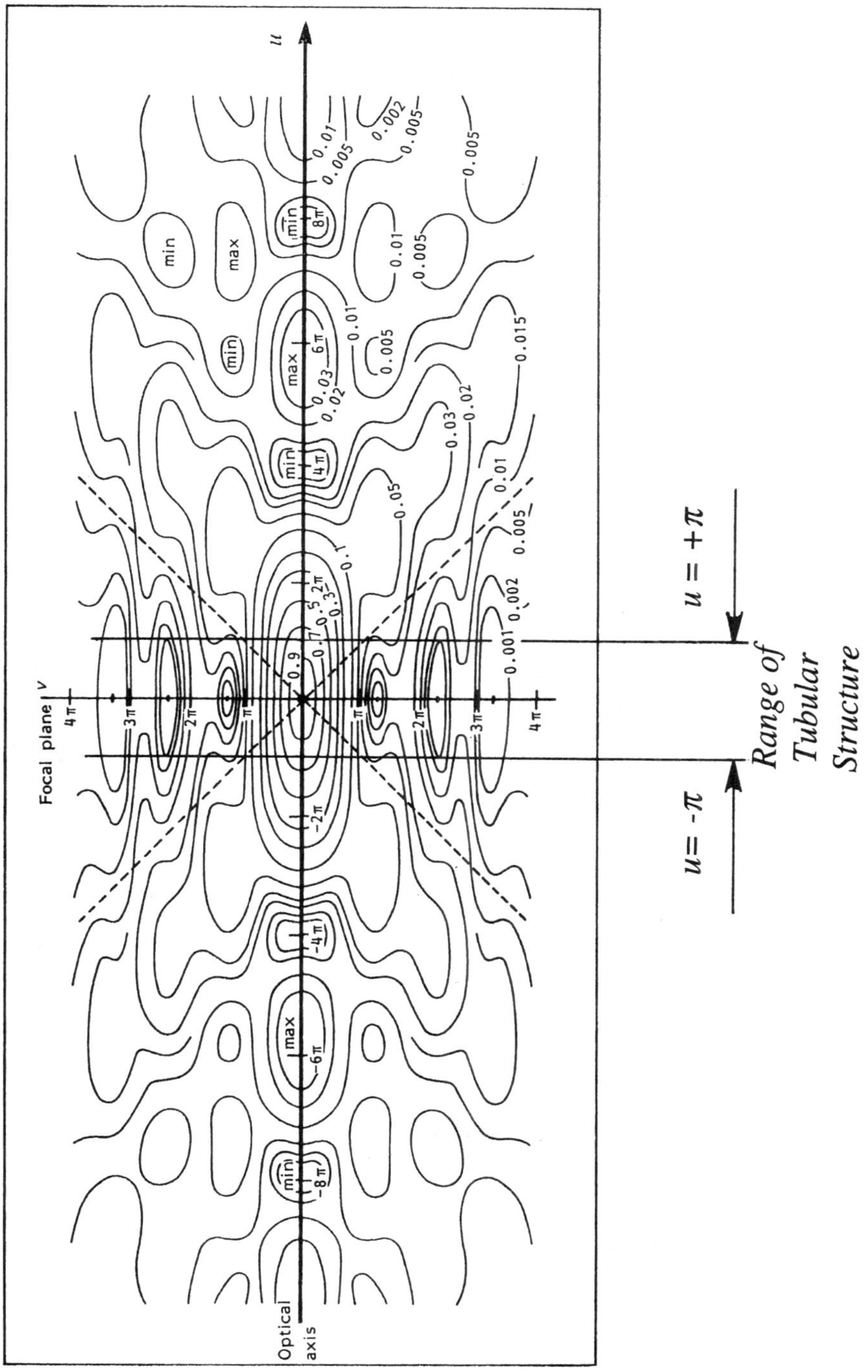

Fig. 2. Isophotes (lines of equal intensity) in the focal region. The tabular structure of the central portion should be noted [adapted from E.H. Linfoot and E. Wolf, Proc. Phys. Soc. B, 69 (1956) 823].

where θ_0, given by Eq. (6), represents the phase shift between the two returned waves. Let us suppose that the photodetector is located at the distance

$$z_0 = -\Delta'/2 \qquad (27)$$

from the point S'_1. It then follows Eqs. (12a) and (13a) that

$$u_1 = -k\left(\frac{a}{R'_1}\right)^2 \frac{\Delta'}{2} \qquad (28a)$$

and

$$u_2 = k\left(\frac{a}{R'_2}\right)^2 \frac{\Delta'}{2}. \qquad (28b)$$

On substituting Eqs. (28a) and (28b) into Eqs. (26a) and Eq. (26b), we obtain for the phase difference $\phi_1 - \phi_2$ the expression

$$\phi_1 - \phi_2 = \left(\frac{R'_1}{a}\right)^2 u_1 - \left(\frac{R'_2}{a}\right)^2 u_2 + \theta_0. \qquad (29)$$

Next, on substituting from Eqs. (4), (6), (28a) and. (28b) into Eq. (29), we find that

$$\phi_1 - \phi_2 = -k\Delta' + k\Delta = k\Delta(1 + M_T^2). \qquad (30)$$

When the laser spot on the disk surface scans over the pit, the phase difference $\phi_1 - \phi_2 = m\pi$, ($m =$ 1,3,5 ...). The intensity in the region of overlap is then a minimum and is given by the expression

$$I_{min}(P) = I_1(P) + I_2(P) - 2\sqrt{I_1(P)I_2(P)}. \qquad (31)$$

On the other hand, when the laser spot scans over the land, the phase difference $\phi_1 - \phi_2 = (m-1)\pi$, ($m = 1,3,5\ldots$). The intensity in the region of overlap is then a maximum and is given by the expression

$$I_{max}(P) = I_1(P) + I_2(P) + 2\sqrt{I_1(P)I_2(P)}. \qquad (32)$$

The intensity distribution in the focal region in systems of large angular aperture is symmetrical about the focal plane (see Refs. [5] or [6]), viz.

$$I_1(-u_1, v_1) = I_1(u_1, v_1) \qquad (33a)$$

and

$$I_2(-u_2, v_2) = I_2(u_2, v_2). \qquad (33b)$$

If we ignore the slight difference between R'_1 and R'_2, we find immediately from Eqs. (28a) and (28b) that

$$u_1 \approx -u_2, \qquad (34a)$$

and from Eqs. (12b) and (13b) that

$$v_1 \approx v_2. \qquad (34b)$$

We can therefore conclude that

$$I_1(P) \approx I_2(P). \qquad (34)$$

The maximum and minimum intensity distributions given by Eqs. (31) and (32) then reduce to

$$I_{max}(P) \approx 4I_1(P) \qquad (36a)$$

and

$$I_{min}(P) \approx 0. \qquad (36b)$$

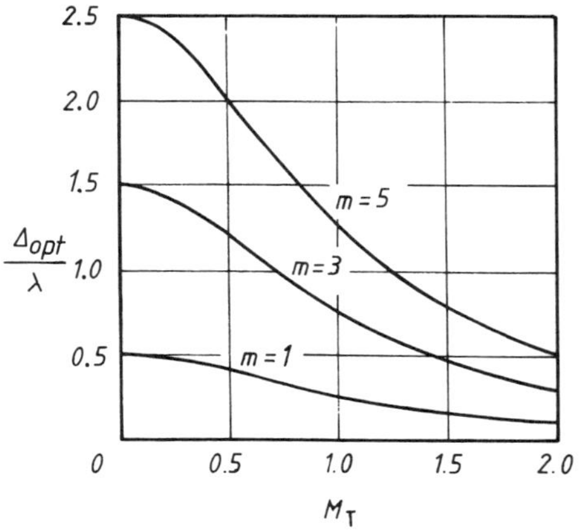

Fig. 3. Optimum pit depth Δ_{opt} as a function of the transverse magnification M_T of the system, when $m = 1$, 3 and 5.

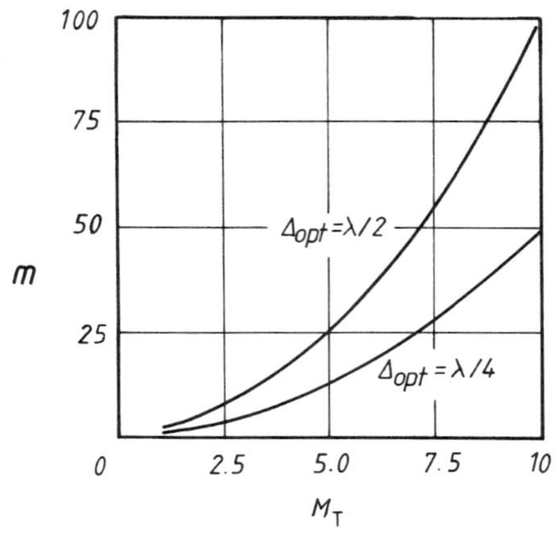

Fig. 4. The order of interference m as a function of the transverse magnification M_T of the system, when the optimum pit depth Δ_{opt} has the values $\lambda/2$ and $\lambda/4$.

The aim of pit-depth optimization is to bring the contrast, C say, of the photodetector output signal to a maximum, namely, to insure that

$$C \equiv \frac{I_{max}(P) - I_{min}(P)}{I_{max}(P) + I_{min}(P)} \simeq 1. \quad (37)$$

This happens when $I_{min}(P) = 0$, i.e., when

$$\phi_1 - \phi_2 = m\pi, \ (m = 1,3,5,\dots). \quad (38)$$

On substituting from Eq. (30) into Eq. (38), one readily obtains for the optimum depth, Δ_{opt} say, of the information pit on the CD data surface the expression

$$\Delta_{opt} = \frac{\lambda}{2} \frac{m}{1 + M_T^2}, \ (m = 1,3,5,\dots). \quad (39)$$

Eq. (39) is the main result of our analysis. In Fig. 3 the optimum pit depth Δ_{opt} is plotted as function of the transverse magnification M_T for the cases when $m = 1$, 3 and 5.

6. Discussion and conclusions

It is seen from Eq. (39) that the optimum depth Δ_{opt} of the information pit is a function of three parameters: the wavelength λ, the transverse magnification M_T of the system and the order of interference m. For systems of low magnification, namely, $M_T \ll 1$, and for the lowest order, ($m = 1$), Eq. (39) gives

$$\Delta_{opt} \simeq \frac{\lambda}{2}. \quad (40)$$

However, the optimum depth Δ_{opt} decreases rapidly when the system magnification M_T increases. The relationship between M_T and the order of interference m is shown in Fig. 4 for the cases when $\Delta_{opt} = \lambda/2$ and $\Delta_{opt} = \lambda/4$.

Finally, we mention that in our analysis we ignored the refraction of the returned waves at the boundary of the plastic substrate of CD. We plan to take it into account in another publication.

We may conclude by saying that we have shown that the optimum pit depth is a function of three parameters: the wavelength λ, the magnification M_T of the system and the order of interference m. For a system of low magnification and for $m = 1$, the optimum pit depth is $\lambda/2$. However, for systems of large magnification, the optimum depth may take other values, including the conventional value $\lambda/4$.

Acknowledgements

Our research was supported by the WEA Manufacturing, Inc.

References

[1] G. Bouwhuis, J. Braat, A. Huijser, J. Pasman, G. van Rosemalen, K. Schouhamer Immint, Principles of Optical Disc Systems, 1st edn., Adam Hilger Ltd., Bristol and Boston, 1985.
[2] K.C. Pohlmann, The Compact Disc, updated edn., A–R Editions, Madison, 1992.
[3] J.G. Dil, B.A.J. Jacobs, J. Opt. Soc. Am. 69 (1979) 950.
[4] US Patent 5,995,481, November 30, 1999.
[5] M. Born, E. Wolf, Principles of Optics, 7th edn., Cambridge University Press, Cambridge, 1999.
[6] E. Collett, E. Wolf, Opt. Lett. 5 (1980) 264.

Section 2 – Radiation Theory and String Excitations

The first two papers in this Section (papers 2.1 and 2.2) are concerned with theoretical models developed in the 1960s and 1970s for the interaction of waves and charged particles with material media. An important role in these theories is played by waves which decay exponentially in amplitude along a certain direction as they propagate. They are known as evanescent waves. Today such waves are well-known, particularly in the area of near-field optics, but at the time when these papers were written such waves were regarded as somewhat of a rarity, except in connection with total internal reflection and the skin effect in metals.

Paper 2.3 deals with an old subject, multipole expansions of electromagnetic fields generated by localized, monochromatic charge-current distributions. The paper is included in this volume because it establishes a connection between the multipole expansion and the so-called angular spectrum representation, which is a generalized plane wave representation of fields containing both ordinary (homogeneous) plane waves and evanescent waves. The connection revealed by this analysis proved useful later, for example in the theory of so-called non-radiating sources. These are localized sources which, although they generate fields within their domain of localization, do not produce any power in the far zone, i.e. they do not radiate. In fact such sources do not produce any field at all outside the source domain. The theory of such sources has been studied since the early years of the last century. The subject has attracted a good deal of attention in the early days of quantum theory, when some physicists, including Ehrenfest, Sommerfeld and Bohm, tried to model electrons and some other elementary particles as sources of this kind. The attempts were, however, not successful. In more recent times, such sources have become of interest again because of the obvious bearing that they have on questions of uniqueness of inverse problems, especially in connection with inverse source problems and inverse scattering problems. If, for example, one tries to reconstruct a source from measurements outside its domain of localization, no information about the non-radiating part of the source distribution can be deduced from such measurements.

In papers 2.4 – 2.6 some properties of non-radiating sources, both scalar and electromagnetic ones, are derived. In particular, in paper 2.6 a new way for characterizing such sources is presented, according to which a non-radiating source and the field which it generates within the source domain satisfy a certain overspecified boundary value problem.

Although there are numerous publications concerning the theory of non-radiating sources, no such source has been produced in a laboratory up to now. In paper 2.7 it is shown that non-radiating sources have analogues in acoustics. Specifically it is demonstrated that when an appropriate force distribution is applied to a finite portion of an infinitely long string, it will produce no displacement of the string outside the region of the applied force. This result was generalized in a later paper, (not included in this volume), by G. Gbur, J.T. Foley and E. Wolf [Wave Motion **30**, 125–134 (1999)] to several more realistic situations, including strings of finite lengths. It seems likely that the existence of non-propagating string excitations could be experimentally verified more easily than oscillations of non-radiating sources. Such experiments would show that non-radiating sources are not just mathematical fictions of theoretical physicists.

Section 2 – Radiation Theory and String Excitations

Reprinted from
VOLUME 9 NUMBER 4 APRIL 1971 THE PHYSICS TEACHER

RESEARCH FRONTIER

THEORY OF ČERENKOVIAN EFFECTS

In recent years new physical models have been proposed for the explanation of the Čerenkov effect and related phenomena.

The Čerenkov effect is named after a Soviet scientist who discovered it in the early thirties while carrying out thesis work at the Lebedev Institute in Moscow. He was investigating problems suggested to him by his teacher, S. I. Vavilov, connected with effects that arise when gamma rays (very short wavelength electromagnetic radiation) from a radium source enter a fluid. Čerenkov observed that as the gamma rays penetrated the liquid, a weak, bluish glow emanated from it. Čerenkov found that the bluish glow was not directly generated by the gamma rays, but was caused by electrons, moving at very high speeds, released from the atoms of the fluid as a result of the gamma-ray bombardment. A theoretical understanding of the phenomenon came a few years later as a result of the work of two other Soviet scientists, I. E. Tamm and I. M. Frank.

The essential condition for the appearance of the phenomenon is the following one: For each medium such as a liquid or a transparent solid, there exists a characteristic velocity $v(\omega)$, known as the phase velocity of the medium for radiation of frequency ω. It is the velocity with which each surface of constant phase of a periodic wave disturbance of frequency ω (e.g., the crests of the wave) will propagate in that medium. When an electron moves in the medium with a uniform velocity V that is greater than the phase velocity $v(\omega)$, light of frequency ω is generated by a subtle mechanism that involves the interaction of the electron with the atoms of the substance.

The traditional more formal explanation of this phenomenon,[1] using Huygens' construction, is illustrated in Fig. 1. At each point A of its paths, the electron emits spherical wavelets which propagate away from it with velocity $v(\omega)$. When the electron has reached a point B, the wavelets emitted from the various intermediate positions between A and B (e.g., P_1, P_2, P_3, . . .) give rise to a common conical wavefront (see Figs. 1 and 2), provided the velocity V of the electron exceeds the phase velocity $v(\omega)$. This conical wavefront is the envelope of the wavelets at a particular instant of time. On this wavefront, the fields associated with the individual wavelets reinforce each other; elsewhere the wavelets interfere destructively and their effects substantially cancel out. The normals to the wavefront make an angle $\theta(\omega)$ with the direction of motion of the electron, given by

$$\cos\theta(\omega) = v(\omega)/V, \quad [V > v(\omega)].$$

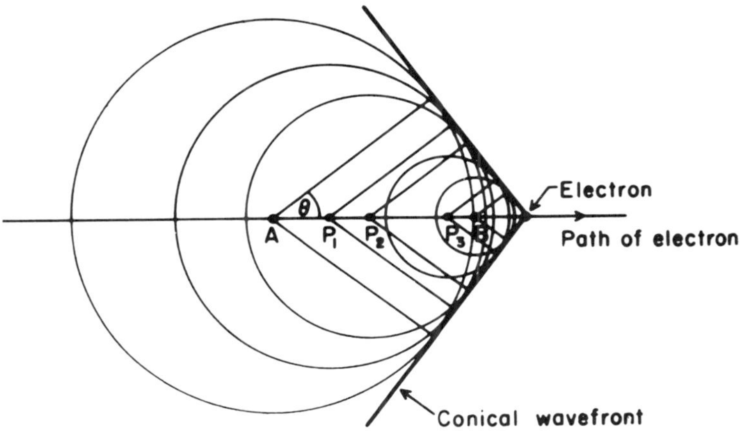

$$\cos\theta(\omega) = \frac{v(\omega)}{V} \qquad (V > v(\omega))$$

Fig. 1. The Huygens' construction of a conical wavefront associated with Cerenkov radiation.

The bluish glow observed by Cerenkov is associated with such a conical light wave. The smaller the ratio v/V (the larger the electron velocity V) the greater is the angle θ at which the Čerenkov radiation is emitted. Although the electron velocity V may exceed the phase velocity in the substance, it cannot, according to Einstein's theory of relativity, exceed the velocity of light in free space (about 300 000 km/sec).

The Čerenkov effect is closely analogous to the formation of a bow wave of a vessel that moves through water at a speed that exceeds the speed of water waves. Another analogous example arises in air, when the speed of an airplane exceeds the speeds of sound waves in air (about 1000 km/h at sea level). When an object crosses the "sound barrier," it produces the well-known sonic boom. Both these effects may be understood with the help of Huygens' construction of the kind we just discussed. The associated conical wavefronts (shock waves) correspond to the appearance of the Čerenkov light cone.

Today the Čerenkov effect has numerous applications in physics.

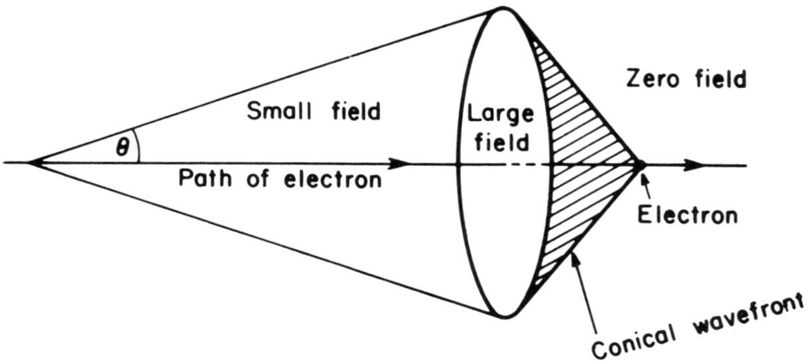

Fig. 2. Illustrating the theory of the Cerenkov effect.

It is used, for example, to determine the speed of very fast moving particles; it is also employed as a kind of filter to distinguish between particles of different types. The importance of Čerenkov's discovery and of the subsequent theoretical interpretation of it was clearly recognized when Čerenkov, Tamm, and Frank were jointly awarded the Nobel Prize for physics in 1958.

A new approach to the analysis of the Čerenkov and related effects was initiated in 1960 by an Italian physicist, G. Toraldo di Francia.[2] This treatment is based on a novel model for the electromagnetic field generated by a uniformly moving electron in free space. According to this model the electromagnetic field associated with the electron may be regarded as being made up of *evanescent plane waves*. These are waves that do not have some of the usual properties of ordinary plane waves; for example their amplitude is not constant but may decrease exponentially along some particular direction in space. The best known phenomenon where evanescent waves arise is that of *total internal reflection* (Fig. 3) : A beam of light is incident on an interface between an optically denser medium and an optically rarer one, for example from glass into air, at an angle ϕ which exceeds the so-called critical angle ϕ_0, given by

$$\sin \phi_0 = n_2/n_1, \quad (n_1 > n_2),$$

where n_1 and n_2 are the refractive indices of the two media. The incident beam is then completely reflected back into the optically denser medium, in contrast with the more commonly encountered situation when a beam incident on an interface is partly reflected and partly transmitted. Surprisingly, when total internal reflection takes place, there is a wave beyond the surface in the optically less dense medium which is just an evanescent wave of the type we are considering. It is not easily detected since its amplitude decays very rapidly with increasing distance from the interface, unless the direction of incidence is extremely close to the critical direction. The existence of such waves has in the past been mainly demonstrated by experiments with wavelengths much longer than optical ones. However, in recent years evanescent waves generated in total reflection have been studied by an ingenious method using monomolecular layers of long-chain fatty acid.[3]

Returning to the uniformly moving electron in free space, let us imagine a plane drawn through its line of motion. The electromagnetic field generated by the electron can be thought of as being made up of evanescent waves of all frequencies, which decay away from the plane on both sides. To explain the Čerenkov effect Toraldo di Francia now assumed the uniformly moving electron to be placed in a narrow gap between two blocks of dielectric material (Fig. 4). The evanescent waves then undergo reflection and refraction at the faces of the blocks. Toraldo di Francia showed that if the velocity of the electron exceeds the phase velocity of light in the dielectric, some of the waves obtained by refraction of the evanescent waves will be ordinary waves, and these carry away radiation from the electron. If now the gap between the blocks is allowed to become vanishingly small, one obtains the situation where an electron is moving uniformly in an infinite dielectric; if its velocity exceeds the phase velocity, the ordinary waves generated from the evanescent waves by refraction add up and they give rise precisely to Čerenkov radiation.

In a somewhat similar way, Toraldo di Francia was also able to explain the Smith-Purcell effect,[4] discovered in 1953. This is the effect observed when an electron moves uniformly in free space close to and parallel to the surface of a

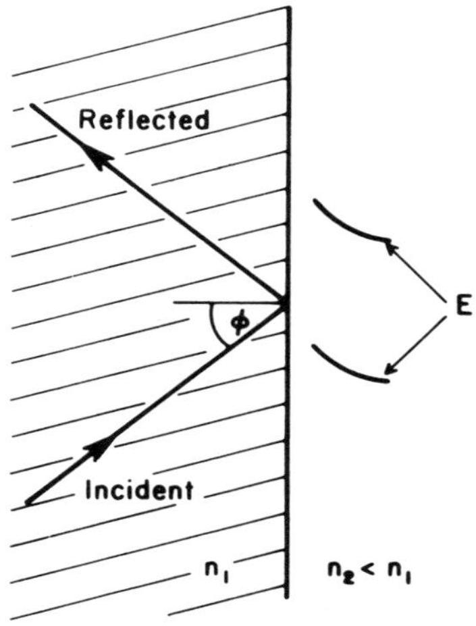

Fig. 3. Total internal reflection: A plane wave is incident from a medium of refractive index n_1 into a medium of refractive index $n_2 < n_1$ at an angle that exceeds the critical angle $\phi_0 = \sin^{-1}(n_2/n_1)$. It is totally reflected at the interface. An evanescent wave is generated in the second medium.

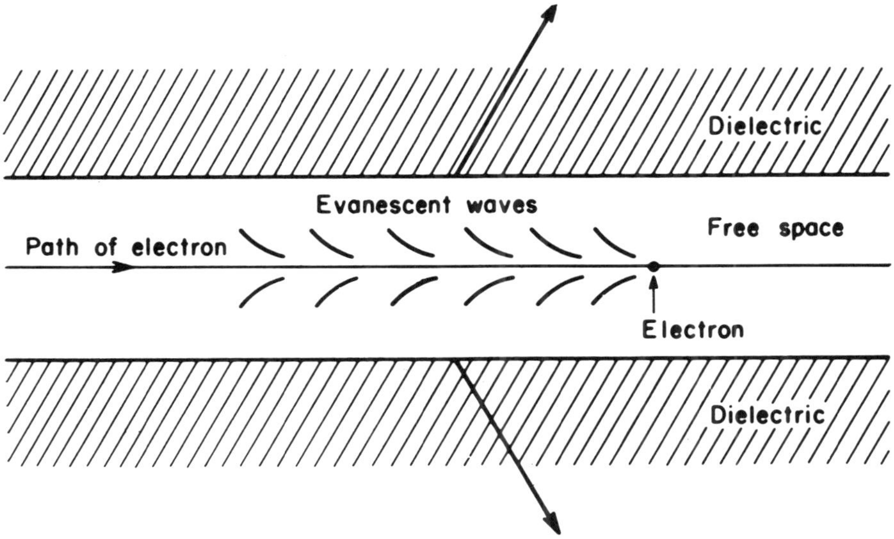

Fig. 4. Toraldo di Francia's model for the explanation of Cerenkov radiation. Evanescent waves created by the uniformly moving electron are refracted at the boundaries of two dielectric blocks, which are eventually brought together. If the velocity of the electron exceeds the phase velocity of light in the dielectric, some of the evanescent waves give rise on refraction to ordinary plane waves. On superposition the ordinary plane waves produce Cerenkov radiation.

metallic grating. Radiation is observed in certain directions, which depend on the speed of the electron and the periodicity of the grating. For an electron moving with speed V, the wavelength λ and the angle θ are related by (Fig. 5)

$$\lambda = d\,(V/c - \cos\theta),$$

where d is the periodicity of the grating and c is the vacuum velocity of light. Toraldo di Francia explained the generation of this radiation as arising from the diffraction of the evanescent waves associated with the freely moving electron on the metallic grating.

Here at the University of Rochester we have taken a somewhat more general approach to problems of this type. Starting with the basic equations govering the electromagnetic field generated by a moving electron, we have developed a theory[5] which describes the radiation arising from any prescribed two-dimensional motion of the electron. The field in each half-space on either side of the plane in which the electron moves, is expressed as superposition of plane monochromatic waves. In general both ordinary waves and evanescent waves are present in this so-called *angular spectrum* representation of the field. The ordinary waves propagate with constant amplitude away from the plane of motion of the electron, while the evanescent waves decay exponentially with increasing distance from this plane.

In the special case when the electron moves uniformly in free space (and under these conditions it does not radiate) the angular spectrum representation consists of evanescent waves only, in agreement with the results of Toraldo di Francia, referred to earlier. Thus the evanescent waves are intimately related to the Coulomb field of the electron and the rate of decay of the evanescent waves is related to the speed of the electron. As the uniform velocity is made larger, the evanescent waves decay more slowly away from the plane in which the electron moves.

If instead of moving in free space the electron moves uniformly in a dielectric medium, the situation is similar, except that the speed of the electron may approach and eventually exceed the phase velocity of the light in the medium. When this happens, ordinary waves are generated, which give rise to Čerenkov radiation. In Toraldo di

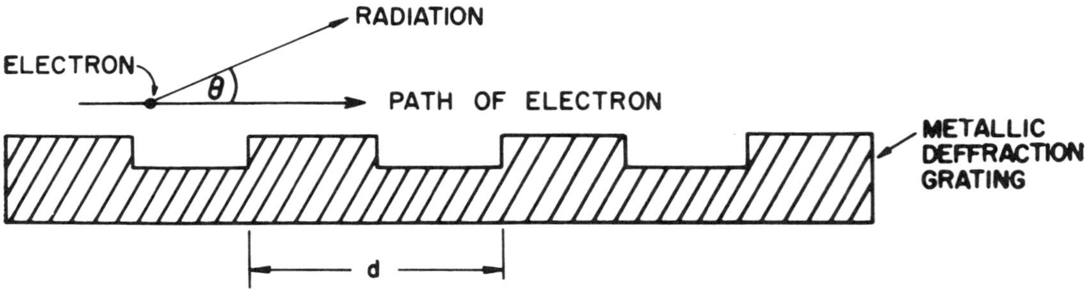

$$\lambda = d\left(\frac{V}{c} - \cos\theta\right)$$

Fig. 5. The Smith–Purcell effect: An electron moves uniformly close to and parallel to the surface of a metallic grating. Radiation is observed in certain directions.

Francia's model, the generation of the ordinary waves was regarded as arising from refraction of evanescent waves at the boundaries of two dielectric blocks that were finally brought into contact. However it is possible to explain the generation of the homogeneous waves on the basis of a molecular model in which the waves in the dielectric are shown to arise from the interaction of the vacuum field of the electron with the molecules that form the dielectric medium. Studies of radiation from electrons moving in dielectric media have been recently made on the basis of this model both in connection with the Čerenkov effect and for more general situations.[6] These investigations have led to a deeper understanding of the nature of the electromagnetic field created by the motion of charged particles both in free space and in material media.

Robin Asby and Emil Wolf
Department of Physics and Astronomy
University of Rochester
Rochester, New York 14627

Footnotes

1. For more details see for example in J. D. Jackson, *Classical Electrodynamics* (Wiley, New York, 1962), Sec. 14.9.
2. G. Toraldo di Francia, Nuovo Cimento **16,** 61 (1960).
3. K. H. Drexhage, Sci. Amer. 222, 108 (March 1970).
4. S. J. Smith and E. M. Purcell, Phys. Rev. **92,** 1069 (1953).
5. R. J. Asby and E. Wolf, J. Opt. Soc. Amer. **61,** 52 (1971).
6. A preliminary account of this work was presented in a talk given at a meeting of the Optical Society of America [E. Lalor and E. Wolf, J. Opt. Soc. Amer. **60,** 722 (1970), Abstract WB 13]. A fuller account of this research is being prepared for publication.

Emil Wolf is Professor of Physics at the University of Rochester, Rochester, New York. He is the author of about ninety articles, chiefly on optics and electromagnetic theory. He is also a coauthor (with Max Born) of the book Principles of Optics, the editor of Progress in Optics, and coeditor (with L. Mandel) of two recently published volumes, Selected Papers on Coherence and Fluctuations of Light (Dover, New York, 1970).

Robin Asby received his B.A. in Mathematics from the University of Cambridge in 1965 and Ph.D. in Mathematical Physics from the University of London in 1968. From September 1968 until August 1970 he was a Research Associate at the University of Rochester, Rochester, New York and is now a Research Fellow at the University of Southampton. His main research interests are in the theory of the interaction of light with matter.

VOLUME 26, NUMBER 20 PHYSICAL REVIEW LETTERS 17 MAY 1971

New Model for the Interaction Between a Moving Charged Particle and a Dielectric, and the Cherenkov Effect*

Éamon Lalor† and Emil Wolf
Department of Physics and Astronomy, University of Rochester, Rochester, New York 14627
(Received 8 April)

A new model is described for the interaction between a moving charged particle executing any two-dimensional motion and an infinite, homogeneous, linear dielectric. The model is based on the exact solution of the self-consistent integro-differential equation of molecular optics. The main results are illustrated by application to the Cherenkov effect.

Consider first a particle of charge e that moves on a trajectory $\vec{r}=\vec{r}_e(t)$ in the plane $z=0$ *in vacuo*. The Fourier frequency transforms [with kernel $(1/2\pi)\exp(i\omega t)$] of the electric and the magnetic fields that are generated by the particle may readily be derived by the use of the retarded potentials. It was shown by Asby and Wolf[1] that in each half-space $z \gtrless 0$ the fields have the following angular spectrum representation:

$$\vec{E}^{(v)}(\vec{r},\omega)=\iint_{-\infty}^{+\infty}\vec{e}^{(v)}(p,q;\omega;\gtrless)\exp[ik(px+qy\pm mz)]dp\,dq, \tag{1}$$

$$\vec{H}^{(v)}(\vec{r},\omega)=\iint_{-\infty}^{+\infty}\vec{h}^{(v)}(p,q;\omega;\gtrless)\exp[ik(px+qy\pm mz)]dp\,dq. \tag{2}$$

Here

$$k=\omega/c, \tag{3}$$

$$m=\begin{cases}+(1-p^2-q^2)^{1/2} \text{ if } p^2+q^2\leqslant 1, & \text{(4a)}\\ +i(p^2+q^2-1)^{1/2} \text{ if } p^2+q^2\geqslant 1; & \text{(4b)}\end{cases}$$

and the complex spectral vector amplitudes are given by

$$\vec{e}^{(v)}(p,q;\omega;\gtrless)=\hat{s}^{(\pm)}\times[\hat{s}^{(\pm)}\times\vec{f}(p,q;\omega)], \tag{5}$$

$$\vec{h}^{(v)}(p,q;\omega;\gtrless)=-\hat{s}^{(\pm)}\times\vec{f}(p,q;\omega), \tag{6}$$

with

$$\vec{f}(p,q;\omega)=(e/cm)(k/2\pi)^2\int_{-\infty}^{+\infty}\vec{V}(t')\exp\{i[\omega t'-k(px_e'+qy_e')]\}dt'. \tag{7}$$

In Eq. (7) x_e', y_e' are the coordinates and $\vec{V}(t')$ the velocity of the charged particle at time t'; c is the speed of light *in vacuo*. In (5) and (6), $\hat{s}^{(\pm)}$ are the unit vectors $(p,q,\pm m)$; the positive or negative sign on $\hat{s}^{(\pm)}$ and in $\pm m$ in (1) and (2), and the symbol > or < in the arguments of $\vec{e}^{(v)}$ and $\vec{h}^{(v)}$, are taken according as the field point $\vec{r}$ is in the half-space $z>0$ or $z<0$, respectively.

VOLUME 26, NUMBER 20 PHYSICAL REVIEW LETTERS 17 MAY 1971

Equations (1) and (2) are exact mode expansions of the fields. Each equation expresses the Fourier frequency transform of the fields as a superposition of plane waves, all of the *same* wave number $k = \omega/c$ appropriate to the particular frequency component ω of the fields. Each mode is labeled by the pair of parameters p, q and the frequency ω, and consists of a plane wave

$$\vec{e}^{(v)}(p, q; \omega; >) \exp[ik(px + qy + mz)] \text{ if } z > 0,$$

or

$$\vec{e}^{(v)}(p, q; \omega; <) \exp[ik(px + qy - mz)] \text{ if } z < 0.$$

In view of the definition of m [Eqs. (4)] we see that if $p^2 + q^2 \leqslant 1$ the waves are ordinary homogeneous waves propagated away from the plane of motion of the particle, and that if $p^2 + q^2 > 1$ they are evanescent waves propagated in directions parallel to the plane of motion of the particle ($z = 0$) and decaying exponentially in amplitude with increasing distance $|z|$ from that plane. It may readily be shown by examining the asymptotic behavior of (1) and (2) as $kr \to \infty$ in any fixed direction that only the homogeneous waves in the angular spectrum give rise to radiation.[2]

Suppose now that the charged particle executes the same motion in an infinite dielectric, which we assume to be linear, homogeneous, isotropic, and nonmagnetic. Then according to molecular optics,[3,4] the average effective fields $\vec{E}'(\vec{r}, \omega)$ and $\vec{H}'(\vec{r}, \omega)$ at the point $\vec{r}$ in either of the two half-spaces $z > 0$ or $z < 0$ may be shown to satisfy the equations

$$\vec{E}'(\vec{r}, \omega) = \vec{E}^{(v)}(\vec{r}, \omega) + N\alpha(\omega)\int \nabla\times\nabla\times[\vec{E}'(\vec{r}\,', \omega)G_\omega(|\vec{r}-\vec{r}\,'|)]d^3r', \tag{8}$$

$$\vec{H}'(\vec{r}, \omega) = \vec{H}^{(v)}(\vec{r}, \omega) - ikN\alpha(\omega)\int \nabla\times[\vec{E}'(\vec{r}\,', \omega)G_\omega(|\vec{r}-\vec{r}\,'|)]d^3r', \tag{9}$$

where the integration extends over the two half-spaces $z > 0$ and $z < 0$, except for a vanishingly small sphere centered on the field point at $\vec{r}$. In these equations $\vec{E}^{(v)}$ and $\vec{H}^{(v)}$ are, of course, the vacuum fields, given by Eqs. (1) and (2); N is the average number of molecules per unit volume; $\alpha(\omega)$ is the average polarizability (at frequency ω) of a molecule; and

$$G_\omega(|\vec{r}-\vec{r}\,'|) = |\vec{r}-\vec{r}\,'|^{-1}\exp(ik|\vec{r}-\vec{r}\,'|) \tag{10}$$

is the outgoing free-space Green's function.

Now according to a mathematical lemma [Ref. 3, Eq. (11) of Appendix V], the integral in (8) may be rewritten as

$$\int \nabla\times\nabla\times[\vec{E}'(\vec{r}\,', \omega)G_\omega(|\vec{r}-\vec{r}\,'|)]d^3r' = \nabla\times\nabla\times\int \vec{E}'(\vec{r}\,', \omega)G_\omega(|\vec{r}-\vec{r}\,'|)d^3r' - (8\pi/3)\vec{E}'(\vec{r}, \omega). \tag{11}$$

On the other hand one can show that in the integral in (9) the operator $\nabla\times$ may be taken outside the integral sign. Further, one has the Lorentz relation between the effective electric field $\vec{E}'$ and the macroscopic Maxwell electric field $\vec{E}^{(d)}$ in the dielectric:

$$\vec{E}'(\vec{r}, \omega) = \vec{E}^{(d)}(\vec{r}, \omega) + (4\pi/3)N\alpha(\omega)\vec{E}'(\vec{r}, \omega). \tag{12}$$

Since the dielectric was assumed to be nonmagnetic, the effective magnetic field is equal to the macroscopic Maxwell field $\vec{H}^{(d)}$ in the dielectric:

$$\vec{H}'(\vec{r}, \omega) = \vec{H}^{(d)}(\vec{r}, \omega). \tag{13}$$

Finally, we also have the Lorentz-Lorenz relation connecting the molecular polarizability per unit volume, $N\alpha$, with the refractive index n of the medium:

$$[n^2(\omega) - 1]/[n^2(\omega) + 2] = (4\pi/3)N\alpha(\omega). \tag{14}$$

Using Eqs. (11)-(14), Eqs. (8) and (9) may be rewritten as the following relations between the Maxwell vacuum fields and the fields in the dielectric:

$$\vec{E}^{(d)}(\vec{r}, \omega) = \frac{1}{n_2}\vec{E}^{(v)}(\vec{r}, \omega) + \frac{1}{4\pi}\left[\frac{n^2(\omega)-1}{n^2(\omega)}\right]\nabla\times\nabla\times\int \vec{E}^{(d)}(\vec{r}\,', \omega)G_\omega(|\vec{r}-\vec{r}\,'|)d^3r', \tag{15}$$

$$\vec{H}^{(d)}(\vec{r}, \omega) = \vec{H}^{(v)}(\vec{r}, \omega) - (ik/4\pi)[n^2(\omega) - 1]\nabla\times\int \vec{E}^{(d)}(\vec{r}\,', \omega)G_\omega(|\vec{r}-\vec{r}\,'|)d^3r'. \tag{16}$$

Equation (15) is an integro-differential equation for the Maxwell electric field in the dielectric in terms of the Maxwell vacuum electric field. We have succeeded in obtaining the exact solution of Eq. (15), and we will describe in another publication the technique we used. Here we only wish to report the physical implications of the solution. For this purpose we represent $\vec{E}^{(d)}(\vec{r}, \omega)$ and $\vec{H}^{(d)}(\vec{r}, \omega)$ as angular spectra of plane waves, all with the wave number $n(\omega)k = n(\omega)\omega/c$ appropriate to waves of frequency ω in the dielectric. In particular, the representation of $\vec{E}^{(d)}$ has the form

$$\vec{E}^{(d)}(\vec{r}, \omega) = \iint_{-\infty}^{+\infty} \vec{e}^{(d)}(p', q'; \omega; \gtrless) \exp[in(\omega)k(p'x + q'y \pm m'z)]dp'dq'. \tag{17}$$

Here m' bears the same relation to p' and q' as m bears to p and q [cf. Eqs. (4)]. The solution of the integro-differential equation (15) then implies the following: *The field $\vec{E}^{(d)}(\vec{r}, \omega)$ in the dielectric is obtained from the vacuum field $\vec{E}^{(v)}(\vec{r}, \omega)$ by a transformation in which each mode (p, q, ω) of the vacuum field is transformed into one and only one mode (p', q', ω) of the field in the dielectric, with*

$$p \to p' = p/n(\omega), \quad q \to q' = q/n(\omega). \tag{18}$$

Moreover, the transformation law for the vector amplitudes $[\vec{e}^{(v)} \to \vec{e}^{(d)}]$ is of a very simple form [given by Eqs. (21) below].

We note that (18) implies that for corresponding modes,

$$\begin{vmatrix} 0 & 0 & 1 \\ p & q & m \\ p' & q' & m' \end{vmatrix} = 0, \tag{19}$$

which shows that the unit (real or complex) vectors (p, q, m) and (p', q', m') of corresponding modes are coplanar, with the normal to the plane of motion of the particle (the plane $z = 0$). If the corresponding modes are homogeneous, i.e., if $p^2 + q^2 \leq 1$ and $p'^2 + q'^2 \leq 1$, then $m = \cos\theta_v$, $m' = \cos\theta_d$, where θ_v and θ_d are the angles which the respective directions of propagation of the waves *in vacuo* and in the dielectric make with the normal to the plane of the particle motion (see Fig. 1). Equation (18) together with Eq. (4a) and a similar formula for m' imply that for corresponding modes

$$\sin\theta_v / \sin\theta_d = n(\omega). \tag{20}$$

The relation (20), together with the complanarity condition expressed by (19), is formally identical with the *law of refraction* (Ref. 3, Sect. 1.5.1). Actually the analogy with the problem of refraction goes further. If we resolve the vector amplitudes $\vec{e}^{(v)}$ and $\vec{e}^{(d)}$ into components parallel and perpendicular to the plane containing corresponding wave normals $(p, q, \pm m)$ and $(p', q', \pm m')$, then the solution of Eq. (15) may be shown to imply that for corresponding modes,

$$\frac{e_{\parallel}^{(d)}}{e_{\parallel}^{(v)}} = \frac{\sin\theta_v}{\sin\theta_d}, \quad \frac{e_{\perp}^{(d)}}{e_{\perp}'^{(v)}} = \frac{\sin 2\theta_v}{\sin 2\theta_d}. \tag{21}$$

These relations are of the same mathematical form as, but simpler than, the Fresnel formulas for refraction and reflection (Ref. 3, Sect. 1.5.2).

The results expressed by Eqs. (18) and (21), which are exact within the domain of classical electrodynamics, express basic laws of interaction between the vacuum field of a charged particle executing any two-dimensional motion, and an infinite dielectric.

Several interesting consequences may readily be derived from these laws. We see from (18) that all those modes of the vacuum field for which

$$p^2 + q^2 \leq n^2(\omega) \tag{22}$$

will be transformed upon interaction with the dielectric into homogeneous modes $(p'^2 + q'^2 \leq 1)$. If $n(\omega) > 1$, the domain (22) is seen to include evanescent modes of the vacuum field, namely those modes labeled by p, q, ω, where

$$1 < p^2 + q^2 \leq n^2(\omega). \tag{23}$$

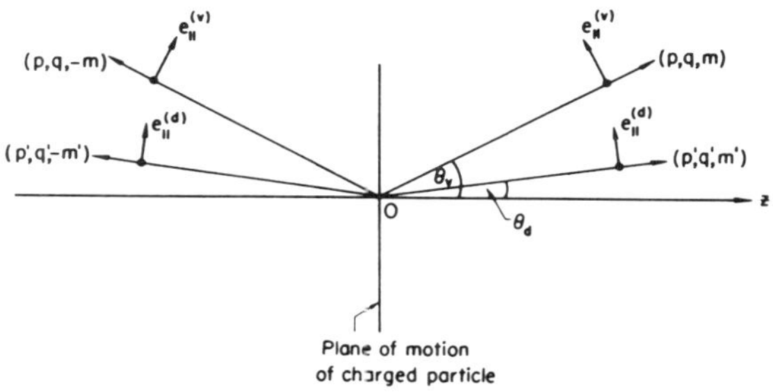

FIG. 1. Illustration of the law of interaction between a moving charged particle and an infinite dielectric. Upon interaction, each (p, q, ω) mode of the vacuum field generated by the moving particle is transformed into a (p', q', ω) mode of the field in the dielectric, where $p' = p/n(\omega)$, $q' = q/n(\omega)$. The components of the electric field amplitudes are transformed according to Eqs. (21). (Only the case when both the modes are homogeneous is illustrated by the figure.)

These evanescent modes will, therefore, upon interaction with the dielectric give rise to radiation. In particular consider a charged particle moving uniformly with speed V in the x direction. From (5) and (7), the spectral vector amplitudes of the vacuum electric field that the charge generates are readily shown to be given by

$$\vec{e}^{(v)}(p, q; \omega; \gtrless) = (ke/2\pi c V m)\hat{s}^{(\pm)} \times [\hat{s}^{(\pm)} \times \vec{V}]\, \delta(p - c/V), \tag{24}$$

where $\delta(x)$ is the Dirac delta function. We see that in this case the parameter p has the sharp value

$$p = p_0 \equiv c/V, \tag{25}$$

but q takes on, of course, all real values. Since the speed V is necessarily smaller than the vacuum speed of light c, p_0 exceeds unity, so that *all* the vacuum modes are now evanescent. The condition (23) for conversion of some of the evanescent modes into homogeneous ones upon interaction with the dielectric now becomes

$$1 \leqslant (c/V)^2 + q^2 \leqslant n^2(\omega). \tag{26}$$

The first inequality is automatically satisfied because $c/V > 1$; the second will be satisfied for some q values if and only if $(c/V)^2 < n^2(\omega)$, i.e., if

$$V > c/n(\omega) \equiv v(\omega), \tag{27}$$

where $v(\omega)$ is the phase velocity of the medium for frequency ω. It follows that *if the velocity of the particle exceeds the phase velocity $v(\omega)$ in the medium, those evanescent modes of frequency ω of the vacuum field for which q lies in the range*

$$-n(\omega)\{1 - [v(\omega)/V]^2\}^{1/2} \leqslant q \leqslant +n(\omega)\{1 - [v(\omega)/V]^2\}^{1/2} \tag{28}$$

(and for which p has, of course, the sharp value $p_0 = c/V$) *will, upon interaction with the dielectric, give rise to radiation.* The radiation field in the dielectric will then, in view of (18), be created by homogeneous plane waves (p', q', ω), where

$$p' = v(\omega)/V, \quad -\{1 - [v(\omega)/V]^2\}^{1/2} \leqslant q' \leqslant \{1 - [v(\omega)/V]^2\}^{1/2}. \tag{29}$$

The first relation of (29) implies that the wave normals of all the homogeneous waves of frequency ω in the dielectric that form the radiation field lie on a cone with semi-angle $\alpha = \cos^{-1}[v(\omega)/V]$ about the direction of propagation of the particle.

Equation (27) is precisely the condition for the generation of Cherenkov radiation,[5] and it is clear that the application of the general results reported in the earlier part of this note would lead to the full description of the Cherenkov field, both above and below threshold.

*Research supported by the U. S. Air Force Office of Scientific Research and the Army Research Office (Durham).

†Present address: Physics Department, University College, Dublin, Ireland.

[1]R. Asby and E. Wolf, J. Opt. Soc. Amer. 61, 52 (1971). This reference also deals with the field due to a charge moving in a dielectric, but the analysis given there is based on the phenomenological Maxwell theory and does not bring out the underlying physical model discussed in the present paper.

[2]For the discussion of the asymptotic behavior of the angular spectrum see, for example, Ref. 1.

[3]M. Born and E. Wolf, *Principles of Optics* (Pergaman, New York, 1969), § 2.4.

[4]L. Rosenfeld, *Theory of Electrons* (North-Holland, Amsterdam, 1951), Chap. VI.

[5]J. V. Jelley, *Cherenkov Radiation and Its Applications* (Pergamon, London, 1958).

Multipole expansions and plane wave representations of the electromagnetic field*

A. J. Devaney and E. Wolf

Department of Physics and Astronomy, University of Rochester, Rochester, New York 14627
(Received 31 May 1973; revised manuscript received 2 August 1973)

A new and conceptually simple derivation is presented of the multipole expansion of an electromagnetic field that is generated by a localized, monochromatic charge-current distribution. The derivation is obtained with the help of a generalized plane wave representation (known also as the angular spectrum representation) of the field. This representation contains both ordinary plane waves, and plane waves that decay exponentially in amplitude as the wave is propagated. The analysis reveals an intimate relationship between the generalized plane wave representation and the multipole expansion of the field and leads to a number of new results. In particular, new expressions are obtained for the electric and magnetic multipole moments in terms of certain components of the spatial Fourier transform of the transverse part of the current distribution. It is shown further that the electromagnetic field at all points outside a sphere that contains the charge-current distribution is completely specified by the radiation pattern (i.e., by the field in the far zone). Explicit formulas are obtained for all the multipole moments in terms of the radiation pattern.

1. INTRODUCTION

Multipole expansions of the electromagnetic field are employed extensively both in classical electrodynamics and in the quantum theory of radiation. Such expansions originated in a restrictive form in the classical theory of diffraction by a sphere, especially in the work of Clebsch,[1] Mie,[2] Debye,[3] and Bromwich,[4] as a "partial wave" expansion.[5] The first fairly general formulation is implicit in the work of Hansen.[6] Important contributions were later made by Heitler,[7] Kramers,[8] Franz,[9] Wallace,[10] Blatt and Weisskopf,[11] Bouwkamp and Casimir,[12] Nisbet,[13] and Wilcox.[14] The methods employed in some of the treatments include tensor analysis, operator calculus, and group theory, though more elementary derivations have been given.[10,12]

In a well-known paper dealing with the differential equations of mathematical physics, Whittaker[15] introduced a multipole expansion of a *source free* scalar wave field. He first expressed the solution of the homogeneous wave equation as a superposition of homogeneous plane waves. On expanding the amplitude function of the plane waves in a series of spherical harmonics he was then lead to a multipole expansion of the free field. Whittaker's analysis is conceptually very simple, and gives a clear understanding of the relationship between the representation of a free field as a superposition of plane waves and its representation in terms of multipole fields.

In the present paper, we show that Whittaker's method may be extended to electromagnetic fields generated by localized charge-current distributions. To illustrate the essential feature of the technique, we first show how a generalized plane wave expansion for a scalar field generated by a localized source distribution may be obtained. The generalized plane wave expansion contains not only ordinary (homogeneous) waves, but also waves that decay in amplitude as the wave is propagated (the so-called evanescent waves, well known from the theory of total internal reflection). Such generalized plane wave expansions have been playing an increasingly important role in recent years in optics and classical electrodynamics, and are generally known as angular spectrum representations.[16] We show that the amplitudes of all the plane waves in this representation are expressible in terms of the spatial Fourier transform, and its analytic continuation, of the source distribution. By expanding the amplitude function of the plane waves in a series of spherical harmonics we are immediately lead to the required multipole expansion. In Sec. 4 we consider the electromagnetic field generated by a localized charge-current distribution. The only essential difference arises from the fact that the amplitudes in the generalized plane wave expansion of the electromagnetic field are now vectors, and the appropriate basic set of functions in terms of which they are expanded are the vector spherical harmonics instead of the ordinary spherical harmonics. In the concluding section (Sec. 5), we show that the angular spectrum representation provides also a new insight into the well-known relationship between the multipole expansion and the so-called Debye potential representation of the electromagnetic field generated by a localized charge-current distribution.

Our analysis not only provides a new derivation of the multipole expansion of the electromagnetic field, but it also reveals an intimate relationship between the multipole expansion and the generalized plane wave representation of the field. From this fact some new results readily follow. In particular, we obtain new expressions for the electric and magnetic multipole moments in terms of certain components of the spatial Fourier transform of the transverse part of the current distribution. We also show that the electromagnetic field at all points outside a sphere that encloses the source distribution is completely specified by the radiation pattern (the far field), and we derive formulas for all the multipole moments in terms of the radiation pattern. Moreover, as we show elsewhere, our results lead readily to some interesting new theorems on properties of fields generated by localized charge-current distributions and on localized charge-current distributions that do not give rise to any radiation.[17]

2. WHITTAKER'S REPRESENTATION OF A SOURCE-FREE MONOCHROMATIC SCALAR WAVE FIELD

We begin with a brief review of Whittaker's derivation of the multipole expansion of a source-free, monochromatic scalar wave field.

Whittaker[15] showed that a wide class of solutions of the reduced wave equation (the Helmholtz equation),

$$(\nabla^2 + k^2)\psi(\mathbf{r}) = 0 \tag{2.1}$$

(k = real constant), may be expressed in the form

$$\psi(\mathbf{r}) = \frac{k}{4\pi} \int_{-\pi}^{\pi} d\beta \int_0^{\pi} d\alpha \, \sin\alpha \hat{\psi}(\mathbf{s}) e^{ik\mathbf{s}\cdot\mathbf{r}}, \tag{2.2}$$

where $\mathbf{s}$ is a real unit vector with Cartesian components

$$s_x = \sin\alpha \cos\beta, \quad s_y = \sin\alpha \sin\beta, \quad s_z = \cos\alpha. \tag{2.3}$$

The right-hand side of (2.2) expresses $\psi(\mathbf{r})$ as a superposition of homogeneous plane waves propagating in all possible directions and, hence, we will refer to it as the homogeneous plane wave expansion of $\psi(\mathbf{r})$.

Whittaker showed further, that the homogeneous plane wave expansion of the multipole field[18]

$$\Lambda_l^m(\mathbf{r}) = j_l(kr) Y_l^m(\theta, \phi), \tag{2.4}$$

where $j_l(kr)$ is the spherical Bessel function of order l and $Y_l^m(\theta, \phi)$ is the spherical harmonic of degree l and order m, is[19]

$$\Lambda_l^m(\mathbf{r}) = (-i)^l \frac{1}{4\pi} \int_{-\pi}^{\pi} d\beta \int_0^{\pi} d\alpha \, \sin\alpha \, Y_l^m(\alpha, \beta) e^{ik\mathbf{s}\cdot\mathbf{r}}. \tag{2.5}$$

Here r, θ, ϕ are the spherical polar coordinates of the field point $\mathbf{r}$, referred to the same system of axes as the unit vector $\mathbf{s}$, so that

$$x = r \sin\theta \cos\phi, \quad y = r \sin\theta \sin\phi, \quad z = r\cos\theta. \tag{2.6}$$

Returning to the general case, Whittaker expanded the plane wave amplitude function $\hat{\psi}(\mathbf{s})$ in a series of spherical harmonics,

$$\hat{\psi}(\mathbf{s}) = \sum_{l=0}^{\infty} \sum_{m=-l}^{l} (-i)^l a_l^m Y_l^m(\alpha, \beta), \tag{2.7}$$

and on substituting from (2.7) into (2.2) and using (2.5) he obtained an alternative representation of $\psi(\mathbf{r})$, namely the *multipole expansion*

$$\psi(\mathbf{r}) = k \sum_{l=0}^{\infty} \sum_{m=-l}^{l} a_l^m \Lambda_l^m(\mathbf{r}). \tag{2.8}$$

The *multipole moments* a_l^m may, of course, be expressed from (2.7) in terms of the plane wave amplitudes $\hat{\psi}(\mathbf{s})$ by using the fact that the spherical harmonics form a orthonormal set over the unit sphere. The result is

$$a_l^m = i^l \int_{-\pi}^{\pi} d\beta \int_0^{\pi} d\alpha \, \sin\alpha \, \hat{\psi}(\mathbf{s}) Y_l^{m*}(\alpha, \beta), \tag{2.9}$$

and shows that the multipole moments are, apart from the trivial factor i^l, simply the projections of the plane wave amplitudes $\hat{\psi}(\mathbf{s})$ onto the set of the spherical harmonics $Y_l^m(\alpha, \beta)$.

It should be noted that since each plane wave in (2.2), and each multipole field in (2.8), obey the Helmholtz equation (2.1) throughout the whole space, each of the two representations is a *mode expansion* of the general solution of that equation. The relations (2.7) and (2.9) establish an intimate connection between the two representations. Unfortunately, Whittaker's elegant results are of limited applicability since they are essentially existence theorems. They leave unanswered the question of how the amplitudes $\hat{\psi}(\mathbf{s})$, or the multipole moments a_l^m, are to be determined in any particular case. Moreover, the representation (2.2) [and, consequently, (2.8)] is valid only for source-free fields. Consequently, Whittaker's derivation of the multipole expansion is of limited use in electromagnetic theory where one must frequently deal with fields generated by sources that are not situated at infinity.

In this paper we will show that the insight gained from Whittaker's analysis leads readily to a new, and basically simple, derivation of the multipole expansion of wavefields generated by a localized source distribution, and also to explicit expressions for the multipole moments. The derivation is based on a generalized plane wave expansion which includes not only homogeneous plane waves, but also evanescent plane waves (i.e., plane waves that decay exponentially in amplitude in a particular direction). In Sec. 3 we consider the scalar field and in Sec. 4 we will treat the electromagnetic field.

3. MULTIPOLE EXPANSION OF A MONOCHROMATIC SCALAR WAVE FIELD GENERATED BY A LOCALIZED SOURCE DISTRIBUTION

Let us consider a real, monochromatic scalar wave field

$$\psi(\mathbf{r}, t) = \mathcal{R}(\psi(\mathbf{r}) e^{-i\omega t}),$$

generated by a source distribution

$$\rho(\mathbf{r}, t) = \mathcal{R}(\rho(\mathbf{r}) e^{-i\omega t}),$$

in the infinite free space. Here, ω is a real positive constant and $\mathcal{R}$ denotes the real part. Then $\psi(\mathbf{r})$ and $\rho(\mathbf{r})$ are related by the differential equation

$$(\nabla^2 + k^2)\psi(\mathbf{r}) = -4\pi\rho(\mathbf{r}), \tag{3.1}$$

where

$$k = \omega/c, \tag{3.2}$$

c being the velocity of propagation. We will assume that the source distribution $\rho(\mathbf{r})$ is a continuous function of position and is confined to a finite region around the origin. Hence, $\rho(\mathbf{r}) \equiv 0$ when $r \equiv |\mathbf{r}| > R$, where R is some real constant.

The field $\psi(\mathbf{r})$ generated by the source distribution $\rho(\mathbf{r})$ is identified with that particular solution of Eq. (3.1) which behaves at infinity as an outgoing spherical wave, and is well known to be given by

$$\psi(\mathbf{r}) = \int_{r' \le R} \rho(\mathbf{r}') \frac{e^{ik|\mathbf{r}-\mathbf{r}'|}}{|\mathbf{r}-\mathbf{r}'|} d^3r'. \tag{3.3}$$

Now, the spherical wave which enters as the kernel of the integral transform (3.3) may be expressed in the following form due to Weyl:[20]

$$\frac{e^{ik|\mathbf{r}-\mathbf{r}'|}}{|\mathbf{r}-\mathbf{r}'|} = \frac{ik}{2\pi} \int_{-\pi}^{\pi} d\beta \int_{C^\pm} d\alpha \, \sin\alpha \, e^{ik\mathbf{s}\cdot(\mathbf{r}-\mathbf{r}')}, \tag{3.4}$$

where $\mathbf{s} = \mathbf{s}(\alpha, \beta)$ is again a unit vector with Cartesian components given by (2.3). However, the polar angle α is no longer necessarily real, but takes on all values on the contours C^+ and C^- in the complex α plane shown in Fig. 1. It is understood that in (3.4) the contour C^+ applies when $z - z' > 0$, and the contour C^- applies when $z - z' < 0$, z and z', being, of course, the Cartesian z coordinates of the field point $\mathbf{r}$ and the integration point $\mathbf{r}'$, respectively.

On substituting from (3.4) into (3.3), and on interchanging the order of integration, we obtain the following representation of $\psi(\mathbf{r})$:[21]

$$\psi(\mathbf{r}) = \frac{ik}{2\pi} \int_{-\pi}^{\pi} d\beta \int_{C^\pm} d\alpha \, \sin\alpha \, \hat{\psi}(\mathbf{s}) e^{ik\mathbf{s}\cdot\mathbf{r}}, \tag{3.5}$$

where the spectral amplitudes $\hat{\psi}(\mathbf{s})$ are given in terms of the source distribution by the formula

$$\hat{\psi}(\mathbf{s}) = \int_{r' \le R} \rho(\mathbf{r}') e^{-ik\mathbf{s}\cdot\mathbf{r}'} d^3r'. \tag{3.6}$$

In (3.5), the contour C^+ is used when $z > R$ and C^- when $z < -R$. It is not difficult to show that the interchange of the order of integration in deriving (3.5) is justified provided $|z| > R$, i.e., provided the field point is outside a strip bounded by planes parallel to the z plane and containing the source. However, since the orientation of our coordinate system is arbitrary, such a representation may be used to represent the field outside *any* strip bounded by two parallel planes tangential to the sphere of radius R, centered at the origin.

The representation (3.5), just like Whittaker's representation (2.2) of the source-free field, expresses $\psi(\mathbf{r})$ as a superposition of plane waves. However, since in (3.5) the unit vector $\mathbf{s}$ is real on a portion of the contour of integration and complex on the rest of the contour, (3.5) includes both *homogeneous* plane waves (corresponding to real $\mathbf{s}$) and *inhomogeneous* (called also evanescent) plane waves (corresponding to complex[22] $\mathbf{s}$). Integrals of this type are said to represent the wave field in the form of an *angular spectrum of plane waves*.

The spectral amplitudes $\hat{\psi}(\mathbf{s})$ as defined in (3.6) are intimately related to the threefold Fourier transform $\tilde{\rho}(\mathbf{K})$ of the source distribution $\rho(\mathbf{r})$:

$$\tilde{\rho}(\mathbf{K}) = \int_{r' \leq R} \rho(\mathbf{r}') e^{-i\mathbf{K}\cdot\mathbf{r}'} d^3r'. \tag{3.7}$$

On comparing (3.6) with (3.7), we conclude that $\hat{\psi}(\mathbf{s})$ is simply $\tilde{\rho}(\mathbf{K})$ with $\mathbf{K} = k\mathbf{s}$; i.e.,

$$\hat{\psi}(\mathbf{s}) = \tilde{\rho}(k\mathbf{s}). \tag{3.8}$$

In the definition of the Fourier transform $\tilde{\rho}(\mathbf{K})$, the variable $\mathbf{K}$ conjugate to $\mathbf{r}'$ is, of course, real. However, since the integral on the right-hand side of (3.7) extends over a finite domain (because, by hypothesis, $\rho \equiv 0$ when $r' > R$) and, moreover, since ρ was assumed to be continuous, it follows that $\tilde{\rho}(\mathbf{K})$ is the boundary value, on the real K_x, K_y, K_z axes, of an entire analytic function of three complex variables[23] K_x, K_y, K_z. Consequently, the relation (3.8) is valid for all (real and complex) unit vectors $\mathbf{s}$ and, moreover, shows that $\hat{\psi}$ is the boundary value (on the contours C^+ and C^-) of an entire analytic function of three complex variables.

Following Whittaker's treatment of the free field, we next expand the spectral amplitudes $\hat{\psi}(\mathbf{s})$ into a series of spherical harmonics, viz.,

$$\hat{\psi}(\mathbf{s}) = \sum_{l=0}^{\infty} \sum_{m=-l}^{l} (-i)^l a_l^m Y_l^m(\alpha, \beta), \tag{3.9}$$

where the expansion coefficients (multipole moments) a_l^m are simply the projections of $\hat{\psi}(\mathbf{s})$ onto the spherical harmonics:

$$a_l^m = i^l \int_{-\pi}^{\pi} d\beta \int_0^{\pi} d\alpha \, \sin\alpha \, \hat{\psi}(\mathbf{s}) Y_l^{m*}(\alpha, \beta). \tag{3.10}$$

It is important to note that although the multipole moments, defined by (3.10), depend explicitly only on those spectral amplitudes $\hat{\psi}(\mathbf{s})$ which are associated with real $\mathbf{s}$ (i.e., those corresponding to homogeneous plane waves in the angular spectrum representation), the expansion (3.9) is valid for all unit vectors associated with the complex contours $C^\pm$; this result is a consequence of the fact that $\hat{\psi}(\mathbf{s})$ is the boundary value of an entire analytic function.

We can also readily express the multipole moments a_l^m in terms of the source distribution $\rho(\mathbf{r})$. To do this, we simply substitute (3.6) into (3.10) and interchange the order of integration. We then obtain the following expressions for the multipole moments:

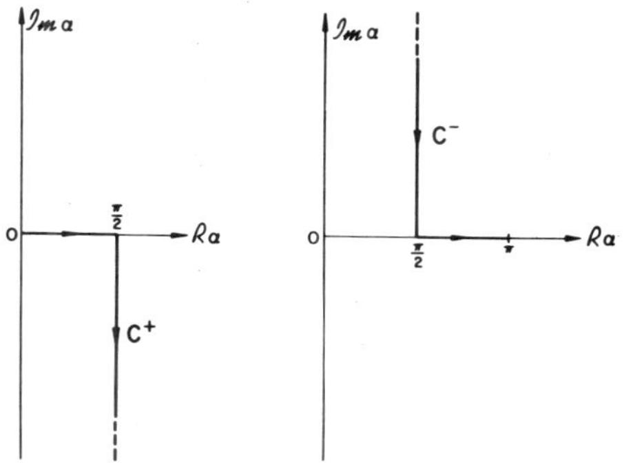

FIG. 1. The α-contours of integration C^+ and C^-.

$$a_l^m = \int_{r' \leq R} d^3r' \rho(\mathbf{r}') i^l \int_{-\pi}^{\pi} d\beta \int_0^{\pi} d\alpha \, \sin\alpha \, Y_l^{m*}(\alpha, \beta) e^{-ik\mathbf{s}\cdot\mathbf{r}'},$$
$$= 4\pi \int_{r' \leq R} \rho(\mathbf{r}') \Lambda_l^{m*}(\mathbf{r}') d^3r'. \tag{3.11}$$

In deriving (3.11) we have made use of (2.5).

Next, we substitute the expansion (3.9) into the representation (3.5) of $\psi(\mathbf{r})$ and interchange the order of integration and summation. We than obtain, in analogy with (2.8), the following series expansion for $\psi(\mathbf{r})$:

$$\psi(\mathbf{r}) = k \sum_{l=0}^{\infty} \sum_{l=-m}^{m} a_l^m \Pi_l^m(\mathbf{r}), \tag{3.12}$$

where

$$\Pi_l^m(\mathbf{r}) = (-i)^l \frac{i}{2\pi} \int_{-\pi}^{\pi} d\beta \int_{C^\pm} d\alpha \, \sin\alpha \, Y_l^m(\alpha, \beta) e^{ik\mathbf{s}\cdot\mathbf{r}}. \tag{3.13}$$

Now, it is not difficult to show that the expression on the right-hand side of (3.13) is precisely the angular spectrum representation of the scalar multipole field of degree l and order m, i.e.,

$$\Pi_l^m(\mathbf{r}) = h_l^{(+)}(kr) Y_l^m(\theta, \phi), \tag{3.14}$$

where (r, θ, ϕ) are the spherical polar coordinates of the field point $\mathbf{r}$ and $h_l^{(+)}$ is the spherical Hankel function[19] of the first kind of order l. This result, which is the counterpart of Whittaker's result expressed by (2.4), appears to have been first stated by Erdélyi,[24] and is proved in Appendix A of the present paper. Thus, we see that (3.12) is the *multipole expansion* of $\psi(\mathbf{r})$ and a_l^m are the corresponding *multipole moments*.

Since the angular spectrum representation (3.5), from which the multipole expansion (3.12) was derived, converges only throughout the two half-spaces $z > R$ and $z < -R$, it would appear that this expansion is valid only at field points $\mathbf{r}$ situated in these two regions. However, as is well known, the multipole expansion (3.12) represents the field correctly at every point outside the source region, i.e., for all values of $r > R$. For the sake of completeness, this result is verified in Appendix C.

The angular spectrum representation also yields readily the far zone approximation for $\psi(\mathbf{r})$. If we let $kr \to \infty$ in a fixed direction specified by a unit vector

$$\mathbf{u}_r = \mathbf{r}/r, \tag{3.15}$$

we have

$$|\mathbf{r} - \mathbf{r}'| \sim r - \mathbf{r}' \cdot \mathbf{u}_r, \tag{3.16}$$

and Eq. (3.3) then gives

$$\psi(r\mathbf{u}_r) \sim \tilde{\rho}(k\mathbf{u}_r) e^{ikr}/r \quad (kr \to \infty), \tag{3.17}$$

where, as before, $\tilde{\rho}(\mathbf{K})$ is the Fourier transform of $\rho(\mathbf{r})$ [Eq. (3.7)]. Now according to (3.8), $\tilde{\rho}(k\mathbf{u}_r)$ is precisely the spectral amplitude $\hat{\psi}(\mathbf{u}_r)$ of ψ, so that (3.17) may be expressed in the form

$$\psi(r\mathbf{u}_r) \sim \hat{\psi}(\mathbf{u}_r)(e^{ikr}/r) \quad (kr \to \infty). \tag{3.18}$$

This formula shows that the *radiation pattern of the field* is precisely given by the spectral amplitude function $\hat{\psi}(\mathbf{u}_r)$ for real unit vectors $\mathbf{u}_r$ (corresponding to points on the contours C^+ and C^- that coincide with portions of the *real* α axis). In other words, the radiation pattern of the field in any particular direction $\mathbf{u}_r$ is equal to the complex amplitude of a certain plane wave in the angular spectrum representation of the field, namely the plane wave propagated in the direction $\mathbf{u}_r$. This result may be understood in physical terms as follows: The angular spectrum representation (3.5) expresses $\psi(\mathbf{r})$ at every point $\mathbf{r}$ outside the source region $r \leqslant R$ as a superposition of plane waves, both homogeneous and evanescent ones. As the field point $\mathbf{r} = r\mathbf{u}_r$ gradually moves further away from the source region in any fixed direction $\mathbf{u}_r$, the evanescent waves gradually die out because of their exponential amplitude decay. The contribution of the homogeneous waves gradually decreases also, but for a different reason, namely because they progressively cancel each other out by destructive phase interference. In the asymptotic limit as $kr \to \infty$, only the single homogeneous plane wave in the angular spectrum that is propagated in the direction $\mathbf{u}_r$ survives, and it is this wave that determines the behavior of $\psi(\mathbf{r})$ in the far zone, as Eq. (3.18) shows. This argument may be made more rigorous with the help of the principle of stationary phase.[25]

If in Eq. (3.18) we express $\hat{\psi}(\mathbf{u}_r)$ in terms of the multipole moments [Eq. (3.9)] we obtain, at once, the asymptotic approximation for ψ, valid in the far zone, (i.e., as $kr \to \infty$):

$$\psi(r\mathbf{u}_r) \sim \frac{e^{ikr}}{r} \sum_{l=0}^{\infty} \sum_{m=-l}^{l} (-i)^l a_l^m Y_l^m(\theta,\phi) \quad (kr \to \infty), \tag{3.19}$$

where (θ, ϕ) are the spherical polar coordinates of the unit direction vector $\mathbf{u}_r$.

Before concluding this section it is worthwhile to stress the following points: The angular spectrum representation (3.5), and the multipole expansion (3.12), are *mode expansions* in the sense that they express the field $\psi(\mathbf{r})$ in terms of certain elementary fields (plane wave fields and multipole fields, respectively), each of which satisfies the same equation as does $\psi(\mathbf{r})$ outside the source region, namely the Helmholtz equation $(\nabla^2 + k^2)\psi = 0$. The range of validity of each of the two expansions is different. The angular spectrum expansion represents $\psi(\mathbf{r})$ outside the strip $|z| \leqslant R$, the multipole expansion represents it outside the sphere $r \leqslant R$. The expansion coefficients in the two representations are related by Eqs. (3.9) and (3.10).

4. MULTIPOLE EXPANSION OF A MONOCHROMATIC ELECTROMAGNETIC FIELD GENERATED BY A LOCALIZED CHARGE-CURRENT DISTRIBUTION

Let us now turn our attention to the electromagnetic field. Just as in the scalar case, two situations are to be distinguished, namely when the field is source free and when it is generated by a localized source distribution. We will discuss here only the second case.[26]

We consider then a real, monochromatic, electromagnetic field

$$\mathbf{E}(\mathbf{r},t) = \mathcal{R}(\mathbf{E}(\mathbf{r})e^{-i\omega t}), \quad \mathbf{H}(\mathbf{r},t) = \mathcal{R}(\mathbf{H}(\mathbf{r})e^{-i\omega t}),$$

generated by a charge-current distribution[27]

$$\rho(\mathbf{r},t) = \mathcal{R}(\rho(\mathbf{r})e^{-i\omega t}), \quad \mathbf{j}(\mathbf{r},t) = \mathcal{R}(\mathbf{j}(\mathbf{r})e^{-i\omega t}),$$

in the infinite free space. As before, ω is a real positive constant and $\mathcal{R}$ denotes the real part. We assume that $\rho(\mathbf{r})$ and $\mathbf{j}(\mathbf{r})$ are continuous and continuously differentiable functions of position and vanish identically when $r > R$, where R is some real constant. From the Maxwell equations it readily follows that $\mathbf{E}(\mathbf{r})$ and $\mathbf{H}(\mathbf{r})$ satisfy (in Gaussian system of units) the equations[28]

$$(\nabla^2 + k^2)\mathbf{E}(\mathbf{r}) = -4\pi[i(k/c)\mathbf{j}(\mathbf{r}) - \nabla\rho(\mathbf{r})], \tag{4.1a}$$

$$(\nabla^2 + k^2)\mathbf{H}(\mathbf{r}) = -4\pi[(1/c)\nabla \times \mathbf{j}(\mathbf{r})]. \tag{4.1b}$$

The fields $\mathbf{E}, \mathbf{H}$ generated by the charge-current distribution $\rho, \mathbf{j}$ are identified with those particular solutions of Eqs. (4.1a) and (4.1b) that behave at infinity as outgoing spherical waves.

It is clear from Eqs. (4.1) that each Cartesian component of the electromagnetic field vectors $\mathbf{E}, \mathbf{H}$ satisfy inhomogeneous Helmholtz equations of the form (3.1). Consequently, we may apply the results established in the preceding section to each of the Cartesian components. In particular, it follows by making use of (3.5) and (3.6) that the field vectors have the following angular spectrum representations,[29] valid throughout the two half-spaces $z < -R$ and $z > R$:

$$\mathbf{E}(\mathbf{r}) = \frac{ik}{2\pi}\int_{-\pi}^{\pi} d\beta \int_{C^\pm} d\alpha \, \sin\alpha \, \hat{\mathbf{E}}(\mathbf{s}) e^{ik\mathbf{s}\cdot\mathbf{r}}, \tag{4.2a}$$

$$\mathbf{H}(\mathbf{r}) = \frac{ik}{2\pi}\int_{-\pi}^{\pi} d\beta \int_{C^\pm} d\alpha \, \sin\alpha \, \hat{\mathbf{H}}(\mathbf{s}) e^{ik\mathbf{s}\cdot\mathbf{r}}, \tag{4.2b}$$

where the spectral amplitude vectors $\hat{\mathbf{E}}(\mathbf{s})$ and $\hat{\mathbf{H}}(\mathbf{s})$ are given by

$$\hat{\mathbf{E}}(\mathbf{s}) = \int_{r' \leq R} [(ik/c)\mathbf{j}(\mathbf{r}') - \nabla\rho(\mathbf{r}')] e^{-ik\mathbf{s}\cdot\mathbf{r}'} d^3r', \tag{4.3a}$$

$$\hat{\mathbf{H}}(\mathbf{s}) = \int_{r' \leq R} [(1/c)\nabla \times \mathbf{j}(\mathbf{r}')] e^{-ik\mathbf{s}\cdot\mathbf{r}'} d^3r'. \tag{4.3b}$$

We may express $\hat{\mathbf{E}}(\mathbf{s})$ and $\hat{\mathbf{H}}(\mathbf{s})$ in simpler forms by introducing the three-dimensional Fourier transforms of $\mathbf{j}$ and ρ:

$$\tilde{\mathbf{j}}(\mathbf{K}) = \int_{r' \leq R} \mathbf{j}(\mathbf{r}') e^{-i\mathbf{K}\cdot\mathbf{r}'} d^3r', \tag{4.4a}$$

$$\tilde{\rho}(\mathbf{K}) = \int_{r' \leq R} \rho(\mathbf{r}') e^{-i\mathbf{K}\cdot\mathbf{r}'} d^3r'. \tag{4.4b}$$

Then (4.3a) and (4.3b) readily give the following expressions for the spectral amplitudes:

$$\hat{\mathbf{E}}(\mathbf{s}) = -(i/c)k\mathbf{s} \times [\mathbf{s} \times \tilde{\mathbf{j}}(k\mathbf{s})], \tag{4.5a}$$

$$\hat{\mathbf{H}}(\mathbf{s}) = (i/c)k\mathbf{s} \times \tilde{\mathbf{j}}(k\mathbf{s}). \tag{4.5b}$$

In deriving Eq. (4.5a) from Eq. (4.3a) we made use of the equation $\nabla \cdot \mathbf{j}(\mathbf{r}) - ick\rho(\mathbf{r}) = 0$ which expresses the conservation of the charge. From Eqs. (4.5) it follows that

$$\hat{\mathbf{E}}(\mathbf{s}) = -\mathbf{s} \times \hat{\mathbf{H}}(\mathbf{s}), \tag{4.6a}$$

$$\mathbf{s} \cdot \hat{\mathbf{E}}(\mathbf{s}) = \mathbf{s} \cdot \hat{\mathbf{H}}(\mathbf{s}) = 0. \tag{4.6b}$$

Consequently, for each $\mathbf{s}$, the terms in the integrands of (4.2a) and (4.2b) are plane waves that satisfy the homogeneous Maxwell equations for a monochromatic field with the frequency $\omega = kc$ throughout the whole space. Thus, the angular spectrum representation (4.2) is, throughout its domain of validity ($|z| > R$), a mode expansion of the electromagnetic field. We note in passing that, in analogy with the scalar case, the spectral amplitudes $\hat{\mathbf{E}}(\mathbf{s})$ and $\hat{\mathbf{H}}(\mathbf{s})$ are, according to Eqs. (4.5), simply related to the spatial Fourier transform, and its analytic continuation, of the current distribution $\mathbf{j}(\mathbf{r})$.

We could now follow the procedure employed in Sec. 3 to expand $\hat{\mathbf{E}}(\mathbf{s})$ and $\hat{\mathbf{H}}(\mathbf{s})$ into a series of spherical harmonics with constant (vector) coefficients. If these expansions were then substituted into Eqs. (4.2), and use was made of Eq. (3.13), we would obtain expansions of the two field vectors as series of the *scalar* multipole fields $\Pi_l^m(\mathbf{r})$. Unfortunately, although such expansions have been discussed in the literature[30] they are not very useful. One reason for this is that the individual terms in the expansion need not have a vanishing divergence and, consequently, such a representation will not provide a mode expansion of the electromagnetic field outside the source region. It is well known that the proper generalization, which is a mode expansion, is an expansion in terms of so-called *electromagnetic multipole fields* [defined by Eqs. (4.15) below], these being the appropriate electromagnetic vector analogs of the scalar multipole fields.

To obtain the expansions for $\mathbf{E}(\mathbf{r})$ and $\mathbf{H}(\mathbf{r})$ in terms of the electromagnetic multipole fields, we first expand the spectral amplitudes $\hat{\mathbf{E}}(\mathbf{s})$ and $\hat{\mathbf{H}}(\mathbf{s})$ in terms of the *vector* spherical harmonics. A vector spherical harmonic $\mathbf{Y}_l^m(\alpha,\beta)$ of degree l and order m may be defined in terms of the ordinary spherical harmonics $Y_l^m(\alpha,\beta)$ by means of the formula

$$\mathbf{Y}_l^m(\alpha,\beta) = \mathcal{L}_s Y_l^m(\alpha,\beta), \tag{4.7}$$

where $\mathcal{L}_s$ is the "orbital angular momentum operator:

$$\mathcal{L}_s = -ik\mathbf{s}\times\nabla_{k\mathbf{s}} = -i\left(\mathbf{u}_\beta \frac{\partial}{\partial\alpha} - \frac{1}{\sin\alpha}\mathbf{u}_\alpha \frac{\partial}{\partial\beta}\right). \tag{4.8}$$

Here $\nabla_{k\mathbf{s}}$ denotes the gradient operator in $k\mathbf{s}$ space, $\mathbf{u}_\alpha$, $\mathbf{u}_\beta$ being unit vectors in the positive α and β directions, respectively. A discussion of the main properties of these functions can be found, in Refs. 11 and 31. Here, we only remark that the vector spherical harmonics $\mathbf{Y}_l^m$ are everywhere tangent to the unit sphere [i.e., $\mathbf{s}\cdot\mathbf{Y}_l^m(\alpha,\beta) = 0$], that they form an orthogonal set in the sense that

$$\int_{-\pi}^{\pi} d\beta \int_0^{\pi} d\alpha \sin\alpha\, \mathbf{Y}_l^{m*}(\alpha,\beta)\cdot\mathbf{Y}_{l'}^{m'}(\alpha,\beta) = l(l+1)\,\delta_{l,l'}\,\delta_{m,m'}, \tag{4.9}$$

and that they, *together with the associated functions* $\mathbf{s}\times\mathbf{Y}_l^m(\alpha,\beta)$, form a complete orthogonal basis[32] for all well-behaved vector functions $\mathbf{F}(\mathbf{s})$ defined on the unit sphere $s^2 = 1$ and tangential to it (i.e., such that $\mathbf{s}\cdot\mathbf{F}(\mathbf{s}) = 0$). Thus, in particular, $\hat{\mathbf{E}}(\mathbf{s})$ and $\hat{\mathbf{H}}(\mathbf{s})$ defined by Eqs. (4.5) may each be expanded in terms of these two sets of vector functions and we have, in analogy with (3.9):

$$\hat{\mathbf{E}}(\mathbf{s}) = \sum_{l=1}^{\infty}\sum_{m=-l}^{l} (-i)^l [a_l^m\, \mathbf{s}\times\mathbf{Y}_l^m(\alpha,\beta) + b_l^m \mathbf{Y}_l^m(\alpha,\beta)], \tag{4.10a}$$

$$\hat{\mathbf{H}}(\mathbf{s}) = \sum_{l=1}^{\infty}\sum_{m=-l}^{l} (-i)^l [-a_l^m \mathbf{Y}_l^m(\alpha,\beta) + b_l^m\, \mathbf{s}\times\mathbf{Y}_l^m(\alpha,\beta)], \tag{4.10b}$$

where we have made use of the relation (4.6a). The factor $(-i)^l$ is included in these expansions to lead to the conventional definition of the expansion coefficients (the multipole moments). The l summations begin now with $l = 1$ rather than with $l = 0$, since there are no vector spherical harmonics of zero degree.

By making use of the orthogonality relation (4.9) we may express the expansion coefficients a_l^m and b_l^m in terms of the spectral amplitudes $\hat{\mathbf{E}}(\mathbf{s})$ and $\hat{\mathbf{H}}(\mathbf{s})$ of the angular spectrum representation (4.2). The result is

$$a_l^m = -\frac{i^l}{l(l+1)} \int_{-\pi}^{\pi} d\beta \int_0^{\pi} d\alpha\, \sin\alpha\, \hat{\mathbf{H}}(\mathbf{s})\cdot\mathbf{Y}_l^{m*}(\alpha,\beta), \tag{4.11a}$$

$$b_l^m = \frac{i^l}{l(l+1)} \int_{-\pi}^{\pi} d\beta \int_0^{\pi} d\alpha\, \sin\alpha\, \hat{\mathbf{E}}(\mathbf{s})\cdot\mathbf{Y}_l^{m*}(\alpha,\beta). \tag{4.11b}$$

We may also express a_l^m and b_l^m in terms of the Fourier transform of the current density itself by substituting into (4.11) from (4.5). We then obtain the following expressions[33] for a_l^m and b_l^m:

$$a_l^m = -\frac{i^{l+1}}{l(l+1)}\left(\frac{k}{c}\right) \int_{-\pi}^{\pi} d\beta \int_0^{\pi} d\alpha\, \sin\alpha [\mathbf{s}\times\tilde{\mathbf{j}}(k\mathbf{s})]\cdot\mathbf{Y}_l^{m*}(\alpha,\beta), \tag{4.12a}$$

$$b_l^m = -\frac{i^{l+1}}{l(l+1)}\left(\frac{k}{c}\right) \int_{-\pi}^{\pi} d\beta \int_0^{\pi} d\alpha\, \sin\alpha \{\mathbf{s}\times[\mathbf{s}\times\tilde{\mathbf{j}}(k\mathbf{s})]\}\cdot\mathbf{Y}_l^{m*}(\alpha,\beta). \tag{4.12b}$$

Again, by analogy with the procedure that we employed in our treatment of the scalar case in Sec. 3, we substitute the expansions (4.10) of the spectral amplitude vectors $\hat{\mathbf{E}}(\mathbf{s})$ and $\hat{\mathbf{H}}(\mathbf{s})$ into the angular spectrum representations (4.2) of the fields, interchange the order of integration and summation and we then obtain the following series expansions for $\mathbf{E}(\mathbf{r})$ and $\mathbf{H}(\mathbf{r})$:

$$\mathbf{E}(\mathbf{r}) = \sum_{l=1}^{\infty}\sum_{m=-l}^{l} [a_l^m \mathbf{E}_{lm}^e(\mathbf{r}) + b_l^m \mathbf{E}_{lm}^h(\mathbf{r})], \tag{4.13a}$$

$$\mathbf{H}(\mathbf{r}) = \sum_{l=1}^{\infty}\sum_{m=-l}^{l} [a_l^m \mathbf{H}_{lm}^e(\mathbf{r}) + b_l^m \mathbf{H}_{lm}^h(\mathbf{r})], \tag{4.13b}$$

where

$$\mathbf{E}_{lm}^e(\mathbf{r}) = \mathbf{H}_{lm}^h(\mathbf{r}) = (-i)^l \frac{ik}{2\pi} \int_{-\pi}^{\pi} d\beta \int_{C\pm} d\alpha\, \sin\alpha [\mathbf{s}\times\mathbf{Y}_l^m(\alpha,\beta)] e^{ik\mathbf{s}\cdot\mathbf{r}}, \tag{4.14a}$$

$$\mathbf{E}_{lm}^h(\mathbf{r}) = -\mathbf{H}_{lm}^e(\mathbf{r}) = (-i)^l \frac{ik}{2\pi} \int_{-\pi}^{\pi} d\beta \int_{C\pm} d\alpha\, \sin\alpha\, \mathbf{Y}_l^m(\alpha,\beta)] e^{ik\mathbf{s}\cdot\mathbf{r}} \tag{4.14b}$$

We show in Appendix B that the integrals on the right-hand side of Eqs. (4.14) are the usual *electromagnetic multipole fields*, i.e.,

$$\mathbf{E}_{lm}^e(\mathbf{r}) = \mathbf{H}_{lm}^h(\mathbf{r}) = \nabla\times\{\nabla\times[\mathbf{r}\,\Pi_l^m(\mathbf{r})]\}, \tag{4.15a}$$

$$\mathbf{E}_{lm}^h(\mathbf{r}) = -\mathbf{H}_{lm}^e(\mathbf{r}) = ik\nabla\times[\mathbf{r}\,\Pi_l^m(\mathbf{r})], \tag{4.15b}$$

where $\Pi_l^m(\mathbf{r})$ is, as before, the scalar multipole field defined by Eq. (3.14). $\mathbf{E}_{lm}^e(\mathbf{r}), \mathbf{H}_{lm}^e(\mathbf{r})$ are the fields generated by an electric multipole and the fields $\mathbf{E}_{lm}^h(\mathbf{r})$, $\mathbf{H}_{lm}^h(\mathbf{r})$ are those generated by a magnetic multipole, each of degree l and order m. Thus, we see that Eqs. (4.13) are the usual *multipole expansions* of the electric and magnetic fields generated by a localized charge-current distribution, the coefficients a_l^m and b_l^m being the electric and magnetic *multipole moments*, respectively. As is well known, each electromagnetic multipole field satisfies the homogeneous Maxwell equations (for a monochromatic field of frequency $\omega = kc$) throughout the whole space, with the exception of the origin. Thus, Eqs. (4.13) are true *mode expansions* of the field **E**, **H** outside the source region. Although we have derived the multipole expansion (4.13) from the angular spectrum representation, which is valid only when $|z| > R$, it is, in fact, valid everywhere outside the source region, i.e., for all $r > R$. The reason for this is essentially the same as in the scalar case. (See Appendix C.)

We see now that the multipole expansion of the electromagnetic field generated by a localized charge-current distribution arises naturally when the spectral amplitudes of the angular spectrum representation of the field are expressed as a series of vector spherical harmonics [Eqs. (4.10)]. The multipole moments are simply the projections of these spectral amplitudes onto the set of vector spherical harmonics [Eqs. (4.11)]. The multipole moments are also seen to be projections of certain components, (namely, those for which $|\mathbf{K}| = k = \omega/c$), of the Fourier transform of the transverse part of the current distribution onto the vector spherical harmonics $\mathbf{Y}_l^m(\alpha,\beta)$ and $\mathbf{s}\times\mathbf{Y}_l^m(\alpha,\beta)$ [Eqs. (4.12a) and (4.12b)].

The formulas (4.12), which express the multipole moments in terms of the transverse components of the Fourier transform $\tilde{\mathbf{j}}$ of the current density appear to be new. It is not difficult to express the multipole moments in a more conventional form involving the current density **j** directly. For this purpose, we first substitute into Eqs. (4.12) from (4.4) and obtain the formulas

$$a_l^m = \frac{i^{l+1}}{l(l+1)}\left(\frac{k}{c}\right)\int_{r'\le R} d^3r'\,\mathbf{j}(\mathbf{r}') \cdot \int_{-\pi}^{\pi} d\beta \int_0^{\pi} d\alpha\,\sin\alpha\,\mathbf{s}\times\mathbf{Y}_l^{m*}(\alpha,\beta)\,e^{-ik\mathbf{s}\cdot\mathbf{r}'}, \tag{4.16a}$$

$$b_l^m = \frac{i^{l+1}}{l(l+1)}\left(\frac{k}{c}\right)\int_{r'\le R} d^3r'\,\mathbf{j}(\mathbf{r}') \cdot \int_{-\pi}^{\pi} d\beta \int_0^{\pi} d\alpha\,\sin\alpha\,\mathbf{Y}_l^{m*}(\alpha,\beta)\,e^{-ik\mathbf{s}\cdot\mathbf{r}'}. \tag{4.16b}$$

In writing down the expression (4.16b) we made use of the relation $\mathbf{s}\times(\mathbf{s}\times\mathbf{Y}_l^m) = -\mathbf{Y}_l^m$, which follows from the fact that the vector spherical harmonics are everywhere tangential to the unit sphere $s^2 = 1$. Except for the nature of the α-contour of integration, the (α,β) integrals in Eqs. (4.16) are very similar to those appearing in the angular spectrum representation of the electromagnetic multipole fields [See Appendix B, Eqs. (B12) and (B13)]. In fact, if one carries out a strictly similar calculation as given in Appendix B, except that instead of the multipole field $\Pi_l^m(\mathbf{r}) = h_l^{(+)}(kr)Y_l^m(\theta,\phi)$ with source at the origin $r = 0$ one considers the source free multipole field $\Lambda_l^m(\mathbf{r}) = j_l(kr)Y_l^m(\theta,\phi)$, then comparison of Eqs. (3.13) with (2.5) shows that, except for a trivial proportionality factor $-i/2$ (which arises from the difference between coefficients on the right-hand sides of the two equations), the only change in the calculation will be the replacement of the α-contours $C^{\pm}$ by the real contour $0 \le \alpha \le \pi$. One then obtains, in place of Eqs. (B12) and (B13), the identities

$$ik\nabla\times[\mathbf{r}\,\Lambda_l^m(\mathbf{r})] = (-i)^l\frac{k}{4\pi}\int_{-\pi}^{\pi} d\beta\int_0^{\pi} d\alpha\,\sin\alpha\,\mathbf{Y}_l^m(\alpha,\beta)\,e^{ik\mathbf{s}\cdot\mathbf{r}}, \tag{4.17a}$$

$$\nabla\times\{\nabla\times[\mathbf{r}\,\Lambda_l^m(\mathbf{r})]\} = (-i)^l\frac{k}{4\pi}\int_{-\pi}^{\pi} d\beta\int_0^{\pi} d\alpha\,\sin\alpha\,\mathbf{s}\times\mathbf{Y}_l^m(\alpha,\beta)\,e^{ik\mathbf{s}\cdot\mathbf{r}}. \tag{4.17b}$$

With the help of these two identities, one readily obtains from Eqs. (4.16) the following, well-known expressions for the multipole moments:

$$a_l^m = \frac{4\pi i}{l(l+1)}\left(\frac{1}{c}\right)\int_{r'\le R}\mathbf{j}(\mathbf{r}')\cdot\{\nabla\times[\nabla\times(\mathbf{r}'\Lambda_l^{m*}(\mathbf{r}'))]\}\,d^3r', \tag{4.18a}$$

$$b_l^m = \frac{4\pi}{l(l+1)}\left(\frac{k}{c}\right)\int_{r'\le R}\mathbf{j}(\mathbf{r}')\cdot\{\nabla\times[\mathbf{r}'\Lambda_l^{m*}(\mathbf{r}')]\}\,d^3r'. \tag{4.18b}$$

Just as in the scalar case, we may readily obtain from the angular spectrum representation the far zone approximation for the electromagnetic fields $\mathbf{E}(\mathbf{r})$ and $\mathbf{H}(\mathbf{r})$. In fact, if we apply the result expressed by Eq. (3.18) to each of the Cartesian components of $\mathbf{E}(\mathbf{r})$ and $\mathbf{H}(\mathbf{r})$ separately, we see at once that

$$\mathbf{E}(r\mathbf{u}_r) \sim \hat{\mathbf{E}}(\mathbf{u}_r)(e^{ikr}/r) \quad (kr\to\infty), \tag{4.19a}$$

$$\mathbf{H}(r\mathbf{u}_r) \sim \hat{\mathbf{H}}(\mathbf{u}_r)(e^{ikr}/r) \quad (kr\to\infty), \tag{4.19b}$$

for any real direction specified by the unit vector $\mathbf{u}_r$. The physical significance of these two relations may be readily understood from similar considerations as given in connection with the corresponding scalar equation (3.18).

If in Eqs. (4.19) we express the spectral amplitudes $\hat{\mathbf{E}}$ and $\hat{\mathbf{H}}$ in terms of the multipole moments by means of Eq. (4.10), we obtain at once the usual asymptotic approximations for the electromagnetic field, valid in the far zone (i.e., as $kr\to\infty$):

$$\mathbf{E}(r\mathbf{u}_r) \sim \frac{e^{ikr}}{r}\sum_{l=1}^{\infty}\sum_{m=-l}^{l}(-i)^l[a_l^m\mathbf{u}_r\times\mathbf{Y}_l^m(\theta,\phi) + b_l^m\mathbf{Y}_l^m(\theta,\phi)], \tag{4.20a}$$

$$\mathbf{H}(r\mathbf{u}_r) \sim \frac{e^{ikr}}{r}\sum_{l=1}^{\infty}\sum_{m=-l}^{l}(-i)^l[-a_l^m\mathbf{Y}_l^m(\theta,\phi) + b_l^m\mathbf{u}_r\times\mathbf{Y}_l^m(\theta,\phi)]. \tag{4.20b}$$

Here, (θ,ϕ) are again the spherical polar coordinates of the unit direction vector $\mathbf{u}_r$.

We note, in passing, that since, according to Eq. (4.19a), the radiation pattern of the field [i.e., the vector function of $\mathbf{u}_r$ that multiplies the scalar field $\exp(ikr)/r$ in the asymptotic expansion of $\mathbf{E}(r\mathbf{u}_r)$ as $kr\to\infty$], is given precisely by the spectral amplitude vector $\hat{\mathbf{E}}(\mathbf{u}_r)$ for all real unit direction vectors $\mathbf{u}_r$, Eqs. (4.11), together with Eqs. (4.6), may also be interpreted as giving all the multipole moments in terms of the radiation pattern. Thus, we see, incidentally, that *all the multipole moments, and hence by Eqs. (4.13) the electromagnetic field at all points outside the sphere $r > R$, are completely specified by the radiation pattern.*

Finally, we may readily deduce from our results expressions for the time averaged power radiated by the source. It is given by the integral of the radial component of the time averaged Poynting vector across a limitingly large sphere Σ of radius r (with $kr\to\infty$):

$$\langle P\rangle = \frac{c}{8\pi}\,\mathcal{R}\int_{\Sigma}[\mathbf{E}(\mathbf{r})\times\mathbf{H}^*(\mathbf{r})]\cdot\mathbf{u}_r\,d\Sigma. \tag{4.21}$$

On substituting from Eqs. (4.19) into (4.21) we obtain the following expression for $\langle P\rangle$:

$$\langle P\rangle = \frac{c}{8\pi}\,\mathcal{R}\int_{-\pi}^{\pi} d\phi\int_0^{\pi} d\theta\,\sin\theta\,[\hat{\mathbf{E}}(\mathbf{u}_r)\times\hat{\mathbf{H}}^*(\mathbf{u}_r)]\cdot\mathbf{u}_r. \tag{4.22}$$

But from the orthogonality relations between the three

vectors $\hat{\mathbf{E}}(\mathbf{u}_r)$, $\hat{\mathbf{H}}(\mathbf{u}_r)$, and $\mathbf{u}_r$, indicated by Eqs. (4.6), we see that

$$[\hat{\mathbf{E}}(\mathbf{u}_r) \times \hat{\mathbf{H}}^*(\mathbf{u}_r)]\cdot\mathbf{u}_r = \hat{\mathbf{E}}^*(\mathbf{u}_r)\cdot\hat{\mathbf{E}}(\mathbf{u}_r), \tag{4.23a}$$

$$= \hat{\mathbf{H}}^*(\mathbf{u}_r)\cdot\hat{\mathbf{H}}(\mathbf{u}_r). \tag{4.23b}$$

If we now substitute from (4.23a) [or (4.23b)] into (4.22), and express $\hat{\mathbf{E}}(\mathbf{u}_r)$ [or $\hat{\mathbf{H}}(\mathbf{u}_r)$] in the series form (4.10a) [or (4.10b)], and if we also make use of the orthogonality relations (4.9) between the vector spherical harmonics we find that

$$\langle P\rangle = \frac{c}{8\pi}\sum_{l=1}^{\infty}\sum_{m=-l}^{l} l(l+1)[|a_l^m|^2 + |b_l^m|^2]. \tag{4.24}$$

Equation (4.24) is the well-known expression for the radiated power in terms of the multipole moments.

5. THE MULTIPOLE EXPANSION AND THE DEBYE POTENTIALS

Many of the existing treatments of multipole expansions and of other problems arising in electromagnetic theory employ a representation of the electromagnetic field in terms of two scalar potentials that was introduced by Debye[3] in a well-known investigation relating to the pressure exerted by light on a homogeneous sphere composed of arbitrary material. In this concluding section we briefly show that the angular spectrum representation gives a new insight into the relation between the multipole expansion, and the Debye representation.[34]

The Debye potentials $\Pi_e(\mathbf{r})$ and $\Pi_h(\mathbf{r})$ are solutions of the scalar Helmholtz equation which yield an electromagnetic field in free space by means of the formulae (the Debye representation):

$$\mathbf{E}(\mathbf{r}) = \nabla\times[\nabla\times(\mathbf{r}\,\Pi_e(\mathbf{r}))] + ik\nabla\times[\mathbf{r}\,\Pi_h(\mathbf{r})], \tag{5.1a}$$

$$\mathbf{H}(\mathbf{r}) = -ik\nabla\times[\mathbf{r}\,\Pi_e(\mathbf{r})] + \nabla\times[\nabla\times(\mathbf{r}\,\Pi_h(\mathbf{r}))]. \tag{5.1b}$$

It is clear that the multipole expansion (4.13) may be expressed in the form (5.1) if we substitute into (4.13) the definitions (4.15) of the electromagnetic multipole fields and interchange the order of differentiation and summations. We then obtain the following expressions for the electric and magnetic Debye potentials:

$$\Pi_e(\mathbf{r}) = \sum_{l=1}^{\infty}\sum_{m=-l}^{l} a_l^m\,\Pi_l^m(\mathbf{r}), \tag{5.2a}$$

$$\Pi_m(\mathbf{r}) = \sum_{l=1}^{\infty}\sum_{m=-l}^{l} b_l^m\,\Pi_l^m(\mathbf{r}). \tag{5.2b}$$

The Debye representation is intimately connected with our decomposition (4.10) of the spectral amplitudes $\hat{\mathbf{E}}(\mathbf{s})$ and $\hat{\mathbf{H}}(\mathbf{s})$ into series of the vector spherical harmonics $\mathbf{Y}_l^m$ and $\mathbf{s}\times\mathbf{Y}_l^m$. To see this, let us introduce in (4.10) the definition (4.7) of the vector spherical harmonic $\mathbf{Y}_l^m$ in terms of the ordinary spherical harmonic $\mathbf{Y}_l^m$. If then we interchange the orders of differentiation and summation we obtain the formulas

$$\hat{\mathbf{E}}(\mathbf{s}) = \mathbf{s}\times\mathcal{L}_s\hat{A}(\mathbf{s}) + \mathcal{L}_s\hat{B}(\mathbf{s}), \tag{5.3a}$$

$$\hat{\mathbf{H}}(\mathbf{s}) = -\mathcal{L}_s\hat{B}(\mathbf{s}) + \mathbf{s}\times\mathcal{L}_s\hat{A}(\mathbf{s}), \tag{5.3b}$$

where

$$\hat{A}(\mathbf{s}) = \sum_{l=1}^{\infty}\sum_{m=-l}^{l}(-i)^l a_l^m Y_l^m(\alpha,\beta), \tag{5.4a}$$

$$\hat{B}(\mathbf{s}) = \sum_{l=1}^{\infty}\sum_{m=-l}^{l}(-i)^l b_l^m Y_l^m(\alpha,\beta). \tag{5.4b}$$

Consider now the scalar fields $A(\mathbf{r})$, $B(\mathbf{r})$, whose spectral amplitudes are $\hat{A}(\mathbf{s})$ and $\hat{B}(\mathbf{s})$, respectively:

$$A(\mathbf{r}) = \frac{ik}{2\pi}\int_{-\pi}^{\pi}d\beta\int_{C\pm}d\alpha\,\sin\alpha\,\hat{A}(\mathbf{s})e^{ik\mathbf{s}\cdot\mathbf{r}}, \tag{5.5a}$$

$$B(\mathbf{r}) = \frac{ik}{2\pi}\int_{-\pi}^{\pi}d\beta\int_{C\pm}d\alpha\,\sin\alpha\,\hat{B}(\mathbf{s})e^{ik\mathbf{s}\cdot\mathbf{r}}. \tag{5.5b}$$

If we substitute from (5.4a) into (5.5a), interchange the order of integration and summation, and recall, also, the angular spectrum representation (3.13) of the multipole field $\Pi_l^m(\mathbf{r})$, we obtain the following expression for $A(\mathbf{r})$:

$$A(\mathbf{r}) = k\sum_{l=1}^{\infty}\sum_{m=-l}^{l}(-i)^l a_l^m\Pi_l^m(\mathbf{r}). \tag{5.6a}$$

In a strictly similar way, we obtain from (5.4b) and (5.5b), if again we use (3.13), the following expression for $B(\mathbf{r})$:

$$B(\mathbf{r}) = k\sum_{l=1}^{\infty}\sum_{m=-l}^{l}(-i)^l b_l^m\Pi_l^m(\mathbf{r}). \tag{5.6b}$$

Comparison of Eqs. (5.2) and (5.6) shows that

$$A(\mathbf{r}) = k\Pi_e(\mathbf{r}),\qquad B(\mathbf{r}) = k\Pi_h(\mathbf{r}), \tag{5.7}$$

i.e., apart from the proportionality factor k, $A(\mathbf{r})$ and $B(\mathbf{r})$ are precisely the Debye potentials and, hence, the *amplitude functions* $\hat{A}(\mathbf{s})$ *and* $\hat{B}(\mathbf{s})$ *introduced in (5.3) are*, apart from the proportionality factor k, *the spectral amplitudes in the angular spectrum representation of the Debye potentials.*[35]

ACKNOWLEDGMENT

We wish to acknowledge our appreciation to Dr. George Sherman for stimulating discussions relating to the analysis presented in this paper.

APPENDIX A: ANGULAR SPECTRUM REPRESENTATION OF A SCALAR MULTIPOLE FIELD

In this Appendix, we show that the angular spectrum representation of the scalar multipole field $\Pi_l^m(\mathbf{r})$, defined by Eq. (3.14), is given by Eq. (3.13). For this purpose we will use the well known result that a multipole field $\Pi_l^m(\mathbf{r})$ of any order $m \geq 0$ may be generated from the spherical wave $\exp(ikr)/kr$, [which apart from a normalization constant is the lowest order scalar multipole field $\Pi_0{}^0(\mathbf{r})$], by means of the following relation[24]:

$$\Pi_l^m(\mathbf{r}) = C_l^m\left\{\left[\frac{1}{ik}\left(\frac{\partial}{\partial x}+i\frac{\partial}{\partial y}\right)\right]^m P_l^{(m)}\left(\frac{1}{ik}\frac{\partial}{\partial z}\right)\right\}\frac{e^{ikr}}{kr}. \tag{A1}$$

Here, the operator

$$P_l^{(m)}\left(\frac{1}{ik}\frac{\partial}{\partial z}\right)$$

is defined by the formula

$$P_l^{(m)}\left(\frac{1}{ik}\frac{\partial}{\partial z}\right) = \frac{d^m}{du^m}P_z(u)\Big|_{u=(1/ik)\partial/\partial l}, \tag{A2}$$

where $P_l(u)$ is the Legendre polynomial of degree of l and C_l^m are normalization constants, defined as

$$C_l^m = (-1)^m(-i)^l\left(\frac{(2l+1)(l-m)!}{4\pi(l+m)!}\right)^{1/2}. \tag{A3}$$

When $m < 0$ we use the identity

$$\Pi_l^{-|m|}(r,\theta,\phi) = (-1)^{|m|}\Pi_l^{|m|}(r,\theta,-\phi), \tag{A4}$$

where, of course, (r,θ,ϕ) are the spherical polar coordinates of $\mathbf{r}$.

We now express the spherical wave $\exp(ikr)/kr$ in (A1) in the form of an angular spectrum representation, given by Weyl's formula (3.4):

$$\frac{e^{ikr}}{kr} = \frac{i}{2\pi} \int_{-\pi}^{\pi} d\beta \int_{C^\pm} d\alpha \, \sin\alpha \, e^{ik\mathbf{s}\cdot\mathbf{r}}, \tag{A5}$$

and interchange the order of integration and differentiation [which may be shown to be justified when $|z| > 0$ because the double integral in (A5) is then uniformly convergent]. We then obtain the following expression for $\Pi_l^m(\mathbf{r})$, valid when $|z| > 0$ and $m \geqslant 0$:

$$\Pi_l^m(\mathbf{r}) = \int_{-\pi}^{\pi} d\beta \int_{C^\pm} d\alpha \, \sin\alpha \, F(\alpha,\beta) e^{ik\mathbf{s}\cdot\mathbf{r}}, \tag{A6}$$

where

$$F(\alpha,\beta) e^{ik\mathbf{s}\cdot\mathbf{r}} = \frac{i}{2\pi} C_l^m \left\{ \left[\frac{1}{ik}\left(\frac{\partial}{\partial x} + i\frac{\partial}{\partial y}\right)\right]^m P_l^{(m)}\left(\frac{1}{ik}\frac{\partial}{\partial z}\right) \right\} e^{ik\mathbf{s}\cdot\mathbf{r}}. \tag{A7}$$

Now, with the Cartesian coordinates of $\mathbf{s}$ given by (2.3), we have

$$\mathbf{s}\cdot\mathbf{r} = x \sin\alpha \cos\beta + y \sin\alpha \sin\beta + z \cos\alpha, \tag{A8}$$

and we then readily obtain from (A7) the following expression for $F(\alpha,\beta)$:

$$F(\alpha,\beta) = \frac{i}{2\pi} C_l^m \sin^m\alpha \, e^{im\beta} P_l^{(m)}(\cos\alpha). \tag{A9}$$

But

$$\sin^m\alpha \, P_l^{(m)}(\cos\alpha) = P_l^m(\cos\alpha), \tag{A10}$$

where P_l^m is the associated Legendre polynomial of degree l and order m. Hence,

$$F(\alpha,\beta) = (i/2\pi) C_l^m P_l^m(\cos\alpha) e^{im\beta} = (i/2\pi)(-i)^l Y_l^m(\alpha,\beta), \tag{A11}$$

where

$$Y_l^m(\alpha,\beta) = i^l C_l^m P_l^m(\cos\alpha) e^{im\beta} \tag{A12}$$

is the spherical harmonic of degree l and order m. Finally, on substituting from (A11) into (A6) we find that

$$\Pi_l^m(\mathbf{r}) = \frac{i}{2\pi}(-i)^l \int_{-\pi}^{\pi} d\beta \int_{C^\pm} d\alpha \, \sin\alpha \, Y_l^m(\alpha,\beta) e^{ik\mathbf{s}\cdot\mathbf{r}}, \tag{A13}$$

which is the desired result valid when $|z| > 0$ and $m \geqslant 0$. Although, as previously mentioned, the interchange the order of integration and differentiation which lead to (A13) is justified only when $|z| > 0$, (A13) can be shown to be valid also when $z = 0$, except at the origin, in the sense of the following limit:

$$\Pi_l^m(\mathbf{r})\big|_{z=0} = \lim_{|z|\to 0} \frac{i}{2\pi}(-i)^l \int_{-\pi}^{\pi} d\beta \int_{C^\pm} d\alpha \, \sin\alpha \, Y_l^m(\alpha,\beta) e^{ik\mathbf{s}\cdot\mathbf{r}}. \tag{A13a}$$

To determine the angular spectrum representation for $\Pi_l^m(\mathbf{r})$ when $m < 0$, we first express the scalar product $\mathbf{s}\cdot\mathbf{r}$ in (A13) in a more explicit form, using Eqs. (A8) and (2.6):

$$\mathbf{s}\cdot\mathbf{r} = r(\sin\theta \sin\alpha \cos(\beta+\phi) + \cos\theta \cos\alpha). \tag{A14}$$

Equation (A13) then becomes, on substitution from (A14):

$$\Pi_l^{|m|}(\mathbf{r}) = \frac{i}{2\pi}(-i)^l \int_{-\pi}^{\pi} d\beta \int_{C^\pm} d\alpha \, \sin\alpha \, Y_l^{|m|}(\alpha,\beta) \times e^{ikr[\sin\theta \sin\alpha \cos(\beta+\phi)+\cos\theta\cos\alpha]}. \tag{A15}$$

From (A14) and (A4) it follows that

$$\Pi_l^{-|m|}(\mathbf{r}) = (-1)^{|m|} \frac{i}{2\pi}(-i)^l \int_{-\pi}^{\pi} d\beta \int_{C^\pm} d\alpha \, \sin\alpha \, Y_l^{|m|}(\alpha,\beta) \times e^{ikr[\sin\theta \sin\alpha \cos(\beta-\phi)+\cos\theta\cos\alpha]}. \tag{A16}$$

If in (A16) we change the variable of integration from β to $-\beta$ and use the relation

$$Y_l^{-|m|}(\alpha,\beta) = (-1)^{|m|} Y_l^{|m|}(\alpha,-\beta),$$

and also the formula (A14), we find that

$$\Pi_l^{-|m|}(\mathbf{r}) = \frac{i}{2\pi}(-i)^l \int_{-\pi}^{\pi} d\beta \int_{C^\pm} d\alpha \, \sin\alpha \, Y_l^{-|m|}(\alpha,\beta) e^{ik\mathbf{s}\cdot\mathbf{r}}. \tag{A17}$$

If now we set $-|m| = m$ where $m < 0$, Eq. (A17) becomes formally identical with (A13). Thus, (A13) is valid for both positive and negative integers m, $(-l \leqslant m \leqslant l)$, as we wished to show.

APPENDIX B: ANGULAR SPECTRUM REPRESENTATION OF AN ELECTROMAGNETIC MULTIPOLE FIELD

We will show in this Appendix that the angular spectrum representation of an electromagnetic multipole field, defined by Eqs. (4.15), is given by Eqs. (4.14), i.e., that

$$\nabla \times \{\nabla \times [\mathbf{r}\Pi_l^m(\mathbf{r})]\} = (-i)^l \frac{ik}{2\pi} \int_{-\pi}^{\pi} d\beta \int_{C^\pm} d\alpha \, \sin\alpha \times [\mathbf{s} \times \mathbf{Y}_l^m(\alpha,\beta)] e^{ik\mathbf{s}\cdot\mathbf{r}}, \tag{B1a}$$

$$ik\nabla \times [\mathbf{r}\Pi_l^m(\mathbf{r})] = (-i)^l \frac{ik}{2\pi} \int_{-\pi}^{\pi} d\beta \int_{C^\pm} d\alpha \, \sin\alpha \times \mathbf{Y}_l^m(\alpha,\beta) e^{ik\mathbf{s}\cdot} . \tag{B1b}$$

We derive first the formula (B1b). We have, by using a standard vector identity,

$$ik\{\nabla \times [\mathbf{r}\Pi_l^m(\mathbf{r})]\} = ik\{\Pi_l^m(\mathbf{r})\nabla \times \mathbf{r} - \mathbf{r} \times \nabla\Pi_l^m(\mathbf{r})\} = k\mathcal{L}_r \Pi_l^m(\mathbf{r}), \tag{B2}$$

where we have used the fact that $\nabla \times \mathbf{r} = 0$. In (B2), $\mathcal{L}_r$ is the "orbital angular momentum operator" in $\mathbf{r}$ space, viz.

$$\mathcal{L}_r = -i\mathbf{r} \times \nabla = -i\left(\mathbf{u}_\phi \frac{\partial}{\partial\theta} - \frac{1}{\sin\theta}\mathbf{u}_\theta \frac{\partial}{\partial\phi}\right), \tag{B3}$$

where $\mathbf{u}_\phi, \mathbf{u}_\theta$ are unit vectors in the positive ϕ and θ directions, respectivply. Now the operator $\mathcal{L}_r$ does not act on the radial coordinate r and, consequently, if we recall the definition (3.14) of the scalar multipole field $\Pi_l^m(\mathbf{r})$, we may express (B2) in the form

$$ik\nabla \times [\mathbf{r}\Pi_l^m(\mathbf{r})] = kh_l^+(kr)\mathbf{Y}_l^m(\theta,\phi), \tag{B4}$$

where

$$\mathbf{Y}_l^m(\theta,\phi) = \mathcal{L}_r Y_l^m(\theta,\phi) \tag{B5}$$

is the vector spherical harmonic of degree l and order m.

Next, we will make use of the following identity, which expresses a vector spherical harmonic as a linear combination of ordinary spherical harmonics:[36]

$$\mathbf{Y}_l^m(\theta,\phi) = a_- Y_l^{m+1}(\theta,\phi)\boldsymbol{\epsilon}_- + a_+ Y_l^{m-1}(\theta,\phi)\boldsymbol{\epsilon}_+ + mY_l^m(\theta,\phi)\mathbf{u}_z, \tag{B6}$$

where

$$a_- = [(l-m)(l+m+1)]^{1/2},$$
$$a_+ = [(l+m)(l-m+1)]^{1/2}, \tag{B7}$$
$$\epsilon_- = \tfrac{1}{2}(\mathbf{u}_x - i\mathbf{u}_y), \qquad \epsilon_+ = \tfrac{1}{2}(\mathbf{u}_x + i\mathbf{u}_y), \tag{B8}$$

and $\mathbf{u}_x, \mathbf{u}_y, \mathbf{u}_z$ are unit vectors in the positive x, y, and z directions, respectively. On substituting from (B6) into (B4), and recalling once again the definition (3.14) of $\Pi_l^m(\mathbf{r})$, we obtain at once the identity

$$ik\nabla \times [\mathbf{r}\,\Pi_l^m(\mathbf{r})] = k[a_-\Pi_l^{m+1}(\mathbf{r})\epsilon_- + a_+\Pi_l^{m-1}(\mathbf{r})\epsilon_+ + m\,\Pi_l^m(\mathbf{r})\mathbf{u}_z]. \tag{B9}$$

We now express each of the three scalar multipole fields appearing on the righthand side of (B9) in terms of the angular spectrum representation (3.13) to obtain the identity

$$ik\nabla \times [\mathbf{r}\,\Pi_l^m(\mathbf{r})] = k(-i)^l \frac{i}{2\pi}\int_{-\pi}^{\pi} d\beta \int_{C\pm} d\alpha\, \sin\alpha\, \mathbf{G}(\alpha,\beta)\, e^{ik\mathbf{s}\cdot\mathbf{r}}, \tag{B10}$$

where

$$\mathbf{G}(\alpha,\beta) = a_- Y_l^{m+1}(\alpha,\beta)\epsilon_- + a_+ Y_l^{m-1}(\alpha,\beta)\epsilon_+ + mY_l^m(\alpha,\beta)\mathbf{u}_z. \tag{B11}$$

But according to (B6), the right-hand side of (B11) is precisely the vector spherical harmonic $\mathbf{Y}_l^m(\alpha,\beta)$, i.e., $\mathbf{G}(\alpha,\beta) = \mathbf{Y}_l^m(\alpha,\beta)$. Hence, (B10) gives

$$ik\nabla \times [\mathbf{r}\,\Pi_l^m(\mathbf{r})] = (-i)^l \frac{ik}{2\pi}\int_{-\pi}^{\pi} d\beta \int_{C\pm} d\alpha\, \sin\alpha\, \mathbf{Y}_l^m(\alpha,\beta)\, e^{ik\mathbf{s}\cdot\mathbf{r}}, \tag{B12}$$

which establishes the representation (B1b).

Next, we apply the curl operator to (B12). The curl operator may be taken under the integral signs on the rhs of (B12) since the double integral may be shown to converge uniformly when $|z| > 0$. We then obtain the formula

$$\nabla \times \{\nabla \times [\mathbf{r}\,\Pi_l^m(\mathbf{r})]\} = (-i)^l \frac{ik}{2\pi}\int_{-\pi}^{\pi} d\beta \int_{C\pm} d\alpha\, \sin\alpha\, \mathbf{s} \times \mathbf{Y}_l^m(\alpha,\beta)\, e^{ik\mathbf{s}\cdot\mathbf{r}}, \tag{B13}$$

which establishes the representation (B1a).

We should note that the two expansions (B1a) and (B1b) are valid for all $z \neq 0$ and, moreover, can be shown to have limiting values as $|z| \to 0$ which correctly represent the electromagnetic multipole fields on the plane $z = 0$, except at the origin [cf. Eq. (A13a) of Appendix A].

APPENDIX: C: DOMAIN OF VALIDITY OF THE MULTIPOLE EXPANSION (3.12)

In this appendix, we verify that the multipole expansion (3.12) is valid not only throughout the two half-spaces $z > R$ and $z > -R$ [where the angular spectrum expansion (3.5) converges], but is, in fact, valid throughout the exterior, $r > R$, of the source region.

As mentioned in Sec. 3, the field $\psi(\mathbf{r})$ can be represented in the form of an angular spectrum of plane waves outside *any* strip bounded by two parallel planes tangential to the sphere of radius R, which surrounds the source. For example, if we choose a new Cartesian coordinate system of axes $O\bar{X}, O\bar{Y}, O\bar{Z}$, obtained from the original system OX, OY, OZ, by a rotation about the origin O, we can represent the field at all points outside the strip $|\bar{z}| < R$ in the form (3.5), where the polar angles α, β of $\mathbf{s}$ and θ, ϕ of the field point $\mathbf{r}$ are referred to the rotated (barred) rather than the original (unbarred) system of axes. Consequently, all the analysis leading to the multipole expansion (3.12) remains valid in the rotated system.

Consider now a field point $\mathbf{r}_0$ that lies outside the source region (i.e., for which $r_0 > R$), but which is situated *within* the strip $|z| < R$ and let us choose the rotated axes $O\bar{X}, O\bar{Y}, O\bar{Z}$ in such a way that $\mathbf{r}_0$ lies outside the strip $|\bar{z}| < R$ (see Fig. 2). Then if $r_0, \bar{\theta}_0, \bar{\phi}_0$ and $r', \bar{\theta}', \bar{\phi}'$ are the spherical polar coordinates of the field point $\mathbf{r}_0$ and of the integration point $\mathbf{r}'$, respectively, referred to the *rotated* system of axes, we have by (3.12)

$$\psi(\mathbf{r}_0) = k \sum_{l=0}^{\infty} \sum_{m=-l}^{l} \bar{a}_l^m h_l^{(+)}(kr_0)\, Y_l^m(\bar{\theta}_0, \bar{\phi}_0), \tag{C1}$$

where [if we also use (2.4)]

$$\bar{a}_l^m = 4\pi \int_{r'\le R} \rho(\mathbf{r}')\, j_l(kr')\, Y_l^{m*}(\bar{\theta}', \bar{\phi}')\, d^3r'. \tag{C2}$$

We can rewrite (C1) in the form

$$\psi(\mathbf{r}_0) = k \sum_{l=0}^{\infty} h_l^{(+)}(kr_0) \left(4\pi \int_{r'\le R} d^3r' \rho(\mathbf{r}')\, j_l(kr') \sum_{m=-l}^{l} \times Y_l^{m*}(\bar{\theta}', \bar{\phi}')\, Y_l^m(\bar{\theta}_0, \bar{\phi}_0)\right). \tag{C3}$$

Now, if r_0, θ_0, ϕ_0 and r', θ', ϕ' are the spherical polar coordinates of the field point $\mathbf{r}_0$ and the integration point $\mathbf{r}'$, respectively, referred to the *original* system of axes we have, from the addition theorem on spherical harmonics (Ref. 37, pp. 290–291)

$$\sum_{m=-l}^{l} Y_l^{m*}(\bar{\theta}', \bar{\phi}')\, Y_l^m(\bar{\theta}_0, \bar{\phi}_0) = \sum_{m=-l}^{l} Y_l^{m*}(\theta', \phi')\, Y_l^m(\theta_0, \phi_0), \tag{C4}$$

each of these two sums being equal to $(2l+1)P_l(\cos\chi)/4\pi$. Here P_l is the Legendre polynomial of degree l, and χ is the angle between the position vectors $\mathbf{r}_0$ and $\mathbf{r}'$. Making use of (C4), Eq. (C3) can be rewritten in the form

$$\psi(\mathbf{r}_0) = k \sum_{l=0}^{\infty} h_l^{(+)}(kr_0) \left(4\pi \int_{r'\le R} d^3r' \rho(\mathbf{r}')\, j_l(kr') \times \sum_{m=-l}^{l} Y_l^{m*}(\theta', \phi')\, Y_l^m(\theta_0, \phi_0)\right), \tag{C5}$$

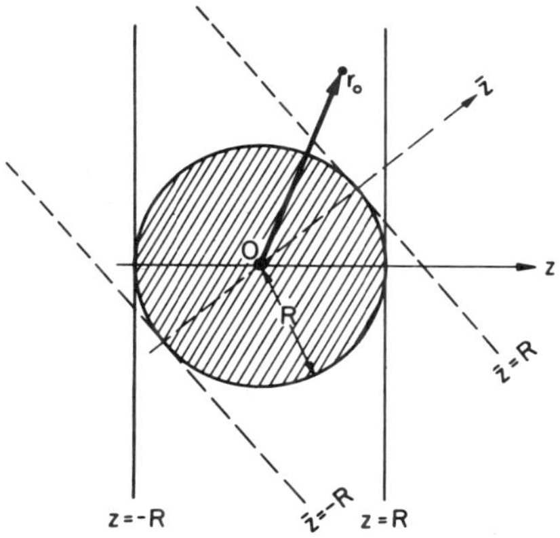

FIG. 2. Notation relating to the proof that the multipole expansion (3.12) is valid throughout the exterior $r > R$ of the source region. The point r_0 is situated in the strip $|z| < R$ (referred to the original coordinate system) but outside the strip $|\bar{z}| < R$ (referred to the rotated system).

or

$$\psi(\mathbf{r}_0) = k \sum_{l=0}^{\infty} \sum_{m=-l}^{l} a_l^m h_l^{(+)}(kr_0) Y_l^m(\theta_0, \phi_0), \tag{C6}$$

with [if use is also made of (2.4)]

$$a_l^m = 4\pi \int_{r' \leq R} \rho(\mathbf{r}') \Lambda_l^{m*}(\mathbf{r}') d^3r'. \tag{C7}$$

Equations (C6) is seen to be precisely Eq. (3.12) evaluated at the point $\mathbf{r}_0$, and referred to the original system of axes and Eq. (C7) is identical with Eq. (3.11). Since $\mathbf{r}_0$ can be taken to be *any* point outside the source region, we conclude that (3.12) is valid at all points $\mathbf{r}$ such that $r > R$.

* Research supported by the Air Force Office of Scientific Research and the Army Research Office (Durham). Some of the results presented here were obtained in the course of doctoral thesis research by one of the authors [A. J. Devaney, "A New Theory of the Debye Representation of Classical and Quantized Electromagnetic Fields", Ph. D. thesis, University of Rochester, Rochester, N.Y. (1971)]. A preliminary account was also presented at the Fall meeting of the Optical Society of America, held in San Francisco, California in October 1972. [Abstract WT17, J. Opt. Soc. Amer., **62**, 1360 (1972).]

[1]A. Clebsch, J. Math. **61**, 195 (1863).
[2]G. Mie, Ann. Phys. (Leipz.) **25**, 377 (1908).
[3]P. Debye, Ann. Phys. (Leipz.) **30**, 57 (1909).
[4]T. J. I'A. Bromwich, Phil. Trans. R. Soc. Lond. **220**, 175 (1920).
[5]For a very readable account of the early studies of scattering of plane waves by spheres see, N. A. Logan, Proc. IEEE **63**, 48 (1965). See also, a note by the same author published in the J. Opt. Soc. Am. **52**, 342 (1962). For a detailed list of references to papers on multipole expansions up to 1953 see M. E. Rose, *Multipole Fields* (Wiley, New York, 1955).
[6]W. W. Hansen, Phys. Rev. **47**, 139 (1935).
[7]W. Heitler, Proc. Camb. Phil. Soc. **32**, 112 (1936).
[8]H. A. Kramers, Physica **10**, 261 (1943).
[9]W. Franz, Z. Phys. **127**, 363 (1950).
[10]P. R. Wallace, Can. J. Phys. **29**, 393 (1951).
[11]J. M. Blatt and V. F. Weisskopf, *Theoretical Nuclear Physics* (Wiley, New York, 1952), App. B.
[12]C. J. Bouwkamp and H. B. G. Casimir, Physica **20**, 539 (1954).
[13]A. Nisbet, Physica **21**, 799 (1955).
[14]C. H. Wilcox, J. Math. Mech. **6**, 167 (1957).
[15]E. T. Whittaker, Math. Ann. **57**, 333 (1902); see also, E. T. Whittaker and G. N. Watson, *A Course of Modern Analysis* (Cambridge U.P., Cambridge, 1963), 4th ed., Sec. 18.6.
[16]For a discussion of this representation see, for example, C. J. Bouwkamp, Rep. Prog. Phys. **17**, 41 (1954) or J. Goodman, *Introduction to Fourier Optics* (McGraw-Hill, New York, 1968), Sec. 3.7. The relationship between homogeneous plane wave expansions and angular spectrum representations is discussed, for certain classes of scalar wave fields, in A. J. Devaney and G. C. Sherman, SIAM Rev. **15**, 765 (1973).
[17]A. J. Devaney and E. Wolf, Phys. Rev. **8**, 1044 (1973).
[18]The customary term "multipole field" for the field $\Lambda_l^m(\mathbf{r})$ defined by Eq. (2.4) is rather inappropriate since this field is everywhere well-behaved. This terminology is undoubtedly responsible for incorrect statements in the literature about the behavior of such a field at the origin. [See, for example, W. Heitler, *The Quantum Theory of Radiation* (Clarendon, Oxford, 1954), 3rd ed., statement following Eq. (9) on p. 402.] It is the scalar field $\Pi_l^m(\mathbf{r})$, defined by Eq. (3.14), and the electromagnetic fields $\mathbf{E}_{lm}^{(e)}$, $\mathbf{H}_{lm}^{(e)}$ and $\mathbf{E}_{lm}^{(h)}$, $\mathbf{H}_{lm}^{(h)}$, defined by Eqs. (4.15), that behave as fields generated by a true multipole at the origin $r = 0$, and hence have an appropriate singularity at that point.
[19]We use the same definitions of the spherical harmonics and of the spherical Bessel and Hankel functions as those given in A. Messiah, *Quantum Mechanics* (North–Holland, Amsterdam, third printing 1965), Vol. I, App. BII and BIV.
[20]H. Weyl, Ann. Phys. (Paris) **60**, 481 (1919). Because the integrand of (3.4) is an entire analytic function of α, the α integration contours $C^{\pm}$ may be deformed in accordance with the rules of integration in the complex plane. When these contours are chosen as shown in Fig. 1, (3.4) corresponds to the following, frequently used, alternative form of the Weyl expansion:

$$\frac{\exp(ik|\mathbf{r}-\mathbf{r}'|)}{|\mathbf{r}-\mathbf{r}'|} = \frac{ik}{2\pi} \iint_{-\infty}^{\infty} \frac{1}{s_z} \exp\{ik[s_x(x-x') + s_y(y-y') + s_z|z-z'|]\} \, ds_x \, ds_y. \tag{3.4a}$$

Here,

$$s_z = +\sqrt{1 - s_x^2 - s_y^2}, \quad \text{if } s_x^2 + s_y^2 \leq 1,$$
$$s_z = +i\sqrt{s_x^2 + s_y^2 - 1}, \quad \text{if } s_x^2 + s_y^2 > 1. \tag{3.4b}$$

The two forms of the Weyl expansion, and the transformations required to pass from one to the other, are discussed in A. Baños, *Dipole Radiation in the Presence of a Conducting Half-Space* (Pergamon, New York, 1966), sec. 2.13.
[21]If the representation (3.4a), rather than (3.4), is employed one obtains, instead of (3.5), the following form of the angular spectrum representation, which is commonly employed in the literature:

$$\psi(\mathbf{r}) = \frac{ik}{2\pi} \iint_{-\infty}^{\infty} \hat{\phi}^{(\pm)}(s) \exp[ik(s_x x + s_y y \pm s_z z)] \, ds_x \, ds_y, \tag{3.5a}$$

where s_z is defined by (3.4b), and the positive signs are used when $z > R$ and the negative signs when $z < -R$. The spectral amplitudes $\phi^{(\pm)}(s)$ are given by

$$\phi^{(\pm)}(\mathbf{s}) = \frac{1}{s_z} \tilde{\rho}(ks_x, ks_y, \pm ks_z), \tag{3.5c}$$

with $\hat{\rho}(\mathbf{K})$ being the threefold Fourier transform of the source distribution, as defined by Eq. (3.7).
[22]The effect of discarding all evanescent plane waves in the angular spectrum representation of the scalar dipole field $\pi_1^0(\mathbf{r})$ [see Eq. (3.14)] has been very clearly analyzed by W. H. Carter, Opt. Commun. 2, 142 (1970).
[23]This result is, essentially, a three-dimensional analogue of a well-known theorem that the Fourier transform of a continuous function which vanishes outside a finite interval is a boundary value of an entire analytic function. This theorem follows at once from a well known result on analytic functions defined by definite integrals [cf., E. T. Copson, *An Introduction to the Theory of Functions of a Complex Variable* (Oxford U.P., London, 1962), Sec. 5.5]. The multidimensional form of the theorem is the well-known Plancherel–Pólya theorem [cf., B. A. Fuks, *Introduction to the Theory of Analytic Functions of Several Complex Variables* (Amer. Math. Soc., Providence, R.I., 1963), p. 352].
[24]A. Erdélyi, Physica **4**, 107 (1937).
[25]See, for example, the Appendix in K. Miyamoto and E. Wolf, J. Opt. Soc. Am. **52**, 615 (1962).
[26]It is planned to present the corresponding analysis for the source-free, classical electromagnetic field in another publication. An analogous treatment of the quantized, free electromagnetic field has been given by A. J. Devaney, *Proceedings of the Third Rochester Conference on Coherence and Quantum Optics*, edited by L. Mandel and E. Wolf (Plenum, New York, 1973), p. 241.
[27]If, in addition to oscillating charges and current densities, there is a contribution from density of magnetization $\mathfrak{M}(\mathbf{r}, t) = \mathfrak{R}(\mathfrak{M}(\mathbf{r}) e^{-i\omega t})$, one must replace $\mathbf{j}(\mathbf{r})$ by $\mathbf{j}(\mathbf{r}) + c\nabla \times \mathfrak{M}(\mathbf{r})$. The introduction of magnetization makes it possible to take into account the effect of spin in the corresponding quantum mechanical formulation.
[28]Conversely, it is not difficult to show that the **E** and **H** fields that are the solutions to Eqs. (4.1) and which behave at infinity as outgoing spherical waves satisfy the full set of Maxwell equations everywhere.
[29]The angular spectrum representation (4.2) may, of course, be expressed in the vectorial generalization of the alternative form discussed in Footnote 21.
[30]See, for example, F. Rohrlich, *Classical Charged Particles* (Addison-Wesley, Reading, Mass., 1965), Sec. 4.3.
[31]E. L. Hill, Am. J. Phys. 22, 211 (1954).
[32]This fact may readily be deduced from the customary form of the completeness theorem (discussed, for example, in Ref. 11, pp. 798–799), which may be stated as follows: An arbitrary, well-behaved vector field $\mathbf{A}(\mathbf{r})$ may be expanded in terms of three types

of vector spherical harmonics $\mathbf{X}_{lm}(\theta,\phi)$, $\mathbf{V}_{lm}(\theta,\phi)$, and $\mathbf{W}_{lm}(\theta,\phi)$ in the form

$$\mathbf{A}(\mathbf{r}) = \sum_{l=0}^{\infty}\sum_{m=-l}^{l}[f_l^m(r)\mathbf{X}_{lm}(\theta,\phi) + g_l^m(r)\mathbf{V}_{lm}(\theta,\phi) + h_l^m(r)\mathbf{W}_{lm}(\theta,\phi)],$$

where f_l^m, g_l^m, and h_l^m are functions of the radial coordinate $r = |\mathbf{r}|$ only. (The vector spherical harmonics $\mathbf{X}_{lm}$, $\mathbf{V}_{lm}$, and $\mathbf{W}_{lm}$ correspond to $\mathbf{Y}^m_{l,l,1}$, $\mathbf{Y}^m_{l,l+1,1}$, and $\mathbf{Y}^m_{l,l-1,1}$ respectively, of Ref. 11.) The vector spherical harmonic $\mathbf{X}_{lm}$ is, except for a normalization factor, the same function that we denoted by $\mathbf{Y}_l^m$; more precisely

$$\mathbf{Y}_l^m = [l(l+1)]^{1/2}\mathbf{X}_{lm}.$$

Moreover, one has the relations [see Eqs. (B3) and (B9) of Ref. 31):

$$\mathbf{s}\times\mathbf{Y}_l^m = i[l(l+1)]^{1/2}\left[\left(\frac{l}{l+1}\right)^{1/2}\mathbf{V}_{lm} + \left(\frac{l+1}{2l+1}\right)^{1/2}\mathbf{W}_{lm}\right],$$

and

$$\mathbf{s}\,Y_l^m = \left[-\left(\frac{l+1}{2l+1}\right)^{1/2}\mathbf{V}_{lm} + \left(\frac{l}{2l+1}\right)^{1/2}\mathbf{W}_{lm}\right],$$

where $\mathbf{s} = \mathbf{r}/r$ is the unit vector in the radial direction. Since, according to these relations, the vector spherical harmonics $\mathbf{Y}_l^m$, $\mathbf{s}\times\mathbf{Y}_l^m$, and $\mathbf{s}\,Y_l^m$ are linearly independent combinations of $\mathbf{X}_{lm}$ $\mathbf{V}_{lm}$, and $\mathbf{W}_{lm}$, it is clear that they too form a complete basis for the expansion of $\mathbf{A}(\mathbf{r})$. Moreover, $\mathbf{Y}_l^m$ and $\mathbf{s}\times\mathbf{Y}_l^m$ are tangential to the unit sphere ($s^2 = 1$) and $\mathbf{s}\,Y_l^m$ is perpendicular to it. Hence, the two types of vector spherical harmonics $\mathbf{Y}_l^m$ and $\mathbf{s}\times\mathbf{Y}_l^m$ form a complete set for arbitrary "tangential vector fields" $\mathbf{A}(\mathbf{r})$, i.e., an arbitrary, well-behaved, vector field $\mathbf{A}(\mathbf{r})$ such that $\mathbf{s}\cdot\mathbf{A}(\mathbf{r}) = 0$, may be expanded in terms of them.

[33]By making use of elementary vector identities and the fact that $\mathbf{s}\cdot\mathbf{Y}_l^m = 0$, one can rewrite (4.12a) in the form

$$a_l^m = -\frac{i^{l+1}}{l(l+1)}\left(\frac{k}{c}\right)\int_{-\pi}^{\pi}d\beta\int_0^{\pi}d\alpha\sin\alpha\,\{\mathbf{s}\times[\mathbf{s}\times\tilde{\mathbf{j}}(k\mathbf{s})]\}\cdot[\mathbf{s}\times\mathbf{Y}_l^{m*}(\alpha,\beta)]. \quad (4.12a')$$

Now, $-\mathbf{k}\times[\mathbf{k}\times\tilde{\mathbf{j}}(\mathbf{k})]/|\mathbf{k}|^2$ is the Fourier transform $\tilde{\mathbf{j}}_T(\mathbf{k})$ of the transverse part $\mathbf{j}_T(\mathbf{r})$ of the current distribution [cf., E. A. Power, *Introductory Quantum Electrodynamics* (American Elsevier, New York, 1964), Sec. 6.3]. Thus, (4.12a') and (4.12b) show that all multipole moments, (and, consequently, the field outside the source region), depend only on those Fourier components $\tilde{\mathbf{j}}_T(\mathbf{k})$ of the transverse part of the current distribution for which $|\mathbf{k}| = k = \omega/c$.

[34]For a discussion of the various aspects of the Debye representation see, for example, the papers by Bouwkamp and Casimir[12], Nisbet[13], and Wilcox[14].

[35]The decomposition (5.3) of the spectral amplitudes $\hat{\mathbf{E}}(\mathbf{s})$ and $\hat{\mathbf{H}}(\mathbf{s})$, which we obtained as a consequence of the completeness of the vector spherical harmonics $\mathbf{Y}_l^m(\alpha,\beta)$ and $\mathbf{s}\times\mathbf{Y}_l^m(\alpha,\beta)$ with respect to all well behaved fields $\mathbf{F}(\mathbf{s})$ that are orthogonal to $\mathbf{s}$ [i.e., such that $\mathbf{s}\cdot\mathbf{F}(\mathbf{s}) \equiv 0$], may also be obtained as a direct consequence of the so-called *Hodge's decomposition theorem* [See, for example, P. Bidal and G. de Rham, Commun. Math. Hel. 19, 1 (1946).] Wilcox[14] employed this theorem in his treatment of the Debye representation, which, however, is quite different from ours.

[36]This identity follows from a well known general expression for vector spherical harmonics in terms of ordinary spherical harmonics. [See, for example, Eq. (1.5), p. 797 in Ref. 11.]

[37]See, for example, B. W. Shore and D. H. Menzel, *Principles of Atomic Spectra* (Wiley, New York, 1968), pp. 290-91.

Reprinted from:
PHYSICAL REVIEW D VOLUME 8, NUMBER 4 15 AUGUST 1973

Radiating and Nonradiating Classical Current Distributions and the Fields They Generate*

A. J. Devaney† and E. Wolf
Department of Physics and Astronomy, University of Rochester, Rochester, New York, 14627
(Received 4 April 1973)

Several general theorems are established relating to well-behaved, localized, monochromatic current distributions and the fields that they generate. In particular, a necessary and sufficient condition for such a current distribution to be nonradiating is established and a general expression for all nonradiating current distributions of this class is obtained.

Oscillating classical charge-current distributions which do not radiate have been studied, from time to time, by various authors. Interest in such distributions arises in connection with extended electron models and models for other elementary particles, and with problems involving electromagnetic self-force and radiation reaction. Early work in this field is associated with the names of Sommerfeld, Herglotz, Hertz, Ehrenfest, and Schott.[1] More recent contributions include those of Bohm and Weinstein,[2] Goedecke,[3] Arnett and Goedecke,[4] and Erber and Prastein.[5] A number of nonradiating distributions have been found, but no general technique for determining such distributions has been discovered, nor are their general properties understood.

We have recently presented a new treatment of multipole expansions of electromagnetic fields (Ref. 6; to be referred to as I), in the course of which we provided some new general expressions for the multipole moments and for the far field generated by a localized charge-current distribution. Here we show that our results lead readily to several general theorems about well-behaved, localized, monochromatic charge-current distributions, both radiating and nonradiating ones, and about the fields that they generate. We begin by summarizing some of the results established in I.

Let $\mathcal{R}\{\rho_\omega(\vec{r})e^{-i\omega t}\}$ and $\mathcal{R}\{\vec{j}_\omega(\vec{r})e^{-i\omega t}\}$ ($\mathcal{R}$ denoting the real part) represent the charge density and the current density, respectively, at the space-time point $\vec{r}, t$, oscillating with frequency ω ($\neq 0$). We assume throughout this paper that $\rho_\omega(\vec{r})$ and $\vec{j}_\omega(\vec{r})$ are continuous and differentiable functions of position, and that they vanish identically outside a sphere σ_R of radius R, about the origin $\vec{r}=0$. We will refer to such source distributions as *well-behaved distributions, localized within* σ_R. Let $\mathcal{R}\{\vec{E}(\vec{r})e^{-i\omega t}\}$ and $\mathcal{R}\{\vec{H}(\vec{r})e^{-i\omega t}\}$ be the electric and magnetic field vectors, respectively, generated by the charge-current distribution. The space-dependent parts[7] $\vec{E}(\vec{r})$ and $\vec{H}(\vec{r})$ of the electromagnetic field vectors may then be represented in the form of multipole expansions, valid when $r>R$:

$$\vec{E}(\vec{r}) = \sum_{l=1}^{\infty} \sum_{m=-l}^{l} [a_l^m \vec{E}^e_{lm}(\vec{r}) + b_l^m \vec{E}^h_{lm}(\vec{r})], \tag{1a}$$

$$\vec{H}(\vec{r}) = \sum_{l=1}^{\infty} \sum_{m=-l}^{l} [a_l^m \vec{H}^e_{lm}(\vec{r}) + b_l^m \vec{H}^h_{lm}(\vec{r})], \tag{1b}$$

where

$$\vec{E}^e_{lm}(\vec{r}) = \vec{H}^h_{lm}(\vec{r}) = \nabla\times\{\nabla\times[\vec{r}h_l^{(+)}(kr)Y_l^m(\theta,\phi)]\}, \tag{2a}$$

$$\vec{E}^{h}_{lm}(\vec{r}) = -\vec{H}^{e}_{lm}(\vec{r}) = ik\nabla\times[\vec{r}h_l^{(+)}(kr)Y_l^m(\theta,\phi)], \quad (2b)$$

and $h_l^{(+)}$ and Y_l^m are the spherical Hankel functions and the spherical harmonics, respectively,[8] r, θ, ϕ being the spherical polar coordinates of $\vec{r}$ and $k = \omega/c$.

Let $\vec{J}_\omega(\vec{K})$ be the Fourier transform of $\vec{j}_\omega(\vec{r})$, i.e.,

$$\vec{J}_\omega(\vec{K}) = \int_{r'\leq R} \vec{j}_\omega(\vec{r}')e^{-i\vec{K}\cdot\vec{r}'}d^3r' \quad (3)$$

and let $\vec{J}^T_\omega(\vec{K})$ be the Fourier transform of the transverse part of $\vec{j}_\omega(\vec{r})$, i.e.,[9]

$$\vec{J}^T_\omega(\vec{K}) = \frac{[\vec{K}\times\vec{J}_\omega(\vec{K})]\times\vec{K}}{\vec{K}^2}. \quad (4)$$

We have shown in I that the multipole moments a_l^m and b_l^m may be expressed in the following form:

$$a_l^m = \frac{i^{l+1}}{l(l+1)}\left(\frac{1}{c}\right) \times \int_{-\pi}^{\pi} d\beta \int_0^{\pi} d\alpha \sin\alpha\, \vec{J}^T_\omega(\vec{k})\cdot[\vec{k}\times\vec{Y}_l^{m*}(\alpha,\beta)], \quad (5a)$$

$$b_l^m = -\frac{i^{l+1}}{l(l+1)}\left(\frac{k}{c}\right) \times \int_{-\pi}^{\pi} d\beta \int_0^{\pi} d\alpha \sin\alpha\, \vec{J}^T_\omega(\vec{k})\cdot\vec{Y}_l^{m*}(\alpha,\beta). \quad (5b)$$

Here

$$\vec{k} = k\vec{s} \quad (k=\omega/c), \quad (6)$$

$\vec{s}$ being a real unit vector, and α and β are the spherical polar angles of $\vec{s}$. Further, $\vec{Y}_l^m(\alpha,\beta)$ are the vector spherical harmonics, defined by the formula $\vec{Y}_l^m(\alpha,\beta) = \mathcal{L}Y_l^m(\alpha,\beta)$ where $Y_l^m(\alpha,\beta)$ is the ordinary spherical harmonic of degree l and order m and $\mathcal{L} = -i\vec{K}\times\nabla_{\vec{K}}$ is the orbital angular momentum operator in $\vec{K}$ space. We also showed in I that $\vec{E}$ and $\vec{H}$ have the following asymptotic behavior as $kr\to\infty$ in a fixed direction specified by a unit vector $\vec{u}_r$:

$$\vec{E}(r\vec{u}_r) \sim \vec{\mathcal{E}}(\vec{u}_r)\frac{e^{ikr}}{r}, \qquad \vec{H}(r\vec{u}_r) \sim \vec{u}_r\times\vec{\mathcal{E}}(\vec{u}_r)\frac{e^{ikr}}{r}; \quad (7)$$

here $\vec{\mathcal{E}}(\vec{u}_r)$, the *radiation pattern* of the field, is given by

$$\vec{\mathcal{E}}(\vec{u}_r) = \frac{ik}{c}\vec{J}^T_\omega(k\vec{u}_r). \quad (8)$$

We will now derive from these results a number of general theorems.

Theorem I. The electromagnetic field at every point $\vec{r}$ *outside* σ_R *generated by a current distribution*[10] $\vec{j}_\omega$, *localized in* σ_R, *is completely and uniquely specified by certain Fourier components* $\vec{J}^T_\omega(\vec{K})$ *of the transverse part of the current density; namely those Fourier components for which* $|\vec{K}| = \omega/c$.

The validity of this theorem is evident by noting that in the expressions (5) for the multipole moments the current enters only through $\vec{J}^T_\omega(\vec{k})$ and that according to (6), $|\vec{k}| = \omega/c$. Hence, the knowledge of the Fourier components $\vec{J}^T_\omega(\vec{K})$ of the transverse part of the current density, for all $\vec{K}$ vectors of magnitude $|\vec{K}| = \omega/c$, allows the determination of all the multipole moments a_l^m and b_l^m, and these, in turn, represent the field uniquely at every point $\vec{r}$ exterior to σ_R, by means of the multipole expansions (1).

We also note, that in view of the relation (8), *the field at every point outside* σ_R *is also uniquely specified by its radiation pattern* $\vec{\mathcal{E}}(\vec{u}_r)$ *and the formulas (8), (5), and (1) allow determination of the field outside* σ_R *in terms of the radiation pattern.* The fact that the radiation pattern uniquely determines the field outside σ_R has been known before (Ref. 11, p. 339); however, no procedure for determining the field from the radiation pattern appears to have been given previously.

We now turn our attention to *nonradiating* current distributions, i.e., to current distributions that generate fields for which the time-averaged power

$$\langle P\rangle = \lim_{kr\to\infty}\frac{c}{8\pi}\,\mathcal{R}\int_\Sigma [\vec{E}^*(r\vec{u}_r)\times\vec{H}(r\vec{u}_r)]\cdot\vec{u}_r\,d\Sigma \quad (9)$$

across a sphere Σ, centered at the origin, and of limitingly large radius vanishes. Now the average power $\langle P\rangle$ may be expressed in the form[6]

$$\langle P\rangle = \frac{c}{8\pi}\sum_{l=1}^{\infty}\sum_{m=-l}^{l} l(l+1)(|a_l^m|^2+|b_l^m|^2). \quad (10)$$

Since each term under the summation sign in (10) is non-negative, $\langle P\rangle$ can vanish only if all the multipole moments a_l^m and b_l^m are zero, and according to Eqs. (1) the field itself then vanishes outside σ_R. Thus, we have established the following theorem:

Theorem II. The electromagnetic field generated by a nonradiating current distribution[10] $\vec{j}_\omega$ *localized in* σ_R, *vanishes identically at every point outside* σ_R.

It should be noted that Theorem II does not imply that the field generated by such a localized current distribution vanishes identically *inside* σ_R. In fact (see Theorem V), one can determine nonradiating current distributions, localized in σ_R, that will generate an electromagnetic field which inside σ_R can be prescribed quite arbitrarily. What Theorem II essentially implies is that a nonradiating current distribution that is localized in σ_R will generate a field that itself is localized in σ_R. In this connection, it should be mentioned that Arnett and Goedecke[4] examined the fields generated by several localized, nonradiating distributions and found that in these special cases the field vanished identically outside σ_R. They conjectured that this property should be true in general.[12]

Theorem III. A necessary and sufficient condition for a current distribution[10] $\vec{j}_\omega$ *localized in* σ_R, *to be nonradiating is the vanishing of certain Fourier components* $\vec{J}^T_\omega(\vec{K})$ *of the transverse part of the current density; namely those components for which* $|\vec{K}| = \omega/c$.

The condition of Theorem III is clearly sufficient. For the vanishing of $\vec{J}^T_\omega(\vec{K})$ for all $\vec{K}$ vectors of magnitude $|\vec{K}| = \omega/c$ implies, just as in connection with Theorem I, that all the multipole moments vanish. According to Eq. (10) the time-averaged radiated power then also vanishes. That this condition is also necessary can be seen as follows: If the current distribution is nonradiating then, according to Theorem II, the field that it generates will vanish outside σ_R and hence the radiation pattern of the field will be identically zero. This result implies, according to Eq. (8) that $\vec{J}^T_\omega(\vec{K}) \equiv 0$ for $|\vec{K}| = k \equiv \omega/c$.

It should be mentioned that Goedecke[3] showed that the condition of our Theorem III is a sufficiency condition for current distributions of a certain class to be nonradiating. (In this connection see also Ref. 5, Eq. R. 3, p. 236.) The condition also resembles a sufficiency condition stated by Bohm and Weinstein[2] for the absence of radiation, to order v/c, from certain rigid charge distribution (v being the instantaneous speed of the charge).

Theorem IV. Any nonradiating current distribution[10] $\vec{j}_\omega$, *localized in* σ_R, *can be represented in the form* $\Re\{\vec{j}_\omega(\vec{r})e^{-i\omega t}\}$, *where*

$$\vec{j}_\omega(\vec{r}) = \frac{1}{4\pi i}\left(\frac{c}{k}\right)\{\nabla \times [\nabla \times \vec{E}(\vec{r})] - k^2\vec{E}(\vec{r})\}, \qquad (11)$$

$k = \omega/c$ *and* $\vec{E}(\vec{r})$ *is a vector field that vanishes at all points outside* σ_R. *Moreover, with every nonradiating current distribution*[10] $\vec{j}_\omega$ *that is localized in* σ_R *there is associated one and only one such field* $\vec{E}(\vec{r})$ *and this field is precisely the electric field that is generated by the current distribution.*

To establish this theorem we observe that if $\vec{E}_0(\vec{r})$ is the electric field generated by a given nonradiating distribution it must, according to Theorem II, vanish identically at all points exterior to σ_R. Moreover, it follows from Maxwell's equations that $\vec{E}_0(\vec{r})$ satisfies the differential equation

$$\vec{j}_\omega(\vec{r}) = \frac{1}{4\pi i}\left(\frac{c}{k}\right)\{\nabla \times [\nabla \times \vec{E}_0(\vec{r})] - k^2\vec{E}_0(\vec{r})\}. \qquad (12)$$

Thus any well-behaved, localized, nonradiating monochromatic current distribution may be represented in the form (12) with $\vec{E}_0(\vec{r})$ being the field that is generated by the current distribution. Suppose now that there is another vector field $\vec{E}(\vec{r})$ that leads to the same current density $\vec{j}_\omega(\vec{r})$ via Eq. (11) and which vanishes at all points outside σ_R. It follows, on substracting Eq. (12) from Eq. (11), that the vector field

$$\vec{G}(\vec{r}) = \vec{E}(\vec{r}) - \vec{E}_0(\vec{r}) \qquad (13)$$

satisfies, for all values of $\vec{r}$, the differential equation

$$\nabla \times [\nabla \times \vec{G}(\vec{r})] - k^2\vec{G}(\vec{r}) = 0\,.$$

Taking the divergence of this equation and using the fact that the div curl $\equiv 0$, we see at once that $\nabla \cdot \vec{G}(\vec{r}) \equiv 0$. Further, using the vector identity

$$\nabla \times (\nabla \times \vec{G}) \equiv \nabla(\nabla \cdot \vec{G}) - \nabla^2\vec{G},$$

we find that $\vec{G}(\vec{r})$ satisfies, for all values of $\vec{r}$, the Helmholtz equation

$$\nabla^2\vec{G}(\vec{r}) + k^2\vec{G}(\vec{r}) = 0\,. \qquad (14)$$

Now, since $\vec{E}$ and $\vec{E}_0$ were assumed to vanish outside σ_R, the same must be true of $\vec{G}(\vec{r})$. It then follows from a theorem (Ref. 11, p. 87–88) on solutions of the Helmholtz equation which decrease sufficiently rapidly at infinity that $\vec{G}(\vec{r}) \equiv 0$ and hence, $\vec{E}(\vec{r}) \equiv \vec{E}_0(\vec{r})$. Thus $\vec{j}_\omega(\vec{r})$ is uniquely represented in the form (11) with $\vec{E}(\vec{r}) = \vec{E}_0(\vec{r})$.

Theorem V. If $\vec{f}(\vec{r})$ *is any vector field that is continuous and has continuous partial derivatives up to the third order and that vanishes at all points outside* σ_R, *then* $\Re\{\vec{j}_\omega(\vec{r})\exp(-i\omega t)\}$, *where*

$$\vec{j}_\omega(\vec{r}) = \frac{1}{4\pi i}\left(\frac{c}{k}\right)\{\nabla \times [\nabla \times \vec{f}(\vec{r})] - k^2\vec{f}(\vec{r})\}, \qquad (k = \omega/c) \qquad (15)$$

is a well-behaved nonradiating current distribution localized in σ_R *and* $\vec{f}(\vec{r})$ *is precisely the space-dependent part of the electric field that is generated by the distribution.*

To establish this result we first note that from the assumption on $\vec{f}(\vec{r})$, $\vec{j}_\omega(\vec{r})$, as defined by Eq. (15), is a well-behaved current distribution localized in σ_R. Next, we take the Fourier transform of Eq. (15). We then find that

$$\vec{K}\times[\vec{K}\times\vec{F}(\vec{K})]+k^2\vec{F}(\vec{K}) = -4\pi i\left(\frac{k}{c}\right)\vec{J}_\omega(\vec{K}), \qquad (16)$$

where $\vec{F}(\vec{K})$ and $\vec{J}_\omega(\vec{K})$ are the Fourier transforms of $\vec{f}(\vec{r})$ and $\vec{j}_\omega(\vec{r})$, respectively. If in Eq. (16) we express the first term in the form $(\vec{K}\cdot\vec{F})\vec{K} - \vec{K}^2\vec{F}$, and then multiply both sides of the equation vectorially by $\vec{K}$, we obtain the relation

$$(k^2-\vec{K}^2)\vec{K}\times\vec{F}(\vec{K}) = (-4\pi ik/c)\vec{K}\times\vec{J}_\omega(\vec{K}).$$

Hence, $\vec{K}\times\vec{J}_\omega(\vec{K})$ and also, therefore, the Fourier transform $\vec{J}_\omega^T(\vec{K})$ of the transverse part of the current [defined by Eq. (4)], vanish for all $\vec{K}$ such that $|\vec{K}| = k \equiv \omega/c$. This result implies, according to Theorem III, that the current distribution is *nonradiating*. Moreover, from Eq. (15) and Theorem IV it follows that $\vec{f}(\vec{r})$ is the space-dependent part of the electric field generated by the current distribution.

For the sake of simplicity we have restricted our discussion here to well-behaved, localized, monochromatic current distributions. We will consider elsewhere generalizations of the results to a wider class of current distributions.

ACKNOWLEDGMENT

We are indebted to Dr. Carlos E. Stroud for stimulating discussions on the subject matter of this note and also for having drawn our attention to several related articles. One of us (E. W.) wishes to express his appreciation to Karl Ruehle, for having provided at his home ideal working conditions during the time when this note was being prepared for publication.

*Work supported by the Army Research Office (Durham).

†Now with the Eikonix Corporation, Burlington, Massachusetts.

[1]References to the relevant papers are given by T. Erber in Fortschr. Phys. 9, 343 (1961) and by G. H. Goedecke in the paper quoted in Ref. 3 below.

[2]D. Bohm and M. Weinstein, Phys. Rev. 74, 1789 (1948).

[3]G. H. Goedecke, Phys. Rev. 135, B281 (1964).

[4]J. B. Arnett and G. H. Goedecke, Phys. Rev. 168, 1424 (1968).

[5]T. Erber and S. M. Prastein, Acta Phys. Austriaca 32, 224 (1970).

[6]A. J. Devaney and E. Wolf (unpublished).

[7]For the sake of brevity we will frequently omit the terms "space-dependent part," i.e., we will refer to $\vec{E}(\vec{r})$ itself as the electric field, etc. Also, since the continuity equation, relating the charge and current densities, implies that $\rho_\omega(\vec{r}) = (1/i\omega)\nabla\cdot\vec{j}_\omega(\vec{r})$ we will from now on regard the source distribution to be completely specified by the current density alone.

[8]We use the definitions of the spherical harmonics and the spherical Hankel functions given in A. Messiah, *Quantum Mechanics* (North-Holland, Amsterdam, 1961), Vol. I, Appendixes BII and BIV.

[9]Cf. E. A. Power, *Introductory Quantum Electrodynamics* (American Elsevier, New York, 1964), Sec. 6.3.

[10]For the sake of brevity "current distribution $\vec{j}_\omega$" is to be understood to mean "well-behaved, monochromatic current distribution of frequency ω."

[11]C. Müller, *Foundations of the Mathematical Theory of Electromagnetic Waves* (Springer, New York, 1969).

[12]Actually the current distributions studied in Ref. 4 are not monochromatic but are represented by a Fourier series. It is not difficult to show that our Theorem II applies separately to each Fourier component.

Volume 59, number 1 OPTICS COMMUNICATIONS 1 August 1986

NON-RADIATING MONOCHROMATIC SOURCES AND THEIR FIELDS ☆

Kisik KIM and Emil WOLF [1]

Department of Physics and Astronomy, University of Rochester, Rochester, NY 14627, USA

Received 24 March 1986

A new theorem is derived from which a number of previously known results relating to non-radiating monochromatic sources readily follow. Expressions are also obtained for the field within any non-radiating spherically symmetric source. As an example explicit expressions are found for the field within non-radiating homogeneous spherical sources and curves are presented which show the behavior of the field in several non-radiating sources of this kind.

1. Introduction

The subject of non-radiating sources has been investigated by many authors in connection with extended electron models and models of other elementary particles and in problems involving electromagnetic self-force and the radiation reaction ‡1. More recently questions relating to the uniqueness of inverse source problems [2,3] and of inverse scattering problems [4, 3] have been shown to be closely related to the presence of non-radiating sources. This subject has also attracted attention in connection with stochastic sources [6,8].

In this paper we obtain a new theorem relating to non-radiating monochromatic source distributions from which some of the previously known results readily follow. We also derive expressions for the field within any spherically symmetric non-radiating source. As an example we calculate the field within a non-radiating homogeneous spherical source and present plots of the field within several sources of this kind.

2. Preliminaries

We begin with the scalar wave equation for the field $V(\boldsymbol{r}, t)$ generated by a source distribution $Q(\boldsymbol{r}, t)$ in free space:

$$(\nabla^2 - c^{-2}\, \partial^2/\partial t^2)\, V(\boldsymbol{r}, t) = -4\pi Q(\boldsymbol{r}, t). \tag{2.1}$$

Here $\boldsymbol{r}$ denotes the position vector of a typical point in space, t denotes the time and c the speed of light. We assume that for all times the source distribution is localized within a finite volume D. We will only consider monochromatic sources and monochromatic fields. Then $V(\boldsymbol{r}, t)$ and $Q(\boldsymbol{r}, t)$ are of the form

$$Q(\boldsymbol{r}, t) = q(\boldsymbol{r}) \exp(-\mathrm{i}\omega t), \quad V(\boldsymbol{r}, t) = v(\boldsymbol{r}) \exp(-\mathrm{i}\omega t) \tag{2.2}$$

and eq. (2.1) yields the reduced wave equation

$$(\nabla^2 + k^2)\, v(\boldsymbol{r}) = -4\pi q(\boldsymbol{r}), \tag{2.3}$$

where

$$k = \omega/c. \tag{2.4}$$

To begin with we will assume that $q(\boldsymbol{r})$ is a continuous function of $\boldsymbol{r}$, both within and on the boundary S of the volume of localization D; it vanishes, of course, outside D.

The outgoing solution of eq. (2.3), which has the asymptotic form

$$v(r\boldsymbol{s}) \sim A(\boldsymbol{s}) \exp(\mathrm{i}kr)/r \quad (\boldsymbol{s}^2 = 1) \tag{2.5}$$

as $kr \to \infty$ along any direction specified by real unit vector $\boldsymbol{s}$, is given by

☆ Research supported by the National Science Foundation.
[1] Also at the Institute of Optics, University of Rochester.
‡1 See ref. [1] for references to pertinent literature.

0 030-4018/86/$03.50 © Elsevier Science Publishers B.V. (North-Holland Physics Publishing Division)

Volume 59, number 1 OPTICS COMMUNICATIONS 1 August 1986

$$v(\mathbf{r}) = \int_D G_0(\mathbf{r} - \mathbf{r}')\, q(\mathbf{r}')\, \mathrm{d}^3 r', \tag{2.6}$$

where

$$G_0(R) = \exp(\mathrm{i}kR)/R \quad (R = |\mathbf{R}|) \tag{2.7}$$

is the outgoing free-space Green's function (for propagation at wave number $k = \omega/c$).

A source is said to be non-radiating if it does not produce any power flow across a sphere of infinite radius, centered on a point in the source region. For such a source the radiation pattern

$$A(\mathbf{s}) \equiv 0 \tag{2.8}$$

for all real unit vectors $\mathbf{s}$, i.e. the radiation pattern $A(\mathbf{s})$ has zero value for all directions. It is known that for such a source [9]

$$v(\mathbf{r}) \equiv 0 \quad \text{for all } \mathbf{r} \notin D. \tag{2.9}$$

It is also known that any of the following three conditions ensures that a monochromatic source is non-radiating [‡2]:

(i) $\tilde{q}(k\mathbf{s}) = 0$ for all real unit vectors $\mathbf{s}$, (2.10)

where $\tilde{q}(\mathbf{K})$ is the Fourier transform of $q(\mathbf{r})$, i.e.

$$\tilde{q}(\mathbf{K}) = \int_D q(\mathbf{r}') \exp(-\mathrm{i}\mathbf{K}\cdot\mathbf{r}')\, \mathrm{d}^3 r'. \tag{2.11}$$

(ii) $q(\mathbf{r})$ is of the form

$$q(\mathbf{r}) = -(1/4\pi)(\nabla^2 + k^2) f(\mathbf{r}), \tag{2.12}$$

where $f(\mathbf{r})$ is any function with continuous second partial derivatives that vanishes identically outside D.

(iii) The retarded and the advanced fields generated by $q(\mathbf{r})$ are equal, i.e.

$$\int_D q(\mathbf{r}')\, G_0(\mathbf{r} - \mathbf{r}')\, \mathrm{d}^3 r' = \int_D q(\mathbf{r}')\, G_0^*(\mathbf{r} - \mathbf{r}')\, \mathrm{d}^3 r', \tag{2.13}$$

where the asterisk denotes the complex conjugate. Eq. (2.13) implies that for a non-radiating source

$$\int_D q(\mathbf{r}')\, j_0(k|\mathbf{r} - \mathbf{r}'|)\, \mathrm{d}^3 r' = 0, \tag{2.14}$$

where $j_0(kr)$ is the spherical Bessel function of order zero, viz.

$$j_0(kR) = (\sin kR)/kR. \tag{2.15}$$

Eq. (2.14) holds for all points $\mathbf{r}$, whether inside or outside the source volume D.

3. A theorem on non-radiating sources

Consider the reduced wave equations for fields $v_1(\mathbf{r})$ and $v_2(\mathbf{r})$ generated by source distributions $q_1(\mathbf{r})$ and $q_2(\mathbf{r})$ localized in domains D_1 and D_2 respectively:

$$(\nabla^2 + k^2)\, v_1(\mathbf{r}) = -4\pi q_1(\mathbf{r}), \tag{3.1a}$$

$$(\nabla^2 + k^2) v_2(\mathbf{r}) = -4\pi q_2(\mathbf{r}). \tag{3.1b}$$

Let D be any volume containing D_1 and D_2 in its interior and let S be the closed surface bounding D (fig. 1). Then according to Green's theorem,

$$\int_D (v_1 \nabla^2 v_2 - v_2 \nabla^2 v_1)\, \mathrm{d}^3 r$$

$$= \int_S (v_1 \partial v_2/\partial n - v_2 \partial v_1/\partial n)\, \mathrm{d}S, \tag{3.2}$$

where $\partial/\partial n$ denotes differentiation along the outward normal to S. If we eliminate the terms $\nabla^2 v_1$ and $\nabla^2 v_2$ in the integral on the left of eq. (3.2) by use of eqs. (3.1) we obtain the identity

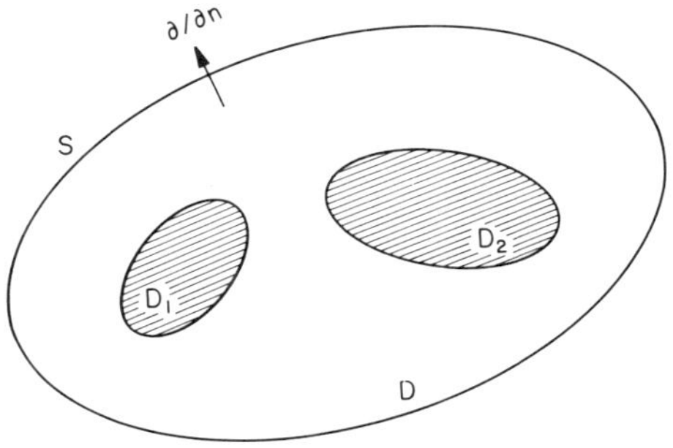

Fig. 1. Illustrating the notation relating to derivation of eq. (3.6)

[‡2] Condition (i) was derived in ref. [2] and its electromagnetic analog was obtained in ref. [1]. Condition (ii) is a scalar analog of a corresponding result derived in ref. [1] for non-radiating charge-current distributions. Condition (iii) follows from a result obtained in ref. [9] when specialized to non-radiating monochromatic sources. The requirement (2.14) was derived in ref. [2].

Volume 59, number 1 OPTICS COMMUNICATIONS 1 August 1986

$$-4\pi \int_D (v_1 q_2 - v_2 q_1)\, d^3r$$

$$= \int_S (v_1 \partial v_2/\partial n - v_2 \partial v_1/\partial n)\, dS. \qquad (3.3)$$

Suppose now that the source distribution q_1 is non-radiating. Then

$$v_1(\boldsymbol{r}) = \nabla v_1(\boldsymbol{r}) = 0 \quad \text{when } \boldsymbol{r} \notin D_1, \qquad (3.4a)$$

and also, of course,

$$q_1(\boldsymbol{r}) = 0 \quad \text{when } \boldsymbol{r} \notin D_1. \qquad (3.4b)$$

If we make use of these results in eq. (3.3) and also use the facts that $v_1(\boldsymbol{r})$ and $\nabla v_1(\boldsymbol{r})$ are continuous across the surface S, the equation reduces to

$$\int_{D_1} (v_1 q_2 - v_2 q_1)\, d^3r = 0. \qquad (3.5)$$

Next suppose that domains D_1 and D_2 are disjoint. Then $q_2(\boldsymbol{r}) \equiv 0$ throughout D_1 and eq. (3.5) implies that

$$\int_{D_1} v_2(\boldsymbol{r})\, q_1(\boldsymbol{r})\, d^3r = 0. \qquad (3.6)$$

We have thus established the following *theorem*: *Any continuous non-radiating monochromatic source distribution $q(\boldsymbol{r})$ localized in a finite closed domain D is orthogonal to every solution $v(\boldsymbol{r})$ in D of the Helmholtz equation* $(\nabla^2 + k^2)v(\boldsymbol{r}) = 0$.

Some of the known results about non-radiating distributions follow at once from this theorem. In particular, with the choice $v_2(\boldsymbol{r}) = \exp(-ik\boldsymbol{s}\cdot\boldsymbol{r})$, eq. (3.6) reduces to eq. (2.10). With the choice $v_2(\boldsymbol{r}) = j_0(k|\boldsymbol{r} - \boldsymbol{r}'|)$, eq. (3.6) reduces to eq. (2.14).

It is to be noted that if we choose

$$v_2(\boldsymbol{r}) = \exp(ik|\boldsymbol{r} - \boldsymbol{r}_0|)/|\boldsymbol{r} - \boldsymbol{r}_0|, \quad \boldsymbol{r}_0 \notin D, \qquad (3.7)$$

eq. (3.6) reduces to

$$\int q_1(\boldsymbol{r}) \frac{\exp(ik|\boldsymbol{r} - \boldsymbol{r}_0|)}{|\boldsymbol{r} - \boldsymbol{r}_0|}\, d^3r = 0, \quad \boldsymbol{r}_0 \notin D, \qquad (3.8)$$

verifying that the source distribution $q_1(\boldsymbol{r})$ does not generate a field outside its domain of localization, as we assumed in deriving the above theorem.

4. Fields within non-radiating, spherically symmetric sources

Let us consider a non-radiating, monochromatic source distribution that is spherically symmetric, i.e. for which q is a function of $r = |\boldsymbol{r}|$ only, localized in a spherical domain of radius a. The condition (2.14) for the source to be non-radiating takes now a simpler form, which can be readily derived with the help of the expansion [‡3]

$$j_0(k|\boldsymbol{r} - \boldsymbol{r}'|)$$

$$= 4\pi \sum_{l=0}^{\infty} \sum_{m=-l}^{l} j_l(kr)\, j_l(kr')\, Y_{lm}(\theta, \varphi)\, Y^*_{lm}(\theta', \varphi'), \qquad (4.1)$$

where the j_l are spherical Bessel functions and the Y_{lm} are spherical harmonics. On substituting from eq. (4.1) into eq. (2.14) and making use of elementary properties of spherical harmonics we readily find that in order for the spherically symmetric source distribution $q(r)$ to be non-radiating it must satisfy the constraint

$$\int_0^a q(r')\, j_0(kr')\, r'^2\, dr' = 0. \qquad (4.2)$$

A relatively simple expression for the field generated by such a source distribution can be obtained from eq. (2.6) with the help of the multipole expansion of the outgoing free-space Green's function, viz. [10]

$$G_0(\boldsymbol{r} - \boldsymbol{r}') = 4\pi i k$$

$$\times \sum_{l=0}^{\infty} \sum_{m=-l}^{l} j_l(kr_<)\, h_l^{(1)}(kr_>)\, Y_{lm}(\theta, \varphi)\, Y^*_{lm}(\theta', \varphi'), \qquad (4.3)$$

where j_l and Y_{lm} have the same meaning as before and $h_l^{(1)}$ is the spherical Hankel function of the first kind and of order l. $r_<$ and $r_>$ denote the smaller and the larger of the distances $r = |\boldsymbol{r}|$ and $r' = |\boldsymbol{r}'|$. On sub-

‡3 This expansion readily follows by taking the imaginary part of the representation (4.3) of the outgoing spherical Green's function, dividing it by k, and making use of the addition theorem for spherical harmonics and of the fact that $j_n(x)$ is the real part of $h_n^{(1)}(x)$.

stituting from eq. (4.3) into eq. (2.6), using the fact that the source distribution is spherically symmetric and using elementary properties of spherical harmonics we readily find that the (spherically symmetric) field generated by the source is given by

$$v(r)=4\pi \mathrm{i}k\int_0^a j_0(kr_<)\,h_0^{(1)}(kr_>)\,q(r')\,r'^2\,\mathrm{d}r'. \qquad (4.4)$$

We will express the right-hand side of eq. (4.4) as the sum of two integrals, one extending over the range $0\leqslant r'\leqslant r$ and the other over the range $r<r'\leqslant a$:

$$v(r)=4\pi \mathrm{i}k\left(h_0^{(1)}(kr)\int_0^r j_0(kr')\,q(r')\,r'^2\,\mathrm{d}r' + j_0(kr)\int_r^a h_0^{(1)}(kr')\,q(r')\,r'^2\,\mathrm{d}r'\right). \qquad (4.5)$$

Now it follows from the requirement (4.2) that

$$\int_0^r j_0(kr')\,q(r')\,r'^2\,\mathrm{d}r' = -\int_r^a j_0(kr')\,q(r')\,r'^2\,\mathrm{d}r', \qquad (4.6)$$

and, using this relation, the expression (4.5) may be re-written in the form

$$v(r)=4\pi \mathrm{i}k\left(j_0(kr)\int_r^a h_0^{(1)}(kr')\,q(r')\,r'^2\,\mathrm{d}r' - h_0^{(1)}(kr)\int_r^a j_0(kr')\,q(r')\,r'^2\,\mathrm{d}r'\right). \qquad (4.7)$$

Next we make use of the relations [11]

$$h_0^{(1)}(kr)=[h_0^{(2)}(kr)]^* = (-\mathrm{i})\exp(\mathrm{i}kr)/kr \qquad (4.8)$$

and

$$j_0(kr)=\tfrac{1}{2}\,[h_0^{(1)}(kr)+h_0^{(2)}(kr)]\,. \qquad (4.9)$$

On substituting for $j_0(kr)$ from eq. (4.9) into eq. (4.7) and making use of eq. (4.8) we finally obtain, after a straightforward calculation, the following expression for the field within a non-radiating spherically symmetric source of radius a:

$$v(r)=A(r)\exp(-\mathrm{i}kr)/r + B(r)\exp(\mathrm{i}kr)/r \quad (r\leqslant a) \qquad (4.10)$$

where

$$A(r)=(2\pi\mathrm{i}/k)\int_r^a r'\,q(r')\exp(\mathrm{i}kr')\,\mathrm{d}r', \qquad (4.11\mathrm{a})$$

$$B(r)=(2\pi\mathrm{i}/k)\int_r^a r'\,q(r')\exp(-\mathrm{i}kr')\,\mathrm{d}r'. \qquad (4.11\mathrm{b})$$

We see from eq. (4.10) that *the field at distance r ($\leqslant a$) from the center of the source is expressed entirely in terms of the source distribution within the shell $r\leqslant r'<a$*. Of course the source distribution in this shell is not entirely independent of the source distribution in the rest of the spherical domain ($0<r'<r$) because of the constraint (4.6).

5. Example: Field within a non-radiating homogeneous spherical source

We will now apply the formulas (4.10) and (4.11) to determine the field within a non-radiating *homogeneous* source of radius a. For such a source [‡4]

$$q(r) = \text{constant } (q_0 \text{ say}) \quad \text{when } r<a$$
$$= 0 \quad \text{when } r>a. \qquad (5.1)$$

The condition (4.2), which expresses the fact that the source is non-radiating, becomes

$$\int_0^a j_0(kr')\,r'^2\,\mathrm{d}r' = 0. \qquad (5.2)$$

On substituting in eq. (5.2) for $j_0(kr')$ its explicit expression $(\sin kr')/kr'$ and integrating by parts we read-

‡4 This source distribution violates our earlier assumption that $q(r)$ is a continuous function of position. However we can expect, by analogy with a well-known theorem relating to the continuity properties of potentials due to volume distribution of piecewise continuous densities [12], that the field $v(r)$ and also its gradient will nevertheless be continuous throughout all space; and that consequently the field generated by this source as well as its normal derivative will vanish on the source boundary. Explicit calculations show that this indeed is the case [see eq. (5.9)].

Volume 59, number 1 OPTICS COMMUNICATIONS 1 August 1986

ily find that eq. (5.2) reduces to ‡5

$$j_1(ka) = 0, \qquad (5.3)$$

where $j_1(x) = (\sin x)/x^2 - (\cos x)/x$ is the spherical Bessel function of order 1.

The three smallest roots of the equation $j_1(x) = 0$ are known to be (ref. [11], p. 467) $x \approx 4.493, 7.725, 10.904$, and hence the three smallest homogeneous non-radiating sources have radii

$$a \approx 0.715\lambda, \quad a \approx 1.571\lambda, \quad a \approx 1.735\lambda,$$

where $\lambda = 2\pi/k = 2\pi c/\omega$ is the wavelength associated with the frequency ω. Roots of the equation $j_1(x) = 0$ which are large compared to unity are given by the asymptotic formula $x = (n + \frac{1}{2})\pi$ where n is a (large) positive integer (ref. [11], p. 371) and hence the radii of large non-radiating homogeneous spherical sources are given by

$$a \sim (n + \tfrac{1}{2})\lambda/2 \quad (n \gg 1, \text{positive integer}).$$

The fields within these sources are given by the expression (4.10), with

$$A(r) = (2\pi \mathrm{i} q_0/k) \int_r^a r' \exp(\mathrm{i}kr')\,\mathrm{d}r', \qquad (5.4a)$$

$$B(r) = -(2\pi \mathrm{i} q_0/k) \int_r^q r' \exp(-\mathrm{i}kr')\,\mathrm{d}r'. \qquad (5.4b)$$

These integrals may be readily evaluated by integrating by parts and one finds that

$$A(r) = (2\pi q_0/k^2)\{[(1/\mathrm{i}k) - r] \exp(\mathrm{i}kr) - [(1/\mathrm{i}k) - a] \exp(\mathrm{i}ka)\} \qquad (5.5a)$$

$$B(r) = -(2\pi q_0/k^2)\{[(1/\mathrm{i}k) + r] \exp(-\mathrm{i}kr) - [(1/\mathrm{i}k) + a] \exp(-\mathrm{i}ka)\}. \qquad (5.5b)$$

On substituting from eqs. (5.5) into eq. (4.10) we obtain, after straightforward calculation, the following expressions for the field within the source:

‡5 The condition (5.3) was previously derived by Carter and Wolf [13], eq. (4.20), in investigations relating to radiation from stochastic sources. They showed that a fully coherent homogeneous source will not radiate at frequency $\omega = kc$ if its radius a is a root of eq. (5.3).

$$v(r) = (4\pi q_0/k^2)\{(a/r) \cos[k(a - r)] - (1/kr) \sin[k(a - r)] - 1\}. \qquad (5.6)$$

The formula (5.6) may be expressed in a simpler form by using the fact that the radius a of the source must satisfy the equation

$$\sin ka = ka \cos ka, \qquad (5.7)$$

as follows at once from eq. (5.3). On making use in eq. (5.6) of elementary trigonometrical identities and the relation (5.7) we finally obtain for the field within a non-radiating homogeneous source of radius a the expression:

$$v(r) = (4\pi q_0/k^2)\{[j_0(kr)/\cos ka] - 1\}. \qquad (5.8)$$

It readily follows from eq. (5.8) that

$$v(a) = [\mathrm{d}v(r)/\mathrm{d}r]_{r=a} = 0, \qquad (5.9)$$

which implies that the field and its normal derivatives are continuous across the source boundary as we suspected (see footnote ‡4).

In fig. 2 the fields within the three smallest spherical homogeneous sources calculated from eq. (5.6), are plotted as functions of kr.

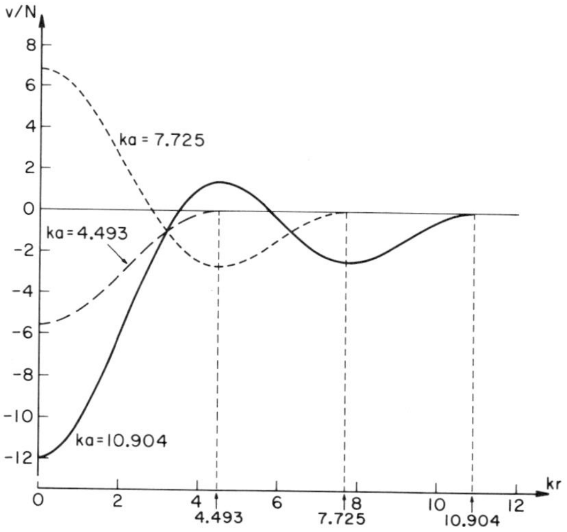

Fig. 2. Normalized fields $v(r)/N$ ($N = 4\pi q_0/k^2$) within the three smallest non-radiating, monochromatic, homogeneous spherical sources.

Volume 59, number 1 OPTICS COMMUNICATIONS 1 August 1986

Acknowledgement

We are obliged to A. Gamliel for assistance with computations relating to fig. 2.

References

[1] A.J. Devaney and E. Wolf, Phys. Rev. D 8 (1973) 1044.
[2] N.J. Bleistein and J.K. Cohen, J. Math. Phys. 18 (1977) 194.
[3] A.J. Devaney and G.C. Sherman, IEEE Trans. Antennas and Propagat. AP-30 (1982) 1034.
[4] A.J. Devaney, J. Math. Phys. 19 (1978) 1526.
[5] A.J. Devaney, J. Math. Phys. 20 (1979) 1687.
[6] B.J. Hoenders and H.P. Baltes, Lett. Nuovo Cimento 25 (1979) 206.
[7] A.J. Devaney and E. Wolf, in Coherence and quantum optics V, eds. L. Mandel and E. Wolf (Plenum Press, New York, 1984) p. 417.
[8] I.J. LaHaie, J. Opt. Soc. Am. 2 (1985) 35.
[9] F.G. Friedlander, Proc. London Math. Soc. 27 (1973) 551.
[10] J.D. Jackson, Classical electrodynamics (2nd Ed., J. Wiley, New York, 1975) p. 742, eq. (16.22).
[11] Handbook of mathematical functions, eds. M. Abramowitz and I.A. Stegun (Dover, New York, 1965), Chapt. 10.
[12] O.D. Kellog, Foundations of potential theory (Dover, New York, 1953) Chapt. VI, Sec. 3.
[13] W.H. Carter and E. Wolf, Optica Acta 28 (1981) 227.

1388 J. Opt. Soc. Am. A/Vol. 6, No. 9/September 1989

A new method for specifying nonradiating, monochromatic, scalar sources and their fields

A. Gamliel

The Institute of Optics, University of Rochester, Rochester, New York 14627

K. Kim

Department of Chemistry, University of Rochester, Rochester, New York 14627

A. I. Nachman

Department of Mathematics, University of Rochester, Rochester, New York 14627

E. Wolf

Department of Physics and Astronomy, University of Rochester, Rochester, New York 14627

Received January 9, 1989; accepted May 5, 1989

Nonradiating sources and the fields that they generate within the source domain are characterized in a novel way, as solutions to an overspecified boundary value problem. This characterization is used to describe a procedure for determining all nonradiating, spherically symmetric sources of a finite radius. An example of a source of this kind is presented and is discussed in detail.

1. INTRODUCTION

As is well known, the subject of nonradiating source distributions is intimately related to questions concerning the uniqueness of various wave-theoretic inverse problems.[1] In spite of this fact, no general technique appears to have been developed so far for specifying all nonradiating sources that can be associated with a given source domain. Moreover, relatively little is known about fields that nonradiating sources generate within the region that they occupy.[2,3]

In the present paper we show that nonradiating sources and the fields that they generate within the source domain are simultaneous solutions of a certain overspecified boundary value problem. Using this fact, a procedure is described for determining all nonradiating spherically symmetric sources of a finite radius. An example of such a source is discussed in detail, and it is verified, by explicit calculations, that the field radiated by the source vanishes at every point outside the source domain as a result of destructive interference.

2. NONRADIATING MONOCHROMATIC SOURCES AND THEIR FIELDS AS SIMULTANEOUS SOLUTIONS TO AN OVERSPECIFIED BOUNDARY VALUE PROBLEM

Consider the reduced wave equation

$$(\nabla^2 + k^2)V(\mathbf{r}) = -4\pi Q(\mathbf{r}), \tag{2.1}$$

where k is a positive real constant and $Q(\mathbf{r})$ is a function that vanishes identically outside a finite simply connected domain D bounded by a closed surface S. We will assume that $Q(\mathbf{r})$ is a bounded function of $\mathbf{r}$ within D. The outgoing solution of Eq. (2.1) has the asymptotic form

$$V(r\hat{s}) \sim A(\hat{s})\,\frac{e^{ikr}}{r} \tag{2.2}$$

as $kr \to \infty$ along any direction specified by a real unit vector $\hat{s}$. The source is said to be nonradiating if the total power generated by it is zero. This requirement is equivalent to the condition that

$$\int_{(4\pi)} |A(\hat{s})|^2 d\Omega = 0, \tag{2.3}$$

where the integral extends over the whole 4π solid angle generated by the unit vector $\hat{s}$. It is known that the condition (2.3) implies that[3,4]

$$V(\mathbf{r}) = 0 \quad \text{when} \quad \mathbf{r} \notin D, \tag{2.4}$$

and consequently also

$$\nabla V(\mathbf{r}) \equiv 0 \quad \text{when} \quad \mathbf{r} \notin D. \tag{2.5}$$

It follows at once from a theorem established in Appendix A that $V(\mathbf{r})$ and $\nabla V(\mathbf{r})$ are necessarily continuous on the surface S bounding the source domain D. Consequently the field $V_{\mathrm{NR}}(\mathbf{r})$, generated by a nonradiating source $Q_{\mathrm{NR}}(\mathbf{r})$, must satisfy the two boundary conditions

$$V_{\mathrm{NR}}(\mathbf{r})\big|_{\mathbf{r}\in S} = 0, \tag{2.6}$$

$$\left.\frac{\partial V_{\mathrm{NR}}(\mathbf{r})}{\partial n}\right|_{\mathbf{r}\in S} = 0, \tag{2.7}$$

0740-3232/89/091388-06$02.00

where $\partial/\partial n$ denotes differentiation along the outward normal to D. We thus have the following result:

Theorem: A bounded nonradiating source distribution $Q_{NR}(\mathbf{r})$ *of finite support and the field* $V_{NR}(\mathbf{r})$ *that it generates are related by the reduced wave equation*

$$(\nabla^2 + k^2)V_{NR}(\mathbf{r}) = -4\pi Q_{NR}(\mathbf{r}), \tag{2.8}$$

subject to the boundary conditions (2.6) and (2.7).

Since the differential equation (2.8) is elliptic, only one of the two boundary conditions (2.6) and (2.7) is required to yield a unique solution for the field if the source function is specified. Thus the above theorem characterizes fields generated by nonradiating sources, and the nonradiating sources themselves as quantities that are related by the reduced wave equation with overspecified boundary conditions.

3. SPHERICALLY SYMMETRIC NONRADIATING SOURCES AND THEIR FIELDS

We will now apply the preceding formulation to obtain expressions for nonradiating spherically symmetric source distributions. In this case $Q_{NR}(\mathbf{r})$ and consequently $V_{NR}(\mathbf{r})$ depend on $\mathbf{r}$ only through the distance $r = |\mathbf{r}|$ from the center of the source domain D, which we take to be a sphere of finite radius a. The reduced wave equation (2.8) now becomes[5]

$$\left[\frac{1}{r^2}\frac{\mathrm{d}}{\mathrm{d}r}\left(r^2\frac{\mathrm{d}}{\mathrm{d}r}\right) + k^2\right]V_{NR}(r) = -4\pi Q_{NR}(r), \tag{3.1}$$

and the boundary conditions (2.6) and (2.7) take the forms

$$V(a) = 0, \tag{3.2}$$

$$\left[\frac{\mathrm{d}V(r)}{\mathrm{d}r}\right]_{r=a} = 0. \tag{3.3}$$

To determine the source and field distributions that are related by Eq. (3.1) we will first expand them formally in terms of the eigenfunctions of an associated Sturm–Liouville system. Let $\{\psi_n(r)\}$ be the set of eigenfunctions and $\{k_n^2\}$ be the set of eigenvalues of the equation

$$\left[\frac{1}{r^2}\frac{\mathrm{d}}{\mathrm{d}r}\left(r^2\frac{\mathrm{d}}{\mathrm{d}r}\right) + k_n^2\right]\psi_n(r) = 0 \qquad (n = 0, 1, 2\ldots), \tag{3.4}$$

$r \in D$, with the boundary condition

$$\psi_n(a) = 0. \tag{3.5}$$

We take the eigenfunctions to be orthonormalized over the spherical domain D:

$$\int_{r\leq a} \psi_n^*(r)\psi_m(r)\mathrm{d}^3r = \delta_{nm}, \tag{3.6}$$

where the asterisk denotes the complex conjugate and δ_{nm} is the Kronecker symbol. From the properties of the solutions of Eq. (3.4) (cf. Ref. 5, Sec. 16.1), one readily finds that

$$\psi_n(r) = \frac{k_n}{\sqrt{2\pi a}} j_0(k_n r) \equiv \frac{k_n}{\sqrt{2\pi a}}\left[\frac{\sin(k_n r)}{k_n r}\right], \tag{3.7}$$

where

$$k_n = \frac{(n+1)\pi}{a}. \tag{3.8}$$

Since the eigenfunctions $\{\psi_n\}$ form a complete set in the Hilbert space of spherically symmetric functions that are square integrable over the domain D, we may represent any field $V_{NR}(r)$ produced by a continuous nonradiating source distribution $Q_{NR}(r)$, localized in the sphere D, in the form

$$V_{NR}(r) = \sum_{n=0}^{\infty} a_n\psi_n(r), \tag{3.9}$$

where the a_n's are constants with $\Sigma_{n=1}^{\infty} |a_n|^2 < \infty$. Since $\nabla^2 V_{NR}$ is also square integrable, we have $\Sigma_{n=1}^{\infty} k_n^4|a_n|^2 < \infty$ or, equivalently, $\Sigma_{n=1}^{\infty} n^4|a_n|^2 < \infty$.

Since the eigenfunctions $\{\psi_n\}$ satisfy the requirement (3.5), the function $V_{NR}(r)$, given by Eq. (3.9), necessarily obeys the boundary condition (3.2). In order that the $V_{NR}(r)$ also obey the boundary condition (3.3), the constants a_n must evidently satisfy the constraint

$$\sum_{n=0}^{\infty} \alpha_n a_n = 0, \tag{3.10}$$

where

$$\alpha_n = \frac{\mathrm{d}}{\mathrm{d}r}\psi_n(r)\bigg|_{r=a}. \tag{3.11}$$

The constants α_n can be evaluated at once. It follows from Eq. (3.7) that

$$\frac{\mathrm{d}\psi_n(r)}{\mathrm{d}r} = \frac{k_n}{a\sqrt{2\pi a}}\left[\cos(k_n a) - \frac{\sin(k_n a)}{k_n a}\right], \tag{3.12}$$

and hence, when the expression (3.8) is used for k_n,

$$\alpha_n \equiv \frac{\mathrm{d}\psi_n(r)}{\mathrm{d}r}\bigg|_{r=a} = \frac{(-1)^{n+1}(n+1)\pi}{a^2\sqrt{2\pi a}}. \tag{3.13}$$

The source distribution $Q_{NR}(r)$ that gives rise to the field distribution (3.9) is readily obtained on substituting the expression (3.9) into Eq. (3.1) and on using Eq. (3.4). We then find that

$$Q_{NR}(r) = -\frac{1}{4\pi}\sum_{n=0}^{\infty}(k^2 - k_n^2)a_n\psi_n(r). \tag{3.14}$$

We have just shown how one may construct spherically symmetric nonradiating source distributions and the corresponding fields that they generate. We will now construct a complete set of such distributions, which can be used as a basis for generating all of them.

Consider the functions

$$V_1(r), V_2(r), V_3(r), \ldots V_m(r), \ldots, \tag{3.15}$$

where

$$V_m(r) = v_0^{(m)}\psi_0(r) + v_1^{(m)}\psi_1(r) + \ldots + v_m^{(m)}\psi_m(r) \tag{3.16}$$

and the $v_j^{(m)}$ $(j = 0, 1, \ldots m)$ are constants to be determined. Irrespective of any particular choice of these constants, all the functions $V_m(r)$ will satisfy the boundary condition (3.2), because all the eigenfunctions $\psi_m(r)$ satisfy it [Eq. (3.5)]. We will now choose the constants $v_j^{(m)}$ so that

(1) Each $V_m(r)$ also satisfies the boundary condition (3.3) and

(2) The set $\{V_m\}$ is orthonormal, i.e.,

$$\int_D V_n^*(r)V_m(r)\mathrm{d}^3r = \delta_{nm}. \tag{3.17}$$

Requirement (1) ensures, in view of the theorem established in Section 2 and because of Eqs. (3.1) and (3.4), that each member $V_m(r)$ of the sequence (3.15) is a field generated by the nonradiating distribution

$$Q_m(r) = -\frac{1}{4\pi}\sum_{j=0}^{m}(k^2 - k_j^2)v_j^{(m)}\psi_j(r). \tag{3.18}$$

We note in passing that the sequence (3.15) does not contain a term, V_0 say, that is just proportional to ψ_0. The reason is that ψ_0 and consequently V_0 cannot be a field produced by a nonradiating source, because ψ_0 satisfies Eq. (3.4) with $n = 0$, subject to the boundary condition (3.2) and, in addition, would also have to satisfy the boundary condition $[\mathrm{d}\psi_0(r)/\mathrm{d}r]_{r=a} = 0$. However, as can readily be shown, for example by the use of Green's theorem, this is impossible unless $\psi_0(r) \equiv 0$ throughout D.

In order to determine the constants $v_j^{(m)}$, subject to the two prescribed conditions on the functions $V_m(r)$, we proceed as follows: We consider the first member of the sequence (3.15), viz.,

$$V_1(r) = v_0^{(1)}\psi_0(r) + v_1^{(1)}\psi_1(r). \tag{3.19}$$

In order that V_1 obey the boundary condition (3.3), we must have

$$\alpha_0 v_0^{(1)} + \alpha_1 v_1^{(1)} = 0, \tag{3.20}$$

where α_0 and α_1 are constants, defined by the formula (3.11). In order that V_1 satisfy the normalization condition (3.17), viz.,

$$\int_D V_1^*(r)V_1(r)\mathrm{d}^3r = 1, \tag{3.21}$$

we must have

$$|v_0^{(1)}|^2 + |v_1^{(1)}|^2 = 1. \tag{3.22}$$

The constraints (3.20) and (3.22) are two simultaneous equations for the two unknown constants $v_0^{(1)}$ and $v_1^{(1)}$. On solving them we find that

$$v_0^{(1)} = -\frac{\alpha_1}{\alpha_0}\left(1 + \left|\frac{\alpha_1}{\alpha_0}\right|^2\right)^{-1/2}\exp[i\phi^{(1)}], \tag{3.23}$$

$$v_1^{(1)} = \left(1 + \left|\frac{\alpha_1}{\alpha_0}\right|^2\right)^{-1/2}\exp[i\phi^{(1)}], \tag{3.24}$$

where $\phi^{(1)}$ is any arbitrary real constant. On substituting for α_0 and α_1 from Eq. (3.13) we obtain for $v_0^{(1)}$ and $v_1^{(1)}$ the values

$$v_0^{(1)} = \frac{2}{\sqrt{5}}\exp[i\phi^{(1)}], \qquad v_1^{(1)} = \frac{1}{\sqrt{5}}\exp[i\phi^{(1)}]. \tag{3.25}$$

The field $V_1(r)$ is now obtained on substituting from Eqs. (3.7), (3.8), and (3.25) into Eq. (3.19). The result is, if we omit the unessential factor $\exp[i\phi^{(1)}]$,

$$V_1(r) = \frac{1}{a}\left(\frac{2\pi}{5a}\right)^{1/2}\left[j_0\left(\frac{\pi r}{a}\right) + j_0\left(\frac{2\pi r}{a}\right)\right]. \tag{3.26}$$

The corresponding nonradiating source distribution $Q_1(r)$ is readily obtained on substituting from Eqs. (3.7), (3.8), and (3.25) into the formula (3.18) with $m = 1$, and one then finds that

$$Q_1(r) = -\frac{k^2}{2a\sqrt{10\pi a}}\left\{\left[1 - \left(\frac{\pi}{ka}\right)^2\right]j_0\left(\frac{\pi r}{a}\right) + \left[1 - \left(\frac{2\pi}{ka}\right)^2\right]j_0\left(\frac{2\pi r}{a}\right)\right\}. \tag{3.27}$$

It is of interest to note that, while $Q_1(r)$ depends on the wave number k, the field $V_1(r)$ that the source generates is independent of k. We also note that when $ka \gg \pi$, the expression (3.27) becomes

$$Q_1(r) \sim -\frac{k^2}{2a\sqrt{10\pi a}}\left[j_0\left(\frac{\pi r}{a}\right) + j_0\left(\frac{2\pi r}{a}\right)\right], \tag{3.28}$$

which is seen to be proportional to $V_1(r)$. Hence when the radius of the source is large compared with the wavelength $\lambda = 2\pi/k$ the field generated within the source domain imitates the behavior of the nonradiating source distributions that generate it.

Curves illustrating the behavior of the nonradiating source distribution $Q_1(r)$ and of the field $V_1(r)$ within the source are shown in Figs. 1 and 2.

It is important to appreciate that each of the two terms within the brackets on the right-hand side of the expression (3.28) does *not* represent a nonradiating source distribution. It is the *sum* of the fields generated by each of them that leads to complete cancellation of the field outside the spherical source domain, as we will now show. For this purpose let us again consider a spherically symmetric source distribution $Q(r)$ in free space occupying a spherical domain D of radius a, centered at the origin. By the use of the multipole expansion and elementary properties of spherical harmonics, one can readily show[6] that at any point outside D

$$V(r > a) = A\frac{e^{ikr}}{r}, \tag{3.29}$$

where

$$A = 4\pi\int_0^a Q(r')j_0(kr')r'^2\mathrm{d}r'. \tag{3.30}$$

In particular, with

$$Q(r) \equiv Q^{(n)}(r) = j_0\left(\frac{n\pi r}{a}\right) \qquad (n = 1, 2, 3, \ldots), \tag{3.31}$$

the field outside D is given by

$$V(r > a) \equiv V^{(n)}(r > a) = A^{(n)}\frac{e^{ikr}}{r}, \tag{3.32}$$

with

$$A^{(n)} = 4\pi\int_0^a j_0\left(\frac{n\pi r'}{a}\right)j_0(kr')r'^2\mathrm{d}r'. \tag{3.33}$$

An integral of the form (3.33) is evaluated in Appendix B, and it follows with the help of Eq. (B7) of that appendix that

$$A^{(n)} = \frac{\sin(ka)}{ka}\,\frac{4\pi(-1)^{n-1}a^3}{(n\pi)^2 - (ka)^2}. \tag{3.34}$$

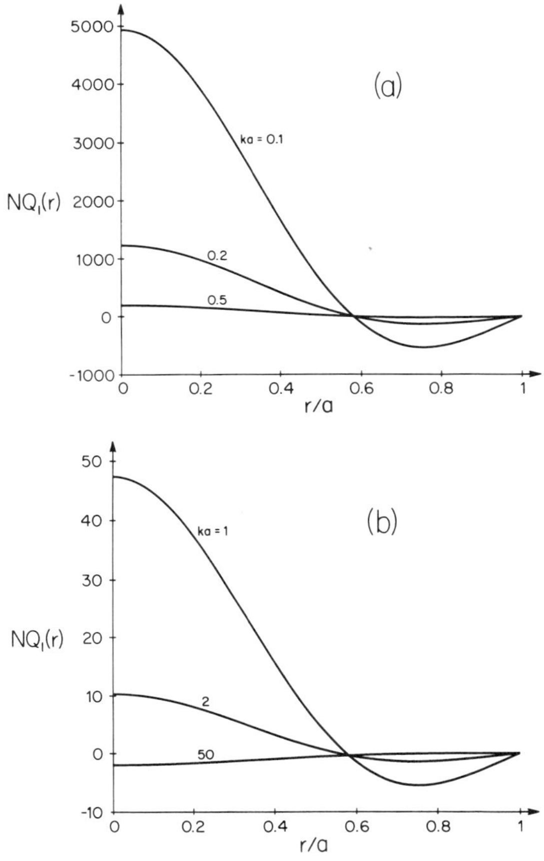

Fig. 1. Normalized nonradiating source distribution $NQ_1(r)$ ($N = 2a\sqrt{10\pi a}/k^2$), for different values of the parameter $ka = 2\pi a/\lambda$.

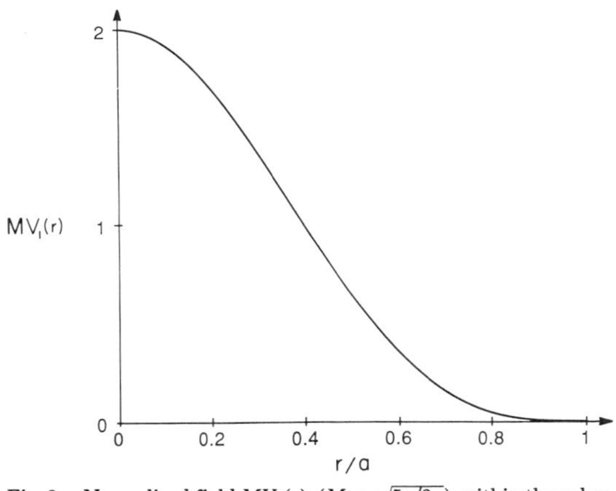

Fig. 2. Normalized field $MV_1(r)$, ($M = a\sqrt{5a/2\pi}$), within the spherical source, generated by the nonradiating distribution $Q_1(r)$, as a function of the normalized distance r/a from the center of the source.

Now, in the notation of Eq. (3.31), the nonradiating source $Q_1(r)$, given by Eq. (3.27), may be expressed in the form

$$Q_1(r) = -\frac{k^2}{2a\sqrt{10\pi a}}\left\{\left[1-\left(\frac{\pi}{ka}\right)^2\right]Q^{(1)}(r) + \left[1-\left(\frac{2\pi}{ka}\right)^2\right]Q^{(2)}(r)\right\}. \tag{3.35}$$

Hence, according to Eqs. (3.32) and (3.34), it will generate the field

$$V_1(r > a) = -\frac{k^2}{2a\sqrt{10\pi a}}\left\{\left[1-\left(\frac{\pi}{ka}\right)^2\right]A^{(1)} + \left[1-\left(\frac{2\pi}{ka}\right)^2\right]A^{(2)}\right\}\frac{e^{ikr}}{r} \tag{3.36}$$

outside the source distribution D. It follows from Eq. (3.34) that

$$\left[1-\left(\frac{\pi}{ka}\right)^2\right]A^{(1)} = -\frac{4\pi a}{k^2}\left[\frac{\sin(ka)}{ka}\right], \tag{3.37a}$$

$$\left[1-\left(\frac{2\pi}{ka}\right)^2\right]A^{(2)} = +\frac{4\pi a}{k^2}\left[\frac{\sin(ka)}{ka}\right]. \tag{3.37b}$$

Since the two expressions on the right-hand sides of these formulas differ in sign, their sum is zero, and hence Eq. (3.36) implies that

$$V_1(r) = 0 \quad \text{for all } r > a. \tag{3.38}$$

Thus the field generated by the source distribution $Q_1(r)$, given by Eq. (3.27), indeed vanishes outside the source distribution, and this fact may be attributed to the destructive interference of the two fields generated by the two source terms whose sum represents the nonradiating source distribution $Q_1(r)$.

We have considered in detail only the first member $V_1(r)$ of the sequence (3.15) of the fields generated by spherically symmetric nonradiating sources. We plan to show elsewhere that the general member of the sequence is given (up to a constant proportionality factor) by

$$V_m(r) = \frac{C_m}{r}\left\{\frac{3}{2}\,\frac{m\sin[\pi(m+1)r/a] + (m+1)\sin(\pi m r/a)}{\cos^2(\pi r/2a)} + m(2m+1)\sin[\pi(m+1)r/a]\right\}, \tag{3.39}$$

where

$$C_m = \frac{1}{[2\pi a m(m+2)(2m+1)(2m+3)]^{1/2}} \tag{3.40}$$

($m = 1, 2, 3, \ldots$) and, moreover, that the set of all nonradiating, spherically symmetric, square-integrable source distributions localized in a sphere of radius a centered at the origin may be represented in the form

$$Q_{NR}(r) = \sum_{m=1}^{\infty} b_m Q_m(r), \tag{3.41}$$

where

$$Q_m(r) = -\frac{1}{4\pi}(\nabla^2 + k^2)V_m(r) \tag{3.42}$$

and b_m are constants that satisfy the constraint

$$\sum_{m=1}^{\infty} m^4|b_m|^2 < \infty. \tag{3.43}$$

APPENDIX A: CONTINUITY OF $V(r)$ and $\nabla V(r)$

In this appendix we show that if the source Q is a bounded (not necessarily continuous) function then the corresponding field is continuous in the whole three-dimensional space $\Re^3$ and has continuous first derivatives. Our proof parallels a classical argument from potential theory (see, for example, Ref. 7).

Theorem

Suppose that D is a bounded domain in $\Re^3$ and Q is a bounded (measurable) function in D. Let

$$V(r) = \int_D \frac{\exp(ik|r - r'|)}{|r - r'|} Q(r')\mathrm{d}^3r' \tag{A1}$$

be the field generated by Q. Then

(a) *V is defined and is continuous at every point in $\Re^3$ and is bounded in every sphere $S_R = \{\mathbf{r}:|\mathbf{r}| \leq R\}$* and

(b) *∇V is defined and is continuous at every point in $\Re^3$.*

Proof

(a) Let R_0 be such that D is contained in the sphere of radius R_0 about the origin. Assume that Q is bounded by M. Then, in view of Eq. (A1), V is bounded by

$$|V(\mathbf{r})| \leq M\int_D \frac{\mathrm{d}^3r'}{|\mathbf{r} - \mathbf{r}'|} \leq M\int_{|\mathbf{r}'|\leq R+R_0} \frac{\mathrm{d}^3r'}{|\mathbf{r}'|} = CM(R + R_0)^2$$

$$\text{for all } \mathbf{r} \text{ in } S_R. \tag{A2}$$

The continuity of V follows from the following inequality: if $\mathbf{r}_1$, $\mathbf{r}_2$, and $\mathbf{r}'$ are points in the sphere S_R, then

$$\left|\frac{\exp(ik|\mathbf{r}_1 - \mathbf{r}'|)}{|\mathbf{r}_1 - \mathbf{r}'|} - \frac{\exp(ik|\mathbf{r}_2 - \mathbf{r}'|)}{|\mathbf{r}_2 - \mathbf{r}'|}\right| \leq C|\mathbf{r}_1 - \mathbf{r}_2|\left(\frac{1}{|\mathbf{r}_1 - \mathbf{r}'|^2} + \frac{1}{|\mathbf{r}_2 - \mathbf{r}'|^2}\right). \tag{A3}$$

The constant in the inequality (A3) depends only on R and k. It is easy to verify the inequality (A3) by using the mean-value theorem.

(b) By using the inequality (A3) it is straightforward to prove that the partial derivatives of V exist at every point and that

$$\nabla V(\mathbf{r}) = \int_D \nabla_{\mathbf{r}}\left(\frac{\exp(ik|\mathbf{r} - \mathbf{r}'|)}{|\mathbf{r} - \mathbf{r}'|}\right)Q(\mathbf{r}')\mathrm{d}^3r'. \tag{A4}$$

To show that ∇V is continuous at an arbitrary point $\mathbf{r}_0$ in $\Re^3$, let $S = S_d(\mathbf{r}_0)$ be the sphere of radius d around $\mathbf{r}_0$. Let us write the integral in Eq. (A4) as

$$\nabla V(\mathbf{r}) = \left(\int_{D\cap S} + \int_{D\setminus S}\right) \times \left[\frac{\mathbf{r} - \mathbf{r}'}{|\mathbf{r} - \mathbf{r}'|^3}(ik|\mathbf{r} - \mathbf{r}'| - 1)\exp(ik|\mathbf{r} - \mathbf{r}'|)Q(\mathbf{r}')\right]\mathrm{d}^3r'. \tag{A5}$$

Note that both integrals are absolutely convergent. When estimating the difference $|\nabla V(\mathbf{r}) - \nabla V(\mathbf{r}_0)|$, one can make the first integral arbitrarily small by choosing d small enough (independently of $\mathbf{r}$ and $\mathbf{r}_0$). In the second integral $\mathbf{r}'$ is outside S, and hence for all $\mathbf{r}$ in $S_{d/2}(\mathbf{r}_0)$ the distance $|\mathbf{r}' - \mathbf{r}|$ is greater than $d/2$; the integrand is then continuous in $\mathbf{r}$, and the result follows.

APPENDIX B: EVALUATION OF AN INTEGRAL

Let us consider the integral

$$I(\kappa, k; a) = \int_0^a j_0(\kappa r')j_0(kr')r'^2\mathrm{d}r'. \tag{B1}$$

Since

$$j_0(x) = \frac{\sin x}{x}, \tag{B2}$$

the integral (B1) can be rewritten in the form

$$I(\kappa, k; a) = \frac{1}{\kappa k}\int_0^a \sin(\kappa r')\sin(kr')\mathrm{d}r'. \tag{B3}$$

If we use an elementary trigonometric identity and recall Eq. (B2) again, we readily find that

$$I(\kappa, k; a) = \frac{a}{2\kappa k}\{j_0[(\kappa - k)a] - j_0[(\kappa + k)a]\}. \tag{B4}$$

Let us now consider the special case when κ is one of the eigenvalues k_n, given by Eq. (3.8), viz.,

$$k_n = \frac{(n + 1)\pi}{a} \qquad (n = 0, 1, 2, \ldots). \tag{B5}$$

Then

$$\begin{aligned} j_0[(k_n \pm k)a] &= j_0[(n + 1)\pi \pm ka] \\ &= \frac{\sin[(n + 1)\pi \pm ka]}{(n + 1)\pi \pm ka} \\ &= \frac{\mp(-1)^n \sin(ka)}{(n + 1)\pi \pm ka}, \end{aligned} \tag{B6}$$

and, on substituting from this formula into Eq. (B4), with $\kappa = k_n$ we readily find that

$$I(k_n, k; a) = \left[\frac{\sin(ka)}{ka}\right]\frac{(-1)^n a^3}{[(n + 1)\pi]^2 - (ka)^2}. \tag{B7}$$

ACKNOWLEDGMENT

This research was supported by the National Science Foundation and by the U.S. Army Research Office.

E. Wolf is also with The Institute of Optics, University of Rochester.

REFERENCES

1. For a review of this subject see, for example, B. Hoenders, "The uniqueness of inverse problems," in *Inverse Source Problems in Optics*, H. P. Baltes, ed. (Springer-Verlag, New York, 1978), Chap. 3, pp. 41–82.
2. The only investigations concerning fields that nonradiating sources generate within the source region that we are aware of are the following: A. J. Devaney and E. Wolf, "Radiating and nonradiating classical current distributions and the fields they generate," Phys. Rev. D **8,** 1044–1047 (1973); Kim and Wolf.[3]
3. K. Kim and E. Wolf, "Non-radiating monochromatic sources and their fields," Opt. Commun. **59,** 1 (1986).
4. F. G. Friedlander, "An inverse problem for radiation fields," Proc. London Math. Soc. **27,** 551–576 (1973).
5. Cf. J. D. Jackson, *Classical Electrodynamics*, 2nd ed. (Wiley, New York, 1975), p. 740, Eq. (16.5) with $l = 0$.
6. See, for example, W. H. Carter, "Band-limited angular-spectrum approximation to a spherical scalar wave field," J. Opt. Soc. Am. **65,** 1054–1058 (1975), appendix.
7. O. D. Kellog, *Foundations of Potential Theory* (Dover, New York, 1953), Chap. VI, Sec. 3.

Nonpropagating string excitations

Michael Berry
H. H. Wills Physics Laboratory, University of Bristol, Tyndall Avenue, Bristol BS8 1TL, United Kingdom

John T. Foley
Department of Physics and Astronomy, Mississippi State University, Mississippi State, Mississippi 39762

Greg Gbur and Emil Wolf
Department of Physics and Astronomy and the Rochester Theory Center for Optical Science and Engineering, University of Rochester, Rochester, New York 14627

(Received 6 August 1997; accepted 3 September 1997)

It is shown that certain force distributions applied to a finite portion of an infinitely long string do not produce any excitation outside the region of the applied force. The existence of such nonpropagating excitations is demonstrated by a simple example, and two general theorems concerning their nature are proven. Some analogies between nonpropagating string excitations and fields produced by nonradiating sources are noted.

It has been known in radiation theory for some time that there are localized source distributions which do not produce a field outside the domain of the source (see, for example, Refs. 1–4). In this paper we show that analogous situations exist for waves on an infinitely long string. Specifically, we show that there are localized force distributions which produce no displacement of the string at any point outside the region of the applied force, and we derive general theorems concerning such situations. These theorems are one-dimensional analogues of theorems encountered with nonradiating sources (usually discussed in three dimensions).

We consider an infinitely long flexible string under tension

T and with mass per unit length μ, undergoing small displacements $y(x,t)$ from the equilibrium position, driven by a force density $f(x,t)$ (force per unit length) and localized in the region $a \leqslant x \leqslant b$. The displacement obeys the wave equation[5]

$$\mu \frac{\partial^2 y(x,t)}{\partial t^2} - T \frac{\partial^2 y(x,t)}{\partial x^2} = f(x,t). \tag{1}$$

Restricting ourselves to simple harmonic driving forces,

$$f(x,t) \equiv \mathrm{Re}\{f(x)e^{-i\omega t}\}, \tag{2}$$

where Re denotes the real part, the steady-state solution $y(x,t)$ of Eq. (1) will have the same time dependence,

$$y(x,t) \equiv \mathrm{Re}\{y(x)e^{-i\omega t}\}, \tag{3}$$

and Eq. (1) then reduces to the one-dimensional inhomogeneous Helmholtz equation,

$$\frac{d^2 y(x)}{dx^2} + k^2 y(x) = q(x), \tag{4}$$

where k is the wave number,

$$k = \frac{\omega}{v}, \quad v = \sqrt{T/\mu} \tag{5a}$$

and

$$q(x) = -f(x)/T. \tag{5b}$$

We will call $q(x)$ the effective force density, or simply the force density.

The outgoing solution of Eq. (4) is well known to be[6]

$$y(x) = \frac{1}{2ik} \int_a^b q(x') e^{ik|x-x'|} dx'. \tag{6}$$

For displacements to the right ($x > b$) and left ($x < a$) of the region of the applied force, $y(x)$ reduces to

$$y(x)|_R = \frac{e^{ikx}}{2ik} \int_a^b q(x') e^{-ikx'} dx' \tag{7}$$

and

$$y(x)|_L = \frac{e^{-ikx}}{2ik} \int_a^b q(x') e^{ikx'} dx'. \tag{8}$$

It is apparent from Eqs. (7) and (8) that the excitations will vanish everywhere outside the force region $a \leqslant x \leqslant b$ if

$$\tilde{q}(k) = 0, \quad \tilde{q}(-k) = 0, \tag{9}$$

with k given by Eq. (5a), and $\tilde{q}(k)$ is the Fourier transform of the force density, i.e.,

$$\tilde{q}(K) = \frac{1}{2\pi} \int_a^b q(x) e^{-iKx} dx. \tag{10}$$

Nontrivial force densities that satisfy Eq. (9) will generate displacements of the string only within the region of the applied force, and will not produce any displacement outside it. We will refer to such a situation as a *nonpropagating string excitation.*

As a simple example of a nonpropagating excitation, let $a = -L$, $b = L$, $L > 0$, and let the force be constant throughout this domain:

$$q(x) = \begin{cases} Q_0 & \text{when } |x| \leqslant L, \\ 0 & \text{when } |x| > L. \end{cases} \tag{11}$$

Upon substituting from Eq. (11) into Eq. (10) and requiring that the two conditions (9) be fulfilled, we find that nontrivial solutions occur if and only if

$$kL = n\pi, \tag{12}$$

where $n = 1,2,\ldots$. This result shows that a *constant* localized force distribution within the region $-L \leqslant x \leqslant L$ produces a nonpropagating excitation only for certain special values of kL. Using this result in the general expression (6) for the displacement, one readily finds that

$$y(x) = \begin{cases} \dfrac{Q_0}{(n\pi/L)^2} \left[1 - (-1)^n \cos \dfrac{n\pi x}{L} \right] & \text{when } |x| \leqslant L \\ 0 \quad \text{when } |x| > L. \end{cases} \tag{13}$$

This displacement, along with the associated force density, is shown in Fig. 1 for the cases $n = 1$ and $n = 2$. We note that the displacement $y(x)$ given by Eq. (13) is continuous everywhere on the string, in particular at the boundary of the region of applied force. One can readily verify that the first derivative $dy(x)/dx$ is also continuous everywhere on the string. In the Appendix, we demonstrate that this behavior is a general property of excitations due to piecewise continuous, localized force distributions. This fact leads to the following theorem about nonpropagating string excitations.

Theorem I: A nonpropagating excitation on an infinitely long string and the piecewise continuous force distribution, assumed to be confined to a finite region $a \leqslant x \leqslant b$, which generates it are related by the inhomogeneous Helmholtz equation (4), subject to the boundary conditions

$$y(a) = y(b) = 0, \quad \left.\frac{dy}{dx}\right|_{x=a} = \left.\frac{dy}{dx}\right|_{x=b} = 0. \tag{14}$$

This theorem implies that the nonpropagating excitations are solutions to an *overspecified* Sturm–Liouville boundary-value problem, because only one of the two sets of boundary conditions (14) is required for a unique solution of the equation. Using Theorem I, one can construct numerous examples of nonpropagating string excitations. The theorem is a one-dimensional analogue of a theorem in radiation theory,[4] as is the following one.[3]

Theorem II: A force distribution, assumed to be confined to a region $a \leqslant x \leqslant b$, which generates a nonpropagating excitation on a infinitely long string, related by Eq. (4), is orthogonal to every solution of the homogeneous Helmholtz equation with wave number k.

To establish this theorem, let $y(x)$ be a nonpropagating excitation and $q(x)$ the force distribution which generates it, and let $u(x)$ be any solution of the homogeneous Helmholtz equation,

$$\frac{d^2 u}{dx^2} + k^2 u = 0. \tag{15}$$

We first multiply Eq. (4) by $u(x)$ and Eq. (15) by $y(x)$ and subtract the equations from each other. We then obtain the identity

$$u \frac{d^2 y}{dx^2} - y \frac{d^2 u}{dx^2} = q(x) u(x). \tag{16}$$

On integrating both sides of Eq. (15) with respect to x over the range $a \leqslant x \leqslant b$ and then integrating by parts on the left we obtain the relation

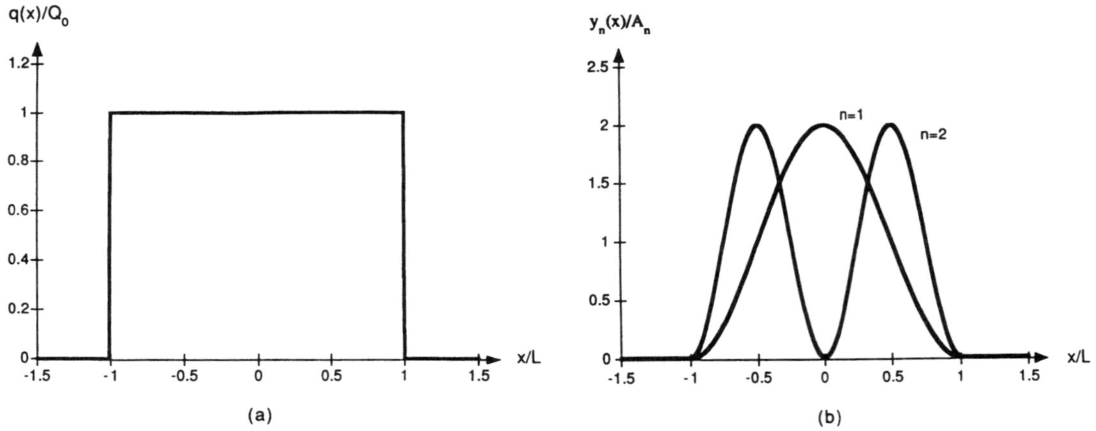

Fig. 1. Force density distribution $q(x)$, shown in (a), which produces nonpropagating excitations (b):

$$y_n(x)=A_n\left[1-(-1)^n \cos\frac{n\pi x}{L}\right], \quad \text{when } |x|\leq L$$
$$=0, \quad \text{when } |x|>L,$$

with $A_n=Q_0/(n\pi/L)^2$, $k=n\pi/L$, for $n=1$ and $n=2$.

$$\left[u\frac{dy}{dx}-y\frac{du}{dx}\right]_a^b=\int_a^b q(x)u(x)dx. \tag{17}$$

According to Theorem I, the left-hand side of Eq. (16) vanishes and consequently

$$\int_a^b q(x)u(x)dx=0, \tag{18}$$

as asserted by Theorem II.

We conclude by noting that in the example that we have presented, the force density which produced the nonpropagating excitation is quite simple [see Eqs. (11) and (12)]. This force density is, of course, not the only one with these properties. Any localized force density which is a piecewise continuous function of x and which obeys the conditions (9) will produce nonpropagating excitations on the string.

ACKNOWLEDGMENTS

This research was supported by the National Science Foundation and by the New York State Foundation for Advanced Technology.

APPENDIX: CONTINUITY OF $y(x)$ AND $dy(x)/dx$

Let $q(x)$ be a piecewise continuous function on a finite interval $a\leq x\leq b$, with its only discontinuities at the points $x_0=a,\ x_1,x_2,\ldots,\ x_n=b$. It is clear from the integral representation (6) that $y(x)$ is continuous at every point on the line, and one can readily deduce from that equation that its first derivative is continuous at every point with the possible exception of the points x_j, $j=0,1,\ldots,n$, where $q(x)$ is discontinuous. Now consider

$$\int_{x_j-\epsilon_1}^{x_j+\epsilon_2} y''(x)dx+k^2\int_{x_j-\epsilon_1}^{x_j+\epsilon_2} y(x)dx=\int_{x_j-\epsilon_1}^{x_j+\epsilon_2} q(x)dx \quad (j=0,1,\ldots,n), \tag{A1}$$

with $\epsilon_1\geq 0$, $\epsilon_2\geq 0$, which follows from the differential equation (4). From this equation it is clear that

$$y'(x)|_{x_j+\epsilon_2}-y'(x)|_{x_j-\epsilon_1} =-k^2\int_{x_j-\epsilon_1}^{x_j+\epsilon_2} y(x)dx+\int_{x_j-\epsilon_1}^{x_j+\epsilon_2} q(x)dx \quad (j=0,1,\ldots,n). \tag{A2}$$

Because $y(x)$ is continuous and $q(x)$ piecewise continuous, it is evident on proceeding to the limits $\epsilon_1\to 0$ and $\epsilon_2\to 0$ that $y'(x)$ is also continuous at the points of discontinuity of $q(x)$.

[1]F. G. Friedlander, "An inverse problem for radiation fields," Proc. London Math. Soc. **27**, 551–576 (1973).
[2]N. Bleistein and J. K. Cohen, "Nonuniqueness in the inverse source problem in acoustics and electromagnetics," J. Math. Phys. **18**, 194–201 (1977).
[3]K. Kim and E. Wolf, "Non-radiating monochromatic sources and their fields," Opt. Commun. **59**, 1–6 (1986).
[4]A. Gamliel, K. Kim, A. I. Nachman, and E. Wolf, "A new method for specifying nonradiating monochromatic sources and their fields," J. Opt. Soc. Am. A **6**, 1388–1393 (1989).
[5]See, for example, P. M. Morse and K. Uno Ingard, *Theoretical Acoustics* (Princeton U.P., Princeton, NJ, 1986), Chap. 4.
[6]See, for example, G. Arfken, *Mathematical Methods for Physicists* (Academic, San Diego, CA, 1985), 3rd ed., Secs. 16.5, 16.6.

Section 3 – Coherence and Statistical Optics

Most of my scientific work has been and still is concerned with coherence properties of light and with related topics in the broad area of statistical optics. The role this research has played in the development of this area of optics is discussed in a lecture "Development of Coherence Theory", of which a shortened version is presented in Sec. 7.3 of this volume. I will, therefore, comment here only briefly on the papers on this subject which are reproduced in Section 3.

Papers 3.1 and, particularly paper 3.2 provide a rigorous basis for the so-called second-order coherence theory of fluctuating scalar wavefields. The main results of the second of these two papers is the introduction of the so-called mutual coherence function, which is a precise measure of the correlation between field fluctuations at two points in space, and the demonstration that in free space the mutual coherence function rigorously obeys two wave equations. Papers 3.3 – 3.5 generalize these results to the full electromagnetic field. In particular, the paper 3.5 formulates the theory of partially polarized electromagnetic radiation in a form which is a natural generalization of the theory of partial coherence of scalar wave fields.

Paper 3.6 deals with the stochastic theory of photoelectric detection of light fluctuations. This subject became of particular interest in the 1960s, in connection with a controversy surrounding certain experiments of Hanbury Brown and Twiss. The controversy is briefly discussed in the article included in Sec. 7.3 of this volume.

Also in the 1960s the so-called phase-space representation of quantized fields become of considerable interest in the rapidly developing field of quantum optics. The phase-space representation of quantized fields is intimately related to the ordering of non-commuting operators and to their mapping onto c-number spaces. A general theory of this subject was developed in three papers by G.S. Agarwal and myself, published in Phys. Rev. D **2**, 2161–2186 (1970), 2187–2205 (1970) and 2206–2225 (1970). Because of limitations of space, they are not reprinted in this volume, but the essential aspects of the theory are outlined in papers 3.7 and 3.8.

Paper 3.9 is concerned with the dependence of the angular distribution of radiant intensity (radiated power per unit solid angle) from planar sources of different states of coherence. An important class of many laboratory sources and sources frequently found in nature are Lambertian sources. A planar source of this type radiates power with an angular dependence proportional to $\cos\theta$, where θ is the angle between the normal of the source and the direction of observation. Although such sources are often regarded as being incoherent this is, in fact, not so. In paper 3.10 it is shown that all planar Lambertian sources have unique spatial coherence properties.

In paper 3.9 a measure of correlations between light vibrations at two points in space, at a particular frequency, was also introduced. It is called the complex degree of spatial coherence or the spectral correlation coefficient. Later, the more appropriate term "degree of spatial coherence" came to be used. This measure of correlations in optical fields has become important in more recent developments of optical coherence theory. In paper 3.11 some properties of this correlation coefficient are discussed and its role in interference experiments is elucidated. In paper 3.15 it is shown that the spectral degree of coherence can be determined by Young's interference experiment and the use of filters.

Because a well-stabilized laser generates a highly directional beam, it has been generally taken for granted that only a source which is spatially highly coherent can produce light which propagates in a narrow solid angle of direction. In paper 3.12 it is shown that this is not so and that some partially coherent sources, including globally incoherent ones, may generate the same far-zone intensity distribution as a completely coherent laser. Later many more such "equivalent" sources were described in the literature.

Optical coherence theory as developed prior to the 1980s was based on a space-time description. In papers 3.13 and 3.14 a new formulation of coherence theory was introduced, which employed a space-frequency representation. This was not a trivial step, because the spectral representation of stationary random fields (i.e. fields whose statistical properties are invariant with respect to the translation of the origin of time) are known not to have a Fourier frequency representation. The new formulation was soon found to be very useful as it simplified the analysis of many problems involving the propagation of light from partially coherent sources and its interaction with various media. It also elucidated the coherence properties of the output of lasers, as shown in paper 3.16. Moreover, it lead to a discovery, reported in paper 3.17, that, in general, the spectrum of light may change on propagation even in free space. Such changes may be of many different kinds: a spectral line may be broadened, for example, or become narrower or it may split into several lines. Particularly surprising is the fact discussed in paper 3.18 and 3.19 that such correlation-induced spectral changes may result in blue shifts or red shifts of spectral lines, i.e. some of them appearing to be shifted towards the shorter or the longer wavelengths. Soon afterwards it was found that similar spectral modifications may be produced when polychromatic light is scattered by a random medium. This discovery and some of its rather interesting implications are discussed in the commentary to the next Section (Sec. 4).*

Young's interference experiment has played an important role in clarifying the role of correlations in optical fields and has provided a way of measuring them. Surprisingly, in recent years, more than 200 years since Thomas Young first described it, some new and interesting aspects of such an experiment have been revealed. In the traditional version of the experiment narrow-band light illuminates the two pinholes. With broad-band light, interference fringes formed by light of different spectral components are displaced relative to each other and, as a result, only an essentially uniform, rather than sinusoidally-modulated, intensity pattern is observed in the region of superposition. In paper 3.20 and 3.22 it is shown that even when broadband light is incident on the pinholes, the intensity in the region of superposition contains, nevertheless, important physical information, namely information about the spectral degree of the light incident on the pinholes. The information may be deduced by spectral analysis rather than from measurement of fringe visibility in the region of superposition. Spectral analysis shows, in agreement with the theory outlined in papers 3.22 and 3.23, that the intensity distribution of each spectral component of the incident light exhibits

*A review of numerous publications on correlation-induced spectral changes, both theoretical and experimental ones, is presented in an article by E. Wolf and D.F.V. James in Rep. Prog. Phys. , (IOP Publishing, Bristol and London), **59**, 771–818 (1996).

sinusoidal modulation, generally with different minima and maxima.† Papers 3.21 and 3.23 describe new methods for determining the spectral degree of coherence and, in addition, paper 3.22 also describes an application of some of the results to synthetic aperture imaging.

The phenomenon of correlation-induced spectral changes has raised questions regarding energy conservation. In papers 3.24 and 3.25, laws of energy conservation are derived for randomly fluctuating fields, both scalar and electromagnetic ones. It is shown that even when the spectra of the radiated fields exhibit shifts of lines or other modifications, such changes do not violate the law of energy conservation.

The problem of storing and retrieving correlation functions of optical fields of any state of coherence has become of some interest in recent years. One such method,‡ proposed in paper 3.26, makes use of an ensemble of instantaneous holograms.

The papers included in this Section are not specifically concerned with coherence effects in scattering. Some papers on this topic are included in the next Section (Sec. 4).

†Somewhat analogous effects have been found in neutron interference experiments. See, for example, H. Rauch, *Phys. Lett.* A **173**, 240 (1993); D.L. Jacobson, S. Werner and H. Rauch, *Phys. Rev.* A **49**, 3196 (1994). See also G.S. Agarwal, *Found. Phys.* **25**, 219 (1995).

‡A recent technique for displaying spatial coherence of optical fields was described by D. Mendlovic, G. Shabtay, A.W. Lohmann and N. Konforti, *Opt. Lett.* **23**, 1084 (1998).

Section 3 – Coherence and Statistical Optics

Reprinted without change of pagination from the
Proceedings of the Royal Society, A, *volume* 225, pp. 96–111, 1954

A macroscopic theory of interference and diffraction of light from finite sources
I. Fields with a narrow spectral range

By E. Wolf
Department of Mathematical Physics, University of Edinburgh

(*Communicated by M. Born, F.R.S.—Received* 27 *January* 1954)

A macroscopic theory of interference and diffraction of light in stationary fields produced by finite sources which emit light within a finite spectral range is formulated. It is shown that a generalized Huygens principle may be obtained for such fields, which involves only observable quantities. The generalized Huygens principle expresses the *intensity* at a typical point of the field in terms of an integral taken twice independently over an arbitrary surface, the integral involving the intensity distribution over the surface and the values of a certain correlation factor, which is found to be the 'degree of coherence' previously introduced by Zernike. Next it is shown that under fairly general conditions, this correlation factor is essentially the normalized integral over the source of the Fourier (frequency) transform of the spectral intensity function of the source, and that it may be determined from simple interference experiments. Further, it is shown that in regions where geometrical optics is a valid approximation, the coherence factor itself then obeys a simple geometrical law of propagation. Several results on partially coherent fields, established previously by Van Cittert, Zernike, Hopkins and Rogers, follow as special cases from these theorems.

The results have a bearing on many optical problems and can also be applied in investigations concerned with other types of radiation.

1. Introduction

In the usual treatments of interference and diffraction of light, the source is assumed to be of vanishingly small dimensions (a point source), emitting strictly monochromate radiation. Such treatments correspond essentially to an idealized wave field created by a (classical) oscillator. Huygens's principle, in the extended formulation of Fresnel, may be regarded as an approximate propagation law for such fields.

97 *Interference and diffraction of light. I*

In recent years remarkable advances have been made in practical optics in connexion with interference and diffraction of light and electrons; in particular, the phase-contrast method (Zernike 1934 *a, b*), the method of the coherent background (Zernike 1948) and the method of reconstructed wave-fronts (Gabor 1949, 1951) must be mentioned. These discoveries, as well as numerous problems in both theoretical and practical optics, make it highly desirable to extend the theory of interference and diffraction to fields produced by an actual source, i.e. a source of finite extension and one which emits light within a finite frequency range.

First steps towards formulating such a theory were made by Berek (1926 *a, b, c, d*), Van Cittert (1934, 1939), Zernike (1938) and Hopkins (1951, 1953). The experimental counterpart of these investigations dates back to Michelson† (1890, 1891 *a, b*, 1892, 1920).‡ In the theoretical papers just referred to correlation factors for light disturbances at two arbitrary points or at two instants of time were introduced and applied to particular problems, notably by Hopkins (1953). However, only a moderate progress was achieved in formulating the general laws relating to such fields and in fact the subject presents to-day a somewhat confused picture. This is mainly because each of the four authors introduced a formally different correlation factor, which in turn led to many disconnected results.

In the present investigation a systematic study is made of interference and diffraction in stationary§ optical fields produced by finite sources which emit light within a finite frequency range. Part I is mainly restricted to the case when the effective frequency range is sufficiently narrow. In § 2 it is shown that a Huygens principle may be formulated for such fields, which involves only observable quantities. In this generalized form, the Huygens principle expresses the *intensity* at an arbitrary point of the field in terms of an integral taken twice independently over an arbitrary surface, the integrand involving (1) the values of the intensity at all points of this surface and (2) a correlation factor, which turns out to be the complex form of the 'degree of coherence' introduced by Zernike. In this formulation the Huygens principle is subject to similar restrictions on its range of validity as encountered in connexion with its usual form, but a rigorous formulation is possible and will be given in part II of this investigation.

In § 3 the significance of the coherence factor is discussed, and it is shown that it may be determined from simple interference experiments. In § 4 it is shown that under fairly general conditions the coherence factor is essentially the normalized integral over the source of the Fourier (frequency) transform of the intensity function $j(\xi, \nu)$ of the source. This relation takes a particularly simple form when the frequency range of the radiation is sufficiently narrow and when some further simplifying conditions are satisfied. Under these restrictions several of the earlier

† See also Michelson & Pease (1921) and Pease (1931).

‡ A fuller historical survey is given in my article in *Vistas in astronomy* (Wolf 1954). In addition to the literature quoted there, reference to a discussion of a more abstract kind may be added: Wiener (1930), chapter III, § 9.

§ By a stationary field we mean here a field of which all observable properties are constant in time. This definition includes as special case the usual case of high frequency sinusoidal time dependence; or the field constituted by the steady flux of (polychromatic) radiation through an optical system. But it excludes fields for which the time average over a macroscopic time-interval of the flux of radiation depends on time.

results of Van Cittert (1934), Zernike (1938) and Hopkins (1951) are then shown to follow as special cases. It is also found that in regions where the approximations of geometrical optics hold, the coherence factor itself then obeys a simple geometrical law of propagation. In § 6 it is shown that when the source is small, the generalized Huygens principle may be expressed in a simple form in which the properties of the source and the transmission properties of the medium are completely separated.

The results have an immediate bearing on many optical problems and can also be applied in investigations concerned with other types of radiation.

2. A generalized Huygens's principle

We shall be concerned with stationary optical fields and begin by considering the propagation of a beam of natural, nearly monochromatic light from a finite source Σ. For reasons of convergence we assume that the radiation field exists only between the instants $t = -T$ and $t = +T$. It is easy to pass to the limit $T \rightarrow \infty$ subsequently.

Let $V(\mathbf{x}, t)$ denote the disturbance at a point specified by the position vector $\mathbf{x}$, at time t. We shall represent V in the form of a Fourier integral:

$$V(\mathbf{x}, t) = \int_{-\infty}^{+\infty} v(\mathbf{x}, \nu)\, \mathrm{e}^{-2\pi i \nu t}\, \mathrm{d}\nu. \tag{2·1}$$

Then

$$v(\mathbf{x}, \nu) = \int_{-T}^{T} V(\mathbf{x}, t)\, \mathrm{e}^{2\pi i \nu t}\, \mathrm{d}t. \tag{2·2}$$

Since the light is assumed to be almost monochromatic, $|\, v(\mathbf{x}, \nu)\,|$ will differ appreciably from zero only in a narrow frequency range $\nu_0 - \Delta\nu \leqslant \nu \leqslant \nu_0 + \Delta\nu$.

Let us take a surface $\mathscr{A}$ cutting across the beam and consider the intensity at a point $P(\mathbf{x})$ on that side of Σ towards which the light is advancing (figure 1).

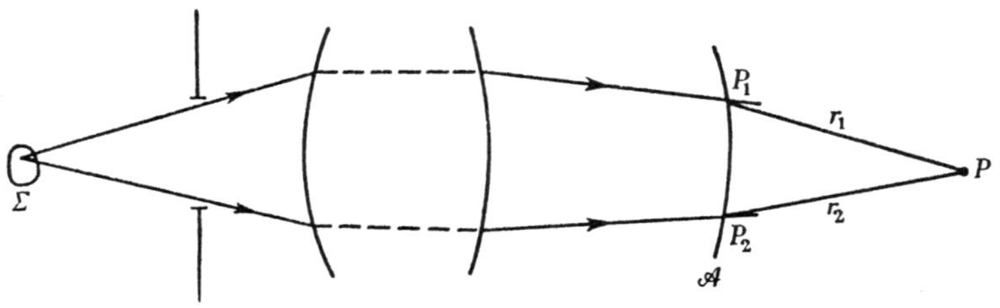

Figure 1

Each Fourier component of (2·1) represents a perfectly monochromatic wave, and therefore (under the usual restrictions on its range of validity) obeys Huygens's principle in the usual form

$$v(\mathbf{x}, \nu) = \int_{\mathscr{A}} v(\mathbf{x}_1, \nu) \frac{\mathrm{e}^{ikr_1}}{r_1} \Lambda_1\, \mathrm{d}\mathbf{x}_1. \tag{2·3}$$

Here $\mathbf{x}_1$ is the position vector of a typical point P_1 on the surface $\mathscr{A}$, r_1 is the distance from P_1 to P, $k = \dfrac{2\pi\nu}{c} = \dfrac{2\pi}{\lambda}$, c being the vacuum velocity of light and λ the wave-length. Further, Λ_1 is the usual inclination factor of Huygens's principle:

$$\Lambda_1 = \frac{\mathrm{i}}{2\lambda}[\cos\phi_1' - \cos\phi_1]; \tag{2·4}$$

the meaning of the angles ϕ_1 and ϕ_1' is shown in figure 2.

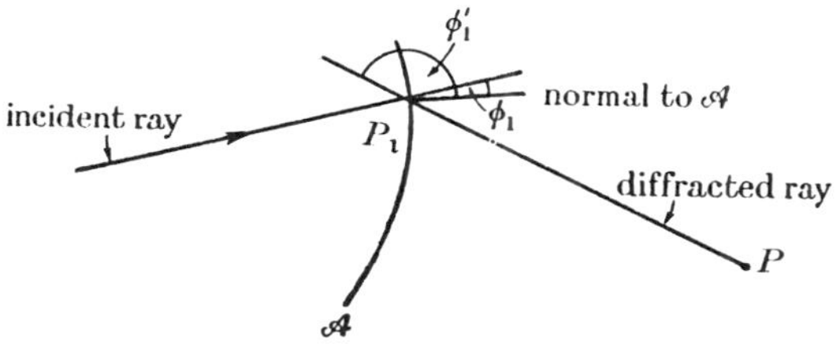

FIGURE 2

From (2·1) and (2·3) it follows that

$$V(\mathbf{x}, t) = \int_{-\infty}^{+\infty} \mathrm{d}\nu\, \mathrm{e}^{-2\pi \mathrm{i}\nu t} \int_{\mathscr{A}} v(\mathbf{x}_1, \nu) \frac{\mathrm{e}^{\mathrm{i}kr_1}}{r_1} \Lambda_1 \,\mathrm{d}\mathbf{x}_1. \tag{2·5}$$

The intensity at P is then given by

$$I(\mathbf{x}) = \langle V(\mathbf{x}, t)\, V^*(\mathbf{x}, t)\rangle, \tag{2·6}$$

where asterisks denote the complex conjugate and the brackets denote average over the interval $-T \leqslant t \leqslant T$. Substituting from (2·5) into (2·6) we obtain

$$I(\mathbf{x}) = \Big\langle \int_{-\infty}^{+\infty}\int_{-\infty}^{+\infty} \mathrm{d}\nu\,\mathrm{d}\nu' \exp\{-2\pi\mathrm{i}(\nu-\nu')t\} \times \int_{\mathscr{A}}\int_{\mathscr{A}} v(\mathbf{x}_1, \nu)\, v^*(\mathbf{x}_2, \nu') \frac{\exp\{\mathrm{i}[kr_1 - k'r_2]\}}{r_1 r_2} \Lambda_1 \Lambda_2'^* \,\mathrm{d}\mathbf{x}_1 \mathrm{d}\mathbf{x}_2 \Big\rangle, \tag{2·7}$$

where $k' = 2\pi\nu'/c$, $\Lambda' = \Lambda(k')$, and the points $P_1(\mathbf{x}_1)$ and $P_2(\mathbf{x}_2)$ explore the surface $\mathscr{A}$ independently.

Since the light is assumed to be nearly monochromatic and of frequency ν_0 the curve $v(\mathbf{x}, \nu)$, considered as a function of ν, will have a peak at $\nu = \nu_0$ and fall off rapidly on both sides, being practically zero outside the range $(\nu_0 - \Delta\nu, \nu_0 + \Delta\nu)$. Under these conditions we may replace both k and k' in the term

$$\exp\{\mathrm{i}[kr_1 - k'r_2]\}\, \Lambda_1 \Lambda_2'^*$$

by $k^{(0)} = 2\pi\nu_0/c$. (2·7) then becomes

$$I(\mathbf{x}) = \int_{\mathscr{A}}\int_{\mathscr{A}} \Gamma(\mathbf{x}_1, \mathbf{x}_2) \frac{\exp\{\mathrm{i}k^{(0)}(r_1 - r_2)\}}{r_1 r_2} \Lambda_1^{(0)} \Lambda_2^{(0)*} \,\mathrm{d}\mathbf{x}_1 \mathrm{d}\mathbf{x}_2, \tag{2·8}$$

where

$$\Gamma(\mathbf{x}_1, \mathbf{x}_2) = \frac{1}{2T}\int_{-T}^{T} \mathrm{d}t \int_{-\infty}^{+\infty}\int_{-\infty}^{+\infty} v(\mathbf{x}_1, \nu)\, v^*(\mathbf{x}_2, \nu') \exp\{-2\pi\mathrm{i}(\nu-\nu')t\}\, \mathrm{d}\nu\, \mathrm{d}\nu'. \tag{2·9}$$

Since
$$\operatorname*{Lim}_{T\to\infty}\int_{-T}^{+T}\exp\{-2\pi i(\nu-\nu')t\}\,\mathrm{d}t=\delta(\nu-\nu'),$$

where δ is the Dirac Delta function, it follows that, for sufficiently large T, (2·9) reduces to

$$\Gamma(\mathbf{x}_1,\mathbf{x}_2)=\frac{1}{2T}\int_{-\infty}^{+\infty}v(\mathbf{x}_1,\nu)\,v^*(\mathbf{x}_2,\nu)\,\mathrm{d}\nu, \tag{2·10}$$

or, using the convolution (Faltung) theorem,

$$\Gamma(\mathbf{x}_1,\mathbf{x}_2)=\langle V(\mathbf{x}_1,t)\,V^*(\mathbf{x}_2,t)\rangle. \tag{2·11}$$

It will be useful to normalize Γ by setting

$$\gamma(\mathbf{x}_1,\mathbf{x}_2)=\frac{\Gamma(\mathbf{x}_1,\mathbf{x}_2)}{\sqrt{\{\Gamma(\mathbf{x}_1,\mathbf{x}_1)\,\Gamma(\mathbf{x}_2,\mathbf{x}_2)\}}}. \tag{2·12}$$

Now $\Gamma(\mathbf{x}_s,\mathbf{x}_s)$ $(s=1,2)$ is nothing but the intensity I_s at the point $\mathbf{x}_s$, so that (2·12) may be written as

$$\gamma(\mathbf{x}_1,\mathbf{x}_2)=\frac{\Gamma(\mathbf{x}_1,\mathbf{x}_2)}{\sqrt{(I_1I_2)}}. \tag{2·13}$$

Substituting from (2·13) into (2·8), we finally obtain

$$I(\mathbf{x})=\int_{\mathscr{A}}\int_{\mathscr{A}}\sqrt{(I_1I_2)}\,\gamma_{12}\frac{\exp\{\mathrm{i}k^{(0)}(r_1-r_2)\}}{r_1r_2}\Lambda_1^{(0)}\Lambda_2^{(0)*}\,\mathrm{d}\mathbf{x}_1\,\mathrm{d}\mathbf{x}_2, \tag{2·14}$$

where γ_{12} has been written for $\gamma(\mathbf{x}_1,\mathbf{x}_2)$.

Equation (2·14) may be regarded as a *generalized Huygens principle*. In the usual formulation (2·3), Huygens's principle applies only to strictly coherent radiation and expresses the (non-observable) *disturbance* at a point in the wave field as sum of contributions from each element $\mathrm{d}\mathbf{x}_1$ of the primary wave (or, more generally, of an arbitrary surface). In the present formulation, the restriction of strict coherence is dropped and the *intensity* is calculated by summing over all products $\mathrm{d}\mathbf{x}_1\mathrm{d}\mathbf{x}_2$ of the surface, each contribution being weighted by the appropriate value of the correlation factor γ_{12}. It will be shown that in any particular case, this factor may be determined from simple experiments, and that under fairly general conditions it may also be calculated from the knowledge of the intensity function of the source and the optical transmission properties of the medium. Hence our generalized Huygens principle involves *observable quantities*† only.

3. Determination of the γ factor from experiment

The relations (2·13) and (2·11) are formally equivalent to relations (4) and (5) of Zernike's (1938) paper and show that $\Gamma(\mathbf{x}_1,\mathbf{x}_2)$ is the *mutual intensity* and $\gamma(\mathbf{x}_1,\mathbf{x}_2)$

† An earlier formulation of Huygens's principle in terms of observable quantities, due to Gabor (1952; Private communication), must also be mentioned: 'Set up a coherent radiation field and apply to it a small perturbation by introducing objects in the path of radiation which do not destroy the coherence. If the *absolute amplitudes* of the perturbed field are known in one cross-section, they are thereby determined in all cross-sections.' This formulation is, however, not sufficiently general, being restricted to strictly coherent radiation.

the complex degree of coherence,† two important concepts introduced by Zernike. The degree of coherence has been previously defined in a different way and under more restrictive conditions by Van Cittert (1934).

In general γ is complex. It is easily seen that its absolute value is less than unity. For one has, using the well-known modulus inequality for integrals,

$$\left|\int_{-T}^{T} V(\mathbf{x}_1, t)\, V^*(\mathbf{x}_2, t)\, \mathrm{d}t\right| \leqslant \left\{\int_{-T}^{T} |\, V(\mathbf{x}_1, t)\, V^*(\mathbf{x}_2, t)\,|\, \mathrm{d}t\right\}. \tag{3·1}$$

Moreover, by Schwarz's inequality,

$$\left\{\int_{-T}^{T} |\, V(\mathbf{x}_1, t)\, V^*(\mathbf{x}_2, t)\,|\, \mathrm{d}t\right\}^2 \leqslant \int_{-T}^{T} |\, V(\mathbf{x}_1, t)\,|^2\, \mathrm{d}t \int_{-T}^{T} |\, V(\mathbf{x}_2, t)\,|^2\, \mathrm{d}t. \tag{3·2}$$

From (3·1) and (3·2),

$$\left|\int_{-T}^{T} V(\mathbf{x}_1, t)\, V^*(\mathbf{x}_2, t)\, \mathrm{d}t\right| \leqslant \left\{\int_{-T}^{T} |\, V(\mathbf{x}_1, t)\,|^2\, \mathrm{d}t\right\}^{\frac{1}{2}} \left\{\int_{-T}^{T} |\, V(\mathbf{x}_2, t)\,|^2\, \mathrm{d}t\right\}^{\frac{1}{2}}, \tag{3·3}$$

or, in terms of Γ, $\quad |\,\Gamma(\mathbf{x}_1, \mathbf{x}_2)\,| \leqslant \sqrt{\{\Gamma(\mathbf{x}_1, \mathbf{x}_1)\, \Gamma(\mathbf{x}_2, \mathbf{x}_2)\}},$

whence

$$|\,\gamma(\mathbf{x}_1, \mathbf{x}_2)\,| \leqslant 1. \tag{3·4}$$

In order to see the significance of γ and also to confirm our earlier statement that it is an observable quantity, we shall apply our generalized Huygens principle to a simple interference experiment.

Figure 3 shows the arrangement. Light from a finite source Σ falls either directly or via an optical system on to a screen $\mathscr{A}$ which has small openings at P_1 and P_2. The resulting interference fringes are observed on a second screen $\mathscr{A}'$.

If $\mathrm{d}\mathscr{A}_1$ and $\mathrm{d}\mathscr{A}_2$ denote the areas of the openings at P_1 and P_2, our generalized Huygens principle (2·14) reduces to the following expression when integration is taken over the non-illuminated side of the screen $\mathscr{A}$:

$$\begin{aligned} I \sim I_1 \gamma_{11} \left(\frac{1}{r_1}\right)^2 \Lambda_1 \Lambda_1^* (\mathrm{d}\mathscr{A}_1)^2 + \sqrt{(I_1 I_2)}\, \gamma_{12} \frac{\exp\{ik(r_1 - r_2)\}}{r_1 r_2} \Lambda_1 \Lambda_2^*\, \mathrm{d}\mathscr{A}_1 \mathrm{d}\mathscr{A}_2 \\ + I_2 \gamma_{22} \left(\frac{1}{r_2}\right)^2 \Lambda_2 \Lambda_2^* (\mathrm{d}\mathscr{A}_2)^2 + \sqrt{(I_2 I_1)}\, \gamma_{21} \frac{\exp\{ik(r_2 - r_1)\}}{r_2 r_1} \Lambda_2 \Lambda_1^*\, \mathrm{d}\mathscr{A}_2 \mathrm{d}\mathscr{A}_1. \end{aligned} \tag{3·5}$$

The upper index zero has now been omitted on k, Λ_1 and Λ_2^*. Since $\gamma_{11} \equiv 1$ it follows that the first term gives precisely the value $I^{(1)}$ of the intensity which would be obtained at P if the opening at P_1 alone was open ($\mathrm{d}\mathscr{A}_2 = 0$), the third term having a similar interpretation:

$$\left.\begin{aligned} I^{(1)}(\mathbf{x}) &= \frac{I_1}{r_1^2} |\, \Lambda_1\,|^2 (\mathrm{d}\mathscr{A}_1)^2, \\ I^{(2)}(\mathbf{x}) &= \frac{I_2}{r_2^2} |\, \Lambda_2\,|^2 (\mathrm{d}\mathscr{A}_2)^2. \end{aligned}\right\} \tag{3·6}$$

† In the first part of this paper, Zernike defined the degree of coherence of two light vibrations as the visibility of the interference fringes that may be obtained from them under the best circumstances, i.e. when both intensities are made equal and only small path-differences introduced. The analytical definition given by equation (4) of his paper [equivalent to (2·12) above] is however more general and applies whether or not the two intensities are equal; nor is it restricted to small path differences.

The second and the fourth term in (3·5) are complex conjugates of each other. Hence using the identity

$$z+z^* = 2\mathscr{R}(z) \tag{3·7}$$

($\mathscr{R}$ denoting the real part), the sum of the two terms may be written as

$$2\sqrt{(I_1 I_2)}\frac{1}{r_1}\frac{1}{r_2}|\Lambda_1||\Lambda_2||\gamma_{12}|\cos[\arg\gamma_{12}+k(r_1-r_2)]\,\mathrm{d}\mathscr{A}_1\mathrm{d}\mathscr{A}_2$$
$$= 2\sqrt{(I^{(1)}I^{(2)})}\,|\gamma_{12}|\cos[\arg\gamma_{12}+k(r_1-r_2)], \tag{3·8}$$

and (3·5) then reduces to

$$I = I^{(1)}+I^{(2)}+2\sqrt{(I^{(1)}I^{(2)})}\,|\gamma_{12}|\cos[\arg\gamma_{12}+k(r_1-r_2)]. \tag{3·9}$$

We see that in the limiting cases $|\gamma_{12}| = 1$ and $|\gamma_{12}| = 0$ (3·9) reduces to the usual laws for the combination of *perfectly coherent* and *completely incoherent* disturbances. Hence (3·9) may be regarded as a *generalized interference law* in which the factor γ_{12} is a measure of the degree of correlation between the disturbances at P_1 and P_2. This law was derived previously in a different manner by Zernike

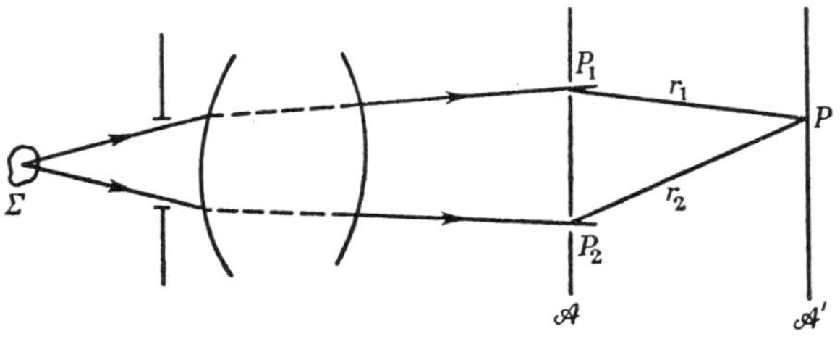

FIGURE 3. An interference experiment with a finite source.

(1938). It has also been derived under a more restrictive definition of the coherence factor by Hopkins (1951).

The generalized interference law (3·9) enables the calculation of $I(\mathbf{x})$ when $I^{(1)}$, $I^{(2)}$ and γ_{12} is known. Conversely when $I^{(1)}$, $I^{(2)}$ and I are known γ_{12} may be determined from measurements of intensities. One has only to measure $I^{(1)}$, $I^{(2)}$ and I for different values of r_1 and r_2 and solve the resulting equations obtained from (3·9) for $|\gamma_{12}|$ and $\arg\gamma_{12}$. Hence the coherence factor may be considered to be an *observable quantity*.

As pointed out by Zernike and Hopkins, the coherence factor is closely related to the *visibility*† and the *position* of the fringes. By (3·9) one has

$$\left.\begin{aligned} I_{\text{max.}} &= I^{(1)}+I^{(2)}+2\sqrt{(I^{(1)}I^{(2)})}\,|\gamma_{12}|, \\ I_{\text{min.}} &= I^{(1)}+I^{(2)}-2\sqrt{(I^{(1)}I^{(2)})}\,|\gamma_{12}|. \end{aligned}\right\} \tag{3·10}$$

† The visibility $\mathscr{V}(\mathbf{x})$ of interference fringes, a concept due to Michelson, is defined by

$$\mathscr{V} = \frac{I_{\text{max.}}-I_{\text{min.}}}{I_{\text{max.}}+I_{\text{min.}}},$$

where $I_{\text{max.}}$ is the maximum of the intensity at the centre of the brightest fringe near $P(\mathbf{x})$ and $I_{\text{min.}}$ is the intensity at the centre of the adjacent dark fringe.

Hence the visibility $\mathscr{V}$ is given by

$$\mathscr{V} = \frac{2}{\sqrt{\frac{I^{(1)}}{I^{(2)}}} + \sqrt{\frac{I^{(2)}}{I^{(1)}}}} |\gamma_{12}|. \tag{3·11}$$

Moreover, it is seen from (3·9) that $\arg \gamma_{12}$ appears formally as the 'phase difference' between the disturbances at the two points; it is equal (in suitable units) to the amount of lateral displacement of the fringe system from the position which it would occupy if $\arg \gamma_{12}$ was equal to zero. In practice one can often set $I^{(1)} = I^{(2)}$. (3·11) shows that the visibility is then simply equal to the absolute value of the coherence factor.

In the paper already referred to, Hopkins (1951) claimed that Zernike's results relating to partially coherent fields may be derived without any resort to statistical analysis. This claim is not justified; Hopkins's analysis, unlike Zernike's, is applicable only to the limiting case of vanishingly narrow frequency range. For a similar reason, a claim by Van Cittert (1939) as to the equivalence of his correlation factor with the degree of coherence of Zernike also does not appear to be justified.

For a full understanding of the coherence problem, it is obviously essential to take the finite frequency range of actual radiation into account. This may be done with the help of Fourier analysis as described in § 2. Alternatively, one may express the disturbance in the form

$$V(\mathbf{x}, t) = U(\mathbf{x}, t) \exp\{-2\pi i \nu_0 t\}, \tag{3·12}$$

where the complex amplitude U is now a function of both position and time. Since the light is assumed to be nearly monochromatic and of frequency ν_0, U will, however, at each point vary slowly (and irregularly) in comparison with the frequency ν_0, and will remain practically constant over an interval of time depending on the coherence length of the light.†

The path differences encountered in instrumental optics are, as a rule, small compared to the coherence length. Hopkins's theory will in these cases lead substantially to the same results as the present analysis. When large path differences are involved Hopkins's theory can, however, no longer be expected to be applicable.

The representation (3·12) was used by Zernike (1938) and also previously by Berek (1926 (*a*)–(*d*)). For a quantitative treatment the Fourier integral approach seems, however, to be more appropriate and has also the attractive feature that it brings the optical coherence problems within the scope of methods well established in connexion with other statistical problems encountered in physics.

Experimental investigations, using small path differences, were recently carried out by Baker (1953) and Arnulf, Dupuy & Flamant (1953). Good agreement with the existing theories was obtained.

† The coherence length may be defined as the maximum path difference between the interfering beams for which interference fringes may be obtained. The finite value of the coherence length arises mainly from (1) natural broadening of spectral lines due to the finite lifetime of atomic states, (2) broadening due to atomic collisions and (3) Doppler broadening due to the thermal motion of the atoms. For an excellent account of the coherence length see Born (1933).

Finally a paper by Duffieux (1953) may also be mentioned. It contains some criticisms of the earlier theories and discusses the connexion between the coherence factor and the transmission functions introduced by him and Lansraux in earlier investigations.

4. Expressions for the mutual intensity function

We shall now derive an explicit expression for the mutual intensity function Γ_{12} in terms of the intensity function of the source and the transmission function of the medium. No restriction on the frequency range will now be imposed.

We assume that the source Σ is a radiating element of a plane surface and divide it into elements $d\Sigma_1, d\Sigma_2, \ldots$, whose linear dimensions are small compared to the wave-length. If $V_m(\mathbf{x}, t)$ denotes the disturbance at the point $P(\mathbf{x})$ and time t due to the mth element of the source, then

$$V(\mathbf{x}, t) = \sum_m V_m(\mathbf{x}, t),$$

and (2·11) becomes

$$\begin{aligned}\Gamma_{12} &= \langle \sum_m V_m(\mathbf{x}_1, t) \sum_n V_n^*(\mathbf{x}_2, t) \rangle \\ &= \langle \sum_m V_m(\mathbf{x}_1, t)\, V_m^*(\mathbf{x}_2, t) \rangle + \langle \sum_{m \neq n} V_m(\mathbf{x}_1, t)\, V_n^*(\mathbf{x}_2, t) \rangle, \qquad (4\cdot1)\end{aligned}$$

where m and n run independently through all possible values. Now to a very good approximation, the disturbances from different elements of the source may be treated as statistically independent, i.e.

$$\langle V_m(\mathbf{x}_1, t)\, V_n^*(\mathbf{x}_2, t) \rangle = 0 \quad \text{when} \quad m \neq n,$$

so that the last term in (4·1) vanishes. If we also introduce the Fourier inverse $v_m(\mathbf{x}, \nu)$ of $V_m(\mathbf{x}, t)$

$$v_m(\mathbf{x}, \nu) = \int_{-T}^{T} V_m(\mathbf{x}, t)\, e^{2\pi i \nu t}\, dt,$$

and use the convolution theorem, we then obtain the following expression for Γ_{12}:

$$\Gamma_{12} = \langle \sum_m V_m(\mathbf{x}_1, t)\, V_m^*(\mathbf{x}_2, t) \rangle \qquad (4\cdot2a)$$

$$= \frac{1}{2T} \sum_m \int_{-\infty}^{+\infty} v_m(\mathbf{x}_1, \nu)\, v_m^*(\mathbf{x}_2, \nu)\, d\nu. \qquad (4\cdot2b)$$

Let

$$K(\mathbf{x}', \mathbf{x}'', \nu) = a(\mathbf{x}', \mathbf{x}'') \exp\{2\pi i (\nu/c) \mathscr{S}(\mathbf{x}', \mathbf{x}'')\} \qquad (4\cdot3)$$

(a and $\mathscr{S}$ real) be the *transmission function* of the medium, defined as the disturbance at $P''(\mathbf{x}'')$ due to a monochromatic point source of frequency ν and of unit strength† at $P'(\mathbf{x}')$. (Strictly a and $\mathscr{S}$ also depend on the frequency, but this dependence may here be neglected.) In particular, if the points are situated in a region where diffraction effects are not dominant, then the phase function $\mathscr{S}$ is simply the

† We define a source of unit strength as one which *in vacuo* would give rise to a disturbance of unit amplitude at a unit distance from it.

Hamilton point characteristic function of the medium, i.e. the optical length of the natural ray joining $\mathbf{x}'$ to $\mathbf{x}''$, and the amplitude a may be obtained from the usual conservation law of geometrical optics. *In vacuo*, for example,

$$K = \frac{\exp\{ik\,|\,\mathbf{x}'-\mathbf{x}''\,|\}}{|\,\mathbf{x}'-\mathbf{x}''\,|}. \tag{4·4}$$

If $\boldsymbol{\xi}_m$ denotes the position vector of the element $d\Sigma_m$ of the source, then the disturbance at a point $P(\mathbf{x})$ due to a typical Fourier component $v_m(\mathbf{x},\nu)$ of $V_m(\mathbf{x},t)$ may be written as

$$v_m(\mathbf{x},\nu) = v_m(\boldsymbol{\xi}_m,\nu)\,K(\boldsymbol{\xi}_m,\mathbf{x},\nu). \tag{4·5}$$

Hence (4·2*b*) becomes

$$\begin{aligned}\Gamma_{12} &= \frac{1}{2T}\sum_m \int_{-\infty}^{+\infty} |\,v_m(\boldsymbol{\xi}_m,\nu)\,|^2\,K(\boldsymbol{\xi}_m,\mathbf{x}_1,\nu)\,K^*(\boldsymbol{\xi}_m,\mathbf{x}_2,\nu)\,d\nu \\ &= \frac{1}{2T}\sum_m a(\boldsymbol{\xi}_m,\mathbf{x}_1)\,a(\boldsymbol{\xi}_m,\mathbf{x}_2)\int_{-\infty}^{+\infty} |\,v_m(\boldsymbol{\xi}_m,\nu)\,|^2 \exp\{2\pi i\nu\tau_{12}(\boldsymbol{\xi}_m)\}\,d\nu,\end{aligned} \tag{4·6}$$

where

$$\tau_{12}(\boldsymbol{\xi}_m) = \frac{1}{c}[\mathscr{S}(\boldsymbol{\xi}_m,\mathbf{x}_1)-\mathscr{S}(\boldsymbol{\xi}_m,\mathbf{x}_2)] \tag{4·7}$$

represents the difference in the time needed for light to travel to $\mathbf{x}_1$ and $\mathbf{x}_2$ from the element $d\Sigma_m$ of the source. Applying the convolution theorem to (4·6) it follows that Γ may also be written in the form

$$\Gamma_{12} = \sum_m a(\boldsymbol{\xi}_m,\mathbf{x}_1)\,a(\boldsymbol{\xi}_m,\mathbf{x}_2)\,\langle V_m(\boldsymbol{\xi}_m,t-\tau_{12})\,V_m^*(\boldsymbol{\xi}_m,t)\rangle. \tag{4·8}$$

Instead of $|\,v_m(\boldsymbol{\xi}_m,\nu)\,|$ which is defined for only a discontinuous set of values $\boldsymbol{\xi}_m$, we may introduce a function $j(\boldsymbol{\xi},\nu)$ which represents the *intensity per unit area* of the source, *per unit frequency range*;† it is defined for the whole continuous set of $\boldsymbol{\xi}$ values. Absorbing the factor $1/2T$ in our definition of j, (4·6) then becomes

$$\Gamma_{12} = \int_\Sigma d\boldsymbol{\xi}\,a(\boldsymbol{\xi},\mathbf{x}_1)\,a(\boldsymbol{\xi},\mathbf{x}_2)\int_{-\infty}^{+\infty} j(\boldsymbol{\xi},\nu)\exp\{2\pi i\nu\tau_{12}(\boldsymbol{\xi})\}\,d\nu. \tag{4·9}$$

The amplitude factor $a(\boldsymbol{\xi},\mathbf{x}_s)$ $(s = 1, 2)$ of the transmission function will as a rule be a slowly varying function of $\boldsymbol{\xi}$. Also in most applications the linear dimensions of the source will be small compared to the distance from the source to $\mathbf{x}_s$. Hence in (4·8) and (4·9) we may as a rule replace $a(\boldsymbol{\xi},\mathbf{x}_s)$ by $a(0,\mathbf{x}_s)$ without introducing an appreciable error. Finally, normalizing Γ as before, we obtain

$$\gamma_{12} = \frac{\sum_m \langle V_m(\boldsymbol{\xi}_m,t-\tau_{12})\,V_m^*(\boldsymbol{\xi}_m,t)\rangle}{\sum_m \langle V_m(\boldsymbol{\xi}_m,t)\,V_m^*(\boldsymbol{\xi}_m,t)\rangle} \tag{4·10a}$$

$$= \frac{\int_\Sigma d\boldsymbol{\xi}\int_{-\infty}^{+\infty} j(\boldsymbol{\xi},\nu)\exp\{2\pi i\nu\tau_{12}(\boldsymbol{\xi})\}\,d\nu}{\int_\Sigma d\boldsymbol{\xi}\int_{-\infty}^{+\infty} j(\boldsymbol{\xi},\nu)\,d\nu}, \tag{4·10b}$$

i.e. *the coherence factor is essentially the normalized integral over the source of the Fourier (frequency) transform of the intensity function of the source.*

† We neglect here the variation of the intensity with direction. The effect of this variation can also be taken into account by introducing a more general intensity function $j(\boldsymbol{\xi},\mathbf{p},\nu)$, $\mathbf{p}$ being a directional variable.

5. Some approximate expressions for the coherence factor

We shall now consider the form which the expressions for γ take when the effective frequency range is sufficiently narrow. We shall show that they lead, in special cases, to several results obtained previously by Van Cittert (1934), Zernike (1938), Hopkins (1951) and Rogers (1953).

Let us assume that the effective frequency range is so narrow that ν may be replaced by ν_0 in the integral in (4·9). This will be permissible, if

$$|\Delta\nu\tau_{12}(\xi)| \ll 1,$$

or, since $\Delta\nu = -c\Delta\lambda/\lambda^2$ and $\tau_{12}(\xi) = \frac{1}{c}[\mathscr{S}(\xi, \mathbf{x}_1) - \mathscr{S}(\xi, \mathbf{x}_2)]$, if

$$\frac{|\Delta\lambda|}{\lambda} \ll \frac{\lambda}{|\mathscr{S}(\xi, \mathbf{x}_1) - \mathscr{S}(\xi, \mathbf{x}_2)|}. \tag{5·1}$$

We also set

$$\int_{-\infty}^{\infty} j(\xi, \nu)\,\mathrm{d}\nu = J(\xi); \tag{5·2}$$

(4·9) then gives on normalizing

$$\gamma_{12} = \frac{1}{\sqrt{(I_1 I_2)}}\int_\Sigma J(\xi)\,a(\xi, \mathbf{x}_1)\,a(\xi, \mathbf{x}_2)\exp\{2\pi i\nu_0\tau_{12}(\xi)\}\,\mathrm{d}\xi, \tag{5·3}$$

where

$$I_s = \Gamma_{ss} = \int_\Sigma J(\xi)\,a^2(\xi, \mathbf{x}_s)\,\mathrm{d}\xi \quad (s = 1, 2). \tag{5·3a}$$

In particular, *in vacuo* (air) one has (cf. (4·4))

$$a(\xi, \mathbf{x}_s) = \frac{1}{R_s}, \quad \mathscr{S}(\xi, \mathbf{x}_s) = R_s,$$

with

$$R_s = |\mathbf{x}_s - \xi|.$$

(5·3) then reduces to

$$\gamma_{12} = \frac{1}{\sqrt{(I_1 I_2)}}\int_\Sigma \frac{J(\xi)}{R_1 R_2}\exp\{ik^{(0)}(R_1 - R_2)\}\,\mathrm{d}\xi, \tag{5·4}$$

where

$$I_s = \int_\Sigma \frac{J(\xi)}{R_s^2}\,\mathrm{d}\xi \quad (s = 1, 2).$$

Integrals of the form (5·4) are well known in optics. They represent the complex amplitude at the point P_2 in a diffraction pattern around P_1, when diffraction takes place at an aperture which is identical in form with Σ, the amplitude in the diffracting aperture being proportional to $J(\xi)$. This expression for the coherence factor was first obtained by Zernike (1938) and later by Hopkins (1951).

If, as before, we neglect the variation of the amplitude factor with ξ, we obtain from (5·3)

$$\gamma_{12} = \frac{\int_\Sigma J(\xi)\exp\{2\pi i\nu_0\tau_{12}(\xi)\}\,\mathrm{d}\xi}{\int_\Sigma J(\xi)\,\mathrm{d}\xi}. \tag{5·5}$$

This relation may be further simplified if either the linear dimensions of the source, or the distance between P_1 and P_2 are small compared to the distance from the source to P_1 and P_2. In the first case one may expand $\tau_{12}(\boldsymbol{\xi})$ at a suitable point O ($\boldsymbol{\xi}=0$) of the source and neglect higher order terms:

$$\tau_{12}(\boldsymbol{\xi}) \sim \frac{1}{c}[\mathscr{S}(0,\mathbf{x}_1)-\mathscr{S}(0,\mathbf{x}_2)]+\frac{1}{c}\boldsymbol{\xi}\cdot\frac{\partial}{\partial\boldsymbol{\xi}}[\mathscr{S}(\boldsymbol{\xi},\mathbf{x}_1)-\mathscr{S}(\boldsymbol{\xi},\mathbf{x}_2)]_{\boldsymbol{\xi}=0}. \tag{5·6}$$

Now by the fundamental property of the $\mathscr{S}$ function,

$$\frac{\partial}{\partial\boldsymbol{\xi}}[\mathscr{S}(\boldsymbol{\xi},\mathbf{x}_s)]_{\boldsymbol{\xi}=0} = -\mathbf{p}_s \quad (s=1,2), \tag{5·7}$$

where $\mathbf{p}_s$ denotes the ray vector† at O of the ray $\overrightarrow{OP_s}$. Hence

$$\tau_{12}(\boldsymbol{\xi}) \sim \frac{1}{c}[\mathscr{S}(0,\mathbf{x}_1)-\mathscr{S}(0,\mathbf{x}_2)]-\frac{1}{c}\boldsymbol{\xi}\cdot(\mathbf{p}_1-\mathbf{p}_2), \tag{5·8}$$

and (5·5) reduces to

$$\gamma_{12} = \frac{\int_\Sigma J(\boldsymbol{\xi})\exp\{-\mathrm{i}k^{(0)}\boldsymbol{\xi}\cdot(\mathbf{p}_1-\mathbf{p}_2)\}\,\mathrm{d}\boldsymbol{\xi}}{\int_\Sigma J(\boldsymbol{\xi})\,\mathrm{d}\boldsymbol{\xi}}\exp\{\mathrm{i}k^{(0)}[\mathscr{S}(0,\mathbf{x}_1)-\mathscr{S}(0,\mathbf{x}_2)]\}. \tag{5·9}$$

Hence *when the effective frequency range is sufficiently narrow and the source small enough, the coherence factor is equal to the product of the term*

$$\exp\{\mathrm{i}k^{(0)}[\mathscr{S}(0,\mathbf{x}_1)-\mathscr{S}(0,\mathbf{x}_2)]\}$$

and the complex amplitude in an associated Fraunhofer diffraction pattern.

Similar analysis may be used when the distance between P_1 and P_2 is small compared to the distance from the source to these points. In place of (5·6) we now have

$$\begin{aligned}\tau_{12}(\boldsymbol{\xi}) &\sim -\frac{1}{c}(\mathbf{x}_2-\mathbf{x}_1)\left[\frac{\partial\mathscr{S}(\boldsymbol{\xi},\mathbf{x})}{\partial\mathbf{x}}\right]_{\mathbf{x}=\mathbf{x}_1}\\ &= -\frac{1}{c}(\mathbf{x}_2-\mathbf{x}_1)\cdot\mathbf{p}_1'(\boldsymbol{\xi}),\end{aligned} \tag{5·10}$$

where $\mathbf{p}_1'(\boldsymbol{\xi})$ is the ray vector at P_1 of the ray $\boldsymbol{\xi}\to\mathbf{x}_1$. (5·5) now reduces to

$$\gamma_{12} = \frac{\int_\Sigma J(\boldsymbol{\xi})\exp\{-\mathrm{i}k^{(0)}(\mathbf{x}_2-\mathbf{x}_1)\cdot\mathbf{p}_1'(\boldsymbol{\xi})\}\,\mathrm{d}\boldsymbol{\xi}}{\int_\Sigma J(\boldsymbol{\xi})\,\mathrm{d}\boldsymbol{\xi}}. \tag{5·11}$$

In particular, assume that P_1 and P_2 are in a plane parallel to the source and illuminated directly by it and that the medium between the source and the plane is homogeneous and of refractive index $n=1$. If we choose as origin of the position vector the point P_1, then $\mathbf{p}_1'(\boldsymbol{\xi}) = -\boldsymbol{\xi}/|\boldsymbol{\xi}|$. Moreover, it will often be permissible to

† A ray vector $\mathbf{p}$ at a point P is defined by the relation $\mathbf{p}=n\mathbf{s}$, $\mathbf{s}$ being the unit vector along the ray at P and n the value of the refractive index at that point.

replace $|\boldsymbol{\xi}|$ by the distance d between the plane of the source and the plane containing P_1 and P_2. (5·11) then reduces to

$$\gamma_{12} = \frac{\int_\Sigma J(\boldsymbol{\xi}) \exp\{ik^{(0)}\mathbf{x}_2 . \boldsymbol{\xi}/d\} \, d\boldsymbol{\xi}}{\int_\Sigma J(\boldsymbol{\xi}) \, d\boldsymbol{\xi}}; \tag{5·12}$$

hence *the coherence factor is again expressed in the form of a complex amplitude in an associated Fraunhofer diffraction pattern.* This result was first obtained, for a source of circular or rectangular form, under a somewhat different definition of the γ factor by Van Cittert (1934). It is of importance in the theory of the stellar interferometer (cf. Hopkins 1951).

From (5·9) one can easily derive a simple '*propagation law*' for the coherence factor, valid (subject to the restrictions mentioned) in regions where diffraction effects are not dominant.† Consider two pairs of points $P_1(\mathbf{x}_1)$, $P_2(\mathbf{x}_2)$ and $P_1'(\mathbf{x}_1')$, $P_2'(\mathbf{x}_2')$ in the field, such that P_1' lies on the ray from O to P_1 and P_2' lies on the ray from O to P_2 (figure 4). Then it immediately follows from (5·9) that

$$\gamma(\mathbf{x}_1', \mathbf{x}_2') = \gamma(\mathbf{x}_1, \mathbf{x}_2) \exp\{ik^{(0)}[\mathscr{S}(\mathbf{x}_1, \mathbf{x}_1') - \mathscr{S}(\mathbf{x}_2, \mathbf{x}_2')]\}. \tag{5·13}$$

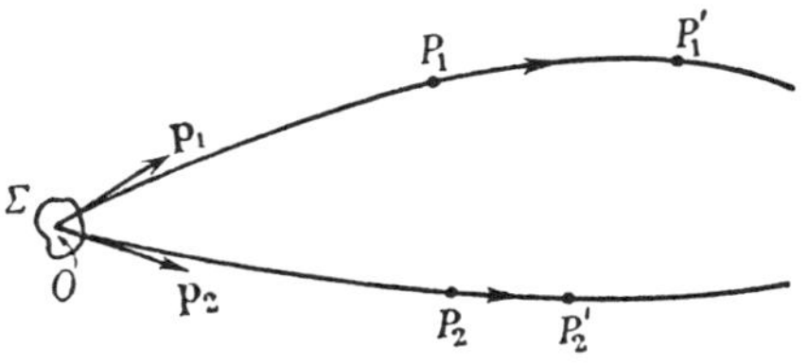

FIGURE 4. Propagation of 'coherence' in regions where geometrical optics is a valid approximation (small source and narrow frequency range assumed).

In particular, if the optical path $[P_1P_1']$ equals the optical path $[P_2P_2']$, e.g. when P_1 and P_2 are one wave-front and P_1' and P_2' on another wave-front, then $\mathscr{S}(\mathbf{x}_1, \mathbf{x}_1') = \mathscr{S}(\mathbf{x}_2, \mathbf{x}_2')$ and (5·13) reduces to

$$\gamma(\mathbf{x}_1', \mathbf{x}_2') = \gamma(\mathbf{x}_1, \mathbf{x}_2). \tag{5·14}$$

We may therefore say that *under the restrictions mentioned, the coherence is propagated in accordance with the laws of geometrical optics.*

It was shown by Hopkins (1951) that the absolute value of the coherence factor for two points in the entry pupil of an optical system and of the corresponding conjugate points in the exit pupil is the same. The result is seen to be an immediate consequence of the theorem just established.

It was also pointed out by Rogers (1953) that the degree of coherence is conserved in a beam of light from image plane to image plane in Gaussian systems. This result too is a special case of the law expressed by (5·13).

† A more general propagation law valid everywhere will be discussed in part II of this investigation.

6. The generalized Huygens principle in the special case of a small source

In this section we shall reformulate the generalized Huygens principle by making use of the relation which expresses the coherence factor in terms of the intensity function of the source, the source being assumed to be sufficiently small. Before doing this, we shall, however, generalize (2·14) by dropping the restriction that the medium between the surface of integration and the point of observation is homogeneous.

If the medium between $\mathscr{A}$ and P is heterogeneous or contains refracting or reflecting surfaces, then clearly the factor $\exp\{ik^{(0)}r_s)/r_s\}$ (with $s = 1, 2$) in (2·14) has to be replaced by the appropriate transmission function $K(\mathbf{x}_s, \mathbf{x}, \nu_0)$, giving

$$I(\mathbf{x}) = \int_{\mathscr{A}}\int_{\mathscr{A}} \sqrt{(I_1 I_2)}\,\gamma_{12} K(\mathbf{x}_1, \mathbf{x}, \nu_0)\, K^*(\mathbf{x}_2, \mathbf{x}, \nu_0)\, \Lambda_1^{(0)} \Lambda_2^{(0)*}\, d\mathbf{x}_1 d\mathbf{x}_2. \tag{6·1}$$

Assume that the conditions under which (5·9) was derived are satisfied. Then, from (5·3a),

$$I_s = \Gamma_{ss} = a^2(0, \mathbf{x}_s) \int_\Sigma J(\boldsymbol{\xi})\, d\boldsymbol{\xi} = a^2(0, \mathbf{x}_s)\, \mathscr{J}, \tag{6·2}$$

where

$$\mathscr{J} = \int_\Sigma J(\boldsymbol{\xi})\, d\boldsymbol{\xi} = \int_\Sigma d\boldsymbol{\xi} \int_{-\infty}^{+\infty} j(\boldsymbol{\xi}, \nu)\, d\nu. \tag{6·3}$$

(6·1) then becomes

$$I(\mathbf{x}) = \mathscr{J} \int_{\mathscr{A}}\int_{\mathscr{A}} \gamma_{12}\, a(0, \mathbf{x}_1)\, a(0, \mathbf{x}_2)\, K(\mathbf{x}_1, \mathbf{x}, \nu_0)\, K^*(\mathbf{x}_2, \mathbf{x}, \nu_0)\, \Lambda_1^{(0)} \Lambda_2^{(0)*}\, d\mathbf{x}_1 d\mathbf{x}_2. \tag{6·4}$$

Substituting from (5·9), setting

$$\frac{\int_\Sigma J(\boldsymbol{\xi}) \exp\{-ik^{(0)}\boldsymbol{\xi} \,.\, (\mathbf{p}_1 - \mathbf{p}_2)\}\, d\boldsymbol{\xi}}{\int_\Sigma J(\boldsymbol{\xi})\, d\boldsymbol{\xi}} = \sigma(\mathbf{p}_1 - \mathbf{p}_2), \tag{6·5}$$

and using (4·3) we obtain

$$I(\mathbf{x}) = \mathscr{J} \int_{\mathscr{A}}\int_{\mathscr{A}} \sigma(\mathbf{p}_1 - \mathbf{p}_2)\, K(0, \mathbf{x}_1, \nu_0)\, K^*(0, \mathbf{x}_2, \nu_0) \times K(\mathbf{x}_1, \mathbf{x}, \nu_0)\, K^*(\mathbf{x}_2, \mathbf{x}, \nu_0)\, \Lambda_1^{(0)} \Lambda_2^{(0)*}\, d\mathbf{x}_1 d\mathbf{x}_2. \tag{6·6}$$

In this formula, *the effect of the source and the transmission properties of the medium are completely separated.* The source is characterized by the factor $\sigma(\mathbf{p}_1 - \mathbf{p}_2)$ which is the normalized Fourier transform of $J(\boldsymbol{\xi})$, and the medium is characterized by the transmission function K which in practice can be calculated by methods well known in optical designing (e.g. from a ray trace). The ray vectors $\mathbf{p}_1$ and $\mathbf{p}_2$ occurring in the source factor σ and the position vectors $\mathbf{x}_1$ and $\mathbf{x}_2$ of points on the surface of integration are connected by the canonical relations (5·7). It is clear that the integration over the surface $\mathscr{A}$ may be replaced by integration over the solid angle which the entry pupil of the instrument subtends at the source. Also instead of integration *twice* over *one* surface (e.g. the exit pupil) one may reformulate (6·6) so as to involve *one* integration over each of *two* different surfaces (e.g. the entry and the exit pupils). Similar formulae were found by Hopkins (1951, 1953).

Within the accuracy of the present analysis (6·6) shows that *in a given medium, sources which have identical σ factors will give rise to the same intensity distribution.* This is probably the main reason why, in Gabor's method of reconstructed wave-fronts (Gabor 1949, 1951), one may use a different source for the reconstruction from that employed in taking the hologram. In practice, the intensity function will be often practically independent of the position of the radiating element, and it then follows that provided the geometrical shapes of the sources are the same the σ factors will be identical.

7. Concluding remarks

Our generalized Huygens principle makes it possible to obtain solutions to a variety of problems encountered in light optics, and with suitable modifications may also be applied in investigations concerned with other kinds of radiation (electron beams, X-rays, micro-waves). Some possible applications of our results to astronomical investigations and their relation to Michelson's pioneering researches on the application of interference methods to astronomy have already been briefly discussed elsewhere (Fürth & Finlay-Freundlich 1954; Wolf 1954).

It is well known that in applications to problems of image formation in optical systems with low numerical aperture, Huygens's principle gives results in excellent agreement with experiments, but it fails at higher apertures. Now (5·9) and (5·11) show that in systems with high aperture the γ factor may vary appreciably over the domain of integration. Consequently it may be expected that our generalized Huygens principle, which takes into account this variation, will have a wider range of validity than Huygens's principle in its usual form. Since in our formulation the Huygens principle involves observable quantities only, it should be possible to determine its range of validity from experiment.

It can be shown that in the limiting case as $|\gamma_{12}| \rightarrow 1$, (6·1) reduces to usual expressions for the intensity due to an ideal monochromatic point source. Thus (6·1) contains practically the whole elementary diffraction theory of image formation as a special case, and together with (4·9) leads, as we have seen, to many of the previously derived results concerning partially coherent fields. However, the limitations of our analysis must also be stressed, by summarizing the main assumptions under which these formulae were derived: The generalized Huygens principle has been established on the assumption that the usual conditions for the validity of Huygens's principle are satisfied and that the effective frequency range is sufficiently narrow; in the derivation of the expression (4·9) for the Γ factor it has been assumed that the effect of the medium is described with a sufficient accuracy by a transmission function of the form (4·3). In (6·6) it was assumed, in addition to both these conditions, that the source is sufficiently small. Some of these restrictions will be removed in part II of this investigation, where it will be shown that the coherence factors of Van Cittert, Zernike and Hopkins are special cases of a more general correlation function which rigorously obeys the wave equation and with the help of which precise propagation laws may be formulated.

In conclusion, I wish to thank Professor Max Born, F.R.S., for stimulating and helpful discussions, and Professor E. Finlay-Freundlich for valuable information concerning possible applications to astronomy. I am also indebted to Dr A. B. Bhatia and Mr G. Weeden for useful suggestions. Finally, I wish to thank Mr and Mrs R. M. Sillitto for having drawn my attention to some important aspects of Michelson's work.

Part of this work was carried out during the tenure of an Imperial Chemical Industries Research Fellowship and was also supported by a Research Grant from the Carnegie Trust for the Universities of Scotland, both of which are gratefully acknowledged.

References

Arnulf, A., Dupuy, O. & Flamant, F. 1953 *Rev. Opt.* (*théor. instrum.*), **32**, 529.
Baker, L. R. 1953 *Proc. Phys. Soc.* B, **66**, 975.
Berek, M. 1926*a* *Z. Phys.* **36**, 675.
Berek, M. 1926*b* *Z. Phys.* **36**, 824.
Berek, M. 1926*c* *Z. Phys.* **37**, 287.
Berek, M. 1926*d* *Z. Phys.* **40**, 420.
Born, M. 1933 *Optik*, §42, 132. Berlin: Springer.
Duffieux, P. 1953 *Rev. Opt.* (*théor. instrum.*), **32**, 129.
Fürth, R. & Finlay-Freundlich, E. 1954 Contribution in *Vistas in astronomy*. London: Pergamon Press.
Gabor, D. 1949 *Proc. Roy. Soc.* A, **197**, 454.
Gabor, D. 1951 *Proc. Phys. Soc.* B, **64**, 449.
Hopkins, H. H. 1951 *Proc. Roy. Soc.* A, **208**, 263.
Hopkins, H. H. 1953 *Proc. Roy. Soc.* A, **217**, 408.
Michelson, A. A. 1890 *Phil. Mag.* **30**, 1.
Michelson, A. A. 1891*a* *Phil. Mag.* **31**, 256.
Michelson, A. A. 1891*b* *Phil. Mag.* **31**, 338.
Michelson, A. A. 1892 *Phil. Mag.* **34**, 280.
Michelson, A. A. 1920 *Astrophys. J.* **51**, 257.
Michelson, A. A. & Pease, F. G. 1921 *Astrophys. J.* **53**, 249.
Pease, F. G. 1931 Contribution to *Ergebnisse der Exacten Naturwissenschaften*. Berlin: Springer, **10**, 84.
Rogers, G. L. 1953 *Nature, Lond.*, **172**, 118.
Van Cittert, P. H. 1934 *Physica*, **1**, 201.
Van Cittert, P. H. 1939 *Physica*, **6**, 1129.
Wiener, N. 1930 *Acta Math.* **55**, 182.
Wolf, E. 1954 Contribution in *Vistas in astronomy*. London: Pergamon Press.
Zernike, F. 1934*a* *Physica*, **1**, 689.
Zernike, F. 1934*b* *Mon. Not. R. Astr. Soc.* **94**, 377.
Zernike, F. 1938 *Physica*, **5**, 785.
Zernike, F. 1948 *Proc. Phys. Soc.* **61**, 158.

PRINTED IN GREAT BRITAIN AT THE UNIVERSITY PRESS, CAMBRIDGE
(BROOKE CRUTCHLEY, UNIVERSITY PRINTER)

Reprinted without change of pagination from the
Proceedings of the Royal Society, A, *volume* 230, pp. 246–265, 1955

A macroscopic theory of interference and diffraction of light from finite sources
II. Fields with a spectral range of arbitrary width

By E. Wolf

The Physical Laboratories, University of Manchester

(*Communicated by M. Born, F.R.S.—Received* 29 *November* 1954)

The results of part I of this investigation are generalized to stationary fields with a spectral range of arbitrary width. For this purpose it is found necessary to introduce in place of the mutual intensity function of Zernike a more general correlation function

$$\hat{\Gamma}(\mathbf{x}_1, \mathbf{x}_2, \tau) = \langle \hat{V}(\mathbf{x}_1, t+\tau)\, \hat{V}^*(\mathbf{x}_2, t)\rangle,$$

which expresses the correlation between disturbances at any two given points $P_1(\mathbf{x}_1)$, $P_2(\mathbf{x}_2)$ in the field, the disturbance at P_1 being considered at a time τ later than at P_2. It is shown that $\hat{\Gamma}$ is an observable quantity. Expressions for $\hat{\Gamma}$ in terms of functions which specify the source and the transmission properties of the medium are derived.

Further, it is shown that *in vacuo* the correlation function obeys rigorously the two wave equations

$$\nabla_s^2 \hat{\Gamma} = \frac{1}{c^2}\frac{\partial^2 \hat{\Gamma}}{\partial \tau^2} \quad (s = 1, 2),$$

where ∇_s^2 is the Laplacian operator with respect to the co-ordinates (x_s, y_s, z_s) of $P_s(\mathbf{x}_s)$. Using this result, a formula is obtained which expresses rigorously the correlation between disturbances at P_1 and P_2 in terms of the values of the correlation and of its derivatives at all pairs of points on an arbitrary closed surface which surrounds P_1 and P_2. A special case of this formula ($P_2 = P_1$, $\tau = 0$) represents a rigorous formulation of the generalized Huygens principle, involving observable quantities only.

1. Introduction

In part I of this investigation (Wolf (1954*a*), to be referred to as I), interference and diffraction of light in stationary fields produced by finite sources which emit light within a small but finite spectral range were studied. Whilst the results obtained are of interest in connexion with a variety of problems, they are inadequate in the treatment of problems where the path differences between the interfering beams are sufficiently large (cf. appendix 2 below). The present paper removes this restriction, and extends the results to stationary fields of any spectral range.

In the investigation of I, the mutual intensity function

$$\hat{\Gamma}(\mathbf{x}_1, \mathbf{x}_2) = \langle \hat{V}(\mathbf{x}_1, t)\, \hat{V}^*(\mathbf{x}_2, t)\rangle \qquad (1\cdot1)$$

of Zernike, played an essential part. In (1·1), $\hat{V}(\mathbf{x}, t)$ represents the complex disturbance at a point specified by the position vector $\mathbf{x}$, at time t, asterisks denoting the complex conjugate and sharp brackets the time average. In order to extend the analysis to fields with a spectral range of arbitrary width, it is found necessary to introduce in place of (1·1) the more general correlation function

$$\hat{\Gamma}(\mathbf{x}_1, \mathbf{x}_2, \tau) = \langle \hat{V}(\mathbf{x}_1, t+\tau)\, \hat{V}^*(\mathbf{x}_2, t)\rangle. \qquad (1\cdot2)$$

With the help of this function which is shown to represent an observable quantity, the generalized Huygens principle derived in I and the generalized interference law (I, (3·9)) of Zernike & Hopkins are extended to the wider class of fields under consideration. Expressions for $\hat{\Gamma}(\mathbf{x}_1, \mathbf{x}_2, \tau)$ in terms of quantities which specify the source and the medium are also derived.

In § 7, it is shown that *in vacuo* the correlation function (1·2) obeys rigorously the wave equations

$$\left.\begin{aligned} \nabla_1^2 \hat{\Gamma} &= \frac{1}{c^2}\frac{\partial^2 \hat{\Gamma}}{\partial \tau^2}, \\ \nabla_2^2 \hat{\Gamma} &= \frac{1}{c^2}\frac{\partial^2 \hat{\Gamma}}{\partial \tau^2}, \end{aligned}\right\} \qquad (1\cdot3)$$

where ∇_1^2 and ∇_2^2 are the Laplacian operators with respect to the co-ordinates of the points specified by the position vectors $\mathbf{x}_1$ and $\mathbf{x}_2$ respectively.

In § 8, a formula is derived which expresses rigorously the correlation between the disturbances at two given points $P_1(\mathbf{x}_1)$ and $P_2(\mathbf{x}_2)$ in the field in terms of the correlation, and its derivatives, between the disturbances at all pairs of points on an arbitrary closed surface surrounding P_1 and P_2. A special case of this formula ($\mathbf{x}_1 = \mathbf{x}_2$, $\tau = 0$) represents a rigorous formulation of our generalized Huygens principle.

As in I, polarization effects are not considered in the present paper. They will be taken into account in part III of this investigation, where it will be shown that a natural generalization of our results leads to a unified treatment of partial coherence and partial polarization and to a formulation of a wide branch of optics in terms of observable quantities only.†

2. Preliminary: a complex representation of real, polychromatic fields

When dealing with a real, monochromatic (or nearly monochromatic) field, it is usual to employ a complex representation, the field variable being identified with the real part of an appropriate complex function.

† A brief preliminary account of some of these results will be found in Wolf (1954*b*).

In the present paper we shall be concerned with polychromatic fields, i.e. with fields which cover a finite spectral range. It will be convenient to use also in this case a complex representation, this being a natural extension of the representation used in connexion with monochromatic fields.

Let $F(t)$ be a real function, defined for all values of t $(-\infty < t < \infty)$ which possess a Fourier integral representation:

$$F(t) = \int_0^\infty (a_\nu \cos 2\pi\nu t + b_\nu \sin 2\pi\nu t)\, d\nu. \tag{2·1}$$

With F we associate the (generally complex) function $\hat{F}$, defined as

$$\hat{F}(t) = \int_0^\infty (a_\nu + ib_\nu)\, e^{-2\pi i\nu t}\, d\nu. \tag{2·2}$$

It is seen that

$$F(t) = \mathscr{R}\hat{F}(t), \tag{2·3}$$

where $\mathscr{R}$ denotes the real part.

$\hat{F}(t)$ will be referred to as the *half-range complex function* associated with the real function $F(t)$; it is characterized by the property that it may be represented by a Fourier integral which contains no terms of negative frequencies.† F defines $\hat{F}$ uniquely and vice versa.

Throughout this paper a real function and the associated half-range complex function will be denoted by the same symbol, the latter being distinguished by a circumflex. The use of half-range complex functions in place of real functions considerably shortens some of our calculations, and enables the resulting formulae to be expressed in a form which closely resembles those obtained in connexion with almost monochromatic fields in I.

3. A space-time correlation function of stationary fields

In order to extend the analysis of I to stationary fields the spectral range of which is arbitrarily wide, it is necessary, as will be seen below, to introduce, in place of the mutual intensity function (1·1) of Zernike, the more general correlation function‡

$$\hat{\Gamma}(\mathbf{x}_1, \mathbf{x}_2, \tau) = \langle \hat{V}(\mathbf{x}_1, t+\tau)\, \hat{V}^*(\mathbf{x}_2, t)\rangle, \tag{3·1}$$

† *Added in proof* 21 *March* 1955: Since this was written I find that the same complex representation of real fields was introduced previously by Gabor (1946, p. 432) in his interesting investigations in communication theory. Gabor also points out that the real and imaginary parts of such 'half-range complex functions' are Hilbert transforms of each other.

‡ The auto-correlation function, which in our notation would be written as $\hat{\Gamma}(\mathbf{x}, \mathbf{x}, \tau)$, was previously employed in optics by a number of authors, e.g. Wiener (1930), Van Cittert (1939) and Parke (1948). Wiener (1930, p. 119) points out that this function played a fundamental part in Schuster's theory of white light.

Added in proof 21 *March* 1955: In a very interesting paper which was published since this was written, Blanc-Lapierre & Dumontet (1955) applied the general theory of random functions to the optical coherence problems. In their treatment which leads to several new results some of which are closely related to ours, the cross-correlation function (3·2) plays also a basic role.

which expresses the correlation between disturbances at the points $P_1(\mathbf{x}_1)$ and $P_2(\mathbf{x}_2)$, the disturbance at $\mathbf{x}_1$ being considered at a time τ later than at $\mathbf{x}_2$. The sharp brackets denote the time average.†

By a straightforward calculation carried out fully in appendix 1, it may be shown that if $\hat{V}$ is a half-range complex function (in the sense defined in the previous section) the correlation function defined by (3·1) is likewise a half-range complex function, the knowledge of $\hat{\Gamma}(\mathbf{x}_1, \mathbf{x}_2, \tau)$ being equivalent to the knowledge of the real function

$$\Gamma(\mathbf{x}_1, \mathbf{x}_2, \tau) = \mathscr{R}\hat{\Gamma}(\mathbf{x}_1, \mathbf{x}_2, \tau) = 2\langle V(\mathbf{x}_1, t+\tau)\, V(\mathbf{x}_2, t)\rangle. \tag{3·2}$$

We shall need an expression for $\hat{\Gamma}$ in terms of the Fourier components of $\hat{V}$. Let

$$\hat{V}(\mathbf{x}, t) = \int_0^\infty v(\mathbf{x}, \nu)\, \mathrm{e}^{-2\pi \mathrm{i}\nu t}\, d\nu;$$

then

$$v(\mathbf{x}, \nu) = \int_{-T}^{T} \hat{V}(\mathbf{x}, t)\, \mathrm{e}^{2\pi \mathrm{i}\nu t}\, \mathrm{d}t. \tag{3·3}$$

It follows from (3·1), on using the convolution theorem, that

$$\hat{\Gamma}(\mathbf{x}_1, \mathbf{x}_2, \tau) = \lim_{T\to\infty} \frac{1}{2T}\int_0^\infty v(\mathbf{x}_1, \nu)\, v^*(\mathbf{x}_2, \nu)\, \mathrm{e}^{-2\pi \mathrm{i}\nu\tau}\, \mathrm{d}\nu. \tag{3·4}$$

We note a useful relation between $\hat{\Gamma}(\mathbf{x}_2, \mathbf{x}_1, -\tau)$ and $\hat{\Gamma}(\mathbf{x}_1, \mathbf{x}_2, \tau)$. From (3·4),

$$\hat{\Gamma}(\mathbf{x}_2, \mathbf{x}_1, -\tau) = \hat{\Gamma}^*(\mathbf{x}_1, \mathbf{x}_2, \tau). \tag{3·5}$$

We shall normalize $\hat{\Gamma}$ by setting

$$\hat{\gamma}(\mathbf{x}_1, \mathbf{x}_2, \tau) = \frac{\hat{\Gamma}(\mathbf{x}_1, \mathbf{x}_2, \tau)}{\sqrt{\hat{\Gamma}(\mathbf{x}_1, \mathbf{x}_1, 0)}\sqrt{\hat{\Gamma}(\mathbf{x}_2, \mathbf{x}_2, 0)}} = \frac{\hat{\Gamma}(\mathbf{x}_1, \mathbf{x}_2, \tau)}{\sqrt{I(\mathbf{x}_1)}\sqrt{I(\mathbf{x}_2)}}, \tag{3·6}$$

where

$$I(\mathbf{x}_s) = \hat{\Gamma}(\mathbf{x}_s, \mathbf{x}_s, 0) = \langle \hat{V}(\mathbf{x}_s, t)\, \hat{V}^*(\mathbf{x}_s, t)\rangle \quad (s = 1, 2) \tag{3·7}$$

is the intensity at the point $P_s(\mathbf{x}_s)$. On applying the modulus inequality for integrals and the Schwarz inequality to (3·4), one has, by a similar argument as in I, § 3,

$$|\hat{\gamma}(\mathbf{x}_1, \mathbf{x}_2, \tau)| \leqslant 1. \tag{3·8}$$

4. An approximate propagation law for $\hat{\Gamma}(\mathbf{x}_1, \mathbf{x}_2, \tau)$ and an extension of the generalized Huygens principle to fields with spectral range of arbitrary width

As in I we consider the propagation of a beam of light from a finite source Σ. The field will again be assumed to be stationary, but no restriction on the width of the spectral range will now be imposed.

† In I, the time average was taken over the finite range $-T \leqslant t \leqslant T$, but it is more convenient mathematically (although it is not more significant physically) to allow the range to become infinite by proceeding to the limit $T\to\infty$, as customary. We shall now understand the time average in this sense:

$$\hat{\Gamma}(\mathbf{x}_1, \mathbf{x}_2, \tau) = \lim_{T\to\infty} \frac{1}{2T}\int_{-T}^{T} \hat{V}(\mathbf{x}_1, t+\tau)\, \hat{V}^*(\mathbf{x}_2, t)\, \mathrm{d}t.$$

The integration over the frequency range was taken in I from $-\infty$ to $+\infty$, but as explained in the preceding section we may set $v(\mathbf{x}, \nu) = 0$ for $\nu < 0$ and hence integrate over the positive range $0 \leqslant \nu < \infty$ only. The quantities which were denoted by $V(\mathbf{x}, t)$, $\Gamma(\mathbf{x}_1, \mathbf{x}_2)$ and $\gamma(\mathbf{x}_1, \mathbf{x}_2)$ in I would in the present notation be therefore written as $\hat{V}(\mathbf{x}, t)$, $\hat{\Gamma}(\mathbf{x}_1, \mathbf{x}_2)$ and $\hat{\gamma}(\mathbf{x}_1, \mathbf{x}_2)$.

Let $\mathscr{A}$ be a surface cutting across the beam (see figure 1) and let $P_1(\mathbf{x}_1)$ and $P_2(\mathbf{x}_2)$ be two points on that side of $\mathscr{A}$ towards which the light is advancing. We shall derive an expression for the correlation $\hat{\Gamma}(\mathbf{x}_1, \mathbf{x}_2, \tau)$ between the disturbances at P_1 and P_2, in terms of the values which $\hat{\Gamma}$ takes at all pairs of points on the surface $\mathscr{A}$.

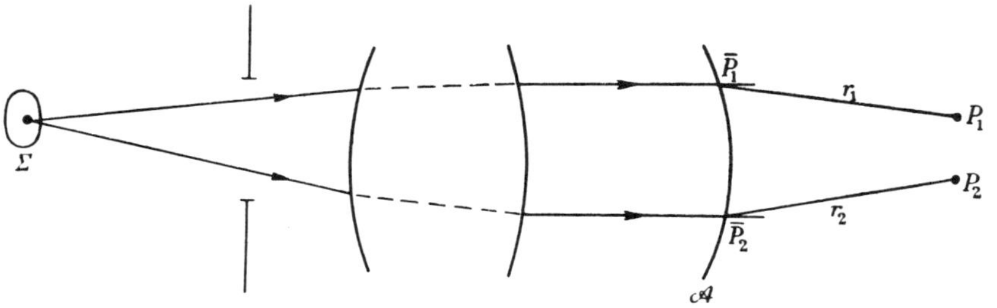

FIGURE 1. Derivation of a propagation law for $\hat{\Gamma}(\mathbf{x}_1, \mathbf{x}_2, \tau)$.

Each Fourier term $v(\mathbf{x}, \nu) \exp\{-\mathrm{i}\,2\pi\nu t\}$ of $\hat{V}$ represents a perfectly monochromatic wave. Hence the value of v at P_1 may be expressed (under the usual restrictions) in terms of the values which v takes at all points on the surface $\mathscr{A}$, by means of the ordinary Huygens principle:

$$v(\mathbf{x}_1, \nu) = \int_{\mathscr{A}} v(\bar{\mathbf{x}}_1, \nu) \frac{\mathrm{e}^{\mathrm{i}kr_1}}{r_1} \Lambda_1 \,\mathrm{d}\bar{\mathbf{x}}_1. \tag{4·1}$$

Here $\bar{\mathbf{x}}_1$ is the position vector at a typical point $\bar{P}_1$ on the surface $\mathscr{A}$, r_1 is the distance from $\bar{P}_1$ to P_1 (see figure 1), $k = 2\pi\nu/c = 2\pi/\lambda$, c being the vacuum velocity of light and λ the wave-length. Λ denotes as before the usual inclination factor (see I, (2·4)).

In a similar way, the values $v(\mathbf{x}_2, \nu)$ at P_2 may be expressed in the form

$$v(\mathbf{x}_2, \nu) = \int_{\mathscr{A}} v(\bar{\mathbf{x}}_2, \nu) \frac{\mathrm{e}^{\mathrm{i}kr_2}}{r_2} \Lambda_2 \,\mathrm{d}\bar{\mathbf{x}}_2. \tag{4·2}$$

From (4·1) and (4·2),

$$v(\mathbf{x}_1, \nu)\, v^*(\mathbf{x}_2, \nu) = \int_{\mathscr{A}}\int_{\mathscr{A}} v(\bar{\mathbf{x}}_1, \nu)\, v^*(\bar{\mathbf{x}}_2, \nu) \frac{\mathrm{e}^{\mathrm{i}k(r_1-r_2)}}{r_1 r_2} \Lambda_1 \Lambda_2^* \,\mathrm{d}\bar{\mathbf{x}}_1 \,\mathrm{d}\bar{\mathbf{x}}_2, \tag{4·3}$$

the points $\bar{P}_1(\bar{\mathbf{x}}_1)$ and $\bar{P}_2(\bar{\mathbf{x}}_2)$ exploring the surface $\mathscr{A}$ independently. Next we multiply (4·3) by $\frac{1}{2T}\exp\{-2\pi\mathrm{i}\nu\tau\}$, integrate over the frequency range and proceed to the limit $T \to \infty$. We then obtain the following expression for $\hat{\Gamma}(\mathbf{x}_1, \mathbf{x}_2, \tau)$:

$$\hat{\Gamma}(\mathbf{x}_1, \mathbf{x}_2, \tau) = \int_{\mathscr{A}}\int_{\mathscr{A}} \left\{ \lim_{T\to\infty} \frac{1}{2T} \int_0^\infty v(\bar{\mathbf{x}}_1, \nu)\, v^*(\bar{\mathbf{x}}_2, \nu) \exp\left\{-2\pi\mathrm{i}\nu\left[\tau - \frac{r_1 - r_2}{c}\right]\right\} \Lambda_1 \Lambda_2^* \,\mathrm{d}\nu \right\} \frac{1}{r_1}\frac{1}{r_2} \,\mathrm{d}\bar{\mathbf{x}}_1 \,\mathrm{d}\bar{\mathbf{x}}_2. \tag{4·4}$$

Now in the integral over the frequency range, the factors Λ_1 and Λ_2^* are well-behaved functions, depending on the frequency only through a multiplicate factor ν.

We shall take Λ_1 and Λ_2^* outside the ν integration in mean values† (denoted by $\bar{\Lambda}_1$ and $\bar{\Lambda}_2^*$). The remaining part of the frequency integral gives, in the limit $T \to \infty$, precisely $\hat{\Gamma}\left(\bar{\mathbf{x}}_1, \bar{\mathbf{x}}_2, \tau - \frac{r_1 - r_2}{c}\right)$. Hence (4·4) reduces to the relatively simple law

$$\hat{\Gamma}(\mathbf{x}_1, \mathbf{x}_2, \tau) = \int_{\mathscr{A}} \int_{\mathscr{A}} \frac{\hat{\Gamma}\left(\bar{\mathbf{x}}_1, \bar{\mathbf{x}}_2, \tau - \frac{r_1 - r_2}{c}\right)}{r_1 r_2} \bar{\Lambda}_1 \bar{\Lambda}_2^* \, d\bar{\mathbf{x}}_1 \, d\bar{\mathbf{x}}_2. \tag{4·5}$$

Equation (4·5) expresses $\hat{\Gamma}(\mathbf{x}_1, \mathbf{x}_2, \tau)$ in terms of the values which this function takes at all pairs of points $\bar{P}_1$ and $\bar{P}_2$ on the surface, the time argument for each pair having the values $\tau - (\tau_1 - \tau_2)$, where $\tau_1 = r_1/c$ and $\tau_2 = r_2/c$ are the times needed for light to travel from $\bar{P}_1$ to P_1 and from $\bar{P}_2$ to P_2 respectively.

Of particular interest is the special case when the points P_1 and P_2 coincide and when, in addition, $\tau = 0$. Denoting the common point by $P(\mathbf{x})$, the left-hand side of (4·5) reduces to the intensity $I(\mathbf{x})$ at P, and, if (3·6) is also used, one obtains

$$I(\mathbf{x}) = \int_{\mathscr{A}} \int_{\mathscr{A}} \frac{\sqrt{I(\bar{\mathbf{x}}_1)}\sqrt{I(\bar{\mathbf{x}}_2)}}{r_1 r_2} \hat{\gamma}\left(\bar{\mathbf{x}}_1, \bar{\mathbf{x}}_2, \frac{r_2 - r_1}{c}\right) \bar{\Lambda}_1 \bar{\Lambda}_2^* \, d\bar{\mathbf{x}}_1 \, d\bar{\mathbf{x}}_2, \tag{4·6}$$

$I(\bar{\mathbf{x}}_1)$ and $I(\bar{\mathbf{x}}_2)$ being the intensities at two typical points $\bar{P}_1$ and $\bar{P}_2$ of the surface $\mathscr{A}$.

In (4·6) $\hat{\gamma}$ may be replaced by $\gamma = \mathscr{R}\hat{\gamma}$, since the imaginary part of $\hat{\gamma}$ contributes nothing to the integral, as I and $\bar{\Lambda}_1 \bar{\Lambda}_2^*$ are real. That the integral is actually real may be shown formally by verifying that it remains unchanged when its complex conjugate is taken. This result follows immediately on using (3·5) and interchanging the independent variables $\bar{\mathbf{x}}_1$ and $\bar{\mathbf{x}}_2$. In place of (4·6) we may therefore write

$$I(\mathbf{x}) = \int_{\mathscr{A}} \int_{\mathscr{A}} \frac{\sqrt{I(\bar{\mathbf{x}}_1)}\sqrt{I(\bar{\mathbf{x}}_2)}}{r_1 r_2} \gamma\left(\bar{\mathbf{x}}_1, \bar{\mathbf{x}}_2, \frac{r_2 - r_1}{c}\right) \bar{\Lambda}_1 \bar{\Lambda}_2^* \, d\bar{\mathbf{x}}_1 \, d\bar{\mathbf{x}}_2. \tag{4·6a}$$

Equation (4·6*a*) (or (4·6)) may be regarded as a *generalized Huygens principle for stationary fields of an arbitrary spectral range*. It expresses the *intensity* at the point $P(\mathbf{x})$ in terms of the intensity distribution over an arbitrary surface $\mathscr{A}$, the contribution from each pair of elements of the surface being weighed by the appropriate value of the correlation factor $\gamma(\mathbf{x}_1, \mathbf{x}_2, \tau)$. This factor, which is a generalization of the degree of coherence of Zernike, may, like the latter, be determined from experiments (cf. § 5 below). It may also be calculated from the knowledge of an (observable) correlation function of the source and of the transmission properties of the medium (cf. § 6 below). Hence our extended formulation of the generalized Huygens principle involves again observable quantities only.

† This step of the analysis, though formally correct, is somewhat unsatisfactory, since the mean values will depend not only on the geometrical situation, but also on the form of v, as function of frequency. It should, however, be borne in mind that the inclination factor Λ of the ordinary Huygens principle represents only a rough approximation, and is, in most practical cases, simply replaced by a constant. The difficulty disappears in the rigorous formulation given in § 8 below.

In order to see more clearly the connexion between the formula just derived, and some of the formulae given previously, consider the special case when the light is practically monochromatic, of frequency ν_0. If we express $\hat{\Gamma}$ and $\hat{\gamma}$ in the form

$$\left.\begin{aligned}\hat{\Gamma}(\mathbf{x}_1,\mathbf{x}_2,\tau) &= G(\mathbf{x}_1,\mathbf{x}_2,\tau)\,\mathrm{e}^{-2\pi \mathrm{i}\nu_0\tau},\\ \hat{\gamma}(\mathbf{x}_1,\mathbf{x}_2,\tau) &= g(\mathbf{x}_1,\mathbf{x}_2,\tau)\,\mathrm{e}^{-2\pi \mathrm{i}\nu_0\tau},\end{aligned}\right\} \tag{4·7}$$

the quantities G and g, considered as functions of τ, will vary slowly compared with the variation of the exponential term. If sufficiently small path differences are involved, the variations of G and g, with τ, may be completely neglected, i.e. we may then write

$$\left.\begin{aligned}G \sim G(\mathbf{x}_1,\mathbf{x}_2) &\sim \hat{\Gamma}(\mathbf{x}_1,\mathbf{x}_2,0),\\ g \sim g(\mathbf{x}_1,\mathbf{x}_2) &\sim \hat{\gamma}(\mathbf{x}_1,\mathbf{x}_2,0).\end{aligned}\right\} \tag{4·8}$$

G is essentially Zernike's mutual intensity (denoted by $\Gamma(\mathbf{x}_1,\mathbf{x}_2)$ in I) and g his complex degree of coherence (denoted by $\gamma(\mathbf{x}_1,\mathbf{x}_2)$ in I). With this substitution, (4·5) reduces to Zernike's propagation law (Zernike 1938, equation (9)).†

$$G(\mathbf{x}_1,\mathbf{x}_2) \sim \int_{\mathscr{A}}\int_{\mathscr{A}} \frac{G(\bar{\mathbf{x}}_1,\bar{\mathbf{x}}_2)}{r_1 r_2}\,\mathrm{e}^{\mathrm{i}k^{(0)}(r_1-r_2)}\,\Lambda_1^{(0)}\Lambda_2^{(0)*}\,\mathrm{d}\bar{\mathbf{x}}_1\,\mathrm{d}\bar{\mathbf{x}}_2, \tag{4·9}$$

where $k^{(0)} = 2\pi\nu_0/c$, $\Lambda^{(0)} = \Lambda(\nu_0)$; and with the same approximation (4·6) reduces to the more restricted formulation of the generalized Huygens principle given in I.

5. The generalized interference law and determination of the correlation function from experiments

5·1. *The case* $\mathbf{x}_1 \neq \mathbf{x}_2$

In order to determine the correlation functions from experiment we may use a procedure similar to that described in I in connexion with the determination of the less general correlation functions of Zernike.

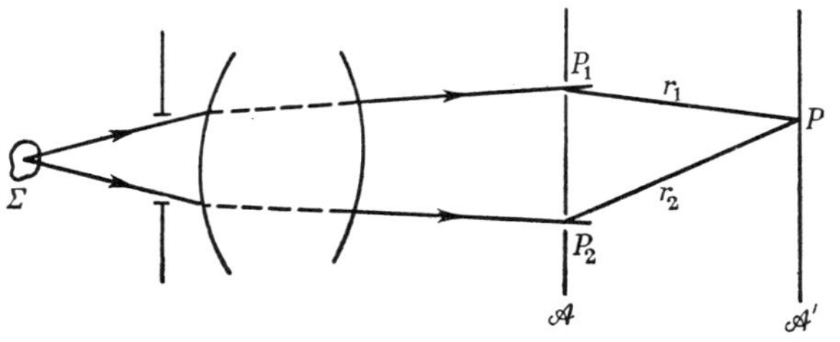

Figure 2. Experimental determination of $\Gamma(\mathbf{x}_1,\mathbf{x}_2,\tau)$.

We place a screen $\mathscr{A}$ across the field so as to pass through the points $P_1(\mathbf{x}_1)$ and $P_2(\mathbf{x}_2)$, small openings $\mathrm{d}\mathscr{A}_1$ and $\mathrm{d}\mathscr{A}_2$ being made at these points. The resulting intensity distribution is observed on a second screen $\mathscr{A}'$ (figure 2).

† Zernike actually neglected the variation of the inclination factor over the surface of integration, setting as usual, $\Lambda = \mathrm{i}/\lambda$.

Taking the integrals (4·6a) over the non-illuminated side of the screen $\mathscr{A}$, we obtain for the intensity $I(\mathbf{x})$ at the point $P(\mathbf{x})$ the expression

$$\begin{aligned} I(\mathbf{x}) \sim & \frac{I(\mathbf{x}_1)}{r_1^2} \,|\, \tilde{\Lambda}_1 \,|^2 (\mathrm{d}\mathscr{A}_1)^2 + \frac{I(\mathbf{x}_2)}{r_2^2} \,|\, \tilde{\Lambda}_2 \,|^2 (\mathrm{d}\mathscr{A}_2)^2 \\ & + \frac{\sqrt{I(\mathbf{x}_1)}\sqrt{I(\mathbf{x}_2)}}{r_1 r_2} \gamma\left(\mathbf{x}_1, \mathbf{x}_2, \frac{r_2 - r_1}{c}\right) \tilde{\Lambda}_1 \tilde{\Lambda}_2^* \mathrm{d}\mathscr{A}_1 \mathrm{d}\mathscr{A}_2 \\ & \qquad + \frac{\sqrt{I(\mathbf{x}_2)}\sqrt{I(\mathbf{x}_1)}}{r_2 r_1} \gamma\left(\mathbf{x}_2, \mathbf{x}_1 \frac{r_1 - r_2}{c}\right) \tilde{\Lambda}_2 \tilde{\Lambda}_1^* \mathrm{d}\mathscr{A}_2 \mathrm{d}\mathscr{A}_1. \end{aligned} \tag{5·1}$$

Now the first term on the right is precisely the intensity $I^{(1)}(\mathbf{x})$ which would be obtained at P if the openings at P_1 alone was open ($\mathrm{d}\mathscr{A}_2 = 0$):

$$I^{(1)}(\mathbf{x}) = \frac{I(\mathbf{x}_1)}{r_1^2} \,|\, \tilde{\Lambda}_1 \,|^2 (\mathrm{d}\mathscr{A}_1)^2, \tag{5·2}$$

the second term

$$I^{(2)}(\mathbf{x}) = \frac{I(\mathbf{x}_2)}{r_2^2} \,|\, \tilde{\Lambda}_2 \,|^2 (\mathrm{d}\mathscr{A}_2)^2 \tag{5·3}$$

having a similar interpretation. Also, on account of (3·5) and remembering that γ is the real part of $\hat{\gamma}$,

$$\gamma\left(\mathbf{x}_2, \mathbf{x}_1, \frac{r_1 - r_2}{c}\right) = \gamma\left(\mathbf{x}_1, \mathbf{x}_2, \frac{r_2 - r_1}{c}\right). \tag{5·4}$$

Further, since $\tilde{\Lambda}$ is purely imaginary, $\tilde{\Lambda}_2 \tilde{\Lambda}_1^* = \tilde{\Lambda}_1 \tilde{\Lambda}_2^* = |\,\tilde{\Lambda}_1\,|\,|\,\tilde{\Lambda}_2\,|$. Hence (5·1) reduces to

$$I(\mathbf{x}) = I^{(1)}(\mathbf{x}) + I^{(2)}(\mathbf{x}) + 2\sqrt{I^{(1)}(\mathbf{x})}\sqrt{I^{(2)}(\mathbf{x})}\gamma\left(\mathbf{x}_1, \mathbf{x}_2, \frac{r_2 - r_1}{c}\right). \tag{5·5}$$

(5·5) represents *a general interference law for stationary fields.*

It is seen that in order to determine $\gamma(\mathbf{x}_1, \mathbf{x}_2, \tau)$ it is only necessary to take the distances r_1 and r_2 such that $\frac{r_2 - r_1}{c} = \tau$, and to measure the intensities $I(\mathbf{x})$, $I^{(1)}(\mathbf{x})$ and $I^{(2)}(\mathbf{x})$. γ is then given by

$$\gamma\left(\mathbf{x}_1, \mathbf{x}_2, \frac{r_2 - r_1}{c}\right) = \frac{I(\mathbf{x}) - I^{(1)}(\mathbf{x}) - I^{(2)}(\mathbf{x})}{2\sqrt{I^{(1)}(\mathbf{x})}\sqrt{I^{(2)}(\mathbf{x})}}. \tag{5·6}$$

If the value of Γ is also required, it is necessary, in addition, to measure the intensities $I(\mathbf{x}_1)$ and $I(\mathbf{x}_2)$ at the points P_1 and P_2. Then from (5·6) and (3·6),

$$\Gamma\left(\mathbf{x}_1, \mathbf{x}_2, \frac{r_2 - r_1}{c}\right) = \frac{1}{2}\sqrt{\frac{I(\mathbf{x}_1)}{I^{(1)}(\mathbf{x})}}\sqrt{\frac{I(\mathbf{x}_2)}{I^{(2)}(\mathbf{x})}}\,[I(\mathbf{x}) - I^{(1)}(\mathbf{x}) - I^{(2)}(\mathbf{x})]. \tag{5·7}$$

In the special case when the effective frequency range is sufficiently narrow (path differences small enough), (5·5) reduces, on substitution from (4·7), to the formula I, (3·9) of Zernike and Hopkins:

$$I(\mathbf{x}) = I^{(1)}(\mathbf{x}) + I^{(2)}(\mathbf{x}) + 2\sqrt{I^{(1)}(\mathbf{x})}\sqrt{I^{(2)}(\mathbf{x})}\,|\,g(\mathbf{x}_1, \mathbf{x}_2)\,|\cos[\arg g(\mathbf{x}_1, \mathbf{x}_2) + k^{(0)}(r_1 - r_2)]. \tag{5·8}$$

5·2. *The case* $\mathbf{x}_1 = \mathbf{x}_2$

It has been assumed so far that the two points P_1 and P_2 are distinct. We must now consider the special case when they coincide.

When $P_1 = P_2$, (3·1) and (3·4) reduce to

$$\hat{\Gamma}(\mathbf{x}_1, \mathbf{x}_1, \tau) = \langle \hat{V}(\mathbf{x}_1, t+\tau)\,\hat{V}^*(\mathbf{x}_1, t)\rangle = \lim_{T\to\infty}\frac{1}{2T}\int_0^\infty |\,v(\mathbf{x}_1,\nu)\,|^2\,\mathrm{e}^{-2\pi i\nu\tau}\,\mathrm{d}\nu. \qquad (5\cdot9)$$

Hence $\hat{\Gamma}$ may now be determined by measuring the spectral intensity function $I(\mathbf{x}_1,\nu) = \lim_{T\to\infty}\frac{1}{2T}|\,v(\mathbf{x}_1,\nu)\,|^2$ at P_1 and evaluating the Fourier integral. (If, in addition to $\mathbf{x}_1 = \mathbf{x}_2$, one also has $\tau = 0$, $\hat{\Gamma}$ reduces, of course, to the intensity at P_1.)

A more direct determination of $\hat{\Gamma}(\mathbf{x}_1, \mathbf{x}_2, \tau)$ appears to be possible, in principle, by the use of some interferometer based on the principle of division of amplitude (cf. Williams 1941).

Suppose that the beam of light is divided at the point $P_1(\mathbf{x}_1)$ into two beams (for example, in the Michelson interferometer), which then proceed via different paths and are reunited again at a point $P(\mathbf{x})$. Let the transmission functions (cf. I, § 4) of the two paths for strictly monochromatic radiation of frequency ν be

$$K(\mathbf{x}_1, \mathbf{x}, \nu) = |\,K_1\,|\,\mathrm{e}^{2\pi i\nu\phi_1}, \quad K(\mathbf{x}_2, \mathbf{x}, \nu) = |\,K_2\,|\,\mathrm{e}^{2\pi i\nu\phi_2}. \qquad (5\cdot10)$$

Then the Fourier component $v(\mathbf{x},\nu)$ at P is related to that at P_1 by

$$v(\mathbf{x},\nu) = (K_1\Lambda_1 + K_2\Lambda_2)\,v(\mathbf{x}_1,\nu)\,\delta\mathscr{A}. \qquad (5\cdot11)$$

where $\delta\mathscr{A}$ is the element of area around P_1 which reflects and transmit the incident light. Hence the intensity at P is given by

$$\begin{aligned} I(\mathbf{x}) &= \lim_{T\to\infty}\frac{1}{2T}\int_0^\infty |\,v(\mathbf{x},\nu)\,|^2\,\mathrm{d}\nu \\ &= I^{(1)}(\mathbf{x}) + I^{(2)}(\mathbf{x}) + \mathscr{J}(\mathbf{x}), \end{aligned} \qquad (5\cdot12)$$

where

$$\left.\begin{aligned} I^{(1)}(\mathbf{x}) &= \lim_{T\to\infty}\frac{1}{2T}\int_0^\infty |\,v(\mathbf{x}_1,\nu)\,|^2\,|\,K_1\,|^2\,|\,\Lambda_1\,|^2(\delta\mathscr{A})^2\,\mathrm{d}\nu, \\ I^{(2)}(\mathbf{x}) &= \lim_{T\to\infty}\frac{1}{2T}\int_0^\infty |\,v(\mathbf{x}_1,\nu)\,|^2\,|\,K_2\,|^2\,|\,\Lambda_2\,|^2(\delta\mathscr{A})^2\,\mathrm{d}\nu, \\ \mathscr{J}(\mathbf{x}) &= \lim_{T\to\infty}\frac{1}{T}\int_0^\infty |\,v(\mathbf{x}_1,\nu)\,|^2\,|\,K_1\,|\,|\,K_2\,|\,|\,\Lambda_1\,|\,|\,\Lambda_2\,|(\delta\mathscr{A})^2\cos 2\pi\nu(\phi_2-\phi_1)\,\mathrm{d}\nu. \end{aligned}\right\} \qquad (5\cdot13)$$

In general, $|\,K_1\,|$, $|\,K_2\,|$, ϕ_1 and ϕ_2 will depend on the frequency, since owing to the refraction at the individual elements of the interferometer, light of different frequencies will proceed along slightly different paths. In many cases, however, this effect will be negligible; $|\,K_1\,|$ and $|\,K_2\,|$ may then be taken outside the integrals in (5·13) and $\mathscr{J}$ becomes, apart from a multiplicative factor, the Fourier cosine transform of the spectral intensity function, this being equal to Γ according to

(5·9). The expression (5·12) for the intensity may then be expressed in a form strictly analogous to (5·5):

$$I(\mathbf{x}) = I^{(1)}(\mathbf{x}) + I^{(2)}(\mathbf{x}) + 2\sqrt{I^{(1)}(\mathbf{x})}\sqrt{I^{(2)}(\mathbf{x})}\,\gamma\left(\mathbf{x}_1, \mathbf{x}_1, \frac{\phi_2-\phi_1}{c}\right), \tag{5·14}$$

with

$$\left.\begin{aligned} I^{(1)}(\mathbf{x}) &= |K_1|^2 |\tilde{\Lambda}_1|^2 (\delta\mathscr{A})^2 I(\mathbf{x}_1), \\ I^{(2)}(\mathbf{x}) &= |K_2|^2 |\tilde{\Lambda}_2|^2 (\delta\mathscr{A})^2 I(\mathbf{x}_1), \\ I(\mathbf{x}_1) &= \lim_{T\to\infty} \frac{1}{2T}\int_0^\infty |v(\mathbf{x}_1,\nu)|^2\,\mathrm{d}\nu = \langle \hat{V}(\mathbf{x}_1,t)\,\hat{V}^*(\mathbf{x}_1,t)\rangle. \end{aligned}\right\} \tag{5·15}$$

The term $I^{(1)}(\mathbf{x})$ represents the intensity which is obtained at the point $P(\mathbf{x})$ if the second beam is excluded ($K_2 = 0$), the term $I^{(2)}(\mathbf{x})$ having a similar interpretation. Hence to measure $\gamma(\mathbf{x}_1, \mathbf{x}_1, \tau)$ it is only necessary (provided $|K_1|$, $|K_2|$, ϕ_1 and ϕ_2 may be treated as independent of ν) to set the interferometer so that $(\phi_2-\phi_1)/c = \tau$. γ is then given by a formula analogous to (5·6):

$$\gamma\left(\mathbf{x}_1, \mathbf{x}_1, \frac{\phi_2-\phi_1}{c}\right) = \frac{I(\mathbf{x}) - I^{(1)}(\mathbf{x}) - I^{(2)}(\mathbf{x})}{2\sqrt{I^{(1)}(\mathbf{x})}\sqrt{I^{(2)}(\mathbf{x})}}. \tag{5·16}$$

And Γ is given by

$$\Gamma\left(\mathbf{x}_1, \mathbf{x}_1, \frac{\phi_2-\phi_1}{c}\right) = I(\mathbf{x}_1)\,\gamma\left(\mathbf{x}_1, \mathbf{x}_1, \frac{\phi_2-\phi_1}{c}\right). \tag{5·17}$$

The investigations of Zernike (1938, 1948; see also I, § 3) brought out the close connexion which exists between the visibility factor of Michelson and the *correlation* (characterized by $\hat{\gamma}(\mathbf{x}_1, \mathbf{x}_2, 0)$) of disturbances *at two points in the field, at the same instant of time*. These investigations interpreted, also from a new point of view, Michelson's method for the determination of the intensity of radiation across a radiating source from the measurements of the visibility of fringes, a method which in recent years has become of fundamental importance in radio astronomy (cf. Ryle 1950; Smith 1952). With the help of the preceding analysis, it will now be shown that there is a complementary relation between the visibility function and the *correlation* (characterized by $\hat{\gamma}(\mathbf{x}_1, \mathbf{x}_1, \tau)$) of disturbances *at two instants of time, at the same point in the field*. This result leads to a new interpretation of Michelson's well-known method (Michelson 1891, 1892) for the determination of the energy distribution in spectral lines from measurements of the visibility.†

Let ν_0 be the mean frequency of the spectral distribution, assumed now to be confined within a narrow frequency range $\nu_0 - \Delta\nu \leqslant \nu \leqslant \nu_0 + \Delta\nu$ and set

$$\hat{\gamma}(\mathbf{x}_1, \mathbf{x}_1, \tau) = h(\mathbf{x}_1, \tau)\,\mathrm{e}^{-2\pi i\nu_0\tau}. \tag{5·18}$$

Further, assume, as is usually the case, that $I^{(1)}(\mathbf{x}) \sim I^{(2)}(\mathbf{x})$. Equation (5·14) then becomes

$$I(\mathbf{x}) = 2I^{(1)}(\mathbf{x})\{1 + |h(\mathbf{x}_1,\tau)|\cos[\arg h(\mathbf{x}_1,\tau) - 2\pi\nu_0\tau]\}. \tag{5·19}$$

Since $\Delta\nu/\nu_0 \ll 1$, h, considered as a function of τ, will change very slowly in comparison with $\cos 2\pi\nu_0\tau$ and $\sin 2\pi\nu_0\tau$, so that the minima and maxima of the intensity are effectively given by τ values which satisfy the relation

$$\sin[\arg h(\mathbf{x}_1,\tau) - 2\pi\nu_0\tau] = 0. \tag{5·20}$$

† In this connexion see also the paper by Van Cittert (1939).

The corresponding maxima and minima have the values

$$\left.\begin{aligned} I_{\text{max.}} &= 2I^{(1)}(\mathbf{x})\,[1+|\,h(\mathbf{x}_1,\tau)\,|],\\ I_{\text{min.}} &= 2I^{(1)}(\mathbf{x})\,[1-|\,h(\mathbf{x}_1,\tau)\,|].\end{aligned}\right\} \tag{5·21}$$

The visibility $\mathscr{V}$ of the fringes is therefore

$$\mathscr{V} = \frac{I_{\text{max.}}-I_{\text{min.}}}{I_{\text{max.}}+I_{\text{min.}}} = |\,h(\mathbf{x}_1,\tau)\,| = |\,\hat{\gamma}(\mathbf{x}_1,\mathbf{x}_1,\tau)\,|, \tag{5·22}$$

i.e. the visibility is equal to the absolute value of the complex correlation factor $\hat{\gamma}(\mathbf{x}_1,\mathbf{x}_1,\tau)$. Hence according to (5·9), using the Fourier inversion formula, it is possible to obtain information about the intensity distribution in the spectrum from measurements of the visibility, provided that suitable assumptions about the associated phase $[\arg\hat{\gamma}(\mathbf{x}_1,\mathbf{x}_1,\tau)]$ are made.†

It is now seen that the two methods of Michelson correspond essentially to the two limiting cases $\tau\to 0$ and $\mathbf{x}_2\to\mathbf{x}_1$ of our theory.

6. Expressions for $\hat{\Gamma}$ in terms of quantities which specify the source and the medium

We shall now derive explicit expressions for the $\hat{\Gamma}$ factor in terms of quantities which specify the source and the transmission properties of the medium.

As in I we assume the source Σ to be a radiating plane area and divide it into elements $\delta\Sigma_1, \delta\Sigma_2, \ldots$, which are small in linear dimensions in comparison with the optical wave-lengths. Let

$$\hat{V}_m(\mathbf{x},t) = \int_0^\infty v_m(\mathbf{x},\nu)\,\mathrm{e}^{-2\pi i\nu t}\,\mathrm{d}\nu \tag{6·1}$$

be the disturbance due to the mth element; the total disturbance $\hat{V}(\mathbf{x},t)$ is then given by

$$\hat{V}(\mathbf{x},t) = \sum_m \hat{V}_m(\mathbf{x},t). \tag{6·2}$$

Hence

$$\begin{aligned}\hat{\Gamma}(\mathbf{x}_1,\mathbf{x}_2,\tau) &= \langle \hat{V}(\mathbf{x}_1,t+\tau)\,\hat{V}^*(\mathbf{x}_2,t)\rangle\\ &= \sum_m\sum_n \hat{\Gamma}_{mn}(\mathbf{x}_1,\mathbf{x}_2,\tau),\end{aligned} \tag{6·3}$$

where

$$\begin{aligned}\hat{\Gamma}_{mn}(\mathbf{x}_1,\mathbf{x}_2,\tau) &= \langle \hat{V}_m(\mathbf{x}_1,t+\tau)\,\hat{V}_n^*(\mathbf{x}_2,t)\rangle\\ &= \lim_{T\to\infty}\frac{1}{2T}\int_0^\infty v_m(\mathbf{x}_1,\nu)\,v_n^*(\mathbf{x}_2,\nu)\,\mathrm{e}^{-2\pi i\nu\tau}\,\mathrm{d}\nu.\end{aligned} \tag{6·4}$$

In most cases of practical interest (e.g. for a gas discharge or incandescent solid) it will be permissible to assume that the radiation from the different elements of the source is mutually incoherent, i.e. that for all values of $\mathbf{x}_1$, $\mathbf{x}_2$ and τ

$$\hat{\Gamma}_{mn}(\mathbf{x}_1,\mathbf{x}_2,\tau) = 0 \quad \text{when} \quad m\neq n. \tag{6·5}$$

† As is evident from (5·19), the phase may in principle be obtained from the measurement of the position of the fringes. This has been pointed out already by Rayleigh (1892) in an open letter to Michelson, in which he discussed the question of a complete determination of the intensity distribution from Michelson's experiments.

There are, however, important cases when this assumption does not hold. For example, if the source is not a 'natural' source, but is a secondary source obtained by imagining a source of natural light by a lens of a finite aperture, then on account of diffraction there will exist a finite degree of correlation in the plane of the secondary source at points which are sufficiently close to each other. Accordingly, we shall first consider the general case when $\hat{\Gamma}_{mn} \neq 0$.

Let $a_m(\nu)$ be the strength of the radiation from the element $\delta\Sigma_m$, at frequency ν, and let $K(\mathbf{x}', \mathbf{x}'', \nu)$ be the transmission function of the medium. Then†

$$v_m(\mathbf{x}, \nu) = a_m(\nu)\, K(\boldsymbol{\xi}_m, \mathbf{x}, \nu), \tag{6·6}$$

where $\boldsymbol{\xi}_m$ denotes the position vector of the mth element of the source. (6·3) then becomes

$$\hat{\Gamma}(\mathbf{x}_1, \mathbf{x}_2, \tau) = \sum_m \sum_n \int_0^\infty J_{mn}(\nu) L(\boldsymbol{\xi}_m, \boldsymbol{\xi}_n;\ \mathbf{x}_1, \mathbf{x}_2;\ \nu)\, \mathrm{e}^{-2\pi i\nu\tau}\, \mathrm{d}\nu, \tag{6·7}$$

with

$$\left.\begin{aligned} J_{mn}(\nu) &= \lim_{T\to\infty} \frac{1}{2T}[a_m(\nu)\, a_n^*(\nu)], \\ L(\boldsymbol{\xi}_m, \boldsymbol{\xi}_n;\ \mathbf{x}_1, \mathbf{x}_2;\ \nu) &= K(\boldsymbol{\xi}_m, \mathbf{x}_1, \nu)\, K^*(\boldsymbol{\xi}_n, \mathbf{x}_2, \nu). \end{aligned}\right\} \tag{6·8}$$

and

Let

$$\int_0^\infty J_{mn}(\nu)\, \mathrm{e}^{-2\pi i\nu u}\, \mathrm{d}\nu = \hat{\Gamma}_{mn}(u). \tag{6·9}$$

We also introduce the frequency transform of L:

$$\int_0^\infty L(\boldsymbol{\xi}_m, \boldsymbol{\xi}_n;\ \mathbf{x}_1, \mathbf{x}_2;\ \nu)\, \mathrm{e}^{-2\pi i\nu u}\, \mathrm{d}\nu = M(\boldsymbol{\xi}_m, \boldsymbol{\xi}_n;\ \mathbf{x}_1, \mathbf{x}_2;\ u). \tag{6·10}$$

The relation (6·7) then becomes, on using the convolution theorem,

$$\hat{\Gamma}(\mathbf{x}_1, \mathbf{x}_2, \tau) = \sum_m \sum_n \int_{-\infty}^{+\infty} \hat{\Gamma}_{mn}(u)\, M(\boldsymbol{\xi}_m, \boldsymbol{\xi}_n;\ \mathbf{x}_1, \mathbf{x}_2;\ \tau - u)\, \mathrm{d}u. \tag{6·11}$$

(6·11) expresses $\hat{\Gamma}(\mathbf{x}_1, \mathbf{x}_2, \tau)$ in terms of $\hat{\Gamma}_{mn}(u)$ and M. The former specifies the source and may be determined from experiments described in § 5. The latter specifies the medium and may be obtained from calculations based on a ray trace.

For incoherent sources (i.e. sources for which (6·5) holds), the double summation in (6·11) reduces to a single summation:

$$\hat{\Gamma}(\mathbf{x}_1, \mathbf{x}_2, \tau) = \sum_m \int_0^\infty J_{mn}(\nu)\, L(\boldsymbol{\xi}_m, \boldsymbol{\xi}_n;\ \mathbf{x}_1, \mathbf{x}_2;\ \nu)\, \mathrm{e}^{-2\pi i\nu\tau}\, \mathrm{d}\nu \tag{6·12}$$

$$= \sum_m \int_{-\infty}^{+\infty} \hat{\Gamma}_{mn}(u)\, M(\boldsymbol{\xi}_m, \boldsymbol{\xi}_n;\ \mathbf{x}_1, \mathbf{x}_2;\ \tau - u)\, \mathrm{d}u. \tag{6·13}$$

If the elements $\delta\Sigma_1, \delta\Sigma_2, \ldots$ are taken small enough, one may replace the summations by integrations over the source, provided that obvious modifications are made: One introduces in place of $\hat{\Gamma}_{mn}$ and J_{mn} which are defined only for the discrete set of $\boldsymbol{\xi}$ values, the functions $\hat{\Omega}$ and j defined for a continuous range of $\boldsymbol{\xi}$ and $\boldsymbol{\xi}'$, such that

$$\hat{\Gamma}_{mn}(u) = \hat{\Omega}(\boldsymbol{\xi}, \boldsymbol{\xi}', u)\, \mathrm{d}\Sigma\, \mathrm{d}\Sigma', \tag{6·14}$$

$$J_{mn}(\nu) = j(\boldsymbol{\xi}, \boldsymbol{\xi}', \nu)\, \mathrm{d}\Sigma\, \mathrm{d}\Sigma', \tag{6·15}$$

† We neglect here the variation of the source strength with direction.

it being assumed that $m \neq n$. $\hat{\Omega}$ and j are, on account of (6·9), Fourier transforms of each other. In place of (6·7) and (6·11) one then obtains

$$\hat{\Gamma}(\mathbf{x}_1, \mathbf{x}_2, \tau) = \int_\Sigma \int_\Sigma \mathrm{d}\boldsymbol{\xi}\,\mathrm{d}\boldsymbol{\xi}' \int_0^\infty j(\boldsymbol{\xi}, \boldsymbol{\xi}', \nu)\, L(\boldsymbol{\xi}, \boldsymbol{\xi}'; \mathbf{x}_1, \mathbf{x}_2; \nu)\, \mathrm{e}^{-2\pi \mathrm{i}\nu\tau}\, \mathrm{d}\nu \tag{6·16}$$

$$= \int_\Sigma \int_\Sigma \mathrm{d}\boldsymbol{\xi}\,\mathrm{d}\boldsymbol{\xi}' \int_{-\infty}^{+\infty} \hat{\Omega}(\boldsymbol{\xi}, \boldsymbol{\xi}', u)\, M(\boldsymbol{\xi}, \boldsymbol{\xi}'; \mathbf{x}_1, \mathbf{x}_2; \tau - u)\, \mathrm{d}u, \tag{6·17}$$

where $\boldsymbol{\xi}$ and $\boldsymbol{\xi}'$ explore the surface Σ of the source independently.

In the case of an incoherent source, we set

$$\hat{\Gamma}_{mm}(u) = \hat{\Omega}(\boldsymbol{\xi}, u)\, \mathrm{d}\Sigma, \tag{6·18}$$

$$J_{mm}(u) = j(\boldsymbol{\xi}, \nu)\, \mathrm{d}\Sigma. \tag{6·19}$$

$j(\boldsymbol{\xi}, \nu)$ is nothing but the spectral intensity function of the source; it represents the intensity per unit area of the source per unit frequency range. In place of (6·12) and (6·13) one then obtains

$$\hat{\Gamma}(\mathbf{x}_1, \mathbf{x}_2, \tau) = \int_\Sigma \mathrm{d}\boldsymbol{\xi} \int_0^\infty j(\boldsymbol{\xi}, \nu)\, L(\boldsymbol{\xi}, \boldsymbol{\xi}; \mathbf{x}_1, \mathbf{x}_2; \nu) \mathrm{e}^{-2\pi \mathrm{i}\nu\tau}\, \mathrm{d}\nu \tag{6·20}$$

$$= \int_\Sigma \mathrm{d}\boldsymbol{\xi} \int_{-\infty}^{+\infty} \hat{\Omega}(\boldsymbol{\xi}, u)\, M(\boldsymbol{\xi}, \boldsymbol{\xi}; \mathbf{x}_1, \mathbf{x}_2; \tau - u)\, \mathrm{d}u. \tag{6·21}$$

As an example, consider the case of an incoherent source in a homogeneous medium. The transmission function of a homogeneous medium is

$$K(\boldsymbol{\xi}, \mathbf{x}, \nu) = \frac{\exp\left\{2\pi \mathrm{i} \frac{\nu}{c} |\mathbf{x} - \boldsymbol{\xi}|\right\}}{|\mathbf{x} - \boldsymbol{\xi}|}, \tag{6·22}$$

so that

$$L(\boldsymbol{\xi}, \boldsymbol{\xi}; \mathbf{x}_1, \mathbf{x}_2, \nu) = \frac{\exp\left\{2\pi \mathrm{i} \frac{\nu}{c} (R_1 - R_2)\right\}}{R_1 R_2}, \tag{6·23}$$

with

$$R_1 = |\mathbf{x}_1 - \boldsymbol{\xi}|, \quad R_2 = |\mathbf{x}_2 - \boldsymbol{\xi}|. \tag{6·24}$$

(6·20) then gives the following expression for $\hat{\Gamma}(\mathbf{x}_1, \mathbf{x}_2, \tau)$:

$$\hat{\Gamma}(\mathbf{x}_1, \mathbf{x}_2, \tau) = \int_\Sigma \mathrm{d}\boldsymbol{\xi} \int_0^\infty \frac{j(\boldsymbol{\xi}, \nu)}{R_1 R_2} \exp\left\{-2\pi \mathrm{i}\nu\left[\tau - \frac{R_1 - R_2}{c}\right]\right\} \mathrm{d}\nu \tag{6·25}$$

$$= \int_\Sigma \frac{\hat{\Omega}\left(\boldsymbol{\xi}, \tau - \frac{R_1 - R_2}{c}\right)}{R_1 R_2} \mathrm{d}\boldsymbol{\xi}. \tag{6·26}$$

7. Differential equations for the correlation function $\hat{\Gamma}$

In its usual form, Huygens's principle describes within a certain degree of accuracy the propagation of the light disturbance $\hat{V}(\mathbf{x}, t)$. As is well known, this principle may be regarded as an approximate formulation of a rigorous theorem due to Kirchhoff (see, for example, Baker & Copson 1950), this theorem being a consequence of the fact that $\hat{V}$ obeys rigorously the wave equation.

In the present investigation we have found a generalization of the Huygens principle which applies to the intensity rather than to the complex disturbance, and, more generally, we have found a kind of Huygens principle for the propagation of the correlation function $\hat{\Gamma}(\mathbf{x}_1, \mathbf{x}_2, \tau)$. These results suggest that $\hat{\Gamma}$ itself obeys certain differential equations and that our propagation laws are essentially some approximate formulations of the associated 'Kirchhoff's theorems'. We shall now show that this indeed is the case.

In vacuo, the complex disturbance $\hat{V}(\mathbf{x}, t)$ satisfies the wave equation

$$\nabla^2 \hat{V} - \frac{1}{c^2}\frac{\partial^2 \hat{V}}{\partial t^2} = 0. \tag{7·1}$$

Consequently each Fourier component $v(\mathbf{x}, \nu)$ obeys the equation

$$\nabla^2 v + \left(\frac{2\pi\nu}{c}\right)^2 v = 0. \tag{7·2}$$

Let

$$\nabla_1^2 \equiv \frac{\partial^2}{\partial x_1^2} + \frac{\partial^2}{\partial y_1^2} + \frac{\partial^2}{\partial z_1^2} \tag{7·3}$$

be the Laplacian operator with respect to the co-ordinates x_1, y_1, z_1 of the point $P_1(\mathbf{x}_1)$. It then follows from (3·4) and (7·2) that

$$\begin{aligned}\nabla_1^2 \hat{\Gamma}(\mathbf{x}_1, \mathbf{x}_2, \tau) &= \lim_{T\to\infty} \frac{1}{2T}\int_0^{+\infty} [\nabla_1^2 v(\mathbf{x}_1, \nu)]\, v^*(\mathbf{x}_2, \nu)\, \mathrm{e}^{-2\pi \mathrm{i}\nu\tau}\, \mathrm{d}\nu \\ &= -\frac{2\pi^2}{c^2}\lim_{T\to\infty}\frac{1}{T}\int_0^{\infty} \nu^2 v(\mathbf{x}_1, \nu)\, v^*(\mathbf{x}_2, \nu)\, \mathrm{e}^{-2\pi \mathrm{i}\nu\tau}\, \mathrm{d}\nu. \end{aligned} \tag{7·4}$$

Also from (3·4),

$$\frac{\partial^2}{\partial \tau^2}\hat{\Gamma}(\mathbf{x}_1, \mathbf{x}_2, \tau) = -2\pi^2 \lim_{T\to\infty}\frac{1}{T}\int_0^{\infty} \nu^2 v(\mathbf{x}_1, \nu)\, v^*(\mathbf{x}_2, \nu)\, \mathrm{e}^{-2\pi \mathrm{i}\nu\tau}\, \mathrm{d}\nu. \tag{7·5}$$

Comparison of (7·4) and (7·5) shows that

$$\nabla_1^2 \hat{\Gamma} - \frac{1}{c^2}\frac{\partial^2 \hat{\Gamma}}{\partial \tau^2} = 0. \tag{7·6}$$

Similarly, if ∇_2^2 denotes the Laplacian operator with respect to the co-ordinates x_2, y_2, z_2 of the point $P_2(\mathbf{x}_2)$, then

$$\nabla_2^2 \hat{\Gamma} - \frac{1}{c^2}\frac{\partial^2 \hat{\Gamma}}{\partial \tau^2} = 0. \tag{7·7}$$

Hence, in vacuo, *the correlation function* $\hat{\Gamma}(\mathbf{x}_1, \mathbf{x}_2, \tau)$ *obeys rigorously the two wave equations* (7·6) *and* (7·7).

Each of the two wave equations describes the variation of the correlation when one of the points (P_2 or P_1) is fixed whilst the other point as well as the parameter τ varies. It will be recalled that τ denotes a time difference; in all experiments it will play the part of the difference in the optical path (divided by c). The 'actual' time makes no appearance in our formulae. This is a most desirable aspect of the theory, since true time variations are not observed in optical fields.

8. A rigorous formulation of the propagation law for $\hat{\Gamma}$ and of the generalized Huygens principle

We are now in a position to derive a rigorous propagation law for $\hat{\Gamma}$ and also to formulate rigorously the generalized Huygens principle.

Let $P(\mathbf{x})$ and $A(\mathbf{a})$ be any two points in the field and let $\mathscr{A}$ be any closed surface surrounding P; the point A may be either inside or outside this surface.

If ∇^2 denotes the Laplacian operator with respect to the co-ordinates of $P(\mathbf{x})$, then, according to (7·6),

$$\nabla^2\hat{\Gamma}(\mathbf{x},\mathbf{a},\tau)-\frac{1}{c^2}\frac{\partial^2\hat{\Gamma}(\mathbf{x},\mathbf{a},\tau)}{\partial\tau^2}=0. \tag{8·1}$$

Hence, using Kirchhoff's integral formula (cf. Baker & Copson 1950, p. 37), we may express $\hat{\Gamma}(\mathbf{x},\mathbf{a},\tau)$ in the following form:

$$\hat{\Gamma}(\mathbf{x},\mathbf{a},\tau)=\frac{1}{4\pi}\int_{\mathscr{A}}\left\{f[\hat{\Gamma}]^-+g\left[\frac{\partial}{\partial\tau}\hat{\Gamma}\right]^-+h\left[\frac{\partial}{\partial n}\hat{\Gamma}\right]^-\right\}\mathrm{d}\bar{\mathbf{x}}, \tag{8·2}$$

where

$$f=\frac{\partial}{\partial n}\left(\frac{1}{r}\right),\quad g=-\frac{1}{cr}\frac{\partial r}{\partial n},\quad h=-\frac{1}{r}, \tag{8·3}$$

r being the distance from a typical point $\bar{P}(\bar{\mathbf{x}})$ on the surface to $P(\mathbf{x})$ (see figure 3a), $\partial/\partial n$ denoting differentiation along the inward normal to $\mathscr{A}$; and the brackets $[\ldots]^-$ denote retarded values, i.e. values obtained by replacing τ by $\tau-r/c$, e.g.

$$[\hat{\Gamma}]^-=\hat{\Gamma}(\bar{\mathbf{x}},\mathbf{a},\tau-r/c). \tag{8·4}$$

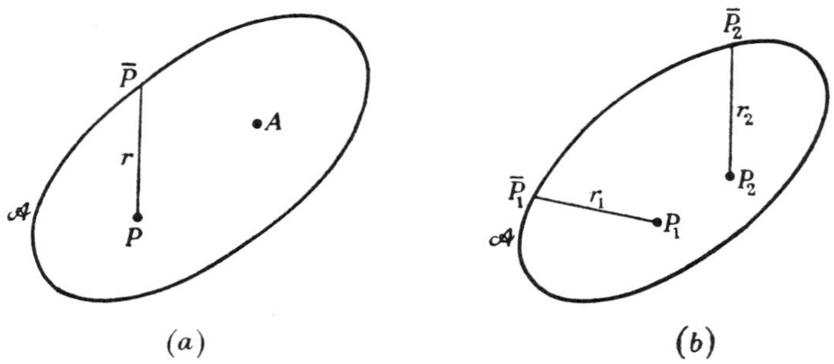

Figure 3. Illustrating Kirchhoff's integral theorem and the 'propagation law' for $\hat{\Gamma}$.

It may be shown by an argument similar to that used to derive Kirchhoff's formula that the integral (8·2) with advanced values ($\tau+r/c$ in place of $\tau-r/c$) also satisfies the wave equation, provided g is replaced by $-g$. If expressions for $\hat{V}$ in terms of retarded terms only are admitted, as it is reasonable to do on physical grounds, then Kirchhoff's integral for $\hat{\Gamma}(\mathbf{x},\mathbf{a},\tau)$ in terms of the retarded values of $\hat{\Gamma}$ clearly represents the physical solution. The situation is, however, different in the case of $\hat{\Gamma}(\mathbf{a},\mathbf{x},\tau)$. For if again only expressions in terms of retarded values are admitted for $\hat{V}$, Kirchhoff's integral for $\hat{\Gamma}(\mathbf{a},\mathbf{x},\tau)$ must involve terms like

$$\begin{aligned}f\langle\hat{V}(\mathbf{a},t+\tau)\,\hat{V}^*(\bar{\mathbf{x}},t-r/c)\rangle&=f\langle\hat{V}(\mathbf{a},t+\tau+r/c)\,\hat{V}^*(\bar{\mathbf{x}},t)\rangle\\&=f\hat{\Gamma}(\mathbf{a},\bar{\mathbf{x}},\tau+r/c),\end{aligned} \tag{8·5}$$

i.e. it must involve 'advanced' values of $\hat{\Gamma}$. The full formula for $\hat{\Gamma}(\mathbf{a},\mathbf{x},\tau)$ therefore is

$$\Gamma(\mathbf{a},\mathbf{x},\tau) = \frac{1}{4\pi}\int_{\mathscr{A}}\left\{f[\hat{\Gamma}]^{+} - g\left[\frac{\partial}{\partial\tau}\hat{\Gamma}\right]^{+} + h\left[\frac{\partial}{\partial n}\hat{\Gamma}\right]^{+}\right\}d\bar{\mathbf{x}}, \tag{8·6}$$

the brackets $[\ldots]^{+}$ denoting the advanced values, e.g.

$$[\hat{\Gamma}]^{+} = \hat{\Gamma}(\mathbf{a},\bar{\mathbf{x}},\tau + r/c). \tag{8·7}$$

We now apply (8·2) and (8·6) to derive an expression for $\hat{\Gamma}(\mathbf{x}_1,\mathbf{x}_2,\tau)$ in terms of an integral which involves the values of $\hat{\Gamma}$ and its derivatives at all pairs of points $\bar{P}_1$, $\bar{P}_2$ on an arbitrary closed surface $\mathscr{A}$ surrounding P_1 and P_2 (see figure 3*b*).

We set $\mathbf{x} = \mathbf{x}_1$, $\mathbf{a} = \mathbf{x}_2$, $\bar{\mathbf{x}} = \bar{\mathbf{x}}_1$ in (8·2). This gives

$$\Gamma(\mathbf{x}_1,\mathbf{x}_2,\tau) = \frac{1}{4\pi}\int_{\mathscr{A}}\left\{f_1[\hat{\Gamma}]_1^{-} + g_1\left[\frac{\partial}{\partial\tau}\hat{\Gamma}\right]_1^{-} + h_1\left[\frac{\partial}{\partial n_1}\hat{\Gamma}\right]_1^{-}\right\}d\bar{\mathbf{x}}_1, \tag{8·8}$$

where the arguments in the brackets $[\ldots]_1^{-}$ are $\bar{\mathbf{x}}_1$, $\mathbf{x}_2$, $\tau - r_1/c$, e.g.

$$[\hat{\Gamma}]_1^{-} = \hat{\Gamma}(\bar{\mathbf{x}}_1,\mathbf{x}_2,\tau - r_1/c); \tag{8·9}$$

r_1 being the distance from $\bar{P}_1$ to P_1, f_1, g_1 and h_1 the appropriate values of f, g and h $\partial/\partial n_1$ denoting differentiation at $\bar{P}_1$ along the inward normal to $\mathscr{A}$.

Next we express each of the retarded terms in (8·8) in terms of Kirchhoff's integral over the surface. Setting $\mathbf{a} = \bar{\mathbf{x}}_1$, $\mathbf{x} = \mathbf{x}_2$, $\bar{\mathbf{x}} = \bar{\mathbf{x}}_2$ in (8·6) and writing τ' in place of τ where τ' is arbitrary for the present, we obtain

$$\hat{\Gamma}(\bar{\mathbf{x}}_1,\mathbf{x}_2,\tau') = \frac{1}{4\pi}\int_{\mathscr{A}}\left\{f_2[\hat{\Gamma}]_2^{+} - g_2\left[\frac{\partial}{\partial\tau'}\hat{\Gamma}\right]_2^{+} + h_2\left[\frac{\partial}{\partial n_2}\hat{\Gamma}\right]_2^{+}\right\}d\bar{\mathbf{x}}_2, \tag{8·10}$$

where the arguments of the terms in the brackets $[\ldots]_2^{+}$ are $\bar{\mathbf{x}}_1$, $\bar{\mathbf{x}}_2$, $\tau' + r_2/c$, e.g.

$$[\hat{\Gamma}]_2^{+} = \hat{\Gamma}(\bar{\mathbf{x}}_1,\bar{\mathbf{x}}_2,\tau' + r_2/c), \tag{8·11}$$

the other symbols having a similar meaning as before. Differentiating (8·10) with respect to τ', we obtain

$$\frac{\partial}{\partial\tau'}\hat{\Gamma}(\bar{\mathbf{x}}_1,\mathbf{x}_2,\tau') = \frac{1}{4\pi}\int_{\mathscr{A}}\left\{f_2\left[\frac{\partial}{\partial\tau'}\hat{\Gamma}\right]_2^{+} - g_2\left[\frac{\partial^2}{\partial\tau'^2}\hat{\Gamma}\right]_2^{+} + h_2\left[\frac{\partial^2}{\partial\tau'\,\partial n_2}\hat{\Gamma}\right]_2^{+}\right\}d\bar{\mathbf{x}}_2. \tag{8·12}$$

Differentiation of (8·10) with respect to n_1 gives

$$\frac{\partial}{\partial n_1}\hat{\Gamma}(\bar{\mathbf{x}}_1,\mathbf{x}_2,\tau') = \frac{1}{4\pi}\int_{\mathscr{A}}\left\{f_2\left[\frac{\partial}{\partial n_1}\hat{\Gamma}\right]_2^{+} - g_2\left[\frac{\partial^2}{\partial n_1\,\partial\tau'}\hat{\Gamma}\right]_2^{+} + h_2\left[\frac{\partial^2}{\partial n_1\,\partial n_2}\hat{\Gamma}\right]_2^{+}\right\}d\bar{\mathbf{x}}_2. \tag{8·13}$$

Setting $\tau' = \tau - r_1/c$ in (8·10) to (8·13) and substituting into (8·8), we finally obtain

$$\begin{aligned}\hat{\Gamma}(\mathbf{x}_1,\mathbf{x}_2\tau) = \frac{1}{(4\pi)^2}\int_{\mathscr{A}}\int_{\mathscr{A}}\Bigg\{&f_1f_2[\hat{\Gamma}] - f_1g_2\left[\frac{\partial}{\partial\tau}\hat{\Gamma}\right] + f_1h_2\left[\frac{\partial}{\partial n_2}\hat{\Gamma}\right]\\ &+ g_1f_2\left[\frac{\partial}{\partial\tau}\hat{\Gamma}\right] - g_1g_2\left[\frac{\partial^2}{\partial\tau^2}\hat{\Gamma}\right] + g_1h_2\left[\frac{\partial^2}{\partial\tau\,\partial n_2}\hat{\Gamma}\right]\\ &+ h_1f_2\left[\frac{\partial}{\partial n_1}\hat{\Gamma}\right] - h_1g_2\left[\frac{\partial^2}{\partial n_1\,\partial\tau}\hat{\Gamma}\right] + h_1h_2\left[\frac{\partial^2}{\partial n_1\,\partial n_2}\hat{\Gamma}\right]\Bigg\}d\bar{\mathbf{x}}_1\,d\bar{\mathbf{x}}_2,\end{aligned} \tag{8·14}$$

where the arguments of the terms in the brackets are $\bar{\mathbf{x}}_1$, $\bar{\mathbf{x}}_2$, $\tau-\frac{r_1-r_2}{c}$, e.g.

$$[\hat{\Gamma}] = \hat{\Gamma}\left(\bar{\mathbf{x}}_1, \bar{\mathbf{x}}_2, \tau-\frac{r_1-r_2}{c}\right). \tag{8·15}$$

The formula (8·14) may be regarded as a rigorous formulation of the propagation law for $\hat{\Gamma}$. It expresses the correlation between disturbances at P_1 and P_2 in terms of the correlations and its derivatives at all pairs of points on an arbitrary closed surface surrounding P_1 and P_2.

In the special case when P_1 and P_2 coincide ($\mathbf{x}_1 = \mathbf{x}_2 = \mathbf{x}$ (say)) and when in addition $\tau = 0$, (8·14) reduces to, when we also substitute from (3·6),

$$\begin{aligned} I(\mathbf{x}) = \frac{1}{(4\pi)^2}\int_{\mathscr{A}}\int_{\mathscr{A}} & \sqrt{\bar{I}_1}\sqrt{\bar{I}_2}\left\{f_1f_2[\hat{\gamma}]+(f_2g_1-f_1g_2)\left[\frac{\partial}{\partial\tau}\hat{\gamma}\right]-g_1g_2\left[\frac{\partial^2}{\partial\tau^2}\hat{\gamma}\right]\right\} \\ & +\sqrt{\bar{I}_1}\left\{f_1h_2\frac{\partial}{\partial n_2}(\sqrt{\bar{I}_2}[\hat{\gamma}])+g_1h_2\frac{\partial}{\partial n_2}\left(\sqrt{\bar{I}_2}\left[\frac{\partial}{\partial\tau}\hat{\gamma}\right]\right)\right\} \\ & +\sqrt{\bar{I}_2}\left\{f_2h_1\frac{\partial}{\partial n_1}(\sqrt{\bar{I}_1}[\hat{\gamma}])-g_2h_1\frac{\partial}{\partial n_1}\left(\sqrt{\bar{I}_1}\left[\frac{\partial}{\partial\tau}\hat{\gamma}\right]\right)\right\} \\ & +h_1h_2\frac{\partial^2}{\partial n_1\partial n_2}(\sqrt{\bar{I}_1}\sqrt{\bar{I}_2}[\hat{\gamma}])\,\mathrm{d}\bar{\mathbf{x}}_1\,\mathrm{d}\bar{\mathbf{x}}_2, \end{aligned} \tag{8·16}$$

where $\bar{I}_1 = I(\bar{\mathbf{x}}_1)$, $\bar{I}_2 = I(\bar{\mathbf{x}}_2)$ and $[\hat{\gamma}] = \hat{\gamma}\,(\bar{\mathbf{x}}_1, \bar{\mathbf{x}}_2, (r_2-r_1)/c)$. By a similar argument to that used in connexion with (4·6), $\hat{\gamma}$ may be replaced in (8·16) by γ. (8·16) may be regarded as a rigorous formulation of the generalized Huygens principle.

Appendix 1. Cross-correlation between real functions in terms of the associated half-range complex functions

We shall establish the following theorem:

Let $F(t)$ and $G(t)$ be any two real functions, which possess Fourier integral representations and let $P(\tau)$ be the cross correlation function

$$P(\tau) = 2\int_{-\infty}^{+\infty} F(t+\tau)\,G(t)\,\mathrm{d}t. \tag{A 1·1}$$

Then the half-range complex function $\hat{P}(\tau)$ associated with $P(\tau)$ is given by

$$\hat{P}(\tau) = \int_{-\infty}^{+\infty} \hat{F}(t+\tau)\,\hat{G}^*(t)\,\mathrm{d}t, \tag{A 1·2}$$

where $\hat{F}$ and $\hat{G}$ are the half-range complex functions associated with F and G respectively.

Proof. Let a_ν and b_ν be the Fourier coefficients of F and c_ν and d_ν the Fourier coefficients of G:

$$F(t) = \int_0^\infty (a_\nu \cos 2\pi\nu t + b_\nu \sin 2\pi\nu t)\,\mathrm{d}\nu, \tag{A 1·3}$$

$$G(t) = \int_0^\infty (c_\nu \cos 2\pi\nu t + d_\nu \sin 2\pi\nu t)\,\mathrm{d}\nu. \tag{A 1·4}$$

In accordance with the definition of § 2, the associated half-range complex functions are then given by

$$\hat{F}(t) = \int_0^\infty (a_\nu + \mathrm{i}b_\nu)\,\mathrm{e}^{-2\pi\mathrm{i}\nu t}\,\mathrm{d}\nu, \tag{A 1·5}$$

$$\hat{G}(t) = \int_0^\infty (c_\nu + \mathrm{i}d_\nu)\,\mathrm{e}^{-2\pi\mathrm{i}\nu t}\,\mathrm{d}\nu. \tag{A 1·6}$$

Consider now the Fourier representations of $P(\tau)$ and $\hat{P}(\tau)$. Substitution from (A 1·4) into (A 1·1) gives

$$P(\tau) = 2\int_{-\infty}^{+\infty} \mathrm{d}t\,F(t+\tau)\int_0^\infty (c_\nu \cos 2\pi\nu t + d_\nu \sin 2\pi\nu t)\,\mathrm{d}\nu. \tag{A 1·7}$$

Now

$$\begin{aligned}\int_{-\infty}^{\infty} F(t+\tau)\cos 2\pi\nu t\,\mathrm{d}t &= \int_{-\infty}^{+\infty} F(u)\cos 2\pi\nu(u-\tau)\,\mathrm{d}u \\ &= \cos 2\pi\nu\tau\int_{-\infty}^{+\infty} F(u)\cos 2\pi\nu u\,\mathrm{d}u + \sin 2\pi\nu\tau\int_{-\infty}^{+\infty} F(u)\sin 2\pi\nu u\,\mathrm{d}u \\ &= \tfrac{1}{2}(a_\nu \cos 2\pi\nu\tau + b_\nu \sin 2\pi\nu\tau),\end{aligned} \tag{A 1·8}$$

where the Fourier inversion formula was used. Similarly

$$\int_{-\infty}^{\infty} F(t+\tau)\sin 2\pi\nu t\,\mathrm{d}t = \tfrac{1}{2}(b_\nu \cos 2\pi\nu\tau - a_\nu \sin 2\pi\nu\tau). \tag{A 1·9}$$

Hence, interchanging the order of integration in (A 1·7) and using the last two relations, one obtains

$$P(\tau) = \int_0^\infty \{[a_\nu c_\nu + b_\nu d_\nu]\cos 2\pi\nu\tau + [b_\nu c_\nu - a_\nu d_\nu]\sin 2\pi\nu\tau\}\,\mathrm{d}\nu. \tag{A 1·10}$$

The associated half-range complex function $\hat{P}(\tau)$ is therefore given by

$$\begin{aligned}\hat{P}(\tau) &= \int_0^\infty \{[a_\nu c_\nu + b_\nu d_\nu] + \mathrm{i}[b_\nu c_\nu - a_\nu d_\nu]\}\,\mathrm{e}^{-2\pi\mathrm{i}\nu\tau}\,\mathrm{d}\nu \\ &= \int_0^\infty (a_\nu + \mathrm{i}b_\nu)(c_\nu - \mathrm{i}d_\nu)\,\mathrm{e}^{-2\pi\mathrm{i}\nu\tau}\,\mathrm{d}\nu.\end{aligned} \tag{A 1·11}$$

Next consider the integral on the right of (A 1·2). Substitution from (A 1·6) gives

$$\int_{-\infty}^{\infty} \hat{F}(t+\tau)\,\hat{G}^*(t)\,\mathrm{d}t = \int_{-\infty}^{+\infty} \mathrm{d}t\,\hat{F}(t+\tau)\int_0^\infty (c_\nu - \mathrm{i}d_\nu)\,\mathrm{e}^{2\pi\mathrm{i}\nu\tau t}\,\mathrm{d}\nu. \tag{A 1·12}$$

But

$$\begin{aligned}\int_{-\infty}^{\infty} \hat{F}(t+\tau)\,\mathrm{e}^{2\pi\mathrm{i}\nu t}\,\mathrm{d}t &= \int_{-\infty}^{+\infty} \hat{F}(u)\,\mathrm{e}^{2\pi\mathrm{i}\nu(u-\tau)}\,\mathrm{d}u \\ &= (a_\nu + \mathrm{i}b_\nu)\,\mathrm{e}^{-2\pi\mathrm{i}\nu\tau},\end{aligned} \tag{A 1·13}$$

where the Fourier inversion formula was used. Interchanging the order of integration in (A 1·12) and using (A 1·13), one obtains

$$\int_{-\infty}^{+\infty} \hat{F}(t+\tau)\,\hat{G}^*(t)\,\mathrm{d}t = \int_0^{\infty} (a_\nu + ib_\nu)(c_\nu - id_\nu)\,\mathrm{e}^{-2\pi i\nu\tau}\,\mathrm{d}\nu. \qquad \text{(A 1·14)}$$

On comparing (A 1·14) with (A 1·11), the theorem follows.

Appendix 2. The range of validity of the restricted theory of paper I

We shall now investigate the range of validity of the restricted formulation of the generalized Huygens principle and of the generalized interference law of Zernike and Hopkins, as given in the preceding paper of this series.

The essential approximation which was made in the derivation of these formulae in I, was the replacement of the exponential term

$$\exp\{\mathrm{i}(kr_1 - k'r_2)\} \quad \text{by} \quad \exp\{\mathrm{i}k^{(0)}(r_1 - r_2)\}$$

in the expression† I, (2·7):

$$I(\mathbf{x}) = \frac{1}{2T}\int_{-T}^{T}\mathrm{d}t\int_0^{\infty}\int_0^{\infty}\mathrm{d}\nu\,\mathrm{d}\nu'\exp\{-2\pi\mathrm{i}(\nu-\nu')\,t\}$$

$$\times\int_{\mathscr{A}}\int_{\mathscr{A}} v(\mathbf{x}_1,\nu)\,v^*(\mathbf{x}_2,\nu')\,\frac{\exp\{\mathrm{i}(kr_1 - k'r_2)\}}{r_1 r_2}\,\Lambda_1\Lambda_2^*\,\mathrm{d}\mathbf{x}_1\,\mathrm{d}\mathbf{x}_2. \qquad \text{(A 2·1)}$$

To see what restriction this implies, we note that

$$kr_1 - k'r_2 = k^{(0)}(r_1 - r_2) + (k - k^{(0)})(r_1 - r_2) + (k - k')\,r_2. \qquad \text{(A 2·2)}$$

(A 2·1) may therefore be written as

$$I(\mathbf{x}) = \int_{\mathscr{A}}\int_{\mathscr{A}} G(\mathbf{x}_1,\mathbf{x}_2,r_2)\,\frac{\exp\{\mathrm{i}[k^{(0)}(r_1 - r_2) + (k - k^{(0)})(r_1 - r_2)]\}}{r_1 r_2}\,\Lambda_1\Lambda_2^*\,\mathrm{d}\mathbf{x}_1\,\mathrm{d}\mathbf{x}_2, \qquad \text{(A 2·3)}$$

where

$$G(\mathbf{x}_1,\mathbf{x}_2,r_2) = \frac{1}{2T}\int_{-T}^{T}\mathrm{d}t\int_0^{\infty}\int_0^{\infty} v(\mathbf{x}_1,\nu)\,v^*(\mathbf{x}_2,\nu')\exp\{-2\pi\mathrm{i}(\nu-\nu')(t - r_2/c)\}\,\mathrm{d}\nu\,\mathrm{d}\nu'. \qquad \text{(A 2·4)}$$

We now change the variable of integration in (A 2·4) from t to $t' = t - r_2/c$. Then, since

$$\lim_{T'\to\infty}\int_{-T'}^{T'}\exp\{-2\pi\mathrm{i}(\nu-\nu')\,t'\}\,\mathrm{d}t' = \delta(\nu-\nu'), \qquad \text{(A 2·5)}$$

where δ is the Dirac delta function, it follows, that in the limit $T\to\infty$, G becomes independent of r_2 and is equal to

$$\Gamma(\mathbf{x}_1,\mathbf{x}_2) = \lim_{T\to\infty} G(\mathbf{x}_1,\mathbf{x}_2,r_2) = \lim_{T\to\infty}\frac{1}{2T}\int_0^{\infty} v(\mathbf{x}_1,\nu)\,v^*(\mathbf{x}_2,\nu)\,\mathrm{d}\nu \qquad \text{(A 2·6)}$$

$$= \lim_{T\to\infty}\frac{1}{2T}\int_{-T}^{T} V(\mathbf{x}_1,t)\,V^*(\mathbf{x}_2,t)\,\mathrm{d}t. \qquad \text{(A 2·7)}$$

† In I, the lower limits of the integrals over the frequency range were actually taken as $-\infty$. We replace them here by 0, since as explained in § 2 above, $v(\mathbf{x},\nu) = 0$ for $\nu < 0$.

Hence in the limit $T \to \infty$, (A 2·3) differs from the approximate expression I, (2·8) by the presence of the term $(k - k^{(0)})(r_1 - r_2)$ in the exponential term and by the factor $\Lambda_1^{(0)}, \Lambda_2^{(0)*}$ in place of $\Lambda_1 \Lambda_2^*$. Setting $k - k^{(0)} = \Delta k$, $r_1 - r_2 = \Delta r$, and replacing the inclination factors by their mean values, it follows that I, (2·8) will be a valid approximation if

$$|\Delta k| \, |\Delta r| \ll 2\pi, \tag{A 2·8}$$

or, since $k = 2\pi/\lambda$, $\Delta k = -2\pi \Delta\lambda/\lambda^2$, and the condition becomes†

$$|\Delta r| \ll \frac{\lambda^2}{|\Delta\lambda|}. \tag{A 2·9}$$

Hence *the more restricted formulation of the generalized Huygens principle and of the generalized interference law given in paper I is applicable if the path differences between the interfering beams are small compared to* $\lambda^2/|\Delta\lambda|$.

The quantity $\lambda^2/|\Delta\lambda|$ has a simple physical interpretation: it represents (apart from a multiplicative factor of the order of unity), the coherence length of the radiation. A proof of this result will be found in Born (1933, p. 137) and Kahan (1952, p. 310).

This work was carried out during the tenure of an Imperial Chemical Industries Research Fellowship, the award of which, by the University of Manchester is gratefully acknowledged. It is also a pleasure to thank Dr F. D. Kahn for some helpful discussions.

References

Baker, B. B. & Copson, E. T. 1950 *The mathematical theory of Huygens's principle*, 2nd ed. Oxford: Clarendon Press.
Blanc-Lapierre, A. & Dumontet, P. 1955 *Rev. Opt.* (*théor. instrum.*), **34**, 1.
Born, M. 1933 *Optik*. Berlin: Springer.
Gabor, D. 1946 *J. Inst. Elect. Engrs.* **93**, Part III, 429.
Kahan, T. 1952 *Suppl. Nuovo Cim.* **9**, 310.
Michelson, A. A. 1891 *Phil. Mag.* **31**, 338.
Michelson, A. A. 1892 *Phil. Mag.* **34**, 280.
Parke III, N. G. 1948 Ph.D. thesis, Massachusetts Institute of Technology. (See also *Tech. Rep. Electron., Mass. Inst. Tech.* no. 95 (1949).)
Rayleigh, Lord 1892 *Phil. Mag.* **34**, 407.
Ryle, M. 1950 *Rep. Progr. Phys.* **13**, 184.
Smith, F. G. 1952 *Mon. Not. R. Astr. Soc.* **112**, 497.
Van Cittert, P. H. 1939 *Physica*, **6**, 1129.
Williams, W. E. 1941 *Applications of interferometry*, 2nd ed. London: Methuen and Co.
Wiener, N. 1930 *Acta Math. Uppsala*, **55**, 117.
Wolf, E. 1954*a* *Proc. Roy. Soc.* A, **225**, 96.
Wolf, E. 1954*b* *Nuovo Cim.* **12**, 884.
Zernike, F. 1938 *Physica*, **5**, 785.
Zernike, F. 1948 *Proc. Phys. Soc.* **61**, 158.

† It is to be noted that this is also essentially the condition I, (5·1) under which the approximate expressions of Van Cittert, Zernike and Hopkins for the space-correlation factor were derived in the previous paper.

PRINTED IN GREAT BRITAIN AT THE UNIVERSITY PRESS, CAMBRIDGE
(BROOKE CRUTCHLEY, UNIVERSITY PRINTER)

IL NUOVO CIMENTO VOL. XII, N. 6 1° Dicembre 1954

Optics in Terms of Observable Quantities.

E. WOLF
Department of Astronomy, The University, Manchester, England

(ricevuto il 26 Settembre 1954)

Summary — Space-time correlation functions are defined which express the correlation between components of the electromagnetic field vectors in stationary fields. These functions form sets of 3×3 matrices, the individual elements of which obey the wave equation. Unlike the field vectors which are not measurable at the high frequencies encountered in Optics our correlation functions may be determined with the help of standard optical instruments. The results enable a unified treatment of theories of partial coherence and partial polarization to be obtained, and suggest a formulation of a wide branch of Optics in terms of observable quantities only.

In all Optical experiments the only quantities which are observable are the averages of certain quadratic functions of the field components. It is therefore tempting to try to formulate the laws of Optical fields directly in terms of such quantities rather than in terms of the unmeasurable field vectors as has been customary in the past.

It was as early as 1852 that STOKES (1) showed that a nearly monochromatic (plane) light wave may be characterized at each point by four parameters which now bear his name (*). If

$$(1) \qquad E_x = a_1(\boldsymbol{x}, t) \cos\{2\pi\nu_0 t - \alpha_1(\boldsymbol{x}, t)\}\,, \qquad E_y = a_2(\boldsymbol{x}, t) \cos\{2\pi\nu_0 t - \alpha_2(\boldsymbol{x}, t)\},$$

(1) G. G. STOKES: *Trans. Camb. Phil. Soc.*, **9**, 399 (1852). Also his *Mathematical and Physical Papers* (Cambridge, 1901), vol. III, p. 233.

(*) Very good accounts of the Stokes parameters may be found in CHANDRASEKHAR (2) and WALKER (3).

are the components of the electric vector of such a wave in two mutually orthogonal directions at right angles to the direction of propagation, the Stokes parameters are defined by

$$\text{(2)} \qquad \begin{cases} P = \langle a_1^2 + a_2^2 \rangle\,, & Q = \langle a_1^2 - a_2^2 \rangle\,, \\ U = \langle 2a_1 a_2 \cos(\alpha_1 - \alpha_2) \rangle\,, & V = \langle 2a_1 a_2 \sin(\alpha_1 - \alpha_2) \rangle\,, \end{cases}$$

the brackets $\langle .... \rangle$ denoting time average. Then the intensity $I(\psi, \varepsilon)$ associated with vibrations in the direction which makes an angle ψ with the x-direction, when retardation ε is introduced between the two components, is given by (see CHANDRASEKHAR ([2]), p. 29)

$$\text{(3)} \qquad I(\psi, \varepsilon) = \tfrac{1}{2}[P + Q \cos 2\psi + (U \cos \varepsilon - V \sin \varepsilon) \sin 2\psi]\,.$$

By measuring I for different values of ψ and ε, the four parameters may be determined.

Let us now introduce in place of E_x and E_y, the associated complex vectors

$$\text{(4)} \qquad \hat{E}_x = a_1(\boldsymbol{x}, t) \exp\{i[2\pi\nu_0 t - \alpha_1(\boldsymbol{x}, t)]\}, \quad \hat{E}_y = a_2(\boldsymbol{x}, t) \exp\{i[2\pi\nu_0 t - \alpha_2(\boldsymbol{x}, t)]\}.$$

Next we construct the four functions

$$\text{(5)} \qquad \mathcal{E}_{ij} = \langle \hat{E}_i(\boldsymbol{x}, t) \hat{E}_j^*(\boldsymbol{x}, t) \rangle,$$

where i and j can each take on the value x or y and asterisk denotes the complex conjugate. The knowledge of these four quantities is equivalent to the knowledge of the four Stokes parameters; in fact the Stokes parameters are simple linear combinations of the $\mathcal{E}_{ij}$'s. If (3) is expressed in terms of these quantities, one finds after a simple calculation that

$$\text{(6)} \qquad I(\psi, \varepsilon) = I_x(\psi) + I_y(\psi) + 2\sqrt{I_x(\psi)}\sqrt{I_y(\psi)}\,|\gamma_{xy}| \cos[\arg \gamma_{xy} + \varepsilon]\,,$$

where

$$\text{(7)} \qquad I_x(\psi) = \mathcal{E}_{xx} \cos^2 \psi\,, \qquad I_y(\psi) = \mathcal{E}_{yy} \sin^2 \psi\,,$$

and

$$\text{(8)} \qquad \gamma_{xy} = \frac{\mathcal{E}_{xy}}{\sqrt{\mathcal{E}_{xx}}\sqrt{\mathcal{E}_{yy}}}\,;$$

([2]) S. CHANDRASEKHAR: *Radiative Transfer* (Oxford, 1950).
([3]) M. J. WALKER: *Amer. Journ. Phys.*, **22**, 170 (1954).

and the well-known inequality $P^2 \geqslant Q^2 + U^2 + V^2$ becomes simply $|\gamma_{xy}| \leqslant 1$. Equation (6) is formally identical with the *generalized interference law* derived in recent years in the theory of partially coherent scalar fields (ZERNIKE (4), HOPKINS (5), WOLF (6)). In the special case when $\gamma = 0$, (6) reduces to the usual law for the combination of completely incoherent fields; when $|\gamma| = 1$ it reduces to the ordinary interference law for fields which are perfectly coherent. Partially coherent and partially polarized fields are characterized by the intermediate values of $|\gamma|$.

Now the Stokes parameters (or the 2 by 2 matrix whose elements are defined by (5)) give the intensity and express the correlation between the components of $\boldsymbol{E}$ at the *same* point in space and at the *same* instant of time. Moreover they are defined only for a plane wave whose effective frequency range is sufficiently narrow. In order to characterize a general stationary field (*), we introduce, by analogy with the scalar case (see WOLF (7)), correlation functions between field components at different points in space and at different instants of time.

Let

$$E_i(\boldsymbol{x}, t) = \int_0^\infty a_{\nu i}(\boldsymbol{x}) \cos \{2\pi\nu t - \alpha_{\nu i}(\boldsymbol{x})\} \, d\nu \,, \qquad (i = x, \, y \text{ or } z) \tag{9}$$

be the Fourier representation of a typical field component over the time interval $-T \leqslant t \leqslant T$, $\boldsymbol{E}$ being formally assumed to be zero outside this range, and define the functions

$$\hat{E}_i(\boldsymbol{x}, t) = \int_0^\infty a_{\nu i}(\boldsymbol{x}) \exp \{i[2\pi\nu t - \alpha_{\nu i}(\boldsymbol{x})\} \, d\nu \,. \tag{10}$$

We now introduce a 3×3 correlation matrix $\mathcal{E}$ whose elements are

$$\mathcal{E}_{ij}(\boldsymbol{x}_1, \boldsymbol{x}_2, \tau) = \langle \hat{E}_i(\boldsymbol{x}_1, t+\tau) \hat{E}_j^*(\boldsymbol{x}_2, t) \rangle \,. \tag{11}$$

Similar considerations to those employed in connection with scalar fields of

(4) F. ZERNIKE: *Physica*, **5**, 785 (1938).

(5) H. H. HOPKINS: *Proc. Roy. Soc.*, A **208**, 263 (1951).

(6) E. WOLF: *Proc. Roy. Soc.*, A **225**, 96 (1954).

(*) By a stationary field we mean here a field of which all *observable* properties are constant in time. This includes as a special case the usual case of high frequency periodic time-dependence; or the field constituted by the steady flux of polychromatic radiation through an optical system. But it excludes fields for which the time average over a macroscopic time interval of the flux of radiation depends on time.

(7) E. WOLF: *Proc. Roy. Soc.*, in press.

arbitrary frequency range and with vector fields characterized by the Stokes parameters indicate, that the expressions for the electric energy density appropriate to various experimental conditions are simple functions of the elements of the $\mathcal{E}$-matrix; and moreover that all these elements may be determined from experiments by means of standard optical instruments.

Since the electric vector satisfies the wave equation, it can readily be shown that each element of the $\mathcal{E}$-matrix satisfies the two equations:

$$(12)\qquad \begin{cases} \nabla_1^2 \mathcal{E}_{ij} = \dfrac{1}{c^2}\dfrac{\partial^2 \mathcal{E}_{ij}}{\partial \tau^2}, \\[2ex] \nabla_2^2 \mathcal{E}_{ij} = \dfrac{1}{c^2}\dfrac{\partial^2 \mathcal{E}_{ij}}{\partial \tau^2}, \end{cases}$$

where ∇_1^2 and ∇_2^2 are the Laplacian operators with respect to the coordinates of $\boldsymbol{x}_1$ and $\boldsymbol{x}_2$, and c is the velocity of light in the vacuum. Thus not only the unmesurable field vectors, but also *the observable correlation functions* here introduced *obey rigorous propagation laws*. This result should prove particularly useful in connection with scattering problems.

In addition to the $\mathcal{E}$-matrix, one can introduce similar matrices involving components of the other field vectors ($\boldsymbol{H}$, $\boldsymbol{D}$ and $\boldsymbol{B}$) and also matrices involving mixed pairs like E_i and H_j. On account of Maxwell's equations, these matrices are related by a set of first order partial differential equations with respect to the variables $\boldsymbol{x}_1$, $\boldsymbol{x}_2$, and τ. In the analysis of all optical experiments $c\tau$ will play the part of an optical path difference. The actual time, like the frequency has been eliminated.

Unlike the $\mathcal{E}$-matrix, the $\mathcal{H}$ matrix is not likely to be of any interest in Optics, since no radiation detectors appear to be available at Optical wavelengths which would respond to the magnetic rather than the electric field. It may, however, prove useful in connection with applications to other types of partially coherent radiation, e.g. in Radio Astronomy. The matrix containing the mixed pairs should prove useful in experiments where the (averaged) flux of energy rather than the energy density is measured.

The matrices here introduced may be expected to play a role in Electromagnetic field theory which is somewhat analogous to that which the Density matrix of von NEUMAN [8] plays in Quantum Mechanics. An analogy between the Stokes parameters and the Density Matrix has been noted previously (PERRIN [9], FALKOFF and MACDONALD [10]; see also FANO [11]). It is, how-

[8] J. v. NEUMANN: *Gött. Nachr.*, 245 (1927).
[9] F. PERRIN, *Journ. Chem. Phys.*, **10**, 415 (1942).
[10] D. L. FALKOFF and J. E. MACDONALD: *Journ. Opt. Soc. Amer.*, **41**, 861 (1951).
[11] U. FANO: *Phys. Rev.*, **93**, 121 (1954).

ever, evident that only by considering more general correlation functions, such as those here introduced, does one obtain an adequate tool for the study of propagation problems in a general stationary electromagnetic field.

A fuller discussion of the subject matter of this note will be published at a later date.

This work was carried out during the tenure of an Imperial Chemical Industries Research Fellowship and was also supported by a grant from the Carnegie Trust for the Universities of Scotland, both of which are gratefully acknowledged.

RIASSUNTO (*)

Si definiscono funzioni di correlazione spazio-tempo che esprimono la correlazione fra componenti dei vettori del campo elettromagnetico in campi stazionari. Queste funzioni formano gruppi di 3×3 matrici i cui elementi individuali soddisfano all'equazione d'onda. A differenza dei vettori di campo che, alle alte frequenze che intervengono in Ottica, non sono misurabili, le nostre funzioni di correlazione possono essere determinate con l'ausilio degli ordinari apparecchi ottici. I risultati consentono un trattamento unificato delle teorie della coerenza parziale e della polarizzazione parziale e suggeriscono una formulazione di un ampio settore dell'Ottica in termini di sole grandezze osservabili.

(*) *Traduzione a cura della Redazione.*

17

THE COHERENCE PROPERTIES OF OPTICAL FIELDS

E. WOLF

DEPARTMENT OF ASTRONOMY, UNIVERSITY OF MANCHESTER

The aim of this introductory lecture is two-fold: first, to summarize the main results obtained in recent years in the theory of partially coherent fields; and secondly to show that an appropriate generalization of these results leads to a formulation of the theories of Interference, Diffraction and Polarization of Light on a much broader basis than has been customary in the past.

* * *

(I) Let us begin by considering the superposition of two beams of light. In the usual treatments one assumes that such superposition is either coherent or incoherent. The former is assumed to arise when both beams originate in the same "point" source, whilst the latter is considered to take place when the beams originate in different sources. This sharp distinction is adequate in connection with many optical problems; but there are other problems, such as those connected with the theory of illumination in the microscope, or with interferometric measurements of stellar diameters, where a more precise distinction must be made. Roughly speaking, it is then necessary to take into account the fact, that, because of the finite extension of any physical source, and because of the finite spectral range of any radiation, coherent as well as incoherent superposition takes place side by side; the vibrations in the two beams will then be partially correlated, one says that the field is *partially coherent.*

The first experimental investigations concerned with partially coherent fields were carried out by Verdet [1], as early as 1860, and later much more extensively by Michelson [2], and Lakeman and Groosmuller [3]. Early theoretical researches are associated with the names of Von Laue [4], Berek [5] and Van Cittert [6, 7].

An important turning point in the development of the subject was the publication of a paper by Zernike [8] in 1938, when he introduced the concepts of *mutual intensity* and of *degree of coherence* of light vibrations at two arbitrary points in a nearly monochromatic, stationary field. Let

$$V(t) = a(t)\, e^{-2\pi i \nu_0 t} \tag{1}$$

be the disturbance at typical point of the field. Because the field is not strictly monochromatic, the amplitude is not constant, but varies irregularly with time. Only certain averages involving $a(t)$ can be measured. If suffices 1 and 2 refer to vibrations at points P_1 and P_2, the *mutual intensity* Γ_{12} may be defined as

$$\Gamma_{12} = \langle V_1(t)\, V_2^*(t) \rangle, \tag{2}$$

the sharp brackets denoting time average; and the *complex degree of coherence* may be defined by

$$\gamma_{12} = \frac{\Gamma_{12}}{\sqrt{\Gamma_{11}}\,\sqrt{\Gamma_{22}}} = \frac{\Gamma_{12}}{\sqrt{I_1}\,\sqrt{I_2}}, \tag{3}$$

I_1 and I_2 being the intensities at the two points. It may be shown that

$$|\gamma_{12}| \leqslant 1. \tag{4}$$

Eq. (2) and (3) define the two quantities in mathematical terms, but both may be determined from simple interference experiments. If the disturbances at P_1 and P_2 are isolated by placing a screen across the field, with small openings at these points, and if the resulting beams are superposed with a path difference δ, then, as has been shown by Hopkins [9] (see also reference [10]), the resulting intensity is given by

$$I = I^{(1)} + I^{(2)} + 2\sqrt{I^{(1)}}\,\sqrt{I^{(2)}}\,|\gamma_{12}| \cos\,[arg\,\gamma_{12} + k_0\delta], \tag{5}$$

$I^{(1)}$ and $I^{(2)}$ being the intensities due to each of the beams and $k_0 = 2\pi\nu_0/c = 2\pi/\lambda_0$ is the wave number. Thus γ may be determined from measurements of intensities. From (5) it readily follows that if $I^{(1)} = I^{(2)}$, the absolute value of the degree of coherence is simply the Michelson visibility of the interference fringes obtained in the region where the beams are superposed. We note that the two extreme cases, namely $\gamma_{12} \doteq 0$ and $|\gamma_{12}| = 1$, characterize complete incoherence and perfect coherence respectively.

Zernike also derived an expression for the degree of coherence between disturbances at any two points on a screen, illuminated by an extended source. It turned out that the variation of the degree of coherence with the distance of separation of the two points is identical with the variation of the complex amplitude in a certain diffraction pattern.

The diffraction pattern being that which would be obtained on replacing the source by a diffracting aperture of the same size and shape as the source, and on filling the aperture by a wave whose complex amplitude is proportional to the *intensity* at the corresponding point of the source. This theorem, which was actually derived earlier by Van Cittert under a different definition of the coherence factor, is of importance to understanding of the role played by a microscope condensor or a condensor in front of a slit of a spectroscope.

Another important theorem (due to Zernike) must be mentioned. It is concerned with the propagation of the mutual intensity and enables the determination of the mutual intensity Γ_{12} for any points P_1 and P_2 in the field, from the knowledge of this function for all pairs of points $\bar{P}_1$ and $\bar{P}_2$ on an arbitrary surface A cutting across the beam (Fig. 1), by means of the formula

$$\Gamma_{12} = \frac{1}{\lambda_0^2} \int_A \int_A \bar{\Gamma}_{12} \frac{e^{-ik_0(r_1 - r_2)}}{r_1 r_2} \, d\bar{\mathbf{x}}_1 \, d\bar{\mathbf{x}}_2. \tag{6}$$

In (6), the points $\bar{P}_1$ and $\bar{P}_2$, specified by the position vectors $\bar{\mathbf{x}}_1$ and $\bar{\mathbf{x}}_2$ explore the surface A independently and $\bar{\Gamma}_{12} = \Gamma(\bar{\mathbf{x}}_1, \bar{\mathbf{x}}_2)$.

The basic conclusions of these theories were verified experimentally by Baker [11] and Arnulf, Dupuy and Flamant [12].

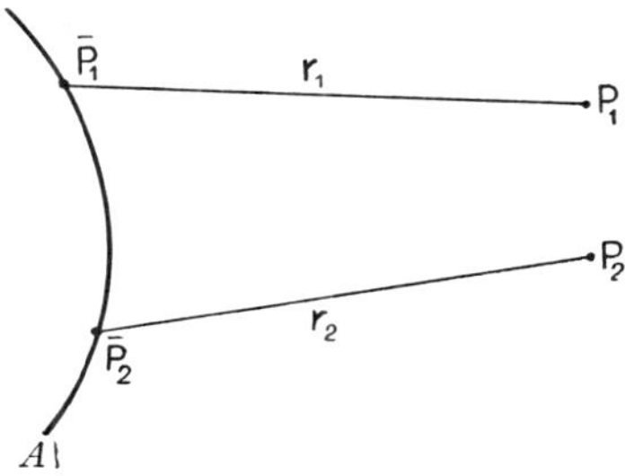

Fig. 1. Illustrating the propagation of the mutual intensity Γ_{12}

* * *

(II). The results which we mentioned so far, were derived on the assumption that the radiation is nearly monochromatic. Now it is well known that with radiation of effective spectral range $\nu_0 - \Delta\nu \leqslant \nu \leqslant \nu_0 + \Delta\nu$, no appreciable interference effects can be obtained if the path difference δ between the beams exceeds the so-called *coherence length* $\bar{\delta} = c\Delta t$, where

$$\Delta\nu \, \Delta t \geqslant 1, \tag{7}$$

c being the velocity of light. The exact value of Δt depends on the energy distribution over the spectral range, but in most cases of practical interest the product on the left of this inequality is of the order of unity. Now the formulae of Van Cittert, Zernike and Hopkins do not contain a parameter which specifies the effective frequency range, and, in consequence, their theories are inadequate when applied to situations which involve path differences that are sufficiently large.

It has been shown recently by Blanc-Lapierre and Dumontet [13], and independently by Wolf [14, 15], that in order to extend Zernike's theory to fields with spectral range of arbitrary width (and therefore to problems involving arbitrarily large path differences), it is necessary to introduce in place of the mutual intensity function of Zernike, the more general correlation function

$$\Gamma_{12}(\tau) = \langle V_1(t+\tau)\, V_2^*(t)\rangle. \tag{8}$$

This function expresses the correlation at two arbitrary points P_1 and P_2 in the field, the disturbance at P_1 being considered at time τ later than at P_2; and instead of (3) one defines

$$\gamma_{12}(\tau) = \frac{\Gamma_{12}(\tau)}{\sqrt{\Gamma_{11}(0)}\,\sqrt{\Gamma_{22}(0)}} = \frac{\Gamma_{12}(\tau)}{\sqrt{I_1}\,\sqrt{I_2}}. \tag{9}$$

In terms of these functions, one obtains in place of the interference law (5) and the propagation law (6) the more general formulae

$$I = I^{(1)} + I^{(2)} + 2\sqrt{I^{(1)}}\,\sqrt{I^{(2)}}\,\gamma_{12}\left(\frac{r_2 - r_1}{c}\right), \tag{10}$$

$$\Gamma_{12}(\tau) = \frac{1}{\lambda_m^2} \int_A \int_A \frac{\overline{\Gamma}_{12}\left(\tau - \dfrac{r_1 - r_2}{c}\right)}{r_1 r_2}\, d\overline{\mathbf{x}}_1\, d\overline{\mathbf{x}}_2, \tag{11}$$

where $r_1 - r_2$ is the path difference introduced between the interfering beams and λ_m is a mean wave-length of the radiation. For small path differences, (10) and (11) may be shown to reduce to the formulae of Zernike and Hopkins.

Eq. (10) and (9) show that $\gamma_{12}(\tau)$ and $\Gamma_{12}(\tau)$ may be determined from the measurements of intensity in suitable interference experiments; both, therefore, represent *observable quantities*.

If the field is created by a finite incoherent source Σ *in vacuo*, the correlation of the disturbances at any two points P_1 and P_2 in the field may be shown to be given (within an accuracy sufficient for the purpose of most applications) by

$$\Gamma_{12}(\tau) = \int_\Sigma d\xi \int_0^\infty \frac{j(\xi, \nu)}{r_1 r_2}\, e^{-2\pi i\nu\left[\tau - \frac{r_1 - r_2}{c}\right]}\, d\nu. \tag{12}$$

Here ξ is a position vector of a typical element of the source, r_1 and r_2 are the distances of P_1 and P_2 from this element and $j(\xi, \nu)$ is the spectral intensity function of the source, i.e. the intensity per unit area, per unit frequency range.

The propagation laws (6) and (11) remind us of Huyghens' principle. Eq. (6) shows, for example, that Γ may be expressed as the sum of contributions due to spherical wavelets originating at all elements of the surface, the contribution from each pair of elements being proportional to the correlation of the vibrations at those two points. Now, as is well known, Huyghens principle may be regarded as an approximate formulation of Kirchhoff's integral theorem, this theorem being a rigorous consequence of the fact that V satisfies the wave equation. These remarks suggest that Γ perhaps also satisfies some differential equation and that Eq. (11) and (6) are simply approximate formulations of some associated "Kirchhoff's theorems". This indeed is the case. I showed recently (Wolf, [14]) that the introduction of the more general correlation function $\Gamma_{12}(\tau)$ is in fact essential for a precise formulation of such laws, and that with this definition of the correlation,

$$\nabla_1^2 \Gamma - \frac{1}{c^2}\frac{\partial^2 \Gamma}{\partial \tau^2} = 0, \qquad \nabla_2^2 \Gamma - \frac{1}{c^2}\frac{\partial^2 \Gamma}{\partial \tau^2} = 0, \tag{13}$$

where ∇_1^2 and ∇_2^2 are the Laplacian operators operating on the co-ordinates of P_1 and P_2 respectively.

Each of the wave equations (13) describes the variation of the correlation when one of the two points is fixed, whilst the other point, as well as the parameter τ varies. In all experiments, $c\tau$ will play the part of an optical path difference. The "actual" time makes no appearance in the theory; this is a most desirable feature, since true time variations are not observed in optical fields.

With the help of the wave equations it is possible, (as shown in [14]), to express rigorously the correlation at any two points P_1 and P_2, in terms of the correlation and its derivatives for all pairs of points on an arbitrary closed surface surrounding P_1 and P_2. A special case of this formula ($P_1 = P_2$, $\tau = 0$) gives the *intensity* at an arbitrary point in the field, in terms of the intensity distribution and its derivatives over the surface, each contribution being weighted by the appropriate value of the correlation factor γ and its derivatives. This formula may be regarded as a rigorous formulation of a generalized Huyghens Principle which involves observable quantities only.

So far our discussion has been simplified by treating light as a scalar phenomenon. It is, however, possible to extend the theory so as to take into account the vectorial character of light, i.e., its polarization properties. Let $E_i(\mathbf{x}, t)$ ($i = x, y$ or z) be the Cartesian rectangular components of the electric vector* at a typical point $P(\mathbf{x})$ at time t. We then define the 3×3 correlation tensor**

$$E_{ij}(\mathbf{x}_1, \mathbf{x}_2, \tau) = \langle E_i(\mathbf{x}_1, t+\tau)\, E_j^*(\mathbf{x}_2, t)\rangle. \tag{14}$$

This tensor, whose components may be shown to be measurable with the help of an interferometer, a polarizer and a phase plate, expresses the correlation between the components of the electric vector at any two given points in the field, and at two instants of time. It may be shown that *in vacuo*, each component of (14) obeys the wave equations

$$\nabla_1^2 E_{ij} - \frac{1}{c^2}\frac{\partial^2 E_{ij}}{\partial \tau^2} = 0, \qquad \nabla_2^2 E_{ij} - \frac{1}{c^2}\frac{\partial^2 E_{ij}}{\partial \tau^2} = 0. \tag{15}$$

For the sake of completeness, we may also introduce correlation tensors involving the components of the magnetic vector $\mathbf{H}(\mathbf{x}, t)$, and correlation tensors involving mixed pairs:

$$\left.\begin{aligned} H_{ij} &= \langle H_i(\mathbf{x}_1, t+\tau)\, H_j^*(\mathbf{x}_2, t)\rangle, \\ G_{ij} &= \langle E_i(\mathbf{x}_1, t+\tau)\, H_j^*(\mathbf{x}_2, t)\rangle. \end{aligned}\right\} \tag{16}$$

It appears that all quantities that are observable in Optical fields are expressible in terms of the components of these correlation tensors. It is, however, not necessarily true that all the components of our correlation tensors are in fact observables, since no radiation detectors appear to be available at optical frequencies which would respond to the magnetic rather than the electric field. The magnetic correlation tensors may, howevers, prove useful in other branches of Electro-

* Strictly speaking, we use here a complex representation of the electric vector. If e_i is the actual (real) component of the electric vector, whose real Fourier representation is

$$e_i = \int_0^\infty (a_\nu \cos 2\pi\nu t + b_\nu \sin 2\pi\nu t)\, d\nu,$$

then E_i is defined by

$$E_i = \int_0^\infty (a_\nu + i b_\nu)\, e^{-2\pi i\nu t}\, d\nu.$$

Such representation of real polychromatic fields was introduced by Gabor [16] in his interesting investigations in communication theory.

** In contrast to (8), subscripts now specify Cartesian components, not points in the field. The points are specified explicitly by their position vectors ($\mathbf{x}_1$, $\mathbf{x}_2$).

magnetic Theory. In terms of E and $\mathcal{G}$ the (time averaged) electric energy density $\langle W^{(e)}\rangle = (8\pi)^{-1}\langle E^2\rangle$ of the field *in vacuo* is given by

$$\langle W^{(e)}(\mathbf{x}, t)\rangle = \frac{1}{16\pi}\mathcal{R}\left[E_{11}(\mathbf{x}, \mathbf{x}, 0) + E_{22}(\mathbf{x}, \mathbf{x}, 0) + E_{33}(\mathbf{x}, \mathbf{x}, 0)\right],$$

and the components of the time averaged energy flux $\mathbf{S} = (c/4\pi)[\mathbf{E}\times\mathbf{H}]$ are

$$\langle S_i(\mathbf{x}, t)\rangle = \frac{c}{8\pi}\mathcal{R}\left[\mathcal{G}_{jk}(\mathbf{x}, \mathbf{x}, 0) - \mathcal{G}_{kj}(\mathbf{x}, \mathbf{x}, 0)\right],$$

$\mathcal{R}$ denoting the real part.

Each of the tensors H_{ij} and $\mathcal{G}_{ij}$ satisfies two wave equations of the form (15). Moreover, on account of Maxwell's equations, the correlation tensors are related by a set of differential equations. Using dyadic notation (e.g. writing $E(\mathbf{x}_1, \mathbf{x}_2, \tau) = \langle\mathbf{E}(\mathbf{x}_1, t+\tau)\,\mathbf{E}^*(\mathbf{x}_2, t)\rangle$ for the (time averaged) dyadic product of the two vectors $\mathbf{E}(\mathbf{x}_1, t+\tau)$ and $\mathbf{E}^*(\mathbf{x}_2, t)$, the differential equations for the propagation of the correlations *in vacuo* may be written as

$$\left.\begin{aligned} &\nabla_1 \wedge E + \frac{1}{c}\frac{\partial}{\partial\tau}\widetilde{\mathcal{G}} = 0, \qquad && \nabla_1 \wedge H - \frac{1}{c}\frac{\partial}{\partial\tau}\mathcal{G} = 0, \\ &\nabla_1 \wedge \widetilde{\mathcal{G}} - \frac{1}{c}\frac{\partial}{\partial\tau}E = 0, && \nabla_1 \wedge \mathcal{G} + \frac{1}{c}\frac{\partial}{\partial\tau}H = 0, \end{aligned}\right\} \tag{17}$$

where ∇_1 denotes the "del" operator operating on the coordinates of $P_1(\mathbf{x}_1)$, and $\widetilde{\mathcal{G}}$ is the "anti-conjugate" of $\mathcal{G}$:

$$\widetilde{\mathcal{G}}_{ij}(\mathbf{x}_1, \mathbf{x}_2, \tau) = \mathcal{G}^*_{ij}(\mathbf{x}_2, \mathbf{x}_1, -\tau). \tag{18}$$

There is a similar set of equations involving the operator ∇_2 which operates on the co-ordinates of $P_2(\mathbf{x}_2)$:

$$\left.\begin{aligned} &\nabla_2 \wedge E - \frac{1}{c}\frac{\partial}{\partial\tau}\mathcal{G} = 0, \qquad && \nabla_2 \wedge H + \frac{1}{c}\frac{\partial}{\partial\tau}\widetilde{\mathcal{G}} = 0, \\ &\nabla_2 \wedge \mathcal{G} + \frac{1}{c}\frac{\partial}{\partial\tau}E = 0, && \nabla_2 \wedge \widetilde{\mathcal{G}} - \frac{1}{c}\frac{\partial}{\partial\tau}H = 0. \end{aligned}\right\} \tag{19}$$

Moreover, from the subsidiary relations $\operatorname{div}\mathbf{E} = \operatorname{div}\mathbf{H} = 0$ of Maxwell's theory, it follows that

$$\nabla_1\cdot E = \nabla_1\cdot H = \nabla_1\cdot \mathcal{G} = \nabla_1\cdot \widetilde{\mathcal{G}} = 0, \tag{20}$$

$$\nabla_2\cdot E = \nabla_2\cdot H = \nabla_2\cdot \mathcal{G} = \nabla_2\cdot \widetilde{\mathcal{G}} = 0. \tag{21}$$

Since in Optics one never measures the instantaneous values of the field vectors, but rather the values of the correlations functions (usually in a somewhat disguised form), the equations (17)–(21) may be regarded as the basic equations which connect the observables of optical fields. Although formally they seem to be similar to Maxwell's equations, they are in fact more complicated as they are differential equations connecting tensor and not vector quantities.

In conclusion we may say that the researches described, which lead from the highly idealized case of a monochromatic field due to an ideal point source, to the polychromatic field due to a finite source, represent an important step towards the treatment of optical problems on a more satisfactory basis than was previously possible. This new development is necessarily a statistical one, in which the observable correlations, rather than the non-measurable field vectors play a central part. This development goes clearly in step with the advances made in recent times in other branches of Physics; and it suggests that statistical methods are likely to play an increasingly important part in optical researches of future years, a conclusion which is also supported by what we already heard in the Session on Information Theory and Optics.

ACKNOWLEDGEMENT

This paper was written during the tenure of an Imperial Chemical Industries Research Fellowship, the award of which by the University of Manchester is gratefully acknowledged.

REFERENCES

1. E. VERDET, Leçons d'Optique Physique (Paris), Tome I, 1869.
2. A. A. MICHELSON, Phil. Mag. **30,** 1, 1890; **31,** 338, 1891; **34,** 280, 1892, Astrophys. J., **51,** 257, 1920.
3. C. LAKEMAN and J. TH. GROOSMULLER, Physica (Gravenhage) **8,** 193, 199, 1928.
4. M. VON LAUE, Ann. Phys., **23,** 1, 1907.
5. M. BEREK, Z. Phys., **36,** 675, 824; **37,** 287; **40,** 420, 1926.
6. P. H. VAN CITTERT, Physica, **1,** 201, 1934.
7. P. H. VAN CITTERT, Physica, **6,** 1129, 1939.
8. F. ZERNIKE, Physica, **5,** 785, 1938; see also Proc. Phys. Soc., **61,** 158, 1948
9. H. H. HOPKINS, Proc. Roy. Soc., A **208,** 263, 1951; see also Proc. Roy. Soc., A, **217,** 408, 1953.
10. E. WOLF, Proc. Roy. Soc., A, **225,** 96, 1954.

11. L. R. Baker, Proc. Phys. Soc., B, **66**, 975, 1953.
12. A. Arnulf, O. Dupuy and F. Flamant, Rev. Opt. (théor. instrum.), **32**, 529, 1953.
13. A. Blanc-Lapierre and P. Dumontet, Rev. Opt. (théor. instrum.), **34**, 1, 1955.
14. E. Wolf, Proc. Roy. Soc., A, **230**, 246, 1955.
15. E. Wolf, Nuovo Cimento, **12**, 884, 1954.
16. D. Gabor, Journ. Inst. Electr. Eng., **93**, Part III, 429, 1946.

Coherence Properties of Partially Polarized Electromagnetic Radiation (*).

E. Wolf (**)

Department of Theoretical Physics - University of Manchester

(ricevuto il 4 Giugno 1959)

Summary. — This paper is concerned with the analysis of partial polarization from the standpoint of coherence theory. After observing that the usual analytic definition of the Stokes parameters of a quasi-monochromatic wave are not unique, a simple experiment is analysed, which brings out clearly the observable parameters of a quasi-monochromatic light wave. The analysis leads to a unique coherency matrix and to a unique set of Stokes parameters, the latter being associated with the representation of the coherency matrix in terms of Pauli's spin matrices. In this analysis the concept of Gabor's analytic signal proves to be basic. The degree of coherence between the electric vibrations in any two mutually orthogonal directions of propagation of the wave depends in general on the choice of the two orthogonal directions. It is shown that its maximum value is equal to the degree of polarization of the wave. It is also shown that the degree of polarization may be determined in a new way from relatively simple experiments which involve a compensator and a polarizer, and that this determination is analogous to the determination of the degree of coherence from Young's interference experiment.

1. – Introduction.

Although it is generally known that there is an intimate connection between partial polarization and partial coherence (see, for example, Wiener (1), p. 187), a systematic analysis of the properties of partially polarized radiation from

(*) The research described in this paper has been partially sponsored by the Air Force Cambridge Research Center of the Air Research and Development Command, United States Air Force, through its European Office, under Contract No. AF 61(052)-169.

(**) Now at the Institute of Optics, University of Rochester, Rochester, N.Y.

(1) N. Wiener: *Acta Math.*, **55**, § 9 (1930).

2737

the standpoint of coherence theory does not appear to have been made so far. There are several reasons why it seems desirable to carry out such an analysis. The numerous investigations made in recent years in connection with partially coherent light (ZERNIKE (2), HOPKINS (3), WOLF (4a,b,c), BLANC-LAPIERRE and DUMONTET (5)) have clearly shown that there is still a great deal to be learnt about the statistical properties of high frequency electromagnetic radiation. Moreover, as we shall briefly indicate, the usual treatments of partial polarization are not entirely satisfactory.

The most systematic treatments of partial polarization utilize the concept of Stokes parameters (introduced by G. G. STOKES in 1852 (6)) which are usually defined as follows: Consider a plane, quasi-monochromatic electromagnetic wave and let the components of the electric vector in two mutually perpendicular directions at right angles to the direction of propagation of the wave be represented in the form

$$(1.1)\qquad \begin{cases} E_x(t) = A_1(t)[\cos \Phi_1(t) - \bar{\omega}t]\,, \\ E_y(t) = A_2(t)[\cos \Phi_2(t) - \bar{\omega}t]\,, \end{cases}$$

where $\bar{\omega}$ denotes the mean frequency and t the time. The Stokes parameters are the four quantities

$$(1.2)\qquad \begin{cases} s_0 = \langle A_1^2\rangle + \langle A_2^2\rangle\,, \\ s_1 = \langle A_1^2\rangle - \langle A_2^2\rangle\,, \\ s_2 = 2\langle A_1 A_2 \cos(\Phi_1 - \Phi_2)\rangle\,, \\ s_3 = 2\langle A_1 A_2 \sin(\Phi_1 - \Phi_2)\rangle\,, \end{cases}$$

where the sharp brackets denote time average. Although to-day there exists an extensive literature in which the Stokes parameters play a central role (see, for example CHANDRASEKHAR (7)), it does not appear to have been noticed that the above relations do not define the parameters uniquely. For in (1.1) only $E_x(t)$ and $E_y(t)$ can be regarded as uniquely associated with the wave, whereas the A's and Φ's may evidently be chosen in many ways, leading to different sets of Stokes parameters. Only a careful analysis of experiment may be expected to lead to a unique set. Such an analysis is carried out in

(2) F. ZERNIKE: *Physica*, **5**, 785 (1938).

(3) H. H. HOPKINS: *Proc. Roy. Soc.*, A **208**, 263 (1951).

(4) E. WOLF: (*a*) *Proc. Roy. Soc.*, A **225**, 96 (1954); (*b*) *Nuovo Cimento*, **12**, 884 (1954); (*c*) *Proc. Roy. Soc.*, A **230**, 96 (1955).

(5) A. BLANC-LAPIERRE and P. DUMONTET: *Revue d'Optique*, **34**, 1 (1955).

(6) G. G. STOKES: *Trans. Camb. Phil. Soc.*, **9**, 399 (1852); also his *Mathematical and Physical Papers*, vol. **3** (Cambridge, 1901), p. 233.

(7) S. CHANDRASEKHAR: *Radiative Transfer* (Oxford, 1950), § 15.

2738

Section **2** of this paper and shows that the unique set obtained is intimately related to the appropriate « degree of coherence » of the electric vibrations in the two orthogonal directions.

Another unsatisfactory feature of the usual treatments of partial polarization has been clearly pointed out in an interesting recent paper by PANCHARANTHAM ([8], expecially p. 339-340): A partially polarized beam is usually described in terms of incoherent superposition of polarized and unpolarized beams and the interference phenomena arising from the superposition of these beams are analysed by using the concept of coherence and incoherence alone. However, the decomposition may be carried out in many different ways and it is by no means evident that the different decompositions will always lead to identical results. In any case this approach masks completely the *invariant characteristics* of the different representations. These unsatisfactory features can only be expected to be removed by the introduction of intermediate states (partial coherence).

In the present paper the basic properties of a quasi-monochromatic partially polarized electromagnetic wave are discussed from the standpoint of coherency theory and the invariant characteristics of such a wave are clearly brought out. The basic tool used for this purpose is the coherency matrix introduced in a previous paper (WOLF ([4b])), specialized to the problem in question; however, unlike in the previous paper, the coherency matrix is introduced here from the analysis of a simple experiment.

2. – The coherency matrix of a plane, quasi-monochromatic electromagnetic wave.

Consider a plane, quasi-monochromatic wave and let $E_x^{(r)}(t)$ and $E_y^{(r)}(t)$ represent (*) the components of the electric vector $\boldsymbol{E}^{(r)}$ at a typical point in the wave field in two mutually orthogonal directions at right angles to the directions of propagation of the wave. We assume that $\boldsymbol{E}^{(r)}$ may be represented as a Fourier integral and write

$$(2.1)\qquad \begin{cases} E_x^{(r)}(t) = \displaystyle\int_0^\infty a_1(\omega)\cos[\varphi_1(\omega) - \omega t]\,\mathrm{d}\omega\,, \\[2ex] E_y^{(r)}(t) = \displaystyle\int_0^\infty a_2(\omega)\cos[\varphi_2(\omega) - \omega t]\,\mathrm{d}\omega\,. \end{cases}$$

([8]) S. PANCHARATNAM: *Proc. Ind. Acad. Sci.*, A **44**, 398 (1956),

(*) The superscript « r » is introduced because the (real) electric wave function will shortly be regarded as the real part of a suitably chosen complex wave function.

2739

Since the wave is assumed to be quasi-monochromatic, the spectral amplitudes $a_1(\omega)$ and $a_2(\omega)$ will be appreciable only in a narrow range

$$\bar{\omega} - \tfrac{1}{2}\Delta\omega \leqslant \omega \leqslant \bar{\omega} + \tfrac{1}{2}\Delta\omega\,, \tag{2.2}$$

where $\Delta\omega$ is small compared with the mean frequency $\bar{\omega}$.

Suppose that the wave is passed through a device (compensator) which introduces retardations in $E_x^{(r)}$ and $E_y^{(r)}$. Let $\varepsilon_1(\omega)$ and $\varepsilon_2(\omega)$ be the phase delays in the Fourier components of frequency ω of $E_x^{(r)}$ and $E_y^{(r)}$ respectively. The electric wave emerging from the compensator has the components

$$\left\{\begin{aligned} \mathscr{E}_x^{(r)}(t) &= \int_0^\infty a_1(\omega)\cos[\varphi_1(\omega) - \varepsilon_1(\omega) - \omega t]\,\mathrm{d}\omega\,, \\ \mathscr{E}_y^{(r)}(t) &= \int_0^\infty a_2(\omega)\cos[\varphi_2(\omega) - \varepsilon_2(\omega) - \omega t]\,\mathrm{d}\omega\,, \end{aligned}\right. \tag{2.3}$$

it being assumed that reflection and absorption losses are negligible. If we use the identity $\cos[A - \varepsilon] = \cos A \cos\varepsilon + \sin A \sin\varepsilon$ in (2.3) and assume that for any two frequencies ω' and ω'' in the range (2.2)

$$|\varepsilon_1(\omega') - \varepsilon_1(\omega'')| \ll 2\pi\,, \qquad |\varepsilon_2(\omega') - \varepsilon_2(\omega'')| \ll 2\pi\,, \tag{2.4}$$

(2.3) may be re-written as

$$\left\{\begin{aligned} \mathscr{E}_x^{(r)} &= E_x^{(r)}(t)\cos\bar{\varepsilon}_1 + E_x^{(i)}(t)\sin\bar{\varepsilon}_1\,, \\ \mathscr{E}_y^{(r)} &= E_y^{(r)}(t)\cos\bar{\varepsilon}_2 + E_y^{(i)}(t)\sin\bar{\varepsilon}_2\,, \end{aligned}\right. \tag{2.5}$$

where $\bar{\varepsilon}_1 = \varepsilon_1(\bar{\omega})$, $\bar{\varepsilon}_2 = \varepsilon_2(\bar{\omega})$ and $E_x^{(i)}$ and $E_y^{(i)}$ are the Fourier integrals *conjugate* to $E_x^{(r)}$ and $E_y^{(r)}$ respectively, *i.e.*

$$\left\{\begin{aligned} E_x^{(i)} &= \int_0^\infty a_1(\omega)\sin[\varphi_1(\omega) - \omega t]\,\mathrm{d}\omega\,, \\ E_y^{(i)} &= \int_0^\infty a_2(\omega)\sin[\varphi_2(\omega) - \omega t]\,\mathrm{d}\omega\,. \end{aligned}\right. \tag{2.6}$$

As is well known (cf. TITCHMARSH [9]), any two conjugate functions are related

[9] E. C. TITCHMARSH: *Introduction to the Theory of Fourier Integrals* (Oxford, 1948), 2nd ed., chap. V.

2740

by Hilbert's reciprocity relations, *i.e.* by relations of the form

$$(2.7)\qquad E_x^{(i)}(t) = \frac{P}{\pi}\int_{-\infty}^{\infty}\frac{E_x^{(r)}(t')}{t'-t}\,\mathrm{d}t'\,,\qquad E_x^{(r)}(t) = -\frac{P}{\pi}\int_{-\infty}^{\infty}\frac{E_x^{(i)}(t')}{t'-t}\,\mathrm{d}t'\,,$$

P denoting the Cauchy principal value at $t'=t$.

Suppose now that the wave emerging from the compensator is sent through a polarizer, which only transmits the component which makes an angle θ with the x-direction. This component is given by

$$(2.8)\qquad \mathscr{E}^{(r)}(t;\,\theta,\,\varepsilon_1,\,\varepsilon_2) = \mathscr{E}_x^{(r)}(t)\cos\theta + \mathscr{E}_y^{(r)}(t)\sin\theta\,,$$

so that the intensity of the light emerging from the polarizer is

$$(2.9)\qquad \begin{aligned} I(\theta_1,\,\varepsilon_1,\,\varepsilon_2) &= 2\langle\mathscr{E}^{(r)^2}(t,\,\theta,\,\varepsilon_1,\,\varepsilon_2)\rangle \\ &= 2\langle\mathscr{E}_x^{(r)^2}\rangle\cos^2\theta + 2\langle\mathscr{E}_y^{(r)^2}\rangle\sin^2\theta + 4\langle\mathscr{E}_x^{(r)}\mathscr{E}_y^{(r)}\rangle\cos\theta\sin\theta\,, \end{aligned}$$

where sharp brackets denote time average. Here the wave field was assumed to be stationary (*), so that the intensity I is independent of the time instant at which the average is taken and the factor 2 on the right of the first equation of (2.9) was introduced to simplify later calculations. Next we substitute from (2.5) into (2.9) and use the following relations which may be readily proved from the properties of Hilbert transforms (**):

$$(2.10)\qquad \left\{\begin{aligned} &\langle E_x^{(r)^2}\rangle = \langle E_x^{(i)^2}\rangle\,,\qquad \langle E_y^{(r)^2}\rangle = \langle E_y^{(i)^2}\rangle\,,\\ &\langle E_x^{(r)}E_y^{(r)}\rangle = \langle E_x^{(i)}E_y^{(i)}\rangle\,,\\ &\langle E_x^{(r)}E_y^{(i)}\rangle = -\langle E_x^{(i)}E_y^{(r)}\rangle\,,\\ &\langle E_x^{(r)}E_x^{(i)}\rangle = \langle E_y^{(i)}E_y^{(r)}\rangle = 0\,. \end{aligned}\right.$$

(*) Stationarity in the strict sense of the theory of random functions would imply that the field vectors are not square integrable and hence we would not be justified in using Fourier integral analysis. This difficulty may be avoided in the usual way by assuming that the field exists only for a finite time interval $-T\leqslant t\leqslant T$ and proceeding to the limit $T\to\infty$ at the end of the calculations. The final results are the same whether or not this refinement is made.

(**) The formulae (2.10) are valid quite generally. When the field is quasi-monochromatic as here assumed, they may be proved in a very simple way by the following

Further, if we set

$$\delta = \bar{\varepsilon}_1 - \bar{\varepsilon}_2 , \tag{2.11}$$

and write $I(\theta, \delta)$ in place of $I(\theta, \bar{\varepsilon}_1, \bar{\varepsilon}_2)$ [since $\bar{\varepsilon}_1$ and $\bar{\varepsilon}_2$ enter the expression for the intensity only through their difference] we obtain the following expression for the time averaged intensity:

$$I(\theta, \delta) = 2\langle E_x^{(r)2}\rangle \cos^2\theta + 2\langle E_y^{(r)2}\rangle \sin^2\theta + \\ + 4\cos\theta\sin\theta\{\cos\delta\langle E_x^{(r)}E_y^{(r)}\rangle - \sin\delta\langle E_x^{(r)}E_y^{(i)}\rangle\} . \tag{2.12}$$

The formulae (2.12) may be expressed in a more convenient form, by using in place of the real wave functions the associated analytic signals of GABOR ([11]),

---

argument due to BRACEWELL [([10]), p. 102]. We set

$$E_x(t) = E_x^{(r)}(t) + iE_x^{(i)}(t) = A_1(t)\exp[-i\bar{\omega}t] ,$$

$$E_y(t) = E_y^{(r)}(t) + iE_y^{(i)}(t) = A_2(t)\exp[-i\bar{\omega}t] .$$

Then, if the field is quasi-monochromatic, the (generally complex) quantities A_1 and A_2 will vary slowly with t in comparison with the periodic term, and we have, for example,

$$\langle E_x^{(r)2}\rangle = \left\langle\left(\frac{A_1\exp[-i\omega t] + A_1^*\exp[i\omega t]}{2}\right)^2\right\rangle =$$

$$= \frac{1}{4}\langle A_1^2\exp[-2i\bar{\omega}t]\rangle + \frac{1}{2}\langle A_1A_1^*\rangle + \frac{1}{4}\langle A_1^{*2}\exp[2i\bar{\omega}t]\rangle .$$

The first and the last term on the right vanish because of the rapidly varying terms $\exp[-2i\bar{\omega}t]$ and $\exp[2i\bar{\omega}t]$, so that

$$\langle E_x^{(r)2}\rangle = \tfrac{1}{2}\langle A_1A_1^*\rangle .$$

Similarly

$$\langle E_x^{(i)2}\rangle = \left\langle\left(\frac{A_1\exp[-i\bar{\omega}t] - A_1^*\exp[i\bar{\omega}t]}{2i}\right)^2\right\rangle = \frac{1}{2}\langle A_1A_1^*\rangle .$$

Comparison of the last two formulae gives the first relation in (2.10); the other relations may be proved in a similar way.

([10]) R. N. BRACEWELL: *Proc. I.R.E.*, **46**, 97 (1958).

([11]) D. GABOR: *Journ. Inst. Elect. Engrs.*, **93**, part III, 429 (1946).

2742

i.e. by using in place of $E_x^{(r)}$ and $E_y^{(r)}$, the functions (*)

$$\text{(2.13)}\qquad \begin{cases} E_x(t) = E_x^{(r)}(t) + i\,E_x^{(i)}(t) = \displaystyle\int_0^\infty a_1(\omega)\exp\left[i[\varphi_1(\omega) - \overline{\omega}t]\right]\mathrm{d}\omega\,, \\ E_y(t) = E_y^{(r)}(t) + i\,E_y^{(i)}(t) = \displaystyle\int_0^\infty a_2(\omega)\exp\left[i[\varphi_2(\omega) - \omega t]\right]\mathrm{d}\omega\,. \end{cases}$$

Using (2.10) we have the relations

$$\text{(2.14}a)\qquad \langle E_x E_x^*\rangle = 2\langle E_x^{(r)2}\rangle = 2\langle E_x^{(i)2}\rangle\,,$$

$$\text{(2.14}b)\qquad \langle E_x E_y^*\rangle = 2\langle E_x^{(r)}E_y^{(r)}\rangle - 2i\langle E_x^{(i)}E_y^{(i)}\rangle\,,$$

etc. With the help of (2.14) the formula (2.12) becomes

$$\text{(2.15)}\qquad I(\theta,\delta) = J_{xx}\cos^2\theta + J_{yy}\sin^2\theta + J_{xy}\cos\theta\sin\theta\exp[-i\delta] + J_{yx}\sin\theta\cos\theta\exp[i\delta]\,,$$

where the J's are the elements of the *coherency matrix*

$$\text{(2.16)}\qquad \boldsymbol{J} = \begin{bmatrix} J_{xx} & J_{xy} \\ J_{yx} & J_{yy} \end{bmatrix} = \begin{bmatrix} \langle E_x E_x^*\rangle & \langle E_x E_y^*\rangle \\ \langle E_y E_x^*\rangle & \langle E_y E_y^*\rangle \end{bmatrix}.$$

The formula (2.15) expresses in a compact form the intensity of the wave after transmission through the compensator (which introduces a phase delay δ) and the polarizer (oriented so as to transmit the component which makes an angle θ with the x-axis) in terms of the coherency matrix $\boldsymbol{J}$ which characterizes the incident wave.

Since $J_{yx} = J_{xy}^*$ the coherency matrix is *Hermitian*. Its trace represents the intensity of the incident wave,

$$\text{(2.17)}\qquad \operatorname{Tr}\boldsymbol{J} = J_{xx} + J_{yy} = \langle E_x E_x^*\rangle + \langle E_y E_y^*\rangle = 2\langle E_x^{(r)2}\rangle + 2\langle E_y^{(r)2}\rangle\,,$$

and its non-diagonal elements express the correlation between the x and y-components of the complex vector $\boldsymbol{E}$. Further it follows from Schwarz' inequality

(*) An analytic signal is a complex function characterized by the property that its Fourier integral contains no spectral components of positive (or negative) frequencies. This fact alone implies that the real and imaginary parts of the signal are conjugate functions and hence Hilbert transforms of each other.

2743

for integrals that $|J_{xy}| \leqslant \sqrt{J_{xx}}\sqrt{J_{yy}}$, $|J_{yx}| \leqslant \sqrt{J_{yy}}\sqrt{J_{xx}}$; hence, since $J_{yx} = J_{xy}$,

$$|\boldsymbol{J}| = J_{xx}J_{yy} - J_{xy}J_{yx} \geqslant 0\,, \tag{2.18}$$

i.e. the discriminant of the coherency matrix is non-negative.

Let $A_1(t)$, $A_2(t)$ be the amplitudes and $\Psi_1(t)$ and $\Psi_2(t)$ the phases of $E_x(t)$ and $E_y(t)$ respectively, *i.e.*

$$E_x(t) = A_1(t)\exp[i\,\Psi_1(t)]\,, \qquad E_y(t) = A_2(t)\exp[i\,\Psi_2(t)]\,. \tag{2.19}$$

Then, from (2.13),

$$\begin{cases} E_x^{(r)}(t) = A_1(t)\cos[\Psi_1(t)]\,, & E_x^{(i)}(t) = A_1(t)\sin[\Psi_1(t)]\,, \\ E_y^{(r)}(t) = A_2(t)\cos[\Psi_2(t)]\,, & E_y^{(i)}(t) = A_2(t)\sin[\Psi_2(t)]\,, \end{cases} \tag{2.20}$$

and, if we introduce quantities $\Phi_1(t)$ and $\Phi_2(t)$ by the relations

$$\Psi_1(t) = \Phi_1(t) - \bar{\omega}t\,, \qquad \Psi_2(t) = \Phi_2(t) - \bar{\omega}t\,, \tag{2.21}$$

where $\bar{\omega}$ is the mean frequency, the components $E_x^{(r)}$, $E_y^{(r)}$ of the (real) electric vector are represented by expressions of the form (1.1), but the representation is now *unique*. In terms of the A's and Ψ's the elements of the coherency matrix are

$$\begin{cases} J_{xx} = \langle A_1^2\rangle\,, \\ J_{yy} = \langle A_2^2\rangle\,, \\ J_{xy} = \langle A_1A_2\exp[i(\Psi_1 - \Psi_2)]\rangle\,, \\ J_{yx} = \langle A_1A_2\exp[-i(\Psi_1 - \Psi_2)\rangle\,. \end{cases} \tag{2.22}$$

We may now introduce a set of Stokes parameters by the relations

$$\begin{cases} s_0 = \langle A_1^2\rangle + \langle A_2^2\rangle & = J_{xx} + J_{yy}\,, \\ s_1 = \langle A_1^2\rangle - \langle A_2^2\rangle & = J_{xx} - J_{yy}\,, \\ s_2 = 2\langle A_1A_2\cos(\Psi_1 - \Psi_2)\rangle & = J_{xy} + J_{yx}\,, \\ s_3 = 2\langle A_1A_2\sin(\Psi_1 - \Psi_2)\rangle & = i(J_{yx} - J_{xy})\,. \end{cases} \tag{2.23}$$

2744

We see that this set of Stokes parameters is unique and that uniqueness has been achieved with the help of analytic signals, the introduction of which was suggested by the appearance of conjugate functions in the analysis of our experiment.

The relation between the Stokes parameters and the coherency matrix may also be expressed in the form

$$\boldsymbol{J} = \tfrac{1}{2}\sum_{i=0}^{3} s_i \boldsymbol{\sigma}_i \,, \tag{2.24}$$

where $\boldsymbol{\sigma}_0$ is the unit matrix

$$\boldsymbol{\sigma}_0 = \begin{bmatrix} 1 & 0 \\ 0 & 1 \end{bmatrix},$$

and $\boldsymbol{\sigma}_1$, $\boldsymbol{\sigma}_2$, $\boldsymbol{\sigma}_3$, are the Pauli spin matrices

$$\boldsymbol{\sigma}_1 = \begin{bmatrix} 1 & 0 \\ 0 & -1 \end{bmatrix}, \qquad \boldsymbol{\sigma}_2 = \begin{bmatrix} 0 & 1 \\ 1 & 0 \end{bmatrix}, \qquad \boldsymbol{\sigma}_3 = \begin{bmatrix} 0 & i \\ -i & 0 \end{bmatrix}. \tag{2.26}$$

The connection between a coherency matrix, Stokes parameters and Pauli's spin matrices has been noted previously (FANO ([12])); however, as already mentioned, the non-uniqueness of the usual analytic definition of the Stokes parameters appears to have escaped attention.

Finally we mention that the coherency matrix (2.16) was introduced in an earlier paper (WOLF ([4b])) from more formal considerations. Our present analysis shows that this matrix appears in a natural way from the analysis of a simple experiment.

3. – Some consequences of the basic intensity formula.

To see the physical significance of the intensity formula (2.15) we re-write it in a somewhat different form. We set

$$\frac{J_{xy}}{\sqrt{J_{xx}}\sqrt{J_{yy}}} = \mu_{xy} = |\mu_{xy}| \exp[i\beta_{xy}] \,. \tag{3.1}$$

([12]) U. FANO: *Phys. Rev.*, **93**, 121 (1954).

2745

It follows from (2.18) that

$$|\mu_{xy}| \leqslant 1 \,.$$

By analogy with the theory of partially coherent scalar fields we may call μ_{xy} *the complex degree of coherence* of the electric vibrations in the x and y directions. It absolute value $|\mu_{xy}|$ is a measure of the degree of correlation of the vibrations and its phase represents their « effective phase difference ».

If we substitute from (3.1) into the intensity formula (2.15) and use the relation $J_{yx} = J^*_{xy}$ we obtain the following expression for the intensity:

$$I(\theta, \delta) = J_{xx} \cos^2\theta + J_{yy} \sin^2\theta + 2\sqrt{J_{xx}}\sqrt{J_{yy}} \cos\theta \sin\theta\, |\mu_{xy}| \cos(\beta_{xy} - \delta) \,. \tag{3.3}$$

This expression is formally identical with the basic interference law of partially coherent fields [Wolf ([4a]), p. 102, Thompson and Wolf ([13]), p. 896]. It shows that the intensity $I(\theta, \delta)$ may be regarded as arising from the interference of two beams of intensities:

$$I^{(1)} = J_{xx} \cos^2\theta \,, \qquad I^{(2)} = J_{yy} \sin^2\theta \,, \tag{3.4}$$

and with complex degree of coherence μ_{xy}, after a phase difference δ has been introduced between them.

Returning to (2.15) we see that the elements of the coherency matrix of a quasi-monochromatic plane wave may be determined from very simple experiments. It is only necessary to measure the intensity for several different values of θ (orientation of polarizer) and δ (delay introduced by a compensator), and solve the corresponding relations obtained from (2.15). Let $\{\theta, \delta\}$, denote the measurement corresponding to a particular pair θ, δ. A convenient set of measurements is the following:

$$\{0°, 0\}\,, \quad \{45°, 0\}\,, \quad \{90°, 0\}\,, \quad \{135°, 0\}\,, \quad \left\{45°, \frac{\pi}{2}\right\}\,, \ \left\{135°, \frac{\pi}{2}\right\} . \tag{3.5}$$

It follows from (2.15) that, in terms of the intensities determined from these six measurements, the elements of the coherency matrix are given by

$$\left\{\begin{aligned} J_{xx} &= I(0°, 0)\,, \\ J_{yy} &= I(90°, 0)\,, \\ J_{xy} &= \frac{1}{2}\{I(45°, 0) - I(135°, 0)\} + \frac{1}{2} i \left\{ I\left(45°, \frac{\pi}{2}\right) - I\left(135°, \frac{\pi}{2}\right)\right\}, \\ J_{yx} &= \frac{1}{2}\{I(45°, 0) - I(135°, 0)\} - \frac{1}{2} i \left\{ I\left(45°, \frac{\pi}{2}\right) - I\left(135°, \frac{\pi}{2}\right)\right\} . \end{aligned}\right. \tag{3.6}$$

([13]) B. J. Thompson and E. Wolf: *Journ. Opt. Soc. Amer.*, **47**, 895 (1957).

Thus we see that the elements of the coherency matrix represent ***measurable*** physical quantities.

In the theory of partially coherent scalar fields, the concept of Michelson's visibility (*) of fringes plays a central role (cf. ZERNIKE [2]). We will now derive an expression for a quantity defined in a similar way and we shall see later that this quantity has a simple physical meaning.

It follows from (2.15) by a straightforward calculation, that the maxima and minima of the intensity (with respect to both θ and δ) are

$$\tag{3.7} \begin{cases} I_{\max} = \dfrac{1}{2}(J_{xx}+J_{yy})\left[1+\sqrt{1-\dfrac{4|\boldsymbol{J}|}{(J_{xx}+J_{yy})^2}}\right], \\ I_{\min} = \dfrac{1}{2}(J_{xx}+J_{yy})\left[1-\sqrt{1-\dfrac{4|\boldsymbol{J}|}{(J_{xx}+J_{yy})^2}}\right]. \end{cases}$$

Hence

$$\tag{3.8} \frac{I_{\max}-I_{\min}}{I_{\max}+I_{\min}} = \sqrt{1-\frac{4|\boldsymbol{J}|}{(J_{xx}+J_{yy})^2}}.$$

Now if the x and y axes are rotated about the direction of propagation of the wave, the coherency matrix will change. There are, however, two invariants for such rotations, namely the discriminant $|\boldsymbol{J}|$ and the trace $\mathrm{Tr}\boldsymbol{J} = J_{xx}+J_{yy}$ of the matrix. Since on the right hand side of (3.8) the elements of $\boldsymbol{J}$ enter only in these combinations, it follows that the expression is invariant with respect to rotations of the axes and hence may be expected to have a physical significance. We shall see shortly (eq. (5.14) below) that it represents the degree of polarization of the wave.

4. – Coherency matrices of natural and of monochromatic radiation.

Light which is most frequently encountered in nature has the property that the intensity of its components in any direction perpendicular to the direction of propagation is the same; and, moreover, the intensity is not af-

(*) The visibility $\mathscr{V}$ of fringes at a point P in the fringe pattern is defined by the formula

$$\mathscr{V} = \frac{I_{\max}-I_{\min}}{I_{\max}+I_{\min}},$$

where $I_{\max}$ and $I_{\min}$ are the maximum and minimum intensities in the immediate neighbourhood of P.

fected by any previous retardation of one of the rectangular components relative to the other, into which the light may have been resolved. In other words

$$I(\theta, \delta) = \text{constant} \tag{4.1}$$

for all values of θ and δ. Such light is called *natural light*; and we may define « natural » electromagnetic radiation of any other spectral range in a strictly similar way.

It is evident from (3.3) that $I(\theta, \delta)$ is independent of δ and θ, if and only if

$$|\mu_{xy}| = 0\,, \qquad \text{and} \qquad J_{xx} = J_{yy}\,. \tag{4.2}$$

The first condition implies that the electric vibrations in the x and y directions are mutually incoherent. According to (3.1) and the relation $J_{yx} = J^*_{yx}$, (4.2) may also be written as

$$J_{xy} = J_{yx} = 0\,, \qquad J_{xx} = J_{yy}\,, \tag{4.3}$$

and it follows that *the coherency matrix of natural radiation* of intensity $J_{xx} + J_{yy} = I$ is

$$\frac{1}{2} I \begin{bmatrix} 1 & 0 \\ 0 & 1 \end{bmatrix}. \tag{4.4}$$

Next let us consider the coherency matrix of monochromatic radiation. In this case the amplitudes A_1 and A_2 and the phases Ψ_1 and Ψ_2 in (2.22) are independent of time and the coherency matrix has the form

$$\begin{bmatrix} A_1^2 & A_1 A_2 \exp[i(\Psi_1 - \Psi_2)] \\ A_1 A_2 \exp[-i(\Psi_1 - \Psi_2)] & A_2^2 \end{bmatrix}. \tag{4.5}$$

We see that in this case

$$|\boldsymbol{J}| = J_{xx} J_{yy} - J_{xy} J_{yx} = 0\,, \tag{4.6}$$

i.e. the discriminant of the coherency matrix is zero. The complex degree of coherence now is

$$\mu_{xy} = \frac{J_{xy}}{\sqrt{J_{xx}}\sqrt{J_{yy}}} = \exp[(\Psi_1 - \Psi_2)\,, \tag{4.7}$$

i.e. its absolute value is unity (complete coherence) and its phase is equal to the difference between the phases of the two components.

5. – The degree of polarization.

Before deriving an expression for the degree of polarization in terms of the coherency matrix we shall establish a simple theorem relating to the coherency matrix of a wave resulting from the superposition of a number of mutually independent waves.

Consider N mutually independent quasi-monochromatic waves propagated in the same direction (z-say) and let $E_x^{(n)}$, $E_y^{(n)}$ $(n = 1, 2, \ldots, N)$ be the analytic signals associated with the components of the electric vibrations of the n-th wave in the directions of the x and y-axes. The components of the resulting wave then are

$$E_x = \sum_{n=1}^{N} E_x^{(n)}, \qquad E_y = \sum_{n=1}^{N} E_y^{(n)}, \tag{5.1}$$

so that the elements of the coherency matrix are

$$\begin{cases} J_{kl} = \langle E_k E_l^* \rangle = \sum_{n=1}^{N} \sum_{m=1}^{N} \langle E_k^{(n)} E_l^{(m)*} \rangle , \\ \quad = \sum_{n=1}^{N} \langle E_k^{(n)} E_l^{(n)*} \rangle + \sum_{n \neq m} \langle E_k^{(n)} E_l^{(m)*} \rangle . \end{cases} \tag{5.2}$$

Since the waves are assumed to be independent, each term under the last summation sign is zero, and it follows that

$$J_{kl} = \sum_{n=1}^{N} J_{kl}^{(n)} , \tag{5.3}$$

where $J_{kl}^{(n)} = \langle E_k^{(n)} E^{(n)*} \rangle$ are the elements of the coherency matrix of the n-th wave. The formula (5.3) shows that the coherency matrix of a wave resulting from the superposition of a number of independent waves is the sum of the coherency matrices of the individual waves.

To find an expression for the degree of polarization of a wave, we first represent the wave as a superposition of a wave of natural radiation and a wave of monochromatic radiation, independent of the former. Let $\boldsymbol{J}$ be the coherency matrix of the given wave and let $\boldsymbol{J}^{(1)}$ and $\boldsymbol{J}^{(2)}$ be the coherency matrices of the two independent waves into which we decompose it. Then according to (4.4) and (4.6) $\boldsymbol{J}^{(1)}$ and $\boldsymbol{J}^{(2)}$ must be of the form

$$\boldsymbol{J}^{(1)} = \begin{bmatrix} A & 0 \\ 0 & A \end{bmatrix}, \qquad \boldsymbol{J}^{(2)} = \begin{bmatrix} B & D \\ D^* & C \end{bmatrix}, \tag{5.4}$$

2749

where $A \geqslant 0$, $B \geqslant 0$, $C \geqslant 0$ and

$$BC - DD^* = 0 . \tag{5.5}$$

In order to show that such a decomposition is possible we must determine quantities A, B, C, D, subject to the above conditions, such that the given coherency matrix $\mathbf{J} = [J_{lk}]$ is equal to the sum of two matrices of the form (5.4),

$$\mathbf{J} = \mathbf{J}^{(1)} + \mathbf{J}^{(2)} . \tag{5.6}$$

The relation (5.6) implies that

$$\begin{cases} J_{xx} = A + B , & J_{xy} = D , \\ J_{yx} = D^* , & J_{yy} = A + C . \end{cases} \tag{5.7}$$

On substituting for B, C, D and D^* from (5.7) into (5.5) we find that (*)

$$A = \tfrac{1}{2}(J_{xx} + J_{yy}) \pm \tfrac{1}{2}\sqrt{(J_{xx} + J_{yy})^2 - 4|\mathbf{J}|} . \tag{5.8}$$

Since $J_{yx} = J^*_{yx}$ the product $J_{xy} J^*_{yx}$ is non-negative, and it follows from (2.18) that

$$|\mathbf{J}| \leqslant J_{xx} J_{yy} \leqslant \tfrac{1}{4}(J_{xx} + J_{yy})^2 , \tag{5.9}$$

so that both the roots (5.8) are non-negative. Consider first the solution with the negative sign in front of the square root. We then have from (5.7),

$$\begin{cases} A = \tfrac{1}{2}(J_{xx} + J_{yy}) - \tfrac{1}{2}\sqrt{(J_{xx} + J_{yy})^2 - 4|\mathbf{J}|} , \\ B = \tfrac{1}{2}(J_{xx} - J_{yy}) + \tfrac{1}{2}\sqrt{(J_{xx} + J_{yy})^2 - 4|\mathbf{J}|} , \\ C = \tfrac{1}{2}(J_{yy} - J_{xx}) + \tfrac{1}{2}\sqrt{(J_{xx} + J_{yy})^2 - 4|\mathbf{J}|} , \\ D = J_{xy} , \\ D^* = J_{yx} . \end{cases} \tag{5.10}$$

Now

$$\sqrt{(J_{xx} + J_{yy})^2 - 4|\mathbf{J}|} = \sqrt{(J_{xx} - J_{yy})^2 + 4J_{xy}J_{yx}} \geqslant |J_{xx} - J_{yy}| . \tag{5.11}$$

Hence B and C are also non-negative as required. The other root given by (5.8) (with the positive sign in front of the square root) leads to negative values

(*) A is seen to be a characteristic root (eigenvalue) of the coherency matrix $\mathbf{J}$.

of B and C and must therefore be rejected. We have thus obtained a unique decomposition of the required kind.

The total intensity of the wave is

$$I_{tot} = \mathrm{Tr}\,\boldsymbol{J} = J_{xx} + J_{yy}\;; \tag{5.12}$$

and the intensity of the monochromatic (and hence *polarized*) part is

$$I_{pol} = \mathrm{Tr}\boldsymbol{J}^{(2)} = B + C = \sqrt{(J_{xx} + J_{yy})^2 - 4\,|\boldsymbol{J}|}\,. \tag{5.13}$$

Hence the *degree of polarization* P of the original wave is

$$P = \frac{I_{pol}}{I_{tot}} = \sqrt{1 - \frac{4\,|\boldsymbol{J}|}{(J_{xx} + J_{yy})^2}}\,. \tag{5.14}$$

Since this expression involves only the two rotational invariants of the coherency matrix $\boldsymbol{J}$, the degree of polarization is independent of the particular choice of the x and y axes, as might have been expected.

Comparison of (5.14) with (3.8) shows that the quantity $(I_{max} - I_{min})/(I_{max} + I_{min})$ is precisely the degree of polarization P of the wave.

Unlike the degree of polarization, the degree of coherence depends on the choice of the x and y directions. We shall not investigate in detail the changes in the degree of coherence as the x, y axes are rotated; we shall only consider an extreme case, which is of special physical interest.

If in the expression (5.14) for the degree of polarization P we write out in full the discriminant $\boldsymbol{J}$, and use the expression (3.1) for the degree of coherence μ_{xy} we find that the following relation holds between P and $|\mu_{xy}|$:

$$1 - P^2 = \frac{J_{xx}\,J_{yy}}{[\frac{1}{2}(J_{xx} + J_{yy})]^2}\,[1 - |\mu_{xy}|^2]\,. \tag{5.15}$$

Since the geometric mean of any two positive numbers cannot exceed their arithmetic mean it follows that $1 - P^2 \leqslant 1 - |\mu_{xy}|^2$, *i.e.*

$$P \geqslant |\mu_{xy}|\,. \tag{5.16}$$

The equality sign in (5.16) will hold if an only if $J_{xx} = J_{yy}$, *i.e.* if the (time averaged) intensities in the two orthogonal directions are equal. We shall now show that a pair of directions always exists for which this is the case.

Suppose that we take a new pair of orthogonal directions x', y' perpendicular to the direction of propagation of the wave and let φ be the angle between x and x'. The components E_x, E_y, of the electric vector (in the complex representations (2.13)) in the new directions are

$$\text{(5.17)} \qquad \begin{cases} E_{x'} = E_x \cos\varphi + E_y \sin\varphi\,, \\ E_{y'} = -E_x \sin\varphi + E_y \cos\varphi\,. \end{cases}$$

Hence the elements of the transformed coherency matrix $\boldsymbol{J}' = [J_{k'l'}] = [\langle E_{k'} E^*_{l'}\rangle]$ are

$$\text{(5.18)} \qquad \begin{cases} J_{x'x'} = J_{xx}c^2 + J_{yy}s^2 + (J_{xy} + J_{yx})cs\,, \\ J_{y'y'} = J_{xx}s^2 + J_{yy}c^2 - (J_{xy} + J_{yx})cs\,, \\ J_{x'y'} = (J_{yy} - J_{xx})cs + J_{xy}c^2 - J_{yx}s^2\,, \\ J_{y'x'} = (J_{yy} - J_{xx})cs + J_{yx}c^2 - J_{xy}s^2\,, \end{cases}$$

where

$$\text{(5.19)} \qquad c = \cos\varphi\,, \qquad s = \sin\varphi\,.$$

The intensities in the x' and y' directions will be equal (*i.e.* $J_{x'x'} = J_{y'y'}$) if

$$J_{xx}c^2 + J_{yy}s^2 + (J_{xy} + J_{yx})cs = J_{xx}s^2 + J_{yy}c^2 - (J_{xy} + J_{yx})cs\,.$$

Solving this equation for φ we obtain

$$\text{(5.20)} \qquad \operatorname{tg} 2\varphi = \frac{J_{yy} - J_{xx}}{J_{xy} + J_{yx}}\,.$$

Since $J_{yx} = J^*_{xy}$ and J_{xx} and J_{yy} are real, this equation always has a real root. Thus *there always exists a pair of directions for which the two intensities are equal. For this pair of directions the degree of coherence* $|\mu_{xy}|$ *of the electric vibrations has its maximum value and this value is equal to the degree of polarization of the wave.*

This special pair of directions has a simple geometrical significance. If, as in (5.6) we represent the wave as incoherent mixture of a wave of natural radiation and a wave of monochromatic (and therefore completely polarized) radiation, the angle χ which the major axes of the vibrational ellipse of the polarized portion makes with the x-direction is given by (see CHANDRASEKHAR ([7]), p. 33, eq. (180))

$$\text{(5.21)} \qquad \operatorname{tg} 2\chi = \frac{s_2}{s_1} = \frac{J_{xy} + J_{yx}}{J_{xx} - J_{yy}}\,.$$

2752

It follows from (5.20) and (5.21) that $(\operatorname{tg} 2\chi)\cdot(\operatorname{tg} 2\varphi) = -1$ so that $\chi - \varphi = 45°$ or $135°$. This implies that *the directions for which $P = |\mu_{xy}|$ are the bisectors of the principal directions (directions of the major and minor axes) of the vibrational ellipse of the polarized portion of the wave.*

It is evident from the foregoing discussion that the introduction of coherence concepts into the theory of partial polarization leads to a clearer understanding of the behaviour of partially polarized radiation and suggests new ways for the measurement of its degree of polarization.

RIASSUNTO (*)

In questo articolo si fa l'analisi della polarizzazione parziale dal punto di vista della teoria della coerenza. Dopo aver rilevato che la usuale definizione analitica dei parametri di Stokes per un'onda quasi monocromatica non è univoca, si prende in esame un semplice esperimento il quale introduce, in maniera chiara, i parametri osservabili per un'onda luminosa quasi monocromatica. L'analisi conduce ad un'unica matrice di coerenza e ad un unico gruppo di parametri di Stokes; quest'ultimo è associato alla rappresentazione della matrice di coerenza in funzione delle matrici di spin di Pauli. Tale analisi mostra la fondamentale importanza del concetto di segnale analitico di Gabor. Il grado di coerenza fra le vibrazioni elettriche in due direzioni qualunque, reciprocamente ortogonali, di propagazione dell'onda dipende, in generale, dalla scelta delle due direzioni ortogonali. Si dimostra che il valore massimo di esso uguaglia il grado di polarizzazione dell'onda. Si dimostra altresì che il grado di polarizzazione può essere dedotto, seguendo una via nuova, da esperimenti relativamente semplici che richiedono l'uso di un compensatore e di un polarizzatore, e che tale determinazione è analoga a quella del grado di coerenza ricavata dall'esperimento di interferenza di Young.

(*) *Traduzione a cura della Redazione.*

2753

PROC. PHYS. SOC., 1964, VOL. 84

Theory of photoelectric detection of light fluctuations†

L. MANDEL‡, E. C. G. SUDARSHAN§ and E. WOLF||

‡ Department of Physics, Imperial College, London
§ Department of Physics, Brandeis University, Waltham, Massachusetts, U.S.A.
|| Department of Physics and Astronomy, University of Rochester, Rochester, New York, U.S.A.

MS. received 1st *June* 1964

Abstract. The basic formulae governing the fluctuations of counts registered by photoelectric detectors in an optical field are derived. The treatment, which has its origin in Purcell's explanation of the Hanbury Brown–Twiss effect, is shown to apply to any quasi-monochromatic light, whether stationary or not, and whether of thermal origin or not. The representation of the classical wave amplitude of the light by Gabor's complex analytic signal appears naturally in this treatment.

It is shown that the correlation of counts registered by N separate photodetectors at N points in space is determined by a $2N$th order correlation function of the complex classical field. The variance of the individual counts is shown to be expressible as the sum of terms representing the effects of classical particles and classical waves, in analogy to a well-known result of Einstein relating to black-body radiation. Since the theory applies to correlation effects obtained with any type of light it applies, in particular, to the output of an optical maser, although, for a maser operating on one mode, correlation effects are likely to be very small.

1. Introduction

In an important note Purcell (1956) has given a very clear explanation of an effect first observed by Brown and Twiss (1956), namely the appearance of correlation in the fluctuations of two photoelectric currents evoked by coherent beams of light. The method employed by Purcell is semi-classical, but it brings out the essence of the phenomenon much more clearly than most other approaches. Purcell's method has been developed further by Mandel (1958, 1959, 1963 a, b) and applied to the analysis of related problems by Alkemade (1959) and by Wolf (1960) (see also Kahn 1958).

In all the publications just referred to, the light incident on the photodetector was implicitly or explicitly assumed to be of thermal origin. However, in view of the development of optical masers the question has been raised in recent months as to the existence of the Hanbury Brown–Twiss effect with light from other sources. In this connection Glauber (1963 a, b) has expressed doubts about the applicability of these stochastic semi-classical methods to the analysis of fluctuation and correlation experiments obtained with light from non-thermal sources. Although partial answers to Glauber's critical remarks have already been given (Mandel and Wolf 1963, Sudarshan 1963 a, b; see also Jaynes and Cummings 1963), it is, of course, desirable to examine this semi-classical approach more closely. This is done in the present paper where new results on fluctuations and correlations are also obtained. The main conclusions are:

† This research was supported in part by the U.S. Army Research Office (Durham).

‡ Now at the Department of Physics and Astronomy, University of Rochester, Rochester, New York, U.S.A.

(i) In the conditions under which light fluctuations are usually measured by photoelectric detectors, the semi-classical treatment applies as readily to light of non-thermal origin as to thermal light, and to non-stationary as well as to stationary fields.

(ii) The representation of the optical field by complex analytic signals (as customary in the classical theory of optical coherence) appears naturally in this treatment.

(iii) There exists a very simple formula for the variance of the number of photoelectrons registered by a photodetector in a given time interval, when it is illuminated by any quasi-monochromatic light beam. This formula expresses the variance as the sum of two terms, one of which can be interpreted as representing the effect of fluctuations in a system of classical particles, and the other as arising from the interference of classical waves. This result, which may also be interpreted as describing the fluctuations of the light itself, is strictly analogous to a well-known result first established by Einstein (1909 a, b) for energy fluctuations in an enclosure containing black-body radiation, under conditions of thermal equilibrium.

(iv) The correlation between the number of photoelectrons registered by N separate photodetectors is expressible in terms of functions of $2N$th order correlations of the classical complex wave amplitudes.

The simplicity of the semi-classical theory and its wide range of validity makes it well suited for the analysis of many problems relating to photoelectric detection of light fluctuations. Our work demonstrates that a full quantum field theoretical treatment is not at all necessary for the analysis of such problems.

2. The probability distribution for photoelectrons ejected from a photo cathode illuminated by a light beam

First, consider an electromagnetic wave interacting with a quantum mechanical system, playing the role of a 'detector', in a bound state $|\psi_b\rangle$. Suppose that $|\psi_b\rangle$, together with a continuum of unbound states $|\psi_\kappa\rangle$, form a complete set of orthonormal eigenstates of the unperturbed time-independent Hamiltonian H_0 of the system, i.e.

$$H_0|\psi_\kappa\rangle = E_\kappa|\psi_\kappa\rangle. \tag{2.1}$$

Here κ stands collectively for all the indices labelling the eigenstates of H_0. If a time-dependent perturbation $H_1(t)$ is applied to the system at time t_0, then the state at time t may be expressed in terms of the set $|\psi_\kappa\rangle$:

$$|\psi(t)\rangle = \int C_\kappa(t)\exp\left(-\frac{iE_\kappa t}{\hbar}\right)|\psi_\kappa\rangle\, d\kappa + C_b(t)\exp\left(-\frac{iE_b t}{\hbar}\right)|\psi_b\rangle. \tag{2.2}$$

The integration with respect to κ is to be interpreted as an integration over ω_k, where

$$\hbar\omega_k = E_\kappa - E_b \geqslant 0, \qquad E_\kappa \equiv E_{k\mu}$$

and a summation over μ, where μ denotes the set of all quantum numbers other than k. The coefficients $C_\kappa(t)$ are given by the familiar formula of first-order perturbation theory

$$C_\kappa(t) = \frac{1}{i\hbar}\int_{t_0}^{t} \langle\psi_\kappa|H_1(t')|\psi_b\rangle \exp(i\omega_k t')\, dt'. \tag{2.3}$$

Let $\rho(\omega_k)\, d\omega_k$ be the number of states $|\psi_\kappa\rangle$ in the energy interval $\hbar\, d\omega_k$. Then the probability that a transition has occurred to any of the unbound states by the time t is

$$\int_0^\infty |C_\kappa(t)|^2\rho(\omega_k)\, d\kappa.$$

If $t-t_0 = \Delta t$ we define a transition probability per unit time by

$$\Pi(t) = \frac{1}{\Delta t}\int_0^\infty |C_\kappa(t)|^2 \rho(\omega_k)\, d\kappa. \tag{2.4}$$

Both $C_\kappa(t)$ and $\Pi(t)$ depend also on the position of the atomic nucleus (to be specified by position vector $\mathbf{R}$ later on).

Let us now consider the interaction between a *single* atom and an incident electromagnetic wave, represented by the vector potential $\mathbf{A}(\mathbf{r}, t)$. Let the momentum of a typical electron of the atom be represented by $\mathbf{p}$. Then the interaction Hamiltonian, in the usual notation is

$$H_1(t) = \frac{e}{mc}\mathbf{A}(\mathbf{r}, t).\mathbf{p}. \tag{2.5}$$

We now express $\mathbf{A}(\mathbf{r}, t)$ in the form of a Fourier integral

$$\mathbf{A}(\mathbf{r}, t) = \int_{-\infty}^{\infty} \mathscr{A}(\mathbf{r}, \omega)\, e^{-i\omega t}\, d\omega. \tag{2.6}$$

Since $\mathbf{A}$ is real,

$$\mathscr{A}(\mathbf{r}, -\omega) = \mathscr{A}^*(\mathbf{r}, \omega). \tag{2.7}$$

Then, from (2.3), (2.5) and (2.6), we obtain

$$\begin{aligned} C_\kappa(t) &= \frac{e}{i\hbar mc}\int_{t_0}^{t_0+\Delta t} dt' \int_{-\infty}^{\infty} d\omega \exp\{i(\omega_k-\omega)t'\} M_\mu(\mathbf{R}, \omega, \omega_k) \\ &= \frac{e\Delta t}{i\hbar mc}\int_{-\infty}^{\infty} d\omega \exp\{i(\omega_k-\omega)(t_0+\tfrac{1}{2}\Delta t)\} \frac{\sin\{\frac{1}{2}(\omega_k-\omega)\Delta t\}}{\frac{1}{2}(\omega_k-\omega)\Delta t} M_\mu(\mathbf{R}, \omega, \omega_k) \end{aligned} \tag{2.8}$$

where

$$M_\mu(\mathbf{R}, \omega, \omega_k) = \langle \psi_\kappa | \mathscr{A}(\mathbf{r}, \omega).\mathbf{p} | \psi_b \rangle \tag{2.9}$$

is the matrix element between the ground state and the continuum state and is a function of $\mathbf{R}$, the position of the atomic nucleus. Now provided $\Delta t \gg 1/\omega_k$ for all k for which $|\mathbf{M}_\mu(\mathbf{R}, \omega, \omega_k)|^2$ is appreciable, the integrand considered as a function of ω_k will be very sharply peaked about ω. In practice the smallest value of ω_k likely to matter when we are dealing with photoelectric transitions is of the order of 10^{14} c/s. The condition $\omega_k \Delta t \gg 1$ is therefore likely to hold for all measurable intervals Δt. The contribution to $C_\kappa(t)$ arises from values of ω in the neighbourhood of ω_k, and since $\omega_k > 0$ we may replace the lower limit in (2.8) by zero. From (2.4) and (2.8),

$$\begin{aligned} \Pi(t) = \frac{e^2\Delta t}{\hbar^2 m^2 c^2}\int_0^\infty\int_0^\infty\int_0^\infty d\omega d\omega' d\omega_k \exp\{i(\omega'-\omega)(t_0+\tfrac{1}{2}\Delta t)\} \frac{\sin\{\frac{1}{2}(\omega_k-\omega)\Delta t\}}{\frac{1}{2}(\omega_k-\omega)\Delta t} \\ \times \frac{\sin\{\frac{1}{2}(\omega_k-\omega')\Delta t\}}{\frac{1}{2}(\omega_k-\omega')\Delta t} \sum_\mu \rho(\omega_k) M_\mu^*(\mathbf{R}, \omega', \omega_k) M_\mu(\mathbf{R}, \omega, \omega_k). \end{aligned} \tag{2.10}$$

It is evident that the integrand effectively vanishes unless ω, ω' and ω_k are nearly equal, to within an amount of the order of $1/\Delta t$. It is reasonable to suppose that $\rho(\omega_k)$ does not vary significantly over such a small range of ω_k and may therefore be replaced by $\{\rho(\omega)\rho(\omega')\}^{1/2}$ in (2.10).

Next let us assume that the incident radiation is quasi-monochromatic, i.e. its effective bandwidth is small compared with the mid-frequency. Then we can choose $1/\Delta t$ small compared with any frequency ω that contributes to (2.10), but large compared with the difference $\omega-\omega'$ between any pair of frequencies ω and ω', and (2.10) then reduces to

$$\Pi(t) = \frac{2\pi e^2}{\hbar^2 m^2 c^2} \int_0^\infty \int_0^\infty d\omega d\omega' \exp\{i(\omega'-\omega)t\}\{\rho(\omega)\}^{1/2}\{\rho(\omega')\}^{1/2} \frac{\sin\{\frac{1}{2}(\omega-\omega')\Delta t\}}{\frac{1}{2}(\omega-\omega')\Delta t} \times \sum_\mu M_\mu^*(\mathbf{R}, \omega', \omega') M_\mu(\mathbf{R}, \omega, \omega). \tag{2.11}$$

Now we have the identity

$$\frac{\sin\{\frac{1}{2}(\omega-\omega')\Delta t\}}{\frac{1}{2}(\omega-\omega')\Delta t} = \frac{1}{\Delta t}\int_{-\Delta t/2}^{\Delta t/2} \exp\{i(\omega'-\omega)\tau\}\, d\tau.$$

Hence the probability of a photoelectric transition per unit time may be expressed in the form

$$\Pi(t) = \frac{2\pi e^2}{\hbar^2 m^2 c^2} \sum_\mu \frac{1}{\Delta t} \int_{-\Delta t/2}^{\Delta t/2} W_\mu^*(\mathbf{R}, t+\tau) W_\mu(\mathbf{R}, t+\tau)\, d\tau \tag{2.12a}$$

where

$$W_\mu(\mathbf{R}, t) = \int_0^\infty d\omega\, e^{-i\omega t}\{\rho(\omega)\}^{1/2} M_\mu(\mathbf{R}, \omega, \omega). \tag{2.12b}$$

Since the Fourier spectrum of $W_\mu(\mathbf{R}, t)$ contains no negative frequencies, W_μ is, according to a well-known theorem (Titchmarsh 1948), analytic and regular in the lower half of the complex t plane.

So far we have been considering the interaction between an incident electromagnetic field and a single atom only. Now suppose that we are dealing with a plane wave incident normally on an extended detector, which is in the form of a thin photoelectric layer containing a large number of electrons in initial states $|\psi_b\rangle$. With the assumption that these atoms may be treated as independent (i.e. that their electron wave functions do not appreciably overlap) and that the states are not appreciably depopulated, we can express the probability $P(t)\Delta t$ of photoelectric detection from any part of the photo-surface by

$$P(t) = \frac{2\pi e^2 N}{\hbar^2 m^2 c^2 \Delta t} \sum_\mu \int_{-\Delta t/2}^{\Delta t/2} W_\mu^*(\mathbf{R}, t+\tau) W_\mu(\mathbf{R}, t+\tau)\, d\tau \tag{2.13}$$

where N is the number of effective electrons. Since we are dealing with an incident plane wave, the right-hand side of (2.13) is independent of $\mathbf{R}$.

In the dipole approximation we may express the matrix element M_μ, given by (2.9), in the form

$$M_\mu(\mathbf{R}, \omega, \omega_k) \sim \mathscr{A}(\mathbf{R}, \omega) . \langle\psi_\kappa|\mathbf{p}|\psi_b\rangle.$$

If $\rho(\omega)$ and the matrix element $\langle\psi_\kappa|\mathbf{p}|\psi_b\rangle$ are effectively independent of ω over the narrow frequency band of the incident light, and if the incident wave is plane as we assumed, then (2.12b) reduces to

$$W_\mu(\mathbf{R}, t) = \{\rho(\omega_0)\}^{1/2} \mathbf{V}(\mathbf{R}, t) . \langle\psi_{\mu,\kappa_0}|\mathbf{p}|\psi_b\rangle \tag{2.14}$$

where $\omega_0 = k_0 c$ is the mid-frequency, and $\mathbf{V}(\mathbf{R}, t)$ is the vector function obtained from $\mathbf{A}(\mathbf{R}, t)$ by suppressing the negative frequency components in the Fourier integral† (2.6):

$$\mathbf{V}(\mathbf{R}, t) = \int_0^\infty \mathscr{A}(\mathbf{R}, \omega)\, e^{-i\omega t}\, d\omega. \tag{2.15}$$

† Such a representation of the field, obtained by suppressing the negative Fourier components is customarily employed in the classical theory of optical coherence, under the name of *analytic signal*, a concept due to D. Gabor (cf. Born and Wolf 1959). The same representation has played an important role already in the early quantum mechanical investigations of radiation and coherence based on the correspondence principle, and is also implicit in some older pioneering researches of von Laue relating to coherence and thermodynamics of light.

From Maxwell's equation $\mathbf{V}(\mathbf{R}, t)$ is transverse, i.e. normal to the direction of propagation of the wave. Let us write

$$\mathbf{V}(\mathbf{R}, t) = V(\mathbf{R}, t)\boldsymbol{\epsilon} \tag{2.16}$$

where $\boldsymbol{\epsilon}$ is a unit (generally complex) vector and $V(\mathbf{R}, t)$ is a complex scalar function. Now one may readily show that for a time interval Δt, which is short compared with the reciprocal of the effective bandwidth of the light (as is here assumed),

$$\frac{1}{\Delta t}\int_{-\Delta t/2}^{\Delta t/2} V^*(\mathbf{R}, t+\tau)V(\mathbf{R}, t+\tau)\,d\tau \simeq V^*(\mathbf{R}, t)V(\mathbf{R}, t).$$

With the aid of this result and (2.14) and (2.16), (2.13) becomes

$$P(t) = \frac{2\pi e^2 N}{\hbar^2 m^2 c^2}\rho(\omega_0)V^*(\mathbf{R}, t)V(\mathbf{R}, t)\sum_{\mu} |\boldsymbol{\epsilon}\,.\,\langle\psi_{\mu,k_0}|\mathbf{p}|\psi_{\mathrm{b}}\rangle|^2. \tag{2.17}$$

Because Σ_μ involves summation over all possible polarizations of the electron the result will be independent of $\boldsymbol{\epsilon}$, so that we may write (2.17) in the form

$$P(t) = \alpha\mathbf{V}^*(\mathbf{R}, t).\mathbf{V}(\mathbf{R}, t) \tag{2.18a}$$

where α represents the quantum efficiency of the photoelectric detector. This result does not depend in an essential way on all the simplifying assumptions made. If there is a whole range of initial electron states $|\psi_{\mathrm{b}}\rangle$, the factor N in (2.17) has to be replaced by a sum over these states. On the other hand, if the electron wave functions overlap, as in a metal, the electron system has to be treated appropriately, and the calculation must be modified. Nevertheless, as long as we are dealing with plane waves falling normally on a thin photoelectric layer, a factorization of the kind embodied in equation (2.14) will still be permissible. The general form of (2.18*a*) will therefore remain valid, although the total cross section for the process, and therefore the constant α, will be affected.

We may identify $\mathbf{V}^*(\mathbf{R}, t)\,.\,\mathbf{V}(\mathbf{R}, t)$ with the instantaneous intensity† $I(t)$ of the classical field, and express (2.18*a*) in the form

$$P(t)\Delta t = \alpha I(t)\Delta t. \tag{2.18b}$$

The probability of photoemission of an electron is therefore proportional to the classical measure of the instantaneous light intensity, defined in terms of the complex analytic signal. In the idealized case of strictly monochromatic radiation this result is, of course, well known, but its generalization to a field which exhibits arbitrary fluctuations is essential for the purposes of the present discussion (see also Brown and Twiss 1957 a).

It should be noted that in equation (2.18) probability enters in two different ways: in the fundamental uncertainties associated with the photoelectric interaction and in the fluctuations of the radiation field itself. This, of course, is a general feature of quantum statistical mechanics (cf. Landau and Lifschitz 1958).

Equation (2.18) shows that the probability of a single photoelectric transition in a small time interval $t, t+\Delta t$ is proportional to Δt. However, we are mainly interested in the probability distribution $p(n, t, T)$ of emission of n photoelectrons in a finite time interval $t, t+T$. If the different photoelectric emissions could be considered as independent statistical events in the sense of classical probability theory, it would follow from (2.18)

† $\mathbf{V}^*\,.\,\mathbf{V}$ is not strictly proportional to the instantaneous energy density; it may be easily shown that $\mathbf{V}^*\,.\,\mathbf{V}$ represents a short-time average of $\mathbf{A}^2$ taken over a time interval of a few mean periods of the light vibrations.

that (see Mandel 1959, 1963 a)

$$p(n, t, T) = \frac{1}{n!}\{\alpha U(t, T)\}^n \exp\{-\alpha U(t, T)\} \tag{2.19}$$

where

$$U(t, T) = \int_t^{t+T} I(t')\, dt'. \tag{2.20}$$

Actually, it is possible to see that it is legitimate to proceed in this way. Consider, for example, a system of two atoms, both of which interact with the incident radiation but do not interact with each other. The product of the unperturbed energy eigenfunctions of the two individual atoms are the energy eigenfunctions of the two-atom system. It is now possible to calculate the probability amplitude for a transition to a final state in which either one or two photoelectrons have been emitted. In particular, the probability amplitude for emission of two photoelectrons in a small time interval $(t, t+\Delta t)$ may be shown to be given by the product of two expressions of the type (2.8). When we take account of the (infinite) degeneracy of the two-atom energy eigenstates, the probability for such a transition in this time interval may then be shown to be given by a product of two factors† of the type (2.10). In any case it seems plausible to look on successive photoelectric emissions from the whole photoelectric surface considered as one system as events which are substantially independent with respect to the electron system, provided the photoelectric layer is not appreciably depopulated. Similar assumptions are implicit in the derivaton of (2.19) from (2.18) by Mandel (1963 a).

Equation (2.19) refers to the photoelectric counting distribution appropriate to a single realization of the incident electromagnetic field. The average of $p(n, t, T)$ over the ensemble of the incident fields is the probability that would normally be derived from counting experiments. If, as is usually the case, $I(t)$ represents a stationary ergodic process, this average will be independent of t. If we denote the ensemble average by $\bar{p}(n, t, T)$, we have from (2.19)

$$\bar{p}(n, t, T) = \frac{1}{n!}\overline{\{\alpha U(t, T)\}^n \exp\{-\alpha U(t, T)\}} \tag{2.21}$$

which, in general, will *not* be a Poisson distribution.

It should be noted that for radiation fields in some states, for example in an eigenstate of the number operator, the probability distribution of U will exhibit somewhat unusual properties, not normally encountered in classical theory. However, in view of a theorem relating to the equivalence of the semi-classical and quantum representations of light beams, established recently by Sudarshan (1963 a, b), such probability distributions can in principle nevertheless always be found.

3. Statistical properties of the counting distribution

Let us next consider the variance of the counts n, recorded in time intervals of duration T. The averages of n and n^2 are given by

$$\bar{n} = \sum_{n=0}^{\infty} n\bar{p}(n, t, T) \tag{3.1}$$

$$\overline{n^2} = \sum_{n=0}^{\infty} n^2\bar{p}(n, t, T). \tag{3.2}$$

† This situation must be contrasted with the situation in which a direct two-electron transition from a single atom takes place.

Using well-known expressions for the first two moments of the Poisson distribution (see, for example, Levy and Roth 1951), we readily find from (3.1), (3.2) and (2.19) that

$$\bar{n} = \alpha \overline{U(t, T)} \tag{3.3}$$

$$\overline{n^2} = \alpha \overline{U(t, T)} + \alpha^2 \overline{\{U(t, T)\}^2} \tag{3.4}$$

so that the variance

$$\overline{(\Delta n)^2} = \overline{(n-\bar{n})^2} = \overline{n^2} - (\bar{n})^2$$

is given by

$$\overline{(\Delta n)^2} = \bar{n} + \alpha^2 \overline{(\Delta U)^2} \tag{3.5}$$

where

$$\overline{(\Delta U)^2} = \overline{(U-\bar{U})^2} = \overline{U^2} - (\bar{U})^2$$

is the variance of $U(t, T)$.

The formula (3.5) has evidently a very simple interpretation. It shows that the variance of the fluctuations in the number of ejected photoelectrons may be regarded as having two separate contributions: (i) from the fluctuations in the number of particles obeying the classical Poisson distribution (term $\bar{n}$), and (ii) from the fluctuations in a classical wave field (wave interference term $\alpha^2 \overline{(\Delta U)^2}$). This result, which holds for any radiation field, is strictly analogous to a celebrated result of Einstein (1909 a, b)† relating to energy fluctuations in an enclosure containing black-body radiation, under conditions of thermal equilibrium. Fürth (1928) has later shown that the same result holds for energy fluctuations of thermal radiation of spectral compositions other than that appropriate to black-body radiation, but like Einstein's analysis, Fürth's considerations apply to closed systems only. We have now shown that a fluctuation formula of this type is also valid for counting fluctuations in *time* intervals, for any light beam (i.e. thermal or non-thermal and stationary and non-stationary), at points that may be situated far away from the sources of the light. Although the result refers to the fluctuations of the photoelectric counts, it can be regarded as reflecting the fluctuation properties of the light itself, in so far as they are accessible to measurement.

Since equation (3.5) applies to any light beam, whether of thermal origin or not, it is likely to be useful in connection with photoelectric experiments relating to fluctuations of light generated by optical masers. In this connection we note that, if an optical maser operates on a single mode and is well stabilized, then the variance $\overline{(\Delta U)^2}$ will be negligible (absence of classical wave intensity fluctuations). In this case (3.5) reduces to‡

$$\overline{(\Delta n)^2} = \bar{n} \tag{3.6}$$

i.e. the variance is the same as for a system of classical particles.

We may draw some further conclusions from the results of § 2. First, let us again consider the case when the light intensity of the classical wave field does not fluctuate significantly. In this case (2.21) becomes

$$\bar{p}(n, t, T) = \frac{1}{n!} \bar{n}^n \exp(-\bar{n}) \tag{3.7}$$

† For a lucid account of the significance of Einstein's result, see Born (1949).

‡ It should be born in mind that the formula was derived on the basis of the first-order perturbation theory. When very intense maser beams are employed the effect of multiple photon interactions might have to be included.

where

$$\bar{n} = \alpha\bar{U} = \alpha\bar{I}T. \tag{3.8}$$

Thus we see that the photoelectrons now obey the Poisson distribution. This situation may be expected to arise in the case already referred to, namely when the photoemission is triggered off by a well-stabilized, single-mode laser beam. That in this case the distribution $\bar{p}(n, t, T)$ will be Poisson's was already noted elsewhere (Mandel 1964, Glauber 1964). This result clearly shows that departure from Poisson's statistics is not a universal consequence of the Bose–Einstein statistics of light quanta as is often erroneously believed to be the case†.

If, on the other hand, the light is of thermal origin, the probability distribution $\bar{p}(n, t, T)$ may be expected to be quite different. For in this case the incident field will fluctuate appreciably and its distribution will as a rule be Gaussian (van Cittert 1934, Blanc-Lapierre and Dumontet 1955, Janossy 1957, 1959). This implies (Mandel 1963 a, p. 191) that for polarized thermal light the probability distribution of the intensity is exponential:

$$p(I) = \frac{1}{\bar{I}}\exp\left(\frac{-I}{\bar{I}}\right). \tag{3.9}$$

It may then be shown from (2.21), (3.9) and (3.8) that, if T is much smaller than the coherence time of the light, the probability distribution of the photoelectrons becomes (Mandel 1958, 1959, 1963 a)

$$\bar{p}(n, t, T) = \frac{\bar{n}^n}{(\bar{n}+1)^{n+1}}. \tag{3.10}$$

This will be recognized as the Bose–Einstein distribution (Morse 1962). It was derived by Bothe (1927) by a somewhat similar argument long ago.

4. Multiple correlations

The one-dimensional counting distributions do not exhaust the range of application of the semi-classical theory, for the relation (2.19) can be applied to any number N of photodetectors, each situated at a different point of the radiation field. Let n_j be the number of counts registered at the jth detector in a time interval t_j, $t_j + T$. Then

$$\overline{n_1 n_2 \dots n_N} = \sum_{n_1=0}^{\infty} \sum_{n_2=0}^{\infty} \dots \sum_{n_N=0}^{\infty} \overline{\{n_1 n_2 \dots n_N \prod_{j=1}^{N} p_j(n_j, t, T)\}} \tag{4.1}$$

where the $p_j(n_j, t, T)$ ($j = 1, 2 \dots N$) are given by expressions such as (2.19). Equation (4.1) may be rewritten in the form

$$\overline{n_1 n_2 \dots n_N} = \overline{\sum_{n_1=0}^{\infty} n_1 p_1(n_1, t_1, T) \sum_{n_2=0}^{\infty} n_2 p_2(n_2, t_2, T) \dots \sum_{n_N=0}^{\infty} n_N p_N(n_N, t_N, T)}. \tag{4.2}$$

Now each of the sums on the right-hand side of (4.2) represents, according to a well-known property of the Poisson distribution (2.19), the parameter of that distribution

† In this connection see the interesting discussion by Rosenfeld (1955, especially pp. 77–78).

$$\alpha_j U_j(t_j, T) = \sum_{n_j=0}^{\infty} n_j p_j(n_j, t_j, T). \tag{4.3}$$

Hence, if we substitute from (4.3) into (4.2), we obtain the formula

$$\overline{n_1 n_2 \ldots n_N} = A\overline{U_1(t_1, T) U_2(t_2, T) \ldots U_N(t_N, T)} \tag{4.4}$$

where

$$A = \alpha_1 \alpha_2 \ldots \alpha_N \tag{4.5}$$

represents the product of the quantum efficiencies of the N detectors. Equation (4.4) shows that *the correlation of the counts registered by the N photodetectors is proportional to the correlation in the integrated intensities* (cf. (2.20))

$$U_j(t_j, T) = \int_{t_j}^{t_j+T} I_j(t')\, dt' \qquad (j = 1, 2, \ldots N) \tag{4.6}$$

of the classical field at the location of the N detectors.

If we substitute from (4.6) into (4.4) and recall that $I_j(t) = \mathbf{V}_j^*(t) . \mathbf{V}_j(t)$ we readily find that

$$\overline{n_1 n_2 \ldots n_N} = A \int_{t_1}^{t_1+T} \int_{t_2}^{t_2+T} \ldots \int_{t_N}^{t_N+T} \Gamma^{(N,N)}(t_1', t_2', \ldots t_N')\, dt_1' dt_2' \ldots dt_N' \tag{4.7}$$

where

$$\Gamma^{(N,N)}(t_1, t_2, \ldots t_N) = \overline{\mathbf{V}_1^*(t_1) . \mathbf{V}_1(t_1) \ldots \mathbf{V}_N^*(t_N) . \mathbf{V}_N(t_N)}. \tag{4.8}$$

Thus the correlation of the counts is completely expressible in terms of the $2N$th order cross-correlation function of the classical field (cf. Mandel 1964, Wolf 1963, 1964). For a stationary field this correlation function will, of course, be independent of the origin of time and if, in addition, ergodicity is assumed, it can also be expressed in the form

$$\tilde{\Gamma}^{(N,N)}(\tau_2, \tau_3, \ldots \tau_N) = \lim_{T\to\infty} \frac{1}{2T} \int_{-T}^{T} \mathbf{V}_1^*(t) . \mathbf{V}_1(t) \mathbf{V}_2^*(t+\tau_2) . \mathbf{V}_2(t+\tau_2) \ldots \mathbf{V}_N^*(t+\tau_N) . \mathbf{V}_N(t+\tau_N)\, dt \tag{4.9}$$

where

$$\tau_j = t_j - t_1 \qquad (j = 2, 3, \ldots N).$$

In a quantized field-theoretical treatment the correlation $\overline{n_1 n_2 \ldots n_N}$ would be expressed in terms of the expectation value of the ordered product of the corresponding creation and annihilation operators (Glauber 1963 b, 1964). This expectation value has already been shown to be equivalent to a cross correlation of the complex classical fields (Sudarshan 1963 a, b, Mandel and Wolf 1965) and the formula (4.7) emphasizes this fact once again.

We can also convert (4.4) into a correlation formula for the fluctuations $\Delta n_j = n_j - \bar{n}_j$. By making a multinomial expansion of the product $\Delta n_1 \Delta n_2 \ldots \Delta n_N$ and applying (4.4) repeatedly, we obtain the formula

$$\overline{\Delta n_1 \Delta n_2 \ldots \Delta n_N} = A\overline{\Delta U_1 \Delta U_2 \ldots \Delta U_N} \tag{4.10}$$

where

$$\Delta U_j = U_j - \bar{U}_j.$$

The value of the correlation depends, of course, on the type of light illuminating the detectors. For light from the usual thermal sources the random process $\mathbf{V}(t)$ will, to a good approximation, be stationary, ergodic and Gaussian, and the correlations appearing on the right-hand side of equations (4.4), (4.7) and (4.10) can then be expressed in terms of second-order cross-correlation functions (Reed 1962). In particular, for

$N = 2$, (4.10) then represents the correlation effect discovered by Brown and Twiss (1956, 1957 a, b). However, from (4.10) it follows that the effect will be small when the fluctuations ΔU in the integrated classical intensity are small, as may be the case for light generated by an optical maser oscillating in a single mode.

Acknowledgments

It is a pleasure to acknowledge helpful discussion with Dr. C. L. Mehta about the subject matter of this paper.

References

ALKEMADE, C. T. J., 1959, *Physica*, **25**, 1145.
BLANC-LAPIERRE, A., and DUMONTET, P., 1955, *Rev. Opt. (Théor. Instrum.)*, **34**, 1.
BORN, M., 1949, *Natural Philosophy of Cause and Chance* (Oxford: Clarendon Press), p. 80.
BORN, M., and WOLF, E., 1959, *Principles of Optics* (London, New York: Pergamon Press), chap. X.
BOTHE, W., 1927, *Z. Phys.*, **41**, 345.
BROWN, R. HANBURY, and TWISS, R. Q., 1956, *Nature, Lond.*, **177**, 27.
—— 1957 a, *Proc. Roy. Soc.* A, **242**, 300.
—— 1957 b, *Proc. Roy. Soc.* A, **243**, 291.
VAN CITTERT, P. H., 1934, *Physica*, **1**, 201.
—— 1939, *Physica*, **6**, 1129.
EINSTEIN, A., 1909 a, *Phys. Z.*, **10**, 185.
—— 1909 b, *Phys. Z.*, **10**, 817.
FÜRTH, R., 1928, *Z. Phys.*, **50**, 310.
GLAUBER, R. J., 1963 a, *Phys. Rev. Letters*, **10**, 84.
—— 1963 b, *Phys. Rev. Letters*, **130**, 2529.
—— 1964, *Proc. 3rd Int. Congress on Quantum Electronics*, Eds N. Bloembergen and P. Grivet (Paris: Dunod; New York: Columbia University Press), p. 111.
JANOSSY, L., 1957, *Nuovo Cim.*, **6**, 14.
—— 1959, *Nuovo Cim.*, **12**, 369.
JAYNES, E. T., and CUMMINGS, F. W., 1963, *Proc. Inst. Elect. Electron. Engrs*, **51**, 89.
KAHN, F. D., 1958, *Optica Acta*, **5**, 93.
LANDAU, L. D., and LIFSHITZ, E. M., 1958, *Statistical Physics* (London: Pergamon Press; Reading, Mass.: Addison Wesley), p. 18.
LEVY, H., and ROTH, L., 1951, *Elements of Probability* (Oxford: Clarendon Press), p. 143.
MANDEL, L., 1958, *Proc. Phys. Soc.*, **72**, 1037.
—— 1959, *Proc. Phys. Soc.*, **74**, 233.
—— 1963 a, *Progress in Optics*, **2**, 181, Ed. E. Wolf (Amsterdam: North-Holland).
—— 1963 b, *Proc. Phys. Soc.*, **81**, 1104.
—— 1964, *Proc. 3rd Int. Congress on Quantum Electronics*, Eds N. Bloembergen and P. Grivet (Paris: Dunod; New York: Columbia University Press), p. 101.
MANDEL, L., and WOLF, E., 1963, *Phys. Rev. Letters*, **10**, 276
MANDEL, L., and WOLF, E., 1965, *Rev. Mod. Phys.*, in the press.
MORSE, P. M., 1962, *Thermal Physics* (New York: Benjamin), p. 218.
PURCELL, E. M., 1956, *Nature, Lond.*, **178**, 1449.
REED, I. S., 1962, *Trans. Inst. Radio Engrs*, **IT-8**, 194.
ROSENFELD, L., 1955, *Niels Bohr and the Development of Physics*, Ed. W. Pauli (London: Pergamon Press).
SUDARSHAN, E. C. G., 1963 a, *Phys. Rev. Letters*, **10**, 277.
—— 1963 b, *Proc. Symp. on Optical Masers* (New York: Brooklyn Polytechnic Press and John Wiley), p. 45.
TITCHMARSH, E. C., 1948, *Introduction to the Theory of Fourier Integrals*, 2nd edn (Oxford: Clarendon Press), p. 128.
WOLF, E., 1960, *Proc. Phys. Soc.*, **76**, 424.
—— 1963, *Proc. Symp. on Optical Masers* (New York: Brooklyn Polytechnic Press and John Wiley), p. 29.
—— 1964, *Proc. 3rd Int. Congress on Quantum Electronics*, Eds N. Bloembergen and P. Grivet (Paris: Dunod; New York: Columbia University Press), p. 13.

VOLUME 21, NUMBER 3 PHYSICAL REVIEW LETTERS 15 JULY 1968

QUANTUM DYNAMICS IN PHASE SPACE*

G. S. Agarwal and E. Wolf
Department of Physics and Astronomy, University of Rochester, Rochester, New York 14627
(Received 22 May 1968)

After a brief summary of recently derived general results relating to the mapping of functions of noncommuting operators on functions of c numbers, equations are given which describe the time evolution of the c-number equivalents (phase-space representations) of the density operator and of a Heisenberg operator. The evaluation of time-ordered functions of operators by c-number techniques is also briefly discussed.

Since the publication of the pioneering papers of Wigner,[1] Groenewold,[2] and Moyal[3] on the representation of quantum-mechanical systems in terms of generalized phase-space distribution functions, a considerable use has been made of such representations in the treatment of various problems. In the last few years, generalized phase-space descriptions have become of central importance in quantum optics, especially in the study of coherence properties of light[4] and in the theory of the laser.[5] As is well known, the phase-space representation of a quantum-mechanical system is not unique; it depends on the rule that is adopted for ordering of functions of noncommuting operators. We have recently developed a general technique for a systematic treatment of problems in this field, based on the use of certain new class of ordering operators. In this note we present the phase-space form of the basic quantum-mechanical equations of motion, which we have derived by the use of this technique.

We will first briefly explain the notation and summarize some of the main results given elsewhere.[6] We consider a correspondence between a function $F(z, z^*)$ of complex c-number vari-

VOLUME 21, NUMBER 3 PHYSICAL REVIEW LETTERS 15 JULY 1968

ables z and z^* (z^* denoting the complex conjugate of z) and a function[7] $G(\hat{a},\hat{a}^\dagger)$ of the boson annihilation and creation operators $\hat{a}$ and $\hat{a}^\dagger$, obeying the commutation relation

$$[\hat{a},\hat{a}^\dagger]=1. \tag{1}$$

The correspondence is characterized by a rule of association between elementary c-number products

$$F_{mn}(z,z^*)=z^m z^{*n} \tag{2}$$

and elementary functions $G_{mn}(\hat{a},\hat{a}^\dagger)$ of the boson operators. The choice of $\hat{G}_{mn}$ depends on the particular choice of the rule of association. For example, for association based on normal ordering,

$$G_{mn}(\hat{a},\hat{a}^\dagger)=\hat{a}^{\dagger n}\hat{a}^m. \tag{3}$$

We introduce a class of linear operators $\hat{\Omega}$, one for each rule of association, which map an arbitrary analytic function F, of the two complex variables $u=z$ and $v=z^*$,

$$F(z,z^*)=\sum_{m=0}^{\infty}\sum_{n=0}^{\infty}c_{mn}F_{mn}(z,z^*), \tag{4}$$

(the c_{mn}'s being c numbers) into the function of the boson operators

$$G(\hat{a},\hat{a}^\dagger)=\hat{\Omega}F(z,z^*), \tag{5}$$

where

$$G(\hat{a},\hat{a}^\dagger)=\sum_{m=0}^{\infty}\sum_{n=0}^{\infty}c_{mn}G_{mn}(\hat{a},\hat{a}^\dagger). \tag{6}$$

For a wide class of rules of association between F_{mn} and $\hat{G}_{mn}$, an explicit form of the mapping $F\to\hat{G}$ and of the inverse mapping $\hat{G}\to F$ may readily be obtained by means of the ordering delta operators[8]

$$\Delta^{(\Omega)}(z_0-\hat{a},z_0^*-\hat{a}^\dagger)=\hat{\Omega}\delta^{(2)}(z_0-z), \tag{7}$$

where $\delta^{(2)}$ is the two-dimensional Dirac delta function $\delta^{(2)}(z_0-z)=\delta(x_0-x)\delta(y_0-y)$, $z_0=x_0+iy_0$, and $z=x+iy$ (x_0,y_0,x,y, real). If one used the Fourier-integral representation of the Dirac delta function and the Baker-Hausdorff identity, one readily finds that for the usual rules of association, and for many others, $\hat{\Delta}^{(\Omega)}$ may be expressed in the following form:

$$\Delta^{(\Omega)}(z_0-\hat{a},z_0^*-\hat{a}^\dagger)=\frac{1}{\pi^2}\int\Omega(\alpha,\alpha^*)\times\exp\{\alpha(z_0^*-\hat{a}^\dagger)-\alpha^*(z_0-\hat{a})\}d^2\alpha, \tag{8}$$

where the integration extends over the whole complex α plane. The function $\Omega(\alpha,\beta)$ in the integrand on the right-hand side of (8), which depends on the particular rule of association, is an entire analytic function of the two complex variables α and β, which has no zeros and is such that $\Omega(-\alpha,-\beta)=\Omega(\alpha,\beta)$ and $\Omega(0,0)=1$. The form of $\Omega(\alpha,\alpha^*)$ for the most commonly used rules of association is given in Table I.

The mapping $F(z,z^*)\to G(\hat{a},\hat{a}^\dagger)$, according to a rule of association characterized by a particular form of the function $\Omega(\alpha,\alpha^*)$ and the inverse mapping $G(\hat{a},\hat{a}^\dagger)\to F(z,z^*)$ may then be shown to be expressible by the two formulas

$$G(\hat{a},\hat{a}^\dagger)=\int F(z_0,z_0^*)\Delta^{(\Omega)}(z_0-\hat{a},z_0^*-\hat{a}^\dagger)d^2z_0, \tag{9a}$$

$$F(z_0,z_0^*)=\pi\,\mathrm{Tr}\{G(\hat{a},\hat{a}^\dagger)\Delta^{(\bar{\Omega})}(z_0-\hat{a},z_0^*-\hat{a}^\dagger)\}. \tag{9b}$$

In Eq. (9a) the integration extends over the whole complex z plane. In (9b), $\hat{\Delta}^{(\bar{\Omega})}$ denotes the $\hat{\Delta}$ operator for ordering reciprocal to that defined by $\hat{\Omega}$:

$$\Delta^{(\bar{\Omega})}(z_0-\hat{a},z_0^*-\hat{a}^\dagger)=\frac{1}{\pi^2}\int[\Omega(\alpha,\alpha^*)]^{-1}\times\exp\{\alpha(z_0^*-\hat{a}^\dagger)-\alpha^*(z_0-\hat{a})\}d^2\alpha. \tag{10}$$

Table I. The form of the function $\Omega(\alpha,\alpha^*)$ [Eq. (8)] and the coefficients μ,ν,λ [Eq. (11)] for commonly used rules of association.

Rule of association	$\Omega(\alpha,\alpha^*)$	μ	ν	λ
Normal	$\exp(\frac{1}{2}\alpha\alpha^*)$	0	0	$\frac{1}{2}$
Antinormal	$\exp(-\frac{1}{2}\alpha\alpha^*)$	0	0	$-\frac{1}{2}$
Weyl	1	0	0	0
Standard	$\exp\{\frac{1}{4}(\alpha^2-\alpha^{*2})\}$	$\frac{1}{4}$	$-\frac{1}{4}$	0
Antistandard	$\exp\{-\frac{1}{4}(\alpha^2-\alpha^{*2})\}$	$-\frac{1}{4}$	$\frac{1}{4}$	0

We shall refer to the c-number function $F(z, z^*)$ corresponding to a given function $G(\hat{a}, \hat{a}^\dagger)$ of the boson operators as the Ω equivalent of $\hat{G}$ or the phase-space representation of $\hat{G}$ for Ω association and will occasionally write $F_G(z, z^*)$ in place of $F(z, z^*)$. $G(\hat{a}, \hat{a}^\dagger)$ may, of course, be expressed in many different, but equivalent, forms with the help of the commutation relation (1). For any particular rule of association there is, however, one and only one Ω equivalent F_G of $\hat{G}$, irrespective of the particular form of $\hat{G}$. We have indicated elsewhere[6,9] how the formulas (9) may be used to give the "Ω-ordered form" of a given function $G(\hat{a}, \hat{a}^\dagger)$ and how the trace of the product of two functions $G_1(\hat{a}, \hat{a}^\dagger)$ and $G_2(\hat{a}, \hat{a}^\dagger)$ may be expressed as an integral in the phase space.

Let us now restrict ourselves to the class of associations for which the functions $\Omega(\alpha, \alpha^*)$ have the form

$$\Omega(\alpha, \alpha^*) = \exp(\mu\alpha^2 + \nu\alpha^{*2} + \lambda\alpha\alpha^*), \tag{11}$$

where μ, ν, and λ are constants. As is seen from Table I, the usual rules of association belong to this class.

We introduce two differential operators

$$\hat{\Lambda}_1 = -2\nu \frac{\overleftarrow{\partial}}{\partial z}\frac{\overrightarrow{\partial}}{\partial z} - 2\mu \frac{\overleftarrow{\partial}}{\partial z^*}\frac{\overrightarrow{\partial}}{\partial z^*} + \lambda\left(\frac{\overleftarrow{\partial}}{\partial z^*}\frac{\overrightarrow{\partial}}{\partial z} + \frac{\overleftarrow{\partial}}{\partial z}\frac{\overrightarrow{\partial}}{\partial z^*}\right), \tag{12}$$

$$\hat{\Lambda}_2 = \frac{1}{2}\left(\frac{\overleftarrow{\partial}}{\partial z}\frac{\overrightarrow{\partial}}{\partial z^*} - \frac{\overleftarrow{\partial}}{\partial z^*}\frac{\overrightarrow{\partial}}{\partial z}\right), \tag{13}$$

where the arrow pointing to the left (right) indicates that the differential operator below it operates on quantities on the left (right) of the $\hat{\Lambda}$ operators. By rather lengthy calculations one can then show that if $F_1(z, z^*)$ and $F_2(z, z^*)$ are the Ω equivalents of $G_1(\hat{a}, \hat{a}^\dagger)$ and $G_2(\hat{a}, \hat{a}^\dagger)$, respectively, the Ω equivalent $F_{12}(z, z^*)$ of the product $G_1(\hat{a}, \hat{a}^\dagger)G_2(\hat{a}, \hat{a}^\dagger)$ may be expressed in the form

$$F_{12}(z, z^*) = F_1(z, z^*)\exp\{\hat{\Lambda}_1 + \hat{\Lambda}_2\}F_2(z, z^*), \tag{14}$$

and the Ω equivalent $F_{21}(z, z^*)$ of the product $G_2(\hat{a}, \hat{a}^\dagger)G_1(\hat{a}, \hat{a}^\dagger)$ may be expressed in the form

$$F_{21}(z, z^*) = F_1(z, z^*)\exp\{\hat{\Lambda}_1 - \hat{\Lambda}_2\}F_2(z, z^*). \tag{15}$$

Consider now the Schrödinger equation of motion for the density operator $\rho(\hat{a}, \hat{a}^\dagger, t)$ of a quantum-mechanical system,

$$i\hbar\partial\hat{\rho}/\partial t = [\hat{H}, \hat{\rho}], \tag{16}$$

where $\hat{H}$ is the Hamiltonian of the system. Taking the Ω equivalent of both sides of (16) and using (14) and (15), we readily obtain the equation

$$i\hbar\partial F_\rho/\partial t = F_H \exp(\hat{\Lambda}_1)\{\exp(\hat{\Lambda}_2) - \exp(-\hat{\Lambda}_2)\}F_\rho. \tag{17}$$

In a similar way we obtain from the Heisenberg equation of motion for an operator $G(\hat{a}, \hat{a}^\dagger)$, assumed to be explicitly independent of time,

$$i\hbar d\hat{G}/dt = -[\hat{H}, \hat{G}], \tag{18}$$

the equation of motion for the Ω equivalent of $\hat{G}$:

$$i\hbar dF_G/dt = -F_H \exp(\hat{\Lambda}_1)\{\exp(\hat{\Lambda}_2) - \exp(-\hat{\Lambda}_2)\}F_G. \tag{19}$$

There are numerous interesting consequences of Eqs. (17) and (19), which will be discussed elsewhere. We only note two immediate consequences of our analysis.

Let $\hat{U}(t, t_0)$ be the unitary evolution operator of a quantum system with a time-dependent Hamiltonian $\hat{H}(t)$. Then $\hat{U}$ obeys the Schrödinger equation[10]:

$$i\hbar\frac{\partial\hat{U}}{\partial t} = \hat{H}\hat{U}. \tag{20}$$

The formal solution of (20), subject to the boundary condition $\hat{U}(t_0, t_0) = \hat{1}$, where $\hat{1}$ is the unit operator, is

$$\hat{U}(t, t_0) = \hat{T}\exp\left\{-\frac{i}{\hbar}\int_{t_0}^{t}\hat{H}(t')dt'\right\}, \tag{21}$$

with $\hat{T}$ denoting the time-ordering operator. Let us now take the Ω equivalent of Eq. (20). With the help of (14) it is seen to be given by

$$i\hbar\frac{\partial F_U}{\partial t} = F_H \exp\{\hat{\Lambda}_1 + \hat{\Lambda}_2\}F_U, \tag{22}$$

which is to be solved for $F_U(t, t_0)$ subject to the boundary condition $F_U(t_0, t_0) = 1$, for all values of z and z^*. Equation (22) is a partial differential equation which does not involve any time ordering. In terms of the solution F_U of (22), the unitary operator U is in accordance with the general correspondence rule (5) given by

$$\hat{U}(t, t_0) = \hat{\Omega}F_U(t, t_0). \tag{23}$$

From (21) and (23) one has the identity

$$\hat{T}\exp\left\{-\frac{i}{\hbar}\int_{t_0}^{t}\hat{H}(t')dt'\right\} = \hat{\Omega}F_U(t, t_0). \tag{24}$$

VOLUME 21, NUMBER 3 PHYSICAL REVIEW LETTERS 15 JULY 1968

In the special case when $\hat{\Omega}$ represents the rule of association based on normal ordering, the identity (24) is at the root of the normal ordering technique developed in recent years for solving quantum-dynamical problems with time-dependent Hamiltonian.[11] It is evident that in this case (with $\hat{\Omega}$ representing the normal ordering association) the identity (24) plays a similar role in the present theory as Wick's theorem plays in quantum field theory.

Returning to the Schrödinger equation (16) for the density operator, consider the case, often encountered in practice, when the Hamiltonian is a quadratic function of $\hat{a}$ and $\hat{a}^\dagger$, i.e., when it is of the form

$$H=\omega\hat{a}^\dagger\hat{a}+\gamma\hat{a}+\gamma^*\hat{a}^\dagger+\delta\hat{a}^2+\delta^*\hat{a}^{\dagger 2}, \tag{25}$$

where ω is a real c number and γ and δ are complex c numbers, all of which may depend on time. The Ω equivalent [within the class of rules of association characterized by Eq. (11)] F_H and $\hat{H}$ may then be shown to be quadratic in z and z^*, and the phase-space equation of motion (17) reduces to

$$i\hbar\frac{\partial F_\rho}{\partial t}=A\frac{\partial^2 F_\rho}{\partial z^2}+B\frac{\partial^2 F_\rho}{\partial z^{*2}}+C\frac{\partial^2 F_\rho}{\partial z\partial z^*}+D\frac{\partial F_\rho}{\partial z}-D^*\frac{\partial F_\rho}{\partial z^*}, \tag{26}$$

where

$$A=2\nu\omega-2\lambda\delta^*,\quad B=-2\mu\omega+2\lambda\delta,$$

$$C=-4\nu\delta+4\mu\delta^*,\quad D=-\omega z-\gamma^*-2\delta^*z^*. \tag{27}$$

Equation (26) has the form of the Fokker-Planck equation. Such equations have recently been encountered in the theory of the laser.[5]

It is also possible to transcribe the Bloch equation[12] for the density operator of a system in thermal equilibrium into an equivalent equation in phase space. All these equations of motion in phase space have the form of a generalized Liouville equation and they can be treated by perturbation techniques (see, for example, Zwanzig[13]). One can also study by the phase-space technique the motion of a system in contact with a reservoir, and one may derive exact master equations for the Ω equivalent of the density operator of the system of interest. Some of these extensions and applications of our theory as well as generalizations of our results to systems with many degrees of freedom will be discussed elsewhere.

---

*Research supported by the U. S. Air Force Office of Scientific Research.

[1]E. Wigner, Phys. Rev. 40, 749 (1932).

[2]H. J. Groenewold, Physica 12, 405 (1946).

[3]J. E. Moyal, Proc. Cambridge Phil. Soc. 45, 99 (1949).

[4]For a review of some of this work see, for example, L. Mandel and E. Wolf, Rev. Mod. Phys. 37, 231 (1965).

[5]See, for example, H. Haken, H. Risken, and W. Weidlich, Z. Physik 206, 355 (1967); and M. Lax and W. H. Louisell, IEEE J. Quantum Electron. QE-3, 47 (1967).

[6]E. Wolf and G. S. Agarwal, to be published.

[7]Throughout this note circumflex denotes an operator. Functions of operators, e.g., $G(\hat{a},\hat{a}^\dagger)$ are sometimes denoted by $\hat{G}$, etc.

[8]These ordering delta operators should be distinguished from those recently introduced by M. Lax, to be published.

[9]G. S. Agarwal and E. Wolf, Phys. Letters 26A, 485 (1968). In this reference we considered $\hat{\Omega}$ to operate on functions of operators, but subsequent work showed that it is preferable to define $\hat{\Omega}$ as operating on functions of c numbers. Also $f^{(\Omega)}(\hat{a},\hat{a}^\dagger)$ of that reference corresponds to $F(z,z^*)$ of the present paper.

[10]It is understood, of course, that in addition to the dependence on time, $\hat{U}$ and $\hat{H}$ are also functions of $\hat{a}$ and $\hat{a}^\dagger$ and F_U and F_H are functions of z and z^*.

[11]See, for example, H. Heffner and W. H. Louisell, J. Math. Phys. 6, 474 (1965); M. Lax, Phys. Rev. 157, 213 (1967).

[12]Cf. T. Matsubara, Progr. Theoret. Phys. (Kyoto) 14, 351 (1955).

[13]R. Zwanzig, in Lectures in Theoretical Physics, edited by W. E. Brittin, B. W. Downs, and J. Downs (Interscience Publishers, Inc., New York, 1961), Vol. 3, p. 106.

Ordering of Operators and Phase Space Descriptions in Quantum Optics*

E. WOLF AND G. S. AGARWAL
Department of Physics and Astronomy,
University of Rochester, Rochester, N. Y., U.S.A.

ABSTRACT

Several new theorems are presented relating to correspondence between functions of *c*-numbers and functions of non-commuting operators, to ordering of operator functions according to prescribed rules and to generalized phase space descriptions. These theorems make it possible to derive in a systematic way many results that have previously been obtained by *ad hoc* methods. As an illustration we derive from first principles the Sudarshan-Glauber diagonal coherent state representation of operators. Some examples are also given concerning the *c*-number mapping of functions of boson operators according to the Weyl, the normal and the anti-normal rules of ordering.

1. **Introduction**

The problem of expressing functions of operators according to some prescribed rule of ordering and the closely related problem of evaluating quantum mechanical expectation values as averages with respect to some generalized phase space distribution function** is of interest in many branches of physics. The first such generalized phase space distribution function was introduced by Wigner [2] in a study of quantum mechanical corrections for thermodynamic equilibrium. Moyal [3] showed

* Research supported by the Air Force Office of Scientific Research.

** By a generalized phase space distribution function we mean here a function which has many of the properties of a true distribution function of classical statistical theory but is not necessarily non-negative and may become singular. Such functions are, therefore, not true probabilities, in general, but they make it possible to treat many quantum mechanical problems by quasi-classical techniques [1].

later in an important paper dealing with the interpretation of quantum mechanics as a statistical theory that Wigner's choice of the generalized phase space distribution function is intimately related to a certain "symmetrization rule of correspondence" between c-numbers and quantum mechanical operators proposed earlier by Weyl [4] from group theoretical considerations [5-7]. Since that time, the Wigner distribution function has been employed in many problems of statistical mechanics [see, for example, ref. 6, 8, 9]. Other rules of correspondence have also been studied [10-12]. More recently problems of c-number descriptions of quantum mechanical systems have become of central importance in studies of coherence properties of light [13-15] and in investigations relating to the theory of laser oscillations [16, 17]. Numerous results in this area and many related results concerning the ordering of operators [17, 18-20] and various phase space descriptions [21-23] have been derived by *ad hoc* methods.

We have recently found a systematic way of treating problems of this type. In this paper we present several general theorems relating to the correspondence between functions of c-numbers and functions of non-commuting operators, to ordering of operators and to generalized phase space descriptions and we show how some of the previously derived results follow readily from these theorems. For the sake of brevity we consider here only systems with one degree of freedom which satisfy equal time commutation relations. However the methods can readily be extended to systems with arbitrary number of degrees of freedom, to dynamical problems [33] and to various problems of multi-time correspondence between quantum mechanical and classical stochastic processes [24]. These extensions as well as the derivation of the theorems given in the present paper will be given elsewhere.

2. Correspondence between c-Number Functions and Functions of Operators

Let p and q be real variables and $\hat{p}$ and $\hat{q}$ Hermitian operators satisfying the commutation relation :

$$[\hat{q}, \hat{p}] = i\hbar. \tag{2.1}$$

We consider some c-number function $F(p, q)$ and will associate with it a function $G(\hat{p}, \hat{q})$ of the operators $\hat{p}$ and $\hat{q}$ according to some prescribed rule. Before we define such an association with some generality, we summarize some of the better known rules of association between elementary c-number functions and operators [4, 5, 10-12] :

Rule of association	Correspondence
Standard	$p^m q^n \to \hat{q}^n \hat{p}^m$
Anti-standard	$p^m q^n \to \hat{p}^m \hat{q}^n$
Weyl's symmetrization	$p^m q^n \to (\hat{p}^m \hat{q}^n)_W$
Rivier's symmetrization	$p^m q^n \to 1/2\{\hat{p}^m \hat{q}^n + \hat{q}^n \hat{p}^m\}$
Born-Jordan	$p^m q^n \to \frac{1}{(m+1)} \sum_{l=0}^{m} \hat{p}^{m-l} \hat{q}^n \hat{p}^l$

The symbol $(\hat{p}^m \hat{q}^n)_W$ in Weyl's rule of association denotes the sum of all possible products involving $m\,\hat{p}$'s and $n\,\hat{q}$'s divided by the total number of terms, e.g. :

$$(\hat{p}^2 \hat{q})_W = 1/3(\hat{p}^2 \hat{q} + \hat{q}\hat{p}^2 + \hat{p}\hat{q}\hat{p}). \tag{2.2}$$

It is often convenient, especially in quantum optics, to consider correspondence between two complex variables z, z^* (z^* denoting the complex conjugate of z) and the boson creation and annihilation operators $\hat{a}$ and $\hat{a}^\dagger$ which obey the commutation relation :

$$[\hat{a}, \hat{a}^\dagger] = 1. \tag{2.3}$$

The most commonly used rules of association in this case are :

Rule of association	Correspondence
Normal	$z^m z^{*n} \to \hat{a}^{\dagger n} \hat{a}^m$
Anti-normal	$z^m z^{*n} \to \hat{a}^m \hat{a}^{\dagger n}$
Weyl's symmetrization	$z^m z^{*n} \to (\hat{a}^m \hat{a}^{\dagger n})_W$
Moment	$z^m z^{*m} \to (\hat{a}^\dagger \hat{a})^m$

The symbol $(\hat{a}^{\dagger n} \hat{a}^m)_W$ is defined in a similar way as before, i.e. it represents the sum of all possible products involving n creations operators $\hat{a}^\dagger$ and m annihilation operators $\hat{a}$, divided by the total number of terms.

With each rule of correspondence we now associate a linear

operator, which transforms any given c-number function into a function of operators. For normal rule of association, for example, this operator, denoted by $\hat{\Omega}_N$, is defined by the following properties* :

(I) $\hat{\Omega}_N \{ z^m z^{*n} \} \equiv \hat{a}^{\dagger n} \hat{a}^m$

(II) $\hat{\Omega}_N \{ cf(z, z^*) \} \equiv c\hat{\Omega}_N \{ f(z, z^*) \}$

(III) $\hat{\Omega}_N \{ f_1(z, z^*) + f_2(z, z^*) \} \equiv \hat{\Omega}_N \{ f_1(z, z^*) \} + \hat{\Omega}_N \{ f_2(z, z^*) \}$

(IV) $\hat{\Omega}_N \{ c \} \equiv c\hat{\mathbf{1}}$.

Here f, f_1 and f_2 denote arbitrary functions, c is a c-number and $\hat{\mathbf{1}}$ the identity operator. The identity sign ($\equiv$) in these formulae and throughout this paper is to be understood as implying that the expressions on the two sides of this sign may be transformed into each other without the use of commutation relations. Thus, for example :

$$\hat{\Omega}_N \{ zz^* \} \equiv \hat{a}^\dagger \hat{a}. \tag{2.4 a}$$

If we use the commutation relation (2.3) we may re-write : $\hat{a}^\dagger a$ as $\hat{a}\hat{a}^\dagger + 1$, so that we also have :

$$\hat{\Omega}_N \{ zz^* \} = \hat{a}\hat{a}^\dagger + 1. \tag{2.4 b}$$

(2.4 a) is an identity, (2.4 b) is a simple equality.

It follows from the defining properties of $\hat{\Omega}_N$, that if a function $F(z, z^*)$ has the power series expansion

$$F(z, z^*) = \sum_{m=0}^{\infty} \sum_{n=0}^{\infty} c_{mn} z^m z^{*n}, \tag{2.5}$$

then

$$\hat{\Omega}_N \{ F(z, z^*) \} \equiv \sum_{m=0}^{\infty} \sum_{n=0}^{\infty} c_{mn} \hat{a}^{\dagger n} \hat{a}^m. \tag{2.6}$$

Thus the operator $\hat{\Omega}_N$ maps each c-number function $F(z, z^*)$ which has a power series expansion onto a function $G(\hat{a}, \hat{a}^\dagger)$ of operators, where

$$G(\hat{a}, \hat{a}^\dagger) = \hat{\Omega}_N \{ F(z, z^*) \}. \tag{2.7}$$

* The operator $\hat{\Omega}_N$ is essentially the same as the normal ordering operator discussed by W. Louisell in ref. [18], § 3.3.

By the use of the commutation relations $G(\hat{a}, \hat{a}^\dagger)$ may, of course, be written in many different ways. The particular form of G (which we denote by $\mathcal{G}$), defined by the relation

$$\mathcal{G}(\hat{a}, \hat{a}^\dagger) \equiv \hat{\Omega}_N \{F(z, z^*)\} \tag{2.8}$$

is of special interest. We call $\mathcal{G}$ the *normally ordered form of* G, or the *principal form of* G *for normal ordering*.

By analogy with $\hat{\Omega}_N$ one may introduce in a strictly similar manner other linear operators, each of which is associated with a particular rule of correspondence. Each such operator is defined by a condition of the type I for the appropriate mapping of an elementary *c*-number product into an elementary function of operators, by linearity conditions of the type II and III (involving $f(z, z^*)$ or $f(p, q)$ and by a condition of the type IV.

We will denote by $\hat{\Omega}$ any one of these "mapping operators". When we wish to stress that the operator $\hat{\Omega}$ is associated with a particular correspondence rule, we will attach an appropriate suffix to $\hat{\Omega}$.

The correspondence between a function F of *c*-numbers and a function G of operators, established according to some prescribed rule of association, viz. the relation

$$G(\hat{a}, \hat{a}^\dagger) = \hat{\Omega}\{F(z, z^*)\}, \tag{2.9}$$

may be expressed in a closed form with the help of the *ordering delta operator* $\Delta^{(\Omega)}$ that we recently introduced [25]. It is defined by the relation

$$\Delta^{(\Omega)}(z_0 - \hat{a}, z_0^* - \hat{a}^\dagger) = \hat{\Omega}\, \delta^{(2)}(z_0 - z), \tag{2.10}$$

where $\delta^{(2)}$ is the two dimensional Dirac delta function :

$$\delta^{(2)}(z - z_0) = \delta(x_0 - x)\, \delta(y_0 - y),$$

$z_0 = x_0 + iy_0$, $z = x + iy$; x_0, y_0, x, y being real parameters. Using the Fourier integral representation :

$$\delta^{(2)}(z_0 - z) = \frac{1}{\pi^2} \int e^{\alpha(z_0^* - z^*) - \alpha^*(z_0 - z)}\, d^2\alpha \tag{2.11}$$

of the $\delta^{(2)}$ function and the Baker-Hausdorff identity [26] one readily finds that for the rules of association which are commonly employed, $\Delta^{(\Omega)}$ may be expressed in the following form :

$$\Delta^{(\Omega)}(z_0 - \hat{a}, z_0^* - \hat{a}^\dagger) = \frac{1}{\pi^2}\int \Omega(\alpha, \alpha^*)\, e^{\alpha(z_0^* - \hat{a}^\dagger) - \alpha^*(z_0 - \hat{a})}\, d^2\alpha. \quad (2.12)$$

Here the integration extends over the whole complex α-plane and $\Omega(\alpha, \alpha^*)$ is a function of α and α^* whose form depends on the particular rule of correspondence. The form of $\Omega(\alpha, \alpha^*)$ for the more important rules of association is given in the following table :

TABLE I

Rule of association	$\Omega(\alpha, \alpha^*)$
Normal	$\exp\{1/2\, \alpha\alpha^*\}$
Anti-normal	$\exp\{-1/2\, \alpha\alpha^*\}$
Weyl.....................	1
Standard	$\exp\{1/4(\alpha^2 - \alpha^{*2})\}$
Anti-standard	$\exp\{-1/4(\alpha^2 - \alpha^{*2})\}$

We note that the functions $\Omega(\alpha, \alpha^*)$ for normal and anti-normal rules of association and also for standard and anti-standard rules are reciprocals of each other; Weyl's rule is self reciprocal.

It is seen from this table that in all the five cases $\Omega(\alpha, \beta)$ has the following properties :

(1) $\Omega(\alpha, \beta)$ is an entire analytic function of two complex variables α and β;
(2) $\Omega(\alpha, \beta) \neq 0$ for all values of α and β;
(3) $\Omega(-\alpha, -\beta) = \Omega(\alpha, \beta)$;
and (4) $\Omega(0, 0) = 1$.

The results that we will present apply not only to the rules of correspondence that are explicitly listed above but they apply, more generally, to any rule of association for which the ordering Δ-operator may be expressed in the form (2.12), *with the function* $\Omega(\alpha, \beta)$ *satisfying the four conditions that we just noted.*

With the help of the ordering Δ-operator one can readily establish the following two theorems* :

Theorem I : The operator function $G(\hat{a}, \hat{a}^\dagger)$, obtained by mapping a *c*-number function $F(z, z^*)$ according to the rule of association characterized by the operator $\hat{\Omega}$, is given by

$$G(\hat{a}, \hat{a}^\dagger) = \int F(z_0, z_0^*)\, \Delta^{(\Omega)}(z_0 - \hat{a}, z_0^* - \hat{a}^\dagger)\, d^2 z_0, \tag{2.13}$$

where the integration extends over the whole complex z_0-plane.

Theorem II : The *c*-number function $F(z_0, z_0^*)$ which corresponds to any given operator function $G(\hat{a}, \hat{a}^\dagger)$ when the rule of association is characterized by the operator $\hat{\Omega}$, is given by

$$F(z_0, z_0^*) = \pi \operatorname{Tr} \{ G(\hat{a}, \hat{a}^\dagger)\, \Delta^{(\overline{\Omega})}(z_0 - \hat{a}, z_0^* - \hat{a}^\dagger) \}, \tag{2.14}$$

where $\Delta^{(\overline{\Omega})}$ denotes the Δ-operator for ordering reciprocal to that defined by $\hat{\Omega}$:

$$\Delta^{(\overline{\Omega})}(z_0 - \hat{a}, z_0^* - \hat{a}^\dagger) = \frac{1}{\pi^2}\int [\Omega(\alpha, \alpha^*)]^{-1} \exp\{\alpha(z_0^* - \hat{a}^\dagger) - \alpha^*(z_0 - \hat{a})\}\, d^2\alpha. \tag{2.15}$$

By analogy with the definition of a normally ordered form given earlier, we define the *$\hat{\Omega}$-ordered form $\mathscr{G}(\hat{a}, \hat{a}^\dagger)$ of an operator function* $G(\hat{a}, \hat{a}^\dagger)$ *(the principal form of* $G(\hat{a}, \hat{a}^\dagger)$ *for $\hat{\Omega}$ ordering)* by the relations** :

$$\mathscr{G}(\hat{a}, \hat{a}^\dagger) = G(\hat{a}, \hat{a}^\dagger) \tag{2.16 a}$$

and

$$\hat{\Omega}\mathscr{G}(z, z^*) \equiv \mathscr{G}(\hat{a}, \hat{a}^\dagger). \tag{2.16 b}$$

* We only display formulae relating to the correspondence between a function of z and z^* and a function of $\hat{a}$ and $\hat{a}^\dagger$. The correspondence between a function of p and q and a function of $\hat{p}$ and $\hat{q}$ is governed by a similar set of formulae. One can readily transfer from one set to the other with the help of the relations :

$$\hat{a} = \left(\frac{1}{2\hbar}\right)^{1/2} (\hat{q} + i\hat{p}) \qquad \hat{a}^\dagger = \left(\frac{1}{2\hbar}\right)^{1/2} (\hat{q} - i\hat{p})$$

$$z = \left(\frac{1}{2\hbar}\right)^{1/2} (q + ip) \qquad z^* = \left(\frac{1}{2\hbar}\right)^{1/2} (q - ip).$$

** $\mathscr{G}(z, z^*)$ is to be understood, of course to be the same function of z and z^* as $\mathscr{G}(\hat{a}^\dagger, \hat{a})$ is of $\hat{a}$ and $\hat{a}^\dagger$.

In view of (2.9) one obtains from these relations

Theorem III : The $\hat{\Omega}$-ordered form $\mathcal{G}(\hat{a}, \hat{a}^\dagger)$ of an operator function $G(\hat{a}, \hat{a}^\dagger)$ *may be obtained from the formula :*

$$\mathcal{G}(\hat{a}, \hat{a}^\dagger) \equiv \hat{\Omega} F(z, z^*), \tag{2.17}$$

where $F(z, z^*)$ *is given by* (2.14).

Theorem III shows that the problem of expressing a given operator function in an ordered form according to some prescribed rule of association* is intimately related to the mapping of operator function onto *c*-number functions. Theorem III provides a systematic way for the solution of this problem.

3. **Phase Space Descriptions**

Let $G^{(1)}(\hat{a}, \hat{a}^\dagger)$ and $G^{(2)}(\hat{a}, \hat{a}^\dagger)$ be two operator functions and let $F_\Omega^{(1)}(z, z^*)$ and $F_{\overline{\Omega}}^{(2)}(z, z^*)$ be the associated *c*-number functions, $G^{(1)}$ and $F_\Omega^{(1)}$ being related by $\hat{\Omega}$-association and $G^{(2)}$ and $F_{\overline{\Omega}}^{(2)}$ via the association reciprocal to the $\hat{\Omega}$-association. According to (2.13) one has :

$$G^{(1)}(\hat{a}, \hat{a}^\dagger) = \int F_\Omega^{(1)}(z_0, z_0^*)\, \Delta^{(\Omega)}(z_0 - \hat{a}, z_0^* - \hat{a}^\dagger)\, d^2 z_0 \tag{3.1 a}$$

$$G^{(2)}(\hat{a}, \hat{a}^\dagger) = \int F_{\overline{\Omega}}^{(2)}(z_0, z_0^*)\, \Delta^{(\overline{\Omega})}(z_0 - \hat{a}, z_0^* - \hat{a}^\dagger)\, d^2 z_0. \tag{3.1 b}$$

Taking the trace of the product of $G^{(1)}$ and $G^{(2)}$, using eq. (3.1) and some properties of the ordering Δ-operators one may establish

Theorem IV :

$$\mathrm{Tr}\,[G^{(1)}(\hat{a}, \hat{a}^\dagger)\, G^{(2)}(\hat{a}, \hat{a}^\dagger)] = \frac{1}{\pi}\int F_\Omega^{(1)}(z_0, z_0^*)\, F_{\overline{\Omega}}^{(2)}(z_0, z_0^*)\, d^2 z_0. \tag{3.2}$$

Let us now take for $G^{(2)}$ the *density operator* ρ,

$$G^{(2)}(\hat{a}, \hat{a}^\dagger) = \rho(\hat{a}, \hat{a}^\dagger), \tag{3.3}$$

and let us set

$$\frac{1}{\pi} F_{\overline{\Omega}}^{(2)}(z_0, z_0^*) = \Phi_{\overline{\Omega}}(z_0, z_0^*). \tag{3.4}$$

* Formula for normally and anti-normally ordered forms which are essentially special cases of eq. (2.17) were recently noted by M. Lax and W. H. Louisell in ref. [17].

Then, dropping the superscript (1), (3.2) gives :

$$\mathrm{Tr}\{ G(\hat{a}, \hat{a}^{\dagger})\ \rho(\hat{a}, \hat{a}^{\dagger}) \} = \int F_{\Omega}(z_0, z_0^*)\ \Phi_{\overline{\Omega}}(z_0, z_0^*)\ d^2 z_0. \tag{3.5}$$

In particular if $G(\hat{a}, \hat{a}^{\dagger})$ is an observable, then $\mathrm{Tr}\{G\rho\}$ is the expectation value $\langle G \rangle$ of G when the system is in the mixed state characterized by the density operator ρ. Eq. (3.5) then shows that this expectation value may be expressed in the form of a phase space average of the associated *c*-number function $F_{\Omega}(z_0, z_0^*)$ with respect to $\Phi_{\overline{\Omega}}(z_0, z_0^*)$, i.e. :

$$\langle G \rangle = \int F_{\Omega}(z_0, z_0^*)\ \Phi_{\overline{\Omega}}(z_0, z_0^*)\ d^2 z_0. \tag{3.6}$$

The function $\Phi_{\overline{\Omega}}(z_0, z_0^*)$, which is proportional to the *c*-number function that corresponds to the density operator via the association reciprocal to $\hat{\Omega}$, is seen to play the role of the distribution function in phase space. In general, however, $\Phi_{\overline{\Omega}}$ is not a true probability distribution function, since it is not necessarily positive and may become singular.

Often, especially in quantum optics, one deals with functions of operators that are already in a form ordered according to one of the rules of correspondence listed earlier, e.g. normally ordered correlations in the quantum theory of coherence [13-15, 21, 29]. In other words $G(\hat{a}, \hat{a}^{\dagger})$ is in its principal form for some particular rule of association $\hat{\Omega}$: $G \equiv \mathscr{G}_{\Omega}(\hat{a}, \hat{a}^{\dagger})$. In this case let us use the same rule of association for the mapping of this operator function onto the *c*-number function. It then follows from eq. (2.16 *b*) and theorem III (eq. (2.17)), that the functional form of the operator $\mathscr{G}_{\Omega}$ and the functional form of its *c*-number representation are the same and eq. (3.5) then becomes :

$$\mathrm{Tr}\{ \mathscr{G}_{\Omega}(\hat{a}, \hat{a}^{\dagger})\ \rho(\hat{a}, \hat{a}^{\dagger}) \} = \int \mathscr{G}_{\Omega}(z_0, z_0^*)\ \Phi_{\overline{\Omega}}(z_0, z_0^*)\ d^2 z_0. \tag{3.7}$$

We will illustrate this result by a simple example. Suppose that $\mathscr{G}_{\Omega}$ is the normally ordered operator

$$\mathscr{G}_{N}(\hat{a}, \hat{a}^{\dagger}) \equiv \hat{a}^{\dagger n} \hat{a}^{m}, \quad (n, m \text{ non-negative integers}). \tag{3.8}$$

Then the *c*-number representation of $\mathscr{G}$ is of the same functional form,

$$F(z_0, z_0^*) = \mathscr{G}_{N}(z_0, z_0^*) = z_0^{*n} z_0^{m}, \tag{3.9}$$

and (3.7) gives

$$\mathrm{Tr}\{\hat{a}^{\dagger n}\hat{a}^{m}\rho(\hat{a},\hat{a}^{\dagger})\} = \int z_0^{*n} z_0^{m} \Phi_A(z_0, z_0^*)\, d^2 z_0, \qquad (3.10)$$

where Φ_A is proportional to the c-number representation of the density operator obtained via the *anti-normal* rule of association.

The result (3.7), with $\hat{\Omega}$ representing normal ordering, is the essence of Sudarshan's theorem [13, 27, 29] on the equivalence of the semi-classical and the quantum theory of optical coherence.

4. **Examples**

We will illustrate our general results by a number of examples.

The Sudarshan-Glauber diagonal representation of operators

Consider the anti-normal rule of association (suffix A) between operator function $G(\hat{a}, \hat{a}^\dagger)$ and a c-number function $F(z_0, z_0^*)$. We have from (2.10) and (2.11) and the properties of $\hat{\Omega}_A$ that

$$\Delta^{(A)}(z_0 - \hat{a}, z_0^* - \hat{a}^\dagger) = \frac{1}{\pi^2}\int e^{\alpha z_0^* - \alpha^* z_0} e^{\alpha^* \hat{a}} e^{-\alpha \hat{a}^\dagger}\, d^2\alpha. \qquad (4.1)$$

Let $| z \rangle$ be an eigenstate of the annihilation operator (known also as a coherent state [14 b]) corresponding to the eigenvalue z :

$$\hat{a}\, | z \rangle = z\, | z \rangle. \qquad (4.2)$$

These states allow the resolution of the identity [28-29] :

$$\frac{1}{\pi}\int | z \rangle \langle z |\, d^2 z = \hat{\mathbf{1}}. \qquad (4.3)$$

Inserting the left hand side of (4.3) between the terms $e^{\alpha^* \hat{a}}$ and $e^{-\alpha \hat{a}^\dagger}$ in (4.1) and using (4.2) and (2.11) one readily finds that :

$$\Delta^{(A)}(z_0 - \hat{a}, z_0^* - \hat{a}^\dagger) = \frac{1}{\pi} | z_0 \rangle \langle z_0 | \qquad (4.4)$$

i.e. $\pi\,\Delta^{(A)}(z_0 - \hat{a}, z_0^* - \hat{a}^\dagger)$ is just the projection operator associated with the eigenstate $| z_0 \rangle$.

On substituting from (4.4) into (2.13) and writing F_A in place of F to indicate that the correspondence anti-normal rule of association, and suppressing the suffix zero on z, we obtain :

$$G(\hat{a}, \hat{a}^\dagger) = \frac{1}{\pi}\int F_A(z, z^*)\, | z \rangle \langle z |\, d^2 z. \qquad (4.5)$$

The possibility of expressing an arbitrary operator G in this form was first noted by Sudarshan [13] and this representation is known as the *diagonal coherent state representation* of the operator G. In a somewhat more restricted form such a representation for the density operator was introduced shortly afterwards by Glauber [14 b], under the name of "P-representation".

Recently Mehta [30] gave an expression for F_A in terms of G, for operator functions G, for which $\langle -z \mid G \mid z \rangle \exp\{\mid z \mid^2\}$ is square integrable*. Mehta's formula may be derived from our theorem II (eq. 2.14), specialized to anti-normal rule of association.

The function $F_N(z, z^*)$

According to (2.14) the *c*-number function $F_N(z, z^*)$ associated with an operator function $G(\hat{a}, \hat{a}^\dagger)$ via the normal rule is given by :

$$F_N(z, z^*) = \pi \operatorname{Tr} \{ G(\hat{a}, \hat{a}^\dagger)\, \Delta^{(A)}(z - \hat{a}, z^* - \hat{a}^\dagger) \}. \quad (4.6)$$

Using the expression (4.4) for $\Delta^{(A)}$, we immediately obtain from (4.6) the following expression for F_N :

$$F_N(z, z^*) = \langle z \mid G(\hat{a}, \hat{a}^\dagger) \mid z \rangle. \quad (4.7)$$

The property of this function** have been studied by Kano [23], Mehta and Sudarshan [22], Glauber [21] and Mandel [31].

Relation between the functions F_Ω *for two different rules of association*

Suppose that F_{Ω_1} and F_{Ω_2} are the *c*-number representations of the same operator function G, according to two different

* In some recent publications, the role of the indices A and N is opposite to that employed in the present paper. Our notation appears to be preferable in view of the general theorem IV (eq. 3.2), which shows, that in the *c*-number evaluation of the trace of a product of two operators, the two *c*-number representations of the operators appear in reciprocal rules of association. In this connection see also footnote 1 in ref. [17] and footnote 14 in M. Lax, *Phys. Rev.*, *157*, 213 (1967).

** In a slightly different context this function was introduced by K. Husimi (*Proc. Physico Math. Soc., Japan*, *22*, 264 (1940)), who studied some of its properties in detail.

rules of association, characterized by $\hat{\Omega}_1$ and $\hat{\Omega}_2$ respectively. Then we have from the basic equations (2.13) and (2.14) :

$$G(\hat{a}, \hat{a}^\dagger) = \int F_{\Omega_1}(z_1, z_1^*)\, \Delta^{(\Omega_1)}(z_1 - \hat{a}, z_1^* - \hat{a}^\dagger)\, d^2 z_1, \quad (4.8)$$

$$F_{\Omega_2}(z_2, z_2^*) = \pi \operatorname{Tr}\{G(\hat{a}, \hat{a}^\dagger)\, \Delta^{(\overline{\Omega}_2)}(z_2 - \hat{a}, z_2^* - \hat{a}^\dagger)\}. \quad (4.9)$$

Substitution from (4.8) into (4.9) gives :

$$F_{\Omega_2}(z_2, z_2^*) = \int F_{\Omega_1}(z_1, z_1^*)\, K_{21}(z_2 - z_1, z_2^* - z_1^*)\, d^2 z_1, \quad (4.10)$$

where

$$K_{21}(z_2 - z_1, z_2^* - z_1^*)$$
$$= \pi \operatorname{Tr}\{\Delta^{(\Omega_1)}(z_1 - \hat{a}, z_1^* - \hat{a}^\dagger)\, \Delta^{(\overline{\Omega}_2)}(z_2 - \hat{a}, z_2^* - \hat{a}^\dagger)\}. \quad (4.11)$$

If one uses the representations (2.12) of $\Delta^{(\Omega_1)}$ and $\Delta^{(\overline{\Omega}_2)}$, one readily finds after a straightforward calculation that (4.11) reduces to

$$K_{21}(\zeta, \zeta^*) = \frac{1}{\pi^2} \int \Omega_1(\alpha, \alpha^*)\, \overline{\Omega}_2(\alpha, \alpha^*)\, e^{-(\alpha\zeta^* - \alpha^*\zeta)}\, d^2\alpha. \quad (4.12)$$

Equation (4.10) expresses the relation between F_{Ω_1} and F_{Ω_2} as a linear transform, with the kernel K_{21} given by (4.12). However a relation of the type (4.10) exists only if the integral on the right hand side of (4.12) converges.

Let suffices W, N and A denote the Weyl, the normal and the anti-normal association respectively. Then, according to Table I,

$$\Omega_W \overline{\Omega}_N = \Omega_A \overline{\Omega}_W = e^{-1/2\alpha\alpha^*} \quad (4.13)$$

$$\Omega_A \overline{\Omega}_N = e^{-\alpha\alpha^*} \quad (4.14)$$

and one obtains the following formulae on substituting from (4.13) and (4.14) into (4.12) and on evaluating the integrals :

$$K_{NW}(\zeta, \zeta^*) = K_{WA}(\zeta, \zeta^*) = \frac{2}{\pi} e^{-2\zeta\zeta^*} \quad (4.15)$$

$$K_{NA}(\zeta, \zeta^*) = \frac{1}{\pi} e^{-\zeta\zeta^*}. \quad (4.16)$$

Hence (4.10) gives* :

$$F_W(z, z^*) = \frac{2}{\pi}\int F_A(z', z'^*)\, e^{-2|z-z'|^2} d^2 z', \tag{4.17}$$

$$F_N(z, z^*) = \frac{2}{\pi}\int F_W(z', z'^*)\, e^{-2|z-z'|^2} d^2 z', \tag{4.18}$$

$$F_N(z, z^*) = \frac{1}{\pi}\int F_A(z', z'^*)\, e^{-|z-z'|^2} d^2 z'. \tag{4.19}$$

Formulae of the type (4.17) and (4.19) for the relation between the phase space representations of the density operator were derived in a different way by Glauber [21] and by Mehta and Sudarshan [22].

The normal ordered form of a function of the number operator

As a last example of the application of our general theorems let us find the normal ordered form of a function $G(\hat{a}, \hat{a}^\dagger)$ which depends on the number operator $N = \hat{a}^\dagger \hat{a}$ only, i.e. of an operator function of the form

$$G(\hat{a}, \hat{a}^\dagger) = f(\hat{a}^\dagger \hat{a}). \tag{4.20}$$

Let $\mathscr{G}_N(\hat{a}, \hat{a}^\dagger)$ denote the required normal ordered form. Then, according to (2.17), (4.7) and (4.20),

$$\mathscr{G}_N(\hat{a}, \hat{a}^\dagger) \equiv \hat{\Omega}_N F_N(z, z^*) \tag{4.21 a}$$

where,

$$F_N(z, z^*) = \langle z \,|\, f(\hat{a}^\dagger \hat{a}) \,|\, z \rangle. \tag{4.21 b}$$

Now let us expand $|z\rangle$ in terms of the Fock states $|n\rangle$ (ref. [29], Chapter 7) :

$$|z\rangle = e^{-1/2|z|^2} \sum_{n=0}^{\infty} \frac{z^n}{\sqrt{n!}} |n\rangle. \tag{4.22}$$

* It should be noted that the inverse kernels K_{AW}, K_{WN} and K_{AN} do not exist. However as each of the relations (4.17)-(4.19) is a Weierstrass transform, it can be inverted by standard mathematical techniques. [See for example : I. I. Hirschman and D. V. Widder, *The Convolution Transform* (Princeton University Press, Princeton, N. J., 1955), Chapter VIII.] The inverse relation involves a differential rather than an integral operator. A formula of this inverse type, expressing a relation between *c*-number representations for standard and Weyl's rule of correspondence was found by N. H. McCoy, *Proc. Nat. Acad. Sci.* (Washington), *18*, 674 (1932), and for some other rules of correspondences by C. L. Mehta in ref. [11].

Then on substituting from (4.22) into (4.21 b) and using the fact that $|n\rangle$ is an eigenstate of the number operator $a^\dagger a$, we obtain, if we also make use of the orthogonality of the Fock states, the following series expansion for F_N :

$$F_N(z, z^*) = e^{-|z|^2} \sum_{n=0}^{\infty} \frac{|z|^{2n}}{n!} f(n). \tag{4.23}$$

We see that in this case F_N is the sum of weighted Poissonian distributions, all with the parameters $|z|^2$, the weighting function being $f(n)$.

In order to evaluate readily the effect of operating by $\hat{\Omega}_N$ on F_N let us expand the exponential term in (4.23) in a power series. We then obtain :

$$\begin{aligned} F_N(z, z^*) &= \sum_{m=0}^{\infty} \sum_{n=0}^{\infty} \frac{(-1)^m f(n)}{m!\, n!} z^{*(m+n)} z^{(m+n)} \\ &= \sum_{r=0}^{\infty} \sum_{s=0}^{r} \frac{(-1)^s f(r-s)}{s!\,(r-s)!} z^{*r} z^r. \end{aligned} \tag{4.24}$$

On going from the first to the second line in (4.24) we made the substitution $m + n = r$, $m = s$. It follows from (4.24) and (4.21 a), if we recall that $\hat{\Omega}_N\{z^{*r} z^r\} \equiv \hat{a}^{\dagger r} \hat{a}^r$:

$$\mathscr{G}_N(\hat{a}, \hat{a}^\dagger) \equiv \sum_{r=0}^{\infty} \sum_{s=0}^{r} \frac{(-1)^s f(r-s)}{s!(r-s)!} \hat{a}^{\dagger r} \hat{a}^r. \tag{4.25}$$

This expression for the normal ordered form of the operator function $f(\hat{a}^\dagger \hat{a})$ has been derived in a somewhat less direct way by Louisell (ref. [18], p. 114).

In the special case when $f(\hat{a}^\dagger \hat{a}) = (\hat{a}^\dagger \hat{a})^p$, where p is a non-negative integer, $f(r-s) = (r-s)^p$ and the infinite series (summation over r in (4.25)) reduces to a finite sum by the use of identity* :

$$\sum_{s=0}^{r} \frac{(-1)^s (r-s)^p}{s!(r-s)!} = 0 \qquad \text{for } r > p \geqslant 0, \qquad r \text{ and } p \text{ integers.} \tag{4.26}$$

In this particular case eq. (4.25) reduces to :

$$\mathscr{G}_N(\hat{a}, \hat{a}^\dagger) = \sum_{r=0}^{p} \sum_{s=0}^{r} \frac{(-1)^s (r-s)^p}{s!(r-s)!} \hat{a}^{\dagger r} \hat{a}^r. \tag{4.27}$$

* We are indebted to Dr. J. H. Eberly for bringing this identity to our notice.

The relation (4.27) expresses the moment of the number operator in terms of normally ordered products and can be immediately used to compute counting moments in terms of counting correlations in photoelectric measurements of light fluctuations [32].

REFERENCES

[1] E. C. G. SUDARSHAN, *Lectures in Theoretical Physics* (Brandeis University, Summer Institute 1961, New York, W. A. Benjamin Co., 1962, Vol. *21*, p. 143); T. F. JORDAN and E. C. G. SUDARSHAN, *Rev. Mod. Phys.*, *33*, 515 (1961).

[2] E. WIGNER, *Phys. Rev.*, *40*, 749 (1932).

[3] J. E. MOYAL, *Proc. Cambridge Phil. Soc.*, *45*, 99 (1949).

[4] H. WEYL, a) *Z. Physik*, *46*, 1 (1927-1928); b) *The Theory of Groups and Quantum Mechanics* (London, Macmillan, 1931; also New York, Dover Publications, Inc., p. 274).

[5] H. J. GROENEWOLD, *Physica*, *12*, 405 (1946).

[6] R. KUBO, *J. Phys. Soc., Japan*, *19*, 2127 (1964).

[7] G. A. BAKER Jr, *Phys. Rev.*, *109*, 2198 (1958).

[8] K. IMRE, E. OZIZMIR, M. ROSENBAUM and P. F. ZWEIFEL, *J. Math. Phys.*, *8*, 1097 (1967).

[9] H. MORI, I. OPPENHEIM and J. ROSS, in *Studies in Statistical Mechanics*, ed. J. de BOER and G. E. UHLENBECK (Amsterdam, North-Holland Publishing Co., 1962), Vol. I, p. 217.

[10] J. R. SHEWELL, *Am. J. Phys.*, *27*, 16 (1959); this paper also contains references to many earlier publications in this field.

[11] C. L. MEHTA, *J. Math. Phys.*, *5*, 677 (1964).

[12] L. COHEN, *J. Math. Phys.*, *7*, 781 (1966).

[13] E. C. G. SUDARSHAN, a) *Phys. Rev. Letters*, *10*, 277 (1963); b) in *Proc. Symp. on Optical Masers* (New York, J. Wiley & Sons, 1963), p. 45.

[14] R. J. GLAUBER, a) *Phys. Rev.*, *130*, 2529 (1963); b) *Phys. Rev.*, *131*, 2766 (1963).

[15] L. MANDEL and E. WOLF, *Rev. Mod. Phys.*, *37*, 231 (1965).

[16] H. HAKEN, H. RISKEN and W. WEIDLICH, *Z. Physik*, *206*, 355 (1967) : see also some of the references quoted in this paper.

[17] M. LAX and W. H. LOUISELL, *I.E.E.E. J. Quant. Electron.*, *QE-3*, 47 (1967), and some of the references given in this paper.

[18] W. H. LOUISELL, *Radiation and Noise in Quantum Electronics* (New York, McGraw-Hill, 1964), Chap. III.

[19] R. M. WILCOX, *J. Math. Phys.*, *8*, 962 (1967).

[20] C. L. MEHTA, *J. Phys.*, *A* (*Proc. Phys. Soc.*, London), ser. 2, *1*, 385 (1968).

[21] R. J. GLAUBER, contribution in *Quantum Optics and Electronics*, Les Houches, 1964; ed. C. de WITT, A. BLANDIN and C. COHEN-TANNOUDJI (New York, Gordon & Breach Publishers, 1965), p. 65.

[22] C. L. MEHTA and E. C. G. SUDARSHAN, *Phys. Rev.*, *138*, B 274 (1965).

[23] Y. KANO, a) *J. Phys. Soc., Japan*, *19*, 1555 (1964); b) *J. Math. Phys.*, *6*, 1913 (1965).

[24] M. LAX, *Quantum Noise*, XI (to be published).

[25] G. S. AGARWAL and E. WOLF, *Phys. Letters*, *26*, A 485 (1968). In the special case when $\hat{\Omega}$ refers to Weyl's rule of correspondence, $\Delta^{(\Omega)}$ is essentially the Δ-operator introduced by R. KUBO in ref. [6].

[26] Ref. [18], p. 102.
[27] J. R. Klauder, *Phys. Rev. Letters*, *16*, 534 (1966).
[28] J. R. Klauder, *Ann. Phys.*, *11*, 123 (1960).
[29] J. R. Klauder and E. C. G. Sudarshan, *Fundamentals of Quantum Optics* (New York, W. A. Benjamin, 1968).
[30] C. L. Mehta, *Phys. Rev. Letters*, *18*, 752 (1967).
[31] L. Mandel, *Phys. Rev.*, *152*, 438 (1966).
[32] L. Mandel, *Phys. Rev. 136*, 1221 (1964).
[33] G. S. Agarwal and E. Wolf, *Phys. Rev. Letters*, *21*, 180 (1968).

Volume 13, number 3 OPTICS COMMUNICATIONS March 1975

ANGULAR DISTRIBUTION OF RADIANT INTENSITY FROM SOURCES OF DIFFERENT DEGREES OF SPATIAL COHERENCE *

E. WOLF **
Department of Physics and Astronomy, University of Rochester, Rochester, New York 14627, USA

and

William H. CARTER
Naval Research Laboratory, Washington, D.C. 20375, USA

Received 20 January 1975

The relationship that exists between the state of coherence of a source and the angular distribution of the radiant intensity is investigated. It is found that the radiant intensity from a large, planar, statistically homogeneous source is related in a simple way to the Fourier transform of its degree of spatial coherence. This result is illustrated by curves that show the angular distribution of radiant intensity for gaussian correlated sources. The curves demonstrate in a striking way how spatial coherence of the source affects the directionality of a light beam that the source generates.

1. Introduction

In spite of the great deal of progress that has been made in recent times in the understanding of coherence properties of light, one important question in this field has so far not been clarified. It concerns the relationship between the coherence properties of a source and the angular distribution of the intensity in the far zone of the field generated by the source. That the angular distribution of the intensity depends on the spatial coherence of the source in a crucial way is evident by reference to two well-known extreme cases. A conventional (thermal) source is spatially almost completely incoherent, exhibiting field correlations over distances of the order of the mean wavelength. Such a source gives rise to an intensity distribution in the far zone that changes slowly with direction, i.e. the field exhibits little dependence on direction. On the other hand a source with high degree of spatial coherence, such as a laser, generates light that is highly directional.

In the present note we report some preliminary results of an investigation concerned with this question. To bring out the essential features of the relationship between spatial coherence of a source and the angular distribution of the intensity in the far zone of the field that the source generates, we will restrict ourselves to large two-dimensional sources that are statistically homogeneous and we will ignore the specific effects that arise from the finite size of the source (i.e. effects that are somewhat analogous to diffraction phenomena with coherent light). We find that under these circumstances the angular distribution of the radiant intensity is related in a simple way to the spatial Fourier transform of the degree of coherence of the source. We illustrate this result by considering in detail radiation from a source whose degree of spatial coherence is given by a gaussian distribution. The angular dependence of the radiant intensity on the variance of the gaussian distribution shows in a striking way the relationship that exists between spatial coherence properties of sources and the directionality of light beams produced by them.

* Preliminary results of this investigation were presented on 18 October 1974 at the Ann. Meeting of the Opt. Soc. of America, held in Houston, Texas [4].

** Currently Visiting Professor, Department of Physics, University of Toronto, Toronto, Canada.

Volume 13, number 3 OPTICS COMMUNICATIONS March 1975

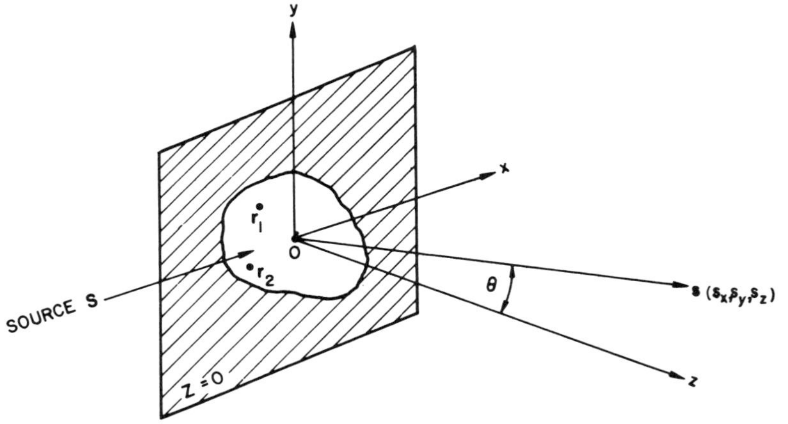

Fig. 1. Illustrating the notation.

2. Angular distribution of radiant intensity from a statistically homogeneous two-dimensional source of any state of spatial coherence

Consider a plane two-dimensional source S, occupying a portion of the plane $z = 0$. We may characterize the (second order) coherence properties of the source by its cross-spectral density function $W(\boldsymbol{r}_1, \boldsymbol{r}_2, \omega)$, (also known as mutual spectral density, cf. ref. [1], § 10.3.2).

It has been shown in a recent paper (ref. [2], eq. (41) *) that the radiant intensity at frequency ω, in the direction specified by a unit vector $\boldsymbol{s}$ (see fig. 1) with cartesian components s_x, s_y, s_z $(s_z \geqslant 0)$ is given by

$$J_\omega(\boldsymbol{s}) = (2\pi k)^2\ \hat{W}(k\boldsymbol{s}_\perp, -k\boldsymbol{s}_\perp, \omega) \cos^2\theta. \tag{1}$$

Here $W(\boldsymbol{f}_1, \boldsymbol{f}_2, \omega)$ is the four-folded spatial Fourier transform of the cross-spectral density function, i.e.

$$W(\boldsymbol{f}_1, \boldsymbol{f}_2, \omega) = \frac{1}{(2\pi)^4} \times \iint_{(z=0)} W(\boldsymbol{r}_1, \boldsymbol{r}_2, \omega) \exp[-\mathrm{i}(\boldsymbol{f}_1 \cdot \boldsymbol{r}_1 + \boldsymbol{f}_2 \cdot \boldsymbol{r}_2)]\, \mathrm{d}^2 r_1\, \mathrm{d}^2 r_2, \tag{2}$$

$\boldsymbol{s}_\perp$ is the two-dimensional vector

$$\boldsymbol{s}_\perp \equiv (s_x, s_y), \tag{3}$$

(i.e., the transverse part, perpendicular to the z-direction, of the unit vector $\boldsymbol{s}$), θ is the angle that the $\boldsymbol{s}$-vector makes with the normal to the source, and $k = \omega/c$.

The integration in (2) extends twice over the plane $z = 0$, once with respect to $\boldsymbol{r}_1$, and once with respect to $\boldsymbol{r}_2$. It follows from (2) that the function $\hat{W}(\boldsymbol{f}_1, \boldsymbol{f}_2, \omega)$, evaluated for the arguments $\boldsymbol{f}_1 = k\boldsymbol{s}_\perp, \boldsymbol{f}_2 = -k\boldsymbol{s}_\perp$ that appear on the r.h.s. of eq. (1), is given by

$$\hat{W}(k\boldsymbol{s}_\perp, -k\boldsymbol{s}_\perp, \omega) = \frac{1}{(2\pi)^4} \times \iint_{(z=0)} W(\boldsymbol{r}_1, \boldsymbol{r}_2, \omega) \exp[-\mathrm{i}k\boldsymbol{s}_\perp \cdot (\boldsymbol{r}_1 - \boldsymbol{r}_2)]\, \mathrm{d}^2 r_1\, \mathrm{d}^2 r_2. \tag{4}$$

We will assume that the source is statistically homogeneous ‡. The cross-spectral density function $W(\boldsymbol{r}_1,$

* In the formula (41), (and in some other formulae of ref. [2]), the notation is somewhat misleading, in that the three-dimensional vector $\boldsymbol{s}$, rather than the two-dimensional vector $\boldsymbol{s}_\perp$ has beean written on the right-hand side of the equation. Consideration of the dimensionality of the vector variables should resolve any confusion as to whether $\boldsymbol{s}$ or $\boldsymbol{s}_\perp$ is meant.

‡ By a statistically homogeneous source one means a source for which all the probability densities that characterize the behavior of the field at an arbitrary number of source points are invariant with respect to translations of all the points on the source plane. This requirement implies, in particular, that the second-order cross-spectral density function is also invariant with respect to such a translation, this being the fact expressed by eq. (5). It is clear that strict statistical homogeneity demands that the source occupies the whole z-plane. Throughout this paper we will use the term "statistical homogeneity" also for a finite source in the approximate sense that eq. (5) holds whenever the points $\boldsymbol{r}_1$ and $\boldsymbol{r}_2$ both lie within the source area.

$\boldsymbol{r}_2, \omega)$ then depends on the two position vectors $\boldsymbol{r}_1$ and $\boldsymbol{r}_2$ only through the difference $\boldsymbol{r}_1 - \boldsymbol{r}_2$, i.e.

$$W(\boldsymbol{r}_1, \boldsymbol{r}_2, \omega) = F_\omega(\boldsymbol{r}_1 - \boldsymbol{r}_2), \tag{5}$$

where F_ω is some function of a single vector variable and of the frequency ω. Let us assume first that the source is infinite in extent, occupying the whole plane $z = 0$. In this case we obtain at once, on substituting from eq. (5) into eq. (4) and changing the variables of integration from $\boldsymbol{r}_1$ and $\boldsymbol{r}_2$ to $\boldsymbol{r}$ and $\boldsymbol{r}'$ where

$$\boldsymbol{r}_1 - \boldsymbol{r}_2 = \boldsymbol{r}', \qquad \boldsymbol{r}_1 + \boldsymbol{r}_2 = 2\boldsymbol{r}, \tag{6}$$

the following expression for $\hat{W}(k\boldsymbol{s}_\perp, -k\boldsymbol{s}_\perp, \omega)$:

$$\hat{W}(k\boldsymbol{s}_\perp, -k\boldsymbol{s}_\perp, \omega) = \frac{1}{(2\pi)^4} \times \iint_{(z=0)} F_\omega(\boldsymbol{r}') \exp(-\mathrm{i}k\boldsymbol{s}_\perp \cdot \boldsymbol{r}') \mathrm{d}^2\boldsymbol{r}' \mathrm{d}^2\boldsymbol{r}. \tag{7}$$

Let us now introduce the two-dimensional spatial Fourier transform of the function $F_\omega(\boldsymbol{r})$ defined as

$$\hat{F}_\omega(\boldsymbol{f}) = \frac{1}{(2\pi)^2} \int_{(z=0)} F_\omega(\boldsymbol{r}') \exp(-\mathrm{i}\boldsymbol{f} \cdot \boldsymbol{r}') \mathrm{d}^2\boldsymbol{r}'. \tag{8}$$

In terms of $\hat{F}_\omega$ eq. (7) may be expressed in the form

$$\hat{W}(k\boldsymbol{s}_\perp, -k\boldsymbol{s}_\perp, \omega) = \frac{1}{(2\pi)^2} \hat{F}_\omega(k\boldsymbol{s}_\perp) \int_{(z=0)} \mathrm{d}^2\boldsymbol{r}. \tag{9}$$

Clearly the integral with respect to $\boldsymbol{r}$ diverges, because it extends over the whole infinite z-plane.

Suppose now that the source is finite, but that for all pairs of points $\boldsymbol{r}_1$ and $\boldsymbol{r}_2$ situated within the area S occupied by the source, the cross-spectral density function is still of the form given by eq. (5). Moreover, if we assume that the source is large compared both with the square of the wavelength $\lambda = 2\pi c/\omega$ and with the coherence area in the source plane (i.e. the effective $\boldsymbol{R}$-domain over which $F_\omega(\boldsymbol{R})$ differs sensibly from zero), it is reasonable to assume that we may obtain a good approximation to $\hat{W}(k\boldsymbol{s}_\perp, -k\boldsymbol{s}_\perp, \omega)$ if we replace the divergent integral in (9) by a corresponding integral extending over the area of the source. We then obtain in place of eq. (9) the approximate relation

$$\hat{W}(k\boldsymbol{s}_\perp, -k\boldsymbol{s}_\perp, \omega) \approx \frac{A}{(2\pi)^2} \hat{F}_\omega(k\boldsymbol{s}_\perp), \tag{10}$$

where A is the source area. It is not difficult to avoid the approximation that we just made, but the exact expression for $\hat{W}$ is too complicated to provide an insight into our problem. We intend to return to this point in another publication.

With the approximation implicit in eq. (10), the expression eq. (1) for the radiant intensity $J_\omega(\boldsymbol{s})$ becomes

$$J_\omega(\boldsymbol{s}) \approx k^2 A \hat{F}_\omega(k\boldsymbol{s}_\perp) \cos^2\theta. \tag{11}$$

We may readily re-write eq. (11) in a form that shows explicitly the dependence of the radiant intensity $J_\omega(\boldsymbol{s})$ on the degree of spatial coherence of the source. The degree of spatial coherence of the source at frequency ω may be defined by the expression

$$\gamma_\omega(\boldsymbol{r}_1, \boldsymbol{r}_2) = \frac{W(\boldsymbol{r}_1, \boldsymbol{r}_2, \omega)}{\sqrt{W(\boldsymbol{r}_1, \boldsymbol{r}_1, \omega)}\sqrt{W(\boldsymbol{r}_2, \boldsymbol{r}_2, \omega)}}, \tag{12}$$

where the normalization ensures that $0 \leqslant |\gamma_\omega(\boldsymbol{r}_1, \boldsymbol{r}_2)| \leqslant 1$. The limiting value $\gamma_\omega = 0$ represents complete spatial incoherence, and the other limiting value $|\gamma_\omega| = 1$ represents complete spatial coherence, (within the framework of second-order coherence theory). In the present case, when the mutual spectral density function has the form indicated by eq. (5), the degree of spatial coherence $\gamma_\omega(\boldsymbol{r}_1, \boldsymbol{r}_2)$ depends on the position vectors $\boldsymbol{r}_1$ and $\boldsymbol{r}_2$ only through their difference $\boldsymbol{r}_1 - \boldsymbol{r}_2$ and is given by

$$\gamma_\omega(\boldsymbol{r}_1 - \boldsymbol{r}_2) = F_\omega(\boldsymbol{r}_1 - \boldsymbol{r}_2)/F_\omega(0). \tag{13}$$

Let $\hat{\gamma}_\omega(\boldsymbol{f})$ denote the two-dimensional spatial Fourier transform of $\gamma_\omega(\boldsymbol{r})$, viz.

$$\hat{\gamma}_\omega(\boldsymbol{f}) = \frac{1}{(2\pi)^2} \int_{(z=0)} \gamma_\omega(\boldsymbol{r}) \exp(-\mathrm{i}\boldsymbol{f} \cdot \boldsymbol{r}) \mathrm{d}^2\boldsymbol{r}. \tag{14}$$

It follows at once, on taking the Fourier transform of eq. (13) that

$$\hat{F}_\omega(\boldsymbol{f}) = F_\omega(0)\, \hat{\gamma}_\omega(\boldsymbol{f}), \tag{15}$$

and, using this relation, (11) becomes

$$J_\omega(\boldsymbol{f}) \approx k^2 A F_\omega(0)\, \hat{\gamma}_\omega(k\boldsymbol{s}_\perp) \cos^2\theta, \tag{16}$$

where, as eq. (5) shows,

$$F_\omega(0) = W(\boldsymbol{r}, \boldsymbol{r}, \omega), \tag{17}$$

represents the spectral density at frequency ω at an arbitrary point $\boldsymbol{r}$ of the source. Although formally the position vector $\boldsymbol{r}$ appears in the expression on the right-hand side of eq. (17), $W(\boldsymbol{r}, \boldsymbol{r}, \omega)$ does not in fact, depend on $\boldsymbol{r}$ because of our assumption that the source is statistically homogeneous.

Eq. (16) is the main formula of this paper. It shows that the angular distribution of the radiant intensity from a large two-dimensional statistically homogeneous source is related in a simple way to the Fourier transform of the degree of spatial coherence of the source. In the next section we will illustrate, by a simple example, some important implications of this formula.

3. Example: gaussian correlated source

Consider a statistically homogeneous source S, whose degree of coherence is given by

$$\gamma_\omega(\boldsymbol{r}_1, \boldsymbol{r}_2) = \exp(-|\boldsymbol{r}_1 - \boldsymbol{r}_2|^2/2\sigma^2) \tag{18}$$

and assume that the linear dimensions of the source are large, compared both to the wavelength $\lambda = 2\pi c/\omega$ and to the root-mean-square width σ of the gaussian distribution in (18). On substituting from eq. (18) into (14) we readily find that in this case

$$\hat{\gamma}_\omega(\boldsymbol{f}) = \frac{\sigma^2}{2\pi} \exp(-\tfrac{1}{2}\sigma^2 f^2), \tag{19}$$

and hence eq. (16) now gives

$$J_\omega(\boldsymbol{s}) \approx \frac{k^2\sigma^2}{2\pi} A F_\omega(0) \times \cos^2\theta \exp[-\tfrac{1}{2}k^2\sigma^2(s_x^2 + s_y^2)], \tag{20}$$

Now $s_x^2 + s_y^2 = \sin^2\theta$ where θ denotes the angle between the $\boldsymbol{s}$-direction and the normal to the source plane. Hence (20) becomes, if we now consider the radiant intensity to be a function of the angle θ rather than of $\boldsymbol{s}$:

$$J_\omega(\theta) \approx \frac{k^2\sigma^2}{2\pi} A F_\omega(0) \times \cos^2\theta \exp(-\tfrac{1}{2}k^2\sigma^2 \sin^2\theta). \tag{21}$$

It will be convenient to normalize the radiant intensity by expressing it as a fraction of the value that it has for the forward direction ($\theta = 0$). Clearly

$$\frac{J_\omega(\theta)}{J_\omega(0)} \approx \cos^2\theta \exp(-\tfrac{1}{2}k^2\sigma^2 \sin^2\theta). \tag{22}$$

From eq. (22) we can see at once how the radiant intensity behaves in two limiting cases. As $\sigma \to 0$, (18) clearly represents a spatially incoherent source (zero correlation distance) and (22) then reduces to

$$\left(\frac{J_\omega(\theta)}{J_\omega(0)}\right)_{\text{incoh}} = \cos^2\theta, \tag{23}$$

i.e. the radiant intensity falls off as $\cos^2\theta$ (not as $\cos\theta$ which is appropriate to a lambertian source), in agreement with a recently derived general result about radiation from spatially incoherent sources (§5 of ref. [2]). In the other extreme case, as $\sigma \to \infty$ (assuming now also that the source occupies the whole z-plane), (18) represents a source that is completely spatially coherent and (22) now reduces to

$$\left(\frac{J_\omega(\theta)}{J_\omega(0)}\right)_{\text{coh}} \begin{array}{ll} = 0, & \text{if } \theta \neq 0; \\ = 1, & \text{if } \theta = 0. \end{array} \tag{24}$$

Thus the source now radiates only in the direction perpendicular to its plane, i.e. the source gives rise to a perfectly collimated uni-directional beam.

Fig. 2 shows the angular distribution of the radiant intensity, computed from eq. (22), for gaussian correlated sources of different correlation distances. Not surprisingly the least directional and the most directional beams correspond to the two extreme cases that we just discussed. For comparison the curve representing the angular distribution of the radiant intensity from a lambertian source is also included. It is evident from these curves that there is a profound modification in the angular distribution of the radiant intensity as one goes from a spatially incoherent source (i.e. one with zero correlation distance), to a partially coherent source with a correlation distance of about one wavelength. In this range a change in the correlation distance in the source plane, by even a fraction of a wavelength, is seen to give rise to a substantial modification in the angular distribution of the energy radiated by the source.

Our main formula (16) may also be used to discuss the inverse problem, namely of determining the coherence properties of a statistically homogeneous source from the knowledge of the angular distribution of the radiant intensity to which the source gives rise. This problem is discussed in another paper [3] and our solution gives, as a by-product, the general form of the degree of spatial coherence of a lambertian source.

Volume 13, number 3 OPTICS COMMUNICATIONS March 1975

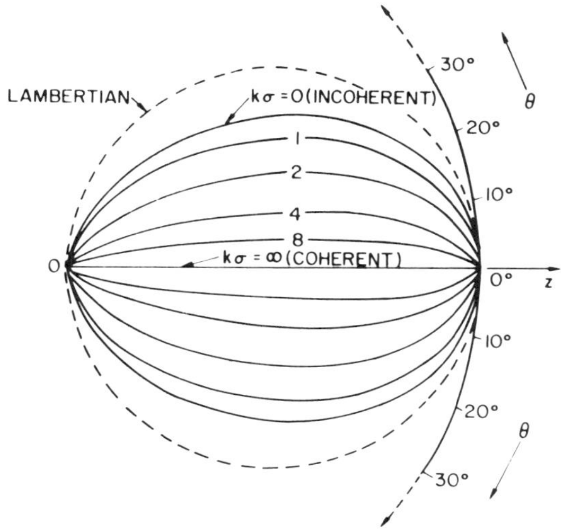

Fig. 2. Angular distribution of the normalized radiant intensity $J_\omega(\theta)/J_\omega(0)$, from a gaussian correlated statistically homogeneous source, computed from eq. (22). The length of the vector pointing from the origin to a typical point on a curve labelled by a particular value of the parameter $k\sigma$ represents the normalized radiant intensity in the direction of that vector, from a source whose degree of spatial coherence is given by eq. (18). The following table gives the root-mean-square values (the effective correlation distance) σ that corresponds to the selected values of $k\sigma$:

$k\sigma$	0	1	2	4	8	∞
σ	0	0.16λ	0.32λ	0.64λ	1.28λ	∞

References

[1] M. Born and E. Wolf, *Principles of Optics*, 4th Ed. (Pergamon Press, Oxford and New York, 1970).
[2] E.W. Marchand and E. Wolf, J. Opt. Soc. Amer. 64 (1974) 1219.
[3] W.H. Carter and E. Wolf, submitted to J. Opt. Soc. Amer.
[4] E. Wolf and W.H. Carter, J. Opt. Soc. Amer. 64 (1974) 1398.

Reprinted from:
JOURNAL OF THE OPTICAL SOCIETY OF AMERICA VOLUME 65, NUMBER 9 SEPTEMBER 1975

Coherence properties of lambertian and non-lambertian sources*

W. H. Carter

Naval Research Laboratory, Washington, D. C. 20375

E. Wolf†

Department of Physics and Astronomy, University of Rochester, Rochester, New York 14627
(Received 6 March 1975)

Some general theorems are derived about spatial coherence properties of a two-dimensional statistically homogeneous source that gives rise to any prescribed angular distribution of radiant intensity. The results are applied to sources that radiate in accordance with Lambert's law. It is found that lambertian sources cannot be spatially strictly incoherent and that they have, in fact, certain unique coherence properties. This result is illustrated by calculations for a blackbody source. The idealized case of a spatially completely incoherent source is discussed for comparison.

Index Headings: Source; Coherence.

We showed in a recent paper[1] that the angular distribution of radiant intensity from a planar statistically homogeneous source depends in a relatively simple way on certain spatial coherence properties of the source. In the present paper, we consider the related problem: Given the angular distribution of the radiant intensity, what conclusions can be drawn about the spatial coherence of the source?

In Sec. I, we establish some general theorems pertaining to this problem. We show, in particular, that from the knowledge of the angular distribution of the radiant intensity the low-frequency part of the cross-spectral density function of the light distribution across the source can be determined, i.e., the part that corresponds to spatial-frequency components f_x, f_y for which $f_x^2+f_y^2\leq k^2$, where k is the wave number of the radiation.

In Sec. II, we analyze, with the help of these results, the coherence properties of lambertian sources and illustrate our conclusions as to the general form of the cross-spectral density function of such sources by explicit calculations for a blackbody source. In Sec. III we discuss, for comparison, the idealized case of a spatially incoherent source. Some of the results that we present in Secs. II and III are in agreement with those obtained previously by other authors,[2-5,13] but go somewhat beyond them.

I. DETERMINATION OF SPATIAL COHERENCE PROPERTIES OF A SOURCE FROM THE ANGULAR DISTRIBUTION OF THE RADIANT INTENSITY

The problem that we examine in this section may be stated thus: A two-dimensional statistically homogeneous source[6] S, which occupies a portion of the plane $z=0$, radiates into the half-space $z>0$. Let $J_\omega(\mathbf{s})$ be the radiant intensity at frequency ω, generated by the source. If $J_\omega(\mathbf{s})$ is known for all $\mathbf{s}$ directions ($\mathbf{s}^2=1$) pointing into the half-space $z>0$ (Fig. 1) what conclusions can be deduced about the spatial coherence properties of the source?

To answer this question, we will make use of the main results, which we now briefly summarize, established in Ref. 1. Let $W(\mathbf{r}_1,\mathbf{r}_2,\omega)$ be the cross-spectral density function at frequency ω, at points of the source specified by position vectors $\mathbf{r}_1$ and $\mathbf{r}_2$. Because of our assumption of statistical homogeneity, W is necessarily of the form

$$W(\mathbf{r}_1,\mathbf{r}_2,\omega)=F_\omega(\mathbf{r}_1-\mathbf{r}_2)\qquad(\mathbf{r}_1\in S,\ \ \mathbf{r}_2\in S)\tag{1}$$

and $W(\mathbf{r}_1,\mathbf{r}_2,\omega)\equiv 0$ if either of the two points lies in the z plane outside the area occupied by the source. The degree of spatial coherence γ_ω at frequency ω, for any pair of points $\mathbf{r}_1$, $\mathbf{r}_2$ on the source area, is then also a function of the difference $\mathbf{r}_1-\mathbf{r}_2$ only, and is given by

$$\gamma_\omega(\mathbf{r}_1-\mathbf{r}_2)\equiv\frac{W(\mathbf{r}_1,\mathbf{r}_2,\omega)}{[W(\mathbf{r}_1,\mathbf{r}_1,\omega)W(\mathbf{r}_2,\mathbf{r}_2,\omega)]^{1/2}}=\frac{F_\omega(\mathbf{r}_1-\mathbf{r}_2)}{F_\omega(0)}\ .\tag{2}$$

We showed in Ref. 1 that when the source area is large compared with the square of the wavelength $\lambda=2\pi c/\omega$ (c = vacuum velocity of light) and also with the coherence area in the source plane (i.e., the effective $\mathbf{r}$ domain throughout which $F_\omega(\mathbf{r})$ differs sensibly from zero), the radiant intensity in the direction specified by the unit vector $\mathbf{s}\equiv(s_x,s_y,s_z)$ may, to a good approximation, be expected to be given by the formula [Eq. (11) of Ref. 1]

$$J_\omega(\mathbf{s})=k^2A\hat{F}_\omega(k\mathbf{s}_\perp)\cos^2\theta\ .\tag{3}$$

In Eq. (3), $\hat{F}_\omega$ is the two-dimensional spatial Fourier transform of F_ω,

$$\hat{F}_\omega(\mathbf{f})=\frac{1}{(2\pi)^2}\int_{(z=0)}F_\omega(\mathbf{r}')\,e^{-i\mathbf{f}\cdot\mathbf{r}'}\,d^2\mathbf{r}'\ ,\tag{4a}$$

$\mathbf{s}_\perp$ is the two-dimensional vector, with components s_x, s_y, θ is the angle that the $\mathbf{s}$ vector makes with the z direction (normal to the source plane), A is the source area, and

$$k=\omega/c\ .\tag{5}$$

Using Eq. (1) we may express $\hat{F}_\omega(\mathbf{f})$ in terms of the cross-spectral density function as

$$\hat{F}_\omega(\mathbf{f})=\frac{1}{(2\pi)^2}\int_{(z=0)}W(\mathbf{r}_1+\mathbf{r}',\mathbf{r}_1)\,e^{-i\mathbf{f}\cdot\mathbf{r}'}\,d^2\mathbf{r}'\ ,\tag{4b}$$

where $\mathbf{r}_1$ is the position vector of any source point.

Equation (3) implies that the radiant intensity in a direction specified by the unit vector $\mathbf{s}$ depends on one

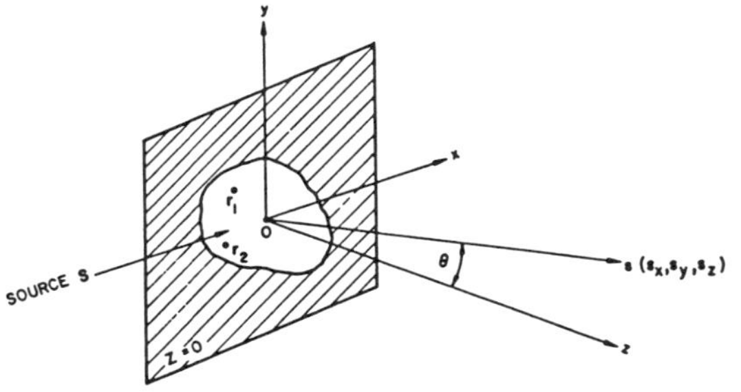

FIG. 1. Illustrating the notation.

and only one (two-dimensional) Fourier spatial-frequency component $\hat{F}_\omega(\mathbf{f})$ of $F_\omega(\mathbf{r})$, namely on the component labeled by $\mathbf{f} \equiv k\mathbf{s}_\perp \equiv (ks_x, ks_y)$. Because

$$\mathbf{s}_\perp^2 = s_x^2 + s_y^2 = 1 - s_z^2 = 1 - \cos^2\theta \le 1 \, , \tag{6}$$

it follows that only those frequency components $\mathbf{f} \equiv (f_x, f_y)$ of $F_\omega(\mathbf{r})$, which lie inside or on the boundary of the circle

$$f_x^2 + f_y^2 = k^2 \tag{7}$$

in the (two-dimensional) spatial-frequency plane f_x, f_y contribute to the radiant intensity $J_\omega(\mathbf{s})$. (The other components are associated with evanescent waves that do not carry energy into the far zone.)

Let us invert Eq. (4a) and separate the inverse into two parts, one containing frequency components $\mathbf{f}$ such that $f^2 \le k^2$, the others containing frequency components such that $f^2 > k^2$,

$$F_\omega(\mathbf{r}) = F_\omega^{\mathrm{LF}}(\mathbf{r}) + F_\omega^{\mathrm{HF}}(\mathbf{r}) \, , \tag{8}$$

where

$$F_\omega^{\mathrm{LF}}(\mathbf{r}) = \int_{f^2 \le k^2} \hat{F}_\omega(\mathbf{f})\, e^{i\mathbf{f}\cdot\mathbf{r}}\, d^2\mathbf{f} \, , \tag{8a}$$

$$F_\omega^{\mathrm{HF}}(\mathbf{r}) = \int_{f^2 > k^2} \hat{F}_\omega(\mathbf{f})\, e^{i\mathbf{f}\cdot\mathbf{r}}\, d^2\mathbf{f} \, . \tag{8b}$$

We will refer to $F_\omega^{\mathrm{LF}}(\mathbf{r})$ and to $F_\omega^{\mathrm{HF}}(\mathbf{r})$ as the *low-frequency part* and the *high-frequency part* of $F_\omega(\mathbf{r})$, respectively. It follows at once, on inverting Eqs. (8a) and (8b), that

$$\int_{(z=0)} F_\omega^{\mathrm{LF}}(\mathbf{r})\, e^{-i\mathbf{f}\cdot\mathbf{r}}\, d^2\mathbf{r} = 0 \quad \text{when } f^2 > k^2 \, , \tag{9a}$$

$$\int_{(z=0)} F_\omega^{\mathrm{HF}}(\mathbf{r})\, e^{-i\mathbf{f}\cdot\mathbf{r}}\, d^2\mathbf{r} = 0 \quad \text{when } f^2 \le k^2 \, . \tag{9b}$$

In view of the Eq. (1), we may also express the cross-spectral density function of the light in the source plane as the sum of low- and high-frequency parts,

$$W(\mathbf{r}_1, \mathbf{r}_2, \omega) = W^{\mathrm{LF}}(\mathbf{r}_1, \mathbf{r}_2, \omega) + W^{\mathrm{HF}}(\mathbf{r}_1, \mathbf{r}_2, \omega) \, , \tag{10}$$

where

$$W^{\mathrm{LF}}(\mathbf{r}_1, \mathbf{r}_2, \omega) = F_\omega^{\mathrm{LF}}(\mathbf{r}_1 - \mathbf{r}_2) \, , \tag{10a}$$

$$W^{\mathrm{HF}}(\mathbf{r}_1, \mathbf{r}_2, \omega) = F_\omega^{\mathrm{HF}}(\mathbf{r}_1 - \mathbf{r}_2) \, . \tag{10b}$$

Let us now change the variable of integration in the integral on the right-hand side (r.h.s.) of Eq. (8a) by setting $\mathbf{f} = k\mathbf{s}_\perp$. Then

$$F_\omega^{\mathrm{LF}}(\mathbf{r}) = k^2 \int_{\mathbf{s}_\perp^2 \le 1} \hat{F}_\omega(k\mathbf{s}_\perp)\, e^{ik\mathbf{s}_\perp\cdot\mathbf{r}}\, d^2\mathbf{s}_\perp \, . \tag{11}$$

Now we have, from Eq. (3)

$$\hat{F}_\omega(k\mathbf{s}_\perp) = \frac{1}{k^2 A}\, \frac{J_\omega(\mathbf{s})}{1 - s_x^2 - s_y^2} \, , \tag{12}$$

where we used the facts that $\mathbf{s}_\perp \equiv (s_x, s_y)$ and that $\cos^2\theta = 1 - s_x^2 - s_y^2$. If we substitute from Eq. (12) into Eq. (11), we obtain the following expression for $F_\omega^{\mathrm{LF}}(\mathbf{r})$:

$$F_\omega^{\mathrm{LF}}(\mathbf{r}) = \frac{1}{A} \iint_{s_x^2 + s_y^2 \le 1} \frac{J_\omega(\mathbf{s})}{1 - s_x^2 - s_y^2}\, e^{ik(s_x x + s_y y)}\, ds_x\, ds_y \, , \tag{13}$$

where $\mathbf{r} \equiv (x, y)$.

From Eqs. (10) and (13), we now obtain the *theorem*: The cross-spectral density function $W(\mathbf{r}_1, \mathbf{r}_2, \omega)$ of a two-dimensional statistically homogeneous source, which is large compared with both the square of the wavelength and the coherence area in the source plane and which generates a radiant intensity distribution $J_\omega(\mathbf{s})$, is necessarily of the form (10) where the low-frequency part W^{LF} of W is given by

$$W^{\mathrm{LF}}(\mathbf{r}_1, \mathbf{r}_2, \omega) = \frac{1}{A} \iint_{s_x^2 + s_y^2 \le 1} \frac{J_\omega(\mathbf{s})}{1 - s_x^2 - s_y^2} \times e^{ik[s_x(x_1 - x_2) + s_y(y_1 - y_2)]}\, ds_x\, ds_y \, , \tag{14a}$$

$[\mathbf{r}_1 \equiv (x_1, y_1);\ \mathbf{r}_2 \equiv (x_2, y_2)]$, and the high-frequency part W^{HF} of W is a function of the form (10b), that satisfied the constraint imposed by Eq. (9b), i.e.,

$$\int_{(z=0)} W^{\mathrm{HF}}(\mathbf{r}_1 + \mathbf{r}', \mathbf{r}_1)\, e^{-i\mathbf{f}\cdot\mathbf{r}'}\, d^2\mathbf{r}' = 0 \tag{14b}$$

for all two-dimensional vectors $\mathbf{f}$ such that $f^2 \le k^2$, $\mathbf{r}_1$ being the position vector of an arbitrary source point.

It seems worthwhile to stress that our theorem implies that, given the angular distribution $J_\omega(\mathbf{s})$ of the radiant intensity, the low-frequency part of the cross-spectral density function of the light distribution across the source is uniquely specified by Eq. (14a). Whether or not the high-frequency part can be chosen quite arbitrarily [subject only to the constraint (14b)] cannot be decided without further analysis, because when the source is finite, some of the Fourier transforms appearing in our analysis have well-known analytic properties that may restrict the (and possibly lead to a unique) high-frequency part as well. We will not pursue this point in the present paper.

We may express our main result in an alternative form. Setting again

$$\mathbf{f} = k\mathbf{s}_\perp \tag{15a}$$

or, in cartesian form,

$$f_x = ks_x \, , \quad f_y = ks_y \, , \tag{15b}$$

Eq. (12) may be rewritten as

$$\hat{F}_\omega(\mathbf{f}) = \frac{1}{A}\, \frac{J_\omega\left(f_x/k,\, f_y/k,\, +\sqrt{k^2 - f_x^2 - f_y^2}/k\right)}{k^2 - f_x^2 - f_y^2} \tag{16}$$

for all two-dimensional vectors $\mathbf{f}$ such that

$$\mathbf{f}^2 \equiv f_x^2 + f_y^2 \le k^2 \, . \tag{17}$$

It is clear now that the result expressed by Eq. (14a) of our theorem is equivalent to the statement that for low spatial frequencies, i.e., for $f^2 \le k^2$, the spatial-frequency spectrum $\hat{F}_\omega(\mathbf{f})$ of the cross-spectral density function is given by Eq. (16).

II. LAMBERTIAN SOURCES

We will now apply the theorem derived in the preceding section to determine the general form of the cross-spectral density function of a two-dimensional statistically homogeneous lambertian source.

For a lambertian source the radiant intensity function is proportional to $\cos\theta$, i.e.,

$$J_\omega(\mathbf{s}) = C\cos\theta \ , \tag{18}$$

where C is a constant. If we substitute from Eq. (18) into Eq. (14a) and use the relation

$$\cos\theta = +\sqrt{1 - s_x^2 - s_y^2} \ , \tag{19}$$

Eq. (14a) becomes

$$W^{\mathrm{LF}}(\mathbf{r}_1, \mathbf{r}_2, \omega) = \frac{C}{A} \iint_{s_x^2+s_y^2 \le 1} \frac{1}{\sqrt{1 - s_x^2 - s_y^2}} \times e^{ik[s_x(x_1-x_2)+s_y(y_1-y_2)]} ds_x\, ds_y \ . \tag{20}$$

The integral on the r.h.s. of Eq. (20) may be readily evaluated. To do so, we change from cartesian to polar variables by means of the transformations

$$s_x = \rho\cos\chi \ , \qquad s_y = \rho\sin\chi \ , \tag{21a}$$

$$x_1 - x_2 = r'\cos\phi \ , \quad y_1 - y_2 = r'\sin\phi \ , \tag{21b}$$

and obtain for W^{LF} the expression

$$W^{\mathrm{LF}}(\mathbf{r}_1, \mathbf{r}_2, \omega) = \frac{C}{A} \int_0^{2\pi} d\chi \int_0^1 \frac{1}{\sqrt{1-\rho^2}}\, e^{ikr'\rho\cos(\phi-\chi)}\rho\, d\rho \ . \tag{22}$$

The integral with respect to χ is well known to be expressible in terms of a Bessel function,[7]

$$\int_0^{2\pi} e^{ikr'\rho\cos(\phi-\chi)} d\chi = 2\pi J_0(kr'\rho) \ , \tag{23}$$

where J_0 is the Bessel function of the first kind and zero order. Hence Eq. (22) reduces to

$$W^{\mathrm{LF}}(\mathbf{r}_1, \mathbf{r}_2, \omega) = \frac{2\pi C}{A} \int_0^1 \frac{1}{\sqrt{1-\rho^2}} J_0(kr'\rho)\rho\, d\rho \ . \tag{24}$$

The integral on the r.h.s. of Eq. (24) may also be evaluated in closed form[8]; we then obtain the simple expression for $W^{\mathrm{LF}}(\mathbf{r}_1, \mathbf{r}_2, \omega)$:

$$W^{\mathrm{LF}}(\mathbf{r}_1, \mathbf{r}_2, \omega) = \frac{2\pi C}{A} \left[\frac{\sin(k|\mathbf{r}_1 - \mathbf{r}_2|)}{k|\mathbf{r}_1 - \mathbf{r}_2|}\right] \ . \tag{25}$$

We have thus established the result that the low-frequency part of the cross-spectral density function of any large[9] two-dimensional statistically homogeneous lambertian source is proportional to the function[10] $\sin(k|\mathbf{r}_1 - \mathbf{r}_2|)/(k|\mathbf{r}_1 - \mathbf{r}_2|)$. Although it may at first seem quite remarkable that all large two-dimensional statistically homogeneous lambertian sources have a cross-spectral density function whose low-frequency parts are of the same form, the result, as we just saw, is an immediate consequence of our general theorem.

We may express the result that we just established in an alternative way by making use of Eq. (16). From Eq. (18), we have, for a lambertian source, if we use Eqs. (19) and (15b)

$$J_\omega\left(f_x/k, f_y/k, \sqrt{k^2 - f_x^2 - f_y^2}/k\right) = C\sqrt{1 - (f_x/k)^2 - (f_y/k)^2} \ , \tag{26}$$

and hence Eq. (16) gives for this case

$$\hat{F}_\omega(\mathbf{f}) = \frac{C}{kA} \frac{1}{\sqrt{k^2 - \mathbf{f}^2}} \ , \tag{27}$$

for $f^2 \le k^2$. Thus we see that in order that a large[9] two-dimensional statistically homogeneous source radiate in accordance with Lambert's law [Eq. (18)], the spatial-frequency spectrum of its cross-spectral density function must be of the form given by Eq. (27) for all low-frequency components, i.e., for all spatial-frequency pairs $\mathbf{f} \equiv (f_x, f_y)$ that satisfy the inequality (17).

Figure 2a illustrates the behavior of the functions $W^{\mathrm{LF}}(\mathbf{r}_1, \mathbf{r}_2, \omega)$, $\hat{F}_\omega(\mathbf{f})$, and $J_\omega(\mathbf{s})$ for a lambertian source, as given by Eqs. (25), (27), and (18).

We conclude this section by demonstrating that a source of blackbody radiation (which is well known to be lambertian) does indeed have a cross-spectral density

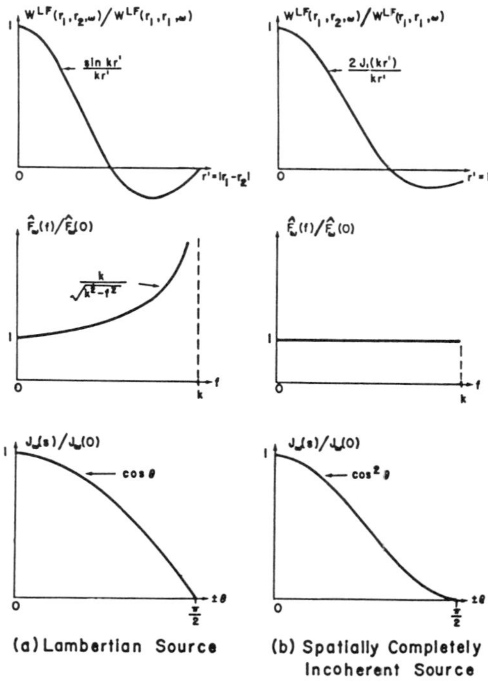

FIG. 2. The low-frequency part $W^{\mathrm{LF}}(\mathbf{r}_1, \mathbf{r}_2, \omega)$ of the cross-spectral density function, the low-frequency part of its spatial-frequency spectrum $F_\omega(\mathbf{f})$, and the angular distribution of radiant intensity $J_\omega(\mathbf{s})$ (a) for a large two-dimensional lambertian source (Ref. 9) and (b) for a large two-dimensional spatially incoherent source. All of the curves are normalized as indicated on the top of the vertical axes.

function of the form predicted by our theorem. Consider a large cavity that contains blackbody radiation at absolute temperature T. It has been shown elsewhere[11] that the 3×3 cross-spectral-correlation tensor of the electric field inside the cavity is given by

$$W_{ij}(\mathbf{r}_1, \mathbf{r}_2, \omega) = \pi A \left\{ \delta_{ij}\left[j_0(kr') - \frac{1}{kr'} j_1(kr') \right] + \frac{r_i' r_j'}{r'^2} j_2(kr') \right\}, \tag{28}$$

where $r' = |\mathbf{r}_1 - \mathbf{r}_2|$ as before. In Eq. (28) the subscripts i, j label cartesian components, A is the blackbody spectrum given by Planck's law,

$$A = \frac{2\hbar w^3}{\pi c^3} \frac{1}{e^{\hbar w/KT} - 1}, \tag{28a}$$

($\hbar = h/2\pi$, h = Planck's constant, K = Boltzmann constant), and j_0, j_1, j_2 are spherical Bessel functions of orders 0, 1, and 2, respectively, i.e.,

$$j_0(kr') = \frac{\sin kr'}{kr'},$$

$$j_1(kr') = \frac{\sin kr'}{(kr')^2} - \frac{\cos kr'}{kr'}, \tag{28b}$$

$$j_2(kr') = \left[\frac{3}{(kr')^3} - \frac{1}{kr'}\right]\sin kr' - \frac{3}{(kr')^2}\cos kr'.$$

Suppose that the cavity that contains the blackbody radiation has the form of a rectangular parallelepiped and suppose that an opening S is made in one of its sides, (which we take to coincide with the plane $z=0$), whose size is large compared to the square of the wavelength $\lambda = 2\pi c/\omega$. S acts as a two-dimensional statistically homogeneous source. If we take $\mathbf{r}_1$ and $\mathbf{r}_2$ to be position vectors of any two points on S, then to a good approximation Eq. (28) represents the cross-spectral density tensor of the radiation leaving S, except for points close to the boundary of S. Now, unlike the cross-spectral density function $W(\mathbf{r}_1, \mathbf{r}_2, \omega)$ that we employed previously, the cross-spectral density tensor $W_{ij}(\mathbf{r}_1, \mathbf{r}_2, \omega)$ takes also into account the polarization properties of the source field. It seems reasonable to assume that the quantity that plays the role of the cross-spectral density function of the scalar theory will be the trace of the cross-spectral density tensor, since the trace is a scalar invariant, independent of the choice of the coordinate axes.[12] Now the trace of the tensor given by Eq. (28) may be evaluated at once and is

$$\mathrm{Tr} W_{ij}(\mathbf{r}_1, \mathbf{r}_2, \omega) = \pi A \left\{ 3j_0(kr') - \frac{3}{kr'} j_1(kr') + j_2(kr') \right\} \tag{29}$$

or, from the explicit expressions (28b) for the spherical Bessel functions,

$$\mathrm{Tr} W_{ij}(\mathbf{r}_1, \mathbf{r}_2, \omega) = C'\left(\frac{\sin kr'}{kr'}\right), \tag{30a}$$

where

$$C' = \frac{4\hbar\omega^3}{c^3} \frac{1}{e^{\hbar\omega/KT} - 1}. \tag{30b}$$

Finally, using the fact that the function $\sin(kr')/(kr')$ has only low-spatial-frequency components ($f^2 \le k^2$) [Eqs. (20), (25), and (15b)], we see at once from Eq. (30a), that for our blackbody source,

$$\mathrm{Tr} W_{ij}(\mathbf{r}_1, \mathbf{r}_2, \omega) = \mathrm{Tr}\, W_{ij}^{\mathrm{LF}}(\mathbf{r}_1, \mathbf{r}_2, \omega) + \mathrm{Tr}\, W_{ij}^{\mathrm{HF}}(\mathbf{r}_1, \mathbf{r}_2, \omega), \tag{31}$$

where

$$\mathrm{Tr} W_{ij}^{\mathrm{LF}}(\mathbf{r}_1, \mathbf{r}_2, \omega) = C' \frac{\sin k|\mathbf{r}_1 - \mathbf{r}_2|}{k|\mathbf{r}_1 - \mathbf{r}_2|}, \tag{32a}$$

$$\mathrm{Tr} W_{ij}^{\mathrm{HF}}(\mathbf{r}_1, \mathbf{r}_2, \omega) = 0, \tag{32b}$$

and C' is given by Eq. (30b). We have thus demonstrated that the trace of the cross-spectral density tensor of our blackbody source is indeed of the form Eq. (25), which we showed to be appropriate to a lambertian source. That in the present case $\mathrm{Tr}\, W_{ij}^{\mathrm{HF}}(\mathbf{r}_1, \mathbf{r}_2, \omega) \equiv 0$ is due to the fact that blackbody radiation field is a free field and hence contains no evanescent-plane-wave components. Of course, if effects that arise from the finite size of the source were taken into account, evanescent waves would necessarily be introduced and the trace of the cross-spectral-density tensor would presumably acquire a nonvanishing high-frequency part.

III. SPATIALLY INCOHERENT SOURCES

There is a widely held belief that lambertian sources are spatially completely incoherent. Actually this is not so. It has been shown by Skinner[3] and by Marchand and Wolf[5,13] that the angular distribution of radiant energy from a spatially incoherent source cannot obey Lambert's law [Eq. (18)], according to which the radiant intensity falls off as $\cos\theta$, but instead falls off as $\cos^2\theta$. To gain a clearer understanding of the differences between lambertian sources and spatially completely incoherent sources, we will now briefly apply our general formulae to the later case.

In the (idealized) case when the source is spatially completely incoherent, the cross-spectral density function of the light distribution across it is of the form

$$W(\mathbf{r}_1, \mathbf{r}_2, \omega) = C\delta^{(2)}(\mathbf{r}_1 - \mathbf{r}_2), \tag{33}$$

where $\delta^{(2)}(\mathbf{r}_1 - \mathbf{r}_2)$ is the two-dimensional Dirac delta function and C is a constant. It follows from Eqs. (33) and (4b) that, in this case,

$$\hat{F}(\mathbf{f}) = \frac{C}{(2\pi)^2} \int_{(z=0)} \delta^{(2)}(\mathbf{r}') e^{-i\mathbf{f}\cdot\mathbf{r}'} d^2\mathbf{r}' = \frac{C}{(2\pi)^2}, \tag{34}$$

and hence, according to Eqs. (3) and (34), the radiant intensity from a spatially incoherent source is given by

$$J_\omega(\mathbf{s}) = \left(\frac{k}{2\pi}\right)^2 AC\cos^2\theta, \tag{35}$$

i.e., it is proportional to $\cos^2\theta$, as we asserted earlier.

The low-frequency part of the cross-spectral density function of a spatially completely incoherent source is according to Eqs. (10a), (8a), and (34) given by

$$W^{LF}(\mathbf{r}_1, \mathbf{r}_2, \omega) = \left(\frac{k}{2\pi}\right)^2 C \times \iint_{s_x^2+s_y^2 \leq 1} e^{ik[s_x(x_1-x_2)+s_y(y_1-y_2)]} ds_x\, ds_y \tag{36}$$

$[\mathbf{r}_1 = (x_1, y_1),\ \mathbf{r}_2 = (x_2, y_2)]$, where in the double integral we changed the variables of integration from f_x, f_y to s_x, s_y by means of the simple transformation (15b). To evaluate the integral on the r.h.s. of Eq. (36) we make a further transformation, from cartesian to polar coordinates, indicated by Eqs. (21). The integral with respect to the polar angle χ is the same as one that we encountered earlier and is given by Eq. (23). Hence Eq. (36) reduces to

$$W^{LF}(\mathbf{r}_1, \mathbf{r}_2, \omega) = \frac{k^2 C}{2\pi} \int_0^1 J_0(kr'\rho)\rho\, d\rho \ . \tag{37}$$

The integral on the r.h.s. may also be readily evaluated[14] and leads to the following expression for W^{LF}:

$$W^{LF}(\mathbf{r}_1, \mathbf{r}_2, \omega) = \frac{k^2 C}{4\pi} \left(\frac{2J_1(k|\mathbf{r}_1 - \mathbf{r}_2|)}{k|\mathbf{r}_1 - \mathbf{r}_2|}\right), \tag{38}$$

where J_1 is the Bessel function of the first kind and of the first order. Thus, we have now shown that the low-frequency part of the cross-spectral density function of a spatially completely incoherent two-dimensional source is necessarily of the form given by Eq. (38).

In connection with the result that we just established, we wish to mention a related investigation of Beran and Parrent.[2] They were led to the conclusion that if the cross-spectral density function, (not just its low-frequency part), of an infinite plane source is of the form given by the r.h.s. of Eq. (38), the source will give rise to the same radiation field as would be generated by a spatially incoherent source. They also asserted that the same result will hold for a finite sufficiently large source. As regards the low-frequency part of the cross-spectral density function, their result is in agreement with ours. However, in view of the remarks made in the paragraph that follows our Eq. (14b) it does not seem to us to be clear without additional considerations whether, for a finite source, their cross-spectral density function represents a limiting case associated with a physically realizable field.

In Fig. 2b, the behavior of the functions $\hat{W}^{LF}(\mathbf{r}_1, \mathbf{r}_2, \omega)$, $\hat{F}(\mathbf{f})$, and $J_\omega(\mathbf{s})$ for a spatially completely incoherent source are shown, as given by Eqs. (38), (34), and (35). Comparison of these curves with those for a lambertian source (Fig. 2a) illustrates a result that seemed apparent from the examination of other special cases in Ref. 1, namely, that changes of the correlation properties in the source plane in intervals as small as a fraction of a wavelength may lead to a substantial modification of the angular distribution of the energy radiated by the source.

*Some of the results of this investigation were presented 24 April 1974 at the Spring Meeting of the Optical Society of America, held in Washington, D.C. (J. Opt. Soc. Am. 64, 545A (1974)).

†Currently Visiting Professor, Department of Physics, University of Toronto, Toronto, Canada.

[1]E. Wolf and W. H. Carter, Opt. Commun. 13, 205 (1975).

[2]M. Beran and G. Parrent, Nuovo Cimento 27, 1049 (1963).

[3]T. J. Skinner, Ph.D. thesis (Boston University, 1965), p. 46.

[4]A. Walther, J. Opt. Soc. Am. 58, 1256 (1968).

[5]E. W. Marchand and E. Wolf, Opt. Commun. 6, 305 (1972).

[6]In this paper, the assumption of statistical homogeneity of the source is to be understood as meaning the property expressed by Eq. (1). For a fuller discussion of this point see footnote 3 of Ref. 1. We implicitly assume that for a finite source, the function $F_\omega(\mathbf{r})$, introduced in Eq. (1), can be defined for all values of $\mathbf{r}$ by analytic continuation; however, within the accuracy of the present theory (which applies only to sufficiently large sources), the exact form of $F_\omega(\mathbf{r})$ is immaterial when $r = |\mathbf{r}|$ is large compared with the linear dimensions of the area of coherence on the source.

[7]G. N. Watson, *A Treatise on the Theory of Bessel Functions* (Cambridge U. P., Cambridge, 1922), p. 26, Eq. (5) (with an obvious substitution).

[8]I. S. Gradshteyn and I. M. Ryzhik, *Tables of Integrals, Series and Products* (Academic, New York, 1965), p. 682, formula 2 of 6.554.

[9]The term "large" is to be understood here in the sense explained in the paragraph that follows Eq. (2).

[10]Somewhat-less-precise versions of this result were obtained previously, in Refs. 2 and 4.

[11]C. L. Mehta and E. Wolf, Phys. Rev. 161, 1328 (1967), Eqs. (3.9)–(3.11).

[12]In this connection, see the papers by B. Karczewski: (a) Phys. Lett. 5, 191 (1963); (b) Nuovo Cimento 30, 906 (1963).

[13]E. W. Marchand and E. Wolf, J. Opt. Soc. Am. 64, 1219 (1974).

[14]M. Born and E. Wolf, *Principles of Optics*, 4th ed. (Pergamon, New York, 1975), Sec. 8.5.2.

Spectral coherence and the concept of cross-spectral purity*

L. Mandel and E. Wolf

Department of Physics and Astronomy, The University of Rochester, Rochester, New York 14627

(Received 26 January 1976)

A new measure of correlations in optical fields, introduced in recent investigations on radiometry with partially coherent sources, is studied and applied to the analysis of interference experiments. This measure, which we call the complex degree of spectral coherence, or the spectral correlation coefficient, characterizes the correlations that exist between the spectral components at a given frequency in the light oscillations at two points in a stationary optical field. A relation between this degree of correlation and the usual degree of coherence is obtained and the role that the complex degree of spectral coherence plays in the spectral structure of a two-beam interference pattern is examined. It is also shown that the complex degree of spectral coherence provides a clear insight into the physical significance of cross-spectral purity. When the optical field at two points is cross-spectrally pure, the absolute value of the complex degree of spectral coherence at these points is found to be the same for every frequency component of the light. This fact is reflected in the visibility of the spectral components of the interference fringes formed by light from these points.

I. INTRODUCTION

Correlations in stationary optical fields are traditionally characterized by the complex degree of coherence, which is defined as the normalized cross-correlation function of the optical field at two points. Thus if $V(\mathbf{r}_1, t)$ and $V(\mathbf{r}_2, t)$ are the analytic signal representations of the light oscillations at two points P_1 and P_2, specified by position vectors $\mathbf{r}_1$ and $\mathbf{r}_2$, respectively, at time t, the complex degree of coherence is given by the formula[1,2]

$$\gamma(\mathbf{r}_1, \mathbf{r}_2, \tau) \equiv \Gamma(\mathbf{r}_1, \mathbf{r}_2, \tau)/[I(\mathbf{r}_1)I(\mathbf{r}_2)]^{1/2} , \tag{1.1}$$

where

$$\Gamma(\mathbf{r}_1, \mathbf{r}_2, \tau) \equiv \langle V^*(\mathbf{r}_1, t)V(\mathbf{r}_2, t+\tau)\rangle \tag{1.2a}$$

is the mutual coherence function of the light, and

$$I(\mathbf{r}) \equiv \langle V^*(\mathbf{r}, t)V(\mathbf{r}, t)\rangle \equiv \Gamma(\mathbf{r}, \mathbf{r}, 0) \tag{1.2b}$$

is the average light intensity. In Eqs. (1.2) the asterisk denotes the complex conjugate and the sharp brackets denote the ensemble average. Under usual circumstances this average will, of course, be equal to the time average, because the field is generally ergodic. As is well known,

$$0 \leqslant |\gamma(\mathbf{r}_1, \mathbf{r}_2, \tau)| \leqslant 1 \tag{1.3}$$

for all possible values of the arguments of γ.

The interpretation of γ as a degree of coherence comes from the role that γ plays in the description of interference fringes formed in a two-beam interference experiment. For example, when the light is passed through two pinholes, located at the points P_1 and P_2 on a screen $\mathcal{A}$, the average light intensity $I(\mathbf{r})$ at a point P specified by the position vector $\mathbf{r}$ on a screen $\mathcal{B}$ some distance away from $\mathcal{B}$ (see Fig. 1), under the usual experimental conditions, is given by

$$I(\mathbf{r}) = I^{(1)}(\mathbf{r}) + I^{(2)}(\mathbf{r}) + 2[I^{(1)}(\mathbf{r})I^{(2)}(\mathbf{r})]^{1/2}\,\mathrm{Re}[\gamma(\mathbf{r}_1, \mathbf{r}_2, \tau_{12})] , \tag{1.4}$$

where

$$\tau_{12} \equiv \frac{s_1 - s_2}{c} = \frac{PP_1 - PP_2}{c} . \tag{1.5}$$

In Eq. (1.4), $I^{(1)}(\mathbf{r}) = \Gamma(\mathbf{r}, \mathbf{r}, 0)$ represents the average intensity of the light at P that would reach that point if only the pinhole at P_1 were open, and $I^{(2)}(\mathbf{r})$ has a similar interpretation. τ_{12} denotes the difference in the times needed for the light to travel from P_1 to P and from P_2 to P, c denotes the vacuum speed of light, and Re the real part.

From Eq. (1.4) one may readily deduce[1,2] that if the two beams are of equal intensities, i.e., if $I^{(1)} = I^{(2)}$, and if the light is quasimonochromatic, the visibility of fringes at the point $P(\mathbf{r})$ on the screen $\mathcal{B}$ is given by

$$\mathcal{V}(\mathbf{r}) \equiv \frac{I_{\max}(\mathbf{r}) - I_{\min}(\mathbf{r})}{I_{\max}(\mathbf{r}) + I_{\min}(\mathbf{r})} = |\gamma(\mathbf{r}_1, \mathbf{r}_2, \tau_{12})| , \tag{1.6}$$

where $I_{\max}(\mathbf{r})$ and $I_{\min}(\mathbf{r})$ denote the maximum and the minimum values of the average intensity in the immediate neighborhood of the point $P(\mathbf{r})$. The formula (1.6) shows clearly that $|\gamma|$, which was introduced by Eqs. (1.1) and (1.2) as a correlation coefficient, is also a measure of the sharpness of the interference fringes to which the light from the pinholes gives rise. It is well known [cf. Ref. 1, Sec. 10.4.1 and Ref. 2, Eq. (3.25)] that the phase of γ also has an operational meaning, being related to the location of the fringe maxima.

The complex degree of coherence $\gamma(\mathbf{r}_1, \mathbf{r}_2, \tau)$ is a measure of the field correlations in the *space–time* domain. However, many physical situations involving coherence properties of light are more naturally described in the *space–frequency* domain. For example, in the study of the mechanism that gives rise to density fluctuations in various substances by means of light scattering experiments, the required information about the scatterer is often most directly described in terms of distribution functions and correlations that depend on position and on the frequency of the light. A measure of correlation in the space–frequency domain has actually been employed in several recent publications[3–5] dealing with radiometry with sources of any state of coherence, but it has not been investigated in any detail so far. It is the purpose of this paper to present some general results relating to this newly defined correlation coefficient, which we call the *complex degree of spectral coherence*, or *the spectral correlation coefficient*, and to show that it leads to an elegant description of second-order coherence phenomena in which the space–frequency dependence is of main physical interest. We illustrate the usefulness of this description by showing that the complex degree of spectral coherence provides a clearer insight into the subject of cross-spectral purity of light[6] that seems not to be well understood even by many workers in optical coherence theory.

II. THE COMPLEX DEGREE OF SPECTRAL COHERENCE

Let $V(\mathbf{r}, t)$ again denote the analytic signal that represents a fluctuating optical field (assumed to be stationary and ergodic) at a point P specified by position vector $\mathbf{r}$, at time t. We will ignore polarization effects, so that we consider $V(\mathbf{r}, t)$ to be a scalar function.

We express $V(\mathbf{r}, t)$ in the form of a generalized Fourier integral with respect to the time variable[7]:

$$V(\mathbf{r}, t) = \int_0^{\infty} v(\mathbf{r}, \nu)\, e^{-2\pi i \nu t}\, d\nu . \tag{2.1}$$

[The lower limit of integration on the right-hand side of Eq. (2.1) is zero rather than minus infinity because $V(\mathbf{r}, t)$ is an analytic signal.] Fourier inversion of Eq. (2.1) then gives

$$v(\mathbf{r}, \nu) = \int_{-\infty}^{\infty} V(\mathbf{r}, t)\, e^{2\pi i \nu t}\, dt . \tag{2.2}$$

We now form the product $v^*(\mathbf{r}_1, \nu)\, v(\mathbf{r}_2, \nu')$ for two different points $\mathbf{r}_1, \mathbf{r}_2$ and two different frequencies ν, ν' and take the average over the ensemble. After interchanging the order of integration and averaging we obtain

$$\begin{aligned}\langle v^*(\mathbf{r}_1, \nu)v(\mathbf{r}_2, \nu')\rangle &= \iint_{-\infty}^{\infty} \langle V^*(\mathbf{r}_1, t)V(\mathbf{r}_2, t')\rangle\, e^{2\pi i(\nu' t' - \nu t)}\, dt\, dt' \\ &= \iint_{-\infty}^{\infty} \langle V^*(\mathbf{r}_1, t)V(\mathbf{r}_2, t+\tau)\rangle\, e^{2\pi i(\nu' - \nu)t} e^{2\pi i \nu' \tau}\, dt\, d\tau ,\end{aligned} \tag{2.3}$$

where the second expression on the right follows from the first by the substitution $t' = t + \tau$. The expression

$\langle V^*(\mathbf{r}_1, t)V(\mathbf{r}_2, t+\tau)\rangle$ in the integrand is just the mutual coherence function $\Gamma(\mathbf{r}_1, \mathbf{r}_2, \tau)$ defined by Eq. (1.2a), and for a stationary optical field it is independent of t. The integration with respect to t can then be carried out immediately, and yields a Dirac δ function, so that we obtain

$$\langle v^*(\mathbf{r}_1, \nu)v(\mathbf{r}_2, \nu')\rangle = \delta(\nu - \nu')W(\mathbf{r}_1, \mathbf{r}_2, \nu) , \tag{2.4}$$

where

$$W(\mathbf{r}_1, \mathbf{r}_2, \nu) = \int_{-\infty}^{\infty} \Gamma(\mathbf{r}_1, \mathbf{r}_2, \tau)\, e^{2\pi i\nu\tau}\, d\tau . \tag{2.5a}$$

The function $W(\mathbf{r}_1, \mathbf{r}_2, \nu)$ is, of course, the cross-spectral density function (also known as the cross-power spectrum) of the optical field at the points $P_1(\mathbf{r}_1)$ and $P_2(\mathbf{r}_2)$. Like $v(\mathbf{r}, \nu)$, $W(\mathbf{r}_1, \mathbf{r}_2, \nu)$ vanishes for $\nu < 0$, as $\Gamma(\mathbf{r}_1, \mathbf{r}_2, \tau)$ is an analytic signal.[1] If we invert the Fourier integral (2.5a), we obtain an expression for the mutual coherence function in terms of the cross-spectral density function,

$$\Gamma(\mathbf{r}_1, \mathbf{r}_2, \tau) = \int_0^{\infty} W(\mathbf{r}_1, \mathbf{r}_2, \nu)\, e^{-2\pi i\nu\tau}\, d\nu . \tag{2.5b}$$

The fact expressed by Eqs. (2.5), namely, that the mutual coherence function and the cross-spectral density function form a Fourier transform pair, is of course the optical analogue of a well known generalization of the Wiener-Khintchine theorem in the theory of stationary random processes.

As is clear from its definition (1.2a), the mutual coherence function $\Gamma(\mathbf{r}_1, \mathbf{r}_2, \tau)$ describes correlations in the space–time domain. The formula (2.4) shows that its Fourier transform, the cross-spectral density function $W(\mathbf{r}_1, \mathbf{r}_2, \nu)$, also characterizes correlations, namely, those between the (generalized) Fourier components of the light oscillations at two points in the field. More specifically, Eq. (2.4) implies that Fourier components of different frequencies ν and ν' are uncorrelated (which is a reflection of the stationarity of the field), while the "strength" of the correlation of Fourier components of the same frequency is characterized by the nonsingular part of the right-hand side of Eq. (2.4), i.e., by the cross-spectral density function $W(\mathbf{r}_1, \mathbf{r}_2, \nu)$. Thus we may say that the cross-spectral density function is a measure of correlations in the space–frequency domain. We will assume that for each pair of values $\mathbf{r}_1$ and $\mathbf{r}_2$, it is a continuous function of ν. This assumption excludes strictly monochromatic waves, but such waves are, of course, not found in nature.

It follows at once from Eq. (2.4) that for each frequency ν, the cross-spectral density function $W(\mathbf{r}_1, \mathbf{r}_2, \nu)$ is Hermitian in the sense that

$$W(\mathbf{r}_2, \mathbf{r}_1, \nu) = W^*(\mathbf{r}_1, \mathbf{r}_2, \nu) . \tag{2.6}$$

Moreover, as we show in the Appendix, the cross-spectral density function is a non-negative definite, continuous matrix,[8] in the sense that, for any positive frequency ν, any set of N position vectors $\mathbf{r}_1, \mathbf{r}_2, \ldots, \mathbf{r}_N$, and any set of N complex numbers $a_1, a_2, \ldots, a_N$,

$$\sum_{i=1}^{N}\sum_{j=1}^{N} a_i^* a_j W(\mathbf{r}_i, \mathbf{r}_j, \nu) \geq 0 . \tag{2.7}$$

With the choice $N=1$, Eq. (2.7) implies that

$$W(\mathbf{r}, \mathbf{r}, \nu) \geq 0 , \tag{2.8}$$

i.e., that the spectral density at any point in the field is non-negative as one would, of course, expect for a power density. Of special interest for the present investigation is the consequence of Eq. (2.7) with the choice $N=2$. In that case Eq. (2.7) is readily seen to imply the additional condition

$$W(\mathbf{r}_1, \mathbf{r}_1, \nu)\, W(\mathbf{r}_2, \mathbf{r}_2, \nu) - W(\mathbf{r}_1, \mathbf{r}_2, \nu)\, W(\mathbf{r}_2, \mathbf{r}_1, \nu) \geq 0 . \tag{2.9}$$

It is clear from this inequality, if use is also made of the Hermiticity relation (2.6), that the quantity

$$\mu(\mathbf{r}_1, \mathbf{r}_2, \nu) \equiv W(\mathbf{r}_1, \mathbf{r}_2, \nu)/[W(\mathbf{r}_1, \mathbf{r}_1, \nu)\, W(\mathbf{r}_2, \mathbf{r}_2, \nu)]^{1/2} \tag{2.10}$$

is normalized so that for all possible values of the arguments $\mathbf{r}_1$, $\mathbf{r}_2$ and ν,

$$0 \leq |\mu(\mathbf{r}_1, \mathbf{r}_2, \nu)| \leq 1 , \tag{2.11}$$

with the understanding that cases in which $W(\mathbf{r}_1, \mathbf{r}_1, \nu)$ or $W(\mathbf{r}_2, \mathbf{r}_2, \nu)$ vanishes are to be interpreted in an appropriate limiting sense. Because the cross-spectral density function $W(\mathbf{r}_1, \mathbf{r}_2, \nu)$ characterizes correlations in the space–frequency domain, as was noted earlier, it seems natural to call the quantity $\mu(\mathbf{r}_1, \mathbf{r}_2, \nu)$, defined by Eq. (2.10), *the complex degree of spectral coherence* or *the spectral correlation coefficient* at frequency ν of the optical field at the points $P_1(\mathbf{r}_1)$ and $P_2(\mathbf{r}_2)$. In Sec. III we show that this quantity plays a similar role in connection with the spectral structure of the fringes obtained when the light from $P_1(\mathbf{r}_1)$ and $P_2(\mathbf{r}_2)$ is allowed to interfere, as does $\gamma(\mathbf{r}_1, \mathbf{r}_2, \tau)$ for the distribution of the average light intensity $I(\mathbf{r})$ in the interference pattern.

Although the cross-spectral density function $W(\mathbf{r}_1, \mathbf{r}_2, \nu)$ and the mutual coherence function $\Gamma(\mathbf{r}_1, \mathbf{r}_2, \tau)$ form a Fourier transform pair, our complex degree of spectral coherence $\mu(\mathbf{r}_1, \mathbf{r}_2, \nu)$ and the usual complex degree of coherence $\gamma(\mathbf{r}_1, \mathbf{r}_2, \tau)$ are related in a slightly more complicated manner. Let us denote the Fourier transform of $\gamma(\mathbf{r}_1, \mathbf{r}_2, \tau)$ by $\phi(\mathbf{r}_1, \mathbf{r}_2, \nu)$:

$$\phi(\mathbf{r}_1, \mathbf{r}_2, \nu) = \int_{-\infty}^{\infty} \gamma(\mathbf{r}_1, \mathbf{r}_2, \tau)\, e^{2\pi i\nu\tau}\, d\tau . \tag{2.12}$$

The quantity $\phi(\mathbf{r}_1, \mathbf{r}_2, \nu)$ is sometimes known as the normalized cross-spectral density, but we must emphasize that it is normalized in a very different sense from the complex degree of spectral coherence $\mu(\mathbf{r}_1, \mathbf{r}_2, \nu)$, and that its modulus is not bounded by unity. If we use the definition (1.1) of γ and the formula (2.5a), we can readily express $\phi(\mathbf{r}_1, \mathbf{r}_2, \nu)$ in terms of the cross-spectral density function $W(\mathbf{r}_1, \mathbf{r}_2, \nu)$ as follows:

$$\phi(\mathbf{r}_1, \mathbf{r}_2, \nu) = W(\mathbf{r}_1, \mathbf{r}_2, \nu)/[I(\mathbf{r}_1)I(\mathbf{r}_2)]^{1/2} , \tag{2.13}$$

where

$$I(\mathbf{r}) = \Gamma(\mathbf{r}, \mathbf{r}, 0) = \int_0^{\infty} W(\mathbf{r}, \mathbf{r}, \nu)\, d\nu . \tag{2.14}$$

In particular when the two points coincide, Eq. (2.13) reduces to

$$\phi(\mathbf{r}, \mathbf{r}, \nu) = W(\mathbf{r}, \mathbf{r}, \nu)/I(\mathbf{r}) , \tag{2.15}$$

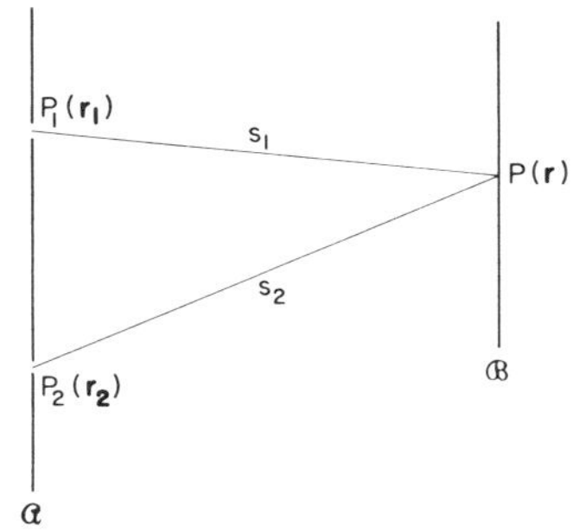

FIG. 1. Outline of the geometry for the interference experiment.

i.e., $\phi(\mathbf{r}, \mathbf{r}, \nu)$ represents the normalized spectral density at the point $\mathbf{r}$, with the normalization

$$\int_0^\infty \phi(\mathbf{r}, \mathbf{r}, \nu)\, d\nu = 1 \,. \tag{2.16}$$

It follows at once from Eqs. (2.13), (2.15), and the definition (2.10), that the complex degree of spectral coherence may be expressed in terms of the Fourier transform $\phi(\mathbf{r}_1, \mathbf{r}_2, \nu)$ of the complex degree of coherence in the form

$$\mu(\mathbf{r}_1, \mathbf{r}_2, \nu) = \phi(\mathbf{r}_1, \mathbf{r}_2, \nu)/[\varphi(\mathbf{r}_1, \mathbf{r}_1, \nu)\varphi(\mathbf{r}_2, \mathbf{r}_2, \nu)]^{1/2} \,, \tag{2.17}$$

provided cases in which $\phi(\mathbf{r}_1, \mathbf{r}_1, \nu)$ or $\phi(\mathbf{r}_2, \mathbf{r}_2, \nu)$ vanish are again interpreted in an appropriate limiting sense.

III. SPECTRAL STRUCTURE OF A TWO-BEAM INTERFERENCE PATTERN

The basic interference law (1.4) for partially coherent beams provides no information about the spectral composition of the light forming the interference pattern. We shall now derive an interference law that provides this information, and show that the complex degree of spectral coherence plays a central role in this formulation.

For this purpose we again consider the two-beam interference experiment illustrated in Fig. 1, and we examine one frequency component of the interference pattern at a time. Because of dispersive effects associated with diffraction, it is also simpler to relate the field at $P(\mathbf{r})$ to the fields at $P_1(\mathbf{r}_1)$ and $P_2(\mathbf{r}_2)$ if we choose one Fourier component at a time. Thus from the Kirchhoff diffraction formula,[1] we may write for the Fourier amplitude at $P(\mathbf{r})$,

$$v(\mathbf{r}, \nu) = K_1(\mathbf{r}_1, \mathbf{r}, \nu)\, e^{2\pi i \nu s_1/c}\, v(\mathbf{r}_1, \nu) + K_2(\mathbf{r}_2, \mathbf{r}, \nu)\, e^{2\pi i \nu s_2/c}\, v(\mathbf{r}_2, \nu) \,, \tag{3.1}$$

where K_1, K_2 are complex numbers that depend chiefly on the position of the points P, P_1, P_2, on the aperture sizes, and on the frequency.[9] In Eq. (3.1) we have separated the oscillatory phase factors from the parameters K_1, K_2, in order to make K_1, K_2 slowly varying functions of position and frequency, although this separation is not essential. By calculating the ensemble average of the product $v^*(\mathbf{r}, \nu)v(\mathbf{r}, \nu')$ for two different frequencies ν, ν' from Eq. (3.1), and making use of Eq. (2.4), we arrive at the relation

$$\begin{aligned} W(\mathbf{r}, \mathbf{r}, \nu) = {} & |K_1|^2 W(\mathbf{r}_1, \mathbf{r}_1, \nu) + |K_2|^2 W(\mathbf{r}_2, \mathbf{r}_2, \nu) \\ & + K_1^* K_2\, e^{-2\pi i \nu (s_1 - s_2)/c}\, W(\mathbf{r}_1, \mathbf{r}_2, \nu) \\ & + K_1 K_2^*\, e^{2\pi i \nu (s_1 - s_2)/c}\, W(\mathbf{r}_2, \mathbf{r}_1, \nu) \,. \end{aligned} \tag{3.2}$$

The first two terms on the right-hand side of Eq. (3.2) have simple meanings. To see this we imagine first that the pinhole at P_2 is closed, so that only light from the pinhole P_1 reaches the plane ℬ of observation. In this case, as is evident from Eq. (3.1), $K_2 = 0$ and it is then clear from Eq. (3.2) that the term

$$|K_1|^2 W(\mathbf{r}_1, \mathbf{r}_1, \nu) \equiv W^{(1)}(\mathbf{r}, \mathbf{r}, \nu) \tag{3.3a}$$

represents the spectral density of the light at the point $P(\mathbf{r})$ that would reach that point from the pinhole P_1 only. Similarly,

$$|K_2|^2 W(\mathbf{r}_2, \mathbf{r}_2, \nu) \equiv W^{(2)}(\mathbf{r}, \mathbf{r}, \nu) \tag{3.3b}$$

represents the spectral density of the light at the point $P(\mathbf{r})$ that would reach that point from the pinhole $P_2(\mathbf{r}_2)$ only.

The last two terms on the right-hand side of Eq. (3.2) may readily be expressed in terms of $W^{(1)}$, $W^{(2)}$, and μ. We have from Eqs. (2.10) and (3.3),

$$\begin{aligned} K_1^* K_2\, W(\mathbf{r}_1, \mathbf{r}_2, \nu)\, e^{-2\pi i \nu (s_1 - s_2)/c} & = [W^{(1)}(\mathbf{r}, \mathbf{r}, \nu) W^{(2)}(\mathbf{r}, \mathbf{r}, \nu)]^{1/2} \mu(\mathbf{r}_1, \mathbf{r}_2, \nu) \\ & \quad \times \exp[i\alpha - 2\pi i \nu (s_1 - s_2/c] \,, \end{aligned} \tag{3.4}$$

where α is the phase of $K_1^*(\mathbf{r}_1, \mathbf{r}, \nu)K_2(\mathbf{r}_2, \mathbf{r}, \nu)$. Under typical conditions the product $K_1^* K_2$ is usually real to a first approximation,[9] but in any case α is only weakly dependent on position and frequency.

If we now substitute from Eqs. (3.3) and (3.4) into Eq. (3.2) and use the fact that, because of the Hermiticity relation (2.6), the last two terms on the right-hand side of Eq. (3.2) are complex conjugates of each other, we finally obtain the following expression for the spectral density of the light at a typical point $\mathbf{r}$ on the screen ℬ:

$$\begin{aligned} W(\mathbf{r}, \mathbf{r}, \nu) = {} & W^{(1)}(\mathbf{r}, \mathbf{r}, \nu) + W^{(2)}(\mathbf{r}, \mathbf{r}, \nu) \\ & + 2[W^{(1)}(\mathbf{r}, \mathbf{r}, \nu) W^{(2)}(\mathbf{r}, \mathbf{r}, \nu)]^{1/2} \\ & \times \mathrm{Re}\{\mu(\mathbf{r}_1, \mathbf{r}_2, \nu) \exp[i\alpha - 2\pi i \nu (s_1 - s_2)/c]\} \,. \end{aligned} \tag{3.5}$$

This equation shows that, in general, the spectral density of the light at $P(\mathbf{r})$ is not just the sum of the spectral densities of the two beams reaching the point $P(\mathbf{r})$ from the two pinholes, but differs from it by a term depending on the complex degree of spectral coherence $\mu(\mathbf{r}_1, \mathbf{r}_2, \nu)$ of the light at the two pinholes. We will refer to Eq. (3.5) as the spectral interference law for partially coherent beams. This law is more general than the usual interference law (1.4), for Eq. (1.4) may readily be deduced from Eq. (3.5) by integration over the complete frequency range $0 \leqslant \nu < \infty$, but it is not possible to deduce Eq. (3.5) from Eq. (1.4). Equation (3.5) implies a spectral modulation of the interference pattern, examples of which have been known before,[6,10] although they were not expressed in terms of the spectral correlation coefficient $\mu(\mathbf{r}_1, \mathbf{r}_2, \nu)$.

It is clear that the spectral interference law may also be considered as expressing the energy distribution as a function of frequency in the interference pattern obtained by superposition of the light from the two pinholes. This could be measured if the fringe pattern were viewed through an ideal filter that passed only the spectral component of frequency ν of the light. However, it is important to appreciate that this filtered pattern would not be sharp, in general, in the sense of having zero minima, as one might naively expect by analogy with the idealized case of strictly coherent monochromatic light. As can be seen from Eq. (3.5), the sharpness of this pattern depends on the absolute value of the complex degree of spectral coherence.

Let us examine the spectral interference law more closely. If we denote the phase of μ by β, i.e., if we write

$$\mu(\mathbf{r}_1, \mathbf{r}_2, \nu) = |\mu(\mathbf{r}_1, \mathbf{r}_2, \nu)| \exp i\beta(\mathbf{r}_1, \mathbf{r}_2, \nu) , \tag{3.6}$$

Eq. (3.5) may be rewritten as

$$W(\mathbf{r}, \mathbf{r}, \nu) = W^{(1)}(\mathbf{r}, \mathbf{r}, \nu) + W^{(2)}(\mathbf{r}, \mathbf{r}, \nu) + 2[W^{(1)}(\mathbf{r}, \mathbf{r}, \nu) W^{(2)}(\mathbf{r}, \mathbf{r}, \nu)]^{1/2} |\mu(\mathbf{r}_1, \mathbf{r}_2, \nu)| \times \cos[\beta(\mathbf{r}_1, \mathbf{r}_2, \nu) - \delta] , \tag{3.7}$$

where

$$\delta \equiv k(s_1 - s_2) - \alpha , \tag{3.8}$$

$$k \equiv 2\pi\nu/c = 2\pi/\lambda . \tag{3.9}$$

Here, δ, which depends on the positions $\mathbf{r}$, $\mathbf{r}_1$, $\mathbf{r}_2$ and on frequency ν, denotes the phase difference introduced between the light from the two pinholes by their passage from the pinholes to P, k is the wave number, and λ the wavelength associated with the frequency ν. If the plane of observation $\mathcal{B}$ is sufficiently far from the plane $\mathcal{A}$ of the two pinholes, the spectral densities $W^{(1)}$ and $W^{(2)}$ will change slowly with the position of the point $P(\mathbf{r})$ in comparison with the last term on the right-hand side of Eq. (3.7). If, as is usually the case, α also varies slowly with position $\mathbf{r}$, then the spectral density $W(\mathbf{r}, \mathbf{r}, \nu)$ of the light in the neighborhood of any point $P(\mathbf{r})$ in the plane $\mathcal{B}$ will vary nearly sinusoidally with position, provided that $\mu(\mathbf{r}_1, \mathbf{r}_2, \nu) \neq 0$, and the extreme values are given by

$$W_{max}(\mathbf{r}, \mathbf{r}, \nu) = W^{(1)}(\mathbf{r}, \mathbf{r}, \nu) + W^{(2)}(\mathbf{r}, \mathbf{r}, \nu) + 2[W^{(1)}(\mathbf{r}, \mathbf{r}, \nu) W^{(2)}(\mathbf{r}, \mathbf{r}, \nu)]^{1/2} |\mu(\mathbf{r}_1, \mathbf{r}_2, \nu)| \tag{3.10a}$$

and

$$W_{min}(\mathbf{r}, \mathbf{r}, \nu) = W^{(1)}(\mathbf{r}, \mathbf{r}, \nu) + W^{(2)}(\mathbf{r}, \mathbf{r}, \nu) - 2[W^{(1)}(\mathbf{r}, \mathbf{r}, \nu) W^{(2)}(\mathbf{r}, \mathbf{r}, \nu)]^{1/2} |\mu(\mathbf{r}_1, \mathbf{r}_2, \nu)| . \tag{3.10b}$$

By analogy with the usual notion of visibility of fringes [defined by the first relation in Eq. (1.6)] we define the *spectral visibility* $\mathcal{V}(\mathbf{r}, \nu)$ at frequency ν at a point $P(\mathbf{r})$ in the fringe pattern by the formula

$$\mathcal{V}(\mathbf{r}, \nu) \equiv \frac{W_{max}(\mathbf{r}, \mathbf{r}, \nu) - W_{min}(\mathbf{r}, \mathbf{r}, \nu)}{W_{max}(\mathbf{r}, \mathbf{r}, \nu) + W_{min}(\mathbf{r}, \mathbf{r}, \nu)} . \tag{3.11}$$

On substitution from Eqs. (3.10) into Eq. (3.11) we readily find that

$$\mathcal{V}(\mathbf{r}, \nu) = \frac{2[W^{(1)}(\mathbf{r}, \mathbf{r}, \nu) W^{(2)}(\mathbf{r}, \mathbf{r}, \nu)]^{1/2}}{W^{(1)}(\mathbf{r}, \mathbf{r}, \nu) + W^{(2)}(\mathbf{r}, \mathbf{r}, \nu)} |\mu(\mathbf{r}_1, \mathbf{r}_2, \nu)| . \tag{3.12}$$

In particular, if the spectral densities of the two beams interfering at $P(\mathbf{r})$ are equal for some frequency ν, i.e., if $W^{(1)}(\mathbf{r}, \mathbf{r}, \nu) = W^{(2)}(\mathbf{r}, \mathbf{r}, \nu)$, Eq. (3.12) reduces to

$$\mathcal{V}(\mathbf{r}, \nu) = |\mu(\mathbf{r}_1, \mathbf{r}_2, \nu)| , \tag{3.13}$$

i.e., the spectral visibility of the fringes is then just equal to the absolute value of the complex degree of spectral coherence.

The phase β of μ can also be directly related to the distribution of the spectral density of the light in the plane of observation. It is clear from Eq. (3.7) that the maxima of W are, to a good approximation, located at points for which

$$\arg\mu(\mathbf{r}_1, \mathbf{r}_2, \nu) \equiv \beta(\mathbf{r}_1, \mathbf{r}_2, \nu) = \delta + 2\pi m, \quad m = 0, \pm 1, \pm 2, \ldots \tag{3.14}$$

or, from Eqs. (3.8) and (3.14), at points for which

$$s_1 - s_2 = m\lambda + \frac{\lambda}{2\pi}[\beta(\mathbf{r}_1, \mathbf{r}_2, \nu) + \alpha], \quad m = 0, \pm 1, \pm 2, \ldots . \tag{3.15}$$

This formula shows that the location of the maxima is the same as would be obtained if the pinholes were illuminated by strictly monochromatic light of frequency ν, with a phase difference of $\beta(\mathbf{r}_1, \mathbf{r}_2, \nu)$ between the waves at the pinholes $P_1(\mathbf{r}_1)$ and $P_2(\mathbf{r}_2)$. Clearly Eq. (3.15) may be used to determine the argument of the complex degree of spectral coherence from measurements of the position of the fringe maxima.

IV. CROSS-SPECTRALLY PURE LIGHT

We shall illustrate the usefulness of some of these results by applying them to light beams that are cross-spectrally pure,[6] and show that they provide additional insight into the physical meaning of cross-spectral purity.

We again consider the interference experiment illustrated in Fig. 1, but we now assume that the normalized spectral density distributions [cf. Eq. (2.15)] at the two pinholes are the same, i.e., that

$$\phi(\mathbf{r}_1, \mathbf{r}_1, \nu) = \phi(\mathbf{r}_2, \mathbf{r}_2, \nu) . \tag{4.1}$$

The light forming the interference fringes on the screen $\mathcal{B}$ (Fig. 1) will, in general, have a different spectral composition. Indeed, the normalized spectral distribution $\phi(\mathbf{r}, \mathbf{r}, \nu)$ in the plane of the screen will, as a rule, vary from point to point; in general $\phi(\mathbf{r}, \mathbf{r}, \nu)$ will not coincide with $\phi(\mathbf{r}_1, \mathbf{r}_1, \nu)$ at any point $P(\mathbf{r})$ of the screen. This effect has been discussed previously[10] in connection with the Alford-Gold experiment.[11] The changes in the spectrum of the light on interference are well known when white light is used; colored fringes then appear in regions of the screen for which the path difference between the interfering beams is sufficiently large.

In the special case when a point P_0 (with position vectors $\mathbf{r}_0$, say) can be found in the interference pattern at which the normalized spectral density distribution $\phi(\mathbf{r}_0, \mathbf{r}_0, \nu)$ coincides with that at the two pinholes, the light at the two pinholes is said to be cross-spectrally pure. It has been shown[6] that for such light the normalized cross-spectral density $\phi(\mathbf{r}_1, \mathbf{r}_2, \nu)$, defined by Eq. (2.12), is necessarily expressible in the form

$$\phi(\mathbf{r}_1, \mathbf{r}_2, \nu) = \phi(\mathbf{r}_j, \mathbf{r}_j, \nu)\gamma(\mathbf{r}_1, \mathbf{r}_2, \tau_0)\, e^{2\pi i\nu\tau_0}, \quad j = 1, 2 \tag{4.2}$$

where

$$\tau_0 = (P_0P_1 - P_0P_2)/c \tag{4.3}$$

is the difference in the times needed for light to travel from the pinholes to P_0. In fact, when a point P_0 exists for which Eq. (4.2) is satisfied, the equation is also satisfied at neighboring points, as $\gamma(\mathbf{r}_1, \mathbf{r}_2, \tau_0) \times \exp(2\pi i\nu\tau_0)$ does not vary with τ_0, to a first approximation, for sufficiently small changes of τ_0. Conditions (4.1) and (4.2) together are known as the conditions for cross-spectral purity, because they ensure that the spectrum can be reproduced as the light propagates.[6] If we take the Fourier transform of Eqs. (4.1) and (4.2) and recall that $\phi(\mathbf{r}_1, \mathbf{r}_2, \nu)$ and the complex degree of coherence $\gamma(\mathbf{r}_1, \mathbf{r}_2, \tau)$ are Fourier transforms of each other [Eq. (2.12)], we find that the cross-spectral purity conditions may be expressed in the equivalent form

$$\gamma(\mathbf{r}_1, \mathbf{r}_1, \tau) = \gamma(\mathbf{r}_2, \mathbf{r}_2, \tau)\ , \tag{4.4a}$$

$$\gamma(\mathbf{r}_1, \mathbf{r}_2, \tau) = \gamma(\mathbf{r}_1, \mathbf{r}_2, \tau_0)\gamma(\mathbf{r}_j, \mathbf{r}_j, \tau - \tau_0), \quad j = 1, 2\ . \tag{4.4b}$$

As $|\gamma(\mathbf{r}_j, \mathbf{r}_j, \tau - \tau_0)|$ has its maximum value unity when $\tau = \tau_0$, we see at once that τ_0 is the value of τ for which the degree of coherence $|\gamma(\mathbf{r}_1, \mathbf{r}_2, \tau)|$ at the two points $P_1(\mathbf{r}_1)$ and $P_2(\mathbf{r}_2)$ is greatest. Equation (4.4b) has been called the *reduction formula* for cross-spectrally pure beams.[6] It is valid for all values of $\tau(-\infty < \tau < \infty)$, and implies that when the light is cross-spectrally pure, the complex degree of coherence of the optical field at the two pinholes is expressible as the product of two correlation coefficients, one of which characterizes spatial coherence and the other one temporal coherence.

We may readily express the condition (4.2) in terms of the complex degree of spectral coherence $\mu(\mathbf{r}_1, \mathbf{r}_2, \nu)$, rather than in terms of the normalized cross-spectral density $\phi(\mathbf{r}_1, \mathbf{r}_2, \nu)$. If we make use of the relation (2.17) and also Eq. (4.1), we see at once that the condition (4.2) for cross-spectrally pure beams is equivalent to the condition

$$\mu(\mathbf{r}_1, \mathbf{r}_2, \nu) = \gamma(\mathbf{r}_1, \mathbf{r}_2, \tau_0)\, e^{2\pi i\nu\tau_0}\ . \tag{4.5}$$

This formula has a clear physical meaning. It implies that when the optical field at the two points $P_1(\mathbf{r}_1)$ and $P_2(\mathbf{r}_2)$ is cross-spectrally pure, the absolute value of the degree of spectral coherence $\mu(\mathbf{r}_1, \mathbf{r}_2, \nu)$ is the same for every frequency component present in the spectrum of the light, and is equal to the absolute value of the complex degree of coherence $|\gamma(\mathbf{r}_1, \mathbf{r}_2, \tau_0)|$, where τ_0 is the value of τ for which $|\gamma(\mathbf{r}_1, \mathbf{r}_2, \tau)|$ is greatest, and moreover, the phases of $\mu(\mathbf{r}_1, \mathbf{r}_2, \nu)$ and of $\gamma(\mathbf{r}_1, \mathbf{r}_2, \tau_0)$ only differ by $2\pi\nu\tau_0$.

Condition (4.5) may be derived directly from the spectral interference law (3.5), if we make use of the relations (2.15), (3.3), and (4.1). We then have

$$\begin{aligned} W(\mathbf{r}, \mathbf{r}, \nu) = \phi(\mathbf{r}_j, \mathbf{r}_j, \nu)\{&|K_1|^2 I(\mathbf{r}_1) + |K_2|^2 I(\mathbf{r}_2) \\ &+ 2|K_1|\,|K_2|[I(\mathbf{r}_1)I(\mathbf{r}_2)]^{1/2} \\ &\times \mathrm{Re}[\mu(\mathbf{r}_1, \mathbf{r}_2, \nu)\exp[i\alpha - 2\pi i\nu(s_1 - s_2)c]]\}\ , \quad j = 1, 2 \end{aligned} \tag{4.6}$$

where $(s_1 - s_2)/c \equiv \tau_{12}$ again is the transit time difference corresponding to the point $\mathbf{r}$. It is clear from this equation that the normalized spectral density at some point $\mathbf{r}$ will coincide with $\phi(\mathbf{r}_j, \mathbf{r}_j, \nu)$, $(j = 1, 2)$, only if the expression within curly brackets is independent of ν. Now $K_1(\mathbf{r}_1, \mathbf{r}, \nu)$ and $K_2(\mathbf{r}_2, \mathbf{r}, \nu)$ and α depend on frequency ν, but they vary only slightly over a small spectral range.[9] The whole expression will therefore be substantially independent of frequency over a range small compared with the midfrequency, at those points $\mathbf{r}$ for which

$$\mu(\mathbf{r}_1, \mathbf{r}_2, \nu)\, e^{-2\pi i\nu\tau_{12}} = f(\mathbf{r}_1, \mathbf{r}_2, \tau_{12})\ , \tag{4.7}$$

where $f(\mathbf{r}_1, \mathbf{r}_2, \tau_{12})$ is a function that does not depend on frequency. $f(\mathbf{r}_1, \mathbf{r}_2, \tau_{12})$ has a simple interpretation. From Eqs. (4.7) and (2.17), with the help of condition (4.1), we have immediately

$$\phi(\mathbf{r}_1, \mathbf{r}_2, \nu)\, e^{-2\pi i\nu\tau_{12}} = \phi(\mathbf{r}_j, \mathbf{r}_j, \nu)\, f(\mathbf{r}_1, \mathbf{r}_2, \tau_{12}); \quad j = 1, 2 \tag{4.8}$$

and integration with respect to ν over the range 0 to ∞ then yields, with the help of Eq. (2.16), and the Fourier inverse of Eq. (2.12),

$$\gamma(\mathbf{r}_1, \mathbf{r}_2, \tau_{12}) = f(\mathbf{r}_1, \mathbf{r}_2, \tau_{12})\ . \tag{4.9}$$

If we substitute this expression for $f(\mathbf{r}_1, \mathbf{r}_2, \tau_{12})$ into Eq. (4.7) we obtain

$$\mu(\mathbf{r}_1, \mathbf{r}_2, \nu)\, e^{-2\pi i\nu\tau_{12}} = \gamma(\mathbf{r}_1, \mathbf{r}_2, \tau_{12})\ . \tag{4.10}$$

In practice, if Eq. (4.10) is satisfied for some point $\mathbf{r}$, then, to a good approximation, it is also satisfied for points within a region in the neighborhood of $\mathbf{r}$. The reason is that, to a first approximation, the modulus of $\gamma(\mathbf{r}_1, \mathbf{r}_2, \tau_{12})$ does not vary with τ_{12}, and the phase varies with τ_{12} as the phase of the factor $\exp(-2\pi i\nu\tau_{12})$ on the left-hand side of Eq. (4.10) (see Ref. 1, Sec. 10.3), provided the range of variation of τ_{12} is much less than the reciprocal bandwidth of the light. However, the position of the region where Eq. (4.10) is satisfied is not arbitrary, as is obvious by closer inspection of Eq. (4.10). For the absolute value of the right-hand side tends to zero as the transit time difference τ_{12} increases, whereas the absolute value of the left-hand side remains constant. If we substitute from Eq. (4.9) into Eq. (4.8), multiply both sides by $\exp[-2\pi i\nu(\tau - \tau_{12})]$, and integrate with respect to ν from 0 to ∞, we obtain

$$\gamma(\mathbf{r}_1, \mathbf{r}_2, \tau) = \gamma(\mathbf{r}_j, \mathbf{r}_j, \tau - \tau_{12})\gamma(\mathbf{r}_1, \mathbf{r}_2, \tau_{12}), \quad (j = 1, 2) \tag{4.11}$$

from which it follows with the help of Eq. (1.3) that, for any τ,

$$|\gamma(\mathbf{r}_1, \mathbf{r}_2, \tau)| \leqslant |\gamma(\mathbf{r}_1, \mathbf{r}_2, \tau_{12})| \ . \tag{4.12}$$

Hence τ_{12} must be that transit time difference, previously denoted by τ_0, for which $|\gamma(\mathbf{r}_1, \mathbf{r}_2, \tau_{12})|$ has its greatest value. With this interpretation of τ_{12}, the conditions (4.10) and (4.11) are seen to be precisely the equivalent conditions (4.5) and (4.4b), respectively, for the reproducibility of the spectral distribution.

The relation (4.5) for cross-spectrally pure beams has implications for the behavior of the spectral visibility of interference fringes formed by cross-spectrally pure beams. To show this we note, first of all, that the condition (4.1) expressing the equality of the normalized spectral densities at the two points $P_1(\mathbf{r}_1)$, $P_2(\mathbf{r}_2)$ implies with the help of Eq. (2.15) that

$$\frac{W(\mathbf{r}_1, \mathbf{r}_1, \nu)}{W(\mathbf{r}_2, \mathbf{r}_2, \nu)} = \frac{I(\mathbf{r}_1)}{I(\mathbf{r}_2)} \ , \tag{4.13}$$

and, because of Eqs. (3.3), that

$$\frac{W^{(1)}(\mathbf{r}, \mathbf{r}, \nu)}{W^{(2)}(\mathbf{r}, \mathbf{r}, \nu)} = \frac{|K_1|^2 I(\mathbf{r}_1)}{|K_2|^2 I(\mathbf{r}_2)} \ . \tag{4.14}$$

On substituting from Eqs. (4.14) and (4.5) into Eq. (3.12) we obtain for the spectral visibility of the interference fringes formed by cross-spectrally pure light, the expression

$$\mathcal{V}(\mathbf{r}, \nu) = \frac{2|K_1||K_2|[I(\mathbf{r}_1)I(\mathbf{r}_2)]^{1/2}}{|K_1|^2 I(\mathbf{r}_1) + |K_2|^2 I(\mathbf{r}_2)} |\gamma(\mathbf{r}_1, \mathbf{r}_2, \tau_0)| \ . \tag{4.15}$$

Now, as we have noted earlier,[9] K_1 and K_2 do not vary significantly with frequency over a range that is small compared with the midfrequency. Hence Eq. (4.15) implies that, to good approximation, the spectral visibility of the interference fringes formed by quasimonochromatic, cross-spectrally pure light beams is the same for all frequency components of the light, and is proportional to the maximum value $|\gamma(\mathbf{r}_1, \mathbf{r}_2, \tau_0)|$ of the degree of coherence at the two pinholes.

APPENDIX: DERIVATION OF THE NON-NEGATIVE DEFINITENESS CONDITION (2.7)

Let us choose a positive integer N, a set of N position vectors $\mathbf{r}_1, \mathbf{r}_2, \ldots, \mathbf{r}_N$, a set of N numbers (real or complex) $a_1, a_2, \ldots, a_N$, a frequency ν_0 and an interval $\nu_0 - \epsilon_1$ to $\nu_0 + \epsilon_2$ containing ν_0 in its interior ($\epsilon_1 > 0$, $\epsilon_2 > 0$), such that $\nu_0 - \epsilon_1 > 0$. Consider the expectation value

$$\mathcal{E}_N \equiv \left\langle \left| \int_{\nu_0-\epsilon_1}^{\nu_0+\epsilon_2} \left[\sum_{j=1}^{N} a_j v(\mathbf{r}_j, \nu) \right] d\nu \right|^2 \right\rangle . \tag{A1}$$

Since $\mathcal{E}_N$ is the expectation value of a non-negative quantity, it must itself be non-negative, i.e., we must have

$$\left\langle \int_{\nu_0-\epsilon_1}^{\nu_0+\epsilon_2} d\nu \int_{\nu_0-\epsilon_1}^{\nu_0+\epsilon_2} d\nu' \left[\sum_{i=1}^{N} \sum_{j=1}^{N} a_i^* a_j v^*(\mathbf{r}_i, \nu) v(\mathbf{r}_j, \nu') \right] \right\rangle \geqslant 0 \ . \tag{A2}$$

If we interchange the orders of averaging and of the integrations and summations, we may rewrite the inequality (A2) in the form

$$\int_{\nu_0-\epsilon_1}^{\nu_0+\epsilon_2} d\nu \int_{\nu_0-\epsilon_1}^{\nu_0+\epsilon_2} d\nu' \left[\sum_{i=1}^{N} \sum_{j=1}^{N} a_i^* a_j \langle v^*(\mathbf{r}_i, \nu) v(\mathbf{r}_j, \nu') \rangle \right] \geqslant 0 \ . \tag{A3}$$

Now according to Eq. (2.4),

$$\langle v^*(\mathbf{r}_i, \nu) v(\mathbf{r}_j, \nu') \rangle = W(\mathbf{r}_i, \mathbf{r}_j, \nu)\delta(\nu - \nu') \ . \tag{A4}$$

If we substitute from Eq. (A4) into Eq. (A3), and carry out the trivial integration with respect to ν', we obtain the inequality

$$\int_{\nu_0-\epsilon_1}^{\nu_0+\epsilon_2} d\nu \left[\sum_{i=1}^{N} \sum_{j=1}^{N} a_i^* a_j W(\mathbf{r}_i, \mathbf{r}_j, \nu) \right] \geqslant 0 \ . \tag{A5}$$

Since ϵ_1 and ϵ_2 are arbitrary non-negative quantities, subject only to the constraint that the range of integration lies wholly in the non-negative frequency range, and since $W(\mathbf{r}_i, \mathbf{r}_j, \nu)$ is assumed to be a continuous function of ν, Eq. (A5) necessarily implies that

$$\sum_{i=1}^{N} \sum_{j=1}^{N} a_i^* a_j W(\mathbf{r}_i, \mathbf{r}_j, \nu_0) \geqslant 0 \ , \tag{A6}$$

which is Eq. (2.7) of the text. This formula shows that for every positive frequency ν, the continuous matrix $W(\mathbf{r}_i, \mathbf{r}_j, \nu)$ is non-negative definite.

*Supported by the Army Research Office and by the National Science Foundation.

The main results of this paper were presented at the Annual Meeting of the Optical Society of America, in Boston, Mass., October 1975 [J. Opt. Soc. Am. **65**, 1198 (1975), Abst. ThB11].

[1] M. Born and E. Wolf, *Principles of Optics*, 5th ed. (Pergamon, Oxford, 1975). It should be noted that the mutual coherence function in this book differs by interchange of the points $\mathbf{r}_1$, $\mathbf{r}_2$ from the definition (1.2a).

[2] L. Mandel and E. Wolf, Rev. Mod. Phys. **37**, 231 (1965).

[3] E. Wolf and W. H. Carter, Opt. Commun. **13**, 205 (1975).

[4] W. H. Carter and E. Wolf, J. Opt. Soc. Am. **65**, 1067 (1975).

[5] E. Wolf and W. H. Carter, Opt. Commun. **16**, 297 (1976).

[6] The concept of cross-spectral purity was introduced by L. Mandel, J. Opt. Soc. Am. **51**, 1342 (1961).

[7] For reasons well known in the theory of stationary random processes, the Fourier representation (2.1) does not exist in the sense of ordinary function theory, and must be understood in terms of the theory of generalized functions [cf. Y. L. Lumley, *Stochastic Tools in Turbulence* (Academic, New York, 1970)]. One may avoid the use of generalized functions by other mathematical refinements, for example with the help of a certain truncation procedure (cf. Ref. 1, Secs. 10.2 and 10.3), or by the use of the Fourier-Stieltjes integral [cf. A. M. Yaglom, *An Introduction to the Theory of Stationary Random Functions* (Prentice-Hall, Englewood Cliffs, N.J., 1962)].

[8] We are dealing here with a cross-spectral density function associated with a stationary random process $V(\mathbf{r}, t)$ labeled by a continuous (vector)parameter $\mathbf{r}$. For a finite-dimensional process labeled by a discrete parameter, a similar non-negative definiteness condition on the cross-spectral density function follows from a theorem due to H. Cramér, Ann. Math. **41**, 215 (1940).

[9] When light of frequency ν is incident on the plane $\mathcal{a}$ of the pinholes along or close to the normal direction, and the angles that the diffracted directions P_1P and P_2P make with the normal to $\mathcal{a}$ are also small, then $K_j \sim (i\nu/cs_j)\delta A_j$ ($j = 1, 2$), where δA_j is the area of the pinhole at r_j (cf. Ref. 1, Secs. 8.2 and 8.3).

[10] L. Mandel, J. Opt. Soc. Am. **52**, 1335 (1962).

[11] W. P. Alford and A. Gold, Am. J. Phys. **26**, 481 (1958).

Volume 25, number 3 OPTICS COMMUNICATIONS June 1978

PARTIALLY COHERENT SOURCES WHICH PRODUCE THE SAME FAR-FIELD INTENSITY DISTRIBUTION AS A LASER *

E. WOLF

Department of Physics and Astronomy and the Institute of Optics, University of Rochester, Rochester, N.Y. 14627, USA

and

E. COLLETT

U.S. Army Electronics Command, Fort Monmouth, N.J. 07003, USA

Received 8 February 1978

It is shown that certain partially coherent model sources whose intensity distribution and degree of coherence are both gaussian will generate the same far-field intensity distributions as a completely coherent laser source.

1. Introduction

In a recent paper [1] we showed that sources of different states of coherence may produce identical distributions of light intensity throughout the far-field and we have demonstrated that complete coherence is not necessary to produce light beams that are as directional as laser beams. In the present note we describe a class of planar sources, each with a different correlation length, which will generate light beams which have the same intensity distributions in the far-zone as a gaussian laser beam. The quasi-homogeneous source that we gave in our previous publication as an example of such equivalence is found to be a limiting case of sources of this type.

2. An equivalence theorem for Schell-model sources with gaussian distributions of intensity and coherence

Let us consider planar sources whose cross-spectral density function †, at some particular frequency ω, is of the form

$$W(\boldsymbol{r}_1, \boldsymbol{r}_2) = \sqrt{I(\boldsymbol{r}_1)}\sqrt{I(\boldsymbol{r}_2)}\, g(\boldsymbol{r}_1 - \boldsymbol{r}_2) \,. \tag{2.1}$$

Here $\boldsymbol{r}_1$ and $\boldsymbol{r}_2$ are the position vectors of two typical points in the source plane, $I(\boldsymbol{r})$ represents the intensity at a point $\boldsymbol{r}$ in this plane and $g(\boldsymbol{r}')$, (with $\boldsymbol{r}' = \boldsymbol{r}_1 - \boldsymbol{r}_2$), is the complex degree of spatial coherence of the light across the source. Sources whose cross-spectral density function is of the form (2.1) appear to have been first considered by Schell [3] ‡ and are usually referred to as *Schell-model sources.* They are a natural generalization of statistically homogeneous sources, but unlike them, the Schell-model sources may be of a finite extent.

* Research supported by the U.S. Army Research Office and the U.S. Army Electronics Command.

† For a definition of the cross-spectral density function see, for example, ref. [2], eqs. (2.4) and (2.5a). The cross-spectral density function as well as the intensity $I(r)$, the degree of spatial coherence $g(r')$ and the radiant intensity $J(\theta)$ are functions of the frequency, but we do not show this dependence explicitly.

‡ Actually Schell assumed that the mutual intensity (cf. ref. [4]) rather than the cross-spectral density function is of the form given by the right-hand side of eq. (2.1). This distinction is of no consequence for the purpose of the present investigation, since we are confining our attention to a single frequency component of the light.

We will consider Schell-model sources whose intensity distribution and degree of spatial coherence are both gaussian, i.e., they may be represented in the form

$$I(\boldsymbol{r}) = A \exp(-r^2/2\sigma_I^2) , \qquad (2.2)$$

$$g(\boldsymbol{r}') = \exp(-r'^2/2\sigma_g^2) , \qquad (2.3)$$

($r = |\boldsymbol{r}|$, $r' = |\boldsymbol{r}'|$), with A, σ_I and σ_g being positive constants. It has recently been shown analytically by Baltes, Steinle and Antes [5] and by Carter and Bertolotti [6] that such sources will generate a field whose radiant intensity #, at frequency ω, in a direction specified by a unit vector $\boldsymbol{s}$ that makes an angle θ with the direction normal to the source plane is given by

$$J(\theta) = J(0) \cos^2\theta \exp(-\sin^2\theta/2\Delta^2) , \qquad (2.4)$$

where

$$\Delta = (1/k\sigma_g)[1 + (\sigma_g/2\sigma_I)^2]^{1/2} , \qquad (2.5)$$

and $k = \omega/c$, c being the speed of light. The radiant intensity $J(0)$ in the direction normal to the source plane, which appears in (2.4), may readily be shown to be given by

$$J(0) = (\sigma_I/\Delta)^2 A . \qquad (2.6)$$

From the formulae (2.4)–(2.6) the following theorem follows at once:

Theorem I: Two Schell-model sources, whose intensity distributions and degree of coherence are both gaussian, will generate fields with identical far-zone intensity distributions, if the r.m.s. widths σ_I and σ_g are such that each of the quantities

$$\Delta^2 \equiv 1/(k\sigma_g)^2 + 1/(2k\sigma_I)^2 \qquad (2.7)$$

and

$$J(0) \equiv (\sigma_I/\Delta)^2 A , \quad [\text{with } \Delta \text{ given by (2.7)}] , \qquad (2.8)$$

are the same for both sources. The radiant intensity generated by each source is then given by eq. (2.4).

The radiant intensity $J(\theta)$ represent the rate at which energy, at the temporal frequency ω, is radiated by the source per unit solid angle around the $\boldsymbol{s}$-direction. It is related to the optical intensity $I(R\boldsymbol{s})$ (the average value of the squared amplitude of the wave function) at the point specified by the position vector $R\boldsymbol{s}$, by the formula $I(R\boldsymbol{s}) = J(\theta)/R^2$, to be understood in the asymptotic sense as $kR \to \infty$.

It follows at once from eq. (2.7) that the r.m.s. widths σ_I and σ_g must necessarily obey the inequalities

$$k\sigma_I \geqslant 1/2\Delta , \quad k\sigma_g \geqslant 1/\Delta . \qquad (2.9)$$

3. Some partially coherent Schell-model sources which generate the same far-zone intensity distribution as a completely coherent laser source

Let us consider the limiting case of eq. (2.3) as

$$\sigma_g \to \infty . \qquad (3.1)$$

In this limit (2.3) becomes

$$g(\boldsymbol{r}') \equiv 1 . \qquad (3.2)$$

It is clear from (3.2) that the light emerging from the source is now completely spatially coherent and cophasal. Moreover, since according to eq. (2.2) its intensity distribution across the source is gaussian it is clear that this limit represents a laser source with a flat output mirror, that operates in its lowest mode, if the effects arising from diffraction at the edge of the mirror are neglected. To distinguish this limiting case from the general case we will modify the notation slightly and we will express the intensity distribution across the exit face of the laser (suffix L) as

$$I_L(\boldsymbol{r}) = A_L \exp(-r^2/2\delta_L^2) . \qquad (3.3)$$

The values of Δ and $J(0)$ appearing in the expression (2.4) for the radiant intensity generated by the laser source are obtained at once from eqs. (2.7) and (2.8) on setting $\sigma_g = \infty$ and replacing σ_I by δ_L and A by A_L. One then obtains

$$\Delta = 1/2k\delta_L , \qquad (3.4)$$

$$J(0) = 4k^2\delta_L^4 A_L . \qquad (3.5)$$

On using the expressions (3.4) and (3.5) we obtain from (2.4) the following expression for the radiant intensity generated by the laser source:

$$J(\theta) = 4k^2\delta_L^4 A_L \cos^2\theta \exp\{-2(k\delta_L)^2 \sin^2\theta\} . \qquad (3.6)$$

With the values of Δ and $J(0)$ given by (3.4) and (3.5) we now obtain from theorem I of sec. 2 the following

Theorem II: Any Schell-model source, whose inten-

Volume 25, number 3 OPTICS COMMUNICATIONS June 1978

sity distribution and degree of coherence are both gaussian and whose parameters σ_g, σ_I and A satisfy the relations

$$1/\sigma_g^2 + 1/(2\sigma_I)^2 = 1/(2\delta_L)^2 \tag{3.7}$$

and

$$A = (\delta_L/\sigma_I)^2 A_L , \tag{3.8}$$

will generate light with the same distribution of radiant intensity $J(\theta)$, [given by eq. (3.6)], as the coherent laser source whose intensity distribution at the output mirror (assumed to be plane) is

$$I_L(r) = A_L \exp(-r^2/2\delta_L^2) . \tag{3.9}$$

For the sake of accuracy we stress once again that the effects arising from diffraction at the edge of the output mirror have been neglected.

While the radiant intensity and hence also the optical intensity in the far-zone, generated by all sources that obey the relations (3.7) and (3.8) will be identical, the fields, just as the different sources, will have different coherence properties. We intend to discuss these differences and related results in another publication.

Finally we note that in view of the inequalities

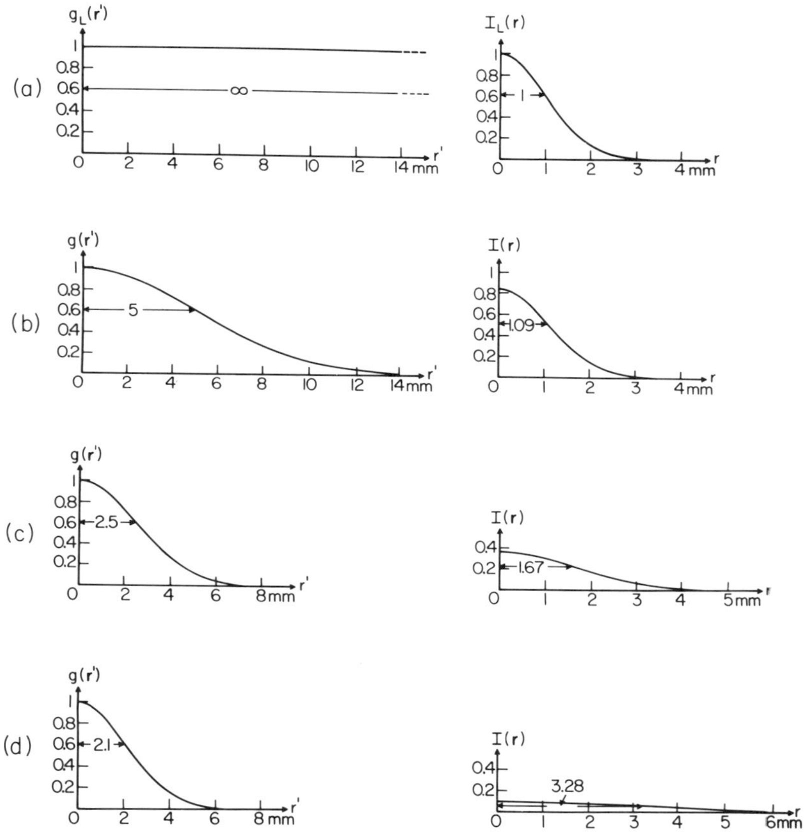

Fig. 1. Illustrating the coherence and the intensity distributions across three partially coherent sources [(b), (c), (d)] which produce fields whose far-zone intensity distributions are the same as that generated by a coherent laser source [(a)]. The parameters characterizing the four sources are:

(a) $\sigma_g = \infty$, $\sigma_I = \delta_L = 1$ mm, $A = 1$ (arbitrary units) (b) $\sigma_g = 5$ mm, $\sigma_I = 1.09$ mm, $A = 0.84$

(c) $\sigma_g = 2.5$ mm, $\sigma_I = 1.67$ mm, $A = 0.36$ (d) $\sigma_g = 2.1$ mm, $\sigma_I = 3.28$ mm, $A = 0.09$.

The normalized radiant intensity generated by all these sources is $J(\theta)/J(0) = \cos^2\theta \exp\{-2(k\delta_L)^2 \sin^2\theta\}$, ($\delta_L = 1$ mm).

Volume 25, number 3 OPTICS COMMUNICATIONS June 1978

(2.9) we must have, because of (3.4),

$$\sigma_I \geqslant \delta_L , \quad \sigma_g \geqslant 2\delta_L . \tag{3.10}$$

The first of these inequalities implies that the r.m.s. widths of the intensity distributions across all the "equivalent" sources cannot be smaller than the r.m.s. width of the intensity distribution across the output mirror of the laser. The second inequality in (3.10) implies that the r.m.s. widths of the degree of coherence of all these sources cannot be smaller than twice the r.m.s. width of the laser source intensity distribution.

The "equivalent" source with the smallest possible value σ_g of its r.m.s. width, namely the value $\sigma_g = 2\delta_L$ represents, according to eqs. (3.7) and (3.8), the unphysical limit of a strictly homogeneous source ($k\sigma_I \to \infty$) of zero intensity ($A \to 0$). However, if

$$\sigma_g \approx 2\delta_L , \tag{3.11}$$

the condition (3.7) will be approximately satisfied with

$$\sigma_I \gg \sigma_g \tag{3.12}$$

which implies that such a source is *quasi-homogeneous* [7]. This is precisely the example of a source given in in ref. [1] that illustrates a basic equivalence theorem that was derived in that reference and which is the root of the two theorems formulated in the present paper.

Our main results are illustrated in fig. 1. The curves indicate the subtle "trade-off" between the source coherence and the source intensity that evidently takes place in order for the four different sources to produce identical distributions of the radiant intensity.

Acknowledgement

We are obliged to Mr. T. Ruehle and Mr. A. Friberg for assistance with the computations relating to fig. 1.

References

[1] E. Collett and E. Wolf, Opt. Lett. 2 (1978) 27.
[2] L. Mandel and E. Wolf, J. Opt. Soc. Am. 66 (1976) 529.
[3] A.C. Schell, The multiple plate antenna, Doctoral Dissertation, Massachusetts Institute of Technology (1961), Sec. 7.5.
[4] M. Born and E. Wolf, Principles of optics, 5th Ed. (Pergamon Press, New York, 1975) Sec. 10.4.1.
[5] H.P. Baltes, B. Steinle and G. Antes, in: Proc. Fourth Rochester Conf. on Coherence and quantum optics, eds. L. Mandel and E. Wolf (Plenum Press, New York, in press).
[6] W.H. Carter and M. Bertolotti, J. Opt. Soc. Am., (in press).
[7] W.H. Carter and E. Wolf, J. Opt. Soc. Am. 67 (1977) 785.

Emil Wolf Vol. 72, No. 3/March 1982/J. Opt. Soc. Am. 343

New theory of partial coherence in the space–frequency domain. Part I: spectra and cross spectra of steady-state sources

Emil Wolf*

Department of Physics and Astronomy, University of Rochester, Rochester, New York 14627

Received July 20, 1981

It is shown that, under very general conditions, the cross-spectral density of a steady-state source of any state of coherence may be expressed in terms of certain new modes of oscillations, each of which represents a completely spatially coherent elementary excitation. Making use of this result, a statistical ensemble of strictly monochromatic oscillations, all of the same temporal frequency, is then introduced that yields the cross-spectral density as a correlation function in the space–frequency domain. From these results two new expressions for the Wiener–Khintchine spectrum of the source and also a new mode representation of the cross-correlation function of the source follow at once.

1. INTRODUCTION

The theory of partial coherence,[1] whether based on classical or on quantum description of the optical field, is generally formulated in terms of space–time correlation functions. In the classical formulation, the simplest correlation function of this type is the *mutual coherence function* $\Gamma(\mathbf{r}_1, \mathbf{r}_2, \tau)$, defined by the formula

$$\Gamma(\mathbf{r}_1, \mathbf{r}_2, \tau) = \langle V^*(\mathbf{r}_1, t) V(\mathbf{r}_2, t + \tau) \rangle. \tag{1.1}$$

Here $V(\mathbf{r}, t)$ is the complex analytic signal representing the fluctuating field (assumed here to be stationary, at least in the wide sense) at a point specified by a position vector $\mathbf{r}$ at time t; the asterisk denotes the complex conjugate, and the angle brackets denote either a time average or an ensemble average. For the sake of simplicity we are ignoring here the polarization features of the field, so we consider $V(\mathbf{r}, t)$ to be a scalar.

Although the mutual coherence function is adequate for the analysis of all the elementary phenomena of interference and diffraction with light of any state of coherence, its use in treatments of problems that involve the interaction of light with matter is mathematically rather unwieldy. This is so because the response of matter to an incident field is most naturally described—at least in the linear regime—not by time-dependent but rather by frequency-dependent response functions, such as the dielectric constant, the refractive index, or the magnetic susceptibility. Moreover, the dominant effects frequently arise from the statistical features of the medium itself rather than of the incident light, for example, when a laser beam propagates through the atmosphere or when it is scattered from a rough surface (giving rise to the well-known speckle phenomenon). In such situations one is basically interested in a single-frequency description, and it seems, therefore, rather inappropriate to analyze such problems with the help of space–time correlation functions of the optical field. Problems of this kind may, of course, be analyzed by employing in place of the mutual coherence function $\Gamma(\mathbf{r}_1, \mathbf{r}_2, \tau)$ its Fourier transform,

$$W(\mathbf{r}_1, \mathbf{r}_2, \omega) = \frac{1}{2\pi} \int_{-\infty}^{\infty} \Gamma(\mathbf{r}_1, \mathbf{r}_2, \tau) e^{i\omega\tau} d\tau, \tag{1.2}$$

known as the *cross-spectral density* or the *cross-power spectrum*. However, it should be noted that the statistical aspects are then still associated with the space–time domain (through Γ) and that the frequency description is introduced at the "level" of second-order correlations rather than at the "field-level."

Attempts to introduce a frequency description of a fluctuating field itself have, in the past, encountered considerable mathematical difficulties because of the well-known fact, briefly discussed in the next section, that a stationary random process does not possess a Fourier representation. In this paper we show that, in spite of this, a rigorous space–frequency representation of stationary random sources can be introduced within the framework of ordinary function theory. A related description of stationary random fields will be described in Part II of this investigation.[2]

The main results that we establish in this paper are the following. We show that under very general conditions, which are likely to hold in most cases of practical interest, one can represent the cross-spectral density of a statistically stationary source of any state of coherence in the form

$$W(\mathbf{r}_1, \mathbf{r}_2, \omega) = \sum_n \lambda_n(\omega) \phi_n^*(\mathbf{r}_1, \omega) \phi_n(\mathbf{r}_2, \omega), \tag{1.3}$$

where the functions $\phi_n(\mathbf{r}, \omega)$ are mutually orthogonal. Each term under the summation sign on the right-hand side of Eq. (1.3) may be regarded as being associated with a natural mode of oscillation of the source, that is, as spatially completely coherent. Using this new expansion, we then show that it is possible to construct an ensemble $\{U(\mathbf{r}, \omega) e^{-i\omega t}\}$ of strictly monochromatic oscillations, all of the same frequency ω, that yields the cross-spectral density of the source as a correlation function in the space–frequency domain, i.e., as

$$W(\mathbf{r}_1, \mathbf{r}_2, \omega) = \langle U^*(\mathbf{r}_1, \omega) U(\mathbf{r}_2, \omega) \rangle, \tag{1.4}$$

where the angle brackets now denote the average over the new ensemble. From representations (1.3) and (1.4) two new expressions for the spectral density of the source in terms of the new modes and the new ensemble of monochromatic oscillations follow at once, viz.,

$$S(\mathbf{r}, \omega) = \sum_n \lambda_n(\omega)|\phi_n(\mathbf{r}, \omega)|^2 \tag{1.5}$$

$$= \langle |U(\mathbf{r}, \omega)|^2 \rangle. \tag{1.6}$$

We believe that our new theory is likely to find useful applications in various problems involving the generation and propagation of partially coherent light and the interaction of light with matter under circumstances when some of the statistical features of the light or of the material medium, or of both, must be taken into account. The theory may also lead to a deeper understanding of the coherence properties of laser modes. It may also elucidate some questions relating to spectroscopy of sources of any state of coherence.

For the sake of simplicity we carry out the analysis within the framework of classical scalar theory. There appear to be no basic difficulties in extending the theory to include also the vectorial features of the source or of the field. Moreover, a corresponding theory can also be formulated within the framework of quantum theory. In fact, the quantum-mechanical results can be written down almost at once from those of the classical theory by making use of the well-known phase–space correspondence between the classical and the quantum theories of coherence [Ref. 1, Sec. 4].

Because the essential feature of the present theory is a new way of representing the spectrum and the cross spectrum of a fluctuating source and of a fluctuating field, we begin with a brief summary of the main classical results relating to the concept of the spectrum of light.

2. CLASSICAL DEFINITIONS OF THE SPECTRUM OF LIGHT

An elementary intuitive way of introducing the spectrum of a field $V(t)$ at some fixed point $\mathbf{r}$ is as follows: We imagine the field to be represented as a Fourier integral

$$V(t) = \int_{-\infty}^{\infty} v(\omega)e^{-i\omega t}d\omega, \tag{2.1}$$

and we assume that Eq. (2.1) may be inverted:

$$v(\omega) = \frac{1}{2\pi}\int_{-\infty}^{\infty} V(t)e^{i\omega t}dt. \tag{2.2}$$

The spectrum $s(\omega)$ of the field at point $\mathbf{r}$ is then identified with the squared modulus of $v(\omega)$, i.e., it is then formally given by

$$s(\omega) = |v(\omega)|^2. \tag{2.3}$$

However, any field found in nature will fluctuate, in the course of time, in a random manner, and hence $V(t)$ and consequently $s(\omega)$ will be random variables. This suggests that the spectrum should be defined as an average of Eq. (2.3) (denoted by angle brackets below), taken over an appropriate ensemble of sample functions:

$$\langle s(\omega)\rangle = \langle |v(\omega)|^2 \rangle. \tag{2.4}$$

However, as is well known, there is a serious difficulty with this definition of the spectrum, which arises from the fact that for a steady-state field the sample functions $V(t)$ cannot tend to zero as t tends to $+\infty$ and to $-\infty$. This fact is quite clear from the standpoint of the theory of random processes for "steady-state" implies that the statistical ensemble is stationary and hence the underlying probability distributions that characterize the fluctuations are invariant with respect to translation of the origin of time. This in turn implies that $V(t)$ cannot behave, in the statistical sense, any differently for large values of t than it behaves for any other values of t and consequently cannot tend to zero at $t \to \pm\infty$. Hence the Fourier transform [Eq. (2.2)] does not exist for such fields.

This difficulty was overcome by Wiener[3] in a classic paper that was the origin of a whole new branch of mathematics, the so-called generalized harmonic analysis. Wiener assumed only that $V(t)$ is measurable (in the sense of Lebesgue) and that the quantity

$$\Gamma(\tau) = \lim_{T\to\infty} \frac{1}{2T}\int_{-T}^{T} V^*(t)V(t+\tau)dt \tag{2.5}$$

exists. He then showed that the function

$$\sigma(\omega) = \frac{1}{2\pi}\int_{-\infty}^{\infty} \Gamma(\tau)\frac{e^{i\omega\tau}-1}{i\tau}d\tau, \tag{2.6}$$

which he called the spectrum of $V(t)$, also exists. We will refer to $\sigma(\omega)$ as the integrated spectrum. Assuming that it is a differentiable function of ω and that one may interchange the orders of differentiation and integration in the derivative of Eq. (2.6), it follows that

$$S(\omega) \equiv \frac{d\sigma(\omega)}{d\omega} = \frac{1}{2\pi}\int_{-\infty}^{\infty} \Gamma(\tau)e^{i\omega\tau}d\tau. \tag{2.7}$$

The expression on the right-hand side of Eq. (2.7) is today generally considered to be a rigorous definition of the spectrum of $V(t)$. It expresses the spectrum as the Fourier transform of the function $\Gamma(\tau)$ introduced by means of Eq. (2.5), i.e., as the Fourier transform of the autocorrelation of $V(t)$, defined as a time average. It should be noted that Wiener's definition of the spectrum is associated with a single function $V(t)$, not with a statistical ensemble of time-dependent functions. In fact, Wiener did not employ any statistical concepts in his analysis.

When one deals with a stationary ensemble $\{V(t)\}$ of random functions rather than with a single function $V(t)$, one defines the autocorrelation function not by the time average [Eq. (2.5)] but rather as an ensemble average (denoted by angle brackets), i.e., as

$$\Gamma(\tau) = \langle V^*(t)V(t+\tau)\rangle. \tag{2.8}$$

Four years after the publication of the classic paper by Wiener, Khintchine[4] showed that in order that a function $\Gamma(\tau)$ be the autocorrelation function [in the sense of definition (2.8)] of a continuous stationary random process, it must be expressible in the form

$$\Gamma(\tau) = \int_{-\infty}^{\infty} e^{i\omega\tau}dF(\omega), \tag{2.9}$$

where $F(\omega)$ is a real, nondecreasing bounded function. This result later played a basic role in the spectral representation that employs the stochastic Fourier–Stieltjes integral of stationary random processes, initiated largely by Kolmogorov.[5] Since it is generally believed that most, if not all, stationary random processes that occur in nature are ergodic, autocorrelation functions (2.8) and (2.5) may be assumed to be equal to each other (with probability one), provided that the function $V(t)$ in Eq. (2.5) is taken to be a sample function of the ensemble $\{V(t)\}$. Under these circumstances the integrated spectrum $\sigma(\omega)$ of Wiener may be identified with the distribution function $F(\omega)$ of the Stieltjes integral representation [Eq. (2.9)] of $\Gamma(\tau)$.

The Wiener–Khintchine spectrum is clearly a much more sophisticated concept than is the physicist's intuitive idea of the spectrum, because, as we noted earlier, the intuitive notion centers around the decomposition of the random process into sinusoidal oscillations [Eqs. (2.1) and (2.4)], whereas the theories of Wiener and Khintchine associate it with the decomposition of the *autocorrelation function* into such oscillatory components. However, it is known that the Wiener–Khintchine spectrum may also be expressed in the form[6]

$$S(\omega) = \lim_{T\to\infty}\left[\frac{1}{2T}\langle|v(\omega;T)|^2\rangle\right], \tag{2.10}$$

where $v(\omega; T)$ is the "finite" Fourier inverse of $V(t)$, viz.,

$$v(\omega;T) = \frac{1}{2\pi}\int_{-T}^{T} V(t)e^{i\omega t}dt. \tag{2.11}$$

Formula (2.10), together with Eq. (2.11), indeed bears some similarity to the mathematically unsound but intuitively appealing expression (2.4) [together with Eq. (2.2)]. It should be noted, however, that Eq. (2.10) involves not only an ensemble average but also a certain limiting procedure. Both are essential: Were the limit $T \to \infty$ omitted, the expression would depend on the choice of the truncation interval in Eq. (2.11) and hence could not represent the spectrum; if, on the other hand, the ensemble averaging were omitted (as is frequently but incorrectly done in the literature), the limit as $T \to \infty$ would in general not exist.[7] It should also be noted that the identification of the expression on the right-hand-side of Eq. (2.10) with the spectrum cannot be made from Wiener's analysis alone because, as was already noted, Wiener's treatment involves a single function, not a statistical ensemble of functions.

It is clear from the preceding remarks that, whereas a formal identification of the spectrum of a steady-state field (or, more generally, of any stationary random process) with the Fourier transform of the autocorrelation function presents no difficulty, its association with sinusoidal oscillatory components of the field is, at best, rather obscure. We show in Section 5, as a consequence of some more general considerations, that the Wiener–Khintchine spectrum of a fluctuating source that is represented by a stationary ensemble may be given, under very general conditions, a new and mathematically rigorous interpretation in terms of strictly sinusoidal oscillations. These oscillations are, however, not related to a Fourier decomposition (which, strictly speaking, does not exist) of the random source variable; it is expressed in terms of the eigenfunctions and the eigenvalues of a Fredholm integral equation, whose kernel is the cross-spectral density of the fluctuating source. In Part II of this investigation similar expressions will be derived for the spectrum of stationary fields.[2]

3. CROSS-SPECTRAL DENSITY AS A HILBERT-SCHMIDT KERNEL

Let $Q(\mathbf{r}, t)$ represent a fluctuating scalar source distribution at a point specified by position vector $\mathbf{r}$ at time t. The source may be either a primary or a secondary one and is assumed to be localized in some finite domain D. For the sake of generality we allow $Q(\mathbf{r}, t)$ to be complex.[8] We assume that the source is a steady-state source; more precisely, we assume that the statistical ensemble that characterizes the fluctuations of $Q(\mathbf{r}, t)$ is stationary, at least in the wide sense, and that it is of zero mean.

Let

$$\Gamma_Q(\mathbf{r}_1, \mathbf{r}_2, \tau) = \langle Q^*(\mathbf{r}_1, t)Q(\mathbf{r}_2, t+\tau)\rangle_t \tag{3.1}$$

be the cross-correlation function of the source distribution. The angle brackets on the right-hand side of Eq. (3.1) denote the ensemble average. We have attached a suffix t to the averaging symbol to stress that the average is taken over an ensemble of *time-dependent* realizations, for we will later also encounter averaging over another ensemble, of frequency-dependent realizations, and we wish to distinguish clearly between the two kinds of averages. If the ensemble of the Q's is ergodic, as will usually be the case, the ensemble average on the right-hand-side of Eq. (3.1) will be equal to the corresponding time average.

We note, in passing, that if the source is a secondary one, $\Gamma_Q(\mathbf{r}_1, \mathbf{r}_2, \tau)$ will be the mutual coherence function of the field distribution across the source (Ref. 1, Sec. 3.1).

We will assume that $|\Gamma_Q(\mathbf{r}_1, \mathbf{r}_2, \tau)|$ falls off sufficiently rapidly with τ as $|\tau| \to \infty$ to ensure that, for all $\mathbf{r}_1$ and $\mathbf{r}_2$ that represent points in the source domain D, Γ_Q is absolutely integrable with respect to τ, i.e., that

$$\int_{-\infty}^{\infty} |\Gamma_Q(\mathbf{r}_1, \mathbf{r}_2, \tau)| d\tau < \infty. \tag{3.2}$$

It then follows from a well-known theorem of Fourier analysis that $\Gamma_Q(\mathbf{r}_1, \mathbf{r}_2, \tau)$ has a Fourier frequency transform

$$W_Q(\mathbf{r}_1, \mathbf{r}_2, \omega) = \frac{1}{2\pi}\int_{-\infty}^{\infty} \Gamma_Q(\mathbf{r}_1, \mathbf{r}_2, \tau)e^{i\omega\tau}d\tau \tag{3.3}$$

and that this transform is a continuous function[9] of ω. $W_Q(\mathbf{r}_1, \mathbf{r}_2, \omega)$ is the cross-spectral density (the cross-power spectrum) of the source distribution $Q(\mathbf{r}, t)$, and its "diagonal" element

$$S_Q(\mathbf{r}, \omega) \equiv W_Q(\mathbf{r}, \mathbf{r}, \omega) \tag{3.4}$$

is the spectral density (the spectrum) of $Q(\mathbf{r}, t)$ at $\mathbf{r}$, in the sense of the definition of Wiener and Khintchine.

Now it is a well-known property of cross-correlation functions that their absolute integrability implies also square-integrability,[10] i.e., expression (3.2) implies that

$$\int_{-\infty}^{\infty} |\Gamma_Q(\mathbf{r}_1, \mathbf{r}_2, \tau)|^2 d\tau < \infty. \tag{3.5}$$

Consequently, by Plancherel's theorem, $W_Q(\mathbf{r}_1, \mathbf{r}_2, \omega)$ is square integrable with respect to ω, i.e.,

$$\int_{-\infty}^{\infty} |W_Q(\mathbf{r}_1, \mathbf{r}_2, \omega)|^2 d\omega < \infty, \tag{3.6}$$

and Eq. (3.3) may, therefore, be inverted:

$$\Gamma_Q(\mathbf{r}_1, \mathbf{r}_2, \tau) = \int_{-\infty}^{\infty} W_Q(\mathbf{r}_1, \mathbf{r}_2, \omega) e^{-i\omega\tau} d\omega. \quad (3.7)$$

Next we assume that $\Gamma_Q(\mathbf{r}_1, \mathbf{r}_2, \tau)$ is a continuous function of $\mathbf{r}_1$ and $\mathbf{r}_2$ and is uniformly bounded throughout the source domain ($\mathbf{r}_1 \in D, \mathbf{r}_2 \in D$) by a function (of τ alone) that is integrable in the range $-\infty < \tau < \infty$. This requirement, together with assumption (3.2), may be shown to imply that $W_Q(\mathbf{r}_1, \mathbf{r}_2, \omega)$ is also a continuous function[12] of $\mathbf{r}_1$ and $\mathbf{r}_2$ throughout D and is consequently bounded in D; hence

$$\iint_{DD} |W_Q(\mathbf{r}_1, \mathbf{r}_2, \omega)|^2 d^3\mathbf{r}_1 d^3\mathbf{r}_2 < \infty. \quad (3.8)$$

The assumptions that the cross-correlation function $\Gamma_Q(\mathbf{r}_1, \mathbf{r}_2, \tau)$ is absolutely integrable with respect to τ, and that it is continuous and uniformly bounded throughout D by an integrable function of τ, are the only assumptions of the present theory. We believe that these requirements will be satisfied in all cases of practical interest; they are stated here explicitly only in order to provide a mathematically rigorous basis for our theory.

The cross-spectral density W_Q has the following properties:

$$W_Q(\mathbf{r}_2, \mathbf{r}_1, \omega) = W_Q^*(\mathbf{r}_1, \mathbf{r}_2, \omega) \quad (3.9)$$

and

$$\iint_{DD} W_Q(\mathbf{r}_1, \mathbf{r}_2, \omega) f^*(\mathbf{r}_1) f(\mathbf{r}_2) d^3r_1 d^3r_2 \geq 0, \quad (3.10)$$

where $f(\mathbf{r})$ is any square-integrable function. Relation (3.9) is nothing but the Fourier inverse of the relation $\Gamma_Q(\mathbf{r}_2, \mathbf{r}_1, -\tau) = \Gamma^*(\mathbf{r}_1, \mathbf{r}_2, \tau)$ that follows at once from the definition [Eq. (3.1)] of Γ_Q. Inequality (3.10), which is intimately related to a generalization of Bochner's theorem [Ref. 9, Chap. 5], is somewhat more difficult to derive and is established in Appendix A.

Functions that satisfy conditions (3.8)–(3.10) are well known in the theory of integral equations and in functional analysis. In particular, expression (3.8) implies that the cross-spectral density is a *Hilbert–Schmidt kernel.* Relation (3.9) states that the kernel is *hermitian*, and inequality (3.10) implies that it is *nonnegative definite.*

We now show that an expansion theorem relating to kernels of this class provides at once a new representation of sources of any state of coherence and that it has some other important implications.

4. NATURAL OSCILLATIONS OF PARTIALLY COHERENT SOURCES

According to Mercer's theorem,[13] any continuous, hermitian, nonnegative definite Hilbert–Schmidt kernel that is not identically zero, and hence, in particular, our cross-spectral density function $W_Q(\mathbf{r}_1, \mathbf{r}_2, \omega)$, may be expressed in the form[14]

$$W_Q(\mathbf{r}_1, \mathbf{r}_2, \omega) = \sum_n \lambda_n(\omega) \phi_n^*(\mathbf{r}_1, \omega) \phi_n(\mathbf{r}_2, \omega), \quad (4.1)$$

the series being absolutely and uniformly convergent. The functions $\phi_n(\mathbf{r}, \omega)$ are the eigenfunctions, and the coefficients $\lambda_n(\omega)$ are the eigenvalues of the homogeneous Fredholm integral equation

$$\int_D W_Q(\mathbf{r}_1, \mathbf{r}_2, \omega) \phi_n(\mathbf{r}_1, \omega) d^3r_1 = \lambda_n(\omega) \phi_n(\mathbf{r}_2, \omega). \quad (4.2)$$

The hermiticity of W_Q ensures that integral Eq. (4.2) has at least one nonzero eigenvalue, and the hermiticity and nonnegative definiteness ensure that all the eigenvalues are real and nonnegative, i.e., that

$$\lambda_n(\omega) \geq 0. \quad (4.3)$$

It is also well known that each eigenfunction ϕ_n is square integrable and that eigenfunctions belonging to different eigenvalues are mutually orthogonal. If there is a degeneracy, i.e., if more than one eigenfunction belongs to a particular eigenvalue, the eigenfunctions may be orthogonalized (if they are not already mutually orthogonal) by the Gram–Schmidt procedure. Hence we may choose the eigenfunctions of integral Eq. (4.2) to form an orthonormal set, i.e.,

$$\int_D \phi_n^*(\mathbf{r}, \omega) \phi_m(\mathbf{r}, \omega) d^3r = \delta_{nm}, \quad (4.4)$$

where δ_{nm} is the Kronecker symbol.

It seems worthwhile to stress that expansion (4.1) holds irrespective of whether the eigenfunctions $\phi_n(\mathbf{r}, \omega)$ associated with nonzero eigenvalues form a complete set in the Hilbert space of square-integrable functions.

We now show that representation (4.1) of the spectral density provides an interesting new model for a finite stationary source of any state of coherence. For this purpose we introduce the degree of spatial coherence[15] at frequency ω of the source distribution by the formula

$$\mu_Q(\mathbf{r}_1, \mathbf{r}_2, \omega) = \frac{W_Q(\mathbf{r}_1, \mathbf{r}_2, \omega)}{[W_Q(\mathbf{r}_1, \mathbf{r}_1, \omega)]^{1/2}[W_Q(\mathbf{r}_2, \mathbf{r}_2, \omega)]^{1/2}}. \quad (4.5)$$

The normalization in Eq. (4.5) ensures that, for all values of its arguments,

$$0 \leq |\mu_Q(\mathbf{r}_1, \mathbf{r}_2, \omega)| \leq 1. \quad (4.6)$$

The extreme case when $\mu_Q(\mathbf{r}_1, \mathbf{r}_2, \omega) = 0$ represents spatial incoherence; the other extreme case, $|\mu_Q(\mathbf{r}_1, \mathbf{r}_2, \omega)| = 1$, represents spatial coherence between the spectral components of frequency ω of the source oscillations at the points $\mathbf{r}_1$ and $\mathbf{r}_2$.

Suppose now that we express Eq. (4.1) in the form

$$W_Q(\mathbf{r}_1, \mathbf{r}_2, \omega) = \sum_n W_Q^{(n)}(\mathbf{r}_1, \mathbf{r}_2, \omega), \quad (4.7)$$

where

$$W_Q^{(n)}(\mathbf{r}_1, \mathbf{r}_2, \omega) = \lambda_n(\omega) \phi_n^*(\mathbf{r}_1, \omega) \phi_n(\mathbf{r}_2, \omega). \quad (4.8)$$

We may regard Eq. (4.7) as representing the cross-spectral density W_Q as the sum of cross-spectral densities $W_Q^{(n)}$, each of which is associated with a *natural mode of oscillation* of the source. From Eqs. (4.8) and (4.5) it follows that the degree of spatial coherence of each mode is given by

$$\mu_Q^{(n)}(\mathbf{r}_1, \mathbf{r}_2, \omega) = \frac{W_Q^{(n)}(\mathbf{r}_1, \mathbf{r}_2, \omega)}{[W_Q^{(n)}(\mathbf{r}_1, \mathbf{r}_1, \omega)]^{1/2}[W_Q^{(n)}(\mathbf{r}_2, \mathbf{r}_2, \omega)]^{1/2}} = \frac{\phi_n^*(\mathbf{r}_1, \omega)\phi_n(\mathbf{r}_2, \omega)}{|\phi_n(\mathbf{r}_1, \omega)||\phi_n(\mathbf{r}_2, \omega)|}. \quad (4.9)$$

Equation (4.9) implies that

$$|\mu_Q^{(n)}(\mathbf{r}_1, \mathbf{r}_2, \omega)| = 1 \quad (4.10)$$

for all values of the arguments $\mathbf{r}_1$, $\mathbf{r}_2$, and ω, i.e., each mode represents a spatially completely coherent contribution. Hence we have established the following result: *The cross-spectal density of any stationary source that obeys the two broad conditions stated in Section 3 may be expressed as the sum of contributions from spatially completely coherent elementary sources.*

In the special case when the integral Eq. (4.2) admits of only one solution, $\phi_1(\mathbf{r}, \omega)$, say, associated with an eigenvalue $\lambda_1(\omega)$, the cross-spectral density of the source will necessarily have the form

$$W_Q(\mathbf{r}_1, \mathbf{r}_2, \omega) = \lambda_1(\omega)\phi_1{}^*(\mathbf{r}_1, \omega)\phi_1(\mathbf{r}_2, \omega), \qquad (4.11)$$

and it is seen at once from Eqs. (4.11) and (4.5) that the source will then be spatially completely coherent at frequency ω. Since the factorization of the cross-spectral density in the form given by Eq. (4.12) is also a necessary and sufficient condition for complete spatial coherence[16] at frequency ω, it follows that *a source will be spatially completely coherent (at a particular frequency ω) if and only if it consists of a single mode.*

We note that according to Eqs. (3.4) and (4.1) the spectrum of the source at point $\mathbf{r}$ may be expressed in terms of contributions from the elementary sources in the form

$$S_Q(\mathbf{r}, \omega) \equiv W_Q(\mathbf{r}, \mathbf{r}, \omega) = \sum_n S_Q{}^{(n)}(\mathbf{r}, \omega), \qquad (4.12)$$

where

$$S_Q{}^{(n)}(\mathbf{r}, \omega) = \lambda_n(\omega)|\phi_n(\mathbf{r}, \omega)|^2. \qquad (4.12a)$$

From the manner in which the λ_n's and the ϕ_n's were introduced, it is clear that *the spectrum of the source distribution depends on its* (second-order) *spatial coherence properties.* This fact is obviously relevant to spectroscopy of partially coherent sources.

If we integrate Eq. (4.12) over the source domain D, interchange the orders of integration and summation, and use Eq. (4.4), we obtain the following expression for the source-integrated spectrum:

$$\int_D S_Q(\mathbf{r}, \omega)\mathrm{d}^3r = \sum_n \int_D S_Q{}^{(n)}(\mathbf{r}, \omega)\mathrm{d}^3r, \qquad (4.13)$$

where

$$\int_D S_Q{}^{(n)}(\mathbf{r}, \omega)\mathrm{d}^3r = \lambda_n(\omega). \qquad (4.14)$$

Equation (4.13) expresses the source-integrated spectrum as a sum of contributions from all the elementary sources (natural modes of oscillation). According to Eq. (4.14), *the contribution from each elementary source to the source-integrated spectrum is equal to the corresponding eigenvalue of integral Eq. (4.2), whose kernel is the cross-spectral density $W_Q(\mathbf{r}_1, \mathbf{r}_2, \omega)$ of the source distribution $Q(\mathbf{r})$.*

The cross-spectral densities $W_Q{}^{(n)}(\mathbf{r}_1, \mathbf{r}_2, \omega)$ of the elementary sources may readily be shown to satisfy the following orthogonality relations:

$$\iint_{DD} W_Q{}^{(n)*}(\mathbf{r}_1, \mathbf{r}_2, \omega)W_Q{}^{(m)}(\mathbf{r}_1, \mathbf{r}_2, \omega)\mathrm{d}^3r_1\mathrm{d}^3r_2 = \lambda_n{}^2(\omega)\delta_{nm}. \qquad (4.15)$$

This relation follows at once from definition (4.8) of $W_Q{}^{(n)}$ and from the orthonormality, expressed by Eq. (4.4), of the eigenfunctions ϕ_n.

We may derive another interesting relation that involves the square of the eigenvalues. We have from Eq. (4.7) the relation

$$|W_Q(\mathbf{r}_1, \mathbf{r}_2, \omega)|^2 = \sum_n\sum_m W_Q{}^{(n)*}(\mathbf{r}_1, \mathbf{r}_2, \omega)W_Q{}^{(m)}(\mathbf{r}_1, \mathbf{r}_2, \omega). \qquad (4.16)$$

On integrating both sides of Eq. (4.16) with respect to $\mathbf{r}_1$ and $\mathbf{r}_2$ over the domain D, interchanging the orders of integration and summation, and on using orthogonality relations (4.15), we find that

$$\iint_{DD} |W_Q(\mathbf{r}_1, \mathbf{r}_2, \omega)|^2\mathrm{d}^3r_1\mathrm{d}^3r_2 = \sum_n \lambda_n{}^2(\omega). \qquad (4.17)$$

This formula, as well as numerous other relations involving the sums of other integral powers of the eigenvalues $\lambda_n(\omega)$, may also be obtained from well-known properties of the kernel and of the iterated kernels associated with basic integral equation (4.2) of the present theory.

So far we have discussed mainly the cross-spectral density. Let us now briefly turn to the cross-correlation function $\Gamma_Q(\mathbf{r}_1, \mathbf{r}_2, \tau)$ of the source. If we substitute for $W_Q(\mathbf{r}_1, \mathbf{r}_2, \omega)$ from Eq. (4.7) into Eq. (3.7) and interchange the orders of integration and summation, we obtain the following new expansion for the cross-correlation function:

$$\Gamma_Q(\mathbf{r}_1, \mathbf{r}_2, \tau) = \sum_n \Gamma_Q{}^{(n)}(\mathbf{r}_1, \mathbf{r}_2, \tau), \qquad (4.18)$$

where

$$\Gamma_Q{}^{(n)}(\mathbf{r}_1, \mathbf{r}_2, \tau) = \int_{-\infty}^{\infty} W_Q{}^{(n)}(\mathbf{r}_1, \mathbf{r}_2, \omega)e^{-i\omega\tau}\mathrm{d}\omega. \qquad (4.19)$$

From Eqs. (4.19) and (4.8) the following expression for $\Gamma_Q{}^{(n)}$ follows at once:

$$\Gamma_Q{}^{(n)}(\mathbf{r}_1, \mathbf{r}_2, \tau) = \int_{-\infty}^{\infty} \lambda_n(\omega)\phi_n{}^*(\mathbf{r}_1, \omega)\phi_n(\mathbf{r}_2, \omega)e^{-i\omega\tau}\mathrm{d}\omega. \qquad (4.20)$$

Formula (4.18), together with Eq. (4.20), is a new mode representation of the cross-correlation function Γ_Q of the source. If we introduce the space–time degree of coherence $\gamma_Q{}^{(n)}(\mathbf{r}_1, \mathbf{r}_2, \tau)$ of each mode $\Gamma_Q{}^{(n)}$, by the usual expression [Ref. 1, Sec. 3.1 and Ref. 17, Sec. 10.3.1],

$$\gamma_Q{}^{(n)}(\mathbf{r}_1, \mathbf{r}_2, \tau) = \frac{\Gamma_Q{}^{(n)}(\mathbf{r}_1, \mathbf{r}_2, \tau)}{[\Gamma_Q{}^{(n)}(\mathbf{r}_1, \mathbf{r}_1, 0)]^{1/2}[\Gamma_Q{}^{(n)}(\mathbf{r}_2, \mathbf{r}_2, 0)]^{1/2}}, \qquad (4.21)$$

we see that $|\gamma_Q{}^{(n)}(\mathbf{r}_1, \mathbf{r}_2, \tau)|$ differs, in general, from unity. This result implies that, whereas each mode in the space–frequency domain is spatially completely coherent, its contribution to the correlation function Γ_Q is, in general, neither spatially nor temporally completely coherent.

The set $\{\Gamma_Q{}^{(n)}(\mathbf{r}_1, \mathbf{r}_2, \tau)\}$ of the modes in the space–time domain may be shown to be orthogonal. More specifically,

$$\int_{-\infty}^{\infty} \mathrm{d}\tau \int_D \mathrm{d}^3r_1 \int_D \mathrm{d}^3r_2\Gamma_Q{}^{(n)*}(\mathbf{r}_1, \mathbf{r}_2, \tau)\,\Gamma_Q{}^{(m)}(\mathbf{r}_1, \mathbf{r}_2, \tau) = 2\pi\delta_{nm}\int_{-\infty}^{\infty} \lambda_n{}^2(\omega)\mathrm{d}\omega. \qquad (4.22)$$

A proof of this result is given in Appendix B.

There are numerous other interesting relations that involve the $\Gamma_Q^{(n)}$'s. We derive only one of them. If we set $\mathbf{r}_1 = \mathbf{r}_2 = \mathbf{r}$, $\tau = 0$ in Eq. (4.19), integrate the resulting expression with respect to $\mathbf{r}$ over the source domain D, and formally interchange the orders of integration, we obtain the formula

$$\int_D \Gamma^{(n)}(\mathbf{r}, \mathbf{r}, 0)\mathrm{d}^3r = \int_{-\infty}^{\infty} \mathrm{d}\omega \int_D \mathrm{d}^3r W_Q^{(n)}(\mathbf{r}, \mathbf{r}, \omega). \tag{4.23}$$

Now $W_Q^{(n)}(\mathbf{r}, \mathbf{r}, \omega)$ is the spectrum $S_Q^{(n)}(\mathbf{r}, \omega)$ at $\mathbf{r}$ of a typical elementary source, and hence, if we use Eq. (4.14), it follows that

$$\int_D \Gamma^{(n)}(\mathbf{r}, \mathbf{r}, 0)\mathrm{d}^3r = \int_{-\infty}^{\infty} \lambda_n(\omega)\mathrm{d}\omega. \tag{4.24}$$

Formula (4.24) shows that the integral over the source domain D of the "intensity" $\Gamma_Q^{(n)}(\mathbf{r}, \mathbf{r}, 0)$ of an elementary source is equal to the frequency integral of the corresponding eigenvalue $\lambda_n(\omega)$.

5. REPRESENTATION OF THE SPECTRUM AND OF THE CROSS SPECTRUM IN TERMS OF AN ENSEMBLE OF MONOCHROMATIC REALIZATIONS

We noted earlier (Section 2) the difficulties that were encountered in previous attempts to introduce the spectral density and the cross-spectral density of a stationary random process through a statistical ensemble of monochromatic oscillations. We saw that the difficulty arose from the fact that a stationary random process does not admit of a Fourier integral decomposition in the context of ordinary function theory. We now show that, in spite of this fact, ensemble representation of the spectral and the cross-spectral densities in terms of monochromatic oscillations can be given, at least in the context of the types of problems that we are considering in this paper, relating to radiation from finite stationary sources of any state of coherence. This new representation is, of course, not based on Fourier integral decomposition (which does not exist) of the sample functions of the process but rather on the natural oscillations of the source that we introduced in the preceding section.

Let us consider an ensemble of functions $\{U_Q(\mathbf{r}, \omega)\}$, each member of which is a linear superposition of the eigenfunctions $\phi_n(\mathbf{r}, \omega)$ of the integral equation (4.2):

$$U_Q(\mathbf{r}, \omega) = \sum_n a_n(\omega)\phi_n(\mathbf{r}, \omega). \tag{5.1}$$

In this expansion the a_n's are random coefficients, whose properties we will specify shortly. Next let us construct the cross-correlation function of the U_Q's at two source points $\mathbf{r}_1$ and $\mathbf{r}_2$, at the same frequency ω. According to Eq. (5.1) it is given by

$$\langle U_Q^*(\mathbf{r}_1, \omega)U_Q(\mathbf{r}_2, \omega)\rangle_\omega = \sum_n \sum_m \langle a_n^*(\omega)a_m(\omega)\rangle_\omega \phi_n^*(\mathbf{r}_1, \omega)\phi_m(\mathbf{r}_2, \omega), \tag{5.2}$$

where the angle brackets denote the ensemble average. We attached the subscript ω to the averaging symbol to stress that the average is now taken over an ensemble of frequency-dependent functions, rather than over the ensemble of time-dependent functions that we encountered earlier, in defining the cross-correlation function $\Gamma_Q(\mathbf{r}_1, \mathbf{r}_2, \tau)$ [Eq. (3.1)]. It is important to distinguish clearly between the two ensembles.

Suppose now that we chose the random coefficients $a_n(\omega)$ so that

$$\langle a_n^*(\omega)a_m(\omega)\rangle_\omega = \lambda_n(\omega)\delta_{nm}, \tag{5.3}$$

where δ_{nm} is the Kronecker symbol, and

$$\sum_n |a_n(\omega)|^2 < \infty. \tag{5.4}$$

Such an ensemble can always be chosen, for example, by taking

$$a_n(\omega) = [\lambda_n(\omega)]^{2/3}e^{i\alpha_n}, \tag{5.5}$$

where, for each n, α_n is a real random variable that is uniformly distributed in the interval $0 \leqslant \alpha_n < 2\pi$ and a_n and α_m are statistically independent when $n \neq m$.

Each of the sample functions $U_Q(\mathbf{r}, \omega)$, given by Eq. (5.1), is readily seen to be square integrable over the source domain D. In fact, the integral of the squared modulus of U_Q, taken over D, then has a simple meaning, as we now show. We have, if we use Eq. (5.1) and interchange the orders of the integration and summation on the right-hand side,

$$\int_D |U_Q(\mathbf{r}, \omega)|^2\mathrm{d}^3r = \sum_n \sum_m a_n^*(\omega)a_m(\omega) \int_D \phi_n^*(\mathbf{r}, \omega)\phi_m(\mathbf{r}, \omega)\mathrm{d}^3r. \tag{5.6}$$

Since, according to Eq. (4.4), the ϕ_n's form an orthonormal set over D, Eq. (5.5) reduces to

$$\int_D |U_Q(\mathbf{r}, \omega)|^2\mathrm{d}^3r = \sum_n |a_n(\omega)|^2. \tag{5.7}$$

Let us now substitute from Eq. (5.3) into Eq. (5.2). We then obtain at once the formula

$$\langle U_Q^*(\mathbf{r}_1, \omega)U_Q(\mathbf{r}_2, \omega)\rangle_\omega = \sum_n \lambda_n(\omega)\phi_n^*(\mathbf{r}_1, \omega)\phi_n(\mathbf{r}_2, \omega). \tag{5.8}$$

On comparing the right-hand side of this equation with the right-hand side of the Mercer expansion [Eq. (4.1)] for the cross-spectral density, we see that they are equal. Hence the left-hand sides must also be equal, i.e.,

$$W_Q(\mathbf{r}_1, \mathbf{r}_2, \omega) = \langle U_Q^*(\mathbf{r}_1, \omega)U_Q(\mathbf{r}_2, \omega)\rangle_\omega. \tag{5.9}$$

Thus we have succeeded in constructing *an ensemble of monochromatic oscillations* $\{U_Q(\mathbf{r}, \omega)e^{-i\omega t}\}$ *that provides a representation of the cross-spectral density* $W_Q(\mathbf{r}_1, \mathbf{r}_2, \omega)$ *of the source as a correlation function in the space–frequency domain.*

Before proceeding further we might mention that the preceding analysis has much in common with that encountered in connection with the well-known Karhunen–Loève orthogonal expansion (Ref. 7, pp. 96 et seq.) of a random process. However, our expansion provides a representation of the cross-spectral density function of a stationary source rather than of the autocorrelation of a random process in terms of a set of orthogonal functions. Moreover, the Karhunen–Loève expansion is not unique, because the orthogonal functions and the associated eigenvalues depend on the choice of the domain over which the decomposition is carried out; on the other hand, the orthogonal functions $\phi_n(\mathbf{r}, \omega)$ and the associated eigenvalues $\lambda_n(\omega)$ of our representations are unique, because the domain is the region of space D occupied by the source. For this reason, our new representation of the cross-spectral density appears to have a deeper physical meaning than the Karhunen–Loève representation of the cross-correlation function.

If in formula (5.9) we let points $\mathbf{r}_1$ and $\mathbf{r}_2$ coincide and use Eq. (3.4), we obtain the following representation of the *spectrum of the fluctuating source distribution* $Q(\mathbf{r}, t)$:

$$S_Q(\mathbf{r}, \omega) = \langle |U_Q(\mathbf{r}, \omega)|^2 \rangle_\omega. \tag{5.10}$$

This formula, together with Eqs. (5.1), (5.3), and (5.4) with the ϕ_n's and λ_n being defined by Eq. (4.2), is a new and mathematically rigorous representation of the spectrum of the fluctuating source distribution $Q(\mathbf{r}, t)$. It agrees with the intuitive concept of a spectrum as an average of the squared modulus of strictly monochromatic oscillations, which has been the basis of many previous attempts (some of which we discussed in Section 2) to provide a satisfactory definition of the spectrum. However, we stress that, unlike in the earlier attempts, our monochromatic oscillations $U_Q(\mathbf{r}, \omega)e^{-i\omega t}$ are associated not with the (nonexistent) Fourier frequency components of the stationary random variable but rather with the eigenfunctions and the eigenvalues of an integral operator, whose kernel is the cross-spectral density [Eq. (4.2)].

Returning to Eq. (5.1), we note that the requirement expressed by Eqs. (5.3) and (5.4) does not fully specify the ensemble of the sample functions $\{U_Q(\mathbf{r}, \omega)\}$. It specifies only their second-order correlation properties by the second-order moments of the expansion coefficients $a_n(\omega)$. We plan to show in another publication that all the moments

$$\langle a_{n_1}^*(\omega_1) a_{n_2}^*(\omega_2) \ldots a_{n_l}^*(\omega_l) a_{n_p}(\omega_p) a_{n_q}(\omega_2) \ldots a_{n_s}(\omega_s) \rangle_\omega$$

may be specified in such a manner that the cross-spectral densities of all orders[18] are given by correlations of the form

$$\langle U_Q^*(\mathbf{r}_1, \omega_1) U_Q^*(\mathbf{r}_2, \omega_2) \ldots U_Q^*(\mathbf{r}_l, \omega_l) \times U_Q(\mathbf{r}_p, \omega_p) U_Q(\mathbf{r}_q, \omega_q) \ldots U_Q(\mathbf{r}_s, \omega_s) \rangle_\omega.$$

Finally, let us compare our new formulation of the second-order correlation theory with the customary one.

In the customary formulation one starts with an ensemble of time-dependent realizations $Q(\mathbf{r}, t)$. One then defines, in terms of them, the cross-correlation function $\Gamma_Q(\mathbf{r}_1, \mathbf{r}_2, \tau)$ [Eq. (3.1)]. The cross-spectral density $W_Q(\mathbf{r}_1, \mathbf{r}_2, \omega)$ may then be obtained by taking its Fourier transform [Eq. (3.3)]. These steps are indicated by arrows in column (a) of Table 1.

In our new formulation, on the other hand, one deals with an ensemble of frequency-dependent realizations $U_Q(\mathbf{r}, \omega)$. The cross-spectral density $W_Q(\mathbf{r}_1, \mathbf{r}_2, \omega)$ can then be identified with their cross-correlation function [Eq. (5.9)]. The cross-correlation function $\Gamma_Q(\mathbf{r}_1, \mathbf{r}_2, \tau)$ may then be obtained by taking its Fourier transform [Eq. (3.7)]. These steps are indicated by arrows in column (b) of Table 1. In neither case can one pass from $Q(\mathbf{r}, t)$ to $U_Q(\mathbf{r}, \omega)$ or vice versa by Fourier transforms.

Apart from introducing greater symmetry between the space–time and the space–frequency descriptions, our new theory provides new insights into properties of optical sources and of optical fields of any state of coherence, as we will demonstrate in other publications.

Table 1. Schematic Comparison of the Usual and the Present Theories[a]

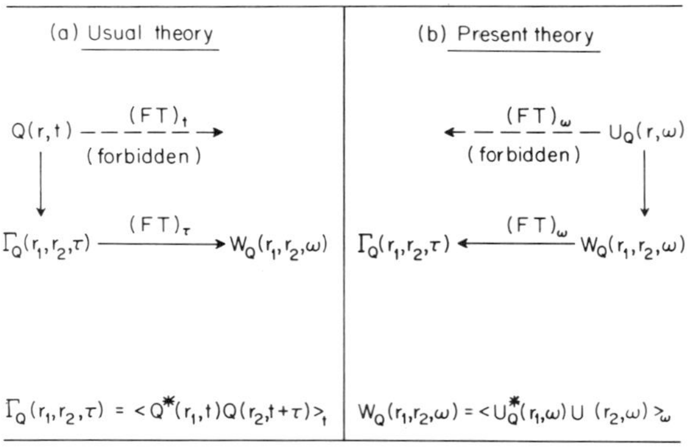

[a] $(FT)_\alpha$ denotes the Fourier transform taken with respect to the variable α.

APPENDIX A: PROOF OF THE NONNEGATIVENESS CONDITION [EQ. (3.10)] SATISFIED BY THE CROSS-SPECTRAL DENSITY

To establish the nonnegativeness condition [Eq. (3.10)] that the cross-spectral density $W_Q(\mathbf{r}_1, \mathbf{r}_2, \omega)$ satisfies,[19] it will be helpful to derive first a corresponding nonnegative definiteness condition obeyed by the cross-correlation function $\Gamma_Q(\mathbf{r}_1, \mathbf{r}_2, \tau)$. To keep the notation simpler we will, throughout this Appendix, suppress the subscript Q on both of these quantities.

Let $Q(\mathbf{r}, t)$ again represent the source function, assumed to be a stationary random function. Then obviously

$$\left\langle \left| \int_D d^3r \int_{-\infty}^{\infty} dt F(\mathbf{r}, t) Q(\mathbf{r}, t) \right|^2 \right\rangle_t \geqslant 0, \tag{A1}$$

where $F(\mathbf{r}, t)$ is any arbitrary well-behaved deterministic function, subject only to the requirement that the integral in inequality (A1) exist. The angle brackets denote, of course, the average taken over the ensemble of $Q(\mathbf{r}, t)$. Written out more explicitly, expression (A1) implies that

$$\left\langle \int_D d^3r_1 \int_D d^3r_2 \int_{-\infty}^{\infty} dt_1 \int_{-\infty}^{\infty} dt_2 F^*(\mathbf{r}_1, t_1) \times F(\mathbf{r}_2, t_2) Q^*(\mathbf{r}_1, t_1) Q(\mathbf{r}_2, t_2) \right\rangle_t \geqslant 0. \tag{A2}$$

We next interchange the orders of the ensemble averaging and of the integrations and use the fact that [Eq. (3.1)]

$$\langle Q^*(\mathbf{r}_1, t_1) Q(\mathbf{r}_2, t_2) \rangle_t = \Gamma(\mathbf{r}_1, \mathbf{r}_2, t_2 - t_1). \tag{A3}$$

Then inequality (A2) becomes the *nonnegative definiteness condition satisfied by the cross-correlation function* Γ:

$$\int_D d^3r_1 \int_D d^3r_2 \int_{-\infty}^{\infty} dt_1 \int_{-\infty}^{\infty} dt_2 F^*(\mathbf{r}_1, t_1) \times F(\mathbf{r}_2, t_2) \Gamma(\mathbf{r}_1, \mathbf{r}_2, t_2 - t_1) \geqslant 0. \tag{A4}$$

Now the cross-correlation function Γ is the Fourier transform of the cross-spectral density W [Eq. (3.7)]:

$$\Gamma(\mathbf{r}_1, \mathbf{r}_2, \tau) = \int_{-\infty}^{\infty} W(\mathbf{r}_1, \mathbf{r}_2, \omega) e^{-i\omega\tau} d\omega. \quad \text{(A5)}$$

On substituting from expression (A5) into expression (A4) we obtain the inequality

$$\int_D d^3r_1 \int_D d^3r_2 \int_{-\infty}^{\infty} dt_1 \int_{-\infty}^{\infty} dt_2 \int_{-\infty}^{\infty} d\omega F^*(\mathbf{r}_1, t_1) \times F(\mathbf{r}_2, t_2) W(\mathbf{r}_1, \mathbf{r}_2, \omega) e^{-i\omega(t_2 - t_1)} \geqslant 0. \quad \text{(A6)}$$

Let us now choose the arbitrary function $F(\mathbf{r}, t)$ to be the product of a function of $\mathbf{r}$ and a function of t:

$$F(\mathbf{r}, t) = f(\mathbf{r}) g(t). \quad \text{(A7)}$$

Then inequality (A6) becomes

$$\int_{-\infty}^{\infty} |G(\omega)|^2 E(\omega) d\omega \geqslant 0, \quad \text{(A8)}$$

where

$$G(\omega) = \int_{-\infty}^{\infty} g(t) e^{i\omega t} dt \quad \text{(A9)}$$

and

$$E(\omega) = \int_D d^3r_1 \int_D d^3r_2 f^*(\mathbf{r}_1) f(\mathbf{r}_2) W(\mathbf{r}_1, \mathbf{r}_2, \omega). \quad \text{(A10)}$$

Now, according to Eq. (3.9), the cross-spectral density $W(\mathbf{r}_1, \mathbf{r}_2, \omega)$ satisfies the hermiticity condition

$$W(\mathbf{r}_2, \mathbf{r}_1, \omega) = W^*(\mathbf{r}_1, \mathbf{r}_2, \omega). \quad \text{(A11)}$$

If we take the complex conjugate of expression (A10) and use expression (A11), we see at once that

$$E^*(\omega) = E(\omega), \quad \text{(A12)}$$

implying that $E(\omega)$ is a *real* function of ω.

Let us now suppose that at some frequency $\omega = \omega_0$,

$$E(\omega_0) < 0. \quad \text{(A13)}$$

Since $W(\mathbf{r}_1, \mathbf{r}_2, \omega)$ is a continuous function of ω [cf. remarks under Eq. (3.3)], it is clear from Eq. (A10) that for any well-behaved function $f(\mathbf{r})$, $E(\omega)$ will also be a continuous function of ω. Hence there will be some neighborhood

$$\Delta(\omega; \omega_0, \epsilon): \ \omega_0 - \epsilon < \omega < \omega_0 + \epsilon, \quad \text{(A14)}$$

($\epsilon > 0$), of ω_0 throughout which

$$E(\omega) < 0. \quad \text{(A15)}$$

Now the function $g(t)$ in Eq. (A9) is any well-behaved function of t. In particular, we may choose it so that the squared modulus $G(\omega)$ of its Fourier transform takes on any prescribed nonnegative values throughout the interval $\Delta(\omega; \omega_0, \epsilon)$ and that it vanishes outside this interval. Then, in view of inequality (A15),

$$\int_{-\infty}^{\infty} |G(\omega)|^2 E(\omega) d\omega = \int_{\Delta(\omega;\omega_0,\epsilon)} |G(\omega)|^2 E(\omega) d\omega < 0. \quad \text{(A16)}$$

Since Eq. (A16) contradicts inequality (A8), it follows that assumption (A13), and consequently inequality (A15), cannot be valid. Hence

$$E(\omega) \geqslant 0. \quad \text{(A17)}$$

Recalling definition (A10) of $E(\omega)$, inequality (A17) implies that, with any well-behaved function $f(\mathbf{r})$,

$$\int_D d^3r_1 \int_D d^3r_2 f^*(\mathbf{r}_1) f(\mathbf{r}_2) W(\mathbf{r}_1, \mathbf{r}_2, \omega) \geqslant 0 \quad \text{(A18)}$$

for each value of ω. This result is the required *nonnegative definiteness condition* (3.10) satisfied by *the cross-spectral density*.

APPENDIX B: PROOF OF THE ORTHOGONALITY RELATIONS [EQ. (4.22)] FOR THE SET OF FUNCTIONS $\Gamma_Q^{(n)}$ ($\mathbf{r}_1, \mathbf{r}_2, \omega$)

Let

$$I_{nm} = \int_{-\infty}^{\infty} d\tau \int_D d^3r_1 \int_D d^3r_2 \times \Gamma_Q^{(n)*}(\mathbf{r}_1, \mathbf{r}_2, \tau) \Gamma_Q^{(m)}(\mathbf{r}_1, \mathbf{r}_2, \tau). \quad \text{(B1)}$$

On substituting on the right-hand side of Eq. (B1) from Eq. (4.19), we obtain for I_{nm} the expression

$$I_{nm} = \int_{-\infty}^{\infty} d\tau \int_D d^3r_1 \int_D d^3r_2 \int_{-\infty}^{\infty} d\omega \int_{-\infty}^{\infty} d\omega' \times W_Q^{(n)*}(\mathbf{r}_1, \mathbf{r}_2, \omega) W_Q^{(m)}(\mathbf{r}_1, \mathbf{r}_2, \omega') e^{i(\omega - \omega')\tau}. \quad \text{(B2)}$$

The integration with respect to τ can be carried out at once and gives $2\pi\delta(\omega - \omega')$, where $\delta(\omega - \omega')$ is the Dirac delta function. The integration with respect to ω' is then trivial and leads to the expression

$$I_{nm} = 2\pi \int_{-\infty}^{\infty} d\omega \int_D d^3r_1 \int_D d^3r_2 \times W_Q^{(n)*}(\mathbf{r}_1, \mathbf{r}_2, \omega) W_Q^{(m)}(\mathbf{r}_1, \mathbf{r}_2, \omega).$$

Finally, if we make use of orthogonality relations (4.15), we obtain the following expression for I_{nm}:

$$I_{nm} = 2\pi\delta_{nm} \int_{-\infty}^{\infty} \lambda_n^2(\omega) d\omega.$$

This formula, together with the defining Eq. (B1) of I_{nm}, is the required orthogonality relation (4.22).

ACKNOWLEDGMENTS

I am obliged to Ari T. Friberg for assistance with the proof given in Appendix A.

This research was supported by the U.S. Army Research Office.

*Also with the Institute of Optics, University of Rochester.

REFERENCES

1. For an account of the theory see, for example, L. Mandel and E. Wolf, "Coherence properties of optical fields," Rev. Mod. Phys. **37**, 231–287 (1965).
2. See also E. Wolf, "A new description of second-order coherence phenomena in the space-frequency domain," in *Optics in Four Dimensions–1980*, M. A. Machado and L. M. Narducci, eds., Conference Proceedings #65 (American Institute of Physics, New York, 1981), pp. 42–48; "New spectral representation of random sources and of the partially coherent fields that they generate," Opt. Commun. **38**, 3–6 (1981).
3. N. Wiener, "Generalized harmonic analysis," Acta Math. **55**, 117–258 (1930). Section 2 of Chap. I of this paper contains a lucid discussion of some of the earlier attempts to provide a satisfactory mathematical description of the spectrum of light. The following

passage seems worth quoting: "... one is astonished by the skill with which authors use clumsy and unsuitable tools to obtain the right results, and one is led to admire the unfailing heuristic insight of a true physicist."

4. A. Y. Khintchine, "Korrelationstheorie der stationären stochastischen Prozesse," Math. Ann. **109,** 605–615 (1934).
5. For pertinent references and a discussion of this point, see A. M. Yaglom, *An Introduction to the Theory of Stationary Random Functions* (Prentice-Hall, Englewood Cliffs, N.J., 1962), especially pp. 35–39.
6. S. Goldman, *Information Theory* (Prentice-Hall, New York, 1953), Sec. 8.4, or D. Middleton, *An Introduction to Statistical Communication Theory* (McGraw-Hill, New York, 1960), Sec. 3.2.
7. W. B. Davenport and W. L. Root, *An Introduction to the Theory of Random Signals and Noise* (McGraw-Hill, New York, 1958), pp. 107–108.
8. In classical wave theory the fluctuating function is usually real, say, $Q^{(r)}(\mathbf{r}, t)$. It is often convenient to associate with $Q^{(r)}(\mathbf{r}, t)$ a certain complex function $Q(\mathbf{r}, t)$, known as the analytic signal (see, for example, Ref. 1, Sec. 3.1, or Ref. 17, Sec. 2). However, when $Q^{(r)}(\mathbf{r}, t)$ is a member of a stationary ensemble, the transition to the complex analytic signal is mathematically unsatisfactory for reasons similar to those that we discussed in a somewhat different context at the beginning of Section 2. This difficulty may be overcome by starting from the real cross-correlation function

$$\Gamma^{(r)}(\mathbf{r}_1, \mathbf{r}_2, \tau) = \langle Q^{(r)}(\mathbf{r}_1, t) Q^{(r)}(\mathbf{r}_2, t + \tau) \rangle_t$$

and defining $\Gamma_Q(\mathbf{r}_1, \mathbf{r}_2, \tau)$ not by Eq. (3.1) but rather as the complex analytic signal associated with $\Gamma^{(r)}(\mathbf{r}_1, \mathbf{r}_2, \tau)$. This procedure is applicable whenever $\Gamma^{(r)}$ is absolutely integrable with respect to τ. If the complex cross-correlation function is introduced in this way, all our subsequent results remain valid. The fact that Γ_Q is an analytic signal implies that the cross-spectral density $W_Q(\mathbf{r}_1, \mathbf{r}_2, \omega)$ will then vanish identically for $\omega < 0$.
9. R. R. Goldberg, *Fourier Transforms* (Cambridge U. Press, Cambridge, 1965), p. 6.
10. A simple proof of this result is as follows: One readily finds from Eq. (3.1), with the help of Schwarz's inequality, that

$$|\Gamma_Q(\mathbf{r}_1, \mathbf{r}_2, \tau)| \leqslant [\Gamma_Q(\mathbf{r}_1, \mathbf{r}_1, 0)]^{1/2} [\Gamma_Q(\mathbf{r}_2, \mathbf{r}_2, 0)]^{1/2}.$$

Hence

$$|\Gamma_Q(\mathbf{r}_1, \mathbf{r}_2, \tau)|^2 \leqslant [\Gamma_Q(\mathbf{r}_1, \mathbf{r}_1, 0)]^{1/2} [\Gamma_Q(\mathbf{r}_2, \mathbf{r}_2, 0)]^{1/2} |\Gamma_Q(\mathbf{r}_1, \mathbf{r}_2, \tau)|,$$

and consequently

$$\int_{-\infty}^{\infty} |\Gamma_Q(\mathbf{r}_1, \mathbf{r}_2, \tau)|^2 d\tau \leqslant [\Gamma_Q(\mathbf{r}_1, \mathbf{r}_1, 0)]^{1/2} [\Gamma_Q(\mathbf{r}_2, \mathbf{r}_2, 0)]^{1/2} \int_{-\infty}^{\infty} |\Gamma_Q(\mathbf{r}_1, \mathbf{r}_2, \tau)| d\tau.$$

Now the first two terms on the right-hand-side of this formula are necessarily finite, as is apparent from Eq. (3.1), and the integral on the right-hand side is finite by assumption [Eq. (3.2)]. Hence the left-hand side must also be finite, implying that Γ_Q is square integrable with respect to τ.
11. More explicitly, this assumption means that there exists a nonnegative function $F(\tau)$ of τ alone such that for all $\mathbf{r}_1 \in D$, $\mathbf{r}_2 \in D$:

$$|\Gamma(\mathbf{r}_1, \mathbf{r}_2, \tau)| < F(\tau)$$

and

$$\int_{-\infty}^{\infty} F(\tau) d\tau < \infty.$$
12. The corresponding result for real functions of a smaller number of variables is established, for example, in E. W. Hobson, *The Theory of Functions of a Real Variable and the Theory of Fourier Series* (Cambridge U. Press, Cambridge, 1926), Vol. II, p. 326.
13. Actually we use here an obvious multidimensional generalization of Mercer's theorem to the case in which the kernel depends on two vector variables $\mathbf{r}_1$ and $\mathbf{r}_2$ rather than on two scalar variables (x_1 and x_2, say). For a discussion of the usual form of Mercer's theorem see, for example, F. Smithies, *Integral Equations* (Cambridge U. Press, Cambridge, 1970), p. 128, or F. Riecz and B. Sz.-Nagy, *Functional Analysis* (Ungar, New York, 1955), p. 245.
14. In two recent papers cited below, Mercer's expansion was used in the treatment of other problems of optical coherence theory. However, the functions that were represented in this way were the equal-time correlation function (the mutual intensity) $\Gamma(\mathbf{r}_1, \mathbf{r}_2, 0)$ and the equal-time degree of coherence $\gamma(\mathbf{r}_1, \mathbf{r}_2, 0)$. Some advantages of basing the theory on the Mercer expansion of the cross-spectral density $W(\mathbf{r}_1, \mathbf{r}_2, \omega)$, as is done in this paper, will become apparent in Part II of this investigation. The two papers are R. Martínez-Herrero, "Expansion of the complex degree of coherence," Nuovo Cimento **54B,** 205–210 (1979); R. Martínez-Herrero and P. M. Majias, "Relation between the expansions of the correlation function at the object and image planes for partially coherent illumination," Opt. Commun. **37,** 234–238 (1981). In this connection see also the early pioneering article by H. Gamo, "Matrix treatment of partial coherence," in *Progress in Optics*, E. Wolf, ed. (North-Holland, Amsterdam, 1964), Vol. 3, pp. 187–336, especially Sec. 4.4.
15. L. Mandel and E. Wolf, "Spectral coherence and the concept of cross-spectral purity," J. Opt. Soc. Am. **66,** 529–535 (1976), Sec. II.
16. L. Mandel and E. Wolf, "Complete coherence in the space-frequency domain," Opt. Commun. **36,** 247–249 (1981).
17. M. Born and E. Wolf, *Principles of Optics*, 6th ed. (Pergamon, Oxford, 1980).
18. C. L. Mehta and L. Mandel, "Some properties of higher order coherence functions," in *Electromagnetic Wave Theory*, Part II, J. Brown, ed. (Pergamon, Oxford, 1967), pp. 1069–1075. See also E. Wolf, "Light fluctuations as a new spectroscopic tool," Jpn. J. Appl. Phys. **4,** Suppl. I, Sec. 6, 1–14 (1965).
19. Our method of proof follows closely an argument given in connection with positive definite functions by S. Bochner in *Lectures on Fourier Integrals* (Princeton U. Press, Princeton, N.J., 1959), pp. 326–327. A nonrigorous proof of a related nonnegative definiteness condition satisfied by the cross-spectral density was given by L. Mandel and E. Wolf in the appendix of Ref. 15.

76 J. Opt. Soc. Am. A/Vol. 3, No. 1/January 1986

New theory of partial coherence in the space-frequency domain. Part II: Steady-state fields and higher-order correlations

Emil Wolf

Department of Physics and Astronomy, University of Rochester, Rochester, New York, 14627

Received March 8, 1985; accepted July 16, 1985

In Part I of this investigation [E. Wolf, J. Opt. Soc. Am. **72,** 343 (1982)] new representations were introduced for the cross-spectral density of a steady-state source of any state of coherence. The central concept in that formulation was the notion of a coherent source mode (a natural mode of oscillation). In the present paper the theory is developed further and new representations are obtained for the cross-spectral densities of all orders, both of the source and of the field that the source generates. These representations involve only the previously introduced coherent source modes and the moments of certain random coefficients that characterize the statistical properties of the source. The results provide a new mathematical framework for analyzing coherence properties of all orders of stationary sources and of stationary fields. Some potential applications of the theory are mentioned.

1. INTRODUCTION

In Part I of this investigation[1] a new representation of fluctuating steady-state primary scalar sources was introduced. One of the main results obtained was the demonstration that under very general conditions that are usually satisfied in practice, the cross-spectral density of such a source may be expressed as a linear superposition of contributions from spatially completely coherent and mutually uncorrelated elementary sources (natural modes of oscillations). It was also shown that an ensemble of strictly monochromatic oscillations, all of the same frequency, may be constructed, which yields rigorously the cross-spectral density of the source as a correlation function in the space–frequency domain. As immediate consequences of these results, two new expressions for the Wiener–Khintchine spectrum of the source were obtained, which are free of the mathematical subtleties involved in its customary definitions. In two brief communications[2,3] similar representations were also introduced for fluctuating steady-state wave fields of any state of coherence in free space. These representations have already found useful applications in studies of radiation fields generated by Gaussian Schell-model sources,[4] in estimations of the degrees of freedom of partially coherent sources,[5] in determining conditions under which a source will produce a partially coherent beam with certain shape-invariant properties,[6,7] in connection with investigations regarding coherence properties of laser modes,[8] in studies relating to foundations of radiometry,[9] and in the analysis of some inverse problems involving partially coherent sources.[10]

In the present paper the theory is developed further. After a brief review in Section 2 of the main results that were obtained in Part I (Ref. 1) we derive, in Section 3, two new representations for the cross-spectral density of the field generated by any steady-state primary source. These representations are analogous to those that we previously obtained for the cross-spectral density of the source, except that each source mode is replaced by the elementary wave field generated by that mode.

In the sections that follow the theory is extended to correlations of all orders. In Section 4 the various higher-order source correlations are defined in both the space–time and the space–frequency domains. In Section 5 new representations of the cross-spectral densities of all orders of a steady-state source are obtained, in terms of the natural modes of oscillations of the source and of moments of certain random coefficients that characterize the statistical properties of the source. In Section 6 analogous representations are derived for the cross-spectral densities of all orders of the field generated by the source. The main results are summarized in the concluding section (Section 7).

Although second-order coherence theory is now a well developed subject that has found many useful applications in connection with numerous optical problems, higher-order coherence phenomena have received appreciably less attention. We believe that the present theory, which provides a new theoretical framework for the analysis of coherence phenomena of all orders in stationary optical fields, will also prove useful, possibly in connection with the theory of speckle patterns, in the analysis of problems relating to propagation of light in the atmosphere, and in the analysis of coherence problems of nonlinear optics. We plan to show in another publication that our theory also clarifies certain questions relating to the concept of complete coherence of arbitrary order.

2. SUMMARY OF THE MAIN RESULTS OF PART I

Let us consider a fluctuating scalar source, occupying a finite domain D. We denote by $Q(\mathbf{r}, t)$ the (generally complex) source distribution at a point $\mathbf{r}$ at time t and assume that the fluctuations can be described by a stationary ensemble and that the mean value of Q is zero. Let

$$\Gamma_Q(\mathbf{r}_1, \mathbf{r}_2, \tau) = \langle Q^*(\mathbf{r}_1, t)Q(\mathbf{r}_2, t + \tau)\rangle \tag{2.1}$$

be the cross-correlation function and

0740-3232/85/010076-10$02.00

$$W_Q(\mathbf{r}_1, \mathbf{r}_2, \omega) = \frac{1}{2\pi}\int_{-\infty}^{\infty} \Gamma_Q(\mathbf{r}_1, \mathbf{r}_2, \tau)e^{i\omega\tau}d\tau \tag{2.2}$$

the cross-spectral density function of the source distribution. In Eq. (2.1) the asterisk denotes the complex conjugate and the angle brackets denote the ensemble average.

It was shown in Ref. 1 that under very general conditions the cross-spectral density may be represented in the form

$$W_Q(\mathbf{r}_1, \mathbf{r}_2, \omega) = \sum_n \lambda_n(\omega)\phi_n{}^*(\mathbf{r}_1, \omega)\phi_n(\mathbf{r}_2, \omega), \tag{2.3}$$

where the ϕ_n are the eigenfunctions and the λ_n are the eigenvalues of the Fredholm integral equation

$$\int_D W_Q(\mathbf{r}_1, \mathbf{r}_2, \omega)\phi_n(\mathbf{r}_1, \omega)d^3\mathbf{r}_1 = \lambda_n(\omega)\phi_n(\mathbf{r}_2, \omega). \tag{2.4}$$

The eigenfunctions may be taken to be orthonormalized over the domain D, i.e.,

$$\int_D \phi_n(\mathbf{r}, \omega)\phi_m(\mathbf{r}, \omega)d^3r = \delta_{nm}, \tag{2.5}$$

where δ_{nm} is the Kronecker symbol and the eigenvalues are real and nonnegative. Because eigenfunctions that are associated with zero eigenvalue do not contribute to the expansion (2.3) they play no role in the present theory. Hence only positive eigenvalues will be included from now on, i.e., we have, for all n,

$$\lambda_n(\omega) > 0. \tag{2.6}$$

If, as we implicitly assumed, the source domain D is three dimensional, the subscript n stands for an ordered triplet (n_1, n_2, n_3) of positive integers. However, strictly similar formulas also apply when the source domain D is two or one dimensional. In general, if D is N dimensional, n will stand for N ordered positive integers $(n_1, n_2 \ldots n_N)$ and the integrals in Eqs. (2.4) and (2.5) will then extend, of course, over the N-dimensional domain D.

Now the degree of spectral coherence of a source (or a field) with cross-spectral density $W(\mathbf{r}_1, \mathbf{r}_2, \omega)$ may be defined by the formula[11]

$$\mu(\mathbf{r}_1, \mathbf{r}_2, \omega) = \frac{W(\mathbf{r}_1, \mathbf{r}_2, \omega)}{[W(\mathbf{r}_1, \mathbf{r}_1, \omega)]^{1/2}[W(\mathbf{r}_2, \mathbf{r}_2, \omega)]^{1/2}}, \tag{2.7}$$

which can be shown to be bounded by zero and by unity in absolute value ($0 \leq |\mu| \leq 1$). The limiting value $\mu = 0$ represents complete spatial incoherence, whereas the limiting value $|\mu| = 1$ represents complete spatial coherence at frequency ω. Since each term

$$W_Q{}^{(n)}(\mathbf{r}_1, \mathbf{r}_2, \omega) = \lambda_n(\omega)\phi_n{}^*(\mathbf{r}_1, \omega)\phi_n(\mathbf{r}_2, \omega) \tag{2.8}$$

under the summation sign in Eq. (2.3) is factorized with respect to the two spatial variables $\mathbf{r}_1$ and $\mathbf{r}_2$, it readily follows that the associated degree of spectral coherence

$$\mu_Q{}^{(n)}(\mathbf{r}_1, \mathbf{r}_2, \omega) = \frac{W_Q{}^{(n)}(\mathbf{r}_1, \mathbf{r}_2, \omega)}{[W_Q{}^{(n)}(\mathbf{r}_1, \mathbf{r}_1, \omega)]^{1/2}[W_Q{}^{(n)}(\mathbf{r}_2, \mathbf{r}_2, \omega)]^{1/2}} \tag{2.9}$$

is unimodular. Hence the expansion (2.3) may be interpreted as representing the cross-spectral density of the source as the sum of contributions from mutually uncorrelated and spatially completely coherent elementary sources (natural modes of oscillations).

It was also shown in Ref. 1 that one may construct an ensemble $\{U_Q(\mathbf{r}, \omega)\exp(-i\omega t)\}$ of monochromatic oscillations, all of the same frequency ω, such that the cross-spectral density $W_Q(\mathbf{r}_1, \mathbf{r}_2, \omega)$ is their cross-correlation function, i.e., that

$$W_Q(\mathbf{r}_1, \mathbf{r}_2, \omega) = \langle U_Q{}^*(\mathbf{r}_1, \omega)U_Q(\mathbf{r}_2, \omega)\rangle_\omega, \tag{2.10}$$

where the angle brackets, with the suffix ω, denote the average taken over this ensemble. The functions $U(\mathbf{r}, \omega)$ were shown to be expressible in the form

$$U_Q(\mathbf{r}, \omega) = \sum_n a_n(\omega)\phi_n(\mathbf{r}, \omega), \tag{2.11}$$

where the a_n's are random coefficients satisfying the requirement that

$$\langle a_n{}^*(\omega)a_m(\omega)\rangle_\omega = \lambda_n(\omega)\delta_{nm}. \tag{2.12}$$

The functions $\phi_n(\mathbf{r}, \omega)$ and the constants $\lambda_n(\omega)$ are, of course, the eigenfunctions and the (nonzero) eigenvalues, respectively, of the integral equation (2.4).

The spectral density $S_Q(\mathbf{r}, \omega)$ (i.e., the Wiener–Khintchine spectrum or the power spectrum) of the source distribution is just the "diagonal element" $W_Q(\mathbf{r}, \mathbf{r}, \omega)$ of the cross-spectral density $W_Q(\mathbf{r}_1, \mathbf{r}_2, \omega)$; hence, according to Eqs. (2.3) and (2.10), it may be expressed in the two equivalent forms

$$S_Q(\mathbf{r}, \omega) = \sum_n \lambda_n(\omega)|\phi_n(\mathbf{r}, \omega)|^2 \tag{2.13}$$

and

$$S_Q(\mathbf{r}, \omega) = \langle|U_Q(\mathbf{r}, \omega)|^2\rangle_\omega. \tag{2.14}$$

Strictly similar results to those that we summarized in this section also apply if instead of a fluctuating source distribution $Q(\mathbf{r}, t)$ one considers a fluctuating field distribution $V(\mathbf{r}, t)$, which is represented by a stationary ensemble, is of zero mean value, and is confined to a finite closed domain D; these results are discussed in detail in Ref. 2.

3. COHERENT-MODE REPRESENTATION OF THE CROSS-SPECTRAL DENSITY OF A FIELD GENERATED BY A FLUCTUATING STEADY-STATE SOURCE

We will now turn our attention to the field $V(\mathbf{r}, t)$, generated by the fluctuating source distribution $Q(\mathbf{r}, t)$. To begin with, we assume that the source is located in free space. Then the field $V(\mathbf{r}, t)$ is coupled to the source $Q(\mathbf{r}, t)$ by the inhomogeneous wave equation

$$\left(\nabla^2 - \frac{1}{c^2}\frac{\partial^2}{\partial t^2}\right)V(\mathbf{r}, t) = -4\pi Q(r, t), \tag{3.1}$$

where c is the speed of light *in vacuo*. Since $Q(\mathbf{r}, t)$ is a stationary random variable and since the relationship between $V(\mathbf{r}, t)$ and $Q(\mathbf{r}, t)$ is linear, $V(\mathbf{r}, t)$ is also a stationary random variable. Let

$$\Gamma_V(\mathbf{r}_1, \mathbf{r}_2, \tau) = \langle V^*(\mathbf{r}_1, t)V(\mathbf{r}_2, t + \tau)\rangle \tag{3.2}$$

be the cross-correlation function (usually called the mutual coherence function) of the field, and let

$$W_V(\mathbf{r}_1, \mathbf{r}_2, \omega) = \frac{1}{2\pi}\int_{-\infty}^{\infty} \Gamma_V(\mathbf{r}_1, \mathbf{r}_2, \tau)e^{i\omega\tau}d\tau \tag{3.3}$$

be its cross-spectral density.[12,13] In Eq. (3.2) the average is taken over the original source ensemble $\{Q(\mathbf{r}, t)\}$. It is shown in Appendix A that, as a consequence of the wave equation (3.1), the cross-spectral density of the field is coupled to the cross-spectral density of the source by the fourth-order linear differential equation

$$\mathcal{M}_2\mathcal{M}_1 W_V(\mathbf{r}_1, \mathbf{r}_2, \omega) = (4\pi)^2 W_Q(\mathbf{r}_1, \mathbf{r}_2, \omega), \tag{3.4}$$

where

$$\mathcal{M}_s \equiv \nabla_s^{\,2} + k^2 \qquad (s = 1, 2); \tag{3.5}$$

($k = \omega/c$) is the Helmholtz operator acting with respect to the coordinates of the point $\mathbf{r}_s$.

We will now make use of the expansion (2.3) for the cross-spectral density W_Q of the source. Since in that expansion $\phi_n(\mathbf{r}, \omega)$ may be regarded as representing the time-independent part of a monochromatic source distribution $\phi_n(\mathbf{r}, \omega) \exp(-i\omega t)$ of frequency ω, $\phi_n(\mathbf{r}, \omega)$ may be expected to generate a monochromatic field of the same frequency ω, whose time-independent part $\psi_n(\mathbf{r}, \omega)$ is the outgoing solution of the Helmholtz equation

$$\mathcal{M}\psi_n(\mathbf{r}, \omega) = -4\pi\phi_n(\mathbf{r}, \omega). \tag{3.6}$$

In Eq. (3.6), $\mathcal{M}$ is, of course, the Helmholtz operator acting with respect to the spatial variable $\mathbf{r}$. The outgoing solution of Eq. (3.6) is well known to be[14]

$$\psi_n(\mathbf{r}, \omega) = \int_D \phi_n(\mathbf{r}', \omega)\frac{\exp(ik|\mathbf{r} - \mathbf{r}'|)}{|\mathbf{r} - \mathbf{r}'|}d^3r'. \tag{3.7}$$

Let us now construct the ensemble of functions

$$U_V(\mathbf{r}, \omega) = \sum_n a_n(\omega)\psi_n(\mathbf{r}, \omega), \tag{3.8}$$

where the a_n's are the same random coefficients as those that appear in Eq. (2.11) and that satisfy the requirements (2.12). Equations (3.8), (3.6), and (2.11) then imply that U_V is the outgoing solution of the Helmholtz equation

$$\mathcal{M}U_V(\mathbf{r}, \omega) = -4\pi U_Q(\mathbf{r}, \omega). \tag{3.9}$$

From Eqs. (3.9) and (2.10) it follows that

$$\mathcal{M}_2\mathcal{M}_1\langle U_V^*(\mathbf{r}_1, \omega)U_V(\mathbf{r}_2, \omega)\rangle_\omega = (4\pi)^2 W_Q(\mathbf{r}_1, \mathbf{r}_2, \omega), \tag{3.10}$$

which implies, in view of Eq. (3.4) and the outgoing behavior of the ψ_n's, that

$$W_V(\mathbf{r}_1, \mathbf{r}_2, \omega) = \langle U_V^*(\mathbf{r}_1, \omega)U_V(\mathbf{r}_2, \omega)\rangle_\omega. \tag{3.11}$$

This formula expresses the cross-spectral density $W_V(\mathbf{r}_1, \mathbf{r}_2, \omega)$ of the field generated by the fluctuating source as a correlation function of the ensemble $\{U_V(\mathbf{r}, \omega)\exp(-i\omega t)\}$ of monochromatic wave fields. The sources of these wavefields are, according to Eq. (3.9), just the monochromatic oscillations $\{U_Q(\mathbf{r}, \omega)\exp(-i\omega t)\}$ introduced through Eqs. (2.11) and (2.12). Moreover, according to Eq. (3.8), members of the ensemble of the fields $\{U_V\}$ are obtained from the members of the ensemble $\{U_Q\}$ of the source oscillations by replacing the eigenfunctions $\phi_n(\mathbf{r}, \omega)$ in the expansion (2.11) by the fields $\psi_n(\mathbf{r}, \omega)$ to which the $\phi_n(\mathbf{r}, \omega)$ give rise [Eq. (3.7)].

Finally, if we substitute from Eq. (3.8) into Eq. (3.11) and use Eq. (2.12), we obtain the following expansion for the cross-spectral density of the field:

$$W_V(\mathbf{r}_1, \mathbf{r}_2, \omega) = \sum_n \lambda_n(\omega)\psi_n^*(\mathbf{r}_1, \omega)\psi_n(\mathbf{r}_2, \omega). \tag{3.12}$$

Since each term

$$W_V^{(n)}(\mathbf{r}_1, \mathbf{r}_2, \omega) = \lambda_n(\omega)\psi_n^*(\mathbf{r}_1, \omega)\psi_n(\mathbf{r}_2, \omega) \tag{3.13}$$

in the expansion (3.12) is factorized with respect to the spatial variables $\mathbf{r}_1$, and $\mathbf{r}_2$, its associated degree of spectral coherence

$$\mu_V^{(n)}(\mathbf{r}_1, \mathbf{r}_2, \omega) = \frac{W_V^{(n)}(\mathbf{r}_1, \mathbf{r}_2, \omega)}{[W_V^{(n)}(\mathbf{r}_1, \mathbf{r}_1, \omega)]^{1/2}[W_V^{(n)}(\mathbf{r}_2, \mathbf{r}_2, \omega)]^{1/2}} \tag{3.14}$$

is unimodular. Hence *the expansion (3.12) represents the cross-spectral density of the field generated by the source distribution $Q(\mathbf{r}, t)$ as a linear superposition of mutually uncorrelated elementary fields that are spatially completely coherent in the space–frequency domain.*

The expansion (3.12) for the cross-spectral density of the field has the same mathematical form as the expansion (2.3) for the cross-spectral density of the source. However, unlike the natural modes $\phi_n(\mathbf{r}, \omega)$ of the source distribution, the elementary fields $\psi_n(\mathbf{r}, \omega)$ are not, in general, mutually orthogonal.

It follows at once from Eqs. (3.12) and (3.11) that the spectral density (the optical intensity at frequency ω) $S_V(\mathbf{r}, \omega) \equiv W_V(\mathbf{r}, \mathbf{r}, \omega)$ of the field may be represented by the following two equivalent formulas:

$$S_V(\mathbf{r}, \omega) = \sum_n \lambda_n(\omega)|\psi_n(\mathbf{r}, \omega)|^2, \tag{3.15}$$

and

$$S_V(\mathbf{r}, \omega) = \langle |U_V(\mathbf{r}, \omega)|^2\rangle_\omega, \tag{3.16}$$

which are strictly analogous to the expressions (2.13) and (2.14) for the spectral density of the source distribution.

The new representations (3.11) and (3.12) for the cross-spectral density of the field lead at once through the Fourier-transform relationship (3.3) (more precisely, through its inverse) to the following two new representations for the mutual coherence function of the field, namely,[15]

$$\Gamma_V(\mathbf{r}_1, \mathbf{r}_1\, \tau) = \int_{-\infty}^{\infty} \langle U_V^*(\mathbf{r}_1, \omega)U_V(\mathbf{r}_2, \omega)\rangle_\omega e^{-i\omega\tau}d\omega \tag{3.17}$$

and (if the order of integration and summation is interchanged)

$$\Gamma_V(\mathbf{r}_1, \mathbf{r}_2, \tau) = \sum_n \int_{-\infty}^{\infty} \lambda_n(\omega)\psi_n^*(\mathbf{r}_1, \omega)\psi_n(\mathbf{r}_2, \omega)e^{-i\omega\tau}d\omega. \tag{3.18}$$

The expression (3.18) corresponds to the expression (4.18), together with Eq. (4.20) of Part I, for the cross-correlation function of the source. Similar remarks apply here as in connection with the formula (4.20) of Part I; namely, that while each term in the expansion (3.12) of the cross-spectral density represents a field that is completely spatially coherent at frequency ω, the corresponding term in the expansion (3.18) represents a field that, in general, is neither spatially nor temporarily fully coherent.

We assumed so far that the source radiates in free space. If the source is embedded in a linear time-independent medium or if the space into which it radiates contains obstacles (e.g., lenses, mirrors, scattering media), whose response to an incident field is linear and time independent, the main formulas that we just derived will hold, provided that the free-space Green function $\exp(ik|\mathbf{r} - \mathbf{r}'|)/|\mathbf{r} - \mathbf{r}'|$ is replaced by the appropriate Green function $G(\mathbf{r}, \mathbf{r}', \omega)$ that represents the field at the point $\mathbf{r}$ produced by a delta-function source of unit strength, located at the point $\mathbf{r}'$. Each elementary field $\psi_n(\mathbf{r}, \omega)$ in the above expressions, given by Eq. (3.7), will then be replaced by the elementary field

$$\Psi_n(\mathbf{r}, \omega) = \int_D \phi_n(\mathbf{r}', \omega) G(\mathbf{r}, \mathbf{r}', \omega) \mathrm{d}^3 r'. \quad (3.19)$$

With this generalization, the present theory can thus be applied to problems involving imaging or scattering of light generated by partially coherent sources.

We have implicitly assumed in this section that the source is a primary one. However, with trivial formal changes, our main results will also apply to fields generated by secondary sources. For example, if a statistically stationary secondary source occupies a finite portion S of a plane $z = 0$ and radiates into an empty half-space $z > 0$, we only need to replace the formula (3.7) by the Rayleigh integral[16]

$$\Psi_n(\mathbf{r}, \omega) = \frac{1}{2\pi} \int_S \phi_n(\mathbf{r}', \omega) \frac{\partial}{\partial z'} \left[\frac{\exp(ik|\mathbf{r} - \mathbf{r}'|)}{|\mathbf{r} - \mathbf{r}'|} \right] \mathrm{d}^2 r', \quad (3.20)$$

where $\phi_n(\mathbf{r}', \omega)$ are now the eigenfunctions of the integral equation

$$\int_S W_V^{(0)}(\mathbf{r}_1, \mathbf{r}_2, \omega) \phi_n(\mathbf{r}_1, \omega) \mathrm{d}^2 r_1 = \lambda_n(\omega) \phi_n(r_2, \omega) \quad (3.21)$$

rather than of the integral equation (2.4). In Eq. (3.21) $W_V^{(0)}(\mathbf{r}_1, \mathbf{r}_2, \omega)$ is, of course, the cross-spectral density of the field in the plane $z = 0$ of the secondary source. With this interpretation of the functions $\phi_n(\mathbf{r}, \omega)$ and $\Psi_n(\mathbf{r}, \omega)$ and with the constants $\lambda_n(\omega)$ now being the eigenvalues of the integral equation (3.21), the main results obtained in this section will again apply.

4. HIGHER-ORDER CORRELATIONS OF A STEADY-STATE SOURCE

Until now we developed a coherent-mode representation of sources and fields within the framework of second-order correlation theory. We will now extend the theory to cover correlations of all orders.

We discussed in Section 2 of Part I certain conceptual difficulties and mathematical subtleties that surround the traditional definition of the spectrum of a stationary random variable. Similar difficulties and subtleties surround the definitions of higher-order spectral densities. In fact, many basic questions in this area are still awaiting solutions. We will begin with the traditional nonrigorous definitions, and we will then show that a natural extension of the theory that we have developed overcomes some of the difficulties, at least in the context of stationary optical sources and stationary optical fields.

Let $Q(\mathbf{r}, t)$ again represent a localized fluctuating source distribution. For the moment we do *not* assume that $Q(\mathbf{r}, t)$ is a stationary random variable. The space–time *cross-correlation function of order* (M, N) of the *source distribution* is defined by the formula[17,18]

$$\begin{aligned} G_Q^{(M,N)}&(\mathbf{r}_1, \mathbf{r}_2, \ldots \mathbf{r}_{M+N}; t_1, t_2, \ldots t_{M+N}) \\ &= \langle Q^*(\mathbf{r}_1, t_1) Q^*(\mathbf{r}_2, t_2) \ldots Q^*(\mathbf{r}_M, t_M) \\ &\times Q(\mathbf{r}_{M+1}, t_{M+1}) Q(\mathbf{r}_{M+2}, t_{M+2}) \ldots Q(\mathbf{r}_{M+N}, t_{M+N}) \rangle. \end{aligned} \quad (4.1)$$

or, written in a more compact form,

$$G_Q^{(M,N)}(\mathbf{r}_1, \mathbf{r}_2, \ldots \mathbf{r}_{M+N}; t_1, t_2, \ldots t_{M+N}) = \left\langle \prod_{j=1}^{M} Q^*(\mathbf{r}_j, t_j) \prod_{k=M+1}^{M+N} Q(\mathbf{r}_k, t_k) \right\rangle, \quad (4.1a)$$

where the angle brackets denote an average taken over the source ensemble.

Let us represent $Q(\mathbf{r}, t)$ as a Fourier integral with respect to the temporal variable:

$$Q(\mathbf{r}, t) = \int_{-\infty}^{\infty} q(\mathbf{r}, \omega) e^{-i\omega t} \mathrm{d}\omega. \quad (4.2)$$

Then, on taking the Fourier inverse, we have

$$q(\mathbf{r}, \omega) = \frac{1}{2\pi} \int_{-\infty}^{\infty} Q(\mathbf{r}, t) e^{i\omega t} \mathrm{d}t. \quad (4.3)$$

If we substitute from Eq. (4.2) in the right-hand side of Eq. (4.1a) and interchange the orders of averaging and integration, we obtain the following representation of the correlation function $G_Q^{(M,N)}$:

$$\begin{aligned} G_Q^{(M,N)}&(\mathbf{r}_1, \mathbf{r}_2, \ldots \mathbf{r}_{M+N}; t_1, t_2, \ldots t_{M+N}) \\ &= \int_{-\infty}^{\infty} \mathrm{d}\omega_1 \int_{-\infty}^{\infty} \mathrm{d}\omega_2 \ldots \int_{-\infty}^{\infty} \mathrm{d}\omega_{M+N} \\ &\times \Phi_Q^{(M,N)}(\mathbf{r}_1, \mathbf{r}_2, \ldots \mathbf{r}_{M+N}; \omega_1, \omega_2, \ldots \omega_{M+N}) \\ &\times \prod_{j=1}^{M} \exp(i\omega_j t_j) \prod_{k=M+1}^{M+N} \exp(-i\omega_k t_k), \end{aligned} \quad (4.4)$$

where

$$\Phi_Q^{(M,N)}(\mathbf{r}_1, \mathbf{r}_2, \ldots \mathbf{r}_{M+N}; \omega_1, \omega_2, \ldots \omega_{M+N}) = \left\langle \prod_{j=1}^{M} q^*(\mathbf{r}_j, \omega_j) \prod_{k=M+1}^{M+N} q(\mathbf{r}_k, \omega_k) \right\rangle. \quad (4.5)$$

We will call the function $\Phi_Q^{(M,N)}$, defined by Eq. (4.5), the *spectral cross-correlation function of order* (M,N) *of the source distribution.*

If we take the Fourier inverse of Eq. (4.4), we obtain the following expression for the spectral cross-correlation function $\Phi_Q^{(M,N)}$ in terms of the space–time correlation function $G_Q^{(M,N)}$:

$$\begin{aligned} \Phi_Q^{(M,N)}&(\mathbf{r}_1, \mathbf{r}_2, \ldots \mathbf{r}_{M+N}; \omega_1, \omega_2, \ldots \omega_{M+N}) \\ &= \frac{1}{(2\pi)^{M+N}} \int_{-\infty}^{\infty} \mathrm{d}t_1 \int_{-\infty}^{\infty} \mathrm{d}t_2 \ldots \int_{-\infty}^{\infty} \mathrm{d}t_{M+N} \\ &\times G_Q^{(M,N)}(\mathbf{r}_1, \mathbf{r}_2, \ldots \mathbf{r}_{M+N}; t_1, t_2, \ldots t_{M+N}) \\ &\times \prod_{j=1}^{M} \exp(-i\omega_j t_j) \prod_{k=M+1}^{M+N} \exp(i\omega_k t_k). \end{aligned} \quad (4.6)$$

Suppose now that the ensemble that represents the fluctuations of $Q(\mathbf{r}, t)$ is stationary. The correlation functions $G_Q^{(M,N)}$ will then be invariant with respect to translation of the origin of time. In particular, if we set

$$t_l = t_1 + \tau_l \qquad (l = 2, 3 \ldots M + N), \qquad (4.7)$$

$G_Q^{(N,N)}$ will be independent of t_1; more precisely, as regards its dependence on the temporal variables, it will be a function of the $M + N - 1$ arguments $\tau_2, \tau_2, \ldots \tau_{M+N}$ only, and we will then write

$$G_Q^{(M,N)}(\mathbf{r}_1, \mathbf{r}_2, \ldots \mathbf{r}_{M+N}; t_1, t_2, \ldots t_{M+N}) \equiv \Gamma_Q^{(M,N)}(\mathbf{r}_1, \mathbf{r}_2, \ldots \mathbf{r}_{M+N}; \tau_2, \tau_3, \ldots \tau_{M+N}). \qquad (4.8)$$

We pointed out in Section 2 of Part I that a stationary random variable [$Q(\mathbf{r}, t)$ in the present case] does not possess a Fourier frequency representation within the framework of ordinary function theory. In that case, Eq. (4.3) must be interpreted in the sense of generalized function theory or one must make use of other sophisticated mathematical techniques, such as the generalized harmonic analysis or the Fourier–Stiltjes integral representation. We will nevertheless still make use of Eqs. (4.3)–(4.6) to obtain certain formal definitions whose significance will be interpreted rigorously later on the basis of our new theory.

On substituting from Eqs. (4.7) into the multiple integral on the right-hand side of Eq. (4.6), making use of Eq. (4.8), and carrying out the (trivial) integration with respect to t_1, we obtain the following expression for the spectral cross-correlation function $\Phi_Q^{(M,N)}$:

$$\Phi_Q^{(M,N)}(\mathbf{r}_1, \mathbf{r}_2, \ldots \mathbf{r}_{M+N}; \omega_1, \omega_2, \ldots \omega_{M+N}) = \delta(\omega_1 + \omega_2 + \ldots \omega_M - \omega_{M+1} - \omega_{M+2} - \ldots \omega_{M+N}) \times W_Q^{(M,N)}(\mathbf{r}_1, \mathbf{r}_2, \ldots \mathbf{r}_{M+N}; \omega_2, \omega_3, \ldots \omega_{M+N}), \qquad (4.9)$$

where $\delta(\omega)$ is the Dirac delta function and[19]

$$W_Q^{(M,N)}(\mathbf{r}_1, \mathbf{r}_2, \ldots \mathbf{r}_{M+N}; \omega_2, \omega_3, \ldots \omega_{M+N}) = \frac{1}{(2\pi)^{M+N-1}} \int_{-\infty}^{\infty} \mathrm{d}\tau_2 \int_{-\infty}^{\infty} \mathrm{d}\tau_3 \ldots \int_{-\infty}^{\infty} \mathrm{d}\tau_{M+N} \times \Gamma_Q^{(M+N)}(\mathbf{r}_1, \mathbf{r}_2, \ldots \mathbf{r}_{M+N}; \tau_2, \tau_3, \ldots \tau_{M+N}) \times \prod_{j=2}^{M} \exp(-i\omega_j \tau_j) \prod_{k=M+1}^{M+N} \exp(i\omega_k \tau_k). \qquad (4.10)$$

The formulas (4.5) and (4.9) imply that any $M + N$ (generalized) Fourier components $q(\mathbf{r}_j, \omega_j)$, $(j = 1, 2, \ldots M + N)$, of a stationary random source distribution $Q(\mathbf{r}, t)$ are δ correlated, or, more precisely, that the $M + N$ components are uncorrelated, i.e., that the spectral cross-correlation function

$$\Phi_Q^{(M,N)}(\mathbf{r}_1, \mathbf{r}_2, \ldots \mathbf{r}_{M+N}; \omega_1, \omega_2, \ldots \omega_{M+N}) = 0 \qquad (4.11)$$

unless

$$\omega_1 + \omega_2 + \ldots \omega_M - \omega_{M+1} - \omega_{M+2} \cdots - \omega_{M+N} = 0; \qquad (4.12)$$

and that when the $M + N$ frequencies are related by Eq. (4.12), the components will, in general, be correlated, the "strength" of the correlation then being characterized by the function $W_Q^{(M,N)}$, given by Eq. (4.10).

In the special case when $M = N = 1$, the formula (4.10) reduces to

$$W_Q^{(1,1)}(\mathbf{r}_1, \mathbf{r}_2; \omega_2) = \frac{1}{2\pi} \int_{-\infty}^{\infty} \Gamma_Q^{(1,1)}(\mathbf{r}_1, \mathbf{r}_2; \tau_2) \exp(i\omega_2 \tau_2) \mathrm{d}\tau_2, \qquad (4.13)$$

where, according to Eqs. (4.8), (4.7), and (4.1),

$$\Gamma_Q^{(1,1)}(\mathbf{r}_1, \mathbf{r}_2; \tau_2) = \langle Q^*(\mathbf{r}_1, t_1) Q(\mathbf{r}_2, t_1 + \tau_2) \rangle. \qquad (4.14)$$

Apart from a trivial change of notation, Eq. (4.13) is nothing but the formula (2.2) for the cross-spectral density in terms of the usual correlation function. The diagonal form ($\mathbf{r}_2 = \mathbf{r}_1$) of Eq. (4.13) is, in fact, just one version of the classic Wiener–Khintchine theorem. It is, therefore, natural to call the function $W_Q^{(M,N)}(\mathbf{r}_1, \mathbf{r}_2, \ldots \mathbf{r}_{M+N}; \omega_2, \omega_3, \ldots \omega_{M+N})$ *the cross-spectral density function of order*[20] (M, N) *of the stationary source distribution* $Q(\mathbf{r}, t)$ and to regard the relationship (4.10) as a *generalization, for higher-order correlations, of the Wiener–Khintchine theorem.*

Although in this section we considered only higher-order correlations of stationary *source* distributions $Q(\mathbf{r}, t)$, it is clear that strictly similar definitions and results apply to stationary *field* distributions $V(\mathbf{r}, t)$.

5. NEW REPRESENTATIONS OF HIGHER-ORDER CROSS-SPECTRAL DENSITIES OF A STEADY-STATE SOURCE

The two main results that we derived in Part I of this investigation[1] were the representations (2.3) and (2.10) of the cross-spectral density $W_Q(\mathbf{r}_1, \mathbf{r}_2, \omega)$. We will now show that there are generalizations of these two representations for cross-spectral density of any order.

By analogy with the formula (2.10) we will try to represent the cross-spectral density of order (M, N) in the form

$$W_Q^{(M,N)}(\mathbf{r}_1, \mathbf{r}_2, \ldots \mathbf{r}_{M+N}; \omega_2, \omega_3, \ldots \omega_{M+N}) = \hat{\sigma}^{(M,N)} \left\langle \prod_{j=1}^{M} U_Q^*(\mathbf{r}_j; \omega_j) \prod_{k=M+1}^{M+N} U_Q(\mathbf{r}_k, \omega_k) \right\rangle_\omega. \qquad (5.1)$$

Here the U_Q are the same random functions as employed in the representation (2.10) of the ordinary cross-spectral density, i.e., the functions

$$U_Q(\mathbf{r}, \omega) = \sum_n a_n(\omega) \phi_n(\mathbf{r}, \omega), \qquad (5.2)$$

where the $a_n(\omega)$ are random coefficients such that [cf. Eq. (2.12)]

$$\langle a_{n_1}^*(\omega) a_{n_2}(\omega) \rangle_\omega = \lambda_{n_1}(\omega) \delta_{n_1 n_2}, \qquad (5.3)$$

$\phi_n(\mathbf{r}, \omega)$ being the eigenfunctions and $\lambda_n(\omega)$ ($\neq 0$) the eigenvalues of the integral equation (2.4). The angle brackets with the subscript ω in Eqs. (5.1) and (5.3) indicate, as before, that the averaging is taken over the ensemble $\{a_n(\omega)\}$ of the random coefficients. The symbol $\hat{\sigma}^{(M,N)}$ on the right-hand side of Eq. (5.1) is a "*substitution operator*" of order (M, N), defined by the property that when it operates on any expression that contains the $M + N$ frequencies ω_1, $\omega_2, \ldots \omega_{M+N}$ one substitutes for ω_1 the value given by the requirement (4.12), i.e., one sets

$$\omega_1 = -\omega_2 - \omega_3 \ldots - \omega_M + \omega_{M+1} + \omega_{M+2} \ldots + \omega_{M+N} \qquad (5.4)$$

that is a consequence of the stationarity of the source distribution.

We note that Eq. (5.3) does not completely specify the ensemble of the random coefficients $a_n(\omega)$. It specifies only their second-order moments, at the chosen frequency ω. Hence the ensemble of the random functions $U_Q(\mathbf{r}, \omega)$ [Eq. (5.2)] is, up to this point, not completely specified either. We will now show that a more detailed specification of the ensemble of the random coefficients $a_n(\omega)$ will make it possible to represent the cross-spectral density $W_Q{}^{(M,N)}$ in the form given by Eq. (5.1).

Let us for the moment assume that $W_Q{}^{(M,N)}$ admits the representation (5.1). It then follows on substituting from Eq. (5.2) in the right-hand side of Eq. (5.1) and on interchanging the order of averaging and summations that

$$W_Q{}^{(M,N)}(\mathbf{r}_1\, \mathbf{r}_2 \ldots \mathbf{r}_{M+N};\, \omega_2\, \omega_3, \ldots \omega_{M+N})$$

$$= \sum_{n_1}\sum_{n_2}\cdots\sum_{n_{M+N}} \hat{\sigma}^{(M,N)}\mathfrak{M}^{(M,N)}_{n_1 n_2 \ldots n_{M+N}}(\omega_1, \omega_2, \ldots \omega_{M+N})$$

$$\times\, \hat{\sigma}^{(M,N)} \prod_{j=1}^{M} \phi_{n_j}{}^*(\mathbf{r}_j, \omega_j) \prod_{k=M+1}^{M+N} \phi_{n_k}(\mathbf{r}_k, \omega_k), \quad (5.5)$$

where

$$\mathfrak{M}^{(M,N)}_{n_1 n_2 \ldots n_{M+N}}(\omega_1, \omega_2, \ldots \omega_{M+N})$$

$$= \left\langle \prod_{j=1}^{M} a_{n_j}{}^*(\omega_j) \prod_{k=M+1}^{M+N} a_{n_k}(\omega_k) \right\rangle_\omega \quad (5.6)$$

is a moment of order (M, N) of the random coefficients $a_n(\omega)$.

Let us now multiply both sides of Eq. (5.5) by the product

$$\hat{\sigma}^{(M,N)} \prod_{j=1}^{M} \phi_{m_j}(\mathbf{r}_j, \omega_j) \prod_{k=M+1}^{M+N} \phi_{m_k}{}^*(\mathbf{r}_k, \omega_k)$$

and integrate both sides of the equation with respect to $\mathbf{r}_1$, $\mathbf{r}_2, \ldots \mathbf{r}_{M+N}$, each integration being taken over the source domain D. If next we interchange the order of integrations and summations and use the orthonormality of the eigenfunctions $\phi_n(\mathbf{r}, \omega)$ expressed by Eq. (2.5) we find (if we finally replace the suffixes $m_1, m_2 \ldots$ by $n_1, n_2 \ldots$)

$$\hat{\sigma}^{(M,N)}\mathfrak{M}^{(M,N)}_{n_1 n_2 \ldots n_{M+N}}(\omega_1, \omega_2, \ldots \omega_{M+N})$$

$$= \int_D \mathrm{d}^3 r_1 \int_D \mathrm{d}^3 r_2 \ldots \int_D \mathrm{d}^3 r_{M+N}$$

$$\times\, W_Q{}^{(M+N)}(\mathbf{r}_1 \mathbf{r}_2, \ldots \mathbf{r}_{M+N};\, \omega_2, \omega_3, \ldots \omega_{M+N})$$

$$\times\, \hat{\sigma}^{(M,N)} \prod_{j=1}^{M} \phi_{n_j}(\mathbf{r}_j, \omega_j) \prod_{k=M+1}^{M+N} \phi_{n_k}{}^*(\mathbf{r}_k, \omega_k). \quad (5.7)$$

We may express the right-hand side of Eq. (5.7) in terms of the cross-correlation function $\Gamma_Q{}^{(M,N)}$ rather than in terms of the cross-spectral density $W_Q{}^{(M,N)}$ by substituting for $W_Q{}^{(M,N)}$ from Eq. (4.10) and interchanging the orders of integration. We then obtain the following expression for the moment $\hat{\sigma}^{(M,N)}\mathfrak{M}^{(M,N)}$:

$$\hat{\sigma}^{(M,N)}\mathfrak{M}^{(M,N)}_{n_1, n_2 \ldots n_{M+N}}(\omega_1, \omega_2, \ldots \omega_{M+N})$$

$$= \frac{1}{(2\pi)^{M+N-1}} \int_{-\infty}^{\infty} \mathrm{d}\tau_2 \int_{-\infty}^{\infty} \mathrm{d}\tau_3 \ldots \int_{-\infty}^{\infty} \mathrm{d}\tau_{M+N}$$

$$\times\, \mathfrak{N}^{(M,N)}_{n_1, n_2 \ldots n_{M+N}}(\tau_2, \tau_3, \ldots \tau_{M+N};\, \omega_2, \omega_3 \ldots \omega_{M+N})$$

$$\times \prod_{j=2}^{M} \exp(-i\omega_j \tau_j) \prod_{k=M+1}^{M+N} \exp(i\omega_k \tau_k), \quad (5.8)$$

where

$$\mathfrak{N}^{(M,N)}_{n_1, n_2 \ldots n_{M+N}}(\tau_2, \tau_3 \ldots \tau_{M+N};\, \omega_2\, \omega_3, \ldots \omega_{M+N})$$

$$= \int_D \mathrm{d}^3 r_1 \int_D \mathrm{d}^3 r_2 \ldots \int_D \mathrm{d}^3 r_{M+N}$$

$$\times\, \Gamma_Q{}^{(M,N)}(\mathbf{r}_1, \mathbf{r}_2, \ldots \mathbf{r}_{M+N};\, \tau_2, \tau_3 \ldots \tau_{M+N})$$

$$\times\, \hat{\sigma}^{(M,N)} \prod_{j=1}^{M} \phi_{n_j}(\mathbf{r}_j, \omega_j) \prod_{k=M+1}^{M+N} \phi_{n_k}{}^*(\mathbf{r}_k, \omega_k). \quad (5.9)$$

Clearly with the choice of the moments $\hat{\sigma}^{(M,N)}\ \mathfrak{M}^{(M,N)}$ the representations (5.1) and (5.5) of the cross-spectral density $W_Q{}^{(M,N)}$ will hold. These two formulas are the required generalizations of the expressions (2.10) and (2.3) for the ordinary cross-spectral density [of order (1, 1)]. The formula (5.1) expresses the cross-spectral density of order (M, N) of the source distribution $Q(\mathbf{r}, t)$ as a correlation function in the space–frequency domain. The formula (5.5) expresses it in terms of the space-dependent part $\phi_n(\mathbf{r}, \omega)$ of the natural modes of the source distribution. It seems worthwhile to stress that these two representations for the cross-spectral density of arbitrary order (m, n) are completely specified by the eigenfunctions $\phi_n(\mathbf{r}, \omega)$ of the lowest-order cross-spectral density $W_Q(\mathbf{r}_1, \mathbf{r}_2, \omega)$ [Eq. (2.4)] and by the moments $\mathfrak{M}_Q{}^{(M,N)}(\omega_1, \omega_2, \ldots \omega_{M+N})$, subject to the constraint (5.4) on the $M + N$ frequencies, of the random coefficients $a_n(\omega)$. The values of these moments, subject to Eq. (5.4), are expressible in terms of the cross-correlation functions $\Gamma_Q(\mathbf{r}_1, \mathbf{r}_2 \ldots \mathbf{r}_{M+N};\, \tau_2, \tau_3, \ldots \tau_{M+N})$ by the formula (5.9).

The constraint (5.4) is a consequence of the assumed stationarity of the source distribution. Now the nonrigorous formula (4.9) for the spectral cross-correlation function implies that when the relation (5.4) is not satisfied, the different frequency components are uncorrelated. As we will see shortly, this fact has an analog, within the framework of our more rigorous theory, if the moments $\mathfrak{M}^{(M,N)}$ are chosen so that

$$\mathfrak{M}^{M,N}_{n_1, n_2 \ldots n_{M+N}}(\omega_1, \omega_2, \ldots \omega_{M+N}) = 0, \quad (5.10)$$

when

$$\omega_1 + \omega_2 + \ldots \omega_M - \omega_{M+1} - \omega_{M+2} - \ldots - \omega_{M+N} \neq 0. \quad (5.11)$$

Before proceeding further, we note that Eqs. (5.10) and (5.11), together with Eqs. (5.8) and (5.9), completely specify all the moments of the random coefficients $a_n(\omega)$ and hence also the ensemble (5.2) of the random functions $U_Q(\mathbf{r}, \omega)$.

Consider now the correlation function

$$\Psi_Q^{(M,N)}(\mathbf{r}_1, \mathbf{r}_2, \ldots \mathbf{r}_{M+N}; \omega_1, \omega_2, \ldots \omega_{M+N}) = \left\langle \prod_{j=1}^{M} U_Q^*(\mathbf{r}_j, \omega_j) \prod_{k=M+1}^{M+N} U_Q(\mathbf{r}_k, \omega_k) \right\rangle_\omega . \quad (5.12)$$

If we substitute from Eq. (5.2) into the right-hand side of Eq. (5.12) and interchange the order of averaging and summations, we obtain the following expression for $\Psi_Q^{(M,N)}$:

$$\Psi_Q^{(M,N)}(\mathbf{r}_1, \mathbf{r}_2, \ldots \mathbf{r}_{M+N}; \omega_1, \omega_2, \ldots \omega_{M+N}) = \sum_{n_1}\sum_{n_2} \ldots \sum_{n_{M+N}} \mathfrak{M}^{M+N}_{n_1 n_2 \ldots n_{M+N}}(\omega_1, \omega_2, \ldots \omega_{M+N}) \times \prod_{j=1}^{M} \phi_{n_j}^*(\mathbf{r}_j, \omega_j) \prod_{k=M+1}^{M+N} \phi_{n_k}(\mathbf{r}_k, \omega_k). \quad (5.13)$$

It follows at once on comparing Eq. (5.13) with Eq. (5.5) and making use of Eq. (5.10) that

$$\Psi_Q^{(M,N)}(\mathbf{r}_1, \mathbf{r}_2, \ldots \mathbf{r}_{M+N}; \omega_1, \omega_2, \ldots \omega_{M+N}) = W_Q^{(M,N)}(\mathbf{r}_1, \mathbf{r}_2, \ldots \mathbf{r}_{M+N}; \omega_2, \omega_3, \ldots \omega_{M+N}) \quad (5.14)$$

when

$$\omega_1 + \omega_2 + \ldots \omega_M - \omega_{M+1} - \omega_{M+2} - \ldots - \omega_{M+N} = 0 \quad (5.14a)$$

and that

$$\Psi_Q^{(M,N)}(\mathbf{r}_1, \mathbf{r}_2, \ldots \mathbf{r}_{M+N}; \omega_1, \omega_2, \ldots \omega_{M+N}) = 0 \quad (5.15)$$

when

$$\omega_1 + \omega_2 + \ldots \omega_M - \omega_{M+1} - \omega_{M+2} - \ldots - \omega_{M+N} \neq 0. \quad (5.15a)$$

The formulas (5.14) and (5.15) may be regarded as a rigorous analog of the formula (4.9) for the spectral cross-correlation function $\Phi_Q^{(M,N)}$. It should be noted, however, that $\Psi^{(M,N)}$ represents correlations between the random functions $U_Q(\mathbf{r}, \omega)$ [Eq. (5.12)], whereas $\Phi_Q^{(M,N)}$ of the usual theory represents correlations between the Fourier components $q(\mathbf{r}, \omega)$ of the random source distribution [Eq. (4.5)]. As we noted earlier, the Fourier components $q(\mathbf{r}, \omega)$ have no real meaning within the framework of ordinary function theory and must be interpreted as generalized functions. On the other hand the random functions $U_Q(\mathbf{r}, \omega)$ are ordinary functions.

6. COHERENT-MODE REPRESENTATION OF THE HIGHER-ORDER CROSS-SPECTRAL DENSITIES OF THE FIELD GENERATED BY A STEADY-STATE SOURCE

We will now generalize the results derived in Section 3 to obtain expressions for higher-order cross-spectral densities of a field generated by a fluctuating source in terms of the natural modes of the source.

In analogy with Eq. (4.1a) we define the space–time *cross-correlation functions of order* (M, N) *of the field* $V(\mathbf{r}, t)$, generated by a fluctuating localized source distribution $Q(\mathbf{r}, t)$ (not necessarily stationary), by the formula

$$G_V^{(M,N)}(\mathbf{r}_1, \mathbf{r}_2, \ldots \mathbf{r}_{M+N}; t_1, t_2, \ldots t_{M+N}) = \left\langle \prod_{j=1}^{M} V^*(\mathbf{r}_j, t_j) \prod_{k=M+1}^{M+N} V(\mathbf{r}_k, t_k) \right\rangle . \quad (6.1)$$

If we introduce the Fourier transform $v(\mathbf{r}, \omega)$ of the field variable $V(\mathbf{r}, t)$, viz.,

$$v(\mathbf{r}, \omega) = \frac{1}{2\pi} \int_{-\infty}^{\infty} V(\mathbf{r}, t) e^{i\omega t} dt, \quad (6.2)$$

we may express $G_V^{(M,N)}$ in terms of the *spectral cross-correlation function of order* (M, N) *of the field*, viz.,

$$\Phi_V^{(M,N)}(\mathbf{r}_1, \mathbf{r}_2, \ldots \mathbf{r}_{M+N}; \omega_1, \omega_2, \ldots \omega_{M+N}) = \left\langle \prod_{j=1}^{M} v^*(\mathbf{r}_j; \omega_j) \prod_{k=M+1}^{M+N} v(\mathbf{r}_k, \omega_k) \right\rangle \quad (6.3)$$

by a formula strictly similar to Eq. (4.4).

If the source fluctuations are statistically stationary, as we will assume from now on, the field fluctuations will also be statistically stationary and the cross-correlation function $G_V^{(M,N)}$ will depend on the $M + N$ temporal arguments $t_1, t_2, \ldots t_{M+N}$ only through the $M + N - 1$ differences $\tau_l = t_l - t_1$ $(l = 2, 3, \ldots M + N)$. We will then write, in analogy with Eq. (4.8),

$$G_V^{(M,N)}(\mathbf{r}_1, \mathbf{r}_2, \ldots \mathbf{r}_{M+N}; t_1, t_2, \ldots t_{M+N}) = \Gamma_V^{(M,N)}\mathbf{r}_1, \mathbf{r}_2, \ldots \mathbf{r}_{M+N}; \tau_2, \tau_3, \ldots \tau_{M+N}). \quad (6.4)$$

By an argument strictly similar to that which led to Eq. (4.9), one finds that the spectral cross-correlation function of order (M, N) of the field has the form

$$\Phi_V^{(M,N)}(\mathbf{r}_1, \mathbf{r}_2, \ldots \mathbf{r}_{M+N}; \omega_1, \omega_2, \ldots \omega_{M+N}) = \delta(\omega_1 + \omega_2 + \ldots \omega_M - \omega_{M+1} - \omega_{M+2} - \ldots - \omega_{M+N}) \times W_V^{(M,N)}(\mathbf{r}_1, \mathbf{r}_2, \ldots \mathbf{r}_{M+N}; \omega_2, \omega_3, \ldots \omega_{M+N}), \quad (6.5)$$

where $W_V^{(M,N)}$ is expressible in terms of $\Gamma_V^{(M,N)}$ by a formula of the form given by Eq. (4.10), viz.,[24]

$$W_V^{(M,N)}(\mathbf{r}_1, \mathbf{r}_2, \ldots \mathbf{r}_{M+N}; \omega_2, \omega_3, \ldots \omega_{M+N}) = \frac{1}{(2\pi)^{M+N-1}} \int_{-\infty}^{\infty} d\tau_2 \int_{-\infty}^{\infty} d\tau_3 \ldots \int_{-\infty}^{\infty} d\tau_{M+N} \times \Gamma_V^{(M,N)}(\mathbf{r}_1, \mathbf{r}_2, \ldots \mathbf{r}_{M+N}; \tau_1, \tau_2, \ldots \tau_{M+N}) \times \prod_{j=2}^{M} \exp(-i\omega_j \tau_j) \prod_{k=M+1}^{M+N} \exp(i\omega_k \tau_k). \quad (6.6)$$

The function $W_V^{(M,N)}(\mathbf{r}_1, \mathbf{r}_2, \ldots \mathbf{r}_{M+N}; \omega_2, \omega_3, \ldots \omega_{M+N})$ is clearly *the cross-spectral density function of order* (M, N) *of the stationary field distribution* $V(\mathbf{r}, t)$.

Suppose that the source is located in free space. We show in Appendix B that as a consequence of the wave equation (3.1), the cross-spectral densities of the field and of the source are then coupled by the following differential equation of order $M + N$:

$$\hat{\sigma}^{(M,N)} \prod_{s=1}^{M+N} \mathcal{M}_s W_V^{(M,N)}(\mathbf{r}_1, \mathbf{r}_2, \ldots \mathbf{r}_{M+N}; \omega_2, \omega_3, \ldots \omega_{M+N})$$
$$= (-4\pi)^{M+N} W_Q^{(M,N)}(\mathbf{r}_1, \mathbf{r}_2, \ldots \mathbf{r}_{M+N}; \omega_2, \omega_3, \ldots \omega_{M+N}), \tag{6.7}$$

where, as before, $\hat{\sigma}^{(M,N)}$ is the substitution operator of order (M, N), and

$$\mathcal{M}_s = \nabla_s^2 + k_s^2 \quad (s = 1, 2, \ldots M + N) \tag{6.8}$$

is the Helmholtz operator associated with the wave number

$$k_s = \frac{\omega_s}{c} \quad (s = 1, 2, \ldots M + N), \tag{6.9}$$

c being the speed of light *in vacuo*.

Let $\psi_s(\mathbf{r}, \omega_s)$ be the outgoing solution of the Helmholtz equation

$$\mathcal{M}_s \psi_s(\mathbf{r}, \omega_s) = -4\pi\phi_s(\mathbf{r}, \omega_s), \tag{6.10}$$

where $\phi_s(\mathbf{r}, \omega_s)$ are, as before, the eigenfunctions of the integral equation (2.4) (with n replaced by s and ω replaced by ω_s). These solutions are represented by integrals of the form (3.7).

Next let us introduce the ensemble of wave functions

$$U_V(\mathbf{r}, \omega) = \sum_n a_n(\omega)\psi_n(\mathbf{r}, \omega), \tag{6.11}$$

where the a_n's are the *same* random coefficients that appear in Eq. (2.11). The moments of these coefficients are defined by Eq. (5.6) and are given by Eqs. (5.7) [or Eqs. (5.8) and (5.9)] and (5.10). It is not difficult to see that the cross-spectral density $W_V^{(M,N)}$ is expressible in the form

$$W_V^{(M,N)}(\mathbf{r}_1, \mathbf{r}_2, \ldots \mathbf{r}_{M+N}; \omega_2, \omega_3, \ldots \omega_{M+N})$$
$$\times \hat{\sigma}^{(M,N)} \left\langle \prod_{j=1}^{M} U_V^*(\mathbf{r}_j; \omega_j) \prod_{k=M+1}^{M+N} U_V(\mathbf{r}_k, \omega_k) \right\rangle_\omega , \tag{6.12}$$

where the angle brackets with suffix ω indicate, as before, that the average is taken over the ensemble of the frequency-dependent random functions, defined by Eq. (6.11). To verify this result we substitute from Eqs. (6.12) and (5.1) into the differential equation (6.7) and make use of Eq. (3.9) and of the outgoing behavior of the wave functions $\psi_n(\mathbf{r}, \omega)$. *The formula* (6.12), which is a generalization of Eq. (3.11), *expresses the cross-spectral density* $W_V^{(M,N)}$ *of the field generated by the fluctuating source as a correlation function of the ensemble* $\{U_V(\mathbf{r}, \omega) \exp(-i\omega t)\}$ *of monochromatic wave fields.*

Finally, if we substitute from Eq. (6.11) into Eq. (6.12) and recall the definition (5.6) of the moments $\mathfrak{M}^{(M,N)}_{n_1 n_2 \ldots n_{M+N}}$, we obtain the following expansion of the cross-spectral density $W_V^{(M,N)}$ of the field:

$$W_V^{(M,N)}(\mathbf{r}_1, \mathbf{r}_2, \ldots \mathbf{r}_{M+N}; \omega_2, \omega_3, \ldots \omega_{M+N})$$
$$= \sum_{n_1} \sum_{n_2} \cdots \sum_{n_{M+N}} \hat{\sigma}^{(M,N)} \mathfrak{M}^{(M,N)}_{n_1, n_2, \ldots n_{M+N}}(\omega_1, \omega_2, \ldots \omega_{M+N})$$
$$\times \hat{\sigma}^{(M,N)} \prod_{j=1}^{M} \psi_{n_j}^*(\mathbf{r}_j; \omega_j) \prod_{k=M+1}^{M+N} \psi_{n_k}(\mathbf{r}_k, \omega_k). \tag{6.13}$$

This formula expresses the cross-spectral density $W_V^{(M,N)}$ of the (stationary) field in terms of the elementary fields $\psi_n(\mathbf{r}, \omega)$ generated by the source modes $\phi_n(\mathbf{r}, \omega)$. It is analogous to the representation (5.5) of the cross-spectral density $W_Q^{(M,N)}$ of the source.

Finally we mention that if the source is not located in free space but is embedded in a linear, time-independent medium or if it radiates into a space that contains obstacles whose response to an incident field is linear and time independent, the main formulas that we derived in this section still hold, provided that the elementary fields $\psi_n(\mathbf{r}, \omega)$ are replaced by the appropriate fields $\Psi_n(\mathbf{r}, \omega)$, as explained in connection with Eq. (3.19). Similar remarks apply if the source is a secondary rather than a primary one.

7. SUMMARY AND DISCUSSION

In spite of the length of some of the formulas derived in this paper, the mathematical structure of this theory is basically fairly simple. After a brief review, presented in Section 2, of the main results derived in Part I of this investigation, we showed in Section 3 that the cross-spectral density of the field generated by a steady-state source may be represented as a correlation function of an ensemble of monochromatic wave functions [Eq. (3.11)] or as a linear superposition of mutually uncorrelated elementary wave fields [Eq. (3.12)]. Each elementary wave field is just the wave field that is generated by the corresponding natural mode of the source [Eqs. (3.7), (3.19), and (3.20)].

In Section 4 we defined cross-spectral densities of arbitrary order of a steady-state source [Eqs. (4.9) and (4.5)], and in Section 5 we derived two new representations for these quantities [Eqs. (5.1) and (5.5)]. These representations are generalizations of the formulas (2.10) and (2.11) for the usual (lowest-order) cross-spectral density. One of them expresses the cross-spectral density of arbitrary order of the source as a correlation function of the ensemble of strictly monochromatic oscillations [Eq. (5.1)]. The other expresses it in terms of the natural modes of oscillation of the source and in terms of the moments of certain random coefficients [Eqs. (5.5) and (5.6)]. Various expressions for the moments are also derived [Eqs (5.7)–(5.11)].

In Section 6 we discussed correlations of arbitrary order of the field generated by the fluctuating source. We first introduced definitions of the various field-correlation functions in strict analogy with the definitions introduced in Section 4 in connection with the source. We then derived two representations of the cross-spectral density of arbitrary order of the field [Eqs. (6.12) and (6.13)], which are generalizations of the corresponding formulas derived in Section 3 for the lowest-order cross-spectral density [Eqs. (3.11) and Eq. (3.12)]. The mathematical and physical significance of these representations is similar to those described in connection with the two new representations of the higher-order cross-spectral density of the source, derived in Section 5. The only difference is that in place of the natural modes of the source, the elementary fields generated by these modes enter the expressions for the cross-spectral densities of the field. The derivation of these two representations makes use of a certain differential equation [Eq. (6.7)], which is derived in Appendix B and which relates the cross-spectral density, of arbitrary order, of the radiated field to the cross-spectral density (of the same order) of the source.

The lowest-order version of this theory[1–3] has already found useful applications, some of which were mentioned in the Introduction. It is likely that the general theory formulated in this paper, which deals with correlations of all orders, will also prove useful. Obvious potential applications are to problems concerning the statistics of speckle patterns, to problems relating to propagation of light in the turbulent atmosphere, and to coherence problems of nonlinear optics.

Finally we might mention that the concept of complete coherence of arbitrary order of steady-state fields has until now presented some difficulties and that the physical significance of this concept has been rather obscure. We plan to show in another publication that our new theory clarifies this question.[25]

APPENDIX A: DERIVATION OF EQ. (3.4)

The field $V(\mathbf{r}, t)$ is coupled to the source distribution $Q(\mathbf{r}, t)$ by the inhomogeneous wave equation (3.1), viz.,

$$\mathcal{L} V(\mathbf{r}, t) = -4\pi Q(\mathbf{r}, t), \tag{A1}$$

where

$$\mathcal{L} \equiv \nabla^2 - \frac{1}{c^2}\frac{\partial^2}{\partial t^2} \tag{A2}$$

is the wave operator. Consequently

$$\mathcal{L}_1 V^*(\mathbf{r}_1, t_1)\mathcal{L}_2 V(\mathbf{r}_2, t_2) = (4\pi)^2 Q^*(\mathbf{r}_1, t_1)Q(\mathbf{r}_2, t_2), \tag{A3}$$

where $\mathcal{L}_1$ and $\mathcal{L}_2$ are the wave operators acting with respect to the variables $(\mathbf{r}_1, t_1)$ and $(\mathbf{r}_2, t_2)$, respectively. If we take the average over the source ensemble of both sides of Eq. (A3) and interchange on the left-hand side the order of the various operations, we obtain the equation

$$\mathcal{L}_2\mathcal{L}_1\langle V^*(\mathbf{r}_1, t_1)V(\mathbf{r}_2, t_2)\rangle = (4\pi)^2\langle Q^*(\mathbf{r}_1, t_1)Q(\mathbf{r}_2, t_2)\rangle. \tag{A4}$$

Since $Q(\mathbf{r}, t)$ and $V(\mathbf{r}, t)$ are stationary random variables, the averages on the two sides of Eq. (A4) depend on t_1 and t_2 only through their difference. In fact, the two averages are just the cross-correlation functions Γ_V and Γ_Q of the field distribution and of the source distribution, respectively [Eq. (3.2) and (2.1)], so that Eq. (A4) may be expressed in the form

$$\mathcal{L}_2\mathcal{L}_1\Gamma_V(\mathbf{r}_1, \mathbf{r}_2, t_2 - t_1) = (4\pi)^2\Gamma_Q(\mathbf{r}_1, \mathbf{r}_2, t_2 - t_1). \tag{A5}$$

If we set

$$t_2 - t_1 = \tau, \tag{A6}$$

it is clear that $\partial^2/\partial t_1^2 = \partial^2/\partial t_2^2 = \partial^2/\partial\tau^2$, and hence Eq. (A5) may be rewritten as

$$L_2L_1\Gamma_V(\mathbf{r}_1, \mathbf{r}_2, \tau) = (4\pi)^2\Gamma_Q(\mathbf{r}_1, \mathbf{r}_2, \tau), \tag{A7}$$

with

$$L_j = \nabla_j^2 - \frac{1}{c^2}\frac{\partial^2}{\partial\tau^2} \qquad (j = 1, 2), \tag{A8}$$

∇_j^2 being, of course, the Laplacian operator acting with respect to the spatial variable $\mathbf{r}_j$.

Let us now express the cross-correlation functions Γ_V and Γ_Q in terms of the cross-spectral densities W_V and W_Q, respectively [the inverses of the Fourier integral relations (3.3) and (2.2)], viz.,

$$\Gamma_V(\mathbf{r}_1, \mathbf{r}_2, \tau) = \int_{-\infty}^{\infty} W_V(\mathbf{r}_1, \mathbf{r}_2, \omega)e^{-i\omega\tau}d\omega, \tag{A9}$$

$$\Gamma_Q(\mathbf{r}_1, \mathbf{r}_2, \tau) = \int_{-\infty}^{\infty} W_Q(\mathbf{r}_1, \mathbf{r}_2, \omega)e^{-i\omega\tau}d\omega. \tag{A10}$$

On substituting from Eqs. (A9) and (A10) into Eq. (A7) and interchanging on the left-hand side the order of the differential operations and of integration, we readily find that

$$\int_{-\infty}^{\infty} \mathcal{M}_2\mathcal{M}_1 W_V(\mathbf{r}_1, \mathbf{r}_2, \omega)e^{-i\omega\tau}d\omega = (4\pi)^2\int_{-\infty}^{\infty} W_Q(\mathbf{r}_1, \mathbf{r}_2, \omega)e^{-i\omega\tau}d\omega, \tag{A11}$$

where

$$\mathcal{M}_j = \nabla_j^2 + k^2 \qquad (j = 1, 2) \tag{A12}$$

($k = \omega/c$). Since Eq. (A11) holds for all values of τ it follows that the cross-spectral densities of the field and of the source distribution are coupled by the equation

$$\mathcal{M}_2\mathcal{M}_1 W_V(\mathbf{r}_1, \mathbf{r}_2, \omega) = (4\pi)^2 W_Q(\mathbf{r}_1, \mathbf{r}_2, \omega), \tag{A13}$$

which is Eq. (3.4) of the text.[26]

APPENDIX B: DERIVATION OF EQ. (6.7)

By an obvious generalization of the argument that led to Eq. (A4) in the preceding appendix, we obtain the equation

$$\prod_{s=1}^{M+N}\mathcal{L}_s G_V^{(M,N)}(\mathbf{r}_1, \mathbf{r}_2, \ldots \mathbf{r}_{M+N}; t_1, t_2, \ldots t_{M+N}) = (-4\pi)^{M+N}G_Q^{(M,N)}(\mathbf{r}_1, \mathbf{r}_2 \ldots \mathbf{r}_{M+N}; t_1, t_2, \ldots t_{M+N}). \tag{B1}$$

Here

$$\mathcal{L}_s \equiv \nabla_s^2 - \frac{1}{c^2}\frac{\partial^2}{\partial t_s^2}, \tag{B2}$$

($s = 1, 2, \ldots M + N$) is the wave operator acting with respect to the variables $(\mathbf{r}_s, t_s)$, and $G_V^{(M,N)}$ and $G_Q^{(M,N)}$ are the cross-correlation functions of the field and of the source, defined by Eqs. (6.1) and (4.1a), respectively.

Let us substitute in Eq. (B1) for the cross-correlation functions their Fourier representations [cf. Eq. (4.4)] and then interchange the orders of the differential operations and of integrations. Since the resulting formula must hold for all values of the $M + N$ temporal arguments, we readily conclude that Eq. (B1) implies the following differential equation that couples the spectral cross-correlation functions $\Phi_V^{(M,N)}$ and $\Phi_Q^{(M,N)}$ of the field and the source distributions, respectively:

$$\prod_{s=1}^{M+N}\mathcal{M}_s\Phi_V^{(M,N)}(\mathbf{r}_1, \mathbf{r}_2, \ldots \mathbf{r}_{M+N}; \omega_1\, \omega_2, \ldots \omega_{M+N}) = (-4\pi)^{(M+N)}\Phi_Q^{(M+N)}(\mathbf{r}_1, \mathbf{r}_2, \ldots \mathbf{r}_{M+N}; t_1, t_2, \ldots t_{M+N}), \tag{B3}$$

where

$$\mathcal{M}_s \equiv \nabla_s^2 + k_s^2 \qquad (s = 1, 2, \ldots M + N) \tag{B4}$$

is the Helmholtz operator associated with the wave number

$$k_s = \frac{\omega_s}{c} \qquad (s = 1, 2, \ldots M + N), \tag{B5}$$

c being the speed of light *in vacuo*.

Suppose now that the source fluctuations and consequently also the field fluctuations are statistically stationary. We may then introduce the cross-spectral densities $W_V^{(M,N)}$ and $W_Q^{(M,N)}$ through the formulas (6.5) and (4.9), and Eq. (B3) implies, after both sides are integrated with respect to ω_1, that they are related by the differential equation

$$\hat{\sigma}^{(M,N)} \prod_{s=1}^{M+N} \mathcal{M}_s W_V^{(M,N)}(\mathbf{r}_1, \mathbf{r}_2, \ldots \mathbf{r}_{M+N}; \omega_2, \omega_3, \ldots \omega_{M+N}) = (-4\pi)^{M+N} W_Q^{(M,N)}(\mathbf{r}_1, \mathbf{r}_2, \ldots \mathbf{r}_{M+N}; \omega_2, \omega_3, \ldots \omega_{M+N}). \tag{B6}$$

Here $\hat{\sigma}^{(M,N)}$ denotes, as before, the substitution operator of order (M, N). The formula (B6) is Eq. (6.7) of the text.

Let us consider the explicit form of Eq. (B6) for some special cases. When $M = N = 1$, Eq. (B6) becomes

$$(\nabla_1^2 + k_2^2)(\nabla_2^2 + k_2^2) W_V^{(1,1)}(\mathbf{r}_1, \mathbf{r}_2, \omega_2) = (4\pi)^2 W_Q^{(1,1)}(\mathbf{r}_1, \mathbf{r}_2, \omega_2) \tag{B7}$$

$(k_2 = \omega_2/c)$, which, except for a trivial change in notation, is Eq. (A13) of Appendix A. When $M = 2$, $N = 1$, Eq. (B6) becomes

$$[\nabla_1^2 + (k_3 - k_2)^2](\nabla_2^2 + k_2^2)(\nabla_3^2 + k_3^2) \times W_V^{(2,1)}(\mathbf{r}_1, \mathbf{r}_2, \mathbf{r}_3; \omega_2, \omega_3) = (4\pi)^3 W_Q^{(2,1)}(\mathbf{r}_1, \mathbf{r}_2, \mathbf{r}_3; \omega_2, \omega_3), \tag{B8}$$

k_2 and k_3 being, of course, given by Eq. (B5) with $s = 2$ and $s = 3$, respectively.

ACKNOWLEDGMENT

This research was supported by the National Science Foundation.

The author is also with the Institute of Optics, University of Rochester.

REFERENCES AND NOTES

1. E. Wolf, "New theory of partial coherence in the space-frequency domain. Part I: Spectra and cross-spectra of steady-state sources," J. Opt. Soc. Am. **72**, 343–351 (1982). There is a misprint in Eq. (5.5) of this reference. The factor $[\lambda_n(\omega)]^{2/3}$ should be replaced by $[\lambda_n(\omega)]^{1/2}$.
2. E. Wolf, "A new description of second-order coherence phenomena in the space-frequency domain," AIP Conf. Proc. **65**, 42–48 (1981).
3. E. Wolf, "New spectral representation of random sources and of the partially coherent fields that they generate," Opt. Commun. **38**, 3–6 (1981).
4. A. Starikov and E. Wolf, "Coherent-mode representation of Gaussian Schell-model sources and of their radiation fields," J. Opt. Soc. Am. **72**, 923–928 (1982).
5. A. Starikov, "Effective number of degrees of freedom of partially coherent sources," J. Opt. Soc. Am. **72**, 1538–1544 (1982).
6. F. Gori, "Mode propagation of the field generated by Collett–Wolf Schell-model sources," Opt. Commun. **46**, 149–154 (1983).
7. F. Gori and R. Grella, "Shape invariant propagation of polychromatic fields," Opt. Commun. **49**, 173–177 (1984).
8. E. Wolf and G. S. Agarwal, "Coherence theory of laser resonator modes," J. Opt. Soc. Am. A **1**, 541–546 (1984).
9. R. Martinez-Herrero and P. M. Mejias, "Radiometric definitions for partially coherent sources," J. Opt. Soc. Am. A **1**, 556–558 (1984); J. T. Foley and M. Nieto-Vesperinas, "Radiance functions that depend nonlinearly on the cross-spectral density," J. Opt. Soc. Am. A **2**, 1446–1447 (1985).
10. I. J. LaHaie, "Inverse source problem for three-dimensional partially coherent sources and fields," J. Opt. Soc. Am. A **2**, 35–45 (1985).
11. L. Mandel and E. Wolf, "Spectral coherence and the concept of cross-spectral purity," J. Opt. Soc. Am. **66**, 529–535 (1976), Sec. II. A relationship between the degree of spectral coherence $\mu(\mathbf{r}_1, \mathbf{r}_2, \omega)$ and the more familiar complex degree of coherence $\gamma(\mathbf{r}_1, \mathbf{r}_2, \tau)$ in the space–time domain, as well as a method for measuring $\mu(\mathbf{r}_1, \mathbf{r}_2, \omega)$, is discussed in E. Wolf, "Young's interference fringes with narrow-band light," Opt. Lett. **8**, 250–252 (1983).
12. M. Born and E. Wolf, *Principles of Optics*, 6th ed. (Pergamon, Oxford, 1980), Sec. 10.3.1.
13. L. Mandel and E. Wolf, "Coherence properties of optical fields," Rev. Mod. Phys. **37**, 231–287 (1965), Sec. 3.1.
14. C. H. Papas, *Theory of Electromagnetic Wave Propagation* (McGraw-Hill, New York, 1965), Sec. 2.1.
15. If $Q(\mathbf{r}, t)$ is the complex analytic signal representation (Ref. 10, Sec. 10.2) of a real source distribution, $V(\mathbf{r}, t)$ will also be an analytic signal. Consequently $\Gamma_Q(\mathbf{r}_1, \mathbf{r}_2, \tau)$ and $\Gamma_V(\mathbf{r}_1, \mathbf{r}_2, \tau)$ will not contain any negative frequency components (Ref. 12, Sec. 10.3.2), and the lower limits in the integrals in Eqs. (3.17) and (3.18) can then be replaced by zero.
16. Lord Rayleigh, *The Theory of Sound* (reprinted by Dover, New York, 1945), Vol. II, Sec. 278 [with a modification appropriate to time dependence $\exp(-i\omega t)$ used in the present paper].
17. Field correlations of arbitrary order appear to have been discussed first in the framework of the classical theory of stationary fields by L. Mandel in Ref. 18. See also E. Wolf, "Basic concepts of optical coherence theory," in *Proceedings of the Symposium on Optical Masers*, J. Fox, ed. (Wiley, New York, 1963), pp. 29–42.
18. L. Mandel, "Some coherence properties of non-Gaussian light," in *Quantum Electronics III*, P. Grivet and N. Bloembergen, eds. (Columbia U. Press, New York, 1964), Vol. I, pp. 101–109.
19. It is implicitly assumed here that $M > 1$. When $M = 1$ one readily finds that the first product term (with the index j) in Eq. (4.10), and in all subsequent formulas where it appears, must be replaced by unity.
20. Cross-spectral density functions of arbitrary orders in classical wave fields were first considered by L. Mandel in Ref. 18. See also Refs. 21–23.
21. E. Wolf, "Light fluctuations as a new spectroscopic tool," Jpn. J. Appl. Phys. Suppl. 1 **4**, 1–14, (1965).
22. C. L. Mehta and L. Mandel, "Some properties of higher order coherence functions," in *Electromagnetic Wave Theory*, J. Brown, ed. (Pergamon, New York, 1967), Part 2, pp. 1069–1075.
23. L. Mandel, "Photoelectric correlations and fourth-order coherence properties of optical fields," AIP Conf. Proc. **65**, 178–195 (1981).
24. Similar remarks apply here to those made in note 19 in connection with Eq. (4.10).
25. A preliminary account of this application of our theory was presented at the 1984 Annual Meeting of the Optical Society of America [J. Opt. Soc. Am. A **1**, 1311 (A) (1984)].
26. Equation (A13) was obtained previously by a somewhat less rigorous argument by W. H. Carter and E. Wolf in "Correlation theory of wave-fields generated by fluctuating, three-dimensional, primary, scalar sources. I: General theory," Opt. Acta **28**, 227–244 (1981), Sec. 2.

250 OPTICS LETTERS / Vol. 8, No. 5 / May 1983

Young's interference fringes with narrow-band light

Emil Wolf*

Schlumberger-Doll Research, P.O. Box 307, Ridgefield, Connecticut 06877

Received January 17, 1983

The changes in the interference pattern in Young's interference experiment, produced by placing two identical narrow-band filters in front of the pinholes, are analyzed. It is shown theoretically that, in general, the fringes will not become sharp (i.e., their maximum visibility will not tend to unity) even when the filters have arbitrarily narrow passbands. The analysis brings out a relationship between the complex degree of coherence in the space–time and the space–frequency domains. When the passbands of the filters are narrow enough, the filtered light is found to be cross-spectrally pure.

1. Introduction

It is well known that the visibility $\mathcal{V}(Q)$ of the fringes at a point Q in a Young's interference pattern is proportional to the absolute value of the complex degree of coherence[1] $\gamma(P_1, P_2, \tau)$ of the light emerging from the pinholes at points P_1 and P_2 (Fig. 1) and that the location of the intensity maxima is expressible in a simple way in terms of the phase of $\gamma(P_1, P_2, \tau)$. The parameter τ represents the difference between the times needed for the light to propagate from the pinholes to point Q.

Suppose that identical filters, with effective passbands $\Delta\omega$, centered on a frequency ω_0, are placed in front of the pinholes. We examine their effect on the interference pattern when $\Delta\omega$ becomes arbitrarily small. Naively one might perhaps expect that when $\Delta\omega$ is decreased the interference fringes will become sharper, with their maximum visibility increasing toward unity. We show that, in general, this is not so and that, when the passbands are sufficiently narrow, the maximum fringe visibility, attained at the center of the pattern (at the symmetric position $\tau = 0$), is equal to the absolute value of the complex degree of spectral coherence[2] $\mu(\mathbf{r}_1, \mathbf{r}_2, \omega_0)$ of the unfiltered light; and that the light transmitted by the two filters then becomes cross-spectrally pure.[2,3]

2. Complex Degree of Coherence of the Filtered Light

Let $V(P_1, t)$ and $V(P_2, t)$ be the analytic signal representations of the fluctuating optical field at the two pinholes at P_1 and P_2 at time t. For simplicity we assume that the field fluctuations are characterized by a statistical ensemble that is stationary and ergodic.[4] The mutual coherence function $\Gamma(\mathbf{r}_1, \mathbf{r}_2, \tau)$ of the light at the two pinholes is then defined in the usual way as

$$\Gamma(P_1, P_2, \tau) = \langle V^*(P_1, t) V(P_2, t + \tau)\rangle, \quad (2.1)$$

where the asterisk denotes the complex conjugate and the angular brackets denote the ensemble average; and the complex degree of coherence of the light at the pinholes is given by

$$\gamma(P_1, P_2, \tau) = \frac{\Gamma(P_1, P_2, \tau)}{[\Gamma(P_1, P_1, 0)]^{1/2}[\Gamma(P_2, P_2, 0)]^{1/2}}, \quad (2.2)$$

where, as is evident from Eq. (2.1), $\Gamma(P_1, P_1, 0)$ and $\Gamma(P_2, P_2, 0)$ are the (averaged) intensities at the two pinholes.

If the light is quasi-monochromatic and the intensities at the two pinholes are equal, as we will assume from now on, the visibility $\mathcal{V}(Q)$ (which is a measure of the fringe contrast) at a point Q in the fringe pattern is equal to the absolute value of the complex degree of coherence:

$$\mathcal{V}(Q) = |\gamma(P_1, P_2, \tau)|. \quad (2.3)$$

Here $\tau = (s_2 - s_1)/c$, where s_1 and s_2 are the distances from P_1 and P_2 to Q (see Fig. 1) and c is the speed of light in vacuum.

Let us represent the mutual coherence function as a Fourier integral:

$$\Gamma(P_1, P_2, \tau) = \int_0^\infty W(P_1, P_2, \omega) e^{-i\omega\tau} d\omega. \quad (2.4)$$

The quantity $W(P_1, P_2, \omega)$ is the cross-spectral density

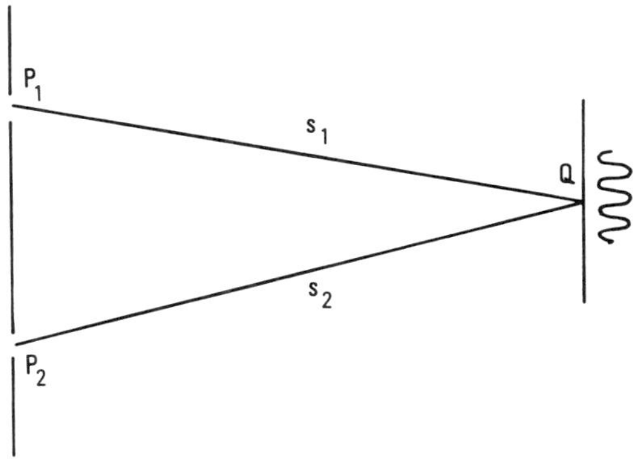

Fig. 1. Illustrating the notation used in the analysis of Young's interference experiment. The parameter τ that occurs in the text is equal to $(s_2 - s_1)/c$, where c is the speed of light *in vacuo*.

of the light at the two pinholes. It was recently shown[5] that it is always possible to construct an ensemble $\{U(P, \omega)\exp(-i\omega t)\}$ of monochromatic oscillations, all of frequency ω, such that

$$W(\mathrm{P}_1, \mathrm{P}_2, \omega) = \langle U^*(\mathrm{P}_1, \omega)U(\mathrm{P}_2, \omega)\rangle_\omega, \qquad (2.5)$$

where the angular brackets, with suffix ω, denote an average over this ensemble.

Suppose now that we place identical filters in front of the two pinholes. Then the light that emerges from the two filters will be represented by the ensembles $\{T(\omega)U(\mathrm{P}_j, \omega)\exp(-i\omega t)\}$, $(j = 1, 2)$, where $T(\omega)$ is the (generally complex) amplitude-transmission function of each filter. The cross-spectral density $W^{(+)}(\mathrm{P}_1, \mathrm{P}_2, \omega)$ of the filtered light at the two pinholes will then be given by a formula of the form of Eq. (2.5) but with $U(\mathrm{P}_j, \omega)$ replaced by $T(\omega)U(\mathrm{P}_j, \omega)$, i.e.,

$$W^{(+)}(\mathrm{P}_1, \mathrm{P}_2, \omega) = \langle T^*(\omega)U^*(\mathrm{P}_1, \omega)T(\omega)U(\mathrm{P}_2, \omega)\rangle_\omega; \qquad (2.6)$$

or, if we take the (deterministic) factor $T^*(\omega)T(\omega)$ outside the averaging and use Eq. (2.5), we find that

$$W^{(+)}(\mathrm{P}_1, \mathrm{P}_2, \omega) = |T(\omega)|^2 W(\mathrm{P}_1, \mathrm{P}_2, \omega). \qquad (2.7)$$

The mutual coherence function $\Gamma^{(+)}(\mathrm{P}_1, \mathrm{P}_2, \omega)$ of the filtered light at the two pinholes is given by the Fourier transform of expression (2.7), i.e., by

$$\Gamma^{(+)}(\mathrm{P}_1, \mathrm{P}_2, \tau) = \int_0^\infty |T(\omega)|^2 W(\mathrm{P}_1, \mathrm{P}_2, \omega)e^{-i\omega\tau}d\omega; \qquad (2.8)$$

and the complex degree of coherence of the filtered light at the pinholes is given by a formula of the form of Eq. (2.2) but with Γ replaced by $\Gamma^{(+)}$, i.e.,

$$\gamma^{(+)}(\mathrm{P}_1, \mathrm{P}_2, \tau) = \frac{\Gamma^{(+)}(\mathrm{P}_1, \mathrm{P}_2, \tau)}{[\Gamma^{(+)}(\mathrm{P}_1, \mathrm{P}_1, 0)]^{1/2}[\Gamma^{(+)}(\mathrm{P}_2, \mathrm{P}_2, 0)]^{1/2}}. \qquad (2.9)$$

Suppose now that each filter has a passband of effective width $\Delta\omega$, centered on a frequency ω_0. If the cross-spectral density $W(\mathrm{P}_1, \mathrm{P}_2, \omega)$ is a continuous function of ω, as we now assume, and if $\Delta\omega$ is so small that $W(\mathrm{P}_1, \mathrm{P}_2, \omega)$ does not appreciably change across the effective passband $\omega_0 - \Delta\omega/2 \leqslant \omega \leqslant \omega_0 + \Delta\omega/2$ of the filters (see Fig. 2), we may replace $W(\mathrm{P}_1, \mathrm{P}_2, \omega)$ in the integral in Eq. (2.8) by $W(\mathrm{P}_1, \mathrm{P}_2, \omega_0)$. We then obtain the following expression for $\Gamma^{(+)}$:

$$\Gamma^{(+)}(\mathrm{P}_1, \mathrm{P}_2, \tau) = W(\mathrm{P}_1, \mathrm{P}_2, \omega_0)\int_0^\infty |T(\omega)|^2 e^{-i\omega\tau}d\omega. \qquad (2.10)$$

In a similar way, it follows that, if the spectral densities $W(\mathrm{P}_1, \mathrm{P}_1, \omega)$ and $W(\mathrm{P}_2, \mathrm{P}_2, \omega)$ are continuous functions of ω and $\Delta\omega$ is small enough, then, to a good approximation,

$$\Gamma^{(+)}(\mathrm{P}_j, \mathrm{P}_j, \tau) = W(\mathrm{P}_j, \mathrm{P}_j, \omega_0)\int_0^\infty |T(\omega)|^2 e^{-i\omega\tau}d\omega, \quad (j = 1, 2). \qquad (2.11)$$

On substituting from Eqs. (2.10) and (2.11) into Eq. (2.9), we obtain the following expression for the complex degree of coherence of the filtered light at the two pinholes:

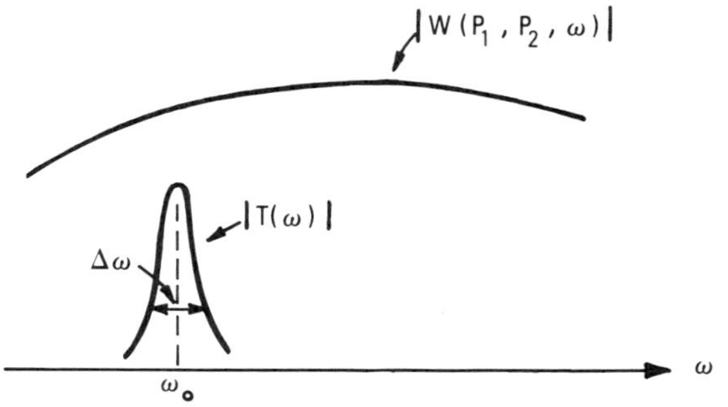

Fig. 2. Schematic illustration of the relative behavior of the modulus of the complex amplitude transmission function $T(\omega)$ of the filters placed in front of the pinholes of Young's interference experiments and of the absolute value of the cross-spectral density $W(\mathrm{P}_1, \mathrm{P}_2, \omega)$ of the unfiltered light at the pinholes. The effective passbands $\omega_0 - \Delta\omega/2 \leqslant \omega \leqslant \omega_0 + \Delta\omega/2$ of the filters are assumed to be so narrow that both the modulus and the phase (not shown) of $W(\mathrm{P}_1, \mathrm{P}_2, \omega)$ and also the spectral densities $W(\mathrm{P}_1, \mathrm{P}_1, \omega)$ and $W(\mathrm{P}_2, \mathrm{P}_2, \omega)$ (with P_1 and P_2 fixed) are substantially constant across it.

$$\gamma^{(+)}(\mathrm{P}_1, \mathrm{P}_2, \tau) = \mu(\mathrm{P}_1, \mathrm{P}_2, \omega_0)\theta(\tau), \qquad (2.12)$$

where

$$\mu(\mathrm{P}_1, \mathrm{P}_2, \omega_0) = \frac{W(\mathrm{P}_1, \mathrm{P}_2, \omega_0)}{[W(\mathrm{P}_1, \mathrm{P}_1, \omega_0)]^{1/2}[W(\mathrm{P}_2, \mathrm{P}_2, \omega_0)]^{1/2}} \qquad (2.13)$$

and

$$\theta(\tau) = \frac{\int_0^\infty |T(\omega)|^2 \exp(-i\omega\tau)d\omega}{\int_0^\infty |T(\omega)|^2 d\omega}. \qquad (2.14)$$

Formula (2.12) is the main result of this Letter. We now show that it has a number of interesting physical implications.

3. Physical Implications of Eq. (2.12) for the Complex Degree of Coherence of the Filtered Light

We note that Eq. (2.12) expresses the complex degree of coherence of the filtered light at the two pinholes as a product of two terms. The first, denoted by $\mu(\mathrm{P}_1, \mathrm{P}_2, \omega_0)$, is precisely the complex degree of spectral coherence[2] at frequency ω_0 of the *unfiltered* light at P_1 and P_2. However, it is also equal to the complex degree of spectral coherence at frequency ω_0 of the filtered light at these two points, viz.,

$$\mu^{(+)}(\mathrm{P}_1, \mathrm{P}_2, \omega_0) = \frac{W^{(+)}(\mathrm{P}_1, \mathrm{P}_2, \omega_0)}{[W^{(+)}(\mathrm{P}_1, \mathrm{P}_1, \omega_0)]^{1/2}[W^{(+)}(\mathrm{P}_2, \mathrm{P}_2, \omega_0)]^{1/2}}, \qquad (3.1)$$

252 OPTICS LETTERS / Vol. 8, No. 5 / May 1983

as is evident on substituting from Eq. (2.7) into the right-hand side of Eq. (3.1) and on comparing the resulting expression with Eq. (2.13). Hence *the degree of spectral coherence remains unchanged by filtering.*

The second term on the right-hand side of Eq. (2.12) is, according to Eq. (2.14), the normalized Fourier transform of the squared modulus of the complex amplitude-transmission function of the filter. This factor alone now specifies the temporal-coherence properties of the filtered light, whereas the first factor on the right-hand side of Eq. (2.1) specifies its spatial-coherence properties at frequency ω_0. It is clear that, unlike the complex degree of spectral coherence μ, the complex degree of (space–time) coherence γ was modified by filtering.

It is also evident from Eq. (2.12) that, considered as a function of τ, $|\gamma^{(+)}|$ attains its maximum when $|\theta(\tau)|$ reaches its maximum, i.e., when $\tau = 0$. Moreover, since $\theta(0) = 1$, it follows that

$$\gamma^{(+)}(P_1, P_2, 0) = \mu(P_1, P_2, \omega_0). \tag{3.2}$$

Hence the "equal-time" complex degree of coherence $\gamma^{(+)}(P_1, P_2, 0)$ of the filtered light at the two pinholes is equal to the degree of spectral coherence $\mu(P_1, P_2, \omega_0)$ of the unfiltered (and also of the filtered) light at the midfrequency ω_0 of the filters. Now $|\gamma^{(+)}(P_1, P_2, 0)|$ is equal to the visibility at the center of the fringe pattern, i.e., at the point Q′ that is equidistant from the two pinholes, and hence, in view of relation (3.2), the fringe visibility at Q′ is also equal to $|\mu(P_1, P_2, \omega_0)|$. It is known that $|\mu|$ may take on any value between zero and unity[2] (with both the extremes being attainable), its actual value depending on the nature of the optical field incident on the pinholes. Hence *the maximum visibility of the interference fringes formed by the filtered light will not, in general, tend to unity as the passbands of the filters decrease.*[6] It is actually not difficult to realize conditions under which the maximum visibility will take on any selected value between (and including) zero and unity, however narrow the effective bandwidth $\Delta\omega$ of the filters may be. One only needs to illuminate the pinholes by a distant, spatially incoherent, uniform, circular, quasi-monochromatic source of mean frequency in the neighborhood of $\omega = \omega_0$, placed symmetrically with respect to the pinholes and separating the pinholes by an appropriate distance that may be readily calculated from the van Cittert–Zernike theorem.

Although, under conditions when Eq. (2.12) applies, the maximum fringe visibility will not change as the bandwidth $\Delta\omega$ of the filter is further reduced, the interference pattern will then, nevertheless, be modified. If $\Delta\omega$ is decreased, but the transmissivity $|T(\omega_0)|$ of the filters at the midfrequency ω_0 is kept fixed, the effective width of $|\theta(\tau)|$ will increase, as one can readily deduce from Eq. (2.14). Hence, according to Eq. (2.12), $|\gamma^{(+)}(P_1, P_2, \tau)|$, and consequently the visibility of the fringes will then fall off more slowly with distance from the centrally located point Q′.

We note that relation (3.2), together with the facts that the modulus of $\gamma^{(+)}(P_1, P_2, 0)$ is equal to the visibility and that the phase of $\gamma^{(+)}(P_1, P_2, 0)$ is directly related to the location of the intensity maximum near the centrally situated point Q′ in the interference pattern formed by the filtered light, indicates a manner in which the complex degree of spectral coherence $\mu(P_1, P_2, \omega)$ can be experimentally determined.

One other conclusion can be drawn from our analysis. According to Eqs. (2.9), (2.11), and (2.14), the complex degree of "self-coherence" of the filtered light at each pinhole is given by

$$\gamma^{(+)}(P_1, P_1, \tau) \equiv \frac{\Gamma^{(+)}(P_1, P_1, \tau)}{\Gamma^{(+)}(P_1, P_1, 0)} = \theta(\tau) \tag{3.3a}$$

and

$$\gamma^{(+)}(P_2, P_2, \tau) \equiv \frac{\Gamma^{(+)}(P_2, P_2, \tau)}{\Gamma^{(+)}(P_2, P_2, 0)} = \theta(\tau). \tag{3.3b}$$

If we make use of Eqs. (3.2) and (3.3a) in Eq. (2.12), we see that

$$\gamma^{(+)}(P_1, P_2, \tau) = \gamma^{(+)}(P_1, P_2, 0)\gamma^{(+)}(P_1, P_1, \tau). \tag{3.4}$$

Moreover, it follows from Eqs. (3.3) that

$$\gamma^{(+)}(P_1, P_1, \tau) = \gamma^{(+)}(P_2, P_2, \tau). \tag{3.5}$$

Formulas (3.4) and (3.5) are special cases of the two requirements for cross-spectral purity [Eqs. (4.4b) and (4.4a) of Ref. 2, with $\tau_0 = 0$]. Hence our analysis also shows that the filtered light that emerges from the pinholes is cross-spectrally pure.

* On leave of absence during the academic year 1982–1983. Permanent address, Department of Physics and Astronomy, University of Rochester, Rochester, New York 14627.

References

1. For explanation of the concepts of coherence theory used in this Letter, see M. Born and E. Wolf, *Principles of Optics*, 6th ed. (Pergamon, Oxford, 1980), Chap. X, or L. Mandel and E. Wolf, "Coherence properties of optical fields," Rev. Mod. Phys. **37**, 231–287 (1965).
2. L. Mandel and E. Wolf, "Spectral coherence and the concept of cross-spectral purity," J. Opt. Soc. Am. **66**, 529–535 (1976).
3. L. Mandel, "Concept of cross-spectral purity in coherence theory," J. Opt. Soc. Am. **51**, 1342–1350 (1961).
4. W. B. Davenport and W. L. Root, *An Introduction to the Theory of Random Signals and Noise* (McGraw-Hill, New York, 1958), pp. 42 and 67.
5. E. Wolf, "A new description of second-order coherence phenomena in the space–frequency domain," AIP Conf. Proc. **65**, 42–48 (1981); "New spectral representation of random sources and of the partially coherent fields that they generate," Opt. Commun. **38**, 3–6 (1981); New theory of partial coherence in the space–frequency domain. Part I: Spectra and cross-spectra of steady-state sources," J. Opt. Soc. Am. **72**, 343–351 (1982).
6. It is, of course, assumed that the measurements are made on a time scale that involves averaging over a time interval that is long compared to the reciprocal bandwidth of the filters and that the detector is sensitive enough to measure the reduced intensities of the filtered light.

Coherence theory of laser resonator modes

Emil Wolf

Department of Physics and Astronomy, University of Rochester, Rochester, New York 14627

G. S. Agarwal

School of Physics, University of Hyderabad, Hyderabad-500 134, India

An analysis of transverse laser resonator modes is presented, based on a recently developed coherence theory in the space-frequency domain. The modes are introduced by means of solutions of an integral equation that expresses a steady-state condition for a second-order correlation function of the field across a mirror of the laser cavity. All solutions of this integral equation are found to be expressible as quadratic forms involving the Fox–Li modes of the conventional theory. If there is no degeneracy, each mode is shown to be necessarily completely spatially coherent, at each frequency, within the framework of second-order correlation theory. It is also shown that, if several transverse modes are excited, the output cannot be completely spatially coherent.

1. INTRODUCTION

Since the publication of a fundamental paper by Fox and Li,[1] studies of laser resonator modes have generally been based on an integral equation that expresses a steady-state condition of the optical field at a mirror of an open resonator. Because these studies idealize the situation by treating the light as being strictly monochromatic, they cannot yield any information about the coherence properties of the output.

An early attempt to elucidate coherence properties of laser resonator modes was made by Wolf.[2] He considered the "transmission line analogue" of a laser resonator that is due to Fox and Li, and he showed that quasi-monochromatic light that is initially partially spatially coherent, or even incoherent, may become completely spatially coherent after a sufficient number of transits. This result implies that spatial coherence of the light in the laser resonator may be generated in the process of propagation and by diffraction at the mirrors of the resonant cavity. Various extensions and generalizations of this analysis were later made by Streifer,[3] Allen *et al.*,[4] and by Gori.[5]

In this paper we present a new theory of laser resonator modes based on an integral equation that expresses a steady-state condition for the cross-spectral density of the field of any spectral composition. It is shown that all the solutions of this equation are expressible as quadratic forms involving the Fox–Li modes of the conventional theory. It is found that if there is no degeneracy, each mode is, within the framework of second-order correlation theory, necessarily completely spatially coherent at each temporal frequency. It is also shown that if several (transverse) modes are excited, the output cannot be spatially completely coherent. These results are in agreement with results of experiments carried out by Bertolotti *et al.* many years ago.[6]

2. COHERENT-MODE REPRESENTATION OF FIELDS OF ANY STATE OF COHERENCE

In the analysis of the main problem that we discuss in this paper we make use of the mode representation of fields of any state of coherence developed in several recent papers. It will therefore be useful to begin by summarizing the main results pertaining to this representation.

Let $V(\mathbf{r}, t)$ be the complex analytic signal representation[7] of a stationary field in some finite closed domain D of free space. Here $\mathbf{r}$ denotes the position vector of a typical point in D and t denotes the time. Let

$$\Gamma(\mathbf{r}_1, \mathbf{r}_2, \tau) = \langle V^*(\mathbf{r}_1, t)V(\mathbf{r}_2, t + \tau)\rangle \tag{2.1}$$

(with the asterisk denoting the complex conjugate) be the mutual coherence function of the field and let

$$W(\mathbf{r}_1, \mathbf{r}_2, \omega) = \frac{1}{2\pi}\int_{-\infty}^{\infty} \Gamma(\mathbf{r}_1, \mathbf{r}_2, \tau)e^{i\omega\tau}\mathrm{d}\tau \tag{2.2}$$

be the cross-spectral density.

It was shown in Ref. 8 that, for a large class of stationary fields, the cross-spectral density is a continuous, Hermitian, nonnegative definite, Hilbert–Schmidt kernel and hence by a three-dimensional generalization of a well-known theorem it admits of a Mercer expansion[11]

$$W(\mathbf{r}_1, \mathbf{r}_2, \omega) = \sum_n \lambda_n(\omega)\psi_n^*(\mathbf{r}_1, \omega)\psi_n(\mathbf{r}_2, \omega). \tag{2.3}$$

Here the ψ_n are the eigenfunctions and the λ_n are the eigenvalues of the Fredholm integral equation

$$\int_D W(\mathbf{r}_1, \mathbf{r}_2, \omega)\psi_n(\mathbf{r}_1, \omega)\mathrm{d}^3r_1 = \lambda_n(\omega)\psi_n(\mathbf{r}_2, \omega). \tag{2.4}$$

The eigenfunctions are orthonormalized over the domain D, i.e.,

$$\int_D \psi_n^*(\mathbf{r}, \omega)\psi_m(\mathbf{r}, \omega)\mathrm{d}^3r = \delta_{nm}, \tag{2.5}$$

where δ_{nm} is the Kronecker symbol. It is known that the integral Eq. (2.4) has at least one nonvanishing eigenvalue and that the expansion (2.3) is absolutely and uniformly convergent.

In the present context the Mercer expansion (2.3) has an interesting physical meaning. It expresses the cross-spectral

0740-3232/84/050541-06$02.00

density $W(\mathbf{r}_1, \mathbf{r}_2, \omega)$ of the field as a linear combination of cross-spectral densities

$$W^{(n)}(\mathbf{r}_1, \mathbf{r}_2, \omega) = \psi_n^*(\mathbf{r}_1, \omega)\psi_n(\mathbf{r}_2, \omega) \quad (2.6)$$

that factorize with respect to the two spatial variables. Such factorized cross-spectral densities represent fields that are completely spatially coherent, because their degree of spectral coherence at frequency ω, viz.,[12]

$$\mu^{(n)}(\mathbf{r}_1, \mathbf{r}_2, \omega) = \frac{W^{(n)}(\mathbf{r}_1,\mathbf{r}_2,\omega)}{[W^{(n)}(\mathbf{r}_1, \mathbf{r}_1, \omega)]^{1/2}[W^{(n)}(\mathbf{r}_2, \mathbf{r}_2, \omega)]^{1/2}}, \quad (2.7)$$

is unimodular:

$$|\mu^{(n)}(\mathbf{r}_1, \mathbf{r}_2, \omega)| = 1. \quad (2.8)$$

Moreover, it follows at once from results of Sec. 3 of Ref. 9 that, throughout D, $W^{(n)}(\mathbf{r}_1, \mathbf{r}_2, \omega)$ satisfies the Helmholtz equations

$$\nabla_j^2 W^{(n)}(\mathbf{r}_1, \mathbf{r}_2, \omega) + k^2 W^{(n)}(\mathbf{r}_1, \mathbf{r}_2, \omega) = 0 \quad (j = 1, 2), \quad (2.9)$$

where ∇_j^2 is the Laplacian operator acting with respect to the variable $\mathbf{r}_j$ and

$$k = \omega/c, \quad (2.10)$$

c being the speed of light in vacuum. Because $W^{(n)}$ satisfies the same differential equations as W (cf. Ref. 13), it is clear that the Mercer expansion (2.3) represents the cross-spatial density of the field as a linear superposition of modes that are completely coherent in the space-frequency domain.

It was shown also in Ref. 10 that one may construct an ensemble of monochromatic wave functions $\{U(\mathbf{r}, \omega) \exp(-i\omega t)\}$, all of the same frequency ω, such that the cross-spatial density of the field is their cross-correlation function, i.e., that

$$W(\mathbf{r}_1, \mathbf{r}_2, \omega) = \langle U^*(\mathbf{r}_1, \omega)U(\mathbf{r}_2, \omega)\rangle_\omega, \quad (2.11)$$

where the angular brackets with the suffix ω denote that the average is taken over this ensemble. The $U(\mathbf{r}, \omega)$ are expressible in the form

$$U(\mathbf{r}, \omega) = \sum a_n(\omega)\psi_n(\mathbf{r}, \omega), \quad (2.12)$$

where the ψ_n are again the eigenfunctions of the integral Eq. (2.4) and the a_n are random coefficients satisfying the requirement that

$$\langle a^*_n(\omega)a_m(\omega)\rangle_\omega = \lambda_n(\omega)\delta_{nm}. \quad (2.13)$$

Moreover, the random functions $U(\mathbf{r}, \omega)$ satisfy the Helmholtz equation

$$\nabla^2 U(\mathbf{r}, \omega) + k^2 U(\mathbf{r}, \omega) = 0 \quad (2.14)$$

throughout D.

Finally we note that, if we substitute for W from Eq. (2.3) into the Fourier inverse of Eq. (2.2), we obtain the following representation of the mutual coherence function of the field:

$$\Gamma(\mathbf{r}_1, \mathbf{r}_2, \tau) = \sum_n \Gamma^{(n)}(\mathbf{r}_1, \mathbf{r}_2, \tau), \quad (2.15)$$

where

$$\Gamma^{(n)}(\mathbf{r}_1, \mathbf{r}_2, \tau) = \int_0^\infty \lambda_n(\omega)\psi_n^*(\mathbf{r}_1, \omega)\psi_n(\mathbf{r}_2, \omega)e^{-i\omega\tau}\mathrm{d}\omega. \quad (2.16)$$

The function $\Gamma^{(n)}(\mathbf{r}_1, \mathbf{r}_2, \tau)$ is the mutual coherence function associated with the mode labeled by the index n.

It should be noted that, unlike the degree of spectral coherence $\mu^{(n)}(\mathbf{r}_1, \mathbf{r}_2, \omega)$, the complex degree of coherence

$$\gamma^{(n)}(\mathbf{r}_1, \mathbf{r}_2, \tau) = \frac{\Gamma^{(n)}(\mathbf{r}_1,\mathbf{r}_2,\tau)}{[\Gamma^{(n)}(\mathbf{r}_1, \mathbf{r}_1, 0)]^{1/2}[\Gamma^{(n)}(\mathbf{r}_2, \mathbf{r}_2, 0]^{1/2}} \quad (2.17)$$

of a mode in the space-time domain is, in general, not unimodular. Hence $\Gamma^{(n)}(\mathbf{r}_1, \mathbf{r}_2, \tau)$ is a mutual coherence function of a field that is not completely coherent in the space-time domain. The distinction between complete coherence in the two domains is rather basic in the context of our subsequent analysis.

3. STEADY-STATE CONDITION FOR SECOND-ORDER COHERENCE OF LIGHT AT A MIRROR OF A LASER RESONATOR

Consider a light distribution across a mirror A. We assume that the distribution is characterized by a statistical ensemble that is stationary, at least in the wide sense. However, we do not impose any restriction on the bandwidth of the light. We denote by $W_0(\boldsymbol{\rho}_1, \boldsymbol{\rho}_2, \omega)$ the cross-spectral density of the light at two points on A, specified by position vectors $\boldsymbol{\rho}_1$ and $\boldsymbol{\rho}_2$. Suppose that some of the light that propagates from A reaches another mirror B, with the two mirrors forming an open, empty, resonant cavity. Let $W_1(\boldsymbol{\rho}_1, \boldsymbol{\rho}_2, \omega)$ be the cross-spectral density of the light returning to A. After reflection at A the process of propagation and reflection continues. We denote by $W_j(\boldsymbol{\rho}_1, \boldsymbol{\rho}_2, \omega)$ the cross-spectral density of the light across the mirror A after completion of j complete cycles $A \to B \to A$.

According to the two-dimensional version of Eq. (2.11), the cross-spectral density W_j may be expressed in the form

$$W_j(\boldsymbol{\rho}_1, \boldsymbol{\rho}_2, \omega) = \langle U_j^*(\boldsymbol{\rho}_1, \omega)U_j(\boldsymbol{\rho}_2, \omega)\rangle_\omega, \quad (3.1)$$

where the average is taken over an appropriate statistical ensemble of random functions U_j. Since each $U_j(\boldsymbol{\rho}, \omega)$ is a boundary value of a field that in the space between the two mirrors satisfies the Helmholtz equation (2.14), U_{j+1} and U_j will be related by a linear transformation, viz.,

$$U_{j+1}(\boldsymbol{\rho}, \omega) = \int_A L(\boldsymbol{\rho}, \boldsymbol{\rho}', \omega)U_j(\boldsymbol{\rho}', \omega)\mathrm{d}^2\rho' \quad (3.2)$$

($j = 0, 1, 2, \ldots$). Here $L(\boldsymbol{\rho}, \boldsymbol{\rho}', \omega)$ is the propagator (independent of j) for propagation of monochromatic light of frequency ω from the point $\boldsymbol{\rho}'$ on the mirror A to the point $\boldsymbol{\rho}$ on the same mirror, after a single reflection at the mirror B. Explicit approximate expressions for this propagator may be obtained with the help of the Huygens–Fresnel principle.[14] On substituting from Eq. (3.2) into Eq. (3.1), we obtain the following relation between the cross-spectral densities W_j and W_{j+1}:

$$W_{j+1}(\boldsymbol{\rho}_1, \boldsymbol{\rho}_2, \omega) = \int_A \int_A L^*(\boldsymbol{\rho}_1, \boldsymbol{\rho}_1', \omega) L(\boldsymbol{\rho}_2, \boldsymbol{\rho}_2', \omega) \times W_j(\boldsymbol{\rho}_1', \boldsymbol{\rho}_2', \omega) \mathrm{d}^2\rho_1' \mathrm{d}^2\rho_2'. \quad (3.3)$$

It seems reasonable to assume that, after a sufficient number of transits between the two mirrors, a steady state is reached in the sense that, for each frequency ω, $W_{j+1}(\boldsymbol{\rho}_1, \boldsymbol{\rho}_2, \omega)$ will be equal to $W_j(\boldsymbol{\rho}_1, \boldsymbol{\rho}_2, \omega)$ up to a proportionality factor, $\sigma(\omega)$, say. The factor $\sigma(\omega)$ represents the loss from diffraction and reflection that occurs in one complete transit. More explicitly, we characterize steady state by the requirement that, for sufficiently large values of j,

$$W_{j+1}(\boldsymbol{\rho}_1, \boldsymbol{\rho}_2, \omega) = \sigma(\omega) W_j(\boldsymbol{\rho}_1, \boldsymbol{\rho}_2, \omega). \quad (3.4)$$

Since the "diagonal values" $W_{j+1}(\boldsymbol{\rho}, \boldsymbol{\rho}, \omega)$ and $W_j(\boldsymbol{\rho}, \boldsymbol{\rho}, \omega)$ represent spectral densities, they are necessarily real and positive, and hence

$$\sigma(\omega) > 0. \quad (3.5)$$

On substituting for W_{j+1} from Eq. (3.4) into Eq. (3.3) and on suppressing the suffix j, we obtain the equation

$$\int_A \int_A W(\boldsymbol{\rho}_1', \boldsymbol{\rho}_2', \omega) L^*(\boldsymbol{\rho}_1, \boldsymbol{\rho}_1', \omega) L(\boldsymbol{\rho}_2, \boldsymbol{\rho}_2', \omega) \mathrm{d}^2\rho_1' \mathrm{d}^2\rho_2' = \sigma(\omega) W(\boldsymbol{\rho}_1, \boldsymbol{\rho}_2, \omega). \quad (3.6)$$

The integral Eq. (3.6) is the basic equation of the present theory. *We identify its solutions* $W(\boldsymbol{\rho}_1, \boldsymbol{\rho}_2, \omega)$ *as the boundary values of the cross-spectral densities on the mirror A of the modes that the cavity can sustain.* It should be noted that, unlike the solutions of the usual integral equation for modes of a laser cavity, the solutions of our integral Eq. (3.6) contain information about their second-order coherence properties.

4. NATURE OF THE SOLUTIONS OF THE INTEGRAL EQUATION (3.6)

As is well known, the propagator $L(\boldsymbol{\rho}_1, \boldsymbol{\rho}_1', \omega)$ that enters our integral Eq. (3.6) is not Hermitian, a fact that already presents a number of difficulties in the conventional (monochromatic) theory of laser modes. In particular, the kernel L cannot be represented in the form of the orthogonal Mercer expansion that is so useful in connection with Fredholm integral equations with Hilbert–Schmidt kernels. There is, however, a biorthogonal generalization of the Mercer expansion, which applies to a class of non-Hermitian kernels and which, to our knowledge, has not been employed previously in the theory of laser modes. We will see that the use of this biorthogonal expansion elucidates the nature of the solutions of our integral Eq. (3.6) and of their relationship to the Fox–Li modes of the conventional theory.

The biorthogonal expansion of a non-Hermitian kernel $L(\boldsymbol{\rho}, \boldsymbol{\rho}', \omega)$ may be introduced in the following manner.[15] Let $\{\alpha_n(\omega)\}$ and $\{\phi_n(\boldsymbol{\rho}, \omega)\}$ be the eigenvalues and the eigenfunctions, respectively, of the Fredholm integral equation

$$\int_A L(\boldsymbol{\rho}_1, \boldsymbol{\rho}_2, \omega) \phi_n(\boldsymbol{\rho}_2, \omega) \mathrm{d}^2\rho_2 = \alpha_n(\omega) \phi_n(\boldsymbol{\rho}_1, \omega), \quad (4.1)$$

and let $\{\beta_n(\omega)\}$ and $\{\chi_n(\boldsymbol{\rho}, \omega)\}$ be the eigenvalues and the eigenfunctions, respectively, of the corresponding integral equation with the adjoint kernel $L^*(\boldsymbol{\rho}_2, \boldsymbol{\rho}_1, \omega)$:

$$\int_A L^*(\boldsymbol{\rho}_2, \boldsymbol{\rho}_1, \omega) \chi_n(\boldsymbol{\rho}_2, \omega) \mathrm{d}^2\rho_2 = \beta_n(\omega) \chi_n(\boldsymbol{\rho}_1, \omega). \quad (4.2)$$

The kernel $L(\boldsymbol{\rho}_1, \boldsymbol{\rho}_2, \omega)$ for laser resonators is defined on a finite domain A and is, for each ω, a continuous function in both the spatial variables $\boldsymbol{\rho}_1$ and $\boldsymbol{\rho}_2$. Such kernels belong to a class of square-integrable kernels for which the following theorems are known to hold[16]:

(1) To each eigenvalue α_n of Eq. (4.1), there corresponds an eigenvalue β_n of Eq. (4.2) and

$$\beta_n = \alpha_n^*. \quad (4.3)$$

Moreover, the ranks (degrees of degeneracy) of α_n and of β_n are the same.

(2) The corresponding eigenfunctions of the two equations are (with suitable normalization) orthonormal over the domain A, i.e.,

$$\int_A \phi_n^*(\boldsymbol{\rho}, \omega) \chi_m(\boldsymbol{\rho}, \omega) \mathrm{d}^2\rho = \delta_{nm}. \quad (4.4)$$

The biorthogonal expansion of the kernel $L(\boldsymbol{\rho}_1, \boldsymbol{\rho}_2, \omega)$ that we employ to study the nature of the solution of the integral Eq. (3.6) is the expansion[15]

$$L(\boldsymbol{\rho}_1, \boldsymbol{\rho}_2, \omega) = \sum_n \alpha_n(\omega) \phi_n(\boldsymbol{\rho}_1, \omega) \chi_n^*(\boldsymbol{\rho}_2, \omega). \quad (4.5)$$

This expansion should not be confused with the Schmidt expansion (which is biorthogonal in a different sense) of non-Hermitian kernels that was used previously in the theory of laser resonator modes.[17–19]

If we substitute from Eq. (4.5) into Eq. (3.6) and interchange the orders of integrations and summations, we obtain the relation

$$\sum_n \sum_m \alpha_n^*(\omega) \alpha_m(\omega) w_{nm}(\omega) \phi_n^*(\boldsymbol{\rho}_1, \omega) \phi_m(\boldsymbol{\rho}_2, \omega) = \sigma(\omega) W(\boldsymbol{\rho}_1, \boldsymbol{\rho}_2, \omega), \quad (4.6)$$

where

$$w_{nm} = \int_A \int_A W(\boldsymbol{\rho}_1', \boldsymbol{\rho}_2', \omega) \chi_n(\boldsymbol{\rho}_1', \omega) \times \chi_m^*(\boldsymbol{\rho}_2', \omega) \mathrm{d}^2\rho_1' \mathrm{d}^2\rho_2'. \quad (4.7)$$

Next let us multiply both sides of Eq. (4.6) by $\chi_N(\boldsymbol{\rho}_1, \omega) \chi_M^*(\boldsymbol{\rho}_2, \omega)$, integrate with respect to $\boldsymbol{\rho}_1$ and $\boldsymbol{\rho}_2$, and make use of biorthogonality relations (4.4) and of expression (4.7) that defines the w_{nm}. We then obtain the relation

$$\sum_n \sum_m \alpha_n^*(\omega) \alpha_m(\omega) w_{nm}(\omega) \delta_{nN} \delta_{mM} = \sigma(\omega) w_{NM}(\omega), \quad (4.8)$$

which implies that

$$[\sigma(\omega) - \alpha_N^*(\omega) \alpha_M(\omega)] w_{NM} = 0 \quad \text{(no summation)}. \quad (4.9)$$

We see from Eq. (4.9) that either $w_{NM}(\omega) = 0$ or $\sigma(\omega) = \alpha_N^*(\omega) \alpha_M(\omega)$. The first case ($w_{NM} = 0$) is of no interest since the corresponding term (with $n = N$, $m = M$) does not contribute to the double sum on the left-hand side of Eq. (4.6).

The other case implies that the eigenvalues of the integral Eq. (3.6) are

$$\sigma_{NM}(\omega) = \alpha_N^*(\omega)\alpha_M(\omega). \tag{4.10}$$

Let us assume, for the moment, that σ_{NM} is nondegenerate in the sense that there are no other pairs $\alpha_{N'}(\omega)$, $\alpha_{M'}(\omega)$ of eigenvalues of integral Eq. (4.1) for which

$$\alpha_N^*(\omega)\alpha_M(\omega) = \alpha_{N'}^*(\omega)\alpha_{M'}(\omega). \tag{4.11}$$

Formula (4.9) implies that with a particular choice

$$\sigma_{kl}(\omega) = \alpha_k^*(\omega)\alpha_l(\omega) \tag{4.12}$$

of a nondegenerate eigenvalue of our integral Eq. (3.6), $w_{NM} = 0$ unless $k = N$, $l = M$, and the expansion (4.6) then reduces the single term

$$W(\boldsymbol{\rho}_1, \boldsymbol{\rho}_2, \omega) = w_{kl}(\omega)\phi_k^*(\boldsymbol{\rho}_1)\phi_l(\boldsymbol{\rho}_2). \tag{4.13}$$

Now the cross-spectral density function $W(\boldsymbol{\rho}_1, \boldsymbol{\rho}_2, \omega)$ is necessarily Hermitian, i.e.,

$$W(\boldsymbol{\rho}_2, \boldsymbol{\rho}_1, \omega) = W^*(\boldsymbol{\rho}_1, \boldsymbol{\rho}_2, \omega), \tag{4.14}$$

as is readily deduced from Eqs. (2.1) and (2.2). Applied to expression (4.13), this condition implies that

$$w_{kl}\phi_k^*(\boldsymbol{\rho}_2, \omega)\phi_l(\boldsymbol{\rho}_1, \omega) = w_{kl}^*\phi_k(\boldsymbol{\rho}_1, \omega)\phi_l^*(\boldsymbol{\rho}_2, \omega), \tag{4.15}$$

i.e.,

$$\frac{\phi_l(\boldsymbol{\rho}_1,\omega)}{\phi_k(\boldsymbol{\rho}_1, \omega)} = \frac{w_{kl}^*}{w_{kl}} \frac{\phi_l^*(\boldsymbol{\rho}_2,\omega)}{\phi_k^*(\boldsymbol{\rho}_2, \omega)}\,. \tag{4.16}$$

With respect to the dependence on the two spatial variables, the left-hand side of Eq. (4.16) is a function of $\boldsymbol{\rho}_1$ only, whereas the right-hand side is a function of $\boldsymbol{\rho}_2$ only. This is possible only if each side is independent of the spatial variables. If we denote each side by $\gamma_{kl}(\omega)$, it then follows that

$$\phi_l(\boldsymbol{\rho}, \omega) = \gamma_{kl}(\omega)\phi_k(\boldsymbol{\rho}, \omega). \tag{4.17}$$

On substituting from Eq. (4.17) into Eq. (4.13), we obtain at once the following expression for W:

$$W(\boldsymbol{\rho}_1, \boldsymbol{\rho}_2, \omega) = w_{kl}(\omega)\gamma_{kl}(\omega)\phi_k^*(\boldsymbol{\rho}_1, \omega)\phi_k(\boldsymbol{\rho}_2, \omega). \tag{4.18}$$

If we substitute from Eq. (4.18) into Eq. (4.7), we obtain the formula

$$w_{nm} = w_{kl}\gamma_{kl} \int_A \phi_k^*(\boldsymbol{\rho}_1', \omega)\chi_n(\boldsymbol{\rho}_1', \omega)\mathrm{d}^2\rho_1' \times \int_A \phi_k(\boldsymbol{\rho}_2', \omega)\chi_m^*(\boldsymbol{\rho}_2'\omega)\mathrm{d}^2\rho_2'. \tag{4.19}$$

If we make again use of the biorthomality relations (4.4), Eq. (4.19) reduces to the formula

$$w_{nm}(\omega) = w_{kl}(\omega)\gamma_{kl}\delta_{kn}\delta_{km}, \tag{4.20}$$

which implies that

$$w_{nm}(\omega) = 0 \quad \text{unless} \quad n = m = k \tag{4.21}$$

and

$$w_{kk}(\omega) = w_{kl}(\omega)\gamma_{kl}(\omega). \tag{4.22}$$

On substituting from Eq. (4.22) into Eq. (4.18), we see that the admissable solutions [which we now denote by[20] $W_k(\boldsymbol{\rho}_1, \boldsymbol{\rho}_2, \omega)$] of our integral Eq. (3.6) are given by

$$W_k(\boldsymbol{\rho}_1, \boldsymbol{\rho}_2, \omega) = w_{kk}(\omega)\phi_k^*(\boldsymbol{\rho}_1, \omega)\phi_k(\boldsymbol{\rho}_2, \omega). \tag{4.23}$$

It is clear from Eq. (4.12) that the corresponding eigenvalues [which we denote by $\sigma_k(\omega)$] are

$$\sigma_k(\omega) = \alpha_k^*(\omega)\alpha_k(\omega). \tag{4.24}$$

The factor $w_{kk}(\omega)$ in Eq. (4.23) depends on normalization. Let us normalize the ϕ_k so that

$$\int_A |\phi_k(\boldsymbol{\rho}, \omega)|^2\mathrm{d}^2\rho = 1. \tag{4.25}$$

Then the right-hand side of Eq. (4.23) can be identified with the two-dimensional version of the Mercer expansion (2.3) of W_k, which now consists of a single term; and one evidently has

$$\phi_k(\boldsymbol{\rho}, \omega) \equiv \psi_k(\boldsymbol{\rho}, \omega), \tag{4.26}$$

$$w_{kk}(\omega) = \lambda_k(\omega). \tag{4.27}$$

Thus each solution of our integral equation (3.6) is also a mode in the sense of the general theory of coherent-mode representation of fields of any state of coherence, discussed in Section 2. We note that Eqs. (4.23), (4.25) and (4.27) imply that

$$\int_A W_k(\boldsymbol{\rho}, \boldsymbol{\rho}, \omega)\mathrm{d}^2\rho = \lambda_k(\omega). \tag{4.28}$$

Since $W_k(\boldsymbol{\rho}, \boldsymbol{\rho}, \omega)$ represents the spectral density at frequency ω at the point $\boldsymbol{\rho}$, Eq. (4.28) implies, roughly speaking, that $\lambda_k(\omega)$ is a measure of the rate at which energy at frequency ω is propagated, in steady state, from the mirror A into the cavity.

The integral equation (4.1) for the functions $\phi_n(\boldsymbol{\rho}, \omega)$ is precisely the integral equation of the elementary (monochromatic) theory of laser resonators modes formulated by Fox and Li.[1,14] We therefore refer to the functions of $\phi_n(\boldsymbol{\rho}, \omega)$ as the Fox–Li modes. The preceding analysis shows that the Fox–Li modes have a broader significance than would appear from the manner in which they were originally introduced.

Since each solution

$$W_k(\boldsymbol{\rho}_1, \boldsymbol{\rho}_2, \omega) = \lambda_k(\omega)\phi_k^*(\boldsymbol{\rho}_1, \omega)\phi_k(\boldsymbol{\rho}_2, \omega) \tag{4.29}$$

of our integral Eq. (3.6) factorizes with respect to the spatial variables $\boldsymbol{\rho}_1$ and $\boldsymbol{\rho}_2$, its degree of spectral coherence at frequency ω is unimodular [cf. Eqs. (2.6)–(2.8)]. Hence the solution (4.29) is the cross-spectral density of a field distribution that is spatially completely coherent at frequency ω over the surface of the mirror A. Stated somewhat differently, *if there is no degeneracy, our integral Eq. (3.6) only admits of solutions that represent light that, at each frequency, is spatially completely coherent within the framework of second-order correlation theory.*

The factorization of the cross-spectral density into a product of a function of $\boldsymbol{\rho}_1$ and a function of $\boldsymbol{\rho}_2$ is known to be both a necessary and a sufficient condition for complete (second-order) spatial coherence.[21] Hence, if the laser operates on more than one transverse mode, the output cannot then be spatially fully coherent across the mirror surface. These conclusions are in agreement with results of experiments.[6]

We have assumed so far that there is no degeneracy in the sense that there are no four eigenvalues of Eq. (4.1) that obey the relation (4.11). If there is a degeneracy, we would find in place of Eq. (4.13) an expression of the form

$$W(\boldsymbol{\rho}_1, \boldsymbol{\rho}_2, \omega) = \sum_k \sum_l c_{kl}(\omega)\phi_k{}^*(\boldsymbol{\rho}_1, \omega)\phi_l(\boldsymbol{\rho}_2, \omega), \quad (4.30)$$

where the c_{kl} are arbitrary coefficients. These coefficients must, however, satisfy the constraint $c_{lk}(\omega) = c_{kl}{}^*(\omega)$, because the cross-spectral density is necessarily Hermitian [Eq. (4.14)]. If we diagonalize the matrix $[c_{kl}(\omega)]$, $c = U^\dagger \Lambda u$, and normalize the $\phi_k(\boldsymbol{\rho}, \omega)$ in accordance with Eq. (4.25), the degenerate solution (4.30) takes the form of its Mercer expansion

$$W(\boldsymbol{\rho}_1, \boldsymbol{\rho}_2, \omega) = \sum_k \Lambda_k(\omega)\, f_k{}^*(\boldsymbol{\rho}_1, \omega) f_k(\boldsymbol{\rho}_2, \omega), \quad (4.31)$$

where

$$f_k(\boldsymbol{\rho}_1, \omega) = \sum_l u_{kl}(\omega)\phi_l(\boldsymbol{\rho}, \omega). \quad (4.32)$$

In such a case, the field across the mirror A is no longer spatially completely coherent.

In our steady-state condition (3.6), the frequency ω is arbitrary. It seems reasonable to assume that, when a laser operates in a steady state, the condition (3.6) will hold for every frequency component that is present in the spectrum of the laser light. It then follows on substituting from Eq. (4.29) into the Fourier inverse of Eq. (2.2) (where $\mathbf{r}_1$ and $\mathbf{r}_2$ are, of course, now replaced by $\boldsymbol{\rho}_1$ and $\boldsymbol{\rho}_2$, respectively) that the mutual-coherence function of a laser resonator mode is given by

$$\Gamma_k(\boldsymbol{\rho}_1, \boldsymbol{\rho}_2, \tau) = \int_0^\infty \lambda_k(\omega)\phi_k{}^*(\boldsymbol{\rho}_1, \omega)\phi_k(\boldsymbol{\rho}_2, \omega)e^{-i\omega\tau}d\omega, \quad (4.33)$$

if there is no degeneracy. If there is a degeneracy, one must use expression (4.31) in place of Eqs. (4.23) and (4.27), and one then obtains for the mutual coherence function of a degenerate mode the formula

$$\Gamma(\boldsymbol{\rho}_1, \boldsymbol{\rho}_2, \tau) = \sum_k \int_0^\infty \Lambda_k(\omega) f^*{}_k(\boldsymbol{\rho}_1, \omega) f_k(\boldsymbol{\rho}_2, \omega)e^{-i\omega\tau}d\omega. \quad (4.34)$$

The spectrum of a laser mode at a point $\boldsymbol{\rho}$ of the mirror A is equal to

$$S_k(\boldsymbol{\rho}, \omega) \equiv W_k(\boldsymbol{\rho}, \boldsymbol{\rho}, \omega) = \lambda_k(\omega)|\phi_k(\boldsymbol{\rho}, \omega)|^2 \quad (4.35)$$

if there is no degeneracy and by

$$S(\boldsymbol{\rho}, \omega) = \sum_k \lambda_k(\omega)|f_k(\boldsymbol{\rho}, \omega)|^2 \quad (4.36)$$

if there is a degeneracy.

5. SUMMARY

In this paper we have introduced transverse modes of an empty open laser cavity as solutions of an integral equation [Eq. (3.6)] that expresses a steady-state condition for the cross-spectral density of the light over a mirror of the cavity. Unlike the usual integral equation for laser resonator modes, the present theory takes into account the frequency spectrum of laser light. We have found that the mutual coherence function of each mode is expressible as a Fourier transform of an expression that involves, in a simple way, the Fox–Li modes of the usual theory [Eqs. (4.33) and (4.34)]. We have also found that if there is no degeneracy, each mode is, within the framework of second-order correlation theory, completely spatially coherent over the mirror surface, at each frequency, and that if several modes are excited, the light is then necessarily partially coherent. We showed further that laser resonator modes are also modes in the sense of a general mode representation of fields of any state of coherence.[8–10]

ACKNOWLEDGMENT

This paper was written when E. Wolf was on leave at Schlumberger-Doll Research in Ridgefield, Connecticut. The hospitality and support received from the management of that laboratory are gratefully acknowledged.

E. Wolf is also with the Institute of Optics, University of Rochester.

REFERENCES

1. A. G. Fox and T. Li, "Resonant modes in a maser interferometer," Bell Syst. Tech. J. **40,** 453–488 (1961).
2. E. Wolf, "Spatial coherence of resonant modes in a maser interferometer," Phys. Lett. **3,** 166–168 (1963).
3. W. Streifer, "Spatial coherence in periodic systems," J. Opt. Soc. Am., **56,** 1481–1489 (1966).
4. L. Allen, S. Gatehouse, and D. G. C. Jones, "Enhancement of optical coherence during light propagation in bounded media," Opt. Commun. **4,** 169–171 (1971).
5. F. Gori, "Propagation of the mutual intensity through a periodic structure," Atti Fond. Giorgio Ronchi **35,** 434–447 (1980).
6. M. Bertolotti, B. Daino, F. Gori, and D. Sette, "Coherence properties of a laser beam," Nuovo Cimento **38,** 1505–1514 (1965).
7. For definition of the analytic signal and for explanation of the concepts of coherence theory used in this paper, see M. Born and E. Wolf, *Principles of Optics*, 6th ed. (Pergamon, Oxford, 1980), Chap. X, or L. Mandel and E. Wolf, "Coherence properties of optical fields," Rev. Mod. Phys. **37,** 231–287 (1965).
8. E. Wolf, "New theory of partial coherence in the space-frequency domain. Part I: spectra and cross spectra of steady-state sources," J. Opt. Soc. Am. **72,** 343–351 (1982). Although this paper deals with stationary primary sources, strictly similar results hold also for stationary secondary sources and for stationary fields. In this connection, see Refs. 9 and 10.
9. E. Wolf, "A new description of second-order coherence phenomena in the space-frequency domain," in *Optics in Four Dimensions—1980*, M. A. Machado and L. M. Narducci, eds., AIP Conf. Proc. **65,** 42–48 (1981).
10. E. Wolf, "New spectral representation of random sources and the partially coherent fields that they generate," Opt. Commun. **38,** 3–6 (1981).
11. Because we assumed the domain D to be three-dimensional, the index n stands for an ordered triplet of integers (n_1, n_2, n_3). We will later need the analogs of some of the formulas for fields in two-dimensional domains. In general, if the domain containing the field is N dimensional, the index n will stand for N ordered integers $(n_1, n_2, \ldots n_N)$, and the integrals in Eqs. (2.4) and (2.5) will, of course, extend over the N-dimensional domain.
12. L. Mandel and E. Wolf, "Spectral coherence and the concept of cross-spectral purity," J. Opt. Soc. Am. **66,** 529–535 (1976), Sec. II.
13. E. Wolf, "Coherence and radiometry," J. Opt. Soc. Am. **68,** 6–17 (1978).

14. G. D. Boyd and H. Kogelnik, "Generalized confocal resonator theory," Bell Syst. Tech. J. **44,** 1347–1369 (1962).
15. P. M. Morse and H. Feshbach, *Methods of Mathematical Physics* (McGraw-Hill, New York (1953)), Part I, pp. 919–920; see also pp. 884–886. Unfortunately neither in this reference nor in any of the well-known texts on integral equations conditions are given for the validity of the biorthogonal expansion [Eq. (4.5) below]. However, $L(\boldsymbol{\rho}_1, \boldsymbol{\rho}_2, \omega)$ belongs to a class of well-behaved kernels to which the important relations (4.3) and (4.4) apply. Moreover, it is known that, for laser cavities of the usual geometries, $L(\boldsymbol{\rho}_1, \boldsymbol{\rho}_2, \omega)$ has a complete set of eigenfunctions in the Hilbert space of square-integrable functions. These facts suggest that for such cavities $L(\boldsymbol{\rho}_1, \boldsymbol{\rho}_2, \omega)$ admits of such an expansion.
16. For proofs of one-dimensional versions of these results, see F. Smithies, *Integral Equations* (Cambridge U. Press, Cambridge, 1970), theorems 6.7.3 and 6.7.4.
17. W. Streifer and H. Gamo, "On the Schmidt expansion for optical resonator modes," in *Proceedings of the Symposium on Quasi-Optics* (Polytechnic, Brooklyn, N.Y., 1964), pp. 351–365.
18. W. Streifer, "Optical resonator modes—rectangular reflectors of spherical curvature," J. Opt. Soc. Am. **55,** 868–877 (1965).
19. J. C. Heurtley and W. Streifer, "Optical resonator modes—circular reflectors of spherical curvature," J. Opt. Soc. Am. **55,** 1472–1479 (1965).
20. It seems worthwhile to stress that the subscript that labels W has here a different meaning from that in Section 3. It now distinguishes the different eigenfunctions of Eq. (3.6), whereas in Section 3 it indicated the number of complete cycles of propagation of light between the two mirrors.
21. L. Mandel and E. Wolf, "Complete coherence in the space-frequency domain," Opt. Commun. **36,** 247–249 (1981).

VOLUME 56, NUMBER 13 PHYSICAL REVIEW LETTERS 31 MARCH 1986

Invariance of the Spectrum of Light on Propagation

Emil Wolf[(a)]
Department of Physics and Astronomy, University of Rochester, Rochester, New York 14627
(Received 27 January 1986)

The question is raised as to whether the normalized spectrum of light remains unchanged on propagation through free space. It is shown that for sources of a certain class that includes the usual thermal sources, the normalized spectrum will, in general, depend on the location of the observation point unless the degree of spectral coherence of the light across the source obeys a certain scaling law. Possible implications of the analysis for astrophysics are mentioned.

PACS numbers: 42.10.Mg, 07.65.−b, 42.68.Hf

Measurements of the spectrum of light are generally made some distance away from its sources and in many cases, as for example in astronomy, they are made exceedingly far away. It is taken for granted that the normalized spectral distribution of the light incident on a detector after propagation from the source through free space is the same as that of the light in the source region. I will refer to this assumption as the assumption of *invariance of the spectrum on propagation.* This assumption, which is implicit in all of spectroscopy, does not appear to have been previously questioned, probably because with light from traditional sources one has never encountered any problems with it. However, with the gradual development of rather unconventional light sources and with the relatively frequent discoveries of stellar objects of an unfamiliar kind, it is obviously desirable to understand whether all such sources generate light whose spectrum is invariant on propagation, and if so, what the reasons for it are. Actually it is not difficult to conceive of sources that generate light whose spectrum is not invariant on propagation. In this note I will show what are the characteristics of a certain class of sources that generate light whose spectrum is invariant, at least in the far zone.

From the standpoint of optical coherence theory, invariance of the spectrum of light on propagation from conventional sources is a rather remarkable fact, as can be seen from the following simple argument. Consider an optical field generated by a stationary source in free space. The basic field variable, say the electric field strength at the space-time point $(\mathbf{r},t)$, may be represented by its complex analytic signal[1,2] $E(\mathbf{r},t)$. According to the Wiener-Khintchine theorem[3] the spectral density of the light at the point $\mathbf{r}$ is then represented by the Fourier transform,

$$S(\mathbf{r},\omega)=\int_{-\infty}^{\infty}\Gamma(\mathbf{r},\tau)e^{i\omega\tau}\,d\tau, \qquad (1)$$

of the autocorrelation function (known in the optical context as the self-coherence function) of the field variable. It is defined as

$$\Gamma(\mathbf{r},\tau)=\langle E^*(\mathbf{r},t)E(\mathbf{r},t+\tau)\rangle, \qquad (2)$$

where the angular brackets denote the ensemble average. Now the spectral density and the self-coherence function are the "diagonal elements" $(\mathbf{r}_2=\mathbf{r}_1=\mathbf{r})$ of two basic optical correlation functions, viz., the cross-spectral density

$$W(\mathbf{r}_1,\mathbf{r}_2,\omega)=\int_{-\infty}^{\infty}\Gamma(\mathbf{r}_1,\mathbf{r}_2,\tau)e^{i\omega\tau}\,d\tau, \qquad (3)$$

and the mutual coherence function

$$\Gamma(\mathbf{r}_1,\mathbf{r}_2,\tau)=\langle E^*(\mathbf{r}_1,t)E(\mathbf{r}_2,t+\tau)\rangle. \qquad (4)$$

It is well known that both the mutual coherence function and the cross-spectral density obey precise propagation laws. For example, in free space[4]

$$(\nabla_j^2+k^2)W(\mathbf{r}_1,\mathbf{r}_2,\omega)=0 \quad (j=1,2), \qquad (5)$$

where

$$k=\omega/c, \qquad (6)$$

with c being the speed of light *in vacuo* and ∇_j^2 being the Laplacian operator acting with respect to the variable $\mathbf{r}_j$. Consequently, both the mutual coherence function and the cross-spectral density and, in fact, also their normalized values change appreciably on propagation. For example, for a spatially incoherent planar source $W(\mathbf{r}_1,\mathbf{r}_2,\omega)$ and $\Gamma(\mathbf{r}_1,\mathbf{r}_2,\tau)$ will be essentially δ correlated with respect to $\mathbf{r}_1$ and $\mathbf{r}_2$ at the source plane but will have nonzero values for widely separated pairs of points which are sufficiently far away from the source. This is the essence of the well known van Cittert–Zernike theorem (Ref. 1, Sect. 10.4.2). In physical terms, the correlation in the field generated by a spatially incoherent source may be shown to have its origin in the process of superposition. We thus have the following rather strange situation: The correlations of the light may change drastically on propagation; yet, under commonly occurring circumstances, their (suitably normalized) diagonal elements, which represent the spectrum of the light or its Fourier transform, remain unchanged.

To obtain some insight into this problem we consider light generated by a very simple model source; namely, a planar source occupying a finite domain D of a plane $z=0$ and radiating into the half space $z>0$, which has the same spectral distribution $S^{(0)}(\omega)$ at each source point $P(\boldsymbol{\rho})$ and whose degree of spectral coherence[5] $\mu^{(0)}(\boldsymbol{\rho}_1,\boldsymbol{\rho}_2,\omega)$ is statistically homogeneous, i.e., has the functional form $\mu^{(0)}(\boldsymbol{\rho}_2-\boldsymbol{\rho}_1,\omega)$. The cross-spectral density of the light across the source plane is then given by

$$W^{(0)}(\boldsymbol{\rho}_1,\boldsymbol{\rho}_2,\omega)=\epsilon(\boldsymbol{\rho}_1)\epsilon(\boldsymbol{\rho}_2)S^{(0)}(\omega)\mu^{(0)}(\boldsymbol{\rho}_2-\boldsymbol{\rho}_1,\omega), \qquad (7)$$

where $\epsilon(\boldsymbol{\rho})=1$ or 0 according to whether the point $P(\boldsymbol{\rho})$ is located within or outside the source area D in the plane $z=0$.

We will also assume that at each effective frequency ω present in the source spectrum, the linear dimensions of the source are much larger than the spectral correlation length [the effective width Δ of $|\mu^{(0)}(\boldsymbol{\rho}',\omega)|$]. Sources of this kind belong to the class of so-called *quasihomogeneous sources*,[6] which have been extensively studied in coherence theory in recent years. Most of the usual thermal sources are of this kind.

The radiant intensity $J_\omega(\mathbf{u})$, i.e., the rate at which energy is radiated at frequency ω per unit solid angle around a direction specified by a unit vector $\mathbf{u}$, is given by the expression [cf. Ref. 6, Eq. (4.8)]

$$J_\omega(\mathbf{u})=k^2AS^{(0)}(\omega)\tilde{\mu}^{(0)}(k\mathbf{u}_\perp,\omega)\cos^2\theta. \qquad (8)$$

In this formula, A is the area of the source,

$$\tilde{\mu}^{(0)}(\mathbf{f},\omega)=\frac{1}{(2\pi)^2}\int\mu^{(0)}(\rho',\omega)e^{-i\mathbf{f}\cdot\boldsymbol{\rho}'}d^2\rho' \qquad (9)$$

is the two-dimensional spatial Fourier transform of the degree of spectral coherence, $\mathbf{u}_\perp$ is the transverse part of the unit vector $\mathbf{u}$, i.e., the component of $\mathbf{u}$ (considered as a two-dimensional vector) perpendicular to the z axis, and θ is the angle between the $\mathbf{u}$ and the z directions (see Fig. 1). Evidently the normalized spectral density $S^{(\infty)}(\mathbf{u},\omega)$ at a point in the far zone, in the direction specified by the unit vector $\mathbf{u}$, is given by

$$S^{(\infty)}(\mathbf{u},\omega)=J_\omega(u)\Big/\int J_\omega(\mathbf{u})d\omega. \qquad (10)$$

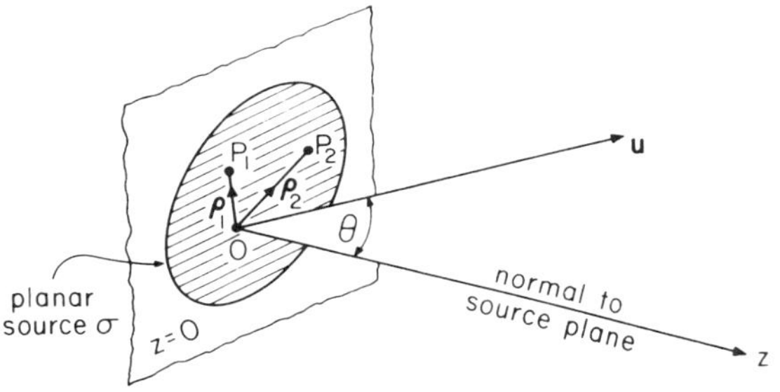

FIG. 1. Illustration of the notation.

On substituting Eq. (8) into Eq. (10) we obtain for the normalized spectrum in the far zone the expression

$$S^{(\infty)}(\mathbf{u},\omega)=\frac{k^2S^{(0)}(\omega)\tilde{\mu}^{(0)}(k\mathbf{u}_\perp,\omega)}{\int k^2S^{(0)}(\omega)\tilde{\mu}^{(0)}(k\mathbf{u}_\perp,\omega)d\omega}. \qquad (11)$$

It is clear from Eq. (11) that the normalized spectrum of the light depends on the direction $\mathbf{u}$; i.e., it is in general not invariant throughout the far zone. However, it is seen at once from Eq. (11) that it will be invariant throughout the far zone if the Fourier transform of the degree of spectral coherence of the light in the source plane is the product of a function of frequency and a function of direction, i.e., it is of the form

$$\tilde{\mu}^{(0)}(k\mathbf{u}_\perp,\omega)=F(\omega)\tilde{H}(\mathbf{u}_\perp). \qquad (12)$$

In this case Eq. (11) reduces to

$$S^{(\infty)}(\mathbf{u},\omega)=\frac{k^2S^{(0)}(\omega)F(\omega)}{\int k^2S^{(0)}(\omega)F(\omega)d\omega}, \qquad (13)$$

and the expression on the right is independent of the direction $\mathbf{u}$.

I will now show that the condition (12) has some interesting implications, which follow from the fact that $\mu^{(0)}$ is a correlation coefficient. Before doing this we note that since $\mathbf{u}$ is a unit vector, $|\mathbf{u}_\perp|<1$. However, we will now assume that the factorization condition (12) holds for all two-dimensional vectors $\mathbf{u}_\perp$ ($0\leq|\mathbf{u}_\perp|<\infty$). This assumption will be trivially satisfied if the degree of spectral coherence $\mu^{(0)}(\boldsymbol{\rho}',\omega)$ is, at each effective temporal frequency ω, band limited in the spatial frequency plane to a circle of radius k about the origin; in more physical terms this condition means that $\mu^{(0)}(\boldsymbol{\rho}',\omega)$ does not vary appreciably over distances of the order of the wavelength $\lambda=2\pi c/\omega$. With this being understood let us take the Fourier transform of Eq. (12). We then find at once that

$$\mu^{(0)}(\boldsymbol{\rho}',\omega)=F(\omega)\int\tilde{H}(\mathbf{u}_\perp)\exp(ik\mathbf{u}_\perp\cdot\boldsymbol{\rho}')\,d^2(k\mathbf{u}_\perp), \qquad (14)$$

i.e.,

$$\mu^{(0)}(\boldsymbol{\rho}',\omega)=k^2F(\omega)H(k\rho'), \qquad (15)$$

where H is, of course, the two-dimensional Fourier transform of $\tilde{H}$. Since $\mu^{(0)}(\boldsymbol{\rho}',\omega)$ is a correlation coefficient it has the value unity when $\rho'=0$, i.e.,

$$\mu^{(0)}(0,\omega)=1, \text{ for all } \omega, \qquad (16)$$

and hence Eq. (15) implies that

$$k^2F(\omega)=[H(0)]^{-1}. \qquad (17)$$

Since the left-hand side of Eq. (17) depends on the frequency but the right-hand side is independent of it,

VOLUME 56, NUMBER 13 PHYSICAL REVIEW LETTERS 31 MARCH 1986

each side must be a constant (α say) and consequently

$$F(\omega) = \alpha/k^2. \tag{18}$$

Two important conclusions follow at once from these results. If we substitute Eq. (18) into Eq. (13) we obtain the following expression for the normalized spectrum of light in the far zone:

$$S^{(\infty)}(\mathbf{u}, \omega) = S^{(\infty)}(\omega) = \frac{S^{(0)}(\omega)}{\int S^{(0)}(\omega)d\omega}. \tag{19}$$

This formula shows that not only is the normalized spectrum of the light now the same throughout the far zone, but it is also equal to the normalized spectrum of the light at each source point.

Next we substitute Eq. (18) into Eq. (15) and set $\alpha H = h$, $\rho' = \rho_2 - \rho_1$. We then obtain for $\mu^{(0)}$ the expression

$$\mu^{(0)}(\boldsymbol{\rho}_2 - \boldsymbol{\rho}_1, \omega) = h[k(\boldsymbol{\rho}_2 - \boldsymbol{\rho}_1)] \quad (k = \omega/c); \tag{20}$$

i.e., the complex degree of spectral coherence is a function of the variable $\xi = k(\boldsymbol{\rho}_2 - \boldsymbol{\rho}_1)$ only. We will refer to Eq. (20) as the *scaling law.* Obviously for a source that satisfies this law, the knowledge of the degree of spectral coherence of the light in the source plane at any particular frequency ω specifies it for all frequencies.

The scaling law (20), which ensures that for sources of the class that we are considering the normalized spectrum of the light is the same throughout the far zone and is equal to the normalized spectrum of the light at each source point [Eq. (19)], is the main result of this note.

It is natural to inquire whether sources are known that obey this scaling law. The answer is affirmative. Many of the commonly occurring sources, including blackbody sources, obey Lambert's radiation law [Ref. 1, Sect. 4.8.1]. It is known[7] that all quasi-homogeneous Lambertian sources have the same degree of spectral coherence, viz.

$$\mu^{(0)}(\boldsymbol{\rho}_2 - \boldsymbol{\rho}_1, \omega) = \sin(k|\boldsymbol{\rho}_2 - \boldsymbol{\rho}_1|)/k|\boldsymbol{\rho}_2 - \boldsymbol{\rho}_1|, \tag{21}$$

which is seen to satisfy the scaling law (20). According to the preceding analysis such sources will generate light whose normalized spectrum is the same throughout the far zone and is equal to the normalized spectrum at each source point. This fact is undoubtedly largely responsible for the commonly held, but nevertheless incorrect, belief that spectral invariance is a general property of light.

This Letter has dealt with what is probably the simplest problem regarding spectral invariance on propagation. It would seem that some significant questions in this area might be profitably studied. Among them are the elucidation of the physical origin of the scaling law, spectral properties of light from a broader class of sources than considered here, the relation between the scaling law and Mandel's results regarding cross-spectrally pure light,[8,9] and relativistic effects. Applications of the results to problems of astrophysics might be of particular interest; at this stage one might only speculate whether source correlations may perhaps not give rise to differences between the spectrum of the emitted light and the spectrum of the detected light that originates in some stellar sources.

It is a pleasure to acknowledge stimulating discussions with Professor Leonard Mandel about the subject matter of this note. This research was supported by the National Science Foundation and by the Air Force Geophysics Laboratory under Air Force Office of Scientific Research Task No. 2310G1.

(a)Also at the Institute of Optics, University of Rochester, Rochester, N. Y. 14627.

1M. Born and E. Wolf, *Principles of Optics* (Pergamon, Oxford and New York, 1980), 6th ed., Sect. 10.2.

2L. Mandel and E. Wolf, Rev. Mod. Phys. **37**, 231 (1965).

3C. Kittel, *Elementary Statistical Physics* (Wiley, New York, 1958), Sect. 28.

4E. Wolf, J. Opt. Soc. Am. **68**, 6 (1978), Eqs. (5.3).

5The degree of spectral coherence is defined by the formula [cf. L. Mandel and E. Wolf, J. Opt. Soc. Am. **66**, 529 (1976)]

$$\mu^{(0)}(\boldsymbol{\rho}_1, \boldsymbol{\rho}_2, \omega) = \frac{W^{(0)}(\boldsymbol{\rho}_1, \boldsymbol{\rho}_2, \omega)}{[W^{(0)}(\boldsymbol{\rho}_1, \boldsymbol{\rho}_1, \omega)]^{1/2}[W^{(0)}(\boldsymbol{\rho}_2, \boldsymbol{\rho}_2, \omega)]^{1/2}}.$$

6W. H. Carter and E. Wolf, J. Opt. Soc. Am. **67**, 785 (1977).

7W. H. Carter and E. Wolf, J. Opt. Soc. Am. **65**, 1067 (1975).

8L. Mandel, J. Opt. Soc. Am. **51**, 1342 (1961).

9See, Mandel and Wolf, Ref. 5.

Reprinted from Nature, Vol. 326, No. 6111, pp. 363-365, 26 March 1987

Non-cosmological redshifts of spectral lines

Emil Wolf

Department of Physics and Astronomy, and Institute of Optics, University of Rochester, Rochester, New York 14627, USA

We showed in a recent report[1] (see also refs 2–4) that the normalized spectrum of light will, in general, change on propagation in free space. We also showed that the normalized spectrum of light emitted by a source of a well-defined class will, however, be the same throughout the far zone if the degree of spectral coherence of the source satisfies a certain scaling law. The usual thermal sources appear to be of this kind. These theoretical predictions were subsequently verified by experiments[5]. Here, we demonstrate that under certain circumstances the modification of the normalized spectrum of the emitted light caused by the correlations between the source fluctuations within the source region can produce redshifts of spectral lines in the emitted light. Our results suggest a possible explanation of various puzzling features of the spectra of some stellar objects, particularly quasars.

To explain why source correlations influence the spectrum of the emitted light consider a very simple example. Suppose that two point sources P_1 and P_2 have identical spectra $S_Q(\omega)$ and that measurements on the emitted field are made at some point P. The sources are assumed at rest relative to an observer at P. Assuming that the source fluctuations can be described by a stationary ensemble, the field at P may be characterized by an ensemble $\{V(P, \omega)\}$ of frequency-dependent realizations[6], each of the form

$$V(P,\omega) = Q(P_1,\omega)\frac{e^{ikR_1}}{R_1} + Q(P_2,\omega)\frac{e^{ikR_2}}{R_2} \qquad (1)$$

where $\{Q(P_j, \omega)\}$, $(j=1, 2)$, characterize the strengths of the two fluctuating point sources, R_1 and R_2 are the distances from P_1 to P and from P_2 to P respectively (see Fig. 1) and $k=\omega/c$, c being the speed of light *in vacuo*. For simplicity polarization effects are ignored and hence V and Q are taken to be scalars. The spectrum of the light at P is then given by

$$S_V(P,\omega) = \langle V^*(P,\omega)V(P,\omega)\rangle \qquad (2)$$

where the asterisk denotes the complex conjugate and the angular brackets denote the ensemble average. On substituting from equation (1) into equation (2) and using the fact that

$$\langle Q^*(P_1,\omega)Q(P_1,\omega)\rangle = \langle Q^*(P_2,\omega)Q(P_2,\omega)\rangle = S_Q(\omega) \qquad (3)$$

the following expression is obtained for the spectrum of the emitted light at P:

$$S_V(P,\omega) = \left(\frac{1}{R_1^2}+\frac{1}{R_2^2}\right)S_Q(\omega) + \left[W_Q(P_1,P_2,\omega)\frac{e^{ik(R_2-R_1)}}{R_1R_2} + \text{c.c.}\right] \qquad (4)$$

Here

$$W_Q(P_1,P_2,\omega) = \langle Q^*(P_1,\omega)Q(P_2,\omega)\rangle \qquad (5)$$

is the so-called cross-spectral density of the source fluctuations and c.c. denotes the complex conjugate.

The formula (4) shows that the spectrum $S_V(P, \omega)$ is, in general, not just proportional to $S_Q(\omega)$ but is modified by the correlation, characterized by $W_Q(P_1, P_2, \omega)$, between the fluctuations of the two source strengths $Q(P_1, \omega)$ and $Q(P_2, \omega)$. Only in some very special cases, for example, when the source fluctuations are uncorrelated $[W_Q(P_1, P_2, \omega)=0]$ will $S_V(P, \omega)$ be proportional to $S_Q(\omega)$. Hence, in general, the spectrum of the light generated by two point sources depends not only on their spectra but also on the correlation between the fluctuations of their strengths.

A generalization of the elementary formula (4) for radiation from three-dimensional steady-state (that is, statistically stationary) sources of any state of coherence is known[7]. Of special interest in the present context is the form that the formula takes when the source has the same normalized spectrum $s_Q(\omega)$, $(\int_0^\infty s_Q(\omega)\,d\omega = 1)$ at each point in the source region and has a degree of spectral coherence[3] (appropriately normalized cross-spectral density) $\mu_Q(\mathbf{r}_1, \mathbf{r}_2, \omega)$ that depends on the position vectors $\mathbf{r}_1$ and $\mathbf{r}_2$ of any source points P_1 and P_2 only through their difference $\mathbf{r}_2-\mathbf{r}_1$. If, in addition, for each frequency that significantly contributes to the source spectrum, the spectral correlation length [the effective spatial width $|\Delta\mathbf{r}'|$ of $|\mu(\mathbf{r}', \omega)|$] is small compared to the linear dimensions of the source, the normalized spectrum $s_V^{(\infty)}(\mathbf{u}, \omega)$ of the emitted light in the far zone, in a direction specified by a unit vector $\boldsymbol{u}$, becomes (see equation (3.11) of ref. 8)

$$s_V^{(\infty)}(\mathbf{u},\omega) = \frac{s_Q(\omega)\tilde{\mu}_Q(k\mathbf{u},\omega)}{\int s_Q(\omega)\tilde{\mu}_Q(k\mathbf{u},\omega)\,d\omega} \qquad (6)$$

where $\tilde{\boldsymbol{\mu}}_Q(\mathbf{K}, \omega)$ is the three-dimensional spatial Fourier transform of the degree of spectral coherence $\mu_Q(\mathbf{r}',\omega) \equiv \mu_Q(\mathbf{r}_2-\mathbf{r}_1, \omega)$.

Let us now choose as the normalized source spectrum $s_Q(\omega)$ a spectral line with a gaussian profile

$$s_Q(\omega) = \frac{1}{\delta\sqrt{2\pi}}\exp[-(\omega-\omega_0)^2/2\delta^2] \quad (\delta \ll \omega_0) \qquad (7)$$

and suppose that at each effective frequency ω, the source correlation decreases with the separation $|\mathbf{r}'| = |\mathbf{r}_2-\mathbf{r}_1|$ of any two source points in a gaussian manner, that is

$$\mu_Q(\mathbf{r}',\omega) = \exp[-r'^2/2\sigma_\mu^2(\omega)] \qquad (8)$$

On taking the Fourier transform of equation (8) and substituting the resulting expression into equation (6) we obtain the following expression for the normalized spectrum of the emitted light in the far zone (see equation (3.21) of ref. 8)

$$s_V^{(\infty)}(\omega) = \frac{s_Q(\omega)\sigma_\mu^3(\omega)\exp\{-\frac{1}{2}[k\sigma_\mu(\omega)]^2\}}{\int_0^\infty s_Q(\omega)\sigma_\mu^3(\omega)\exp\{-\frac{1}{2}[k\sigma_\mu(\omega)]^2\}\,d\omega} \qquad (9)$$

Here, $s_V^{(\infty)}(\omega)$ is written in place of $s_V^{(\infty)}(\mathbf{u}, \omega)$, because the spectrum of the far field is now independent of $\mathbf{u}$, as a consequence of the assumed isotropy of μ_Q (see equation (8)).

The formula (9) shows that the spectrum of the emitted light in the far zone depends both on the spectrum of the source fluctuations and on the manner in which the effective source correlation length $\sigma_\mu(\omega)$ depends on the frequency ω.

Let us consider two particular cases. (1) Suppose first that $\sigma_\mu(\omega)$ is independent of ω. Letting ζ denote the (now constant) value of σ_μ and with $s_Q(\omega)$ given by equation (7), one can readily evaluate the integral in the denominator on the right of equation (9) and one then finds that

$$s_V^{(\infty)}(\omega) \approx \frac{\alpha}{\delta\sqrt{2\pi}}\exp\left[-\left(\omega-\frac{\omega_0}{\alpha^2}\right)^2\Big/2(\delta/\alpha)^2\right] \qquad (10)$$

where

$$\alpha^2 = 1+\left(\frac{\delta}{\Delta}\right)^2 \qquad (11a)$$

and

$$\frac{1}{\Delta} = \frac{\zeta}{c} \qquad (11b)$$

When the source is effectively spatially incoherent, $\zeta \to 0$. Then according to equation (11) $\Delta \to \infty$ and $\alpha \to 1$ and it follows from equations (10) and (7) that in this case

$$s_V^{(\infty)}(\omega) \to s_Q(\omega) \qquad (12)$$

Hence, in the limiting case of a completely incoherent source of the class that is considered here, the normalized spectrum of the emitted light in the far zone is identical with the normalized

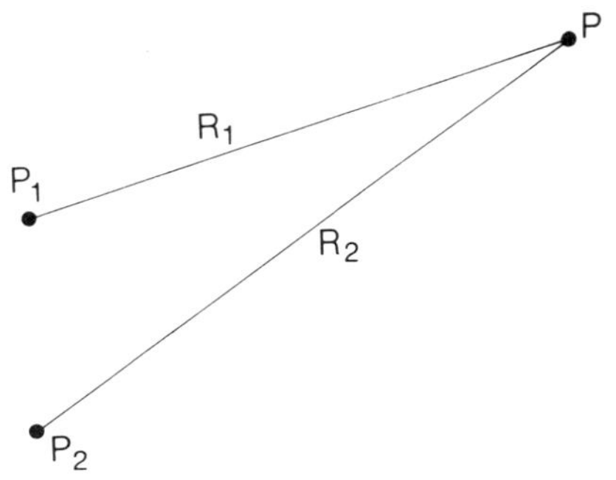

Fig. 1 Illustrating the notation relating to derivation of the formula (4).

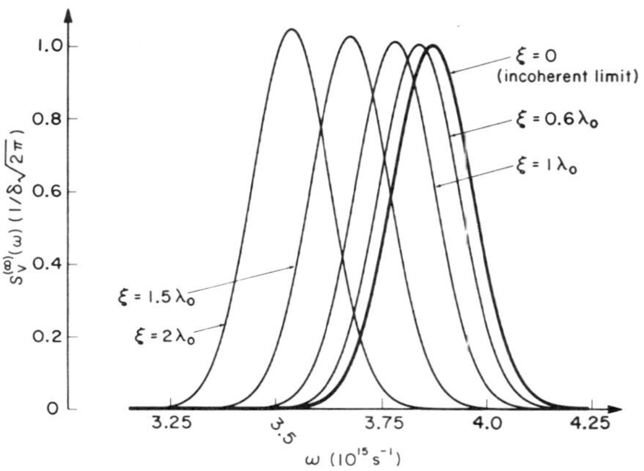

Fig. 2 Spectra $s_V^{(\infty)}(\omega)$ of the far field from sources with spectrum $s_Q(\omega)=(\delta\sqrt{2\pi})^{-1}\exp[-(\omega-\omega_0)^2/2\delta^2]$ and degree of spectral coherence $\mu_Q(\mathbf{r}',\omega)=\exp(-r'^2/2\zeta^2)$, with $\omega_0=3.887\times10^{15}\,\mathrm{s}^{-1}$ ($\lambda_0=4{,}861$ Å) and $\delta=9.57\times10^{13}\,\mathrm{s}^{-1}$, for several selected values of the effective source-correlation length ζ. The solid curve ($\zeta\to0$) also represents the source spectrum $s_Q(\omega)$.

spectrum of the source fluctuations.

However, when the source fluctuations are correlated over an effective distance $\zeta>0$, equation (10) shows that the spectrum $s_V^{(\infty)}(\omega)$, although it is also a line with a gaussian profile, is centred at a lower frequency $\omega_0'\approx\omega_0/\alpha^2<\omega_0$. Hence the source correlations give rise to a spectral line $s_V^{(\infty)}(\omega)$ that is redshifted with respect to the spectral line produced by the completely spatially incoherent source with the source spectrum $s_Q(\omega)$. The shifted line is narrower, having root-mean-square width $\delta'=\delta/\alpha<\delta$ and has α-times greater height. Examples of spectra of light in the far zone, produced by several sources which emit the same spectral line but which have different correlation lengths are shown in Fig. 2. From the formula (10) one can readily deduce that the relative shift of the line, namely,

$$z\equiv\frac{\lambda_0-\lambda_0'}{\lambda_0}=-\frac{\omega_0-\omega_0'}{\omega_0'} \tag{13}$$

($\lambda_0=2\pi c/\omega_0$, $\lambda_0'=2\pi c/\omega_0'$) is given by

$$z=\left(\frac{\delta}{\Delta}\right)^2=\left(\frac{\delta}{c}\right)^2\zeta^2 \tag{14}$$

which shows that in this case the redshift increases quadratically with the spectral source-correlation length ζ. (2) Next consider the situation when $\sigma_\mu(\omega)=a/\omega$ where a is a positive constant. The expression (9) for the normalized spectrum of the emitted light in the far zone now reduces to

$$s_V^{(\infty)}(\omega)=\frac{s_Q(\omega)/\omega^3}{\int_0^\infty[s_Q(\omega)/\omega^3]\,\mathrm{d}\omega} \tag{15}$$

Note that this expression is independent of the value of the constant a.

When $s_Q(\omega)$ is a line with a gaussian profile, given by equation (7), the spectrum $s_V^{(\infty)}(\omega)$, given by equation (15) is no longer strictly gaussian but it can be closely approximated by a gaussian and can be shown to be redshifted with respect to $s_Q(\omega)$ by the relative amount

$$z\approx3\left(\frac{\delta}{\omega_0}\right)^2. \tag{16}$$

An example of this situation is illustrated in Fig. 3.

This case $[\sigma_\mu(\omega)=a/\omega]$ is of special interest because, according to equation (8), the degree of spectral coherence is now given by

$$\mu_Q(\mathbf{r}',\omega)=\exp[-(kr')^2/2(a/c)^2], \tag{17}$$

that is, it has the functional form

$$\mu_Q(\mathbf{r}',\omega)=f(kr')\quad(k=\omega/c=2\pi/\lambda) \tag{18}$$

Thus the degree of spectral coherence of the source distribution now satisfies the three-dimensional analogue of a requirement (called the scaling law) derived in ref. 1, as a sufficient condition for the spectrum of the light emitted by a planar secondary source of a well-defined class to have certain invariance properties on propagation. It will be shown in another publication (J. T. Foley and E. Wolf, in preparation) that for three-dimensional primary sources of an analogous class, whose degree of coherence satisfies this law, the spectrum of the emitted light has similar invariance properties. We conjecture that the usual thermal sources obey such a scaling law.

Now briefly consider the question of a physical mechanism for producing source correlations. Such correlations must clearly be manifestations of some cooperative phenomena. At the atomic level possible candidates may perhaps be superradiance and superfluorescence[9]. An effect of this kind was first predicted by Dicke in 1954 when he showed[10] that under certain circumstances energy from excited atoms may be released cooperatively in a much shorter time than the natural lifetime of the excited states of the atoms and with much larger emission intensity than would be obtained were the atoms radiating independently.

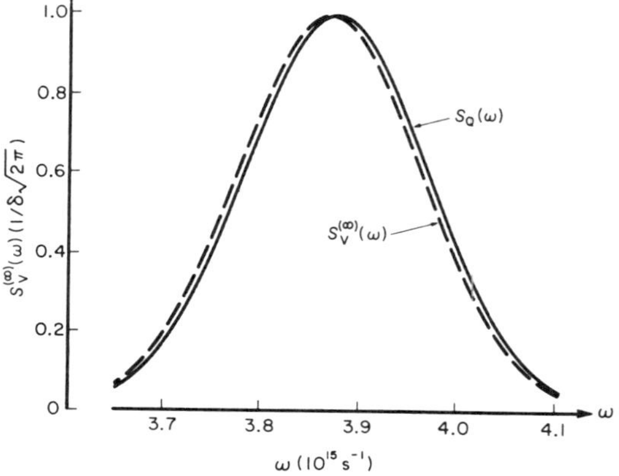

Fig. 3 The spectrum $s_V^{(\infty)}(\omega)$ of the far field from a source with source spectrum $s_Q(\omega)=(\delta\sqrt{2\pi})^{-1}\exp[-(\omega-\omega_0)^2/2\delta^2]$ and degree of spectral coherence $\mu_Q(\mathbf{r}',\omega)=\exp[-(kr')^2/2(a/c)^2]$ ($a=$ an arbitrary constant), with $\omega_0=3.887\times10^{15}\,\mathrm{s}^{-1}$ ($\lambda_0=4{,}861$ Å) and $\delta=9.57\times10^{13}\,\mathrm{s}^{-1}$. The source spectrum $s_Q(\omega)$ is shown for comparison. Note that $\mu_Q(\mathbf{r}',\omega)$ now obeys the scaling law.

However not enough is known at present about the coherence properties of large three-dimensional systems of this kind to make it possible to determine whether superradiance and superfluorescence might involve correlations that could give rise to spectral line shifts.

There is, however, quite a different mechanism, which can be described at the macroscopic level, and which can imitate effects of source correlations; namely effects of correlations between the refractive index at pairs of points in a spatially random but statistically homogeneous, time-invariant medium. If a wave illuminates such a medium, say a dilute gas, then, as is well known, the medium acts as a secondary source, namely as a set of oscillating charges set in motion by the incident wave. The secondary waves produced by the oscillating charges then combine with each other and with the incident wave and generate the scattered field. If the gas is not too dilute the collective response of the microscopic charges to the incident field can be described by macroscopic parameters such as the dielectric susceptibility or the refractive index. Now within the accuracy of the first Born approximation the basic equation for scattering is of the same form as the basic equation for radiation from primary sources, the 'equivalent source' for scattering being the product of the scattering potential (which is a simple function of the refractive index) and of the incident wave. This correspondence clearly implies that our results regarding the effects of source correlations on the spectrum of the emitted light must have analogues regarding the effects of a spatially random medium with correlated refractive index distribution on the spectrum of the light that is scattered by it. This topic will be discussed elsewhere.

Let us now consider some implications of this analysis. Using equation (14), the spectral line in Fig. 2, produced by the source whose correlation length $\zeta = \lambda_0$ is readily found to have a redshift given by $z = 0.0241$ with respect to the source spectrum. It is of interest to note that if an observer detected such a redshift unaware of its true origin and interpreted it on the basis of the Doppler shift formula $v/c = \Delta\lambda/\lambda_0 = z$ he would incorrectly conclude that the source was receeding from him with a speed $v = 0.0241c \approx 7{,}230 \text{ km s}^{-1}$.

It seems worthwhile to note that there is a maximum line shift that can be produced by source correlations. This can be seen from the basic formula (6) which indicates that $s_V^{(\infty)}(\mathbf{u}, \omega) = 0$ when $s_Q(\omega) = 0$, implying that the spectrum of the far field can only contain those frequencies that are already present in the source spectrum. Consequently the maximum attainable frequency shift of the line cannot exceed its effective frequency range. However, any frequency contribution from the source spectrum to the normalized spectrum of the far field can be greatly magnified or greatly reduced, as is evident from equation (6) and from Fig. 2.

We have mainly considered effects of source correlations under circumstances when the source spectrum consists of a single line and when the degree of spectral coherence μ_Q that characterizes the source correlations depends on a single parameter. Preliminary calculations show that with a suitably chosen μ_Q which depends on a larger number of parameters, redshifts of several lines may be produced, all of which will have approximately the same z-values.

In this article we have considered redshifts of spectral lines. However, it is not difficult to specify source correlations which will produce blueshifts. Examples of this kind are given in a forthcoming publication[11].

It seems plausible that the mechanism discussed in this article may be responsible for some of the so far unexplained features of quasar spectra, including line asymmetries and small differences in the observed redshifts of different lines. In this connection it is of interest to recall that the role of coherence in the emission of radiation from quasars was stressed by Hoyle, Burbidge and Sargent in a well-known article[12].

I thank Mr A. Gamliel and Mr K. Kim for carrying out computations relating to the analysis presented in this article. The fact that scattering can also produce shifts of spectral lines was noted independently by Professor Franco Gori, who informed me of this result when commenting on an early version of the manuscript of this article. This investigation was supported by the NSF and by the US Air Force Geophysical Laboratory.

Received 10 September 1986; accepted 3 February 1987.

1. Wolf, E. *Phys. Rev. Lett.* **56,** 1370–1372 (1986).
2. Mandel, L. *J. opt. Soc. Am.* **51,** 1342–1350 (1961).
3. Mandel, L. & Wolf, E. *J. opt. Soc. Am.* **66,** 529–535 (1976).
4. Gori, F. & Grella, R. *Optics Commun.* **49,** 173–177 (1984).
5. Morris, G. M. & Faklis, D. *Optics Commun.* (in the press).
6. Wolf, E. *J. opt. Soc. Am.* **72,** 343–351 (1982); *J. opt. Soc. Am.* A **3,** 76–85 (1986).
7. Carter, W. H. & Wolf, E. *Optica Acta* **28,** 227–244 (1981).
8. Carter, W. H. & Wolf, E. *Optica Acta* **28,** 245–259 (1981).
9. Schuurmans, M. F. H., Vrehen, Q. H. F., Polder, D. & Gibbs, H. M. in *Advances in Atomic and Molecular Physics* Vol. 17 (eds Bates, D. & Bederson, B.) 167–228 (Academic, New York, 1981).
10. Dicke, R. H. *Phys. Rev.* **93,** 99–110 (1954).
11. Wolf, E. *Optics Commun.* (in the press).
12. Hoyle, F., Burbidge, G. R. & Sargent, W. L. W. *Nature* **209,** 751–753 (1966).

Volume 62, number 1 OPTICS COMMUNICATIONS 1 April 1987

REDSHIFTS AND BLUESHIFTS OF SPECTRAL LINES CAUSED BY SOURCE CORRELATIONS☆

Emil WOLF[1]

Department of Physics and Astronomy, University of Rochester, Rochester, NY 14627, USA

Received 24 November 1986

We recently showed that the spectrum of light emitted by a source depends not only on the spectrum of the source distribution but also on the degree of spectral coherence of the source fluctuations. In this note we show that with a degree of spectral coherence of certain kind, specified by two parameters, the spectrum of the emitted light will be displaced relative to the source spectrum. The displacement will be either toward the lower or toward the higher frequencies, depending on the choice of the parameters.

1. Introduction

It has been known for some time that the spectrum of light generally changes on propagation, even in free space [1,2]. Such changes are basically due to correlation properties of the source. Recently we derived a condition for the normalized spectrum of light generated by a planar, secondary, quasi-homogeneous source to be the same throughout the far zone and in the source plane [3]. We referred to this condition, which is a requirement on the functional form of its degree of spectral coherence, as the scaling law and we noted that all quasi-homogeneous lambertian sources satisfy this law. We have also shown that when the scaling law is not satisfied the spectrum of the emitted light will, in general, no longer be invariant on propagation. These theoretical predictions have been recently verified by experiments [4].

In another recent paper [5] we considered radiation from three-dimensional, quasi-homogeneous sources and we showed that if the source spectrum consists of a line with a gaussian profile and if the degree of spectral coherence of the source is appropriately chosen, the spectrum of the emitted light will also consist of a line with gaussian profile, but this line will be redshifted with respect to the spectral line of the source distribution. The amount of the redshift depends on the spectral correlation length of the source. This result has important implications for astrophysics, some of which were briefly mentioned in ref. [5].

In the present note we again consider a source whose spectrum consists of a single line with a gaussian profile but we assume somewhat different correlation properties of the source. More specifically we choose a degree of spectral coherence of the source distribution which depends on two parameters rather than on a single parameter as we have done previously. The spectrum of the emitted light is again found to be a line with gaussian profile, but this line may be redshifted or blueshifted relative to the spectral line of the source distribution, depending on the choice of the parameters.

2. The spectrum of light produced by a three-dimensional quasi-homogeneous source

Let us consider a fluctuating source-distribution $Q(\mathbf{r}, t)$ occupying a finite domain of volume D in free space and let $V(\mathbf{r}, t)$ denote the field generated by the source. Here $\mathbf{r}$ denotes the position vector of a typical point and t the time. Both $V(\mathbf{r}, t)$ and $Q(\mathbf{r}, t)$ are taken to be analytic signals [6]. They are related by the inhomogeneous wave equation

☆ Research supported by the National Science Foundation under Grant PHY-8314626 and the Air Force Geophysics Laboratory under AFOSR Task 2310G1.

[1] Also at the Institute of Optics, University of Rochester.

$$\nabla^2 V(\boldsymbol{r}, t) - c^{-2}(\partial^2/\partial t^2)\, V(\boldsymbol{r}, t) = -4\pi\, Q(\boldsymbol{r}, t)\,. \tag{2.1}$$

We will assume that the statistical ensembles that characterize the source fluctuations are stationary. Let $W_Q(\boldsymbol{r}_1, \boldsymbol{r}_2, \omega)$ and $W_V(\boldsymbol{r}_1, \boldsymbol{r}_2, \omega)$ be the cross-spectral densities of the source distribution and of the field distribution respectively. They may be represented in the form [7]

$$W_Q(\boldsymbol{r}_1, \boldsymbol{r}_2, \omega) = \langle U_Q^*(\boldsymbol{r}_1, \omega)\, U_Q(\boldsymbol{r}_2, \omega)\rangle\,, \tag{2.2a}$$

$$W_V(\boldsymbol{r}_1, \boldsymbol{r}_2, \omega) = \langle U_V^*(\boldsymbol{r}_1, \omega)\, U_V(\boldsymbol{r}_2, \omega)\rangle\,, \tag{2.2b}$$

where $\{U_Q(\boldsymbol{r}, \omega)\}$ and $\{U_V(\boldsymbol{r}, \omega)\}$ are ensembles of suitably chosen realizations, angular brackets denote averages taken over these ensembles and the asterisk denotes the complex conjugate. As consequence of the wave equation (2.1) the two cross-spectral densities may be shown to be related by the equation [‡1] [ref. [7], eq. (3.10); ref. [8a], eq. (2.11)]

$$(\nabla_1^2 + k^2)\,(\nabla_2^2 + k^2)\, W_V(\boldsymbol{r}_1, \boldsymbol{r}_2, \omega) = (4\pi)^2\, W_Q(\boldsymbol{r}_1, \boldsymbol{r}_2, \omega)\,, \tag{2.3}$$

where ∇_1^2 and ∇_2^2 are the laplacian operators acting with respect to the coordinates of the points $\boldsymbol{r}_1$ and $\boldsymbol{r}_2$ respectively and

$$k = \omega / c \tag{2.4}$$

is the wave number associated with the frequency ω, c being the speed of light in vacuo.

Using eq. (2.3) one can show that the radiant intensity $J_\omega(\boldsymbol{u})$ generated by the source, i.e. the rate at which energy is radiated at frequency ω per unit solid angle around a direction specified by a unit vector $\boldsymbol{u}$ is given by [ref. [8a], eq. (3.9)]

$$J_\omega(\boldsymbol{u}) = (2\pi)^6\, \tilde{W}_Q(-k\boldsymbol{u}, k\boldsymbol{u}, \omega)\,, \tag{2.5}$$

where

$$\tilde{W}_Q(\boldsymbol{K}_1, \boldsymbol{K}_2, \omega) = (2\pi)^{-6} \int_D \int_D W_Q(\boldsymbol{r}_1, \boldsymbol{r}_2, \omega) \times \exp[-\mathrm{i}(\boldsymbol{K}_1 \cdot \boldsymbol{r}_1 + \boldsymbol{K}_2 \cdot \boldsymbol{r}_2)]\, \mathrm{d}^3 r_1\, \mathrm{d}^3 r_2 \tag{2.6}$$

‡1 The definition of the cross-spectral densities employed in refs. [7] and [8] differ by complex conjugation. Throughout this note we employ those of ref. [7]; hence some of the formulas we now use [e.g. eq. (2.5) below] differ trivially from the corresponding formulas of refs. [8].

is the six-dimensional Fourier transform of W_Q.

We will restrict our attention to quasi-homogeneous sources. For such sources one has, to a good approximation,

$$W_Q(\boldsymbol{r}_1, \boldsymbol{r}_2, \omega) = S_Q[(\boldsymbol{r}_1 + \boldsymbol{r}_2)/2, \omega]\, \mu_Q(\boldsymbol{r}_2 - \boldsymbol{r}_1, \omega)\,, \tag{2.7}$$

where

$$S_Q(\boldsymbol{r}, \omega) \equiv W_Q(\boldsymbol{r}, \boldsymbol{r}, \omega) = \langle U_Q^*(\boldsymbol{r}, \omega)\, U_Q(\boldsymbol{r}, \omega)\rangle \tag{2.8}$$

is the source spectrum and

$$\mu_Q(\boldsymbol{r}_2 - \boldsymbol{r}_1, \omega) \equiv W_Q(\boldsymbol{r}_1, \boldsymbol{r}_2, \omega) \times [S_Q(\boldsymbol{r}_1, \omega)]^{-1/2}\, [S_Q(\boldsymbol{r}_2, \omega)]^{-1/2} \tag{2.9}$$

is the degree of spectral coherence of the source distribution. Moreover, for each effective frequency ω contained in the source spectrum, $S_Q(\boldsymbol{r}, \omega)$ varies much more slowly with $\boldsymbol{r}$ than $\mu_Q(\boldsymbol{r}', \omega)$ varies with $\boldsymbol{r}'$. With sources of this class eq. (2.5) takes the form [ref. [8b], eq. (3.11)]

$$J_\omega(\boldsymbol{u}) = (2\pi)^6 \tilde{S}_Q(0, \omega) \tilde{\mu}_Q(k\boldsymbol{u}, \omega)\,, \tag{2.10}$$

where the tilde now denotes three-dimensional Fourier transforms.

Let us next assume that the source spectrum is the same at each source point. We will then write $S_Q(\omega)$ in place of $S_Q(\boldsymbol{r},\omega)$. In this case $\tilde{S}_Q(0, \omega) = D S_Q(\omega)/(2\pi)^3$ and the formula (2.10) becomes (D again denoting the source volume)

$$J_\omega(\boldsymbol{u}) = (2\pi)^3 D S_Q(\omega) \tilde{\mu}_Q(k\boldsymbol{u}, \omega)\,. \tag{2.11}$$

Now the radiant intensity $J_\omega(\boldsymbol{u})$ is trivially related to the spectrum $S_V^{(\infty)}(R\boldsymbol{u}, \omega) \equiv W_V^{(\infty)}(R\boldsymbol{u}, R\boldsymbol{u}, \omega)$ of the far field by the formula [9] $S_V^{(\infty)}(R\boldsymbol{u}, R\boldsymbol{u}, \omega) \sim J_\omega(\boldsymbol{u})/R^2$ as $kR \to \infty$, with the unit vector $\boldsymbol{u}$ fixed. Hence we obtain at once from eq. (2.11) the following expression for the spectrum of the emitted light in the far zone:

$$S_V^{(\infty)}(\boldsymbol{u}, \omega) = (2\pi)^3 (D/R^2)\, S_Q(\omega)\, \tilde{\mu}_Q(k\boldsymbol{u}, \omega)\,. \tag{2.12}$$

This formula shows that the spectrum $S_V^{(\infty)}(\boldsymbol{u}, \omega)$ of the emitted light in the far zone depends, in general, not only on the source spectrum $S_Q(\omega)$ but also on

Volume 62, number 1 OPTICS COMMUNICATIONS 1 April 1987

the degree of spectral coherence of the source distribution. It seems worthwhile to note that the dimensions of $S_V^{(\infty)}$ and of S_Q are different. Since $\tilde{\mu}_Q$ is the three-dimensional Fourier transform of μ_Q, $[\tilde{\mu}_Q]=L^3$ (brackets denoting dimensions and L denotes length). Hence eq. (2.12) implies that $[S_V^{(\infty)}]=[S_Q]L^4$, in agreement with eq. (2.3).

3. A class of source correlations that generate lineshifts

In ref. [5] we considered quasi-homogeneous sources whose spectrum was a line of gaussian profile,

$$S_Q(\omega)=A\exp[-(\omega-\omega_0)^2/2\delta_0^2]\ , \quad (\delta_0/\omega_0\ll 1)\ , \tag{3.1}$$

and whose degree of spectral coherence was also gaussian viz.,

$$\mu_Q(\boldsymbol{r}',\omega)=\exp[-r'^2/2\sigma^2(\omega)]\ , \tag{3.2}$$

where $r'=|\boldsymbol{r}'|$. The three-dimensional Fourier transform of μ_Q is then given by

$$\tilde{\mu}_Q(\boldsymbol{K},\omega)=[\sigma(\omega)/\sqrt{2\pi}]^3\exp[-\tfrac{1}{2}K^2\sigma^2(\omega)], \tag{3.3}$$

($K=|\boldsymbol{K}|$). In particular we showed that if $\sigma(\omega)$ is constant (ζ say) such a source will emit light whose spectrum in the far zone is redshifted with respect to the source spectrum, the amount of the shift depending on the effective source correlation length ζ.

The degrees of spectral coherence of the form (3.2), with $\sigma(\omega)=\zeta$ (constant) form a one-parameter family. In this note we will consider quasi-homogeneous sources whose degrees of coherence are of a somewhat more general form. Specifically we assume that for these sources

$$\tilde{\mu}_Q(\boldsymbol{K})=B\exp[-\tfrac{1}{2}(K-K_1)^2\zeta^2]\ , \tag{3.4}$$

where B, K_1 and ζ are positive constants. We have written $\tilde{\mu}_Q(\boldsymbol{K})$ rather than $\tilde{\mu}_Q(\boldsymbol{K},\omega)$ on the left-hand side of eq. (3.4), because $\tilde{\mu}_Q$ is now independent of ω. Only two of the three constants in the expression (3.4) are independent, because the Fourier transform $\mu_Q(\boldsymbol{r}')$ of $\tilde{\mu}_Q(\boldsymbol{K})$ satisfies the requirement that $\mu_Q(0)=1$, which is a necessary condition for $\mu_Q(\boldsymbol{r}')$ to be a correlation coefficient.

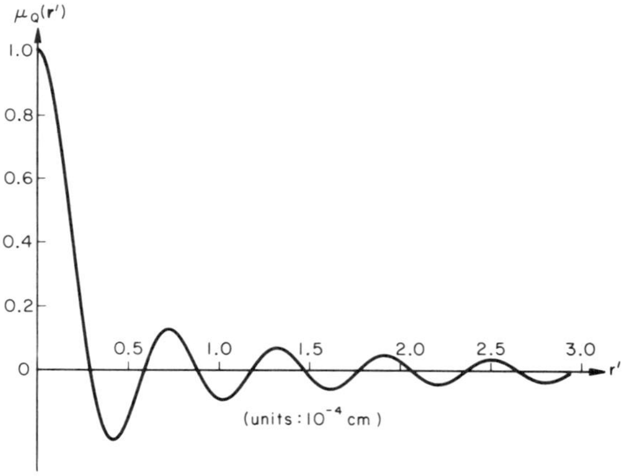

Fig. 1. The behaviour of the correlation coefficient $\mu_Q(\boldsymbol{r}')=[(\sin K_1r')/K_1r']\exp(-r'^2/2\zeta^2)$, with $K_1=1.07\times10^5\ \mathrm{cm}^{-1}$, $\zeta=1.5$ cm [associated with curve (d) in fig. 2].

It can be shown by a long but straightforward calculation (which we omit because of limitation of space) that if

$$K_1\zeta\gg 1 \tag{3.5}$$

the degree of spectral coherence, whose Fourier transform is given by eq. (3.4), is

$$\mu_Q(\boldsymbol{r}')=[(\sin K_1r')/K_1r']\ \exp(-r'^2/2\zeta^2) \tag{3.6}$$

and that the constant B in eq. (3.4) is given in terms of the two other parameters by the formula

$$B=\zeta/2(2\pi)^{3/2}K_1^2\ . \tag{3.7}$$

From now on we will only consider situations for which the constraint (3.5) holds. Eq. (3.6) then shows that the degree of spectral coherence has the form of the sinc function $(\sin K_1r')/K_1r'$, modulated by the gaussian function $\exp(-r'^2/2\zeta^2)$. The behaviour of such a two-parameter correlation coefficient is shown in fig. 1.

It follows on substituting from eqs. (3.1) and (3.4) into eq. (2.12) that the spectrum of the light in the far zone, generated by such a source, is given by

$$S_V^{(\infty)}(\omega)=(2\pi)^3(D/R^2)AB\exp[-(\omega-\omega_0)^2/2\delta_0^2]\times\exp[-(\omega-\omega_1)^2/2\delta_1^2]\ , \tag{3.8}$$

where $\omega = kc$ as before and

$$\omega_1 = K_1 c, \quad \delta_1 = c/\zeta . \tag{3.9}$$

We have written $S_V^{(\infty)}(\omega)$ rather than $S_V^{(\infty)}(\boldsymbol{u}, \omega)$ on the left-hand side of eq. (3.8) since, because of the assumed isotropy of the source, $S_V^{(\infty)}$ is now independent of $\boldsymbol{u}$. In terms of the parameters ω_1 and δ_1 the factor B, given by eq. (3.7), becomes

$$B = c^3/2(2\pi)^{3/2}\omega_1^2\delta_1 . \tag{3.10}$$

Let us now consider the expression (3.8) more closely. For this purpose it is convenient to set

$$\alpha_0 = 1/2\delta_0^2, \quad \alpha_1 = 1/2\delta_1^2 . \tag{3.11}$$

One then finds after a straightforward calculation that eq. (3.8) may be expressed in the form

$$S_V^{(\infty)}(\omega) = AC \exp[-(\omega - \omega_{01})^2/2\delta_{01}^2] , \tag{3.12}$$

where

$$\omega_{01} = (\alpha_0\omega_0 + \alpha_1\omega_1)/(\alpha_0 + \alpha_1) , \tag{3.13}$$

$$1/\delta_{01}^2 = 2(\alpha_0 + \alpha_1) = (1/\delta_0^2) + (1/\delta_1^2) , \tag{3.14}$$

and

$$C = (2\pi)^3(DB/R^2) \times \exp\{-[\alpha_0\alpha_1/(\alpha_0 + \alpha_1)](\omega_1 - \omega_0)^2\} . \tag{3.15}$$

The formula (3.12) shows that the spectrum of the emitted light in the far zone is also a line with gaussian profile, but it is not centered on the frequency ω_0 of the source spectrum [cf. eq. (3.1)] but rather on the frequency ω_{01}, given by eq. (3.13). Since according to eqs. (3.11) α_0 and α_1 are positive constants one can readily deduce from the expression (3.13) that

$$\omega_{01} < \omega_0 \quad \text{when } \omega_1 < \omega_0 ,$$

and that

$$\omega_{01} > \omega_0 \quad \text{when } \omega_1 > \omega_0 .$$

Since according to eq. (3.9) $\omega_1 = K_1 c$, this result implies that *if the parameter K_1 of the degree of spectral coherence (3.6) is smaller than the wavenumber $k_0 = \omega_0/c$ associated with the source spectrum $S_Q(\omega)$, the spectrum $S_V^{(\infty)}(\omega)$ of the emitted light is redshifted with respect to $S_Q(\omega)$; and that if K_1 is greater than k_0 it is blueshifted with respect to it.* We also see from eq. (3.14) that $1/\delta_{01}^2 > 1/\delta_0^2$ i.e. that $\delta_{01} < \delta_0$. Hence in either case the spectral line of the emitted light is narrower than the spectral line of the source distribution.

4. Examples

To illustrate the preceding analysis we consider a few examples. For simplicity we will choose

$$\delta_1 = \delta_0 . \tag{4.1}$$

Then, according to eq. (3.11), $\alpha_1 = \alpha_0$ and the expression (3.12) becomes

$$S_V^{(\infty)}(\omega) = A\bar{C} \exp[-(\omega - \bar{\omega})^2/\delta_0^2] , \tag{4.2}$$

where

$$\bar{\omega} = \tfrac{1}{2}(\omega_0 + \omega_1) , \tag{4.3}$$

$$\bar{C} = (2\pi)^3(DB/R)\exp[-(\omega_1 - \omega_0)^2/4\delta_0^2]. \tag{4.4}$$

We see that the spectral line of the emitted light is now centered on the average value $\bar{\omega}$ of the frequencies ω_0 and ω_1.

Let us consider the normalized spectrum

$$s_V^{(\infty)}(\omega) = S_V^{(\infty)}(\omega) \Big/ \int_0^\infty S_V^{(\infty)}(\omega)\, d\omega \tag{4.5}$$

of the emitted light. On substituting from eq. (4.2) into eq. (4.5) and on using eq. (4.3) we obtain the following expression for $s_V^{(\infty)}(\omega)$:

$$s_V^{(\infty)}(\omega) = (1/\delta_0\sqrt{\pi}) \times \exp\{-[\omega - \tfrac{1}{2}(\omega_0 + \omega_1)]^2/\delta_0^2\} . \tag{4.6}$$

In fig. 2 curves are plotted showing the normalized source spectrum

$$s_Q(\omega) = (1/\delta_0\sqrt{2\pi}) \exp[-(\omega - \omega_0)^2/2\delta_0^2] , \tag{4.7}$$

Volume 62, number 1 OPTICS COMMUNICATIONS 1 April 1987

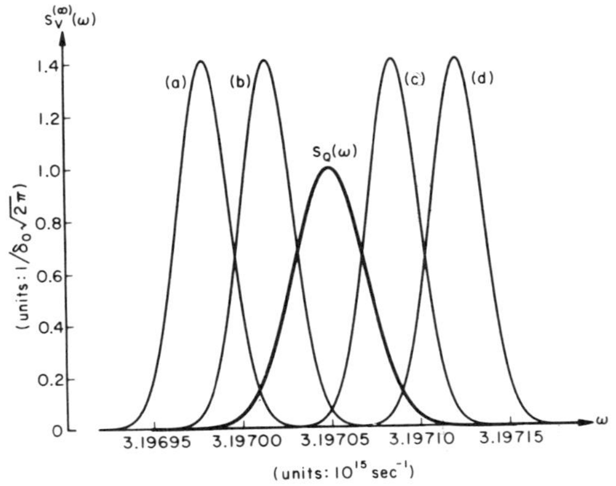

Fig. 2. Redshifts and blueshifts of spectral lines caused by source correlations. The normalized spectrum $s_Q(\omega)$ of the source distribution is a line of gaussian profile [given by eq. (4.7)], with $\omega_0=3.197049\times10^{15}$ s^{-1} (sodium line of wavelength $\lambda_0=5895.924$ Å) and rms width $\delta_0=2\times10^{10}$ s^{-1}. Curves (a) – (d) show the normalized spectra of the emitted light [lines with gaussian profiles given by eq. (4.6)], generated by the source distribution, each with $\delta_1=\delta_0$ ($\zeta=c/\delta_1=1.5$ cm) and with $\omega_1=\omega_0-1.4\times10^{11}$ s^{-1} (a), $\omega_1=\omega_0-0.7\times10^{11}$ s^{-1} (b), $\omega_1=\omega_0+0.7\times10^{11}$ s^{-1} (c) and $\omega_1=\omega_0+1.4\times10^{11}$ s^{-1} (d).

taken to be one of the sodium lines, as well as a number of emitted lines for different values of the parameter $\omega_1=K_1c$, of the degree of spectral coherence of the source; the other parameter, ζ, is kept fixed and chosen so that $\delta_1=c/\zeta$ is equal to δ_0. It is seen that with increasing values of the difference $|\omega_0-\omega_1|$ the shift of the emitted spectral line also increases. This, of course, is to be expected since when $\delta_1=\delta_0$, the shift is given by $|\bar{\omega}-\omega_0|=\frac{1}{2}|\omega_0-\omega_1|$.

Acknowledgements

I am obliged to Prof. G.S. Agarwal and to Mr. K. Kim for helpful discussions and to Mr. A. Gamliel for carrying out the computations relating to figs. 1 and 2.

References

[1] (a) L. Mandel, J. Opt. Soc. Am. 51 (1961) 1342;
(b) L. Mandel and E. Wolf, J. Opt. Soc. Am. 66 (1976) 529.
[2] F. Gori and R. Grella, Optics Comm. 49 (1984) 173.
[3] E. Wolf, Phys. Rev. Lett. 56 (1986) 1370.
[4] G.M. Morris and D. Faklis, Optics Comm. 62 (1987) 5.
[5] E. Wolf, Nature, 326 (1987) 363.
[6] M. Born and E. Wolf, Principles of optics (Pergamon Press, Oxford and New York, 6th ed., 1980), sec. 10.2.
[7] E. Wolf, J. Opt. Soc. Am. A 3 (1986) 76, eqs. (2.10) and (3.11).
[8] W.H. Carter and E. Wolf, (a) Optica Acta 28 (1981) 227; (b) Optica Acta 28 (1981) 245.
[9] E. Wolf, J. Opt. Soc. Am. 68 (1978) 1597, eq. (B16).

Volume 81, number 3,4 OPTICS COMMUNICATIONS 15 February 1991

Spectral changes produced in Young's interference experiment

Daniel F.V. James

The Institute of Optics, University of Rochester, Rochester, NY 14627, USA

and

Emil Wolf [1]

Department of Physics and Astronomy, University of Rochester, Rochester, NY 14627, USA

Received 20 June 1990

It has been known for some time that the spectrum of radiation generated by partially coherent sources may be different from the source spectrum. In this letter we study the spectral changes produced on superposing partially coherent light emerging from two pinholes in Young's interference experiment. We find that whilst only small changes take place with narrow-band light, drastic modifications in the spectrum can take place when the bandwidth of the incident light is sufficiently broad. Our analysis distinguishes clearly the contributions due to diffraction and due to the state of coherence of the light incident on the pinholes.

The interference experiment of Thomas Young [1], performed in 1801, was one of the pillars on which the wave theory of light was erected. More than a century later Zernike [2] showed theoretically that the Young interference experiment also reveals more subtle features of light, namely some of its spatial coherence properties, a prediction which was later confirmed by experiment [3]. Zernike's work also provided a new and physically more basic interpretation of Michelson's classic technique for measuring stellar diameters [4]. It was later found that Young's and Zernike's investigations do not exhaust the amount of information which can be deduced from the analysis of the light produced by superposing two beams. In particular the spectral analysis of the light in the region of superposition was found to provide information about the *spectral* coherence properties of the light incident on the two pinholes [5,6]. In the present paper we show, in agreement with earlier work by Mandel on cross-spectral purity [7] and by Wolf on the effect of source correlations [8], that the spectrum of light changes when two beams are superposed. We find that the circumstances in which appreciable spectral changes are produced in such experiments are to some extent complementary to those which govern the formation of a good-quality interference pattern. To obtain an interference pattern having many fringes, narrow-band light must be used. On the other hand, to obtain substantial spectral changes, the bandwidth of the incident light must be broad. We illustrate our analysis by examining the changes in the spectrum produced on interference of narrow-band light which propagates from a spatially incoherent source and also of broad-band radiation from a blackbody source. In the later case drastic spectral changes, due both to diffraction at the pinholes and due to the coherence properties of the incident light are found to occur.

Let us first consider Young's interference experiment with quasi-monochromatic light of mean frequency $\bar{\omega}$. The light is incident from the left onto an opaque screen containing two small pinholes at points P_1 and P_2 (fig. 1). Let $V(P, t)$ denote the complex field amplitude at an arbitrary point P, and at time t. We assume that its fluctuations are statistically stationary, at least in the wide sense. The expectation

[1] Also at The Institute of Optics, University of Rochester.

Volume 81, number 3,4 OPTICS COMMUNICATIONS 15 February 1991

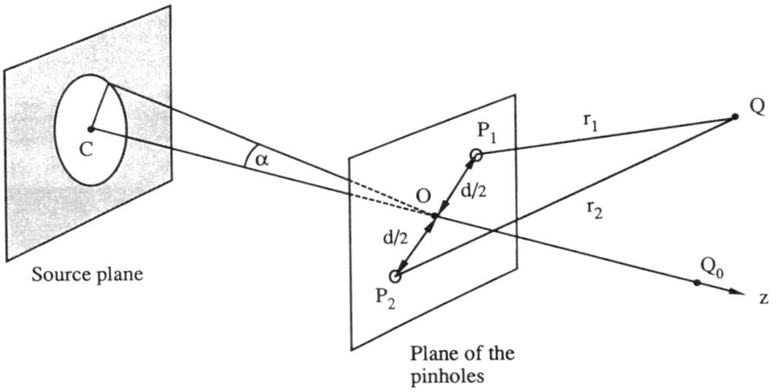

Fig. 1. Illustration of the geometry and notation. The distances and angles are exaggerated to show details.

value [#1] (denoted by angular brackets) of the intensity at P is then given by the usual expression $I(P)=\langle V^*(P,t)\;V(P,t)\rangle$.

Let Q be a point in the half-space on the right of the screen, in a region where the light from the two pinholes is superposed. Under fairly general conditions the expectation value of the intensity at Q is given by the well-known interference law [#2]

$$I(Q)=I^{(1)}(Q)+I^{(2)}(Q)+2\sqrt{I^{(1)}(Q)}\,\sqrt{I^{(2)}(Q)}\times|\gamma_{12}[(r_1-r_2)/c]|\times\cos\{\alpha_{12}[(r_1-r_2)/c]-\bar{k}(r_1-r_2)\}\,.$$

In this formula $I^{(1)}(Q)$ is the expectation value of the intensity which would be observed at Q if the pinhole at P_1 alone were open, $I^{(2)}(Q)$ having a similar meaning with respect to the other pinhole, r_1 and r_2 are the distances between P_1 and Q and between P_2 and Q respectively and $\bar{k}=\bar{\omega}/c$ is the wavenumber associated with the mean frequency $\bar{\omega}$, c being the speed of light in free space. Further

$$\gamma_{12}(\tau)\equiv|\gamma_{12}(\tau)|\exp\{\mathrm{i}[\alpha_{12}(\tau)-\bar{\omega}\tau]\} \tag{2}$$

is the complex degree of coherence of the light fluctuations at the two pinholes. It is defined in terms of the mutual coherence function

$$\Gamma_{12}(\tau)=\langle V^*(P_1,t)\;V(P_2,t+\tau)\rangle \tag{3}$$

by the expression

$$\gamma_{12}(\tau)=\frac{\Gamma_{12}(\tau)}{\sqrt{\Gamma_{11}(0)}\,\sqrt{\Gamma_{22}(0)}}\,. \tag{4}$$

Suppose now that instead of the intensity we wish to determine the spectrum (or more precisely the spectral density) $S(Q,\omega)$ of the light at Q, without restricting ourselves to quasi-monochromatic light. According to the Wiener-Khintchine theorem

$$S(Q,\omega)=\frac{1}{2\pi}\int_{-\infty}^{\infty}\langle V^*(Q,t)\;V(Q,t+\tau)\rangle\times\exp(\mathrm{i}\omega\tau)\,\mathrm{d}\tau\,. \tag{5}$$

The spectrum $S(Q,\omega)$ cannot be determined from the formula (1). Instead we must use the more general spectral interference law [#3]

$$S(Q,\omega)=S^{(1)}(Q,\omega)+S^{(2)}(Q,\omega)+2\sqrt{S^{(1)}(Q,\omega)}\sqrt{S^{(2)}(Q,\omega)}\times|\mu_{12}(\omega)|\cos[\beta_{12}(\omega)-k(r_1-r_2)]\,. \tag{6}$$

Here $S^{(1)}(Q,\omega)$ is the spectral density that would be observed at Q if the pinhole at P_1 alone were open, with $S^{(2)}(Q,\omega)$ having a similar meaning. Further

$$\mu_{12}(\omega)=|\mu_{12}(\omega)|\exp[\mathrm{i}\beta_{12}(\omega)] \tag{7}$$

[#1] Ensemble averages rather than time averages are used throughout this paper. If the field fluctuations are ergodic, as is usually the case, the two averages are equal to each other.

[#2] Ref. [9], sec. 10.3.1, eq. (20). We use here slightly different definitions of $\gamma_{12}(\tau)$ and $\Gamma_{12}(\tau)$.

[#3] Ref. [5], eq. (3.7). We have omitted here an unessential phase term.

Volume 81, number 3,4 OPTICS COMMUNICATIONS 15 February 1991

is the complex degree of spectral coherence (also known as the degree of spatial coherence at frequency ω) of the light at the two pinholes. It is defined in terms of the cross-spectral density function

$$W_{12}(\omega)=\frac{1}{2\pi}\int_{-\infty}^{\infty}\Gamma_{12}(\tau)\exp(\mathrm{i}\omega\tau)\,\mathrm{d}\tau \tag{8}$$

by the expression

$$\mu_{12}(\omega)=\frac{W_{12}(\omega)}{\sqrt{S(P_1,\omega)}\sqrt{S(P_2,\omega)}}, \tag{9}$$

where $S(P_1,\omega)$ and $S(P_2,\omega)$ are the spectral densities at the two pinholes. In eq. (6), r_1 and r_2 have the same meanings as before and

$$k=\omega/c \tag{10}$$

is the wavenumber associated with the frequency ω.

It is important to appreciate that although the cross-spectral density $W_{12}(\omega)$ and the mutual coherence function $\Gamma_{12}(\tau)$ form a Fourier transform pair, the two degrees of coherence, $\mu_{12}(\omega)$ and $\gamma_{12}(\tau)$ do not form such a pair. Each of the degrees of coherence carries different physical information about the correlation properties of the field.

Let us now consider the spectrum at Q when the pinholes are illuminated by a uniform, circular, quasi-homogeneous [10] secondary source centered at a point C on the normal to the screen, passing through the midpoint O of P_1 and P_2 (the z axis in fig. 1) and located in a plane parallel to the screen. (By uniform we mean here that the spectrum of light is the same at every source point.) Further we assume that the pinholes are in the far zone of the source. It readily follows from one of the reciprocity relations for fields generated by quasi-homogeneous sources [10], which is analogous to the far zone form of the Van-Cittert–Zernike theorem [ref. 9, §10.4.2], that

$$\mu_{12}(\omega)=\frac{2J_1(\omega\alpha d/c)}{\omega\alpha d/c}, \tag{11}$$

where $J_1(x)$ is the first-order Bessel function of the first kind, α is the angular semi-diameter which the source subtends at O and d is the distance (assumed to be sufficiently small) between the two pinholes.

Let us suppose that the point Q is on the axis [#4], so that $r_1=r_2$ $(=r$ say). We denote this point by Q_0. If $S^{(\mathrm{i})}(\omega)$ is the spectral density of the light incident on each of the two pinholes, then, provided that the angles of incidence and diffraction are small,

$$S^{(1)}(Q_0,\omega)=S^{(2)}(Q_0,\omega)\approx\frac{A^2}{(2\pi)^2c^2r^2}\,\omega^2S^{(\mathrm{i})}(\omega), \tag{12}$$

where A is the area of either of the two (identical) pinholes [#5]. On substituting from eqs. (11) and (12) into the spectral interference law (6) we obtain the following expression for the spectral density at Q_0:

$$S(Q_0,\omega)=2\left(\frac{A}{2\pi cr}\right)^2\omega^2\left[1+\frac{2J_1(\omega\alpha d/c)}{\omega\alpha d/c}\right]S^{(\mathrm{i})}(\omega). \tag{13}$$

This formula shows that, apart from geometrical factors, the spectrum of the light at the point Q_0 differs for the spectrum $S^{(\mathrm{i})}(\omega)$ of light incident on the pinholes for two reasons: Because of the presence of the multiplicative factor ω^2 due to diffraction at the pinholes and also because of the partial coherence of the incident light, characterized by the second term in the square bracket. This latter contribution is seen to depend on the angular semi-diameter α of the source viewed from O and on the distance d between the pinholes.

It is of interest to note that when $d\ll c/\omega\alpha$ or $d\gg c/\omega\alpha$ for all effective frequencies contained in the spectrum of the incident field, the spectral density $S(Q_0,\omega)$ is, according to eqs. (12) and (13), proportional to the spectral density $S^{(1)}(Q_0,\omega)$, which differs from the spectral density $S^{(\mathrm{i})}(\omega)$ of the incident field by a multiplicative factor proportional to ω^2, due to diffraction. In the first case the

[#4] This simplifying assumption is made here to illustrate the essential feature of this phenomenon. For points Q not on the axis we would have, in place of eq. (13) the more general formula $S(Q,\omega)=2(A/2\pi cr)^2\omega^2\{1+[2J_1(\omega\alpha d/c)/(\omega\alpha d/c)]\cos[\omega(r_1-r_2)/c]\}S^{(\mathrm{i})}(\omega)$. This spectrum differs from the axial spectrum $S(Q_0,\omega)$ by the presence of rapid oscillatory modulations arising from the frequency-dependent factor $\cos[\omega(r_1-r_2)/c]$. Because the point Q is chosen to be on axis, colour effects, which are usually produced in interference experiments with broadband light, are absent.

[#5] Eq. (12) is a good approximation if the pinholes are sufficiently small.

radiation emerging from the pinholes is essentially mutually completely coherent, whereas in the other it is mutually completely incoherent.

We will now illustrate our analysis by two examples. In both cases we assume that the source is uniform and that the geometry of fig. 1 applies.

Example 1: Narrow-band spectrum.

Let us assume that the light incident on the aperture in the source plane is filtered thermal light and that its spectrum consists of a single line of lorentzian profile, centered at frequency ω_0. The spectrum of the light incident on each of the pinholes will then be given by the expression [#6]

$$S^{(i)}(\omega)=B\frac{1}{\Gamma^2+(\omega-\omega_0)^2}, \tag{14}$$

where B and Γ are positive constants and $\Gamma/\omega_0 \ll 1$. On substituting from eq. (14) into eq. (13) we obtain for the spectral density at Q_0 the expression

$$S(Q_0,\omega)=C\frac{\omega^2}{\Gamma^2+(\omega-\omega_0)^2}\left[1+\frac{2J_1(\omega\alpha d/c)}{\omega\alpha d/c}\right], \tag{15}$$

where C is a positive constant, independent of frequency. Straightforward but somewhat lengthy calculation shows that eq. (15) represents a spectral line whose maximum is not at the central frequency ω_0 of the incident light but at a slightly higher frequency

$$\omega_0' \approx \omega_0\left\{1+\left(\frac{\Gamma}{\omega_0}\right)^2\left[\frac{1+J_0(\omega_0\alpha d/c)}{1+2J_1(\omega_0\alpha d/c)/(\omega_0\alpha d/c)}\right]\right\}. \tag{16}$$

In fig. 2 the relative frequency shift is plotted as a function of the pinhole separation d, for selected values of ω_0 and Γ. We see that the shift is very small and that it is an oscillatory function of the distance between the two pinholes.

Example 2: Broad-band spectrum.

Suppose now that the source is a secondary, thermal, quasi-homogeneous source and that its spectrum is given by Planck's law. The spectrum of the radiation incident on the pinholes will then also be given by Planck's law [#6], i.e.

$$S^{(i)}(\omega)=B'\frac{\omega^3}{\exp(\hbar\omega/k_B T)-1}, \tag{17}$$

($\hbar$ is the Planck constant divided by 2π, k_B is the Boltzmann constant, T is the temperature and B' is

[#6] That the normalized spectrum of the radiation incident on the pinholes is the same as the normalized source spectrum is a consequence of the fact that a quasi-homogeneous thermal source satisfies the so-called scaling law (ref. [11]).

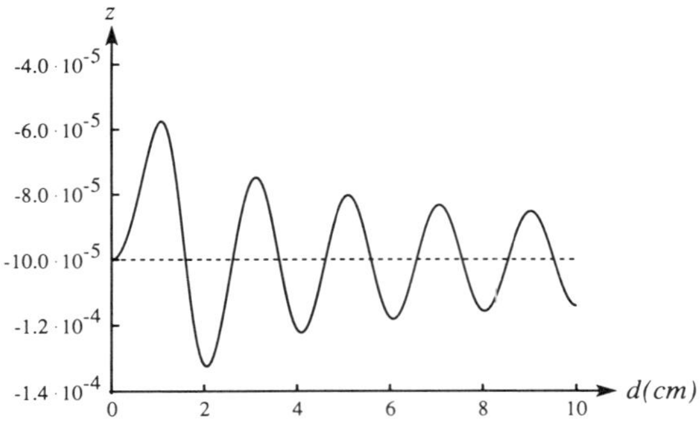

Fig. 2. The relative frequency shift $z=(\omega_0-\omega_0')/\omega_0'$ produced by interference, as a function of the separation d of the two pinholes. The spectrum of the incident light consists of a single spectral line of lorentzian profile, with central frequency $\omega_0=3.25\times10^{15}\,s^{-1}$ and width $\Gamma=3.25\times10^{13}\,s^{-1}$. The angular semi-diameter of the source was taken to be 2.96×10^{-5} radians.

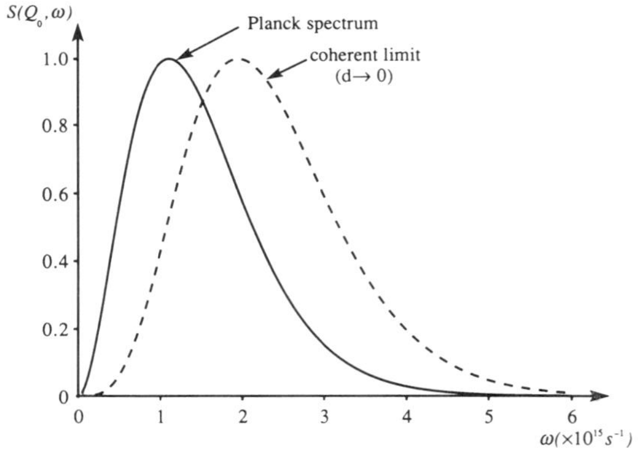

Fig. 3. The Planck spectrum of a source at temperature $T=3000$ K (solid line) and the spectrum at Q_0 obtained in the coherent limit, i.e. when the pinholes are very close to each other (dashed line). Each spectrum is normalized to have the value unity at the peak.

Volume 81, number 3,4 OPTICS COMMUNICATIONS 15 February 1991

a positive constant). On substituting from eq. (17) into eq. (13) we obtain the following expression for the spectrum at Q_0:

$$S(Q_0, \omega) = C' \frac{\omega^5}{\exp(\hbar\omega/k_B T) - 1} \left[1 + \frac{2J_1(\omega\alpha d/c)}{\omega\alpha d/c} \right], \quad (18)$$

with C' being another positive constant.

Fig. 3 shows the Planck spectrum of the source and the spectrum $S(Q_0, \omega)$ that would be observed in the coherent limit, i.e. when the radiation from the source propagates through pinholes when they are very close to each other. The difference between the two curves is due to diffraction. In fig. 4 the corresponding spectra are drawn for larger separation d between the pinholes. The difference between these spectra is due to the dependence of the degree of spatial coherence $\mu_{12}(\omega)$ on the separation d between the pinholes.

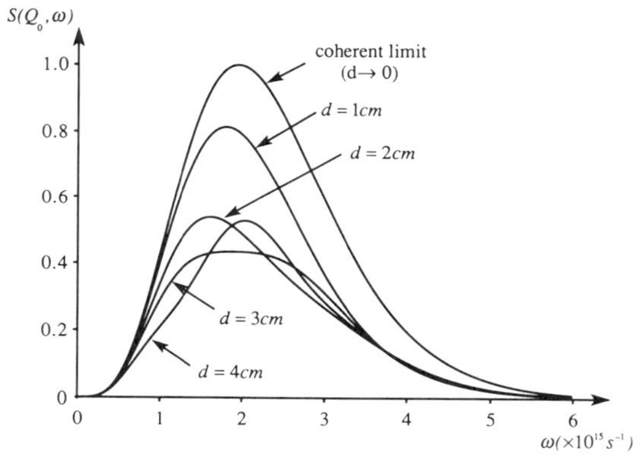

Fig. 4. The changes in the Planck spectrum by interference for different pinhole separations d. The source was assumed to be at temperature $T = 3000$ K and to subtend an angular semi-diameter of $\alpha = 2.96 \times 10^{-5}$ radians at O. The units on the vertical axis are arbitrary, but are the same for all values of d.

It is seen from eq. (13) or from the more general equation given in footnote #4, that when the incident light has sufficiently broad band, spectral measurements with a fixed value of d allow the value of α to be deduced. Potential application of this result to the development of a new interferometric technique will be described elsewhere.

This research was supported by the Department of Energy under grant DE-FG02-90ER 14119. The views expressed in this article do not constitute an endorsement by the Department of Energy.

References

[1] T. Young, Phil. Trans. Roy. Soc. (London), XCII 12 (1802) 387; Young's Works, Vol. 1, pp. 140, 170.
[2] F. Zernike, Physica 5 (1938) 785.
[3] B.J. Thompson and E. Wolf, J. Opt. Soc. Am. 47 (1957) 895.
[4] A.A. Michelson, Phil. Mag. (5) 30 (1890) 1; Studies in Optics (University of Chicago Press, Chicago, Third impression, 1962), Chap. XI.
[5] L. Mandel and E. Wolf, J. Opt. Soc. Am. 66 (1976) 529.
[6] E. Wolf, Optics Lett. 8 (1983) 250.
[7] L. Mandel, J. Opt. Soc. Am. 51 (1961) 1342.
[8] E. Wolf, Phys. Rev. Lett. 58 (1987) 2646.
[9] M. Born and E. Wolf, Principles of Optics (Pergamon, Oxford, 6th ed., 1980).
[10] W.H. Carter and E. Wolf, J. Opt. Soc. Am. 67 (1977) 785.
[11] E. Wolf, Phys. Rev. Lett. 56 (1986) 1370.

Radio Science, Volume 26, Number 5, Pages 1239–1243, September–October 1991

Determination of field correlations from spectral measurements with application to synthetic aperture imaging

Daniel F. V. James

The Institute of Optics, University of Rochester, Rochester, New York

Emil Wolf[1]

Department of Physics and Astronomy, University of Rochester, Rochester, New York

(Received September 9, 1990; accepted March 19, 1991.)

By making use of some new results in optical coherence theory we show that spectral analysis, rather than fringe visibility measurements, can be used to determine correlation properties of the field. A principle is also formulated, according to which, in many interferometric measurements a trade-off can be made between the length of the baseline and the frequency at which measurements are made. This principle is a rigorous generalization of the so-called space-frequency equivalence theorem for two antenna detection systems, and it applies to radiation from a broader class of sources than previously considered.

INTRODUCTION

The use of multiple-element interferometers for synthetic aperture mapping of astronomical radio sources is one of the most successful techniques in modern astronomy (for a description, see, for example, *Thompson et al.* [1986]). These instruments measure the complex degree of coherence of the field produced by a distant radio source, for several baseline lengths simultaneously. The detection system filters the incoming signal to become quasi-monochromatic, i.e., narrow band around a known central frequency. Interference fringes are then produced and the complex degree of coherence is determined from visibility measurements. Assuming that the source is spatially incoherent, the intensity distribution across the source is then determined by Fourier inversion, in accordance with the far-zone form of a classic result of coherence theory, the van Cittert-Zernike theorem [*Born and Wolf*, 1980, section 10.4.2].

Recent research has shown that the state of coherence of a wave field affects not only the fringe visibility but also the spectrum of the field in the fringe pattern [*James and Wolf*, 1991]. This result is intimately related to a prediction made a few years ago [*Wolf*, 1986; 1987*a*, *b*, *c*] and since then confirmed by several experiments [*Morris and Faklis*, 1987; *Bocko et al.*, 1987; *Faklis and Morris*, 1988; *Gori et al.*, 1988; *Indebetouw*, 1989; *Kandpal et al.*, 1989], according to which the spectrum of a field generated by a fluctuating source depends not only on the source spectrum but also on the correlation properties of the source. Some implications of this phenomenon for cosmology [*Wolf*, 1987*a*; *Naklikar*, 1989], signal modulation [*Gamliel and Wolf*, 1988; *Indebetouw*, 1989], spectroradiometry standards specifications [*Kandpal et al.*, 1989; *Foley*, 1990], enhanced backscattering [*Lagendijk*, 1990], and other fields have been considered.

In this paper we propose another application of this effect, namely determining the degree of spectral coherence (defined by (3) and (4) below) from measurements of the changes of the spectrum produced on interference. We also show that for a broad class of sources the degree of spectral coherence at any two points in the radiation field depends on the separation of the two points and on the frequency only through their product, and we discuss a potential application of this result to remote sensing.

[1] Also at the Institute of Optics, University of Rochester, Rochester, New York.

Paper number 91RS01032.
0048-6604/91/91RS-01032$08.00

DETERMINATION OF CORRELATIONS FROM SPECTRAL MEASUREMENTS

To illustrate the procedure, we consider a two-element interferometer, with identical antennas at points P_1 and P_2, exposed to incident signals $V_1(t)$ and $V_2(t)$, respectively, t denoting the time. These quantities are the complex analytic signals associated with the real electromagnetic field [*Born and Wolf*, 1980, section 10.2]. We assume that the electronic signals produced by the detectors are proportional to these quantities, ignoring here detector and antenna effects. Further, we assume that the fluctuations of the incident field are characterized by a statistical ensemble which is stationary, at least in the wide sense [*Davenport and Root*, 1958] and ergodic, and that the power spectra of the two signals are the same. We denote each of the power spectra by $S^{(i)}(\omega)$ ("i" standing for "incident"), ω denoting frequency.

Suppose that the two signals are added electronically, with zero path difference for simplicity. According to the Wiener-Khintchine theorem [*Kittel*, 1958], the spectrum $S(\omega)$ of the combined signal is then given by the formula

$$S(\omega) = \frac{1}{2\pi}\int_{-\infty}^{\infty} \langle\{V_1^*(t) + V_2^*(t)\}\{V_1(t+\tau) + V_2(t+\tau)\}\rangle \cdot \exp(i\omega\tau)\, d\tau, \tag{1}$$

the angular brackets denoting the ensemble average. Because we assumed the field to be ergodic, this ensemble average is, of course, equal to the corresponding time average. From (1) we readily deduce that

$$S(\omega) = 2S^{(i)}(\omega)\{1 + \mathrm{Re}\,[\mu_{12}(\omega)]\}, \tag{2}$$

where

$$\mu_{12}(\omega) = \frac{W_{12}(\omega)}{S^{(i)}(\omega)} \tag{3}$$

is the complex degree of spectral (or spatial) coherence [*Mandel and Wolf*, 1976] at frequency ω,

$$W_{12}(\omega) = \frac{1}{2\pi}\int_{-\infty}^{\infty} \langle V_1^*(t)V_2(t+\tau)\rangle \exp(i\omega\tau)\, d\tau \tag{4}$$

is the cross-spectral density function of the two signals $V_1(t)$ and $V_2(t)$, and Re in (2) denotes the real part. The complex degree of spectral coherence $\mu_{12}(\omega)$ should not be confused with the better known complex degree of coherence, usually denoted by $\gamma_{12}(\tau)$, used in traditional coherence theory. The two degrees of coherence provide different information about the correlation properties of the field, and no simple relation exists between them. However, when the bandwidth is sufficiently narrow and is centered at frequency ω_0, $\gamma_{12}(0) \approx \mu_{12}(\omega_0)$ [*Wolf*, 1983, equation (3.2)].

Because the spectrum $S^{(i)}(\omega)$ can be measured by simply disconnecting one of the antennas, we see at once from (2) that the real part of the degree of spectral coherence can be determined from the spectral measurements by means of the formula

$$\mathrm{Re}\,\{\mu_{12}(\omega)\} = \frac{S(\omega)}{2S^{(i)}(\omega)} - 1. \tag{5}$$

More generally, when the signals from the two antennas have a path difference Δs, one has instead of (2), the more general formula

$$S(\omega) = 2S^{(i)}(\omega)\{1 + |\mu_{12}(\omega)| \cos\,[\alpha_{12}(\omega) - \omega\Delta s/c]\}, \tag{6}$$

where c is the speed of light [*Mandel and Wolf*, 1976, equation (3.7)] and $\alpha_{12}(\omega)$ is the phase of $\mu_{12}(\omega)$. By taking advantage of the fact that Δs is a parameter controlled by the observer it is possible to recover the entire function $\mu_{12}(\omega)$. We will, however, only consider the case where $\mu_{12}(\omega)$ is real and $\Delta s = 0$.

In radio astronomy the observable sky extends over a broad frequency range ($0.1 \times 10^{-9}\ \mathrm{s}^{-1} \lesssim \omega \lesssim 140 \times 10^{-9}\ \mathrm{s}^{-1}$), the observability being limited at the lower end by plasma absorption in the ionosphere and at the upper end by absorption in the rotational bands of atmospheric molecules [*Rohlfs*, 1986, section 1.2]. Over this broad range it is possible to measure the spectra using ground-based antennas. Then by the use of (5) one can calculate the real part of the degree of spectral coherence over that range. For rotationally symmetric sources the knowledge of this quantity is sufficient to determine the normalized intensity distribution across the source, as can be readily deduced from the Fourier inverse of (18) below.

SPACE-FREQUENCY EQUIVALENCE

We stress that the preceding analysis applies to radiation from any planar source, whose fluctuations are statistically stationary, at least in the wide sense. A possibility of substituting, in some cases, frequency sampling for spatial sampling when a detection system consisting of two widely spaced

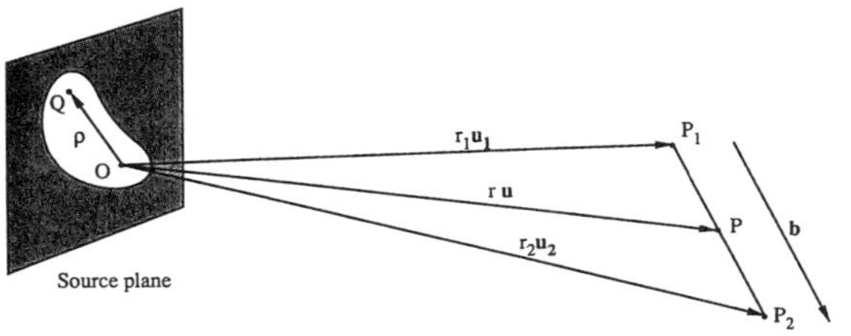

Fig. 1. Illustrating the notation relating to the determination of the degree of spectral coherence $\mu_{12}(\omega)$ of the signals received at antennas located at points P_1 and P_2, from measurements of power spectra.

antennas is used has been noted by *Kock and Stone* [1959] (see also *Cole* [1979]), and the underlying principle is known as space-frequency equivalence. It forms the basis of a technique in radio astronomy known as bandwidth synthesis mapping. We might mention that in Kock and Stone's and in Cole's papers the type of sources to which the space-frequency equivalence applies was not actually specified; we believe that the analysis given in both these papers was restricted to spatially incoherent sources which have the same normalized spectrum at every source point. We will now derive a more general and mathematically more precise theorem which applies to radiation fields produced by sources of a broader class.

Suppose that the source of the radiation field is a planar, secondary, quasi-homogeneous source [*Carter and Wolf*, 1977]. This class of sources includes both fully incoherent sources and also some partially coherent ones. We assume that the normal to plane of the distant source is directed to the midpoint between the two antennas of the detection system. Let $S^{(0)}(\boldsymbol{\rho}, \omega)$ be the spectral density at a typical point Q in the source. Further let $r_1\boldsymbol{u}_1$ and $r_2\boldsymbol{u}_2$, ($\boldsymbol{u}_1^2 = \boldsymbol{u}_2^2 = 1$), be the position vectors of the points P_1 and P_2 where the antennas are located, referred to some origin O in the source (see Figure 1). According to the second reciprocity theorem for quasi-homogeneous sources [*Carter and Wolf*, 1977, equation (4.13)] (with slightly different definitions and notation), the degree of spectral coherence of the field at the points P_1 and P_2 can be expressed in terms of the two-dimensional Fourier transform

$$\tilde{S}^{(0)}(\mathbf{f}, \omega) = \frac{1}{(2\pi)^2} \int_\sigma S^{(0)}(\boldsymbol{\rho}, \omega) e^{-i\mathbf{f}\cdot\boldsymbol{\rho}}\, d^2\rho \qquad (7)$$

of the spectral density $S^{(0)}(\boldsymbol{\rho}, \omega)$ in the form

$$\mu_{12}(\omega) = \frac{\tilde{S}^{(0)}[k(\mathbf{u}_{2\perp} - \mathbf{u}_{1\perp}), \omega]}{\tilde{S}^{(0)}(0, \omega)} e^{ik(r_2 - r_1)}. \qquad (8)$$

Here $\boldsymbol{u}_{1\perp}$ and $\boldsymbol{u}_{2\perp}$ are the projections, considered as two-dimensional vectors, of the unit vectors $\boldsymbol{u}_1$ and $\boldsymbol{u}_2$ onto the source plane and

$$k = \frac{\omega}{c}. \qquad (9)$$

Let us assume that the source has the same normalized spectrum, $s^{(0)}(\omega)$ say, at every source point. Evidently,

$$s^{(0)}(\omega) = \frac{S^{(0)}(\boldsymbol{\rho}, \omega)}{I^{(0)}(\boldsymbol{\rho})}, \qquad (10)$$

where

$$I^{(0)}(\boldsymbol{\rho}) = \int_0^\infty S^{(0)}(\boldsymbol{\rho}, \omega)\, d\omega \qquad (11)$$

represents the spatial intensity distribution across the source. It is this quantity that one usually wishes to measure. On substituting from (10) for $S^{(10)}(\boldsymbol{\rho}, \omega)$ into (8) we obtain for $\mu_{12}(\omega)$ the expression

$$\mu_{12}(\omega) = \frac{\tilde{I}^{(0)}[k(\mathbf{u}_{2\perp} - \mathbf{u}_{1\perp})]}{\tilde{I}^{(0)}(0)} e^{ik(r_2 - r_1)}, \qquad (12)$$

where $\tilde{I}^{(0)}(\boldsymbol{f})$ is the two-dimensional Fourier transform of $I^{(0)}(\boldsymbol{\rho})$, defined by a formula of the same form as used in (7).

Let us introduce the baseline vector

$$\mathbf{b} = r_2\mathbf{u}_2 - r_1\mathbf{u}_1 \qquad (13)$$

and the mean of the position vectors of the antennas

$$r\mathbf{u} = \tfrac{1}{2}(r_1\mathbf{u}_1 + r_2\mathbf{u}_2), \qquad (\mathbf{u}^2 = 1). \qquad (14)$$

Note that $\boldsymbol{u}$ is normal to the source plane. Assuming that $|\boldsymbol{b}| << r_1$ and that $|\boldsymbol{b}| << r_2$, one may readily show that

$$k(\mathbf{u}_{2\perp} - \mathbf{u}_{1\perp}) \approx \frac{k}{r}\mathbf{b}_\perp \qquad (15)$$

and that

$$r_2 - r_1 \approx \mathbf{b} \cdot \mathbf{u}. \tag{16}$$

In (15),

$$\mathbf{b}_\perp = \mathbf{u} \times (\mathbf{b} \times \mathbf{u}) \tag{17}$$

represents the projection, considered as a two-dimensional vector, of the baseline vector $\boldsymbol{b}$ onto the source plane. On substituting from (15) and (16) in the formula (12) we find that

$$\mu_{12}(\omega) = \frac{\tilde{I}^{(0)}[k\mathbf{b}_\perp/r]}{\tilde{I}^{(0)}(0)} e^{ik\mathbf{b}\cdot\mathbf{u}}. \tag{18}$$

Thus we have obtained the result that with the class of sources we are considering the degree of spectral coherence of the far field has the functional form

$$u_{12}(\omega) = g(k\mathbf{b}; \mathbf{u}, r), \tag{19}$$

i.e., it depends on the frequency ω only through the product $\boldsymbol{kb} = \boldsymbol{\omega b}/c$. Now a planar, secondary, quasi-homogeneous source which has the same spectrum at each source point and whose degree of spectral coherence $\mu^{(0)}(\boldsymbol{\rho}_1, \boldsymbol{\rho}_2, \omega)$ has the functional form $g(k\boldsymbol{\rho}')$, $\boldsymbol{\rho}' = \boldsymbol{\rho}_2 - \boldsymbol{\rho}_1$, is said to obey the scaling law [*Wolf*, 1986]. Radiation from sources of this kind have been shown to have an important property regarding the invariance of the normalized spectrum on propagation. Because the degree of spectral coherence of the far field, given by (19), depends on the frequency in a similar manner, we will refer to such a radiation field as a scaling law field.

To bring into evidence an important implication of (19), we will write from now on $\mu(b, \omega; \boldsymbol{u}, \boldsymbol{n}, r)$ instead of $\mu_{12}(\omega)$, with $b = |\boldsymbol{b}|$ denoting the length of the baseline and $\boldsymbol{n} = \boldsymbol{b}/b$ the unit vector along $\boldsymbol{b}$. Suppose now that we keep $\boldsymbol{u}$, $\boldsymbol{n}$, and r fixed and vary b and ω. Equation (19) then implies, if we recall (9), that

$$\mu(b, \omega; \mathbf{u}, \mathbf{n}; r) = \mu(\beta b, \omega/\beta; \mathbf{u}, \mathbf{n}; r), \tag{20}$$

where β is any positive number. This formula shows that with the directions $\boldsymbol{u}$ and $\boldsymbol{n}$ and the distance r kept fixed, the degree of spectral coherence at pairs of points separated by a distance b in the far field, produced by a source of the class which we are considering, can be determined with an interferometer having a fixed baseline of length b_0, provided the measurements are made at frequency $\omega_0 b_0/b$. We will refer to this result as the interferometric equivalence principle for scaling law fields. This principle lends great power to the proposed method of measurement of the intensity distribution across the source. Using the formula (5), we can determine $Re\{\mu_{12}(\omega)\}$ over the broad range of frequencies at which radiation is received, from a single spectral measurement at a fixed baseline length. Then, according to the interferometric equivalence principle, this spectral information can be traded for information one would normally deduce from several measurements, each involving filtered radiation and a broad range of baseline lengths. Thus, provided the source one wishes to map belongs to the class which we are considering and provided also that it has broad enough spectrum, inversion can be made with high accuracy, using only two antennas at fixed separation.

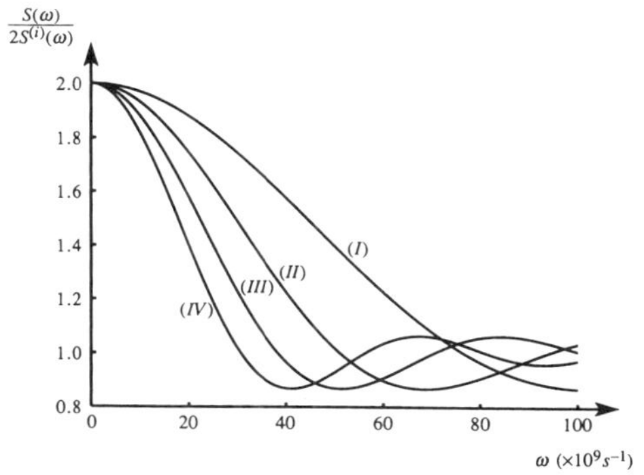

Fig. 2. The ratio $S(\omega)/2S^{(i)}(\omega)$ of the power spectrum of the combined signal, $S(\omega)$, to the sum, $2S^{(i)}(\omega)$, of the power spectra of each signal, plotted as a function of the frequency ω, for several values of the baseline length b, for a radiation field generated by a uniform, circular, incoherent planar source whose diameter subtends an angle 2α at the point P midway between the antennas. The midpoint is assumed to be located on the normal to the source plane, passing through the center O of the source, and the baseline vector $\boldsymbol{b} = b\boldsymbol{n}$ is taken to be perpendicular to this normal. The chosen baseline lengths are (line I) $b = (0.05 \times 10^{-9}\ s)c/\alpha$, (line II) $b = (0.075 \times 10^{-9}\ s)c/\alpha$, (line III) $b = (0.1 \times 10^{-9}\ s)c/\alpha$, and (line IV) $b = (0.125 \times 10^{-9}\ s)c/\alpha$.

As an illustration of this principle let us consider the degree of spectral coherence generated in the far zone by a uniform, circular source, with centre at the origin O, which subtends an angular diameter 2α at the midpoint P between the two antennas, it being assumed that P lies on the normal to the source plane through O., Assuming that $\boldsymbol{b}$ is perpendicular to $\boldsymbol{u}$, one readily finds from (8) that, in this case,

$$\mu(b, \omega; \mathbf{u}, \mathbf{n}; r) = \frac{2J_1(b\omega\alpha/c)}{b\omega\alpha/c}. \tag{21}$$

Evidently, the expression (21) satisfies the interferometric equivalence principle. Figure 2 shows, for this case, the ratio of the power spectrum, $S(\omega)$, of the combined signal to the sum of the power spectra, $2S^{(i)}(\omega)$, of the signals received at each of the two antenna, as a function of frequency, for several different baseline lengths.

CONCLUSION

We conclude by stressing that (5) together with the Fourier inverse of (18) and the interferometric equivalence principle for scaling law fields expressed by (20) provide the basis for a new and potentially powerful interferometric technique.

Acknowledgments. This research was supported by the National Science Foundation and by the Army Research Office.

REFERENCES

Bocko, M. F., D. H. Douglass, and R. S. Knox, Observation of frequency shifts of spectral lines due to source correlations, *Phys. Rev. Lett.*, *58*, 2649–2651, 1987.

Born, M., and E. Wolf, *Principles of Optics*, 6th ed., Pergamon, New York, 1980.

Carter, W. H., and E. Wolf, Coherence and radiometry with quasihomogeneous planar sources, *J. Opt. Soc. Am.*, *67*, 785–796, 1977.

Cole, T. W., Mutual coherence function in radio astronomy, *J. Opt. Soc. Am.*, *69*, 554–557, 1979.

Davenport, W. B., and W. L. Root, *Random Signals and Noise*, p. 60, McGraw-Hill, New York, 1958.

Faklis, D., and G. M. Morris, Spectral shifts produced by source correlations, *Opt. Lett.*, *13*, 4–6, 1988.

Foley, J. T., The effect of an aperture on the spectrum of partially coherent light, *Opt. Commun.*, *75*, 347–352, 1990.

Gamliel, A., and E. Wolf, Spectral modulation by control of source correlations, *Opt. Commun.*, *65*, 91–96, 1988.

Gori, F., G. Guattari, C. Palma, and C. Padovani, Observation of optical redshifts and blueshifts produced by source correlations, *Opt. Commun.*, *67*, 1–4, 1988.

Indebetouw, G., Synthesis of polychromatic light sources with arbitrary degrees of coherence: Some experiments, *J. Mod. Opt.*, *36*, 251–259, 1989.

James, D. F. V., and E. Wolf, Spectral changes produced in Young's interference experiment, *Opt. Commun.*, *81*, 150–154, 1991.

Kandpal, H. C., J. S. Vaishya, and K. C. Joshi, Wolf shift and its application in spectroradiometry, *Opt. Commun.*, *73*, 169–172, 1989.

Kittel, C., *Elementary Statistical Physics*, p. 133, John Wiley, New York, 1958.

Kock, W. E., and J. L. Stone, Space-frequency equivalence, *Proc. I/R/E*, *46*, 449–450, 1958.

Lagendijk, A., Terestial redshifts from a diffuse light source, *Phys. Lett. A.*, *147*, 389–392, 1990.

Mandel, L., and E. Wolf, Spectral coherence and the concept of cross-spectral purity, *J. Opt. Soc. Am.*, *66*, 529–535, 1976.

Morris, G. M., and D. Faklis, Effects of source correlations on the spectrum of light, *Opt. Commun.*, *62*, 5–11, 1987.

Narlikar, J. V., Noncosmological redshifts, *Space Sci. Rev.*, *50*, 523–614, sect. 6.3 and 7.2, 1989.

Rohlfs, K., *Tools of Radio Astronomy*, Springer-Verlag, New York, 1986.

Thompson, A. R., J. M. Moran, and G. W. Swenson, *Interferometry and Synthesis in Radio Astronomy*, John Wiley, New York, 1986.

Wolf, E., Young's interference fringes with narrow-band light, *Opt. Lett.*, *8*, 250–252, 1983.

Wolf, E., Invariance of spectrum of light on propagation, *Phys. Rev. Lett.*, *56*, 1370–1372, 1986.

Wolf, E., Non-cosmological redshifts of spectral lines, *Nature*, *326*, 363–365, 1987*a*.

Wolf, E., Redshifts and blueshifts of spectral lines caused by source correlations, *Opt. Commun.*, *62*, 12–16, 1987*b*.

Wolf, E., Red shifts and blue shifts of spectral lines emitted by two correlated sources, *Phys. Rev. Lett.*, *58*, 2646–2648, 1987*c*.

D. F. V. James, The Institute of Optics, University of Rochester, Rochester, NY 14627.

E. Wolf, Department of Physics and Astronomy, University of Rochester, NY 14627.

Volume 157, number 1 PHYSICS LETTERS A 15 July 1991

Some new aspects of Young's interference experiment

Daniel F.V. James [a] and Emil Wolf [a,b]

[a] *The Institute of Optics, University of Rochester, Rochester, NY 14627, USA*
[b] *Department of Physics and Astronomy, University of Rochester, Rochester, NY 14627, USA*

Received 5 April 1991; accepted for publication 10 April 1991
Communicated by J.P. Vigier

It was first predicted by Zernike in a classic paper that the fringes formed in Young's interference experiment reveal some of the spatial coherence properties of the light incident on the two pinholes. More recently it was found that the interference fringes also provide information about the *spectral* coherence properties of the incident light, because its degree of spatial coherence affects the spectrum of the light in the fringe pattern. The present paper is concerned with elucidating the spectral changes which take place in such superposition experiments, when the light incident on the two small apertures has appreciable bandwidth.

Young's interference experiment, which was performed in 1801 [1], is one of the most fundamental experiments of all of physics. Fresnel made use of it in 1818 in his classic memoir in which he placed the wave theory of light on a sound foundation [2]; and Zernike in 1938 in a basic paper in which he introduced the concept of the degree of coherence of light [3]. Moreover, since quantum mechanics was formulated in the 1920's, the experiment has played an important role in the elucidation of the wave–particle duality of both light and matter. Numerous books and papers dealing with the foundation of quantum mechanics, which continue to be published in large number to this day, also pay a good deal of attention to this experiment (see for example refs. [4,5]).

In the last few years several investigations have been made concerning the effects of spatial coherence on the spectra of optical fields. Mandel was first to show that when two light beams with the same spectrum are superposed the spectrum of light in the region of superposition will, in general, differ from the spectrum of the two beams [6]. More recently it was demonstrated both theoretically and experimentally that coherence properties of a source affect the spectrum of the emitted light, even on propagation in free space [#1].

The spectral changes produced in Young's interference experiment are of special interest not only because of the conceptual simplicity of the experimental arrangement, but also because of the role that this experiment has played and is still playing in clarifying some aspects of quantum physics.

In a recent paper [12] we discussed the spectral changes that are produced at axial points in Young's interference experiment, when the light incident on the pinholes originates in a planar, secondary, circular, black-body source, under symmetrical illumination. In the present paper we extend the analysis to the whole fringe pattern. Our results reveal that when the spectrum of the incident light is broad, the spectrum at an off-axis point in the region of superposition is appreciably modified and exhibits oscillatory behaviour which is not present at points on the axis. We begin by summarizing the main formulas which we will need.

Let us consider a partially coherent light field incident from the left onto an opaque screen $\mathscr{B}$ containing two small identical apertures at points P_1 and P_2 (fig. 1). We represent the fields at these points by ensembles of frequency-dependent realizations [13] $\{U(P_1, \omega)\}$ and $\{U(P_2, \omega)\}$ respectively. We assume that the apertures are sufficiently small so that the amplitude of the field is effectively constant over each of them and

[#1] See, for example refs. [7–9] and also the review articles by Milonni and Singh [10] and by Wolf [11].

Volume 157, number 1 PHYSICS LETTERS A 15 July 1991

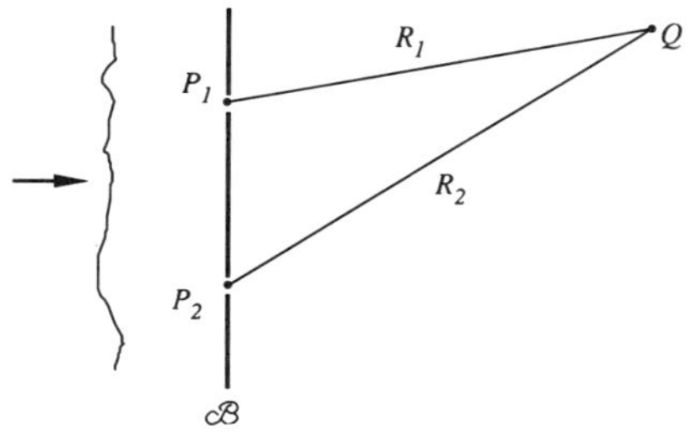

Fig. 1. Illustrating the geometry and notation relating to eq. (1).

that the angles of incidence and diffraction are also small. Then the field at a point Q beyond the screen is, to a good approximation given by

$$U(Q,\omega)=\mathrm{i}kA\left(U(P_1,\omega)\frac{\mathrm{e}^{\mathrm{i}kR_1}}{R_1}+U(P_2,\omega)\frac{\mathrm{e}^{\mathrm{i}kR_2}}{R_2}\right). \tag{1}$$

Here A is the area of either aperture, R_1 and R_2 are the distances from each of the pinholes to the point Q, $k=\omega/c$ is the wavenumber and c is the speed of light.

The spectrum at the point Q is given by the expression

$$S(Q,\omega)=\langle U^*(Q,\omega)\,U(Q,\omega)\rangle\,, \tag{2}$$

where the angular brackets denote an ensemble average. On substituting from eq. (1) into eq. (2), we obtain for the spectrum at the point Q the expression

$$S(Q,\omega)=S^{(1)}(Q,\omega)+S^{(2)}(Q,\omega)+2\sqrt{S^{(1)}(Q,\omega)}\,\sqrt{S^{(2)}(Q,\omega)}\;|\mu_{12}(\omega)|\,\cos[\beta_{12}(\omega)+\omega(R_2-R_1)/c]\,. \tag{3}$$

Here $S^{(1)}(Q,\omega)$ is the spectral density which would be observed at Q if the aperture at P_1 alone was open and $S^{(2)}(Q,\omega)$ has a similar meaning with regard to the pinhole at P_2. Further,

$$\mu_{12}(\omega)\equiv|\mu_{12}(\omega)|\,\exp[\mathrm{i}\beta_{12}(\omega)]=\frac{\langle U^*(P_1,\omega)\,U(P_2,\omega)\rangle}{\sqrt{\langle U^*(P_1,\omega)\,U(P_1,\omega)\rangle}\sqrt{\langle U^*(P_2,\omega)\,U(P_2,\omega)\rangle}} \tag{4}$$

is the degree of spectral coherence (also called the degree of spatial coherence) of the light at the two pinholes.

Formula (3) is the so-called *spectral interference law* for beams of any state of coherence [14]. In most cases of interest one may assume that $S^{(1)}(Q,\omega)\approx S^{(2)}(Q,\omega)$. Eq. (3) then reduces to

$$S(Q,\omega)=2S^{(1)}(Q,\omega)\,\{1+|\mu_{12}(\omega)|\,\cos[\beta_{12}(\omega)+\omega(R_2-R_1)/c]\}\,. \tag{5}$$

This formula implies the following two results: (1) At any fixed frequency ω, the spectral density varies sinusoidally with the position of the point Q across the observation plane, with the amplitude and the phase of the variation depending on the value of the degree of the spectral coherence $\mu_{12}(\omega)$; and (2) at any fixed point Q in the observation plane the spectrum $S(Q,\omega)$ will differ from the spectrum $S^{(1)}(Q,\omega)$, the change depending on the degree of spectral coherence of the light at the two pinholes. Such a spectral modification arising on interference was first observed with coherent pulses by Alford and Gold [15]. A qualitative study of this effect, for the special case of cross-spectrally pure beams, was later made by Mandel [16].

In order to elucidate the spectral changes that can be produced on interference, let us consider a simple example. Suppose that the pinholes are separated by a distance d and are illuminated symmetrically by (spatially) effectively incoherent light emerging from a circular aperture located in the plane $\mathscr{A}$; and that the spectral density in the region of superposition is measured in the immediate neighbourhood of a point Q, at a distance x

Volume 157, number 1 PHYSICS LETTERS A 15 July 1991

from the axis of symmetry on a screen $\mathscr{C}$ at distance R behind the plane $\mathscr{B}$ containing the pinholes (see fig. 2). We assume that

$$R_1 \approx R_2 \approx R \tag{6}$$

and that $x/R \ll 1$, so that we may make the approximation

$$R_2 - R_1 \approx xd/R . \tag{7}$$

We also assume that the pinholes are in the far zone of the source. The degree of spectral coherence of the light incident on the pinholes may be readily determined by the use of one of the reciprocity relations for fields generated by a planar quasi-homogeneous source [17], which is closely analogous to the far-zone form of the Van Cittert–Zernike theorem (ref. [18], section 10.4.2). It is found to be given by the expression

$$\mu_{12}(\omega) = \frac{2J_1(\omega\alpha d/c)}{\omega\alpha d/c} . \tag{8}$$

Here J_1 is the Bessel function of the first kind and first order and α is the angular radius which the circular aperture in the plane $\mathscr{A}$ subtends at a point midway between the pinholes. Under these circumstances expression (5) for the spectrum of the light in the interference pattern becomes

$$S(Q,\omega) = 2S^{(1)}(Q,\omega)\left(1 + \left|\frac{2J_1(\omega\alpha d/c)}{\omega\alpha d/c}\right| \cos[\beta_{12}(\omega) + \omega xd/Rc]\right), \tag{9}$$

where $\beta_{12}(\omega) = 0$ or π according to whether $J_1(\omega\alpha d/c)$ is positive or negative.

Suppose now that the spectrum of the light incident on the pinholes is given by Planck's law:

$$S_{\mathrm{Pl}}(\omega) = B\frac{\omega^3}{\exp(\hbar\omega/k_{\mathrm{B}}T) - 1} \tag{10}$$

($\hbar$ is Planck's constant divided by 2π, k_{B} is Boltzmann's constant, T is the absolute temperature and B is a positive constant). In this case, taking into account the effects of diffraction at the pinholes and of propagation from the pinholes to the point Q in the observation plane, one finds that

$$S^{(1)}(Q,\omega) = C\frac{\omega^5}{\exp(\hbar\omega/k_{\mathrm{B}}T) - 1} , \tag{11}$$

where $C = BA^2/c^2R^2$ and A, as before, denotes the area of each pinhole. Because we assumed that the angles of incidence and diffraction are small, the spectrum $S^{(1)}(Q,\omega)$ is independent of the precise position of the point of observation Q. Two spectra, given by eqs. (10) and (11), are plotted in fig. 3.

In fig. 4 spectra are plotted as calculated by the use of formulas (9) and (11), at different points in the plane of observation and for various separations of the pinholes, when the blackbody source is at temperature $T = 3000$ K. On comparing figs. 3 and 4 we see that the spectrum changes drastically on interference and that, with increasing distance x from the axis, it exhibits more and more rapid oscillations. These modifications of the spec-

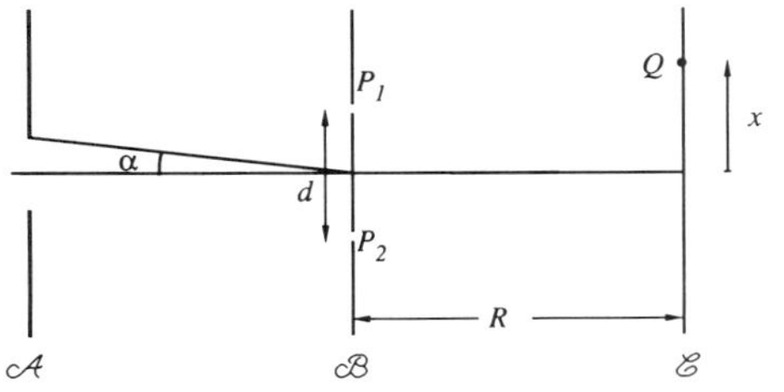

Fig. 2. Notation relating to Young's interference experiment and illustrating the geometry of our example. The value of d was taken as 0.1 cm, with $R = 150$ cm.

Volume 157, number 1 PHYSICS LETTERS A 15 July 1991

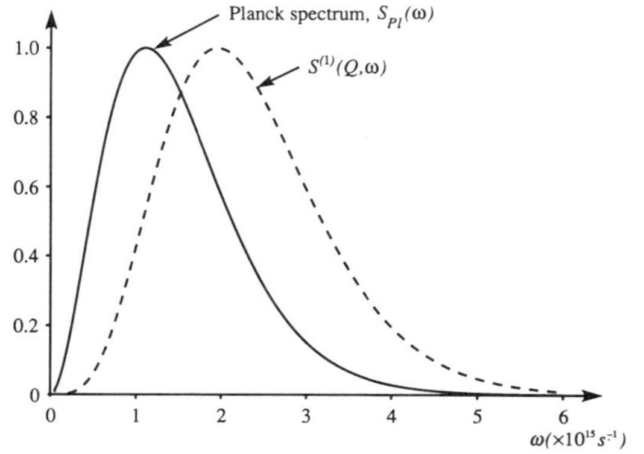

Fig. 3. The Planck spectrum of a source, $S_{Pl}(\omega)$, at temperature $T=3000$ K (solid line) and the spectrum of light observed with only one pinhole open, $S^{(1)}(Q,\omega)$ (dashed line). Each spectrum is normalized to have the value unity at the peak.

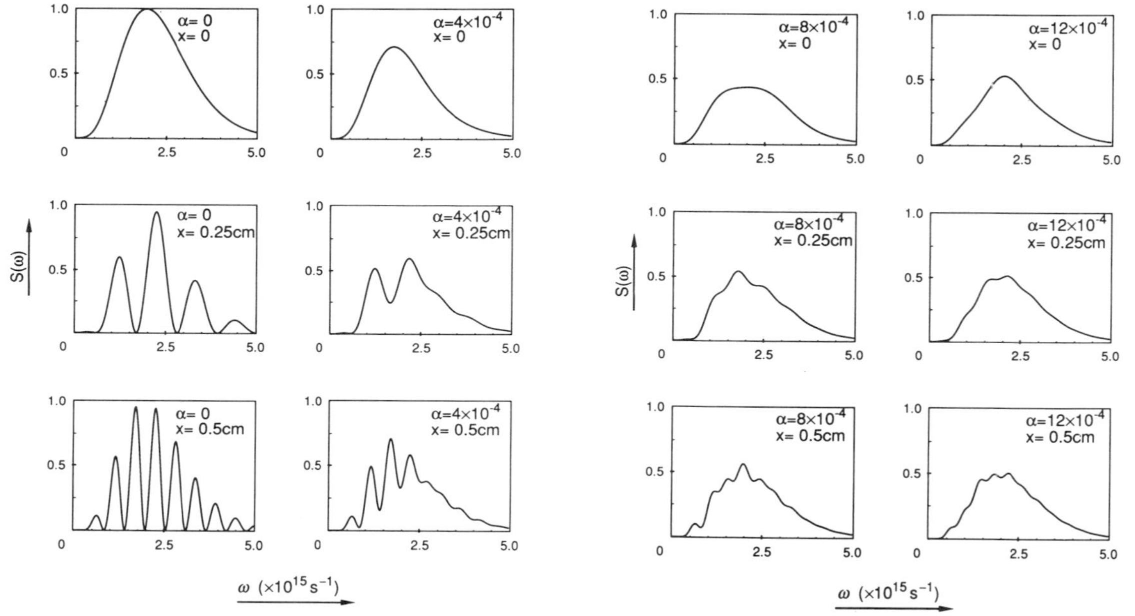

Fig. 4. The spectra at different points Q located at off-axis distance x, when angular radius of the source subtended at the pinholes is α (in radian). The spectrum of the incident light was assumed to be a blackbody spectrum at temperature 3000 K. The calculations were performed with the choice $d=0.1$ cm and $R=150$ cm.

trum are caused both by diffraction and by the partially coherent superposition (characterized by the degree of coherence given by eq. (8)) of the light emerging from the pinholes.

It is evident from eq. (3) that from measurements of spectral distributions in the plane of observation, one can determine the degree of spectral coherence $\mu_{12}(\omega)$ of the light incident on the pinholes. In particular this information could be deduced from measurements of the fringe visibility and the location of the fringe maxima, after the light at the two pinholes, or in the region of superposition, is passed through filters which are sharply tuned to the particular frequency ω [19]. In this connection it is of interest to recall that measurement of the

Volume 157, number 1 PHYSICS LETTERS A 15 July 1991

degree of spectral coherence is at the root of the important technique of aperture synthesis in radio astronomy. Usually such measurements are carried out at a fixed frequency ω for different separations d of two antennas (which correspond to the pinholes of our analysis). In order to obtain an appreciable range of different separations without resorting to physically moving the large antennas, arrays of telescopes, such as the Very Large Array at Socorro, New Mexico, have been constructed [20]. However, it is known that in certain, fairly common situations, measurement at a single frequency for different range of separations d is equivalent to measurements over a range of frequencies, obtained with a *fixed* separation d of the antenna [21] [#2]. In such a situation one can thus avoid the use of a large array of radio telescopes and deduce the same information from measurements with just two antennas, provided that the bandwidth of the incident radiation is broad enough.

It is clear from eq. (5) that appreciable spectral modification takes place only when the degree of spectral coherence changes substantially over the bandwidth of the incident light. Generally this will be the case when the spectrum of the light incident on the aperture is sufficiently broad. Under these circumstances no interference fringes will be formed. On the other hand when the spectrum of the incident light is narrow, interference fringes will, in general, be formed, but no appreciable spectral changes will then be produced. Thus our analysis brings into evidence a kind of complementarity involving fringe formation and spectral changes produced on superposing two beams.

This research was supported by the Department of Energy under grant DE-FG02-90ER 14119. The views expressed in this article do not constitute an endorsement by the Department of Energy.

[#2] A related technique, known as multi-frequency synthesis, used to enhance the resolving power of conventional radio telescope arrays, has recently been reported [22].

References

[1] T. Young, Philos. Trans. R. Soc. XCII 12 (1802) 387; Young's works, Vol. 1, pp. 140, 170.
[2] A.J. Fresnel, Oeuvres completes d'Augustin Fresnel, Vol. 1, eds. H. de Senarmont, E. Verdet and L. Fresnel (Imprimerie Imperiale, Paris, 1866) pp. 89, 129.
[3] F. Zernike, Physica 5 (1938) 785.
[4] P.A.M. Dirac, The principles of quantum mechanics, 4th Ed. (Oxford Univ. Press, Oxford, 1958).
[5] L. Mandel, in: Progress in optics, Vol. 13, ed. E. Wolf (North-Holland, Amsterdam, 1976) ch. 2, section 9, pp. 61–65.
[6] L. Mandel, J. Opt. Soc. Am. 51 (1961) 1342.
[7] E. Wolf, Phys. Rev. Lett. 56 (1986) 1370.
[8] G.M. Morris and D. Faklis, Opt. Commun. 62 (1987) 5.
[9] F. Gori, G. Guattari, C. Palma and G. Padovani, Opt. Commun. 67 (1988) 1.
[10] P.W. Milonni and S. Singh, in: Advances in atomic, molecular and optical physics, Vol. 28, eds. D. Bates and B. Bederson (Academic Press, New York, 1991) section 8, pp. 127–137.
[11] E. Wolf, in: International trends in optics, ed. J.W. Goodman (Academic Press, New York, 1990) ch. 16, pp. 221–232.
[12] D.F.V. James and E. Wolf, Opt. Commun. 81 (1991) 150.
[13] E. Wolf, Opt. Commun. 38 (1981) 3.
[14] L. Mandel and E. Wolf, J. Opt. Soc. Am. 66 (1976) 529.
[15] W.P. Alford and A. Gold, Am. J. Phys. 26 (1958) 481.
[16] L. Mandel, J. Opt. Soc. Am. 52 (1962) 1335.
[17] W.H. Carter and E. Wolf, J. Opt. Soc. Am. 67 (1977) 785.
[18] M. Born and E. Wolf, Principles of optics, 6th Ed. (Pergamon, Oxford, 1980).
[19] E. Wolf, Opt. Lett. 8 (1983) 250.
[20] A.R. Thompson, J.M. Moran and G.W. Swenson, Interferometry and synthesis in radio astronomy (Wiley, New York, 1986).
[21] D.F.V. James and E. Wolf, Radio Sci., in press.
[22] J.E. Conway, T.J. Cornwall and P.N. Wilkinson, Mon. Not. R. Astron. Soc. 246 (1990) 490.

1 January 1997

OPTICS COMMUNICATIONS

ELSEVIER

Optics Communications 145 (1998) 1–4

Determination of the degree of coherence of light from spectroscopic measurements

Daniel F.V. James [a], Emil Wolf [b,1]

[a] Theoretical Division T-4, Mail Stop B-268, Los Alamos National Laboratory, Los Alamos, NM 87545, USA

[b] Department of Physics and Astronomy and Rochester Theory Center for Optical Science and Engineering, University of Rochester, Rochester, NY 14627, USA

Received 9 May 1997; accepted 24 June 1997

Abstract

A method is proposed for determination of the amplitude and the phase of the spectral degree of coherence from spectroscopic measurements. The technique is illustrated by a numerical example. © 1998 Elsevier Science B.V.

In traditional coherence theory, correlation of the fluctuating field at two points in space is characterized by a complex degree of coherence, which is a measure of the "statistical similarity" of light vibrations in the space-time domain [1]. In recent years a different measure of correlations has begun to be used, the so-called spectral degree of coherence [2]. It is a measure of correlations in the space-frequency domain. This measure is somewhat more subtle because, as is well-known, a frequency representation of stationary optical fields does not exist in the sense of ordinary function theory. This fact lead Norbert Wiener [3] to formulate a new branch of mathematics, the so-called generalized harmonic analysis, which provided the first satisfactory theoretical definition of the spectrum of light.

The concept of the spectral degree of coherence has led to several important recent developments [2]. Among them was the discovery that the spectrum of a field radiated by a partially coherent source differs, in general, from the spectrum of the source. This fact has suggested new methods for determining field correlations and for obtaining information about sources from measurements of spectra of the radiated fields in suitable interference experiments. In spite of these developments little attention has been paid to the question of how to measure the spectral degree of coherence. In this Letter a method for doing so is proposed.

Let us consider a Young's interference experiment, with partially coherent light, illustrated in Fig. 1. The light is incident from the left upon an opaque screen $\mathscr{A}$ containing two small pinholes located at points P_1 and P_2. The optical field at these points it may be represented by the ensembles of frequency-dependent realizations $\{U(P_1,\omega)\}$ and $\{U(P_2,\omega)\}$, respectively [1]. The light at some point P beyond the screen is then represented by the ensemble of frequency-dependent realizations $\{U(P,\omega)\}$, where $U(P,\omega)$ is related to $U(P_1,\omega)$ and $U(P_2,\omega)$ by the formula

$$U(P,\omega)=\frac{\mathrm{i}A\omega}{c}\left\{U(P_1,\omega)\frac{\exp(\mathrm{i}\omega R_1/c)}{R_1}+U(P_2,\omega)\frac{\exp(\mathrm{i}\omega R_2/c)}{R_2}\right\}. \quad (1)$$

Here R_1 and R_2 are the distances from the points P_1 and P_2 to the point P, respectively, A is the area of either of the two pinholes (assumed to be identical) and c is the speed of light. In Eq. (1) it is implicitely assumed that the pinholes are so small that the amplitude and the phase of the field are effectively constant over each of them and that the angles of incidence and diffraction are small.

The spectrum of the light at the point P is given by the

[1] Also at the Institute of Optics, University of Rochester, Rochester, NY 14627, USA.

[2] They are reviewed in Ref. [4].

PII S0030-4018(97)00369-6

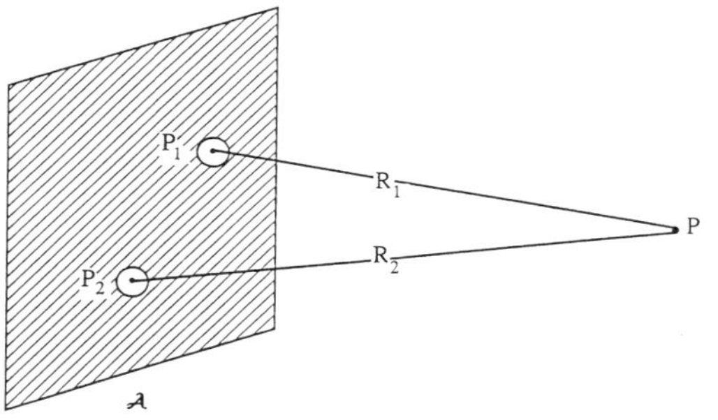

Fig. 1. A schematic illustration of the two-slit interference experiment.

expression $S(P,\omega)=\langle U^*(P,\omega)U(P,\omega)\rangle$, where the angular brackets denote an average over the ensembles of frequency-dependent realizations. The spectral degree of coherence, $\mu_{12}(\omega)$, of the light at P_1 and P_2 is defined by the formula [2]

$$\mu_{12}(\omega)\equiv|\mu_{12}(\omega)|\exp|\beta_{12}(\omega)| = \frac{\langle U^*(P_1,\omega)U(P_2,\omega)\rangle}{\sqrt{S(P_1,\omega)S(P_2,\omega)}}. \quad (2)$$

In this Letter we will address the problem of deducing both the modulus $|\mu_{12}(\omega)|$ and the phase $\beta_{12}(\omega)$ of the spectral degree of coherence from measurements of the spectrum $S(P,\omega)$ of the light at a point of superposition of the two beams emerging from the pinholes.

We assume that the spectrum of the light which is incident on the two pinholes is the same, i.e. that $S(P_1,\omega)=S(P_2,\omega)$. Further, we take $R_1\approx R_2\approx R$, where R is the distance from the point P to a point midway between P_1 and P_2. In this case the spectrum at the point P is given by the expression {cf. [2], Eq. (4.3-54)}

$$S(P,\omega)=2S^{(1)}(P,\omega)\times\{1+|\mu_{12}(\omega)|\cos[\omega T-\beta_{12}(\omega)]\}, \quad (3)$$

where $T=(R_1-R_2)/c$ is the time delay between the light reaching P via the two paths from the openings at P_1 and P_2. In Eq. (3), $S^{(1)}(P,\omega)=(A^2\omega^2/c^2R^2)\ S(P_1,\omega)$ is the spectrum of the light which would be observed at the point P if only one of the pinholes was opened.

It will be convenient to introduce a function $f(T,\omega)$, defined by the expression

$$f(T,\omega)=\frac{S(P,\omega)}{2S^{(1)}(P,\omega)}-1. \quad (4)$$

Because both $S(P,\omega)$ and $S^{(1)}(P,\omega)$, can be obtained by spectral measurements, the function $f(T,\omega)$ can be experimentally determined. We will see that in order to deduce the spectral degree of coherence $\mu_{12}(\omega)$, one only needs to measure $f(T,\omega)$ for several values of the parameter T. From Eqs. (3) and (4), one can readily deduce that $f(T,\omega)$ can be expressed in terms of the real and imaginary parts of the spectral degree of coherence $\mu_{12}(\omega)$ by the formula

$$f(T,\omega)=C_{12}(\omega)\cos(\omega T)-S_{12}(\omega)\sin(\omega T), \quad (5)$$

where

$$C_{12}(\omega)\equiv\mathrm{Re}\{\mu_{12}(\omega)\}=|\mu_{12}(\omega)|\cos[\beta_{12}(\omega)], \quad (6a)$$

$$S_{12}(\omega)\equiv\mathrm{Im}\{\mu_{12}(\omega)\}=|\mu_{12}(\omega)|\sin[\beta_{12}(\omega)]. \quad (6b)$$

Let us suppose that $f(T,\omega)$ is measured over a broad spectral bandwidth for two different values of the time delay, T_1 and T_2. From Eq. (5) we then have

$$f(T_1,\omega)=C_{12}(\omega)\cos(\omega T_1)+S_{12}(\omega)\sin(\omega T_1), \quad (7a)$$

,

$$f(T_2,\omega)=C_{12}(\omega)\cos(\omega T_2)+S_{12}(\omega)\sin(\omega T_2). \quad (7b)$$

Eqs. (7a) and (7b) are a pair of coupled linear equations for the unknown values of the real and imaginary parts $C_{12}(\omega)$ and $S_{12}(\omega)$ of the spectral degree of coherence. The solution of these equations is readily found to be

$$C_{12}(\omega)\equiv\frac{\sin(\omega T_2)f(T_1,\omega)-\sin(\omega T_1)f(T_2,\omega)}{\sin[\omega(T_2-T_1)]}, \quad (8a)$$

$$S_{12}(\omega)\equiv-\frac{\cos(\omega T_2)f(T_1,\omega)-\cos(\omega T_1)f(T_2,\omega)}{\sin[\omega(T_2-T_1)]}, \quad (8b)$$

provided that

$$\sin[\omega(T_2-T_1)]\neq 0. \quad (9)$$

It follows from Eqs. (6a) and (6b) that, in terms of $C_{12}(\omega)$

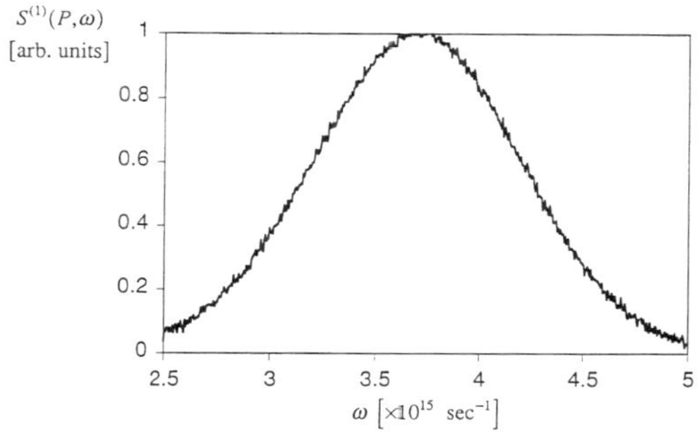

Fig. 2. The spectrum $S^{(1)}(P,\omega)$, which would be observed at P when one of the pinholes was open. In order to simulate actual experimental data, random noise has been added (for details see the text).

and $S_{12}(\omega)$, the modulus $|\mu_{12}(\omega)|$ of the spectral degree of coherence is given by the expression

$$|\mu_{12}(\omega)| = \sqrt{[C_{12}(\omega)]^2 + [S_{12}(\omega)]^2}, \qquad (10)$$

and its phase $\beta_{12}(\omega)$ by the formulas

$$\cos[\beta_{12}(\omega)] = \frac{C_{12}(\omega)}{\sqrt{[C_{12}(\omega)]^2 + [S_{12}(\omega)]^2}}, \qquad (11a)$$

$$\sin[\beta_{12}(\omega)] = \frac{S_{12}(\omega)}{\sqrt{[C_{12}(\omega)]^2 + [S_{12}(\omega)]^2}}. \qquad (11b)$$

For any selected pair of time delays T_1 and T_2, one might expect to have certain frequencies ω_0 in the spectral band which violate the condition (9), i.e. frequencies for which $\sin[\omega_0(T_2 - T_1)] = 0$. It is clear from Eqs. (8) that for such frequencies Eqs. (10) and (11) do not hold. However one can avoid this difficulty by measuring the function $f(T,\omega)$ for another value of the time delay T_3, for which the condition (9) holds.

To illustrate this technique, we have performed a numerical simulation of such experiments. We have taken the spectrum to be a broad Gaussian given by $S^{(1)}(\Gamma,\omega) = \exp[-(\omega-\omega_0)^2/2\Gamma^2]$ with $\omega_0 = 3.7 \times 10^{15}$ s^{-1} and $\Gamma = 0.5 \times 10^{15}$ s^{-1} (see Fig. 2). In order to simulate the effect of experimental noise, we have added to the spectrum a zero-mean random signal with Gaussian statistics and a variance corresponding to a peak noise-to-signal ratio of 1%. The spectral degree of coherence was assumed to have the functional form

$$\mu_{12}(\omega) = sinc(\omega/\Omega)\exp[i(a\omega^2 + b\omega^3)], \qquad (12)$$

where $\mathrm{sinc}(x) \equiv \sin(x)/x$, $\Omega = 2 \times 10^{15}$ s^{-1}, $a = 7.3 \times 10^{-32}$ s^2 and $b = 2 \times 10^{-47}$ s^3. Fig. 3a and Fig. 3b show the spectra at the point P corresponding to $T_1 = 10^{-14}$ s and $T_2 = 2 \times 10^{-14}$ s respectively. Noise has again been added. Using this simulated experimental data and the formulas (10) and (11), the modulus and the cosine and sine of the phase of the spectral degree of coherence $\mu_{12}(\omega)$ were calculated. The results are shown in Fig. 4a and Fig. 4b. We see that there is good agreement between the calculated and the assumed theoretical quantities, except close to the frequencies at which $\sin[\omega(T_2 - T_1)] = 0$ (the frequencies $\omega = 3.14 \times 10^{15}$ s, 3.77×10^{15} s and 4.08×10^{15} s^{-1}), as one would expect.

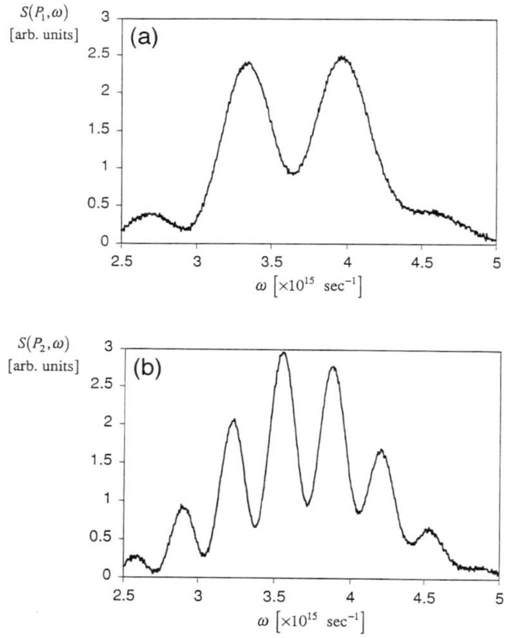

Fig. 3. The spectrum $S(P,\omega)$ which would be observed at P with both pinholes illuminated and with (a) $T_1 = 10^{-14}$ s and (b) $T_2 = 2\times 10^{-14}$ s when the spectral degree of coherence is given by Eq. (12) with $\Omega = 2\times10^{15}$ s^{-1}, $a = 7.3\times10^{-32}$ s and $b = 2 \times 10^{-47}$ s.

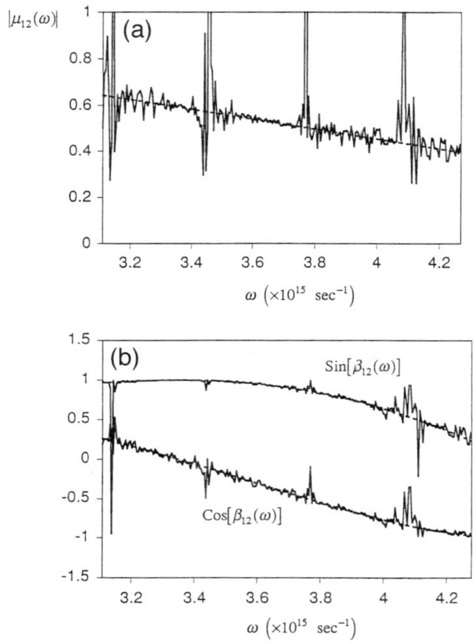

Fig. 4. (a) The modulus and (b) the cosine and the sine of the phase of $\mu_{12}(\omega)$ as deduced from the simulated experimental data shown in Figs. 2 and 3. The dashed lines shows the actual values calculated using Eq. (11). The regions where the technique breaks down are around the frequencies at which $\sin[\omega(T_2 - T_1)] = 0$ (i.e. at $\omega = 3.14\times10^{15}$ s^{-1}, 3.46×10^{15} s^{-1}, 3.77×10^{15} s^{-1} and 4.08×10^{15} s^{-1}).

Acknowledgements

This work was supported by the Department of Energy, under grant DE-FG02-90ER 14119 and by the Air Force Office of Scientific Research under grant F49620-96-1-0460. The views expressed in this article do not constitute an endorsement by the Department of Energy.

References

[1] M. Born, E. Wolf, Principles of Optics, Pergamon Press, Oxford, sixth ed., 1980, Sec. 10.3.1.

[2] L. Mandel, E. Wolf, Optical Coherence and Quantum Optics, Cambridge University Press, Cambridge, 1995, Sec. 4.3.2.

[3] N. Wiener, Acta Math. (Uppsala) 55 (1930) 117.

[4] E. Wolf, D.F.V. James, Rep. Progr. Phys. 59 (1996) 771.

PHYSICAL REVIEW A VOLUME 54, NUMBER 5 NOVEMBER 1996

Correlation-induced spectral changes and energy conservation

Girish S. Agarwal[1] and Emil Wolf[2]
[1]*Physical Research Laboratory, Navrangpura, Ahmedabad, 380 009, India*
[2]*Department of Physics and Astronomy and Rochester Theory Center for Optical Science and Engineering, University of Rochester, Rochester, New York 14627*

(Received 15 April 1996)

An energy conservation law is derived for fields generated by random, statistically stationary, scalar sources of any state of coherence. It is shown that correlation-induced spectral changes are in strict agreement with this law and that, basic to the understanding of such changes, is a distinction that must be made between the spectrum of a source and the spectrum of the field that the source generates. This distinction, which is obviously relevant for spectroscopy, does not appear to have been previously recognized. [S1050-2947(96)06610-3]

PACS number(s): 42.50.Ar

I. INTRODUCTION

It has been predicted theoretically some years ago [1,2] that the spectrum of light and of other radiation may change on propagation, even in free space. This phenomenon, which was soon verified experimentally [3–5], has attracted a good deal of attention and has resulted in the publication of about 100 papers on this subject [6]. The spectral changes, which have their origin in spatial correlation properties of sources, may be of very different kinds. They may, for example, consist of redshifts or blueshifts of spectral lines, narrowing or broadening of the lines, and generation of new lines. Moreover, different spectral changes may occur in different directions of observation.

In spite of the considerable interest that has been shown in this effect, an explicit demonstration that such spectral changes do not violate energy conservation has, up to now, been demonstrated only under somewhat restricted circumstances [12–14]. In this paper we derive an energy conservation law for fields produced by random, statistically stationary sources of any state of coherence and we demonstrate that correlation-induced spectral changes do not violate this law. Our analysis, which is valid for both classical and quantum sources, shows also that basic for the understanding of this phenomenon is a distinction that must be made between the spectrum of a source and the spectrum of the field that the source generates. This distinction, which is obviously very relevant to the interpretation of spectroscopic data, does not appear to have been previously appreciated.

II. THE SOURCE SPECTRUM AND THE SPECTRUM OF THE RADIATED FIELD

We consider radiation from a scalar source, localized in a finite region D of space (Fig. 1). Let $p(\mathbf{r},t)$ denote the source density distribution and $E(\mathbf{r},t)$ the field generated by the source, at a point $\mathbf{r}$, at time t. We take both $p(\mathbf{r},t)$ and $E(\mathbf{r},t)$ to be the complex analytic signal representations [15] of a real source variable and a real field variable. In order to illustrate the main aspects of the theory we will ignore the vectorial nature of the problem and take $p(\mathbf{r},t)$ and $E(\mathbf{r},t)$ to be scalars. We may think of p and E as representing Cartesian components of the polarization vector and of one of the electromagnetic field vectors, respectively, although this, of course, is only a rough analogy. We will show elsewhere that the main conclusions of our analysis hold when the full electromagnetic nature of the source and of the field are taken into account.

Since we are interested in spectral properties it is convenient to deal with the Fourier transforms

$$\tilde{p}(\mathbf{r},\omega)=\frac{1}{2\pi}\int_{-\infty}^{\infty}p(\mathbf{r},t)e^{i\omega t}dt \tag{2.1}$$

and

$$\tilde{E}(\mathbf{r},\omega)=\frac{1}{2\pi}\int_{-\infty}^{\infty}E(\mathbf{r},t)e^{i\omega t}dt \tag{2.2}$$

of $p(\mathbf{r},t)$ and $E(\mathbf{r},t)$, respectively. They are related by the inhomogeneous Helmholtz equation

$$(\nabla^2+k^2)\tilde{E}(\mathbf{r},\omega)=-4\pi k^2\tilde{p}(\mathbf{r};\omega), \tag{2.3}$$

where

$$k=\omega/c \tag{2.4}$$

is the free-space wave number associated with the frequency ω, c being the speed of light *in vacuo*. The outgoing solution of Eq. (2.3) is

$$\tilde{E}(\mathbf{r},\omega)=k^2\int_D\frac{e^{ik|\mathbf{r}-\mathbf{r}'|}}{|\mathbf{r}-\mathbf{r}'|}\tilde{p}(\mathbf{r}',\omega)d^3r'. \tag{2.5}$$

The field $\tilde{E}^{(\infty)}(r\mathbf{u},\omega)$ at a point $\mathbf{r}=r\mathbf{u}$ ($\mathbf{u}^2=1$) in the far zone (the radiation field) is obtained at once from Eq. (2.5) by making use of the asymptotic approximation (see Fig. 1)

$$\frac{e^{ik|\mathbf{r}-\mathbf{r}'|}}{|\mathbf{r}-\mathbf{r}'|}\sim e^{-ik\mathbf{u}\cdot\mathbf{r}'}\frac{e^{ikr}}{r} \tag{2.6}$$

($kr\rightarrow\infty$ with $\mathbf{u}$ fixed) and hence

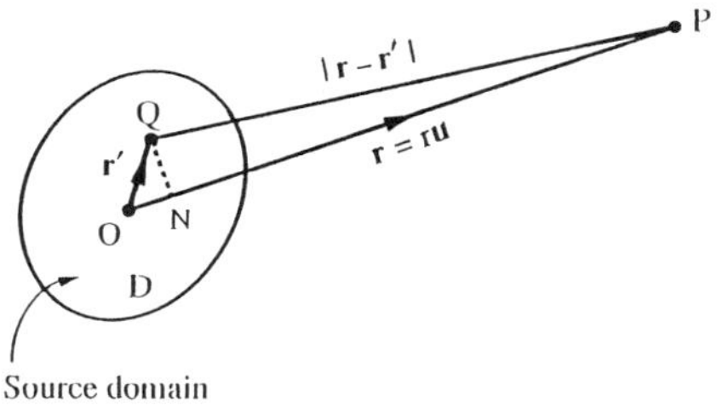

FIG. 1. Illustrating notation relating to Eq. (2.6). P is a point in the far zone of the source, **u** is a unit vector, and $\mathbf{u}\cdot\mathbf{r}'=\overline{ON}$.

$$\widetilde{E}^{(\infty)}(r\mathbf{u},\omega)=k^2\frac{e^{ikr}}{r}\int_D \widetilde{p}(\mathbf{r}',\omega)e^{-ik\mathbf{u}\cdot\mathbf{r}'}d^3r' . \quad (2.7)$$

In any realistic situation the source distribution and, consequently, the field distribution are not deterministic but are random functions of time. We assume that the ensembles that characterize the temporal fluctuations of $p(\mathbf{r},t)$ and $E(\mathbf{r},t)$ are statistically stationary, at least in the wide sense [16]. The spectrum $S_p(\mathbf{r},\omega)$ of the source distribution and the spectrum $S_E(\mathbf{r},\omega)$ of the field distribution are then given by the formulas

$$\langle \widetilde{p}^*(\mathbf{r},\omega)\widetilde{p}(\mathbf{r},\omega')\rangle = S_p(\mathbf{r},\omega)\,\delta(\omega-\omega') , \quad (2.8a)$$

$$\langle \widetilde{E}^*(\mathbf{r},\omega)\widetilde{E}(\mathbf{r},\omega')\rangle = S_E(\mathbf{r},\omega)\,\delta(\omega-\omega') , \quad (2.8b)$$

where the angular brackets denote the ensemble average and δ is the Dirac delta function.

On substituting from Eq. (2.7) into Eq. (2.8b) we obtain, for the spectrum of the radiation field, the expression

$$S_E^{(\infty)}(r\mathbf{u},\omega)=\frac{k^4}{r^2}\int_D\int_D W_p(\mathbf{r}',\mathbf{r}'',\omega)e^{-ik\mathbf{u}\cdot(\mathbf{r}''-\mathbf{r}')}d^3r'\,d^3r'' , \quad (2.9)$$

where $W_p(\mathbf{r}',\mathbf{r}'',\omega)$ is the cross-spectral density of the polarization density defined by the formula [18]

$$\langle p^*(\mathbf{r}',\omega)p(\mathbf{r}'',\omega')\rangle = W_p(\mathbf{r}',\mathbf{r}'',\omega)\,\delta(\omega-\omega') . \quad (2.10)$$

It is convenient to express the right-hand side of Eq. (2.9) in a somewhat different form. For this purpose we introduce the spectral degree of coherence [19] of the polarization density

$$\mu_p(\mathbf{r}',\mathbf{r}'',\omega)=\frac{W_p(\mathbf{r}',\mathbf{r}'',\omega)}{\sqrt{S_p(\mathbf{r}',\omega)}\sqrt{S_p(\mathbf{r}'',\omega)}} . \quad (2.11)$$

On substituting for W_p from Eq. (2.11) into the integral in Eq. (2.9) we see at once that

$$S_E^{(\infty)}(r\mathbf{u},\omega)=\frac{k^4}{r^2}\int_D\int_D \sqrt{S_p(\mathbf{r}',\omega)}\sqrt{S_p(\mathbf{r}'',\omega)}\,\mu_p(\mathbf{r}',\mathbf{r}'',\omega)\times e^{-ik\mathbf{u}\cdot(\mathbf{r}''-\mathbf{r}')}d^3r'\,d^3r'' . \quad (2.12)$$

If, as is often the case, the source is homogeneous in the sense that the source spectrum $S_p(\mathbf{r},\omega)$ is the same at every source point, i.e., if

$$S_p(r\mathbf{u},\omega)\equiv S_p(\omega) \quad \text{for all } \mathbf{r}\in D , \quad (2.13)$$

formula (2.12) reduces to

$$S_E^{(\infty)}(\mathbf{r},\omega)=M(\omega,\mathbf{u},r)S_p(\omega) , \quad (2.14)$$

where

$$M(\omega,\mathbf{u},r)=\frac{k^4}{r^2}\int_D\int_D \mu_p(\mathbf{r}',\mathbf{r}'',\omega)e^{-ik\mathbf{u}\cdot(\mathbf{r}''-\mathbf{r}')}d^3r'\,d^3r'' . \quad (2.15)$$

Formula (2.14), together with Eq. (2.15), confirms again the result established in several previous publications, that the spectrum of the radiation field depends, in general, not only on the spectrum of the source but also on its correlation properties, represented by the spectral degree of coherence $\mu_p(\mathbf{r}',\mathbf{r}'',\omega)$. Consequently, the two spectra will, in general, differ from each other. It is not difficult to show that this is so even when the source is spherically symmetric, as has already been previously demonstrated [12–14]. Because the proportionality factor M in formula (2.14) depends not only on the frequency ω but also on the unit vector **u**, the spectrum of the radiation will be different, in general, in different directions of observation. These conclusions have also been confirmed by quantum-mechanical calculations relating to simple atomic systems [20,21].

III. ENERGY CONSERVATION IN PARTIALLY COHERENT FIELDS

One of several misconceptions surrounding the subject of correlation-induced spectral changes concerns the question of energy conservation. In order to show that energy is indeed conserved in such situations, we will first derive an energy conservation law that holds for statistically stationary fields of any state of coherence.

With a suitable choice of units, the average energy flux vector $\mathbf{F}_\omega(\mathbf{r})$ at frequency ω, in a stationary optical field, is given by the formula [cf. Eq. (5.7-13), Ref. [11]]

$$\mathbf{F}_\omega(\mathbf{r})\,\delta(\omega-\omega') = -\frac{i}{2k}[\langle \widetilde{E}^*(\mathbf{r},\omega)\nabla\widetilde{E}(\mathbf{r},\omega')-\widetilde{E}(\mathbf{r},\omega')\nabla\widetilde{E}^*(\mathbf{r},\omega)\rangle] . \quad (3.1)$$

If we use an elementary vector identity we readily find that

$$\nabla\cdot\mathbf{F}_\omega(\mathbf{r})\,\delta(\omega-\omega') = -\frac{i}{2k}[\langle \widetilde{E}^*(\mathbf{r},\omega)\nabla^2\widetilde{E}(\mathbf{r},\omega')-\widetilde{E}(\mathbf{r},\omega')\nabla^2\widetilde{E}^*(\mathbf{r},\omega)\rangle] . \quad (3.2)$$

Let us eliminate $\nabla^2\widetilde{E}$ and $\nabla^2\widetilde{E}^*$ on the right-hand side by the use of Eq. (2.3). We then obtain, for $\nabla\cdot\mathbf{F}_\omega(\mathbf{r})$, the expression

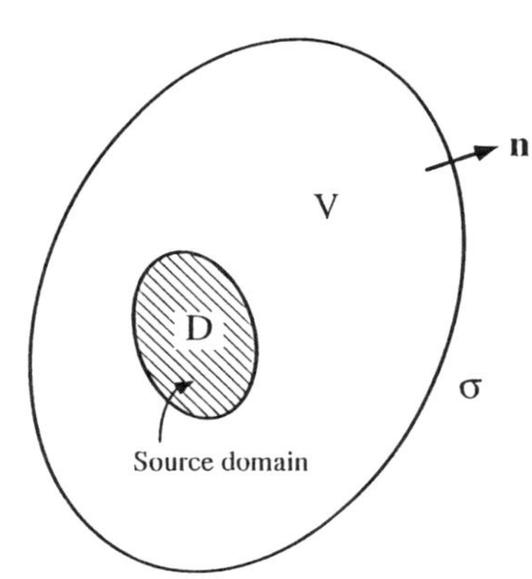

FIG. 2. Illustrating notation relating to Eq. (3.5).

$$\nabla \cdot \mathbf{F}_\omega(\mathbf{r})\,\delta(\omega-\omega')$$
$$=2\pi i k[\langle \widetilde{p}(\mathbf{r},\omega)\widetilde{E}^*(\mathbf{r},\omega')-\widetilde{p}^*(\mathbf{r},\omega)\widetilde{E}(\mathbf{r},\omega')\rangle]. \quad (3.3)$$

Next we eliminate $\widetilde{E}$ and $\widetilde{E}^*$ by the use of Eq. (2.5) and find that

$$\nabla \cdot \mathbf{F}_\omega(\mathbf{r})=4\pi k^3 \text{ Im} \int_D W_p^*(\mathbf{r},\mathbf{r}',\omega)\,\frac{e^{ik|\mathbf{r}-\mathbf{r}'|}}{|\mathbf{r}-\mathbf{r}'|}\,d^3r', \quad (3.4)$$

where $W_p(\mathbf{r},\mathbf{r}',\omega)$ is the cross-spectral density of the polarization defined by Eq. (2.10) and Im denotes the imaginary part.

Formula (3.4) expresses an *energy conservation law* for stationary fields of any state of coherence. When the point $\mathbf{r}$ is outside the source domain D, $W_p(\mathbf{r},\mathbf{r}',\omega)=0$ and Eq. (3.4) reduces to

$$\nabla \cdot \mathbf{F}_\omega(\mathbf{r})=0. \quad (3.4')$$

The physical significance of formula (3.4) becomes more apparent if we convert it into integral form. Let us integrate both sides of that equation throughout a domain V, bounded by a closed surface σ, which contains the source domain D in its interior (see Fig. 2). On using the Gauss theorem and the fact that $W_p(\mathbf{r},\mathbf{r}',\omega)=0$ for all points $\mathbf{r}$ located outside D, we find that

$$\int_\sigma \mathbf{F}_\omega(\mathbf{r})\cdot\mathbf{n}\,d\sigma=4\pi k^3 \text{ Im} \int_D\int_D W_p^*(\mathbf{r},\mathbf{r}',\omega)\,\frac{e^{ik|\mathbf{r}-\mathbf{r}'|}}{|\mathbf{r}-\mathbf{r}'|}\times d^3r\,d^3r', \quad (3.5)$$

where $\mathbf{n}$ is the unit outward normal to σ. Finally, if we use the fact that the cross-spectral density is Hermitian $[W^*(\mathbf{r},\mathbf{r}',\omega)=W(\mathbf{r}',\mathbf{r},\omega)]$ and also that the free-space Green's function $\exp[ik|\mathbf{r}-\mathbf{r}'|]/|\mathbf{r}-\mathbf{r}'|$ is symmetric with respect to $\mathbf{r}$ and $\mathbf{r}'$, formula (17) may be expressed in the form [22]

$$\int_\sigma \mathbf{F}_\omega(\mathbf{r})\cdot\mathbf{n}\,d\sigma=4\pi k^4\int_D\int_D W_p(\mathbf{r},\mathbf{r}',\omega)\,\frac{\sin(k|\mathbf{r}-\mathbf{r}'|)}{k|\mathbf{r}-\mathbf{r}'|}\times d^3r\,d^3r'. \quad (3.6)$$

Since the left-hand side of Eq. (3.6) represents the rate at which energy emerges from the volume V, the right-hand side evidently represents the rate at which the source radiates energy. This term is seen to depend on the second-order correlation properties of the polarization, represented by the cross-spectral density $W_p(\mathbf{r},\mathbf{r}',\omega)$.

Formula (3.6) is *the integral form of the energy conservation law for statistically stationary fields of any state of coherence*. It is to be noted that it *holds at each frequency*. This fact is a consequence of the assumed stationarity of the source because different frequency components of members of a statistically stationary ensemble are uncorrelated [23].

IV. CONSISTENCY OF THE PHENOMENON OF SPECTRAL CHANGES IN FREE PROPAGATION WITH ENERGY CONSERVATION

We will now show that neither the difference between the spectra of the source polarization and of the radiated field nor the dependence of the field spectrum on the direction of observation noted in Sec. II [cf. Eq. (2.14)] violates the law of energy conservation.

The energy flux vector $\mathbf{F}_\omega^{(\infty)}(r\mathbf{u})$ and the spectral density $S_E^{(\infty)}(r\mathbf{u},\omega)$ of the radiation field are, for a suitable choice of units, simply related [24]:

$$\mathbf{F}_\omega^{(\infty)}(r\mathbf{u})=S_E^{(\infty)}(r\mathbf{u},\omega)\mathbf{u}. \quad (4.1)$$

If we substitute for $S_E^{(\infty)}$ expression (2.9), Eq. (4.1) gives

$$\mathbf{F}_\omega^{(\infty)}(r\mathbf{u})=\mathbf{u}\,\frac{k^4}{r^2}\int_D\int_D W_p(\mathbf{r}',\mathbf{r}'',\omega)e^{-ik\mathbf{u}\cdot(\mathbf{r}''-\mathbf{r}')}d^3r'\,d^3r''. \quad (4.2)$$

Hence the total energy flux radiated by the source is given by the expression

$$\int_{4\pi}\mathbf{F}_\omega^{(\infty)}(r\mathbf{u})\cdot\mathbf{u}r^2 d\Omega=k^4\int_{4\pi}d\Omega\int_D\int_D W_p(\mathbf{r}',\mathbf{r}'',\omega)\times e^{-ik\mathbf{u}\cdot(\mathbf{r}''-\mathbf{r}')}d^3r'\,d^3r'' \quad (4.3)$$

$(kr\to\infty)$, where $d\Omega$ is the element of solid angle generated by the real unit vector $\mathbf{u}$ and the Ω integration extends over the whole 4π solid angle. Let us interchange, on the right-hand side of Eq. (4.3), the order of the angular and the spatial integrations and use the identity [25]

$$\int_{4\pi}d\Omega\;e^{-ik\mathbf{u}\cdot(\mathbf{r}-\mathbf{r})}=4\pi\,\frac{\sin(k|\mathbf{r}'-\mathbf{r}''|)}{k|\mathbf{r}'-\mathbf{r}''|}. \quad (4.4)$$

Equation (4.3) then becomes

$$\int_{4\pi}\mathbf{F}_\omega^{(\infty)}(r\mathbf{u})\cdot\mathbf{u}r^2 d\Omega$$
$$=4\pi k^4\int_D\int_D W_p(\mathbf{r}',\mathbf{r}'',\omega)\,\frac{\sin(k|\mathbf{r}'-\mathbf{r}''|)}{k|\mathbf{r}'-\mathbf{r}''|}\,d^3r'\,d^3r''. \quad (4.5)$$

Relation (4.5), together with the fact that in the far zone the flux vector is in the outward radial direction [as seen from Eq. (4.1)], is precisely the energy conservation law (3.6), specialized to the situation where the surface σ is taken to be a sphere of infinitely large radius centered on a point in the source region. We have thus demonstrated that expression (2.9) and, consequently, also expression (2.14) for the spectrum of the radiated field do not violate energy conservation. This conclusion confirms the correctness of the prediction evident from Eqs. (2.14) and (2.15) that, in general, the spectrum of the radiation field differs from the source spectrum and that it may be different at different points of observation.

V. QUANTUM FORMULATION

The preceding analysis was based entirely on the statistical theory of classical fields. However, it can readily be seen that the same conclusions also follow when the field is quantized. One only needs to replace the classical field variable $\tilde{E}(\mathbf{r},\omega)$ by the positive frequency part $\hat{\tilde{E}}^{(+)}(\mathbf{r},\omega)$ of the electric field operator and the polarization $\tilde{p}(\mathbf{r},\omega)$ of the source by the positive frequency part $\hat{\tilde{p}}^{(+)}(\mathbf{r},\omega)$ of the polarization operator. In place of Eqs. (2.8) and (2.10) we then have

$$\langle \hat{\tilde{p}}^{(-)}(\mathbf{r},\omega)\hat{\tilde{p}}^{(+)}(\mathbf{r},\omega')\rangle = S_p(\mathbf{r},\omega)\,\delta(\omega-\omega'), \quad (5.1a)$$

$$\langle \hat{\tilde{E}}^{(-)}(\mathbf{r},\omega)\hat{\tilde{E}}^{(+)}(\mathbf{r},\omega')\rangle = S_E(\mathbf{r},\omega)\,\delta(\omega-\omega'), \quad (5.1b)$$

$$\langle \hat{\tilde{p}}^{(-)}(\mathbf{r},\omega)\hat{\tilde{p}}^{(+)}(\mathbf{r}',\omega')\rangle = W_p(\mathbf{r},\mathbf{r}',\omega)\,\delta(\omega-\omega'), \quad (5.2)$$

where the angular brackets now denote the quantum-mechanical expectation value. Similarly, Eq. (3.1) for the average flux vector will now be replaced by the formula

$$F_\omega(\mathbf{r})\,\delta(\omega-\omega') = -\frac{i}{2k}\langle \hat{\tilde{E}}^{(-)}(\mathbf{r},\omega)\nabla\hat{\tilde{E}}^{(+)}(\mathbf{r},\omega') - \hat{\tilde{E}}^{(+)}(\mathbf{r},\omega')\nabla\hat{\tilde{E}}^{(-)}(\mathbf{r},\omega)\rangle. \quad (5.3)$$

With these definitions our basic conversation law (3.6) holds.

ACKNOWLEDGMENTS

This research was supported by the U.S. Department of Energy under Grant No. DE-FG02-90ER 14119 and partially supported by the National Science Foundation, Grant No. PHY94-15583.

[1] E. Wolf, Phys. Rev. Lett. **56**, 1370 (1986).
[2] E. Wolf, Nature **326**, 363 (1987).
[3] G. Morris and D. Faklis, Opt. Commun. **62**, 5 (1987).
[4] M. F. Bocko, D. H. Douglass, and R. S. Knox, Phys. Rev. Lett. **58**, 2649 (1987).
[5] D. Faklis and G. M. Morris, Opt. Lett. **13**, 4 (1988).
[6] A comprehensive review of this research is given by E. Wolf and D. F. V. James, Rep. Prog. Phys. **59**, 77 (1996); see also Refs. [7–11].
[7] E. Wolf, in *International Trends in Optics*, edited by J. W. Goodman (Academic, San Diego, 1991), p. 221.
[8] P. W. Milonni and S. Singh, Adv. At. Mol. Opt. Phys. **28**, 127 (1991).
[9] E. Wolf, in *Recent Developments in Quantum Optics*, edited by R. Inguva (Plenum, New York, 1993), p. 369.
[10] H. C. Kandpal, J. S. Vaishya, and K. C. Joshi, Opt. Eng. (Bellingham) **33**, 1996 (1994).
[11] L. Mandel and E. Wolf, *Optical Coherence and Quantum Optics* (Cambridge University Press, Cambridge, 1995), Sec. 5.8.
[12] E. Wolf and A. Gamliel, J. Mod. Opt. **39**, 927 (1992).
[13] M. Dušek, Opt. Commun. **100**, 24 (1993).
[14] G. Hazak and R. Zamir, J. Mod. Opt. **41**, 1653 (1994).
[15] L. Mandel and E. Wolf, *Optical Coherence and Quantum Optics* (Ref. [11]), Sec. 3.1.
[16] L. Mandel and E. Wolf, *Optical Coherence and Quantum Optics* (Ref. [11]), p. 47. Strictly speaking the Fourier transforms $\tilde{p}(\mathbf{r},\omega)$ and $\tilde{E}(\mathbf{r},\omega)$ must then be interpreted as generalized functions. One may, however, avoid this complication by employing somewhat different random functions that are square integrable, using the methods of coherence theory in the space frequency domain (see Ref. [17] and Ref. [11], Sec. 4.7).
[17] E. Wolf, J. Opt. Soc. Am. **72**, 343 (1982); J. Opt. Soc. Am. A **3**, 76 (1986).
[18] L. Mandel and E. Wolf, *Optical Coherence and Quantum Optics* (Ref. [11]), Eq. (4.3-39).
[19] L. Mandel and E. Wolf, *Optical Coherence and Quantum Optics* (Ref. [11]), Eq. (4.3-47b).
[20] G. V. Varada and G. S. Agarwal, Phys. Rev. A **44**, 7626 (1992).
[21] D. F. V. James, Phys. Rev. A **47**, 1336 (1993).
[22] Because the left-hand side of Eq. (3.6) is real and so is the function $\sin(k|\mathbf{r}-\mathbf{r}'|)/(k|\mathbf{r}-\mathbf{r}'|)$, $W_p(\mathbf{r},\mathbf{r}',\omega)$ may evidently be replaced by $\mathrm{Re}W_p(\mathbf{r},\mathbf{r}',\omega)$ on the right-hand side of Eq. (18), with Re denoting the real part.
[23] L. Mandel and E. Wolf, *Optical Coherence and Quantum Optics* (Ref. [11]), Eq. (2.4-10).
[24] L. Mandel and E. Wolf, *Optical Coherence and Quantum Optics* (Ref. [11]), Eq. (5.7-32).
[25] R. Courant and D. Hilbert, *Methods of Mathematical Physics* (Interscience, New York, 1962), Vol. II, p. 195.

PHYSICAL REVIEW E VOLUME 59, NUMBER 4 APRIL 1999

Energy conservation law for randomly fluctuating electromagnetic fields

Greg Gbur,[1] Daniel James,[2,*] and Emil Wolf[1]
[1]Department of Physics and Astronomy and Rochester Theory Center for Optical Science and Engineering, University of Rochester, Rochester, New York 14627
[2]Theoretical Division T-4, Mail Stop B268, Los Alamos National Laboratory, Los Alamos, New Mexico 87545
(Received 29 May 1998)

An energy conservation law is derived for electromagnetic fields generated by any random, statistically stationary, source distribution. It is shown to provide insight into the phenomenon of correlation-induced spectral changes. The results are illustrated by an example. [S1063-651X(99)01403-8]

PACS number(s): 42.25.Kb, 03.50.De

I. INTRODUCTION

Classical electromagnetic theory deals with deterministic sources and deterministic fields. It follows from Maxwell's equations that such fields obey well-known conservation laws for energy, linear momentum, and angular momentum. The situation regarding conservation laws is rather different when the sources and the fields fluctuate randomly either in space or in time. Such situations are actually very common and are also more realistic, because sources found in nature or produced in laboratories undergo some irregular, unpredictable, fluctuations.

Around 1960, after the rigorous laws of coherence theory of the electromagnetic field had been formulated, various conservation laws for such fields were derived [1]. They turned out to be rather complicated and, probably because of this, little use has been made of them.

About ten years ago the phenomenon of correlation-induced spectral changes was discovered, and it has been extensively studied since then, both theoretically and experimentally [2]. This phenomenon is characterized by changes in the spectrum of the field on propagation, as a consequence of source correlations. In particular the field spectrum may differ from the spectrum of the source, and may be different at different points in space. The source correlations may give rise to shifts of spectral lines, or to broadening or narrowing of the lines, or they may generate much more drastic changes, e.g., producing new lines or suppressing some of the lines present in the source spectrum.

It might appear at first sight that correlation-induced spectral changes violate energy conservation. That this is not so was demonstrated, under somewhat special circumstances, in several papers [3], and this question was examined under more general conditions in Ref. [4], within the framework of scalar theory.

In the present paper we generalize the results of Ref. [4], and we derive an energy conservation law which is valid for all statistically stationary fluctuating electromagnetic fields. We further show that correlation-induced changes of spectra of electromagnetic fields of any state of coherence are consistent with this conservation law, and we illustrate the results by an example.

*Electronic address: dfvj@t4.lanl.gov

II. ENERGY CONSERVATION IN RANDOMLY FLUCTUATING ELECTROMAGNETIC FIELDS

We begin by deriving an energy conservation law for an electromagnetic field generated by a randomly fluctuating statistically stationary source occupying a domain D. Let $\langle \mathbf{F}(\mathbf{r},\omega)\rangle$ represent the expectation value of the flux density vector (the Poynting vector) at frequency ω, at an arbitrary point $\mathbf{r}$ in the field. It is given by the expression (using coherence theory in the space-frequency domain—see Sec. 4.7 of Ref. [5])

$$\langle \mathbf{F}(\mathbf{r},\omega)\rangle = \frac{c}{8\pi}\mathrm{Re}\langle \mathbf{E}^*(\mathbf{r},\omega)\times\mathbf{H}(\mathbf{r},\omega)\rangle, \tag{2.1}$$

where Re denotes the real part, and the asterisk denotes the complex conjugate. On taking the divergence of this expression and on using the vector identity

$$\nabla\cdot(\mathbf{a}\times\mathbf{b}) = \mathbf{b}\cdot(\nabla\times\mathbf{a}) - \mathbf{a}\cdot(\nabla\times\mathbf{b}), \tag{2.2}$$

it follows that

$$\nabla\cdot\langle \mathbf{F}(\mathbf{r},\omega)\rangle = \frac{c}{8\pi}\mathrm{Re}\{\langle \mathbf{H}^*(\mathbf{r},\omega)\cdot[\nabla\times\mathbf{E}(\mathbf{r},\omega)]\rangle - \langle \mathbf{E}^*(\mathbf{r},\omega)\cdot[\nabla\times\mathbf{H}(\mathbf{r},\omega)]\rangle\}. \tag{2.3}$$

The right-hand side of Eq. (2.3) may be simplified by making use of the relations

$$\nabla\times\mathbf{E}(\mathbf{r},\omega) = ik\mathbf{H}(\mathbf{r},\omega), \tag{2.4a}$$

$$\nabla\times\mathbf{H}(\mathbf{r},\omega) = -ik\mathbf{E}(\mathbf{r},\omega) - 4\pi ik\mathbf{P}(\mathbf{r},\omega), \tag{2.4b}$$

which follow from Maxwell's equations. We have assumed that the source is nonmagnetic. Using Eqs. (2.4) in Eq. (2.3), one finds that

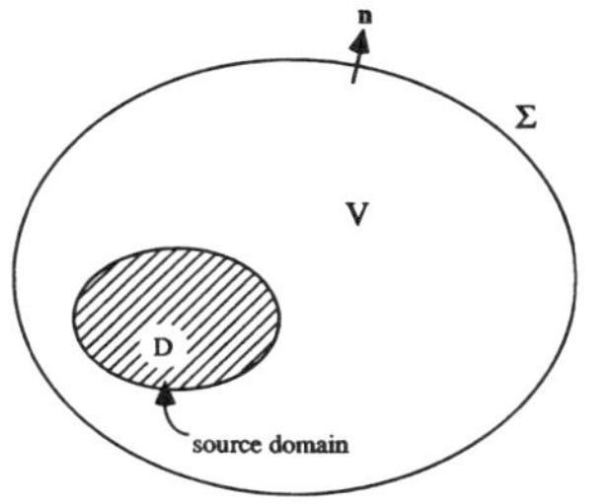

FIG. 1. Illustrating notation relating to the integral form (2.13) of the energy conservation law for fluctuating, statistically stationary, electromagnetic fields.

$$\nabla\cdot\langle\mathbf{F}(\mathbf{r},\omega)\rangle=\frac{kc}{8\pi}\mathrm{Re}\{i\langle\mathbf{H}^*(\mathbf{r},\omega)\cdot\mathbf{H}(\mathbf{r},\omega)\rangle+i\langle\mathbf{E}^*(\mathbf{r},\omega)\cdot\mathbf{E}(\mathbf{r},\omega)\rangle+4\pi i\langle\mathbf{E}^*(\mathbf{r},\omega)\cdot\mathbf{P}(\mathbf{r},\omega)\rangle\}. \tag{2.5}$$

The first two terms on the right of Eq. (2.5) are purely imaginary, and hence do not contribute to the left-hand side. Equation (2.5) therefore reduces to

$$\nabla\cdot\langle\mathbf{F}(\mathbf{r},\omega)\rangle=-\frac{kc}{2}\mathrm{Im}\langle\mathbf{E}^*(\mathbf{r},\omega)\cdot\mathbf{P}(\mathbf{r},\omega)\rangle. \tag{2.6}$$

On eliminating the magnetic field from Eqs. (2.4a) and (2.4b), we can solve the resulting equation for the electric field subject to the requirement that it is outgoing at infinity, and we find that

$$\mathbf{E}(\mathbf{r},\omega)=[k^2+\nabla(\nabla\cdot)]\int_D\mathbf{P}(\mathbf{r}',\omega)\frac{e^{ik|\mathbf{r}-\mathbf{r}'|}}{|\mathbf{r}-\mathbf{r}'|}d^3\mathbf{r}'. \tag{2.7}$$

Next we substitute from Eq. (2.7) into Eq. (2.6), and obtain the formula

$$\nabla\cdot\langle\mathbf{F}(\mathbf{r},\omega)\rangle=-\frac{kc}{2}\mathrm{Im}\left\{\left\langle k^2\int_D\mathbf{P}(\mathbf{r},\omega)\cdot\mathbf{P}^*(\mathbf{r}',\omega)\frac{e^{-ik|\mathbf{r}-\mathbf{r}'|}}{|\mathbf{r}-\mathbf{r}'|}d^3r'\right\rangle+\left\langle\mathbf{P}(\mathbf{r},\omega)\cdot\nabla\int_D\mathbf{P}^*(\mathbf{r}',\omega)\cdot\nabla\frac{e^{-ik|\mathbf{r}-\mathbf{r}'|}}{|\mathbf{r}-\mathbf{r}'|}d^3r'\right\rangle\right\}. \tag{2.8}$$

Let us now introduce the cross-spectral density tensor $W_{ij}^{(P)}(\mathbf{r}_1,\mathbf{r}_2,\omega)$ of the source polarization, defined by the formula

$$W_{ij}^{(P)}(\mathbf{r}_1,\mathbf{r}_2,\omega)=\langle P_i^*(\mathbf{r}_1,\omega)P_j(\mathbf{r}_2,\omega)\rangle, \tag{2.9}$$

where the angular brackets denote averages over the ensemble of the space-frequency realization of the source polarization $\mathbf{P}(\mathbf{r},\omega)$, and the suffixes i and j label Cartesian components. The tensor $W_{ij}^{(P)}(\mathbf{r}_1,\mathbf{r}_2,\omega)$ is a measure of the correlations of the polarization at pairs of points in the source, at frequency ω. On interchanging the order of the various operations on the right-hand side of Eq. (2.8), the formula may be expressed in the more compact form

$$\nabla\cdot\langle\mathbf{F}(\mathbf{r},\omega)\rangle=-\frac{kc}{2}\mathrm{Im}\int_D W_{ij}^{(P)}(\mathbf{r}',\mathbf{r},\omega)(k^2\delta_{ij}+\partial_i\partial_j)\frac{e^{-ik|\mathbf{r}-\mathbf{r}'|}}{|\mathbf{r}-\mathbf{r}'|}d^3r', \tag{2.10}$$

where summation over repeated indices is to be taken.

Equation (2.10) is the *differential form* of an energy conservation law for statistically stationary random electromagnetic fields. We note that when the point $\mathbf{r}$ is outside the source domain D, $W_{ij}^{(P)}(\mathbf{r}',\mathbf{r},\omega)=0$, and Eq. (2.10) reduces to the simple form

$$\nabla\cdot\langle\mathbf{F}(\mathbf{r},\omega)\rangle=0. \tag{2.11}$$

The physical significance of formula (2.10) becomes more apparent if one converts it into integral form. Let us, therefore, integrate both sides of Eq. (2.10) over a volume V, bounded by a surface Σ, which completely encloses the source domain D. Making use of the divergence theorem of vector calculus and of the fact that $W_{ij}^{(P)}(\mathbf{r}',\mathbf{r},\omega)=0$ for all points $\mathbf{r}$ located outside the domain D, it follows that

$$\int_\Sigma\langle\mathbf{F}(\mathbf{r},\omega)\rangle\cdot\mathbf{n}\,d\Sigma=-\frac{kc}{2}\mathrm{Im}\int_D\int_D W_{ij}^{(P)}(\mathbf{r}',\mathbf{r},\omega)(k^2\delta_{ij}+\partial_i\partial_j)\frac{e^{-ik|\mathbf{r}-\mathbf{r}'|}}{|\mathbf{r}-\mathbf{r}'|}d^3r\,d^3r', \tag{2.12}$$

where $\mathbf{n}$ denotes the unit outward normal to Σ at the point $\mathbf{r}$ (see Fig. 1). Noting that $W_{ij}^{(P)}(\mathbf{r}',\mathbf{r},\omega)$, summed over the subscripts i and j, is Hermitian, and that the expression $e^{-ik|\mathbf{r}-\mathbf{r}'|}/|\mathbf{r}-\mathbf{r}'|$ is symmetric with respect to $\mathbf{r}$ and $\mathbf{r}'$, Eq. (2.12) may be rewritten in the form

$$\int_\Sigma\langle\mathbf{F}(\mathbf{r},\omega)\rangle\cdot\mathbf{n}d\Sigma=\frac{k^2c}{2}\int_D\int_D W_{ij}^{(P)}(\mathbf{r}',\mathbf{r},\omega)(k^2\delta_{ij}+\partial_i\partial_j)\frac{\sin k|\mathbf{r}-\mathbf{r}'|}{k|\mathbf{r}-\mathbf{r}'|}d^3r\,d^3r'. \tag{2.13}$$

This formula is the *integral form* of the conservation law. It shows that the rate at which the source radiates energy across any surface Σ which completely encloses the source domain D depends on the second-order correlation properties of the source polarization, represented by the cross-spectral density tensor $W_{ij}^{(P)}(\mathbf{r}',\mathbf{r},\omega)$. The conservation laws (2.10) and (2.13) are generalizations to electromagnetic fields of energy conservation laws derived not long ago for fluctuating scalar fields [Ref. [4], Eqs. (3.4) and (3.6)].

III. SOURCE SPECTRUM AND THE SPECTRUM OF THE RADIATED FIELD

We now apply the energy conservation law to elucidate the phenomenon of correlation-induced spectral changes [2]. Let us consider the field in the far zone of the source, at a point specified by the position vector $R\mathbf{u}$, $(\mathbf{u}^2=1)$. The electric and the magnetic fields are given by the expressions [6]

$$\mathbf{E}(R\mathbf{u},\omega)\sim(2\pi)^3k^2\frac{e^{ikR}}{R}\{[\mathbf{u}\times\widetilde{\mathbf{P}}(k\mathbf{u},\omega)]\times\mathbf{u}\} \quad (3.1a)$$

and

$$\mathbf{H}(R\mathbf{u},\omega)\sim(2\pi)^3k^2\frac{e^{ikR}}{R}[\mathbf{u}\times\widetilde{\mathbf{P}}(k\mathbf{u},\omega)], \quad (3.1b)$$

where

$$\widetilde{\mathbf{P}}(\mathbf{k},\omega)=\frac{1}{(2\pi)^3}\int_D \mathbf{P}(\mathbf{r},\omega)e^{-i\mathbf{k}\cdot\mathbf{r}}d^3r \quad (3.2)$$

is the spatial Fourier transform of the source polarization [7]. In tensor notation, Eqs. (3.1a) and (3.1b) take the forms

$$E_i(R\mathbf{u},\omega)\sim(2\pi)^3k^2\frac{e^{ikR}}{R}(\delta_{ij}-u_iu_j)\widetilde{\mathbf{P}}_j(k\mathbf{u},\omega), \quad (3.3a)$$

$$H_i(R\mathbf{u},\omega)\sim(2\pi)^3k^2\frac{e^{ikR}}{R}\varepsilon_{ijk}u_j\widetilde{\mathbf{P}}_k(k\mathbf{u},\omega), \quad (3.3b)$$

where δ_{ij} is the Kroenecker delta symbol, and ε_{ijk} is the completely antisymmetric unit tensor of Levi-Civita.

Let us now define the cross-spectral density tensors $W_{ij}^{(E)}$ and $W_{ij}^{(H)}$ of the field by formulas analogous to that by which the polarization tensor was introduced [Eq. (2.9)], viz.

$$W_{ij}^{(E)}(\mathbf{r}_1,\mathbf{r}_2,\omega)=\langle E_i^*(\mathbf{r}_1,\omega)E_j(\mathbf{r}_2,\omega)\rangle, \quad (3.4a)$$

$$W_{ij}^{(H)}(\mathbf{r}_1,\mathbf{r}_2,\omega)=\langle H_i^*(\mathbf{r}_1,\omega)H_j(\mathbf{r}_2,\omega)\rangle. \quad (3.4b)$$

Using Eqs. (3.3) in Eqs. (3.4), we find that at points in the far zone of the source the field correlation tensors are given by the expressions

$$W_{ij}^{(E)}(R\mathbf{u}_1,R\mathbf{u}_2,\omega)=\frac{(2\pi)^6k^4}{R^2}(\delta_{im}-u_{1i}u_{1m})\times(\delta_{jn}-u_{2j}u_{2n})\widetilde{W}_{mn}^{(P)}(-k\mathbf{u}_1,k\mathbf{u}_2,\omega), \quad (3.5a)$$

$$W_{ij}^{(H)}(R\mathbf{u}_1,R\mathbf{u}_2,\omega)=\frac{(2\pi)^6k^4}{R^2}\varepsilon_{imn}\varepsilon_{jpq}u_{1m}u_{2p}\times\widetilde{W}_{nq}^{(P)}(-k\mathbf{u}_1,k\mathbf{u}_2,\omega), \quad (3.5b)$$

where $u_{\alpha i}$, $(i=1,2,3)$, is the ith component of the unit vector $\mathbf{u}_\alpha$, and

$$\widetilde{W}_{ij}^{(P)}(\mathbf{k}_1,\mathbf{k}_2,\omega)=\frac{1}{(2\pi)^6}\int_D\int_D W_{ij}^{(P)}(\mathbf{r}_1,\mathbf{r}_2,\omega)e^{-i(\mathbf{k}_1\cdot\mathbf{r}_1+\mathbf{k}_2\cdot\mathbf{r}_2)}d^3r_1\,d^3r_2 \quad (3.6)$$

is the six-dimensional Fourier transform of the cross-spectral density of the source polarization.

Let us now determine the field spectrum in the far zone. The power spectrum $S^{(\infty)}(R\mathbf{u},\omega)$ of the field in the far zone at distance R from the source, in a direction specified by a unit vector $\mathbf{u}$, may be identified with the ensemble average of the energy density multiplied by the speed of light, [see Ref. [5], Eqs. (5.7-31)] viz.

$$S^{(\infty)}(R\mathbf{u},\omega)\equiv c\langle U^{(\infty)}(R\mathbf{u},\omega)\rangle=\frac{c}{16\pi}\langle E_i^*(R\mathbf{u},\omega)E_i(R\mathbf{u},\omega)\rangle+\frac{c}{16\pi}\langle H_i^*(R\mathbf{u},\omega)H_i(R\mathbf{u},\omega)\rangle$$
$$=\frac{c}{16\pi}[W_{ii}^{(E)}(R\mathbf{u},R\mathbf{u},\omega)+W_{ii}^{(H)}(R\mathbf{u},R\mathbf{u},\omega)]. \quad (3.7)$$

On making use of Eqs. (3.5) we obtain for the spectrum of the field in the far zone expression [8]

$$S^{(\infty)}(R\mathbf{u},\omega)=\frac{8\pi^5k^4c}{R^2}[(\delta_{ij}-u_iu_j)\widetilde{W}_{ij}^{(P)}(-k\mathbf{u},k\mathbf{u},\omega)]. \quad (3.8)$$

The spectrum of each Cartesian component of the source polarization may be defined by the expression

$$S_i^{(P)}(\mathbf{r},\omega)\equiv W_{ii}^{(P)}(\mathbf{r},\mathbf{r},\omega)\quad \text{(no summation)}. \quad (3.9)$$

Let us define the spectral degree of coherence of the source polarization by the formula

$$\mu_{ij}^{(P)}(\mathbf{r}_1,\mathbf{r}_2,\omega)=\frac{W_{ij}^{(P)}(\mathbf{r}_1,\mathbf{r}_2,\omega)}{\sqrt{S_i^{(P)}(\mathbf{r}_1,\omega)}\sqrt{S_j^{(P)}(\mathbf{r}_2,\omega)}}. \tag{3.10}$$

Using elementary properties of the source polarization tensor and the Schwarz inequality, it is not difficult to show that

$$0\leqslant|\mu_{ij}^{(P)}(\mathbf{r}_1,\mathbf{r}_2,\omega)|\leqslant 1. \tag{3.11}$$

Evidently $\mu_{ij}^{(P)}$ represents the correlation between Cartesian components of the polarization.

If we substitute for $\tilde{W}_{ij}^{(P)}$ in Eq. (3.8) from Eq. (3.6), we find that

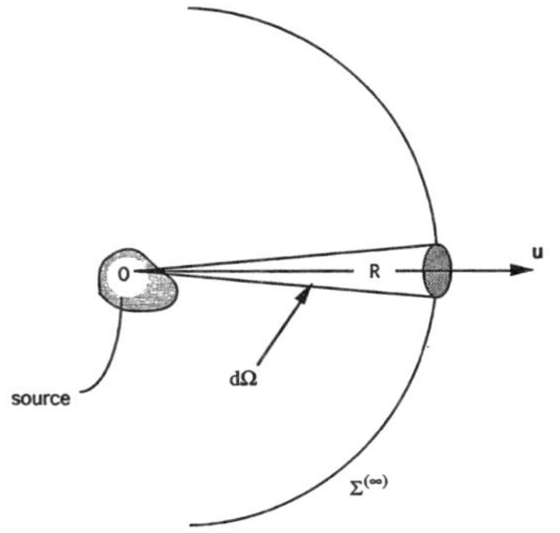

FIG. 2. Illustrating notation relating to the spectrum of the radiated field in the far zone of a fluctuating source polarization.

$$S^{(\infty)}(R\mathbf{u},\omega)=\frac{1}{8\pi}\frac{k^4c}{R^2}\left[(\delta_{ij}-u_iu_j)\int_D\int_D W_{ij}^{(P)}(\mathbf{r}',\mathbf{r},\omega)e^{-ik\mathbf{u}\cdot(\mathbf{r}-\mathbf{r}')}d^3r\,d^3r'\right]. \tag{3.12}$$

If we then express $W_{ij}^{(P)}$ in Eq. (3.12) in terms of the spatial degree of coherence and the spectral densities by the use of Eq. (3.10), we finally obtain for the spectrum of the field in the far zone the expression

$$S^{(\infty)}(R\mathbf{u},\omega)=\frac{1}{8\pi}\frac{k^4c}{R^2}(\delta_{ij}-u_iu_j)\int_D\int_D\sqrt{S_i^{(P)}(\mathbf{r}',\omega)}\sqrt{S_j^{(P)}(\mathbf{r},\omega)}\mu_{ij}^{(P)}(\mathbf{r}',\mathbf{r},\omega)e^{-ik\mathbf{u}\cdot(\mathbf{r}-\mathbf{r}')}d^3r\,d^3r'. \tag{3.13}$$

It is evident from this equation that the spectrum of the far field depends not only on the source spectrum, but also on the correlations between Cartesian components of the polarization. Hence, except perhaps in some special cases, the spectrum of the far field will differ from the source spectrum, and will also depend upon the direction of observation **u**.

We will now show that in spite of the fact that source correlations induce spectral changes in the far field, formula (3.12) is consistent with our new energy conservation law (2.13). For this purpose we integrate both sides of Eq. (3.12) over all directions **u**, and multiply them by R^2. We then obtain the formula

$$\int_{\Sigma^{(\infty)}}S^{(\infty)}(R\mathbf{u},\omega)d\Sigma^{(\infty)}=\frac{1}{8\pi}k^4c\int_{(4\pi)}d\Omega(\delta_{ij}-u_iu_j)\int_D\int_D W_{ij}^{(P)}(\mathbf{r}',\mathbf{r},\omega)e^{-ik\mathbf{u}\cdot(\mathbf{r}-\mathbf{r}')}d^3r\,d^3r', \tag{3.14}$$

where we used the fact that $R^2d\Omega=d\Sigma^{(\infty)}$ is the differential surface element of a large sphere $\Sigma^{(\infty)}$ centered in the source region (see Fig 2). The product u_iu_j on the right side of Eq. (3.14) may be expressed as a differential operator acting on the exponent, and Eq. (3.14) then becomes

$$\int_{\Sigma^{(\infty)}}S^{(\infty)}(R\mathbf{u},\omega)d\Sigma^{(\infty)}=\frac{1}{8\pi}k^2c\int_{(4\pi)}d\Omega\int_D\int_D W_{ij}^{(P)}(\mathbf{r}',\mathbf{r},\omega)(k^2\delta_{ij}+\partial_i\partial_j)e^{-ik\mathbf{u}\cdot(\mathbf{r}-\mathbf{r}')}d^3r\,d^3r', \tag{3.15}$$

where the integral with respect to Ω is taken over the whole 4π solid angle generated by the real unit vector **u**. On making use of the identity (see the footnote on p. 123 of Ref. [5])

$$\frac{\sin k|\mathbf{r}-\mathbf{r}'|}{k|\mathbf{r}-\mathbf{r}'|}=\frac{1}{4\pi}\int_{(4\pi)}e^{-ik\mathbf{u}\cdot(\mathbf{r}-\mathbf{r}')}d\Omega, \tag{3.16}$$

formula (3.15) may be rewritten as

$$\int_{\Sigma^{(\infty)}}S^{(\infty)}(R\mathbf{u},\omega)d\Sigma^{(\infty)}=\frac{k^2c}{2}\int_D\int_D W_{ij}^{(P)}(\mathbf{r}',\mathbf{r},\omega)[k^2\delta_{ij}+\partial_i\partial_j]\frac{\sin k|\mathbf{r}-\mathbf{r}'|}{k|\mathbf{r}-\mathbf{r}'|}d^3r\,d^3r'. \tag{3.17}$$

The right-hand side of this equation is identical to the right-hand side of the integral form of the energy conservation law (2.13). The left-hand sides are also equal to each other because of the well-known relations between the average flux vector $\langle \mathbf{F}^{(\infty)} \rangle$ and the spectral density $\langle S^{(\infty)} \rangle$ in the far field viz. $\langle \mathbf{F}^{(\infty)}(R\mathbf{u},\omega) \rangle = S^{(\infty)}(R\mathbf{u},\omega)\mathbf{u}$ (see, for instance, Eqs. (5.7-32) of Ref. [5]). Hence the two equations (2.13) and (3.17) are equivalent and consequently correlation-induced spectral changes are consistent with energy conservation.

IV. EXAMPLE

We will illustrate our main results by considering a quasi-homogeneous, isotropic source with a source spectrum which is taken to be scalar. For such a source the cross-spectral density tensor can be well approximated by (cf. Ref. [5], Sec. 5.2.2)

$$W_{ij}^{(P)}(\mathbf{r}_1,\mathbf{r}_2,\omega) \approx S((\mathbf{r}_1+\mathbf{r}_2)/2,\omega)\mu_{ij}(\mathbf{r}_2-\mathbf{r}_1,\omega), \tag{4.1}$$

where $S(\mathbf{r},\omega)$ is assumed to vary much more slowly with $\mathbf{r}$ than $\mu_{ij}(\mathbf{r}',\omega)$ varies with $\mathbf{r}'$. Because the source is assumed to be isotropic, it must have the form (cf. Ref. [9])

$$\mu_{ij}(\mathbf{r},\omega) = \delta_{ij}A(r,\omega) + B(r,\omega)r_i r_j, \tag{4.2}$$

where r_i is the ith component of the vector $\mathbf{r}$. The normalization $\mu_{ii}(0,\omega)=1$ (no summation) implies that

$$A(0,\omega)=1, \tag{4.3a}$$

$$r^2 B(r,\omega) \to 0 \quad \text{as } r \to 0. \tag{4.3b}$$

In this case the six-dimensional Fourier transform of the source polarization tensor (4.1) is given by the expression

$$\tilde{W}_{ij}^{(P)}(-k\mathbf{u},k\mathbf{u},\omega) = \frac{1}{(2\pi)^3}\int d^3r[\delta_{ij}A(r,\omega) + r_i r_j B(r,\omega)]e^{-ik\mathbf{u}\cdot\mathbf{r}} \times \frac{1}{(2\pi)^3}\int S(\mathbf{R},\omega)d^3R. \tag{4.4}$$

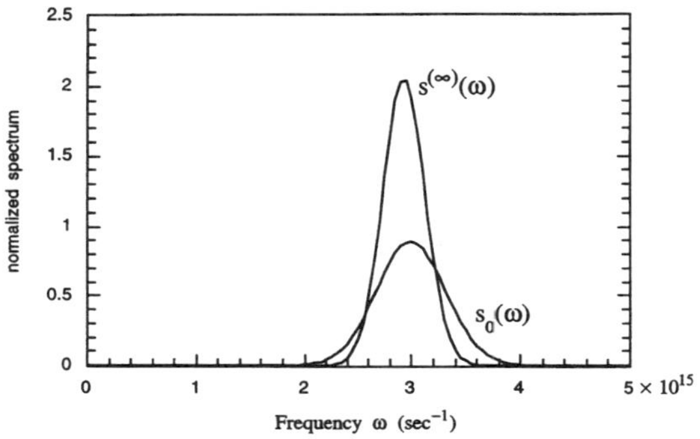

FIG. 3. Normalized spectrum $s_0(\omega) \equiv S_0(\omega)/\int_0^\infty S_0(\omega')d\omega'$ of a homogeneous, isotropic source [represented by Eqs. (4.1), (4.2), and (4.9)] and the normalized spectrum $s^{(\infty)}(\omega) \equiv S^{(\infty)}(\omega)/\int_0^\infty S^{(\infty)} \times(\omega')d\omega'$ of the far field generated by the source, when

$$S(\mathbf{R},\omega) \equiv S_0(\omega) = \frac{I_0}{\sqrt{2\pi}\delta}\exp[-(\omega-\omega_0)^2/2\delta^2]$$

with $\sigma/c = 10^{-15}$ sec, $\omega_0 = 3\times 10^{15}$ sec^{-1}, and $\delta = 2\times 10^{14}$ sec^{-1}.

If $\tilde{A}(q,\omega)$ and $\tilde{B}(q,\omega)$ denote the Fourier transforms of $A(r,\omega)$ and $B(r,\omega)$, respectively, i.e.,

$$\tilde{A}(q,\omega) = \frac{1}{(2\pi)^3}\int A(r,\omega)e^{-i\mathbf{q}\cdot\mathbf{r}}d^3r,$$

$$\tilde{B}(q,\omega) = \frac{1}{(2\pi)^3}\int B(r,\omega)e^{-i\mathbf{q}\cdot\mathbf{r}}d^3r, \tag{4.5}$$

and we make use of the identity

$$\frac{1}{(2\pi)^3}\int r_i r_j B(r,\omega)e^{-i\mathbf{q}\cdot\mathbf{r}}d^3r = -\frac{\partial^2}{\partial q_i \partial q_j}\tilde{B}(q,\omega) = -\left(\delta_{ij} - \frac{q_i q_j}{q^2}\right)\frac{1}{q}\frac{d}{dq}\tilde{B}(q,\omega) - \frac{q_i q_j}{q^2}\frac{d^2}{dq^2}\tilde{B}(q,\omega), \tag{4.6}$$

formula (4.4) becomes

$$\tilde{W}_{ij}^{(P)}(-k\mathbf{u},k\mathbf{u},\omega) = \tilde{S}(0,\omega)\left\{\delta_{ij}\left[\tilde{A}(k,\omega) - \frac{1}{k}\frac{d}{dk}\tilde{B}(k,\omega)\right] + u_i u_j\left[\frac{1}{k}\frac{d}{dk}\tilde{B}(k,\omega) - \frac{d^2}{dk^2}\tilde{B}(k,\omega)\right]\right\}. \tag{4.7}$$

On substituting from Eq. (4.7) into Eq. (3.8), and carrying out the summations, we find that

$$S^{(\infty)}(R\mathbf{u},\omega) = \frac{8\pi^5 k^4 c}{R^2}2\left[\tilde{A}(k,\omega) - \frac{1}{k}\frac{d}{dk}\tilde{B}(k,\omega)\right]\tilde{S}(0,\omega). \tag{4.8}$$

Formula (4.8) shows that the spectrum of the field produced by a source of the kind we are considering is independent of the direction of observation $\mathbf{u}$.

As a specific example, let us choose

$$A(r,\omega) = e^{-r^2/2\sigma^2}, \tag{4.9a}$$

$$B(r,\omega)=\frac{1}{\sigma^2}e^{-r^2/2\sigma^2}, \tag{4.9b}$$

where σ is a positive constant, assumed to be independent of ω. In this case,

$$\tilde{A}(k,\omega)=\frac{\sigma^3}{(2\pi)^{3/2}}e^{-\sigma^2k^2/2}, \tag{4.10a}$$

$$\tilde{B}(k,\omega)=\frac{1}{\sigma^2}\tilde{A}(k,\omega). \tag{4.10b}$$

If we assume that the source spectrum is the same at each source point, i.e., that

$$\begin{aligned} S(\mathbf{r},\omega)&\equiv S_0(\omega), \quad \mathbf{r}\in D\\ &=0, \quad \mathbf{r}\notin D, \end{aligned} \tag{4.11}$$

the formula (4.8) becomes

$$S^{(\infty)}(R\mathbf{u},\omega)=\frac{\sqrt{2\pi}k^4c}{R^2}\sigma^3e^{-\sigma^2k^2/2}V_0S_0(\omega), \tag{4.12}$$

where V_0 is the volume of the domain D occupied by the source.

We see that the normalized spectrum $S^{(\infty)}(R\mathbf{u},\omega)$ of the field in the far zone differs from the source spectrum $S^{(0)}(\omega)$. This is illustrated for a specific case in Fig. 3. In spite of the difference between the two spectra, the result is consistent with the law of conservation of energy, as we showed earlier on general grounds.

ACKNOWLEDGMENTS

This research was supported by the U.S. Air Force Office of Scientific Research under Grant Nos. F 49620-96-1-0400 and F 49620-97-1-0482, and by the U.S. Department of Energy under Grant No. DE-FG02-90 ER 14119.

[1] P. Roman and E. Wolf, Nuovo Cimento **17**, 462 (1960); P. Roman, *ibid.* **22**, 1005 (1961); M. Beran and G. Parrent, J. Opt. Soc. Am. **52**, 48 (1962).

[2] For a discussion of this effect and a review of the publications on this subject see E. Wolf and D. F. V. James, Rep. Prog. Phys. **59**, 771 (1996).

[3] E. Wolf and A. Gamliel, J. Mod. Opt. **39**, 927 (1992); M. Dusek, Opt. Commun. **100**, 24 (1993); G. Hazak and R. Zamir, J. Mod. Opt. **41**, 1653 (1994).

[4] G. S. Agarwal and E. Wolf, Phys. Rev. A **54**, 4424 (1996).

[5] L. Mandel and E. Wolf, *Optical Coherence and Quantum Optics* (Cambridge University Press, Cambridge, 1995).

[6] See W. H. Carter and E. Wolf, Phys. Rev. A **36**, 1258 (1987) where the current density **J** rather than the polarization density **P** was used. These two quantities are related by the continuity equation which, in the space-frequency domain, takes the form $\mathbf{J}(\mathbf{r},\omega)=i\omega\mathbf{P}(\mathbf{r},\omega)$.

[7] Formula (3.1a) is sometimes expressed in the more compact form

$$E_i(R\mathbf{u},\omega)\sim(2\pi)^3k^2\frac{e^{ikR}}{R}\tilde{P}_i^{(t)}(k\mathbf{u},\omega),$$

where

$$\tilde{P}_i^{(t)}(k\mathbf{u},\omega)\equiv([\mathbf{u}\times\tilde{\mathbf{P}}(k\mathbf{u},\omega)]\times\mathbf{u})_i=(\delta_{ij}-u_iu_j)\tilde{P}_j(k\mathbf{u},\omega)$$

are components of the transverse polarization. (cf. Ref. [6]).

[8] The left-hand side of Eq. (3.8) is invariant with respect to a rotation of axes, and therefore so must be the right-hand side. That this is so follows at once from the following relation involving the cross-spectral density tensor of the polarization W_{ij}^P and the cross-spectral density tensor of the *transverse polarization* $W_{ij}^{P^{(t)}}$

$$(\delta_{ij}-u_iu_j)\tilde{W}_{ij}^P(-k\mathbf{u},k\mathbf{u},\omega)=\mathrm{Tr}\{\tilde{W}_{ij}^{P^{(t)}}(-k\mathbf{u},k\mathbf{u},\omega)\},$$

where Tr denotes the trace. [cf. Ref. [6], Eq. (D7)].

[9] G. K. Batchelor, *The Theory of Homogeneous Turbulence* (Cambridge University Press, Cambridge, 1986), Secs. 3.3 and 3.4.

March 15, 1999 / Vol. 24, No. 6 / OPTICS LETTERS 367

Storage and retrieval of correlation functions of partially coherent fields

Emil Wolf

Department of Physics and Astronomy and Rochester Theory Center for Optical Science and Engineering, University of Rochester, Rochester, New York 14627

Tomohiro Shirai

Mechanical Engineering Laboratory, Ministry of International Trade and Industry, Tsukuba 305-8564, Japan

Girish Agarwal

Physical Research Laboratory, Navrangpura, Ahmedabad 380 009, India

Leonard Mandel

Department of Physics and Astronomy and Rochester Theory Center for Optical Science and Engineering, University of Rochester, Rochester, New York 14627

Received November 30, 1998

A new method is described for determining the two-point equal-time coherence function (the mutual intensity) and the two-point equal-time intensity correlation function of partially coherent fields. The method is reminiscent of conventional holography but differs from it in several important respects. © 1999 Optical Society of America

OCIS code: 090.0090.

It is well known that coherence functions of partially coherent fields and, in particular, the degree of coherence, can carry important physical information. For example, they can contain information about the distribution of intensity across sources, about the nature of rough surfaces, or about the structure of random media. Measurements of coherence functions are normally performed in real time by means of interference experiments. It would be useful to have a method available that makes it possible to store the correlation functions and retrieve them at a later time. In this Letter we propose such a method.

The possibility of storing and retrieving the degree of coherence by holography was demonstrated many years ago by Lurie.[1] His theoretical analysis and experiments were, however, restricted to the correlation between just two light beams. In this Letter we consider the broader problem of storage and retrieval of the degree of coherence of light oscillations at any pair of points in a cross section of a light beam by a technique that resembles holography but differs from it in several important respects.

Consider a statistically stationary quasi-monochromatic field of any state of coherence incident on a photographic emulsion located in the plane $z = 0$. Let this field be represented by an ensemble with complex envelopes $U^{(i)}(\mathbf{r}, t)$ at position $\mathbf{r}$ at time t. Similarly, let $U^{(R)}(\mathbf{r}, t)$ represent an ensemble of stationary reference fields at $\mathbf{r}$, t (see Fig. 1). Then the total field $U^{(T)}(\mathbf{r}, t)$ at the space–time point $\mathbf{r}$, t is given by the sum

$$U^{(T)}(\mathbf{r}, t) = U^{(i)}(\mathbf{r}, t) + U^{(R)}(\mathbf{r}, t), \tag{1}$$

and the instantaneous light intensity at $\mathbf{r}$, t is obtained by taking the squared modulus of this quantity,

$$\begin{aligned} I^{(T)}(\mathbf{r}, t) &= |U^{(T)}(\mathbf{r}, t)|^2 \\ &= I^{(i)}(\mathbf{r}, t) + I^{(R)}(\mathbf{r}, t) \\ &\quad + U^{(R)*}(\mathbf{r}, t)U^{(i)}(\mathbf{r}, t) + \text{c.c.}, \end{aligned} \tag{2}$$

where the asterisk and c.c. denote the complex conjugate.

In conventional holography the incident field $U^{(i)}$ and the reference field $U^{(R)}$ are mutually coherent, and they give rise to an interference pattern in the $z = 0$ plane that is recorded in a time interval that is very long compared with the optical coherence time τ_c of the light. This procedure yields information about the ensemble average $\langle I^{(T)}(\mathbf{r}, t)\rangle$ of the instantaneous light

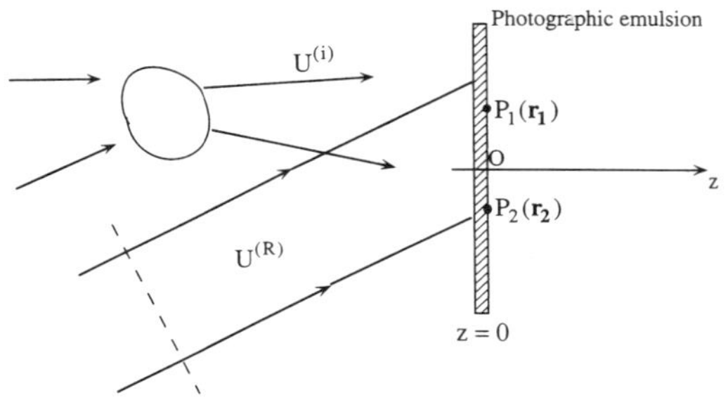

Fig. 1. Formation of the instantaneous hologram. A photographic film is located in the plane $z = 0$, which is illuminated by a quasi-monochromatic field of any state of coherence, with complex envelope $U^{(i)}(\mathbf{r}, t)$ and by a reference beam $U^{(R)}(\mathbf{r}, t)$. $U^{(i)}$ and $U^{(R)}$ are mutually incoherent, and the photographic exposure time is made short compared with the coherence time of the light.

0146-9592/99/060367-03$15.00/0

intensity at $\mathbf{r}$. In the present case we assume the opposite, viz., that the incident field $U^{(i)}$ and the reference field $U^{(R)}$ are mutually incoherent. Moreover, by recording the light intensity in the plane $z = 0$ over a time interval that is short compared with τ_c, we obtain a record not of $\langle I^{(T)}(\mathbf{r}, t)\rangle$ but of the instantaneous light intensity $I^{(T)}(\mathbf{r}, t)$. For want of a better name we shall refer to this record as the "instantaneous hologram," which is quite different from what is recorded in conventional holography. We shall assume that the reference beam is derived from a well-stabilized laser and can be treated as essentially a plane wave, free from intensity fluctuations.

The ensemble of many instantaneous holograms can be used to reproduce the ensemble of instantaneous light intensities $I^{(T)}(\mathbf{r}, t)$. The instantaneous hologram is obtained by processing the photographic emulsion so that its amplitude transmittance $\mathcal{T}(\mathbf{r}, t)$, at $\mathbf{r}, t$ is proportional to $I^{(T)}(\mathbf{r}, t)$, i.e.,

$$\mathcal{T}(\mathbf{r}, t) = \beta I^{(T)}(\mathbf{r}, t), \tag{3}$$

where β is a real constant.

For the reconstruction or retrieval phase of the procedure the processed instantaneous hologram is illuminated by a stationary quasi-monochromatic plane wave $V(\mathbf{r}, t)$ (see Fig. 2) that is mutually *incoherent* with all the other waves. Then the field $U^{(E)}(\mathbf{r}, t)$ at the exit face of the instantaneous hologram is represented by

$$U^{(E)}(\mathbf{r}, t) = \beta V(\mathbf{r}, t) I^{(T)}(\mathbf{r}, t), \tag{4}$$

and, with the help of Eq. (2), we obtain

$$U^{(E)}(\mathbf{r}, t) = \beta V(\mathbf{r}, t)[I^{(i)}(\mathbf{r}, t) + I^{(R)}(\mathbf{r}, t) + U^{(R)*}(\mathbf{r}, t)U^{(i)}(\mathbf{r}, t) + \text{c.c.}]. \tag{5}$$

The assumed character of $U^{(R)}(\mathbf{r}, t)$ makes $I^{(R)}(\mathbf{r}, t)$ independent of $\mathbf{r}$ and the ensemble average $\langle I^{(R)}(\mathbf{r}, t)\rangle$ independent of both $\mathbf{r}$ and t. We now use Eq. (5) to calculate the mutual intensity

$$J^{(E)}(\mathbf{r}_1, \mathbf{r}_2) \equiv \langle U^{(E)*}(\mathbf{r}_1, t)U^{(E)}(\mathbf{r}_2, t)\rangle \tag{6}$$

(Ref. 2, p. 168) at two points, $\mathbf{r}_1$ and $\mathbf{r}_2$, at the exit face of the instantaneous hologram. $J^{(E)}(\mathbf{r}_1, \mathbf{r}_2)$ can be determined experimentally in the usual way by letting the light from the two points $\mathbf{r}_1$, $\mathbf{r}_2$ come together in another plane and making measurements of the resulting interference pattern. From Eqs. (5) and (6) we have, after averaging over the ensemble,

$$\begin{aligned} J^{(E)}(\mathbf{r}_1, \mathbf{r}_2) = \beta^2 J^{(V)}(\mathbf{r}_1, \mathbf{r}_2)\{&\langle I^{(i)}(\mathbf{r}_1, t)I^{(i)}(\mathbf{r}_2, t)\rangle \\ &+ I^{(R)}[\langle I^{(i)}(\mathbf{r}_1, t)\rangle + \langle I^{(i)}(\mathbf{r}_2, t)\rangle] \\ &+ I^{(R)2} + J^{(R)}(\mathbf{r}_1, \mathbf{r}_2)J^{(i)*}(\mathbf{r}_1, \mathbf{r}_2) + \text{c.c.}\}. \end{aligned} \tag{7}$$

where $J^{(X)}(\mathbf{r}_1, \mathbf{r}_2) \equiv \langle X^*(\mathbf{r}_1, t)X(\mathbf{r}_2, t)\rangle$ (with $X = E$ or V or i) are the mutual intensities of the E or V or i fields. In deriving Eq. (7) we made use of the fact that a number of terms vanish because of the assumed stationarity and zero means of the fields $U^{(i)}$, $U^{(R)}$ and V. In particular, we have used a theorem on expectation values of products of analytic signals representing stationary random processes (Ref. 2, theorem II, p. 105). It follows that the unknown mutual intensity $J^{(i)}(\mathbf{r}_1, \mathbf{r}_2)$ of the incident field $U^{(i)}$ can be obtained from a knowledge of the mutual intensities $J^{(E)}(\mathbf{r}_1, \mathbf{r}_2)$, $J^{(V)}(\mathbf{r}_1, \mathbf{r}_2)$ and $J^{(R)}(\mathbf{r}_1, \mathbf{r}_2)$, together with the intensity correlation function $\langle I^{(i)}(\mathbf{r}_1, t)I^{(i)}(\mathbf{r}_2, t)\rangle$. $J^{(R)}(\mathbf{r}_1, \mathbf{r}_2)$ and $J^{(V)}(\mathbf{r}_1, \mathbf{r}_2)$ may be taken as known (see Figs. 1 and 2), and $J^{(E)}(\mathbf{r}_1, \mathbf{r}_2)$ is derivable from measurements of the interference pattern via the ensemble of instantaneous holograms, as described above. Finally, the intensity correlation $\langle I^{(i)}(\mathbf{r}_1, t)I^{(i)}(\mathbf{r}_2, t)\rangle$ on the right-hand side of Eq. (7) is obtainable from measurements of $J^{(E)}(\mathbf{r}_1, \mathbf{r}_2)$ made with the reference field $U^{(R)}$ tuned off, in which case it follows from Eq. (7) that

$$J^{(E)}(\mathbf{r}_1, \mathbf{r}_2) = \beta^2 J^{(V)}(\mathbf{r}_1, \mathbf{r}_2)\langle I^{(i)}(\mathbf{r}_1, t)I^{(i)}(\mathbf{r}_2, t)\rangle. \tag{8}$$

It is worth noting that Eq. (8) forms the basis for an interesting new procedure for determining two-point equal-time intensity correlation functions of an optical field from interferometric measurements of the average light intensity at various points. Unlike conventional holography, the new method works best when the $E^{(i)}(\mathbf{r}, t)$ and $E^{(R)}(\mathbf{r}, t)$ fields are comparable in strength. When $\langle I^{(R)}\rangle \gg \langle I^{(i)}\rangle$, it is apparent from Eq. (7) that the term in $J^{(E)}(\mathbf{r}_1, \mathbf{r}_2)$ becomes negligible compared with $I^{(R)2}$.

In practice it is often the equal-time complex degree of coherence $\gamma(\mathbf{r}_1, \mathbf{r}_2)$ rather than the mutual intensity that is of chief interest. It is defined by the formula

$$\gamma(\mathbf{r}_1, \mathbf{r}_2) = \frac{J(\mathbf{r}_1, \mathbf{r}_2)}{\sqrt{J(\mathbf{r}_1, \mathbf{r}_1)}\sqrt{J(\mathbf{r}_2, \mathbf{r}_2)}}, \tag{9}$$

where $J(\mathbf{r}, \mathbf{r})$ represents the averaged intensity $\langle I(\mathbf{r})\rangle$ of the light at point $\mathbf{r}$.

To be sure, the implied procedure for measuring the mutual intensity $J^{(E)}(\mathbf{r}_1, \mathbf{r}_2)$ of the field $U^{(E)}$ is considerably more complicated than typical measurements in conventional holography. The light from points $\mathbf{r}_1$

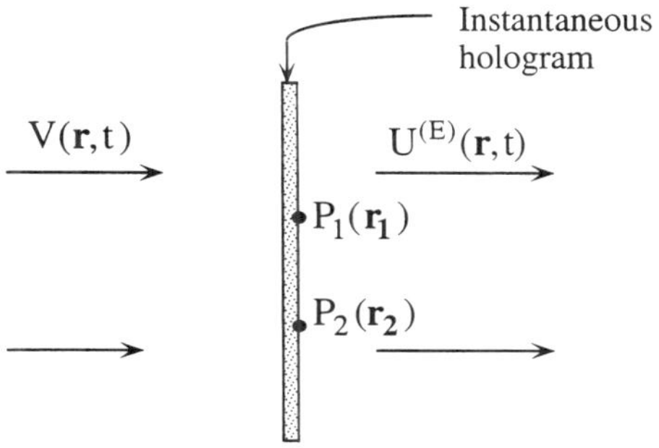

Fig. 2. Retrieval of the mutual intensity function $J^{(i)}(\mathbf{r}_1, \mathbf{r}_2) = \langle U^{(i)*}(\mathbf{r}_1, t)U^{(i)}(\mathbf{r}_2, t)\rangle$ of the field $U^{(i)}$. Each instantaneous hologram is illuminated by a quasi-monochromatic plane wave $V(\mathbf{r}, t)$.

and $\mathbf{r}_2$ immediately behind the instantaneous hologram must be allowed to come together at various points $\mathbf{r}$ in another plane, such that the path difference $||\mathbf{r}_1 - \mathbf{r}| - |\mathbf{r}_2 - \mathbf{r}|| \ll c\tau_c$, and the instantaneous light intensity at $\mathbf{r}$ must be measured and recorded. This measurement then has to be repeated many (perhaps thousands of) times for each different realization of the instantaneous hologram, and the results have to be averaged for each pair of points $\mathbf{r}_1$, $\mathbf{r}_2$.

In principle, a large number of similar photographic plates could be used to record the ensemble of instantaneous holograms, but this is not very practical. It is probably more convenient to make use of a continuous photographic film, as is used for movies, for example, to record the succession of instantaneous holograms. After development the successive frames of the film serve as the ensemble of instantaneous holograms and can be used for the reconstruction. Instead of exposing each frame of the film for a brief time much shorter than τ_c, it would probably be more convenient to make use of pulsed laser light sources, both for the reference field $U^{(R)}$ and for the illuminating field that gives rise to $U^{(i)}$. Such lasers can easily be made to generate subpicosecond pulses.

If the practical problems can be solved, holographic procedures of the type we have described would seem to be even more powerful and more versatile than is generally realized.

One of us (E. Wolf) wishes to acknowledge stimulating discussions with W. H. Carter about holography with partially coherent light. This research was supported by the U.S. Air Force Office of Scientific Research under grant F496W-96-1-0400, by the U.S. Department of Energy under grant DE-FG02-90ER 14119, by the U.S. Office of Naval Research, and by the National Science Foundation. Girish Agarwal wishes to express his thanks for support from National Science Foundation grant INT97-12760, which has made possible his participation in this project.

References

1. M. Lurie, J. Opt. Soc. Am. **56,** 1369 (1966); **58,** 614 (1968).
2. See, for example, L. Mandel and E. Wolf, *Optical Coherence and Quantum Optics* (Cambridge University Press, Cambridge, 1995).

Section 4 – Scattering

The first paper in this Section, paper 4.1, is entitled "Three-dimensional structure determination of semi-transparent objects from holographic data". The reference to holography was included in the title because the reconstruction technique proposed in that paper requires the knowledge not only of the amplitude, but also of the phase of the scattered field, which is generally difficult to measure at optical wavelengths; and I showed in paper 4.2 written around the same time how both the amplitude and the phase of the scattered field can be determined by means of holography. Of course today there are other techniques for measuring the phase. The inverse method proposed in paper 4.1 has developed into a rather successful inversion technique known today as diffraction tomography. It provides an improvement, for some applications, over the well-known computed axial tomography, or CAT, which is an important medical diagnostic tool. Today, both CAT and diffraction tomography find many uses also outside the field of medicine.*

Very recently a different tomographic method for determining the structure of an object was proposed in papers 4.17, which avoids the need for phase measurements. The theory makes use of a generalization of the optical cross-section theorem, derived in a paper (not included in this volume) by P.S. Carney, E. Wolf and G.S. Agarwal, *J. Opt. Soc. Am. A* **14**, 3366-3371 (1997). Information about the structure of the object is deduced from knowledge of the power which is extinguished by the object when it is illuminated by two plane waves whose relative phase is controlled.

Paper 4.3 presents a review of researches relating to a rather poorly understood subject in the theory of scattering, the so-called Ewald-Oseen extinction theorem. In its traditional form, first derived in molecular optics, the theorem expresses the extinction of a field incident on a scattering object at every point within the scatter in terms of a surface integral which contains the induced field on the boundary of the scattering object. The physical significance of the theorem has been subject to a good deal of controversy. The paper summarizes researches in this subject and suggests a new interpretation of the theorem; namely that it represents a non-local boundary condition for the solution of the interior scattering problem. Subsequent research, described in papers 4.4 and 4.5 verified the correctness of this hypothesis, within the framework of non-relativistic quantum-mechanical theory of potential scattering. The impossibility of formulating problems of potential scattering as local boundary value problems, in general, has been demonstrated in paper 4.16.

An interesting development in non-linear optics was the discovery made in the 1970s of a method for generating a wave which is the "phase conjugated version" of a given wave. One of the main uses of phase-conjugation is to correct distortions which may be imparted on an incident wave by its interactions with a scattering medium. In paper 4.6 the theory of such a "healing process" was developed, both for scalar and for electromagnetic fields.

Paper 4.7 is concerned with the analytic properties of general plane wave representations of scattered fields, which contains both ordinary (homogeneous) plane waves, as well as evanescent (inhomogeneous) waves that were briefly mentioned in commentary on papers 2.1 and 2.2. A con-

*A review of diffraction tomography is presented in E. Wolf, "Principles and development of diffraction tomography" in *Trends in Optics*, A. Consortini, ed. (Academic Press, San Diego, 1996), pp. 83–110.

sequence of the analytic properties established in this paper is the fact that a field scattered from a finite object with a continuous scattering potential must contain both homogeneous and evanescent waves unless it vanishes at every point outside the scattering volume. This result has become of considerable interest in recent years in the field of near-field optics.

It was mentioned in the commentary to Section 3, in connection with the discussion of possible changes in the spectrum of a field arising from correlation properties of its source, that similar spectral modifications may arise on scattering of polychromatic light on a random medium. The simplest changes are shifts of spectral lines. In paper 4.8 such changes are discussed for scattering from a random static medium, i.e. a medium whose macroscopic properties, assumed to be characterized by a continuous random response function, do not change in time. In paper 4.9 spectral changes induced by static scattering on a system of particles are discussed. It is also shown in that paper that this effect might be used to determine the structure factor of some scattering systems.

In paper 4.10, spectral changes produced by dynamic scattering on media whose dielectric susceptibility varies randomly both in space and in time are discussed. In paper 4.11, a special situation is considered when this process generates shifts of spectral lines. In a later paper, 4.12, it was shown that in dynamic scattering from a medium with suitable space-time correlation functions, shifts of spectral lines may be generated which are indistinguishable from the Doppler effect, even though the source, the scattering medium and the observer are all at rest with respect to each other. In paper 4.13 a model scatterer of this kind is discussed and the possible relevance of this effect to the origin of discrepancies observed in some quasars spectra is mentioned. A broader class of scattering media which generates Doppler-like shifts of spectral lines is discussed in paper 4.14. The relevance of the phenomenon of correlation-induced spectral changes to spectroscopy of astronomical sources has been the subject of some controversy, which is discussed in a review article included as paper 4.15.

Section 4 – Scattering

Volume 1, number 4 OPTICS COMMUNICATIONS September/October 1969

THREE-DIMENSIONAL STRUCTURE DETERMINATION OF SEMI-TRANSPARENT OBJECTS FROM HOLOGRAPHIC DATA *

Emil WOLF
Department of Physics and Astronomy, University of Rochester, Rochester, N.Y. 14627, USA

Received 11 August 1969

A solution is presented to an inverse scattering problem that arises in the application of holography to the determination of the three-dimensional structure of weakly scattering semi-transparent objects. This solution, together with a result obtained in another recent publication, relating to the determination of the complex amplitude distribution of scattered fields from measurements of the intensity transmission functions of holograms, makes it possible to calculate the distribution of the (generally complex) refractive index throughout the object. In general many holograms are needed, corresponding to different directions of illumination of the object.

One of the very attractive features of holography is the fact that it produces a three-dimensional image of a three-dimensional object. So far this aspect of holography has been mainly utilized for visual displays.

It is obvious that a hologram stores in some complicated way information about the three-dimensional structure of a scattering object. It therefore ought to be possible to determine this structure computationally, from measurements performed on the hologram. Such a technique would be of value in many fields, particularly in biology.

The determination from holograms, of the three-dimensional strtucture of an object, i.e. the determination of the three-dimensional distribution of the (generally complex) refractive index throughout the object falls naturally into two parts:

(1) Calculation of the distribution of the amplitude and the phase of the field scattered by the object from measurements of the transmissivity of the hologram.

(2) Computational reconstruction from these data of the distribution of the refractive index throughout the object.

A solution to the first problem was presented in another recent publication [1], where it was shown how both the amplitude and the phase of the scattered field may be obtained from measurements of the intensity distribution of the field emerging from a transilluminated hologram. A solution to the second problem is presented in the present note.

Let a plane monochromatic wave ‡

$$U^{(i)}(\boldsymbol{R}) = \exp(ik_0 s_0 \cdot \boldsymbol{R}) , \tag{1}$$

propagated in the direction specified by the unit vector s_0 be incident on a semi-transparent weakly scattering object situated in free space (fig. 1). Let

$$U(\boldsymbol{R}) = U^{(i)}(\boldsymbol{R}) + U^{(s)}(\boldsymbol{R}) \tag{2}$$

represent the resulting field at a point specified by the position vector $\boldsymbol{R}$. In (2), $U^{(s)}(\boldsymbol{R})$ represents, of course, the scattered field.

The field $U(\boldsymbol{R})$ satisfies the equation

$$\nabla^2 U(\boldsymbol{R}) + k_0^2 n^2(\boldsymbol{R})\, U(\boldsymbol{R}) = 0 , \tag{3}$$

where $n(\boldsymbol{R})$ is the (possibly complex) refractive index at the point $\boldsymbol{R}$. Since we have assumed that the object is situated in free space, $n(\boldsymbol{R}) \equiv 1$ outside the region of space occupied by the object. From (2) and (3) and the fact that the incident

* Research supported by the Air Force Cambridge Research Laboratories. An account of the main results reported in this note was presented at the Spring Meeting of the Optical Society of America, held in San Diego, March 12, 1969 (Abstract WB 20; J. Opt. Soc. Amer. 59 (1969) 482) and at a symposium on Coherent Optics and Instrumentation held at Oakland University, Rochester, Michigan, April 18, 1969.

‡ For the sake of simplicity we ignore polarization effects and treat the field as a scalar. A fuller analysis, which will take into account the vectorial features of the field will be given in a later publication. As customary a time dependent periodic factor $\exp(-i\omega t)$ [$\omega = k_0 c$; c = vacuum velocity of light] is omitted throughout.

Volume 1, number 4 OPTICS COMMUNICATIONS September/October 1969

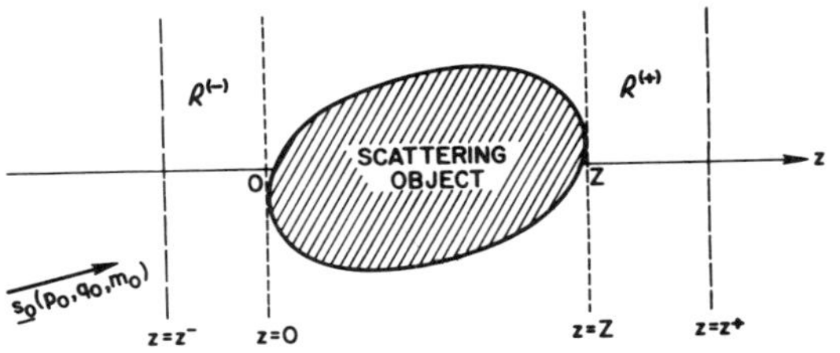

Fig. 1. Illustrating the notation. $\mathcal{R}^{(-)}$ denotes the region $z < 0$, $\mathcal{R}^{(+)}$ denotes the region $z > Z$.

field satisfies the equation $(\nabla^2 + k_0^2)U^{(i)} = 0$ it follows that the scattered field obeys the equation

$$(\nabla^2 + k_0^2)\,U^{(s)}(\boldsymbol{R}) = F(\boldsymbol{R})\,U(\boldsymbol{R})\,, \tag{4}$$

where

$$F(\boldsymbol{R}) = -k_0^2[n^2(\boldsymbol{R}) - 1]\,. \tag{5}$$

We will refer to the function $F(\boldsymbol{R})$ as the *scattering potential* of the object. It is evidently zero at all points outside the object. Our aim is to show how to determine $F(\boldsymbol{R})$ from the knowledge of the scattered field; for according to (5) the knowledge of $F(\boldsymbol{R})$ is equivalent to the knowledge of the refractive index function $n(\boldsymbol{R})$.

Since we have assumed that the object scatterers weakly $[\,|U^{(s)}| \ll |U^{(i)}|\,]$, a good approximation to the solution of eq. (4) for the scattered field is given by the first Born approximation [2]:

$$U^{(s)}(\boldsymbol{R}) = -\frac{1}{4\pi}\int F(\boldsymbol{R}')\exp(\mathrm{i}k_0 \boldsymbol{s}_0\cdot\boldsymbol{R}')\,G(|\boldsymbol{R}-\boldsymbol{R}'|)\,\mathrm{d}^3\boldsymbol{R}'\,, \tag{6}$$

where $G(r) = \exp(\mathrm{i}k_0 r)/r$. In eq. (6) the integral extends only formally over the whole infinite space, since the scattering potential $F(\boldsymbol{R}')$ vanishes outside the region occupied by the object. We now make use of the following representation of the spherical wave *:

$$G(|\boldsymbol{R}-\boldsymbol{R}'|) = \frac{\mathrm{i}k_0}{2\pi}\iint_{-\infty}^{\infty}\frac{1}{m} \times \exp\{\mathrm{i}k_0[p(x-x') + q(y-y') + m|z-z'|]\}\,\mathrm{d}p\,\mathrm{d}q\,, \tag{7}$$

where

$$m = +(1-p^2-q^2)^{\frac{1}{2}} \quad \text{when} \quad p^2+q^2 \leq 1 \tag{8a}$$

$$= +\mathrm{i}(p^2+q^2-1)^{\frac{1}{2}} \quad \text{when} \quad p^2+q^2 > 1 \tag{8b}$$

and $\boldsymbol{R} \equiv (x,y,z)$, $\boldsymbol{R}' \equiv (x',y',z')$.

Suppose now that the point $\boldsymbol{R}$ lies either in the region $\mathcal{R}^{(+)}$ throughout which $z - z'$ is positive ($z > Z$ in fig. 1) or in the region $\mathcal{R}^{(-)}$ throughout which $z - z'$ is negative ($z < 0$ in fig. 1), irrespective of the location of the point $\boldsymbol{R}'$ in the object. If we substitute from (7) into (6), we obtain the following expressions for the scattered field, valid throughout the regions $\mathcal{R}^{(+)}$ (upper sign) and $\mathcal{R}^{(-)}$ (lower sign):

$$U^{(s)}(\boldsymbol{R}) = \iint_{-\infty}^{\infty} A^{(\pm)}(p,q;p_0,q_0) \times \exp[\mathrm{i}k_0(px+qy\pm mz)]\,\mathrm{d}p\,\mathrm{d}q\,, \tag{9}$$

where

$$A^{(\pm)}(p,q;p_0,q_0) = -\frac{\mathrm{i}k_0}{8\pi^2 m}\int F(\boldsymbol{R}') \times \exp\{-\mathrm{i}k_0[(p-p_0)x' + (q-q_0)y' + (\pm m - m_0)z']\}\,\mathrm{d}^3\boldsymbol{R}'. \tag{10}$$

Equation (9) represents the scattered field throughout each of the regions $\mathcal{R}^{(+)}$ and $\mathcal{R}^{(-)}$ in the form of an angular spectrum of plane waves [5]. The "spectral amplitude function" $A^{(\pm)}(p,q;p_0,q_0)$ is expressed in terms of the scattering potential $F(\boldsymbol{R})$ by eq. (10). According to (8) the plane waves in the angular spectrum are *homogeneous* if $p^2+q^2 \leq 1$ and *evanescent* if $p^2+q^2 > 1$. Since for the homogeneous waves m is real, we obtain at once from (10), the following relation:

$$A_h^{(\pm)}(p,q;p_0,q_0) = -\frac{\mathrm{i}\pi k_0}{m} \times \hat{F}[k_0(p-p_0),\ k_0(q-q_0),\ k_0(\pm m - m_0)]\,, \tag{11}$$

where $\hat{F}$ is the Fourier inverse of F:

$$\hat{F}(u,v,w) = \frac{1}{(2\pi)^3}\iiint_{-\infty}^{\infty} F(x,y,z) \times \exp[-\mathrm{i}(ux+vy+wz)]\,\mathrm{d}x\,\mathrm{d}y\,\mathrm{d}z\,. \tag{12}$$

On the left-hand side of (11) we have written $A_h^{(\pm)}$ in place of $A^{(\pm)}$ to remind us that the relation (11) holds only for the homogeneous part of the angular spectrum, i.e. for each pair of values of p and q such that $p^2+q^2 \leq 1$.

Consider now the scattered field $U^{(s)}$ in two fixed planes $z = z^+$ and $z = z^-$, the first one being

* This formula follows by a straightforward modification of a formula due to Weyl [3]. For a direct derivation of it, see, for example [4].

situated in the region $\mathcal{R}^{(+)}$, the second in the region $\mathcal{R}^{(-)}$ (see fig. 1). It follows at once on taking the Fourier inverse of (9) with respect to the variables x and y, with z having the fixed values z^+ and z^-, that

$$A^{(\pm)}(p,q;p_0,q_0) = k_0^2 \exp(\mp ik_0 mz^\pm)\,\hat{U}^{(s)}(k_0 p, k_0 q; z^\pm)\ , \tag{13}$$

where

$$\hat{U}^{(s)}(u,v;z^\pm) = \frac{1}{(2\pi)^2}\iint_{-\infty}^{\infty} U^{(s)}(x,y,z^\pm)\exp[-i(ux+vy)]\,dx\,dy \tag{14}$$

is the Fourier inverse of $U^{(s)}$ with respect to the variables x and y.

On comparing eqs. (11) and (13) and on using (8a) we obtain the important relation

$$\hat{F}(U,V,W^\pm) = \frac{iw}{\pi}\exp(\mp iwz^\pm)\,\hat{U}_h^{(s)}(u,v;z^\pm)\ , \tag{15}$$

where

$$\left.\begin{aligned} U &= u - k_0 p_0\ ,\\ V &= v - k_0 q_0\ ,\\ W^\pm &= \pm w - k_0 m_0\ ,\end{aligned}\right\} \tag{16}$$

and

$$w = (k_0^2 - u^2 - v^2)^{\frac{1}{2}}\ . \tag{17}$$

The subscript h on $U_h^{(s)}$ on the right-hand side of eq. (15) indicates that the relation is valid only for those two-dimensional Fourier components $\hat{U}^{(s)}$ of $U^{(s)}$ about which the information is carried by homogeneous waves [6], i.e. for those components for which

$$u^2 + v^2 \leqslant k_0^2\ . \tag{18}$$

In words eq. (15) shows that some of the *three-dimensional Fourier components of the scattering potential* may be immediately determined from the knowledge of the *two-dimensional Fourier components of the scattered field* in the two planes $z = z^+ > Z$ and $z = z^- < 0$. It is now clear that eq. (15), together with the result of ref. [1] relating to the determination of the complex scattered field from measurements of the *intensity transmission function* of the hologram, provides a method for determining the three-dimensional structure of weakly scattering semi-transparent objects.

The restriction imposed by the inequality (18) on the range of accessible Fourier components of the scattering potential may be readily deduced. For this purpose we re-write eqs. (16) in vector form,

$$\boldsymbol{K}^\pm = \boldsymbol{k}^\pm - \boldsymbol{k}_0\ , \tag{19}$$

where $\boldsymbol{K}^\pm \equiv (U,V,W^\pm)$, $\boldsymbol{k}^\pm \equiv (u,v,\pm w)$, $\boldsymbol{k}_0 \equiv (k_0 p_0, k_0 q_0, k_0 m_0)$. Eqs. (8a) and (17) imply that $|\boldsymbol{k}_0|^2 = |\boldsymbol{k}|^2 = k_0^2$. By allowing the propagation vector $\boldsymbol{k}_0$ of the plane wave incident on the object to take on different directions in successive experiments it is clear that all $\boldsymbol{K}^\pm$-vectors are, in principle, accessible for which

$$|\boldsymbol{K}^\pm| \leqslant 2k_0 = 4\pi/\lambda_0\ , \tag{20}$$

where $\lambda_0 = 2\pi/k_0$ is the wavelength of the incident plane wave. If we set $K_x^\pm \equiv U = 2\pi/\Delta x$, $K_y^\pm \equiv V = 2\pi/\Delta y$, $K_z^\pm = W^\pm = 2\pi/\Delta z$, (20) implies that the formula (15) allows us to determine all those three-dimensional Fourier components of the scattering potential, for which the spatial periods Δx, Δy, Δz satisfy the inequality

$$\frac{1}{(\Delta x)^2} + \frac{1}{(\Delta y)^2} + \frac{1}{(\Delta z)^2} \leqslant \left(\frac{2}{\lambda_0}\right)^2\ . \tag{21}$$

If we denote the left-hand side of (21) by $1/L^2$, (21) gives

$$L \geqslant \lambda_0/2\ . \tag{22}$$

The quantity L, which has the dimension of length, may be taken as a rough measure of the limit of resolution down to which the details of the structure could be determined with the help of formula (15). According to eq. (22) this limit is about half a wavelength. However, as was shown in ref. [1], the measurements of the intensity transmission function of a side-band hologram can only yield information about the scattered field down to details of the order of about $9\lambda_0$. For this reason * the theoretical limit indicated by the inequality (22) cannot be expected to be attainable when the scattered field is determined by side-band holography.

* *Note added in proof*: Detailed analysis shows that if, in spite of this limitation, the scattered field is determined not only for different directions of incidence (p_0, q_0, m_0), but also for different orientations of the observing plane (plane of the hologram), it is possible, in principle, to reconstruct all the three-dimensional Fourier components of the scattering potential down to the theoretical limit indicated by eq. (22). Such experiments would, however, involve difficult practical problems of proper alignment, etc.

I am obliged to Mr. E. Lalor and Dr. W. H. Carter for some very helpful discussions of this point.

Volume 1, number 4 OPTICS COMMUNICATIONS September/October 1969

Finally we note that eqs. (16) and (19) are of the same form as the von Laue equations of X-ray structure analysis and the momentum transfer equation of scattering theory respectively. From eq. (19) one may readily derive an equation that is formally identical with Bragg's law. The full connection between our equations and these classic equations, as well as a more complete analysis of the problem discussed in this note will be given in a forthcoming publication.

REFERENCES

[1] E. Wolf. Determination of the Amplitude and the Phase of Scattered Fields by Holography. J. Opt. Soc. Amer., in press.
[2] See, for example, P. Roman, Advanced Quantum Theory (Addison-Wesley Publishing Company, Inc., 1965) p. 155.
[3] H. Weyl, Ann. Physik 60 (1919) 481.
[4] A. Banos. Dipole radiation in the presence of a conducting half-space (Pergamon Press. Oxford, 1966), eq. (2.19).
[5] See, for example, C. J. Bouwkamp, in: Rep. Progr. Phys. (Physical Society, London) 74 (1954) 41.
[6] In this connection see § III of the paper by J. R. Shewell and E. Wolf, J. Opt. Soc. Amer. 58 (1968) 1598 (especially pp. 1599-1600).

Reprinted from JOURNAL OF THE OPTICAL SOCIETY OF AMERICA, Vol. 60, No. 1, 18–20, January 1970
Printed in U. S. A.

Determination of the Amplitude and the Phase of Scattered Fields by Holography*

EMIL WOLF
Department of Physics and Astronomy, University of Rochester, Rochester, New York 14627
(Received 23 May 1969)

It is shown how both the amplitude and the phase of a scattered field may be determined by holography. It is estimated that information about details down to about nine wavelengths of light can be obtained by this technique. The result is of importance for unambiguous determination of the three-dimensional structure of semitransparent objects, such as are frequently encountered, for example, in biology.
INDEX HEADINGS: Holography; Diffraction; Fourier transform; Scattering.

In conventional holography, a light wave scattered by an object and a reference wave that is coherent with respect to the scattered wave are superimposed on a photographic plate. When the plate is suitably processed and is then illuminated by the reference wave alone, an image of the original object is obtained. It is clear that the processed photographic plate (the hologram) stores in some way information about both the amplitude and the phase of the scattered field, in spite of the fact that the plate is affected only by the intensity of the wave that is incident on it. In the present note we present a solution to the decoding problem of holography, i.e., to the problem of extracting the amplitude and the phase of the scattered field from measurements of the intensity distribution of the light emerging from a transilluminated hologram. This information is needed for determining the three-dimensional structure of semitransparent objects. The solution to this later problem, which opens up a new field of applications of holography, particularly for the study of biological systems, is presented in another publication.[1]

Let a plane monochromatic light wave $U^{(0)}(\mathbf{R})\times\exp(-i\omega t)$, propagated in the direction specified by a unit vector $\mathbf{s}_0$, be incident on a scattering object (Fig. 1). Let $U^{(s)}(\mathbf{R})$ represent the scattered wave, assumed to be of the same frequency as the incident wave, and let

$$U(\mathbf{R})=U^{(0)}(\mathbf{R})+U^{(s)}(\mathbf{R}) \tag{1}$$

be the total field. We wish to determine the complex field U at all points in a selected plane $z=z_1$ outside the scatterer, with the help of side-band holography.[2]

Let us superimpose on the wavefield $U(\mathbf{R})$ a plane monochromatic reference wave

$$U^{(r)}(\mathbf{R})=A^{(r)}\exp(ik\mathbf{s}_r\cdot\mathbf{R})$$

($k=\omega/c$, c=velocity of light) of the same frequency ω propagated in the direction specified by a unit vector $\mathbf{s}_r(p_r,q_r,m_r)$ and let us place a photographic plate in the plane $z=z_1$ (Fig. 2). The total field incident on the photographic plate at $\mathbf{R}_1(x,y,z_1)$ is

$$U^{(t)}(\mathbf{R}_1)=U(\mathbf{R}_1)+U^{(r)}(\mathbf{R}_1). \tag{2a}$$

It will be convenient to change the notation slightly, writing in place of Eq. (2a)

$$U_1^{(t)}(\mathbf{r})=U_1(\mathbf{r})+U_1^{(r)}(\mathbf{r}), \tag{2b}$$

where $\mathbf{r}$ denotes the two-dimensional position vector $\mathbf{r}(x,y)$ specifying the location of the point on the photo-

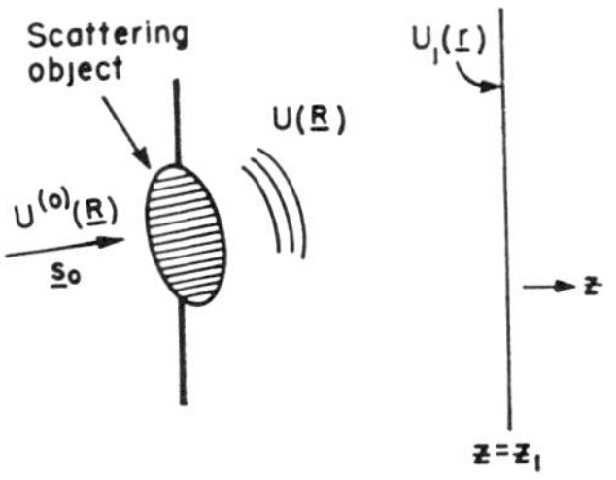

FIG. 1. Illustrating the notation relating to scattering. $U^{(0)}(\mathbf{R})$ = incident field (plane wave propagated in the direction specified by the unit vector $\mathbf{s}_0$); $U(\mathbf{R})=U^{(0)}(\mathbf{R})+U^{(s)}(\mathbf{R})$; $U^{(s)}(\mathbf{R})$ = scattered field; $\mathbf{r}$ = two-dimensional position vector of a typical point (x,y) in the plane $z=z_1$; $U_1(\mathbf{r})\equiv U(x,y,z_1)$.

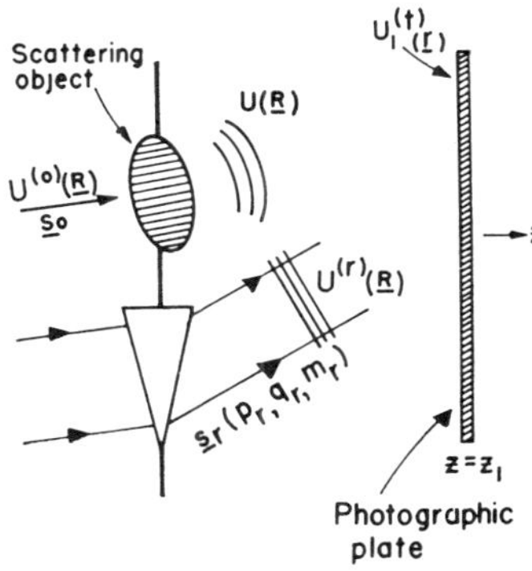

FIG. 2. Illustrating the notation relating to the formation of the hologram. $U^{(r)}(\mathbf{R})$ = reference field [plane wave propagated in the direction specified by the unit vector $\mathbf{s}_r(p_r,q_r,m_r)$]; $U^{(t)}(\mathbf{R})$ = total field incident on the photographic plate in the plane $z=z_1$; $U_1^{(r)}(\mathbf{r})\equiv U^{(r)}(x,y,z_1)$; $U_1^{(t)}(\mathbf{r})\equiv U^{(t)}(x,y,z_1)$. Other symbols have same meaning as in Fig. 1.

* Research supported by the Air Force Cambridge Research Laboratories. An account of the main results reported in this note was presented at the Spring Meeting of the Optical Society of America, held in San Diego, March 12, 1969 [J. Opt. Soc. Am. 59, 482A (1969)] and at a symposium on Coherent Optics and Instrumentation held at Oakland University, Rochester, Michigan, 18 April 1969.

[1] E. Wolf, Optics Communications 1, No. 4, 153 (1969).

[2] E. N. Leith and J. Upatnieks, J. Opt. Soc. Am. 53, 1377 (1963).

graphic plate, with respect to the origin $(0,0,z_1)$ in the plane of the photographic plate; the suffix 1 on $U_1^{(t)}$, etc., indicates that the point is situated in that plane.

Suppose that the photographic plate is processed so that it will have an amplitude transmission function that is linear in the intensity $I^{(t)}=U_1^{(t)*}U_1^{(t)}$. If the plate so processed (the hologram) is now illuminated by a plane monochromatic wave of frequency ω, incident normally on the plate (Fig. 3), the field distribution on the exit face of the hologram is given by

$$V_H(\mathbf{r})=V_H^{(\mathrm{I})}(\mathbf{r})+V_H^{(\mathrm{II})}(\mathbf{r})+V_H^{(\mathrm{III})}(\mathbf{r}), \tag{3}$$

where

$$V_H^{(\mathrm{I})}(\mathbf{r})=CU_1(\mathbf{r})\exp[-ik(p_1x+q_ry)], \tag{4a}$$

$$V_H^{(\mathrm{II})}(\mathbf{r})=C^*U_1^*(\mathbf{r})\exp[ik(p_rx+q_ry)], \tag{4b}$$

$$V_H^{(\mathrm{III})}(\mathbf{r})=D+FU_1^*(\mathbf{r})U_1(\mathbf{r}). \tag{4c}$$

Here C, D, and F are constants and asterisks denote complex conjugates. With a plate processed in the usual way and neglecting effects arising from possible nonuniformities in the thickness of the plate, the constants D and F are real. We see from Eqs. (3) and (4) that $V_H(\mathbf{r})$ is then a *real* function of $\mathbf{r}$. Hence if $I_H(r)=V_H^*(\mathbf{r})V_H(\mathbf{r})$ is the intensity distribution on the exit face of the transilluminated hologram (see Fig. 3),

$$V_H(\mathbf{r})=\mu(\mathbf{r})[I_H(\mathbf{r})]^{\frac{1}{2}}, \tag{5}$$

where, for every $\mathbf{r}$, $\mu(\mathbf{r})$ is either $+1$ or -1. From the continuity of $V_H(\mathbf{r})$ it follows that $\mu(\mathbf{r})$ can only switch values at points where $I_H(\mathbf{r})$ is zero. However, the transmission function of a photographic plate, processed in the usual way, is necessarily non-negative, and so is, therefore, $V_H(\mathbf{r})$. Hence we may take $\mu(\mathbf{r})\equiv 1$ and it follows that $V_H(\mathbf{r})$ may be unambiguously determined from measurement of the intensity transmission of the transilluminated hologram. Next we will show how from the knowledge of $V_H(\mathbf{r})$ we may determine the unknown (generally complex) field $U_1(\mathbf{r})$.

Let us take the two-dimensional Fourier transform $\hat{V}_H$ of V_H:

$$\hat{V}_H(u,v)=\frac{1}{(2\pi)^2}\int_{-\infty}^{\infty}\!\!\int V_H(x,y)\exp[-i(ux+vy)]dxdy. \tag{6}$$

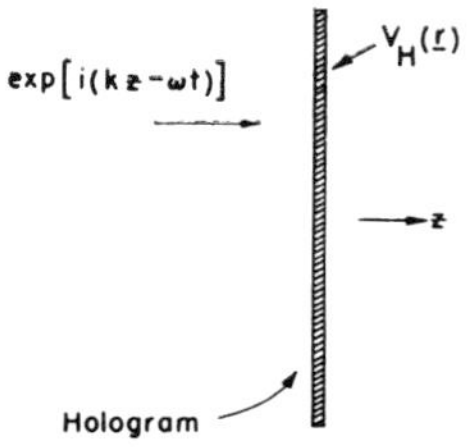

Fig. 3. Illustrating the notation relating to the transilluminated hologram: $V_H(\mathbf{r})$ represents the distribution of the field on the exit face of the transilluminated hologram, when the hologram is illuminated by a normally incident plane monochromatic wave.

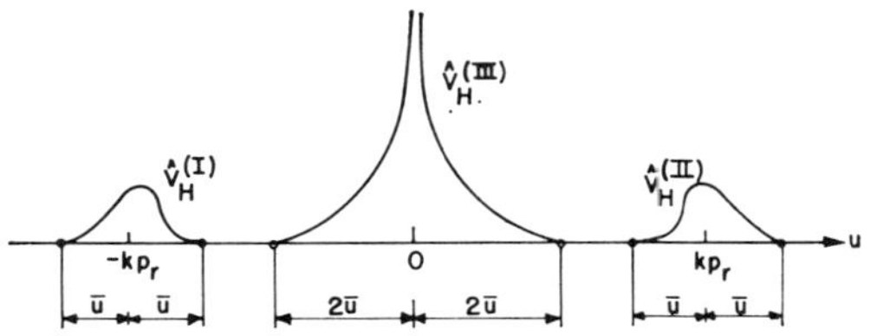

Fig. 4. The general form of the spatial-frequency spectrum $\hat{V}_H(u)$ of a (one-dimensional) transilluminated hologram. The scattered field in the plane $z=z_1$ is assumed to be band limited to the range $u\leq k|p_r|/3$. $\hat{V}_H(u)$ is, in general, complex, but for the purpose of illustration is considered here to be real.

From Eqs. (6), (3), and (4) it follows with the help of the convolution theorem that

$$\hat{V}_H(u,v)=\hat{V}_H^{(\mathrm{I})}(u,v)+\hat{V}_H^{(\mathrm{II})}(u,v)+\hat{V}_H^{(\mathrm{III})}(u,v), \tag{7}$$

where

$$\hat{V}_H^{(\mathrm{I})}(u,v)=C\hat{U}_1(u+kp_r,\ v+kq_r), \tag{8a}$$

$$\hat{V}_H^{(\mathrm{II})}(u,v)=C^*\hat{U}_1^*(-u+kp_r,\ -v+kq_r), \tag{8b}$$

$$\hat{V}_H^{(\mathrm{III})}(u,v)=D\delta^{(2)}(u,v)+F\int_{-\infty}^{\infty}\!\!\int\hat{U}_1(u',v')\times\hat{U}_1^*(u'-u,\ v'-v)du'dv', \tag{8c}$$

and

$$\hat{U}_1(u,v)=\frac{1}{(2\pi)^2}\int_{-\infty}^{\infty}\!\!\int U_1(x,y)\exp[-i(ux+vy)]dxdy \tag{9}$$

is the Fourier transform of $U_1,(x,y)$.

Now although, in principle, $U_1(u,v)$ may contain all possible spatial-frequency components (u,v), it is reasonable to assume that components of sufficiently high frequencies will not give significant contributions to the scattered field. We may therefore assume that U_1 is effectively band limited to some range

$$-\bar{u}\leqslant u\leqslant\bar{u},\quad -\bar{v}\leqslant v\leqslant\bar{v}. \tag{10}$$

We will also assume that[3]

$$\bar{u}\leqslant k(p_r^2+q_r^2)^{\frac{1}{2}}/3\sqrt{2},\qquad \bar{v}\leqslant k(p_r^2+q_r^2)^{\frac{1}{2}}/3\sqrt{2}. \tag{11}$$

Under these circumstances, the spatial spectrum $\hat{V}_H(u,v)$ of the transilluminated hologram will consist of a sharp central peak $(\hat{V}_H^{(\mathrm{III})})$ at $u=v=0$ and two side bands $\hat{V}_H^{(\mathrm{I})}$ and $\hat{V}_H^{(\mathrm{II})}$ around the spatial-frequency pairs $u=-kp_r$, $v=-kq_r$ and $u=kp_r$, $v=kq_r$, respectively. The general form of V_H is illustrated in Fig. 4.

[3] If the field distribution $U_1(x,y)$ does not satisfy this band-limitation condition, the unwanted spatial-frequency components could be eliminated by using standard optical spatial-filtering techniques. Such spatial-frequency components cannot, of course, be reconstructed at a later stage, i.e., only the filtered field can be reconstructed.

The conditions (11) ensure that $\hat{V}_H^{(\mathrm{I})}$, $\hat{V}_H^{(\mathrm{II})}$, and $\hat{V}_H^{(\mathrm{III})}$ do not overlap.

It is evident from the two dimensional analog of Fig. 4 that

$$\hat{V}_H^{(\mathrm{I})}(u,v)=\hat{V}_H(u,v)\hat{\epsilon}(u,v), \tag{12}$$

where

$$\left.\begin{aligned}\hat{\epsilon}(u,v)=1 \quad\text{if}\quad & -kp_r-\bar{u}\leqslant u\leqslant -kp_r+\bar{u},\\ & -kp_r-\bar{v}\leqslant v\leqslant -kp_r+\bar{v},\\ =0 \quad & \text{otherwise.}\end{aligned}\right\} \tag{13}$$

From Eqs. (12) and (8a) it follows that

$$\hat{U}_1(u+kp_r,\ v+kq_r)=(1/C)\hat{V}_H(u,v)\hat{\epsilon}(u,v). \tag{14}$$

Next we take the Fourier transform of Eq. (14) and make use of the convolution theorem and of the shift theorem on Fourier transforms. We then obtain a closed expression for the unknown field U_1:

$$U_1(x,y)=\frac{\bar{u}\bar{v}}{\pi^2C}\int\!\!\int_{-\infty}^{\infty}[I_H(x',y')]^{\frac{1}{2}}\times\mathrm{sinc}\left[\frac{\bar{u}(x-x')}{\pi},\ \frac{\bar{v}(y-y')}{\pi}\right]\times\exp[ik(p_rx'+q_ry')]dx'dy', \tag{15}$$

where

$$\mathrm{sinc}[\xi,\eta]=(\sin\pi\xi/\pi\xi)(\sin\pi\eta/\pi\eta). \tag{16}$$

It is evident from our method of derivation, that in Eq. (15) $\bar{u}$ and $\bar{v}$ may be replaced by any pair of positive numbers U and V such that $U\geqslant\bar{u}$, $V\geqslant\bar{v}$. In view of Eq. (11) we may choose $U=V=k(p_r^2+q_r^2)^{\frac{1}{2}}/3\sqrt{2}$ and in place of Eq. (15) we then obtain the formula

$$U_1(x,y)=\frac{k^2(p_r^2+q_r^2)}{18\pi^2C}\int\!\!\int_{-\infty}^{\infty}\sqrt{I_H(x',y')}\times\mathrm{sinc}\left[\frac{k(p_r^2+q_r^2)^{\frac{1}{2}}(x-x')}{3\sqrt{2}\pi},\ \frac{k(p_r^2+q_r^2)^{\frac{1}{2}}(y-y')}{3\sqrt{2}\pi}\right]\times\exp[ik(p_rx'+q_ry')]dx'dy'. \tag{15a}$$

This form of our solution has the advantage that it does not explicitly involve the effective spatial-frequency range of the scattered field.

The formula (15) [or (15a)] allows the determination of the unknown scattered field—both in amplitude and in phase—from the measurements of the *intensity* distribution of the field emerging from the hologram.[4]

As pointed out in the introduction, our results may be used to determine the three-dimensional structure of semitransparent objects. For this purpose it is of importance to have an estimate of the limit of resolution down to which the scattered field may be determined by the present method. To obtain such an estimate, let Δx and Δy be the spatial periods associated with the maximum spatial frequencies $\bar{u}$, $\bar{v}$, i.e.,

$$\bar{u}=2\pi/\Delta x,\quad \bar{v}=2\pi/\Delta y. \tag{17}$$

Then from Eqs. (17) and (11),

$$p_r^2+q_r^2\geqslant\left(\frac{2\pi}{k}\right)^2\frac{18}{L^2}, \tag{18}$$

where

$$\frac{1}{L^2}=\frac{1}{(\Delta x)^2}+\frac{1}{(\Delta y)^2}. \tag{19}$$

The quantity L, defined by Eq. (19), may be regarded as a rough measure of the limit of the resolution. Now

$$p_r^2+q_r^2=\sin^2\theta_r, \tag{20}$$

where θ_r is the angle that the reference wave makes with the normal to the photographic plate. Hence the inequality (18) may be expressed in the form

$$L\geqslant 3\sqrt{2}\lambda/\sin\theta_r, \tag{21}$$

where $\lambda=2\pi/k$ is the wavelength of the light.

A realistic figure[5] for θ_r is about 30°. With this choice of θ_r, the inequality (20) gives

$$L\gtrsim 6\sqrt{2}\lambda. \tag{22}$$

Thus we may expect that this technique makes it possible to determine the complex scattered field down to details of the order of about nine wavelengths of the light that illuminates the object.

[4] Because of the finite size of the hologram, the intensity distribution $I_H(x,y)$ cannot be determined throughout the whole infinite plane $z=z_1$, as formally needed in formula (15). However, because the sinc function in Eq. (15) effectively vanishes when $|x-x'|$ and $|y-y'|$ exceed a moderate multiple of $\pi/\bar{u}$ and $\pi/\bar{v}$, respectively, the integration in Eq. (15) needs only be extended over such a finite x',y' domain. This result implies, incidently, that information about the scattered field U_1 at the point (x,y) is stored in a region of the hologram, centered on the point (x,y), whose linear dimensions are moderate multiples of $\pi/\bar{u}$ and $\pi/\bar{v}$ in the x and y directions, respectively.

[5] High-resolution photographic plates are available that would appear, at first sight, to make it profitable to employ reference beams with appreciably larger values of θ_r. It seems doubtful, however, that this would lead to a significant improvement of the resolution limit of this technique, because of the difficulty of measuring, to sufficient accuracy, the intensity distribution $I_H(x,y)$ of the transilluminated hologram.

Reprinted from: COHERENCE AND QUANTUM OPTICS
Edited by L. Mandel and E. Wolf
Book available from: Plenum Publishing Corporation
227 West 17th Street, New York, N. Y. 10011

A GENERALIZED EXTINCTION THEOREM AND ITS ROLE IN SCATTERING THEORY*

Emil Wolf

University of Rochester, Rochester, N.Y.

When an electromagnetic wave is incident on a homogeneous medium with a sharp boundary, it is extinguished inside the medium in the process of interaction and is replaced by a wave propagated in the medium with a velocity different from that of the incident wave. A classic theorem of molecular optics due to P.P. Ewald (1912) and C.W. Oseen (1915) expresses the extinction of the incident wave in terms of an integral relation, that involves the induced field on the boundary of the medium. Various generalizations of this theorem have recently been proposed and it was also shown that the customary physical interpretation of the theorem is incorrect.

In this paper results of a recent investigation carried out in collaboration with D.N. Pattanayak are presented, which provide a generalization of the extinction theorem to any medium. Like the recent generalization due to J.J. Sein our derivation is based entirely on Maxwell's theory and not on molecular optics. A hypothesis is put forward as to the true physical significance of the extinction theorem and it is shown how the theorem may be used to solve scattering problems in a novel way. An analogous extinction theorem for non-relativistic quantum mechanics is also presented.

***

One of the most poorly understood theorems of classical electrodynamics is undoubtedly the so-called *extinction theorem* first formulated by P.P. Ewald [1] in 1912 in his basic investigations on the foundations of crystal optics and later by C.W. Oseen [2] in 1915 in his studies of dispersion of light in material media.

Let me first say a few words about the usual formulation of the theorem.

Suppose that a plane electromagnetic wave is incident from vacuo on a material medium with a sharp boundary. The medium will for the moment be assumed to be of the simplest kind - a linear, homogeneous, isotropic, non-magnetic dielectric.

We know that under the influence of the incident electromagnetic field another field will be generated inside the dielectric, which will have a different wave number and hence a different phase velocity. We may, therefore, say that inside the medium, the incident wave, propagated with the vacuum velocity of light c, is somehow *extinguished* by the interaction with the medium and is replaced by a new wave propagated with the velocity c/n, where n is the refractive index of the medium. The question then arises: how does the extinguishing of the incident wave come about? The Ewald-Oseen theorem provides an answer to this question. In mathematical terms the theorem may be expressed in the form: [3]

$$\underline{E}^{(i)}(\underline{r}_<) + \frac{1}{4\pi k^2} \nabla\times\nabla\times \int_S \{\underline{E}(\underline{r}') \frac{\partial}{\partial n} G_o(\underline{r}_<, \underline{r}') - G_o(\underline{r}_<, \underline{r}') \frac{\partial}{\partial n} \underline{E}(\underline{r}')\} \, dS = 0, \qquad (1)$$

valid at every point $\underline{r}_<$ inside the volume V bounded by a surface S (see Fig. 1) occupied by the medium. Here $\underline{E}^{(i)}$ and $\underline{E}$ represent the Fourier transforms (for frequency components $\omega = kc$, c being the vacuum velocity of light) of the incident electric field and of the total electric field generated inside the medium respectively (and taken in the integral in (1) in the limit as the surface is approached from inside V),

$$G_o(\underline{r}, \underline{r}') = \exp\{ik|\underline{r} - \underline{r}'|\}/|\underline{r} - \underline{r}'| \qquad (2)$$

is the outgoing free-space Green's function of the Helmholtz equation and $\partial/\partial n$ denotes differentiation along the outward normal to the boundary surface.

The relation (1), which is essentially in the form as formulated by Oseen, was originally derived not from the macroscopic Maxwell theory, but rather from molecular optics, which is a microscopic theory. In this later theory the response of the medium to the incident field is expressed in terms of elementary dipole fields, generated by the interaction of the incident wave with the individual molecules of the medium. We note that in (1), the second term formally cancels the incident electric field at every point inside the medium. Since the second term involves the values of the total field $\underline{E}$ on the boundary surface S only it has been generally asserted that Eq. (1) implies that the incident field is extinguished entirely by those molecular dipoles that are situated on the boundary S of the dielectric. This is the original formulation of the Ewald-Oseen extinction theorem.

In the last few years the Ewald-Oseen extinction theorem has attracted a good deal of attention and various modifications and generalizations of it for more complicated media have been proposed and applied to numerous problems of current research interest. Here is a partial listing of the relevant publications, indicating the authors, year of publication and topics. The complete references are given in footnote 5.

(a)	A. Wierzbicki	(1961,1962)	Quadrupole radiation, reflection, refraction
(b)	N. Bloembergen and P.S. Pershan	(1962)	Non-linear optics
(c)	B.A. Sotskii	(1963)	Metals, optically active media
(d)	R.K. Bullough	(1968)	Many-body optics
(e)	J.J. Sein	(1969,1970)	Spatial dispersion, excitons
(f)	É. Lalor	(1969)	New formulation
(g)	É. Lalor and E.Wolf	(1971)	Interaction of charged particle with a dielectric
(h)	T. Suzuki (1971)		Diffraction
(i)	G.S. Agarwal, D.Pattanayak and E. Wolf	(1971	Spatial Dispersion
(j)	É. Lalor and E. Wolf	(1972)	Refraction and reflection
(k)	J.R. Birman and J.J. Sein	(1972)	Polaritons in bounded media

It is thus clear that the extinction theorem is playing an increasingly greater role in widely different areas. Of the numerous investigations those of J.J. Sein [4,5e] are of particular relevance to the subject matter of this talk. Sein showed that the extinction theorem which, as I already noted, was derived originally from molecular optics may also be derived from Maxwell's theory and he also showed that the traditional interpretation of the theorem is incorrect [4].

In this talk I will present results of a recent investigation that I carried out in collaboration with D. Pattanayak [6], (see also [7]), which we have attempted to answer the following two questions:

(1) Can the extinction theorem be generalized within the framework of Maxwell's theory to a medium of any kind, i.e. with arbitrary response?

(2) What is the true meaning of the theorem?

The first question has been partially answered already by Sein, but we will present quite a general and rigorous answer to it. Let me add that Sein's recognition that the extinction theorem follows also from Maxwell's theory represents an important contribution, since attempts to generalize it within the framework of molecular optics encounters formidable difficulties because of the local field corrections (associated with the Lorentz internal field).

As regards the second question - namely what is the true meaning of the theorem - we will put forward a hypothesis, supported by a few explicit solutions that we obtained with the help of the theorem.

We will also show that the extinction theorem has a strict analogue in potential scattering in non-relativistic quantum mechanics.

The full derivation of our main results is rather lengthy and I will only indicate the main steps.

Let us then consider the scattering of monochromatic electromagnetic wave incident from vacuo on a body with a sharp boundary. We assume the body to be of arbitrary kind; its response could be, for example, non-linear or non-local as in the case of spatial dispersion.

From Maxwell's equations for monochromatic fields, on eliminating, the electric displacement vector $\underline{D}$ and the magnetic induction vector $\underline{B}$ via the relations

$$\underline{D} = \underline{E} + 4\pi\, \underline{P}, \qquad \underline{B} = \underline{H} + 4\pi\, \underline{M}, \tag{3}$$

where $\underline{P}$ and $\underline{M}$ denote the polarization and magnetization vectors respectively, we obtain the four equations

$$\hat{L}\, \underline{E} = \underline{F}_e, \tag{4a}$$

$$\hat{L}\, \underline{H} = \underline{F}_h, \tag{4b}$$

$$\nabla.\underline{E} = 4\pi(\rho - \nabla.\underline{P}) \tag{4c}$$

$$\nabla.\underline{H} = -4\pi\ \nabla.\underline{M}, \tag{4d}$$

where $\hat{L}$ is the operator

$$\hat{L} = -k^2 + \nabla x \nabla x \tag{5}$$

and the source terms $\underline{F}_e$ and $\underline{F}_h$ are given by

$$\underline{F}_e = 4\pi[\frac{ik}{c}\underline{j} + k^2\underline{P} + ik\ \nabla x\ \underline{M}] \quad , \tag{6a}$$

$$\underline{F}_h = 4\pi[\frac{1}{c}\nabla x\underline{j} - ik\ \nabla x\ \underline{P} + k^2\underline{M}] \ . \tag{6b}$$

The vectors $\underline{E}$, $\underline{H}$, $\underline{j}$, $\underline{P}$ and $\underline{M}$ and the scalar ρ are, of course, functions of position ($\underline{r}$) and are taken at a fixed frequency ω.

The sources of the incident field are assumed to be in the domain $\tilde{V}$ exterior to V (see Fig. 1) and we take them to be at a finite distance [8] from V. It is clear then, since the exterior $\tilde{V}$ of V is vacuo, that

$$\underline{F}_e(\underline{r}) = 4\pi[\frac{ik}{c}\underline{j}_c + k^2\underline{P} + ik\ \nabla x\ \underline{M}\ ] \qquad \text{if } \underline{r}\ \varepsilon\ V \tag{7a}$$

$$= \frac{4\pi ik}{c}\underline{j}_{ext} \qquad \text{if } \underline{r}\ \varepsilon\ \tilde{V}\ , \tag{7b}$$

$$\underline{F}_h(\underline{r}) = 4\pi[\frac{1}{c}\nabla x\ \underline{j}_c - ik\ \nabla x\ \underline{P} + k^2\ \underline{M}\ ] \qquad \text{if } \underline{r}\ \varepsilon\ V \tag{8a}$$

$$= \frac{4\pi}{c}\ \nabla x\ \underline{j}_{ext} \qquad \text{if } \underline{r}\ \varepsilon\ \tilde{V}\ . \tag{8b}$$

In Eqs. (7) and (8), $\underline{j}_c$ denotes the conduction current density and $\underline{j}_{ext}$ denotes the external current density (representing the source). The corresponding charge densities are, of course, related to the current densities by the continuity equation. Further the polarization vector $\underline{P}$ and the magnetization vector $\underline{M}$ in (7a) and (8a) are assumed to be given functions of the electromagnetic field vectors $\underline{E}$ and $\underline{H}$, whose exact form depends on the nature of the medium. It is to be noted that because of this fact, the equations (4a) and (4b) are in general *coupled* to each other.

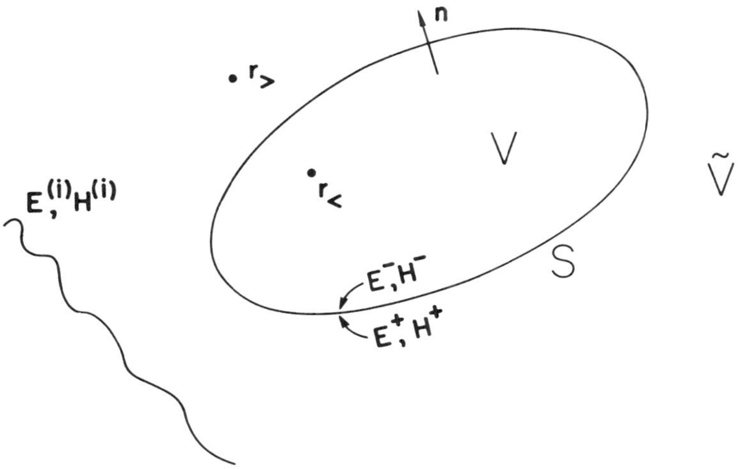

Fig. 1 Notation relating to scattering of an electromagnetic wave on a material medium.

To complete the formulation we must specify the behavior of the fields at the boundary. As is well known Maxwell's equations imply that across the boundary

$$\underline{n} \times (\underline{E}^{+} - \underline{E}^{-}) = 0, \qquad \underline{n} \times (\underline{H}^{+} - \underline{H}^{-}) = \frac{4\pi}{c} \underline{K} , \qquad (9)$$

where the superscripts plus and minus denote limiting values as the boundary surface S is approached from outside and inside of the medium respectively (see Figure 1), $\underline{n}$ is the unit outward normal to the boundary surface S and $\underline{K}$ represents the surface current density which will be non-zero only for a perfect conductor. Although these conditions are generally referred to as *boundary conditions*, they are, actually jump conditions - or *saltus* conditions as they are called in the older literature. A clear appreciation of the difference between true boundary conditions and the saltus conditions is, as we shall see later, at the heart of the proper interpretation of the extinction theorem.

For later purpose we also note that the corresponding problem in non-relativistic quantum mechanics, namely the scattering from a finite step potential is mathematically much simpler since in place of several coupled equations involving the vector fields

$\underline{E}$ and $\underline{H}$, the quantum mechanical problem involves only a single equation for the Schrödinger scalar wave function $\psi(\underline{r})$. Moreover, since ψ and its normal $\partial\psi/\partial n$ must, according to basic postulates of quantum mechanics be *continuous* across a finite potential step we now have in place of the conditions (9) the conditions

$$\psi^+ - \psi^- = 0, \qquad \left(\frac{\partial\psi}{\partial n}\right)^+ - \left(\frac{\partial\psi}{\partial n}\right)^- = 0. \tag{10}$$

These are *continuity conditions* and not true boundary conditions either.

Returning to the electromagnetic problem we introduce a dyadic Green's function associated with the $\hat{L}$-operator,

$$\hat{L}\,\underline{\underline{G}} = 4\pi\delta(\underline{r}-\underline{r}')\underline{\underline{U}}, \tag{11}$$

($\underline{\underline{U}}$ = unit dyadic), which obeys the vectorial form of the Sommerfeld radiation condition at infinity:

$$\lim_{r\to\infty} r\,[\nabla\times\underline{\underline{G}} - ik\,\hat{\underline{r}}\times\underline{\underline{G}}\,] = 0, \tag{12}$$

where $\hat{\underline{r}}$ is the unit vector in the direction of $\underline{r}$ and $r = |\underline{r}|$. It is known that [9]

$$\underline{\underline{G}}(\underline{r},\underline{r}') = (\underline{\underline{U}} + \frac{1}{k^2}\nabla\nabla)\,G_o(\underline{r},\underline{r}') , \tag{13a}$$

where G_o is the outgoing free-space Green's function of the Helmholtz equation, viz.

$$G_o(\underline{r},\underline{r}') = \frac{e^{ik|\underline{r}-\underline{r}'|}}{|\underline{r}-\underline{r}'|} . \tag{13b}$$

Now our equation involving the electric field is of the form

$$\hat{L}\,\underline{E} = \underline{F}_e , \tag{14}$$

with a similar equation involving $\underline{H}$. From (11) and (14) one obtains, if one also uses the vectorial form of Green's theorem, the following identity valid for integration through any domain V' bounded by a closed surface S':

$$\int_{V'} E(\underline{r}')\delta(\underline{r}-\underline{r}')d^3\underline{r}' = \frac{1}{4\pi}\int_{V'} \underline{F}_e(\underline{r}') \cdot \underline{\underline{G}}(\underline{r},\underline{r}')d^3\underline{r}' - \frac{1}{4\pi}\sum_e(\underline{r}), \tag{15}$$

where

$$\sum_e(\underline{r}) = \int_{S'} \{[\underline{n}\times \nabla\times\underline{E}(\underline{r}')]\cdot\underline{\underline{G}}(\underline{r},\underline{r}')+[\underline{n}\times \underline{E}(\underline{r}')]\cdot \nabla\times\underline{\underline{G}}\}dS' \tag{16}$$

and n is the unit normal to S' pointing outward from the volume V'.

Let us now take the volume V' to coincide either with the scattering volume V or with the exterior $\tilde{V}$ of it. Also we can take the field point $\underline{r}$ to be either in V or in $\tilde{V}$. Applying the theorem (15) separately to each of these four cases we obtain the following four relations:

(a) $\underline{r}\varepsilon V,\ \underline{r}'\varepsilon V$:

$$\underline{E}(\underline{r}_<) = \frac{1}{4\pi}\int_V \underline{F}_e\cdot\underline{\underline{G}}\, d^3\underline{r}' - \frac{1}{4\pi}\sum_e^{(-)}(\underline{r}_<), \tag{17a}$$

(b) $\underline{r}\varepsilon V,\ \underline{r}'\varepsilon\tilde{V}$:

$$0 = \frac{ik}{c}\int_{\tilde{V}} \underline{j}_{ext}\cdot\underline{\underline{G}}\, d^3\underline{r}' + \frac{1}{4\pi}\cdot\sum_e^{(+)}(\underline{r}_<), \tag{17b}$$

(c) $\underline{r}\varepsilon\tilde{V},\ \underline{r}'\varepsilon\tilde{V}$:

$$\underline{E}(r_>) = \frac{ik}{c}\int_{\tilde{V}} \underline{j}_{ext}\cdot\underline{\underline{G}}\, d^3\underline{r}' + \frac{1}{4\pi}\sum_e^{(+)}(r_>), \tag{17c}$$

(d) $\underline{r}\varepsilon\tilde{V},\ \underline{r}'\varepsilon V$:

$$0 = \frac{1}{4\pi}\int_V \underline{F}_e\cdot\underline{\underline{G}}\, d^3\underline{r}' - \frac{1}{4\pi}\sum_e^{(-)}(\underline{r}_>). \tag{17d}$$

Here

$$\sum_e^{(+)} = \int_{S^\pm} \{[n\times \nabla\times\underline{E}]\cdot\underline{\underline{G}} + [\underline{n}\times\underline{E}]\cdot\nabla\times\underline{\underline{G}}\}dS, \tag{18}$$

and the upper or lower signs are taken on $\Sigma_e^{\pm}$ and on $S^{\pm}$ according as the limiting values are taken from outside (S^+) or inside (S^-) of the scattering volume. In deriving (17b) and (17c) we also used the radiation condition (12) which ensures that there is no contribution from a sphere of infinitely large radius (the outer boundary of the volume V).

Now the integral containing the external current, taken over the exterior $\tilde{V}$ of our scattering volume has a clear physical meaning. From the significance of $\underline{\underline{G}}$ as the outgoing free-space Green's function of the $\tilde{L}$-operator, this integral must evidently represent the unperturbed incident field, i.e.

$$\frac{ik}{c}\int_{\tilde{V}} \underline{j}_{ext}(\underline{r}')\cdot\underline{\underline{G}}(\underline{r},\underline{r}')d^3\underline{r}' = \underline{E}^{(i)}(\underline{r}), \tag{19}$$

irrespective whether the point $\underline{r}$ is situated inside or outside the scattering volume. Hence the equations (17b) and (17c) may be expressed in the compact form

$$\underline{E}^{(i)}(\underline{r}_<) + \frac{1}{4\pi}\sum\nolimits_e^{(+)}(\underline{r}_<) = 0, \tag{20a}$$

$$\underline{E}(\underline{r}_>) = \underline{E}^{(i)}(\underline{r}_>) + \frac{1}{4\pi}\sum\nolimits_e^{(+)}(\underline{r}_>) . \tag{20b}$$

We note that (20a) has some resemblance to the Ewald-Oseen extinction theorem, since it expressed the cancellation of the incident field at every point $\underline{r}$ inside the scattering medium V in terms of an integral involving the field on the boundary of the medium only. However because $\Sigma_e^{(+)}$ rather than $\Sigma_e^{(-)}$ appears in this surface integral, the integral involves the limiting values of the field taken from the outside, rather than from the inside of the scattering volume; but one can easily transform (20a) and also(20b) so as to involve the limiting values from the inside, since from the definition (18) of $\Sigma_e^{(-)}$ and $\Sigma_e^{(+)}$ and from the saltus conditions (9) one easily finds that

$$\frac{1}{4\pi}\left[\sum\nolimits_e^{(+)}(\underline{r}) - \sum\nolimits_e^{(-)}(\underline{r})\right] = -\,ik\int_{S^-}\left(\underline{n}\times\underline{M} - \frac{1}{c}\underline{K}\right)\cdot\underline{\underline{G}}ds \tag{21}$$

Using this result in (20a) and (20b) and the expressions for $\Sigma_e^{(+)}$ one then obtains the following two relations:

$$\underline{E}^{(i)}(\underline{r}_<) + \frac{1}{4\pi}\underline{S}_e(\underline{r}_<) = 0, \tag{22}$$

$$\underline{E}(\underline{r}_>) = \underline{E}^{(i)}(\underline{r}_>) + \frac{1}{4\pi}\underline{S}_e(\underline{r}_>), \tag{23}$$

where

$$\underline{S}_e(\underline{r}) = \int_{S^-} \{[\underline{n}\times(\nabla\times\underline{E} - 4\pi ik\underline{M}) + \frac{4\pi ik}{c}\underline{K}].\underline{\underline{G}}(\underline{r}.\underline{r}') + [\underline{n}\times\underline{E}].\nabla\times\underline{\underline{G}}(\underline{r},\underline{r}')\}dS. \tag{24}$$

The relation (22) must be satisfied at each point $\underline{r}_<$ inside the scattering volume V. It is one form of our *generalized* Ewald-Oseen theorem, valid for scattering by *any* medium, irrespective of the nature of the constitutive relations. I will indicate shortly how it reduces to the usual form of the Ewald-Oseen theorem when the medium is of the simplest kind. But first I want to say a little about what I believe is the true meaning of the theorem and also discuss briefly the significance of the complementary relation (23). For this purpose we also must note there is an analogous set of relations to (22) and (23), involving the *magnetic* rather than the electric field. They can be derived in a similar way and are

$$\underline{H}^{(i)}(\underline{r}_<) + \frac{1}{4\pi}\underline{S}_h(\underline{r}_<) = 0, \tag{25}$$

$$\underline{H}(\underline{r}_>) = \underline{H}^{(i)}(\underline{r}_>) + \frac{1}{4\pi}\underline{S}_h(\underline{r}_>), \tag{26}$$

where

$$\underline{S}_h(\underline{r}) = \int_{S^-} \{[\underline{n}\times(\nabla\times\underline{H} + 4\pi ik\underline{P} - (4\pi/c)\underline{j})].\underline{\underline{G}} + [\underline{n}\times\underline{H} + (4\pi/c)\underline{K}].\nabla\times\underline{\underline{G}}\}\, dS. \tag{27}$$

We know that inside the medium, the $\underline{E}$ and $\underline{H}$ fields obey the equations (4a) and (4b),

$$\hat{L}\,\underline{E}(\underline{r}_<) = \underline{F}_e(\underline{r}_<) , \qquad \hat{L}\,\underline{H}(\underline{r}_<) = \underline{F}_h(\underline{r}_<) .$$

However these are *general field equations* valid inside the medium. They do not completely specify the scattered field since they involve neither the incident field, nor any boundary conditions. Our hypothesis is that the *two extinction theorems (22) and (25) represent boundary conditions subject to which the (generally coupled) field equations (4) provide unique solution for the fields* $\underline{E}$, $\underline{H}$ *inside the scattering medium (i.e. inside the volume* V), *when an electromagnetic field* $\underline{E}^{(i)}$, $\underline{H}^{(i)}$ *is incident on the medium.* Thus, according to this hypothesis, the two extinction theorems allow us to replace the original saltus problem - involving the solution both inside and outside the medium - by a *boundary value problem* for determining the field inside the scattering medium. The boundary conditions for this later problem are of a somewhat unusual kind, having the form of *non-local* relations. Once the solution inside the scattering medium has been obtained, the solution outside it may be determined from Eqs. (23) and (26) by substituting the boundary values into the surface integrals occuring in these formulae. Our new interpretation of the extinction theorems is supported by explicit solutions that were obtained for several special cases [10].

Up to this point we have considered only two of the four relations (17), namely (17b) and (17c). If the other two relations, viz. (17a) and (17d) are also used, as well as the relations (19) and (21) and the corresponding formulae involving the magnetic fields one obtains an alternative set of equations in the form of *integro-differential equations* valid both inside and outside the medium - for the unknown electromagnetic fields $\underline{E}$ and $\underline{H}$. Lack of time prevents a discussion of this point here but we will later consider briefly the analogous situation for the case of quantum mechanical potential scattering.

With the help of various vector identities and Maxwell's equations, our general extinction theorems may be expressed in many alternative but equivalent forms. The extinction theorem for the electric field may, for example, be transformed into the form

$$\underline{E}^{(i)}(\underline{r}_<) + \frac{1}{k^2}\nabla\times\nabla\times\,[\,\underline{I}^{(E)}(\underline{r}_<) + \underline{I}^{(P)}(\underline{r}_<) + \underline{I}^{(M)}(\underline{r}_<) + \underline{I}^{(J)}(\underline{r}_<)] = 0 , \tag{28}$$

where

$$\underline{I}^{(E)}(\underline{r}_<) = \frac{1}{4\pi}\int_{S^-}\{\underline{E}(\underline{r}')\frac{\partial G_o(\underline{r}_<,\underline{r}')}{\partial n} - G_o(\underline{r}_<,\underline{r}')\frac{\partial \underline{E}(\underline{r}')}{\partial n}\}\, dS, \quad (29)$$

$$\underline{I}^{(P)}(\underline{r}_<) = -\int_{S^-}[\underline{n}\ \nabla .\ \underline{P}(\underline{r}')]\ G_o(\underline{r}_<,\underline{r}')\, dS, \quad (30)$$

$$\underline{I}^{(M)}(\underline{r}_<) = -ik\int_{S^-}[\underline{n}\times\underline{M}(\underline{r}')]\ G_o(\underline{r}_<,\underline{r}')\, dS, \quad (31)$$

$$\underline{I}^{(J)}(\underline{r}_<) = -\frac{i}{kc}\int_{S^-}[\underline{n}\ \nabla .\ \underline{j}(\underline{r}')-\underline{K}(\underline{r}')]\ G_o(\underline{r}_<,\underline{r}')\, dS. \quad (32)$$

Clearly if the medium is non-magnetic, $\underline{I}^{(M)} \equiv 0$, if it is a non-conductor $\underline{I}^{(J)} \equiv 0$. For a linear, homogeneous ,spatially non-dispersive non-magnetic dielectric, not only do these two terms vanish, but so does also the term $\underline{I}^{(P)}$, unless the frequency of the incident field coincides with a frequency at which the dielectric constant vanishes; for except in this case the polarization field is necessarily transverse [11] (i.e. $\nabla.\underline{P} = 0$). If we exclude this exceptional case, (28) reduces to

$$\underline{E}^{(i)}(\underline{r}_<) + \frac{1}{4\pi k^2}\ \nabla\times\nabla\times\int_{S^-}\{\underline{E}(\underline{r}')\frac{\partial G_o(\underline{r}_<,\underline{r}')}{\partial n} - G_o(\underline{r}_<,\underline{r}')\frac{\partial \underline{E}(\underline{r}')}{\partial n}\}dS=0, \quad (33)$$

which is seen to be identical with the Oseen formulation (1) of the extinction theorem. In this case ($\underline{M} = \nabla.\underline{P} = \underline{j} = \underline{K} = 0$, $\underline{P} = \chi\underline{E}$, χ being a constant), the equation of motion (4a) is not coupled to (4b) (since no magnetic term now occurs on the r.h.s. of (6a)) and reduces to

$$\nabla^2\underline{E}(\underline{r}_<) + n^2k^2\underline{E}(\underline{r}_<) = 0, \quad (34)$$

where

$$n^2 = 1 + 4\pi\chi . \quad (35)$$

According to our hypothesis the electric field $\underline{E}$ inside the medium is that solution of (34), which obeys the condition (33) at every point $\underline{r}$ inside the medium. Once this solution has been found, the field outside the medium is obtained from the formula

$$\underline{E}(\underline{r}_>)=\underline{E}^{(i)}(\underline{r}_>)+\frac{1}{4\pi k^2}\nabla\times\nabla\times\int_{S^-}\{\underline{E}(\underline{r}')\frac{\partial G_o(\underline{r}_>,\underline{r}')}{\partial n}-G_o(\underline{r}_>,\underline{r}')\frac{\partial\underline{E}(\underline{r}')}{\partial n}\}dS, \tag{36}$$

to which Eq. (23) may be shown to reduce in the present case.

One can also show that in some other special cases, our general extinction theorem (22) for the electric field reduces to various extinction theorems derived in recent years by other authors. Moreover, one can readily show that in the special case when no scattering medium is present at all our general extinction theorem for the electric field reduces to

$$\underline{E}^{(i)}(\underline{r}_<)+\frac{1}{4\pi}\int_{S^-}\{\underline{E}^{(i)}(\underline{r}')\frac{\partial G_o(\underline{r}_<,\underline{r}')}{\partial n}-G_o(\underline{r}_<,\underline{r}')\frac{\partial\underline{E}^{(i)}(\underline{r}')}{\partial n}\}dS=0, \tag{37}$$

which will be recognized as the classic *integral theorem of Helmholtz and Kirchhoff* (cf. for example, ref. 3a, p.377).

Returning to the general forms (22) and (25) of the extinction theorems it seems quite remarkable that for a completely arbitrary medium - e.g. an inhomogeneous, anisotropic, non-linear or spatially dispersive medium - the cancelation of the incident field inside the medium is expressible entirely by the values that the field takes at the boundary of the medium

Finally I will show that the main results that we obtained for electromagnetic scattering have a strict analogue in non-relativistic quantum-mechanical potential scattering. Consider scattering of a free particle of momentum $\underline{p}$ on a three-dimensional potential barrier or potential well, characterized by a potential $\mathcal{V}(\underline{r})$ that vanishes outside a finite volume V, bounded by a surface S. (Fig. 2). For simplicity we assume that the potential $\mathcal{V}(\underline{r})$ has at most a finite discontinuity on S.

The Schrodinger equation for this problem may be written in the form

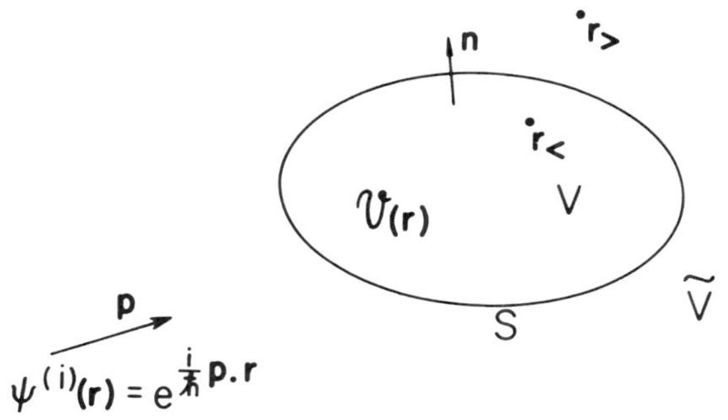

Fig. 2 Notation relating to quantum mechanical potential scattering.

$$(\nabla^2 + k^2)\ \psi(\underline{r}) = U(\underline{r})\psi(\underline{r}), \tag{38}$$

where

$$k^2 = \frac{2m}{\hbar^2} E, \qquad U(\underline{r}) = \frac{2m}{k^2}\mathcal{V}(\underline{r}) \quad \text{if } r \in V$$
$$= 0 \quad \text{if } r \in \tilde{V}. \tag{39}$$

Here m denotes the mass of the particle and $E = p^2/2m$ its energy, $\hbar$ is the Planck constant divided by 2π and $\tilde{V}$ denotes the (infinite) domain outside V.

The equation for the associated Green's function is

$$(\nabla^2 + k^2)\ G_o(\underline{r},\underline{r}') = -\ 4\pi\delta(\underline{r}-\underline{r}'), \tag{40}$$

the Green's function being, of course, the outgoing spherical wave [Eq. (13b) above].

From (38) and (39) we obtain, if we also use Green's theorem, the following identity valid for integration throughout any domain V' bounded by a closed surface S' :

$$\int_{V'} \psi(\underline{r}')\delta(\underline{r}-\underline{r}')d^3\underline{r}' = -\frac{1}{4\pi}\int_{V'} U(\underline{r}')\psi(\underline{r}')G_o(\underline{r},\underline{r}')d^3\underline{r}' - \frac{1}{4\pi}\sum(\underline{r}), \quad (41)$$

where

$$\sum(\underline{r}) = \int_{S'} \{\psi(r') \frac{\partial G_o(\underline{r},\underline{r}')}{\partial n} - G_o(\underline{r},\underline{r}') \frac{\partial\psi(\underline{r}')}{\partial n}\} dS \quad (43)$$

and $\partial/\partial n$ denotes differentiation along the outward normal to S'.

Let us now take the volume V' to coincide either with the scattering volume V or with the exterior $\tilde{V}$ of it. Again the field point $\underline{r}$ may be taken to be either in V or in $\tilde{V}$. Thus, in analogy with the electromagnetic case, we obtain four formulae:

(a) $\underline{r}\epsilon V$, $\underline{r}'\epsilon V$

$$\psi(\underline{r}_<) = -\frac{1}{4\pi}\int_V \psi(\underline{r}')U(\underline{r}')G_o(\underline{r}_<,\underline{r}')d^3\underline{r}' - \frac{1}{4\pi}\sum(\underline{r}_<), \quad (43a)$$

(b) $\underline{r}\epsilon V$, $\underline{r}'\epsilon\tilde{V}$

$$0 = -\frac{1}{4\pi}\int_{\tilde{V}} \psi(\underline{r}')U(\underline{r}')G_o(\underline{r}_<,\underline{r}')d^3r') + \frac{1}{4\pi}\sum(\underline{r}_<) - \frac{1}{4\pi}\sum^{(\infty)}(\underline{r}_<), \quad (43b)$$

(c) $\underline{r}\epsilon\tilde{V}$, $\underline{r}'\epsilon\tilde{V}$

$$\psi(\underline{r}_>) = -\frac{1}{4\pi}\int_{\tilde{V}} \psi(\underline{r}')U(\underline{r}')G_o(\underline{r}_>,\underline{r}')d^3\underline{r}' + \frac{1}{4\pi}\sum(\underline{r}_>) - \frac{1}{4\pi}\sum^{(\infty)}(\underline{r}_>), \quad (43c)$$

(d) $\underline{r}\epsilon\tilde{V}$, $\underline{r}'\epsilon V$

$$0 = -\frac{1}{4\pi}\int_V \psi(\underline{r}')U(\underline{r}')G_o(\underline{r}_>,\underline{r}')d^3\underline{r}' - \frac{1}{4\pi}\sum(\underline{r}_>), \quad (43d)$$

where

$$\sum(\underline{r}) = \int_S \{\psi(\underline{r}') \frac{\partial G_o(\underline{r},\underline{r}')}{\partial n} - G_o(\underline{r},\underline{r}') \frac{\partial\psi(\underline{r}')}{\partial n}\} dS, \quad (44a)$$

$$\sum^{(\infty)}(\underline{r}) = \lim_{R\to\infty} \int_{S_R} \{\psi(\underline{r}') \frac{\partial G_o(\underline{r},\underline{r}')}{\partial n} - G_o(\underline{r},\underline{r}') \frac{\partial \psi(\underline{r}')}{\partial n}\} dS, \tag{44b}$$

$\Sigma^{(\infty)}$ being a contribution from a sphere of limitingly large radius $R\to\infty$ and $\partial/\partial n$ denotes differentiation along the outward normals to the respective volume regions. In the integral (44a) for $\Sigma(\underline{r})$ we need not distinguish between limits from inside and outside of V since ψ and $\partial\psi/\partial n$ must be continuous at the surface S [cf. Eq. (10) above].

The integrals over V in (43b) and (43c) vanish since $U(\underline{r}) = 0$ in $\tilde{V}$. Also the contribution from the surface at infinity must evidently represent the incident wave, i.e.

$$-\frac{1}{4\pi}\sum^{(\infty)}(\underline{r}) = \psi^{(i)}(\underline{r}). \tag{45}$$

Hence (40b) and (40c) reduce to

$$\psi^{(i)}(\underline{r}_<) + \frac{1}{4\pi}\sum(\underline{r}_<) = 0, \tag{46}$$

$$\psi(\underline{r}_>) = \psi^{(i)}(\underline{r}_>) + \frac{1}{4\pi}\sum(\underline{r}_>). \tag{47}$$

Further from (43a) we obtain if we also use (46)

$$\psi(\underline{r}_<) = \psi^{(i)}(\underline{r}_<) - \frac{1}{4\pi}\int_V \psi(\underline{r}')U(\underline{r}')G_o(\underline{r}_<,\underline{r}')d^3\underline{r}', \tag{48}$$

and from (43d) if we use (47)

$$\psi(\underline{r}_>) = \psi^{(i)}(\underline{r}_>) - \frac{1}{4\pi}\int_V \psi(\underline{r}')U(\underline{r}')G_o(\underline{r}_>,\underline{r}')d^3\underline{r}'. \tag{49}$$

The relations (46) and (47) are strictly analogous to those that we found for the electromagnetic case. In particular (46) represents *an extinction theorem* expressing the cancellation of the incident wave function at every point inside the potential barrier or potential well in terms of the values of the total wave function ψ and its normal derivative $\partial\psi/\partial n$ at all points on the boundary of the potential barrier or potential well. We assert that this theorem has the same kind of significance as we postulated for the electromagnetic extinction theorem: It is a (non-local) boundary condition subject to which the Schrödinger equation (38) has to be solved inside the scattering volume V. Once this solution is known the wave function outside V can be determined from (47) by substitution. Note that (47) involves only a surface integral $\Sigma(\underline{r}_>)$, not a volume integral as it does in the usual formulation. The other two equations (47) and (48), which are seen to be of the same form irrespective whether the field point is inside or outside V will be recognized as the *usual integral equations for potential scattering*. Mathematically they are equivalent to the Schrödinger equation (38) together with the two equations (46) and (47).

We made use of the quantum mechanical extinction theorem (46) and the associated formula (47) to solve simple scattering problems, in order to verify the correctness of this new formulation of scattering. The results agree with those obtained by conventional methods based on the integral equations (48) and (49). I might add that the extinction theorem also leads correctly to *bound states* under appropriate conditions. One finds in these cases that the Schrödinger equation can then be solved subject to our non-local boundary condition (expressed by the extinction theorem (46))only when $\psi^{(i)}(\underline{r}) \equiv 0$.

Let me remark here that a quantum-mechanical extinction theorem, was also derived, from a different approach by Melvin Lax [12] in 1952 in his treatment of multiple scattering.

I will end by summarizing our main conclusions:

(1) We obtained on the basis of Maxwell's theory generalization of the classic Ewald-Oseen extinction theorem, valid rigorously for scattering from medium of any prescribed macroscopic response.

(2) We have put forward a hypothesis as to the true meaning of the theorem: it is a non-local boundary condition for the solution of the equation of motion for the *interior* scattering problem.

(3) We have shown that once the boundary values have been determined the *exterior* scattering problem can be solved in a novel way, involving only surface integrations.

(4) We have shown that these results have strict analogues in non-relativistic quantum-mechanical potential scattering and provide a new approach to solving such problems.

Finally let me say that scattering problems involving sharp boundaries, such as we have considered here are, in general, hard to solve even approximately, since the Born approximation cannot be used in such cases. It is possible that our new formulation might provide a basis for the development of useful approximate techniques for solving problems of this kind. For this new formulation takes very explicitly into account the sharp boundary, the very presence of which makes the usual perturbation methods inapplicable. We plan to discuss this and related problems in other publications.

References

* Research supported by the Air Force Office of Scientific Research and the Army Research Office (Durham).

1. P.P. Ewald, (a) Dissertation, Univ. of Munich, 1912;
   (b) Ann. Phys. *49*, 1 (1915).
2. C.W. Oseen, Ann. Phys. *48*, 1 (1915).
3. For a detailed account of the theorem see
   (a) M. Born and E. Wolf, *Principles of Optics* (Pergamon Press, Oxford and New York, 1970) 4th ed., §2.4,

   or

   (b) L. Rosenfeld, *Theory of Electrons* (North-Holland Publishing Co., Amsterdam, 1951) Chapt. VI, §4.
4. J.J. Sein, *An Integral-Equation Formulation of the Optics of Spatially-Dispersive Media*, Ph.D. Dissertation, New York University, 1969, Appendix III.
5. (a) A. Wierzbicki, Bul. Acad. Polonaise des Sciences *9*, 833 (1961); Acta Phys. Pol. *21*, 557 (1962), ibid *21*, 575 (1962).
   (b) N. Bloembergen and P.S. Pershan, Phys. Rev. *127*, 206 (1962).
   (c) B.A. Sotskii, Opt. Spectro. *14*, 57 (1963).
   (d) R.K. Bullough, J. Phys. A. (Proc. Phys. Soc.) Ser. 2, *1*, 409 (1968).
   (e) J.J. Sein, ref. 4 above and Opt. Comm. *2*, 170 (1970).
   (f) É. Lalor, Opt. Comm. *1*, 50 (1969).
   (g) É. Lalor and E. Wolf, Phys. Rev. Lett. *26*, 1274 (1971).
   (h) T. Suzuki, J. Opt. Soc. Amer. *61*, 1029 (1971).
   (i) G.S. Agarwal, D.N. Pattanayak and E. Wolf, Opt. Comm. *4*, 260 (1971).
   (j) É. Lalor and E. Wolf, J. Opt. Soc. Amer. *62*, 1165 (1972).
   (k) J.L. Birman and J.J. Sein, Phys. Rev. *B6*, 2482 (1972).

6. Preliminary results of this investigation were presented in a lecture at the annual meeting of the Optical Society of America held in Ottawa in October 1971 (Abstr. WC16, J.Opt.Soc.Amer., *61*, 1560 (1971)) and in a note published in Optics Commun. *6*, 217 (1972).
7. While this manuscript was being prepared for publication a paper reporting some closely related results was published by J. de Goede and P. Mazur, Physica *58*, 568 (1972). This paper also contains some additional references to publications concerning the extinction theorem.
8. A slightly different argument to that given below is needed if the incident field is a plane wave (i.e. if the source is at infinity), but the final formulae remain the same. The case of plane wave incidence is discussed explicitly in connection with the quantum mechanical extinction theorem, in the last part of this paper.
9. See, for example, Chen-To Tai, *Dyadic Green's Functions in Electromagnetic Theory* (In-text Educational Publishers, Scranton and San Francisco, 1971).
10. One of them was presented in reference 5j .
11. That this is so follows at once from the Maxwell equation $\nabla.\underline{D} = 0$. For this implies that $0 = \nabla.(\varepsilon\,\underline{E}) = \varepsilon\,\nabla.\underline{E}$ (ε = dielectric constant), so that $\nabla.E = 0$ and hence also $\nabla.\underline{P} = 0$ (because of the linearity of the medium), unless $\varepsilon = 0$.
12. M. Lax, Phys. Rev. *85*, 646 (1952).

PHYSICAL REVIEW D VOLUME 13, NUMBER 4 15 FEBRUARY 1976

Scattering states and bound states as solutions of the Schrödinger equation with nonlocal boundary conditions*

D. N. Pattanayak
Department of Physics, University of Toronto, Toronto M5S 1A7 Canada

E. Wolf†
Department of Physics and Astronomy, University of Rochester, Rochester, New York 14627
(Received 21 July 1975)

The problem of determining the Schrödinger wave function of a nonrelativistic particle that is either scattered by a potential of a finite range or that is bound to it is reformulated in a novel way. It is shown that in either case the wave function must satisfy a certain boundary condition on the surface that delimits the effective range of the potential. For scattering states the boundary condition is analogous to the mathematical formulation of the Ewald-Oseen extinction theorem of classical electromagnetic theory. The new formulation is illustrated by determining the scattering states and the bound states for a central potential. It is also shown that a boundary condition that is used in band-structure calculations in solids is an immediate consequence of our quantum-mechanical extinction theorem for bound states.

I. INTRODUCTION

One of the standard procedures for determining the wave function of a nonrelativistic particle scattered by a potential of a finite range or bound to it consists of the following: One writes down the general solution of the Schrödinger equation with the given potential for both the interior and the exterior of the volume V throughout which the potential is effective. One then determines the values of the coefficients that enter the general solution by invoking the continuity of the wave function and of its normal derivative across the surface S bounding V and by taking into account the postulated asymptotic behavior of the wave function at infinity. For more complicated potentials, where the general form of the solution cannot be found, one usually proceeds in another way: One recasts the Schrödinger equation into an integral equation and solves it either in the Born approximation or in the form of a perturbation series. These procedures have been employed since the days when wave mechanics was formulated, and the majority of physicists have obtained their first acquaintance with some of the basic features of quantized systems from solutions derived in this way.

In neither of these procedures does one impose any constraints on the values of the wave function on the surface S of the domain throughout which the potential is effective, though it is clear that a boundary condition on the wave function at this surface (to be distinguished from a mere "continuity requirement") is implicitly contained in the integral equation formulation. Although this latter fact is sometimes alluded to in the literature,[1] the boundary conditions to our knowledge have never been "extracted" from the integral equation and their general form appears to have remained unknown up to now. A similar situation existed for a very long time in the classical theory of scattering of electromagnetic waves on media with sharp boundaries,[2] where only very recently has this situation been clarified.[3] It turned out that in this case the "missing" boundary condition is nothing else but the mathematical formulation of the so-called Ewald-Oseen extinction theorem, whose fundamental role (but not its full understanding) in problems of interaction between an electromagnetic field and a material medium was already apparent about 60 years ago, from Ewald's pioneering research on the foundation of crystal optics.[4] We refer elsewhere for a discussion of this theorem.[5] Here we only mention that it elucidates the manner in which an electromagnetic wave that is incident on a material medium with vacuum velocity of light is extinguished at all points inside the medium in the process of interacting with it, so that another wave, propagated with a different velocity (appropriate to the refractive index of the medium), can be generated there (see also footnote 11 below).

In the present paper we first apply, in Sec. II, a similar procedure to the one we used in our discussion of the electromagnetic case[3a,b] to reformulate the problem of nonrelativistic quantum-mechanical scattering on a potential of a finite range. We derive a certain integral relation that involves the values of the wave function and of its normal derivative on the surface S that delimits the range of the potential. We postulate that the solution of the interior problem, i.e., the wave function at all points inside the scattering volume, is that solution of the Schrödinger equation which satisfies this constraint. This new boundary condition resembles the Ewald-Oseen extinction theorem for scattering of electromagnetic waves. In the general case this boundary condition is nonlocal, i.e., it involves the values of the wave function and also of its normal derivative at every point on the surface S. The solution of the interior problem provides as a by-product these boundary values. Once they have been determined, the solution of the exterior problem (determination of the wave function at points outside V) reduces to surface integration. In Sec. III we illustrate these results by applying them to scattering from a central potential. We find that in this case the extinction theorem gives rise to a set of *local* boundary conditions, one for each partial wave into which the wave function may be decomposed.

In Sec. IV we show how a straightforward modification of the theory of Sec. II for scattering states leads to a corresponding formulation for bound states. In Sec. V we illustrate these results by showing how the bound states for a central potential can be determined from our theory. In Sec. VI we show that a well-known boundary condition employed in band-structure calculations in solids notably by Kohn and Rostoker[6] and Ziman[7] is a straightforward consequence of our bound-state form of the extinction theorem, specialized to the muffin-tin potential.

We will show in another publication[16] that our theory can also be extended to resonance states and that it leads to a new derivation and to a generalization to noncentral force fields of the boundary conditions introduced into nuclear resonance theory by Kapur and Peierls,[8a] Siegert,[8b] Humblet,[8c] and Humblet and Rosenfeld[8d] in well-known papers.

II. POTENTIAL SCATTERING AS A NONLOCAL BOUNDARY-VALUE PROBLEM

We consider nonrelativistic scattering of a free particle of mass m and momentum $\vec{p}$ on a three-dimensional potential $\mathcal{V}(\vec{r})$. We assume that $\mathcal{V}(\vec{r})$ vanishes outside a finite volume V bounded by a closed surface S and takes on finite values on the boundary S. The time-independent part $\psi(\vec{r})$ of the wave function $\Psi(\vec{r},t)=\psi(\vec{r})e^{-iEt/\hbar}$ of the particle then satisfies throughout the whole space the Schrödinger equation

$$(\nabla^2+k^2)\psi(\vec{r})=U(\vec{r})\psi(\vec{r}), \tag{2.1}$$

where

$$k^2=\frac{2m}{\hbar^2}E, \tag{2.2a}$$

and

$$U(\vec{r})=\frac{2m}{\hbar^2}\mathcal{V}(\vec{r}) \quad \text{if } \vec{r}\in V$$
$$=0 \quad \text{if } \vec{r}\in\tilde{V}, \tag{2.2b}$$

and $\tilde{V}$ denotes the infinite volume exterior to S (see Fig. 1). $E=p^2/2m$ is, of course, the energy of the particle and $\hbar$ is the Planck constant divided by 2π.

The Green's function $G(\vec{r},\vec{r}\,')$ associated with the operator (∇^2+k^2) that appears on the left-hand side of the Schrödinger equation (2.1) satisfies the differential equation

$$(\nabla^2+k^2)G(\vec{r},\vec{r}\,')=-4\pi\delta(\vec{r}-\vec{r}\,'), \tag{2.3}$$

where δ is the three-dimensional Dirac δ function. Of the possible solutions of (2.3) we choose the one that behaves at infinity as an outgoing wave, i.e.,

$$G(\vec{r},\vec{r}\,')=\frac{e^{ik|\vec{r}-\vec{r}\,'|}}{|\vec{r}-\vec{r}\,'|}, \tag{2.4}$$

where k is taken to be the positive square root of the expression on the right-hand side of Eq. (2.2a).

From Eqs. (2.1) and (2.3) we obtain in the usual way, with the help of Green's theorem, the identity valid for integration throughout an arbitrary domain V' bounded by a closed surface S'

$$\int_{V'}\psi(\vec{r}\,')\delta(\vec{r}-\vec{r}\,')d^3r'=-\frac{1}{4\pi}\int_{V'}U(\vec{r}\,')\psi(\vec{r}\,')G(\vec{r},\vec{r}\,')d^3r' - \frac{1}{4\pi}\Sigma'(\vec{r}), \tag{2.5}$$

where

$$\Sigma'(\vec{r})=\int_{S'}\left[\psi(\vec{r}\,')\frac{\partial G(\vec{r},\vec{r}\,')}{\partial n'}-G(\vec{r},\vec{r}\,')\frac{\partial\psi(\vec{r}\,')}{\partial n'}\right]dS' \tag{2.6}$$

and $\partial/\partial n'$ denotes differentiation along the outward

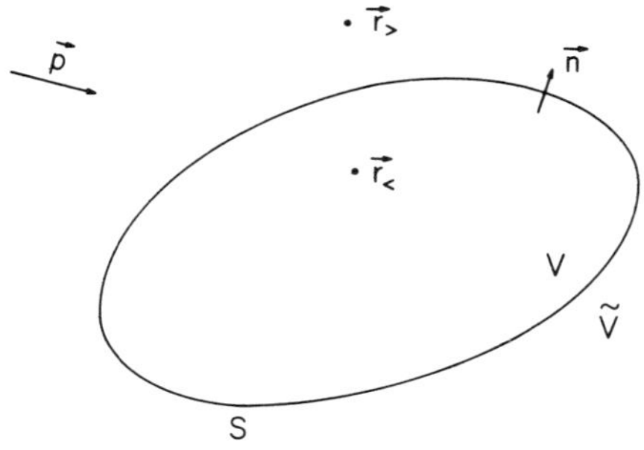

FIG. 1. Illustration of the notation relating to potential scattering.

normal to S'.

Let us now take the volume V' to coincide either with the scattering volume V or with the exterior volume $\tilde{V}$ (Fig. 1). The field point $\vec{r}$ may be taken to be situated either in V (in which case we denote it by $\vec{r}_<$) or in $\tilde{V}$ (in which case we denote it by $\vec{r}_>$). If we also use the fact that $U \equiv 0$ in $\tilde{V}$ we obtain from (2.5) the following four identities appropriate to the four possible cases:

(a) $\vec{r} \in V$, $\vec{r}' \in V$:

$$\psi(\vec{r}_<) = -\frac{1}{4\pi}\int_V \psi(\vec{r}')U(\vec{r}')G(\vec{r}_<, \vec{r}')d^3r' - \frac{1}{4\pi}\Sigma(\vec{r}_<); \tag{2.7a}$$

(b) $\vec{r} \in V$, $\vec{r}' \in \tilde{V}$:

$$0 = \frac{1}{4\pi}\Sigma(\vec{r}_<) - \frac{1}{4\pi}\Sigma^{(\infty)}(\vec{r}_<); \tag{2.7b}$$

(c) $\vec{r} \in \tilde{V}$, $\vec{r}' \in \tilde{V}$:

$$\psi(\vec{r}_>) = \frac{1}{4\pi}\Sigma(\vec{r}_>) - \frac{1}{4\pi}\Sigma^{(\infty)}(\vec{r}_>); \tag{2.7c}$$

(d) $\vec{r} \in \tilde{V}$, $\vec{r}' \in V$:

$$0 = -\frac{1}{4\pi}\int_V \psi(\vec{r}')U(\vec{r}')G(\vec{r}_>, \vec{r}')d^3r' - \frac{1}{4\pi}\Sigma(\vec{r}_>), \tag{2.7d}$$

where

$$\Sigma(\vec{r}) = \int_S \left[\psi(\vec{r}')\frac{\partial G(\vec{r}, \vec{r}')}{\partial n} - G(\vec{r}, \vec{r}')\frac{\partial \psi(\vec{r}')}{\partial n}\right]dS \tag{2.8a}$$

and

$$\Sigma^{(\infty)}(\vec{r}) = \lim_{R\to\infty}\int_{S_R}\left[\psi(\vec{r}')\frac{\partial G(\vec{r}, \vec{r}')}{\partial n_R} - G(\vec{r}, \vec{r}')\frac{\partial \psi(\vec{r}')}{\partial n_R}\right]dS. \tag{2.8b}$$

In (2.8b), S_R denotes the outer boundary of the exterior region $\tilde{V}$ surrounding the scatterer, chosen for convenience to be a large sphere of radius R that is taken in the limit $R \to \infty$. In (2.8a) $\partial/\partial n$ denotes differentiation along the outward normal to S; in (2.8b) $\partial/\partial n_R$ denotes differentiation along the outward radial direction.

The surface integral $\Sigma^{(\infty)}$ may be readily evaluated and its value is independent of the form of the potential. To see this let us express the wave function ψ that appears in the integrand of $\Sigma^{(\infty)}$ as the sum of the incident wave $\psi^{(i)}(\vec{r}) = e^{i(\vec{p}\cdot\vec{r})/\hbar}$ and of the scattered wave $\psi^{(s)}(\vec{r})$:

$$\psi(\vec{r}) = \psi^{(i)}(\vec{r}) + \psi^{(s)}(\vec{r}). \tag{2.9}$$

Since the scattered wave $\psi^{(s)}$ must behave at infinity as an outgoing wave, it does not contribute to $\Sigma^{(\infty)}$, i.e., for any point[9] $\vec{r}$,

$$\lim_{R\to\infty}\int_{S_R}\left[\psi^{(s)}(\vec{r}')\frac{\partial G(\vec{r}, \vec{r}')}{\partial n_R} - G(\vec{r}, \vec{r}')\frac{\partial \psi^{(s)}(\vec{r}')}{\partial n_R}\right]dS = 0. \tag{2.10}$$

On the other hand, since the incident wave $\psi^{(i)}(\vec{r})$ obeys the Helmholtz equation, $(\nabla^2 + k^2)\psi^{(i)} = 0$, throughout the whole space it follows at once from the Helmholtz-Kirchhoff integral theorem [Ref. 5(a), Sec. 8.3.1 or Ref. 9(a), p.24] that

$$\lim_{R\to\infty}\int_{S_R}\left[\psi^{(i)}(\vec{r}')\frac{\partial G(\vec{r}, \vec{r}')}{\partial n_R} - G(\vec{r}, \vec{r}')\frac{\partial \psi^{(i)}(\vec{r}')}{\partial n_R}\right]dS = -4\pi\psi^{(i)}(\vec{r}). \tag{2.11}$$

On substituting from (2.9) into (2.8b) and on using the results expressed by Eqs. (2.10) and (2.11) it follows that

$$-\frac{1}{4\pi}\Sigma^{(\infty)}(\vec{r}) = \psi^{(i)}(\vec{r}). \tag{2.12}$$

On making use of this result in Eqs. (2.7b) and (2.7c) we obtain the following two basic equations of our theory:

$$\psi^{(i)}(\vec{r}_<) + \frac{1}{4\pi}\Sigma(\vec{r}_<) = 0, \tag{2.13}$$

and

$$\psi(\vec{r}_>) = \psi^{(i)}(\vec{r}_>) + \frac{1}{4\pi}\Sigma(\vec{r}_>). \tag{2.14}$$

Before discussing the significance of these two equations we note, in passing, that if we substitute $\Sigma(\vec{r}_<)$ from (2.13) into (2.7a) we obtain the formula

$$\psi(\vec{r}_<) = \psi^{(i)}(\vec{r}_<) - \frac{1}{4\pi}\int_V \psi(\vec{r}')U(\vec{r}')G(\vec{r}_<, \vec{r}')d^3r', \tag{2.15}$$

and that if we substitute for $\Sigma(\vec{r}_>)$ from (2.14) into (2.7d) we obtain a formula of the same form, viz.,

$$\psi(\vec{r}_>) = \psi^{(i)}(\vec{r}_>) - \frac{1}{4\pi}\int_V \psi(\vec{r}')U(\vec{r}')G(\vec{r}_>, \vec{r}')d^3r'. \tag{2.16}$$

Equations (2.15) and (2.16) are the usual integral equations for potential scattering. Equation (2.15) is an integral equation for the wave function at points inside the volume V. Equation (2.16) gives the wave function throughout the exterior of the volume V in terms of the solution of the interior problem.

To discuss the significance of Eqs. (2.13) and (2.14) let us rewrite them in a more explicit form by substituting for Σ from (2.8a):

$$\psi^{(i)}(\vec{r}_<) + \frac{1}{4\pi}\int_S \left[\psi(\vec{r}')\frac{\partial G(\vec{r}_<, \vec{r}')}{\partial n} - G(\vec{r}_<, \vec{r}')\frac{\partial\psi(\vec{r}')}{\partial n}\right] dS = 0, \quad (2.13')$$

$$\psi(\vec{r}_>) = \psi^{(i)}(\vec{r}_>) + \frac{1}{4\pi}\int_S \left[\psi(\vec{r}')\frac{\partial G(\vec{r}_>, \vec{r}')}{\partial n} - G(\vec{r}_>, \vec{r}')\frac{\partial\psi(\vec{r}')}{\partial n}\right] dS. \quad (2.14')$$

Equation (2.13′), which must be satisfied for every point $\vec{r}_<$ inside the domain V occupied by the scattering potential, is of the same mathematical form (apart from a simplification arising from the scalar nature of the present problem), as the classic Ewald-Oseen extinction theorem for electromagnetic waves.[5] Although it has been realized for many years that this theorem is rather basic for elucidating some aspects of the interaction between an electromagnetic wave and a material medium, its true mathematical significance has been clarified only recently.[3a,b] In analogy with this recent development we postulate that Eq. (2.13′), which may be called a *quantum-mechanical extinction theorem*,[10] is a *boundary condition for the solution of the interior scattering problem*; more precisely that *the solution* $\psi(\vec{r})$ *of our scattering problem valid at all points that are situated inside the volume V occupied by the scattering potential is that solution of the Schrödinger equation (2.1) that satisfies the constraint expressed by the extinction theorem (2.13′)*. We stress that *(2.13′) has to be satisfied for every point*[11] $\vec{r}_<$ *in V*. Since (2.13′) involves the values of ψ (and also of its normal derivative) at every point on the boundary surface S of the volume V, it is a *nonlocal* boundary condition. However, we will see later that in the special but important case of a central potential, (2.13′) reduces to a set of local boundary conditions, one for each partial wave into which ψ may be decomposed.

The meaning of the other equation, (2.14′), is quite clear. It represents the *solution to the exterior scattering problem* [i.e., the determination of $\psi(\vec{r})$ at points outside V] in terms of the boundary values on the surface S of ψ and its normal derivative $\partial\psi/\partial n$. These boundary values may be obtained from the solution of the interior problem.

Our theory separates clearly the interior and the exterior problems. It allows the solution of the interior problem to be obtained without any reference to the scattered field outside the domain containing the scattering potential. This theory may readily be extended to the problem of determining bound states. Before showing this we will illustrate our new formulation by an example.

III. EXAMPLE: SCATTERING FROM A CENTRAL POTENTIAL

Consider the scattering of a nonrelativistic particle on a central potential $\mathcal{V}(r)$ $(r = |\vec{r}|)$, which vanishes outside a sphere of radius a centered at the origin $r = 0$. To apply our technique to this problem, we will represent each term that occurs in the extinction theorem (2.13′) as an expansion in terms of separable solutions of the Schrödinger equation for the central-potential problem in spherical polar coordinates. We choose the polar axis $\theta = 0$ in the direction of the momentum of the incident particle, whose wave function is

$$\psi^{(i)}(\vec{r}) = e^{i(\vec{p}\cdot\vec{r})/\hbar}. \quad (3.1)$$

The expansion of $\psi^{(i)}$ is well known to be[12]

$$\psi^{(i)}(\vec{r}) = \sum_{l=0}^{\infty} i^l(2l+1)P_l(\cos\theta)j_l(kr), \quad (3.2)$$

where $k = p/\hbar$, $j_l(kr)$ is the spherical Bessel function of order l, and $P_l(\cos\theta)$ is the Legendre polynomial of this order. The wave function $\psi(\vec{r}_<)$ that enters the surface integral in (2.13′) has the general expansion (Ref. 12, p.385)

$$\psi(\vec{r}_<) = \sum_{l=0}^{\infty} \chi_l(\vec{r}_<)P_l(\cos\theta), \quad (3.3)$$

where

$$\chi_l(r) = \frac{u_l(r)}{r}, \quad (3.4)$$

and u_l is a regular solution of the radial equation

$$\left\{\frac{d^2}{dr^2} + \left[\epsilon - U(r) - \frac{l(l+1)}{r^2}\right]\right\}u_l(r) = 0, \quad (3.5)$$

with

$$\epsilon = k^2 = \frac{2m}{\hbar^2}E, \quad (3.6a)$$

$$U(r) = \begin{cases} \frac{2m}{\hbar^2}\mathcal{V}(r) & \text{if } r \le a \\ 0 & \text{if } r > a. \end{cases} \quad (3.6b)$$

It is known that the physically acceptable solution of (3.5) has an expansion of the form (Ref. 12, p.351)

$$u_l(r) = A_l r^{l+1}(1 + C_{1l}r + C_{2l}r^2 + \cdots). \quad (3.7)$$

The successive coefficients $C_{1l}, C_{2l}, \ldots$ may be obtained by substituting in (3.5) the power-series expansion for $U(r)$ and the expansion (3.7) for $u_l(r)$ and equating coefficients of equal powers. The normalization constants A_l will be determined shortly with the help of our boundary condition (the quantum-mechanical extinction theorem), (2.13′).

For the Green's function we have the expansion[13]

$$G(\vec{r}_<, \vec{r}') = k \sum_{l=0}^{\infty} (2l+1) j_l(kr_<) h_l^{(+)}(kr') P_l(\cos\Theta), \tag{3.8}$$

valid with $r_< < r'$, where $h_l^{(+)}$ is the spherical Hankel function of the first kind and order l and Θ is the angle between the directions of the vectors $\vec{r}_<$ and $\vec{r}'$.

We now substitute the expansion for $\psi(\vec{r}_<)$ [from Eqs. (3.3), (3.4), and (3.7)] and that for $G(\vec{r}_<, \vec{r}')$ [Eq. (3.8)] in the surface integral $\Sigma(\vec{r}_<)$, defined by Eq. (2.8a), which occurs in our extinction theorem (2.13). After a straightforward calculation that is carried out in Appendix A we find that

$$\frac{1}{4\pi}\Sigma(\vec{r}_<) = \sum_{l=0}^{\infty} \alpha_l j_l(kr_<) P_l(\cos\theta), \tag{3.9}$$

where

$$\alpha_l = (ka)^2[\chi_l(a) h_l^{(+)\prime}(ka) - (1/k)\chi_l'(a) h_l^{(+)}(ka)] \tag{3.10}$$

and prime denotes differentiation with respect to the argument.

Now according to the extinction theorem (2.13), the sum of the expansions that occurs on the right-hand side of Eqs. (3.2) and (3.9) must vanish at every point $\vec{r}_<$ inside the scattering volume. This clearly is only possible if

$$\alpha_l = i^l(2l+1), \tag{3.11}$$

or, more explicitly, using (3.10), if

$$(ka)^2[\chi_l(a) h_l^{(+)\prime}(ka) - (1/k)\chi_l'(a) h_l^{(+)}(ka)] = i^l(2l+1). \tag{3.11′}$$

Equation (3.11′) is nothing but a set of boundary conditions, by means of which the normalization constants A_l that appear in the expansion (3.7) may be determined. We have from (3.4) and (3.7)

$$\chi_l(r) = A_l r^l \sum_{n=0}^{\infty} C_{n,l} r^n \tag{3.12}$$

(with $C_{0,l} = 1$ for all l). On setting $r = a$ in (3.12) and in the expansion for $\chi_l'(r)$ obtained from (3.12) by term-by-term differentiation and then substituting the resulting expansions into (3.11′) we readily obtain from that equation the following expressions for the normalization constants A_l:

$$A_l = \frac{i^l(2l+1)}{k^2 a^{l+2}} \left\{ \sum_{n=0}^{\infty} C_{n,l} a^n [h_l^{(+)\prime}(ka) - (n+l)(1/ka) h_l^{(+)}(ka)] \right\}^{-1}. \tag{3.13}$$

Thus we have now obtained the complete solution for the wave function inside the domain $r \le a$. It is given by Eq. (3.3), with $\chi_l(r)$ given by (3.12) and the A_l's given by (3.13). We stress that this solution was obtained without any reference to the scattered wave outside the domain containing the scattering potential.

It may be worthwhile to point out that the extinction theorem (2.13), which according to our theory plays, in general, the role of a nonlocal boundary condition, reduced in the present case (central potential) to a set of *local* boundary conditions, given by Eqs. (3.11), there being one such boundary condition for each of the partial solutions labeled by the suffix l.

Let us now determine the wave function $\psi(\vec{r})$ for points situated in the exterior domain ($r > a$). For this purpose, we must, according to Eq. (2.14) evaluate the surface integral $\Sigma(\vec{r})$, defined by Eq. (2.8a) for values of r such that $r > a$. This evaluation can be carried out in a similar way as the evaluation of $\Sigma(\vec{r})$ with $r < a$, except that we now make explicit use of the values of the normalization constants A_l [Eq. (3.13)]. The calculation is carried out in Appendix B and leads to the result that

$$\frac{1}{4\pi}\Sigma(\vec{r}_>) = \sum_{l=0}^{\infty} \beta_l h_l^{(+)}(kr_>) P_l(\cos\theta), \tag{3.14}$$

where

$$\beta_l = i^l(2l+1) \frac{\sum_{n=0}^{\infty} C_{n,l} a^n [j_l'(ka) - (n+l)(1/ka) j_l(ka)]}{\sum_{n=0}^{\infty} C_{n,l} a^n [h_l^{(+)\prime}(ka) - (n+l)(1/ka) h_l^{(+)}(ka)]}, \tag{3.15}$$

and the $C_{n,l}$ are, of course, the same constants as those that occur in the expansion (3.12). Finally on substituting from (3.14) into (2.14), we obtain the solution to the exterior scattering problem from a central potential, i.e., the solution valid at all points outside the spherical domain of radius a containing the potential:

$$\psi(\vec{r}_>) = \psi^{(i)}(\vec{r}_>) + \sum_{l=0}^{\infty} \beta_l h_l^{(+)}(kr_>) P_l(\cos\theta). \tag{3.16}$$

IV. BOUND STATES

The technique that we have described in the preceding sections in connection with scattering can also be readily applied to the problem of determining bound states of the particle. The corresponding equations may, in fact, be determined almost at once from the basic equations of Sec. II by a simple formal modification.

For bound states the equations (2.1)–(2.8) retain their validity, m now denoting the mass of the particle bound to the potential. The succeeding equations must, however, be modified, because the asymptotic behavior of the wave function at large distance from the domain V containing the potential is quite different now. The wave function for a bound state decreases exponentially with the distance r from the domain V as $r \to \infty$ in any given direction (θ, ϕ); more precisely,

$$\psi(\vec{r}) \sim A(\theta, \phi)\frac{e^{ikr}}{r} \quad (r \to \infty), \tag{4.1}$$

where

$$k = iK, \quad K > 0. \tag{4.2}$$

Equation (2.2a) may now be expressed in the form

$$K^2 = -\frac{2m}{\hbar^2}E, \tag{4.3}$$

which implies the well-known criterion, namely $E < 0$, for bound states.

When k is of the form (4.2), the Green's function (2.4) becomes

$$G(\vec{r}, \vec{r}\,') = \frac{e^{-K|\vec{r} - \vec{r}\,'|}}{|\vec{r} - \vec{r}\,'|}, \tag{4.4}$$

and it is readily seen that, because of the asymptotic behavior of the wave function and of the Green's function, the integral on the right-hand side of Eq. (2.8b) vanishes when ψ represents a bound state, i.e., we now have

$$\Sigma^{(\infty)}(\vec{r}) = 0. \tag{4.5}$$

Comparison of (4.5) with (2.12) shows that in the present case we must have

$$\psi^{(i)}(\vec{r}) \equiv 0, \tag{4.6}$$

which expresses the well-known fact that a bound state cannot be generated by a particle that is incident externally on the potential.

The equations for bound states that correspond to the equations (2.13) and (2.14) for scattering states are obtained at once from these equations by making use of (4.6), i.e., we now have

$$\frac{1}{4\pi}\Sigma(\vec{r}_<) = 0 \tag{4.7}$$

and

$$\psi(\vec{r}_>) = \frac{1}{4\pi}\Sigma(\vec{r}_>), \tag{4.8}$$

or, more explicitly,

$$\int_S \left[\psi(r')\frac{\partial G(\vec{r}_<, \vec{r}\,')}{\partial n} - G(\vec{r}_<, \vec{r}\,')\frac{\partial \psi(\vec{r}\,')}{\partial n}\right] dS = 0 \tag{4.7'}$$

and

$$\psi(\vec{r}_>) = \frac{1}{4\pi}\int_S \left[\psi(\vec{r}\,')\frac{\partial G(\vec{r}_>, \vec{r}\,')}{\partial n} - G(\vec{r}_>, \vec{r}\,')\frac{\partial \psi(\vec{r}\,')}{\partial n}\right] dS, \tag{4.8'}$$

where the Green's function G is given by (4.4).

The formula (4.7) [or (4.7′)], which we will refer to as the *quantum-mechanical extinction theorem for bound states,* must be satisfied at every point $\vec{r}_<$ inside the volume V occupied by the potential. It may be regarded as a *boundary condition* on the bound-state wave function $\psi(\vec{r})$. More precisely, *the bound-state wave functions are those solutions of the Schrödinger equation (2.1) [with k of the form (4.2)] which for every point $\vec{r}_<$ inside the domain V occupied by the potential satisfy the relation (4.7′).* Only for certain values of K, and hence of the energy E, will such solutions exist. Thus the boundary condition (4.7′) implicitly specifies the energies of the bound states. The other formula, (4.8′), expresses the wave functions of a bound state at all points exterior to the volume V, in terms of the boundary values of the wave function and of its normal derivatives on the surface S bounding the volume. These boundary values may be determined from solution of the interior problem.

It is of interest to note that if we substitute from (4.7) into (2.7a) we obtain the formula

$$\psi(\vec{r}_<) = -\frac{1}{4\pi}\int_V \psi(\vec{r}\,')U(\vec{r}\,')G(\vec{r}_<, \vec{r}\,')d^3r', \tag{4.9}$$

and, if we substitute for $\Sigma(\vec{r}_>)$ from (4.8) into (2.7d) we find that

$$\psi(\vec{r}_>) = -\frac{1}{4\pi}\int_V \psi(\vec{r}\,')U(\vec{r}\,')G(\vec{r}_>, \vec{r}\,')d^3r'. \tag{4.10}$$

Equations (4.9) and (4.10) are the usual equations for bound states.[14] They may, of course, also be obtained from Eqs. (2.15) and (2.16) on making use of (4.6).

V. EXAMPLE. BOUND STATES FOR A CENTRAL POTENTIAL

To illustrate the results of the preceding section we will now show how they may be used to determine the bound states associated with a central potential $\mathcal{V}(r)$ ($r=|\vec{r}|$) that vanishes outside a sphere of radius a, centered at the origin.

We may shorten the analysis by making use of the formulas derived in Sec. III in connection with scattering from a central potential, with appropriate modifications. It is clear that Eqs. (3.3)–(3.10) still apply, provided that we make use of the fact that k is now of the form indicated by Eq. (4.2). From (3.9) and our extinction theorem (4.7) for bound states we obtain the result that

$$\sum_{l=0}^{\infty} \alpha_l j_l(iKr_<)P_l(\cos\theta)=0 \tag{5.1}$$

for all values $r_< \leq a$. Because of the orthogonality of the Legendre polynomials $P_l(\cos\theta)$, it follows from (5.1) that we now have

$$\alpha_l=0 \quad (l=0,1,2,\ldots) \tag{5.2}$$

or, on making use of (3.10) and (4.2),

$$\chi_l(a)h_l^{(+)\prime}(iKa)-(1/iK)\chi_l{}'(a)h_l^{(+)}(iKa)=0 \quad (l=0,1,2,\ldots). \tag{5.3}$$

Now χ_l has a series expansion of the form (3.12) and it is convenient to express it in the form

$$\chi_l(r)=A_l\phi_l(r;\,iK), \tag{5.4}$$

where

$$\phi_l(r;\,iK)=r^l\sum_{n=0}^{\infty}C_{n,l}r^n. \tag{5.5}$$

The second argument iK is indicated explicitly in ϕ_l to remind us of this dependence (which is a consequence of the fact that the $C_{n,l}$'s themselves obviously depend on the parameter iK, because it appears in the Schrödinger equation). Hence (5.3) may be expressed in the form

$$A_l[\phi_l(a;iK)h_l^{(+)\prime}(iKa)-(1/iK)\phi_l{}'(a;iK)h_l^{(+)}(iKa)]=0 \quad (l=0,1,\ldots). \tag{5.6}$$

This equation implies that either

$$A_l=0, \tag{5.7a}$$

or that

$$\phi_l(a;iK)h_l^{(+)\prime}(iKa)-(1/iK)\phi_l{}'(a;iK)h_l^{(+)}(iKa)=0. \tag{5.7b}$$

The first case, $A_l=0$, is trivial, of no physical interest. The required solutions are obviously those that are associated with the second condition (5.7b). It represents a set of equations for the values of K for which bound states exist. We will denote the K roots of Eq. (5.7b), with l fixed, by $K_l^{(j)}$, where the superscript j distinguishes the different roots of that equation (if any). The wave function $\psi_l^{(j)}$ of the bound state that corresponds to any particular root $K_l^{(j)}$ is for all interior points ($r\leq a$) given by an expression that has the form of a single term under the summation sign on the right-hand side of (3.3); in view of (5.4) it may be written as

$$\psi_l^{(j)}(\vec{r}_<)=A_l^{(j)}\phi_l(r_<;\,iK_l^{(j)})P_l(\cos\theta), \tag{5.8}$$

where $A_l^{(j)}$ is a normalization constant. To obtain an expression for the wave function $\psi_l^{(j)}$ valid at exterior points ($r>a$), we only need, according to (4.8), to substitute into the surface integral $\Sigma(r_>)$, defined by (2.8a), the values that $\psi_l^{(j)}$ and its normal derivative take on the boundary surface $r=a$. The values may be determined from (5.8). The surface integral is of the same form as that which is evaluated in Appendix B, in connection with the case of scattering. Strictly similar calculations [or, more simply, an obvious modification of a typical term that occurs on the right-hand side of Eq. (B3)] lead to the result that

$$\frac{1}{4\pi}\Sigma(r_>)=A_l^{(j)}(iK_l^{(j)}a)^2[\phi_l(a;iK_l^{(j)})j_l{}'(iK_l^{(j)}a)-(1/iK_l^{(j)})\phi_l{}'(a;iK_l^{(j)})j_l(iK_l^{(j)}a)]h_l^{(+)}(iK_l^{(j)}r_>)P_l(\cos\theta). \tag{5.9}$$

Using this result in (4.8) we obtain the following expression for the bound-state wave function, valid at points exterior to the spherical domain occupied by the potential:

$$\psi_l^{(j)}(\vec{r}_>)=A_l^{(j)}F_l^{(j)}(a)h_l^{(+)}(iK_l^{(j)}r_>)P_l(\cos\theta), \tag{5.10}$$

where

$$F_l^{(j)}(a)=(iK_l^{(j)}a)^2[\phi_l(a;iK_l^{(j)})j_l{}'(iK_l^{(j)}a)-(1/iK_l^{(j)})\phi_l{}'(a;iK_l^{(j)})j_l(iK_l^{(j)}a)]. \tag{5.11}$$

If we recall the asymptotic behavior of the spherical Hankel function $h_l^{(+)}(iKr)$, it is clear from (5.10) that the bound-state wave functions decay exponentially as $r_>\to\infty$, as required. The values of the constants $A_l^{(j)}$ that appear in the expressions (5.8) and (5.10) may be obtained (up to an arbitrary phase factor) from the normalization requirement that the integral of $\psi_l^{(j)*}(\vec{r})\psi_l^{(j)}(\vec{r})$

taken over the whole space should equal to unity.

VI. THE EXTINCTION THEOREM FOR BOUND STATES AND THE KOHN-ROSTOKER-ZIMAN BOUNDARY CONDITIONS EMPLOYED IN BAND-STRUCTURE CALCULATIONS IN SOLIDS

We will now show that a certain boundary condition, frequently employed in band-structure calculations in solids, is an immediate consequence of our extinction theorem for bound states.

Consider a crystal in the muffin-tin potential model.[15] In this model the potential is periodic and is spherically symmetric within the sphere inscribed into each Wigner-Seitz cell and is constant in the interstitial region D (see Fig. 2). We take this constant to be zero by shifting the zero of the energy. To determine the bound states we must, according to the considerations of Sec. IV, first solve the Schrödinger equation with this potential, subject to the extinction theorem (4.7′) for bound states, i.e., subject to the constraint

$$\int_S \left[\psi(\vec{r}\,')\frac{\partial G(\vec{r}_<,\vec{r}\,')}{\partial n} - G(\vec{r}_<,\vec{r}\,')\frac{\partial\psi(\vec{r}\,')}{\partial n}\right] dS = 0\,, \tag{6.1}$$

for every point $\vec{r}_<$ within the crystal. In (6.1)

$$G(\vec{r},\vec{r}\,') = \frac{e^{-K|\vec{r}-\vec{r}\,'|}}{|\vec{r}-\vec{r}\,'|}\,, \tag{6.2}$$

where $K=(-2mE)^{1/2}/\hbar$, E being the energy of a bound state. The integration in (6.2) extends over the surface S of the crystal.

We may rewrite the integral in Eq. (6.1) as the sum of integrals over the surfaces $S_{l'}$ of all the individual cells,

$$\sum_{l'}\int_{S_{l'}} \left[\psi(\vec{r}\,')\frac{\partial G(\vec{r}_<,\vec{r}\,')}{\partial n} - G(\vec{r}_<,\vec{r}\,')\frac{\partial\psi(\vec{r}\,')}{\partial n}\right] dS_{l'} = 0\,, \tag{6.3}$$

where l' are lattice vectors. That Eqs. (6.1) and (6.3) are equivalent is clear if we note that the contributions from any two surface areas that are common to neighboring cells cancel out in the summation in (6.3) because the corresponding surface normals point in opposite directions. Let us set $\vec{r}\,'=\vec{\rho}\,'+\vec{l}\,'$, where $\vec{\rho}\,'$ represents points on the surface S' of a particular cell. If we label the admissible wave functions by the wave vector $\vec{k}$, we have by Bloch's theorem

$$\psi_{\vec{k}}(\vec{\rho}\,'+\vec{l}\,') = \psi_{\vec{k}}(\vec{\rho}\,')e^{i\vec{k}\cdot\vec{l}\,'}\,, \tag{6.4}$$

and (6.3) may now be expressed in the form

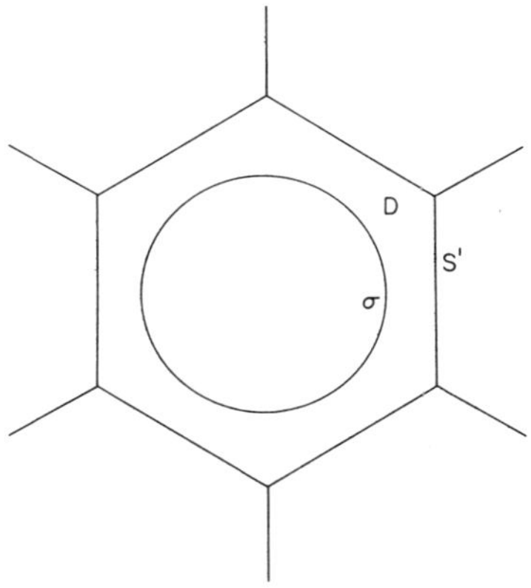

FIG. 2. Illustration of the notation relating to derivation of the Kohn-Rostoker-Ziman boundary condition (6.10). S' = boundary of a Wigner-Seitz cell, σ = surface of the inscribed sphere, D = interstitial region.

$$\int_{S'} \left[\psi_{\vec{k}}(\vec{\rho}\,')\frac{\partial g_{\vec{k}}(\vec{r}_<,\vec{\rho}\,')}{\partial n} - g_{\vec{k}}(\vec{r}_<,\vec{\rho}\,')\frac{\partial\psi_k(\vec{\rho}\,')}{\partial n}\right] dS' = 0\,, \tag{6.5}$$

where

$$g_{\vec{k}}(\vec{r}_<,\vec{\rho}\,') = \sum_{l'} G(\vec{r}_<,\vec{\rho}\,'+\vec{l}\,')e^{i\vec{k}\cdot\vec{l}\,'} = \sum_{l'} \frac{e^{-K|\vec{r}_<-\vec{\rho}\,'-\vec{l}\,'|}}{|\vec{r}_<-\vec{\rho}\,'-\vec{l}\,'|}e^{i\vec{k}\cdot\vec{l}\,'}\,. \tag{6.6}$$

$\partial/\partial n'$ denotes differentiation along the outward normal to S'.

We will now show that the condition (6.5), which involves integration over the surface S' of a Wigner-Seitz cell, implies a similar condition involving integration over the surface σ of the inscribed sphere. To see this, we use the fact that for all points $\vec{r}\,'$ in the interstitial region D

$$(\nabla^2 - K^2)\psi_{\vec{k}}(\vec{r}\,') = 0 \tag{6.7}$$

and that the function $g_{\vec{k}}(\vec{r}_<,\vec{r}\,')$ satisfies the equation

$$(\nabla^2 - K^2)g_{\vec{k}}(\vec{r}_<,\vec{r}\,') = -4\pi\sum_{l'}\delta(\vec{r}_< - \vec{r}\,' - \vec{l}\,')\,. \tag{6.8}$$

If we multiply (6.7) by $g_{\vec{k}}(\vec{r}_<,\vec{r}\,')$, (6.8) by $\psi_{\vec{k}}(\vec{r}\,')$, subtract the equations from each other, integrate the resulting equation throughout D, and apply Green's theorem we readily find that for all points $\vec{r}_<$ inside the inscribed sphere

$$\int_{S'}\left[\psi_{\vec{k}}(\vec{\rho}\,')\frac{\partial g_{\vec{k}}(\vec{r}_<,\vec{\rho}\,')}{\partial n'}-g_{\vec{k}}(\vec{r}_<,\vec{\rho}\,')\frac{\partial\psi_{\vec{k}}(\vec{\rho}\,')}{\partial n'}\right]dS'$$

$$-\int_{\sigma}\left[\psi_{\vec{k}}(\vec{\rho})\frac{\partial g_{\vec{k}}(\vec{r}_<,\vec{\rho})}{\partial n}-g_{\vec{k}}(\vec{r}_<,\vec{\rho})\frac{\partial\psi_{\vec{k}}(\vec{\rho})}{\partial n}\right]d\sigma=0\,. \tag{6.9}$$

In the second integral in (6.9), $\vec{\rho}$ denotes a typical point on the inscribed sphere σ and $\partial/\partial n$ denotes differentiation in the outward radial direction. According to (6.5) the first integral in (6.9) vanishes and we obtain the condition

$$\int_{\sigma}\left[\psi_{\vec{k}}(\vec{\rho})\frac{\partial g_{\vec{k}}(\vec{r}_<,\rho)}{\partial n}-g_{\vec{k}}(\vec{r}_<,\vec{\rho})\frac{\partial\psi_{\vec{k}}(\vec{\rho})}{\partial n}\right]d\sigma=0 \tag{6.10}$$

that must be valid for all points $\vec{r}_<$ inside the sphere σ. Equation (6.10), together with (6.6), is the boundary condition employed by Kohn and Rostoker,[6] Ziman,[7] and others in band-structure calculations in solids.

VII. SUMMARY AND DISCUSSION

It is clear that the method developed in this paper provides an alternative approach to that traditionally used for the analysis of problems of nonrelativistic one-particle scattering and for bound-state problems with finite range potentials. Our analysis brings out clearly the nature of the boundary conditions that the Schrödinger wave function must satisfy on the surface S that delimits the effective range of the potential. These boundary conditions [Eqs. (2.13) and (4.7)], which turn out to be analogous to the mathematical formulation of the Ewald-Oseen extinction theorem for electromagnetic fields, are nonlocal in general, i.e., they involve the values of the wave function and of its normal derivative at every point on the surface S. However, in the special case of a central potential, they reduce to a set of local boundary conditions, one for each of the partial waves [Eqs. (3.11′) and (5.3)]. We have also seen (Sec. VI) that our extinction theorem for bound states provides a new insight into the origin of the boundary condition that is used in calculations of energy of Bloch waves in solids.

Our analysis may also be extended to resonance states. We will discuss such an extension in another publication.[16] Here we only mention that our theory leads in a natural way to the boundary condition introduced by Kapur and Peierls,[8a] Siegert,[8b] Humblet,[8c] and Humblet and Rosenfeld[8d] in well-known papers dealing with nuclear resonance theory, and that it also provides a generalization of the boundary conditions to noncentral fields of force.

APPENDIX A: DERIVATION OF THE EXPANSION (3.9) FOR $\Sigma(\vec{r}_<)$

The surface integral $\Sigma(\vec{r}_<)$ is defined by Eq. (2.8a) as

$$\Sigma(\vec{r}_<)=\int_S\left[\psi(\vec{r}\,')\frac{\partial G(\vec{r}_<,\vec{r}\,')}{\partial n}-G(\vec{r}_<,\vec{r}\,')\frac{\partial\psi(\vec{r}\,')}{\partial n}\right]dS\,. \tag{A1}$$

On making use of the expansion (3.3) for $\psi(\vec{r}_<)$ and of the expansion (3.8) for $G(\vec{r}_<,\vec{r}\,')$, (A1) becomes

$$\Sigma(\vec{r}_<)=(ka)^2\sum_{l=0}^{\infty}\sum_{l'=0}^{\infty}(2l+1)[\chi_{l'}(a)h_l^{(+)\prime}(ka)-(1/k)\chi_{l'}{}'(a)h_l^{(+)}(ka)]\times I_{ll'}(kr_<,\theta)\,, \tag{A2}$$

where

$$I_{ll'}(kr_<,\theta)=\int_0^{2\pi}\int_0^{\pi}P_{l'}(\cos\theta')P_l(\cos\Theta)j_l(kr_<)\times\sin\theta'd\theta'd\phi'\,. \tag{A3}$$

Now Θ in (A3) represents the angle between the vectors $\vec{r}_<$ and $\vec{r}\,'$, i.e., between the directions (θ,φ) and (θ',ϕ') and there exists a well-known expansion for $P_l(\cos\Theta)$ in terms of spherical harmonics with arguments (θ,ϕ) and (θ',ϕ'), viz. (Ref. 13, p. 68),

$$P_l(\cos\Theta)=\frac{4\pi}{2l+1}\sum_{m=-l}^{l}Y_l^{m*}(\theta,\phi)Y_l^m(\theta',\phi')\,. \tag{A4}$$

On using the expansion (A4) in (A3) and carrying out the angular integrations, we obtain for $I_{ll'}(kr_<,\theta)$ the value

$$I_{ll'}(kr_<,\theta)=\left(\frac{4\pi}{2l+1}\right)\delta_{ll'}j_l(kr_<)P_l(\cos\theta)\,, \tag{A5}$$

where $\delta_{ll'}$ is the Kronecker symbol. On substituting from (A5) into (A2) we obtain the result that

$$\frac{1}{4\pi}\Sigma(\vec{r}_<)=(ka)^2\sum_{l=0}[\chi_l(a)h_l^{(+)\prime}(ka)-(1/k)\chi_l{}'(a)h_l^{(+)}(ka)]\times j_l(kr_<)P_l(\cos\theta)\,, \tag{A6}$$

and this is Eq. (3.9) of the text.

APPENDIX B: DERIVATION OF THE EXPANSION (3.14) FOR $\Sigma(\vec{r}_>)$

We recall that according to Eq. (2.8a) the surface integral $\Sigma(\vec{r}_>)$ is defined as

$$\Sigma(\vec{r}_>) = \int_S \left[\psi(\vec{r}')\frac{\partial G(\vec{r}_>,\vec{r}')}{\partial n} - G(\vec{r}_>,\vec{r}')\frac{\partial\psi(\vec{r}')}{\partial n}\right]dS\,. \tag{B1}$$

The values of $\psi(\vec{r}')$ and $\partial\psi(\vec{r}')/\partial n$ which enter on the right-hand side of (B1) are now the limiting values as $r_< \to a$ of the wave function (3.3) and of its normal derivative, with χ_l in (3.3) being given by Eqs. (3.12) and (3.13), respectively. For $G(\vec{r}_>,\vec{r}')$ we have an expansion of the form (3.8), viz.,

$$G(\vec{r}_>,\vec{r}') = k\sum_{l=0}^{\infty}(2l+1)j_l(kr')h_l^{(+)}(kr_>)P_l(\cos\Theta)\,, \tag{B2}$$

valid with $r_> > r'$ as appropriate here, Θ denoting the angle between the vectors $\vec{r}_>$ and $\vec{r}'$. On substituting into (B1) for $\psi(\vec{r}')$ from (3.3), with $r_< \to r' = a$ and for $G(\vec{r}_>,\vec{r}')$ from (B2) and proceeding in a strictly similar way as in the evaluation of $\Sigma(\vec{r}_<)$ carried out in Appendix A, we obtain the formula

$$\frac{1}{4\pi}\Sigma(\vec{r}_>) = (ka)^2\sum_{l=0}^{\infty}[\chi_l(a)j_l'(ka) - (1/k)\chi_l'(a)j_l(ka)] \times h_l^{(+)}(kr_>)P_l(\cos\theta)\,, \tag{B3}$$

where, of course, θ is the angle that the vector $\vec{r}_>$ makes with the direction of the momentum of the incident particle.

Next we substitute into (B3) the series expansions that are obtained on letting $r \to a$ in the expansion (3.12) for $\chi_l(r)$ and in the corresponding expansion for $\chi_l'(r)$ [obtained by term-by-term differentiation of (3.12)]. We then find after a straightforward calculation that

$$\frac{1}{4\pi}\Sigma(\vec{r}_>) = \sum_{l=0}^{\infty}\beta_l h_l^{(+)}(kr_>)P_l(\cos\theta)\,, \tag{B4}$$

where

$$\beta_l = i^l(2l+1) \times \frac{\sum_{n=0}^{\infty} C_{n,l}a^n[j_l'(ka) - (n+l)(1/ka)j_l(ka)]}{\sum_{n=0}^{\infty} C_{n,l}a^n[h_l^{(+)\prime}(ka) - (n+l)(1/ka)h_l^{(+)}(ka)]}\,. \tag{B5}$$

The formulas (B4) and (B5) are Eqs. (3.14) and (3.15) of the text.

---

*Research supported by the Army Research Office.

†At the Department of Physics, University of Toronto, Toronto, Canada, during the academic year 1974–1975.

[1]See, for example, P. Roman, *Advanced Quantum Theory* (Addison-Wesley, Reading, Mass., 1965), p. 150.

[2]An exception to this statement is the idealized case of scattering on a perfect conductor, where the boundary conditions have long been known.

[3](a) D. N. Pattanayak and E. Wolf, Opt. Commun. 3, 217 (1972). (b) E. Wolf, in *Coherence and Quantum Optics*, edited by L. Mandel and E. Wolf (Plenum, New York, 1973), p. 339. The following publications have much in common with these two papers but do not provide a precise boundary-value formulation of the scattering problem: J. J. Sein, Ph.D. thesis, New York University, 1969 (unpublished); Opt. Commun. 2, 170 (1970); J. D. Goede and P. Mazur, Physica 58, 568 (1972).

[4]P. P. Ewald, Ph.D. dissertation, Univ. of Munich, 1912 (unpublished); Ann. Phys. (Leipzig) 49, 1 (1916).

[5](a) M. Born and E. Wolf, *Principles of Optics* (Pergamon, New York, 1975), 5th ed., Sec. 2.4.2. (b) L. Rosenfeld, *Theory of Electrons* (North-Holland, Amsterdam, 1951), Chap. VI, Sec. 4.

[6]W. Kohn and N. Rostoker, Phys. Rev. 94, 1111 (1954), Eq. (A.1.3).

[7]J. M. Ziman, in *Theory of Condensed Matter*, lectures presented at an international course held at the International Centre for Theoretical Physics, Trieste, 1967 (International Atomic Energy Agency, Vienna, 1968), p. 3, Eq. (36).

[8](a) P. L. Kapur and R. Peierls, Proc. R. Soc. London A166, 277 (1938); (b) A. J. F. Siegert, Phys. Rev. 56, 750 (1939); (c) J. Humblet, Mém. Soc. Roy. Sci. de Liège 12, No. 4 (1952); (d) J. Humblet and L. Rosenfeld, Nucl. Phys. 26, 529 (1961).

[9]This fact is the essence of the well-known Sommerfeld's radiation condition of classical diffraction theory. [Cf. (a) B. B. Baker and E. T. Copson, *The Mathematical Theory of Huygens' Principle* (Clarendon Press, Oxford, 1950), 2nd ed., pp. 25 and 28, or (b) A. Sommerfeld, *Partial Differential Equations of Physics* (Academic, New York, 1949), p. 189].

[10]A quantum-mechanical extinction theorem was briefly mentioned by M. Lax [Phys. Rev. 85, 646 (1952), passage following his Eq. (6.10)] in the context of the theory of multiple scattering, but was not interpreted as a boundary condition.

[11]The fact that Eq. (2.13′) has to be satisfied for every interior point $\vec{r}_<$ makes it a rather unconventional type of a boundary condition. In molecular optics, where the classic electromagnetic form of the extinction theorem originated, it has the following basis: The response of the medium to the incident field may be described in terms of elementary induced dipole fields, whose sources are the molecules of the medium. The field in the medium that is generated by all these dipoles may be shown to be expressible as the sum of two terms. One of them precisely cancels the incident field—propagated with vacuum velocity c—at every point inside the medium. The other term gives rise to the actual macroscopic field in the medium,

propagated with a different velocity, c/n, where n is the refractive index of the medium. The cancellation of the incident field at every point in the medium by a part of the induced field is mathematically expressed by the extinction theorem and hence the theorem includes in its formulation all the interior points. In the introduction to Chapter VI of Ref. 5b, where the classic form of the extinction theorem is derived from microscopic considerations, Rosenfeld writes: "··· the physical relations disclosed by the analysis are such as to make the theory one of the most beautiful of natural philosophy." Unfortunately some of the beauty seems to be lacking in the rather formal Green's function derivations of the extinction theorems of macroscopic electromagnetic theory and of quantum mechanics. Of course, an aesthetic criterion may not be the most appropriate one to indicate the ultimate value of the extinction theorems in these fields.

[12]A. Messiah, *Quantum Mechanics* (North-Holland, Amsterdam, 1961), Vol. I, p. 358.

[13]J. D. Jackson, *Classical Electrodynamics* (Wiley, New York, 1967), pp. 541 and 68. The function $h_l^{(1)}$ of that reference corresponds to $ih_l^{(+)}$ of the present paper.

[14]K. Gottfried, *Quantum Mechanics* (Benjamin, New York, 1966), p. 98.

[15]J. M. Ziman, *The Principles of the Theory of Solids* (Cambridge University Press, England, 1972), p. 103.

[16]D. N. Pattanayak and E. Wolf, Phys. Rev. D (to be published).

PHYSICAL REVIEW D VOLUME 13, NUMBER 8 15 APRIL 1976

Resonance states as solutions of the Schrödinger equation with a nonlocal boundary condition*

D. N. Pattanayak
Department of Physics, University of Toronto, Toronto M5S 1A7 Canada

E. Wolf
Department of Physics and Astronomy, University of Rochester, Rochester, New York 14627

(Received 17 November 1975)

Resonance states of a system consisting of a particle interacting with a finite-range potential are introduced in a novel way, independent of the notion of scattering. It is shown that a resonance wave function satisfies a certain nonlocal boundary condition on the surface that delimits the range of the potential. With a central potential our general boundary condition reduces to a set of local ones that are identical with those obtained previously by Humblet and Rosenfeld by different methods.

There exist several definitions of resonance parameters and resonance states of a system consisting of a particle interacting with a potential. Some of them are related to singularities of the scattering cross section, of a Green's function, or of the scattering matrix.[1] When the potential is spherically symmetric and of finite range, some of the definitions have been shown to be equivalent to a boundary-value formulation that involves the boundary values of the radial wave function of a resonance state and of its normal derivative on a spherical surface delimiting the domain throughout which the potential is effective.[2-4] This boundary-value formulation is a refinement of a procedure introduced by Kapur and Peierls in a well-known paper[5] dealing with the dispersion formula for nuclear reactions. All these definitions have one feature in common, namely that they introduce the resonance state as a kind of extrapolation or as a limit of scattering states. Mittleman[6] has stressed the desirability of defining resonance states without using a "preconceived notion" of its structure.

In a recent paper[7] we have reformulated in a novel way the problem of determining the scattering states and the bound states of a nonrelativistic particle interacting with a finite-range potential. We showed that such states are represented by solutions of the Schrödinger equation that satisfy a certain boundary condition, which is nonlocal in general, on the surface that delimits the effective range of the potential. In the present paper we use the same approach to introduce resonance states in a new way. Our definition of a resonance state will be seen to be entirely independent of the notion of scattering and applies to a particle interacting with any finite-range potential, central as well as noncentral ones. With a central potential our definition will be shown to reduce to the boundary-value formulation of Siegert,[2] Humblet,[3] and Humblet and Rosenfeld.[4]

Consider a particle of mass m interacting with a finite-range potential $\mathcal{V}(\vec{r})$, which vanishes outside a volume V bounded by a closed surface S. The time-independent part $\psi_E(\vec{r})$ of the wave function $\psi_E(\vec{r},t)=\psi_E(\vec{r})\exp(-iEt/\hbar)$ then satisfies, throughout the whole space, the time-independent Schrödinger equation

$$(\nabla^2+k^2)\psi_E(\vec{r})=U(\vec{r})\psi_E(\vec{r}), \tag{1}$$

where

$$k^2=\frac{2m}{\hbar^2}E \tag{2}$$

and

$$U(\vec{r})=\frac{2m}{\hbar^2}\mathcal{V}(\vec{r}). \tag{3}$$

We have shown in Ref. 7 by an elementary argument involving nothing more than the use of Green's theorem that the following four equations (not all independent of each other) are a necessary consequence of the fact that $\psi_E(\vec{r})$ satisfies the Schrödinger equation:

$$\psi_E(\vec{r}_<)=-\frac{1}{4\pi}\int_V \psi_E(\vec{r}\,')U(\vec{r}\,')G_E(\vec{r}_<,\vec{r}\,')d^3r' - \frac{1}{4\pi}\Sigma(\vec{r}_<), \tag{4}$$

$$0=\frac{1}{4\pi}\Sigma(\vec{r}_<)-\frac{1}{4\pi}\Sigma^{(\infty)}(\vec{r}_<), \tag{5}$$

$$\psi_E(\vec{r}_>)=\frac{1}{4\pi}\Sigma(\vec{r}_>)-\frac{1}{4\pi}\Sigma^{(\infty)}(\vec{r}_>), \tag{6}$$

$$0=-\frac{1}{4\pi}\int_V \psi_E(\vec{r}\,')U(\vec{r}\,')G(\vec{r}_>,\vec{r}\,')d^3r' - \frac{1}{4\pi}\Sigma(\vec{r}_>). \tag{7}$$

In these formulas $\vec{r}_<$ represents any point inside V, $\vec{r}_>$ represents any point outside V, and

$$G_E(\vec{r}, \vec{r}') = \frac{e^{ik|\vec{r}-\vec{r}'|}}{|\vec{r}-\vec{r}'|}, \tag{8}$$

$$\Sigma(\vec{r}) = \int_S \left[\psi_E(\vec{r}') \frac{\partial G_E(\vec{r}, \vec{r}')}{\partial n} - G_E(\vec{r}, \vec{r}') \frac{\partial \psi(\vec{r}')}{\partial n}\right] dS, \tag{9}$$

$$\Sigma^{(\infty)}(\vec{r}) = \lim_{R\to\infty} \int_{S_R} \left[\psi_E(\vec{r}') \frac{\partial G_E(\vec{r}, \vec{r}')}{\partial n_R} - G_E(\vec{r}, \vec{r}') \frac{\partial \psi(\vec{r}')}{\partial n_R}\right] dS, \tag{10}$$

with S_R in (10) denoting a sphere of radius R centered at any convenient point in the region V. Further, $\partial/\partial n$ in Eq. (9) denotes differentiation along the outward normal to S and $\partial/\partial n_R$ in (10) denotes differentiation along the outward radial direction. In Ref. 7, where the energy E was assumed to be real, k in Eq. (8) represents the positive root of the right-hand side of Eq. (2) when $E > 0$ and the root with positive imaginary part when $E < 0$. However, the relations (4)–(7) retain their validity whether or not E is real. When E is complex we interpret k in Eq. (8) as that root of the expression on the right-hand side of (2) that has a non-negative real part.

By analogy with classical electromagnetic theory we may define *natural modes* of our system (particle + potential) as those solutions of the Schrödinger equation that are well behaved throughout the whole space and that are outgoing at infinity. The natural modes may be either radiative (in which case E is complex) or nonradiative (in which case E is real). We will identify the radiative natural modes with *resonances*[8] and the nonradiative ones with *bound states* of the system. Let us consider the implications of Eqs. (4)–(7) when $\psi_E(\vec{r})$ is a natural mode.

The requirement that natural modes are solutions of the Schrödinger equation that are outgoing at infinity implies that $\Sigma^{(\infty)}(\vec{r})$, defined by Eq. (10), identically vanishes (cf. Ref. 7), i.e., that for all points $\vec{r}$

$$\Sigma^{(\infty)}(\vec{r}) = 0. \tag{11}$$

The relation (5) then reduces to

$$\Sigma(\vec{r}_<) = 0, \tag{12a}$$

or more explicitly

$$\int_S \left[\psi_E(\vec{r}') \frac{\partial G_E(\vec{r}_<, \vec{r}')}{\partial n} - G_E(\vec{r}_<, \vec{r}') \frac{\partial \psi_E(\vec{r}')}{\partial n}\right] dS = 0. \tag{12b}$$

We interpret the formula (12b), which must be satisfied for every point $\vec{r}_<$ situated in the volume V, as a *boundary condition for natural modes*. More precisely *the natural modes of our system are those well-behaved solutions of the Schrödinger equation that satisfy the constraint expressed by Eq. (12b) for every point $\vec{r}_<$ in the volume V*. For similar reasons that we gave in Ref. 7 in connection with a boundary condition of this kind, we may refer to Eq. (12b) as the *extinction theorem for natural modes*.

We will see shortly that the Schrödinger equation (1) can be solved subject to our (nonlocal) boundary condition (12b) only for certain values of E, i.e., we are dealing with an *eigenvalue problem*. According to our classification, those eigenvalues E that are real are associated with bound states and those that are complex are associated with resonances. Since we have already discussed bound states from the present standpoint in Ref. 7 we may from now on confine our attention to resonance states, though our main results apply to both cases.

Our boundary condition (12b) may be used to solve the interior problem, i.e., to determine the (unnormalized) wave functions at all points $\vec{r}_<$ in V. The solution will provide, as a by-product, the complex energies of the resonance states.[9] The solution to the "exterior problem," i.e., determination of the wave function at points $\vec{r}_>$ outside V may then be written down at once in a closed form by the use of Eq. (6), as simplified by the requirement (11)

$$\psi_E(\vec{r}_>) = \frac{1}{4\pi} \Sigma(\vec{r}_>), \tag{13a}$$

or, more explicitly,

$$\psi_E(\vec{r}_>) = \frac{1}{4\pi} \int_S \left[\psi_E(\vec{r}') \frac{\partial G_E(\vec{r}_>, \vec{r}')}{\partial n} - G_E(\vec{r}_>, \vec{r}') \frac{\partial \psi_E(\vec{r}')}{\partial n}\right] dS. \tag{13b}$$

It is clear that the boundary values $\psi_E(\vec{r}')$ and $\partial\psi_E(\vec{r}')/\partial n$ on S that enter the integrand on the right-hand side of (13b) may be obtained from the solution of the interior problem by letting $\vec{r}_<$ move to the boundary surface S.

Up to now we have only made use of two of our four relations (4)–(7), namely of Eqs. (5) and (6) and we have seen that, together with the time-independent Schrödinger equation (1), they provide the complete specification of the (unnormalized) wave functions of resonance states and of bound states, as defined in the present paper. We will now make use of some of the implications of Eqs. (5) and (6) in the remaining two relations (4) and (7) and will see that they reduce to a more familiar pair of equations that our wave functions must satisfy. On substituting from (12a) into (4) we obtain the equation

$$\psi_E(\vec{r}_<) = -\frac{1}{4\pi}\int_V \psi_E(\vec{r}\,')U(\vec{r}\,')G_E(\vec{r}_<, \vec{r}\,')d^3r' , \quad (14)$$

and on substituting from (13a) into (7) we find that

$$\psi_E(\vec{r}_>) = -\frac{1}{4\pi}\int_V \psi_E(\vec{r}\,')U(\vec{r}\,')G_E(\vec{r}_>, \vec{r}\,')d^3r' . \quad (15)$$

These two formulas will be recognized as the "interior" and the "exterior" forms of the equation to which the integral equation of the conventional theory of nonrelativistic potential scattering reduces in the absence of the incoming wave.[10] This is as it should be, since the property of "no incoming wave" is considered to be characteristic of resonance states and of bound states.[11]

We will next show that when the particle interacts with a central potential $[U(\vec{r}) \equiv U(r), r = |\vec{r}|]$ which vanishes outside a sphere of radius $r = a$, our nonlocal boundary condition (12b) reduces to a set of local boundary conditions previously obtained for this case by other authors.

With a central potential the wave function $\psi(\vec{r}_<)$ may be expanded in a series of partial waves[12]

$$\psi_E(\vec{r}_<) = \sum_{l=0}^{\infty} \chi_l(\vec{r}_<; k)P_l(\cos\theta) , \quad (16)$$

where

$$\chi_l(r; k) = \frac{u_l(r; k)}{r} , \quad (17)$$

and $u_l(r; k)$ is a regular solution of the radial equation

$$\left[\frac{d^2}{dr^2} + \left(\epsilon - U(r) - \frac{l(l+1)}{r^2}\right)\right]u_l(r; k) = 0 \quad (18)$$

$$(l = 0, 1, 2, \ldots)$$

$(\epsilon = k^2 = 2mE/\hbar)$. The Green's function may be expanded as (cf. Ref. 13)

$$G_E(\vec{r}_<, \vec{r}\,') = k \sum_{l=0}^{\infty} (2l+1)j_l(kr_<)h_l^{(+)}(kr') \times P_l(\cos\Theta) , \quad (19)$$

valid with $r_< < r'$, where $h_l^{(+)}$ is the spherical Hankel function of the first kind and order l and Θ is the angle between the directions of the vectors $\vec{r}_<$ and $\vec{r}\,'$. If we now substitute from (16) and (19) into our boundary condition (12b) we find (cf. Appendix A of Ref. 7) that it implies that

$$\sum_{l=0}^{\infty} \alpha_l j_l(kr_<)P_l(\cos\theta) = 0 , \quad (20)$$

where

$$\alpha_l = (ka)^2[\phi_l(a; k)h_l^{(+)\prime}(ka) - (1/k)\phi_l'(a; k)h_l^{(+)}(ka)] , \quad (21)$$

the prime denoting differentiation with respect to the radial argument. Because of the orthogonality of the Legendre polynomials $P_l(\cos\theta)$, it follows from (20) that we must have $\alpha_l = 0$ for all l. If we express these conditions in terms of the functions $u_l(r; k)$, they are seen to imply that

$$[h_l^{(+)\prime}(ka) + (1/ka)h_l^{(+)}(ka)]u_l(a; k) - (1/k)h_l^{(+)}(ka)u_l'(a; k) = 0 \quad (l = 0, 1, 2, \ldots) . \quad (22)$$

Equation (22) is a set of (local) boundary conditions on the radial wave functions u_l of the resonance states. These boundary condtions can only be satisfied for certain values of k [that then give the complex resonance energies via the relation (2)], so that we are dealing with an eigenvalue problem. More specifically, the eigenvalue problem is that of determining, for each l, the solutions of Eq. (18) which are well behaved throughout the range $0 \le r \le a$ and which satisfy the boundary conditions (22) when $r = a$. These solutions must, of course, also satisfy the usual boundary conditions $u_l(0; k) = 0$ $(l = 0, 1, 2, \ldots)$ when $r = 0$.

The set of equations (22) is, except for notation, precisely the set of boundary conditions for the radial wave functions of resonance states obtained previously by Humblet [Ref. 3, Eq. (11.1)] and Humblet and Rosenfeld [Ref. 4, Eq. (1.33)] by entirely different methods. For the special case when $l = 0$ the boundary condition was derived eariler by Siegert[2] and it is also implicit in the well-known derivation of Kapur and Peierls,[5] of the dispersion formula for nuclear reactions, by the technique of perturbation of boundary conditions.

We may summarize our analysis by saying that we derived, on the basis of a new definition of a resonance state (of a system consisting of a particle interacting with a finite-range potential) a general boundary condition that the wave function of such a state must satisfy. Our definition of a resonance state is independent of the notion of scattering, and our derivation of the boundary condition is entirely based on two identities that follow from the Schrödinger equation. We also showed that our definition of a resonance implies that the wave function of such a state is a solution of the homogeneous integral equation to which the usual integral equation of potential scattering reduces in the absence of any incoming wave. With a central potential our nonlocal boundary condition reduces to a set of local ones for the radial wave functions of the resonance states, which are in agreement with those obtained previously for this special case by other techniques.

We wish to acknowledge helpful discussions with J. T. Foley, concerning the definition and classification of natural modes.

*Work supported by the Army Research Office and the National Research Council of Canada. This investigation was carried out during the academic year 1974-1975 when one of the authors (E.W.) was at the Department of Physics, University of Toronto, Toronto, Canada.

[1]For a review of some of the literature on this subject see R. M. More and E. Gerjuoy, Phys. Rev. A 7, 1288 (1973).

[2]A. J. Siegert, Phys. Rev. 56, 750 (1939).

[3]J. Humblet, Mém. Soc. R. Sc. Liège 12, No. 4 (1952).

[4]J. Humblet and L. Rosenfeld, Nucl. Phys. 26, 529 (1961).

[5]P. L. Kapur and R. Peierls, Proc. R. Soc. London A166, 277 (1938).

[6]M. H. Mittleman, Phys. Rev. 182, 128 (1969).

[7]D. N. Pattanayak and E. Wolf, Phys. Rev. D 13, 913 (1976).

[8]The resonances so defined may be subdivided into "proper resonances" and "virtual states," according to whether the real part of E is positive or negative.

[9]As is customary, we speak of a solution of the Schrödinger equation as representing a state of the system, even when the energy eigenvalue is complex. However, such language must be used with caution [cf. H. M. Nussenzveig, *Causality and Dispersion Relations* (Academic, New York, 1972), p. 160].

[10]A. Messiah, *Quantum Mechanics* (North-Holland, Amsterdam, 1962), Vol. 2, p. 811.

[11]Ref. 4, p. 541.

[12]A. Messiah, *Quantum Mechanics* (North-Holland, Amsterdam, 1961), Vol. 1, p. 385.

[13]J. D. Jackson, *Classical Electrodynamics* (Wiley, New York, 1962), p. 541 and p. 68. The function $h_l^{(1)}$ of that reference corresponds to $-ih_l^{(+)}$ of the present paper.

Scattering theory of distortion correction by phase conjugation

G. S. Agarwal*

Joint Institute for Laboratory Astrophysics, University of Colorado and National Bureau of Standards, Boulder, Colorado

Ari T. Friberg and E. Wolf †

Department of Physics and Astronomy, University of Rochester, Rochester, New York 14627

Received September 1, 1982

The correction of wave distortions by the technique of optical phase conjugation is examined first on the basis of a newly derived integral equation for scattering of monochromatic scalar waves in the presence of a phase-conjugate mirror. The solution is developed in an iterative series, and the first- and second-order terms are analyzed and illustrated diagrammatically. A generalization of the integral equation is then presented, which takes into account the electromagnetic nature of light. It is also shown that, if the conjugate wave is generated without losses or gains and with a complete reversal of polarization, a total elimination of distortions may be achieved by this technique under circumstances that frequently occur in practice.

1. INTRODUCTION

One of the main applications envisaged for the rapidly developing technique of optical phase conjugation is the correction of distortions that are imparted on a wave field by its interaction with a scattering medium.[1] A number of experiments have been carried out that demonstrate the possibility of such a "healing" process,[2] but no satisfactory theory of this phenomenon has been developed so far. This is undoubtedly so because the physical processes that are involved are very complex, a fact that does not appear to be generally appreciated; they include two, conceptually distinct, scattering processes, namely, scattering of the original wave and of the conjugated wave, as well as a nonlinear interaction. The simple, intuitive arguments that have been put forward so far to explain the success of this technique have greatly oversimplified the problem. Only some special cases have been treated adequately until now, namely, corrections of distortions produced by weak scatterers[3,4] and corrections under conditions when the phase-conjugate wave is generated without losses or gains.[5]

In Ref. 5 we put forward an integral equation for scattering of monochromatic waves in the presence of a phase-conjugate mirror (PCM), from which the degree of corrections attainable by this technique can, in principle, be deduced. The equation was obtained within the framework of scalar wave theory by a plausibility argument, on the assumptions that the incident field contains no evanescent components, that the scatterer is nonabsorbing, and that the effects of the evanescent waves are negligible at the PCM—a condition that is likely to hold in most cases of practical interest. In the present paper some implications of this integral equation are studied, and a generalization of the equation is obtained that takes into account the electromagnetic nature of light.

In Section 2 we recall the (scalar) integral equation for the scattered conjugated field. In Section 3 we present a formal solution of this equation in a form of an iterative series. In Sections 4 and 5 we analyze the first- and second-order contributions to the conjugated field and illustrate the results diagrammatically. In Section 6 we examine the correction of distortions for the case when the conjugated wave is generated without losses or gains, and we show that under these circumstances a complete correction is obtained to all orders of scattering. In Section 7 we present a generalization of our basic integral equation within the framework of Maxwell's electromagnetic theory, under the assumption that the conjugated wave is generated with a complete reversal of polarization.[6] We find that if, in addition, there are no losses or gains on phase conjugation, a complete correction of distortions produced by nonabsorbing scatterers will again be achieved. In the concluding section (Section 8), the main assumptions implicit in our theory are summarized, and some possible generalizations are mentioned.

2. BASIC INTEGRAL EQUATION FOR THE SCATTERED FIELD IN THE PRESENCE OF A PHASE-CONJUGATE MIRROR

We consider a monochromatic scalar wave field

$$U^{(i)}(\mathbf{r}, t) = U^{(i)}(\mathbf{r})e^{-i\omega t} \tag{2.1}$$

that is incident upon a scattering medium occupying a finite volume $\mathcal{V}$ in free space. In Eq. (2.1), $\mathbf{r}$ denotes the position vector of a typical field point, t denotes the time, and ω denotes a (real) frequency. The wave field is taken to be incident from a half-space $\mathcal{R}^-$ on one side of the scatterer [see Fig. 1(a)]. The scattering medium is assumed to be linear, time independent, and nonabsorbing. We denote by $n(\mathbf{r})$ the (real) distribution of the refractive index throughout the scattering volume.

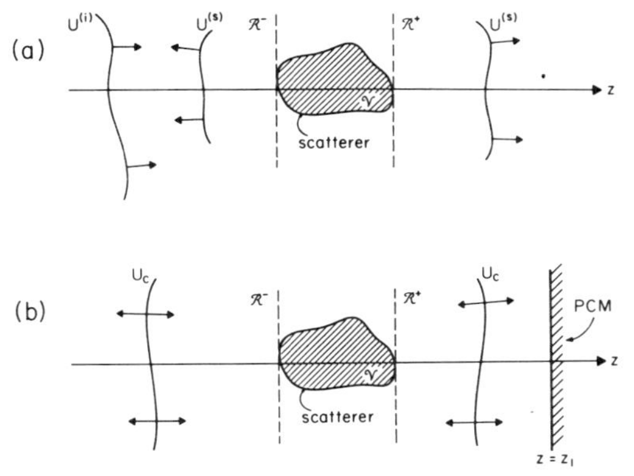

Fig. 1. Schematic diagram illustrating the notation. (a) In the absence of the PCM, the total field U is given by the sum of the incident field $U^{(i)}$ and the scattered field $U^{(s)}$. (b) In the presence of the PCM in the plane $z = z_1$, the total field throughout the domain $z < z_1$ is denoted by U_c.

When the field interacts with the scatterer, a new field [with the time-dependent factor $\exp(-i\omega t)$ omitted from now on]

$$U(\mathbf{r}) = U^{(i)}(\mathbf{r}) + U^{(s)}(\mathbf{r}) \tag{2.2}$$

is generated, where $U^{(s)}(\mathbf{r})$ represents the scattered field. It is well known that $U(\mathbf{r})$ satisfies the integral equation[7]

$$U(\mathbf{r}) = U^{(i)}(\mathbf{r}) - \frac{1}{4\pi}\int_{\mathcal{V}} G(\mathbf{r}, \mathbf{r}')F(\mathbf{r}')U(\mathbf{r}')\mathrm{d}^3r', \tag{2.3}$$

where

$$G(\mathbf{r}, \mathbf{r}') = \frac{\exp(ik|\mathbf{r} - \mathbf{r}'|)}{|\mathbf{r} - \mathbf{r}'|} \tag{2.4}$$

($k = \omega/c$, c being the speed of light in vacuum) is the outgoing free-space Green's function and

$$F(\mathbf{r}) = -k^2[n^2(\mathbf{r}) - 1] \tag{2.5}$$

represents the scattering potential.

Suppose now that a PCM,[8] assumed for simplicity to be infinite, is placed in a plane $z = z_1$ in the half-space $\mathcal{R}^+$ on the side opposite that from which the field $U^{(i)}(\mathbf{r})$ is incident [see Fig. 1(b)]. Were the scatterer absent, the PCM would replace the field distribution $U^{(i)}(\mathbf{r})|_{z=z_1}$ in that plane with the distribution $\mu U^{(i)*}(\mathbf{r})|_{z=z_1}$ that would give rise to an additional contribution to the total field in the half-space $z < z_1$. Here the asterisk denotes the complex conjugate and μ is a (generally complex) constant that accounts for the losses ($|\mu| < 1$) or gains ($|\mu| > 1$) that arise in the process of phase conjugation. The assumption of constant μ represents an idealization; in practice the PCM may respond differently to plane waves that propagate in different directions.

When both the scatterer and the PCM are present, a new field distribution is generated throughout the domain $z < z_1$ that we will denote by $U_c(\mathbf{r})$. It represents the total field after phase conjugation; for the sake of brevity, we will refer to it as the conjugate field.[9] If the incident field $U^{(i)}(\mathbf{r})$ contains no evanescent components, and if, in addition, the effects of the evanescent waves that might be created in the scattering process are negligible[10] at the PCM, then the field $U_c(\mathbf{r})$ satisfies the integral equation[5,11]

$$U_c(\mathbf{r}) = U_c{}^{(0)}(\mathbf{r}) - \frac{1}{4\pi}\int_{\mathcal{V}} \hat{G}_c(\mathbf{r}, \mathbf{r}')F(\mathbf{r}')U_c(\mathbf{r}')\mathrm{d}^3r', \tag{2.6}$$

where

$$U_c{}^{(0)}(\mathbf{r}) = U^{(i)}(\mathbf{r}) + \mu U^{(i)*}(\mathbf{r}) \tag{2.7}$$

and $\hat{G}_c(\mathbf{r}, \mathbf{r}')$ is an operator Green's function[12] that takes into account the presence of the PCM.

The operator Green's function $\hat{G}_c(\mathbf{r}, \mathbf{r}')$ was shown[5,11] to be expressible in terms of the following four quantities:

(1) The complex "reflection coefficient" μ of the PCM.

(2) The complex-conjugation operator $\hat{C}$, defined by the property that

$$\hat{C}f(\mathbf{r}) \equiv f^*(\mathbf{r}), \tag{2.8}$$

where $f(\mathbf{r})$ is an arbitrary function.

(3) The free-space Green's function $G(\mathbf{r}, \mathbf{r}')$.

(4) The function $G_>{}^{(H)}(\mathbf{r}, \mathbf{r}')$ associated with the "homogeneous" part of the free-space Green's function $G(\mathbf{r}, \mathbf{r}')$.

More specifically, the function $G_>{}^{(H)}(\mathbf{r}, \mathbf{r}')$ is related to the free-space Green's function $G(\mathbf{r}, \mathbf{r}')$ by the formula

$$G(\mathbf{r}, \mathbf{r}') = \Theta(z - z')G_>{}^{(H)}(\mathbf{r}, \mathbf{r}') + \Theta(z' - z)G_<{}^{(H)}(\mathbf{r}, \mathbf{r}') + G^{(I)}(\mathbf{r}, \mathbf{r}'), \tag{2.9}$$

where $G_>{}^{(H)}(\mathbf{r}, \mathbf{r}')$ and $G_<{}^{(H)}(\mathbf{r}, \mathbf{r}')$ represent the contributions to $G(\mathbf{r}, \mathbf{r}')$ from all homogeneous plane waves, $G^{(I)}(\mathbf{r}, \mathbf{r}')$ is the contribution from all inhomogeneous (evanescent) plane waves, and $\Theta(\zeta)$ is the unit step function, viz.,

$$\begin{aligned}\Theta(\zeta) &= 1 \quad \text{if } \zeta > 0,\\ &= 0 \quad \text{if } \zeta < 0.\end{aligned} \tag{2.10}$$

By using the Weyl representation of a spherical wave,[13] the functions $G_>{}^{(H)}(\mathbf{r}, \mathbf{r}')$, $G_<{}^{(H)}(\mathbf{r}, \mathbf{r}')$, and $G^{(I)}(\mathbf{r}, \mathbf{r}')$ may be shown to be expressible explicitly as

$$G_>{}^{(H)}(\mathbf{r}, \mathbf{r}') = \frac{i}{2\pi}\iint_{|\boldsymbol{\kappa}|\leqslant k} \frac{1}{w}\exp\{i[\boldsymbol{\kappa}\cdot(\boldsymbol{\rho} - \boldsymbol{\rho}') + w(z - z')]\}\mathrm{d}^2\kappa, \tag{2.11a}$$

$$G_<{}^{(H)}(\mathbf{r}, \mathbf{r}') = \frac{i}{2\pi}\iint_{|\boldsymbol{\kappa}|\leqslant k} \frac{1}{w}\exp\{i[\boldsymbol{\kappa}\cdot(\boldsymbol{\rho} - \boldsymbol{\rho}') - w(z - z')]\}\mathrm{d}^2\kappa, \tag{2.11b}$$

and

$$G^{(I)}(\mathbf{r}, \mathbf{r}') = \frac{i}{2\pi}\iint_{|\boldsymbol{\kappa}|>k} \frac{1}{w}\exp\{i[\boldsymbol{\kappa}\cdot(\boldsymbol{\rho} - \boldsymbol{\rho}') + w|z - z'|]\}\mathrm{d}^2\kappa, \tag{2.12}$$

respectively. In the integrals on the right-hand sides of Eqs. (2.11) and (2.12), $\mathbf{r} = (\boldsymbol{\rho}, z)$ and $\mathbf{r}' = (\boldsymbol{\rho}', z')$ (with $\boldsymbol{\rho}$ and $\boldsymbol{\rho}'$ being two-dimensional vectors orthogonal to the z axis), and

$$\begin{aligned} w &= +(k^2 - \kappa^2)^{1/2} \quad \text{when } |\boldsymbol{\kappa}| \leqslant k, &(2.13a)\\ &= +i\,(\kappa^2 - k^2)^{1/2} \quad \text{when } |\boldsymbol{\kappa}| > k. &(2.13b)\end{aligned}$$

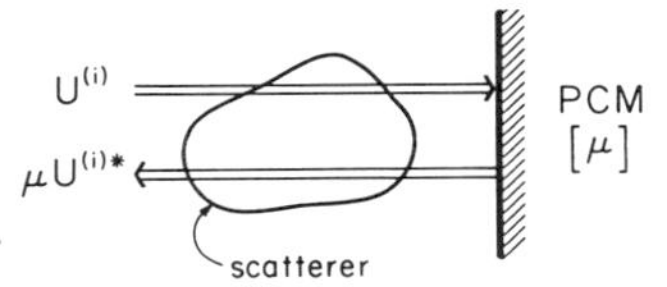

Fig. 2. Diagrammatic illustration of the zeroth-order term $U_c^{(0)}$ in iterative expansion (3.1) for the conjugate field U_c. $U^{(i)}$ denotes the incident wave, and $\mu U^{(i)*}$ denotes the wave that is generated by the PCM in the absence of the scatterer.

In terms of the four quantities listed as (1)–(4) above, the operator Green's function $\hat{G}_c(\mathbf{r}, \mathbf{r}')$ that appears in the basic integral equation (2.6) may then be expressed in the form

$$\hat{G}_c(\mathbf{r}, \mathbf{r}') = G(\mathbf{r}, \mathbf{r}') + \mu G_>^{(H)*}(\mathbf{r}, \mathbf{r}')\hat{C}. \quad (2.14)$$

Here the first term on the right-hand side represents the propagator for scattering in the absence of the PCM; the presence of the PCM in taken into account by the second term.

Making use of Eq. (2.9) and the relation

$$G_>^{(H)*}(\mathbf{r}, \mathbf{r}') = -G_<^{(H)}(\mathbf{r}, \mathbf{r}') \quad (2.15)$$

that readily follows from Eqs. (2.11), we may write the operator Green's function $\hat{G}_c(\mathbf{r}, \mathbf{r}')$ given by Eq. (2.14) alternatively as

$$\hat{G}_c(\mathbf{r}, \mathbf{r}') = \Theta(z - z')(1 + \mu\hat{C})G_>^{(H)}(\mathbf{r}, \mathbf{r}') + \Theta(z' - z)G_<^{(H)}(\mathbf{r}, \mathbf{r}')(1 - \mu\hat{C}) + G^{(I)}(\mathbf{r}, \mathbf{r}'). \quad (2.16)$$

This form of the operator Green's function $\hat{G}_c(\mathbf{r}, \mathbf{r}')$ will prove convenient later (Sections 6 and 7).

We note that the removal of the PCM is equivalent to letting $\mu \to 0$. We see from Eq. (2.14) [or Eq. (2.16)] that in this limit

$$\hat{G}_c(\mathbf{r}, \mathbf{r}') = G(\mathbf{r}, \mathbf{r}') \qquad (\mu = 0), \quad (2.17)$$

and, according to Eq. (2.7), one then also has

$$U_c^{(0)}(\mathbf{r}) = U^{(i)}(\mathbf{r}) \qquad (\mu = 0). \quad (2.18)$$

Since, in the absence of the PCM, $U_c(\mathbf{r}) \to U(\mathbf{r})$, we see at once, on using Eqs. (2.17) and (2.18), that our integral equation (2.6) then correctly reduces to the usual integral equation (2.3).

3. ITERATIVE EXPANSION OF THE SOLUTION TO THE BASIC INTEGRAL EQUATION

The basic integral equation for the scattered field in the presence of a PCM [Eq. (2.6)] may be formally solved in the form of an iterative expansion[14]

$$U_c(\mathbf{r}) = \sum_{n=0}^{\infty} U_c^{(n)}(\mathbf{r}), \quad (3.1)$$

where $U_c^{(0)}(\mathbf{r})$ is given by Eq. (2.7) and

$$U_c^{(n)}(\mathbf{r}) = -\frac{1}{4\pi}\int_{\mathcal{V}} \hat{G}_c(\mathbf{r}, \mathbf{r}')F(\mathbf{r}')U_c^{(n-1)}(\mathbf{r}')\mathrm{d}^3r' \qquad (n = 1, 2, 3, \ldots). \quad (3.2)$$

We note that, in the absence of the scatterer, $F(\mathbf{r}) \equiv 0$ and all the terms $U_c^{(n)}(\mathbf{r})$ in series (3.1) then vanish, except the first one (i.e., the term $n = 0$). Hence in this case we have

$$U_c(\mathbf{r}) \to U_c^{(0)}(\mathbf{r}) = U^{(i)}(\mathbf{r}) + \mu U^{(i)*}(\mathbf{r}) = (1 + \mu\hat{C})U^{(i)}(\mathbf{r}), \quad (3.3)$$

as expected. We emphasize that the result expressed by Eq. (3.3) holds only when there are no evanescent waves associated with the incident field[15] $U^{(i)}(\mathbf{r})$. The physical meaning of the right-hand side of Eq. (3.3) is illustrated schematically in Fig. 2. We give later (in Sections 4 and 5) similar diagrammatic representations for the next two terms ($n = 1$ and $n = 2$) of iterative expansion (3.1).

In applying the technique of optical phase conjugation to the correction of wave-front distortions suffered by a light wave on interaction with the scattering medium, one is usually interested in the field $U_c(\mathbf{r})$ in the half-space $\mathcal{R}^-$ only, i.e., on that side of the scatterer from which the wave $U^{(i)}(\mathbf{r})$ is incident. It follows from Eqs. (2.14) and (2.9) that, if the evanescent waves are neglected in the half-space $\mathcal{R}^-$, the operator Green's function $\hat{G}_c(\mathbf{r}_<, \mathbf{r}')$, with $\mathbf{r}_< \in \mathcal{R}^-$ and $\mathbf{r}' \in \mathcal{V}$, then reduces to [cf. also Eq. (2.16)]

$$\hat{G}_c^{(H)}(\mathbf{r}_<, \mathbf{r}') = G_<^{(H)}(\mathbf{r}_<, \mathbf{r}') + \mu G_>^{(H)*}(\mathbf{r}_<, \mathbf{r}')\hat{C}. \quad (3.4)$$

Hence, in this approximation, each term $U_c^{(n)}(\mathbf{r}_<)$ in expansion (3.1) with $n > 1$ may be represented, throughout the half-space $\mathcal{R}^-$, in the following form:

$$U_c^{(n)}(\mathbf{r}_<) = -\frac{1}{4\pi}\int_{\mathcal{V}} \hat{G}_c^{(H)}(\mathbf{r}_<, \mathbf{r}')F(\mathbf{r}')U_c^{(n-1)}(\mathbf{r}')\mathrm{d}^3r', \quad (3.5)$$

where $\hat{G}_c^{(H)}(\mathbf{r}_<, \mathbf{r}')$ is given by Eq. (3.4). The field $U_c^{(n-1)}(\mathbf{r}')$ under the integral sign on the right-hand side of Eq. (3.5) must be calculated by the use of recursion formula (3.2), which is valid throughout the domain $z < z_1$. It should be noted that, although here the effects of the evanescent waves are neglected at the PCM and in the half-space $\mathcal{R}^-$, they are included inside the scattering medium.

4. DISTORTION CORRECTION IN THE FIRST BORN APPROXIMATION

We now analyze in some detail a few of the lowest-order terms of iterative expansion (3.1) in the special case when the observation point $\mathbf{r}_<$ is located in the half-space $\mathcal{R}^-$ and the evanescent waves are omitted in that half-space. If the scatterer is sufficiently weak and also thin enough, the total field $U_c(\mathbf{r}_<)$ in the half-space $\mathcal{R}^-$ may then be adequately approximated by the first two terms in expansion (3.1), i.e., by

$$U_c(\mathbf{r}_<) \cong U_c^{(\mathrm{I})}(\mathbf{r}_<) = U_c^{(0)}(\mathbf{r}_<) + U_c^{(1)}(\mathbf{r}_<), \quad (4.1)$$

where $U_c^{(0)}(\mathbf{r}_<)$ is given by Eq. (3.3) (with $\mathbf{r} = \mathbf{r}_<$) and $U_c^{(1)}(\mathbf{r}_<)$ is obtained from Eq. (3.5) with $n = 1$, viz.,

$$U_c^{(1)}(\mathbf{r}_<) = -\frac{1}{4\pi}\int_{\mathcal{V}} \hat{G}_c^{(H)}(\mathbf{r}_<, \mathbf{r}')F(\mathbf{r}')U_c^{(0)}(\mathbf{r}')\mathrm{d}^3r'. \quad (4.2)$$

This approximation represents the first Born approximation (superscript I) to the field $U_c(\mathbf{r}_<)$ in the presence of the PCM.

If we substitute from Eqs. (3.3) and (3.4) into Eq. (4.2), we obtain at once the formula

$$U_c^{(1)}(\mathbf{r}_<) = -\frac{1}{4\pi}\int_{\mathcal{V}} [G_<^{(H)}(\mathbf{r}_<, \mathbf{r}') + \mu G_>^{(H)*}(\mathbf{r}_<, \mathbf{r}')\hat{C}]\, F(\mathbf{r}')(1 + \mu\hat{C})U^{(i)}(\mathbf{r}')\mathrm{d}^3r'. \quad (4.3)$$

On expanding the product under the integral sign, we obtain for $U_c^{(1)}(\mathbf{r}_<)$ the expression

$$\begin{aligned} U_c^{(1)}(\mathbf{r}_<) = &-\frac{1}{4\pi}\int_{\mathcal{V}} G_<^{(H)}(\mathbf{r}_<, \mathbf{r}')F(\mathbf{r}')U^{(i)}(\mathbf{r}')\mathrm{d}^3r' && \text{(a)} \\ &-\frac{\mu}{4\pi}\int_{\mathcal{V}} G_<^{(H)}(\mathbf{r}_<, \mathbf{r}')F(\mathbf{r}')U^{(i)*}(\mathbf{r}')\mathrm{d}^2r' && \text{(b)} \\ &-\frac{\mu}{4\pi}\int_{\mathcal{V}} G_>^{(H)*}(\mathbf{r}_<, \mathbf{r}')F(\mathbf{r}')U^{(i)*}(\mathbf{r}')\mathrm{d}^3r' && \text{(c)} \\ &-\frac{|\mu|^2}{4\pi}\int_{\mathcal{V}} G_>^{(H)*}(\mathbf{r}_<, \mathbf{r}')F(\mathbf{r}')U^{(i)}(\mathbf{r}')\mathrm{d}^3r'. && \text{(d)} \end{aligned} \quad (4.4)$$

Here we made use of the fact that the scatterer was assumed to be nonabsorbing and that, consequently, the scattering potential $F(\mathbf{r})$ is a real function of position throughout the volume $\mathcal{V}$.

We note that each term on the right-hand side of Eq. (4.4) involves either the function $G_>^{(H)}(\mathbf{r}_<, \mathbf{r}')$ or the function $G_<^{(H)}(\mathbf{r}_<, \mathbf{r}')$, both of which are associated with the "homogeneous" part of the free-space Green's function $G(\mathbf{r}, \mathbf{r}')$ [cf. Eqs. (2.9) and (2.11)]. This situation is a direct consequence of the fact that we neglected the evanescent waves both at the PCM and in the half-space $\mathcal{R}^-$. With this neglection it is clear that, in the first Born approximation, the conjugate field $U_c(\mathbf{r}_<)$ in the half-space $\mathcal{R}^-$ contains no effects of the evanescent waves that may have been created on scattering inside the volume $\mathcal{V}$.

In the approximation that the evanescent waves are neglected in the half-space $\mathcal{R}^-$, the quantity $G_<^{(H)}(\mathbf{r}_<, \mathbf{r}')$ represents the propagator from point $\mathbf{r}'$ inside the scattering volume $\mathcal{V}$ to point $\mathbf{r}_<$ in the half-space $\mathcal{R}^-$. Similarly, with the effects of the evanescent waves neglected at the PCM, the quantity $\mu G_>^{(H)*}(\mathbf{r}_<, \mathbf{r}')\hat{C}$ in Eq. (4.3), which gives rise to the last two terms in Eq. (4.4), may be interpreted as the propagator from $\mathbf{r}'$ to $\mathbf{r}_<$ via the PCM. Hence each of the four terms (a)–(d) in Eq. (4.4) specifies a well-defined elementary event associated with scattering in the presence of a PCM, namely, first-order scattering, which may or may not be followed or preceded by phase conjugation. These elementary events are illustrated diagrammatically in Fig. 3. The scattering taking place at a typical point within the volume $\mathcal{V}$ is denoted by a dot in each of the diagrams.

In view of relation (2.15), the terms (b) and (c) in Eq. (4.4) are seen to cancel each other. The remaining two terms, viz., (a) and (d), represent the effects of backscattering[4] in the first Born approximation; the term (a) arises from the backscattering of the incident wave $U^{(i)}$, whereas the term (d) arises from the backscattering of the wave $\mu U^{(i)*}$. By making use of relation (2.15), Eq. (4.4) may be expressed in the form

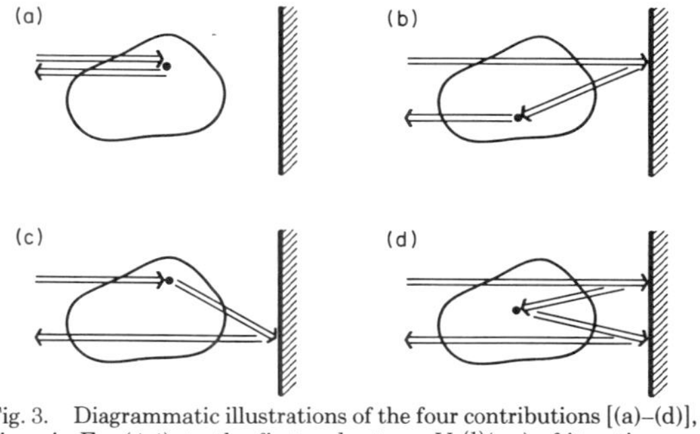

Fig. 3. Diagrammatic illustrations of the four contributions [(a)–(d)], given in Eq. (4.4), to the first-order term $U_c^{(1)}(\mathbf{r}_<)$ of iterative expansion (3.1).

$$U_c^{(1)}(\mathbf{r}_<) = (1 - |\mu|^2)U_{\mathrm{bs}}^{(1)}(\mathbf{r}_<), \quad (4.5)$$

where

$$U_{\mathrm{bs}}^{(1)}(\mathbf{r}_<) = -\frac{1}{4\pi}\int_{\mathcal{V}} G_<^{(H)}(\mathbf{r}_<, \mathbf{r}')F(\mathbf{r}')U^{(i)}(\mathbf{r}')\mathrm{d}^3r' \quad (4.6)$$

is the (first-order) backscattered contribution to the total field in the half-space $\mathcal{R}^-$ [represented by diagram (a) in Fig. 3]. We note that the complex "reflection coefficient" μ of the PCM enters expression (4.5) for the field $U_c^{(1)}(\mathbf{r}_<)$ only through its modulus.

According to Eqs. (4.1) and (4.5), the effect of backscattering prevents, in general, a complete cancellation of distortions by phase conjugation within the accuracy of the first Born approximation. However, distortion correction may be achieved in the first Born approximation if either

(1) $U_{\mathrm{bs}}^{(1)}(\mathbf{r}_<) = 0$, i.e., there is no backscattering into the half-space $\mathcal{R}^-$, or

(2) $|\mu| = 1$, i.e., there are no losses or gains in the process of phase conjugation.

Under the usual circumstances $|\mu| \ll 1$, and Eqs. (4.1) and (4.5) then yield

$$U_c^{(\mathrm{I})}(\mathbf{r}_<) \cong U_c^{(0)}(\mathbf{r}_<) + U_{\mathrm{bs}}^{(1)}(\mathbf{r}_<) \qquad (|\mu| \ll 1). \quad (4.7)$$

Hence, if the backscattered field $U_{\mathrm{bs}}^{(1)}(\mathbf{r}_<)$ is physically removed by some experimental means, the wave-front distortions resulting from the interaction of the incident field with a nonabsorbing scatterer can be corrected in the first Born approximation by the use of a PCM for which $|\mu| \ll 1$.

5. DISTORTION CORRECTION IN THE SECOND BORN APPROXIMATION

We now go one step beyond the first Born approximation by including the next higher term in iterative expansion (3.1), i.e., we approximate the field $U_c(\mathbf{r}_<)$ in the half-space $\mathcal{R}^-$ by

$$U_c(\mathbf{r}_<) \cong U_c^{(\mathrm{II})}(\mathbf{r}_<) = U_c^{(0)}(\mathbf{r}_<) + U_c^{(1)}(\mathbf{r}_<) + U_c^{(2)}(\mathbf{r}_<). \quad (5.1)$$

The first two terms on the right-hand side of approximation (5.1) were discussed in Sections 3 and 4 above. The term $U_c^{(2)}(\mathbf{r}_<)$ is obtained from Eq. (3.5) with $n = 2$. Approximation (5.1) represents the second Born approximation (superscript II) to the conjugate field $U_c(\mathbf{r}_<)$ in the half-space $\mathcal{R}^-$.

Setting $n = 2$ in Eq. (3.5) and making use of Eq. (3.2) to find an expression for the field $U_c{}^{(1)}(\mathbf{r}')$ under the integral sign, we obtain the formula

$$U_c{}^{(2)}(\mathbf{r}_<) = \left(-\frac{1}{4\pi}\right)^2 \int_{\mathcal{V}} \int_{\mathcal{V}} \hat{G}_c{}^{(H)}(\mathbf{r}_<, \mathbf{r}'') \times F(\mathbf{r}'')\hat{G}_c(\mathbf{r}'', \mathbf{r}')F(\mathbf{r}')U_c{}^{(0)}(\mathbf{r}')\mathrm{d}^3r'\mathrm{d}^3r''. \quad (5.2)$$

We note that, unlike in the expression for $U_c{}^{(1)}(\mathbf{r}_<)$, the "total" free-space Green's function G [contained in the operator Green's function $\hat{G}_c(\mathbf{r}'', \mathbf{r}')$ in the integrand—cf. Eq. (2.14)] appears in expression (5.2) for $U_c{}^{(2)}(\mathbf{r}_<)$. This fact implies that, within the scattering volume $\mathcal{V}$, contributions carried by both the homogeneous waves and the evanescent waves are now included [cf. Eq. (2.9)]. However, outside the scatterer the effects of the evanescent waves have again been neglected.

If we now substitute for the operator Green's function $\hat{G}_c(\mathbf{r}'', \mathbf{r}')$ from Eq. (2.14) and for the zeroth-order field $U_c{}^{(0)}(\mathbf{r}')$ from Eq. (3.3), Eq. (5.2) becomes

$$\begin{aligned} U_c{}^{(2)}(\mathbf{r}_<) = \left(\frac{1}{4\pi}\right)^2 \int_{\mathcal{V}} \int_{\mathcal{V}} & [G_<{}^{(H)}(\mathbf{r}_<, \mathbf{r}'') + \mu G_>{}^{(H)*}(\mathbf{r}_<, \mathbf{r}'')\hat{C}]F(\mathbf{r}'') \\ & \times [G(\mathbf{r}'', \mathbf{r}') + \mu G_>{}^{(H)*}(\mathbf{r}'', \mathbf{r}')\hat{C}]\, F(\mathbf{r}') \\ & \times (1 + \mu\hat{C})U^{(i)}(\mathbf{r}')\mathrm{d}^3r'\mathrm{d}^3r''. \end{aligned} \quad (5.3)$$

On expanding the product under the integral signs on the right-hand side of Eq. (5.3), we obtain the following explicit expression for the second-order contribution $U_c{}^{(2)}(\mathbf{r}_<)$ to the total field in the half-space $\mathcal{R}^-$:

$$\begin{aligned} U_c{}^{(2)}(\mathbf{r}_<) = &+ \left(\frac{1}{4\pi}\right)^2 \int_{\mathcal{V}} \int_{\mathcal{V}} G_<{}^{(H)}(\mathbf{r}_<, \mathbf{r}'')F(\mathbf{r}'')G(\mathbf{r}'', \mathbf{r}')F(\mathbf{r}')U^{(i)}(\mathbf{r}')\mathrm{d}^3r'\mathrm{d}^3r'' && \text{(a)} \\ &+ \mu\left(\frac{1}{4\pi}\right)^2 \int_{\mathcal{V}} \int_{\mathcal{V}} G_<{}^{(H)}(\mathbf{r}_<, \mathbf{r}'')F(\mathbf{r}'')G(\mathbf{r}'', \mathbf{r}')F(\mathbf{r}')U^{(i)*}(\mathbf{r}')\mathrm{d}^3r'\mathrm{d}^3r'' && \text{(b)} \\ &+ \mu\left(\frac{1}{4\pi}\right)^2 \int_{\mathcal{V}} \int_{\mathcal{V}} G_<{}^{(H)}(\mathbf{r}_<, \mathbf{r}'')F(\mathbf{r}'')G_>{}^{(H)*}(\mathbf{r}'', \mathbf{r}')F(\mathbf{r}')U^{(i)*}(\mathbf{r}')\mathrm{d}^3r'd^3r'' && \text{(c)} \\ &+ \mu\left(\frac{1}{4\pi}\right)^2 \int_{\mathcal{V}} \int_{\mathcal{V}} G_>{}^{(H)*}(\mathbf{r}_<, \mathbf{r}'')F(\mathbf{r}'')G^*(\mathbf{r}'', r')F(\mathbf{r}')U^{(i)*}(\mathbf{r}')d^3r'd^3r'' && \text{(d)} \\ &+ |\mu|^2\left(\frac{1}{4\pi}\right)^2 \int_{\mathcal{V}} \int_{\mathcal{V}} G_<{}^{(H)}(\mathbf{r}_<, \mathbf{r}'')F(\mathbf{r}'')G_>{}^{(H)*}(\mathbf{r}'', \mathbf{r}')F(\mathbf{r}')U^{(i)}(\mathbf{r}')\mathrm{d}^3r'\mathrm{d}^3r'' && \text{(e)} \\ &+ |\mu|^2\left(\frac{1}{4\pi}\right)^2 \int_{\mathcal{V}} \int_{\mathcal{V}} G_>{}^{(H)*}(\mathbf{r}_<, \mathbf{r}'')F(\mathbf{r}'')G^*(\mathbf{r}'', \mathbf{r}')F(\mathbf{r}')U^{(i)}(\mathbf{r}')\mathrm{d}^3r'\mathrm{d}^3r'' && \text{(f)} \\ &+ |\mu|^2\left(\frac{1}{4\pi}\right)^2 \int_{\mathcal{V}} \int_{\mathcal{V}} G_>{}^{(H)*}(\mathbf{r}_<, \mathbf{r}'')F(\mathbf{r}'')G_>{}^{(H)}(\mathbf{r}'', \mathbf{r}')F(\mathbf{r}')U^{(i)}(\mathbf{r}')\mathrm{d}^3r'\mathrm{d}^3r'' && \text{(g)} \\ &+ \mu|\mu|^2\left(\frac{1}{4\pi}\right)^2 \int_{\mathcal{V}} \int_{\mathcal{V}} G_>{}^{(H)*}(\mathbf{r}_<, \mathbf{r}'')F(\mathbf{r}'')G_>{}^{(H)}(\mathbf{r}'', \mathbf{r}')F(\mathbf{r}')U^{(i)*}(\mathbf{r}')\mathrm{d}^3r'\mathrm{d}^3r''. && \text{(h)} \end{aligned} \quad (5.4)$$

Here we again made use of the fact that the scattering potential $F(\mathbf{r})$ is assumed to be real.

The eight terms (a)–(h) of Eq. (5.4) are illustrated in Fig. 4 in a diagrammatic form. Each diagram represents a single elementary event consisting of two scattering processes (denoted by dots, as before) and m phase conjugations at the PCM, where $m = 0, 1, 2,$ or 3. We see that each term in Eq. (5.4) contains the function $G_>{}^{(H)}$ or the function $G_<{}^{(H)}$ (depicted, as before, by hollow arrows in the diagrams). Some of the terms contain, in addition, the free-space Green's function $G(\mathbf{r}'', \mathbf{r}')$ (depicted by a solid arrow), which, unlike $G_>{}^{(H)}$ and $G_<{}^{(H)}$, includes contributions from the evanescent waves; the appearance of $G(\mathbf{r}'', \mathbf{r}')$ in these terms allows for the possibility of a conversion of an evanescent wave into a homogeneous one, which subsequently contributes to the field $U_c{}^{(2)}(\mathbf{r}_<)$ in the half-space $\mathcal{R}^-$ [diagrams (a), (b), (d), and (f) of Fig. 4].

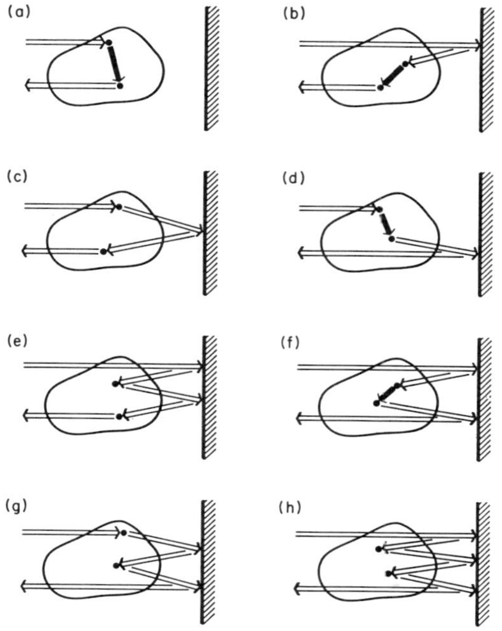

Fig. 4. Diagrammatic illustrations of the eight contributions [(a)–(h)], given in Eq. (5.4), to the second-order term $U_c{}^{(2)}(\mathbf{r}_<)$ of iterative expansion (3.1).

If we make use of Eq. (2.15) and also of the relation

$$G^{(I)*}(\mathbf{r}, \mathbf{r}') = G^{(I)}(\mathbf{r}, \mathbf{r}') \tag{5.5}$$

that readily follows from Eq. (2.12), we find that many of the terms in Eq. (5.4) cancel one another. The formula (5.4) for the field $U_c^{(2)}(\mathbf{r}_<)$ may then be expressed in the following compact form:

$$U_c^{(2)}(\mathbf{r}_<) = (1 - |\mu|^2)[U_{\rm bs}^{(2)}(\mathbf{r}_<) + {}'U^{(2)}(\mathbf{r}_<)], \tag{5.6}$$

where

$$U_{\rm bs}^{(2)}(\mathbf{r}_<) = \left(\frac{1}{4\pi}\right)^2 \int_{\mathcal{V}} \int_{\mathcal{V}} G_<^{(H)}(\mathbf{r}_<, \mathbf{r}'')F(\mathbf{r}'') \times G(\mathbf{r}'', \mathbf{r}')F(\mathbf{r}')U^{(i)}(\mathbf{r}')\mathrm{d}^3r'\mathrm{d}^3r'' \tag{5.7}$$

is the (second-order) backscattered contribution to the conjugate field in the half-space $\mathcal{R}^-$ [diagram (a) in Fig. 4] and

$${}'U^{(2)}(\mathbf{r}_<) = -\mu\left(\frac{1}{4\pi}\right)^2 \int_{\mathcal{V}} \int_{\mathcal{V}} G_>^{(H)*}(\mathbf{r}_<, \mathbf{r}'')F(\mathbf{r}'') \times G_>^{(H)}(\mathbf{r}'', \mathbf{r}')F(\mathbf{r}')U^{(i)*}(\mathbf{r}')\mathrm{d}^3r'\mathrm{d}^3r'' \tag{5.8}$$

represents the (second-order) field contribution resulting from the sum of the three elementary events, each of which contains only a single phase conjugation at the PCM [diagrams (b)–(d) in Fig. 4]. It is seen that, in general, the field $U_c^{(2)}(\mathbf{r}_<)$ does not vanish throughout the half-space $\mathcal{R}^-$ and that it contains contributions of the evanescent waves created inside the scatterer through the backscattered term $U_{\rm bs}^{(2)}(\mathbf{r}_<)$. However, in the special case when the conjugated waves are generated without losses or gains ($|\mu| = 1$), the second-order field contribution $U_c^{(2)}(\mathbf{r}_<)$ to the total field vanishes identically in the half-space $\mathcal{R}^-$ [cf. Eq. (5.6)].

If $|\mu| \ll 1$, as is usually the case, Eq. (5.6) yields

$$U_c^{(2)}(\mathbf{r}_<) \cong U_{\rm bs}^{(2)}(\mathbf{r}_<) + {}'U^{(2)}(\mathbf{r}_<) \qquad (|\mu| \ll 1), \tag{5.9}$$

where the directly backscattered contribution $U_{\rm bs}^{(2)}(\mathbf{r}_<)$ is independent of μ and the term ${}'U^{(2)}(\mathbf{r}_<)$, given by Eq. (5.8), is proportional to μ. It can be seen, on comparing Eq. (5.8) with the expression for term (h) in Eq. (5.4), that the contribution ${}'U^{(2)}(\mathbf{r}_<)$ is equal to $-1/|\mu|^2$ times the contribution shown schematically in diagram (h) of Fig. 4.

6. DISTORTION CORRECTION IN THE ABSENCE OF LOSSES OR GAINS ON PHASE CONJUGATION

In view of the complexity of the physical processes implicit in the technique of distortion correction by phase conjugation, the theoretical analysis concerning the attainable degree of correction becomes prohibitive when one tries to carry out the calculations explicitly beyond the first few terms in the iterative expansion. However, it was found recently[5] that, in the special case when $|\mu| = 1$, i.e., when there are no losses or gains on phase conjugation at the PCM, the analysis can be carried out to all orders, and it leads to an interesting result. We now briefly discuss this case.

Let us examine a typical term $U_c^{(n)}(\mathbf{r})$ with $n \geq 1$ in the iterative expansion (3.1), making use of expression (2.16) for the Green's function $\hat{G}_c(\mathbf{r}, \mathbf{r}')$. With the field point $\mathbf{r} = \mathbf{r}_<$ situated in the half-space $\mathcal{R}^-$ and with the evanescent waves omitted in that half-space, an expression for the term $U_c^{(n)}(\mathbf{r}_<)$ is given by Eq. (3.5). One finds that, irrespective of the exact value of the complex "reflectance" μ of the PCM, the nth-order field contribution $U_c^{(n)}(\mathbf{r}_<)$ may be expressed in the form (cf. Ref. 5)

$$U_c^{(n)}(\mathbf{r}_<) = \left(-\frac{1}{4\pi}\right)^n \int_{\mathcal{V}} \mathrm{d}^3r_1 F(\mathbf{r}_1) \ldots \int_{\mathcal{V}} \mathrm{d}^3r_n F(\mathbf{r}_n) \times G_<^{(H)}(\mathbf{r}_<, \mathbf{r}_1)(1 - \mu\hat{C})\hat{P}_{1,2} \ldots \hat{P}_{n-1,n}(1 + \mu\hat{C})U^{(i)}(\mathbf{r}_n), \tag{6.1}$$

where

$$\hat{P}_{j,j+1} \equiv \hat{G}_c(\mathbf{r}_j, \mathbf{r}_{j+1}) = (1 + \mu\hat{C})A_{j,j+1} + B_{j,j+1}(1 - \mu\hat{C}) + G^{(I)}(\mathbf{r}_j, \mathbf{r}_{j+1}), \tag{6.2}$$

with

$$A_{j,k} = \Theta(z_j - z_k)G_>^{(H)}(\mathbf{r}_j, \mathbf{r}_k) \tag{6.3a}$$

and

$$B_{j,k} = \Theta(z_k - z_j)G_<^{(H)}(\mathbf{r}_j, \mathbf{r}_k). \tag{6.3b}$$

We recall that the expressions (6.1)–(6.3) are applicable only when the scatterer is nonabsorbing. It can be shown that the nth-order term $U_c^{(n)}(\mathbf{r}_<)$ represents the sum of contributions from 2^{n+1} elementary events. Each of these events involves n scattering processes and m phase conjugations, with m taking on the values 0, 1, 2, ..., $n + 1$; the number of events containing m phase conjugations is given by the binomial coefficient $\binom{n+1}{m}$.

It was further shown in Ref. 5 that, in the special case when $|\mu| = 1$, the following identity holds:

$$(1 - \mu\hat{C})\hat{P}_{1,2}\hat{P}_{2,3} \ldots \hat{P}_{n-1,n} = \hat{Q}_{1,2}\hat{Q}_{2,3} \ldots \hat{Q}_{n-1,n}(1 - \mu\hat{C}), \tag{6.4}$$

where

$$\hat{Q}_{j,j+1} = (1 - \mu\hat{C})B_{j,j+1} + G^{(I)}(\mathbf{r}_j, \mathbf{r}_{j+1}). \tag{6.5}$$

In deriving this identity, use was made of relation (5.5) and of the operator identity

$$(1 - \mu\hat{C})(1 + \mu\hat{C}) = (1 - |\mu|^2), \tag{6.6}$$

which follows readily from definition (2.8) of the complex-conjugation operator $\hat{C}$.

On substituting from Eq. (6.4) into Eq. (6.1), we obtain the following expression for $U_c^{(n)}(\mathbf{r}_<)$, valid when $|\mu| = 1$:

$$U_c^{(n)}(\mathbf{r}_<) = \left(-\frac{1}{4\pi}\right)^n \int_{\mathcal{V}} \mathrm{d}^3r_1 F(\mathbf{r}_1) \ldots \int_{\mathcal{V}} \mathrm{d}^3r_n F(\mathbf{r}_n) \times G_<^{(H)}(\mathbf{r}_<, \mathbf{r}_1)\hat{Q}_{1,2} \ldots \hat{Q}_{n-1,n} \times (1 - \mu\hat{C})(1 + \mu\hat{C})U^{(i)}(\mathbf{r}_n) \quad (n = 1, 2, 3, \ldots). \tag{6.7}$$

If we make use once more of identity (6.6) and recall that $|\mu|$ is now assumed to be unity, we see at once from Eq. (6.7) that

$$U_c^{(n)}(\mathbf{r}_<) = 0, \qquad n = 1, 2, 3, \ldots. \tag{6.8}$$

Hence, with $\mathbf{r} = \mathbf{r}_<$, the right-hand side of Eq. (3.1) for the field $U_c(\mathbf{r}_<)$ in the half-space $\mathcal{R}^-$ reduces, in this case, to the single term $U_c^{(0)}(\mathbf{r}_<)$ given by Eq. (3.3), i.e.,

$$U_c(\mathbf{r}_<) = U^{(i)}(\mathbf{r}_<) + e^{i\phi}U^{(i)*}(\mathbf{r}_<) \qquad (|\mu| = 1), \tag{6.9}$$

where ϕ is the argument (phase) of μ. This result implies that, *if there are no losses or gains on phase conjugation at the PCM* (i.e., if $|\mu| = 1$), *the field* $U_c(\mathbf{r}_<)$ *in the half-space* $\mathcal{R}^-$ *does not depend at all on the scatterer, and, consequently, distortions introduced by the scatterer on the incident field* $U^{(i)}$ *are now completely eliminated by the technique of phase conjugation*, it being assumed, as before, that the scatterer is nonabsorbing and that the effects of the evanescent waves outside the scatterer are negligible.

7. ELECTROMAGNETIC THEORY OF DISTORTION CORRECTION BY PHASE CONJUGATION

Until now we represented the (monochromatic) optical field by a scalar function of position, and hence we completely ignored its polarization properties. We now briefly discuss some generalizations of our main results, which take the polarization properties into account. For this purpose we must first know what changes are introduced into the state of an electromagnetic field on interaction with a PCM.

Let $\mathbf{E}^{(i)}(\mathbf{r})$ be the electric field vector [with the time-dependent part $\exp(-i\omega t)$ omitted] of an electromagnetic field incident in free space (i.e., in the absence of the scatterer) upon an infinite PCM located in the plane $z = z_1$. The response of the PCM will depend on the experimental technique that is used to generate the conjugate field.[6] We consider only the case when the PCM produces, in the plane of the mirror, the transformation

$$\mathbf{E}^{(i)}(\mathbf{r})|_{z=z_1} \to \mu\mathbf{E}^{(i)*}(\mathbf{r})|_{z=z_1}, \tag{7.1}$$

where μ is a constant that represents the "reflectivity" of the PCM. Formula (7.1) implies that the state of polarization of light is completely reversed on interaction with the PCM. The assumption of constant "reflection coefficient" μ implies that the response of the PCM to a plane wave is independent of both the angle of incidence and the state of polarization of the wave. This, of course, is an idealization. However, we wish to mention that experimental techniques have been developed, for example, by Zel'dovich and co-workers[16] for producing PCM's that give rise to complete reversal of the state of polarization, at least in the paraxial regime.

One can show by an argument similar to that leading to Theorem VI of Ref. 15 that, if the distribution $\mathbf{E}^{(i)}(\mathbf{r})|_{z=z_1}$ contains no evanescent contributions, the distribution $\mu\mathbf{E}^{(i)*}(\mathbf{r})|_{z=z_1}$ on the right-hand side of formula (7.1) will generate, in the absence of the scatterer, an electric field $\mu\mathbf{E}^{(i)*}(\mathbf{r})$ throughout the domain $z < z_1$.

Suppose now that the scatterer is present in the half-space $z < z_1$. For simplicity we assume that the scatterer is linear, time independent, isotropic, nonabsorbing, nonmagnetic, and spatially nondispersive and that it occupies a finite volume $\mathcal{V}$. It then follows from Maxwell's equations that the electric field $\mathbf{E}(\mathbf{r})$ satisfies (in the Gaussian system of units) the differential equation[17]

$$\nabla \times \nabla \times \mathbf{E}(\mathbf{r}) - k^2 n^2(\mathbf{r})\mathbf{E}(\mathbf{r}) = 0, \tag{7.2}$$

where

$$n^2(\mathbf{r}) = 1 + 4\pi\chi(\mathbf{r}) \tag{7.3}$$

and $\chi(\mathbf{r})$ is the dielectric susceptibility, defined by the formula

$$\mathbf{P}(\mathbf{r}) = \chi(\mathbf{r})\mathbf{E}(\mathbf{r}), \tag{7.4}$$

with $\mathbf{P}(\mathbf{r})$ denoting the induced polarization. We may rewrite Eq. (7.2) in the form

$$\nabla \times \nabla \times \mathbf{E}(\mathbf{r}) - k^2\mathbf{E}(\mathbf{r}) = -F(\mathbf{r})\mathbf{E}(\mathbf{r}), \tag{7.5}$$

where

$$F(\mathbf{r}) = -k^2[n^2(\mathbf{r}) - 1] = -4\pi k^2\chi(\mathbf{r}) \tag{7.6}$$

is the scattering potential associated with the distribution $\chi(\mathbf{r})$ of the dielectric susceptibility throughout the volume $\mathcal{V}$. Since we assumed that the scatterer is nonabsorbing, the scattering potential $F(\mathbf{r})$ is a real function of position.

It is convenient to introduce the dyadic Green's function $\overleftrightarrow{G}(\mathbf{r}, \mathbf{r}')$, defined as the outgoing free-space solution to the differential equation[18]

$$\nabla \times \nabla \times \overleftrightarrow{G}(\mathbf{r}, \mathbf{r}') - k^2\overleftrightarrow{G}(\mathbf{r}, \mathbf{r}') = 4\pi\delta^{(3)}(\mathbf{r} - \mathbf{r}')\overleftrightarrow{I}, \tag{7.7}$$

where $\delta^{(3)}(\mathbf{r} - \mathbf{r}')$ is the three-dimensional Dirac delta function and $\overleftrightarrow{I}$ denotes the unit dyadic. It is well known that $\overleftrightarrow{G}(\mathbf{r}, \mathbf{r}')$ is related to the scalar Green's function $G(\mathbf{r}, \mathbf{r}')$ [Eq. (2.4)] by the formula

$$\overleftrightarrow{G}(\mathbf{r}, \mathbf{r}') = \left(\overleftrightarrow{I} + \frac{1}{k^2}\nabla\nabla\right) G(\mathbf{r}, \mathbf{r}'). \tag{7.8}$$

Let us first consider the case when an electric field $\mathbf{E}^{(i)}(\mathbf{r})$ is incident upon the scatterer $\mathcal{V}$ in the absence of the PCM. When the field interacts with the scatterer, a new field $\mathbf{E}(\mathbf{r})$ is generated that can be shown, on using Eqs. (7.5) and (7.7), to satisfy the integral equation

$$\mathbf{E}(\mathbf{r}) = \mathbf{E}^{(i)}(\mathbf{r}) - \frac{1}{4\pi}\int_{\mathcal{V}} \overleftrightarrow{G}(\mathbf{r}, \mathbf{r}') \cdot F(\mathbf{r}')\mathbf{E}(\mathbf{r}')\mathrm{d}^3r'. \tag{7.9}$$

This equation represents an electromagnetic analog to the usual integral equation (2.3) for scattering of scalar waves.

Next, suppose that the electric field $\mathbf{E}^{(i)}(\mathbf{r})$ is incident upon the scatterer with the PCM being present. We denote by $\mathbf{E}_c(\mathbf{r})$ the total electric field that is now generated. In analogy with the scalar case, we refer to $\mathbf{E}_c(\mathbf{r})$ as the conjugate electric field [cf. Ref. 9]. By arguments similar to those used in deriving the corresponding integral equation (2.6) for the conjugate scalar field $U_c(\mathbf{r})$, one can show that the conjugate electric field $\mathbf{E}_c(\mathbf{r})$ satisfies the integral equation

$$\mathbf{E}_c(\mathbf{r}) = \mathbf{E}_c^{(0)}(\mathbf{r}) - \frac{1}{4\pi}\int_{\mathcal{V}} \overleftrightarrow{G}_c(\mathbf{r}, \mathbf{r}') \cdot F(\mathbf{r}')\mathbf{E}_c(\mathbf{r}')\mathrm{d}^3r', \tag{7.10}$$

where $\overleftrightarrow{G}_c(\mathbf{r}, \mathbf{r}')$ is a dyadic operator Green's function, to be discussed shortly, which takes into account the presence of the PCM. The field $\mathbf{E}_c^{(0)}(\mathbf{r})$ is related to the incident field $\mathbf{E}^{(i)}(\mathbf{r})$ by the formula

$$\begin{aligned}\mathbf{E}_c^{(0)}(\mathbf{r}) &= \mathbf{E}^{(i)}(\mathbf{r}) + \mu\mathbf{E}^{(i)*}(\mathbf{r}) \\ &= (1 + \mu\hat{C})\mathbf{E}^{(i)}(\mathbf{r}),\end{aligned} \tag{7.11}$$

where $\hat{C}$ denotes, as before, the complex-conjugation operator defined by Eq. (2.8). Formula (7.11) is strictly analogous to Eq. (3.3), and it represents the total electric field in the absence of the scatterer [$F(\mathbf{r}) \equiv 0$] but with the PCM present.

The dyadic operator Green's function $\overleftrightarrow{G}_c(\mathbf{r}, \mathbf{r}')$ may be constructed in a manner similar to that employed in the der-

536 J. Opt. Soc. Am./Vol. 73, No. 5/May 1983

ivation[5,11] of the operator Green's function $\hat{G}_c(\mathbf{r}, \mathbf{r}')$ [see also Ref. 19], and it can be expressed in the following analogous form [cf. Eq. (2.16)]:

$$\ddot{G}_c(\mathbf{r}, \mathbf{r}') = \Theta(z - z')(1 + \mu\hat{C})\ddot{G}_>^{(H)}(\mathbf{r}, \mathbf{r}') + \Theta(z' - z)\ddot{G}_<^{(H)}(\mathbf{r}, \mathbf{r}')(1 - \mu\hat{C}) + \ddot{G}^{(I)}(\mathbf{r}, \mathbf{r}'). \quad (7.12)$$

Here $\Theta(\zeta)$ is again the unit step function [Eq. (2.10)]

$$\ddot{G}_>^{(H)}(\mathbf{r}, \mathbf{r}') = \left(\ddot{I} + \frac{1}{k^2}\nabla\nabla\right) G_>^{(H)}(\mathbf{r}, \mathbf{r}'), \quad (7.13a)$$

$$\ddot{G}_<^{(H)}(\mathbf{r}, \mathbf{r}') = \left(\ddot{I} + \frac{1}{k^2}\nabla\nabla\right) G_<^{(H)}(\mathbf{r}, \mathbf{r}'), \quad (7.13b)$$

and

$$\ddot{G}^{(I)}(\mathbf{r}, \mathbf{r}') = \left(\ddot{I} + \frac{1}{k^2}\nabla\nabla\right) G^{(I)}(\mathbf{r}, \mathbf{r}'), \quad (7.14)$$

where the functions $G_>^{(H)}(\mathbf{r}, \mathbf{r}')$, $G_<^{(H)}(\mathbf{r}, \mathbf{r}')$, and $G^{(I)}(\mathbf{r}, \mathbf{r}')$ are given by Eqs. (2.11a), (2.11b), and (2.12), respectively. We emphasize that the vectorial integral equation (7.10) is valid only for PCM's that give rise to a complete reversal of polarization in the plane of the mirror, as implied by Eq. (7.1). The incident field was again assumed to contain no evanescent components, and the effects of the evanescent waves have been neglected at the PCM.

Just as in the scalar case, integral equation (7.10) may be formally solved by iteration. We express the conjugate electric field $E_c(\mathbf{r})$ in the form

$$\mathbf{E}_c(\mathbf{r}) = \sum_{n=0}^{\infty} \mathbf{E}_c^{(n)}(\mathbf{r}), \quad (7.15)$$

where the zeroth-order term $\mathbf{E}_c^{(0)}(\mathbf{r})$ is given by Eq. (7.11) and the higher-order terms are obtained from the recursion formula

$$\mathbf{E}_c^{(n)}(\mathbf{r}) = -\frac{1}{4\pi}\int_V \ddot{G}_c(\mathbf{r}, \mathbf{r}') \cdot F(\mathbf{r}')\mathbf{E}_c^{(n-1)}(\mathbf{r}')\mathrm{d}^3r' \quad (n = 1, 2, 3, \ldots). \quad (7.16)$$

Because the mathematical structures of Eqs. (7.11) and (7.16) and of the dyadic operator Green's function (7.12) are similar to the structures of the corresponding formulas for the scalar case [Eqs. (3.3), (3.2), and (2.16), respectively], one may expect that many of the results that we obtained for the conjugate scalar field $U_c(\mathbf{r})$ will have strict analogs for the conjugate electric field $\mathbf{E}_c(\mathbf{r})$. In particular, one can show that, if $|\mu| = 1$, i.e., if the conjugate electric field is generated without losses or gains, and if, in addition, the effects of the evanescent waves are negligible at the PCM and at the detector in the half-space $\mathcal{R}^-$, then (with the detector located at $\mathbf{r}_< \in \mathcal{R}^-$)

$$\mathbf{E}_c^{(n)}(\mathbf{r}_<) = 0 \quad \text{for } n = 1, 2, 3, \ldots \quad (|\mu| = 1). \quad (7.17)$$

Using Eqs. (7.11), (7.15), and (7.17), it follows at once that (with $\mathbf{r}_< \in \mathcal{R}^-$)

$$\mathbf{E}_c(\mathbf{r}_<) = \mathbf{E}^{(i)}(\mathbf{r}_<) + e^{i\phi}\mathbf{E}^{(i)*}(\mathbf{r}_<) \quad (|\mu| = 1), \quad (7.18)$$

where ϕ denotes the phase of the complex "reflectance" μ of the PCM.

Equation (7.18) represents an electromagnetic generalization of the corresponding result derived in Section 6 on the basis of scalar wave theory [cf. Eq. (6.9)]. It implies that, within the framework of the electromagnetic theory, a complete correction of distortions is obtained if the following conditions are satisfied:

(1) The phase conjugation at the PCM gives rise to a complete reversal of polarization [see formula (7.1)].

(2) There are no losses or gains on phase conjugation (i.e., $|\mu| = 1$).

(3) The scatterer is linear, time independent, isotropic, nonmagnetic, and spatially nondispersive and occupies a finite volume.

(4) The effects of the evanescent waves are negligible at the PCM and in the region of the half-space $\mathcal{R}^-$ where the detector is situated.

If $|\mu| \neq 1$ and the scatterer is sufficiently weak, the electric field $\mathbf{E}_c(\mathbf{r}_<)$ in the half-space $\mathcal{R}^-$ will, to a good approximation, be given by the first few terms of iterative expansion (7.15). For example, within the accuracy of the first Born approximation, one may deduce from our integral equation (7.10) that

$$\mathbf{E}_c(\mathbf{r}_<) \cong \mathbf{E}_c^{(0)}(\mathbf{r}_<) + \mathbf{E}_c^{(1)}(\mathbf{r}_<), \quad (7.19)$$

where $\mathbf{E}_c^{(0)}(\mathbf{r}_<)$ is given by Eq. (7.11) and

$$\mathbf{E}_c^{(1)}(\mathbf{r}_<) = -(1 - |\mu|^2)\frac{1}{4\pi}\int\left(\ddot{I} + \frac{1}{k^2}\nabla\nabla\right) G^{(H)}(\mathbf{r}_<, \mathbf{r}') \cdot F(\mathbf{r}')\mathbf{E}^{(i)}(\mathbf{r}')\mathrm{d}^3r'. \quad (7.20)$$

The term $\mathbf{E}_c^{(1)}(\mathbf{r}_<)$ represents the contribution to the conjugate electric field $\mathbf{E}_c(\mathbf{r}_<)$ in the half-space $\mathcal{R}^-$ that arises from backscattering, in the first Born approximation [cf. Eqs. (4.5) and (4.6)]. If the first-order backscattering is negligible, then Eq. (7.19) gives $\mathbf{E}_c(\mathbf{r}_<) \cong \mathbf{E}_c^{(0)}(\mathbf{r}_<)$, showing, in view of Eq. (7.11), that in this approximation the scatterer has no effect on the conjugated electric field in the half-space $\mathcal{R}^-$, i.e., that the distortions in the incident field produced by a nonabsorbing scatterer are then eliminated by phase conjugation.

8. DISCUSSION

In this paper we have developed a systematic approach, both within the framework of the scalar wave theory and within the framework of Maxwell's electromagnetic theory, to the problem of determining the degree of distortion correction that may be achieved by the technique of optical phase conjugation. This approach is based on new integral equations for scattering of monochromatic waves in the presence of a phase-conjugate mirror (PCM). We treated only situations in which the distorting medium is linear, time independent, and nonabsorbing, even though the basic integral equations may be shown to apply also to absorbing media. It was found that, if the PCM produces in its plane a complete reversal of polarization, essentially the same conclusions follow from the scalar wave theory and from the electromagnetic theory.

In the derivation of the basic integral equations [Eqs. (2.6) and (7.10)], we assumed that the effects of the evanescent waves are negligible at the PCM. We also assumed that the

PCM is infinite in extent and that it is characterized by a constant "reflectivity" μ. In practice, the PCM will be finite, and this may be expected to lead to a reduction in its ability to produce a field that would compensate for high-spatial-frequency components of the distortions imparted on the incident wave by the scatterer. Moreover, μ may be a function of position across the PCM and will, in general, also depend on the angles of incidence of the plane-wave components that are present in the angular spectrum representation of the field. These more-complex situations could be treated by appropriately modifying our integral equations, but such modifications are not discussed in the present paper.

Perhaps the most striking feature of the present approach is that the contributions that arise from elementary events consisting of scattering and phase-conjugation processes of all orders appear explicitly in the expressions for the conjugated fields. Such contributions must, in general, be included when the wave fields are monochromatic. The situation may be different when pulsed waves are used or when the macroscopic properties of the scatterer vary with time; however, the analysis of such cases would require a separate investigation.

Finally, we wish to mention that, although we assumed throughout this paper that the scatterer is "deterministic," our integral equations may be used even when the scattering medium is random in nature. In such a case the scattering potential and the conjugate field become random processes, and our integral equations then refer to single realizations. Quantities of physical interest, such as field correlations, can then be obtained from the solutions of these equations by taking appropriate ensemble averages.[20]

ACKNOWLEDGMENT

This research was supported by the U.S. Army Research Office.

* G. S. Agarwal was a Joint Institute for Laboratory Astrophysics Visiting Fellow, 1981–1982; permanent address, School of Physics, University of Hyderabad, Hyderabad, India.

† Also at the Institute of Optics, University of Rochester; during the academic year 1982–1983, at Schlumberger-Doll Research, Ridgefield, Connecticut 06877.

REFERENCES

1. For reviews of the technique of optical phase conjugation see, for example, A. Yariv, "Phase conjugate optics and real-time holography," IEEE J. Quantum Electron. **QE-14,** 650–660 (1978); D. M. Pepper, "Nonlinear optical phase conjugation," Opt. Eng. **21,** 156–183 (1982).
2. See, for example, B. Ya. Zel'dovich, V. I. Popovichev, V. V. Ragul'skii, and F. S. Faizullov, "Connection between the wave fronts of the reflected and exciting light in stimulated Mandel'shtam–Brillouin scattering," Sov. Phys. JETP **15,** 109–113 (1972); O. Yu. Nosach, V. I. Popovichev, V. V. Ragul'skii, and F. S. Faizullov, "Cancellation of phase distortions in an amplifying medium with a 'Brillouin mirror,' " Sov. Phys. JETP **16,** 435–438 (1972); D. N. Bloom and G. C. Bjorklund, "Conjugate wave-front generation and image reconstruction by four-wave mixing," Appl. Phys. Lett. **31,** 592–594 (1977); and V. Wang and C. R. Guiliano, "Correction of phase aberrations via stimulated Brillouin scattering," Opt. Lett. **2,** 4–6 (1978).
3. G. S. Agarwal and E. Wolf, "Theory of phase conjugation with weak scatterers," J. Opt. Soc. Am. **72,** 321–326 (1982).
4. G. S. Agarwal, A. T. Friberg, and E. Wolf, "Effect of backscattering in phase conjugation with weak scatterers," J. Opt. Soc. Am. **72,** 861–863 (1982).
5. G. S. Agarwal, A. T. Friberg, and E. Wolf, "Elimination of distortions by phase conjugation without losses or gains," Opt. Commun. **43,** 446–450 (1982).
6. Several schemes, such as those based on three-wave mixing [A. Yariv, "Three-dimensional pictorial transmission in optical fibers," Appl. Phys. Lett. **28,** 88–89 (1976)], four-wave mixing [R. W. Hellwarth, "Generation of time-reversed wave fronts by nonlinear refraction," J. Opt. Soc. Am. **67,** 1–3 (1977)], stimulated Brillouin scattering (Zel'dovich *et al.*[2]), etc. have been proposed and used for the generation of the conjugated field. The polarization properties of the conjugated field will, in general, be different from those of the incident (probe) field. A complete reversal of polarization can be achieved by suitably arranging the experimental geometry and by choosing the polarization properties of the various interacting waves appropriately [see, for example, Ref. 16 below and J. F. Lam, D. G. Steel, R. A. McFarlane, and R. C. Lind, "Atomic coherence effects in resonant degenerate four-wave mixing," Appl. Phys. Lett. **38,** 977–979 (1981)].
7. This integral equation is derived, in the context of time-independent quantum-mechanical potential scattering, for example in P. Roman, *Advanced Quantum Theory* (Addison-Wesley, Reading, Mass., 1965), Sec. 3.2.
8. The concept of a PCM is a convenient idealization that describes the effect of a true physical device located beyond the plane $z = z_1$, by means of which a field distribution U is replaced by a new field distribution μU^* in that plane. The transformation $U \to \mu U^*$ is usually achieved by nonlinear optical interactions, such as stimulated scattering processes or optical parametric interactions [see Ref. 1].
9. The term "conjugate" (or "conjugated") field is somewhat ambigous and must be interpreted with caution. The field $U_c(\mathbf{r})$ is generated as a result of the interaction of the incident field $U^{(i)}(\mathbf{r})$ with the scattering medium in the presence of the PCM. In general, it will include, in addition to $U^{(i)}(\mathbf{r})$, also contributions (usually ignored) arising from backscattering of the incident field and from successive conjugations of waves backscattered onto the PCM [see, for example, Figs. 3(a), 3(d), 4(a), and 4(h)].
10. For a discussion of some of the complications that arise when the evanescent waves are taken into account, see E. Wolf and W. H. Carter, "Comments on the theory of phase-conjugated waves," Opt. Commun. **40,** 397–400 (1982). The approximation resulting from the neglection of the evanescent waves in the scattered field is examined, within the accuracy of the first Born approximation, in App. A of Ref. 3.
11. A. T. Friberg, "Integral equation of the scattered field in the presence of a phase-conjugate mirror" J. Opt. Soc. Am. (to be published.)
12. Throughout this paper, a caret above a symbol denotes an operator.
13. See, for example, A. Baños, *Dipole Radiation in the Presence of a Conducting Half-Space* (Pergamon, Oxford, 1966), Eq. (2.19).
14. We will not discuss here the conditions under which series (3.1) will converge, a subject that would require a separate investigation.
15. E. Wolf, "Phase conjugacy and symmetries in spatially bandlimited wavefields containing no evanescent components," J. Opt. Soc. Am. **70,** 1311–1319 (1980). Equation (2.1) of this reference contains a misprint. $U^{(2)}(x, y, z)e^{i\omega t}$ should be replaced with $U^{(2)}(x, y, z)e^{-i\omega t}$. Also, Eq. (1.8) should read $A(u/k, v/k) = k^2\tilde{U}(u, v; z)e^{-iwz}$. These corrections do not affect any other equations or conclusions of that paper.
16. B. Ya. Zel'dovich and V. V. Shkunov, "Spatial-polarization wavefront reversal in four-photon interaction," Sov. J. Quantum Electron. **9,** 379–381 (1979); B. Ya. Zel'dovich and T. V. Yakovleva, "Spatial-polarization wavefront reversal in stimulated scattering of the Rayleigh line wing," Sov. J. Quantum Electron. **10,** 501–505 (1980).
17. Alternatively, by using the Maxwell equation $\nabla \cdot \mathbf{D} = 0$ (where

$\mathbf{D} \equiv \mathbf{E} + 4\pi\mathbf{P}$ is the electric displacement vector), the constitutive relation $\mathbf{D} = n^2\mathbf{E}$ and the vector identity $\nabla \cdot (n^2\mathbf{E}) = n^2\nabla \cdot \mathbf{E} + \mathbf{E} \cdot \nabla n^2$, Eq. (7.2) may be rewritten in the familiar form

$$\nabla^2\mathbf{E}(\mathbf{r}) + k^2n^2(\mathbf{r})\mathbf{E}(\mathbf{r}) + \nabla\{\mathbf{E}(\mathbf{r}) \cdot \nabla \log[n^2(\mathbf{r})]\} = 0.$$

However, form (7.2) is more convenient for the present purposes.

18. See, for example, C.-T. Tai, *Dyadic Green's Functions in Electromagnetic Theory* (Intext, Scranton, Pa., 1971), Sec. 14.
19. G. S. Agarwal, "Dipole radiation in the presence of a phase conjugate mirror," Opt. Commun. **42,** 205–207 (1982).
20. In this connection, see V. I. Tatarskii, *The Effects of the Turbulent Atmosphere on Wave Propagation* (National Technical Information Service, Springfield, Va., 1971); E. Wolf, "New theory of partial coherence in the space-frequency domain. Part I: spectra and cross spectra of steady-state sources," J. Opt. Soc. Am. **72,** 343–351 (1982).

886 J. Opt. Soc. Am. A/Vol. 2, No. 6/June 1985

Analyticity of the angular spectrum amplitude of scattered fields and some of its consequences

Emil Wolf and Manuel Nieto-Vesperinas*

Department of Physics and Astronomy, University of Rochester, Rochester, New York 14627

Received September 27, 1984; accepted December 3, 1984

Analytic properties of the angular spectrum amplitude of the field scattered from a finite object with a continuous scattering potential are established, and some of its consequences are discussed. In particular, it is shown that the scattered field must contain both homogeneous and evanescent waves, unless it vanishes everywhere outside the scattering volume.

1. INTRODUCTION

The angular spectrum representation of wave fields is a powerful mathematical tool for the treatment of various radiation and interaction problems. It has found useful applications in connection with scattering of radio waves from the ionosphere,[1] in the analysis of radiation patterns of antennas,[2,3] in the theory of multipole radiation[4] and Čerenkov radiation,[5] in the theory of image formation,[6,7] self-imaging,[8] and focusing,[9] in connection with diffraction,[10,11] inverse diffraction,[12] and inverse scattering,[13,14] in the theory of rough surface scattering,[15] in holography,[16] in studies of electromagnetic radiation in spatially dispersive media,[17] in investigations relating to the foundations of radiometry,[18] in phase-conjugate optics,[19] and in many other fields.

In the present paper we consider an important question relating to this representation when applied to scattering of scalar waves from a finite object, namely, the question of the analytic behavior of the angular spectrum amplitude of the scattered field. We show that when the scattering potential is a continuous function throughout the (finite) domain occupied by the scatterer, the angular spectrum amplitude of the scattered field may be extended into an entire function of two complex variables. From this result we derive a number of new theorems. In particular, we show that the scattered field must always contain both homogeneous and evanescent waves, unless it vanishes everywhere outside the scattering volume.

2. INTEGRAL REPRESENTATION OF THE ANGULAR SPECTRUM AMPLITUDE OF SCATTERED FIELDS

Let $V^{(i)}(\mathbf{r}, t) = U^{(i)}(\mathbf{r}) \exp(-i\omega t)$ be a monochromatic scalar wave field that is incident upon a scatterer that occupies a finite domain D in free space. Here $\mathbf{r}$ denotes the position vector of a typical point in space, and t denotes the time. We assume that throughout the whole space the incident field satisfies the source-free wave equation. Consequently $U^{(i)}(\mathbf{r})$ will satisfy, throughout the whole space, the Helmholtz equation

$$(\nabla^2 + k^2)U^{(i)}(\mathbf{r}) = 0, \tag{2.1}$$

where $k = \omega/c$, c being the speed of light *in vacuo*.

We assume that the response of the scattering medium to the incident field is time independent and linear. The time-independent part of the total field $V(\mathbf{r}, t) = U(\mathbf{r}) \exp(-i\omega t)$ generated by the interaction between the incident field and the scatterer will then satisfy the equation

$$[\nabla^2 + k^2 n^2(\mathbf{r})]U(\mathbf{r}) = 0, \tag{2.2}$$

where $n(\mathbf{r})$ is the refractive index of the scattering medium. Since the scatterer is assumed to be situated in free space, $n(\mathbf{r}) = 1$ at every point outside the domain D. If we express $U(\mathbf{r})$ as the sum of the incident field $U^{(i)}(\mathbf{r})$ and the scattered field $U^{(s)}(\mathbf{r})$, *i.e.*,

$$U(\mathbf{r}) = U^{(i)}(\mathbf{r}) + U^{(s)}(\mathbf{r}), \tag{2.3}$$

it follows at once on subtracting Eq. (2.1) from Eq. (2.2) and making use of Eq. (2.3) that scattered field obeys the equation

$$(\nabla^2 + k^2)U^{(s)}(\mathbf{r}) = F(\mathbf{r})U(\mathbf{r}), \tag{2.4}$$

where

$$F(\mathbf{r}) = -k^2[n^2(\mathbf{r}) - 1] \tag{2.5}$$

represents the scattering potential. Evidently $F(\mathbf{r}) \equiv 0$ at all points outside the scattering domain D.

By using standard Green's function techniques Eq. (2.4) may be transformed into the basic integral equation of potential scattering, viz.,

$$U^{(s)}(\mathbf{r}) = -\frac{1}{4\pi}\int_D U(\mathbf{r}')F(\mathbf{r}')G(\mathbf{r}, \mathbf{r}')\mathrm{d}^3r', \tag{2.6}$$

where $G(\mathbf{r}, \mathbf{r}')$ is the free-space Green's function

$$G(\mathbf{r}, \mathbf{r}') = \frac{\exp(ik|\mathbf{r} - \mathbf{r}'|)}{|\mathbf{r} - \mathbf{r}'|}. \tag{2.7}$$

Suppose that the scattering domain D is located in the strip $0 \le z \le L$, and let $\mathcal{R}^-$ and $\mathcal{R}^+$ be the half-spaces $z < 0$ and $z > L$, respectively (see Fig. 1). The scattered field in these two half-spaces may be readily expressed in the form of the angular spectrum of plane waves. For this purpose we make use of the well-known Weyl representation of the free-space Green's function, viz.,[20]

0740-3232/85/060886-05$02.00

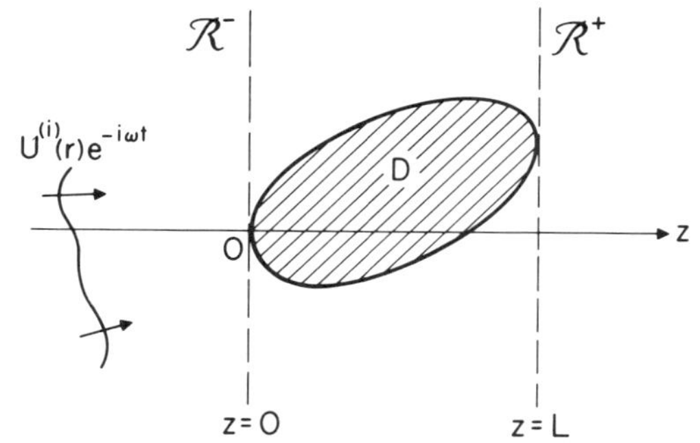

Fig. 1. Illustrating the notation.

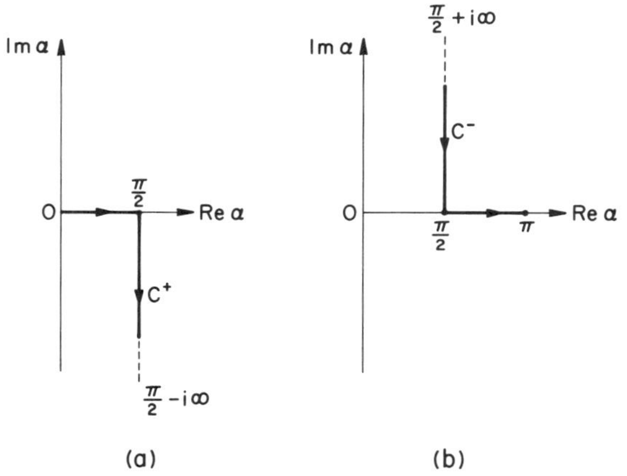

Fig. 2. The α contours C^+ and C^-.

$$G(\mathbf{r}, \mathbf{r}') = \frac{ik}{2\pi} \int_{-\pi}^{\pi} \mathrm{d}\beta \int_{C^\pm} \mathrm{d}\alpha \sin\alpha \exp[ik\mathbf{s} \cdot (\mathbf{r} - \mathbf{r}')], \tag{2.8}$$

where $\mathbf{s} \equiv \mathbf{s}(\alpha, \beta)$ is a unit vector with Cartesian components

$$s_x = \sin\alpha\cos\beta, \qquad s_y = \sin\alpha\sin\beta, \qquad s_z = \cos\alpha. \tag{2.9}$$

The polar angle β $(0 \leq \beta < 2\pi)$ is real, but the other polar angle, α, is either real or pure imaginary. When $z - z' > 0$ it takes on all values on the contour C^+ shown in Fig. 2(a). When $z - z' < 0$ it takes on all values on the contour C^- shown in Fig. 2(b).

We now substitute in Eq. (2.6) the integral representation (2.8) of the free-space Green's function, and we interchange the orders of integration. The resulting formula may be written in the form[21]

$$U^{(s)}(\mathbf{r}) = \frac{ik}{2\pi} \int_{-\pi}^{\pi} \mathrm{d}\beta \int_{C^\pm} \mathrm{d}\alpha \sin\alpha A(\alpha, \beta) \exp(ik\mathbf{s} \cdot \mathbf{r}), \tag{2.10}$$

where

$$A(\alpha, \beta) = -\frac{1}{4\pi} \int_D U(\mathbf{r}')F(\mathbf{r}') \exp(-ik\mathbf{s} \cdot \mathbf{r}')\, \mathrm{d}^3r'. \tag{2.11}$$

In Eq. (2.10) the α integration extends over the contour C^+ if the field point $\mathbf{r}$ is situated in the half-space $\mathcal{R}^+$, and it extends over the contour C^- if $\mathbf{r}$ is situated in the half-space $\mathcal{R}^-$ (Figs. 1 and 2).

Equation (2.10) expresses the scattered field throughout the half-spaces $\mathcal{R}^+$ and $\mathcal{R}^-$ in the form of angular spectra of plane waves, with the angular spectrum amplitude $A(\alpha, \beta)$ being given by expression (2.11). Although Eq. (2.11) contains the unknown total field $U(\mathbf{r})$ under the integral sign, one can nevertheless draw a number of important consequences from that equation, as we will show.

It is to be noted that along the horizontal portions of the contour C^+ and C^-, α takes on real values. The corresponding contributions $A(\alpha, \beta)\exp(ik\mathbf{s} \cdot \mathbf{r})$ to the integral on the right-hand side of Eq. (2.10) then represent homogeneous plane waves that propagate away from the scatterer. Along the vertical portions of the contours, the values of α are pure imaginary, and the corresponding contributions may then readily be shown to represent evanescent plane waves whose amplitudes decay exponentially with increasing distance, projected along the z direction, from the scatterer. Specifically, $A(\alpha, \beta) \exp(ik\mathbf{s} \cdot \mathbf{r})$ represents

in $\mathcal{R}^+$:
- a homogeneous wave if $0 \leq \alpha \leq \pi/2$, $\quad 0 \leq \beta < 2\pi$,
- an evanescent wave if $\alpha = (\pi/2) - i\gamma$, $\quad 0 \leq \gamma \leq \infty$, $0 \leq \beta < 2\pi$;

in $\mathcal{R}^-$:
- a homogeneous wave if $(\pi/2) < \alpha \leq \pi$, $\quad 0 \leq \beta < 2\pi$,
- an evanescent wave if $\alpha = (\pi/2) + i\gamma$, $\quad \infty > \gamma \geq 0$, $0 \leq \beta < 2\pi$.

For the real values of α, the spectral amplitude $A(\alpha, \beta)$ has a simple meaning. To see this, let us consider the scattered field at points in the far zone. More precisely, let us consider the asymptotic limit of expression (2.10) as $kr \to \infty$, i.e., as the distance of the field point from the scatterer increases without limit, in a fixed direction specified by a unit vector

$$\mathbf{u} = \mathbf{r}/r \tag{2.12}$$

$(r = |\mathbf{r}|)$. Let (θ, φ) be the (real) spherical polar angles of $\mathbf{u}$:

$$\mathbf{u} \equiv (\sin\theta\cos\varphi, \quad \sin\theta\sin\varphi, \quad \cos\theta). \tag{2.13}$$

We have, by elementary geometry (Fig. 3) $|\mathbf{r} - \mathbf{r}'| \sim r - \mathbf{r}' \cdot \mathbf{u}$, $1/|\mathbf{r} - \mathbf{r}'| \sim 1/r$ as $kr \to \infty$. Making use of these approximations in Eq. (2.7), we obtain at once the asymptotic approxi-

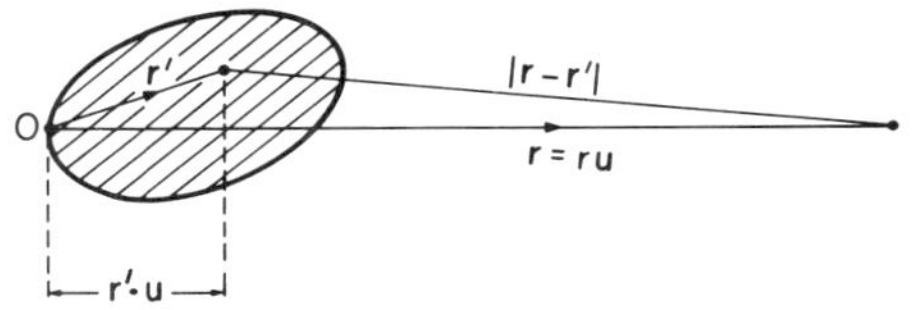

Fig. 3. Notation relating to the derivation of the asymptotic approximation (2.14) to the free-space Green's function $G(\mathbf{r}, \mathbf{r}')$.

mation for the free-space Green's function $G(\mathbf{r}, \mathbf{r}')$, viz.,

$$G(\mathbf{r}, \mathbf{r}') \sim \frac{\exp(ikr)}{r} \exp(-ik\mathbf{u} \cdot \mathbf{r}') \qquad (kr \to \infty). \tag{2.14}$$

On substituting for $G(\mathbf{r}, \mathbf{r}')$ from the formula (2.14) into Eq. (2.6), we find that the scattered field throughout the far zone is given by

$$U^{(s)}(r\mathbf{u}) \sim -\frac{1}{4\pi} \frac{\exp(ikr)}{r} \int_D U(\mathbf{r}')F(\mathbf{r}') \exp(-i\mathbf{u} \cdot \mathbf{r}')\mathrm{d}^3r' \qquad (kr \to \infty). \tag{2.15}$$

Finally, if we recall the expression (2.11) for the spectral amplitude, it follows at once from formula (2.15) that

$$U^{(s)}(r\mathbf{u}) \sim A(\theta, \varphi) \frac{\exp(ikr)}{r} \qquad (kr \to \infty). \tag{2.16}$$

This formula shows that for a pair of real values (θ, φ) of its arguments, the spectral amplitude function $A(\theta, \varphi)$ is just the *scattering amplitude* for the direction $\mathbf{u}$ specified by polar angles (θ, φ).

3. ANALYTIC PROPERTIES OF THE SPECTRAL AMPLITUDE FUNCTION $A(\alpha, \beta)$ AND SOME OF ITS CONSEQUENCES

According to Eqs. (2.11) and (2.9), the angular spectrum amplitude function $A(\alpha, \beta)$ may be represented in the form

$$A(\alpha, \beta) = -\frac{1}{4\pi} \int_D U(\mathbf{r}')F(\mathbf{r}') \exp[-ik(x' \sin\alpha \cos\beta + y' \sin\alpha \sin\beta + z' \cos\alpha)]\mathrm{d}^3\mathbf{r}' \tag{3.1}$$

$[\mathbf{r}' \equiv (x', y', z')]$. We have introduced the angular spectrum amplitude function (to be referred to from now on as the spectral amplitude function for short) for values of α on the contours C^+ and C^- (Fig. 2) in the complex α plane and for real values of β in the range $0 \le \beta < 2\pi$. We will now broaden the definition of this function. Assuming that the scattering potential $F(\mathbf{r}')$ is a continuous function of $\mathbf{r}'$ throughout the domain D, the integrand of the integral on the right-hand side of Eq. (3.1) is then also continuous in $\mathbf{r}'$ throughout D; moreover, it is also a continuous function of α and β for all values of these variables, real or complex. We also note that the integrand is, for each $\mathbf{r}' \in D$ and for any value α, an entire function of β and that for each $\mathbf{r}' \in D$ and for any value of β it is an entire function of α. Hence according to a well-known theorem on functions represented by definite integrals,[22] $A(\alpha, \beta)$ is, for each value of β, an entire function of α, and for each value of α it is an entire function of β. This result implies, according to a theorem of Hartogs,[23] the following important theorem:

Theorem I: *The function $A(\alpha, \beta)$, defined by Eq. (3.1) with any potential $F(\mathbf{r})$ that is continuous in a finite scattering volume D, is an entire function of two complex variables.*

This result has a number of interesting consequences. In particular, it follows at once from Theorem I and formula (2.16) that the following is also true:

Theorem II: *The scattering amplitude $A(\theta, \varphi)$ of a field scattered from a finite domain D throughout which the potential $F(\mathbf{r})$ is continuous is the boundary value on the real unit sphere ($\sin\theta \cos\varphi, \sin\theta \sin\varphi, \cos\theta, 0 \le \theta \le \pi, 0 \le \varphi < 2\pi$) of an entire function $A(\alpha, \beta)$ of two complex variables.*

The results expressed by Theorems I and II are analogous for the scattering amplitude of results established previously for the radiation pattern of a field generated by any monochromatic source distribution of finite support (Ref. 4, Sec. 3).

From Theorem II one obtains at once by the principle of analytic continuation

Theorem III: *If the scattering amplitude $A(\theta, \varphi)$ of a field scattered from a finite domain D throughout which the potential $F(\mathbf{r})$ is continuous vanishes for all directions $\mathbf{u}(\theta, \varphi)$ within a finite solid angle $\delta\Omega$, however small, it vanishes for all directions; and, more generally, one then has $A(\alpha, \beta) \equiv 0$ for all values α, β, real or complex.*

The first part of Theorem III was established previously in a different manner by Müller.[24]

The analyticity of $A(\alpha, \beta)$, expressed by Theorem I, also has some important consequences regarding the structure of the scattered field at arbitrary distance from the scatterer. To show this we recall that Eq. (2.10) represents the scattered field throughout the half-spaces $\mathcal{R}^+$ and $\mathcal{R}^-$ as superposition of homogeneous and evanescent plane waves. Let us examine whether it is possible to generate a scattered field that does not contain any homogeneous waves or that does not contain any evanescent waves.

Suppose first that the scattered field does not contain any homogeneous waves either in the half-space $\mathcal{R}^+$ or in the half-space $\mathcal{R}^-$. Then $A(\alpha, \beta) = 0$ for all α, β such that

$$0 \le \alpha \le \pi/2, \qquad 0 \le \beta < 2\pi \tag{3.2a}$$

if there are no homogeneous waves in $\mathcal{R}^+$ and $A(\alpha, \beta) = 0$ for all α, β such that

$$\frac{\pi}{2} < \alpha \le \pi, \qquad 0 \le \beta < 2\pi \tag{3.2b}$$

if there are no homogeneous waves in $\mathcal{R}^-$. In either case it follows at once from Theorem III that $A(\alpha, \beta) \equiv 0$ for all values of α and β, and hence, according to Eq. (2.10) the scattered field $U^{(s)}(\mathbf{r})$ vanishes identically throughout both the half-spaces $\mathcal{R}^+$ and $\mathcal{R}^-$. From this conclusion and from well-known analytic properties of solutions of the Helmholtz equation[25] a sharper result follows, namely, that the scattered field $U^{(s)}(\mathbf{r})$ then vanishes everywhere outside the scattering volume.

Next suppose that the scattered field does not contain any evanescent waves either in the half-space $\mathcal{R}^+$ or in the half-space $\mathcal{R}^-$. Then $A(\alpha, \beta) = 0$ for all α, β such that

$$\alpha = \frac{\pi}{2} - i\gamma, \qquad 0 \le \gamma < \infty, \qquad 0 \le \beta < 2\pi \tag{3.3a}$$

if there are no evanescent waves in $\mathcal{R}^+$ and $A(\alpha, \beta) = 0$ for all α, β such that

$$\alpha = \frac{\pi}{2} + i\gamma, \qquad \infty > \gamma \ge 0, \qquad 0 \le \beta < 2\pi \tag{3.3b}$$

if there are no evanescent waves in the half-space $\mathcal{R}^-$. Since according to Theorem I $A(\alpha, \beta)$ is an entire function of α and β, it follows at once from Eqs. (3.3) by the principle of analytic continuation that when either of the conditions (3.3) holds, $A(\alpha, \beta) \equiv 0$ for all α and β. Hence by the same argument as given in connection with the previous case (absence of homogeneous waves), the scattered field $U^{(s)}(\mathbf{r})$ vanishes everywhere outside the scattering volume.

These results may be summarized in the following theorem:

Theorem IV: The scattered field $U^{(s)}(\mathbf{r})$ *generated from a scatterer that occupies a finite domain D throughout which the scattering potential* $F(\mathbf{r})$ *is continuous and that is located in the strip* $0 \le z \le L(<\infty)$ *contains necessarily both homogeneous and evanescent waves in both half-spaces* $\mathcal{R}^-(z<0)$ *and* $\mathcal{R}^+(z>L)$*, unless the scattered field vanishes identically everywhere outside D.*[26]

ACKNOWLEDGMENTS

We are obliged to D. Colton for helpful advice in connection with the proof of Theorem I. This research was supported by the National Science Foundation.

The authors are also with the Institute of Optics, University of Rochester.

* M. Nieto-Vesperinas is on leave from Instituto de Optica, CSIC, Serrano 121, 28006 Madrid, Spain.

REFERENCES

1. H. G. Booker, J. A. Ratcliffe, and D. H. Shinn, "Diffraction from an irregular screen with applications to ionospheric problems," Phil. Trans. R. Soc. London Ser. A **242**, 579–609 (1950).
2. D. R. Rhodes, "On a fundamental principle in the theory of planar antennas," Proc. IEEE **52**, 1013–1021 (1964).
3. D. M. Kerns, "Scattering matrix description and nearfield measurements of electroacoustic tranducers," J. Acoust. Soc. Am. **57**, 497–507 (1975).
4. A. J. Devaney and E. Wolf, "Multipole expansions and plane wave representations of the electromagnetic field," J. Math. Phys. **15**, 234–244 (1974).
5. E. Lalor and E. Wolf, "New model for the interaction between a moving charged particle and a dielectric, and the Čerenkov effect," Phys. Rev. Lett. **26**, 1274–1277 (1971).
6. A. Walther, "Gabor's theorem and energy transfer through lenses," J. Opt. Soc. Am. **57**, 639–644 (1967); also "Systematic approach to the teaching of lens theory," Am. J. Phys. **35**, 808–816 (1967).
7. R. Mittra and P. L. Ransom, "Imaging with coherent fields" in *Proceedings of the Symposium on Modern Optics*, J. Fox, ed. (Wiley, New York, 1967), pp. 619–647.
8. W. D. Montgomery, "Self-imaging objects of infinite aperture," J. Opt. Soc. Am. **57**, 772–778 (1967).
9. B. Richards and E. Wolf, "Electromagnetics diffraction in optical systems II. Structure of the image field in an aplanatic system," Proc. R. Soc. London Ser. A **253**, 358–379 (1959).
10. D. Gabor, "Light and information" in *Progress in Optics*, E. Wolf, ed. (North-Holland, Amsterdam, 1961), Vol. 1, pp. 109–153, App. I.
11. G. C. Sherman, "Integral-transform formulation of diffraction theory," J. Opt. Soc. Am. **57**, 1490–1498 (1967).
12. J. R. Shewell and E. Wolf, "Inverse diffraction and a new reciprocity theorem," J. Opt. Soc. Am. **58**, 1596–1603 (1968).
13. E. Wolf, "Three-dimensional structure determination of semitransparent objects from holographic data," Opt. Commun. **1**, 153–156 (1969).
14. M. Nieto-Vesperinas, "Inverse scattering problems: A study in terms of the zeros of entire functions," J. Math. Phys. **25**, 2109–2115 (1984).
15. G. S. Agarwal, "Interaction of electromagnetic waves at rough dielectric surfaces," Phys. Rev. B **15**, 2371–2382 (1977).
16. E. Wolf and J. R. Shewell, "Diffraction theory of holography," J. Math. Phys. **11**, 2254–2267 (1970).
17. G. S. Agarwal, D. N. Pattanayak, and E. Wolf, "Electromagnetic fields in spatially dispersive media" Phys. Rev. B **10**, 1447–1475 (1974).
18. A. Walther, "Radiometry and coherence," J. Opt. Soc. Am. **58**, 1256–1259 (1968).
19. G. S. Agarwal, A. T. Friberg, and E. Wolf, "Scattering theory of distortion correction by phase conjugation," J. Opt. Soc. Am. **73**, 529–538 (1983).
20. H. Weyl, "Ausbreitung elektromagnetischer Wellen über einem ebenen Leiter," Ann. Phys. (Leipzig) **60**, 481–500 (1919). Because the integrand of Eq. (2.8) is an entire function of α, the α integration contours $C^{\pm}$ may be deformed in accordance with the rules of integration in the complex plane. When these contours are chosen as shown in Fig. 2, Eq. (2.8) transforms to the following, frequently used, alternative form of the Weyl representation:
$$\frac{\exp(ik|\mathbf{r}-\mathbf{r}'|)}{|\mathbf{r}-\mathbf{r}'|} = \frac{ik}{2\pi}\int_{-\infty}^{\infty}\frac{1}{s_z} \times \exp\{ik[s_x(x-x')+s_y(y-y')+s_z|z-z'|]\}\,ds_x ds_y. \quad \text{(R1)}$$
Here,
$$s_z = +(1-s_x^2-s_y^2)^{1/2} \quad \text{when } s_x^2+s_y^2 \le 1,$$
$$= +i(s_x^2+s_y^2-1)^{1/2} \quad \text{when } s_x^2+s_y^2 > 1. \quad \text{(R2)}$$
The two forms of the Weyl expansion, and the transformation required to pass from one to the other, are discussed in A. Baños, *Dipole Radiation in the Presence of a Conducting Half-Space* (Pergamon, New York, 1966), Sec. 2.13.
21. If instead of Eq. (2.8) we make use in Eq. (2.6) of the alternative representation (R1) of $G(\mathbf{r}, \mathbf{r}')$, we readily obtain the following alternative form of the angular spectrum representation of the scattered fields:
$$U^{(s)}(r) = \iint_{-\infty}^{\infty} a^{(\pm)}(s_x, s_y)\exp[ik(s_x x + s_y y \pm s_z z)]ds_x ds_y. \quad \text{(R3)}$$
Here s_z is given by Eq. (R2), and
$$a^{(\pm)}(s_x, s_y) = \frac{ik}{2\pi}A(\alpha, \beta). \quad \text{(R4)}$$
In Eq. (R3) the upper or the lower sign is taken according to whether the point $\mathbf{r} \equiv (x, y, z)$ is situated in the half-space $\mathcal{R}^+$ or $\mathcal{R}^-$, respectively. In Eq. (R4) the upper or lower sign is associated with polar angle α that is on the contour C^+ or C^-, respectively. Angular spectrum representations of the form (R3) are frequently used in optics.
22. E. T. Copson, *Theory of Functions of a Complex Variable* (Oxford U. Press, Oxford, 1970), pp. 107–108.
23. W. F. Osgood, *Topics in the Theory of Functions of Several Complex Variables* (Dover, New York, 1966), p. 33.
24. C. Müller, "Radiation patterns and radiation fields," J. Rat. Mech. Anal. **4**, 235–246 (1955), especially pp. 237–238.
25. D. Colton and R. Kress, *Integral Equation Methods in Scattering Theory* (Wiley, New York, 1983), Theorem 3.5, p. 72.
26. Whether there are finite range potentials that do not give rise to scattering [i.e., for which $U^{(s)}(\mathbf{r}) \equiv 0$ everywhere outside the scattering volume D] does not appear to be known. However, the answer to the corresponding question for the radiation problem is known: There are monochromatic source distributions of finite support D that do not generate any fields outside D [see, for example, A. J. Devaney and E. Wolf, "Radiating and non-radiating classical current distributions and the fields they generate," Phys. Rev. **D 8**, 1044–1047 (1973)].

(see overleaf)

Emil Wolf

Emil Wolf is Professor of Physics and Professor of Optics at the University of Rochester. His main areas of research are physical optics and the theory of partial coherence. He is the coauthor, with Max Born, of the text *Principles of Optics* and is the Editor of *Progress in Optics*. He was the President of the Optical Society of America in 1978 and received the Society's Frederic Ives Medal in 1977. He is also the 1980 recipient of the Albert A. Michelson Medal of the Franklin Institute. In 1981 he was elected an Honorary Member of the Optical Society of India.

Manuel Nieto-Vesperinas

Manuel Nieto-Vesperinas was born in Madrid, Spain, on August 4, 1950. He received the B.Sc. and M.Sc. degrees from Universidad Autonoma de Madrid in 1973 and the Ph.D. degree in physics from the University of London, U.K., in 1978. He has been a Senior Scientist at the Instituto de Optica, C.S.I.C. (National Research Council, Madrid) since 1981. Between 1978 and 1981 he was a Visiting Scientist at Queen Elizabeth College, University of London, and at the Physikalisches Institut der Universität Erlangen-Nürnberg, West Germany. During the academic years 1983–1985 he is a Visiting Scientist at the Institute of Optics and Department of Physics and Astronomy, University of Rochester, New York. His research interests lie in physical optics, including coherence theory, wave propagation and scattering in random media and rough surfaces, inverse scattering, and mathematical methods for processing of scattering data (phase problems and resolution problems), to which he has contributed about 44 papers and several conference communications. He is an advisory editor to *Optics Communications* and a member of the Optical Society of America.

Frequency shifts of spectral lines produced by scattering from spatially random media

E. Wolf

Department of Physics and Astronomy, University of Rochester, Rochester, New York 14627

J. T. Foley

Department of Physics and Astronomy, Mississippi State University, Mississippi State, Mississippi 39762

F. Gori

Departimento di Fisica, Università di Roma "La Sapienza," P. le Aldo Moro 2-00185, Rome, Italy

Received July 29, 1988; accepted February 23, 1989

Scattering of polychromatic light by a medium whose dielectric susceptibility is a random function of position is considered within the accuracy of the first Born approximation. It is shown, in particular, that if the two-point spatial correlation function of the dielectric susceptibility has Gaussian form and the spectrum of the incident light has a Gaussian profile, the spectrum of the scattered light may be shifted toward the shorter or the longer wavelengths, depending on the angle of scattering. The results are analogous to those derived recently in connection with radiation from partially coherent sources [Nature (London) **326**, 363 (1987)].

I. INTRODUCTION

It was predicted theoretically not long ago, that, in general, correlations between source fluctuations produce changes in the spectrum of the emitted light.[1,2] For radiation from planar secondary sources this prediction was subsequently verified by experiment.[3] It was also shown theoretically that the changes may be such as to produce red shifts or blue shifts of spectral lines,[4–6] and this prediction too has been verified.[7–9] A similar effect can also be expected to arise with acoustical waves and was, in fact, observed not long ago.[10] Some other modifications of spectra due to source correlations have also been discussed.[11]

Because of the well-known analogy that exists between the processes of radiation and scattering, one might expect that similar phenomena will arise when a polychromatic wave is scattered by a medium whose dielectric susceptibility is a random function of position. We show, in the present paper, that this indeed is the case. First we derive an expression, valid within the accuracy of the first Born approximation, for the spectrum of the scattered light in terms of the spectrum of the incident light and the two-point spatial correlation function of the dielectric susceptibility of the medium. We then show that if the spectrum of the incident light consists of a single line of Gaussian profile and the correlation function of the dielectric susceptibility is also a Gaussian function, the spectrum of the scattered field will consist of a line that has approximately a Gaussian profile. However, this line is, in general, red shifted or blue shifted with respect to that of the incident light, depending on the angle of scattering. This result may appear, at first sight, to contradict the well-known fact that there is no frequency change in linear scattering on a time-invariant medium and that different frequency components of the incident and also of the scattered light are uncorrelated. The resolution of this apparent paradox is briefly discussed in the concluding section.

2. EXPRESSION FOR THE SPECTRUM OF THE SCATTERED FIELD

Let us consider a field incident upon a scattering medium occupying a finite volume V. Suppose that the incident field propagates in a direction specified by a real unit vector $\hat{s}_0$. We do *not* assume, however, that the field is monochromatic; rather we consider it to fluctuate at each point, generally in a random manner, characterized by an ensemble that is statistically stationary. The cross-spectral density $W^{(i)}(\mathbf{r}_1, \mathbf{r}_2, \omega)$ of the incident field at points whose location is specified by position vectors $\mathbf{r}_1$ and $\mathbf{r}_2$ may be expressed in the form[12]

$$W^{(i)}(\mathbf{r}_1, \mathbf{r}_2, \omega) = \langle U^{(i)*}(\mathbf{r}_1, \omega) U^{(i)}(\mathbf{r}_2, \omega)\rangle, \tag{2.1}$$

where $\{U^{(i)}(\mathbf{r}, \omega)\}$ represents a statistical ensemble of random functions, all of the form

$$U^{(i)}(\mathbf{r}, \omega) = a(\omega)\exp(ik\hat{s}_0 \cdot \mathbf{r}), \tag{2.2}$$

with

$$k = \frac{\omega}{c}, \tag{2.3}$$

c being the speed of light *in vacuo*. In Eq. (2.2) $a(\omega)$ are (generally complex) frequency-dependent random variables, and the angle brackets in Eq. (2.1) denote the expectation value, taken over the ensemble of the incident field.

0740-3232/89/081142-08$02.00

The spectrum

$$S^{(i)}(\mathbf{r}, \omega) \equiv W^{(i)}(\mathbf{r}, \mathbf{r}, \omega) = \langle U^{(i)*}(\mathbf{r}, \omega) U^{(i)}(\mathbf{r}, \omega)\rangle \quad (2.4)$$

of the incident field is then, according to Eq. (2.2), given by

$$S^{(i)}(\omega) \equiv S^{(i)}(\mathbf{r}, \omega) = \langle a^*(\omega) a(\omega)\rangle \quad (2.5)$$

and is seen to be independent of position.

We assume that the scatterer is weak in the sense that the amplitude of the scattered field $U^{(s)}(\mathbf{r}, \omega)$ is small compared with the amplitude of the incident field $[|U^{(s)}| \ll |U^{(i)}|]$. We may then calculate $U^{(s)}$ on the basis of the first Born approximation, which yields[13]

$$U^{(s)}(\mathbf{r}, \omega) = a(\omega) \int_V F(\mathbf{r}', \omega) G(\mathbf{r}, \mathbf{r}', \omega) \exp(ik\hat{s}_0 \cdot \mathbf{r}') \mathrm{d}^3 r', \quad (2.6)$$

where

$$F(\mathbf{r}, \omega) = (\omega/c)^2 \eta(\mathbf{r}, \omega), \quad (2.7)$$

$\eta(\mathbf{r}, \omega) = [n^2(\mathbf{r}, \omega) - 1]/4\pi$ being the dielectric susceptibility of the scattering medium, $n(\mathbf{r}, \omega)$ its refractive index, and

$$G(\mathbf{r}, \mathbf{r}', \omega) = \frac{\exp[ik(\mathbf{r} - \mathbf{r}')]}{|\mathbf{r} - \mathbf{r}'|} \quad (2.8)$$

the outgoing free-space Green's function.

We will consider only the scattered field in the far zone. If we set $\mathbf{r} = r\hat{s}$, $(\hat{s}^2 = 1)$, the Green's function may then be approximated by its asymptotic form (see Fig. 1)

$$G(\mathbf{r}, \mathbf{r}', \omega) \sim \frac{e^{ikr}}{r} \exp(-ik\hat{s} \cdot \mathbf{r}') \qquad (kr \to \infty). \quad (2.9)$$

On substituting from the formula (2.9) into Eq. (2.6), we find at once that

$$U^{(\infty)}(r\hat{s}, \omega) = a(\omega) \frac{e^{ikr}}{r} \int_V F(\mathbf{r}', \omega) \exp[-ik(\hat{s} - \hat{s}_0) \cdot \mathbf{r}'] \mathrm{d}^3 r', \quad (2.10)$$

where we have written $U^{(\infty)}$ rather than $U^{(s)}$ to stress that the expression on the right-hand side of Eq. (2.10) represents the scattered field in the far zone. If we introduce the Fourier transform of the scattering potential by the formula

$$\tilde{F}(\mathbf{K}, \omega) = \int_V F(\mathbf{r}', \omega) \exp(-i\mathbf{K} \cdot \mathbf{r}') \mathrm{d}^3 r', \quad (2.11)$$

Eq. (2.10) becomes

$$U^{(\infty)}(r\hat{s}, \omega) = a(\omega) \tilde{F}[k(\hat{s} - \hat{s}_0), \omega] \frac{e^{ikr}}{r}, \quad (2.12)$$

or, using Eq. (2.7),

$$U^{(\infty)}(r\hat{s}, \omega) = a(\omega) \left(\frac{\omega}{c}\right)^2 \tilde{\eta}[k(\hat{s} - \hat{s}_0), \omega] \frac{e^{ikr}}{r}, \quad (2.12a)$$

where $\tilde{\eta}(\mathbf{K}, \omega)$ is the Fourier transform, defined by a formula of the form (2.11), of the dielectric susceptibility $\eta(\mathbf{r}', \omega)$.

Let us next consider the spectrum of the scattered field in the far zone. It can be determined from the following formula analogous to Eq. (2.4), which is

$$S^{(\infty)}(r\hat{s}, \omega) = \langle U^{(\infty)*}(r\hat{s}, \omega) U^{(\infty)}(r\hat{s}, \omega)\rangle. \quad (2.13)$$

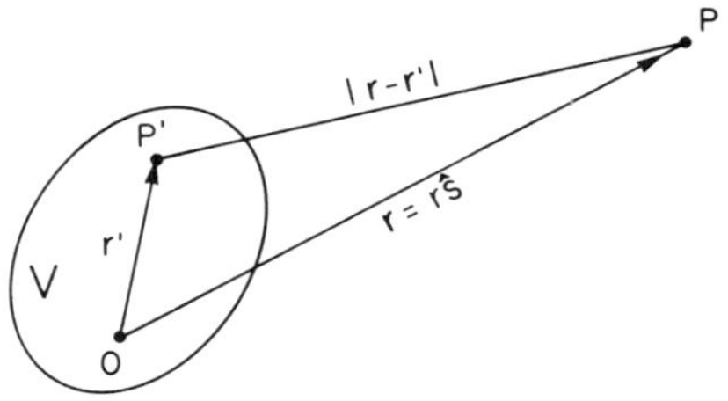

Fig. 1. Illustrating the notation relating to the asymptotic approximation (2.9).

On substituting from Eq. (2.12a) into Eq. (2.13) and on using Eq. (2.5), we find at once that

$$S^{(\infty)}(r\hat{s}, \omega) = \frac{1}{r^2} \left(\frac{\omega}{c}\right)^4 \tilde{\eta}^*[k(\hat{s} - \hat{s}_0), \omega] \tilde{\eta}[k(\hat{s} - \hat{s}_0), \omega] S^{(i)}(\omega). \quad (2.14)$$

Up to now we have assumed that the scattering medium was deterministic. Let us now assume, instead, that its physical properties are characterized by random functions of position but are independent of time. An example of such a medium would be the atmosphere under conditions when its temporal fluctuations are slow enough to be ignored. Then, for each ω, the dielectric susceptibility $\eta(\mathbf{r}, \omega)$ will be a random function of $\mathbf{r}$, and consequently $\tilde{\eta}(\mathbf{K}, \omega)$ will be a random function of $\mathbf{K}$. A meaningful measure of the spectrum of the scattered field in the far zone is then the expectation value of the right-hand side of Eq. (2.14), taken over an ensemble of different realizations of the scattering medium; i.e., it is given by the expression

$$S^{(\infty)}(r\hat{s}, \omega) = \frac{1}{r^2} \left(\frac{\omega}{c}\right)^4 \langle \tilde{\eta}^*[k(\hat{s} - \hat{s}_0), \omega] \tilde{\eta}[k(\hat{s} - \hat{s}_0), \omega]\rangle_\eta S^{(i)}(\omega). \quad (2.15)$$

Here the angle brackets, with the subscript η, denote the average value taken over this ensemble. It is to be noted that the expression (2.15) involves two averaging procedures, one over the ensemble of the incident field [cf. Eq. (2.5)] and the other over an ensemble of scatterers. We implicitly assume here that these two sources of randomness are mutually independent.

Although the formula (2.15) for the far-field spectrum involves an average, obtained from experiments with macroscopically similar but microscopically different scatters, one can in some cases deduce the value of this average, at least to a good approximation, from experiments performed with a single scatterer. For example, often the necessarily finite size of the detector aperture will provide spatial averaging, which is essentially equivalent to ensemble averaging.[14,15] The value of the ensemble average may also be obtained sometimes by the use of a moving aperture in front of the scatterer.[16]

If we express $\tilde{\eta}$ in terms of its Fourier inverse η and interchange the order of averaging and integrations, we find that

$$\langle \tilde{\eta}^*(\mathbf{K}_1, \omega) \tilde{\eta}(\mathbf{K}_2, \omega)\rangle_\eta = \int_V \int_V C_\eta(\mathbf{r}_1', \mathbf{r}_2', \omega) \times \exp[-i(\mathbf{K}_2 \cdot \mathbf{r}_2' - \mathbf{K}_1 \cdot \mathbf{r}_1')] \mathrm{d}^3 r_1' \mathrm{d}^3 r_2', \quad (2.16)$$

where

$$C_\eta(\mathbf{r}_1', \mathbf{r}_2', \omega) = \langle \eta^*(\mathbf{r}_1', \omega)\eta(\mathbf{r}_2', \omega)\rangle_\eta \tag{2.17}$$

is the spatial correlation function of the dielectric susceptibility. If we introduce its six-dimensional Fourier transform

$$\tilde{C}_\eta(\mathbf{K}_1, \mathbf{K}_2, \omega) = \int_V \int_V C_\eta(\mathbf{r}_1', \mathbf{r}_2', \omega) \times \exp[-i(\mathbf{K}_1 \cdot \mathbf{r}_1' + \mathbf{K}_2 \cdot \mathbf{r}_2')]\mathrm{d}^3r_1'\mathrm{d}^3r_2', \tag{2.18}$$

Eq. (2.16) becomes

$$\langle \tilde{\eta}^*(\mathbf{K}_1, \omega)\tilde{\eta}(\mathbf{K}_2, \omega)\rangle_\eta = \tilde{C}_\eta(-\mathbf{K}_1, \mathbf{K}_2, \omega). \tag{2.19}$$

Finally, on substituting from Eq. (2.19) into Eq. (2.15), we obtain the following expression for the spectrum of the scattered field in the far zone:

$$S^{(\infty)}(r\hat{s}, \omega) = \frac{1}{r^2}(\omega/c)^4\tilde{C}_\eta[-k(\hat{s} - \hat{s}_0), k(\hat{s} - \hat{s}_0), \omega]S^{(i)}(\omega). \tag{2.20}$$

The formula (2.20) shows that the spectrum of the scattered field in the far zone differs, in general, from the spectrum $S^{(i)}(\omega)$ of the incident field by the effect of two multiplicative factors, namely, a factor proportional to ω^4 and the factor $\tilde{C}_\eta$. The ω^4 factor is a reflection of the fact that on the microscopic level the medium responds to the incident field as a set of dipole oscillators [cf. Ref. 17, Secs. 2.2.1 and 2.2.3]. The other factor, $\tilde{C}_\eta$, may be regarded as representing the correlation that exists between them. Because $\tilde{C}_\eta$ in Eq. (2.20) depends on the momentum transfer vector $\mathbf{K} = k(\hat{s} - \hat{s}_0)$, it is a function of the angle of scattering, and hence, in general, the far-field spectrum will be different in different directions of observation.

The situation is somewhat different in the idealized special case when the dielectric susceptibility of the scattering medium is completely uncorrelated, i.e., when

$$\begin{aligned} C_\eta(\mathbf{r}_1', \mathbf{r}_2', \omega) &= I(\omega)\delta^{(3)}(\mathbf{r}_2' - \mathbf{r}_1') \quad && \text{when } \mathbf{r}_1' \in V, \quad \mathbf{r}_2' \in V \\ &= 0 && \text{otherwise,} \end{aligned} \tag{2.21}$$

where $\delta^{(3)}(\mathbf{r}')$ is the three-dimensional Dirac delta function and $I(\omega)$ is a nonnegative function of frequency. We then have, according to Eqs. (2.18) and (2.21),

$$\tilde{C}_\eta[-k(\hat{s} - \hat{s}_0), k(\hat{s} - \hat{s}_0), \omega] = I(\omega)V, \tag{2.22}$$

and the formula (2.20) reduces to

$$[S^{(\infty)}(r\hat{s}, \omega)]_{\text{uncorr}} = \frac{V}{r^2}\left(\frac{\omega}{c}\right)^4 I(\omega)S^{(i)}(\omega). \tag{2.23}$$

We see that in this special case the spectrum of the scattered field is the same for all angles of scattering and that it differs from the source spectrum $I(\omega)$ only by a multiplicative factor that is proportional to the dipole term ω^4.

It seems worthwhile to stress that in deriving Eq. (2.20) no assumption was made regarding homogeneity or isotropy of the scatterer. When the scatterer is statistically homogeneous, as is often the case, the formula (2.20) considerably simplifies, as we will now show.

3. SPECTRUM OF THE FIELD SCATTERED BY A STATISTICALLY HOMOGENEOUS MEDIUM

When the medium is statistically homogeneous, the correlation function $C_\eta(\mathbf{r}_1', \mathbf{r}_2', \omega)$ will depend on $\mathbf{r}_1'$ and $\mathbf{r}_2'$ only through the difference $\mathbf{r}_2' - \mathbf{r}_1'$, and we will then write

$$\begin{aligned} C_\eta(\mathbf{r}_1', \mathbf{r}_2', \omega) &= C_\eta(\mathbf{r}_2' - \mathbf{r}_1', \omega) \quad && \text{when } \mathbf{r}_1' \in V, \quad \mathbf{r}_2' \in V \\ &= 0 && \text{otherwise.} \end{aligned} \tag{3.1}$$

In this case Eq. (2.18) becomes

$$\tilde{C}_\eta(\mathbf{K}_1, \mathbf{K}_2, \omega) = \int_V \int_V C_\eta(\mathbf{r}_2' - \mathbf{r}_1', \omega)\exp[-i(\mathbf{K}_1 \cdot \mathbf{r}_1' + \mathbf{K}_2 \cdot \mathbf{r}_2')]\mathrm{d}^3\mathbf{r}_1'\mathrm{d}^3r_2', \tag{3.2}$$

and hence

$$\tilde{C}_\eta[-k(\hat{s} - \hat{s}_0), k(\hat{s} - \hat{s}_0), \omega] = \int_V \int_V C_\eta(\mathbf{r}_2' - \mathbf{r}_1', \omega)\exp[-ik(\hat{s} - \hat{s}_0) \cdot (\mathbf{r}_2' - \mathbf{r}_1')]\mathrm{d}^3r_1'\mathrm{d}^3r_2'. \tag{3.3}$$

If we change the variables of integration from $\mathbf{r}_1'$, $\mathbf{r}_2'$ to $\mathbf{r}$, $\mathbf{r}'$ by setting

$$\mathbf{r} = (\mathbf{r}_1' + \mathbf{r}_2')/2, \qquad \mathbf{r}' = (\mathbf{r}_2' - \mathbf{r}_1') \tag{3.4}$$

and assume that the linear dimensions of the scattering volume are large compared both with the correlation length of the dielectric susceptibility [the effective $\mathbf{r}'$ range of $|C_\eta(\mathbf{r}', \omega)|$] and with the wavelength $\lambda = 2\pi c/\omega$, Eq. (3.3) gives

$$\tilde{C}_\eta[-k(\hat{s} - \hat{s}_0), k(\hat{s} - \hat{s}_0), \omega] \approx V\tilde{C}_\eta[k(\hat{s} - \hat{s}_0), \omega], \tag{3.5}$$

where $\tilde{C}_\eta(\mathbf{K}, \omega)$ is the three-dimensional Fourier transform of $C_\eta(\mathbf{r}', \omega)$, viz.,

$$\tilde{C}_\eta(\mathbf{K}, \omega) = \int_V C_\eta(\mathbf{r}', \omega)\exp(-i\mathbf{K} \cdot \mathbf{r}')\mathrm{d}^3r'. \tag{3.6}$$

On substituting from formula (3.5) into the general formula (2.20) we obtain the following simpler formula for the spectrum of the scattered field:

$$S^{(\infty)}(r\hat{s}, \omega) \approx \frac{V}{r^2}\left(\frac{\omega}{c}\right)^4 \tilde{C}_\eta[k(\hat{s} - \hat{s}_0), \omega]S^{(i)}(\omega). \tag{3.7}$$

This formula again shows that the spectrum of the incident field is, in general, modified by interaction with the scattering medium. The modification arises from a dipole contribution that is proportional to the fourth power of the frequency and by the influence of the dielectric susceptibility correlations, characterized by the correlation function $C_\eta(\mathbf{r}', \omega)$.

4. FREQUENCY SHIFTS GENERATED BY SCATTERING

To illustrate the effect of the dielectric susceptibility correlations of the scattering medium, let us consider the situation when the correlation function is a three-dimensional Gaussian distribution, i.e., when

$$C_\eta(\mathbf{r}', \omega) = \frac{A}{(2\pi\sigma^2)^{3/2}} \exp(-r'^2/2\sigma^2), \tag{4.1}$$

where A and σ are positive constants, i.e., they are independent of both $\mathbf{r}'$ and ω.

The three-dimensional Fourier transform [defined by Eq. (3.6)] of the expression (4.1) is

$$\tilde{C}_\eta(\mathbf{K}, \omega) = A \exp[-(K\sigma)^2/2]. \tag{4.2}$$

Now with $\mathbf{K} = k(\hat{s} - \hat{s}_0)$ we have, since $\hat{s}$ and $\hat{s}_0$ are unit vectors and $k = \omega/c$,

$$K^2 = 4(\omega/c)^2 \sin^2(\theta/2), \tag{4.3}$$

where θ is the scattering angle ($\hat{s} \cdot \hat{s}_0 = \cos\theta$). Hence it follows that, in this case,

$$\tilde{C}_\eta[k(\hat{s} - \hat{s}_0), \omega] = A \exp\left[-2\left(\frac{\omega}{c}\right)^2 \sigma^2 \sin^2\left(\frac{\theta}{2}\right)\right], \tag{4.4}$$

and the expression (3.7) for the spectrum of the scattered field becomes

$$S^{(\infty)}(r\hat{s}, \omega) = \frac{AV}{r^2}\left(\frac{\omega}{c}\right)^4 \exp\left[-2\left(\frac{\omega}{c}\right)^2 \sigma^2 \sin^2\left(\frac{\theta}{2}\right)\right] S^{(i)}(\omega). \tag{4.5}$$

Suppose that the spectrum of the incident field is a Gaussian function of rms width Γ_0, centered at frequency ω_0, i.e., that

$$S^{(i)}(\omega) = B \exp\left[\frac{-(\omega - \omega_0)^2}{2\Gamma_0^2}\right], \tag{4.6}$$

where B, ω_0 and Γ_0 are positive constants. On substituting from Eq. (4.6) into Eq. (4.5), we see that the right-hand side contains the product of two Gaussian functions. With the help of a product theorem for Gaussian functions established in Appendix A, we show in Appendix B that the resulting expression for the spectrum of the scattered field can be written in the form

$$S^{(\infty)}(r, \theta; \omega) = N(r)H(\theta)\omega^4 \exp\left\{-\frac{1}{2}\left[\frac{\omega - \tilde{\omega}(\theta)}{\tilde{\Gamma}(\theta)}\right]^2\right\}, \tag{4.7}$$

where we have now written $S^{(\infty)}(r, \theta; \omega)$ rather than $S^{(\infty)}(r\hat{s}, \omega)$. The various quantities that appear on the right-hand side of Eq. (4.7) are defined by the following formulas:

$$N(r) = \frac{VAB}{c^4 r^2}, \tag{4.8a}$$

$$H(\theta) = \exp\left[-\frac{2}{\alpha^2(\theta)}\left(\frac{\sigma}{\bar{\lambda}_0}\right)^2 \sin^2\left(\frac{\theta}{2}\right)\right], \tag{4.8b}$$

$$\tilde{\omega}(\theta) = \frac{\omega_0}{\alpha^2(\theta)}, \tag{4.8c}$$

$$\tilde{\Gamma}(\theta) = \frac{\Gamma_0}{\alpha(\theta)}, \tag{4.8d}$$

where

$$\alpha(\theta) = \left\{1 + \left[2\left(\frac{\sigma}{\bar{\lambda}_0}\right)\left(\frac{\Gamma_0}{\omega_0}\right)\sin\left(\frac{\theta}{2}\right)\right]^2\right\}^{1/2} \tag{4.8e}$$

and

$$\bar{\lambda}_0 = \frac{\lambda_0}{2\pi}. \tag{4.8f}$$

We see from Eq. (4.8e) that $\alpha(\theta) \geq 1$ and that the equality holds only [i.e., that $\alpha(\theta) = 1$] when either $\sigma = 0$ (completely uncorrelated scatterer) or $\Gamma_0 = 0$ (monochromatic light) or when $\theta = 0$ (forward scattering). Hence except in these special cases

$$\tilde{\omega}(\theta) < \omega_0, \qquad \tilde{\Gamma}(\theta) < \Gamma_0, \tag{4.9}$$

implying that the Gaussian function in Eq. (4.7) is centered on a frequency that is lower than the central frequency ω_0 of the incident light and its rms width is smaller. The fact that $\tilde{\omega}(\theta) < \omega_0$ implies that the Gaussian function in the expression (4.7) is centered at a lower frequency (i.e., is red shifted) with respect to the Gaussian spectral line of the incident light, the magnitude of the shift depending on the angle of scattering,[18] θ. On the other hand, the factor ω^4 in the expression (4.7) is an increasing function of the frequency and hence will produce a shift toward the higher frequencies (i.e., a blue shift). Consequently the spectrum of the scattered field will be either red shifted or blue shifted with respect to the spectrum of the incident light, depending on the magnitudes of these two contributions. It is shown in Appendix C that the maximum of the spectrum [Eq. (4.7)] of the scattered field at scattering angle θ occurs at frequency $\omega = \omega_0'(\theta)$, where

$$\omega_0'(\theta) = \frac{\omega_0}{2\alpha^2(\theta)}\left\{1 + \left[1 + \alpha^2(\theta)\left(\frac{4\Gamma_0}{\omega_0}\right)^2\right]^{1/2}\right\}. \tag{4.10}$$

It is customary, especially in astronomy, to specify frequency shifts by the quantity

$$z = \frac{\lambda_0' - \lambda_0}{\lambda_0} = \frac{\omega_0 - \omega_0'}{\omega_0'}, \tag{4.11}$$

where $\lambda_0 = 2\pi c/\omega_0$ is the original wavelength and $\lambda_0' = 2\pi c/\omega_0'$ is the corresponding shifted wavelength. Evidently $z > 0$ when the line is red shifted and $z < 0$ when it is blue shifted. On substituting from Eq. (4.10) into Eq. (4.11) we readily find that in the present case

$$z(\theta) = \frac{2\alpha^2(\theta) - \left\{1 + \left[1 + \alpha^2(\theta)\left(\frac{4\Gamma_0}{\omega_0}\right)^2\right]^{1/2}\right\}}{1 + \left[1 + \alpha^2(\theta)\left(\frac{4\Gamma_0}{\omega_0}\right)^2\right]^{1/2}}. \tag{4.12}$$

Were the scatterer completely uncorrelated ($\sigma = 0$), we would have, according to Eq. (4.8c), $\alpha = 1$ for all values of θ, and Eq. (4.12) would reduce to

$$z_{\text{uncorr}} = \frac{1 - \left[1 + \left(\frac{4\Gamma_0}{\omega_0}\right)^2\right]^{1/2}}{1 + \left[1 + \left(\frac{4\Gamma_0}{\omega_0}\right)^2\right]^{1/2}}. \tag{4.13}$$

In Figs. 2–4 our main results are illustrated by a number of computed curves. Figure 2 shows spectra of the scattered field at different angles of scattering for some selected values of the parameters. Figure 3 illustrates the behavior of the height factor $H(\theta)$ defined by Eq. (4.8b). The behavior of

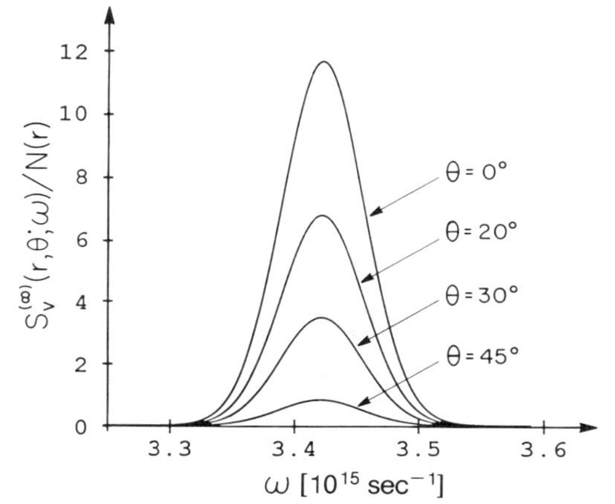

Fig. 2. Spectrum [in units of $N(r)$] of light scattered at various angles θ, from a Gaussian correlated medium with rms width $\sigma = 3\lambda_0$, $\lambda_0 = 5500$Å ($\omega_0 = 3.427 \times 10^{15}$ sec^{-1}). The spectrum of the incident light is a line with Gaussian profile of rms width $\Gamma_0 = 10^{-2}$ ω_0). The curve labeled $\theta = 0$ also represents the spectrum [in units of $N(r)$] for the case when the scatterer is completely uncorrelated ($\sigma = 0$) [see Eq. (4.4)].

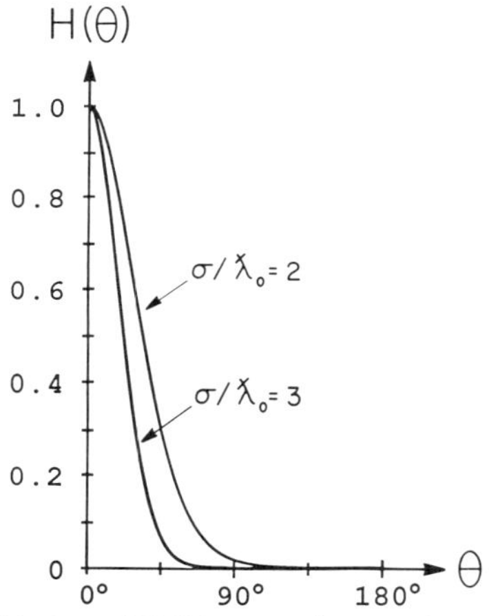

Fig. 3. Height factor [$H(\theta)$] [Eq. (4.8b)], with $\Gamma_0/\omega_0 = 10^{-2}$. The values of the correlation parameter σ/λ_0 are indicated on the curves.

the relative frequency shift as a function of the angle of scattering is shown in Fig. 4. We see that for a certain range of directions around the forward direction $\theta = 0$, the line is blue shifted ($z < 0$), but the magnitude of the shift decreases with increasing θ, and eventually red-shifted lines ($z > 0$) are produced. The lower curves in Fig. 4 show the difference between the values pertaining to scattering from a spatially random medium of finite (nonzero) correlation length ($\sigma > 0$) and from one that is completely uncorrelated ($\sigma = 0$). This difference is seen to be positive for all angles of scattering $\theta \neq 0$, indicating that when the scatterer has a finite (nonzero) correlation length the spectrum is always red shifted with respect to the spectrum that is produced with a completely uncorrelated scatterer.

In the case when

$$(4\Gamma_0/\omega_0)^2 \ll 1, \qquad \sigma/\lambda_0 = O(1) \tag{4.14}$$

(conditions that were satisfied in connection with the computed curves), $z(\theta)$ as given by Eq. (4.12) can be approximated by an expression that clearly indicates the roles of the physical parameters. One finds, after a straightforward calculation, that

$$z(\theta) \approx 4\left(\frac{\Gamma_0}{\omega_0}\right)^2\left[\left(\frac{c}{\lambda_0}\right)^2 \sin^2(\theta/2) - 1\right]$$
$$= 4\left[\left(\frac{\sigma}{L_0}\right)^2 \sin^2\left(\frac{\theta}{2}\right) - \left(\frac{\Gamma_0}{\omega_0}\right)^2\right], \tag{4.15}$$

where $L_0 = c/\Gamma_0$ is the coherence length of the incident light (cf. Note 18). Further, when the conditions (4.14) are fulfilled, the fractional shift of the center frequency of the Gaussian factor in Eq. (4.7), defined by the expression

$$z_G(\theta) \equiv \frac{\omega_0 - \tilde{\omega}(\theta)}{\tilde{\omega}(\theta)}, \tag{4.16}$$

is readily found to be approximately equal to

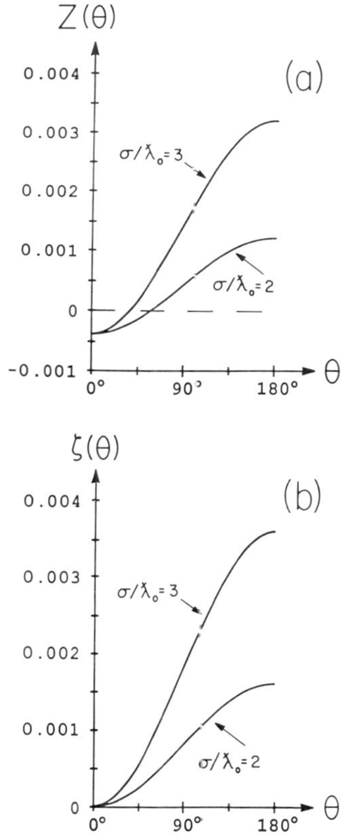

Fig. 4. Relative frequency shift $z(\theta)$, (a), and the difference $\zeta(\theta) = z(\theta) - z_{\text{uncorr}}$, (b), of the spectral line of light scattered at various angles θ from a Gaussian-correlated medium with rms width σ. The spectrum of the incident light is a line with a Gaussian profile of rms width $\Gamma_0 = 10^{-2}\,\omega_0$.

$$z_G(\theta) \approx 4\left(\frac{\sigma}{L_0}\right)^2 \sin^2\left(\frac{\theta}{2}\right). \tag{4.17}$$

Hence *the first term in the formula (4.15) is a red shift due to the spatial correlations in the physical properties of the medium. The second term is a blue shift due to the dipole factor* ω^4.

5. CONCLUDING REMARKS

We have considered in this paper the effect of a random medium whose physical properties do not change in time on the spectrum of the light scattered by it. It is commonly stated that under these circumstances the frequency of the incident light will not change by the process of scattering. How is it possible then that our analysis indicates that such scattering may produce frequency shifts of the spectral lines? The resolution of this apparent paradox lies in the fact that most previous work was concerned with the scattering of essentially *monochromatic* light, of well-defined frequency, ω_0, say. The spectrum of the scattered light will then necessarily also be essentially monochromatic and of the same frequency ω_0. However, when the incident light is *polychromatic*, the different frequency components of the incident light, which are scattered in any particular direction, will be scattered with different strengths. As a result, the spectrum of the scattered light will differ from that of the incident light, even though the different frequency components are uncorrelated. As we showed, the difference may manifest itself as a frequency shift. This possibility of generating frequency shifts by scattering is analogous to that discovered not long ago in connection with effects of source correlations on the spectrum of the emitted light.[4–9] The only difference is that we now deal with secondary sources, namely, with the polarization induced in the scattering medium by the incident wave. The induced polarization will, in general, be correlated over finite distances of the scattering medium and will thus imitate correlations in primary sources. It was this analogy between scattering and radiation that, in fact, led to the present analysis.

Finally we might mention that our results suggest a new method for determining correlation properties of refractive-index distribution in spatially random media, based on the analysis of the spectral changes produced by a spatially random medium when illuminated by polychromatic light of known spectral composition.

APPENDIX A: PRODUCT THEOREM FOR GAUSSIAN FUNCTIONS

In this appendix we will establish the following theorem.

Theorem: If

$$G(\omega - \omega_j; \Gamma_j) = \exp[-(\omega - \omega_j)^2/2\Gamma_j^2] \qquad (j = 1, 2), \tag{A1}$$

then

$$G(\omega - \omega_1; \Gamma_1)G(\omega - \omega_2; \Gamma_2) = G[\omega_1 - \omega_2; (\Gamma_1^2 + \Gamma_2^2)^{1/2}]G(\omega - \tilde{\omega}; \tilde{\Gamma}), \tag{A2}$$

where

$$\tilde{\omega} = \frac{\omega_1\Gamma_2^2 + \omega_2\Gamma_1^2}{\Gamma_1^2 + \Gamma_2^2}, \tag{A3}$$

$$\frac{1}{\tilde{\Gamma}^2} = \frac{1}{\Gamma_1^2} + \frac{1}{\Gamma_2^2}. \tag{A4}$$

To establish this theorem we multiply together the two formulas (A1) and express the product in the form

$$G(\omega - \omega_1; \Gamma_1)G(\omega - \omega_2; \Gamma_2) = \exp[-g(\omega)], \tag{A5}$$

where

$$g(\omega) = \frac{1}{2\Gamma_1^2\Gamma_2^2}[\Gamma_2^2(\omega - \omega_1)^2 + \Gamma_1^2(\omega - \omega_2)^2]$$
$$= \frac{1}{2\Gamma_1^2\Gamma_2^2}(a^2\omega^2 - 2b\omega + c) \tag{A6}$$

and

$$a^2 = \Gamma_1^2 + \Gamma_2^2, \tag{A7a}$$

$$b = \omega_1\Gamma_2^2 + \omega_2\Gamma_1^2, \tag{A7b}$$

$$c = \omega_1^2\Gamma_2^2 + \omega_2^2\Gamma_1^2. \tag{A7c}$$

On completing the square in Eq. (A6), we find that

$$g(\omega) = \frac{a^2}{2\Gamma_1^2\Gamma_2^2}\left(\omega - \frac{b}{a^2}\right)^2 + \frac{1}{2\Gamma_1^2\Gamma_2^2}\left(c - \frac{b^2}{a^2}\right)$$
$$= \frac{1}{2\tilde{\Gamma}^2}(\omega - \tilde{\omega})^2 + \frac{1}{2\Gamma_1^2\Gamma_2^2}\left(c - \frac{b^2}{a^2}\right), \tag{A8}$$

where

$$\frac{1}{\tilde{\Gamma}^2} = \frac{a^2}{\Gamma_1^2\Gamma_2^2} = \frac{1}{\Gamma_1^2} + \frac{1}{\Gamma_2^2}, \tag{A9}$$

$$\tilde{\omega} = \frac{b}{a^2} = \frac{\omega_1\Gamma_2^2 + \omega_2\Gamma_1^2}{\Gamma_1^2 + \Gamma_2^2}. \tag{A10}$$

By using Eqs. (A7a)–(A7c), one can readily show that

$$c - \frac{b^2}{a^2} = \frac{\Gamma_1^2\Gamma_2^2}{\Gamma_1^2 + \Gamma_2^2}(\omega_1 - \omega_2)^2. \tag{A11}$$

On substituting from Eq. (A11) into Eq. (A8) and from Eq. (A8) into Eq. (A5), we find that

$$G(\omega - \omega_1; \Gamma_1)G(\omega - \omega_2; \Gamma_2)$$
$$= \exp\left[-\frac{(\omega_1 - \omega_2)^2}{2(\Gamma_1^2 + \Gamma_2^2)}\right]\exp\left[-\frac{(\omega - \tilde{\omega})^2}{2\tilde{\Gamma}^2}\right]$$
$$= G[\omega_1 - \omega_2; (\Gamma_1^2 + \Gamma_2^2)^{1/2}]G(\omega - \tilde{\omega}; \tilde{\Gamma}), \tag{A12}$$

where the last step follows from Eq. (A1). This completes the proof of the theorem.

Some remarks about the implications of this theorem are in order. Equation (A2) shows that *the product of two Gaussian functions is proportional to a new Gaussian function*, $G(\omega - \tilde{\omega}; \tilde{\Gamma})$, the constant of proportionality being $G[\omega_1 - \omega_2; (\Gamma_1^2 + \Gamma_2^2)^{1/2}]$. The center frequency and width of the new Gaussian function are $\tilde{\omega}$ and $\tilde{\Gamma}$, given by the formulas (A10) and (A9), respectively.

The relationship between $\tilde{\omega}$ and the two original center frequencies ω_1 and ω_2 can be expressed in a different form. It follows from Eq. (A3) that

$$\tilde{\omega} = f_1\omega_1 + f_2\omega_2, \tag{A13}$$

where

$$f_1 = \frac{\Gamma_2^2}{\Gamma_1^2 + \Gamma_2^2}, \qquad f_2 = \frac{\Gamma_1^2}{\Gamma_1^2 + \Gamma_2^2}. \tag{A14}$$

Since $f_1 + f_2 = 1$, *$\tilde{\omega}$ is a weighted average of the two original center frequencies.* In particular, if $0 < \omega_1 < \omega_2$, and neither Γ_1 nor Γ_2 is zero,

$$\omega_1 < \tilde{\omega} < \omega_2. \tag{A15}$$

The relationship between $\tilde{\Gamma}$ and the original widths, Γ_1 and Γ_2, can be deduced from Eq. (A4). If neither Γ_1 nor Γ_2 is zero, we find that *the new Gaussian is narrower than either of the two original Gaussians*, i.e., that

$$\tilde{\Gamma} < \Gamma_1, \qquad \tilde{\Gamma} < \Gamma_2. \tag{A16}$$

Finally, it follows directly from Eq. (A12) that the maximum value (as a function of ω) of the product $G(\omega - \omega_1; \Gamma_1)G(\omega - \omega_2; \Gamma_2)$ occurs at the frequency $\omega = \tilde{\omega}$ and is equal to $G[(\omega_1 - \omega_2; (\Gamma_1^2 + \Gamma_2^2)^{1/2}]$, whereas the maximum values of each of the original Gaussians occur at $\omega = \omega_j$ and are each equal to unity. Therefore, as long as $\omega_1 \neq \omega_2$, *the maximum value of the product of two Gaussians is smaller than the maximum value of either of the two original Gaussians.*

APPENDIX B: DERIVATION OF EXPRESSION (4.7) FOR SPECTRUM OF THE SCATTERED FIELD

We have, according to Eqs. (4.5) and (4.6),

$$S^{(\infty)}(r\hat{s}, \omega) = \frac{AV}{r^2}\left(\frac{\omega}{c}\right)^4 \exp\left(-\frac{\omega^2}{2\Gamma_1^2}\right) B \exp\left[-\frac{(\omega - \omega_0)^2}{2\Gamma_0^2}\right], \tag{B1}$$

where[19]

$$\frac{1}{\Gamma_1^2} = \frac{4}{c^2}\sigma^2 \sin^2\left(\frac{\theta}{2}\right). \tag{B2}$$

If we set

$$G(\omega - \omega_j; \Gamma_j) = \exp\left[-\frac{(\omega - \omega_j)^2}{2\Gamma_j^2}\right], \tag{B3}$$

$$N = \frac{VAB}{c^4 r^2}, \tag{B4}$$

the expression (B1) becomes

$$S^{(\infty)}(r\hat{s}, \omega) = N\omega^4 G(\omega; \Gamma_1)G(\omega - \omega_0; \Gamma_0). \tag{B5}$$

Now according to the product theorem for Gaussian functions established in Appendix A,

$$G(\omega; \Gamma_1)G(\omega - \omega_0; \Gamma_0) = G[\omega_0; (\Gamma_1^2 + \Gamma_2^2)^{1/2}]G(\omega - \tilde{\omega}; \tilde{\Gamma}), \tag{B6}$$

where

$$\tilde{\omega} = \frac{\omega_0 \Gamma_1^2}{\Gamma_0^2 + \Gamma_1^2} \tag{B7}$$

and

$$\frac{1}{\tilde{\Gamma}^2} = \frac{1}{\Gamma_0^2} + \frac{1}{\Gamma_1^2}. \tag{B8}$$

It will be convenient to rewrite Eqs. (B7) and (B8) as

$$\frac{\tilde{\omega}}{\omega_0} = \frac{1}{1 + (\Gamma_0/\Gamma_1)^2}, \tag{B9}$$

$$\frac{\tilde{\Gamma}}{\Gamma_0} = \frac{1}{[1 + (\Gamma_0/\Gamma_1)^2]^{1/2}}. \tag{B10}$$

Now from Eq. (B2) it follows that

$$\frac{\Gamma_0}{\Gamma_1} = \frac{2}{c}\sigma\Gamma_0 \sin\left(\frac{\theta}{2}\right), \tag{B11}$$

or, when the relation $\lambda_0\omega_0 = c$, ($\lambda_0 = \lambda_0/2\pi$) is used, Eq. (B11) gives

$$\frac{\Gamma_0}{\Gamma_1} = 2\left(\frac{\sigma}{\lambda_0}\right)\left(\frac{\Gamma_0}{\omega_0}\right)\sin\left(\frac{\theta}{2}\right). \tag{B12}$$

On substituting from Eq. (B12) into Eqs. (B9) and (B10) we find that

$$\frac{\tilde{\omega}}{\omega_0} = \frac{1}{\alpha^2}, \tag{B13}$$

$$\frac{\tilde{\Gamma}}{\Gamma_0} = \frac{1}{\alpha}, \tag{B14}$$

where

$$\alpha = \left\{1 + \left[2\left(\frac{\sigma}{\lambda_0}\right)\left(\frac{\Gamma_0}{\omega_0}\right)\sin\left(\frac{\theta}{2}\right)\right]^2\right\}^{1/2}. \tag{B15}$$

Further we have, if we use Eqs. (B10),

$$\Gamma_0^2 + \Gamma_1^2 = \Gamma_1^2\left[1 + \left(\frac{\Gamma_0}{\Gamma_1}\right)^2\right] = \Gamma_1^2\left(\frac{\Gamma_0}{\tilde{\Gamma}}\right)^2$$

or, if we also use Eq. (B14),

$$\Gamma_0^2 + \Gamma_1^2 = \alpha^2\Gamma_1^2. \tag{B16}$$

Hence the first factor on the right-hand side of the identity (B6) has the explicit form

$$G[\omega_0; (\Gamma_0^2 + \Gamma_1^2)^{1/2}] = \exp\left[-\frac{\omega_0^2}{2\alpha^2\Gamma_1^2}\right]$$

or, if we make use of Eq. (B12),

$$G[\omega_0; (\Gamma_0^2 + \Gamma_1^2)^{1/2}] = \exp\left[-\frac{2}{\alpha^2}\left(\frac{\sigma}{\lambda_0}\right)^2 \sin^2\left(\frac{\theta}{2}\right)\right]. \tag{B17}$$

On substituting from Eq. (B17) into Eq. (B6) we find that

$$G(\omega; \Gamma_1)G(\omega - \omega_0; \Gamma_0) = \exp\left[-\frac{2}{\alpha^2}\left(\frac{\sigma}{\lambda_0}\right)^2 \sin^2\left(\frac{\theta}{2}\right)\right]G(\omega - \tilde{\omega}; \tilde{\Gamma}). \tag{B18}$$

Finally, on substituting from Eq. (B18) into the expression (B5) we obtain the following expression for the spectrum of the scattered field:

$$S^{(\infty)}(r\hat{s}, \omega) = N(r)H(\theta)\omega^4 \exp\left\{-\frac{1}{2}\left[\frac{\omega - \tilde{\omega}(\theta)}{\tilde{\Gamma}(\theta)}\right]^2\right\}, \tag{B19}$$

where

$$H(\theta) = \exp\left[-\frac{2}{\alpha^2(\theta)}\left(\frac{\sigma}{\lambda_0}\right)^2 \sin^2\left(\frac{\theta}{2}\right)\right]. \tag{B20}$$

The quantities $\tilde{\omega}(\theta)$ and $\tilde{\Gamma}(\theta)$, which appear on the right-hand side of Eq. (B19), are given by the formulas (B13) and (B14), respectively, with $\alpha(\theta)$ given by Eq. (B15). The factor $N(r)$ is defined by the formula (B4).

APPENDIX C: DERIVATION OF THE FORMULA (4.10) FOR THE FREQUENCY AT WHICH THE SHIFTED LINE ATTAINS ITS MAXIMUM

The formula (4.7) for the spectrum of the scattered field may be written in the form

$$S^{(\infty)}(r, \theta; \omega) = N(r)H(\theta)f(\omega, \theta), \tag{C1}$$

where

$$f(\omega, \theta) = \omega^4 \exp\left\{-\frac{1}{2}\left[\frac{\omega - \tilde{\omega}(\theta)}{\tilde{\Gamma}(\theta)}\right]^2\right\}, \tag{C2}$$

with $\tilde{\omega}(\theta)$ and $\tilde{\Gamma}(\theta)$ given by the formulas (4.8c) and (4.8d), respectively.

To determine the maximum of $f(\omega, \theta)$ as a function of ω (with θ fixed), we differentiate Eq. (C2) with respect to ω. We find that

$$\frac{\partial f(\omega, \theta)}{\partial \omega} = \left[4\omega^3 - \frac{\omega - \tilde{\omega}}{\tilde{\Gamma}^2}\omega^4\right]\exp\left[-\frac{1}{2}\left(\frac{\omega - \tilde{\omega}}{\tilde{\Gamma}}\right)^2\right]. \tag{C3}$$

Obviously

$$\frac{\partial f(\omega, \theta)}{\partial \omega} = 0$$

when $\omega = 0$ or when $\omega = \omega_\pm$, where

$$4 - \frac{\omega_\pm - \tilde{\omega}}{\tilde{\Gamma}^2}\omega_\pm = 0,$$

i.e., when

$$\omega_\pm = \frac{1}{2}\tilde{\omega}\left\{1 \pm \left[1 + \left(\frac{4\tilde{\Gamma}}{\tilde{\omega}}\right)^2\right]^{1/2}\right\}. \tag{C4}$$

If we make use of Eqs. (4.8c) and (4.8d), it follows that

$$\omega_\pm = \frac{\omega_0}{2\alpha^2}\left\{1 \pm \left[1 + \alpha^2\left(\frac{4\Gamma_0}{\omega_0}\right)^2\right]^{1/2}\right\}. \tag{C5}$$

Both of these roots can readily be shown to be frequencies at which $f(\omega, \theta)$ and consequently $S^{(\infty)}(r, \theta; \omega)$ attain maximum values. Only the one with the positive sign is relevant because of our use of the complex analytic signal representation of the field[12] [see also Ref. 17, Sec. 10.2]. Denoting this root by $\omega_0'(\theta)$, we have

$$\omega_0'(\theta) = \frac{\omega_0}{2\alpha^2}\left\{1 + \left[1 + \alpha^2\left(\frac{4\Gamma_0}{\omega_0}\right)^2\right]^{1/2}\right\}. \tag{C6}$$

ACKNOWLEDGMENTS

We wish to express our thanks to Avshalom Gamliel for assistance with computations relating to Figs. 2–4.

This research was supported by the National Science Foundation and by the U.S. Army Research Office.

E. Wolf is also with the Institute of Optics, University of Rochester.

REFERENCES AND NOTES

1. E. Wolf, "Invariance of spectrum of light on propagation," Phys. Rev. Lett. **56,** 1370–1372 (1986).
2. See also L. Mandel, "Concept of cross-spectral purity in coherence theory," J. Opt. Soc. Am. **51,** 1342–1350 (1961); L. Mandel and E. Wolf, "Spectral coherence and the concept of cross-spectral purity," J. Opt. Soc. Am. **66,** 529–535 (1976); F. Gori and R. Grella, "Shape invariant propagation of polychromatic fields," Opt. Commun. **49,** 173–177 (1984).
3. G. M. Morris and D. Faklis, "Effects of source correlation on the spectrum of light," Opt. Commun. **62,** 5–11 (1987).
4. E. Wolf, "Non-cosmological redshifts of spectral lines," Nature (London) **326,** 363–365 (1987).
5. E. Wolf, "Red shifts and blue shifts of spectral lines caused by source correlations," Opt. Commun. **62,** 12–16 (1987).
6. Z. Dacic and E. Wolf, "Changes in the spectrum of partially coherent light beam propagating in free space," J. Opt. Soc. Am. A **5,** 1118–1126 (1988).
7. D. Faklis and G. M. Morris, "Spectral shifts produced by source correlations," Opt. Lett. **13,** 4–6 (1988).
8. F. Gori, G. Guattari, C. Palma, and G. Padovani, "Observation of optical redshifts and blueshifts produced by source correlations," Opt. Commun. **67,** 1–4 (1988).
9. W. H. Knox and R. S. Knox, "Direct observation of the optical Wolf shift using white-light interferometry," J. Opt. Soc. Am. A **4**(13), P131 (1987).
10. M. F. Bocko, D. H. Douglass, and R. S. Knox, "Observation of frequency shifts of spectral lines due to source correlations," Phys. Rev. Lett. **58,** 2649–2651 (1987).
11. A. Gamliel and E. Wolf, "Spectral modulation by control of source correlations," Opt. Commun. **65,** 91–96 (1988); see also E. Wolf, "Red shifts and blue shifts of spectral lines emitted by two correlated sources," Phys. Rev. Lett. **58,** 2646–2648 (1987).
12. E. Wolf, "New Theory of partial coherence in the space-frequency domain. Part I: spectra and cross-spectra of steady state sources," J. Opt. Soc. Am. **72,** 343–351 (1982); "Part II: steady-state fields and higher-order correlations," J. Opt. Soc. Am. A **3,** 76–85 (1986).
13. See, for example, P. Roman, *Advanced Quantum Theory* (Addison-Wesley, Reading, Mass., 1965), Sec. 3.2.
14. L. G. Shirley and N. George, "Diffuser radiation patterns over a large dynamic range. 1: Strong diffusers," Appl. Opt. **27,** 1850–1861 (1988), Sec. II.
15. See also J. W. Goodman, "Statistical properties of laser speckle patterns," in *Laser Speckle and Related Phenomena*, 2nd ed., J. C. Dainty ed. (Springer, New York, 1984), Sec. 2.6.1.
16. T. S. McKechnie, "Speckle reduction," in *Laser Speckle and Related Phenomena*, 2nd ed., J. C. Dainty ed. (Springer, New York, 1984), Sec. 4.3.
17. M. Born and E. Wolf, *Principles of Optics*, 6th ed. (Pergamon, Oxford, 1980), Secs. 2.2.1 and 2.2.3.
18. It follows from Eqs. (4.8e) and (4.8c) that for a given value of θ the magnitude of this red shift depends on the product of the two ratios σ/λ_0 (the correlation length of the medium in units of λ_0) and Γ_0/ω_0 (the linewidth of the incident light in units of ω_0). Since the linewidth of the incident light is Γ_0, its coherence length, L_0, may be defined by the expression $L_0 = c/\Gamma_0$. Therefore $(\sigma/\lambda_0)(\Gamma_0/\omega_0) = \sigma/L_0$, and the expression (4.8e) may be rewritten as

$$\alpha(\theta) = \left\{1 + \left[2\left(\frac{\sigma}{L_0}\right)\sin\left(\frac{\theta}{2}\right)\right]^2\right\}^{1/2}.$$

Hence, for fixed θ, $\alpha(\theta)$ is a monotonically increasing function of the ratio σ/L_0; consequently Eq. (4.8c) implies that the center frequency $\tilde{\omega}$ of the Gaussian factor in Eq. (4.7) is a monotonically decreasing function of this ratio.
19. For the sake of simplicity we do not show explicitly the dependence of Γ_1 (nor of the quantities $\tilde{\omega}$, $\tilde{\Gamma}$, and α defined below) on θ throughout the main part of this appendix.

JOSA Errata Vol. 7, No. 1/January 1990/J. Opt. Soc. Am. A 173

Frequency shifts of spectral lines produced by scattering from spatially random media: errata

E. Wolf

Department of Physics and Astronomy, University of Rochester, Rochester, New York 14627

J. T. Foley

Department of Physics and Astronomy, Mississippi State University, Mississippi State, Mississippi 39762

F. Gori

Departimento di Fisica, Università di Roma "La Sapienza," P. le Aldo Moro 2-00185, Rome, Italy

Owing to an editorial error, incorrect characters appeared in two equations in our paper.[1] The equations are printed correctly below.

$$\tilde{C}_\eta(\mathbf{K}_1, \mathbf{K}_2, \omega) = \int_V \int_V C_\eta(\mathbf{r}_2' - \mathbf{r}_1', \omega) \exp[-i(\mathbf{K}_1 \cdot \mathbf{r}_1' + \mathbf{K}_2 \cdot \mathbf{r}_2')] \mathrm{d}^3\mathbf{r}_1' \mathrm{d}^3 r_2', \tag{3.2}$$

$$S^{(\infty)}(r\hat{s}, \omega) \approx \frac{V}{r^2} \left(\frac{\omega}{c}\right)^4 \tilde{C}_\eta[k(\hat{s} - \hat{s}_0), \omega] S^{(i)}(\omega). \tag{3.7}$$

The sentence containing Eq. (2.13) was printed incorrectly and should read as follows:

It can be determined from the following formula, which is analogous to Eq. (2.4):

$$S^{(\infty)}(r\hat{s}, \omega) = \langle U^{(\infty)*}(r\hat{s}, \omega) U^{(\infty)}(r\hat{s}, \omega) \rangle. \tag{2.13}$$

REFERENCE

1. E. Wolf, J. T. Foley, and F. Gori, "Frequency shifts of spectral lines produced by scattering from spatially random media," J. Opt. Soc. Am. A 6, 1142–1149 (1989).

0740-3232/90/010173-01$02.00

1340 OPTICS LETTERS / Vol. 23, No. 17 / September 1, 1998

Spectral changes produced by static scattering on a system of particles

Aristide Dogariu

Center for Research and Education in Optics and Lasers, University of Central Florida, Orlando, Florida 32816

Emil Wolf

Department of Physics and Astronomy, and Rochester Theory Center for Optical Science and Engineering, University of Rochester, Rochester, New York 14627

Received May 4, 1998

A formula of considerable generality is derived for the spectrum of light produced by static scattering of polychromatic light on a system of particles. It applies to deterministic as well as to random distributions of monodispersed or polydispersed particles. We illustrate it by examples. It is suggested that spectral changes produced by scattering could be used for determining the structure of some scattering systems.

OCIS codes: 290.0290, 300.0300.

During the past few years a class of phenomena was discovered that is manifested by changes in the spectrum of light on propagation, even in free space. It arises from correlations that are in general present both in the source distribution and in the distribution of the refractive index in a scattering medium. Since the discovery of this effect[1] in 1986 more than 100 papers on this subject have been published, both theoretical and experimental ones.[2] Several applications of this effect have been considered, for example, in connection with remote sensing[3-5] and in the design of novel types of filters[6,7] that could be used, for example, in optical information processing and for signal coding in optical cryptography.

Spectral changes that arise on scattering of polychromatic light have been considered both for static scattering (e.g., Refs. 8–11) and for dynamic scattering (e.g., Refs. 12–16). In all these investigations scattering from a continuous distribution of matter was considered. In the present Letter we derive a formula for the spectrum of light generated on scattering of polychromatic light on a system of particles. The formula is of considerable generality; it applies to a deterministic as well as to a random distribution of particles, both monodispersed and polydispersed. We illustrate the formula by examples. We suggest that spectral changes produced by scattering could be used as a novel tool for determining the structure of some scattering systems.

It will be useful to begin by recalling some results obtained in Ref. 8 relating to frequency shifts of spectral lines produced by scattering on spatially random media whose response to an incident field is represented by a continuous model.

Let us consider a polychromatic plane wave with spectral density $S^{(i)}(\omega)$, ω denoting frequency, propagating in a direction specified by a unit vector $\mathbf{u}_0$, incident on a random medium occupying a finite volume V. We assume that the macroscopic properties of the medium do not change in time, i.e., that we are dealing with static scattering. Let $F(\mathbf{r}, \omega)$ be the scattering potential of the medium. It is given in terms of the dielectric susceptibility $\eta(\mathbf{r}, \omega)$ and the refractive index $n(\mathbf{r}, \omega)$ by the formula

$$F(\mathbf{r}, \omega) = \left(\frac{\omega}{c}\right)^2 \eta(\mathbf{r}, \omega) = \frac{1}{4\pi}\left(\frac{\omega}{c}\right)^2 [n^2(\mathbf{r}, \omega) - 1]. \tag{1}$$

In these expressions $\mathbf{r}$ denotes the position vector of a point in the scatterer and c the speed of light in vacuum. The refractive index, the dielectric susceptibility, and, consequently, the scattering potential are assumed to be random functions of position.

Let

$$C_F(\mathbf{r}_1, \mathbf{r}_2, \omega) = \langle F^*(\mathbf{r}_1, \omega) F(\mathbf{r}_2, \omega)\rangle \tag{2}$$

be the spatial correlation function of the scattering potential, the angle brackets denoting an average taken over the statistical ensemble of realizations of the scatterer. It was shown in Ref. 8 that, within the accuracy of the first Born approximation, the spectral density $S^{(\infty)}(r\mathbf{u}, \omega)$ of the scattered field in the far zone, at a point at distance r from an origin located in a region of the scatterer, in a direction specified by a unit vector $\mathbf{u}$ is given by the expression [Ref. 8, Eq. (2.20), expressed in terms of F rather than η],

$$S^{(\infty)}(r\mathbf{u}, \omega) = \frac{1}{r^2}\tilde{C}_F[-k(\mathbf{u} - \mathbf{u}_0), k(\mathbf{u} - \mathbf{u}_0), \omega] S^{(i)}(\omega), \tag{3}$$

where

$$\tilde{C}_F(\mathbf{K}_1, \mathbf{K}_2, \omega) = \int_V \int_V C_F(\mathbf{r}_1', \mathbf{r}_2', \omega) \times \exp[-i(\mathbf{K}_1 \cdot \mathbf{r}_1' + \mathbf{K}_2 \cdot \mathbf{r}_2')] \mathrm{d}^3\mathbf{r}_1' \mathrm{d}^3\mathbf{r}_2' \tag{4}$$

is the six-dimensional spatial Fourier transform of the correlation function (2) of the scattering potential. It

0146-9592/98/171340-03$15.00/0

also follows from Eq. (2.19) of Ref. 8 that $\tilde{C}_F$ may be expressed in the form

$$\tilde{C}_F(\mathbf{K}_1, \mathbf{K}_2, \omega) = \langle \tilde{F}^*(-\mathbf{K}_1, \omega)\tilde{F}(\mathbf{K}_2, \omega)\rangle, \quad (5)$$

where

$$\tilde{F}(\mathbf{K}, \omega) = \int_V F(\mathbf{r}', \omega)\exp(-i\mathbf{K}\cdot\mathbf{r}')\mathrm{d}^3\mathbf{r}' \quad (6)$$

is the three-dimensional spatial Fourier transform of the scattering potential.

Let us now assume that the incident field is scattered by a system of particles rather than by a continuous medium. Suppose that there are L types of particles forming this system, $m(l)$ of each type, $(l = 1, 2, \ldots L)$, located at points specified by position vectors $\mathbf{r}_{lm}$. We may characterize the response of each of the particles to an incident field by a scattering potential $U_l(\mathbf{r}, \omega)$. The scattering potential of the whole system of the particles is represented by the expression

$$F(\mathbf{r}, \omega) = \sum_{l=1}^{L}\sum_{m(l)} U_l(\mathbf{r} - \mathbf{r}_{lm}, \omega). \quad (7)$$

The three-dimensional Fourier transform of $F(\mathbf{r}, \omega)$ is then given by the formula

$$\tilde{F}(\mathbf{K}, \omega) = \sum_{l=1}^{L}\sum_{m(l)}\int_V U_l(\mathbf{r}' - \mathbf{r}_{lm}, \omega)\exp(-i\mathbf{K}\cdot\mathbf{r}')\mathrm{d}^3r'. \quad (8)$$

From Eq. (8), we readily find that $\tilde{F}(\mathbf{K}, \omega)$ may be expressed in the form

$$\tilde{F}(\mathbf{K}, \omega) = \sum_{l=1}^{L}\sum_{m(l)} \tilde{U}_l(\mathbf{K}, \omega)\exp(-i\mathbf{K}\cdot\mathbf{r}_{lm}), \quad (9)$$

$\tilde{U}_l$ being the three-dimensional spatial Fourier transform of U_l. In Eq. (9) the function $\tilde{U}_l(\mathbf{K}, \omega)$ is a deterministic function but the sum is, in general, a random function, because the positions of the particles may be random. Taking the square modulus of Eq. (9), averaged over the ensemble of the particle distributions, we obtain the formula

$$\langle|\tilde{F}[\mathbf{K}, \omega]|^2\rangle = \sum_{l=1}^{L}\sum_{l'=1}^{L}\tilde{U}_l^*(\mathbf{K}, \omega)\tilde{U}_{l'}(\mathbf{K}, \omega)S_{ll'}(\mathbf{K}), \quad (10)$$

where

$$S_{ll'}(\mathbf{K}) = \left\langle \sum_{m(l)}\sum_{m(l')}\exp[-i\mathbf{K}\cdot(\mathbf{r}_{ml} - \mathbf{r}_{ml'})]\right\rangle \quad (11)$$

is a generalized structure function of the particle system. It is clear from Eq. (5) that $\langle|\tilde{F}[\mathbf{K}, \omega]|^2\rangle$ is equal to $\tilde{C}_F(-\mathbf{K}, \mathbf{K}, \omega)$ and hence, setting $\mathbf{K} = k(\mathbf{u} - \mathbf{u}_0)$ and using Eq. (10), we finally obtain from Eq. (3) the following expression for the spectrum of the scattered light in the far zone:

$$S^{(\infty)}(r\mathbf{u}, \omega) = \frac{1}{r^2}\sum_{l=1}^{L}\sum_{l'=1}^{L}\tilde{U}_l^*[k(\mathbf{u} - \mathbf{u}_0), \omega] \times \tilde{U}_{l'}[k(\mathbf{u} - \mathbf{u}_0), \omega]S_{ll'}[k(\mathbf{u} - \mathbf{u}_0)]S^{(i)}(\omega). \quad (12)$$

Formula (12) is the main result of this Letter. It shows how the spectrum of the scattered field in the far zone differs, in general, from the spectrum $S^{(i)}(\omega)$ of the incident field and that the difference arises from several distinct causes, namely, (1) the frequency dependence of the generalized structure function $S_{ll'}$ and of the Fourier transform term $\tilde{U}_l$ of the individual potentials on the wave number $k = \omega/c$ via the momentum transfer vector $\mathbf{q} = k(\mathbf{u} - \mathbf{u}_0)$ and (2) the second argument, ω in the potentials $U_l(\mathbf{r}, \omega)$, which represents the effect of dispersion and absorption by the particles. Formula (12) also shows that because the spectrum $S^{(\infty)}(r\mathbf{u}, \omega)$ of the scattered field in the far zone depends on the direction $\mathbf{u}$ of scattering it will, in general, be different in different directions of observation. In this connection we might mention that formula (12) may give new insight into the observations of Čerňanský[17] relating to broadening and shifts of spectral lines in some x-ray diffraction experiments.[18]

Formula (12) might provide a new approach to the solution of inverse scattering problems, because it shows how changes of spectra induced by scattering depend on several features of the system of the scattering particles.

To illustrate the use of formula (12) let us consider a system of identical, spherical particles $(L = 1)$, each of radius a, and with the Fourier transform [with argument $\mathbf{K} = k(\mathbf{u} - \mathbf{u}_0)$] of the scattering potential described by the Rayleigh–Gans approximation.[19] The interparticle interactions can be easily accounted for by consideration of a system of identical hard spheres. For the present purpose, we chose the Percus–Yevick approximation because of its simple closed-form expression.[20] In addition, the structure function for this case has been successfully used to interpret light-scattering measurements for monodispersed suspensions of colloidal particles with solid loads up to 54%.[21] Figure 1 presents the results of calculations for the far-field spectrum $S^{(\infty)}(r\mathbf{u}, \omega)$ produced by scattering on media with 45% volume fraction of the spheres. More extensive calculations reveal that the angular behavior of the scattered spectra depends strongly on the overall density of particles, i.e., on the form of the structure function.

It is well known that the theory of fractals provides a powerful tool for analyzing disordered systems. We therefore take as our second example scattering of polychromatic light from a fractal aggregate consisting of identical particles of radius a. It can be shown from scale invariance arguments that the scattering characteristics of fractal clusters have scaling properties extended over a domain limited by the size of an individual particle and the overall size of an aggregate. In

1342 OPTICS LETTERS / Vol. 23, No. 17 / September 1, 1998

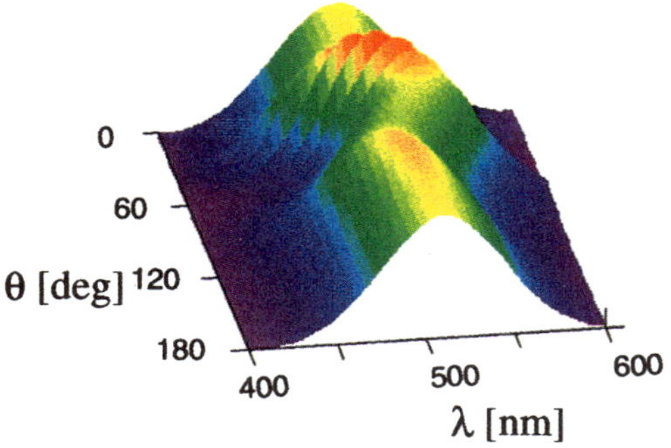

Fig. 1. Spectrum of the scattered light in the far field at different angles of scattering, calculated for monodispersed systems of spherical particles with radius $a = 150$ nm and volume fraction 45%. The spectrum of the incident field has a Gaussian shape with 70-nm FWHM, centered at $\lambda = 500$ nm.

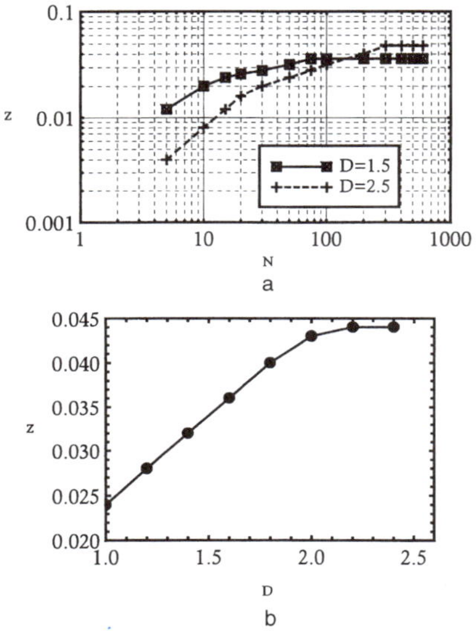

Fig. 2. (a) Relative wavelength shifts $z = (\lambda_0' - \lambda_0)/\lambda_0$ for scattering on fractal aggregates with increasing number of monomers and different fractal dimension. (b) Relative wavelength shifts for scattering from a cluster containing 1000 particles and different fractal dimensions. The spectrum of the incident light is the same as that shown in Fig. 1a, and the relative wavelength shifts are calculated for scattering angle $\theta = \pi/6$.

this case, the structure function is well approximated by the Fisher–Burford formula[22]:

$$S(\mathbf{K}) = 1 + \frac{D\Gamma(D-1)[1 + (\mathbf{K}R)^{-2}]^{(1-D)/2}}{(\mathbf{K}a)^D} \times \sin\left[\frac{D-1}{\tan(\mathbf{K}a)}\right], \tag{13}$$

where $R = aN^{1/D}$ is the radius of gyration of an aggregate with fractal dimension D containing N monomers.

Figure 2a shows the values of the relative wavelength shift $z = (\lambda_0' - \lambda_0)/\lambda_0$, where $\lambda_0 = 2\pi c/\omega$ is the original wavelength and λ_0' is the corresponding shifted wavelength. As can be seen, monotonic dependence is obtained when the number of particles per aggregate increases, confirming the power law behavior of the structure function. It is also to be noted that the scaling behavior depends on the aggregate morphology as exemplified in Fig. 2b, where plots of the z numbers are displayed for the case of a cluster with a constant number of particles but with different fractal dimensions.

The results presented in this Letter suggest that structural information about the scattering system distribution might be obtained from the knowledge of the changes in the spectrum of light generated on scattering of polychromatic light.

We are grateful to Daniel James and Greg Gbur for some helpful comments. This research was supported by the Department of Energy under grant DE-FG02-90ER 14119, by the Air Force Office of Scientific research under grant F49620-96-1-0400, and by the National Science Foundation under grant ERC-94-02989.

References

1. E. Wolf, Phys. Rev. Lett. **56,** 1370 (1986).
2. For a review of these investigations see E. Wolf and D. F. V. James, Rep. Progr. Phys. **59,** 771 (1996).
3. D. F. V. James, H. C. Kandpal, and E. Wolf, Astrophys. J. **445,** 406 (1995).
4. H. C. Kandpal, K. Saxena, D. S. Mehta, J. S. Vaishya, and K. C. Joshi, J. Mod. Opt. **42,** 447 (1995).
5. S. Vicalvi, G. Shirripa Spagnolo, and M. Santarsiero, Opt. Commun. **130,** 241 (1996).
6. E. Wolf, T. Shirai, H. Chen, and W. Wang, J. Mod. Opt. **44,** 1345 (1997).
7. T. Shirai, E. Wolf, H. Chen, and W. Wang, J. Mod. Opt. **45,** 799 (1998).
8. E. Wolf, J. T. Foley, and F. Gori, J. Opt. Soc. Am. A **6,** 1142 (1989); erratum, **7,** 173 (1990).
9. T. Shirai and T. Asakura, J. Opt. Soc. Am. **12,** 1354 (1995).
10. T. Shirai and T. Asakura, Opt. Commun. **123,** 234 (1996).
11. T. A. Leskova, A. A. Maradudin, A. V. Shchegrov, and E. R. Méndez, Phys. Rev. Lett. **79,** 1010 (1997).
12. J. T. Foley and E. Wolf, Phys. Rev. A **40,** 588 (1989).
13. E. Wolf, Phys. Rev. Lett. **63,** 2220 (1989).
14. D. F. V. James, M. P. Savedoff, and E. Wolf, Astrophys. J. **359,** 67 (1990).
15. D. F. V. James and E. Wolf, Phys. Lett. A **146,** 167 (1990).
16. D. F. V. James and E. Wolf, Phys. Lett. A **188,** 239 (1994).
17. M. Čerňanský, Phys. Stat. Sol. (b) **114,** 365 (1982).
18. In this connection see also M. Dušek, Phys. Rev. E **52,** 6833 (1995).
19. H. C. van de Hulst, *Light Scattering by Small Particles* (Dover, New York, 1981), Chap. 7.
20. N. W. Ashcroft and D. C. Langreth, Phys. Rev. **159,** 500 (1967).
21. A. Dogariu, K. Kutsche, P. Likamwa, G. Boreman, and B. Moudgil, Opt. Lett. **22,** 585 (1997).
22. M. E. Fisher and R. J. Burford, Phys. Rev. **156,** 583 (1967).

PHYSICAL REVIEW A VOLUME 40, NUMBER 2 JULY 15, 1989

Scattering of electromagnetic fields of any state of coherence from space-time fluctuations

Emil Wolf*
Department of Physics and Astronomy, University of Rochester, Rochester, New York 14627

John T. Foley
Department of Physics and Astronomy, Mississippi State University, Mississippi State, Mississippi 39762
(Received 23 January 1989)

A theory is developed, valid within the accuracy of the first-order Born approximation, of the scattering of electromagnetic fields that applies under much more general conditions than do the current theories. The incident field is assumed to be statistically homogeneous and stationary, of arbitrary state of coherence, of arbitrary state of polarization, and to have arbitrary spectrum. The medium is assumed to be linear and spatially and temporally random, with the randomness being characterized by an ensemble that is statistically homogeneous, isotropic, and stationary. The analysis is not restricted to scattering from a medium in thermal equilibrium and applies in both off-resonance and near-resonance regions of the medium.

I. INTRODUCTION

As is well known, light scattering phenomena are of great importance in physics, astronomy, chemistry, meteorology, biology, and in other fields. Systematic studies of light scattering have spanned a long period of time, dating back to the pioneering researches of Tyndall and Rayleigh, made around 1870. Since then, other distinguished scientists, such as Smoluchowski, Einstein, Brillouin, Raman, and many others, have made substantial contributions to this field.[1] A new chapter in the development of the subject came with the development of the laser in 1960, which has produced light whose spectrum is exceedingly narrow and which is highly directional. These properties of laser light have made it possible, in recent years, to obtain important detailed information not previously accessible to experiment about the interaction of light with liquid, solids, and gases and about microscopic properties of matter.[2]

Much of the existing literature on the theory of light scattering is restricted to scattering of monochromatic light[3] and assumes that the density fluctuations of the scattering medium may be described by equilibrium thermodynamics. The first of these restrictions makes it impossible to explain some phenomena, such as the recently discovered frequency shifts of spectral lines that may be produced under appropriate circumstances by the scattering of partially coherent light from a spatially random medium;[4] and the assumption of thermodynamic equilibrium has in some cases led to disagreement of at least one order of magnitude between theory and experiment.[5]

In this paper we present a theory of scattering of electromagnetic waves, valid within the accuracy of the first-order Born approximation, that applies under much more general conditions than do the current theories. In particular, the incident field can be any statistically homogeneous and stationary field and be of arbitrary state of coherence, of arbitrary state of polarization, and can have arbitrary spectrum. The medium is assumed to be linear and spatially and temporally random, with the randomness being characterized by a statistical ensemble that is homogeneous, isotropic, and stationary. Our analysis is based on statistical continuum theory and is not restricted to scattering under equilibrium conditions. Moreover, the results apply in both off-resonance and near-resonance regions of the medium.[6] The central quantity of this theory is a generalized analogue of the Van Hove correlation function, well known in the theory of neutron scattering.

We illustrate the generality of our theory by showing that, when appropriate assumptions are made about the incident field and the scattering medium, our main formula for the spectrum of the scattered field yields several well-known results of conventional scattering theory. In the accompanying paper[7] we show that, under appropriate circumstances, our theory also predicts frequency shifts that depend on the state of coherence of the incident light and on the correlation properties of the fluctuating medium, a fact that may be of particular interest for astronomy.

II. DIELECTRIC RESPONSE OF A LINEAR, INHOMOGENEOUS, ISOTROPIC MEDIUM

It will be useful to begin by recalling some standard results relating to an electromagnetic field in a linear, inhomogeneous, isotropic, and nonmagnetic medium, whose macroscopic properties do not depend on time. In such a medium the Fourier transforms

$$\tilde{\mathbf{E}}(\mathbf{r},\omega)=\frac{1}{2\pi}\int_{-\infty}^{\infty}\mathbf{E}(\mathbf{r},t)e^{i\omega t}dt \; , \tag{2.1a}$$

$$\tilde{\mathbf{P}}(\mathbf{r},\omega)=\frac{1}{2\pi}\int_{-\infty}^{\infty}\mathbf{P}(\mathbf{r},t)e^{i\omega t}dt \tag{2.1b}$$

of the (real) electric field $\mathbf{E}(\mathbf{r},t)$ and the induced polarization $\mathbf{P}(\mathbf{r},t)$, respectively,[8] ($\mathbf{r}$ denotes a position vector of a field point and t denotes the time) are connected by the simple constitutive relation

$$\tilde{\mathbf{P}}(\mathbf{r},\omega)=\tilde{\eta}(\mathbf{r},\omega)\tilde{\mathbf{E}}(\mathbf{r},\omega)\ , \tag{2.2}$$

where $\tilde{\eta}(\mathbf{r},\omega)$ is the dielectric susceptibility that describes the response of the medium at frequency ω.

It is known from microscopic theory that for many media $\tilde{\eta}(\mathbf{r},\omega)$ may be expressed in terms of the average number $N(\mathbf{r})$ of molecules per unit volume and the mean polarizability $\alpha(\omega)$ of each molecule by the formula

$$\tilde{\eta}(\mathbf{r},\omega)=\frac{N(\mathbf{r})\alpha(\omega)}{1-(4\pi/3)N(\mathbf{r})\alpha(\omega)}\ . \tag{2.3}$$

Equation (2.3) is one form of the well-known Lorentz-Lorenz relation (Ref. 9, Sec. 2.3.3).

On taking the Fourier transform of Eq. (2.2) and making use of the fact that the response of the medium must necessarily be causal, we obtain the following constitutive relation valid in the space-time domain:

$$\mathbf{P}(\mathbf{r},t)=\frac{1}{2\pi}\int_0^\infty \eta(\mathbf{r},t')\mathbf{E}(\mathbf{r},t-t')dt'\ , \tag{2.4}$$

where

$$\eta(\mathbf{r},t)=\int_{-\infty}^{\infty}\tilde{\eta}(\mathbf{r},\omega)e^{-i\omega t}d\omega\ . \tag{2.5}$$

Suppose next that the macroscopic properties of the medium change in the course of time, in either a deterministic or a random way, but that the response is still linear and isotropic. Then, in place of Eq. (2.4) we have the more general causal, linear relationship[10]

$$\mathbf{P}(\mathbf{r},t)=\frac{1}{2\pi}\int_0^\infty \eta(\mathbf{r},t;t')\mathbf{E}(\mathbf{r},t-t')dt' \tag{2.6}$$

that contains a generalized dielectric susceptibility function $\eta(\mathbf{r},t;t')$ which depends on two time arguments. Its dependence on the second temporal argument t' characterizes the response of the medium to sufficiently short pulses.

To obtain some insight into the behavior of the generalized dielectric susceptibility function let us take its Fourier transform with respect to its second argument:

$$\hat{\eta}(\mathbf{r},t;\omega')=\frac{1}{2\pi}\int_0^\infty \eta(\mathbf{r},t;t')e^{i\omega't'}dt'\ . \tag{2.7}$$

If the response of the medium is time independent, $\hat{\eta}=\tilde{\eta}$, and for many media $\tilde{\eta}$ is related to the time-independent number density $N(\mathbf{r})$ of the molecules by Eq. (2.3). However, if the response of the medium changes in time, the number density also becomes a function of time, $N(\mathbf{r},t)$, say. If the temporal variations are not too rapid, we may expect that the following generalization of Eq. (2.3) will hold:

$$\hat{\eta}(\mathbf{r},t;\omega')=\frac{N(\mathbf{r},t)\alpha(\omega')}{1-\dfrac{4\pi}{3}N(\mathbf{r},t)\alpha(\omega')}\ . \tag{2.8}$$

Although the precise range of validity of this equation can only be determined from detailed microscopic considerations, these remarks indicate the physical origin of the generalized dielectric susceptibility, which depends [via Eqs. (2.8) and (2.7)] on two, rather than one, temporal arguments.

When the effective frequencies of the electric field are not too close to any of the resonance frequencies of the medium, the constitutive relation (2.6) may be aproximated by a more familiar formula, as we will now show. For this purpose we first rewrite Eq. (2.6) in the following form that readily follows by the use of Fourier integral representations of $\eta(\mathbf{r},t;t')$ and $\mathbf{E}(\mathbf{r},t-t')$ in the integral of Eq. (2.6):

$$\mathbf{P}(\mathbf{r},t)=\int_{-\infty}^{\infty}\hat{\eta}(\mathbf{r},t;\omega')\tilde{\mathbf{E}}(\mathbf{r},\omega')e^{-i\omega't}d\omega'\ . \tag{2.9}$$

Suppose that $|\tilde{\mathbf{E}}(\mathbf{r},\omega)|$ is appreciable only in the neighborhood of frequencies $\pm\omega_0$. If these frequencies are not close to any of the resonance frequencies of the medium, the variation of $\hat{\eta}(\mathbf{r},t;\omega')$ with ω' in the neighborhood of $\omega'=\pm\omega_0$ $(\omega_0>0)$ may be neglected and Eq. (2.9) then gives

$$\mathbf{P}(\mathbf{r},t)\approx\hat{\eta}(\mathbf{r},t;+\omega_0)\int_0^\infty\tilde{\mathbf{E}}(\mathbf{r},\omega')e^{-i\omega't}d\omega' + \hat{\eta}(\mathbf{r},t;-\omega_0)\int_{-\infty}^0\tilde{\mathbf{E}}(\mathbf{r},\omega')e^{-i\omega't}d\omega'\ . \tag{2.10}$$

Now, since $\eta(\mathbf{r},t;t')$ and $\mathbf{E}(\mathbf{r},t)$ are real, it follows at once from Eqs. (2.7) and (2.1a) that

$$\hat{\eta}(\mathbf{r},t;-\omega_0)=\hat{\eta}^*(\mathbf{r},t;\omega_0) \tag{2.11}$$

and

$$\tilde{\mathbf{E}}(\mathbf{r},\omega')=\tilde{\mathbf{E}}^*(\mathbf{r},-\omega')\ . \tag{2.12}$$

Hence, Eq. (2.10) may be rewritten as

$$\mathbf{P}(\mathbf{r},t)\approx 2\,\mathrm{Re}\left[\hat{\eta}(\mathbf{r},t;\omega_0)\int_0^\infty\tilde{\mathbf{E}}(\mathbf{r},\omega')e^{-i\omega't}d\omega'\right]\ , \tag{2.13}$$

where Re denotes the real part. Now, since we assumed that ω_0 is not too close to any of the resonance frequencies of the medium, the imaginary part of $\hat{\eta}(\mathbf{r},t;\omega_0)$ will be negligible and hence Eq. (2.13) may be rewritten as

$$\mathbf{P}(\mathbf{r},t)=\hat{\eta}(\mathbf{r},t;\omega_0)\left[2\,\mathrm{Re}\left[\int_0^\infty\tilde{\mathbf{E}}(\mathbf{r},\omega')e^{-i\omega't}d\omega'\right]\right]\ , \tag{2.14a}$$

or, using Eq. (2.12) again,

$$\mathbf{P}(\mathbf{r},t)=\hat{\eta}(\mathbf{r},t;\omega_0)\mathbf{E}(\mathbf{r},t)\ . \tag{2.14b}$$

Frequently the constitutive relation of a time-dependent medium is written, without any justification, in the form $\mathbf{P}(\mathbf{r},t)=\eta(\mathbf{r},t)\mathbf{E}(\mathbf{r},t)$. Comparison of this formula with Eq. (2.14b) reveals the real significance and the approximate nature of the time-dependent response function $\eta(\mathbf{r},t)$. Moreover, our analysis indicates that, unlike Eq. (2.6), such a constitutive relation will not describe adequately the response of a medium under circumstances when the field contains frequencies that are in the resonance region of the medium.

III. DETERMINISTIC SCATTERING IN THE FIRST-ORDER BORN APPROXIMATION

In Sec. IV we will investigate the scattering of electromagnetic waves of arbitrary state of coherence from a medium which fluctuates both in space and in time. The incident and the scattered fields, and the dielectric susceptibility, will then be random functions of position and time. It will be useful, however, to derive first some general formulas that we will then need, pertaining to situations where the fields and the response of the medium are deterministic.

Let $\mathbf{E}^{(i)}(\mathbf{r},t)$ and $\mathbf{H}^{(i)}(\mathbf{r},t)$ denote the electric and the magnetic field vectors of a deterministic electromagnetic field incident on a deterministic medium that occupies a finite volume V in free space. We assume that the medium is of the kind considered in Sec. II. Its dielectric response is then characterized by formula (2.6). Further, let the scattered field vectors produced by the interaction of the incident field with the medium be denoted by $\mathbf{E}^{(s)}(\mathbf{r},t)$ and $\mathbf{H}^{(s)}(\mathbf{r},t)$. The total field (inside and outside the medium) may then be written as

$$\mathbf{E}(\mathbf{r},t)=\mathbf{E}^{(i)}(\mathbf{r},t)+\mathbf{E}^{(s)}(\mathbf{r},t)\ , \tag{3.1a}$$

$$\mathbf{H}(\mathbf{r},t)=\mathbf{H}^{(i)}(\mathbf{r},t)+\mathbf{H}^{(s)}(\mathbf{r},t)\ . \tag{3.1b}$$

It will satisfy Maxwell's equations

$$\nabla\times\mathbf{E}(\mathbf{r},t)=-\frac{1}{c}\frac{\partial\mathbf{H}(\mathbf{r},t)}{\partial t}\ , \tag{3.2}$$

$$\nabla\times\mathbf{H}(\mathbf{r},t)=\frac{1}{c}\frac{\partial\mathbf{E}(\mathbf{r},t)}{\partial t}+\frac{4\pi}{c}\frac{\partial\mathbf{P}(\mathbf{r},t)}{\partial t}\ , \tag{3.3}$$

$$\nabla\cdot\mathbf{E}(\mathbf{r},t)=-4\pi\nabla\cdot\mathbf{P}(\mathbf{r},t)\ , \tag{3.4}$$

$$\nabla\cdot\mathbf{H}(\mathbf{r},t)=0\ , \tag{3.5}$$

where c is the speed of light *in vacuo*. If we make use of the fact that the incident field satisfies the source-free Maxwell equations [Eqs. (3.2)–(3.5) with $\mathbf{P}\equiv\mathbf{0}$], subtract these equations from Eqs. (3.2)–(3.5) and use Eqs. (2.6) and (3.1), we obtain the equations

$$\nabla\times\mathbf{E}^{(s)}(\mathbf{r},t)=-\frac{1}{c}\frac{\partial\mathbf{H}^{(s)}(\mathbf{r},t)}{\partial t}\ , \tag{3.6}$$

$$\nabla\times\mathbf{H}^{(s)}(\mathbf{r},t)=\frac{1}{c}\frac{\partial\mathbf{E}^{(s)}(\mathbf{r},t)}{\partial t}+\frac{4\pi}{c}\frac{\partial[\mathbf{P}_1(\mathbf{r},t)+\mathbf{P}_2(\mathbf{r},t)]}{\partial t}\ , \tag{3.7}$$

$$\nabla\cdot\mathbf{E}^{(s)}(\mathbf{r},t)=-4\pi\nabla\cdot[\mathbf{P}_1(\mathbf{r},t)+\mathbf{P}_2(\mathbf{r},t)]\ , \tag{3.8}$$

$$\nabla\cdot\mathbf{H}^{(s)}(\mathbf{r},t)=0\ , \tag{3.9}$$

where

$$\mathbf{P}_1(\mathbf{r},t)=\frac{1}{2\pi}\int_0^\infty\eta(\mathbf{r},t;t')\mathbf{E}^{(i)}(\mathbf{r},t-t')dt'\ , \tag{3.10}$$

$$\mathbf{P}_2(\mathbf{r},t)=\frac{1}{2\pi}\int_0^\infty\eta(\mathbf{r},t;t')\mathbf{E}^{(s)}(\mathbf{r},t-t')dt'\ . \tag{3.11}$$

Suppose now that the scattering medium is weak in the sense that for all values of its arguments the polarization induced by the scattered field is much smaller than the polarization induced by the incident field, i.e., that

$$|\mathbf{P}_2(\mathbf{r},t)|\ll|\mathbf{P}_1(\mathbf{r},t)|\ , \tag{3.12}$$

and also that

$$\left|\frac{\partial\mathbf{P}_2(\mathbf{r},t)}{\partial t}\right|\ll\left|\frac{\partial\mathbf{P}_1(\mathbf{r},t)}{\partial t}\right|\ , \tag{3.13a}$$

$$|\nabla\cdot\mathbf{P}_2(\mathbf{r},t)|\ll|\nabla\cdot\mathbf{P}_1(\mathbf{r},t)|\ . \tag{3.13b}$$

Under these circumstances, Eqs. (3.6)–(3.9) give the following equations for the scattered field:

$$\nabla\times\mathbf{E}^{(s)}(\mathbf{r},t)=-\frac{1}{c}\frac{\partial\mathbf{H}^{(s)}(\mathbf{r},t)}{\partial t}\ , \tag{3.14}$$

$$\nabla\times\mathbf{H}^{(s)}(\mathbf{r},t)=\frac{1}{c}\frac{\partial\mathbf{E}^{(s)}(\mathbf{r},t)}{\partial t}+\frac{4\pi}{c}\frac{\partial\mathbf{P}_1(\mathbf{r},t)}{\partial t}\ , \tag{3.15}$$

$$\nabla\cdot\mathbf{E}^{(s)}(\mathbf{r},t)=-4\pi\nabla\cdot\mathbf{P}_1(\mathbf{r},t)\ , \tag{3.16}$$

$$\nabla\cdot\mathbf{H}^{(s)}(\mathbf{r},t)=0\ , \tag{3.17}$$

where $\mathbf{P}_1(\mathbf{r},t)$ is given by Eq. (3.10). Equations (3.14)–(3.17), together with Eq. (3.10), characterize the behavior of the scattered electromagnetic field with the same kind of accuracy as the first-order Born approximation characterizes the scattered field in quantum collision theory.

The solutions of Eqs. (3.14)–(3.17), which behave as outgoing waves at infinity, can be expressed in the form (Ref. 9, Sec. 2.2.2)

$$\mathbf{E}^{(s)}(\mathbf{r},t)=\nabla\times\nabla\times\mathbf{\Pi}(\mathbf{r},t)\ , \tag{3.18a}$$

$$\mathbf{H}^{(s)}(\mathbf{r},t)=\frac{1}{c}\nabla\times\frac{\partial\mathbf{\Pi}(\mathbf{r},t)}{\partial t}\ , \tag{3.18b}$$

where $\mathbf{\Pi}(\mathbf{r},t)$ is the Hertz vector

$$\mathbf{\Pi}(\mathbf{r},t)=\int_V\frac{\mathbf{P}_1(\mathbf{r}',t-R/c)}{R}d^3r'\ , \tag{3.19}$$

with

$$R=|\mathbf{r}-\mathbf{r}'|\ . \tag{3.20}$$

For later purposes it will be more useful to consider the fields in the space-frequency domain, rather than in the space-time domain. We, therefore, introduce Fourier transforms, defined by formulas of the form

$$\tilde{f}(\mathbf{r},\omega)=\frac{1}{2\pi}\int_{-\infty}^{\infty}f(\mathbf{r},t)e^{i\omega t}dt\ . \tag{3.21}$$

Equations (3.18) then imply that

$$\tilde{\mathbf{E}}^{(s)}(\mathbf{r},\omega)=\nabla\times\nabla\times\tilde{\mathbf{\Pi}}(\mathbf{r},\omega)\ , \tag{3.22a}$$

$$\tilde{\mathbf{H}}^{(s)}(\mathbf{r},\omega)=-ik\nabla\times\tilde{\mathbf{\Pi}}(\mathbf{r},\omega)\ , \tag{3.22b}$$

where

$$\tilde{\mathbf{\Pi}}(\mathbf{r},\omega)=\int_V\tilde{\mathbf{P}}_1(\mathbf{r}',\omega)\frac{e^{ikR}}{R}d^3r' \tag{3.23}$$

and

$$k=\omega/c\ . \tag{3.24}$$

We are mainly interested in the far field. Therefore, we will set

$$\mathbf{r}=r\mathbf{u},\quad \mathbf{k}\equiv k\mathbf{u}=(\omega/c)\mathbf{u},\quad |\mathbf{u}|=1\ , \tag{3.25}$$

and consider the asymptotic behavior of Eqs. (3.22) as $kr\to\infty$, with $\mathbf{u}$ being kept fixed. By elementary geometry, we have, in this limit, $R\sim r-\mathbf{u}\cdot\mathbf{r}'$ and, with this approximation, the expression (3.23) becomes

$$\tilde{\mathbf{\Pi}}(r\mathbf{u},\omega)\sim(2\pi)^3\mathcal{P}_1(\mathbf{k},\omega)\frac{e^{ikr}}{r},\quad kr\to\infty \tag{3.26}$$

where

$$\mathcal{P}_1(\mathbf{k},\omega)=\frac{1}{(2\pi)^3}\int_V\tilde{\mathbf{P}}_1(\mathbf{r}',\omega)e^{-i\mathbf{k}\cdot\mathbf{r}'}d^3r'\ , \tag{3.27a}$$

or, in terms of $\mathbf{P}_1(\mathbf{r},t)$ rather than $\tilde{\mathbf{P}}_1(\mathbf{r},\omega)$,

$$\mathcal{P}_1(\mathbf{k},\omega)=\frac{1}{(2\pi)^4}\int_V d^3r' e^{-i\mathbf{k}\cdot\mathbf{r}'}\int_{-\infty}^{\infty}\mathbf{P}_1(\mathbf{r}',t')e^{i\omega t'}dt'\ . \tag{3.27b}$$

With $\tilde{\mathbf{\Pi}}$ given by Eq. (3.26), Eqs. (3.22) may be shown to give the following expressions for the scattered field in the far zone:[11]

$$\tilde{\mathbf{E}}^{(s)}(r\mathbf{u},\omega)\sim-(2\pi)^3k^2\mathbf{u}\times[\mathbf{u}\times\mathcal{P}_1(\mathbf{k},\omega)]\frac{e^{ikr}}{r}\ , \tag{3.28a}$$

$$\tilde{\mathbf{H}}^{(s)}(r\mathbf{u},\omega)\sim(2\pi)^3k^2[\mathbf{u}\times\mathcal{P}_1(\mathbf{k},\omega)]\frac{e^{ikr}}{r}\ . \tag{3.28b}$$

For later purposes it will be useful to express the quantity $\mathcal{P}_1(\mathbf{k},\omega)$ in a different form. We find, after using the Fourier integral representations of $\eta(\mathbf{r},t;t')$ and $\mathbf{E}^{(i)}(\mathbf{r},t-t')$ in Eq. (3.10), that

$$\mathbf{P}_1(\mathbf{r},t)=\int_{-\infty}^{\infty}\hat{\eta}(\mathbf{r},t;\omega')\tilde{\mathbf{E}}^{(i)}(\mathbf{r},\omega')e^{-i\omega' t}d\omega'\ , \tag{3.29}$$

and hence, on taking the Fourier transform, that

$$\tilde{\mathbf{P}}_1(\mathbf{r},\omega)=\int_{-\infty}^{\infty}\bar{\eta}(\mathbf{r},\omega-\omega';\omega')\tilde{\mathbf{E}}^{(i)}(\mathbf{r},\omega')d\omega'\ , \tag{3.30}$$

where

$$\bar{\eta}(\mathbf{r},\omega;\omega')=\frac{1}{2\pi}\int_{-\infty}^{\infty}\hat{\eta}(\mathbf{r},t;\omega')e^{i\omega t}dt\ . \tag{3.31}$$

On substituting from Eq. (3.30) into Eq. (3.27a) we obtain the following expression for the vector function $\mathcal{P}_1(\mathbf{k},\omega)$ that enters the expression (3.28) for the scattered field in the far zone, within the accuracy of the first-order Born approximation:

$$\mathcal{P}_1(\mathbf{k},\omega)=\frac{1}{(2\pi)^3}\int_V d^3r' e^{-i\mathbf{k}\cdot\mathbf{r}'}\times\int_{-\infty}^{\infty}\bar{\eta}(\mathbf{r}',\omega-\omega';\omega')\tilde{\mathbf{E}}^{(i)}(\mathbf{r}',\omega')d\omega'\ . \tag{3.32}$$

IV. SCATTERING FROM SPACE-TIME FLUCTUATIONS IN THE FIRST-ORDER BORN APPROXIMATION

We will now consider the problem of the scattering of a randomly fluctuating electromagnetic field from a randomly fluctuating medium. We will retain the assumptions that the medium is linear, isotropic, and nonmagnetic, but we will allow its generalized dielectric susceptibility function $\hat{\eta}(\mathbf{r},t;\omega')$ [cf. Eqs. (2.6) and (2.7)] to be, at each frequency ω', a random function of both position ($\mathbf{r}$) and time (t). More specifically we will assume the following.

(i) The dielectric susceptibility function is statistically homogeneous and stationary, at least in the wide sense,[12] and of approximately zero mean. Then the correlation function $\langle\hat{\eta}^*(\mathbf{r}_1,t_1;\omega')\hat{\eta}(\mathbf{r}_2,t_2,\omega')\rangle$, where the angular brackets denote average over an ensemble of the random medium, depends on $\mathbf{r}_1$ and $\mathbf{r}_2$ and on t_1 and t_2 only through the differences $\mathbf{r}_2-\mathbf{r}_1$ and t_2-t_1, respectively, i.e., it has the form

$$\langle\hat{\eta}^*(\mathbf{r}_1,t_1;\omega')\hat{\eta}(\mathbf{r}_2,t_2;\omega')\rangle=G(\mathbf{r}_2-\mathbf{r}_1,t_2-t_1;\omega')\ . \tag{4.1}$$

(ii) The spatial extent of the susceptibility correlation is small compared to the size of the scattering volume V, i.e., the separation distances $R=|\mathbf{R}|=|\mathbf{r}_2-\mathbf{r}_1|$ for which $|G(\mathbf{R},T;\omega'|$ has appreciable values are much smaller than the linear dimensions of V.

(iii) The scattering is so weak that, to a good approximation, it may be described within the framework of the first-order Born approximation.

As regards the incident electric field we will assume that it is statistically homogeneous and stationary, at least in the wide sense,[12] which implies that the correlation tensor $\langle E_l^{(i)}(\mathbf{r}_1,t_1)E_m^{(i)}(\mathbf{r}_2,t_2)\rangle$ (the subscripts l and m labeling Cartesian components) is a function of the differences $\mathbf{r}_2-\mathbf{r}_1$ and of t_2-t_1 only. Hence the correlation tensor of the incident electric field has the form

$$\langle E_l^{(i)}(\mathbf{r}_1,t_1)E_m^{(i)}(\mathbf{r}_2,t_2)\rangle=\mathcal{E}_{lm}^{(i)}(\mathbf{r}_2-\mathbf{r}_1,t_2-t_1)\ . \tag{4.2}$$

Here the angular brackets denote the average taken over the ensemble that characterizes the statistical properties of the incident field. We will also need the cross-spectral density tensor $W_{lm}^{(i)}(\mathbf{r}_2-\mathbf{r}_1,\omega)$ of the incident electric field. It may be defined formally by the equation

$$\langle[\tilde{E}_l^{(i)}(\mathbf{r}_1,\omega)]^*\tilde{E}_m^{(i)}(\mathbf{r}_2,\omega')\rangle=\mathbf{W}_{lm}^{(i)}(\mathbf{r}_2-\mathbf{r}_1,\omega)\delta(\omega-\omega')\ , \tag{4.3}$$

where $\tilde{E}_l^{(i)}$ and $\tilde{E}_m^{(i)}$ are the Fourier transforms of $E_l^{(i)}$ and $E_m^{(i)}$, respectively,[13] and δ is the Dirac delta functon. According to the Wiener-Khintchine theorem[14] the electric cross-spectral density tensor is the Fourier transform of the electric correlation tensor, i.e.,

$$W_{lm}^{(i)}(\mathbf{R},\omega)=\frac{1}{2\pi}\int_{-\infty}^{\infty}\mathcal{E}_{lm}^{(i)}(\mathbf{R},T)e^{i\omega T}dT\ . \tag{4.4}$$

We stress that the expectation values in Eqs. (4.1) and (4.2) [or (4.3)] are taken over two different ensembles. In Eq. (4.1) it is taken over the ensemble that characterizes the fluctuations of the dielectric susceptibility, whereas in Eq. (4.2) it is taken over the ensemble that characterizes the fluctuations of the incident field. We will assume that these two kinds of fluctuations are statistically independent. The assumption is evidently reasonable if the incident field is not exceptionally strong.

We will now derive expressions for the angular and spectral distribution of energy of the scattered field in the far zone. For this purpose we first note that the quantity $\mathcal{P}_1(\mathbf{k},\omega)$, given by Eq. (3.32), is now a random variable because both $\tilde{\eta}$ and $\tilde{\mathbf{E}}^{(i)}$ are themselves random variables. Hence the (generalized) Fourier transforms $\tilde{\mathbf{E}}^{(s)}$ and $\tilde{\mathbf{H}}^{(s)}$ of the scattered field in the far zone, given by Eqs. (3.28), are also random variables. For each realization we have from Eq. (3.28a),

$$[\tilde{\mathbf{E}}^{(s)}(r\mathbf{u},\omega)]^*\cdot\tilde{\mathbf{E}}^{(s)}(r\mathbf{u},\omega')=(2\pi)^6\frac{k^2k'^2}{r^2}e^{i(k'-k)r}\{\mathbf{u}\times[\mathbf{u}\times\mathcal{P}_1^*(\mathbf{k},\omega)]\}\cdot\{\mathbf{u}\times[\mathbf{u}\times\mathcal{P}_1(\mathbf{k}',\omega')]\} , \tag{4.5}$$

where

$$\mathbf{k}=k\mathbf{u}=\frac{\omega}{c}\mathbf{u},\quad \mathbf{k}'=k'\mathbf{u}=\frac{\omega'}{c}\mathbf{u} . \tag{4.6}$$

It is shown in Appendix A that the right-hand side of Eq. (4.5) may be simplified by the use of an elementary vector identity. Equation (4.5) then reduces to

$$[\tilde{\mathbf{E}}^{(s)}(r\mathbf{u},\omega)]^*\cdot\tilde{\mathbf{E}}^{(s)}(r\mathbf{u},\omega') =(2\pi)^6\frac{k^2k'^2}{r^2}e^{i(k'-k)r} \times(\delta_{lm}-u_lu_m)\mathcal{P}_{1l}^*(\mathbf{k},\omega)\mathcal{P}_{1m}(\mathbf{k}',\omega') , \tag{4.7}$$

where δ_{lm} denotes the Kronecker symbol, and summation over repeated suffices is implied.

Let us now take averages of Eq. (4.7) over the ensembles of the incident field and of the fluctuating medium. Denoting this double average by double angular brackets, we have at once from Eq. (4.7) that

$$\langle\langle[\tilde{\mathbf{E}}^{(s)}(r\mathbf{u},\omega]^*\cdot\tilde{\mathbf{E}}^{(s)}(r\mathbf{u},\omega')\rangle\rangle =(2\pi)^6\frac{k^2k'^2}{r^2}e^{i(k'-k)r}(\delta_{lm}-u_lu_m) \times\langle\langle\mathcal{P}_{1l}^*(\mathbf{k},\omega)\mathcal{P}_{1m}(\mathbf{k}',\omega')\rangle\rangle . \tag{4.8}$$

We show in Appendix B that under the assumptions stated at the beginning of this section the double average on the right-hand side of Eq. (4.8) is given by

$$\langle\langle\mathcal{P}_{1l}^*(\mathbf{k},\omega)\mathcal{P}_{1m}(\mathbf{k}',\omega')\rangle\rangle =\frac{V\delta(\omega-\omega')}{(2\pi)^6}\int_V d^3Re^{-i\mathbf{k}\cdot\mathbf{R}} \times\int_{-\infty}^{\infty}\bar{G}(\mathbf{R},\omega-\omega_1;\omega_1) \times W_{lm}^{(i)}(\mathbf{R},\omega_1)d\omega_1 , \tag{4.9}$$

where $\bar{G}$ is the Fourier transform, defined by the formula

$$\bar{G}(\mathbf{R},\Omega;\omega')=\frac{1}{2\pi}\int_{-\infty}^{\infty}G(\mathbf{R},T;\omega')e^{i\Omega T}dT , \tag{4.10}$$

of the correlation function (4.1) of the generalized susceptibility function $\hat{\eta}(\mathbf{r},t;\omega')$ of the scattering medium.

Finally, on substituting from Eq. (4.9) into Eq. (4.8) we find that

$$\langle\langle[\tilde{\mathbf{E}}^{(s)}(r\mathbf{u},\omega)]^*\cdot\tilde{\mathbf{E}}^{(s)}(r\mathbf{u},\omega')\rangle\rangle=S^{(s)}(r\mathbf{u},\omega)\delta(\omega-\omega') , \tag{4.11}$$

where

$$S^{(s)}(r\mathbf{u},\omega)=\frac{Vk^4}{r^2}(\delta_{lm}-u_lu_m)\int_{-\infty}^{\infty}d\omega_1\int_V\bar{G}(\mathbf{R},\omega-\omega_1;\omega_1)W_{lm}^{(i)}(\mathbf{R},\omega_1)e^{-i\mathbf{k}\cdot\mathbf{R}}d^3R , \tag{4.12}$$

with k and $\mathbf{k}$ being defined in Eq. (4.6).

It is clear from (4.11) that the function $S^{(s)}(r\mathbf{u},\omega)$ is proportional to the expectation value of the electric energy density at frequency ω, of the scattered electric field at a typical point $r\mathbf{u}$ in the far zone. According to well-known properties of the far field [cf. Ref. 11, Eqs. (4.10) and (4.11)] it is also proportional to the expectation values of the magnetic energy density and of the magnitude of the Poynting vector at frequency ω in the far zone. We will therefore refer to $S^{(s)}(r\mathbf{u},\omega)$ as the *spectral density (spectrum) of the scattered field.*

Formula (4.12) is the main result of this investigation. It expresses the spectral density of the scattered field throughout the far zone as a linear transform of the cross-spectral density tensor $W_{lm}^{(i)}$ [defined by Eq. (4.3)] of the fluctuating incident field. The kernel of the transform is, apart from a simple geometrical factor, the Fourier transform [Eq. (4.10)] of the two-point correlation function G [defined by Eq. (4.1)] of the generalized dielectric susceptibility of the scattering medium. This two-point correlation function is somewhat analogous to the well-known Van Hove time-dependent two-particle correlation function[15] (known also as the pair distribution function), frequently employed in the theory of neutron scattering. We will now specialize formula (4.12) to some special cases of practical interest.

V. SOME SPECIAL CASES

A. Plane, polychromatic, linearly polarized incident wave

Suppose that the incident field is a fluctuating polychromatic plane wave, propagating in a direction specified by a real unit vector $\mathbf{u}_0$, with its electric vector linearly polarized along a direction specified by a unit vector $\mathbf{e}_0$ ($\mathbf{u}_0\cdot\mathbf{e}_0=0$). Then each realization of the incident electric field may be represented in the form

$$\mathbf{E}^{(i)}(\mathbf{r},t)=\mathbf{e}_0\int_{-\infty}^{\infty}A(\omega)e^{i(k\mathbf{u}_0\cdot\mathbf{r}-\omega t)}d\omega , \tag{5.1}$$

where $A(\omega)$ is, for each frequency ω, a random variable.[13]

The Fourier transform of $\mathbf{E}^{(i)}(\mathbf{r},t)$ evidently is

$$\tilde{\mathbf{E}}^{(i)}(\mathbf{r},\omega)=A(\omega)e^{ik\mathbf{u}_0\cdot\mathbf{r}}\mathbf{e}_0 . \tag{5.2}$$

Hence, Eq. (4.3) gives, in this case,

$$\langle A^*(\omega)A(\omega')\rangle e^{i\mathbf{u}_0\cdot(k'\mathbf{r}_2-k\mathbf{r}_1)}e_{0l}e_{0m} = W_{lm}^{(i)}(\mathbf{r}_2-\mathbf{r}_1,\omega)\delta(\omega-\omega') . \tag{5.3}$$

Now the spectral density $S^{(i)}(\omega)$, say, of the incident field is just the trace of $W_{lm}^{(i)}(0,\omega)$, and hence we have from Eq. (5.3)

$$\langle A^*(\omega)A(\omega')\rangle = S^{(i)}(\omega)\delta(\omega-\omega') . \tag{5.4}$$

It follows from Eqs. (5.3) and (5.4) that

$$W_{lm}^{(i)}(\mathbf{R},\omega)=S^{(i)}(\omega)e^{ik\mathbf{u}_0\cdot\mathbf{R}}e_{0l}e_{0m} . \tag{5.5}$$

On substituting from Eq. (5.5) into Eq. (4.12) we obtain for the expectation value of the spectral density of the scattered field the expression

$$S^{(s)}(r\mathbf{u},\omega)=\frac{Vk^4\sin^2\psi}{r^2}\int_{-\infty}^{\infty}d\omega'\int_V \bar{G}(\mathbf{R},\omega-\omega';\omega')S^{(i)}(\omega')e^{-i(\mathbf{k}-\mathbf{k}_0')\cdot\mathbf{R}}d^3R , \tag{5.6}$$

where

$$\mathbf{k}=k\mathbf{u}=\frac{\omega}{c}\mathbf{u} \tag{5.7a}$$

is the wave vector of the ω component of the scattered field,

$$\mathbf{k}_0'=k'\mathbf{u}_0=\frac{\omega'}{c}\mathbf{u}_0 \tag{5.7b}$$

is the wave vector of the ω' component of the incident field, and ψ is the angle between the direction of observation ($\mathbf{u}$) and the direction of polarization ($\mathbf{e}_0$) of the incident electric field, i.e., $\cos\psi=\mathbf{u}\cdot\mathbf{e}_0$. We have also made use here of the identity

$$(\delta_{lm}-u_lu_m)e_{0l}e_{0m}=1-(\mathbf{u}\cdot\mathbf{e}_0)^2=1-\cos^2\psi=\sin^2\psi . \tag{5.8}$$

If we define a function $\mathcal{S}(\mathbf{K},\Omega;\omega')$ by the formula

$$\mathcal{S}(\mathbf{K},\Omega;\omega')\equiv\frac{1}{(2\pi)^3}\int_V \bar{G}(\mathbf{R},\Omega;\omega')e^{-i\mathbf{K}\cdot\mathbf{R}}d^3R , \tag{5.9}$$

Eq. (5.6) reduces to

$$S^{(s)}(r\mathbf{u},\omega)=\frac{(2\pi)^3Vk^4\sin^2\psi}{r^2}\times\int_{-\infty}^{\infty}\mathcal{S}(\mathbf{k}-\mathbf{k}_0',\omega-\omega';\omega')S^{(i)}(\omega')d\omega' . \tag{5.10}$$

The function $\mathcal{S}(\mathbf{K},\Omega;\omega')$, introduced by the formula (5.9), has a simple meaning. If in Eq. (5.9) we substitute for $\bar{G}$ from Eq. (4.10) we see at once that

$$\mathcal{S}(\mathbf{K},\Omega;\omega')=\frac{1}{(2\pi)^4}\int_V d^3R\int_{-\infty}^{\infty}G(\mathbf{R},T;\omega')\times e^{-i(\mathbf{K}\cdot\mathbf{R}-\Omega T)}dT , \tag{5.11a}$$

or, more explicitly, using Eq. (4.1)

$$\mathcal{S}(\mathbf{K},\Omega;\omega')=\frac{1}{(2\pi)^4}\int_V d^3R\times\int_{-\infty}^{\infty}\langle\hat{\eta}^*(\mathbf{r},t;\omega')\hat{\eta}(\mathbf{r}+\mathbf{R},t+T;\omega')\rangle\times e^{-i(\mathbf{K}\cdot\mathbf{R}-\Omega T)}dT , \tag{5.11b}$$

i.e., it is the space-time Fourier transform of the two-point correlation function of the dielectric susceptibility of the scattering medium. We may thus regard the function $\mathcal{S}(\mathbf{K},\Omega;\omega')$ as the *generalized structure function* of the medium.[16] Hence we see from Eq. (5.10) that when the incident field is a linearly polarized polychromatic plane wave, the spectral density of the scattered field is equal, apart from simple geometrical factors, to a "weighted integral" taken over the spectrum of the incident field, the weighting factor being the generalized structure function of the medium.

B. Plane, monochromatic, linearly polarized incident wave

Suppose that the incident field is again a linearly polarized plane wave, but that it is monochromatic. Then the spectrum $S^{(i)}(\omega)$ has the form

$$S^{(i)}(\omega)=\tfrac{1}{2}I_0[\delta(\omega-\omega_0)+\delta(\omega+\omega_0)] , \tag{5.12}$$

where ω_0 and I_0 are positive constants. The formula (5.10) now reduces to

$$S^{(s)}(r\mathbf{u},\omega)=\frac{(2\pi)^3I_0Vk^4\sin^2\psi}{2r^2}\times[\mathcal{S}(\mathbf{k}-\mathbf{k}_0,\omega-\omega_0;\omega_0)+\mathcal{S}(\mathbf{k}+\mathbf{k}_0,\omega+\omega_0;-\omega_0)] , \tag{5.13}$$

where

$$\mathbf{k}_0=\frac{\omega_0}{c}\mathbf{u}_0 . \tag{5.14}$$

If the fluctuations of the dielectric susceptibility are slow compared to the optical period $2\pi/\omega_0$, then for $\omega>0$,

$$\mathscr{S}(\mathbf{k}+\mathbf{k}_0,\omega+\omega_0;-\omega_0)\cong 0\ , \tag{5.15}$$

and Eq. (5.13) reduces to

$$S^{(s)}(r\mathbf{u},\omega)\approx\frac{(2\pi)^3 I_0 V k^4 \sin^2\psi}{2r^2}\mathscr{S}(\mathbf{k}-\mathbf{k}_0,\omega-\omega_0;\omega_0)\ . \tag{5.16}$$

This formula is essentially a well-known expression of classical scattering theory for the intensity of the scattered light.[17-19]

C. Static limit

Finally, let us consider the special case when the physical properties of the scattering medium do not change in the course of time but they still change randomly with position. $\hat{\eta}(\mathbf{r},t;\omega')$ will then be independent of t [in which case we will write $\tilde{\eta}(\mathbf{r},\omega')$ in place of $\hat{\eta}(\mathbf{r},t;\omega')$ as we did before—cf. Eq. (2.2)]. The correlation function $G(\mathbf{R},T;\omega')$ defined by Eq. (4.1), will then be independent of the temporal argument and we will denote it by $g(\mathbf{R},\omega')$. Instead of Eq. (4.1) we now have[20]

$$\langle\tilde{\eta}^*(\mathbf{r}_1,\omega')\tilde{\eta}(\mathbf{r}_2,\omega')\rangle=g(\mathbf{r}_2-\mathbf{r}_1,\omega')\ , \tag{5.17}$$

and Eq. (4.10) gives

$$\bar{G}(\mathbf{R},\Omega;\omega')=g(\mathbf{R},\omega')\delta(\Omega)\ . \tag{5.18}$$

On substituting from this equation into the general formula (4.12) we obtain the following expression for the spectral density of the scattered field in the static limit:

$$S^{(s)}(r\mathbf{u},\omega)=\frac{Vk^4}{r^2}(\delta_{lm}-u_l u_m)\times\int_V g(\mathbf{R},\omega)W_{lm}^{(i)}(\mathbf{R},\omega)e^{-i\mathbf{k}\cdot\mathbf{R}}d^3R\ . \tag{5.19}$$

Two special cases of this formula are of particular interest. When the incident field is a linearly polarized polychromatic plane wave [Eq. (5.1)], the cross-spectral density tensor of the incident field is given by Eq. (5.5), and, if we also make use of the identity (5.8), Eq. (5.19) reduces to

$$S^{(s)}(r\mathbf{u},\omega)=\frac{(2\pi)^3 V k^4\sin^2\psi}{r^2}\tilde{g}(\mathbf{k}-\mathbf{k}_0,\omega)S^{(i)}(\omega)\ , \tag{5.20}$$

where, as before, $\mathbf{k}=(\omega/c)\mathbf{u}$, $\mathbf{k}_0=(\omega/c)\mathbf{u}_0$, and $\tilde{g}(\mathbf{k},\omega)$ is the three-dimensional Fourier transform of the function $g(\mathbf{R},\omega)$, i.e.,

$$\tilde{g}(\mathbf{K},\omega)=\frac{1}{(2\pi)^3}\int_V g(\mathbf{R},\omega)e^{-i\mathbf{K}\cdot\mathbf{R}}d^3R\ , \tag{5.21}$$

or, more explicitly, using Eq. (5.17),

$$\tilde{g}(\mathbf{K},\omega)=\frac{1}{(2\pi)^3}\int_V\langle\tilde{\eta}^*(\mathbf{r},\omega')\tilde{\eta}(\mathbf{r}+\mathbf{R},\omega')\rangle e^{-i\mathbf{K}\cdot\mathbf{R}}d^3R\ . \tag{5.22}$$

The formula (5.20) is the electromagnetic analogue of a formula for scalar scattering, derived not long ago[4] in a different manner.

Finally, let us suppose that the field incident on the scatterer is monochromatic. Then $S^{(i)}(\omega)$ is given by Eq. (5.12) and Eq. (5.20) reduces, when $\omega>0$, to

$$S^{(s)}(r\mathbf{u},\omega)=\frac{(2\pi)^3 I_0 V k^4\sin^2\psi}{2r^2}\tilde{g}(\mathbf{k}-\mathbf{k}_0,\omega)\delta(\omega-\omega_0)\ . \tag{5.23}$$

A formula of this kind was first derived by Einstein[21] in a well-known investigation that was the starting point of the statistical theory of light scattering.

VI. CONCLUDING REMARKS

We have developed, in this paper, a statistical continuum theory of scattering of electromagnetic fields, valid within the accuracy of the first-order Born approximation. The theory has a much wider range of validity than those that are currently available. In particular the incident field may be of any state of coherence and polarization and have arbitrary spectrum, provided only that it is statistically stationary and homogeneous. The medium is assumed to be linear, statistically homogeneous, isotropic, and nonmagnetic, and of linear dimensions that are large compared with the spatial correlation lengths over which the random variation of its physical properties are correlated at each effective frequency contained in the spectrum of the incident field. No assumption is made regarding the thermodynamic state of the scattering medium.

The response of the scattering medium is described by a generalized "two-time" dielectric susceptibility function, which takes into account the effects of both its spatial and its temporal variations. Our analysis elucidates, as a by-product, the physical significance and the approximate nature of the "one-time" response function that is employed in the usual theories.

Our main formula [Eq. (4.12)] expresses the spectrum of the scattered field as a linear transform of the cross-spectral density tensor of the fluctuating incident field. The kernel of the transform is, apart from a simple geometrical factor, a Fourier transform of the two-point correlation function of the generalized dielectric susceptibility of the scattering medium. We show that many of the well-known formulas of the usual scattering theories readily follow from it as special or limiting cases. In particular, Eq. (4.12) yields a well-known expression derived by Einstein in a classic paper that was the starting point of the statistical theory of light scattering, as well as various formulas that are frequently in the analysis of modern scattering experiments with laser light.

In general, the interaction of an electromagnetic field with a random medium produces changes in the spectrum of the field. In an accompanying paper[7] we show on the basis of the present theory that the modification may be such as to produce frequency shifts of spectral lines.

ACKNOWLEDGMENTS

This research was supported by the National Science Foundation and by the U.S. Army Research Office.

APPENDIX A: A VECTOR IDENTITY USED IN THE DERIVATION OF EQ. (4.7)

We start with the vector identity

$$(\mathbf{A}\times\mathbf{B})\cdot(\mathbf{C}\times\mathbf{D})=(\mathbf{A}\cdot\mathbf{C})(\mathbf{B}\cdot\mathbf{D})-(\mathbf{A}\cdot\mathbf{D})(\mathbf{B}\cdot\mathbf{C}) . \tag{A1}$$

It follows at once from this identity that

$$\begin{aligned}\{\mathbf{u}\times[\mathbf{u}\times\mathcal{P}_1^*(\mathbf{k},\omega)]\}\cdot\{\mathbf{u}\times[\mathbf{u}\times\mathcal{P}_1(\mathbf{k}',\omega')]\}&=(\mathbf{u}\cdot\mathbf{u})[\mathbf{u}\times\mathcal{P}_1^*(\mathbf{k},\omega)]\cdot[\mathbf{u}\times\mathcal{P}_1(\mathbf{k}',\omega')]\\&\quad-\mathbf{u}\cdot[\mathbf{u}\times\mathcal{P}_1(\mathbf{k}',\omega')]\{[\mathbf{u}\times\mathcal{P}_1^*(\mathbf{k},\omega)]\cdot\mathbf{u}\}\\&=[\mathbf{u}\times\mathcal{P}_1^*(\mathbf{k},\omega)]\cdot[\mathbf{u}\times\mathcal{P}_1(\mathbf{k}',\omega')] .\end{aligned} \tag{A2}$$

On using the identity (A1) once again we obtain from Eq. (A2) the identities

$$\begin{aligned}\{\mathbf{u}\times[\mathbf{u}\times\mathcal{P}_1^*(\mathbf{k},\omega)]\}\cdot\{\mathbf{u}\times[\mathbf{u}\times\mathcal{P}_1(\mathbf{k}',\omega')]\}&=\mathcal{P}_1^*(\mathbf{k},\omega)\cdot\mathcal{P}_1(\mathbf{k}',\omega')-[\mathbf{u}\cdot\mathcal{P}_1^*(\mathbf{k},\omega)][\mathbf{u}\cdot\mathcal{P}_1(\mathbf{k}',\omega')]\\&=(\delta_{lm}-u_l u_m)\mathcal{P}_{1l}^*(\mathbf{k},\omega)\mathcal{P}_{1m}(\mathbf{k}',\omega') ,\end{aligned} \tag{A3}$$

where $\mathcal{P}_{1l}$ and $\mathcal{P}_{1m}$ denote the lth and the mth component, respectively, of $\mathcal{P}_1$, δ_{lm} denotes the Kronecker symbol, and summation over repeated suffixes is implied.

APPENDIX B: DERIVATION OF FORMULA (4.9)

It follows from Eq. (3.32) that

$$\begin{aligned}\langle\langle\mathcal{P}_{1l}^*(\mathbf{k},\omega)\mathcal{P}_{1m}(\mathbf{k}',\omega')\rangle\rangle=\frac{1}{(2\pi)^6}\int_V d^3r_1\int_V d^3r_2\int_{-\infty}^{\infty}d\omega_1\int_{-\infty}^{\infty}d\omega_2\, e^{i(\mathbf{k}\cdot\mathbf{r}_1-\mathbf{k}'\cdot\mathbf{r}_2)}\\ \times\langle\bar{\eta}^*(\mathbf{r}_1,\omega-\omega_1;\omega_1)\bar{\eta}(\mathbf{r}_2,\omega'-\omega_2;\omega_2)\rangle\\ \times\langle[\tilde{E}_l^{(i)}(\mathbf{r}_1,\omega_1)]^*\tilde{E}_m^{(i)}(\mathbf{r}_2,\omega_2)\rangle ,\end{aligned} \tag{B1}$$

where we have made use of the assumption that the fluctuations of the medium and of the incident field are statistically independent. Now according to Eq. (4.3) the second expectation value that occurs on the right-hand side of Eq. (B1) is given by

$$\langle[\tilde{E}_l^{(i)}(\mathbf{r}_1,\omega_1)]^*\tilde{E}_m^{(i)}(\mathbf{r}_2,\omega_2)\rangle=W_{lm}^{(i)}(\mathbf{r}_2-\mathbf{r}_1,\omega_1)\delta(\omega_2-\omega_1) . \tag{B2}$$

On substituting from Eq. (B2) into Eq. (B1) and on carrying out the trivial integration with respect to ω_2, we find that

$$\begin{aligned}\langle\langle\mathcal{P}_{1l}^*(\mathbf{k},\omega)\mathcal{P}_{1m}(\mathbf{k}',\omega')\rangle\rangle=\frac{1}{(2\pi)^6}\int_V d^3r_1\int_V d^3r_2\int_{-\infty}^{\infty}d\omega_1\, e^{i(\mathbf{k}\cdot\mathbf{r}_1-\mathbf{k}'\cdot\mathbf{r}_2)}\\ \times\langle\bar{\eta}^*(\mathbf{r}_1,\omega-\omega_1;\omega_1)\bar{\eta}(\mathbf{r}_2,\omega'-\omega_1;\omega_1\rangle\\ \times W_{lm}^{(i)}(\mathbf{r}_2-\mathbf{r}_1,\omega_1) .\end{aligned} \tag{B3}$$

Now the expectation value on the right-hand side of Eq. (B3), which involves the dielectric susceptibility, may be expressed in a simpler form. We find from Eqs. (3.31), (4.1), and the Wiener-Khintchine theorem that

$$\langle\bar{\eta}^*(\mathbf{r}_1,\Omega;\omega')\bar{\eta}(\mathbf{r}_2,\Omega';\omega')\rangle=\bar{G}(\mathbf{r}_2-\mathbf{r}_1,\Omega;\omega')\delta(\Omega-\Omega') , \tag{B4}$$

where

$$\bar{G}(\mathbf{R},\Omega,\omega')=\frac{1}{2\pi}\int_{-\infty}^{\infty}G(\mathbf{R},T;\omega')e^{i\Omega T}dT , \tag{B5}$$

$G(\mathbf{R},T;\omega')$ being the correlation function defined by Eq. (4.1), viz.,

$$G(\mathbf{R},T;\omega')=\langle\hat{\eta}^*(\mathbf{r},t;\omega')\hat{\eta}(\mathbf{r}+\mathbf{R},t+T;\omega')\rangle . \tag{B6}$$

On substituting from Eq. (B4) into Eq. (B3) we obtain the formula

$$\begin{aligned}\langle\langle\mathcal{P}_{1l}^*(\mathbf{k},\omega)\mathcal{P}_{1m}(\mathbf{k}',\omega')\rangle\rangle=\frac{\delta(\omega-\omega')}{(2\pi)^6}\int_V d^3r_1\int_V d^3r_2\, e^{-i\mathbf{k}\cdot(\mathbf{r}_2-\mathbf{r}_1)}\\ \times\int_{-\infty}^{\infty}d\omega_1\,\bar{G}(\mathbf{r}_2-\mathbf{r}_1,\omega-\omega_1;\omega_1)W_{lm}^{(i)}(\mathbf{r}_2-\mathbf{r}_1,\omega_1) .\end{aligned} \tag{B7}$$

Because of our assumption that the spatial correlation length of the dielectric susceptibility fluctuations is small compared with the linear dimensions of the scattering volume, the above expression can be readily shown to reduce to

$$\langle\langle\mathcal{P}_{1l}^*(\mathbf{k},\omega)\mathcal{P}_{1m}(\mathbf{k}',\omega')\rangle\rangle=\frac{V\delta(\omega-\omega')}{(2\pi)^6}\int_V d^3R\, e^{-i\mathbf{k}\cdot\mathbf{R}}\int_{-\infty}^{\infty}d\omega_1\,\bar{G}(\mathbf{R},\omega-\omega_1;\omega_1)W_{lm}^{(i)}(\mathbf{R},\omega_1) , \tag{B8}$$

where V is the volume occupied by the scattering medium.

*Also at the Institute of Optics, University of Rochester.

[1]For reviews of some of these contributions, see, for example, L. Fabelinskii, *Molecular Scattering of Light* (Plenum, New York, 1968); G. B. Benedek, in *Brandeis University Summer Institute on Theoretical Physics, Boston, 1966,* edited by M. Chretien, E. P. Gross, and S. Desser (Gordon and Breach, New York, 1968), Vol. 2, pp. 5–98; B. Crosignani, P. Di Porto, and M. Bertolotti, *Statistical Properties of Scattered Light* (Academic, New York, 1975); B. J. Berne and R. Pecora, *Dynamic Light Scattering* (Wiley, New York, 1976).

[2]See, for example, B. Chu, *Laser Light Scattering* (Academic, New York, 1974); or B. J. Berne and R. Pecora, cited in Ref. 1.

[3]Notable exceptions are (a) L. Mandel, Phys. Rev. **181**, 75 (1969); (b) P. N. Pusey, in *Photon Correlation Spectroscopy and Velocimetry,* edited by H. A. Cummins and E. R. Pike (Plenum, New York, 1977), pp. 45–141.

[4]E. Wolf, J. T. Foley, and F. Gori, J. Opt. Soc. Am. A (to be published).

[5]See, for example, J. Schroder, in *Treatise on Material Science and Technology* (Academic, New York, 1977), Vol. 12, p. 158, and references given therein.

[6]This theory will, however, not apply when the effective frequencies of the incident field are very close to any of the resonance frequencies of the medium because the first-order Born approximation will then no longer adequately describe the scattering process.

[7]J. T. Foley and E. Wolf, following paper, Phys. Rev. A **40**, 588 (1989).

[8]It is customary in optical coherence theory to represent the real fields (such as the electric field **E** or the polarization **P**) by so-called analytic signals (see, for example, Ref. 9, Sec. 10.2). We do *not* adopt this procedure here.

[9]M. Born and W. Wolf, *Principles of Optics,* 6th ed. (Pergamon, Oxford, England, 1980).

[10]Cf. L. Mandel and E. Wolf, Opt. Commun. **8**, 95 (1973).

[11]A. T. Friberg and E. Wolf, J. Opt. Soc. Am. **73**, 26 (1983), Eq. (4.5).

[12]W. B. Davenport and W. L. Root, *Random Signals and Noise* (McGraw-Hill, New York, 1958), p. 60.

[13]These quantities must be interpreted in the sense of the theory of generalized functions because, as is well known, the sample functions of a stationary random process do not possess Fourier transforms within the framework of ordinary function theory.

[14]See, for example, C. Kittel, *Elementary Statistical Physics* (Wiley, New York, 1958), Sec. 28.

[15]L. Van Hove, Phys. Rev. **95**, 249 (1954); Physica, **24**, 404 (1958); see also R. J. Glauber, in *Lectures on Theoretical Physics,* edited by W. E. Brittin, B. W. Downs, and J. Downs (Interscience, New York, 1962), Vol IV, Sec. V, p. 571.

[16]When $\mathcal{S}(\mathbf{K},\Omega;\omega')$ is independent of ω' it becomes the analogue (for fluctuations of the dielectric susceptibility) of the usual dynamical structure factor for particle-density fluctuations.

[17]L. I. Komarov and I. Z. Fisher, Zh. Eksp. Teor. Fiz. **43**, 1927 (1962) [Sov. Phys.—JETP **16**, 1358 (1963)], Eq. (24).

[18]N. G. Van Kampen, in *Quantum Optics,* Proceedings of the International School of Physics "Enrico Fermi," Course XLII, Varenna, 1967, edited by R. J. Glauber (Academic, New York, 1969), p. 235, Eq. (7a).

[19]R. Pecora, J. Chem. Phys. **40**, 1604 (1964), Eq. (30).

[20]The function $g(\mathbf{R},\omega')$ is analogous to the pair distribution function introduced by F. Zernike and J. A. Prins [Z. Phys. **41**, 184 (1927)], in their well-known study of x-ray scattering from liquids.

[21]A. Einstein, Ann. Phys. **33**, 1275 (1910). An English translation of this paper was published in *Colloid Chemistry,* edited by J. Alexander (Reinhold, New York, 1926), Vol. 1, pp. 323–339.

PHYSICAL REVIEW A VOLUME 40, NUMBER 2 JULY 15, 1989

Frequency shifts of spectral lines generated by scattering from space-time fluctuations

John T. Foley
Department of Physics and Astronomy, Mississippi State University, Mississippi State, Mississippi 39762

Emil Wolf*
Department of Physics and Astronomy, University of Rochester, Rochester, New York 14627

(Received 23 January 1989)

The scattering of a class of partially coherent electromagnetic fields by a model medium whose dielectric susceptibility fluctuates in space and in time is considered within the accuracy of the first-order Born approximation. It is shown that if the incident field is a linearly polarized, polychromatic plane wave whose spectrum is a line of Gaussian profile, the spectrum of the scattered light is, in some cases, approximately Gaussian and is shifted towards the shorter or the longer wavelengths, depending upon the angle of scattering.

I. INTRODUCTION

It was predicted not long ago that, in general, correlations between source fluctuations produce changes in the spectrum of the emitted light.[1,2] For radiation from planar secondary sources this prediction was subsequently verified by experiment.[3] It was also shown theoretically that the changes may be such as to produce red shifts or blue shifts of spectral lines,[4-6] and this prediction has also been verified.[7-9] A similar effect can also be expected to arise with acoustical waves and was, in fact, observed recently.[10]

Because of the well-known analogy that exists between the processes of radiation and scattering, one might expect that similar phenomena will arise when a polychromatic wave is scattered by a medium whose dielectric susceptibility fluctuates in space and in time. In fact, it was shown recently[11] that, in some cases, when a polychromatic plane wave field is scattered from a medium whose dielectric susceptibility is a static, random function of position, the spectrum of the scattered light will have approximately the same profile as the incident light, but be blue shifted or red shifted, depending upon the angle of scattering.

In the present paper we investigate the scattering, within the accuracy of the first-order Born approximation, of a linearly polarized, electromagnetic polychromatic plane wave from a model medium whose dielectric susceptibility fluctuates both in space and in time. We utilize the theory developed in the accompanying paper,[12] which applies to a very wide class of problems of this kind. We will show, in particular, that when the spectrum of the incident field is a line of Gaussian profile, and the correlation function of the dielectric susceptibility fluctuations at any two space-time points is a Gaussian function of the spatial and temporal variables, the spectrum of the scattered light is approximately also Gaussian, but is blue shifted or red shifted, depending upon the angle of scattering. An approximate formula for the line shift, which emphasizes the roles of the physical parameters, is developed, and some numerical results are presented.

II. FORMULATION OF THE PROBLEM

Let us begin with a brief outline of the problem that is studied in this paper. We consider the scattering of a fluctuating, polychromatic, linearly polarized electromagnetic plane wave with the electric field $\mathbf{E}^{(i)}(\mathbf{r},t)$ and magnetic field $\mathbf{H}^{(i)}(\mathbf{r},t)$, propagating in free space in the direction specified by the real unit vector $\mathbf{u}_0$. The wave is incident upon a medium which occupies a finite volume V and whose dielectric susceptibility fluctuates both in space and in time (see Fig. 1). We denote by $\mathbf{E}^{(s)}(\mathbf{r},t)$ and $\mathbf{H}^{(s)}(\mathbf{r},t)$ the scattered (i.e., total minus incident) electric and magnetic fields, respectively. We wish to determine the spectrum of the scattered light in the far zone when the spectrum of the incident light consists of a single line of Gaussian profile and the correlation function of the

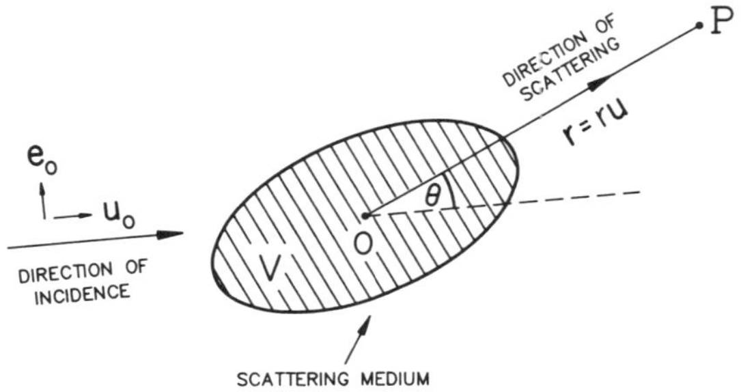

FIG. 1. Notation relating to the scattering geometry. $\mathbf{u}_0$ and $\mathbf{u}$ are real unit vectors in the direction of propagation of the incident field $\mathbf{E}^{(i)},\mathbf{H}^{(i)}$ and the scattered field $\mathbf{E}^{(s)},\mathbf{H}^{(s)}$, respectively. $\mathbf{e}_0$ is a real unit vector in the direction of polarization of the incident electric field. P is a typical observation point in the far zone.

dielectric susceptibility fluctuations is a Gaussian function both in the spatial and temporal variables.

A. The incident field

Let

$$\mathbf{E}^{(i)}(\mathbf{r},t)=\mathbf{e}_0\int_{-\infty}^{\infty} A(\omega)e^{i(k\mathbf{u}_0\cdot\mathbf{r}-\omega t)}d\omega \ , \qquad (2.1a)$$

$$\mathbf{H}^{(i)}(\mathbf{r},t)=\mathbf{u}_0\times\mathbf{E}^{(i)}(\mathbf{r},t) \qquad (2.1b)$$

be the (real[13]) electric and magnetic vectors of the incident wave. Here $\mathbf{u}_0$ is a real unit vector in the direction of propagation of the incident wave, $\mathbf{e}_0$ is a (real) unit vector in the direction of polarization of the incident electric field ($\mathbf{e}_0\cdot\mathbf{u}_0=0$), and

$$k=\omega/c \ , \qquad (2.2)$$

c being the speed of light in a vacuum. We also assume that the field is stationary, at least in the wide sense,[14] as is well known; the representation (2.1a) must then be interpreted in terms of generalized functions. Under these circumstances the cross-spectral density tensor of the incident electric field is given by (cf. Sec. V A of Ref. 12)

$$W_{lm}^{(i)}(\mathbf{R},\omega)=S^{(i)}(\omega)e^{ik\mathbf{u}_0\cdot\mathbf{R}}e_{0l}e_{0m} \ , \qquad (2.3)$$

where $S^{(i)}(\omega)$ is the spectrum of the incident field and the subscripts l and m label Cartesian components. In terms of the spectral amplitude $A(\omega)$ the spectrum $S^{(i)}(\omega)$ is given by

$$\langle A^*(\omega)A(\omega')\rangle=S^{(i)}(\omega)\delta(\omega'-\omega) \ , \qquad (2.4)$$

where the angular brackets denote an ensemble average.

We will only consider the case where the spectrum is a line of Gaussian profile, centered on frequency ω_0 and of rms width Γ_0, i.e.,

$$S^{(i)}(\omega)=\frac{A}{\sqrt{2\pi}\Gamma_0}[g(\omega-\omega_0;\Gamma_0)+g(\omega+\omega_0;\Gamma_0)] \ , \qquad (2.5)$$

where A, ω_0, and Γ_0 are positive constants and

$$g(\omega-\omega_0;\Gamma_0)=e^{-(\omega-\omega_0)^2/2\Gamma_0^2} \ . \qquad (2.6)$$

For later purposes we note some coherence properties of the incident field. The space-time correlation tensor of the field is just the Fourier transform of the cross-spectral density tensor (2.3), i.e.,

$$\mathcal{E}_{lm}^{(i)}(\mathbf{R},T)=e_{0l}e_{0m}\int_{-\infty}^{\infty} S^{(i)}(\omega)e^{i(k\mathbf{u}_0\cdot\mathbf{R}-\omega T)}d\omega \ . \qquad (2.7a)$$

If we choose the axes of our Cartesian coordinate system so that one of them coincides with the direction of the real unit vector $\mathbf{e}_0$, (i.e., along the direction of polarization of the incident electric field), then only one component of $\mathcal{E}_{lm}(\mathbf{R},T)$ will be nonzero and it will be given by

$$\mathcal{E}(\mathbf{R},T)=\int_{-\infty}^{\infty} S^{(i)}(\omega)e^{i(k\mathbf{u}_0\cdot\mathbf{R}-\omega T)}d\omega \ . \qquad (2.7b)$$

The degree of coherence of the incident field is then given by

$$\gamma^{(i)}(\mathbf{R},T)\equiv\frac{\mathcal{E}^{(i)}(\mathbf{R},T)}{\mathcal{E}^{(i)}(0,0)} =\frac{\int_{-\infty}^{\infty} S^{(i)}(\omega)e^{i(k\mathbf{u}_0\cdot\mathbf{R}-\omega T)}d\omega}{\int_{-\infty}^{\infty} S^{(i)}(\omega)d\omega} \ . \qquad (2.8)$$

In particular, with $T=0$,

$$\gamma^{(i)}(\mathbf{R},0)=\frac{\int_{-\infty}^{\infty} S^{(i)}(\omega)e^{ik\mathbf{u}_0\cdot\mathbf{R}}d\omega}{\int_{-\infty}^{\infty} S^{(i)}(\omega)d\omega} \ . \qquad (2.9)$$

We see from Eq. (2.9) that when the two field points $\mathbf{r}_1$ and $\mathbf{r}_2$ ($\mathbf{R}=\mathbf{r}_2-\mathbf{r}_1$) are located in a plane perpendicular to the direction of propagation, i.e., when $\mathbf{u}_0\cdot\mathbf{R}=0$,

$$\gamma^{(i)}(\mathbf{R},0)=1 \ . \qquad (2.10)$$

We may express this result by saying that the incident field has *complete transverse coherence.* On the other hand, when the two points are located along the direction of propagation, then $\mathbf{u}_0\cdot\mathbf{R}=R$, and if, in addition, the spectrum is given by Eq. (2.5) the effective width of $|\gamma^{(i)}(\mathbf{R},0)|$ is readily seen to be given by

$$L_0\sim\frac{c}{\Gamma_0} \ . \qquad (2.11)$$

We will refer to this quantity as the *longitudinal coherence length* of the incident field.

B. The scattering medium

We will assume that the scattering medium is linear, isotropic, and nonmagnetic, and that the fluctuations of its generalized dielectric susceptibility $\hat{\eta}(\mathbf{r},t;\omega')$ (cf. Ref. 12, Sec. II) are statistically homogeneous and stationary, at least in the wide sense. As before, we will denote the space-time correlation function of the medium by $G(\mathbf{R},T;\omega')$:

$$\langle\hat{\eta}^*(\mathbf{r},t;\omega')\hat{\eta}(\mathbf{r}+\mathbf{R},t+T;\omega')\rangle=G(\mathbf{R},T;\omega') \ . \qquad (2.12)$$

Here the asterisk denotes the complex conjugate. We will also assume that the scattering is so weak that, to a good accuracy, it may be described within the framework of the first-order Born approximation.

The resonance frequencies of the medium, i.e., the frequencies of its atomic or molecular transitions, will be assumed not to be in the immediate vicinity of the center frequencies $\pm\omega_0$ of the incident light. Under these circumstances the correlation function $G(\mathbf{R},T;\omega')$ may be approximated, over the effective frequency range (the spectral line) of the incident light, by $G(\mathbf{R},T;\omega_0)$ when $\omega'>0$ and by $G(\mathbf{R},T;-\omega_0)$ when $\omega'<0$.

We will take as our model scatterer one that satisfies the above requirements and has a space-time correlation function that is the product of a Gaussian function of $\mathbf{R}$ of rms width σ and a Gaussian function of T of rms width τ, i.e., that

$$G(\mathbf{R},T;\pm\omega_0)=\frac{B}{(2\pi\sigma^2)^{3/2}}e^{-R^2/2\sigma^2}e^{-T^2/2\tau^2} \ , \qquad (2.13)$$

where B, σ, and τ are positive constants. Evidently, τ is the effective time duration over which the fluctuations of the dielectric susceptibility are correlated, i.e., it is the *correlation time of the medium.* The constant σ is clearly the effective distance over which the fluctuations of the dielectric constant are correlated, i.e., it is the *spatial correlation length* of the medium. We will assume that σ is much smaller than the linear dimensions of the scattering volume. Finally, we will assume that the fluctuations of the medium and the fluctuations of the incident field are statistically independent.

III. SPECTRUM OF THE SCATTERED LIGHT IN THE FAR ZONE

When the incident field is of the type described in Sec. II A and the medium fulfills the assumptions stated above, the spectrum of the scattered light at the position $\mathbf{r}=r\mathbf{u}$ $(\mathbf{u}\cdot\mathbf{u}=1)$ in the far zone, $S^{(s)}(r\mathbf{u},\omega)$, is to a good approximation, given by Eq. (5.10) of Ref. 12, viz.,

$$S^{(s)}(r\mathbf{u},\omega)=\frac{(2\pi)^3\omega^4 V\sin^2\psi}{c^4r^2}\times\int_{-\infty}^{\infty}\mathcal{S}(\mathbf{k}-\mathbf{k}',\omega-\omega';\omega')S^{(i)}(\omega')d\omega' \ , \tag{3.1}$$

where ψ is the angle between $\mathbf{u}$ and $\mathbf{e}_0$, $\mathbf{k}$ is the wave vector of the scattered light,

$$\mathbf{k}=\frac{\omega}{c}\mathbf{u} \ , \tag{3.2}$$

$\mathbf{k}'$ is the wave vector of the ω' component of the incident field [cf. Eq. (2.1a)],

$$\mathbf{k}'=\frac{\omega'}{c}\mathbf{u}_0 \ , \tag{3.3}$$

and $\mathcal{S}(\mathbf{K},\Omega;\omega')$ is the *generalized structure function* of the medium [Ref. 12, Eq. (5.11a)]

$$\mathcal{S}(\mathbf{K},\Omega;\omega')=\frac{1}{(2\pi)^4}\int_V\int_{-\infty}^{\infty}G(\mathbf{R},T;\omega')e^{-i(\mathbf{K}\cdot\mathbf{R}-\Omega T)}d^3R\,dT \ . \tag{3.4}$$

The formula (3.1) shows that the spectrum of the scattered light in the far zone differs from the spectrum of the incident light $S^{(i)}(\omega)$ by the effect of two factors: namely, a factor proportional to ω^4 and a factor which is a linear transform of $S^{(i)}(\omega)$, the kernel of the transform being the generalized structure function of the medium. The ω^4 factor is a reflection of the fact that on the microscopic level the medium responds to the incident field as a set of dipole oscillators (cf. Ref. 15, Secs. 2.2.1 and 2.2.3). The generalized structure factor $\mathcal{S}$ depends upon the correlation between the oscillators. Consequently, the linear transform represents the effect of the interaction of the incident light with these correlated oscillators. Since $\mathcal{S}$ depends upon the "momentum transfer vector" $\mathbf{K}=\mathbf{k}-\mathbf{k}'$, it depends on the angle of scattering, and hence the spectrum of the scattered light in the far field will not only differ, in general, from the spectrum of the incident light, but will also be different in different directions of observation.

It is useful, for the purposes of calculation, to divide $S^{(s)}(r\mathbf{u},\omega)$ into two parts: the part denoted by $S_+^{(s)}(r\mathbf{u},\omega)$, generated by the positive frequency part of the spectrum of the incident light, and the part denoted by $S_-^{(s)}(r\mathbf{u},\omega)$, generated by the negative frequency part of the incident spectrum. Upon substituting Eq. (2.5) into Eq. (3.1) and assuming that $\Gamma_0/\omega_0\ll 1$ we find that, to a good approximation,

$$S^{(s)}(r\mathbf{u},\omega)=S_+^{(s)}(r\mathbf{u},\omega)+S_-^{(s)}(r\mathbf{u},\omega) \ , \tag{3.5}$$

where

$$S_\pm^{(s)}(r\mathbf{u},\omega)=\frac{(2\pi)^3\omega^4 V\sin^2\psi}{c^4r^2}\times\int_{-\infty}^{\infty}\mathcal{S}(\mathbf{k}-\mathbf{k}',\omega-\omega';\pm\omega_0)S_\pm^{(i)}(\omega')d\omega' \ , \tag{3.6}$$

with

$$S_\pm^{(i)}(\omega)=\frac{A}{(2\pi\Gamma_0^2)^{1/2}}g(\omega\mp\omega_0;\Gamma_0) \ . \tag{3.7}$$

We will now calculate the spectrum of the scattered light in the far zone for the case when the correlation function of the dielectric susceptibility is given by Eq. (2.13).

A. Evaluation of the generalized structure function $\mathcal{S}(\mathbf{k}-\mathbf{k}',\omega-\omega';\pm\omega_0)$

Upon substituting Eq. (2.13) into Eq. (3.4) and extending the spatial integration over all space (which is justified since σ was assumed to be small compared to the linear dimensions of V), we find that

$$\mathcal{S}(\mathbf{K},\Omega;\pm\omega_0)=\frac{B(2\pi\tau^2)^{1/2}}{(2\pi)^4}e^{-K^2\sigma^2/2}e^{-\Omega^2\tau^2/2} \ . \tag{3.8}$$

It will be useful to introduce the parameters

$$\Gamma_\sigma=c/\sigma \ , \tag{3.9}$$

$$\Gamma_\tau=1/\tau \ . \tag{3.10}$$

Each of these parameters has a clear physical significance. Γ_σ is the reciprocal of the time it takes light to cross a spatial correlation length σ. Γ_τ is the bandwidth of the temporal fluctuations of the medium.

Upon using Eqs. (3.9), (3.10), and (2.6), Eq. (3.8) can be rewritten as

$$\mathcal{S}(\mathbf{K},\Omega;\pm\omega_0)=\frac{B}{(2\pi)^4}\left[\frac{2\pi}{\Gamma_\tau^2}\right]^{1/2}g(Kc;\Gamma_\sigma)g(\Omega;\Gamma_\tau) \ . \tag{3.11}$$

Since $\mathbf{u}$ and $\mathbf{u}_0$ are unit vectors,

$$|\mathbf{k}-\mathbf{k}'|^2=\left|\frac{\omega}{c}\mathbf{u}-\frac{\omega'}{c}\mathbf{u}_0\right|^2 ,$$
$$=\frac{1}{c^2}(\omega^2-2\omega\omega'\cos\theta+\omega'^2) ,$$
$$=\frac{1}{c^2}\omega^2\sin^2\theta+\frac{1}{c^2}(\omega'-\omega\cos\theta)^2 , \quad (3.12)$$

where θ is the angle of scattering ($\mathbf{u}\cdot\mathbf{u}_0=\cos\theta$). It follows from Eqs. (3.11) and (3.12) that

$$\mathcal{S}(\mathbf{k}-\mathbf{k}',\omega-\omega';\pm\omega_0)=\frac{B}{(2\pi)^4}\left[\frac{2\pi}{\Gamma_\tau^2}\right]^{1/2} g(\omega\sin\theta;\Gamma_\sigma)$$
$$\times g(\omega'-\omega\cos\theta;\Gamma_\sigma)g(\omega-\omega';\Gamma_\tau) . \quad (3.13)$$

B. Evaluation of the spectrum of the scattered field $S^{(s)}(r\mathbf{u},\omega)$

It is shown in Appendix A that upon substituting from Eqs. (3.13) and (3.7) into Eq. (3.6) we obtain, after a straightforward calculation, the following expression for the positive frequency part of the spectrum of the scattered field:

$$S_+^{(s)}(r\mathbf{u},\omega)=N(r)H(\theta,\psi)\omega^4\exp\left[-\frac{1}{2}\left[\frac{\omega-\hat{\omega}(\theta)}{\hat{\Gamma}(\theta)}\right]^2\right] . \quad (3.14)$$

Here

$$N(r)=\frac{VAB}{2\pi c^4 r^2}\left[\frac{2\pi}{\Gamma_\tau^2+\alpha_\tau^2\Gamma_0^2}\right]^{1/2} , \quad (3.15)$$

$$H(\theta,\psi)=\sin^2\psi\exp\left[-\frac{1}{2}\left[\frac{\omega_0}{\Gamma_\sigma}\right]^2\frac{\alpha_\tau^2}{\alpha^2(\theta)}\times\left[\sin^2\theta+\frac{(1-\cos\theta)^2}{\alpha_\tau^2}\right]\right] , \quad (3.16)$$

$$\hat{\omega}(\theta)=\frac{\omega_0}{\alpha^2(\theta)}\left[1+\frac{\Gamma_\tau^2}{\Gamma_\sigma^2}\cos\theta\right] , \quad (3.17)$$

$$\hat{\Gamma}(\theta)=\frac{(\Gamma_\tau^2+\alpha_\tau^2\Gamma_0^2)^{1/2}}{\alpha(\theta)} , \quad (3.18)$$

and

$$\alpha_\tau^2=1+\frac{\Gamma_\tau^2}{\Gamma_\sigma^2} , \quad (3.19)$$

$$\alpha^2(\theta)=1+4\left[\frac{\Gamma_0}{\Gamma_\sigma}\right]^2\sin^2(\theta/2)+\frac{\Gamma_\tau^2}{\Gamma_\sigma^2}\left[1+\frac{\Gamma_0^2}{\Gamma_\sigma^2}\sin^2\theta\right] . \quad (3.20)$$

In a similar way, one can show from Eqs. (3.13), (3.7), and (3.6) that the negative frequency part of the scattered spectrum is given by

$$S_-^{(s)}(r\mathbf{u},\omega)=N(r)H(\theta,\psi)\omega^4\exp\left[-\frac{1}{2}\left[\frac{\omega+\hat{\omega}(\theta)}{\hat{\Gamma}(\theta)}\right]^2\right] . \quad (3.21)$$

For physically reasonable values of the parameters ω_0, Γ_0, σ, and τ, the frequency $\hat{\omega}(\theta)$ is close to ω_0 and $\hat{\Gamma}(\theta)$ is much smaller than ω_0. Under these circumstances,

$$S_-^{(s)}(r\mathbf{u},\omega)\approx 0 \quad \text{when } \omega>0 \quad (3.22)$$

for all positions in the far zone $\mathbf{r}=r\mathbf{u}$. It then follows from Eqs. (3.22), (3.5), and (3.14) that for $\omega>0$ the spectrum of the scattered light in the far zone is given by, to a good approximation,

$$S^{(s)}(r\mathbf{u},\omega)=S_+^{(s)}(r\mathbf{u},\omega)$$
$$=N(r)H(\theta,\psi)\omega^4\exp\left[-\frac{1}{2}\left[\frac{\omega-\hat{\omega}(\theta)}{\hat{\Gamma}(\theta)}\right]^2\right] , \quad (3.23)$$

where $N(r)$, $H(\theta,\psi)$, $\hat{\omega}(\theta)$, and $\hat{\Gamma}(\theta)$ are given by Eqs. (3.15)–(3.18).

It is evident from Eq. (3.23) that, in terms of its behavior as a function of frequency, the spectrum of the scattered light is a product of two factors: ω^4 and a Gaussian function of rms width $\hat{\Gamma}(\theta)$ and center frequency $\hat{\omega}(\theta)$. It is shown in Appendix C that for all values of the parameters Γ_0, ω_0, σ, and τ, and for all scattering angles θ ($0\le\theta\le\pi$),

$$\hat{\omega}(\theta)\le\omega_0 . \quad (3.24)$$

It is also shown there that the equality holds only in the following three special cases: (a) $\theta=0$, i.e., for forward scattering; (b) $\Gamma_\sigma\to\infty$ ($\sigma\to 0$), i.e., when the dielectric susceptibility fluctuations are spatially uncorrelated; and (c) $\Gamma_0\to 0$ and $\Gamma_\tau\to 0$ ($\tau\to\infty$), i.e., when the incident light is monochromatic and the dielectric susceptibility fluctuations are static.

The fact that, except in these special cases, $\hat{\omega}(\theta)<\omega_0$ implies that the Gaussian function in expression (3.23) is centered at a lower frequency (i.e., is red shifted) with respect to the Gaussian spectral line which produced it $[S_+^{(i)}(\omega)]$, the magnitude of the shift depending on angle of scattering θ. On the other hand, the factor ω^4 in the expression (3.23) is an increasing function of frequency and hence will produce a shift towards the higher frequencies (i.e., a blue shift). Consequently, the spectrum of the scattered light will be either red shifted or blue shifted with respect to $S_+^{(i)}(\omega)$, depending on the relative magnitudes of these two contributions.

IV. FREQUENCY SHIFT OF THE SPECTRUM OF THE SCATTERED LIGHT

A. General form

By a straightforward calculation (cf. Ref. 11, Appendix C), one can show that the spectrum of the scattered light, given by Eq. (3.23), is maximum as a function of ω when $\omega=\omega_0'(\theta)$, where

$$\omega_0'(\theta)=\frac{\hat{\omega}(\theta)}{2}\left\{1+\left[1+\left[\frac{4\hat{\Gamma}(\theta)}{\hat{\omega}(\theta)}\right]^2\right]^{1/2}\right\}. \tag{4.1}$$

For the purposes of the present calculation, we will rewrite Eq. (3.17) as

$$\hat{\omega}(\theta)=\omega_0\frac{f(\theta)}{\alpha^2(\theta)}, \tag{4.2}$$

where

$$f(\theta)=1+\left[\frac{\Gamma_\tau}{\Gamma_\sigma}\right]^2\cos\theta . \tag{4.3}$$

It follows from Eqs. (3.18) and (4.2) that

$$\frac{\hat{\Gamma}(\theta)}{\hat{\omega}(\theta)}=\frac{\alpha(\theta)}{f(\theta)}\left[\frac{\Gamma_\tau^2+\alpha_\tau^2\Gamma_0^2}{\omega_0^2}\right]^{1/2}. \tag{4.4}$$

Upon substituting Eqs. (4.2) and (4.4) into Eq. (4.1), we find, after some algebra, that

$$\omega_0'(\theta)=\frac{\omega_0}{2\alpha^2(\theta)}\left\{f(\theta)+\left[f^2(\theta)+16\alpha^2(\theta)\left[\frac{\Gamma_\tau^2+\alpha_\tau^2\Gamma_0^2}{\omega_0^2}\right]\right]^{1/2}\right\}. \tag{4.5}$$

It is customary, especially in astronomy, to specify frequency shifts by the quantity

$$z=\frac{\lambda_0'-\lambda_0}{\lambda_0}=\frac{\omega_0-\omega_0'}{\omega_0'}, \tag{4.6}$$

where $\lambda_0=2\pi c/\omega_0$ is the original wavelength and $\lambda_0'=2\pi c/\omega_0'$ is the corresponding "shifted" wavelength. Evidently $z>0$ when the line is red shifted and $z<0$ when it is blue shifted. Upon substituting from Eq. (4.5) into Eq. (4.6), we find that in the present case

$$z(\theta)=\frac{2\alpha^2(\theta)-\left\{f(\theta)+\left[f^2(\theta)+16\alpha^2(\theta)\left[\frac{\Gamma_\tau^2+\alpha_\tau^2\Gamma_0^2}{\omega_0^2}\right]\right]^{1/2}\right\}}{f(\theta)+\left[f^2(\theta)+16\alpha^2(\theta)\left[\frac{\Gamma_\tau^2+\alpha_\tau^2\Gamma_0^2}{\omega_0^2}\right]\right]^{1/2}}. \tag{4.7}$$

B. Approximate forms for $z(\theta)$ and $H(\theta,\psi)$

The expression (4.7) for $z(\theta)$ is too complicated to make it possible to draw any conclusions about the roles that the various physical mechanisms play in generating the line shift. Similar remarks apply to the expression (3.16) for $H(\theta,\psi)$, which describes the strength of the scattered light. However, if

$$\frac{\Gamma_\tau^2}{\Gamma_\sigma^2}\ll 1, \tag{4.8a}$$

$$\frac{\Gamma_0^2}{\Gamma_\sigma^2}\ll 1, \tag{4.8b}$$

$$16\left[\frac{\Gamma_\tau^2+\Gamma_0^2}{\omega_0^2}\right]\ll 1 \tag{4.8c}$$

(conditions which are usually fulfilled in practice), $z(\theta)$ as given by Eq. (4.7) and $H(\theta,\psi)$ as given by Eq. (3.16) can be approximated by expressions which clearly indicate the roles of the various physical parameters, as we will now show.

It follows from Eq. (4.7) that $z(\theta)$ can be rewritten as

$$z(\theta)=\frac{2\alpha^2(\theta)-\{f(\theta)+[1+q(\theta)]^{1/2}\}}{f(\theta)+[1+q(\theta)]^{1/2}}, \tag{4.9}$$

where

$$\begin{aligned} q(\theta)&=f^2(\theta)-1+16\alpha^2(\theta)\left[\frac{\Gamma_\tau^2+\alpha_\tau^2\Gamma_0^2}{\omega_0^2}\right]\\ &=2\left[\frac{\Gamma_\tau}{\Gamma_\sigma}\right]^2\cos\theta+\left[\frac{\Gamma_\tau}{\Gamma_\sigma}\right]^4\cos^2\theta\\ &\quad+16\alpha^2(\theta)\left[\frac{\Gamma_\tau^2+\alpha_\tau^2\Gamma_0^2}{\omega_0^2}\right]. \end{aligned} \tag{4.10}$$

It follows from Eq. (3.20) that to first order in the small quantities, which appear in the inequalities (4.8),

$$\alpha^2(\theta)\approx 1+\left[\frac{\Gamma_\tau}{\Gamma_\sigma}\right]^2+4\frac{\Gamma_0^2}{\Gamma_\sigma^2}\sin^2(\theta/2), \tag{4.11}$$

and if we use this approximation, we deduce from Eq. (4.10) that

$$\sqrt{1+q(\theta)}\approx 1+\left[\frac{\Gamma_\tau}{\Gamma_\sigma}\right]^2\cos\theta+8\left[\frac{\Gamma_\tau^2+\Gamma_0^2}{\omega_0^2}\right]. \tag{4.12}$$

Upon substituting from Eqs. (4.11) and (4.12) into Eq. (4.9), we find that, to first order in small quantities,

$$z(\theta)\approx 2\left[\frac{\Gamma_\tau^2}{\Gamma_\sigma^2}+2\frac{\Gamma_0^2}{\Gamma_\sigma^2}\right]\sin^2(\theta/2)-4\left[\frac{\Gamma_\tau^2+\Gamma_0^2}{\omega_C^2}\right]. \tag{4.13}$$

We note that in the static limit ($\Gamma_\tau\to 0$), Eq. (4.13) has the same form as the corresponding expression of scalar scattering theory [Ref. 11, Eq. (4.15)]. The two terms on the right-hand side of Eq. (4.13) have a simple physical interpretation as regards to the shift they provide: since $\sin^2(\theta/2) \geq 0$, *the first term represents a red shift* (as long as $\theta \neq 0$); on the other hand, *the second term represents a blue shift.*

We will now show that each of the terms can be related to a particular physical mechanism. Let us recall from Sec. III B, that the total shift of the line is comprised of two parts: the shift of the center frequency $\hat{\omega}(\theta)$ of the Gaussian term in Eq. (3.23) from the center frequency ω_0 of the incident line, and a shift due to the ω^4 factor. The shift of the Gaussian term can be described by the corresponding shift parameter

$$z_G(\theta) = \frac{\omega_0 - \hat{\omega}(\theta)}{\hat{\omega}(\theta)} ,$$

$$= \frac{\alpha^2(\theta) - f(\theta)}{f(\theta)} , \qquad (4.14)$$

where Eq. (4.2) was used above. Upon using Eq. (4.11), we find that to first order in small parameters,

$$z_G(\theta) \approx 2 \left[\frac{\Gamma_\tau^2}{\Gamma_\sigma^2} + 2 \frac{\Gamma_0^2}{\Gamma_\sigma^2} \right] \sin^2(\theta/2) . \qquad (4.15)$$

Upon comparing Eqs. (4.15) and (4.13), we see the first term in Eq. (4.13) represents a *red shift* due to *correlations in the fluctuations of the physical properties of the medium and the finite bandwidth of the incident light* and the second term represents a *blue shift* due to the *dipole factor* ω^4.

In order to discuss the details of the dependences of the frequency shifts upon the physical parameters, we will rewrite Eq. (4.13) in a slightly different form. It follows from Eqs. (3.9) and (3.10) that the ratio $\Gamma_\tau/\Gamma_\sigma$ may be expressed as

$$\frac{\Gamma_\tau}{\Gamma_\sigma} = \frac{\sigma/c}{\tau} , \qquad (4.16)$$

i.e., it is the ratio of the time light takes to cross the spatial correlation length of the medium to the correlation time of the medium. In view of Eqs. (2.11) and (3.9), the ratio Γ_0/Γ_σ may be expressed as

$$\frac{\Gamma_0}{\Gamma_\sigma} = \frac{\sigma}{L_0} , \qquad (4.17)$$

i.e., it is the ratio of the spatial correlation length of the medium to the coherence length of the incident light. Upon substituting from Eqs. (4.16) and (4.17) into Eq. (4.13), we find that the frequency shift may be expressed as

$$z(\theta) \approx 2 \left[\left[\frac{\sigma/c}{\tau} \right]^2 + 2 \left[\frac{\sigma}{L_0} \right]^2 \right] \sin^2(\theta/2)$$

$$-4 \left[\left[\frac{\Gamma_\tau}{\omega_0} \right]^2 + \left[\frac{\Gamma_0}{\omega_0} \right]^2 \right] \qquad (4.18)$$

It follows from Eq. (4.18) that the red-shift term is a monotonically increasing function of the two ratios mentioned above and the blue-shift term is a monotonically increasing function of the bandwidth of the temporal fluctuations, as measured in units of ω_0, and the bandwidth of the incident light, as measured in units of ω_0. Moreover, Eq. (4.18) shows that for appropriate values of the parameters it contains, the spectrum of the scattered light will be blue shifted $[z(\theta) < 0]$ for small angles of scattering θ and will become red shifted $[z(\theta) > 0]$ for larger values of θ.

We conclude this section with a brief discussion of the approximate form for the factor $H(\theta, \psi)$. Since $c = \omega_0 \lambda\!\!\!^{-}_0$ ($\lambda\!\!\!^{-}_0 = \lambda_0/2\pi$), it follows from Eq. (3.9) that $\omega_0/\Gamma_\sigma = \sigma/\lambda\!\!\!^{-}_0$. Therefore Eq. (3.16) may be rewritten as

$$H(\theta, \psi) = \sin^2\psi \exp \left[-\frac{1}{2} \left[\frac{\sigma}{\lambda\!\!\!^{-}_0} \right]^2 \frac{\alpha_\tau^2}{\alpha^2(\theta)} \times \left[\sin^2\theta + \frac{(1-\cos\theta)^2}{\alpha_\tau^2} \right] \right] . \qquad (4.19)$$

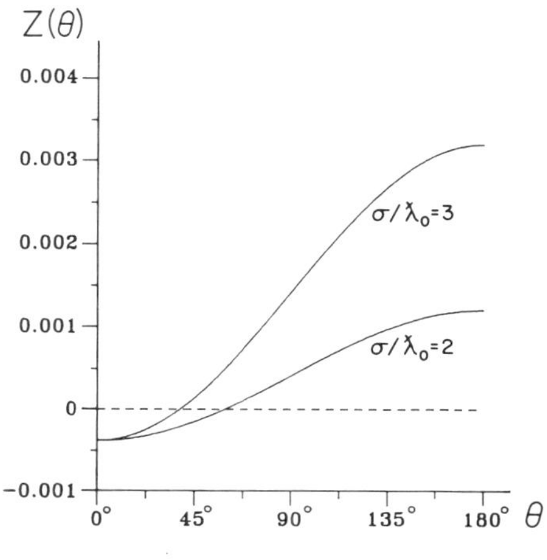

FIG. 2. Plots of the frequency shift parameter $z(\theta)$ with $\Gamma_0/\omega_0 = 0.01$ and $\Gamma_\tau/\Gamma_\sigma = 0$, for two selected values of $\sigma/\lambda\!\!\!^{-}_0$.

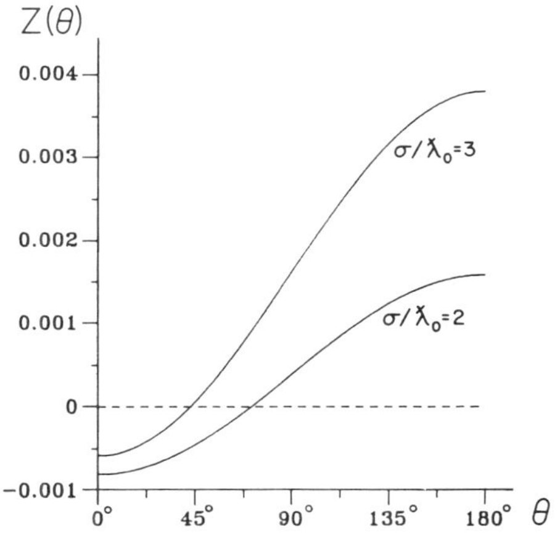

FIG. 3. Plots of the frequency shift parameter $z(\theta)$ with $\Gamma_0/\omega_0 = 0.01$ and $\Gamma_\tau/\Gamma_\sigma = 0.02$, for two selected values of $\sigma/\lambda\!\!\!^{-}_0$.

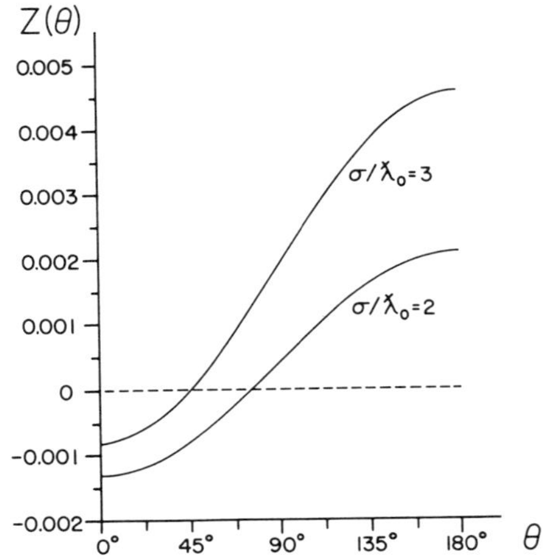

FIG. 4. Plots of the frequency shift parameter $z(\theta)$ with $\Gamma_0/\omega_0=0.01$ and $\Gamma_\tau/\Gamma_\sigma=0.03$, for two selected values of $\sigma/ƛ_0$.

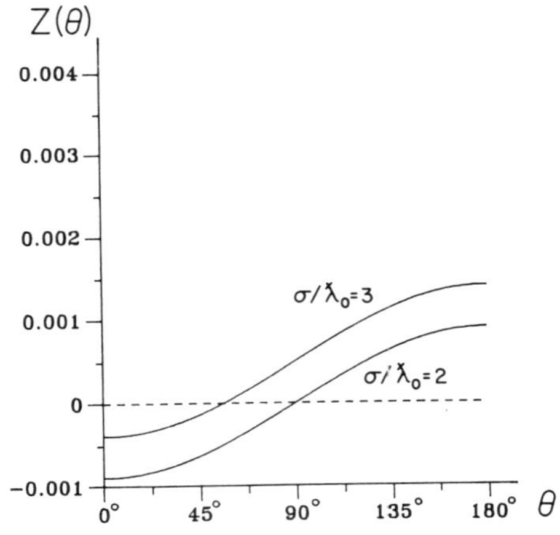

FIG. 6. Plots of the frequency shift parameter $z(\theta)$ with $\Gamma_0/\omega_0=0$ and $\Gamma_\tau/\Gamma_\sigma=0.03$, for two selected values of $\sigma/ƛ_0$.

It then follows from Eqs. (3.19) and (3.20) that[16]

$$H(\theta,\psi)\approx\sin^2\psi\exp\left[-2\left[\frac{\sigma}{ƛ_0}\right]^2\sin^2(\theta/2)\right]. \tag{4.20}$$

Equation (4.20) shows that, for a fixed value of ψ, the strength of the scattered light decreases with increasing θ, and may be small at the values of θ where large red shifts occur. Equation (4.20) also shows that, for fixed value of ψ, the decrease of the strength of the scattered light with increasing θ is larger for more highly spatially correlated media (media with larger σ).

C. Numerical results

Figures 2–6 and Fig. 7 show plots of $z(\theta)$ and $H(\theta,90°)$, respectively, as functions of θ, for some selected values of the parameters. They were calculated from Eqs. (4.7) and (3.16). In all the cases presented below, conditions (4.8) are fulfilled, and hence $z(\theta)$ and $H(\theta,\psi)$ are well described by expressions (4.13) [or (4.18)] and (4.20), respectively.

In Fig. 2, $z(\theta)$ versus θ is plotted for the cases[17] $\Gamma_0/\omega_0=0.01$, $\Gamma_\tau/\Gamma_\sigma=0$, and $\sigma/ƛ_0=2,3$. The expected $\sin^2(\theta/2)$ behavior [cf. Eq. (4.13)] is evident, and it is clear from the curves that an increase of the spatial correlation length of the medium produces larger shifts in the spectrum of the scattered light. In Figs. 3 and 4, $\Gamma_0/\omega_0=0.01$, $\sigma/ƛ_0=2,3$, and $\Gamma_\tau/\Gamma_\sigma$ is increased to 0.02, 0.03. It is clear from the figures that this increase of the ratio $\Gamma_\tau/\Gamma_\sigma$ produces larger shifts.

In Fig. 5, $z(\theta)$ is plotted as a function of θ for the cases $\Gamma_0/\omega_0=0$ (monochromatic incident light), $\Gamma_\tau/\Gamma_\sigma=0.02$,

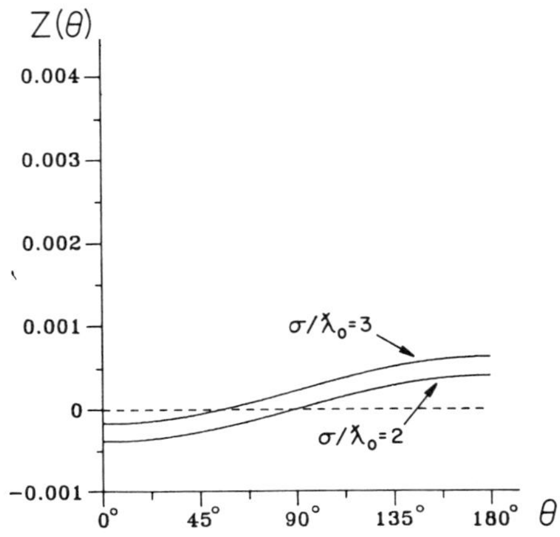

FIG. 5. Plots of the frequency shift parameter $z(\theta)$ with $\Gamma_0/\omega_0=0$ and $\Gamma_\tau/\Gamma_\sigma=0.02$, for two selected values of $\sigma/ƛ_0$.

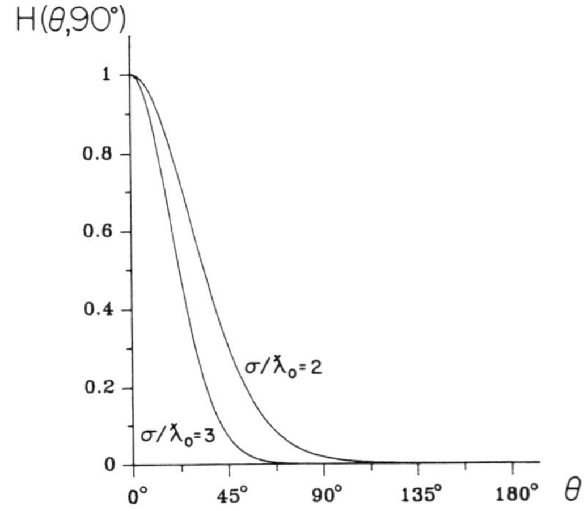

FIG. 7. Plots of the factor $H(\theta,90°)$ with $\Gamma_0/\omega_0=0.01$ and $\Gamma_\tau/\Gamma_\sigma=0.02$, for two selected values of $\sigma/ƛ_0$.

and $\sigma/\lambda_0=2,3$. Upon comparing Fig. 5 to Fig. 3, we see that the shifts are greatly reduced when the linewidth of the incident light is reduced. In Fig. 6, $\Gamma_0/\omega_0=0$, $\sigma/\lambda_0=2,3$, and $\Gamma_\tau/\Gamma_\sigma$ is increased to 0.03. Upon comparing this figure to Fig. 5, it is again clear that an increase in the ratio $\Gamma_\tau/\Gamma_\sigma$ produces larger shifts. Furthermore, if we compare Fig. 6 to Fig. 4, we see again that the shift is significantly smaller when the linewidth of the incident light is reduced.

In Fig. 7 the factor $H(\theta,90°)$ is plotted versus θ for the same values of the parameters as were used in Fig. 3. It is evident from this curve that the scattering is strongest in the forward direction and is weak when $90°<\theta\leq 180°$. Since, for all the sets of parameter values used in Figs. 2–6, $H(\theta,\psi)$ is approximately given by Eq. (4.20), the plots of $H(\theta,90°)$ for the parameter sets used in Fig. 2 and Figs. 4–6 are indistinguishable from those of Fig. 7 and are not presented.

ACKNOWLEDGMENTS

This research was supported by the National Science Foundation and the U.S. Army Research Office.

APPENDIX A: EVALUATION OF $S_+^{(s)}(r\mathbf{u},\omega)$

In the subsequent calculations presented in this Appendix, the following product theorem for Gaussian functions (Ref. 11, Appendix A) will be needed.

Theorem. If

$$g(\omega-\omega_j;\Gamma_j)=e^{-(\omega-\omega_j)^2/2\Gamma_j^2}\quad (j=1,2)\ , \tag{A1}$$

then

$$g(\omega-\omega_1;\Gamma_1)g(\omega-\omega_2;\Gamma_2)$$
$$=g[\omega_1-\omega_2;(\Gamma_1^2+\Gamma_2^2)^{1/2}]g(\omega-\tilde{\omega};\tilde{\Gamma})\ , \tag{A2}$$

where

$$\tilde{\omega}=\frac{\omega_1\Gamma_2^2+\omega_2\Gamma_1^2}{\Gamma_1^2+\Gamma_2^2}\ , \tag{A3}$$

$$\frac{1}{\tilde{\Gamma}^2}=\frac{1}{\Gamma_1^2}+\frac{1}{\Gamma_2^2}\ . \tag{A4}$$

1. Derivation of an alternative form for the generalized structure function (3.13)

It follows from the theorem we just stated that

$$g(\omega'-\omega\cos\theta;\Gamma_\sigma)g(\omega'-\omega;\Gamma_\tau)$$
$$=g[\omega(1-\cos\theta);(\Gamma_\sigma^2+\Gamma_\tau^2)^{1/2}]g(\omega'-\tilde{\omega};\tilde{\Gamma})\ , \tag{A5}$$

where[18]

$$\tilde{\omega}=\omega\frac{1+(\Gamma_\tau/\Gamma_\sigma)^2\cos\theta}{1+(\Gamma_\tau/\Gamma_\sigma)^2}\ , \tag{A6}$$

$$\frac{1}{\tilde{\Gamma}^2}=\frac{1}{\Gamma_\sigma^2}+\frac{1}{\Gamma_\tau^2}\ . \tag{A7}$$

For later reference we note that Eq. (A6) can be rewritten as

$$\tilde{\omega}=\omega/\beta\ , \tag{A8}$$

where

$$\beta=\frac{\alpha_\tau^2}{f}\ , \tag{A9}$$

$$\alpha_\tau^2=1+(\Gamma_\tau/\Gamma_\sigma)^2\ , \tag{A10}$$

$$f=1+(\Gamma_\tau/\Gamma_\sigma)^2\cos\theta\ . \tag{A11}$$

Upon substituting Eq. (A5) into Eq. (3.13) we find that

$$\mathcal{S}(\mathbf{k}-\mathbf{k}',\omega-\omega';\omega_0)$$
$$=\frac{B}{(2\pi)^4}(2\pi/\Gamma_\tau^2)^{1/2}g(\omega\sin\theta;\Gamma_\sigma)$$
$$\times g[\omega(1-\cos\theta);(\Gamma_\sigma^2+\Gamma_\tau^2)^{1/2}]g(\omega'-\tilde{\omega};\tilde{\Gamma})\ . \tag{A12}$$

By using the definition of g [cf. Eq. (A1)], the first two Gaussian functions on the right-hand side of Eq. (A12) can be combined, and we find that

$$\mathcal{S}(\mathbf{k}-\mathbf{k}',\omega-\omega';\omega_0)$$
$$=\frac{B}{(2\pi)^4}(2\pi/\Gamma_\tau^2)^{1/2}g(\omega;\Gamma_1)g(\omega'-\tilde{\omega};\tilde{\Gamma})\ , \tag{A13}$$

where

$$\frac{1}{\Gamma_1^2}=\frac{\sin^2\theta}{\Gamma_\sigma^2}+\frac{(1-\cos\theta)^2}{\Gamma_\sigma^2+\Gamma_\tau^2}\ ,$$
$$=\frac{\sin^2\theta}{\Gamma_\sigma^2}+\frac{(1-\cos\theta)^2}{\Gamma_\sigma^2\alpha_\tau^2}\ . \tag{A14}$$

2. Calculation of $S_+^{(s)}(r\mathbf{u},\omega)$

Upon substituting Eqs. (3.7) and (A13) into Eq. (3.6), we find that

$$S_+^{(s)}(r\mathbf{u},\omega)=M(r)(\sin^2\psi)\omega^4 g(\omega;\Gamma_1)$$
$$\times\int_{-\infty}^{\infty}g(\omega'-\tilde{\omega};\tilde{\Gamma})g(\omega'-\omega_0;\Gamma_0)d\omega'\ , \tag{A15}$$

with

$$M(r)=\frac{V}{2\pi c^4r^2}\frac{AB}{\Gamma_0\Gamma_\tau}\ . \tag{A16}$$

By a straightforward calculation, one can show that

$$\int_{-\infty}^{\infty}g(\omega'-\tilde{\omega};\tilde{\Gamma})g(\omega'-\omega_0;\Gamma_0)d\omega'$$
$$=\left[\frac{2\pi\tilde{\Gamma}^2\Gamma_0^2}{\tilde{\Gamma}^2+\Gamma_0^2}\right]^{1/2}g[\tilde{\omega}-\omega_0;(\tilde{\Gamma}^2+\Gamma_0^2)^{1/2}]$$
$$=\left[\frac{2\pi\tilde{\Gamma}^2\Gamma_0^2}{\tilde{\Gamma}^2+\Gamma_0^2}\right]^{1/2}g(\omega-\omega_0\beta;\Gamma_2)\ , \tag{A17}$$

where Eq. (A8) was used in the last step and

$$\Gamma_2=\beta(\tilde{\Gamma}^2+\Gamma_0^2)^{1/2} . \tag{A18}$$

Upon substituting Eq. (A17) into Eq. (A15) we find that

$$S_+^{(s)}(r\mathbf{u},\omega)=N(r)(\sin^2\psi)\omega^4 g(\omega;\Gamma_1)g(\omega-\omega_0\beta;\Gamma_2) , \tag{A19}$$

where

$$N(r)=\frac{ABV}{2\pi c^4 r^2}\left[\frac{2\pi\tilde{\Gamma}^2}{\Gamma_\tau^2(\tilde{\Gamma}^2+\Gamma_0^2)}\right]^{1/2} . \tag{A20}$$

It follows from Eqs. (A7) and (A10) that

$$\tilde{\Gamma}=\Gamma_\tau/\alpha_\tau . \tag{A21}$$

Straightforward algebraic manipulations show that Eq. (A20) can be rewritten as

$$N(r)=\frac{ABV}{2\pi c^4 r^2}\left[\frac{2\pi}{\Gamma_\tau^2+\Gamma_0^2\alpha_\tau^2}\right]^{1/2} . \tag{A22}$$

Upon again using the product theorem for Gaussian functions [cf. Eqs. (A1)–(A4)], we find that Eq. (A19) can be rewritten as

$$\begin{aligned} S_+^{(s)}(r\mathbf{u},\omega)&=N(r)(\sin^2\psi)g[\omega_0\beta;(\Gamma_1^2+\Gamma_2^2)^{1/2}]\\ &\quad\times\omega^4 g[\omega-\hat{\omega}(\theta);\hat{\Gamma}(\theta)]\\ &=N(r)(\sin^2\psi)g[\omega_0;\Gamma_d(\theta)]\\ &\quad\times\omega^4 g[\omega-\hat{\omega}(\theta);\hat{\Gamma}(\theta)] , \end{aligned} \tag{A23}$$

where

$$\frac{1}{\Gamma_d^2(\theta)}=\frac{\beta^2(\theta)}{\Gamma_1^2(\theta)+\Gamma_2^2(\theta)} , \tag{A24}$$

$$\hat{\omega}(\theta)=\omega_0\frac{\beta(\theta)\Gamma_1^2(\theta)}{\Gamma_1^2(\theta)+\Gamma_2^2(\theta)} , \tag{A25}$$

$$\frac{1}{\hat{\Gamma}^2(\theta)}=\frac{1}{\Gamma_1^2(\theta)}+\frac{1}{\Gamma_2^2(\theta)} . \tag{A26}$$

The function $\beta(\theta)$ is given by Eqs. (A9)–(A11), $\Gamma_1(\theta)$ by Eq. (A14), and $\Gamma_2(\theta)$ by Eq. (A18).

It is shown in Appendix B that $\hat{\omega}(\theta)$, $\hat{\Gamma}(\theta)$, and $\Gamma_d(\theta)$ can be rewritten as [see Eqs. (B7), (B11), and (B13)],

$$\hat{\omega}(\theta)=\frac{\omega_0}{\alpha^2(\theta)}\left[1+\frac{\Gamma_\tau^2}{\Gamma_\sigma^2}\cos\theta\right] , \tag{A27}$$

$$\hat{\Gamma}(\theta)=\frac{(\Gamma_\tau^2+\alpha_\tau^2\Gamma_0^2)^{1/2}}{\alpha(\theta)} , \tag{A28}$$

$$\frac{1}{\Gamma_d^2(\theta)}=\frac{\alpha_\tau^2}{\alpha^2(\theta)}\frac{1}{\Gamma_\sigma^2}\left[\sin^2\theta+\frac{(1-\cos\theta)^2}{\alpha_\tau^2}\right] , \tag{A29}$$

where $\alpha^2(\theta)$ is as given in Eq. (B8). Using Eq. (A29) and the definition of g in Eq. (A23) we find that

$$S_+^{(s)}(r\mathbf{u},\omega)=N(r)H(\theta,\psi)\omega^4\exp\left[-\frac{1}{2}\left[\frac{\omega-\hat{\omega}(\theta)}{\hat{\Gamma}(\theta)}\right]^2\right] \tag{A30}$$

where

$$\begin{aligned} H(\theta,\psi)=\sin^2\psi\exp\Bigg[-\frac{1}{2}\left[\frac{\omega_0}{\Gamma_\sigma}\right]^2\frac{\alpha_\tau^2}{\alpha^2(\theta)}\\ \times\left[\sin^2\theta+\frac{(1-\cos\theta)^2}{\alpha_\tau^2}\right]\Bigg] . \end{aligned} \tag{A31}$$

APPENDIX B: SIMPLIFICATION OF THE EXPRESSIONS FOR $\hat{\omega}(\theta)$, $\hat{\Gamma}(\theta)$, and $\Gamma_d(\theta)$

1. Simplification of $\hat{\omega}(\theta)$

Equation (A25) can be rewritten as

$$\hat{\omega}(\theta)=\frac{\omega_0\beta(\theta)}{1+[\Gamma_2(\theta)/\Gamma_1(\theta)]^2} . \tag{B1}$$

Upon substituting Eqs. (A9), (A14), (A18), and (A21) into Eq. (B1) we find, after some algebra, that

$$\hat{\omega}(\theta)=\frac{\omega_0\alpha_\tau^2 f(\theta)}{h(\theta)+(\alpha_\tau\Gamma_0/\Gamma_\sigma)^2[\alpha_\tau^2\sin^2\theta+(1-\cos\theta)^2]} , \tag{B2}$$

where

$$h(\theta)=f^2(\theta)+\left[\frac{\Gamma_\tau}{\Gamma_\sigma}\right]^2[\alpha_\tau^2\sin^2\theta+(1-\cos\theta)^2] . \tag{B3}$$

It follows directly from Eq. (A11) that

$$f^2(\theta)=1+2\left[\frac{\Gamma_\tau}{\Gamma_\sigma}\right]^2\cos\theta+\left[\frac{\Gamma_\tau}{\Gamma_\sigma}\right]^4\cos^2\theta . \tag{B4}$$

Furthermore, upon using Eq. (A10) we find that

$$\begin{aligned} &\left[\frac{\Gamma_\tau}{\Gamma_\sigma}\right]^2[\alpha_\tau^2\sin^2\theta+(1-\cos\theta)^2]\\ &\quad=2\left[\frac{\Gamma_\tau}{\Gamma_\sigma}\right]^2-2\left[\frac{\Gamma_\tau}{\Gamma_\sigma}\right]^2\cos\theta+\left[\frac{\Gamma_\tau}{\Gamma_\sigma}\right]^4\sin^2\theta . \end{aligned} \tag{B5}$$

Upon substituting Eqs. (B4) and (B5) into Eq. (B3) we obtain, after straightforward calculations, the following expression for $h(\theta)$:

$$h(\theta)=\alpha_\tau^4 , \tag{B6}$$

where α_τ^2 is given by Eq. (A10). It follows from Eq. (B2) that

$$\hat{\omega}(\theta)=\omega_0\frac{f(\theta)}{\alpha^2(\theta)} , \tag{B7}$$

where $f(\theta)$ is given by Eq. (A11) and

$$\alpha^2(\theta)=\alpha_\tau^2+\left[\frac{\Gamma_0}{\Gamma_\sigma}\right]^2[\alpha_\tau^2\sin^2\theta+(1-\cos\theta)^2]$$

$$=1+\left[2\frac{\Gamma_0}{\Gamma_\sigma}\sin(\theta/2)\right]^2+\left[\frac{\Gamma_\tau}{\Gamma_\sigma}\right]^2\left[1+\frac{\Gamma_0^2}{\Gamma_\sigma^2}\sin^2\theta\right] . \quad \text{(B8)}$$

2. Simplification of the expression for $\hat{\Gamma}(\theta)$

It follows from Eq. (A26) that

$$\hat{\Gamma}^2(\theta)=\frac{\Gamma_1^2(\theta)\Gamma_2^2(\theta)}{\Gamma_1^2(\theta)+\Gamma_2^2(\theta)} , \quad \text{(B9)}$$

and from Eq. (A25) that Eq. (B9) can be rewritten as

$$\hat{\Gamma}^2(\theta)=\frac{\hat{\omega}(\theta)}{\omega_0\beta(\theta)}\Gamma_2^2(\theta) . \quad \text{(B10)}$$

We then find from Eqs. (B7) and (A18) that Eq. (B10) can be rewritten as

$$\hat{\Gamma}^2(\theta)=\frac{f(\theta)}{\alpha^2(\theta)\beta(\theta)}\beta^2(\theta)(\tilde{\Gamma}^2+\Gamma_0^2) ,$$

$$=\frac{f(\theta)}{\alpha^2(\theta)}\beta(\theta)\left[\frac{\Gamma_\tau^2}{\alpha_\tau^2}+\Gamma_0^2\right]$$

$$=\frac{f(\theta)}{\alpha^2(\theta)}\frac{\alpha_\tau^2}{f(\theta)}\left[\frac{\Gamma_\tau^2}{\alpha_\tau^2}+\Gamma_0^2\right] ,$$

$$=\frac{\Gamma_\tau^2+\alpha_\tau^2\Gamma_0^2}{\alpha^2(\theta)} .$$

Therefore

$$\hat{\Gamma}(\theta)=\frac{(\Gamma_\tau^2+\alpha_\tau^2\Gamma_0^2)^{1/2}}{\alpha(\theta)} . \quad \text{(B11)}$$

3. Simplification of the expression for $\Gamma_d(\theta)$

From Eqs. (A24) and (A25), it follows that

$$\frac{1}{\Gamma_d^2(\theta)}=\frac{\hat{\omega}(\theta)}{\omega_0\Gamma_1^2(\theta)}\beta(\theta) . \quad \text{(B12)}$$

Upon using Eqs. (B7) and (A9) in Eq. (B12), we find that

$$\frac{1}{\Gamma_d^2(\theta)}=\frac{f(\theta)}{\alpha^2(\theta)}\frac{1}{\Gamma_1^2(\theta)}\frac{\alpha_\tau^2}{f(\theta)}$$

$$=\frac{\alpha_\tau^2}{\alpha^2(\theta)}\frac{1}{\Gamma_\sigma^2}\left[\sin^2\theta+\frac{(1-\cos\theta)^2}{\alpha_\tau^2}\right] , \quad \text{(B13)}$$

where Eq. (A13) was used in the last step.

APPENDIX C: THE BEHAVIOR OF $\hat{\omega}(\theta)$

It follows from Eqs. (3.17) and (3.20) that

$$\omega_0-\hat{\omega}(\theta)=\frac{\omega_0}{\alpha^2(\theta)}\left[\alpha^2(\theta)-\left[1+\frac{\Gamma_\tau^2}{\Gamma_\sigma^2}\cos\theta\right]\right]$$

$$=\frac{\omega_0}{\alpha^2(\theta)}\left[4\left[\frac{\Gamma_0}{\Gamma_\sigma}\right]^2\sin^2(\theta/2)+\left[\frac{\Gamma_\tau}{\Gamma_\sigma}\right]^2\left[1+\frac{\Gamma_0^2}{\Gamma_\sigma^2}\sin^2\theta\right]-\left[\frac{\Gamma_\tau}{\Gamma_\sigma}\right]^2\cos\theta\right] . \quad \text{(C1)}$$

Since $1-\cos\theta=2\sin^2(\theta/2)$, Eq. (C1) can be rewritten as

$$\omega_0-\hat{\omega}(\theta)=\frac{\omega_0}{\alpha^2(\theta)}\left\{\left[4\left[\frac{\Gamma_0}{\Gamma_\sigma}\right]^2+2\left[\frac{\Gamma_\tau}{\Gamma_\sigma}\right]^2\right]\sin^2(\theta/2)+\left[\frac{\Gamma_0}{\Gamma_\sigma}\right]^2\left[\frac{\Gamma_\tau}{\Gamma_\sigma}\right]^2\sin^2\theta\right\} . \quad \text{(C2)}$$

Since according to Eq. (3.20) $\alpha^2(\theta)\geq 1$, it follows from Eq. (C2) that

$$\hat{\omega}(\theta)\leq\omega_0 , \quad \text{(C3)}$$

with the equality holding only in the following three special cases: (a) $\theta=0$, i.e., for forward scattering; (b) $\Gamma_\sigma\to\infty$ ($\sigma\to 0$), i.e., when the dielectric susceptibility fluctuations are spatially uncorrelated; (c) $\Gamma_0\to 0$ and $\Gamma_\tau\to 0$ ($\tau\to\infty$), i.e., when the incident light monochromatic and the dielectric susceptibility fluctuations are static.

---

*Also at the Institute of Optics, University of Rochester.

[1]E. Wolf, Phys. Rev. Lett. **56**, 1370 (1986).
[2]See also L. Mandel, J. Opt. Soc. Am. **51**, 1342 (1961); L. Mandel and E. Wolf, *ibid.* **66**, 529 (1976); F. Gori and R. Grella, Opt. Commun. **49**, 173 (1984).
[3]G. M. Morris and D. Faklis, Opt. Commun. **62**, 5 (1987).
[4]E. Wolf, Nature **326**, 363 (1987).
[5]E. Wolf, Opt. Commun. **62**, 12 (1987).
[6]Z. Dacic and E. Wolf, J. Opt. Soc. Am. A **5**, 1118 (1988).
[7]D. Faklis and G. M. Morris, Opt. Lett. **13**, 4 (1988).
[8]F. Gori, G. Guattari, C. Palma, and G. Padovani, Opt. Commun. **67**, 1 (1988).
[9]W. H. Knox and R. S. Knox, J. Opt. Soc. Am. A **4(13)**, P131 (1987).
[10]M. F. Bocko, D. H. Douglass, and R. S. Knox, Phys. Rev. Lett. **58**, 2649 (1987).
[11]E. Wolf, J. T. Foley, and F. Gori, J. Opt. Soc. Am. A (to be published).
[12]E. Wolf and J. T. Foley, preceding paper, Phys. Rev. A **40**, 579 (1989).
[13]It is customary in optical coherence theory to represent real fields by complex ones by the so-called analytic signals (see, for example, Ref. 15, Sec. 10.2). We do *not* adopt this procedure here.
[14]W. B. Davenport and W. L. Root, *Random Signals and Noise* (McGraw-Hill, New York, 1958), p. 60.
[15]M. Born and E. Wolf, *Principles of Optics,* 6th ed. (Pergamon, Oxford, England, 1980).

[16]The ratio σ/λ_0 is not necessarily small.

[17]The expressions (4.7) and (3.16) for $z(\theta)$ and $H(\theta,\psi)$ can be rewritten entirely in terms of θ, ψ, Γ_0/ω_0, $\Gamma_\tau/\Gamma_\sigma$, and σ/λ_0. For the sake of brevity we do not display such forms for $z(\theta)$ and $H(\theta,\psi)$; however, the three ratios are used to label the curves in Figs. 2–7.

[18]For the sake of economy of notation we do not show explicitly the dependence of $\bar{\omega}$ (and also of the quantities α, β, f, Γ_1, and Γ_2 defined below) on θ throughout the main part of this appendix.

VOLUME 63, NUMBER 20 PHYSICAL REVIEW LETTERS 13 NOVEMBER 1989

Correlation-Induced Doppler-Like Frequency Shifts of Spectral Lines

Emil Wolf[(a)]
Department of Physics and Astronomy, University of Rochester, Rochester, New York 14627
(Received 7 February 1989)

The question is examined whether there may be scattering media which could generate spectral frequency shifts in radiation from sources that are at rest relative to the observer and yet would imitate Doppler shifts. Scattering kernels of media whose physical properties fluctuate randomly in both space and time are presented which achieve this to a good approximation. The frequency shifts produced in this manner are not necessarily small. The results might be of particular interest in connection with the long-standing controversy about the origin of some discrepancies observed in the spectra of quasars.

PACS numbers: 42.50.Ar, 03.80.+r, 05.40.+j, 98.50.−v

It has been predicted theoretically[1-3] and confirmed experimentally[4-7] not long ago that correlations in the fluctuations of a source distribution can give rise to frequency shifts of lines in the spectrum of the emitted radiation, even when the source is at rest relative to the observer. Because of the close analogy that exists between the processes of radiation and scattering, a similar effect may be expected to occur when radiation is scattered by a "static" medium, i.e., a medium whose constitutive parameters (e.g., the dielectric susceptibility) are independent of time but are random functions of position, with appropriate correlation properties. Indeed, a very recent theoretical analysis[8] has shown this to be the case.

The possibility of generating frequency shifts of spectral lines by a mechanism other than the motion of the source relative to the observer or by gravitation might be of particular interest for astronomy, especially in connection with the long-standing controversy surrounding quasars (see, for example, Refs. 9–14). However, source correlations and correlations produced by static scattering do not imitate, in all respects, frequency shifts produced by the motion of a source relative to the observer (i.e., Doppler shifts). The relative frequency shifts[15]

$$z=\frac{\bar{\lambda}'-\bar{\lambda}}{\bar{\lambda}}=\frac{\bar{\omega}-\bar{\omega}'}{\bar{\omega}'} \tag{1}$$

are, in general, frequency dependent when they are generated by such correlations, whereas they are frequency independent when they are manifestations of the Doppler effect. Moreover, frequency shifts which are generated by source correlations or by correlations in the constitutive parameters of static scatterers are restricted in magnitude to values that are of the order of or smaller than the effective widths of the spectral lines[16] [see the discussion following Eq. (10) below].

In the present Letter we investigate whether a more general correlation mechanism, involving dynamic scattering, could give rise to relative frequency shifts that are essentially the same for all the lines in the spectrum of radiation originating in a (stationary) source and whose magnitude could be arbitrarily large.

Consider a linearly polarized plane electromagnetic wave of spectral profile $S^{(i)}(\omega)$ that propagates in the direction specified by a unit vector $\mathbf{u}$, incident on a linear medium, localized in space, whose dielectric susceptibility[17] $\hat{\eta}(\mathbf{r},t;\omega)$ is a random function of both position ($\mathbf{r}$) and time (t). We assume that the ensemble which characterizes the statistical behavior of the medium is homogeneous, isotropic, and stationary, at least in the wide sense.[18] It has recently been shown that, within the accuracy of the first Born approximation, the spectrum of the scattered field at a point $\mathbf{r}=r\mathbf{u}'$ ($|\mathbf{u}'|=1$) in the far zone of the scatterer is given by[19]

$$S^{(\infty)}(\omega';r\mathbf{u}',\mathbf{u})=A\omega'^4\int_{-\infty}^{\infty}\mathcal{S}\left(\frac{\omega'}{c}\mathbf{u}'-\frac{\omega}{c}\mathbf{u},\omega'-\omega;\omega\right)S^{(i)}(\omega)d\omega\,, \tag{2}$$

where

$$A=\frac{(2\pi)^3V\sin^2\psi}{c^4r^2}. \tag{3}$$

In Eq. (2)

$$\mathcal{S}(\mathbf{K},\Omega;\omega)=\frac{1}{(2\pi)^4}\int_V d^3R\int_{-\infty}^{\infty}dT\,G(\mathbf{R},T;\omega)e^{-i(\mathbf{K}\cdot\mathbf{R}-\Omega T)} \tag{4}$$

is the generalized structure function of the scattering medium, being the four-dimensional Fourier transform of the correlation function[20]

$$G(\mathbf{R},T;\omega)=\langle\hat{\eta}^*(\mathbf{r},t;\omega)\hat{\eta}(\mathbf{r}+\mathbf{R},t+T;\omega)\rangle \tag{5}$$

of the dielectric susceptibility and the angular brackets denote the average, taken over the ensemble of the random medium. In Eq. (3) V represents the volume of the scatterer (whose linear dimensions are assumed to be large compared to the correlation distance of $\hat{\eta}$), ψ denotes the angle between the direction of the electric vector of the incident wave and the direction $\mathbf{u}'$ of scattering, and c is the speed of light *in vacuo*.

For our purpose it is convenient to rewrite Eq. (2) in the form

$$S^{(\infty)}(\omega';r\mathbf{u}',\mathbf{u})=A\omega'^4\int_{-\infty}^{\infty}\mathcal{K}(\omega',\omega;\mathbf{u}',\mathbf{u})S^{(i)}(\omega)d\omega\,, \tag{6}$$

where

$$\mathcal{K}(\omega',\omega;\mathbf{u}',\mathbf{u})=\mathcal{S}\left[\frac{\omega'}{c}\mathbf{u}'-\frac{\omega}{c}\mathbf{u},\omega'-\omega;\omega\right]. \tag{7}$$

We will refer to the function $\mathcal{K}(\omega',\omega;\mathbf{u}',\mathbf{u})$ as the *scattering kernel.*

In the special case of static scattering the correlation function $G(\mathbf{R},T;\omega)$ will be independent of the temporal argument T and we will then denote it by $g(\mathbf{R};\omega)$. In this case Eqs. (4) and (7) imply that the scattering kernel is given by

$$\mathcal{K}(\omega',\omega;\mathbf{u}',\mathbf{u})=\tilde{g}\left[\frac{\omega'}{c}\mathbf{u}'-\frac{\omega}{c}\mathbf{u},\omega\right]\delta(\omega'-\omega)\,, \tag{8}$$

where

$$\tilde{g}(\mathbf{K};\omega)=\frac{1}{(2\pi)^3}\int_V g(\mathbf{R};\omega)e^{-i\mathbf{K}\cdot\mathbf{R}}d^3R \tag{9}$$

is the three-dimensional spatial Fourier transform of $g(\mathbf{R},\omega)$ and δ is the Dirac delta function. On substituting from Eq. (8) into Eq. (6) we see that the spectrum of the scattered radiation is given by

$$S^{(\infty)}(\omega';r\mathbf{u}',\mathbf{u})=A\omega'^4\tilde{g}\left[\frac{\omega'}{c}(\mathbf{u}'-\mathbf{u});\omega'\right]S^{(i)}(\omega')\,. \tag{10}$$

This expression, which is the electromagnetic analog of the main result of Ref. 8 relating to scattering of scalar waves by static media, shows that spatial correlation of the dielectric susceptibility, represented by the two-point correlation function $g(\mathbf{R};\omega)$, modifies the spectrum of the radiation incident on the scatterer. However, because the influence of the correlations is manifested in Eq. (10) only through a multiplicative factor, no new frequency components are generated by static scattering [i.e., $S^{(\infty)}(\omega';r\mathbf{u}',\mathbf{u})=0$ whenever $S^{(i)}(\omega')=0$]. This conclusion is, of course, an immediate consequence of the presence of the factor $\delta(\omega'-\omega)$ in the expression (8), which implies that each frequency component ω of the incident radiation gives rise to one and only one frequency ω' in the scattered radiation and that, moreover, $\omega'=\omega$.

Let us now turn to the more general case of a medium whose dielectric susceptibility varies randomly not only in space but also in time. If $S^{(i)}(\omega)=\delta(\omega-\omega_0)$, Eq. (6) gives

$$S^{(\infty)}(\omega';r\mathbf{u}',\mathbf{u})=A\omega'^4\mathcal{K}(\omega',\omega_0;\mathbf{u}',\mathbf{u})\,. \tag{11}$$

This formula implies that a frequency component $\omega=\omega_0$ of the incident radiation will give rise to a frequency component ω' in the scattered radiation and that, in general, $\omega'\neq\omega_0$; and, moreover, there may be several values of ω', possibly a continuous range of them, associated with any particular value of ω_0. An example of the latter situation is provided by Brillouin scattering from a simple fluid under equilibrium conditions. The scattering kernel is then proportional to the sum of three Lorentzian distributions, centered at frequencies $\omega'=\omega_0$, $\omega_0[1+2(v/c)\sin(\theta/2)]$, and $\omega_0[1-2(v/c)\sin(\theta/2)]$, where v is the speed of sound in the fluid and θ is the angle of scattering. Another example is discussed in Ref. 21.

As one of the simplest generalizations of the formula (8) to media which vary randomly both in space and in time let us suppose that the scattering kernel has the form

$$\mathcal{K}(\omega',\omega;\mathbf{u}',\mathbf{u})=f(\omega',\omega;\mathbf{u}',\mathbf{u})\delta(a\omega'-\omega)\,, \tag{12}$$

where a is independent of the frequencies but may depend on $\mathbf{u}$ and $\mathbf{u}'$. In this case the formula (6) gives the following expression for the spectrum of the scattered radiation:

$$S^{(\infty)}(\omega';r\mathbf{u}',\mathbf{u})=A\omega'^4f(\omega',a\omega';\mathbf{u}',\mathbf{u})S^{(i)}(a\omega')\,. \tag{13}$$

Suppose further that the spectrum $S^{(i)}(\omega)$ of the incident radiation consists of a single line centered on the frequency $\omega=\bar{\omega}$ and that over the width of the line the factor $\omega'^4f(\omega',a\omega';\mathbf{u},\mathbf{u}')$ does not change appreciably with ω'. Then, according to Eq. (13) the spectrum $S^{(\infty)}(\omega';r\mathbf{u}',\mathbf{u})$ of the scattered radiation will also consist of a single line, but this line will be centered close to the frequency $\omega'=\bar{\omega}'$, where

$$\bar{\omega}'=\bar{\omega}/a\,. \tag{14}$$

On substituting from Eq. (14) into the formula (1) we see that the spectral line has been shifted in frequency by the relative amount

$$z=\frac{\bar{\omega}-\bar{\omega}/a}{\bar{\omega}/a}=a-1\,. \tag{15}$$

Thus apart from a small contribution arising from the factor $\omega'^4f(\omega',a\omega';\mathbf{u}',\mathbf{u})$, *the relative frequency shift is independent of the central frequency $\bar{\omega}$ of the spectral line of the incident radiation and of the width of the line.*

Evidently, if instead of a single line the spectrum of the incident radiation consisted of several lines, each would be shifted by essentially the same amount $z = a - 1$, thus imitating a Doppler shift. The shift will be towards the lower frequencies (redshift) if $a > 1$ and towards higher frequencies (blueshift) if $a < 1$ and can, in principle, have any magnitude.

Next let us now consider a more general kernel, of the form

$$\mathcal{K}(\omega',\omega;\mathbf{u}',\mathbf{u}) = f(\omega',\omega;\mathbf{u}',\mathbf{u})e^{-(a\omega'-\omega)^2/2\sigma^2}, \quad (16)$$

where a and σ are positive quantities that are independent of the frequencies. Suppose that the spectrum of the incident radiation is the line

$$S^{(i)}(\omega) = Be^{-(\omega-\bar{\omega})^2/2\Gamma^2}, \quad (17)$$

where B, $\bar{\omega}$, and Γ are positive constants. On substituting from Eqs. (16) and (17) into the formula (6) and assuming that the factor $f(\omega',\omega;\mathbf{u}',\mathbf{u})$ does not vary appreciably with ω over the effective width of the spectral line (17), we obtain for $S^{(\infty)}(\omega';r\mathbf{u}',\mathbf{u})$ the expression

$$S^{(\infty)}(\omega';r\mathbf{u}',\mathbf{u}) = AB\omega'^4 f(\omega',\bar{\omega};\mathbf{u}',\mathbf{u})\int_{-\infty}^{\infty} e^{-(a\omega'-\omega)^2/2\sigma^2} e^{-(\omega-\bar{\omega})^2/2\Gamma^2} d\omega. \quad (18)$$

The integral in Eq. (18) may be readily evaluated [most simply by making use of the so-called product theorem for Gaussian functions (cf. Appendix A of Ref. 8)] and one then finds that

$$S^{(\infty)}(\omega';r\mathbf{u}',\mathbf{u}) = AB\tilde{\Gamma}(2\pi)^{1/2}\omega'^4 f(\omega',\bar{\omega};\mathbf{u}',\mathbf{u}) \times e^{-(\omega'-\bar{\omega}/a)^2/2\Gamma'^2}, \quad (19)$$

where

$$\Gamma' = \frac{1}{a}(\sigma^2+\Gamma^2)^{1/2}, \quad \tilde{\Gamma} = \left(\frac{1}{\sigma^2}+\frac{1}{\Gamma^2}\right)^{-1/2}. \quad (20)$$

The formula (19) shows that the spectrum $S^{(\infty)}(\omega';r\mathbf{u}',\mathbf{u})$ of the scattered radiation is proportional to the product of the factor $\omega'^4 f(\omega',\bar{\omega};\mathbf{u}',\mathbf{u})$ and a Gaussian function centered at the frequency $\omega' = \bar{\omega}/a$ and of rms width Γ'. This Gaussian function is shifted with respect to the Gaussian function (17) by an amount whose relative value is again given by the expression (15) and is, therefore, independent of the mean frequency $\bar{\omega}$ of the incident radiation. The widths of the two lines are not the same, however. Moreover, the factor $\omega'^4 f(\omega',\bar{\omega};\mathbf{u}',\mathbf{u})$ in Eq. (19) will produce a distortion of the line and possibly an additional frequency shift, whose relative value may or may not be small compared to the value $z = a - 1$, depending on the exact form of the function $f(\omega'\omega;\mathbf{u}',\mathbf{u})$.

It is evident from the preceding analysis that there is a possibility that scattering from some media whose macroscopic properties fluctuate randomly both in space and in time may generate frequency shifts of spectral lines which closely resemble Doppler shifts. As mentioned earlier, this possibility might be of particular interest in connection with the long-standing quasar controversy. However, in order to apply the present theory to quasar problems it would be necessary to have a reliable model available for the medium through which radiation originating in these astronomical sources passes, and no models that have been proposed so far can be accepted with full confidence. Moreover, the very process of generation of the radiation that reaches us from quasars is not currently understood. One should also bear in mind that throughout the preceding analysis the incident field was assumed to be a plane wave. For applications to quasars and possibly other astronomical objects, additional averaging over a range of directions of incidence would also have to be performed. Moreover, extension of the analysis beyond the first Born approximation and also to anisotropic media might be required.[22]

There is a widely held opinion that no other mechanism except the Doppler effect and gravitation exist within the framework of present-day physics, which can account even for parts of the redshifts observed in the spectra of radiation that reaches us from astronomical sources. The analysis presented in this Letter raises a question about the correctness of this opinion.

This research was supported by the National Science Foundation and by the Army Research Office. I wish to thank Professor Jack W. Sulentic for making available to me a preprint of his paper (cited in Ref. 16) prior to publication and for stimulating discussions about quasar astronomy.

[(a)]Also at the Institute of Optics, University of Rochester, Rochester, NY 14627.

[1]E. Wolf, Nature (London) **326**, 363 (1987).

[2]E. Wolf, Opt. Commun. **62**, 12 (1987).

[3]E. Wolf, Phys. Rev. Lett. **58**, 2646 (1987).

[4]D. Faklis and G. M. Morris, Opt. Lett. **13**, 4 (1988).

[5]F. Gori, G. Guattari, C. Palma, and G. Padovani, Opt. Commun. **67**, 1 (1988).

[6]G. Indebetouw, J. Mod. Opt. **36**, 251 (1989).

[7]Similar observations of frequency shifts with acoustical rather than with optical sources have been made by M. F. Bocko, D. H. Douglass, and R. S. Knox, Phys. Rev. Lett. **58**, 2649 (1987).

[8]E. Wolf, J. T. Foley, and F. Gori, J. Opt. Soc. Am. A, **6**, 1142 (1989).

[9]G. B. Field, H. Arp, and J. N. Bahcall, *The Redshift Controversy* (Benjamin, Reading, MA, 1973).

[10]G. Burbidge, in *Objects of High Redshift,* edited by G. O. Abell and P. J. E. Peebles (D. Reidel, Boston, 1980), p. 99.

[11]J. W. Narlikar, in *Quasars,* edited by G. Swarup and V. K. Kapahi (D. Reidel, Boston, 1986), p. 463.

[12]H. Arp, *Quasars, Redshifts and Controversies* (Interstellar Media, Berkeley, CA, 1987). See also a review of this book and related discussions in Sky Telescope **75**, 38–43 (1988).

[13]J. W. Sulentic, in *New Ideas in Astronomy,* edited by F. Bertola, J. W. Sulentic, and B. F. Madore (Cambridge Univ. Press, Cambridge, 1988), p. 123.

[14]G. Burbidge, Mercury **17**, 136 (1988).

[15]We use here the standard notation: $\bar{\lambda},\bar{\omega}$ are the wavelength and the frequency of the line center that would be measured in a reference frame of the source and $\bar{\lambda}',\bar{\omega}'$ are the corresponding quantities that would be measured in the observer's frame.

[16]Such frequency shifts may nevertheless have relevance to quasars because different emission lines in many quasar spectra exhibit small but physically significant differences in their z numbers [C. M. Gaskell, Astrophys. J. **263**, 79 (1982); in *Quasars and Gravitational Lenses,* Proceedings of the Twenty-Fourth Liège International Astrophysics Colloquium (Institute d'Astrophysique, Université de Liège, Liege, 1983), p. 473; J. Sulentic, Astrophys. J. **343**, 54 (1989)].

[17]The use of response functions which depend on both time and frequency is, in general, necessary when the macroscopic physical properties of the medium change in time [cf. L. Mandel and E. Wolf, Opt. Comm. **8**, 95 (1973)].

[18]W. B. Davenport and W. L. Root, *Random Signals and Noise* (McGraw-Hill, New York, 1958), p. 60.

[19]E. Wolf and J. T. Foley, Phys. Rev. A **40**, 579 (1989), Eq. (5.10), with obvious changes in notation.

[20]This correlation function is analogous to the well-known Van Hove two-particle correlation function frequently employed in neutron scattering [L. Van Hove, Phys. Rev. **95**, 249 (1954)].

[21]J. T. Foley and E. Wolf, Phys. Rev. A **40**, 588 (1989).

[22]A correlation function $G(\mathbf{R},T;\omega)$ for a class of anisotropic media which will generate Doppler-like frequency shifts is considered in a paper by D. F. V. James, M. P. Savedoff, and E. Wolf, submitted recently to the *Astrophysical Journal.* It is also shown in that paper that the usual quasar models imply characteristic anisotropies which are consistent with correlation functions of this class.

The Astrophysical Journal, 359:67–71, 1990 August 10

SHIFTS OF SPECTRAL LINES CAUSED BY SCATTERING FROM FLUCTUATING RANDOM MEDIA

Daniel F. V. James,[1] Malcolm P. Savedoff,[2] and Emil Wolf[1,2]
University of Rochester, Rochester, New York
Received 1989 October 13; accepted 1990 February 12

ABSTRACT

A model scatterer is introduced, whose dielectric response function is a random function of space and time and which produces frequency shifts of spectral lines that imitate the Doppler effect in its main features. The possible relevance of this effect to the origin of discrepancies observed in some quasar spectra is discussed.

Subject headings: galaxies: redshifts — quasars — radiation mechanisms

I. INTRODUCTION

In the last few years several new closely related processes have been discovered that can generate frequency shifts of spectral lines. This development followed the theoretical prediction made by Wolf (1986) and since then verified by experiment (Morris and Faklis 1987) that, contrary to a commonly held belief, the spectrum of light is, in general, not invariant on propagation in free space. For radiation from primary sources, the spectral changes are induced by correlations between fluctuations of the source distribution at different points in the source region (Wolf 1987*a*, *b*; see also James and Wolf 1989); for radiation from secondary sources, such as an illuminated aperture, the changes are induced by the correlations between the field fluctuations at pairs of points in the aperture plane (Wolf 1986; Dacic and Wolf 1988; Gamliel 1989). In scattering processes the changes in the spectrum are induced by correlations between fluctuating response functions of the scattering medium, e.g., its dielectric susceptibility, either at different points in the scatterer, when the frequency-dependent macroscopic response is time-independent (Wolf, Foley, and Gori 1989), or at different spacetime points, when it is time-dependent (Foley and Wolf 1989). In all these cases the spectral changes are the consequence of correlations involving an appropriate variable characterizing the source, the field, or the response of a scatterer. With appropriate correlations, the changes are manifested as frequency shifts of spectral lines. This possibility has been predicted theoretically (Wolf 1987*a*, *b*; Wolf, Foley, and Gori 1989; Foley and Wolf 1989) and has been confirmed by a number of laboratory experiments using sources with controllable coherence properties, both optically (Faklis and Morris 1988; Gori *et al.* 1988; Indebetouw 1989), and also acoustically (Bocko, Douglass, and Knox 1987) for cases when the shifts are induced by source correlations.

Very recently it was predicted (Wolf 1989*a*, *b*) that the frequency shifts induced by scattering from time-dependent random media with suitable correlation properties may imitate a Doppler shift of any magnitude, even though the source, the medium, and the observer are all at rest with respect to each other. It seems therefore possible, in principle, that this effect may contribute toward the redshifts observed in the spectra of some astronomical objects. In this connection we recall that the long-standing controversy over the interpretation of quasar redshifts continues despite the pronouncement that "the 'evidence' for non-cosmological redshifts is a collection of unrelated curiosities having no predictive power" (Weedman 1986, p. 37). The contrary view was ably argued by Arp (1987). Even Hubble (1936) questioned the validity, in all cases, of the velocity interpretation. Most non-Doppler hypotheses proposed in the past violated established physical laws in some way, e.g., the principle of conservation of momentum and energy.

It is important to appreciate that there are two quite distinct aspects to the redshift controversy, which have not always been kept in focus in the lengthy dispute, namely, the questions (1) whether most quasars are at cosmological distances and (2) whether *some* quasars are associated with objects of lower redshifts (e.g., Markarian 205 with NGC 4319). We wish to make it quite clear at the outset that, in this paper, we are not questioning whether most quasars are at cosmological distances. However, we show that scattering from suitably correlated fluctuating random media such as those discussed in this paper provides a possible additional mechanism for generating redshifts.

In the present paper we consider an explicit form for a correlation function of the dielectric susceptibility of a fluctuating random scattering medium which will generate frequency shifts that are essentially indistinguishable from those that might be caused by the motion of a source relative to the observer or by gravitation; and we show that the usual quasar models imply characteristic anisotropies which are consistent with our mechanism.

II. THE CORRELATION FUNCTION OF A SCATTERING MEDIUM WHICH WILL GENERATE DOPPLER-LIKE FREQUENCY SHIFTS

Consider a linearly polarized plane wave, with spectrum $S^{(i)}(\omega)$, incident in a direction specified by a unit vector $\boldsymbol{u}$ on a fluctuating random scattering medium, occupying a volume V. It was recently shown (Wolf and Foley 1989) that, under fairly general conditions, the spectrum of the scattered radiation at a point $\boldsymbol{r} = r\boldsymbol{u}'$ ($|\boldsymbol{u}'| = 1$) in the far zone of the scatterer is given by the following expression, valid within the accuracy of the first-order Born approximation:

$$S^{(\infty)}(r\boldsymbol{u}', \omega') = A\omega'^4 \int_{-\infty}^{\infty} \mathscr{S}\left(\frac{\omega'}{c}\boldsymbol{u}' - \frac{\omega}{c}\boldsymbol{u}, \omega' - \omega; \omega\right) S^{(i)}(\omega) d\omega . \tag{1}$$

[1] The Institute of Optics.
[2] Department of Physics and Astronomy.

The function

$$\mathscr{S}(\boldsymbol{K}, \Omega; \omega) = \frac{1}{(2\pi)^4} \int_V d^3R \int_{-\infty}^{\infty} dT\, G(\boldsymbol{R}, T; \omega) e^{-i(\boldsymbol{K}\cdot\boldsymbol{R} - \Omega T)} \quad (2)$$

is the generalized structure function of the medium, being the four-dimensional Fourier transform of the correlation function

$$G(\boldsymbol{R}, T; \omega) = \langle \hat{\eta}^*(\boldsymbol{r} + \boldsymbol{R}, t + T; \omega)\hat{\eta}(\boldsymbol{r}, t; \omega)\rangle \quad (3)$$

of the generalized dielectric susceptibility $\hat{\eta}(\boldsymbol{r}, t; \omega)$ of the scattering medium (Mandel and Wolf 1973). The angular brackets in equation (3) denote the ensemble average. The factor A in equation (1) is given by the expression

$$A = \frac{(2\pi)^3 V \sin^2 \psi}{c^4 r^2}, \quad (4)$$

where ψ is the angle between the electric vector of the incident field and the direction of scattering and c is the speed of light in a vacuum.

As in earlier papers (Wolf 1989*a*, *b*), it is convenient to introduce a scattering kernel $\mathscr{K}(\omega', \omega; \boldsymbol{u}', \boldsymbol{u})$ by the formula

$$\mathscr{K}(\omega', \omega; \boldsymbol{u}', \boldsymbol{u}) \equiv \mathscr{S}\left(\frac{\omega'}{c}\boldsymbol{u}' - \frac{\omega}{c}\boldsymbol{u}, \omega' - \omega; \omega\right). \quad (5)$$

Expression (1) for the far-zone spectrum then becomes

$$S^{(\infty)}(r\boldsymbol{u}', \omega') = A\omega'^4 \int_{-\infty}^{\infty} \mathscr{K}(\omega', \omega; \boldsymbol{u}', \boldsymbol{u}) S^{(i)}(\omega) d\omega. \quad (6)$$

Suppose now that the correlation properties of the fluctuating medium are characterized by an anisotropic Gaussian function, viz.,

$$G(\boldsymbol{R}, T; \omega) = G_0 \exp\left[-\left(\frac{X^2}{2\sigma_x^2} + \frac{Y^2}{2\sigma_y^2} + \frac{Z^2}{2\sigma_z^2} + \frac{T^2}{2\tau^2}\right)\right], \quad (7)$$

where $\sigma_x, \sigma_y, \sigma_z, \tau$, and G_0 are positive constants and (X, Y, Z) are the components of the vector $\boldsymbol{R}$ with respect to a suitably chosen set of Cartesian coordinate axes. The correlation function (7) is a natural generalization of those considered by Foley and Wolf (1989) and by Ishimaru (1978). On substituting from equation (7) into equation (2), we readily find that the generalized structure function of the medium is given by

$$\mathscr{S}(\boldsymbol{K}, \Omega; \omega) = \frac{G_0}{(2\pi)^2} \sigma_x \sigma_y \sigma_z \tau \times \exp\left[-\frac{1}{2}\left(\sigma_x^2 K_x^2 + \sigma_y^2 K_y^2 + \sigma_z^2 K_z^2 + \tau^2 \Omega^2\right)\right], \quad (8)$$

where (K_x, K_y, K_z) are the Cartesian components of the vector $\boldsymbol{K}$ with respect to the same coordinate axes.

It follows from equations (8) and (5) that when the correlation function $G(\boldsymbol{R}, T; \omega)$ of the scattering medium is the anisotropic Gaussian distribution (7), the scattering kernel becomes

$$\mathscr{K}(\omega', \omega; \boldsymbol{u}', \boldsymbol{u}) = \frac{G_0}{(2\pi)^2 c} \sigma_x \sigma_y \sigma_z \sigma_\tau \exp\left(-\frac{1}{2}\Delta\right), \quad (9)$$

where

$$\Delta = \frac{1}{c^2}\left[\sigma_x^2(\omega' u_x' - \omega u_x)^2 + \sigma_y^2(\omega' u_y' - \omega u_y)^2 + \sigma_z^2(\omega' u_z' - \omega u_z)^2 + \sigma_\tau^2(\omega' - \omega)^2\right], \quad (10)$$

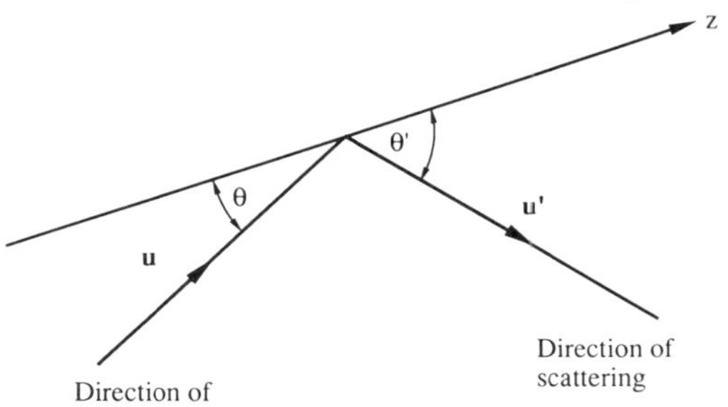

FIG. 1.—Diagram illustrating the significance of the angles θ and θ'

u_x, u_y, u_z and u_x', u_y', u_z' are the components of the unit vectors $\boldsymbol{u}$ and $\boldsymbol{u}'$, respectively, with respect to the same coordinate axes used in connection with equations (7) and (8), and

$$\sigma_\tau \equiv c\tau. \quad (11)$$

Suppose now that the anisotropy of the correlation function $G(\boldsymbol{R}, T; \omega)$ is strong in one particular direction, say in the z-direction, in the sense that

$$\sigma_z \gg \sigma_x, \quad \sigma_z \gg \sigma_y, \quad \text{and } \sigma_z \gg \sigma_\tau. \quad (12)$$

Then, with some additional inequalities being satisfied,[3] equation (10) may be approximated as

$$\Delta \approx \frac{\sigma_z^2}{c^2}(\omega' \cos\theta' - \omega\cos\theta)^2, \quad (13)$$

where θ and θ' are the angles which the unit vectors $\boldsymbol{u}$ (direction of incidence) and $\boldsymbol{u}'$ (direction of scattering) make with the z-axis (see Fig. 1).

With the approximation (13), the scattering kernel (eq. [9]) becomes

$$\mathscr{K}(\omega', \omega; \boldsymbol{u}', \boldsymbol{u}) \approx \frac{\sigma_x\sigma_y\sigma_z\sigma_\tau G_0}{(2\pi)^2} \times \exp\left[-\frac{\sigma_z^2}{2c^2}(\omega'\cos\theta' - \omega\cos\theta)^2\right]. \quad (14)$$

It is convenient to express this formula in the form

$$\mathscr{K}(\omega', \omega; \boldsymbol{u}', \boldsymbol{u}) \approx \frac{\sigma_x\sigma_y\sigma_z\sigma_\tau G_0}{c(2\pi)^2} \exp\left[-\frac{(\alpha\omega' - \omega)^2}{2\Gamma_1^2}\right], \quad (15)$$

where

$$\alpha = \frac{\cos\theta'}{\cos\theta} \quad (16)$$

and

$$\Gamma_1 = \frac{c}{\sigma_z \cos\theta}. \quad (17)$$

[3] If we assume that $\sigma_x = \sigma_y$, then a set of sufficiency, but not necessary, conditions for approximation (13) to hold is $\cos^2\theta' \gg (\sigma_x/\sigma_z)^2$, $\cos^2\theta' \gg (\sigma_\tau/\sigma_z)^2$, $\cos\theta\cos\theta' \gg (\sigma_x/\sigma_z)^2$, $\cos\theta\cos\theta' \gg (\sigma_\tau/\sigma_z)^2$, $\cos^2\theta \gg (\sigma_x/\sigma_z)^2$, and $\cos^2\theta \gg (\sigma_\tau/\sigma_z)^2$.

Expression (15) is precisely the form obtained not long ago (Wolf 1989*a*, eq. [16], and 1989*b*, eq. [6]) for scattering kernels of media which generate frequency shifts that imitate the Doppler shift in its main features. We will now verify this by explicit calculation.

Suppose that the spectrum of the incident light consists of a single line of Gaussian profile,

$$S^{(i)}(\omega) = I_0 \exp\left[-\frac{(\omega-\omega_0)^2}{2\Gamma_0^2}\right], \tag{18}$$

where ω_0, Γ_0, and I_0 are positive constants. This particular choice of the profile is not essential, but it simplifies the calculation. In order to determine the far-zone spectrum of the scattered light, we must evaluate, according to equation (6), the product $\mathscr{K}(\omega', \omega; \boldsymbol{u}', \boldsymbol{u})S^{(i)}(\omega)$. We readily find from equations (15) and (18), after a straightforward but long calculation most easily performed with the help of the so-called product theorem for Gaussian functions (Wolf, Foley, and Gori 1989, Appendix A), that

$$\mathscr{K}(\omega', \omega; \boldsymbol{u}', \boldsymbol{u})S^{(i)}(\omega) = B \exp\left[-\frac{(\alpha\omega'-\omega_0)^2}{2(\Gamma_0^2+\Gamma_1^2)}\right] \times \exp\left[-\frac{(\omega-\tilde{\omega})^2}{2\tilde{\Gamma}^2}\right], \tag{19}$$

where

$$B = \frac{1}{c(2\pi)^2} G_0 \sigma_x \sigma_y \sigma_z \sigma_\tau I_0 , \tag{20}$$

$$\tilde{\omega} = \frac{\alpha\omega'\Gamma_0^2 + \omega_0 \Gamma_1^2}{\Gamma_0^2+\Gamma_1^2}, \tag{21}$$

and

$$\frac{1}{\tilde{\Gamma}^2} = \frac{1}{\Gamma_0^2} + \frac{1}{\Gamma_1^2} . \tag{22}$$

On substituting from equation (19) in the equation (6) and performing the integration, we readily find that

$$S^{(\infty)}(r\boldsymbol{u}', \omega') = C\omega'^4 \exp\left[-\frac{(\omega'-\omega_0')^2}{2(\Gamma_0^2+\Gamma_1^2)/\alpha^2}\right], \tag{23}$$

where

$$C = \frac{(2\pi)^{3/2} V G_0 I_0 \sigma_x \sigma_y \sigma_z \sigma_\tau \tilde{\Gamma} \sin^2\psi}{c^5 r^2}, \tag{24}$$

and[4]

$$\omega_0' = \frac{\omega_0}{|\alpha|} . \tag{25}$$

Formula (23) shows that the spectrum of the scattered field in the far zone is proportional to the product of the factor ω'^4 and a Gaussian function. This function is, however, not centered on the mean frequency ω_0 of the incident light but rather on the frequency ω_0' given by equations (25) and (16). This line is therefore shifted with respect to the spectral line of the incident light (eq. [18]) by the relative amount

$$z \equiv \frac{\omega_0-\omega_0'}{\omega_0'} = |\alpha| - 1 \tag{26}$$

or, more explicitly, if equation (16) is used, by the relative amount

$$z = \left|\frac{\cos\theta'}{\cos\theta}\right| - 1 . \tag{27}$$

The rms width of this spectral line differs, however, from the rms width Γ_0 of the line of the incident light, being given by

$$\Gamma' = \frac{(\Gamma_0^2+\Gamma_1^2)^{1/2}}{|\alpha|} . \tag{28}$$

If for a moment we ignore the proportionality factor ω'^4 in equation (23), we see from equation (26) that the *relative frequency shift, which can take any value in the range* $-1 \le z \le \infty$, *is independent of the central frequency* ω_0 *of the incident light* and thus imitates the Doppler shift. Evidently the spectral line of the scattered light is redshifted ($z > 0$) when $\theta' < \theta$ and is blueshifted ($z < 0$) when $\theta' > \theta$, with respect to the spectral line of the incident light.

The multiplicative factor ω'^4, just as in the case of Rayleigh scattering, produces a small amount of blueshift which is frequency-dependent; it also produces a frequency-dependent change of the intensity of the line, which has the effect of making the source appear to be bluer. In Figure 2 we illustrate our analysis by an example which shows frequency shifts of a pair of spectral lines due both to this mechanism and to the Doppler effect.

As may be seen from equations (25) and (28), the fractional widths Γ'/ω_0' of the lines in the spectrum of scattered light are different from the fractional widths Γ/ω_0 of the lines in the unshifted spectrum. Such a change does not occur for the Doppler shift in radiation from a source of any state of coherence, as was shown recently (James 1989). However, changes of this ratio occur in other types of correlation-induced spectral shifts (see § I). Further, if more complicated correlation functions than the Gaussian are used, more complicated line profile changes are likely to be produced. However, there may be other physical processes, unrelated to the statistical properties of the fluctuating medium, which can give rise to line irregularities. In this connection it is interesting to recall that Gaskell (1982) discovered systematic differences in redshifts of lines in the same spectrum originating in strongly and weakly ionized emission regions in some quasars; emission lines of highly ionized regions appear to be blueshifted by a relative amount of the order of $z = -0.002$ with respect to the redshift of the narrow spectral lines. In a recent paper (Sulentic 1989) a classification scheme was proposed for the shape and wavelength displacement of the broad emission lines of Hβ with respect to the adjacent forbidden line of O III in active galactic nuclei. This investigation showed that over half of the active galactic nuclei considered had a measurable shift of spectral lines in the broad-line region with respect to those of the narrow-line region. A possibility exists that such irregularities in individual quasar spectra and also discordant redshifts in some quasar-galaxy pairs which appear to be connected may be due to the coherence effect discussed in the present paper.

[4] We have written here $|\alpha|$ rather than α, because when $\alpha < 0$ the negative rather than the positive frequency part of the spectrum of the incident field is relevant (see Foley and Wolf 1989, § III, especially eqs. [3.5]–[3.7]).

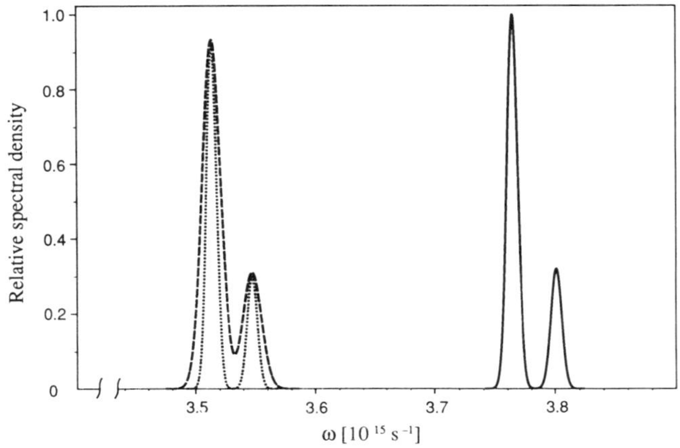

FIG. 2.—Two O III lines ($\lambda = 4959$ Å and $\lambda = 5007$ Å) as seen at rest (*solid line*), Doppler-shifted (*dotted line*), and shifted by the process described in this paper (*dashed line*), both by a relative amount $z = 0.0714$. The FWHM of both lines was taken as 84 km s^{-1}. The constant C in eq. (23) was chosen so that the height of the stronger shifted line is the same as for the Doppler-shifted line. For shifts induced by the correlation mechanism and shown in the figure, $\sigma_z = 50$ μm, $\theta = 30°$, and $\theta' = 21.°89$.

III. DISCUSSION

We have shown that frequency shifts of spectral lines which mimic the Doppler effect can be generated by scattering from a random medium which has a physically plausible correlation function for the dielectric fluctuations. We emphasize that this mechanism cannot be explained either by naive considerations involving photon fluxes or by radiative transfer or coherent wave propagation. We now briefly examine whether strong anisotropies such as are introduced in § II may perhaps be present in or near the envelopes of some quasars. Our discussion will necessarily be only qualitative, because little information exists at present on the detailed structure and geometry of the inner regions of quasars.

The required anisotropy is not incompatible with the physical parameters for active galactic nuclei (AGNs) derived from the usual synchrotron radiation model. The relative sizes of the parameters required for our analysis may be estimated from conditions in AGNs as discussed by Begelman, Blandford, and Rees (1984), in connection with radio galaxies and their jets; by Weedman (1986), who justifies placing the quasars as an energetic extension of the Seyfert 1 galaxies; and by Osterbrock (1989, pp. 59–88). If the scattering medium is located outside the region in which the broad-line emission takes place, it may be assumed to have similar properties to the narrow-line emitting region (NLR). In the vicinity of the NLR, i.e., at a radius of 100 pc, as modeled by Osterbrock, we have a temperature of 10^4 K, an electron density $N_e \sim 10^4$ cm^{-3}, and a magnetic field (derived by assuming the equipartition of thermal and magnetic pressures) of 8×10^{-4} G. This value for the magnetic field is compatible with the value estimated for the nucleus of the radio galaxy NGC 6251 by Begelman, Blandford, and Rees (1984).

The correlation function introduced in § II is required to have one correlation length, σ_z, significantly larger than the others, i.e., the symmetry has to be strongly broken in one direction, taken to be the z-direction. We assume that this direction is along the jet. (Although Weedman estimates that only 0.1% of all quasars have jets, we assume that jets are characteristic of conditions in these regions.) At a radius of 100 pc it is hard to conceive of structures small compared with the intrinsic size of the "central engine." Thus we take as a characteristic length along the jet a scale comparable to the Schwarzschild radius associated with masses of 10^6–10^9 $M_\odot$, i.e., larger than 2.95×10^{11} cm. As noted by Begelman, Blandford, and Rees (1984), polarization of the observed radio emission from these regions suggests that the magnetic field is oriented predominantly in the direction along the jet. For cyclotron motion of charged particles in this magnetic field we associate characteristic transverse lengths with the gyration radius r_g and a characteristic time with the inverse of the gyration frequency, $2\pi/\omega_g$.

Simple theory for the motion of charged particles in a magnetic field (see, for example, Landau and Lifshitz 1971, § 21) gives the following formulae for these parameters:

$$r_g = \frac{p}{qB} \quad \text{and} \quad \frac{c}{\omega_g} = \frac{\gamma mc}{qB}, \tag{29}$$

where p is the relativistic momentum of the particle perpendicular to the magnetic field, q is its charge, m is the particle mass, v is its velocity, and $\gamma = (1 - v^2/c^2)^{-1/2}$. A high degree of anisotropy will exist if r_g and c/ω_g are both less than, say, 3×10^9 cm. The r_g condition therefore implies that the momentum $p \lesssim 750$ MeV/c for both protons and electrons. The ω_g condition will be fulfilled for electrons only if $p \lesssim 1400$ mc. Since most relativistic gases are characterized by a phase-space density proportional to p^β, where β is in the range -4 to -5, the macroscopic properties are dominated by the lower momentum particles. This suggests that the medium is characterized by $p \lesssim mc$ and our conditions are satisfied. Hence it seems that there exist natural anisotropies in some quasar atmospheres appropriate to the scattering mechanism described in § II.

There may be other sources of anisotropy present—for example, those associated with turbulence. However, we wish to mention that anisotropy of the correlation function of the scattering medium is not a necessary requirement for generating Doppler-like frequency shifts of spectral lines. An example of an isotropic correlation function which will generate such shifts was found by James and Wolf (1990) after the completion of this paper.

Our analysis is rather incomplete for several reasons. In particular, since no adequate knowledge exists at the present time about correlations in susceptibility fluctuations in the vicinity of quasars, our model can only be considered as indicative of some of the possibilities. Further, we only considered scattering of an incident plane wave. To draw realistic conclusions, additional averaging, involving a range of directions of incidence, would have to be performed. Moreover, analysis beyond the accuracy of the first-order Born approximation, on which our calculations are based, may be required. Also, we did not treat unscattered radiation. Nevertheless, our analysis clearly indicates that some observed redshifts may contain contributions which arise from the intrinsic properties of the medium surrounding the radiating sources, in addition to those due to the Doppler effect or gravitation. It is thus possible, as we noted in § I, that the long-standing controversy relating to pairs of objects of different redshifts which appear to be physically connected might be resolved by taking into account the correlation mechanism discussed in this paper.

The authors are indebted to Professor Jack W. Sulentic for helpful advice and useful comments. This work was supported by the National Science Foundation and by the Army Research Office.

REFERENCES

Arp, H. 1987, *Quasars, Redshifts and Controversies* (Berkeley: Interstellar Media).
Begelman, M. C., Blandford, R. D., and Rees, M. J. 1984, *Rev. Mod. Phys.*, **56**, 255.
Bocko, M. F., Douglass, D. H., and Knox, R. S. 1987, *Phys. Rev. Letters*, **58**, 2649.
Dacic, Z., and Wolf, E. 1988, *J. Opt. Soc. Am. A*, **5**, 1118.
Faklis, D., and Morris, G. M. 1988, *Optics Letters*, **13**, 4.
Foley, J. T., and Wolf, E. 1989, *Phys. Rev. A*, **40**, 588.
Gamliel, A. 1989 *J. Opt. Soc. Am. A*, in press.
Gaskell, C. M. 1982, *Ap. J.*, **263**, 79.
Gori, F., Guattari, G., Palma, C., and Padovani, C. 1988, *Optics Comm.*, **67**, 1.
Hubble, E. P. 1936, *The Realm of the Nebulae* (New Haven: Yale University Press).
Indebetouw, G. 1989, *J. Mod. Optics*, **36**, 251.
Ishimaru, A. 1978, *Wave Propagation and Scattering in Random Media* Vol. **2** (New York: Academic).
James, D. F. V. 1989, *Phys. Letters A*, **140**, 213.
James, D. F. V., and Wolf, E. 1989, *Optics Comm.*, **72**, 1.
James, D. F. V., and Wolf, E. 1990, *Phys. Letters A*, in press.
Landau, L. D., and Lifshitz, E. M. 1971, *The Classical Theory of Fields* (3d ed.; Oxford: Pergamon).
Mandel, L., and Wolf, E. 1973, *Optics Comm.*, **8**, 95.
Morris, G. M., and Faklis, D. 1987, *Optics Comm.*, **62**, 5.
Osterbrock, D. E. 1989, *Astrophysics of Gaseous Nebulae and Active Galactic Nuclei* (Mill Valley: University Science Books).
Sulentic, J. W. 1989, *Ap. J.*, **343**, 54.
Weedman, D. W. 1986, *Quasar Astronomy* (Cambridge: Cambridge University Press).
Wolf, E. 1986, *Phys. Rev. Letters*, **56**, 1370.
———. 1987*a*, *Nature*, **326**, 363.
———. 1987*b*, *Optics Comm.*, **62**, 12.
———. 1989*a*, *Phys. Rev. Letters*, **63**, 2220.
———. 1989*b*, in *Proc. Sixth Rochester Conf. on Coherence and Quantum Optics*, ed. J. H. Eberly, L. Mandel, and E. Wolf (New York: Plenum), in press.
Wolf, E., and Foley, J. T. 1989, *Phys. Rev. A*, **40**, 579.
Wolf, E., Foley, J. T., and Gori, F. 1989, *J. Opt. Soc. Am. A*, **6**, 1142; **7**, 173.

Daniel F. V. James: University of Rochester, Institute of Optics, Rochester, NY 14627

Malcolm P. Savedoff and Emil Wolf: University of Rochester, Department of Physics and Astronomy, Rochester, NY 14627

ELSEVIER

23 May 1994

Physics Letters A 188 (1994) 239–244

PHYSICS LETTERS A

A class of scattering media which generate Doppler-like frequency shifts of spectral lines

Daniel F.V. James [a], Emil Wolf [b,a]

[a] *Institute of Optics, University of Rochester, Rochester, NY 14627, USA*

[b] *Department of Physics and Astronomy, University of Rochester, Rochester, NY 14627, USA*

Received 18 February 1994; accepted for publication 3 March 1994
Communicated by J.P. Vigier

Abstract

It has been known for some time that under certain circumstances scattering from fluctuating media can generate shifts of spectral lines which are indistinguishable from those due to the Doppler effect, even though the source of the radiation, the scatterer and the observer are at rest with respect to each other. In this Letter a class of correlation functions of scattering media is introduced, each of which can generate such Doppler-like shifts. Red shifts as well as blue shifts can be produced by this mechanism, depending on the scattering geometry. We illustrate the results by an example involving diffusion in the presence of a random source.

It was predicted not long ago [1] that dynamic scattering on a random medium whose dielectric response function is suitably correlated in space and time can give rise to shifts of spectral lines that are indistinguishable from those which would be caused by the Doppler effect, even though the scattering medium is localized in a finite volume region which, for all times, is at rest both with respect to the source of the radiation and with respect to the observer.

Scattering from some model media which generate such Doppler-like shifts of spectral lines [#1] was discussed in Refs. [2,3]. In these two papers the correlation functions of the dielectric susceptibility of the scattering media were assumed to be Gaussian in the spatial as well as in the temporal variables. The question arises as to whether there are media with other kinds of correlations which would produce Doppler-like shifts of spectral lines. In this Letter we show that this is so. More specifically we show that there is a broad class of scattering media, with very different correlation functions, which can produce Doppler-like shifts.

Let us consider a plane monochromatic wave of frequency ω_0 and wave vector $\boldsymbol{k}_0 = k_0 \boldsymbol{u}_0$ ($k_0 = \omega_0/c$, $\boldsymbol{u}_0^2 = 1$, c is the speed of light in vacuum), incident on some fluctuating random medium occupying a finite domain D (see Fig. 1). The medium is assumed to be statistically stationary and homogeneous [#2]. It has been shown [4] that under these circumstances, and within the accuracy of the first Born approximation, the spectrum $S^{(\infty)}(r\boldsymbol{u},$

[#1] By Doppler-like shifts $\Delta\lambda$ we mean that all lines present in the spectrum have the same fractional shift $\Delta\lambda/\lambda$, where λ is the central wavelength of the line, i.e. this ratio is independent of λ.

[#2] Strict homogeneity requires, of course, that the medium occupies all space. We assume, as is customary, that the linear dimensions of the domain D are very large compared to the distances over which the correlations of the fluctuations in the macroscopic physical properties of the medium are non-negligible.

SSDI 0375-9601(94)00174-N

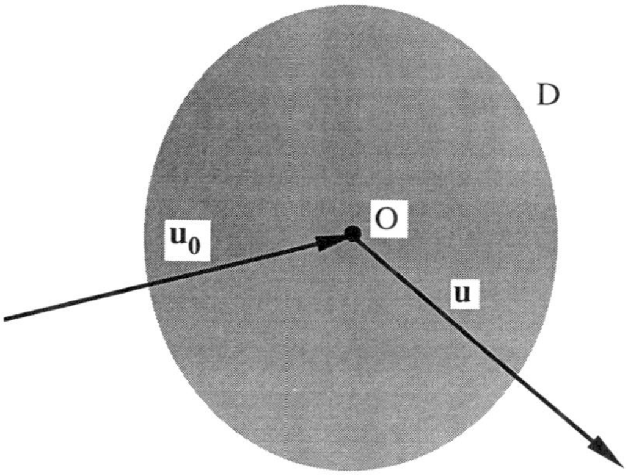

Fig. 1. Illustration of the geometry.

ω) of the scattered field at a point $\boldsymbol{r}=r\boldsymbol{u}$ ($\boldsymbol{u}^2=1$) in the far zone is given by the expression

$$S^{(\infty)}(r\boldsymbol{u},\omega)=A[\mathscr{S}(\boldsymbol{k}-\boldsymbol{k}_0,\omega-\omega_0;\omega_0)+\mathscr{S}(\boldsymbol{k}+\boldsymbol{k}_0,\omega+\omega_0;\omega_0)]\,, \tag{1}$$

where

$$\boldsymbol{k}=\omega\boldsymbol{u}/c \qquad (\boldsymbol{u}^2=1)\,, \tag{2}$$

and

$$\mathscr{S}(\boldsymbol{K},\Omega;\omega_0)=\frac{1}{(2\pi)^4}\int \mathrm{d}^3R\int \mathrm{d}T\,G(\boldsymbol{R},T;\omega_0)\exp[-\mathrm{i}(\boldsymbol{K}\cdot\boldsymbol{R}-\Omega T)] \tag{3}$$

is a structure function of the medium, i.e. the four-dimensional Fourier transform of the correlation function of the generalized time-dependent susceptibility $\hat{\eta}(\boldsymbol{r},t;\omega_0)$ [5] of the medium, viz.,

$$G(\boldsymbol{R},T;\omega_0)=\langle\hat{\eta}^*(\boldsymbol{r},t;\omega_0)\hat{\eta}(\boldsymbol{r}+\boldsymbol{R},t+T;\omega_0)\rangle\,, \tag{4}$$

the angular brackets denoting the ensemble average. The factor A on the right-hand side of Eq. (1) will be regarded to be a constant, although it may depend weakly on ω.

Let us first consider a simple model medium for which the generalized time-dependent susceptibility has the form

$$\hat{\eta}(\boldsymbol{r},t;\omega_0)=\alpha(\omega_0)n(\boldsymbol{r},t)\,, \tag{5}$$

where $\alpha(\omega)$ is a causal, complex function of frequency (and consequently its real and imaginary parts are coupled by dispersion relations [6]) and $n(\boldsymbol{r},t)$ is a real, zero-mean random function of position and time.

On substituting Eq. (5) into Eq. (4) we obtain for G the expression

$$G(\boldsymbol{R},T;\omega_0)=|\alpha(\omega_0)|^2n_0^2g(\boldsymbol{R},T)\,, \tag{6}$$

where

$$n_0=\langle[n(\boldsymbol{r},t)]^2\rangle^{1/2} \tag{7}$$

is the root-mean square value of $n(\boldsymbol{r},t)$ and

$$g(\boldsymbol{R},T)=\frac{\langle n(\boldsymbol{r}+\boldsymbol{R},t+T)n(\boldsymbol{r},t)\rangle}{\langle[n(\boldsymbol{r},t)]^2\rangle} \tag{8}$$

is the normalized correlation function (the correlation coefficient) of $n(\boldsymbol{r}, t)$. From Eqs. (1), (3) and (6) it is clear that the spectrum of the scattered field in the far zone is, in this case, given by the formula

$$S^{(\infty)}(r\boldsymbol{u}, \omega) = A|\alpha(\omega_0)|^2 n_0^2[\tilde{g}(\boldsymbol{k}-\boldsymbol{k}_0, \omega-\omega_0) + \tilde{g}(\boldsymbol{k}+\boldsymbol{k}_0, \omega+\omega_0)] , \tag{9}$$

where

$$\tilde{g}(\boldsymbol{K}, \Omega) = \frac{1}{(2\pi)^4} \int d^3 R \int dT\, g(\boldsymbol{R}, T) \exp[-i(\boldsymbol{K}\cdot\boldsymbol{R} - \Omega T)] \tag{10}$$

is the four-dimensional Fourier transform of the correlation coefficient $g(\boldsymbol{R}, T)$ of $n(\boldsymbol{r}, t)$.

In order to demonstrate that under certain circumstances the far-zone spectrum $S^{(\infty)}(r\boldsymbol{u}, \omega)$, given by formula (9), will be a shifted Doppler-like version of the spectrum of the incident field, we will first introduce a useful representation of the functions $g(\boldsymbol{R}, T)$ and $\tilde{g}(\boldsymbol{K}, \Omega)$. For this purpose we recall that because $g(\boldsymbol{R}, T)$ is a normalized real correlation function, it is both symmetric ($g(-\boldsymbol{R}, -T) = g(\boldsymbol{R}, T)$) and non-negative definite. From these two properties one can readily show that $g(\boldsymbol{R}, T)$ has a maximum with respect to all its variables when $\boldsymbol{R} = T = 0$. Let us next introduce a five-dimensional X, Y, Z, T, W space ($\boldsymbol{R} = (X, Y, Z)$) and consider the hypersurface

$$W = g(X, Y, Z, T) \tag{11}$$

in this space. We choose a rectangular, Cartesian coordinate system with its axes along the four principal directions of curvature [#3] at the origin $X = Y = Z = T = 0$ [#4]. Let ρ_X, ρ_Y, ρ_Z and ρ_T be the principal radii of curvature at the origin. Without loss of generality we may represent g in the form

$$g(\boldsymbol{R}, T) = f\left(\frac{X}{\rho_X}, \frac{Y}{\rho_Y}, \frac{Z}{\rho_Z}, \frac{T}{\rho_T/c}\right), \tag{12}$$

where the function $f(\zeta_1, \zeta_2, \zeta_3, \zeta_4)$ is such that

$$\left(\frac{\partial^2 f}{\partial\zeta_i \partial\zeta_j}\right)_0 = -\delta_{ij} \qquad (i, j = 1, 2, 3, 4) , \tag{13}$$

δ_{ij} being the Kronecker symbol and the subscript zero on the left indicates that all the derivatives are taken at the origin $\zeta_1 = \zeta_2 = \zeta_3 = \zeta_4 = 0$ [#5].

The preceding considerations apply generally. We will restrict our further analysis to a particular class of correlation functions $g(\boldsymbol{R}, T)$, namely to those whose Fourier transforms $\tilde{g}(\boldsymbol{K}, \Omega)$ have a global maximum at the origin $\boldsymbol{K} = \Omega = 0$ [#6].

From expression (12) it follows that

$$\tilde{g}(\boldsymbol{K}, \Omega) = \frac{\rho_X \rho_Y \rho_Z \rho_T}{c} \tilde{f}(\rho_X K_x, \rho_Y K_y, \rho_Z K_z, \rho_T \Omega/c) , \tag{14}$$

[#3] For a discussion of the analogous three-dimensional case see Ref. [7].

[#4] Since the Maxwell equations are not invariant under four-dimensional rotations, we assume a priori that $(\partial^2 g/\partial X_i \partial T)_0 = 0$ $(i = 1, 2, 3)$, where X_i are the components of $\boldsymbol{R}$. If we interpret $n(\boldsymbol{r}, t)$ as the density fluctuations of some scattering medium, this assumption is equivalent to the assumption that the density flux is zero in the given coordinate frame.

[#5] An equivalent method for demonstrating the validity of Eqs. (12) and (13), without introducing a fifth dimension, is to consider the Hessian matrix of the function $g(\boldsymbol{R}, T)$ at the origin (see Ref. [8]). The coordinate axes are chosen so that the Hessian is diagonal, with eigenvalues $-1/\rho_X^2$, $-1/\rho_X^2$, $-1/\rho_Z^2$ and $-1/\rho_T^2$.

[#6] One can show by the use of the multi-dimensional version of Bochner's theorem (see Ref. [9]) that a sufficiency condition for this to be so is that $g(\boldsymbol{R}, T) \geqslant 0$ for all values of the arguments of g. For under these circumstances $\tilde{g}(\boldsymbol{K}, \Omega)$ will be non-negative definite and since also $\tilde{g}(-\boldsymbol{K}, -\Omega) = \tilde{g}(\boldsymbol{K}, \Omega)$, the origin $\boldsymbol{K} = \Omega = 0$ will evidently be a global maximum of $\tilde{g}(\boldsymbol{K}, \Omega)$.

where $\tilde{f}$ is, of course, the four-dimensional Fourier transform of f. In view of Eq. (13), the function $\tilde{f}(q_1, q_2, q_3, q_4)$ has r.m.s. widths unity in all directions and, moreover, by our earlier assumption, it has a maximum at $q_1=q_2=q_3=q_4$. On substituting from Eq. (10) into Eq. (9) we obtain the following expression for the far-zone spectrum of the scattered light:

$$S^{(\infty)}(r\boldsymbol{u}, \omega)=A|\alpha(\omega_0)|^2 n_0^2 \frac{\rho_X\rho_Y\rho_Z\rho_T}{c}$$
$$\times\left[\tilde{f}\left(\frac{\rho_X}{c}(\omega u_x-\omega_0 u_{0x}), \frac{\rho_Y}{c}(\omega u_y-\omega_0 u_{0y}), \frac{\rho_Z}{c}(\omega u_z-\omega_0 u_{0z}), \frac{\rho_T}{c}(\omega-\omega_0)\right)\right.$$
$$\left.+\tilde{f}\left(\frac{\rho_X}{c}(\omega u_x+\omega_0 u_{0x}), \frac{\rho_Y}{c}(\omega u_y+\omega_0 u_{0y}), \frac{\rho_Z}{c}(\omega u_z+\omega_0 u_{0z}), \frac{\rho_T}{c}(\omega+\omega_0)\right)\right]. \tag{15}$$

This result has the following interpretation: Since, as already mentioned, $\tilde{f}$ has an r.m.s. width of unity in all the four principal directions it is clear that, in order for the spectral density $S^{(\infty)}(r\boldsymbol{u}, \omega)$ to have a value that differs significantly from zero in any particular direction, $\boldsymbol{u}$, and at any particular frequency ω, the following inequalities must hold:

$$\frac{\rho_i}{c}|\omega u_i-\omega_0 u_{0i}|\lesssim 1 \qquad (i=1, 2, 3, 4) \tag{16a}$$

or

$$\frac{\rho_i}{c}|\omega u_i+\omega_0 u_{0i}|\lesssim 1 \qquad (i=1, 2, 3, 4)\,. \tag{16b}$$

Here $u_1=u_x$, $u_2=u_y$, $u_3=u_z$, $u_4=1$ and $u_{01}=u_{0x}$, $u_{02}=u_{0y}$, $u_{03}=u_{0z}$, $u_{04}=1$. From inequalities (16) one can readily deduce that the far-zone spectral density will only have significant values for frequencies ω which lie in the four ranges

$$\frac{\omega_0}{\alpha_i}-\frac{c}{\rho_i|u_i|}\leqslant\omega\leqslant\frac{\omega_0}{\alpha_i}+\frac{c}{\rho_i|u_i|} \qquad (i=1, 2, 3, 4)\,, \tag{17}$$

where

$$\alpha_i=\left|\frac{u_i}{u_{0i}}\right|. \tag{18}$$

Suppose now that the normalized correlation function $g(\boldsymbol{R}, T)$ is strongly anisotropic in one of the principal directions, e.g. that

$$\rho_Z\gg\rho_X, \qquad \rho_Z\gg\rho_Y, \qquad \rho_Z\gg\rho_T\,. \tag{19}$$

Inequalities (17) then imply that one of the four frequency ranges is much narrower than the other three, so that the spectral density will have significant values only when

$$\omega\approx\frac{\omega_0}{\alpha_z}\,. \tag{20}$$

Recalling definition (18) of the α, this result implies that the spectrum of the scattered light will be centered at the frequency $\omega_0'=|u_{0z}/u_z|\omega_0$ or, more explicitly, at the frequency

$$\omega_0'=\left|\frac{\cos\theta_0}{\cos\theta}\right|\omega_0\,, \tag{21}$$

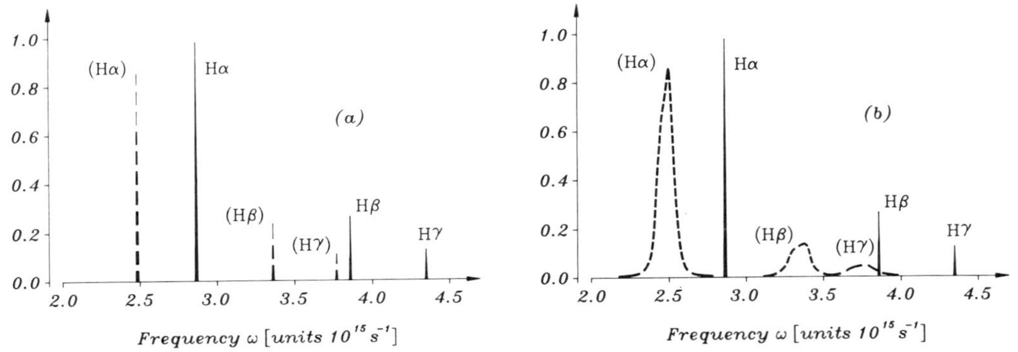

Fig. 2. The three narrow hydrogen Balmer lines Hα, $\omega_0=2.8705\times10^{15}$ s^{-1}; Hβ, $\omega_0=3.875\times10^{15}$ s^{-1} and Hγ, $\omega_0=4.3402\times10^{15}$ s^{-1}, as they would appear in the rest frame of their source (solid lines) and Doppler shifted (dashed lines), due to motion at velocity $0.145c$ away from the observer (a). The same lines shifted and broadened by scattering on a medium at rest with respect to the source and the observer, which has a correlation coefficient whose Fourier transform is given by Eq. (24), with the relevant components of the unit vectors $\boldsymbol{u}$ and $\boldsymbol{u}_0$ specified by $u_{0z}=0.8$, $u_z=0.9211$ and $\boldsymbol{u}\cdot\boldsymbol{u}_0=0.965$. The value of $D_\perp$ was chosen as 3×10^{-2} m^2 s^{-1} (corresponding to the diffusion constant for atomic hydrogen at 10^4 K, with particle density of 10^{16} m^{-3}, and a collison cross-section equal to the area of the first Bohr orbit.) The anisotropy parameter was taken to be $D_\parallel/D_\perp=500$.

where θ_0 and θ are the angles which the directions of incidence and scattering make with the z-direction, i.e. with the direction of the strong anisotropy. The relative frequency shift induced by scattering from the anisotropic medium is evidently given by

$$z\equiv\frac{\omega_0-\omega_0'}{\omega_0'}=\left|\frac{\cos\theta}{\cos\theta_0}\right|-1\,. \tag{22}$$

Since the expression is independent of ω_0, *the frequency shift will be the same for every line present in the spectrum of the incident light*, just as is the case when the shifts are generated by the Doppler effect. The z-number is seen to depend on the directions of incidence and scattering; it is positive (representing a red shift) when $\theta<\theta_0$ and is negative (representing a blue shift) when $\theta>\theta_0$.

We will illustrate the preceding analysis by an example. Suppose that the random field $n(\boldsymbol{r},t)$, which may be interpreted as density in Eq. (5), obeys the anisotropic diffusion equation

$$\left[D_\parallel\frac{\partial^2}{\partial z^2}+D_\perp\left(\frac{\partial^2}{\partial x^2}+\frac{\partial^2}{\partial y^2}\right)-\frac{\partial}{\partial t}\right]n(\boldsymbol{r},t)=\sigma(\boldsymbol{r},t)\,, \tag{23}$$

where $\sigma(\boldsymbol{r},t)$ is some random, highly uncorrelated source of the fluctuations and the diffusion coefficients $D_\parallel$ and $D_\perp$ are assumed to be constants. Using standard methods for calculating structure functions [10–12] one can show that, in this case, the Fourier transform of the normalized correlation function has the form

$$\tilde{g}(\boldsymbol{K},\Omega)=\frac{\tilde{f}_\sigma(\boldsymbol{K},\Omega)}{[D_\parallel K_z^2+D_\perp(K_x^2+K_y^2)]^2+\Omega^2}\,, \tag{24}$$

where $\tilde{f}_\sigma(\boldsymbol{K},\Omega)$ is a slowly varying function of $\boldsymbol{K}$ and Ω. The far-zone spectrum that would result from scattering from such a medium in the case of strong anisotropy ($D_\parallel\gg D_\perp$) is illustrated in Fig. 2. The figure shows that the spectrum shifted by this mechanism has very broad, asymmetric spectral lines, and that it is Doppler like (i.e. that the relative shift $z=\Delta\lambda/\lambda=-\Delta\omega/\omega$ is the same for each line).

This research was supported by the Department of Energy under grant DE-FG02-90ER 14119. The views expressed in this article do not constitute an endorsement by the Department of Energy.

References

[1] E. Wolf, Phys. Rev. Lett. 63 (1989) 2220.
[2] D.F.V. James, M.P. Savedoff and E. Wolf, Astrophys. J. 359 (1990) 67.
[3] D.F.V. James and E. Wolf, Phys. Lett. A 146 (1990) 167.
[4] E. Wolf and J.T. Foley, Phys. Rev. A 40 (1989) 579, Eq. (5.13).
[5] J. Mandel and E. Wolf, Opt. Commun. 8 (1973) 95.
[6] H.M. Nussenzveig, Causality and dispersion relations (Academic Press, New York, 1972) ch. 1.
[7] L.P. Eisenhart, Introduction to differential geometry (Princeton Univ. Press, Princeton, 1947) § 40, pp. 222–227.
[8] M.R. Hestenes, Calculus of variations and optimal control theory (Wiley, New York, 1966) pp. 20,21.
[9] R.R. Goldberg, Fourier transforms (Cambridge Univ. Press, Cambridge, 1961).
[10] R.D. Mountain, Rev. Mod. Phys. 38 (1966) 205.
[11] B.J. Berne and R. Pecora, Dynamic light scattering (Wiley, New York, 1975) ch. 10.
[12] L.D. Landau and E.M. Lifshits, Fluid mechanics (Pergamon, London, 1959) ch. 17.

The Redshift Controversy and a New Mechanism for Generating Frequency Shifts of Spectral Lines

Compiled from a lecture presented by EMIL WOLF of the University of Rochester, Rochester, New York at the National Physical Laboratory in New Delhi, India on January 11, 1991

My talk will be partly concerned with astronomy. You may wonder why anyone should speak about astronomy in a Laboratory whose chief mission is to maintain and realize units based on the International System of units. I will answer this question later in this talk and I will then try to show you that research in a subject which appears to be far removed from the mission of your laboratory has a strong bearing on the standards activities in which some of you are engaged.

Most of you know that the lines observed in the spectra of radiation that reaches us from astronomical sources are shifted relative to those observed in the spectra from the same elements on earth. Usually the lines are shifted towards the longer wavelengths and one then speaks of a redshift. In a few cases, they are shifted towards the shorter wavelengths and then one speaks of a blueshift. From the observed shifts, astronomers draw conclusions which have far reaching consequences for our understanding of the structure of the Universe.

I will begin by briefly reviewing the conventional interpretation of the physical origin of this effect and I will indicate why it is of basic importance for astronomy.

Consider a source which is moving with respect to an observer. For simplicity we will consider only radial motion. Let v be the velocity of the source relative to the observer. The observer then detects a shift $\Delta\lambda$, of a line centered on a wave length λ and according to Doppler's formula

$$z \equiv \frac{\Delta\lambda}{\lambda} = \frac{1+v/c}{[1-(v/c)^2]^{1/2}} - 1$$

$$\approx v/c \quad \text{if} \quad v/c \ll 1,$$

where c is the velocity of light. Fig. 1 shows the variation of z with v/c. For sufficiently small velocities v, z is seen to vary linearly with v. For positive values of v, i.e. when source is moving away from the observer, z can have any value between 0 and ∞ and the shift is then towards the longer wavelengths. When the source is moving towards the observer, z becomes negative and can have any value between 0 and -1.

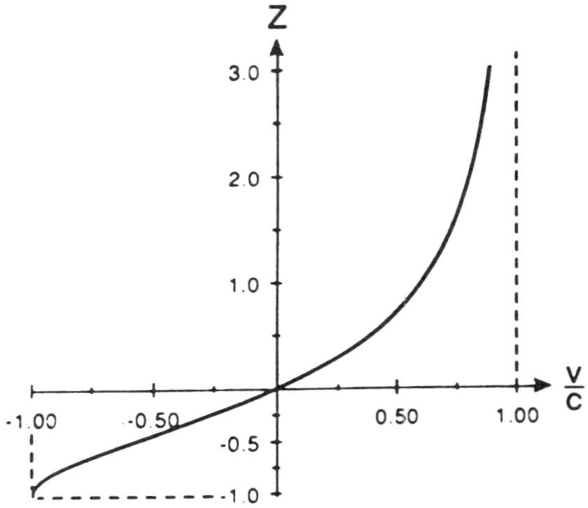

Fig. 1 — The Doppler effect : variation of z with v/c.

One of the first observations of the redshift was made by Huggins in 1868. Twenty years later Vogel observed it in the light from the star Capella. He found that its redshift $z \simeq 10^{-4}$, which corresponds to a speed of recession of about 30 km/sec. During the period 1912-1925, first high quality photographs were obtained by Vesto Slipher of Lowell Observatory and they could be used to determine accurately the velocities. Slipher analyzed spectra of about 40 galaxies and found that thirty eight of them were redshifted and two were blueshifted. The largest speed which was determined from these observations, using Doppler's formula, was 1,800 km/sec (about 4 million miles/hour). From these and from other results, Hubble postulated in 1929 that the speed of recession v is proportional to distance d of the astronomical source from the observer, i.e. that

$$v = Hd, \qquad \ldots(2)$$

where H is now called Hubble's constant. This is one of the basic laws of modern astrophysics. There is some uncertainty about the exact value of the Hubble constant, but it is often assumed to have the approximate value 50 km/sec/megaparsec. (1 megaparsec = 10^6 parsec, 1 parsec. = 3.26 light years).

Fig. 2 shows a set of photographs of some nebulae, with their spectra on the right hand side. It was from the study of such photographs, that Hubble was led to the law expressed by Eq. (2). The actual observed spectra are the cigar-shaped traces. They are shown between the corresponding laboratory spectra. We note the two calcium lines H and K in the spectra. The arrows indicate how far these lines have shifted from their original position. As we go

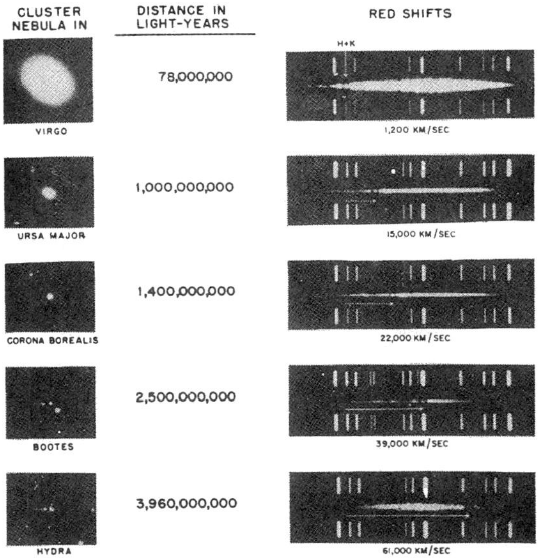

Fig. 2—Photographs of some Nebulae and their spectra. (By Courtesy Hale Observatory, USA)

through the sequence of photographs the observed lines are seen to be shifted further and further to the right. These photographs were taken with the same exposure time and on the same scale. We see that the nebulae are becoming smaller and smaller as we go down this sequence of the pictures, suggesting that they are further and further away. On the average we can assume that the nebulae have similar physical attributes. We may then conclude from these photographs that the larger the redshift, the further are these objects away from us. This is the basis of the origin of Hubble's law. Clearly this law provides supporting evidence for the Big Bang theory, because it indicates that with greater redshifts the recession velocity is also greater and the object is further away. In this case, instead of speaking of the Doppler redshift, one often speaks of the cosmological redshift, because it is related to the expansion of the whole universe.

In the 1950's supporting evidence for Hubble's law came from radio astronomy. One found that the 21 cm hydrogen line had exactly the same z number as the lines in the optical region of the spectrum obtained from the same source. The fact that the z-numbers of the spectral lines of very different wavelengths are the same strongly suggests that the observed redshift is indeed in agreement with the Doppler-effect interpretation.

However, as time went on some problems started to appear. In 1960 it was found that certain kind of objects have very different properties from those one is familiar with from observations of stars and galaxies. Astronomers called these objects quasi-stellar objects. Today they are called quasars. Their spectral lines were found at unusual wavelengths and at first the lines could not be identified. They were very broad, much broader than one finds in the spectra of galaxies. They had few emission lines unlike the spectra of radiation from galaxies which typically have many lines. In 1963 Marteen Schmidt identified four hydrogen lines H_α, H_β, H_γ, and H_δ in the spectrum of one of the quasars called 3C273 (Fig. 3), by noting that these lines have exactly the same separation from each other as the four hydrogen lines observed in the terrestrial spectrum of hydrogen. The lines had extremely large z number by the standards of those days namely z = 0.16, corresponding to the speed of recession of 45,000 km/

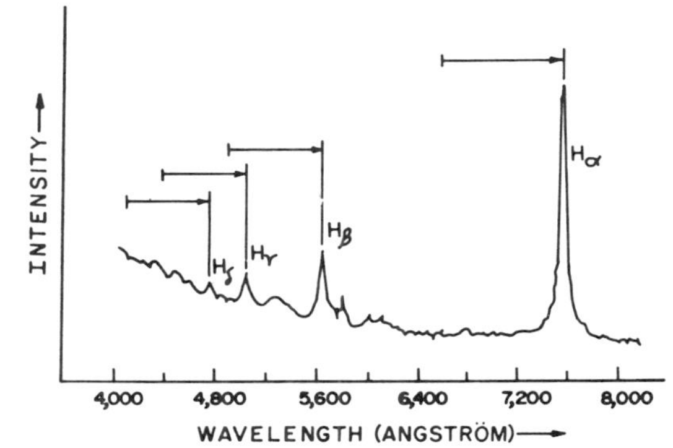

Fig. 3—Spectrum from the quasar 3C 273, showing shifted H_α, H_β, H_γ and H_δ emission lines.

sec. i.e. 15 percent of the speed of light. According to Hubble's law, this places the quasar at a distance of about 3 billion light years. It is clear now why astronomers had problems in identifying these lines, because the shifts were extremely large, orders of magnitude larger than astronomers had observed previously. Soon afterwards lines in the spectra of other quasars were identified and some were found to have even larger redshifts than those found by Schmidt.

There are some strange things about quasars. If they are at the distances deduced from the Doppler interpretation of their redshifts. and from Hubble's law they have very high intrinsic luminosities—they are several trillion times more luminous than the sun and are typically 1000 times more luminous than an entire galaxy, consisting of many billions of stars. Yet one knows that quasars are isolated objects, not composite ones like galaxies, which consist of very many stars. This can be deduced from the fact that the luminosities of quasars vary in time, with periodicities of the order of a week or a month and the theory of relativity puts some restrictions on the size of such objects. As time went by, quasars with larger and larger redshifts were discovered. Quasar with the largest z-number observed up to now, namely 4.7, would have, according to the Doppler formula the speed of recession $v = 0.94c$. According to Hubble's law this implies that the quasar is about 19 billion light years away. The exact age of the universe is not known but, according to the Big Bang theory, it is about 17 to 20 billion years old. (These different estimates of the age of the universe are related to the uncertainty about the value of Hubble's constant). This would mean that radiation from this quasar must have originated when the universe was only a few percent of its present age. The estimated speeds of recession of these quasars pose other problems, connected with relativistic bulk motion, about which very little is known. One has considerable knowledge about relativistic motion of elementary particles but bulk motion at such speeds is a different story. There are other problems, one of which is indicated in the photographs shown in Figures 4 and 5. Fig. 4 is a famous photograph of Markarian 205 which shows a quasar interacting with the NGC galaxy. This redshift of the quasar is twelve times larger than that of the galaxy. Obviously if the connection is real there is something wrong with the current interpretation of the redshift data. According to the Doppler formula and Hubble's law, this quasar must be twelve times futher away than the galaxy, so the two could not be interacting, as appears to be the case from the photograph. This is an example of the sort of problems leading to the redshift controversy. It has been very much argued about. Fig. 5 shows an even

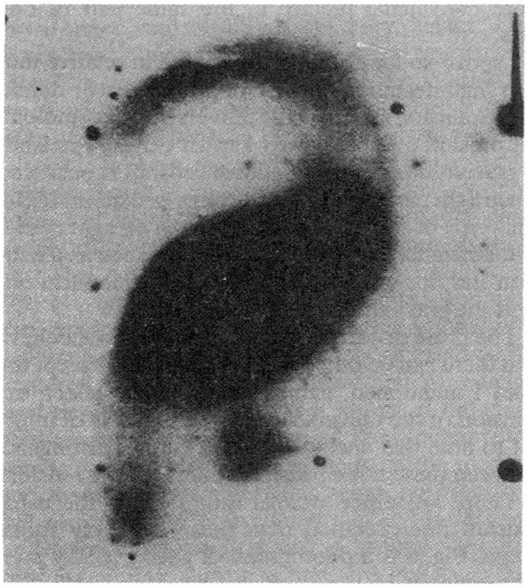

Fig. 4—NGC 4319 galaxy (big oval blob) and the quasar Markarian 205 (smaller blob below it). They appear to be connected, even though according to the usual interpretation the quasar's redshift is much greater than that of the galaxy, indicating that it is much further away. [From W. J. Kaufmann II ***Galaxies and Quasars*** *(W. H. Freeman and Co., San Francisco, 1979), 146]. [By Courstesy H. Arp, Hale Obervatory, USA]*

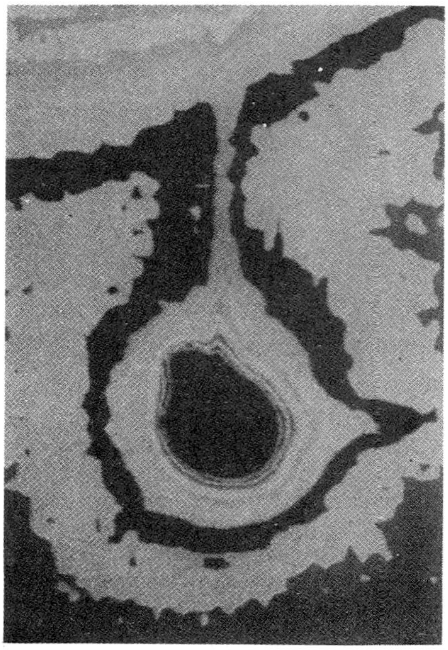

Fig. 5—Another picture of NGC 4319 and Markarian 205 [From G. Bur-bidge, Mercury **17**, 136 (1988)].

more convincing picture of the same pair of objects.

I spoke so far about redshifts due to relative motion (manifestation of the Doppler effect) or due to the expansion of the whole universe (the cosmological redshift). Another well-known cause of redshift is gravitation. The gravitational redshift is observed when light passes through a strong gravitational field. This type of shift is only produced when light passes in the neighbourhood of a dense star such as a neutron star, for example. With most astronomical objects the gravitational shift is negligible.

The great majority of astronomers do not believe that there can be other causes of redshifts except the ones I mentioned, namely the Doppler effect, expansion of the universe or gravitation. It is of interest to note that spectral analysis cannot distinguish between these three possible sources of the redshift.

There are other strange things about the usual redshift interpretation than those I already spoke about. Fig. 6 is a photograph of galaxy-galaxy pair. The redshift of the companion galaxy is twice as large as for the main galaxy and yet they appear to be connected. There are many examples of this kind which have been disputed and argued about for several years.

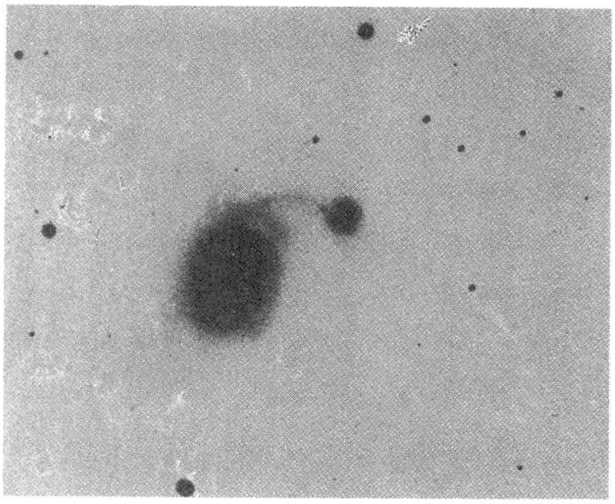

*Fig. 6—NGC7603. A light print to show disturbance of inner part of main galary. Darker print show strong bridge between main galaxy at cz = 8800 km/sec. and companion galaxy at cz = 16900 km/sec. [H. Arp in **Confrontation of Cosmological Theories and Observational Data**, M. S. Langair, ed., (1974), p. 61[By courtesy Reidel Publshers, Holland]*

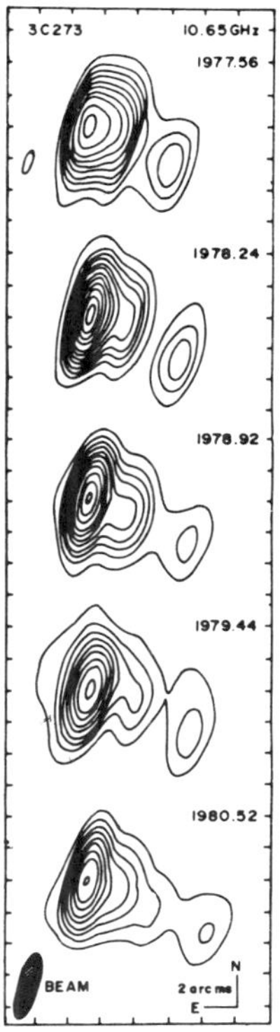

Fig. 7—Radio maps made by VLBI technique from mid-1977 to mid-1980 indicating that the quasar 3C 273 is undergoing a superluminal expansion. [Pearson et al., Nature **290**, 365 (1981)].

Fig. 7 illustrates so called superluminal expansion involving two radio sources, whose positions were recorded from mid 1977 to mid 1980. Assuming the usual redshift interpretation to be correct and that Hubble's law applies, the measured angular separation of the two radio sources implies that their centers are separating with transverse speed of about ten times the speed of light. There are many other examples of such superluminous sources. In some case their speed of separation is even greater. According to the theory of relativity this, of course, cannot happen. Hence one has to find some other explanation. Two explanations have been put forward to solve such puzzles. One is based on the theory of relativity which indicates that instead of these objects being aligned across the sky as they appear to be, they are aligned radially with very small angular separation when viewed from the earth. By applying relativistic corrections to this model, one can resolve the paradox. But it is rather puzzling why so many of these sources would be so well aligned. The other model is based on the concept of a gravitational lens. As radiation passes through a strong gravitational field it becomes focused and hence the objects may actually be at quite different positions than they appear to be. None of these arguments seems to me to be convincing and one might wonder whether a different interpretation of the origin of the red-shifts in the spectra of these objects would not lead to a more plausible resolution of the paradox.

Nevertheless most astronomers do not consider these problems as contradicting the current theories in any way. Here are some typical comments on the situation:

"Standard processes in physics cannot explain how nearby quasar can have a high redshift" [W.J. Kaufmann-III: *Galaxies and Quasars*, (1979), p. 145].

"But photographs of quasarlike objects that are apparently connected to galaxies of much lower redshifts still present puzzles. Either these photographs are coincidences or new laws of physics are needed to explain the radiation from the quasarlike objects" [H.L. Shipman: *Black Holes, Quasars and the Universe*, (1980), p. 239].

These are the kind of statements which are made by the vast majority of astronomers. Only very few astronomers are willing to consider the possibility that there may be some other explanation of the puzzles. They form a very small minority, but they include eminent scientists such as Hoyle, Burbidge, Arp and the well-known Indian astrophysicist Narlikar. These few scientists are of the opinion that, at least in some cases, the usual redshift interpretation is incorrect and that we really do not understand the ture origin of all the observed redshifts.

In the next part of this talk I will show that the statments I just cited are misleading and that there is a mechanism, deeply rooted in present day physics, which has nothing to do with the Doppler effect, the expanding universe or gravitation and which can nevertheless generate redshifts of spectral lines.

Let us consider the electromagnetic field generated by some source. For simplicity we will ignore complications that may arise from the polarization properties of the field. Let $Q(\mathbf{r}, t)$ be a source variable which depends on position $\mathbf{r}$ and time t. It generates a field $V(\mathbf{r}, t)$. These two quantities are related by the inhomogeneous wave equation

$$\nabla^2 V(\mathbf{r}, t) - \frac{1}{c^2}\frac{\partial^2}{\partial t^2} V(\mathbf{r}, t) = -4\pi Q(\mathbf{r}, t) \qquad \ldots(3)$$

or, in the space-frequency domain, by the equation

$$\nabla^2 U(\mathbf{r}, \omega) + k^2 U(\mathbf{r}, \omega) = -4\pi q(\mathbf{r}, \omega), \qquad \ldots(4)$$

where

$$k = \omega/c. \qquad \ldots(5)$$

Since we are interested in the space-frequency description, let us start with Eq. (4). We will deal with realistic sources which fluctuate in a random manner. The fluctuations are largely caused by spontaneous emission. Even in the case of a laser there are some fluctuations because spontaneous emission cannot be totally suppressed and there are other contribution to the randomness, e.g. mechanical vibrations of mirrors at the ends of the resonant cavity, fluctuations in pressure and temperature, etc. Thus we cannot regard $q(\mathbf{r}, \omega)$ as a deterministic function but we must consider it to be a member of a statistical ensemble which characterizes the fluctuations. Because the source variable $q(\mathbf{r}, \omega)$ is a random variable, so is, of course, the field variable $U(\mathbf{r}, \omega)$.

There are two different types of spectra—the spectrum of the source, $S_Q(\mathbf{r}, \omega)$, and the spectrum of the field, $S_V(\mathbf{r}, \omega)$, defined by the formulas

$$S_Q(\mathbf{r}, \omega) = \langle q^*(\mathbf{r}, \omega) q(\mathbf{r}, \omega)\rangle, \qquad \ldots(6a)$$

$$S_V(\mathbf{r}, \omega) = \langle U^*(\mathbf{r}, \omega) U(\mathbf{r}, \omega)\rangle, \qquad \ldots(6b)$$

where the angular brackets denote the ensemble average.

In statistical optics one deals with correlations, called cross-spectral densities, which are generalization of spectra. The generalization consists in replacing the single position variable $\mathbf{r}$ by two position variables $\mathbf{r}_1$ and $\mathbf{r}_2$, i.e. one defines the cross spectral densities by the formulas

$$W_Q(\mathbf{r}_1, \mathbf{r}_2; \omega) = \langle q^*(\mathbf{r}_1, \omega) q(\mathbf{r}_2, \omega)\rangle, \qquad \ldots(7a)$$

$$W_V(\mathbf{r}_1, \mathbf{r}_2; \omega) = \langle U^*(\mathbf{r}_1, \omega) U(\mathbf{r}_2, \omega)\rangle. \qquad \ldots(7b)$$

In general the fluctuations will exhibit some similarities over certain region of space. The cross-spectral densities characterize the *statistical similarity* of the fluctuating quantities at the points $\mathbf{r}_1$ and $\mathbf{r}_2$. By replacing $\mathbf{r}_1$ by $\mathbf{r}$ and $\mathbf{r}_2$ by $\mathbf{r}$, one obtains the source spectrum $S_Q(\mathbf{r}, \omega)$ and the field spectrum $S_V(\mathbf{r}, \omega)$:

$$S_Q(\mathbf{r}, \omega) = W_Q(\mathbf{r}_1, \mathbf{r}_2; \omega)\Big|_{\substack{\mathbf{r}_1 \to \mathbf{r} \\ \mathbf{r}_2 \to \mathbf{r}}}, \qquad \ldots(8a)$$

$$S_V(\mathbf{r}, \omega) = W_V(\mathbf{r}_1, \mathbf{r}_2; \omega)\Big|_{\substack{\mathbf{r}_1 \to \mathbf{r} \\ \mathbf{r}_2 \to \mathbf{r}}}. \qquad \ldots(8b)$$

Since the field propagates in the form of waves which satisfy the inhomogeneous wave equation, one can show that the cross spectral densities of the field and of the source are related by the equation

$$(\nabla_2^2 + k^2)(\nabla_1^2 + k^2) W_V(\mathbf{r}_1, \mathbf{r}_2; \omega) = (4\pi)^2 W_Q(\mathbf{r}_1, \mathbf{r}_2; \omega). \qquad \ldots(9)$$

This equation shows that there is a linear transform relationship from the cross-spectral density of the source to the cross-spectral density of the field. However, there is no such relationship between the *spectra* of the field and of the source. If we wish to determine the spectrum of the field, we have to start from the cross-spectral density of the source (which tells us about correlation i.e. about statistical similarity of fluctuations in the source at different points) and then determine the field spectrum by the following sequence:

$$W_Q(\mathbf{r}_1, \mathbf{r}_2; \omega) \rightarrow W_V(\mathbf{r}_1, \mathbf{r}_2; \omega) \rightarrow W_V(\mathbf{r}, \mathbf{r}; \omega) \equiv S_V(\mathbf{r}, \omega). \quad \ldots(10)$$

Let me illustrate this result by a simple example, involving two small fluctuating sources, located in the neighborhood of points P_1 and P_2 (Fig. 8). Since the sources fluctuate, they must be described, for each frequency ω, by statistical ensembles $\{Q(P_1, \omega)\}$ and $\{Q(P_2, \omega)\}$. The field which these

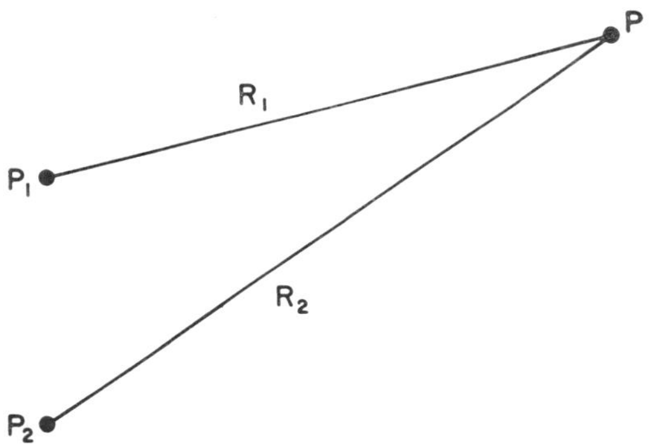

Fig. 8—Illustrating the geometry relating to the calculation of spectra from two correlated sources.

sources generate at a point P will also fluctuate. We may characterize the field fluctuations by a field ensemble $\{U(P, \omega)\}$. A typical realization of the field at point P, at frequency ω, is given by:

$$U(P, \omega) = Q(P_1, \omega)\frac{e^{ikR_1}}{R_1} + Q(P_2, \omega)\frac{e^{ikR_2}}{R_2}, \quad \ldots(11)$$

where R_1 and R_2 are the distances from P_1 to P and P_2 to P respectively. The spectrum of the field at P, is given by

$$S_V(P, \omega) = \langle U^*(P, \omega)U(P, \omega)\rangle, \quad \ldots(12)$$

where the angular brackets denote the ensemble average. Substituting for $U(P, \omega)$ from Eq. (11) into Eq. (12) we see that

$$S_V(P, \omega) = S_Q(\omega)\left[\frac{1}{R_1^2} + \frac{1}{R_2^2}\right] + \left[W_Q(P_1, P_2, \omega)\frac{e^{ik(R_2 - R_1)}}{R_1 R_2} + cc\right], \quad \ldots(13)$$

where cc denotes the complex conjugate. We assumed here that the two sources, at P_1 and P_2, have the same spectrum, $S_Q(\omega)$.

The crucial fact to note in Eq. (13) is that in addition to the source spectrum $S_Q(\omega)$ we have a contribution to the field spectrum which involves $W_Q(P_1, P_2, \omega)$, i.e. which contains a function that characterizes the correlation of the fluctuations at the two source points.

Equation (13) shows that, in general, the spectrum of the field is not proportional to the spectrum of the source but that it depends also on the correlation of the fluctuations of the two small sources. There are, however, two exceptional cases:

(1) When $W_Q(P_1, P_2, \omega) \equiv 0$, i.e. when the two sources are completely uncorrelated, $S_V(\omega)$ is seen to be proportional to $S_Q(\omega)$. This is the case of two *mutually incoherent* sources.

(2) Suppose that $R_2 = R_1$. If it happens that $W_Q(P_1, P_2, \omega)$ is proportional to $S_Q(\omega)$, then again $S_V(\omega)$ will be proportional to $S_Q(\omega)$. This happens when the two sources are completely correlated, i.e. when they are *mutually completely coherent.*

These are precisely the most common situations. However, we see that, in general, the theory predicts that the spectrum of the field is different from the spectrum of the source. As we just saw in the two most common situations namely when the sources are either mutually incoherent or are mutually completely coherent, this effect is absent. Because incoherent and coherent sources are the ones most commonly encountered, this is probably one of the reasons (but not the only one) why this effect has not been observed previously.

The theoretical prediction that the spectrum of the field is, in general, different from the spectrum of the source has been verified experimentally with different types of sources. Before discussing some of these experiments, let me recall what is meant by a primary and by a secondary source. A primary source is a set of radiating atoms or radiating molecules. If we have such a source and allow the radiation to pass through an aperture i.e. through an opening in an opaque screen in some optical system (see Fig. 9), the illuminated aperture behaves essentially as a primary source. The illuminated aperture is an example of a secondary source. We will mainly

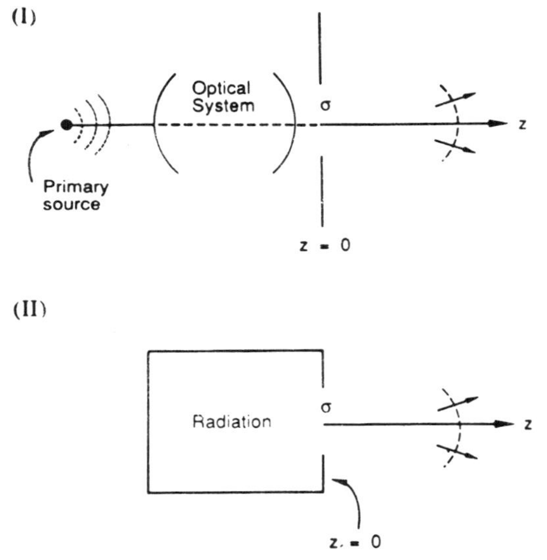

Fig. 9 —Examples of a planar secondary source.

examine spectra of radiation propagating from such secondary sources. There are, of course, strong similarities between primary and secondary sources. In the case of a primary source the randomness comes from true source fluctuations, while in the case of a secondary source it comes from fluctuating boundary conditions of the field in the secondary source plane.

A few years ago I examined the conditions under which the spectrum of light from a certain class of a secondary source does not change in the course of propagation to the far zone [*Phys. Rev. Lett.*, **56**, 1370 (1986)]. I found a sufficiency condition, which I called the *scaling law*, for the invariance of spectrum; more precisely a condition for the far-zone spectrum of the radiated field to be the same as the source spectrum, apart from trivial geometrical factors. Before stating this law let me define the degree of spectral coherence at frequency ω. It is given by the formula

$$\mu_A(\mathbf{r}_1, \mathbf{r}_2, \omega) = \frac{W_A(\mathbf{r}_1, \mathbf{r}_2, \omega)}{[W_A(\mathbf{r}_1, \mathbf{r}_1, \omega) W_A(\mathbf{r}_2, \mathbf{r}_2, \omega)]^{1/2}}, \quad \ldots(14)$$

where the subscript A stands for the source variable or for the field variable. The scaling law asserts that when the normalized spectrum of the field is the same throughout the far zone and is equal to the normalized spectrum of the source then the degree of spectral coherence of a source of a certain class, must have the following functional form:

$$\mu_Q(\mathbf{r}_1 - \mathbf{r}_2, \omega) = f[k(\mathbf{r}_1 - \mathbf{r}_2)]. \quad \ldots(15)$$

This formula shows that if one changes the wavelength, one can compensate for the change in μ_Q by an appropriate change in $(\mathbf{r}_2 - \mathbf{r}_1)$.

Do we know any source of this kind? The answer is Yes. Most common thermal laboratory sources, such as blackbody sources and other lambertian sources are of this type. It is known that such sources have a degree of spectral coherence given by the expression

$$\mu(\mathbf{r}_1 - \mathbf{r}_2, \omega) = \frac{\sin k|\mathbf{r}_1 - \mathbf{r}_2|}{k\,|\mathbf{r}_1 - \mathbf{r}_2|} \quad \ldots(16)$$

Evidently this degree of coherence satisfies the scaling law (15). These remarks indicate that the most common thermal sources obey the scaling law and, therefore, they will generate light whose spectrum does not change on propagation to the far zone.

Two of my colleagues at the University of Rochester, Professor Morris and Dr. Faklis, tested experimentally some of these predictions. They measured the spectra of light produced by two planar secondary sources. Each of these sources was obtained by letting light from a primary thermal source (tungsten lamp), located directly in front of the aperture in plane I, produce a secondary source in plane II after passing through an optical system and they measured the spectra both in these planes and in the plane III located in the far zone.

The first system was a conventional lens [Fig. (10a)]. it can be shown that in this case the secondary source in the plane II obeys the scaling law.

The second lens system is a so-called Fourier achromat [Fig. (10b)]. This is a combination of lenses which is free of chromatic aberration, i.e. free of the detrimental effect caused by the fact that glass dis-

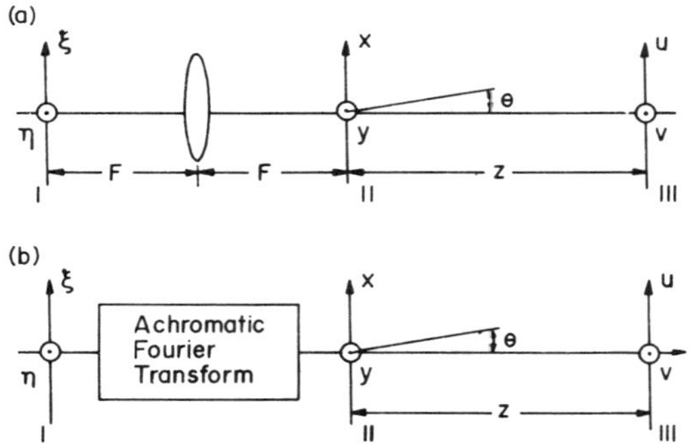

Fig. 10 —Schematics of an experimental set up used for realization of secondary source with controlled degree of spectral coherence that (a) satisfies the scaling law (b) violates the scaling law. [*G. M. Morris and D. Faklis, Opt. Commun,* **62**, 5 (1987)].

perses light of different wavelengths along different paths. Now because of the manner in which this system has been designed, the secondary source in the plane II no longer obeys the scaling law; in fact the degree of spectral coherence of the light in that plane can be shown to be effectively independent of the frequency over the whole frequency-range for which the lens has been achromatized.

According to our theoretical results we would expect that in case (a) the spectrum will be the same in all θ-directions, whereas in case (b) it will depend on θ, i.e. on the location of the point of observation in the plane III.

The experimental results are shown in Figure 11. Figure 11a shows the measured spectrum at some point in the far zone, obtained in experiments with the conventional lens. This is the situation when the scaling law is satisfied (providedthe apertures are not too small) and, in agreement with our theoretical predictions, measurements at other points in the far zone gave identical spectra. The curves corresponding to measurements at different points coincided, so they cannot be distinguished on this diagram. Morris and Faklis also measured the spectrum at points in the plane II of the secondary source and they found that it was indeed identical with the measured far-zone spectrum, when properly normalized

Figure 11b shows the measured far-zone spectrum at points located in different directions, in experiments using the Fourier achromat. This is the case when the secondary source does not obey the scaling law. As the theory predicts, the measured spectra at points in different directions were now found to be different.

These experiments confirm the theoretical prediction that, in general, the spectrum of light changes on propagation.

Let me now return to the example of a system consisting of two small sources, discussed earlier (Fig. 8). The spectrum of the field produced by these sources is given by Eq. (13). The equation can be rewritten (with the choice $R_2 = R_1 = R$) as

$$S_V(\omega) = 2S_Q(\omega)[1 + Re\, \mu_Q(\omega)] \quad \ldots (17)$$

(*Re* denoting the real part), where $S_V(\omega) = (R^2/2)S_v(P, \omega)$ and $\mu_Q(\omega)$ is the degree of spectral correlation between the sources at P_1 and P_2:

$$\mu_Q(\omega) = \frac{W_Q(P_1, P_2; \omega)}{S_Q(\omega)}. \quad \ldots (18)$$

One can show that

$$0 \leq |\mu_Q(\omega)| \leq 1. \quad \ldots (19)$$

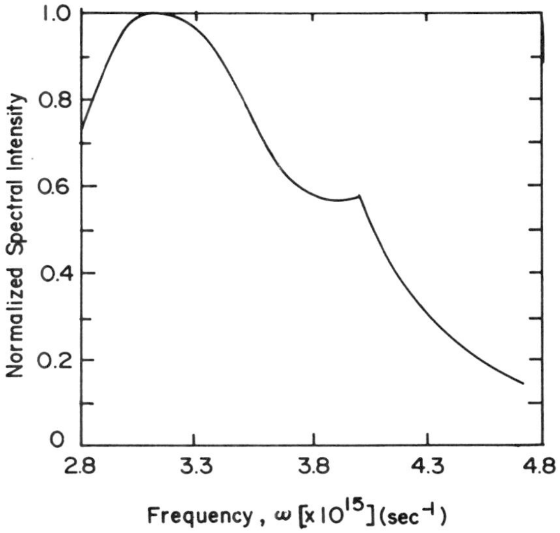

Fig. 11a—Normalized far field spectrum when the scaling law is satisfied. It was found to be the same at all off-axis angles [After Morris and Faklis, loc. cit.].

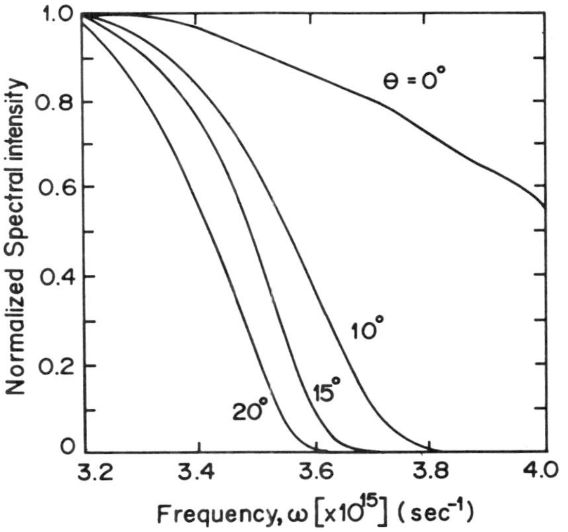

Fig. 11b—Normalized far field spectrum at different off-axis angles when the scaling law is not satisfied [After Morris and Faklis, loc. cit.].

The extreme condition $\mu_Q(\omega) = 0$ applies to uncorrelated sources; whilst the other extreme $|\mu_Q(\omega)| = 1$ applies to fully correlated sources.

We see from Eq. (17) that the spectrum of the field differs from the spectrum of the source, depending on the degree of correlation. Without going into details, one can easily see that with appropriate choices of the degree of correlation many different types of spectral changes can be produced. Suppose, for example, that the source spectrum $S_Q(\omega)$ consists of a line with a Gaussian profile, centered on frequency ω_0, and that the degree of spectral coherence $\mu_Q(\omega)$ is also a Gaussian function, centered at a different frequency, ω_1. The product of two Gaussian is again a Gaussian but the new Gaussian function will not have the same central frequency as the original spectral line. It will be shifted towards the lower or the higher frequencies depending on whether $\omega_1 < \omega_0$ or $\omega_1 > \omega_0$. Fig. 12 shows results of simple computer simulations of such situations. The curve (a) is the original spectral line. Choosing suitable Gaussian correlations one obtains either the curve (b) which is a redshifted line or the curve (c) which is a blueshifted line, depending on the choice of the central frequency ω_1 of the degree of coherence. When I speak here of shifts I mean the change in the location of the maximum. The lines are actually not simply shifted, but they are slightly distorted and their heights and widths also change. This is similar to what happens with the Doppler shift.

The possibility of producing shifts of spectral lines by this method was tested in Rochester by my colleagues Professors Bocko, Douglass and Knox, using acoustic waves. In their experiment (Fig. 13) the acoustical sources $Q_1(t)$ and $Q_2(t)$ were produced by mixing randomly generated and mutually uncorrelated signals $X(t)$ and $Y(t)$ in such a manner that

$$Q_1(t) = X(t) + Y(t), \qquad \text{...(20a)}$$

$$Q_2(t) = X(t) - Y(t). \qquad \text{...(20b)}$$

The assumption that $X(t)$ and $Y(t)$ are uncorrelated is expressed by the equation

$$\langle X(t)Y(t+\tau)\rangle = 0 \text{ for all } \tau. \qquad \text{...(21)}$$

One can show that the two sources Q_1 and Q_2 will have the same spectrum i.e. that

$$S_2(\omega) = S_1(\omega). \qquad \text{...(22)}$$

But the condition (21) does not mean that source Q_1 and Q_2 are independent. They are correlated, i.e. $\langle Q_1(t)Q_2(t+\tau)\rangle \neq 0$, in general, because each of them has contributions from both X and Y, but in different ratios. For example if $Y=0$, then the two sources would be perfectly correlated. If $X=0$ they would be completely anticorrelated. In between as the ratio of X to Y is changed, one obtains different correlations between the two sources. Thus we have here a situation in which we have two sources which have the same spectrum but they are partially correlated and the amount of correlation depends on the ratio of X to Y, which can be controlled. The spectra at the detector were analyzed. With a single radiating source they observed the spectrum represented by the curve A from one of the sources and the spectrum represented by the curve B from the other (Fig. 14a). When both the sources radiated simultaneously the spectrum of the detector was indeed found to depend on the correlation existing between them

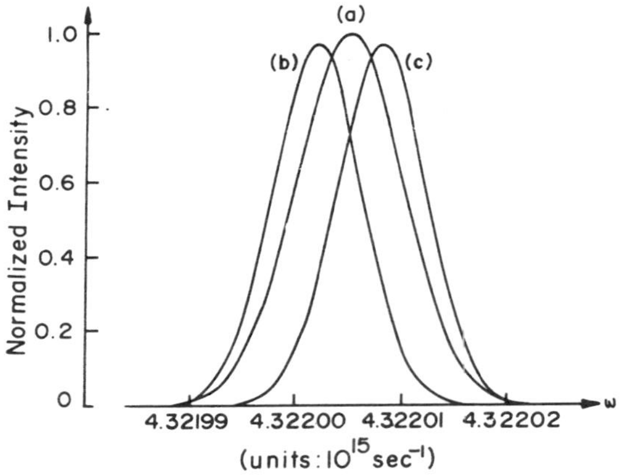

Fig. 12—Redshifts and blueshifts of spectral lines induced by suitable correlation between two small sources: (a) source spectrum; (b) and (c) field spectra, depending on the choice of correlation. [*After E. Wolf, Phys. Rev. Lett.* **58**, 2646 (1987)].

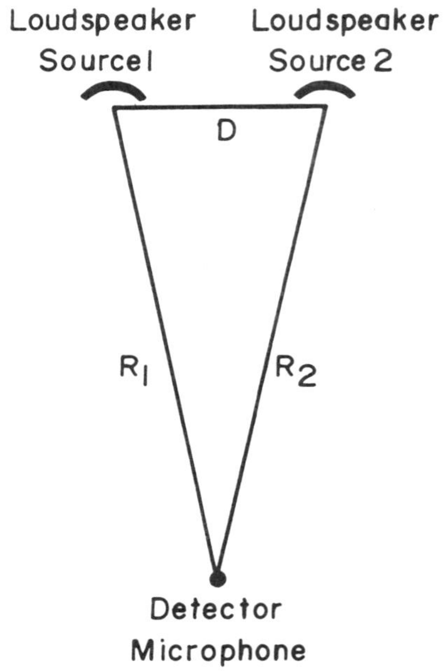

Fig. 13—Arrangement used for an experimental test of spectral shifts with acoustic sources. [*After M. F. Bocko, D. H. Douglass and R. S. Knox, Phys. Rev. Lett.* **58**, 2649 (1987)].

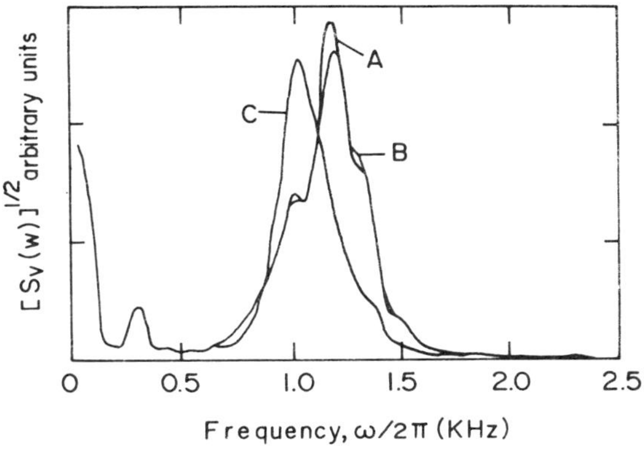

(a)

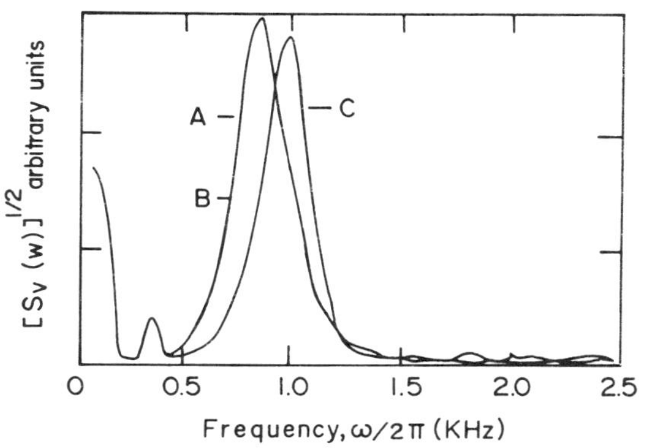

(b)

Fig. 14(a & b)—Curve labeled A and B are the results of surements at the microphone when only one source, at a time, is switched on. Curve C represents the red-shifted or the blue-shifted acoustic spectral line when both sources are switched on, to act as a correlated pair [After Bocko et al., loc. cit.].

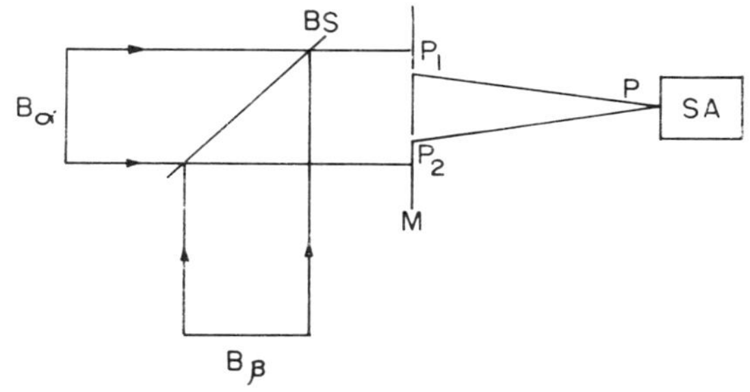

Fig. 15—The experimental layout of an experiment of F. Gori, C. Palma and C. Padovani, Opt. commun, **67**, 1 (1988) *to demonstrate generation of shifts of spectral lines by source correlations.*

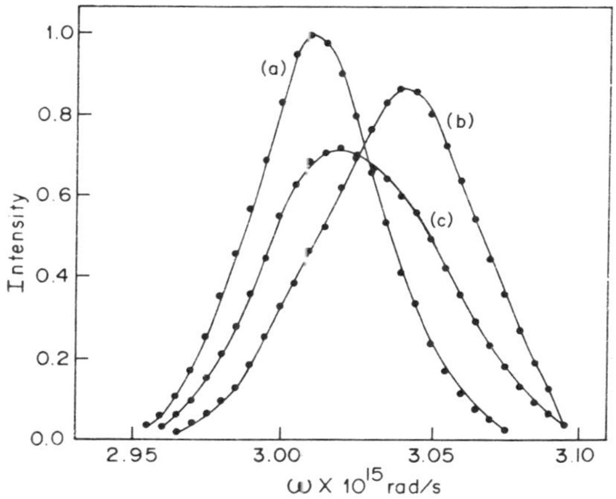

Fig. 16—Curve (c) represents the spectrum (multiplied by a factor 2) produced by a single source. Curve (a) and curve (b) represent the spectrum when both the beams B_1 and B_β are used. Curve (b) represents the spectra when the beam B_α and B_β are interchanged. [After F. Gori, et al., loc. cit.]

(curve C in Fig. 14a), characterized by the ratio X(t)/Y(t), as the theory predicts. The observed spectrum C was redshifted in comparison with the curves A and B. The experiment was repeated by changing the ratio of X to Y suitably. The observed spectrum, curve C in Fig. 14b, was then blueshifted.

Later several experiments with optical rather than with acoustical waves were performed. Fig. 15 shows the layout of an optical experiment carried out by Professor Gori and his co-workers at the University of Rome in Italy. Two mutually independent beams with nearly the same spectra, were brought together via a beam splitter at two pinholes P_1 and P_2. Light from P_1 and P_2 was partially correlated, because each of the two openings received contribution from both beams B_α and B_β. This is similar to the acoustic experiment described earlier, where the two sources had contributions from two signals X and Y. The contributions at P_1 and P_2 can be easily changed by changing the orientation of the beam splitter. The results of their experiments are shown in Fig. 16. It is seen that with appropriate orientations of the beam splitter, a redshift (curve a) or a blueshift (curve b) were produced. Curve c is the spectrum when only one of the two beams (B_α or B_β) illuminates the two pinholes.

I will now briefly discuss some applications of this effect. One of the most important applications is connected with spectroradiometric scales and a good deal of work in this area has been done here at the National Physical Laboratory in New Delhi. Some of you probably know that there are considerable discrepancies in the standards of spectroradiometric scales which are maintained by different laboratories. Fig. 17 shows the intercomparison of spectral irradiance scales maintained at eight National Standards Laboratories viz. NPL (U.K.), ASMW (G.D.R.), NML (Australia), PTB (Germany), ETL (Japan), NBS (U.S.A.), VNIIOFI (U.S.S.R.) and NRC (Canada). We see that there is a scatter of the order of a few percent between these scales. There have been many arguments about the origin of such large discrepancies.

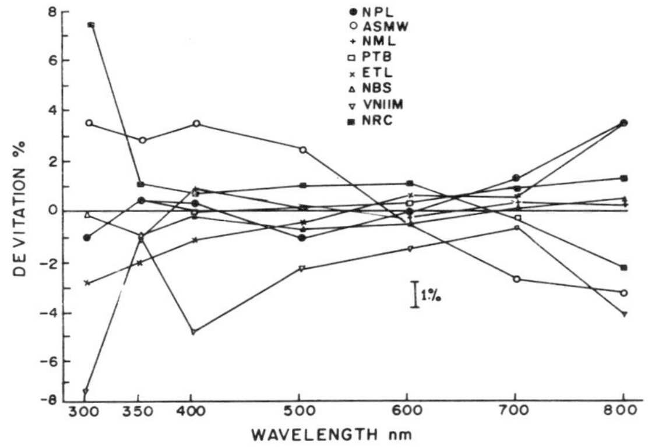

Fig. 17—Comparison of spectral irradiance scales maintained at eight National Standards Laboratories. [After J. R. Moore, Lighting Res. and Techn. **12**, 213 (1980)].

Let us consider how such standards are determined. Typically radiation from a source is randomized in an integrating sphere (Fig. 18), to suppress polarization and directional effects. The light which emerges from the optical system is detected by a monochromator or a spectrum analyzer. Now, from

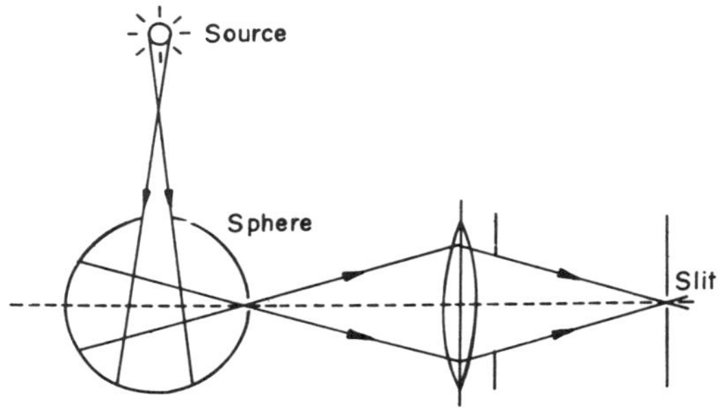

Fig. 18—Experimental arrangement showing the use of an integrating sphere for the measurement of spectral irradiance. [After J. R. Moore, loc..cit.].

our point of view, the crucial thing is that apertures of different sizes and also lenses of different focal lengths for focussing the radiation on a detector or on a monochromator slit are used. It has been realized by Kandpal, Vaishya and Joshi at this laboratory that when the sizes of the apertures and the focal lengths are changed, the coherence properties of the secondary sources are also changed. For example, if we make the aperture sufficiently small, the radiation which emerges from it will be essentially completely spatially coherent. But as the aperture sizes are increased, the secondary source becomes less and less coherent and finally becomes effectively incoherent. This means that by changing the sizes of the aperture and the focal lengths of the lenses one is changing the coherence properties of the secondary source. According to the theory which I have outlined, if the degree of coherence is changed, the spectrum is modified. This is exactly what Kandpal, Vaishya and Joshi have observed. They made measuements using apertures of different diameters. The Table below shows that there is a significant dif-

Spectral Shift Variation with Different-size Apertures

Peak Transmission Wavelength (nm)	*Half Band-width (nm)*	*Shifted Peak Transmission Wavelength (nm) with Secondary Aperture* dia. 2.4 mm	dia. 10 mm
422.0	9	421.0	422.0
484.1	9	483.6	484.1
512.4	5	514.1	512.4
566.1	13	564.1	566.1
609.1	8	610.3	609.1
652.0	8	653.2	652.0

From H. C. Kandpal, J. S. Vaishya and K. C. Joshi, *Opt. Commun*, **73**, 169 (1989).

ference in the peak transmission wavelength with a small and large aperture. Fig. 19 shows that with different sizes of apertures, one obtains different normalized spectral distribution. I believe that this work, which was done here at this laboratory, is of considerable significance and is likely to lead to the resolution of the problems concerning the origin of the differences in the spectroradiometric scales maintained by the different National Laboratories.

Other experiments on the effects of source correlations on spectra of emitted radiation have been carried out in many laboratories. Particularly interesting is the work of Indebetouw of Virginia Polytechnic State University in USA, who found a pow-

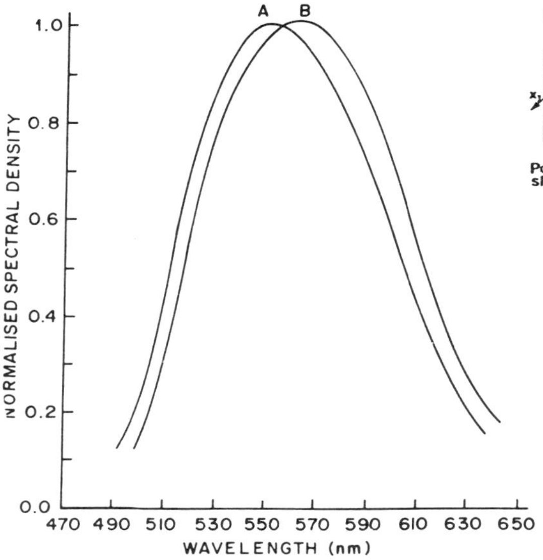

Fig. 19 — Curve labeled A and B are results of measurements of spectral distribution of a lamp — V(λ) filter combination in the presence and the absence of a secondary aperture respectively. [After H. C. Kandpal, J. S. Vaishya and K. C. Joshi, Opt. Commun. **73**, 163 (1989)].

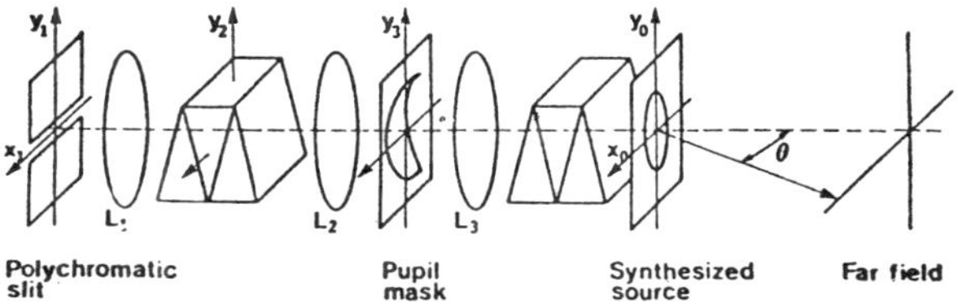

Fig. 20 — Layout of the optical system used for synthesizing secondary sources of prescribed degrees of spectral degrees of coherence. [After G. Indebetouw, J. Mod. Opt. **36**, 251 (1989)].

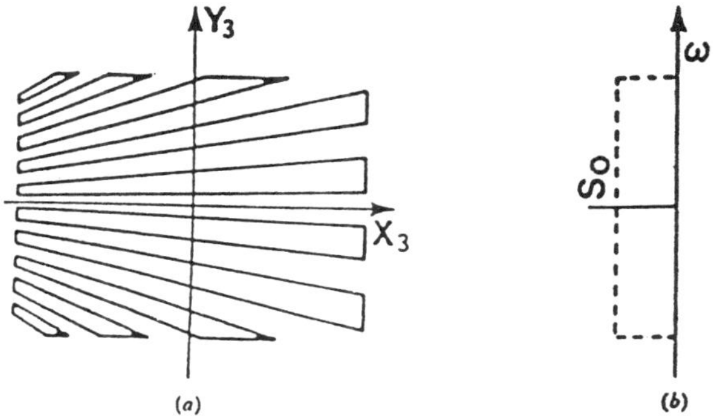

Fig. 21 — A pupil mask [After G. Indebetouw, loc. cit.].

erful way for producing one-dimensional sources of prescribed degrees of coherence. The experimental setup of Indebetouw is indicated in Fig. 20. Light emerging from an incoherently illuminated slit was dispersed by a prism in the x, y-plane, with a pupil mask located at the focal plane of a lens L_2. Images of the slit in different colors are formed at different positions in the plane of the pupil mask. Detailed calculations show that the degree of coherence in the plane of the synthesized source is proportional to the Fourier transform of the square of the transmission function of a mask placed in the pupil plane. Thus an appropriately chosen pupil mask can produce any prescribed degree of coherence at each frequency of the dispersed spectrum. In general, such a source does not satisfy the scaling law [Eq. (15)] and consequently the light which it generates will exhibit a spectral energy redistribution upon propagation. Let us consider as an example a slit aperture illuminated by light with uniform spectrum. It was shown by Indebetouw that by choosing a pupil mask illustrated in Fig. 21 [coded 0, 1, 0, 1 etc] the spectrum of the light emerging from the system will be completely different, showing many maxima and minima (Fig. 22).

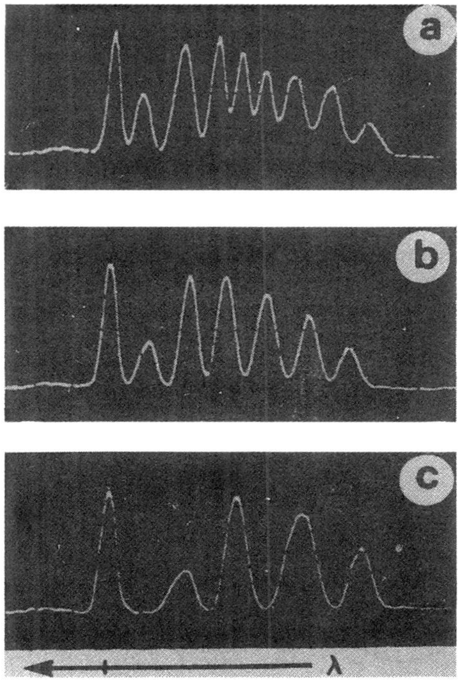

Fig. 22 — Far field spectra of the source synthesized with the pupil mask of Fig. 21, showing a modulation frequency increasing monotonically with the direction of observation θ = 3° (a), θ = 0° (b) and θ = − 3° (c). [After G. Indebetouw, loc. cit].

We have considered so far radiation from planar, secondary sources. Similar results also hold for radiation from three-dimensional primary sources. One can show that the far-zone spectrum $S_V^{(\infty)}(\mathbf{ru}, \omega)$, $(\mathbf{u}^2 = 1)$, of the field is given by

$$S_V^{(\infty)}(\mathbf{r}\mathbf{u},\omega)=\frac{(2\pi)^3}{r^2}VS_Q(\omega)\tilde{\mu}_Q(k\mathbf{u},\omega), \qquad \ldots(23)$$

where $S_Q(\omega)$ is the source spectrum, V is the volume of the source and

$$\tilde{\mu}_Q(\mathbf{K},\omega)=\frac{1}{(2\pi)^3}\int_V \mu_Q(\mathbf{r}',\omega)e^{-i\mathbf{K}\cdot\mathbf{r}'}d^3r'. \qquad \ldots(24)$$

It is seen from equation (23) that the far-zone spectrum again depends on the degree of coherence $\mu_Q(\mathbf{r}',\omega)$ of the source.

Let me illustrate the formula (23) by a simple example, namely by radiation from a source with a line of Gaussian profile, having a Gaussian degree of coherence. The far field spectrum will depend on the spectral coherence length ζ of the source, as shown in Fig. 23. All these spectra have been normalized.

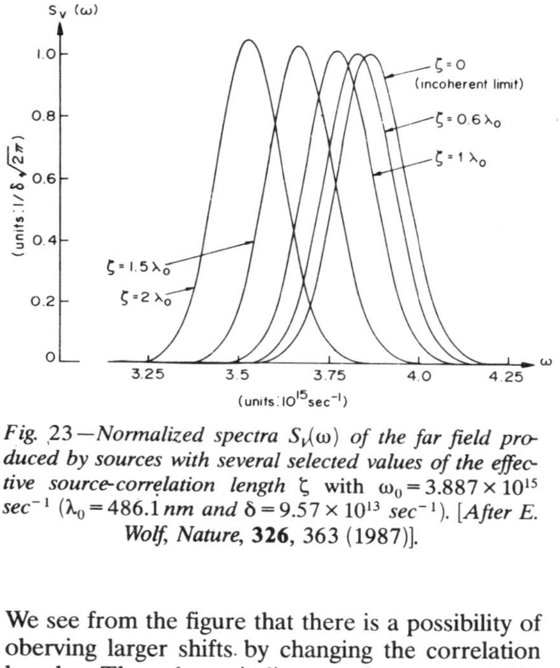

Fig. 23—Normalized spectra $S_V(\omega)$ of the far field produced by sources with several selected values of the effective source-correlation length ζ with $\omega_0=3.887\times10^{15}$ sec^{-1} ($\lambda_0=486.1$ nm and $\delta=9.57\times10^{13}$ sec^{-1}). [After E. Wolf, Nature, **326**, 363 (1987)].

We see from the figure that there is a possibility of oberving larger shifts by changing the correlation length. The data indicates that a redshift $z=\Delta\lambda/\lambda=0.054$ will occur for a soruce having correlation length of $\zeta=1.5\lambda$. Now we may ask how fast would an incoherent source have to recede from the observer in order to produce such a redshift. According to the Doppler formula, the speed of recession v, when $z=0.054$, would be 16,200 km/sec. By Hubble's law this source would then be at a distance of about 324 megaparsec. So we have a situation here where the source is completely at rest with respect to the observer; and yet, because of its correlation properties, gives rise to a shift in its spectrum, which will be exactly the same as if the source was completely incoherent and was moving away at the above speed, and would also be very distant. This shows that it may be necessary in some cases to know the coherence properties of the source for proper interpretation of observations relating to the shifts of spectral lines.

So far I have talked mainly about secondary sources, usual types essentially illuminated apertures. Let me now say a few words about completely different types of secondary sources. Suppose that we have a medium whose refractive index or, equivalently, its dielectric susceptibility, vary randomly in space. Suppose that the medium is illuminated by a plane wave. Let us examine the scattered field in differnet directions. We can regard the illuminated body as a source of radiation, because under the influence of the incident electric field the electrons in the atoms in the medium are displaced from their equilibrium positions and will oscillate. The polarization induced in this way in the scattering medium will, in general, be correlated over finite distances and will imitate correlations in primary or in secondary sources. So this is another secondary source, of a very different kind than we have considered so far. If one works out the theory in detail one finds that the spectrum in the far field, $S_V^{(\infty)}(\mathbf{r}\mathbf{u},\omega)$, depends on the spectrum of the incident light $S_V^{(i)}(\omega)$ and on the correlation function $C_\eta(\mathbf{R},\omega)$ of the dielectric susceptibility, $\eta(\mathbf{r},\omega)$, according to the formula

$$S_V^{(\infty)}(\mathbf{r}\mathbf{u},\omega)=\frac{k^4\sin^2\theta}{\mathbf{r}^2}\tilde{C}_\eta(k\mathbf{u}-k\mathbf{u}_0,\omega)S_V^{(i)}(\omega), \qquad \ldots(25)$$

where

$$\tilde{C}_\eta(\mathbf{K},\omega)=\frac{1}{(2\pi)^3}\int_V C_\eta(\mathbf{R},\omega)e^{-\mathbf{K}\cdot\mathbf{R}}d^3R \qquad \ldots(26)$$

and

$$C_\eta(\mathbf{R},\omega)=\langle\eta^*(\mathbf{r},\omega)\eta(\mathbf{r}+\mathbf{R},\omega)\rangle. \qquad \ldots(27)$$

Here $\mathbf{u}_0$ and $\mathbf{u}$ are real unit vectors in the direction of propagation of the incident and the scattered field respectively. We may call this type of scattering *static scattering*, because nothing moves at the macroscopic level. The scattering arises from the inhomogeneities in the macroscopic physical properties of the medium.

The formula (25) shows that even on static scattering the spectrum of the scattered light differs, in general, from the spectrum of the incident light.

Let us compare the frequency shifts which can be produced by source correlation and by static scattering. The structures of the two formulas Eqs. (23) and (25) are seen to be similar. If we look at the former [the spectrum of the field generated by a fluctuating source—Eq. (23)] we see that the spectrum of the far field is modified by the correlation function of the source fluctuations. According to Eq. (25) pertaining to scattering, the spectrum of the scattered field is modified by the presence of correlations of the dielectrc susceptibility at different points in the scattering medium. Since both are product relations, this mechanism does not introduce new frequencies (i.e. $S_V^{(\infty)}(\mathbf{ru}, \omega) = 0$ whenever $S_V^{(i)}(\omega) = 0$). Moreover, the induced spectral shifts cannot exceed the effective width of the line. It is also clear that if the spectrum has several lines, their z numbers would, in general, be different i.e. they would be shifted by different relative amounts (different z-numbers).

The situation is different with dynamic scattering. For example for Brillouin scattering (Fig. 24) a spectral line of the incident light splits into three lines. In addition to the original frequency ω_0, we have also, two other lines, called the Brillouin doublet, at frequencies $\omega_0 \pm \Delta\omega$, where

$$\frac{\Delta\omega}{\omega} = 2n\left(\frac{v}{c}\right)\sin\frac{\theta}{2}, \qquad \ldots(28)$$

n being the refractive index, v the speed of the sound wave and θ the angle of scattering. One can produce other and possibly rather drastic spectral changes by other types of dynamic scattering. For our purposes the essential thing to note is that as a result of interaction of radiation with a medium whose physical parameters vary with time new frequencies can be generated. The theory of scattering of light of any state of coherence by media whose physical properties vary randomly both in space and in time has been worked out only quite recently. It was found that for a broad class of scatterers, the far field spectrum is given by the following expression:

$$S^{(\infty)}(\omega'; \mathbf{ru}, \mathbf{u}') = A\omega'^4 \int_{-\infty}^{\infty} S\left(\frac{\omega'}{c}\mathbf{u}' - \frac{\omega}{c}\mathbf{u}, \omega' - \omega; \omega\right) S^{(i)}(\omega)\, d\omega, \qquad \ldots(29)$$

where

$$S(\mathbf{K}, \Omega; \omega) = \frac{1}{(2\pi)^4}\int d^3R \int_{-\infty}^{\infty} dT\, C_\eta(\mathbf{R}, T; \omega) e^{-i(\mathbf{K}.\mathbf{R} - \Omega T)} \qquad \ldots(30)$$

and

$$C_\eta(\mathbf{R}, T; \omega) = \langle \eta^*(\mathbf{r}, t; \omega)\eta(\mathbf{r} + \mathbf{R}, t + T; \omega)\rangle \qquad \ldots(31)$$

Here $S^{(i)}(\omega)$ is the spectrum of the incident wave in a direction specified by a unit vector **u** and **u**′ is the unit vector of the direction of the scattered wave. C_η is the correlation function of the dielectric susceptibility at two space-time points; it is an analogue of the so-called van Hove correlation function encountered in neutron scattering and in plasma physics. We see from Eq. (29) that the far field spectrum depends on the correlation of the dielectric susceptibility at two space-time points. It is to be also noted that Eq. (29) is neither a product relation nor a convolution. Consequently one can show that this type of scattering can modify the spectrum of the scattered field very drastically, in some cases. In Fig. 25, a laboratory spectrum which consists of a doublet and the corresponding spectrum calculated with Doppler's formula are shown, assuming that the source is moving away from the observer with a speed of v = 0.0714c. It is compared with the spectrum obtained by scattering on a fluctuating medium having suitable space-time correlations. We see that

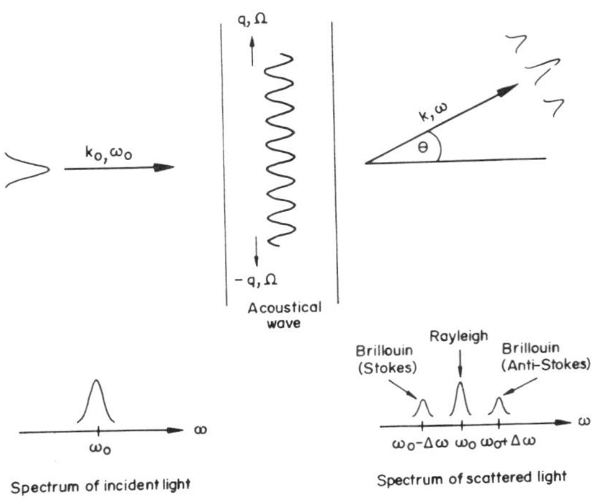

Fig. 24—Brillouin Scattering

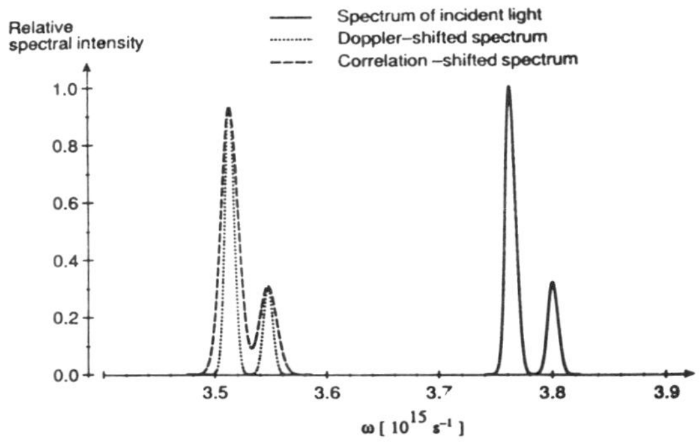

Fig. 25—Two O_{III} lines (λ = 4959 Å and 5007 Å) as seen at rest (solid line), Doppler shifted (dotted line) and shifted due to scattering on suitably correlated media (dashed line). [After D. F. V. James, M. P. Savedoff and E. Wolf, The Astrophys J. **359**, 67 (1990)]

in this case the scattering completely imitates the Doppler effect, with both shifted lines having the same z-number, even though the source, the scattering medium and the observer are at rest relative to each other. We also note that the shift is many times larger than the width of the lines.

As we have seen spectral changes may be generated as a result of correlations in primary sources and also in secondary sources such as illuminated apertures. We have also seen that spectral changes may be produced by scattering on suitably correlated media. A number of applications based on this phenomenon have already been noted, such as the possibility of modulating signals by controlling source correlations, and are pointing the way towards improvement in the accuracy of spectroradiometric scales. Applications in the area of enhanced backscattering are also currently being considered. In view of the results we have discussed the usual procedure of satellite tracking may perhaps require more careful analysis than is customarily made. In the prevailing practice, the satellites are tracked by monitoring laser light which is reflected from them and one estimates the speed of the satellite from the observed spectral shift by the use of the Doppler formula. However, the effect of correlations which might be induced in the reflected light by surface roughness could also cause spectral changes, which are not currently taken into account. Another potential application is in the field of atmospheric turbulence. One could measure the spectral changes produced in single or double passage of light through the turbulent medium; by inversion one would be able to obtain some information about the correlation function of the medium.

It is, of course, tempting to reflect on the possibility that this new mechanism for producing shifts of spectral lines might perhaps lead to the solution of some of the astronomical puzzles which are at the center of the redshift controversy. However, much more work needs to be done before one could claim with confidence that the theory I described may significantly contribute to the resolution of this controversy.

7 December 1998

PHYSICS LETTERS A

ELSEVIER

Physics Letters A 249 (1998) 243–247

Remarks on boundary conditions for scalar scattering

T.D. Visser [1], P.S. Carney, E. Wolf

Department of Physics and Astronomy and Rochester Theory Center for Optical Science and Engineering, University of Rochester, Rochester, NY 14627, USA

Received 6 August 1998; accepted for publication 8 September 1998
Communicated by P.R. Holland

Abstract

The question is discussed whether potential scattering problems can be treated as boundary value problems associated with differential equations, as is sometimes suggested in the literature. We show that, except in some very special cases, this is not possible. The values of the wave function and its normal derivative on the boundary of a finite-range potential cannot be prescribed arbitrarily but are implicit in the integral equation of potential scattering. We derive two coupled singular integral equations for the boundary values for the case when the scattering potential is homogeneous. © 1998 Published by Elsevier Science B.V.

PACS: 03.65.Nk; 03.80.+r; 41.20.Cv; 42.25.Fx
Keywords: Potential scattering; Nonrelativistic scattering; Boundary values in scattering; Surface integral equations in scattering

1. Introduction

Potential scattering problems, whether in quantum mechanics, optics, acoustics or in other fields are generally treated by means of integral equations. For relatively simple situations, such as for scattering of a plane wave by a homogeneous sphere, alternative methods are available, e.g. expanding the solution formally both inside and outside the scatterer in terms of modes and determining the coefficients of the modes by matching the two expansions at the boundary, making use of the (assumed) continuity of the wave function and of its normal derivative.

In contrast to these well-known standard approaches, several problems have been treated in the literature as true boundary value problems of differential equations. Such problems are basically limited to scattering by a hard sphere and a soft sphere in acoustics. The hard sphere can be regarded as the limiting case of an infinitely strong scattering potential. For the soft sphere, however, it is not clear whether there exists an equivalent scattering potential. Perhaps because of the success of the boundary value approach in these two special cases, the impression has been given in the literature that more general scattering problems can also be treated in this way [1,2]. However, it is well known that the values of the wavefunction on the boundary of a finite-range potential are implicitly contained in the integral equation of potential scattering and they cannot, therefore, be specified a priori.

In this Letter we show that for the case of scattering from a homogeneous finite-range potential, the values of the wavefunction and of its normal derivative on

[1] On leave from Department of Physics and Astronomy, Free University, Amsterdam, The Netherlands.

PII S0375-9601(98)00705-1

the boundary of the scatterer satisfy a coupled pair of singular integral equations, indicating more explicitly than is apparent from the integral equation that the boundary values cannot be prescribed arbitrarily.

2. Potential scattering of scalar waves

Let us consider a monochromatic plane wave of unit amplitude and frequency ω, propagating in the direction specified by a unit vector s_0,

$$V^{(i)}(\boldsymbol{r},t) = U^{(i)}(\boldsymbol{r})\exp(-\mathrm{i}\omega t), \tag{2.1}$$

$$U^{(i)}(\boldsymbol{r}) = \mathrm{e}^{\mathrm{i}k\boldsymbol{s}_0\cdot\boldsymbol{r}}, \tag{2.2}$$

incident on a scattering medium occupying a volume V, bounded by a closed surface S in free space (see Fig. 1). In Eq. (2.2) $k = \omega/c = 2\pi/\lambda_0$ is the wavenumber, c being the speed of light in vacuum and λ_0 the vacuum wavelength. We assume that the macroscopic physical properties of the scatterer are independent of time. We denote by $n(\boldsymbol{r})$ the refractive index of the medium at frequency ω. In general it is a function of position, specified by the position vector $\boldsymbol{r}$. The total field (i.e. the sum of the incident and the scattered field) generated by the interaction of the incident field with the scatterer satisfies the basic equation of potential theory [3,4]

$$(\nabla^2 + k^2)U(\boldsymbol{r}) = -4\pi F(\boldsymbol{r})U(\boldsymbol{r}), \tag{2.3}$$

where

$$F(\boldsymbol{r}) \equiv \frac{k^2}{4\pi}[n^2(\boldsymbol{r}) - 1] \tag{2.4}$$

is the so-called scattering potential. The solution of Eq. (2.3) for the total field must behave far away from

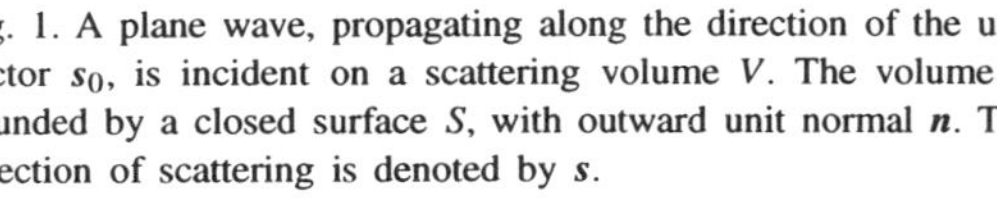

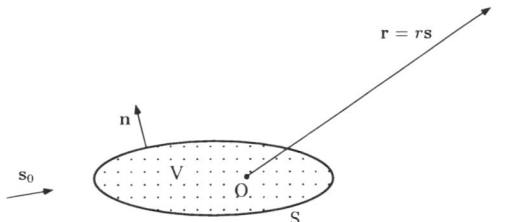

Fig. 1. A plane wave, propagating along the direction of the unit vector $\boldsymbol{s}_0$, is incident on a scattering volume V. The volume is bounded by a closed surface S, with outward unit normal $\boldsymbol{n}$. The direction of scattering is denoted by $\boldsymbol{s}$.

the scatterer as the sum of the incident plane wave and an outgoing spherical wave, i.e. it must have the asymptotic behavior

$$U(r\boldsymbol{s}) \sim \mathrm{e}^{\mathrm{i}k\boldsymbol{s}_0\cdot\boldsymbol{r}} + f(\boldsymbol{s},\boldsymbol{s}_0)\frac{\mathrm{e}^{\mathrm{i}kr}}{r}, \tag{2.5}$$

as $kr \to \infty$, with the direction of scattering, characterized by the unit vector $\boldsymbol{s}$, kept fixed, f being the scattering amplitude.

The solution of Eq. (2.3), subject to the usual assumption that $U(\boldsymbol{r})$ and its normal derivative $\partial U/\partial n \equiv \boldsymbol{n}\cdot\boldsymbol{\nabla}U$ at the boundary S of the scattering volume are continuous and subject to the asymptotic condition (2.5) represents the total field $U(\boldsymbol{r})$ generated by scattering. (A generalization of the theory which takes into account possible discontinuities at the boundaries has recently been discussed in Ref. [5]. It indicates that under certain circumstances such discontinuities can have significant effects on the far field.)

As is well known, Eq. (2.3) subject to the continuity conditions, may be recast into the integral equation [4,6]

$$U(\boldsymbol{r}) = U^{(i)}(\boldsymbol{r}) + \int_V F(\boldsymbol{r}')U(\boldsymbol{r}')\frac{\mathrm{e}^{\mathrm{i}k|\boldsymbol{r}-\boldsymbol{r}'|}}{|\boldsymbol{r}-\boldsymbol{r}'|}\,\mathrm{d}^3r', \tag{2.6}$$

where

$$G(\boldsymbol{r}-\boldsymbol{r}') = \frac{\mathrm{e}^{\mathrm{i}k|\boldsymbol{r}-\boldsymbol{r}'|}}{|\boldsymbol{r}-\boldsymbol{r}'|}. \tag{2.7}$$

is the outgoing free space Green's function of the Helmholtz operator $\mathcal{L} \equiv \nabla^2 + k^2$.

There is a different class of scattering problems which is sometimes discussed in the literature (see, for example, Ref. [1]). For such problems one *prescribes* boundary conditions on the surface S of the scatterer and one seeks a solution which behaves as an outgoing spherical wave at infinity. Problems of this kind are treated most frequently for a "hard" and a "soft" sphere [7]. In the former case one imposes the Dirichlet boundary condition $U(\boldsymbol{r}) = 0$ on S; in the latter case one imposes the Neumann boundary condition $\partial U(\boldsymbol{r})/\partial n = 0$ on S [8]. It is sometimes suggested that many more problems of scattering from a bounded medium may also be treated as boundary value problems [2,7]. However, the physical significance and even the possibility of such an approach is obscure at best.

The solutions to the potential scattering problem and to a boundary value problem have certain features in common such as the asymptotic behavior expressed by relation (2.5) and results such as certain reciprocity theorems and the optical cross-section theorem. However, the boundary value formulation does not explicitly take into account the nature of the medium, characterized by the scattering potential. Although, in principle, boundary values are, of course, associated with any potential scattering problem, knowledge of them can only be obtained by solving them first by other methods, e.g. by the use of the integral equation of potential scattering or by solving a pair of singular integral equations for the boundary values and their normal derivatives as we will now show.

3. Boundary values for the field and its normal derivative

In this section we derive integral equations for the values of the field and its normal derivative on the boundary of the scatterer. Our starting point is the scalar version of the so-called Ewald–Oseen extinction theorem (see, for instance, Refs. [9,10]).

We consider a homogeneous scattering volume V with constant refractive index n. The scalar version of the extinction theorem is expressed by the formula (see Ref. [11] or Eq. (A.4) of Appendix A of this Letter)

$$U^{(i)}(\boldsymbol{r}_<) = -\frac{1}{4\pi}\int_S \left(U(\boldsymbol{r}')\frac{\partial G(\boldsymbol{r}_< - \boldsymbol{r}')}{\partial n'} - G(\boldsymbol{r}_< - \boldsymbol{r}')\frac{\partial U(\boldsymbol{r}')}{\partial n'} \right) \mathrm{d}S. \tag{3.1}$$

Here $\boldsymbol{r}_<$ is any point inside the scatterer, and G is the outgoing free space Green's function, given by Eq (2.7). As the point $\boldsymbol{r}_<$ approaches the boundary S, with $\boldsymbol{r}_s$ denoting a point on the surface S, the first term of the integrand of Eq. (3.1) becomes singular when $\boldsymbol{r}' = \boldsymbol{r}_s$. A similar expression occurs also in potential theory. The result is [12,13]

$$U^{(i)}(\boldsymbol{r}_s) = \tfrac{1}{2}U(\boldsymbol{r}_s) - \frac{1}{4\pi}\mathrm{P}\int_S \left(U(\boldsymbol{r}')\frac{\partial G(\boldsymbol{r}_s - \boldsymbol{r}')}{\partial n'} - G(\boldsymbol{r}_s - \boldsymbol{r}')\frac{\partial U(\boldsymbol{r}')}{\partial n'} \right) \mathrm{d}S, \tag{3.2}$$

where P denotes the Cauchy principal value taken at $\boldsymbol{r}' = \boldsymbol{r}_s$.

The field inside the scatterer satisfies the equation

$$(\nabla^2 + k_1^2)U(\boldsymbol{r}) = 0, \tag{3.3}$$

where $k_1 = 2\pi n/\lambda_0$ (with λ_0 denoting the wavelength in vacuo) is the wave number of the field inside the scatterer. Let G_1 be a Green's function of the operator $\mathcal{L}_1 \equiv \nabla^2 + k_1^2$. It satisfies the differential equation

$$(\nabla^2 + k_1^2)G_1(\boldsymbol{r} - \boldsymbol{r}') = -4\pi\delta^{(3)}(\boldsymbol{r} - \boldsymbol{r}'). \tag{3.4}$$

By a trivial generalization of the Helmholtz–Kirchhoff integral theorem for free space (see Section 8.3 of Ref. [4]) to a homogeneous material medium we obtain the formula

$$U(\boldsymbol{r}_<) = -\frac{1}{4\pi}\int_S \left(U(\boldsymbol{r}')\frac{\partial G_1(\boldsymbol{r}_< - \boldsymbol{r}')}{\partial n'} - G_1(\boldsymbol{r}_< - \boldsymbol{r}')\frac{\partial U(\boldsymbol{r}')}{\partial n'} \right) \mathrm{d}S. \tag{3.5}$$

As the point $\boldsymbol{r}_<$ approaches the boundary surface S, one obtains in a similar manner as in connection with Eq. (3.2)

$$U(\boldsymbol{r}_s) = \tfrac{1}{2}U(\boldsymbol{r}_s) - \frac{1}{4\pi}\mathrm{P}\int_S \left(U(\boldsymbol{r}')\frac{\partial G_1(\boldsymbol{r}_s - \boldsymbol{r}')}{\partial n'} - G_1(\boldsymbol{r}_s - \boldsymbol{r}')\frac{\partial U(\boldsymbol{r}')}{\partial n'} \right) \mathrm{d}S. \tag{3.6}$$

Eqs. (3.2) and (3.6) are two coupled singular integral equations for the field U and its normal derivative $\partial U/\partial n$ on the boundary of the scatterer. We note that these equations imply, if we recall that the scattered field $U^{(s)} = U - U^{(i)}$, that at any point $\boldsymbol{r}_s$ on the boundary of the scatterer,

$$U^{(s)}(\boldsymbol{r}_s) = \frac{1}{4\pi}\mathrm{P}\int_S \left(U(\boldsymbol{r}')\frac{\partial \mathcal{G}(\boldsymbol{r}_s - \boldsymbol{r}')}{\partial n'} - \mathcal{G}(\boldsymbol{r}_s - \boldsymbol{r}')\frac{\partial U(\boldsymbol{r}')}{\partial n'} \right) \mathrm{d}S, \tag{3.7}$$

where

$$\mathcal{G}(\boldsymbol{r}-\boldsymbol{r}') = G(\boldsymbol{r}-\boldsymbol{r}') - G_1(\boldsymbol{r}-\boldsymbol{r}'). \tag{3.8}$$

This is a necessary condition which the scattered field on the boundary must satisfy.

From Eq. (3.6) or (3.7) it is seen that neither the field nor its normal derivative can be prescribed arbitrarily at the boundary of the scatterer.

Finally, we mention that a pair of singular integral equations for the value of an electromagnetic field and of its normal derivative at the boundary, rather than for a scalar field, was derived by Pattanayak [14].

4. Conclusions

We have shown that, contrary to suggestions often made in the literature, potential scattering problems involving finite-range potentials cannot be formulated as boundary value problems of differential equations, except in some very special cases. In fact, to determine the boundary values which the field and its normal derivative take on the boundary of the scatterer, one must first solve the scattering problem by other methods (e.g. by using the integral equation of potential scattering) which yields the boundary values as a byproduct. We have derived a coupled pair of singular integral equations (Eqs. (3.2) and (3.6)) which the field and its normal derivative must satisfy on the boundary of a homogeneous scatterer.

Acknowledgement

This research was supported by the National Science Foundation and the New York State Foundation for Science and Technology.

Appendix A. The extinction theorem for scalar fields

In this appendix we present a simple derivation of the scalar version of the extinction theorem in the form originally derived by Pattanayak and Wolf [11]. We recall Eq. (2.3),

$$(\nabla^2 + k^2)U(\boldsymbol{r}) = -4\pi F(\boldsymbol{r})U(\boldsymbol{r}), \tag{A.1}$$

and the fact that the outgoing free-space Green's function G satisfies the equation

$$(\nabla^2 + k^2)G(\boldsymbol{r}-\boldsymbol{r}') = -4\pi\delta^{(3)}(\boldsymbol{r}-\boldsymbol{r}'). \tag{A.2}$$

We interchange the role of $\boldsymbol{r}$ and $\boldsymbol{r}'$, multiply Eq. (A.1) by G, Eq. (A.2) by U, subtract the resulting equations, integrate over the scattering volume V and use Green's theorem. We then find that if the point $\boldsymbol{r}$ is within the scattering volume (in which case we again write $\boldsymbol{r}_<$ rather than $\boldsymbol{r}$)

$$\int_S \left(U(\boldsymbol{r}') \frac{\partial G(\boldsymbol{r}_< - \boldsymbol{r}')}{\partial n'} - G(\boldsymbol{r}_< - \boldsymbol{r}') \frac{\partial U(\boldsymbol{r}')}{\partial n'} \right) \mathrm{d}S$$
$$= -4\pi U(\boldsymbol{r}_<) + 4\pi \int_V F(\boldsymbol{r}')U(\boldsymbol{r}') \frac{\mathrm{e}^{\mathrm{i}k|\boldsymbol{r}_< - \boldsymbol{r}'|}}{|\boldsymbol{r}_< - \boldsymbol{r}'|} \mathrm{d}^3\boldsymbol{r}'. \tag{A.3}$$

The volume integral on the right of Eq. (A.3) represents $U^{(s)}(\boldsymbol{r}_<) = U(\boldsymbol{r}_<) - U^{(i)}(\boldsymbol{r}_<)$, where $U^{(s)}(\boldsymbol{r}_<)$ is the scattered field (cf. Eq. (2.6)). Hence Eq. (A.3) may be expressed in the form

$$U^{(i)}(\boldsymbol{r}_<) = -\frac{1}{4\pi} \int_S \left(U(\boldsymbol{r}') \frac{\partial G(\boldsymbol{r}_< - \boldsymbol{r}')}{\partial n'} \right.$$
$$\left. - G(\boldsymbol{r}_< - \boldsymbol{r}') \frac{\partial U(\boldsymbol{r}')}{\partial n'} \right) \mathrm{d}S, \tag{A.4}$$

which is the extinction theorem, Eq. (3.1). It shows that the values which U and $\partial U/\partial n$ take on the boundary are such that the incident field is extinguished at every point $\boldsymbol{r}_<$ inside the scattering volume.

References

[1] P.M. Morse, K.U. Ingard, Theoretical Acoustics (Princeton Univ. Press, Princeton, 1968) pp. 410, 419.
[2] M. Kerker, The Scattering of Light and other Electromagnetic Radiation (Academic Press, New York, 1969) p. 430.
[3] P. Roman, Advanced Quantum Theory, Ch. 3 (Addison-Wesley, Reading, 1965).
[4] M. Born, E. Wolf, Principles of Optics, 7th Ed., Section 13.1 (Cambridge Univ. Press, Cambridge), in press.
[5] T.D. Visser, E. Wolf, Phys. Lett. A 234 (1997) 1; 237 (1998) 389 (E).
[6] A. Sommerfeld, Wave Mechanics (Dutton, New York, 1929) pp. 43, 195.

[7] H.W. Wyld, Mathematical Methods for Physics (Addison-Wesley, Reading, 1976) pp. 197–203.

[8] E. Butkov, Mathematical Physics (Addison-Wesley, Reading, 1968) p. 323.

[9] E. Wolf, in: Coherence and Quantum Optics, Proc. of the Third Rochester Conference on Coherence and Quantum Optics, eds. L. Mandel, E. Wolf (Plenum Press, New York, 1973) p. 339.

[10] H. Fearn, D.V.F. James, P.W. Milonni, Am. J. Phys. 64 (1996) 986.

[11] D.N. Pattanayak, E. Wolf, Phys. Rev. D 13 (1976) 913.

[12] O.D. Kellogg, Foundations of Potential Theory (Springer, Berlin, 1967) pp. 215, 160–172.

[13] H. Hönl, A.W. Maue, K. Westpfahl, in: Handbuch der Physik, Band XXV/1 (Springer, Berlin, 1961) pp. 233–236.

[14] D.N. Pattanayak, Opt. Comm. 15 (1975) 335.

Diffraction tomography using power extinction measurements

P. Scott Carney and E. Wolf

Department of Physics and Astronomy and the Rochester Theory Center for Optical Science and Engineering, University of Rochester, Rochester, New York 14627

G. S. Agarwal

Physical Research Laboratory, Ahmedabad 38009, India

Received February 3, 1999; revised manuscript received June 7, 1999; accepted June 24, 1999

We propose a new method for determining structures of semitransparent media from measurements of the extinguished power in scattering experiments. The method circumvents the problem of measuring the phase of the scattered field. We illustrate how this technique may be used to reconstruct both deterministic and random scatterers. © 1999 Optical Society of America [S0740-3232(99)00211-2]

OCIS codes: 290.3200, 110.6960.

1. INTRODUCTION

In the short-wavelength limit, tomographic reconstruction of two- and three-dimensional media has long been carried out from intensity measurements. More accurate methods of reconstruction that take into account diffraction[1] require knowledge of both the field amplitude and the phase of the scattered field. For rapidly varying fields such as optical fields the phase may be prohibitively difficult to measure and presents, at best, a technical challenge at lower frequencies. In this paper we propose a method to circumvent the phase problem. We will show that one can determine a function that is related to the scattering amplitude and makes it possible to reconstruct the scattering object for certain model media.

We begin by recalling a well-known result in scattering theory, the optical cross-section theorem.[2] It relates the total power extinguished from a plane wave on scattering to the scattering amplitude in the forward (incident) direction. More explicitly, let

$$\Psi^{(i)}(\mathbf{r}, t) = \psi^{(i)}(\mathbf{r})\exp(-i\omega t) \tag{1.1}$$

be a monochromatic field incident on the scatterer. We assume that it is a plane wave that propagates in the direction of a unit vector $\mathbf{s}_0$,

$$\psi^{(i)}(\mathbf{r}) = a\exp(ik\mathbf{r}\cdot\mathbf{s}_0), \tag{1.2}$$

with $k = \omega/c$, c being the speed of light in vacuum. Let

$$\Psi^{(s)}(\mathbf{r}, t) = \psi^{(s)}(\mathbf{r})\exp(-i\omega t) \tag{1.3}$$

represent the scattered wave. The total field [with time dependence $\exp(-i\omega t)$ being omitted from now on] is then given by the expression

$$\psi(\mathbf{r}) = \psi^{(i)}(\mathbf{r}) + \psi^{(s)}(\mathbf{r}). \tag{1.4}$$

In the far zone in a direction specified by the unit vector $\mathbf{s}$, the scattered field has the asymptotic form

$$\psi^{(s)}(r\mathbf{s}) \sim a\frac{\exp(ikr)}{r}f(\mathbf{s}, \mathbf{s}_0), \tag{1.5}$$

$f(\mathbf{s}, \mathbf{s}_0)$ being the so-called scattering amplitude.

The total power extinguished from the incident field as a result of scattering and absorption is given by the formula

$$P^{(e)} = |a|^2\frac{4\pi}{k}\Im f(\mathbf{s}_0, \mathbf{s}_0), \tag{1.6}$$

where $\Im$ denotes the imaginary part. In general, one needs to know the scattering amplitude for all directions of incidence and scattering in order to reconstruct the low-pass-filtered version of the scattering object; however, equation (1.6) gives information only about the imaginary part of the scattering amplitude $f(\mathbf{s}, \mathbf{s}_0)$ in the forward direction $\mathbf{s} = \mathbf{s}_0$. Within the accuracy of the first-order Born approximation, $f(\mathbf{s}_1, \mathbf{s}_2)$ is related to the Fourier transform of the susceptibility $\eta(\mathbf{r})$ of the medium by the formula[1]

$$f(\mathbf{s}_1, \mathbf{s}_2) = k^2\int \eta(\mathbf{r})\exp[-ik\mathbf{r}\cdot(\mathbf{s}_2 - \mathbf{s}_1)]\mathrm{d}^3r, \tag{1.7}$$

and consequently Eq. (1.6) yields information only about the volume integral of the imaginary part of the susceptibility of the scattering object.

We will make use of a recent generalization of the optical cross-section theorem to introduce a method of determining a complex function that is related to the scattering amplitude of the object whose structure is to be determined. It is possible to determine this function experimentally from measurements of power alone. In many cases this function is simply related to the structure of the object.

2. THE DATA FUNCTION

Let us consider the power extinguished from a coherent beam consisting of two monochromatic plane waves,

$$\psi^{(i)}(\mathbf{r}) = a_1 \exp(ik\mathbf{r}\cdot\mathbf{s}_1) + a_2 \exp(ik\mathbf{r}\cdot\mathbf{s}_2), \quad (2.1)$$

propagating in the directions specified by the unit vectors $\mathbf{s}_1$ and $\mathbf{s}_2$. In this case the extinguished power, $P^{(e)}(a_1, a_2)$, is given by the expression[3]

$$P^{(e)}(a_1, a_2) = \frac{4\pi}{k}\Im[|a_1|^2 f(\mathbf{s}_1, \mathbf{s}_1) + a_1^* a_2 f(\mathbf{s}_1, \mathbf{s}_2) + a_2^* a_1 f(\mathbf{s}_2, \mathbf{s}_1) + |a_2|^2 f(\mathbf{s}_2, \mathbf{s}_2)]. \quad (2.2)$$

By making two measurements of the total extinguished power with different relative phases between the two incident waves, one can determine the cross terms by using the formula

$$P^{(e)}(a_1, a_2) - P^{(e)}(a_1, -a_2) = \frac{8\pi}{k}\Im[a_1^* a_2 f(\mathbf{s}_1, \mathbf{s}_2) + a_2^* a_1 f(\mathbf{s}_2, \mathbf{s}_1)]. \quad (2.3)$$

By making two additional measurements with plane waves of different relative phases, one can determine the quantity

$$D(\mathbf{s}_1, \mathbf{s}_2) = \frac{k}{8\pi a_1^* a_2}\{P^{(e)}(a_1, ia_2) - P^{(e)}(a_1, -ia_2) + i[P^{(e)}(a_1, a_2) - P^{(e)}(a_1, -a_2)]\}. \quad (2.4)$$

We will refer to $D(\mathbf{s}_1, \mathbf{s}_2)$ as the data function. It may be seen that the data function is related to the scattering amplitude by the expression

$$D(\mathbf{s}_1, \mathbf{s}_2) = f(\mathbf{s}_1, \mathbf{s}_2) - f^*(\mathbf{s}_2, \mathbf{s}_1). \quad (2.5)$$

3. DETERMINATION OF OBJECT STRUCTURE FROM THE DATA FUNCTION

We will now describe a method for determining the structure of the object from knowledge of the data function. While it is, in principle, possible to determine the exact structure of the object from complete knowledge of the scattering amplitude (or even from knowledge of the scattering amplitude over a continuous segment of the scattering angles) e.g., by use of the three-dimensional Marchenko method,[4] to do so presents a computational challenge in the volume of calculations required as well as in the regularization of the data and convergence problems. It seems unlikely that such a technique would be useful in practice. However, as we will see, the data function provides sufficient information to calculate the structure function, at least for some scattering media.

We assume that the data function has been determined by continuous sampling of all available real directions of propagation, i.e., that $D(\mathbf{s}_1, \mathbf{s}_2)$ is known for all values of the real unit vectors $\mathbf{s}_j$ ($s_{jx}^2 + s_{jy}^2 + s_{jz}^2 = 1, j = 1, 2$, the subscript labeling the Cartesian components).

A. Absorptive Part of the Susceptibility in the First-Order Born Approximation

We consider scattering on a medium with complex dielectric susceptibility $\eta(\mathbf{r})$. The total field (incident plus scattered) satisfies the equation

$$\nabla^2\psi(\mathbf{r}) + k^2\psi(\mathbf{r}) = -4\pi k^2 \eta(\mathbf{r})\psi(\mathbf{r}). \quad (3.1)$$

Within the accuracy of the first-order Born approximation the scattering amplitude is given by the expression

$$f(\mathbf{s}_1, \mathbf{s}_2) = k^2\tilde{\eta}[k(\mathbf{s}_1 - \mathbf{s}_2)], \quad (3.2)$$

where

$$\tilde{\eta}(\mathbf{K}) = \int \eta(\mathbf{r})\exp(-i\mathbf{K}\cdot\mathbf{r})\mathrm{d}^3 r \quad (3.3)$$

is the three-dimensional spatial Fourier transform of the dielectric susceptibility.

Let α be the imaginary part of the generally complex dielectric susceptibility,

$$\alpha(\mathbf{r}) \equiv \Im\eta(\mathbf{r}). \quad (3.4)$$

The absorbed power, often characterized by the absorption cross section, is proportional to the volume integral of $\alpha(\mathbf{r})$, and so we will refer to α as the absorptive part of the susceptibility (Ref. 5, p. 219). The data function is related to the Fourier transform of α by the simple formula

$$D(\mathbf{s}_1, \mathbf{s}_2) = 2ik^2\tilde{\alpha}[k(\mathbf{s}_1 - \mathbf{s}_2)], \quad (3.5)$$

valid for real unit vectors $\mathbf{s}_1$ and $\mathbf{s}_2$, $\tilde{\alpha}(\mathbf{K})$ denoting the three-dimensional spatial Fourier transform of $\alpha(\mathbf{r})$.

We may now reconstruct a low-pass-filtered version of $\alpha(\mathbf{r})$ from the data available within the sphere of radius $2k$ in the Fourier space, centered on the origin. Since the data function D is a function of only the difference of two unit vectors, we will formally integrate out an unnecessary variable. More precisely, we integrate D over the average vector variable $\mathbf{S} = (\mathbf{s}_1 + \mathbf{s}_2)/2$ of the two unit vectors $\mathbf{s}_1$ and $\mathbf{s}_2$, which is orthogonal to the difference vector $\mathbf{s} = \mathbf{s}_1 - \mathbf{s}_2$:

$$\mathcal{D}(\mathbf{s}) = \frac{1}{2\pi}\int_{(2\pi)} \mathrm{d}\phi D[\mathbf{S}(\phi) + \mathbf{s}/2, \mathbf{S}(\phi) - \mathbf{s}/2]. \quad (3.6)$$

Here $\mathbf{S}$ is a vector of length $\sqrt{1 - \mathbf{s}^2/4}$ lying in the plane perpendicular to $\mathbf{s}$ and making an angle ϕ with respect to some arbitrary reference line in the plane.[6] It follows from Eqs. (3.5) and (3.6) that the low-pass (subscript LP) reconstruction of $\alpha(\mathbf{r})$ is given by the formula

$$\alpha_{LP}(\mathbf{r}) = \frac{k}{i16\pi^3}\int_{|\mathbf{s}|\leq 2} \exp(ik\mathbf{r}\cdot\mathbf{s})\mathcal{D}(\mathbf{s})\mathrm{d}^3 s. \quad (3.7)$$

To demonstrate the feasibility of this inverse technique, we first calculate the solution of the direct (forward) problem without any approximation (i.e., to all orders of perturbation) in the form of an infinite series. We then apply the inverse technique described above to that solution.

In Fig. 1, computer simulations of the low-pass reconstruction of the absorptive part of the susceptibility of a homogeneous sphere are shown. We first calculated the scattering amplitude for spheres of various radii and susceptibilities by the method of partial waves, with 35 terms in the series, thus including the effects of multiple scattering (see Appendix A). We then applied the algorithm described above to the calculated data in order to reconstruct the potential. It is seen that for media that are

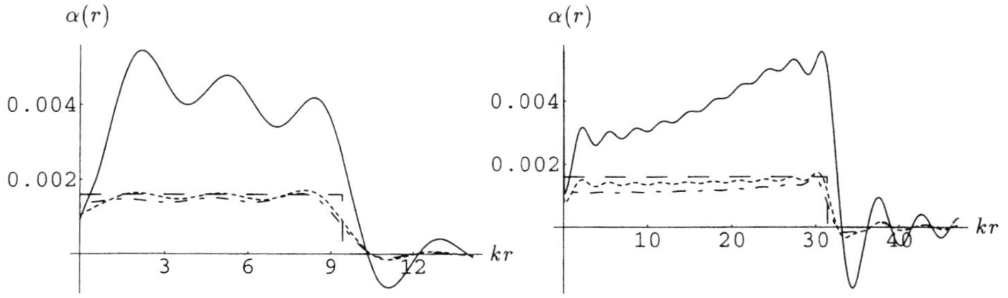

Fig. 1. Left, sphere with size parameter $ka = 3\pi$ (radius $a = 1.5\lambda$); right, sphere of radius $ka = 10\pi$ ($a = 5\lambda$). In all cases $\alpha = 0.01/2\pi$; the original profile is shown by long-dashed curves. Short–long-dashed curves, reconstruction of spheres with susceptibility $\eta = 0.01i/2\pi$; short-dashed curves, reconstruction of spheres with susceptibility $\eta = (0.01 + 0.01i)/2\pi$; solid curves, reconstruction of spheres with susceptibility $\eta = (0.0512 + 0.01i)/2\pi$.

purely absorptive (real part of $\eta = 0$) the reconstruction is quite accurate. The effects of multiple scattering begin to distort the reconstruction as the real part of the susceptibility (representing dispersion) increases. As can be seen by comparing the reconstructions, the size of the scatterer as well as the refractive index determine whether the single-scattering approximation (i.e., the first-order Born approximation) gives reasonably accurate results. This was perhaps to be expected.

B. Spatial Variations of a Dielectric Random Scatterer

Suppose that the scatterer is dielectric (i.e., is nonabsorbing) such that

$$\langle \Im \eta(\mathbf{r}) \rangle = 0, \tag{3.8}$$

the angle brackets denoting the ensemble average. We assume that the fluctuations of the medium are characterized by correlations that have a very short range in comparison with the scale of variations of the average susceptibility. We may then model the scatterer by the quasi-homogeneous approximation.[7–9] The correlation function of the dielectric susceptibility can then be taken to have the form

$$\langle \eta(\mathbf{r}_1)\eta(\mathbf{r}_2)\rangle \equiv C(\mathbf{r}_1, \mathbf{r}_2) \approx \Gamma\left(\frac{\mathbf{r}_1 + \mathbf{r}_2}{2}\right) g(\mathbf{r}_2 - \mathbf{r}_1), \tag{3.9}$$

where g is the degree of correlation normalized so that $g(0) = 1$. We will refer to Γ as the intensity (strength) of the susceptibility, and it may be seen that when $\mathbf{r}_1 = \mathbf{r}_2 = \mathbf{r}$, Γ is given exactly by the expression $\Gamma(\mathbf{r}) = \langle \eta(\mathbf{r})\eta(\mathbf{r})\rangle$. Moreover, $\Gamma(\mathbf{R})$ is assumed to vary much more slowly with $\mathbf{R}$ than $g(\mathbf{r})$ varies with $\mathbf{r}$. In this case the scattering amplitude is given, to leading order in the susceptibility, by the formula[3]

$$f(\mathbf{s}_1, \mathbf{s}_2) = k^4 \tilde{\Gamma}[k(\mathbf{s}_1 - \mathbf{s}_2)] \tilde{\mathcal{G}}\left[-k\left(\frac{\mathbf{s}_1 + \mathbf{s}_2}{2}\right)\right], \tag{3.10}$$

where

$$\mathcal{G}(\mathbf{r}) \equiv g(\mathbf{r}) G(\mathbf{r}), \tag{3.11}$$

$\tilde{\mathcal{G}}(\mathbf{K})$ being the spatial Fourier transform of $\mathcal{G}(\mathbf{r})$, and $G(\mathbf{r})$ is the out-going free-space Green function, viz.,

$$G(\mathbf{r}) = \frac{\exp(ikr)}{r}. \tag{3.12}$$

In this case the data function is given by the expression

$$D(\mathbf{s}_1, \mathbf{s}_2) = 2ik^4 \tilde{\Gamma}[k(\mathbf{s}_1 - \mathbf{s}_2)] \Im \tilde{\mathcal{G}}\left(-k\frac{\mathbf{s}_1 + \mathbf{s}_2}{2}\right). \tag{3.13}$$

By making use of the Fourier representation of the Green function and the convolution theorem, it can be seen that

$$\tilde{\mathcal{G}}(\boldsymbol{\kappa}) = \frac{1}{2\pi^2} \int \frac{\tilde{g}(\boldsymbol{\kappa} - \boldsymbol{\kappa}')}{\kappa'^2 - (k + i\epsilon)^2} \mathrm{d}^3\kappa', \tag{3.14}$$

it being understood that $\epsilon \to 0^+$ on the right-hand side of this equation.

If $\tilde{g}(\boldsymbol{\kappa})$ is isotropic, i.e., if $\tilde{g}(\boldsymbol{\kappa}) \equiv \tilde{g}(|\boldsymbol{\kappa}|)$, one finds that the quantity $\Im \tilde{\mathcal{G}}[-k(\mathbf{s}_2 + \mathbf{s}_2)/2]$ is a simple multiplicative factor in the data function at a fixed frequency and depends only on k. Explicitly,

$$\tilde{\mathcal{G}}(-k\mathbf{s}) = \frac{1}{\pi} \int_0^1 \mathrm{d}x \int_{-\infty}^{\infty} \mathrm{d}\kappa' \kappa'^2 \times \frac{\tilde{g}[(k^2 + \kappa'^2 + 2xk\kappa')^{1/2}]}{\kappa'^2 - (k + i\epsilon)^2}. \tag{3.15}$$

If, in addition, $\tilde{g}[(k^2 + \kappa'^2 + 2xk\kappa')^{1/2}]$ is analytic in κ' in the upper half of the complex κ' plane[10] for all values of x in the range $0 \leqslant x \leqslant 1$, then

$$\tilde{\mathcal{G}}(-k\mathbf{s}) = i \int_{\sqrt{2}}^{2} \mathrm{d}y k y \tilde{g}(ky), \tag{3.16}$$

and, consequently,

$$D(\mathbf{s}_1, \mathbf{s}_2) = 2ik^5 \tilde{\Gamma}[k(\mathbf{s}_1 - \mathbf{s}_2)] \int_{\sqrt{2}}^{2} \tilde{g}(ky) y \mathrm{d}y. \tag{3.17}$$

Let us compare this situation with the case when the medium is delta correlated, i.e., when

$$g(\mathbf{r}) = \delta^{(3)}(\mathbf{r}). \tag{3.18a}$$

Then

$$\tilde{g}(\boldsymbol{\kappa}) = 1. \tag{3.18b}$$

In this case data function can again be shown to be related to a Fourier component of the object. Specifically,

$$D(\mathbf{s}_1, \mathbf{s}_2) = 2ik^5\tilde{\Gamma}[k(\mathbf{s}_1 - \mathbf{s}_2)], \tag{3.19}$$

for all real unit vectors $\mathbf{s}_1$ and $\mathbf{s}_2$. We see that as long as the magnitude of the degree of correlation, $|g(\mathbf{r})|$, is narrow enough for the quasi-homogeneous approximation to apply, the data function represents Fourier components of the intensity $\Gamma(\mathbf{r})$ of the susceptibility. The effect of a finite but small correlation length in the random medium is to multiply the Fourier components by some constant, as can be seen on comparing Eqs. (3.17) and (3.19). This effectively amplifies or deamplifies the intensity of the object on reconstruction but otherwise has no significant effect.

A low-pass-filtered version of the intensity (strength of the susceptibility) of the object can now be determined. It is given by the formula

$$\Gamma_{LP}(\mathbf{r}) = \frac{1}{i16\pi^3k^2}\int_{|\mathbf{s}|\leq 2} \exp(ik\mathbf{r}\cdot\mathbf{s})\mathcal{D}(\mathbf{s})\mathrm{d}^3s, \tag{3.20}$$

where $\mathcal{D}(\mathbf{s})$ is as defined in Eq. (3.6).

As an example, consider a model scatterer characterized by an ensemble of independent homogeneous spheres. The spheres are assumed to have the same radius a and the same susceptibility η_0 and are centered at a point specified by the position vector $\mathbf{r}_0$, distributed with a probability density $p(\mathbf{r}_0)$. In this case the correlation function of the susceptibility of the medium is given by the expression

$$C(\mathbf{r}_1, \mathbf{r}_2) = \eta_0^2\int B_a(\mathbf{r}_1 - \mathbf{r}_0)B_a(\mathbf{r}_2 - \mathbf{r}_0) \times p(\mathbf{r}_0)\mathrm{d}^3r_0, \tag{3.21}$$

where

$$B_a(\mathbf{r}) = \begin{cases} 1 & \text{if } |\mathbf{r}| \leq a \\ 0 & \text{otherwise} \end{cases}, \tag{3.22}$$

and the integration is taken over all space. Equation (3.21) can be expressed in terms of the average and difference coordinates, $\mathbf{R} = (\mathbf{r}_1 + \mathbf{r}_2)/2$ and $\mathbf{r} = \mathbf{r}_1 - \mathbf{r}_2$, respectively, as

$$C(\mathbf{r}_1, \mathbf{r}_2) = \eta_0^2\int B_a(\mathbf{r}' + \mathbf{r}/2)B_a(\mathbf{r}' - \mathbf{r}/2) \times p(\mathbf{r}' + \mathbf{R})\mathrm{d}^3r'. \tag{3.23}$$

If $p(\mathbf{r})$ varies slowly with $\mathbf{r}$ on the scale of the radius a, one can replace $p(\mathbf{r}' + \mathbf{R})$ by the first terms of its Taylor expansion about $\mathbf{R}$, and one finds that

$$C(\mathbf{r}_1, \mathbf{r}_2) \approx \Gamma(\mathbf{R})g(\mathbf{r}) + \eta_0^2\int B_a(\mathbf{r}' + \mathbf{r}/2) \times B_a(\mathbf{r}' - \mathbf{r}/2)\mathbf{r}'\cdot\nabla p(\mathbf{R})\mathrm{d}^3r' + \ldots, \tag{3.24}$$

where

$$\Gamma(\mathbf{R}) = \frac{4\pi}{3}a^3\eta_0^2 p(\mathbf{R}). \tag{3.25}$$

The degree of correlation, g, is related to the geometrical overlap of two spheres[11] and is given by the expression

$$g(\mathbf{r}) = \begin{cases} 1 - \dfrac{3r}{4a} + \dfrac{r^3}{16a^3} & \text{for } r \leq 2a \\ 0 & \text{for } r > 2a \end{cases} \tag{3.26}$$

Retaining only the first term in Eq. (3.24), one obtains the quasi-homogeneous approximation for the correlation function of the random scatterer.

The scattering amplitude of a sphere of radius a centered on the point $\mathbf{r}_0$, $f(\mathbf{s}_1, \mathbf{s}_2; \mathbf{r}_0)$, is given in terms of the scattering amplitude $f_a(\mathbf{s}_1 - \mathbf{s}_2)$ of a sphere of radius a centered on the origin by the expression

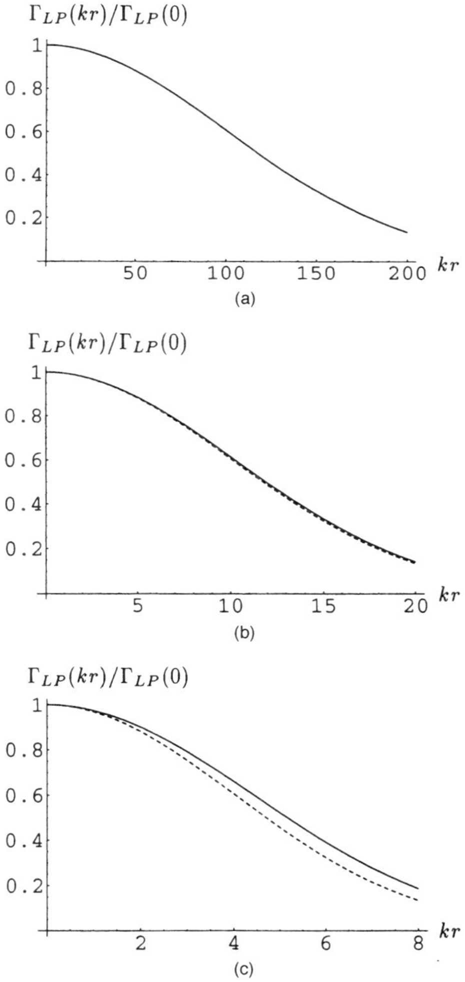

Fig. 2. Susceptibility intensity function Γ for a medium characterized by an ensemble of independent homogeneous spheres each with susceptibility $\eta = 0.21/4\pi$ and size parameter $ka = 4$. Dashed curves, original probability distribution, with (a) $k\sigma = 100$, (b) $k\sigma = 10$, and (c) $k\sigma = 4$. Solid curves, reconstructed fluctuation structure function. Both types of curves have been normalized so that the peak values are unity.

$$f(\mathbf{s}_1, \mathbf{s}_2; \mathbf{r}_0) = \exp[ik\mathbf{r}_0 \cdot (\mathbf{s}_2 - \mathbf{s}_1)] f_a(\mathbf{s}_1 - \mathbf{s}_2). \tag{3.27}$$

Thus the ensemble-average scattering amplitude for such a scatterer is given by the formula

$$\langle f(\mathbf{s}_1, \mathbf{s}_2; \mathbf{r}_0)\rangle = f_a(\mathbf{s}_1 - \mathbf{s}_2)\tilde{p}[k(\mathbf{s}_1 - \mathbf{s}_2)], \tag{3.28}$$

where the tilde indicates, as usual, the three-dimensional spatial Fourier transform. It follows directly from Eq. (3.28) that the data function is given by the expression

$$D(\mathbf{s}_1, \mathbf{s}_2) = [f_a(\mathbf{s}_1 - \mathbf{s}_2) - f_a^*(\mathbf{s}_2 - \mathbf{s}_1)] \times \tilde{p}[k(\mathbf{s}_1 - \mathbf{s}_2)]. \tag{3.29}$$

In this case the function $\mathcal{D}(\mathbf{s})$, defined by Eq. (3.6) has the value

$$\mathcal{D}(\mathbf{s}) = [f_a(\mathbf{s}) - f_a^*(-\mathbf{s})]\tilde{p}(k\mathbf{s}). \tag{3.30}$$

According to Eq. (3.20) the reconstructed intensity of the susceptibility is then given by the expression

$$\Gamma_{LP}(\mathbf{r}) = \frac{1}{k^3}\int \alpha_{LP}(\mathbf{r} - \mathbf{r}')p(\mathbf{r}')\mathrm{d}^3r', \tag{3.31}$$

where $\alpha_{LP}(\mathbf{r})$ is the reconstructed absorptive part of the susceptibility of the homogeneous sphere of radius a and susceptibility η_0 given in Eq. (A5) of Appendix A.

Figure 2 shows the numerically reconstructed intensity of the susceptibility Γ obtained by the use of Eq. (3.31) with a Gaussian probability distribution $p(\mathbf{r})$, viz.,

$$p(\mathbf{r}) = (\pi\sigma^2/2)^{-3/2}\exp(-r^2/2\sigma^2). \tag{3.32}$$

In Fig. 2(a), which corresponds to the case $k\sigma = 100$, the reconstruction is so accurate that the curves are indistinguishable. In Fig. 2(b) ($k\sigma = 10$), some deviation from the original profile is evident. In Fig. 2(c) ($k\sigma = 4$), the width of the probability distribution is equal to the radius of the spheres. One might expect that the quasi-homogeneous approximation will break down in this limit; indeed the figure shows that in this case the reconstruction differs significantly from the original function.

4. CONCLUSION

We have proposed a new technique for diffraction tomography that makes use of a generalization of the optical cross-section theorem and avoids the problems of measuring the phase of the scattered field as well as measuring the directional dependence of the scattered field. We have also shown that by using this method, one can determine the absorptive part of the susceptibility of a deterministic scatterer and the fluctuation strength function of a random scatterer. We have demonstrated the feasibility of the proposed method by numerical simulations.

APPENDIX A: THE HOMOGENEOUS SPHERE

In this appendix, formulas are given that relate the forward and inverse problems. We first determine the scattering amplitude associated with a homogeneous sphere of refractive index $N = (4\pi\eta + 1)^{1/2}$ and of radius a in the form of an infinite series. The formulas for the inverse problem in the single-scattering approximation are then expressed in terms of the (exact) series for the scattering amplitude. That is, the low-pass reconstructed object function α_{LP} and the averaged data function $\mathcal{D}$ are then expressed in terms of the series for the scattering amplitude.

The calculation of the scattering amplitude (usually referred to as the forward problem) can be performed by the method of partial waves (Ref. 12, p. 932). The scattering amplitude is then expressed in series form as

$$f(\mathbf{s}_1, \mathbf{s}_2) = \frac{1}{k}\sum_{m=0}^{\infty}(2m + 1)\frac{i\beta_m}{\beta_m - i\gamma_m}P_m(\mathbf{s}_1 \cdot \mathbf{s}_2), \tag{A1}$$

where

$$\beta_m = kj_m(Nka)j'(ka) - nkj_m'(Nka)j_m(ka), \tag{A2}$$

$$\gamma_m = nkj_m'(Nka)n_m(ka) - kj_m(Nka)n_m'(ka). \tag{A3}$$

In these formulas n_m and j_m are the spherical Neumann functions and the spherical Bessel functions, respectively, of order m, and the prime indicates the derivative with respect to the argument. Further, P_m is the Legendre polynomial of order m. The series converges rapidly when the number of terms retained exceeds the size parameter ka. In this case the averaged data function can be expressed in the form

$$\mathcal{D}(\mathbf{s}) = \frac{2i}{k}\sum_{m=1}^{\infty}(2m + 1)\Re\left(\frac{\beta_m}{\beta_m - i\gamma_m}\right)P_m(1 - s^2/2). \tag{A4}$$

Using formula (3.7), one finds that

$$\alpha_{LP}(\mathbf{r}) = \sum_{m=0}^{\infty}\frac{(2m + 1)}{\pi^2kr}\Re\left(\frac{\beta_m}{\beta_m - i\gamma_m}\right) \times \frac{\mathrm{d}}{\mathrm{d}(kr)}[krj_m(kr)n_m(kr)]. \tag{A5}$$

ACKNOWLEDGMENTS

This research was supported by the U.S. Department of Energy under grant DE-FG02-90ER 14119, by the U.S. Air Force Office of Scientific Research under grant F49620-96-1-0400, and by the U.S. Air Force Research Laboratory under grant F4162299WS014. G. S. Agarwal acknowledges National Science Foundation grant INT97-12760, which made his collaboration on this project possible. P. S. Carney is obliged to Greg Gbur for helpful discussions concerning the quasi-homogeneous approximation.

Address correspondence by fax to E. Wolf: 716-473-6087.

REFERENCES AND NOTES

1. See, for example, E. Wolf, "Principles and development of diffraction tomography," in *Trends in Optics*, A. Consortini, ed. (Academic, San Diego, Calif., 1996), pp. 83–110.

2. H. C. van de Hulst, "On the attenuation of plane waves by obstacles of arbitrary size and form," Physica **15**, 740–746 (1949).
3. P. S. Carney, E. Wolf, and G. S. Agarwal, "Statistical generalizations of the optical theorem with applications to inverse scattering," J. Opt. Soc. Am. A **14**, 3366–3371 (1997).
4. R. G. Newton, "Present status of the generalized Marchenko method for the solution of the inverse scattering problem in three dimensions," in *Inverse Problems in Mathematical Physics*, L. Päivärinta and E. Somersalo, eds. (Springer-Verlag, Berlin, 1993).
5. M. Born and E. Wolf, *Principles of Optics*, 7th ed. (Cambridge U. Press, Cambridge, UK, 1999).
6. Within the accuracy of the first-order Born approximation, D is constant over these values of $\mathbf{S}$. However experimental values of D may require the averaging in Eq. (3.6) owing to noise and multiple scattering.
7. W. H. Carter and E. Wolf, "Coherence and radiometry with quasihomogeneous planar sources," J. Opt. Soc. Am. **67**, 785–796 (1977).
8. R. A. Silverman, "Locally stationary random processes," IRE Trans. Inf. Theory **3**, 182–187 (1957).
9. G. Gbur and K. Kim, "The quasi-homogeneous approximation for a class of three-dimensional primary sources," Opt. Commun. **163**, 20–23 (1999). This paper is soon to be reprinted in Opt. Commun. owing to a large number of printer's errors in the original publication.
10. In order that $\tilde{g}$ satisfy requirements of analyticity in κ', $\tilde{g}(|\boldsymbol{\kappa}|)$ must be expressible as an analytic function of $|\boldsymbol{\kappa}|^2$.
11. See Appendix in G. Gbur and E. Wolf, "Determination of the density correlation function from scattering with polychromatic light," Opt. Commun. (to be published).
12. C. Cohen-Tannoudji, B. Diu, and F. Laloe, *Quantum Mechanics* (Hermann, Paris, 1977).

Section 5 – Foundations of Radiometry

The principles of radiometry are based on one of the oldest models for energy propagation in optical fields. It is a very useful model, used to describe, for example, the distribution of the field intensity from sources of different kinds, usually thermal ones, or to evaluate the performance of various radiation detectors. In its generalized form, known as the theory of radiative energy transfer, it plays a basic role in astrophysics in connection with investigations of stellar atmospheres and the interior of stars. However, except for some special situations, the foundations of radiometry, and of radiative energy transfer have so far not been clarified. In paper 5.4 a review is presented of various investigations which showed that a close relationship exists between radiometry and the theory of partial coherence.

Traditional radiometry has been developed for sources which are essentially spatially incoherent. In paper 5.1 the radiometric laws are generalized to sources of any state of coherence and to the fields which such sources generate. An interesting and a rather useful class of partially coherent sources are the so-called quasi-homogeneous sources,* introduced in paper 5.2. The paper includes the formulation of a pair of reciprocity relations pertaining to such sources and to the radiation field which they produce. One of these relations is a generalization of a basic theorem of coherence theory, the van Cittert-Zernike theorem, the other may be regarded as a natural counterpart of it, which was not previously known. Paper 5.3 is concerned with radiance functions of partially coherent fields which are produced by planar secondary, quasi-homogeneous sources in free space. The paper shows how the coherence properties of the source influence the radiometric behavior of the field.

Traditional radiometry is concerned with energy propagation. In paper 5.5 it is shown that is also possible to describe the propagation of a second-order correlation function of the field by a radiometric model, at least when the source is quasi-homogeneous.

Paper 5.6 is concerned with a much more difficult problem, namely with the derivation of the free-space transport equation of radiometry from statistical wave theory. The field is assumed to be generated by a quasi-homogeneous source, again radiating in free space, and it is shown that under these circumstances, the free-space transport equation follows from statistical wave theory in the short wavelength limit.

Two papers include in Section 3 also have bearing on the foundations of radiometry. They are papers 3.9 and 3.10. References to additional papers concerned with foundations of radiometry and radiative energy transfer, not included in this Section, are cited in the list of references in Sec. 8 of this volume. They are papers 124, 129, 137, 179, 192 and 243.

---

*There are analogous concepts, known as local stationarity, in the theory of non-stationary random processes [R.A. Silverman, *Trans. Inst. Radio Engrs.* **IT-3**, 182 (1957)], and local homogeneity in the theory of scattering on random media [R.A. Silverman, *Proc. Camb. Phil. Soc.* **54**, 530 (1958)].

Section 5 – Foundations of Radiometry

Reprinted from:

JOURNAL OF THE OPTICAL SOCIETY OF AMERICA VOLUME 64, NUMBER 9 SEPTEMBER 1974

Radiometry with sources of any state of coherence*

E. W. Marchand

Research Laboratories, Eastman Kodak Co., Rochester, New York 14650

E. Wolf

Department of Physics and Astronomy, University of Rochester, Rochester, New York 14627

(Received 20 February 1974)

The basic laws of radiometry are generalized to fields generated by a two-dimensional stationary source of any state of coherence. Important in this analysis is the concept of the generalized radiance function, introduced by Walther in 1968. The concepts of generalized radiant emittance and of generalized radiant intensity are introduced and it is shown how all these quantities may be expressed in terms of coherence functions of the source. Both the generalized radiance of Walther and the generalized radiant emittance may take on negative values, indicating that these quantities have, in general, a less-direct physical meaning than have the corresponding quantities of traditional radiometry (which presumably represents the incoherent limit of the present theory). The generalized radiant intensity is, however, always found to be non-negative and, just as in the incoherent limit, represents the angular distribution of the energy flux in the far zone.

Index Headings: Radiometry; Coherence.

Although the concepts of radiometry are familiar to anyone who has even the most rudimentary knowledge of optics, it does not appear to be generally realized that the fundamental laws of radiometry have never been systematically derived from the basic theories of light. Radiometry appears to have been systematized around the turn of the century in connection with the theory of heat radiation, and a well-known book by Max Planck,[1] the first edition of which appeared in 1906, remains probably the most comprehensive account to this day of the fundamentals of the subject. It is taken for granted that traditional radiometry describes, in some approximation, the behavior of radiation fields created by incoherent sources, and it is known that blackbody sources, with which the early investigations on heat radiation were chiefly concerned, are among the most incoherent ones found in nature. Yet modern researches in coherence theory have shown that even such sources can generate fields that are highly correlated (i.e., spatially highly coherent) throughout arbitrarily large regions of space, a result expressed quantitatively by the so-called VanCittert–Zernike theorem.[2] This fact raises serious questions as to the range of validity of even the most basic laws of radiometry and underscores the need for clarifying how, and under what conditions, these laws can be deduced from the presently accepted theories of light. Closely related to these questions is the problem, of considerable practical importance at the present time, of the possibility of extending the concepts and the laws of traditional radiometry to fields produced by laser sources and, more generally, by sources of arbitrary states of coherence.

In the present paper, we take a step towards solving these problems and present a generalization of some of the basic concepts of radiometry to fields generated by any statistically stationary two-dimensional source.[3] Basic in our treatment is the concept of generalized radiance for a source of any state of coherence, introduced a few years ago by Walther[4] in an important investigation. In terms of it, we define generalizations of the concepts of radiant emittance and of radiant intensity and examine these quantities in some detail. We show how the generalized radiance, the generalized radiant emittance, and the generalized radiant intensity may be expressed in terms of various second-order correlation functions of the fluctuating source. A result emerges that may appear somewhat surprising without closer examination: Both the generalized radiance of Walther and our generalized radiant emittance may take on negative values, and they also exhibit some other apparently unphysical features. These facts indicate that neither of these two quantities is, in general, measurable; but it will become clear that they are nevertheless useful mathematical constructs in terms of which considerable insight can be obtained into the behavior of partially coherent fields. The generalized radiant intensity is found to be always non-negative and to represent correctly the (measurable) angular distribution of the energy flux in the far zone. We also examine the incoherent and the coherent limits of our theory. In the limiting case of incoherence both the radiance and the radiant emittance are always found to be non-negative, and this limit obviously represents traditional radiometry. However, we also find that a strictly incoherent finite source cannot radiate in accordance with Lambert's law.

Only the basic mathematical structure of the generalized radiometry is discussed in this paper. We plan to present, in a subsequent paper, numerical results illustrating radiation properties of sources of different geometries and different states of coherence.

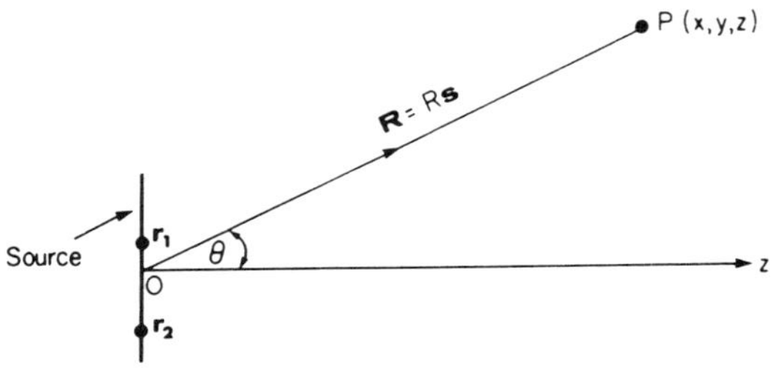

FIG. 1. Illustration of the notation.

I. GENERALIZATION OF BASIC CONCEPTS OF RADIOMETRY TO FIELDS CREATED BY SOURCES OF ANY STATE OF COHERENCE

We consider an optical field generated by a source occupying a portion of the plane $z=0$, around the origin O of our coordinate system. The source is assumed to be statistically stationary but of any state of coherence and to emit light into the half-space $z>0$. We have shown in an earlier paper[5] that the intensity $I(\mathbf{R})$ at a point $P(x,y,z)$ in the far zone $(k|\mathbf{R}|\rightarrow\infty)$ may be expressed in the form (see Fig. 1)

$$I(\mathbf{R})=\int_0^\infty I_\omega(\mathbf{R})\,d\omega, \tag{1}$$

with

$$I_\omega(\mathbf{R})\sim\left(\frac{2\pi}{kR}\right)^2\cos^2\theta\,\mathcal{A}(\mathbf{s},\mathbf{s};\omega). \tag{2}$$

Here

$$k=\omega/c=2\pi/\lambda \tag{3}$$

is the wave number, ω the frequency, λ the wavelength, c the vacuum velocity of light, and $\mathcal{A}(\mathbf{s},\mathbf{s};\omega)$ is the angular self-correlation function of the field in the half-space $z>0$ [cf. Eq. (12) below]. In Eq. (2), $\mathbf{s}$ is the real unit vector in the $\mathbf{R}$ direction,

$$\mathbf{s}=\frac{\mathbf{R}}{R}, \tag{4}$$

with components

$$s_x=\frac{x}{R},\quad s_y=\frac{y}{R},\quad s_z=\frac{z}{R}=(1-s_x^2-s_y^2)^{\frac{1}{2}}=\cos\theta, \tag{5}$$

where θ is the angle between the position vector $\mathbf{R}$ and the z axis.

The intensity I appearing in Eq. (1) was defined in the manner customary in physical optics, i.e., as the average of the square of the absolute value of a complex scalar field variable. It is usually assumed (Ref. 2, §8.4) that this quantity is equal, in appropriate units, to the absolute value of the average Poynting vector $\mathbf{S}$, i.e.,

$$I(\mathbf{R})=|\langle\mathbf{S}(\mathbf{R})\rangle|. \tag{6}$$

However, in radiometry, the basic quantity is the radiant flux. Now a surface element $d\sigma$ around a point P in the far zone, at distance R from the origin, at right angles to the direction OP may be expressed in the form

$$d\sigma = R^2\,d\Omega, \tag{7}$$

where

$$d\Omega=\frac{ds_x\,ds_y}{s_z} \tag{8}$$

is the element of the solid angle that the surface element subtends at the origin O. Hence, from the physical significance of the Poynting vector (Ref. 2, p. 9) and from the relations (6) and (7), it follows that the flux traversing the area element $d\sigma$ and the intensity I are related by the formula

$$dF=IR^2\,d\Omega. \tag{9}$$

Hence Eqs. (1) and (2) may be rewritten in the following form, which is more appropriate to the purposes of radiometry,

$$dF=\int_0^\infty dF_\omega\,d\omega, \tag{10}$$

where

$$dF_\omega\sim\left(\frac{2\pi}{k}\right)^2\cos^2\theta\,\mathcal{A}(\mathbf{s},\mathbf{s};\omega)\,d\Omega. \tag{11}$$

We take Eqs. (10) and (11) as the starting point of our analysis, bearing in mind that we are considering light created by a stationary plane source of any state of coherence.

We will now formally express the angular self-correlation function $\mathcal{A}(\mathbf{s},\mathbf{s};\omega)$ that occurs in Eq. (11) as an integral over the source plane. According to Eq. (17) of Ref. 5 and the Fourier inverse of Eq. (15) of Ref. 5,

$$\mathcal{A}(\mathbf{s},\mathbf{s};\omega)=\left(\frac{k}{2\pi}\right)^4\iint_{(z=0)}W(\mathbf{r}_1,\mathbf{r}_2;\omega)\times\exp\{-ik[(\mathbf{r}_1-\mathbf{r}_2)\cdot\mathbf{s}]\}\,d^2\mathbf{r}_1\,d^2\mathbf{r}_2, \tag{12}$$

where $W(\mathbf{r}_1,\mathbf{r}_2;\omega)$ is the cross-spectral density function for the pair of points $\mathbf{r}_1$ and $\mathbf{r}_2$ in the source plane $z=0$. In the double integral on the right-hand side of Eq. (12), it is to be understood that the points $\mathbf{r}_1$ and $\mathbf{r}_2$ take independently all possible positions in the source plane $z=0$. Let us introduce new variables of integration in Eq. (12) by means of the formulas

$$(\mathbf{r}_1-\mathbf{r}_2)=\mathbf{r}',\quad \tfrac{1}{2}(\mathbf{r}_1+\mathbf{r}_2)=\mathbf{r}. \tag{13}$$

Then Eq. (12) can readily be shown to be expressible in the form

$$\mathcal{A}(\mathbf{s},\mathbf{s};\omega)=\left(\frac{k}{2\pi}\right)^2\frac{1}{\cos\theta}\int_{(z=0)}B_\omega(\mathbf{r},\mathbf{s})\,d^2\mathbf{r}, \tag{14}$$

where

$$B_\omega(\mathbf{r},\mathbf{s})=\left(\frac{k}{2\pi}\right)^2\cos\theta\int_{(z'=0)}W(\mathbf{r}+\tfrac{1}{2}\mathbf{r}',\mathbf{r}-\tfrac{1}{2}\mathbf{r}';\omega)\times\exp(-ik\mathbf{s}\cdot\mathbf{r}')\,d^2\mathbf{r}' \quad (15)$$

and the integrals in Eqs. (14) and (15) extend over the plane $z=0$ and $z'=0$, respectively. On substituting from Eq. (14) into (11) and recalling that according to Eq. (5) $s_z=\cos\theta$, we find that

$$dF_\omega\sim\cos\theta\,d\Omega\int_{(z=0)}B_\omega(\mathbf{r},\mathbf{s})\,d^2\mathbf{r}, \quad (16)$$

so that the flux at frequency ω, radiated by the source across the hemisphere at infinity in the half-space $z>0$, is given by

$$F_\omega\sim\int_{(2\pi)}d\Omega\cos\theta\int_{(z=0)}B_\omega(\mathbf{r},\mathbf{s})\,d^2\mathbf{r}, \quad (17)$$

where the integration extends over the source plane $z=0$ and over the solid angle subtended by the hemisphere in the half-space $z>0$ and centered at the origin.

Formally Eq. (17) is of the same mathematical form as a basic law of radiometry, for the flux generated by a spatially incoherent source with radiance function $B_\omega(\mathbf{r},\mathbf{s})$. However, Eq. (17), with $B_\omega(\mathbf{r},\mathbf{s})$ defined by Eq. (15), now gives the flux generated by a stationary source of any state of coherence. We shall refer to the function $B_\omega(\mathbf{r},\mathbf{s})$, defined by Eq. (15), as *generalized radiance*. It is the same function as that introduced not long ago by Walther.[4] However, this quantity must be interpreted with some caution, because, as we shall see later, it does not possess all the properties normally attributed to radiance. In particular, as we show in Sec. VI, the generalized radiance may be negative for some values of its arguments and does not, in general, vanish for all points $\mathbf{r}$ in the source plane that lie outside the domain occupied by the source. Nevertheless, as we shall see, the generalized radiance is a useful quantity, because, in terms of it, the concepts of radiant emittance and of radiant intensity can readily be generalized to fields created by sources of any state of coherence. For this purpose, we need only to rewrite Eq. (17) in two different ways,

$$F_\omega=\int_{(z=0)}E_\omega(\mathbf{r})\,d^2\mathbf{r} \quad (18)$$

and

$$F_\omega=\int_{(2\pi)}J_\omega(\mathbf{s})\,d\Omega, \quad (19)$$

where

$$E_\omega(\mathbf{r})=\int_{(2\pi)}B_\omega(\mathbf{r},\mathbf{s})\cos\theta\,d\Omega \quad (20)$$

and

$$J_\omega(\mathbf{s})=\cos\theta\int_{(z=0)}B_\omega(\mathbf{r},\mathbf{s})\,d^2\mathbf{r}. \quad (21)$$

The expressions (20) and (21) are formally identical with the usual expressions of traditional radiometry for the radiant emittance and the radiant intensity, respectively, from a spatially incoherent source. Hence we shall refer to the quantity $E_\omega(\mathbf{r})$, defined by Eq. (20), as the *generalized radiant emittance* (at frequency ω) at the point $\mathbf{r}$ of the source and to the quantity $J_\omega(\mathbf{s})$, defined by Eq. (21), as the *generalized radiant intensity* (at frequency ω) in the $\mathbf{s}$ direction. To avoid possible misunderstanding, we stress that the radiometric term "radiant intensity" [defined by Eq. (21)] and the physical-optics term "intensity" [defined by Eq. (6)] have different meanings.

We now examine briefly some of the properties of these quantities and show that they may be expressed in several alternative forms involving various correlation functions of the optical field.

II. THE GENERALIZED RADIANCE

We defined the generalized radiance by the formula (15), viz.,

$$B_\omega(\mathbf{r},\mathbf{s})=\left(\frac{k}{2\pi}\right)^2\cos\theta\int_{(z'=0)}W(\mathbf{r}+\tfrac{1}{2}\mathbf{r}',\mathbf{r}-\tfrac{1}{2}\mathbf{r}';\omega)\times\exp(-ik\mathbf{s}\cdot\mathbf{r}')\,d^2\mathbf{r}', \quad (22)$$

W being the cross-spectral density function. We show in Appendix A that B_ω may also be expressed in the form

$$B_\omega(\mathbf{r},\mathbf{s})=k^2\cos\theta\int_{(\mathbf{f}\text{ plane})}\hat{W}(k\mathbf{s}+\tfrac{1}{2}\mathbf{f},-k\mathbf{s}+\tfrac{1}{2}\mathbf{f};\omega)\times\exp(i\mathbf{f}\cdot\mathbf{r})\,d^2\mathbf{f}, \quad (23)$$

where

$$\hat{W}(\mathbf{f}_1,\mathbf{f}_2;\omega)=\frac{1}{(2\pi)^4}\int\int_{(z=0)}W(\mathbf{r}_1,\mathbf{r}_2;\omega)\times\exp\{-i(\mathbf{f}_1\cdot\mathbf{r}_1+\mathbf{f}_2\cdot\mathbf{r}_2)\}\,d^2\mathbf{r}_1\,d^2\mathbf{r}_2 \quad (24)$$

is the four-dimensional spatial Fourier transform of the cross-spectral density function $W(\mathbf{r}_1,\mathbf{r}_2;\omega)$.

Alternatively, if we recall that, according to Eq. (17) of Ref. 5, $\hat{W}$ and the angular correlation function $\mathcal{A}$ are connected by the relation

$$\mathcal{A}(\mathbf{s}_1,\mathbf{s}_2;\omega)=k^4\hat{W}(k\mathbf{s}_1,-k\mathbf{s}_2;\omega), \quad (25)$$

we see from Eqs. (23) and (25), if we change the

variables of integration from $\mathbf{f}$ to $\mathbf{s}'=\mathbf{f}/k$, that the generalized radiance may also be expressed in the form

$$B_\omega(\mathbf{r},\mathbf{s})=\cos\theta\iint_{-\infty}^{\infty}\mathcal{A}(\mathbf{s}+\tfrac{1}{2}\mathbf{s}',\mathbf{s}-\tfrac{1}{2}\mathbf{s}';\omega)\times\exp(ik\mathbf{s}'\cdot\mathbf{r})\,ds_x'\,ds_y'. \quad (26)$$

The formal similarity in the structure of the integrals (22) and (26) should be noted. Equation (22) expresses the generalized radiance in terms of the cross-spectral density function $W(\mathbf{r}_1,\mathbf{r}_2;\omega)$, i.e., in terms of correlations of the field at pairs of points on the source plane. Equation (26) expresses the generalized radiance in terms of the angular correlation function $\mathcal{A}(\mathbf{s}_1,\mathbf{s}_2;\omega)$, i.e., in terms of correlations of the plane waves in the angular spectrum representation that are propagated in pairs of (possibly complex) directions.

Although for each point $\mathbf{r}$ the domain of integration of the right-hand side of Eq. (22) is formally infinite, the integration extends, actually, over a finite domain of the $\mathbf{r}'$ plane. This follows from the fact that the source was assumed to be finite and hence the integrand in Eq. (22) vanishes when either or both of the points $(\mathbf{r}+\frac{1}{2}\mathbf{r}')$ and $(\mathbf{r}-\frac{1}{2}\mathbf{r}')$ falls outside the area of the plane $z=0$ occupied by the source.

In Appendix B we show that the generalized radiance $B_\omega(\mathbf{r},\mathbf{s})$ is always real. However, we shall see later (end of Sec. VI) that it may take on negative values in some cases. Hence the generalized radiance cannot, in general, be interpreted as a true flux density. This fact, which may appear somewhat surprising at first, can be appreciated readily if we recall that the generalized radiance was introduced by means of the relation [Eq. (16)]

$$dF_\omega\sim\cos\theta\,d\Omega\int_{(z=0)}B_\omega(\mathbf{r},\mathbf{s})\,d^2\mathbf{r}, \quad (27)$$

where dF_ω represents the flux at frequency ω, radiated by the source into the solid angle $d\Omega$ about the direction $\mathbf{s}$. Thus, although dF_ω is necessarily non-negative, there is no reason to expect that B_ω will also have this property. In fact, because B_ω is a function of conjugate Fourier variables, namely, $\mathbf{r}$ and $\mathbf{s}$, we cannot prescribe the values of B_ω in Eq. (16) arbitrarily for all values of its arguments. This situation is strictly analogous to that encountered in connection with the Wigner distribution function in quantum statistics.[6] The Wigner distribution function is also a function of conjugate Fourier variables and is well known to take on negative values, in general. Nevertheless, this distribution function is a useful mathematical tool, as it allows calculation of various expectation values to be carried out by means of formulas that are strictly similar to those employed in ordinary probability theory. We might expect the generalized radiance to share this property with the Wigner distribution function, but we shall not study this question in the present paper.

Although, as already pointed out, the generalized radiance can take on negative values, it can be shown to be non-negative for a spatially incoherent source. This result will be established in Sec. VI.

Irrespective of the state of coherence of the source, the following result holds:

$$\int_{(z=0)}B_\omega(\mathbf{r},\mathbf{s})\,d^2\mathbf{r}=(2\pi k)^2\cos\theta\hat{W}(k\mathbf{s},-k\mathbf{s};\omega). \quad (28)$$

This identity follows when the integral of the expression (23) for B_ω is taken with respect to $\mathbf{r}$ over the whole z plane. If the order of the $\mathbf{r}$ and $\mathbf{f}$ integrations is then interchanged and use is made of the relation

$$\int_{(\mathbf{f}\text{ plane})}\exp(i\mathbf{f}\cdot\mathbf{r})\,d^2\mathbf{r}=(2\pi)^2\delta^{(2)}(\mathbf{f}), \quad (29)$$

where $\delta^{(2)}(\mathbf{f})$ is the two-dimensional Dirac δ function, the relation (28) is then immediately obtained.

III. THE GENERALIZED RADIANT EMITTANCE

The generalized radiant emittance was defined by formula (20) as

$$E_\omega(\mathbf{r})=\int_{(2\pi)}B_\omega(\mathbf{r},\mathbf{s})\cos\theta\,d\Omega. \quad (30)$$

If we use expression (15) for B_ω, Eq. (30) becomes

$$E_\omega(\mathbf{r})=\int_{(z'=0)}W(\mathbf{r}+\tfrac{1}{2}\mathbf{r}',\mathbf{r}-\tfrac{1}{2}\mathbf{r}';\omega)K_\omega(\mathbf{r}')\,d^2\mathbf{r}', \quad (31)$$

where

$$K_\omega(\mathbf{r}')=\left(\frac{k}{2\pi}\right)^2\iint_{s_x^2+s_y^2\leqslant 1}\cos\theta\exp(-ik\mathbf{s}\cdot\mathbf{r}')\,ds_x\,ds_y. \quad (32)$$

The integral on the right-hand side of Eq. (32) may be evaluated readily. If we change to polar coordinates via the relations

$$s_x=\rho\cos\phi,\qquad s_y=\rho\sin\phi, \quad (33a)$$

$$x'=r'\cos\phi',\qquad y'=r'\sin\phi', \quad (33b)$$

and use the fact that $\cos\theta=(1-s_x^2-s_y^2)^{\frac{1}{2}}$, Eq. (32) becomes

$$K_\omega(\mathbf{r}')=\left(\frac{k}{2\pi}\right)^2\int_0^1\rho(1-\rho^2)^{\frac{1}{2}}\,d\rho\times\int_0^{2\pi}\exp[-ik\rho r'\cos(\phi-\phi')]\,d\phi, \quad (34)$$

$$r'=|\mathbf{r}'|=(x'^2+y'^2)^{\frac{1}{2}}.$$

The integral with respect to ϕ is well known to have the value[7] $2\pi J_0(k\rho r')$ where J_0 is the Bessel function of the first kind and zero order. Hence

$$K_\omega(\mathbf{r}')=\frac{k^2}{2\pi}\int_0^1 \rho(1-\rho^2)^{\frac{1}{2}}J_0(k\rho r')\,d\rho. \tag{35}$$

The integral on the right-hand side may be evaluated in terms of a spherical Bessel function. We have[8]

$$\int_0^1 \rho(1-\rho^2)^{\frac{1}{2}}J_0(k\rho r')\,d\rho=\left(\frac{\pi}{2}\right)^{\frac{1}{2}}\frac{J_{\frac{3}{2}}(kr')}{(kr')^{\frac{3}{2}}}, \tag{36}$$

where $J_{\frac{3}{2}}(x)$ is the Bessel function of the first kind and order $\frac{3}{2}$, whose explicit form, in terms of trigonometric functions, is

$$J_{\frac{3}{2}}(x)=\left(\frac{2}{\pi x}\right)^{\frac{1}{2}}\left[\frac{\sin x}{x}-\cos x\right]. \tag{37}$$

From Eqs. (35) and (36), we obtain the following expression for K,

$$K_\omega(\mathbf{r}')=\frac{k^2}{2(2\pi)^{\frac{1}{2}}}\frac{J_{\frac{3}{2}}(kr')}{(kr')^{\frac{3}{2}}}. \tag{38}$$

Using this result in Eq. (31), we finally obtain the expression for the generalized radiant emittance in terms of the cross-spectral density function W of the source,

$$E_\omega(\mathbf{r})=\frac{k^2}{2(2\pi)^{\frac{1}{2}}}\int_{(z'=0)} W(\mathbf{r}+\tfrac{1}{2}\mathbf{r}',\mathbf{r}-\tfrac{1}{2}\mathbf{r}';\omega)\frac{J_{\frac{3}{2}}(kr')}{(kr')^{\frac{3}{2}}}\,d^2\mathbf{r}'. \tag{39}$$

The similarity, in the mathematical structure, of expression (15) for the generalized radiance B_ω and expression (39) for the generalized radiant emittance E_ω should be noted.

We shall demonstrate in Sec. VI that the generalized radiant emittance, like the generalized radiance, may take on negative values. However, as shown in Sec. V, the generalized radiant emittance is non-negative for a spatially incoherent source.

IV. THE GENERALIZED RADIANT INTENSITY

We introduced the generalized radiant intensity for a two-dimensional source of any state of coherence by formula (21), viz.,

$$J_\omega(\mathbf{s})=\cos\theta\int_{(z=0)} B_\omega(\mathbf{r},\mathbf{s})\,d^2\mathbf{r}, \tag{40}$$

where $B_\omega(\mathbf{r},\mathbf{s})$ is the generalized radiance. We now express J_ω in several alternative forms and show that it represents the angular distribution of the flux density in the far zone.

From Eqs. (40) and (28), we obtain an expression for the generalized radiant intensity in terms of the four-dimensional spatial Fourier transform $\hat{W}$ [Eq. (24)] of the cross-spectral density function in the source plane,

$$J_\omega(\mathbf{s})=(2\pi k)^2\hat{W}(k\mathbf{s},-k\mathbf{s};\omega)\cos^2\theta. \tag{41}$$

Alternatively, we may express J_ω in terms of the spectral density function W of the field in the far zone, using the asymptotic relation (34) of Ref. 5. We then obtain from Eq. (41) the formula

$$J_\omega(\mathbf{s})\simeq R^2W(R\mathbf{s},R\mathbf{s};\omega),\quad kR\gg 1 \tag{42}$$

where, as before, $\mathbf{R}$ is a point in the direction $\mathbf{s}$ in the far zone (see Fig. 1). Because the spectral density function $W(\mathbf{R},\mathbf{R};\omega)$ is necessarily non-negative, so is the generalized radiant intensity $J_\omega(\mathbf{s})$, i.e.,

$$J_\omega(\mathbf{s})\geq 0, \tag{43}$$

irrespective of the state of coherence of the source.

Alternatively, using the relation (25) we may also express J_ω in terms of the angular self-correlation function $\mathcal{A}(\mathbf{s},\mathbf{s};\omega)$. From Eq. (41) we then obtain

$$J_\omega(\mathbf{s})=\left(\frac{2\pi}{k}\right)^2\cos^2\theta\,\mathcal{A}(\mathbf{s},\mathbf{s};\omega). \tag{44}$$

Now, if we compare Eq. (44) with Eq. (2) and also use Eq. (6), we see that

$$J_\omega(\mathbf{s})=R^2|\langle\mathbf{S}_\omega(R\mathbf{s})\rangle|,\quad kR\gg 1 \tag{45}$$

where $\langle\mathbf{S}_\omega(R\mathbf{s})\rangle$ is the averaged Poynting vector at the point $\mathbf{R}=R\mathbf{s}$.

Multiplying both sides of Eq. (45) by the element $d\Omega$ of the solid angle around the $\mathbf{s}$ direction and using formula (7) for the corresponding element $d\sigma$ of the surface of a large sphere of radius R centered at the source, we obtain

$$J_\omega(\mathbf{s})d\Omega\sim|\langle\mathbf{S}_\omega(R\mathbf{s})\rangle|\,d\sigma,\quad kR\gg 1. \tag{46}$$

This formula shows that our generalized radiant intensity represents the flux per unit solid angle around the direction $\mathbf{s}$, emitted by the source into the far zone. Thus, the generalized radiant intensity introduced by Eqs. (21) and (15) for radiation from a source of any state of coherence has the same physical significance as the radiant intensity of traditional radiometry, even though the generalized radiance $B_\omega(\mathbf{r},\mathbf{s})$ and the generalized radiant emittance $E_\omega(\mathbf{r})$ may take on negative values.

V. SPECIAL CASE (A), INCOHERENT SOURCES

In the idealized case when the source is spatially strictly incoherent, the cross-spectral density function W for points in the source plane may be represented by a two-dimensional Dirac δ function,[9]

$$W(\mathbf{r}_1,\mathbf{r}_2;\omega)=w(\mathbf{r}_1,\omega)\delta^{(2)}(\mathbf{r}_1-\mathbf{r}_2), \tag{47}$$

where

$$w(\mathbf{r}_1,\omega)\geq 0, \tag{48}$$

and, of course, $w(\mathbf{r}_1,\omega)\equiv 0$ if the point $\mathbf{r}_1$ is outside the source area. In this case the expression (15) for the generalized radiance reduces to

$$B_\omega(\mathbf{r},\mathbf{s})=\left(\frac{k}{2\pi}\right)^2 \cos\theta\, w(\mathbf{r},\omega). \tag{49}$$

With B_ω given by Eq. (49), the expressions (20) and (21) become

$$E_\omega(\mathbf{r})=\left(\frac{k}{2\pi}\right)^2 w(\mathbf{r},\omega)\int_{(2\pi)} \cos^2\theta\, d\Omega, \tag{50}$$

$$J_\omega(\mathbf{s})=\left(\frac{k}{2\pi}\right)^2 \cos^2\theta \int_{(z=0)} w(\mathbf{r},\omega)\, d^2\mathbf{r}. \tag{51}$$

The integral in Eq. (50) may be evaluated readily by expressing $d\Omega$ in spherical polar coordinates, and is found to have the value $2\pi/3$. Hence

$$E_\omega(\mathbf{r})=\frac{k^2}{6\pi}w(\mathbf{r},\omega). \tag{52}$$

It is useful to express B_ω and J_ω in terms of E_ω rather than in terms of $w(r;\omega)$. From Eqs. (49) and (52), we see that

$$B_\omega(\mathbf{r},\mathbf{s})=\frac{3}{2\pi}E_\omega(\mathbf{r})\cos\theta, \tag{53}$$

and, in view of Eq. (52), Eq. (51) may be expressed in the form

$$J_\omega(\mathbf{s})=\frac{3}{2\pi}\cos^2\theta\int_{(z=0)} E_\omega(\mathbf{r})\, d^2\mathbf{r}. \tag{54}$$

For points of observation in the direction at right angles to the plane of the source, $s_x=s_y=0$ and $\theta=0$, and Eq. (54) gives $J_\omega=J_\omega{}^{(0)}$, where

$$J_\omega{}^{(0)}=\frac{3}{2\pi}\int_{(z=0)} E_\omega(\mathbf{r})\, d^2\mathbf{r}. \tag{55}$$

From Eqs. (54) and (55), it follows that

$$J_\omega(\mathbf{s})=J_\omega{}^{(0)}\cos^2\theta. \tag{56}$$

According to Eq. (56), the radiant intensity now decreases in proportion to $\cos^2\theta$ as θ increases—not in proportion to $\cos\theta$ characteristic of an isotropically radiating source. This result implies that a spatially incoherent plane source of finite extent cannot radiate isotropically into the half-space $z>0$, irrespective of the exact form of its radiant-emittance function $E_\omega(\mathbf{r})$. This result is not contradicted by blackbody sources which, as is well known, radiate isotropically and are often incorrectly considered to be spatially incoherent. A careful analysis shows that blackbody-radiation sources exhibit field correlations over distances of the order of the mean wavelength of the radiation.[10]

We see from Eqs. (49) and (52) that, in view of Eq. (48), the generalized radiance and the generalized radiant emittance of a plane incoherent source are necessarily non-negative. The generalized radiant intensity is, of course, also non-negative.

VI. SPECIAL CASE (B), COHERENT SOURCES

For a coherent source the cross-spectral density function W factorizes in the form[11]

$$W(\mathbf{r}_1,\mathbf{r}_2;\omega)=v(\mathbf{r}_1)v^*(\mathbf{r}_2). \tag{57}$$

Of course, $v(\mathbf{r})=0$ when the point $\mathbf{r}$ lies outside the source area. On substituting from Eq. (57) into Eq. (15), we obtain, for the generalized radiance of a coherent source, the expression

$$B_\omega(\mathbf{r},\mathbf{s})=\left(\frac{k}{2\pi}\right)^2\cos\theta\int_{(z'=0)} v(\mathbf{r}+\tfrac{1}{2}\mathbf{r}')v^*(\mathbf{r}-\tfrac{1}{2}\mathbf{r}') \times\exp(-ik\mathbf{s}\cdot\mathbf{r}')\, d^2\mathbf{r}'. \tag{58}$$

The generalized radiant emittance of a coherent source may be determined by substituting from Eq. (57) into Eq. (39). We then obtain

$$E_\omega(\mathbf{r})=\frac{k^2}{2(2\pi)^{\frac{1}{2}}}\int_{(z=0)} v(\mathbf{r}+\tfrac{1}{2}\mathbf{r}')v^*(\mathbf{r}-\tfrac{1}{2}\mathbf{r}')\frac{J_{\frac{3}{2}}(kr')}{(kr')^{\frac{3}{2}}}\, d^2\mathbf{r}'. \tag{59}$$

To determine the generalized radiant intensity of a coherent source, we shall use Eq. (41) and recall that, according to Eq. (24), $\hat{W}(\mathbf{f}_1,\mathbf{f}_2;\omega)$ is the four-fold Fourier inverse of the cross-spectral density function $W(\mathbf{r}_1,\mathbf{r}_2;\omega)$. In the present case, when W is given by Eq. (57), $\hat{W}$ clearly has the form

$$\hat{W}(\mathbf{f}_1,\mathbf{f}_2;\omega)=\hat{v}(\mathbf{f}_1)\hat{v}^*(-\mathbf{f}_2), \tag{60}$$

where $\hat{v}(\mathbf{f})$ is the two-fold Fourier transform of $v(\mathbf{r})$, i.e.,

$$\hat{v}(\mathbf{f})=\frac{1}{(2\pi)^2}\int_{(z=0)} v(\mathbf{r})\exp(-i\mathbf{f}\cdot\mathbf{r})\, d^2\mathbf{r}. \tag{61}$$

From Eqs. (41), (60), and (61), it follows that the generalized radiant intensity of a coherent source is given by

$$J_\omega(\mathbf{s})=(2\pi k)^2\cos^2\theta\,|\hat{v}(k\mathbf{s})|^2. \tag{62}$$

From Eqs. (58) and (59), we can readily verify our earlier assertions that both the generalized radiance and the generalized radiant emittance may take on negative values. To see this, consider, for example, the values of these quantities at a point of the source that coincides with the origin $\mathbf{r}=0$ of our coordinate system. The expressions (58) and (59) then reduce to

$$B_\omega(0,\mathbf{s})=\left(\frac{k}{2\pi}\right)^2\cos\theta\int_{(z'=0)} v(\tfrac{1}{2}\mathbf{r}')v^*(-\tfrac{1}{2}\mathbf{r}')\times\exp(-ik\mathbf{s}\cdot\mathbf{r}')\,d^2\mathbf{r}', \quad (63)$$

$$E_\omega(0)=\frac{k^2}{2(2\pi)^{\frac{1}{2}}}\int_{(z'=0)} v(\tfrac{1}{2}\mathbf{r}')v^*(-\tfrac{1}{2}\mathbf{r}')\frac{J_{\frac{3}{2}}(kr')}{(kr')^{\frac{3}{2}}}\,d^2\mathbf{r}'. \quad (64)$$

Suppose, next, that the source is uniform, i.e., $v(\mathbf{r})=v_0=$ const when the point $\mathbf{r}$ is within the area of the plane $z=0$ occupied by the source and is zero elsewhere on that plane. Moreover, suppose that the source has the form of a square of sides a, parallel to the x and y axes and centered at the origin. Equation (63) then may readily be shown to reduce to

$$B_\omega(0,\mathbf{s})=\frac{|v_0|^2(ka)^2}{\pi^2}\left(\frac{\sin(kas_x)}{kas_x}\right)\left(\frac{\sin(kas_y)}{kas_y}\right)\cos\theta. \quad (65)$$

If we now choose the $\mathbf{s}$ direction to be in the xz plane, then $s_y=0$, $s_x=\sin\theta$, and Eq. (65) gives

$$B_\omega(0,\mathbf{s})=\frac{|v_0|^2(ka)^2}{\pi^2}\left[\frac{\sin(ka\sin\theta)}{ka\sin\theta}\right]\cos\theta. \quad (66)$$

Clearly, B_ω will be negative, for example, for directions for which $\pi<ka\sin\theta<2\pi$. If we choose the side of the square $a=10^5\lambda$, we see from Eq. (66) that $B_\omega(0,\mathbf{s})$ will be negative for θ values such that $5\times10^{-6}<\sin\theta<10^{-5}$, i.e., 2.85×10^{-4} degrees $<\theta<5.7\times10^{-4}$ degrees.

An example of a situation where the generalized radiant emittance becomes negative is provided by a uniform coherent source that has the form of an annulus about the origin $\mathbf{r}=0$. If r_1 and r_2 are the inner and outer radii of the annulus, Eq. (64) can readily be shown to reduce to

$$E_\omega(0)=|v_0|^2\left[\frac{\sin(2kr_1)}{2kr_1}-\frac{\sin(2kr_2)}{2kr_2}\right]. \quad (67)$$

This formula shows, incidently, that the generalized radiant emittance at the center of the annulus, i.e., in the source plane outside the domain occupied by the source itself has a nonzero value in general. Moreover, if we choose suitable values for the two radii, the generalized radiant emittance at the origin becomes negative. For example, with the choice $r_1=10^5\lambda$, $r_2=(10^5+\frac{1}{8})\lambda$, Eq. (67) gives $E_\omega(0)=-8\times10^{-7}|v_0|^2$.

The peculiar features of the generalized radiance and of the generalized radiant emittance, illustrated by the two examples just considered, underscore our contention that, in general, these two quantities are not measurable. On the other hand, the generalized radiant intensity defined by Eq. (21) may be shown to be non-negative in both of these cases, in agreement with the general conclusion expressed by Eq. (43). Clearly, the generalized radiant intensity is the basic measurable quantity that characterizes some of the radiation properties of a source of any state of coherence.

APPENDIX A: DERIVATION OF THE EXPRESSION (23) FOR THE GENERALIZED RADIANCE

The generalized radiance was defined by the formula (15), viz.,

$$B_\omega(\mathbf{r},\mathbf{s})=\left(\frac{k}{2\pi}\right)^2\cos\theta\int_{(z'=0)} W(\mathbf{r}+\tfrac{1}{2}\mathbf{r}',\mathbf{r}-\tfrac{1}{2}\mathbf{r}';\omega)\times\exp(-ik\mathbf{s}\cdot\mathbf{r}')\,d^2\mathbf{r}', \quad (\text{A}1)$$

where $W(\mathbf{r}_1,\mathbf{r}_2;\omega)$ is the cross-spectral density function.

Let us represent W as a four-dimensional Fourier integral [given by the inverse of Eq. (24)]

$$W(\mathbf{r}_1,\mathbf{r}_2;\omega)=\iint_{(\mathbf{f}_1,\mathbf{f}_2\text{ planes})}\hat{W}(\mathbf{f}_1,\mathbf{f}_2;\omega)\times\exp[i(\mathbf{f}_1\cdot\mathbf{r}_1+\mathbf{f}_2\cdot\mathbf{r}_2)]\,d^2\mathbf{f}_1\,d^2\mathbf{f}_2. \quad (\text{A}2)$$

If we substitute from Eq. (A2) into Eq. (A1), rearrange some of the terms in the integrand, and invert the order of integration, we find that

$$B_\omega(\mathbf{r},\mathbf{s})=\left(\frac{k}{2\pi}\right)^2\cos\theta\iint_{(\mathbf{f}_1,\mathbf{f}_2\text{ planes})}\hat{W}(\mathbf{f}_1,\mathbf{f}_2;\omega)\times\exp[i(\mathbf{f}_1+\mathbf{f}_2)\cdot\mathbf{r}]G(\tfrac{1}{2}\mathbf{f}_1-\tfrac{1}{2}\mathbf{f}_2-k\mathbf{s})\,d^2\mathbf{f}_1\,d^2\mathbf{f}_2, \quad (\text{A}3)$$

where

$$\begin{aligned}G(\tfrac{1}{2}\mathbf{f}_1-\tfrac{1}{2}\mathbf{f}_2-k\mathbf{s})&=\int_{(z'=0)}\exp[i\mathbf{r}'\cdot(\tfrac{1}{2}\mathbf{f}_1-\tfrac{1}{2}\mathbf{f}_2-k\mathbf{s})]\,d^2\mathbf{r}'\\&=(2\pi)^2\delta^{(2)}(\tfrac{1}{2}\mathbf{f}_1-\tfrac{1}{2}\mathbf{f}_2-k\mathbf{s})\\&=4(2\pi)^2\delta^{(2)}(\mathbf{f}_1-\mathbf{f}_2-2k\mathbf{s}),\end{aligned} \quad (\text{A}4)$$

and $\delta^{(2)}$ is again the two-dimensional Dirac δ function. Integration in Eq. (A3) with respect to $\mathbf{f}_2$ is trivial; we obtain for B_ω

$$B_\omega(\mathbf{r},\mathbf{s})=4k^2\cos\theta\int_{(\mathbf{f}\text{ plane})}\hat{W}(\mathbf{f}_1,\mathbf{f}_1-2k\mathbf{s};\omega)\times\exp[2i(\mathbf{f}_1-k\mathbf{s})\cdot\mathbf{r}]\,d^2\mathbf{f}_1. \quad (\text{A}5)$$

Let us now introduce in place of $\mathbf{f}_1$ a new variable

$$\mathbf{f}=2(\mathbf{f}_1-k\mathbf{s}). \quad (\text{A}6)$$

Formula (A5) then becomes

$$B_\omega(\mathbf{r},\mathbf{s})=k^2\cos\theta\int_{(\mathbf{f}\text{ plane})}\hat{W}(k\mathbf{s}+\tfrac{1}{2}\mathbf{f},\,-k\mathbf{s}+\tfrac{1}{2}\mathbf{f};\omega)\times\exp(i\mathbf{f}\cdot\mathbf{r})\,d^2\mathbf{f},\quad\text{(A7)}$$

which is Eq. (23).

APPENDIX B: REALITY OF THE GENERALIZED RADIANCE

In this Appendix we show that the generalized radiance, defined by Eq. (15), viz.,

$$B_\omega(\mathbf{r},\mathbf{s})=\left(\frac{k}{2\pi}\right)^2\cos\theta\int_{(z'=0)}W(\mathbf{r}+\tfrac{1}{2}\mathbf{r}',\,\mathbf{r}-\tfrac{1}{2}\mathbf{r}';\omega)\times\exp(-ik\mathbf{s}\cdot\mathbf{r}')\,d^2\mathbf{r}'\quad\text{(B1)}$$

is real. We first take the complex conjugate (denoted by an asterisk) of Eq. (B1) and obtain

$$B^*_\omega(\mathbf{r},\mathbf{s})=\left(\frac{k}{2\pi}\right)^2\cos\theta\int_{(z'=0)}W^*(\mathbf{r}+\tfrac{1}{2}\mathbf{r}',\,\mathbf{r}-\tfrac{1}{2}\mathbf{r}';\omega)\times\exp(ik\mathbf{s}\cdot\mathbf{r}')\,d^2\mathbf{r}'.\quad\text{(B2)}$$

Now, from the definition of the cross-spectral density [cf. Eq. (6) of Ref. 5], it follows that

$$W^*(\mathbf{r}_1,\mathbf{r}_2;\omega)=W(\mathbf{r}_2,\mathbf{r}_1;\omega),\quad\text{(B3)}$$

so that Eq. (B2) can be expressed in the form

$$B^*_\omega(\mathbf{r},\mathbf{s})=\left(\frac{k}{2\pi}\right)^2\cos\theta\int_{(z'=0)}W(\mathbf{r}-\tfrac{1}{2}\mathbf{r}',\,\mathbf{r}+\tfrac{1}{2}\mathbf{r}';\omega)\times\exp(ik\mathbf{s}\cdot\mathbf{r}')\,d^2\mathbf{r}'.\quad\text{(B4)}$$

Next we change the variables of integration on the right-hand side of Eq. (B4) from $\mathbf{r}'$ to $-\mathbf{r}'$ and obtain

$$B^*_\omega(\mathbf{r},\mathbf{s})=\left(\frac{k}{2\pi}\right)^2\cos\theta\int_{(z'=0)}W(\mathbf{r}+\tfrac{1}{2}\mathbf{r}',\,\mathbf{r}-\tfrac{1}{2}\mathbf{r}';\omega)\times\exp(-ik\mathbf{s}\cdot\mathbf{r}')\,d^2\mathbf{r}'.\quad\text{(B5)}$$

Comparison of Eq. (B5) with Eq. (B1) shows that

$$B^*_\omega(\mathbf{r},\mathbf{s})=B_\omega(\mathbf{r},\mathbf{s}),\quad\text{(B6)}$$

implying that the generalized radiance $B_\omega(\mathbf{r},\mathbf{s})$ is real.

REFERENCES

*Research supported in part by the Air Force Office of Scientific Research.

[1]M. Planck, *The Theory of Heat Radiation*, translation from the Second Edition (Dover, New York, 1959).

[2]M. Born and E. Wolf, *Principles of Optics, 4th ed.* (Pergamon, Oxford and New York, 1970), §10.4.2.

[3]A preliminary account of our main results was published in Opt. Commun. **6**, 305 (1972) and was also presented at a meeting of the Optical Society of America held in Rochester, N. Y., 9–12 October 1973 [J. Opt. Soc. Am. **63**, 1285A (1973)]. In Eq. (12) of the published paper, a factor $(2\pi)^{-2}$ should be omitted.

[4]A. Walther, J. Opt. Soc. Am. **58**, 1256 (1968). In a recent paper [J. Opt. Soc. Am. **63**, 1622 (1973)] Walther modified his original definition of the generalized radiance after asserting that it depends on the choice of the coordinate system. This assertion, however, is misleading in the context of his earlier paper relating to radiation from planar sources. For, as we show in a Letter on p. 1273 in the present issue, the generalized radiance, as originally defined by Walther, is, in fact, invariant with respect to an arbitrary displacement of the origin of coordinates in the source plane and is also invariant with respect to rotation of axes about the normal to the plane of the source. In any case, as will be clear from the discussion in the present paper, it is not the generalized radiance, but rather the generalized radiant intensity that has a direct physical significance.

[5]E. W. Marchand and E. Wolf, J. Opt. Soc. Am. **62**, 379 (1972). In Eq. (10) and in some of the subsequent equations of this reference there is an error: m_2 should be replaced by its complex conjugate m_2^*. This error does not, however, affect the main results.

[6]E. Wigner, Phys. Rev. **40**, 749 (1932). For a good discussion of some of the properties of the Wigner distribution function, see K. Imre, E. Ozizmir, M. Rosenbaum, and P. F. Zweifel, J. Math. Phys. **8**, 1907 (1967).

[7]See, for example, G. N. Watson, *A Treatise on the Theory of Bessel Functions* (Cambridge U. P., 1922), p. 20, Eq. (5) (with an obvious substitution).

[8]I. S. Gradsteyn and I. M. Ryzhik, *Tables of Integrals, Series and Products* (Academic, New York, 1965), p. 688, formula 1 of §6.567, with $\nu=0$, $\mu=\frac{1}{2}$.

[9]Alternative representations of spatially incoherent sources are discussed in a paper by M. Beran and G. Parrent, Nuovo Cimento **27**, 1049 (1963).

[10](a) C. L. Mehta, Nuovo Cimento **28**, 401 (1963); (b) R. C. Bourret, Nuovo Cimento **18**, 347 (1960); (c) C. L. Mehta and E. Wolf, Phys. Rev. **134**, A1143 (1964); Phys. Rev. **134**, A1149 (1964).

[11]J. Peřina, *Coherence of Light* (Van Nostrand, London, 1972), §4.2.

Coherence and radiometry with quasihomogeneous planar sources*

W. H. Carter†
Naval Research Laboratory, Washington, D.C. 20375

E. Wolf
Department of Physics and Astronomy, University of Rochester, Rochester, New York 14627
(Received 18 November 1976)

The concept of a quasihomogeneous source is introduced. Unlike a source that is strictly homogeneous in its statistical properties, a quasihomogeneous source may be finite. Many physical sources, both primary and secondary ones, are adequately approximated by this model. Coherence and radiometric properties of light generated by such sources (assumed, for simplicity, to be planar) are discussed and an important reciprocity relation is shown to exist between light in the far zone and in the source plane. This relation implies that the degree of coherence in the far zone is given by the classic form of the van Cittert-Zernike theorem, even though the source may have a high degree of spatial coherence over arbitrarily large areas. The reciprocity relation also provides a generalization of a recently derived result that expresses the angular dependence of the radiant intensity in terms of the degree of spatial coherence of light in the source plane. The dependence of all the basic radiometric quantities on the distribution of the optical intensity across the source and on the degree of spatial coherence of the light emerging from the source is discussed and is illustrated, for some typical sources, by computed curves.

I. INTRODUCTION

In a recent paper,[1] the angular distribution of the radiant intensity from sources of different degrees of spatial coherence was studied. The analysis was restricted to radiation from large statistically homogeneous planar sources; it was found that the radiant intensity function is related in a very simple manner to the two-dimensional spatial Fourier transform of the degree of coherence across the source.[2,3] No finite source can, of course, be homogeneous in the strict statistical sense and it is evidently desirable to refine the analysis to take into account the finite size of a real physical source.

In the present paper we relax some of the conditions that are characteristic of statistical homogeneity and we are then led to a class of sources that we call *quasihomogeneous*. Unlike for a homogeneous source the intensity distribution of light across a quasihomogeneous source is no longer necessarily constant but can vary slowly with position, remaining sensibly constant over distances of the order of the correlation length of the light vibrations in the source plane. We show that the light from such sources reflects the source properties in a very direct manner. In particular we find that while the angular distribution of the radiant intensity depends in a simple manner on the two-dimensional spatial Fourier transform of the (spectral) degree of spatial coherence[4] of the light emerging from the source, the degree of spatial coherence of the far field is proportional (apart from a trivial phase factor), to the normalized two-dimensional spatial Fourier transform of the intensity distribution across the source. The second part of this new reciprocity theorem[5] is formally identical with the far-zone form of the classic theorem of van Cittert and Zernike,[6] that predicts the degree of coherence of light from a spatially incoherent source. This result implies that the far-zone form of the van Cittert–Zernike theorem applies to light from any quasihomogeneous source. Such a source may possess a high degree of coherence over arbitrarily large areas of the source plane.

We also obtain expressions for the radiance and the radiant emittance of a quasihomogeneous source. We find that the radiance factorizes into the product of two functions, one of which is a function of position, whereas the other is a function of direction and both have simple significance in terms of quantities that characterize the source. We also find that the radiant emittance of a quasihomogeneous source is proportional to the optical intensity, the proportionality factor depending on the degree of coherence of the source. Finally, we obtain a simple expression for the total radiant flux generated by a source of this type. This expression makes it possible to compare the radiation efficiency of quasihomogeneous sources of the same intensity distribution but of different states of coherence. We illustrate our main results by a number of curves computed for typical model sources.

The analysis of this paper not only provides a real insight into the far-zone behavior of light from quasihomogeneous sources, but it also completely clarifies, for sources of this class, the relationship between radiometry and physical optics. The concluding section (Sec. VIII) presents our main results in a compact form.

II. CONCEPT OF A QUASIHOMOGENEOUS SOURCE

Consider a statistically stationary light distribution (cf. Ref. 6, p. 498) in some plane $z=0$. Let $W(\mathbf{r}_1, \mathbf{r}_2)$ be the cross-spectral density function (also called the mutual spectral density) of the light, at some chosen frequency ω, at two points Q_1 and Q_2 in that plane, the points being specified by position vectors $\mathbf{r}_1$ and $\mathbf{r}_2$, respectively. Further let

$$I(\mathbf{r}) = W(\mathbf{r}, \mathbf{r}) \tag{2.1}$$

represent the optical intensity[7] of the light at a point P, specified by position vector $\mathbf{r}$ in the source plane and let

$$\mu(\mathbf{r}_1, \mathbf{r}_2) = W(\mathbf{r}_1, \mathbf{r}_2)/[I(\mathbf{r}_1)]^{1/2}[I(\mathbf{r}_2)]^{1/2} \tag{2.2a}$$

be the complex degree of spatial coherence[4] at the particular frequency ω of the light vibrations at the two points P_1 and P_2. All these quantities, i.e., W, I, and μ depend also on the frequency ω, but we do not indicate this dependence explicitly. It will be convenient to rewrite Eq. (2.2a) in the form

$$W(\mathbf{r}_1, \mathbf{r}_2) = [I(\mathbf{r}_1)]^{1/2}[I(\mathbf{r}_2)]^{1/2}\mu(\mathbf{r}_1, \mathbf{r}_2) . \tag{2.2b}$$

Suppose now that the light distribution in the plane $z=0$ is statistically homogeneous,[8] at least in the sense of the second-order correlation theory. The cross-spectral density function $W(\mathbf{r}_1, \mathbf{r}_2)$ then depends on $\mathbf{r}_1$ and $\mathbf{r}_2$ only through the difference $\mathbf{r}_1 - \mathbf{r}_2$, i.e.,

$$W(\mathbf{r}_1, \mathbf{r}_2) = F(\mathbf{r}_1 - \mathbf{r}_2) , \tag{2.3}$$

where $F(\mathbf{r})$ is a function of a single two-dimensional vector variable $\mathbf{r}$. It follows from (2.3) and (2.1) that we now have

$$I(\mathbf{r}) = F(0) , \tag{2.4}$$

and we see from (2.2a), (2.3), and (2.4) that in this case

$$\mu(\mathbf{r}_1, \mathbf{r}_2) = F(\mathbf{r}_1 - \mathbf{r}_2)/F(0) . \tag{2.5}$$

Formulas (2.4) and (2.5) show that statistical homogeneity implies that the optical intensity of the light is the same at every point in the plane $z=0$ and that the complex degree of spatial coherence $\mu(\mathbf{r}_1, \mathbf{r}_2)$ of the light depends on $\mathbf{r}_1$ and $\mathbf{r}_2$ only through the difference $\mathbf{r}_1 - \mathbf{r}_2$.

Conversely, it is clear that if (a) $I(\mathbf{r})$ is independent of $\mathbf{r}$ and (b) $\mu(\mathbf{r}_1, \mathbf{r}_2)$ is a function of $\mathbf{r}_1 - \mathbf{r}_2$, only then the light distribution in the plane $z=0$ is necessarily statistically homogeneous, at least in the sense of the second-order correlation theory. For when both the conditions (a) and (b) are satisfied, we see from Eq. (2.2b) that the cross-spectral density is necessarily of the form (2.3).

The definition of statistical homogeneity demands that a relation of the form (2.3), or equivalently, the conditions (a) and (b) above, are satisfied for all position vectors $\mathbf{r}_1$, $\mathbf{r}_2$, and $\mathbf{r}$ that represent points in the plane $z=0$. Hence, if this plane is the source plane (the source being either a true primary source or a secondary one), the source must necessarily occupy the whole infinite plane. Obviously that requirement is never satisfied in practice.

It is clear that if we drop the requirement (a) above, we may represent a source of a finite extent. We will also weaken the requirement (b) by demanding that only the complex degree of spatial coherence be of the form

$$\mu(\mathbf{r}_1, \mathbf{r}_2) = g(\mathbf{r}_1 - \mathbf{r}_2) \tag{2.6}$$

for points $\mathbf{r}_1$, and $\mathbf{r}_2$ within the source area in the plane $z=0$. In (2.6), $g(\mathbf{r}')$ is a function of a single (two-dimensional) vector variable $\mathbf{r}'$. According to (2.6) and (2.2b), the cross-spectral density function of the light in the plane $z=0$ is then of the form

$$W(\mathbf{r}_1, \mathbf{r}_2) = [I(\mathbf{r}_1)]^{1/2}[I(\mathbf{r}_2)]^{1/2} g(\mathbf{r}_1 - \mathbf{r}_2) , \tag{2.7}$$

where $I(\mathbf{r})$ is a non-negative function of $\mathbf{r}$ that is assumed to vanish identically when $\mathbf{r}$ represents points in the source plane $z=0$ that are situated outside the source area.

Some far-zone properties of light generated by a finite (secondary) source distribution whose cross-spectral density function is of the form (2.7) appear to have been first investigated by Schell and others.[9,10] Unfortunately, the formulas that describe the behavior of such optical fields are too complicated to provide a simple intuitive insight into their properties. For this reason we will now make several simplifying assumptions about the behavior of $g(\mathbf{r}_1 - \mathbf{r}_2)$ and $I(\mathbf{r})$ that are nevertheless realistic enough to apply to many sources of practical interest. We will also assume from now on that the linear dimensions of the source are large compared with the wavelength of the light.

The function $g(\mathbf{r}_1 - \mathbf{r}_2)$ is a measure of spatial coherence that exists in the light vibrations at the points Q_1 and Q_2 in the source area. We will assume that $g(\mathbf{r}')$, considered a function of $\mathbf{r}' = \mathbf{r}_1 - \mathbf{r}_2$, is appreciably different from zero only within some $\mathbf{r}'$ domain whose linear dimensions are much smaller then those of the source. We also assume that the optical intensity $I(\mathbf{r})$ changes so slowly with position across the source, that it is sensibly constant over regions whose linear dimensions are of the order of the correlation distance [the effective range of $g(\mathbf{r}')$] of the light in the source plane. Many sources encountered, in practice, may be expected to satisfy these conditions. Obvious examples are a blackbody source, or a secondary source obtained by imaging a primary thermal source by a system whose impulse response (the spread function) is only appreciable over areas of the image plane throughout which the optical intensity is essentially constant.[11] Under these circumstances we may approximate in (2.7) by setting

$$[I(\mathbf{r}_1)]^{1/2}[I(\mathbf{r}_2)]^{1/2} = I[\tfrac{1}{2}(\mathbf{r}_1, \mathbf{r}_2)] , \tag{2.8}$$

and we then obtain in place of (2.7) the simpler expression

$$W(\mathbf{r}_1, \mathbf{r}_2) \approx I[\tfrac{1}{2}(\mathbf{r}_1 + \mathbf{r}_2)] g(\mathbf{r}_1 - \mathbf{r}_2) . \tag{2.9}$$

When the cross-spectral density function of the light across the source may be approximated in the form (2.9), where $I(\mathbf{r})$ and $g(\mathbf{r}')$ are functions with the properties that we just stated, the source will be said to be *quasihomogeneous*.[12,13]

We will express the basic features of the relative behavior that we just discussed, of the functions $I(\mathbf{r})$ and $g(\mathbf{r}')$, by saying that $I(\mathbf{r})$ is a *slow* function of $\mathbf{r}$, and $g(\mathbf{r}')$ is a *fast* function of $\mathbf{r}'$. It should be noted that our requirement that $I(\mathbf{r})$ vary slowly with r is violated at points that are situated on the boundary of the source, unless $I(\mathbf{r})$ decreases to zero continuously and slowly enough as the point $\mathbf{r}$ approaches the boundary from within the source area. This requirement will, as a rule, not be satisfied in practice. Nor will $\mu(\mathbf{r}_1, \mathbf{r}_2)$ be generally of the form $g(\mathbf{r}_1 - \mathbf{r}_2)$ when $\mathbf{r}_1$ and $\mathbf{r}_2$ represent points that are close to the edge of the source (cf. Ref. 11). Nevertheless we will treat such sources as if they were quasihomogeneous, provided that the postulated conditions on $I(\mathbf{r})$ and $g(\mathbf{r}')$ are, to a good approximation, satisfied elsewhere within the source area. For under these circumstances the effect, on the general behavior of the field, of the departure from true quasihomogeneity may be expected to be negligible.[14]

III. SPATIAL-FREQUENCY TRANSFORM OF THE CROSS-SPECTRAL DENSITY FUNCTION OF A QUASIHOMOGENEOUS SOURCE

For the purpose of later analysis we will now derive an expression for the four-dimensional spatial Fourier transform of the cross-spectral density function $W(\mathbf{r}_1, \mathbf{r}_2)$ of the light distribution in a quasihomogeneous planar source. We will see in Sec. IV that this expression is basic for the elucidation of some important properties of fields generated by sources of this type.

Let $\tilde{W}(\mathbf{f}_1, \mathbf{f}_2)$ be the (four-dimensional) Fourier spectrum of the cross-spectral density function $W(\mathbf{r}_1, \mathbf{r}_2)$, i.e.,

$$W(\mathbf{r}_1, \mathbf{r}_2) = \iint \tilde{W}(\mathbf{f}_1, \mathbf{f}_2)\, e^{i(\mathbf{f}_1\cdot\mathbf{r}_1+\mathbf{f}_2\cdot\mathbf{r}_2)}\, d^2f_1\, d^2f_2 . \tag{3.1a}$$

We assume that (3.1a) may be inverted, so that

$$\tilde{W}(\mathbf{f}_1, \mathbf{f}_2) = \frac{1}{(2\pi)^4} \iint W(\mathbf{r}_1, \mathbf{r}_2)\, e^{-i(\mathbf{f}_1\cdot\mathbf{r}_1+\mathbf{f}_2\cdot\mathbf{r}_2)}\, d^2r_1\, d^2r_2 . \tag{3.1b}$$

The integrations in (3.1a) extend, of course, over the complete $\mathbf{f}_1$ and $\mathbf{f}_2$ planes. In (3.1b) the integrations are formally taken over the complete $\mathbf{r}_1$ and $\mathbf{r}_2$ planes (i.e., twice independently over the full source plane $z=0$), it being understood that $W(\mathbf{r}_1, \mathbf{r}_2)$ has zero value when either or both of the points represented by the vectors $\mathbf{r}_1$ and $\mathbf{r}_2$ are situated outside the area occupied by the source. If we substitute in Eq. (3.1b) for $W(\mathbf{r}_1, \mathbf{r}_2)$ the expression (2.9) appropriate for a quasihomogeneous source, we obtain the following expression for $\tilde{W}$:

$$\tilde{W}(\mathbf{f}_1, \mathbf{f}_2) = \frac{1}{(2\pi)^4} \iint I\left(\frac{\mathbf{r}_1+\mathbf{r}_2}{2}\right) g(\mathbf{r}_1 - \mathbf{r}_2) \times e^{-i(\mathbf{f}_1\cdot\mathbf{r}_1+\mathbf{f}_2\cdot\mathbf{r}_2)}\, d^2r_1\, d^2r_2 . \tag{3.2}$$

Next let us introduce new variables of integration $\mathbf{r}, \mathbf{r}'$, defined by the formulas

$$\mathbf{r} = \tfrac{1}{2}(\mathbf{r}_1 + \mathbf{r}_2), \qquad \mathbf{r}' = (\mathbf{r}_1 - \mathbf{r}_2) . \tag{3.3}$$

We then have the inverse relations

$$\mathbf{r}_1 = \mathbf{r} + \tfrac{1}{2}\mathbf{r}', \qquad \mathbf{r}_2 = \mathbf{r} - \tfrac{1}{2}\mathbf{r}' . \tag{3.4}$$

The argument in the exponential term in Eq. (3.2) becomes

$$\mathbf{f}_1 \cdot \mathbf{r}_1 + \mathbf{f}_2 \cdot \mathbf{r}_2 = \mathbf{r} \cdot (\mathbf{f}_1 + \mathbf{f}_2) + \mathbf{r}' \cdot [\tfrac{1}{2}(\mathbf{f}_1 - \mathbf{f}_2)] . \tag{3.5}$$

The Jacobian of the transformation $(\mathbf{r}_1, \mathbf{r}_2) \to (\mathbf{r}, \mathbf{r}')$ is readily seen to be unity. Using this fact and relations (3.3) and (3.5), the expression (3.2) for $\tilde{W}$ becomes

$$\tilde{W}(\mathbf{f}_1, \mathbf{f}_2) = \frac{1}{(2\pi)^4} \iint I(\mathbf{r}) g(\mathbf{r}') \times e^{-i\{\mathbf{r}\cdot(\mathbf{f}_1+\mathbf{f}_2)+\mathbf{r}'\cdot[(\mathbf{f}_1-\mathbf{f}_2)/2]\}}\, d^2r\, d^2r' , \tag{3.6}$$

the integrations extending over the complete $\mathbf{r}$ and $\mathbf{r}'$ planes (i.e., twice independently over the whole source plane $z=0$). If we introduce the two-dimensional spatial Fourier spectra of $I(\mathbf{r})$ and of $g(\mathbf{r}')$, i.e.,

$$\tilde{I}(\mathbf{f}) = \frac{1}{(2\pi)^2} \int I(\mathbf{r})\, e^{-i\mathbf{f}\cdot\mathbf{r}}\, d^2r , \tag{3.7}$$

$$\tilde{g}(\mathbf{f}) = \frac{1}{(2\pi)^2} \int g(\mathbf{r}')\, e^{-i\mathbf{f}\cdot\mathbf{r}'}\, d^2r' , \tag{3.8}$$

the expression (3.6) takes the simple form

$$\tilde{W}(\mathbf{f}_1, \mathbf{f}_2) = \tilde{I}(\mathbf{f}_1 + \mathbf{f}_2)\tilde{g}[\tfrac{1}{2}(\mathbf{f}_1 - \mathbf{f}_2)] . \tag{3.9}$$

From Eqs. (3.9) and (2.9) we see that when the cross-spectral density function $W(\mathbf{r}_1, \mathbf{r}_2)$ factorizes into the product of the intensity $I(\mathbf{r})$ and the correlation $g(\mathbf{r}')$, with the arguments $\mathbf{r}$ and $\mathbf{r}'$ defined by Eqs. (3.3), the four-dimensional spatial Fourier spectrum $\tilde{W}(\mathbf{f}_1, \mathbf{f}_2)$ of the cross-spectral density function $W(\mathbf{r}_1, \mathbf{r}_2)$ also factorizes. Moreover the two factors of $\tilde{W}(\mathbf{f}_1, \mathbf{f}_2)$ are the two-dimensional Fourier spectra $\tilde{I}(\mathbf{f}')$ and $\tilde{g}(\mathbf{f})$ of $I(\mathbf{r})$ and of $g(\mathbf{r}')$, respectively, with the arguments

$$\mathbf{f}' = \mathbf{f}_1 + \mathbf{f}_2, \qquad \mathbf{f} = \tfrac{1}{2}(\mathbf{f}_1 - \mathbf{f}_2) . \tag{3.10}$$

The result that we just established holds whenever $W(\mathbf{r}_1, \mathbf{r}_2)$ factorizes in the form (2.9). If the source is quasihomogeneous then, in addition, $I(\mathbf{r})$ is a "slow" function of $\mathbf{r}$ and $g(\mathbf{r}')$ is a "fast" function of $\mathbf{r}'$. In view of the well known reciprocity inequality[15] involving the effective widths of Fourier transform pairs, it follows that for a quasihomogeneous source not only does $\tilde{W}(\mathbf{f}_1, \mathbf{f}_2)$ factorize in the form (3.9) but the function $\tilde{I}(\mathbf{f}')$ is a "fast" function of $\mathbf{f}'$ and $\tilde{g}(\mathbf{f})$ is a "slow" function of $\mathbf{f}$. We will see shortly that this result implies an interesting reciprocity between the light distribution in the source plane and in the far zone of a quasihomogeneous planar source.

IV. FAR-ZONE BEHAVIOR OF A FIELD GENERATED BY A QUASIHOMOGENEOUS SOURCE

We will now consider the far-zone behavior of a field generated by a quasihomogeneous source, located in the plane $z=0$ and radiating into the half-space $z>0$. Let $\mathbf{R}_1$ and $\mathbf{R}_2$ denote the position vectors of two points P_1 and P_2 in that half-space and let us set

$$\mathbf{R}_1 = R_1\mathbf{s}_1 , \qquad \mathbf{R}_2 = R_2\mathbf{s}_2 , \tag{4.1}$$

where $\mathbf{s}_1$ and $\mathbf{s}_2$ are unit vectors pointing from the origin 0 (chosen in the source area) to the points P_1 and P_2, respectively (see Fig. 1). We will examine first the behavior of the cross-spectral density function $W(R_1\mathbf{s}_1, R_2\mathbf{s}_2)$ in the asymptotic limit as $kR_1 \to \infty$, $kR_2 \to \infty$, with $\mathbf{s}_1$ and $\mathbf{s}_2$ being kept fixed. We will denote this asymptotic limit by $W^{(\infty)}(R_1\mathbf{s}_1, R_2\mathbf{s}_2)$. Moreover, to bring out clearly an important reciprocity relation between light

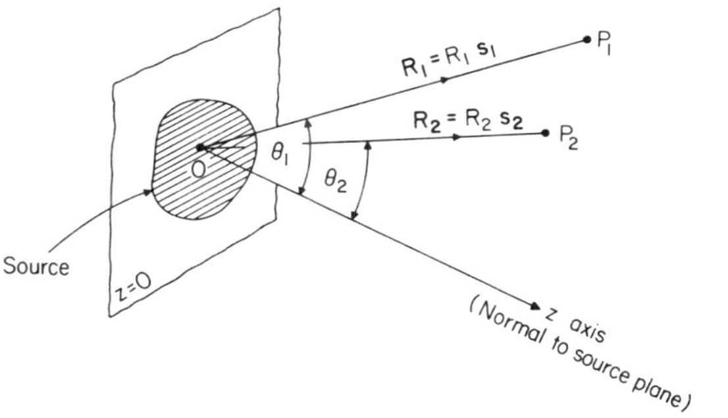

FIG. 1. Illustrating the notation relating to the far-zone behavior of the field.

in the source plane and in the far zone we will, from now on, denote the cross-spectral density function of light in the source plane $z=0$ by $W^{(0)}(\mathbf{r}_1,\mathbf{r}_2)$ rather than by $W(\mathbf{r}_1,\mathbf{r}_2)$ as we have done in Secs. II and III.

It has been shown elsewhere[16] that $W^{(\infty)}$ may be expressed in terms of the four-dimensional spatial Fourier inverse $\tilde{W}^{(0)}$ of $W^{(0)}$ by the formula

$$W^{(\infty)}(R_1\mathbf{s}_1,R_2\mathbf{s}_2)=(2\pi k)^2\cos\theta_1\cos\theta_2 \times\tilde{W}^{(0)}(k\mathbf{s}_{1\perp},-k\mathbf{s}_{2\perp})\,e^{ik(R_1-R_2)}/R_1R_2. \quad (4.2)$$

Here θ_1 and θ_2 are the angles that the unit vectors $\mathbf{s}_1$ and $\mathbf{s}_2$ make with the normal to the source plane (see Fig. 1), $\mathbf{s}_{1\perp}$ and $\mathbf{s}_{2\perp}$ are the projections (considered as vectors) of $\mathbf{s}_1$ and $\mathbf{s}_2$ onto the source plane, and

$$k=\omega/c \quad (4.3)$$

(c is the vacuum speed of light) is the wave number associated with the frequency component ω under consideration. Now, for a quasihomogeneous source we have, according to Eq. (3.9),

$$\tilde{W}^{(0)}(k\mathbf{s}_{1\perp},-k\mathbf{s}_{2\perp})=\tilde{I}^{(0)}[k(\mathbf{s}_{1\perp}-\mathbf{s}_{2\perp})]\tilde{g}^{(0)}[\tfrac{1}{2}k(\mathbf{s}_{1\perp}+\mathbf{s}_{2\perp})], \quad (4.4)$$

where we have written $\tilde{I}^{(0)}$ for $\tilde{I}$ and $\tilde{g}^{(0)}$ for $\tilde{g}$, to remind us that these quantities pertain to light in the source plane $z=0$. On substituting from (4.4) into (4.2) we finally obtain the following expression for the cross-spectral density function of the light in the far zone of a quasihomogeneous source:

$$W^{(\infty)}(R_1\mathbf{s}_1,R_2\mathbf{s}_2)=(2\pi k)^2\cos\theta_1\cos\theta_2\tilde{I}^{(0)}[k(\mathbf{s}_{1\perp}-\mathbf{s}_{2\perp})] \times\tilde{g}^{(0)}[\tfrac{1}{2}k(\mathbf{s}_{1\perp}+\mathbf{s}_{2\perp})]\,e^{ik(R_1-R_2)}/R_1R_2. \quad (4.5)$$

The implications of this formula can most easily be understood in physical terms by considering separately the intensity distribution and the degree of coherence of the light in the far zone.

The optical intensity[7] $I^{(\infty)}(R\mathbf{s})$ at a typical point in the far zone is, according to (4.5), given by

$$I^{(\infty)}(R\mathbf{s})\equiv W^{(\infty)}(R\mathbf{s},R\mathbf{s})=(2\pi k)^2\tilde{I}^{(0)}(0)\tilde{g}^{(0)}(k\mathbf{s}_\perp)\cos^2\theta/R^2. \quad (4.6)$$

Hence the radiant intensity $J(\mathbf{s})$, which is known to be related to the optical intensity in the far zone by the formula[17]

$$J(\mathbf{s})\equiv R^2 I^{(\infty)}(R\mathbf{s}) \quad (kR\to\infty), \quad (4.7)$$

is given by

$$J(\mathbf{s})=(2\pi k)^2\tilde{I}^{(0)}(0)\tilde{g}^{(0)}(k\mathbf{s}_\perp)\cos^2\theta. \quad (4.8)$$

The complex degree of spatial coherence of the light in the far zone is defined by the expression[4]

$$\mu^{(\infty)}(R_1\mathbf{s}_1,R_2\mathbf{s}_2)\equiv\frac{W^{(\infty)}(R_1\mathbf{s}_1,R_2\mathbf{s}_2)}{[I^{(\infty)}(R_1\mathbf{s}_1)]^{1/2}[I^{(\infty)}(R_2\mathbf{s}_2)]^{1/2}}. \quad (4.9)$$

On substituting on the right-hand side of Eq. (4.9) from (4.5) and (4.6) we deduce at once that, in the present case,

$$\mu^{(\infty)}(R_1\mathbf{s}_1,R_2\mathbf{s}_2)=\frac{\tilde{I}^{(0)}[k(\mathbf{s}_{1\perp}-\mathbf{s}_{2\perp})]}{\tilde{I}^{(0)}(0)} \times\tilde{G}^{(0)}(k\mathbf{s}_{1\perp},k\mathbf{s}_{2\perp})\,e^{ik(R_1-R_2)}, \quad (4.10)$$

where

$$\tilde{G}^{(0)}(k\mathbf{s}_{1\perp},k\mathbf{s}_{2\perp})=\frac{\tilde{g}^{(0)}[\tfrac{1}{2}k(\mathbf{s}_{1\perp}+\mathbf{s}_{2\perp})]}{[\tilde{g}^{(0)}(k\mathbf{s}_{1\perp})]^{1/2}[\tilde{g}^{(0)}(k\mathbf{s}_{2\perp})]^{1/2}}. \quad (4.11)$$

Now it was pointed out at the end of Sec. III that for a quasihomogeneous source, $\tilde{I}^{(0)}(\mathbf{f}')$ is a "fast" function of $\mathbf{f}'$, whereas $\tilde{g}^{(0)}(\mathbf{f})$ is a "slow" function of $\mathbf{f}$. It is thus clear that for arguments $k(\mathbf{s}_{1\perp}-\mathbf{s}_{2\perp})$ for which the factor $\tilde{I}^{(0)}[k(\mathbf{s}_{1\perp}-\mathbf{s}_{2\perp})]/\tilde{I}^{(0)}(0)$ in (4.10) differs appreciably from zero, $\mathbf{s}_{2\perp}$ may be replaced by $\mathbf{s}_{1\perp}$ on the right-hand side of (4.11) so that we then have

$$\tilde{G}^{(0)}(k\mathbf{s}_{1\perp},k\mathbf{s}_{2\perp})\approx\tilde{G}^{(0)}(k\mathbf{s}_{1\perp},k\mathbf{s}_{2\perp})\equiv 1. \quad (4.12)$$

Using this result in (4.10) it follows that

$$\mu^{(\infty)}(R_1\mathbf{s}_1,R_2\mathbf{s}_2)\approx\frac{\tilde{I}^{(0)}[k(\mathbf{s}_{1\perp}-\mathbf{s}_{2\perp})]}{\tilde{I}^{(0)}(0)}\,e^{ik(R_1-R_2)}. \quad (4.13)$$

Formulas (4.8) and (4.13), which are among the main results of this paper, bring into evidence a remarkable *reciprocity theorem*[5] which may be expressed as follows.

(*a*) *The angular distribution of the radiant intensity* $J(\mathbf{s})$ *from a quasihomogeneous planar source is proportional to the product of the two-dimensional spatial Fourier transform of the complex degree of spatial coherence of the light in the source plane and the square of the cosine of the angle that the* $\mathbf{s}$ *direction makes with the normal to the source plane* [Eq. (4.8)].

(*b*) *The complex degree of spatial coherence of the light in the far zone of a quasihomogeneous planar source is, apart from a simple geometrical phase factor, equal to the normalized Fourier transform of the distribution of the optical intensity across the source* [Eq. (4.13)].

It should be noted that these results imply that *with a quasihomogeneous source the effects on the far field of the optical intensity distribution across the source and of the coherence properties of the source are quite distinct*. The distribution of the optical intensity $I^{(0)}(\mathbf{r})$ across the source determines completely the (second-order) coherence properties of the far field. The complex degree of spatial coherence $g^{(0)}(\mathbf{r}')$ of the light in the source plane determines the distribution of the optical intensity throughout the far field; or, in what amounts to the same thing, it completely determines the relative angular distribution of the radiant intensity. More specifically, the relative distribution of the radiant intensity, $J(\mathbf{s})/J(0)$ is, according to Eq. (4.8), given by

$$J(\mathbf{s})/J(0)=[\tilde{g}^{(0)}(k\mathbf{s}_\perp)/\tilde{g}^{(0)}(0)]\cos^2\theta, \quad (4.14)$$

where $J(0)$ is the radiant intensity in the direction $s_x=s_y=0$, normal to the plane of the source. We see, in particular, that the relative angular distribution of *the radiant intensity is independent of the size of the source.*

The result (*a*) [Eq. (4.8)] is a generalization to quasihomogeneous sources of a result that was recently established for a large "homogeneous" source.[18]

The result (*b*) [Eq. (4.13)] *is formally identical with the far-zone form of a classic theorem of coherence theory, due to van Cittert and Zernike* (Ref. 6, Sec. 10.4.2). However, while in the classic formulation the source is assumed to be spatially incoherent, we have now shown that the theorem (in its far-zone form), holds much more generally, namely for radiation from any quasihomogeneous source. From our definition of quasihomogeneity (Sec. II) it is clear that such a source may have a high degree of coherence over large areas, possibly of many wavelengths in linear dimensions. In fact the definition of a quasihomogeneous source does not impose any restrictions whatsoever on the size of its "area of coherence," provided only that the source is sufficiently large. It seems remarkable that the van Cittert–Zernike theorem should apply also under conditions that are so different from those assumed in its original derivation.

We will illustrate our results by an example. Suppose that the (quasihomogeneous) source is circular and of radius $a \gg \lambda$ and that both the intensity distribution across the source and its degree of spatial coherence are given by Gaussian functions of zero mean and of variances σ_I and σ_g respectively, i.e.,

$$I^{(0)}(\mathbf{r}) = \begin{cases} A\,e^{-r^2/2\sigma_I^2} & \text{if } r \le a\,, \\ 0 & \text{if } r > a \end{cases} \tag{4.15}$$

and

$$g^{(0)}(\mathbf{r}') = e^{-r'^2/2\sigma_g^2}\,, \tag{4.16}$$

where A is a positive constant. There is no proportionality constant on the right-hand side of Eq. (4.16), since the degree of spatial coherence is normalized so that $g^{(0)}(0) = 1$. Now since the source is assumed to be quasihomogeneous we also have the constraints

$$\sigma_g \ll \sigma_I\,, \quad \sigma_g \ll a\,, \quad \lambda \ll a\,. \tag{4.17a}$$

To simplify the calculations we also assume that

$$\sigma_I \ll a\,. \tag{4.17b}$$

To determine the far-zone behavior of the field generated by this source we first determine the Fourier transforms defined by Eqs. (3.7) and (3.8) of the distributions (4.15) and (4.16). We readily find that

$$\tilde{I}^{(0)}(\mathbf{f}) \approx (A\sigma_I^2/2\pi)\,e^{-\sigma_I^2 f^2/2}\,, \tag{4.18a}$$

$$\tilde{g}^{(0)}(\mathbf{f}) = (\sigma_g^2/2\pi)\,e^{-\sigma_g^2 f^2/2}\,. \tag{4.18b}$$

In calculating $\tilde{I}^{(0)}(\mathbf{f})$ we made use of the constraint (4.17b), which enabled us to replace the truncated Gaussian distribution (4.15) by the complete Gaussian distribution. The required expression for the radiant intensity is now obtained at once on substituting from Eqs. (4.18) into Eq. (4.8). We then find that

$$J(\mathbf{s}) = J(0)\cos^2\theta\, e^{-k^2\sigma_g^2 \sin^2\theta/2}\,, \tag{4.19a}$$

where

$$J(0) = k^2 A \sigma_I^2 \sigma_g^2\,. \tag{4.19b}$$

In deriving formula (4.19a) we used the fact that

$$s_\perp^2 = s_x^2 + s_y^2 = \sin^2\theta\,. \tag{4.20}$$

Expression (4.19a), for the radiant intensity, is formally identical with Eq. (22) of Ref. 1 for the radiant intensity from a large "homogeneous" Gaussian-correlated source. This fact illustrates a result that we derived earlier in this paper, namely, that the relative distribution $J(\mathbf{s})/J(0)$ of the radiant intensity generated by a quasihomogeneous source is independent of the size of the source. Formula (4.19a) also illustrates another of our general results, namely that the relative distribution $J(\mathbf{s})/J(0)$ is independent of the distribution of the optical intensity $I^{(0)}(\mathbf{r})$ across the source. Since some other important consequences of Eq. (4.19a) have already been discussed in Ref. 1, we will not analyze it here any further.

Next let us consider the complex degree of spatial coherence $\mu^{(\infty)}$ of light in the far field generated by the source under consideration. We find at once on substituting from Eq. (4.18a) into Eq. (4.13) that

$$\mu^{(\infty)}(R_1\mathbf{s}_1, R_2\mathbf{s}_2) = e^{-k^2\sigma_I^2 u_{12}^2/2}\, e^{-ik(R_1-R_2)}\,, \tag{4.21}$$

where

$$u_{12}^2 = (s_{1x} - s_{2x})^2 + (s_{1y} - s_{2y})^2\,, \tag{4.22}$$

s_{jx}, s_{jy} $(j = 1, 2)$ being the x and y components of the unit vectors $\mathbf{s}_j$ $(j = 1, 2)$, specifying the directions from the origin to the two points P_1 and P_2 respectively, in the far zone (Fig. 1).

In Fig. 2 we display curves, computed from formulas (4.19a) and (4.21), showing the behavior of the radiant intensity and of the absolute value of the degree of spatial coherence of the light in the far zone of a quasihomogeneous source, whose optical intensity distribution and degree of spatial coherence are given by Eqs. (4.15) and (4.16), respectively, subject to the constraints expressed by Eqs. (4.17). The behavior of the radiant intensity is also displayed in a different manner in Fig. 3 (which is substantially identical with Fig. 2 of Ref. 1). We see that as the correlation length [characterized by the root-mean-square (rms) width σ_g] of the light in the source plane increases, i.e., as the source becomes more and more spatially coherent, the angular distribution of the radiant intensity becomes more and more peaked around the forward direction ($\theta = 0$).

To appreciate the implications of Fig. 2(d), let us choose the points P_1 and P_2 in the far zone at the same distance from the origin ($R_2 = R_1$). Under these circumstances we readily see from Eq. (4.22) that the variable u_{12} is a measure of the angular separation between the two points when viewed from the source, at least for points that are situated sufficiently close to the forward direction. Figure 2(d) then shows that as the rms width σ_I of the intensity across the source increases (i.e., as the effective source size becomes larger and larger) the angular separation of the points in the far zone at which the light has appreciable correlation becomes smaller and smaller.

V. RADIANCE OF A QUASIHOMOGENEOUS SOURCE

The radiance (at frequency $\omega = kc$) of a planar source may be defined by the formula [Ref. 17, Eq. (22)]

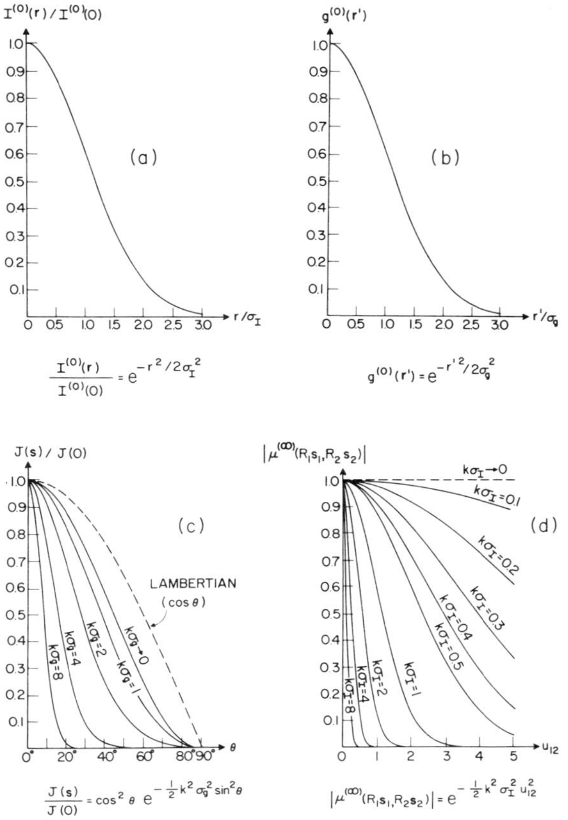

FIG. 2. Radiation from a planar quasihomogeneous source with Gaussian dependence of both (a) the optical intensity and (b) of the degree of spatial coherence. It is assumed that the linear dimensions of the source are large compared to the rms width σ_I of the optical-intensity distribution $I^{(0)}(\mathbf{r})$. (c) Behavior of the normalized radiant intensity and (d) of the absolute value of the degree of spatial coherence in the far zone. For comparison the behavior of the radiant intensity of a Lambertian source is also shown in (c).

$$B(\mathbf{r},\mathbf{s}) = \left(\frac{k}{2\pi}\right)^2 \cos\theta \int_{(z'=0)} W^{(0)}(\mathbf{r}+\tfrac{1}{2}\mathbf{r}', \mathbf{r}-\tfrac{1}{2}\mathbf{r}') \times e^{-ik\mathbf{s}_\perp\cdot\mathbf{r}'}\, d^2r' \tag{5.1}$$

where, as before, θ represents the angle which the $\mathbf{s}$ direction makes with the normal to the source plane[19] (Fig. 4).

For a quasihomogeneous source we have, according to Eq. (2.9),

$$W^{(0)}(\mathbf{r}+\tfrac{1}{2}\mathbf{r}', \mathbf{r}-\tfrac{1}{2}\mathbf{r}') = I^{(0)}(\mathbf{r})g^{(0)}(\mathbf{r}') \,. \tag{5.2}$$

On substituting from (5.2) into (5.1) we obtain for the radiance the following simple expression:

$$B(\mathbf{r},\mathbf{s}) = k^2 I^{(0)}(\mathbf{r})\tilde{g}^{(0)}(k\mathbf{s}_\perp)\cos\theta \ , \tag{5.3}$$

where $\tilde{g}(\mathbf{f})$ is, of course, the two-dimensional spatial Fourier transform of $\tilde{g}(r')$, defined by Eq. (3.8). Equation (5.3) shows that *for a quasihomogeneous source the*

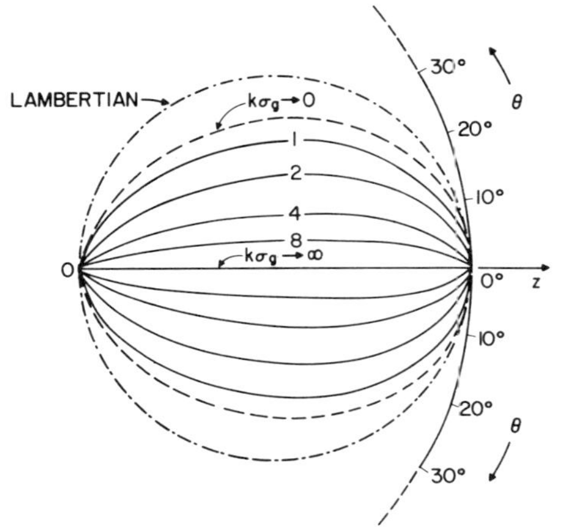

FIG. 3. Polar diagrams of the normalized radiant intensity $J(\mathbf{s})/J0)$ of Fig. 2(c). The length of the vector pointing from the origin to a typical point on a curve labeled by a particular value of the parameter $k\sigma_g$ represents the normalized radiant intensity in the direction of that vector, generated by a Gaussian-correlated planar quasihomogeneous source whose degree of spatial coherence is given by $g^{(0)}(\mathbf{r}') = e^{-r'^2/2\sigma_g^2}$.

radiance $B(\mathbf{r},\mathbf{s})$ is proportional to the product of a function of position and a function of direction. The function of position is just the optical intensity $I^{(0)}(\mathbf{r})$, at the point $\mathbf{r}$ of the source. The function of direction is the product of $\cos\theta$ and of the two-dimensional spatial Fourier transform $\tilde{g}^{(0)}(\mathbf{f})$ of the degree of spatial coherence of the source, evaluated at the (two-dimensional) spatial frequency $k\mathbf{s}_\perp$.

It was pointed out in Ref. 17 that the radiance function, defined by Eq. (5.1) may, in general, acquire negative values. However, it is easy to see that this cannot occur when the source is quasihomogeneous. For in the first place the factor $I^{(0)}(\mathbf{r})$ on the right-hand side of Eq. (5.3) is non-negative, since it represents the optical intensity. Further, $\tilde{g}^{(0)}(\mathbf{f})$ is the Fourier transform of a function that is non-negative definite,[20] and hence, according to a two-dimensional generalization of a classic theorem of Bochner,[21] $\tilde{g}^{(0)}(\mathbf{f})$ is necessarily non-negative for all $\mathbf{f}$. Moreover, since $-\frac{1}{2}\pi \le \theta \le \frac{1}{2}\pi$, we also have $\cos\theta \ge 0$. Thus we see that all the factors on the right-hand side of Eq. (5.3) are non-

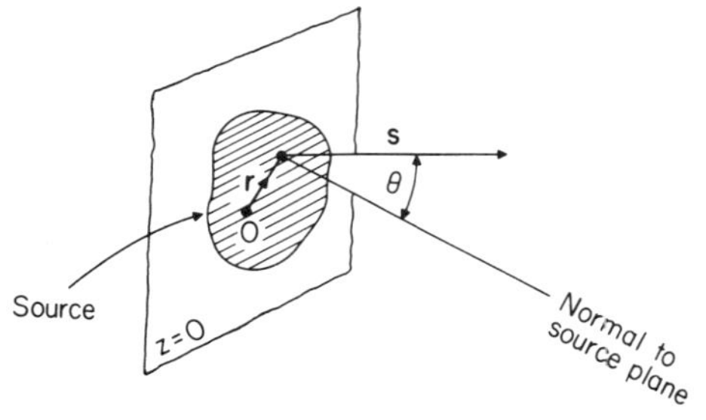

FIG. 4. Illustrating the notation relating to the definition of the radiance $B(\mathbf{r},\mathbf{s})$.

negative and hence we have shown that *for a quasihomogeneous source*, we have

$$B(\mathbf{r},\mathbf{s})\geq 0 \tag{5.4}$$

for all possible values of the arguments **r** and **s**.

As an illustration let us consider the source that we discussed in Sec. IV. The distribution of the optical intensity across the source and the degree of spatial coherence are then given by Eqs. (4.15) and (4.16), respectively, subject to the constraints expressed by Eqs. (4.17). In this case we obtain at once, on substituting from Eqs. (4.15) and (4.18b) into (5.3) and on making use of (4.20), the following expression for the radiance:

$$B(\mathbf{r},\mathbf{s})=\begin{cases}(k^2A\sigma_g^2/2\pi)\cos\theta\, e^{-r^2/2\sigma_I^2}e^{-k^2\sigma_g^2\sin^2\theta/2}, & \text{when } r\leq a\\ 0, & \text{when } r>a.\end{cases} \tag{5.5}$$

This expression is indeed non-negative for all values of **r** and θ, in agreement with (5.4).

Returning to a general quasihomogeneous source we recall that we derived, in Sec. IV, an expression for the radiant intensity $J(\mathbf{s})$ of a quasihomogeneous source by making use of a relation between $J(\mathbf{s})$ and the optical intensity $I^{(\infty)}(R\mathbf{s})$ of the far field [Eq. (4.7)]. We will now verify, for the sake of completeness, that the same expression is obtained from the radiometric formula [Ref. 17, Eq. (21)]:

$$J(\mathbf{s})=\cos\theta\int_{(z=0)}B(\mathbf{r},\mathbf{s})\,d^2r\,. \tag{5.6}$$

We have, on substituting from (5.3) into (5.6),

$$J(\mathbf{s})=k^2\cos^2\theta\tilde{g}^{(0)}(k\mathbf{s}_\perp)\int_{(z=0)}I^{(0)}(\mathbf{r})\,d^2r\,. \tag{5.7}$$

Now according to (3.7) the integral on the right-hand side of (5.7) has the value $(2\pi)^2\tilde{I}(0)$. On using this fact, Eq. (5.7) reduces to

$$J(\mathbf{s})=(2\pi k)^2\tilde{I}^{(0)}(0)\tilde{g}^{(0)}(k\mathbf{s}_\perp)\cos^2\theta\,, \tag{5.8}$$

in agreement with (4.8).

Finally, we recall that in place of his original expression (5.1) for the radiance, Walther[22] proposed, more recently, an alternative expression which in our notation may be written

$$'B(\mathbf{r},\mathbf{s})=\left(\frac{k}{2\pi}\right)^2\cos\theta\,\Re\int_{(z'=0)}W^{(0)}(\mathbf{r}',\mathbf{r})\,e^{ik\mathbf{s}\cdot(\mathbf{r}'-\mathbf{r})}\,d^2r'\,, \tag{5.9}$$

where $\Re$ denotes the real part. The expressions (5.1) and (5.9) may be regarded as two different definitions of the radiance function on the basis of physical optics. There has been a short polemic[23,24] about the relative merits of these two definitions. We show in Appendix A that within the accuracy implicit in the model of a quasihomogeneous source, the two definitions are essentially equivalent, i.e., for a quasihomogeneous source

$$'B(\mathbf{r},\mathbf{s})\approx B(\mathbf{r},\mathbf{s}) \tag{5.10}$$

for all values of the arguments **r** and **s**. Thus for such a source $'B(\mathbf{r},\mathbf{s})$, just like $B(\mathbf{r},\mathbf{s})$, is given by the right-hand side of Eq. (5.3).

VI. RADIANT EMITTANCE OF A QUASIHOMOGENEOUS SOURCE

In terms of the radiance $B(\mathbf{r},\mathbf{s})$ the radiant emittance $E(\mathbf{r})$ of a planar source of any state of coherence is defined by the usual expression [Ref. 17, Eq. (20)]

$$E(\mathbf{r})=\int_{(2\pi)}B(\mathbf{r},\mathbf{s})\cos\theta\,d\Omega, \tag{6.1}$$

where the integration extends over the solid angle 2π, formed by all the **s** directions that point into the half-space $z>0$. For a quasihomogeneous source, $B(\mathbf{r},\mathbf{s})$ was seen to be given by the expression (5.3). On substituting from (5.3) into (6.1) we obtain the following expression for the radiant emittance of a quasihomogeneous source:

$$E(\mathbf{r})=C_g I^{(0)}(\mathbf{r}), \tag{6.2}$$

where

$$C_g=k^2\int_{(2\pi)}\tilde{g}^{(0)}(k\mathbf{s}_\perp)\cos^2\theta\,d\Omega. \tag{6.3}$$

The factor C_g, defined by (6.3), is independent of **r** and depends only on the degree of correlation $g^{(0)}$. Hence we see that *for a quasihomogeneous source the radiant emittance $E(r)$ is proportional to the optical intensity $I(\mathbf{r})$, the proportionality constant, C_g, depending on the degree of spatial coherence of the light distribution across the source.*

We noted in Sec. V that $\tilde{g}^{(0)}(\mathbf{f})$ is necessarily non-negative for all values of its argument and hence we see from (6.3) that $C_g\geq 0$. Moreover, in Appendix B we show that $C_g\leq 1$. Hence

$$0\leq C_g\leq 1; \tag{6.4}$$

and using this result and the fact that the optical intensity $I(\mathbf{r})$ is necessarily non-negative, we see from (6.2) that

$$0\leq E(\mathbf{r})\leq I^{(0)}(\mathbf{r}). \tag{6.5}$$

Hence *the radiant emittance $E(\mathbf{r})$ at any point **r** of a planar quasihomogeneous source is necessarily non-negative*[25] *and cannot exceed the value of the optical intensity $I^{(0)}(\mathbf{r})$ at that point.*

The formula (6.3) expresses C_g as an integral over directions. It is not difficult to express C_g as an integral over the source plane. For this purpose we will make use, in place of (6.1), of the following expression for the radiant emittance, derived elsewhere [Ref. 17, Eq. (39)]:

$$E(\mathbf{r})=\frac{k^2}{2(2\pi)^{1/2}}\int_{(z'=0)}W^{(0)}(\mathbf{r}+\tfrac{1}{2}\mathbf{r}',\mathbf{r}-\tfrac{1}{2}\mathbf{r}')\times\frac{J_{3/2}(kr')}{(kr')^{3/2}}\,d^2r', \tag{6.6}$$

where

$$J_{3/2}(x)=(2/\pi x)^{3/2}(\sin x/x-\cos x) \tag{6.7}$$

is the Bessel function of the first kind and order $\frac{3}{2}$. When the source is quasihomogeneous, $W^{(0)}$ has the form given by Eq. (2.9), viz.,

$$W^{(0)}(\mathbf{r}_1,\mathbf{r}_2)=I^{(0)}[\tfrac{1}{2}(\mathbf{r}_1+\mathbf{r}_2)]\,g^{(0)}(\mathbf{r}_1-\mathbf{r}_2), \tag{6.8a}$$

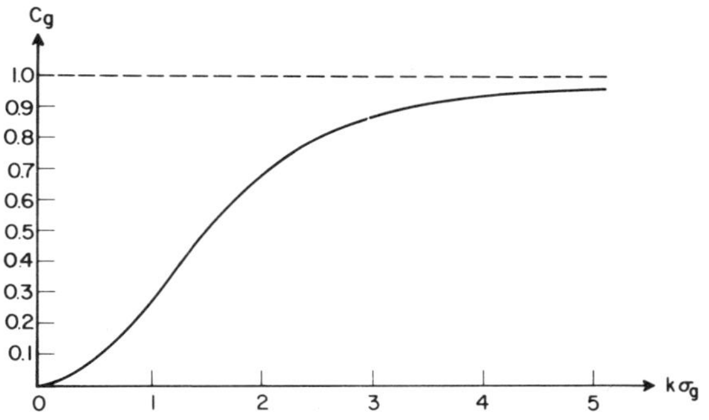

FIG. 5. Factor C_g [given by Eq. (6.14)], for a planar quasi-homogeneous source with a Gaussian dependence of the degree of spatial coherence, viz., $g^{(0)}(\mathbf{r}') = e^{-r'^2/2\sigma_g^2}$, as a function of the normalized rms width $k\sigma_g$.

so that

$$W^{(0)}(\mathbf{r} + \tfrac{1}{2}\mathbf{r}', \mathbf{r} - \tfrac{1}{2}\mathbf{r}') = I^{(0)}(\mathbf{r}) g^{(0)}(\mathbf{r}'), \tag{6.8b}$$

and the expression (6.6) for the radiant emittance then reduces to (6.2), with the following alternative but equivalent expression for the proportionality constant C_g:

$$C_g = \frac{k^2}{2(2\pi)^{1/2}} \int_{(z'=0)} g^{(0)}(\mathbf{r}') \frac{J_{3/2}(kr')}{(kr')^{3/2}} d^2r'. \tag{6.9}$$

As an example of the dependence of the factor C_g on the degree of spatial coherence of the light in the source plane let us again consider a quasihomogeneous source that is "Gaussian correlated." In this case we have

$$g^{(0)}(\mathbf{r}') = e^{-r'^2/2\sigma_g^2}, \tag{6.10}$$

and hence, according to (4.18b) and (4.20),

$$\tilde{g}^{(0)}(k\mathbf{s}_\perp) = (\sigma_g^2/2\pi) e^{-k^2\sigma_g^2 \sin^2\theta/2}. \tag{6.11}$$

In this case, expression (6.3) becomes

$$C_g = \frac{k^2\sigma_g^2}{2\pi} \int_0^{2\pi} d\phi \int_0^{\pi/2} e^{-k^2\sigma_g^2 \sin^2\theta/2} \cos^2\theta \sin\theta \, d\theta, \tag{6.12}$$

where we have used the fact that $d\Omega = \sin\theta \, d\theta \, d\phi$. The integration with respect to ϕ can be carried out at once and gives the value of 2π so that

$$C_g = k^2\sigma_g^2 \int_0^{\pi/2} e^{-k^2\sigma_g^2 \sin^2\theta/2} \cos^2\theta \sin\theta \, d\theta. \tag{6.13}$$

It is shown in Appendix C that the expression (6.13) for C_g may be expressed in the simpler form

$$C_g = 1 - \frac{F(k\sigma_g/\sqrt{2})}{k\sigma_g/\sqrt{2}}, \tag{6.14}$$

where $F(a)$ is the Dawson integral

$$F(a) = e^{-a^2} \int_0^a e^{t^2} dt. \tag{6.15}$$

From available tables of the Dawson integral[26] one can readily plot C_g as a function of the parameter $k\sigma_g$, i.e., as a function of the effective correlation length in the source plane expressed in units of $\lambda/2\pi = 1/k$. The resulting curve is displayed in Fig. 5. The figure implies, in view of the relation (6.2), the following: The radiant emittance $E(\mathbf{r})$ at any point $\mathbf{r}$ of a Gaussian correlated quasihomogeneous planar source increases monotonically from zero value to the maximum possible value $I^{(0)}(\mathbf{r})$ as the rms width of the degree of coherence increases from $k\sigma_g = 0$ to $k\sigma_g = \infty$; these two extreme cases represent complete spatial incoherence and complete spatial coherence, respectively.

VII. RADIANT FLUX FROM A QUASIHOMOGENEOUS SOURCE

The total flux F of radiant energy, at frequency ω, radiated into the half-space $z > 0$ by a source located in the plane $z = 0$, is given by either of the two formulas [cf. Ref. 17, Eqs. (18) and (19)]

$$F = \int_{(z=0)} E(\mathbf{r}) d^2r, \tag{7.1a}$$

$$F = \int_{(2\pi)} J(\mathbf{s}) d\Omega, \tag{7.1b}$$

where, as before, $E(\mathbf{r})$ denotes the radiant emittance and $J(\mathbf{s})$ the radiant intensity. For a quasihomogeneous source we have, according to Eq. (6.2),

$$E(\mathbf{r}) = C_g I^{(0)}(\mathbf{r}), \tag{7.2}$$

where the factor C_g, which depends entirely on the degree of spatial coherence of the source, is given by Eqs. (6.3) [or (6.9)]. On substituting from Eq. (7.2) into Eq. (7.1a) we obtain the following expression for the total radiant flux[27]:

$$F = C_g \int_{(z=0)} I^{(0)}(\mathbf{r}) d^2r. \tag{7.3}$$

This formula shows how the total radiant flux depends on the optical intensity and the degree of spatial coherence of the light emerging from the source. In particular we see that two sources with identical distributions of the optical intensity but with different degrees of spatial coherence $g_1(\mathbf{r}')$ and $g_2(\mathbf{r}')$, give rise to radiant fluxes whose relative ratio is

$$F_2/F_1 = C_{g_2}/C_{g_1}. \tag{7.4}$$

The integral that occurs in the expression (7.3) for the radiant flux may, of course, be expressed in terms of the two-dimensional spatial Fourier transform $\tilde{I}^{(0)}$ of the optical intensity $I^{(0)}$ [cf. Eq. (3.7)] and one obtains for F the equivalent expression

$$F = (2\pi)^2 \tilde{I}(0) C_g. \tag{7.5}$$

This formula also readily follows from Eq. (7.1b) for the radiant flux if we substitute for $J(\mathbf{s})$ the expression (4.8).

Since according to Eq. (6.4), we have $C_g \le 1$, we see from (7.3) that

$$F \le \int_{(z=0)} I^{(0)}(\mathbf{r}) d^2r, \tag{7.6}$$

i.e., the total radiant flux cannot exceed the integral of the optical intensity taken over the source area.

It is clear that the behavior of the factor C_g is fundamental in clarifying the dependence of the total radiant flux generated by a quasihomogeneous source on the degree of spatial coherence of the source. For the special case of a Gaussian correlated source, we have already seen (Fig. 5) that

$$C_g \to 0 \quad \text{as } k\sigma_g \to 0 \tag{7.7a}$$

and

$$C_g \to 1 \quad \text{as } k\sigma_g \to \infty. \tag{7.7b}$$

Equation (7.7a), together with Eq. (7.3) implies that with a Gaussian correlated source of finite optical intensity no net flux is generated at all as the correlation distance of the light across the source tends to zero (spatially incoherent source). Equation (7.7b) implies that in the other extreme case as the correlation distance tends to infinity[28] the source (which now becomes spatially completely coherent), with the same total integrated optical intensity, will generate maximum radiant flux [cf. Eq. (7.6)].

VIII. SUMMARY

For convenience we summarize below the main formulas that we obtained in this paper.

First we recall the basic definitions. The equation numbers below refer to equations with the same number of the text.

We defined *a quasihomogeneous planar source* confined to a portion of the plane $z=0$ by the property that the cross-spectral density of the light (at some particular frequency $\omega = kc$) in the source area is, to a good approximation, of the form

$$W^{(0)}(\mathbf{r}_1, \mathbf{r}_2) = I^{(0)}[\tfrac{1}{2}(\mathbf{r}_1+\mathbf{r}_2)]\, g^{(0)}(\mathbf{r}_1-\mathbf{r}_2), \tag{2.9}$$

where $I^{(0)}(\mathbf{r}) \equiv W^{(0)}(\mathbf{r},\mathbf{r})$ is the optical intensity at the point $\mathbf{r}$ of the source and $g^{(0)}(\mathbf{r}_1-\mathbf{r}_2) = \mu^{(0)}(\mathbf{r}_1,\mathbf{r}_2)$ [Eq. (2.6)], is the complex degree of spatial coherence of the light in the source plane. $I^{(0)}(\mathbf{r})$ is assumed to be a "slow" function of $\mathbf{r}$ and $g^{(0)}(\mathbf{r}')$ is assumed to be a "fast" function of $\mathbf{r}'$. It is further assumed that the linear dimensions of the source are large compared with both the effective range of $g^{(0)}(r')$ and with the wavelength $\lambda = 2\pi c/\omega = 2\pi/k$.

The convention employed in the definitions of two-dimensional Fourier transforms is as follows:

$$\tilde{I}^{(0)}(\mathbf{f}) = \frac{1}{(2\pi)^2}\int I^{(0)}(\mathbf{r})\, e^{-i\mathbf{f}\cdot\mathbf{r}}\, d^2r, \tag{3.7}$$

$$\tilde{g}^{(0)}(\mathbf{f}) = \frac{1}{(2\pi)^2}\int g^{(0)}(\mathbf{r}')\, e^{-i\mathbf{f}\cdot\mathbf{r}'}\, d^2r'. \tag{3.8}$$

The superscripts (0) and (∞) label quantities pertaining to the source plane $z=0$ and to the far zone, respectively.

With these assumptions and definitions we obtained the following results.

A. Far-zone behavior

Cross-spectral density:

$$W^{(\infty)}(R_1\mathbf{s}_1, R_2\mathbf{s}_2) = [I^{(\infty)}(R_1\mathbf{s}_1)]^{1/2}[I^{(\infty)}(R_1\mathbf{s}_2)]^{1/2} \times \mu^{(\infty)}(R_1\mathbf{s}_1, R_2\mathbf{s}_2).$$

Optical intensity:

$$I^{(\infty)}(R\mathbf{s}) = (2\pi k)^2\, \tilde{I}^{(0)}(0)\, \tilde{g}^{(0)}(k\mathbf{s}_\perp)\cos^2\theta/R^2. \tag{4.6}$$

Complex degree of spatial coherence:

$$\mu^{(\infty)}(R_1\mathbf{s}_1, R_2\mathbf{s}_2) = \frac{\tilde{I}^{(0)}[k(\mathbf{s}_{1\perp}-\mathbf{s}_{2\perp})]}{\tilde{I}^{(0)}(0)}\, e^{ik(R_1-R_2)}. \tag{4.13}$$

B. Radiometric quantities

Radiance:

$$B(\mathbf{r},\mathbf{s}) = k^2 I^{(0)}(\mathbf{r})\, \tilde{g}^{(0)}(k\mathbf{s}_\perp)\cos\theta. \tag{5.3}$$

Radiant intensity:

$$J(\mathbf{s}) = (2\pi k)^2\, \tilde{I}^{(0)}(0)\, g^{(0)}(k\mathbf{s}_\perp)\cos^2\theta. \tag{4.8}$$

Radiant emittance:

$$E(\mathbf{r}) = C_g I^{(0)}(\mathbf{r}), \tag{6.2}$$

with

$$C_g = k^2\int_{(2\pi)} \tilde{g}^{(0)}(k\mathbf{s}_\perp)\cos^2\theta\, d\Omega = \frac{k^2}{2(2\pi)^{1/2}}\int_{(z'=0)} g^{(0)}(\mathbf{r}')\,\frac{J_{3/2}(kr')}{(kr')^{3/2}}\, d^2r'. \tag{6.3), (6.9}$$

C. Total radiant flux

$$F = C_g\int_{(z'=0)} I^{(0)}(\mathbf{r}')\, d^2r' = (2\pi)^2\, \tilde{I}^{(0)}(0)\, C_g. \tag{7.3), (7.5}$$

ACKNOWLEDGMENT

We wish to acknowledge our appreciation to Mr. J. T. Foley for much assistance with the computations relating to some of the figures that appear in this paper.

APPENDIX A: EQUIVALENCE OF WALTHER'S TWO DEFINITIONS OF THE RADIANCE WHEN THE SOURCE IS QUASIHOMOGENEOUS

In place of his original definition (5.1) of the radiance, in terms of the cross-spectral density function of the light in the source plane, viz.,

$$B(\mathbf{r},\mathbf{s}) = \left(\frac{k}{2\pi}\right)^2 \cos\theta\int_{(z'=0)} W^{(0)}(\mathbf{r}+\tfrac{1}{2}\mathbf{r}', \mathbf{r}-\tfrac{1}{2}\mathbf{r}') \times e^{-ik\mathbf{s}\cdot\mathbf{r}'}\, d^2r', \tag{A1}$$

Walther proposed in a later paper an alternative definition, which may be expressed in the form [Ref. 22, Eq. (11); see also Refs. 23 and 24]

$$'B(\mathbf{r},\mathbf{s}) = \left(\frac{k}{2\pi}\right)^2 \cos\theta\, \mathcal{R}\int_{(z'=0)} W^{(0)}(\mathbf{r}',\mathbf{r}) \times e^{-ik\mathbf{s}\cdot(\mathbf{r}'-\mathbf{r})}\, d^2r', \tag{A2}$$

where $\mathcal{R}$ denotes the real part.

Let us substitute for $W^{(0)}(\mathbf{r}',\mathbf{r})$ into (A2) its Fourier integral representation [cf. (3.1a)]. We then obtain for $'B(\mathbf{r},\mathbf{s})$ the expression

$$'B(\mathbf{r},\mathbf{s})=\left(\frac{k}{2\pi}\right)^2\cos\theta\,\Re\int_{(z'=0)} d^2r'\,e^{-ik\mathbf{s}_\perp\cdot(\mathbf{r}'-\mathbf{r})}$$
$$\times\iint \tilde{W}^{(0)}(\mathbf{f}_1,\mathbf{f}_2)\,e^{i(\mathbf{f}_1\cdot\mathbf{r}'+\mathbf{f}_2\cdot\mathbf{r}_2)}\,d^2f_1\,d^2f_2, \quad \text{(A3)}$$

where we have made use of the fact that $\mathbf{s}\circ(\mathbf{r}'-\mathbf{r})=\mathbf{s}_\perp\cdot(\mathbf{r}'-\mathbf{r})$ because $\mathbf{r}$ and $\mathbf{r}'$ are position vectors of points in the source plane $z'=0$. The $\mathbf{r}'$ integration on the right-hand side of (A3) extends over the plane $z'=0$ and the other two integrations extend over the complete $\mathbf{f}_1$ and $\mathbf{f}_2$ planes. We may readily reduce the threefold double integrals to a single double integral by straightforward manipulation:

$$'B(\mathbf{r},\mathbf{s})=\left(\frac{k}{2\pi}\right)^2\cos\theta\,\Re\iiint\tilde{W}^{(0)}(\mathbf{f}_1,\mathbf{f}_2)$$
$$\times e^{i\mathbf{r}'\cdot(\mathbf{f}_1-k\mathbf{s}_\perp)}\,e^{i\mathbf{r}\cdot(\mathbf{f}_2+k\mathbf{s}_\perp)}\,d^2r'\,d^2f_1\,d^2f_2$$
$$=k^2\cos\theta\,\Re\iint\tilde{W}^{(0)}(\mathbf{f}_1,\mathbf{f}_2)\,\delta^{(2)}(\mathbf{f}_1-k\mathbf{s}_\perp)$$
$$\times e^{i\mathbf{r}\cdot(\mathbf{f}_2+k\mathbf{s}_\perp)}\,d^2f_1\,d^2f_2$$
$$=k^2\cos\theta\,\Re\int\tilde{W}^{(0)}(k\mathbf{s}_\perp,\mathbf{f}_2)\,e^{i\mathbf{r}\cdot(\mathbf{f}_2+k\mathbf{s}_\perp)}\,d^2f_2. \quad \text{(A4)}$$

Now for a quasihomogeneous source we have, according to (3.9),

$$\tilde{W}^{(0)}(k\mathbf{s}_\perp,\mathbf{f}_2)=\tilde{I}^{(0)}(k\mathbf{s}_\perp+\mathbf{f}_2)\,\tilde{g}^{(0)}[\tfrac{1}{2}(k\mathbf{s}_\perp-\mathbf{f}_2)], \quad \text{(A5)}$$

and using this expression, (A4) becomes

$$'B(\mathbf{r},\mathbf{s})=k^2\cos\theta\,\Re\int\tilde{I}^{(0)}(k\mathbf{s}_\perp+\mathbf{f}_2)$$
$$\times\tilde{g}^{(0)}\left(\frac{k\mathbf{s}_\perp-\mathbf{f}_2}{2}\right)e^{i(k\mathbf{s}_\perp+\mathbf{f}_2)\cdot\mathbf{r}}\,d^2f_2. \quad \text{(A6)}$$

Let us change the variable of integration on the right-hand side of (A6) from $\mathbf{f}_2$ to $\mathbf{f}'=k\mathbf{s}_\perp+\mathbf{f}_2$. Then (A6) becomes

$$'B(\mathbf{r},\mathbf{s})=k^2\cos\theta\,\Re\int\tilde{I}^{(0)}(\mathbf{f}')$$
$$\times\tilde{g}^{(0)}(k\mathbf{s}_\perp-\tfrac{1}{2}\mathbf{f}')\,e^{i\mathbf{f}'\cdot\mathbf{r}}\,d^2f', \quad \text{(A7)}$$

where the integration extends over the complete $\mathbf{f}'$ plane. Now as pointed out in the last paragraph of Sec. III, $\tilde{I}^0(\mathbf{f}')$ is a "fast" function of $\mathbf{f}'$, and $\tilde{g}^{(0)}(\mathbf{f})$ is a "slow" function of $\mathbf{f}$, because the source is assumed to be quasihomogeneous. Under these circumstances one may obviously replace $\tilde{g}^0(k\mathbf{s}_\perp-\frac{1}{2}\mathbf{f}')$ by $\tilde{g}^0(k\mathbf{s}_\perp)$ on the right-hand side of (A7), without introducing an appreciable error and one then obtains for $'B(\mathbf{r},\mathbf{s})$ the expression

$$'B(\mathbf{r},\mathbf{s})\approx k^2\cos\theta\,\Re\left(\tilde{g}^{(0)}(k\mathbf{s}_\perp)\int\tilde{I}^{(0)}(\mathbf{f}')\,e^{i\mathbf{f}'\cdot\mathbf{r}}\,d^2f'\right). \quad \text{(A8)}$$

The integral on the right-hand side of (A8) is precisely the Fourier representation of the optical intensity $I^{(0)}(\mathbf{r})$ at the point $\mathbf{r}$ of the source, so that (A7) implies that

$$'B(\mathbf{r},\mathbf{s})\approx k^2\cos\theta\,\Re[I^{(0)}(\mathbf{r})\,\tilde{g}^{(0)}(k\mathbf{s}_\perp)]. \quad \text{(A9)}$$

Now the optical intensity $I^{(0)}(\mathbf{r})$ is real and, for reasons explained in the paragraph that precedes Eq. (5.4), $\tilde{g}^0(k\mathbf{s}_\perp)$ is also real. Hence we may omit the symbol $\Re$ on the right-hand side of Eq. (A9) and we finally obtain for $'B(\mathbf{r},\mathbf{s})$ the expression

$$'B(\mathbf{r},\mathbf{s})\approx k^2 I^{(0)}(\mathbf{r})\,\tilde{g}^{(0)}(k\mathbf{s}_\perp)\cos\theta. \quad \text{(A10)}$$

The right-hand side of Eq. (A10) is precisely the expression (5.3) for $B(\mathbf{r},\mathbf{s})$ and hence we have established the result that

$$'B(\mathbf{r},\mathbf{s})\approx B(\mathbf{r},\mathbf{s}), \quad \text{(A11)}$$

i.e., *for a quasihomogeneous source, Walther's two expressions* (A1) *and* (A2) *for the radiance are essentially equivalent to each other.*

APPENDIX B: PROOF THAT $C_g \leqslant 1$ [EQ. (6.4)]

The factor C_g is defined by Eq. (6.3) as

$$C_g=k^2\int_{(2\pi)} g^{(0)}(k\mathbf{s}_\perp)\cos^2\theta\,d\Omega. \quad \text{(B1)}$$

Now $\mathbf{s}_\perp$ is the vector with components s_x, s_y (projections of the unit vector $\mathbf{s}$ onto two mutually orthogonal axes in the source plane $z=0$). We also have [cf. (4.20)]

$$\cos^2\theta=1-s_x^2-s_y^2=s_z^2, \quad \text{(B2)}$$

and the element $d\Omega$ of the solid angle may readily be shown to be expressible in the form

$$d\Omega=\frac{ds_x\,ds_y}{s_z}. \quad \text{(B3)}$$

If we make use of relations (B2) and (B3), (B1) may be rewritten in the form

$$C_g=k^2\iint_{s_x^2+s_y^2\leqslant 1}\tilde{g}^{(0)}(k\mathbf{s}_\perp)(1-s_x^2-s_y^2)^{1/2}\,ds_x\,ds_y. \quad \text{(B4)}$$

Now if we represent $g^{(0)}(\mathbf{r})$ as a Fourier integral, we have

$$g^{(0)}(\mathbf{r})=\int\tilde{g}^{(0)}(\mathbf{f})\,e^{i\mathbf{f}\cdot\mathbf{r}}\,d^2f; \quad \text{(B5)}$$

and then setting $\mathbf{r}=0$, we see that

$$g^{(0)}(0)=\int\tilde{g}^{(0)}(\mathbf{f})\,d^2f. \quad \text{(B6)}$$

Since $g(\mathbf{r})$ is the complex degree of spatial coherence it follows that [cf. Eqs. (2.5) and (2.6)] $g^{(0)}(0)=1$, and hence (B6) implies that

$$\int\tilde{g}^{(0)}(\mathbf{f})\,d^2f=1\,. \quad \text{(B7)}$$

The integration on the left-hand side of (B7) extends over the complete $\mathbf{f}$ plane. Let us divide the integral into contributions from the interior and the exterior of the circle $\mathbf{f}^2=k^2$ in the $\mathbf{f}$ plane:

$$\int_{f^2\leqslant k^2}\tilde{g}^{(0)}(\mathbf{f})\,d^2f+\int_{f^2>k^2}\tilde{g}^{(0)}(\mathbf{f})\,d^2f=1\,. \quad \text{(B8)}$$

Now, in deriving the inequality (5.4), we pointed out that $\tilde{g}^0(\mathbf{f})$ is non-negative for all $\mathbf{f}$. Hence both the integrals on the left-hand side of (B8) are non-negative and consequently

$$\int_{f^2 \leq k^2} \tilde{g}^{(0)}(\mathbf{f})\, d^2 f \leq 1 \,. \tag{B9}$$

If we change the variable of integration by setting $\mathbf{f} = k\mathbf{s}_\perp \equiv ks_x,\ ks_y$, (B9) becomes

$$k^2 \iint_{s_x^2 + s_y^2 \leq 1} \tilde{g}^{(0)}(k\mathbf{s}_\perp)\, ds_x\, ds_y \leq 1 \,. \tag{B10}$$

Returning now to Eq. (B4) and using the facts that $\tilde{g}^0$ is non-negative and that $(1 - s_x^2 - s_y^2)^{1/2} \leq 1$ throughout the domain of integration we see that

$$C_g \leq k^2 \int_{s_x^2 + s_y^2 \leq 1} \tilde{g}^{(0)}(k\mathbf{s}_\perp)\, ds_x\, ds_y \,. \tag{B11}$$

From Eqs. (B11) and (B10) it follows that

$$C_g \leq 1 \,. \tag{B12}$$

APPENDIX C: DERIVATION OF EXPRESSION (6.14) FOR C_g

According to (6.13) we have for a Gaussian correlated source

$$C_g = k^2 \sigma_g^2 \int_0^{\pi/2} e^{-k^2 \sigma_g^2 \sin^2\theta/2} \cos^2\theta \sin\theta\, d\theta \,. \tag{C1}$$

If we change the variable of integration by setting $x = \cos\theta$, (C1) becomes

$$C_g = 2a^2 e^{-a^2} \int_0^1 x^2 e^{a^2 x^2}\, dx \,, \tag{C2}$$

where

$$a = k\sigma_g/\sqrt{2} \,. \tag{C3}$$

By partial integration, we obtain, from (C2),

$$C_g = 2a^2 e^{-a^2}\left(\frac{1}{2a^2} e^{a^2} - \frac{1}{2a^2}\int_0^1 e^{a^2 x^2}\, dx\right)$$
$$= 1 - e^{-a^2} \int_0^1 e^{a^2 x^2}\, dx \,. \tag{C4}$$

Next we change the variable of integration from x to $t = ax$. Then (C4) gives

$$C_g = 1 - F(a)/a \,, \tag{C5}$$

where $F(a)$ is the Dawson integral,[26] defined

$$F(a) = e^{-a^2} \int_0^a e^{t^2}\, dt \,. \tag{C6}$$

Formula (C5), with the parameter a being defined by Eq. (C3), is Eq. (6.14) of the text.

*Preliminary results of this investigation were presented at the 1976 Annual Meeting of the Optical Society of America, held in Tucson, Arizona, October 18–22, 1976 [Abstract TuE13, J. Opt. Soc. Am. 66, 1075 (1976)].

†Visiting Research Fellow, Physics Dept., University of Reading, Reading, England, during the academic year 1976–1977.

[1]E. Wolf and W. H. Carter, "Angular Distribution of Radiant Intensity from Sources of Different Degrees of Spatial Coherence," Opt. Commun. 13, 205–209 (1975).

[2]The converse problem of determining the degree of coherence in the source plane from the knowledge of the angular distribution of the radiant intensity was studied by W. H. Carter and E. Wolf in, "Coherence Properties of Lambertian and Non-Lambertian Sources," J. Opt. Soc. Am. 65, 1067–1071 (1975).

[3]Some special cases were also treated by H. P. Baltes, B. Steinle, and G. Antes in, "Spectral Coherence Area and the Radiant Intensity from Statistically Homogeneous and Isotropic Planar Sources," Opt. Commun. 18, 242–246 (1976).

[4]The properties of this correlation coefficient (also called the complex degree of spectral coherence or the spectral correlation coefficient) were recently studied by L. Mandel and E. Wolf in, "Spectral coherence and the concept of cross-spectral purity," J. Opt. Soc. Am. 66, 529–535 (1976). This correlation coefficient must be distinguished from the more familiar complex degree of coherence (cf. Ref. 6, Sec. 10.3.1), often denoted by $\gamma(\mathbf{r}_1, \mathbf{r}_2, \tau)$ [or $\gamma_{12}(\tau)$], even though $\gamma(\mathbf{r}_1, \mathbf{r}_2, 0)$ is also a measure of spatial coherence.

[5]Such a theorem was alluded to by A. Walther in his important paper, "Radiometry and Coherence," J. Opt. Soc. Am. 58, 1256–1259 (1968). However Walther did not formulate the theorem in precise mathematical terms, nor did he state to what class of sources it applies.

[6]M. Born and E. Wolf, *Principles of Optics*, 5th ed. (Pergamon, New York, 1975), Sec. 10.4.2.

[7]In physical optics the quantity $I(\mathbf{r})$ defined by Eq. (2.1), is usually referred to simply as intensity (at frequency ω). We use here the adjective "optical" to distinguish it clearly from the radiometric concept of radiant intensity, that we will also encounter in the present paper.

[8]V. I. Tatarskii, *Wave Propagation in a Turbulent Medium* (McGraw-Hill, New York, 1961) (reprinted by Dover, New York, 1967), Sec. 1.3.

[9]A. C. Schell, "The Multiple Plate Antenna," Doctoral dissertation (Massachusetts Institute of Technology, 1961) (unpublished), Sec. 7.5.

[10]See also R. A. Shore, "Partially Coherent Diffraction by a Circular Aperture," in *Electromagnetic Theory and Antennas*, edited by E. C. Jordan (Pergamon, London, 1963), Part 2, pp. 787–795; and A. K. Jaiswal, G. P. Agrawal, and C. L. Mehta, "Coherence Functions in the Far-field Diffraction Plane," Nuovo Cimento B 15, 295–307 (1973).

[11]Strictly speaking this condition only ensures that $|\mu(\mathbf{r}_1, \mathbf{r}_2)|$ has appreciable values when the separation $|\mathbf{r}_1 - \mathbf{r}_2|$ is small compared with the linear dimensions of the source. This condition does not ensure that $\mu(\mathbf{r}_1, \mathbf{r}_2)$ is a function of $\mathbf{r}_1 - \mathbf{r}_2$ only over the whole domain occupied by the secondary source. The size of the domain where $\mu(\mathbf{r}_1, \mathbf{r}_2)$ is only a function of $\mathbf{r}_1 - \mathbf{r}_2$ depends on the imaging properties of the optical system and must be separately examined in each particular case. In any case $\mu(\mathbf{r}_1, \mathbf{r}_2)$ cannot be expected to be a function of $\mathbf{r}_1 - \mathbf{r}_2$ only when either of the two points is close to the boundary of the source, whether the source is a secondary or a primary one. This fact is, however, of no great practical consequence if the source is sufficiently large.

[12]Our concept of a quasihomogeneity bears some similarity to that of local homogeneity, well known in the statistical theory of turbulence (cf. Ref. 8, Secs. 5 and 7), but the two concepts are not equivalent. A factorization (of the mutual intensity rather than of the cross-spectral density function) of the form (2.9) was assumed in a recent paper by S. Wadaka and T. Sato, "Merits of symmetric scanning for detection of coherence function in incoherent imaging system," J. Opt. Soc. Am. 66, 145–147 (1976). However in this work no motivation for this factorization is given, nor are any restrictions stated on the behavior of the two factors.

[13]Equation (2.9) is also analogous to a basic relation in the theory of locally stationary random processes, established by R. A. Silverman, in "Locally Stationary Random Processes," IRE Trans. Information Theory 3, 182–187 (1957), Eq. (11).

[14]The situation here is somewhat similar to that encountered in connection with the Kirchhoff approximation in elementary diffraction theory (cf. Ref. 6, Sec. 8.3.2).

[15]See, for example R. Bracewell, ***The Fourier Transform and Its Applications*** (McGraw-Hill, New York, 1961), pp. 160–161.

[16]E. W. Marchand and E. Wolf, "Angular Correlation and the Far-Zone Behavior of Partially Coherent Fields," J. Opt. Soc. Am. **62**, 379–385 (1972), Eq. (34).

[17]E. W. Marchand and E. Wolf, "Radiometry with Sources of Any State of Coherence," J. Opt. Soc. Am. **64**, 1219–1226 (1974), Eq. (42).

[18]Equation (16) of Ref. 1. There is a misprint in that equation: $J_\omega(\mathbf{f})$ should be replaced by $J_\omega(\mathbf{s})$ on the left-hand side of that equation.

[19]The expression (5.1) for the radiance was proposed by A. Walther in the paper quoted in Ref. 5. Essentially the same expression for the radiance was proposed in a somewhat different context by L. S. Dolin, in Izv. VUZov: Radiophys. 7, 559–563 (1964), whose title, in English translation, is "Description in Weakly Inhomogeneous Wave Fields."

[20]A function $g(\mathbf{r})$ is said to be non-negative definite if for any non-negative integer N, any set of N vectors $\mathbf{r}_i$, and any set of N (real or complex) numbers a_i,

$$\sum_{i=1}^{N}\sum_{j=1}^{N} a_i^* a_j g(\mathbf{r}_i - \mathbf{r}_j) \geq 0.$$

That this condition is obeyed by our degree of spatial coherence follows at once from the non-negative definiteness of the cross-spectral density function $W(\mathbf{r}_1, \mathbf{r}_2)$, established in the Appendix of Ref. 4, and from Eq. (2.7) above if we take into account the fact that $I(\mathbf{r}) \geq 0$.

[21]For an account of Bochner's theorem see, for example, R. R. Goldberg, *Fourier Transforms* (Cambridge U. P., Cambridge, England, 1965), Chap. 5.

[22]A. Walther, "Radiometry and Coherence," J. Opt. Soc. Am. **63**, 1622–1623 (1973).

[23]E. W. Marchand and E. Wolf, "Walther's Definition of Generalized Radiance," J. Opt. Soc. Am. **64**, 1273–1274 (1974).

[24]A. Walther, "Reply to Marchand and Wolf," J. Opt. Soc. Am. **64**, 1275 (1974).

[25]It has been shown in Ref. 17 that, in general, the radiant emittance of a partially coherent source acquires negative values.

[26]M. Abramowitz and I. A. Stegun, *Handbook of Mathematical Functions*, Natl. Bur. Stds. (U.S.GPO, Washington, D. C., 1964), p. 319; also (Dover, New York, 1965); W. L. Miller and A. R. Gordon, J. Phys. Chem. **35**, 2785 (1931).

[27]The integral in Eq. (7.3) extends only formally over the complete plane $z = 0$, since $I^{(0)}(\mathbf{r})$ vanishes at point $\mathbf{r}$ outside the domain occupied by the source.

[28]The limit $k\sigma_g \to \infty$ must be interpreted with some caution, since for a quasihomogeneous source we must have $\sigma_g \ll l$, where l represents a typical linear dimension of the source. Hence in proceeding to the limit $k\sigma_g \to \infty$ we must allow $kl \to \infty$ in such a way that σ_g/l tends to some number that is much smaller than unity.

JOURNAL OF MODERN OPTICS, 1991, VOL. 38, NO. 10, 2053–2068

Radiance functions of partially coherent fields

JOHN T. FOLEY

Department of Physics, Mississippi State University, Mississippi State, MS 39762, USA

and EMIL WOLF

Department of Physics and Astronomy, University of Rochester, Rochester, NY 14627, USA

(*Received 14 April 1991*)

Abstract. The behaviour of the spectral radiance in fields generated by planar, secondary, quasi-homogeneous sources is investigated. Examples are presented which show how both the spectral intensity and the degree of coherence of the source affect the spatial and the angular distribution of the spectral radiance. These examples clearly show the influence of source coherence on the radiometric behaviour of partially coherent optical fields.

1. Introduction

In the last few years considerable progress has been made towards clarifying the relationship between the basic quantities of traditional radiometry and those of statistical wave theory (see, for example [1]). In particular, an explicit expression was obtained for the spectral radiance of a field generated in free space by any planar, secondary, quasi-homogeneous source. It was found that the spectral radiance depends in a relatively simple manner on the spectral intensity distribution of the light across the source plane and on its degree of spectral coherence [2, 3]. The dependence of the spectral radiance on the state of coherence of the source may appear to be somewhat surprising at first, because coherence concepts do not enter the traditional radiometric description of energy propagation in optical fields. To some extent this is because most traditional light sources generate radiation that obeys Lambert's law, and Lambertian sources have certain unique coherence properties which, under ordinary circumstances, are not readily noticeable. With other types of sources, coherence concepts cannot, as a rule, be ignored, a point that was first clearly brought out in important papers by Walther [4, 5].

In this paper we investigate the behaviour of the spectral radiance of fields produced by some model sources. We show by means of polar diagrams how both the spectral intensity and the degree of coherence of the source affect the spatial and the angular distributions of the spectral radiance. These examples clearly show the influence of source coherence on the radiometric behaviour of partially coherent optical fields.

2. The radiance function of a field produced by a quasi-homogeneous source

We consider first a field produced by a planar, secondary source that occupies a finite domain σ in the plane $z=0$ and radiates into the half-space $z>0$. We assume that the source fluctuations are characterized by a statistical ensemble that is statistically stationary, at least in the wide sense [6]. We denote by $U(\mathbf{r}, \nu)$ a typical realization, at frequency ν, of the ensemble of the field which the source generates at a point P, whose location is specified by a position vector $\mathbf{r}$, in the half-space into which it radiates. Under very general conditions, $U(\mathbf{r}, \nu)$ may be represented in the form of an angular spectrum of plane waves; namely, [7],

$$U(\mathbf{r}, \nu) = \int a(\mathbf{s}_\perp, \nu) \exp(\mathrm{i}k\mathbf{s}\cdot\mathbf{r})\, \mathrm{d}^2 s_\perp, \tag{1}$$

where

$$\mathbf{s} \equiv (s_x, s_y, s_z), \qquad \mathbf{s}_\perp \equiv (s_x, s_y, 0), \tag{2}$$

$$s_z = \begin{cases} +(1-\mathbf{s}_\perp^2)^{1/2}, & \text{if } |\mathbf{s}_\perp| \leqslant 1, \\ +\mathrm{i}(\mathbf{s}_\perp^2 - 1)^{1/2}, & \text{if } |\mathbf{s}_\perp| > 1, \end{cases} \tag{3}$$

and the integration on the right-hand side extends over the whole $\mathbf{s}_\perp$-plane.

In the above notation, the complex version of the generalized radiance introduced by Walther [5] may be written in the form

$$\mathscr{B}_\nu(\mathbf{r}, \mathbf{s}) = s_z \langle [U(\mathbf{r}, \nu)]^* a(\mathbf{s}_\perp, \nu) \rangle \exp(\mathrm{i}k\mathbf{s}\cdot\mathbf{r}), \tag{4}$$

where

$$k = \frac{2\pi\nu}{c}, \tag{5}$$

c is the speed of light in a vacuum and the angular brackets denote the ensemble average. It was shown in [2] that, in terms of its boundary value,

$$\mathscr{B}_\nu^{(0)}(\boldsymbol{\rho}, \mathbf{s}) = s_z \langle [U(\boldsymbol{\rho}, \nu)]^* a(\mathbf{s}_\perp, \nu) \rangle \exp(\mathrm{i}k\mathbf{s}_\perp\cdot\boldsymbol{\rho}) \tag{6}$$

at points $\boldsymbol{\rho}$ in the source plane $\mathscr{B}_\nu(\mathbf{r}, \mathbf{s})$ may be expressed in the form

$$\mathscr{B}_\nu(\mathbf{r}, \mathbf{s}) = \exp(\mathrm{i}k\mathbf{s}\cdot\mathbf{r}) \int_\sigma G^*(\mathbf{R}, \nu) \mathscr{B}_\nu^{(0)}(\boldsymbol{\rho}, \mathbf{s}) \exp(-\mathrm{i}k\mathbf{s}_\perp\cdot\boldsymbol{\rho})\, \mathrm{d}^2\rho, \tag{7}$$

where

$$G(\mathbf{R}, \nu) = -\frac{1}{2\pi}\frac{\partial}{\partial z}[\exp(\mathrm{i}kR)/R] \tag{8}$$

and

$$\mathbf{R} = \mathbf{r} - \boldsymbol{\rho}, \qquad R = |\mathbf{R}|. \tag{9}$$

It was noted in [2] that the function $\mathscr{B}_\nu(\mathbf{r}, \mathbf{s})$ does not, in general, have all the properties attributed to the radiance in traditional radiometry, but that it acquires them in the asymptotic limit as the wavenumber $k \equiv 2\pi/\lambda \to \infty$, provided that the field is produced by a quasi-homogeneous source. The derivation of this result, which was only briefly sketched out in [2], will be discussed more fully in this paper.

A planar, secondary, quasi-homogeneous source [8] is characterized by the properties that its degree of spectral coherence [9] $\mu^{(0)}(\boldsymbol{\rho}_1, \boldsymbol{\rho}_2, \nu)$ depends on the position vectors $\boldsymbol{\rho}_1$ and $\boldsymbol{\rho}_2$ of any two points in the source plane only through the difference $\boldsymbol{\rho}' = \boldsymbol{\rho}_2 - \boldsymbol{\rho}_1$ and that it changes much more rapidly with $\boldsymbol{\rho}'$ than the spectral density $S^{(0)}(\boldsymbol{\rho}, \nu)$ changes with $\boldsymbol{\rho}$, it being assumed that the dimensions of the source at large compared with the effective wavelengths $\boldsymbol{\lambda}=c/\nu$ for which the source spectrum has non-negligible values. To stress the dependence of the degree of spectral coherence on the difference $\boldsymbol{\rho}_2-\boldsymbol{\rho}_1$ rather than on $\boldsymbol{\rho}_1$ and $\boldsymbol{\rho}_2$ separately, we will write $g^{(0)}(\boldsymbol{\rho}_2-\boldsymbol{\rho}_1, \nu)$ in place of $\mu^{(0)}(\boldsymbol{\rho}_1, \boldsymbol{\rho}_2, \nu)$.

For a field generated by a planar, secondary, quasi-homogeneous source the expression (7) was shown to take the form ([2], equations (3.6) and (3.7))

$$\mathscr{B}_\nu(\mathbf{r}, \mathbf{s})=k^2 s_z \tilde{g}^{(0)}(k\mathbf{s}_\perp, \nu) C_\nu^*(\mathbf{r}, \mathbf{s}_\perp) \exp(\mathrm{i}k\mathbf{s}\cdot\mathbf{r}), \tag{10}$$

where

$$\tilde{g}^{(0)}(\mathbf{f}, \nu)=\frac{1}{(2\pi)^2}\int g^{(0)}(\boldsymbol{\rho}', \nu)\exp(-\mathrm{i}\mathbf{f}\cdot\boldsymbol{\rho}')\,\mathrm{d}^2\rho' \tag{11}$$

is the two-dimensional spatial Fourier transform of the degree of spectral coherence of the source and

$$C_\nu(\mathbf{r}, \mathbf{s}_\perp)=\int_\sigma G(\mathbf{R}, \nu) S^{(0)}(\boldsymbol{\rho}, \nu)\exp(\mathrm{i}k\mathbf{s}_\perp\cdot\boldsymbol{\rho})\,\mathrm{d}^2\rho. \tag{12}$$

To determine the asymptotic behaviour of the expression (10) as $k\to\infty$, we first rewrite the integral (11) in the more explicit form

$$C_\nu(\mathbf{r}, \mathbf{s}_\perp)=C_\nu^{(1)}(\mathbf{r}, \mathbf{s}_\perp)+C_\nu^{(2)}(\mathbf{r}, \mathbf{s}_\perp), \tag{13}$$

where

$$C_\nu^{(1)}(\mathbf{r}, \mathbf{s}_\perp)=\frac{kz}{2\pi\mathrm{i}}\int_\sigma S^{(0)}(\boldsymbol{\rho}, \nu)\frac{\exp[\mathrm{i}k\phi(R, \boldsymbol{\rho})]}{R^2}\,\mathrm{d}^2\rho, \tag{14}$$

$$C_\nu^{(2)}(\mathbf{r}, \mathbf{s}_\perp)=\frac{z}{2\pi}\int_\sigma S^{(0)}(\boldsymbol{\rho}, \nu)\frac{\exp[\mathrm{i}k\phi(R, \boldsymbol{\rho})]}{R^3}\,\mathrm{d}^2\rho, \tag{15}$$

and

$$\phi(R, \boldsymbol{\rho})=R+\mathbf{s}_\perp\cdot\boldsymbol{\rho}. \tag{16}$$

On comparing the expressions (14) and (15) it becomes evident that, except perhaps for some cases involving special symmetry, we can omit the contribution $C_\nu^{(2)}$ in comparison to $C_\nu^{(1)}$ in equation (13) when k is large enough; i.e. we then have

$$C_\nu(\mathbf{r}, \mathbf{s}_\perp)\sim C_\nu^{(1)}(\mathbf{r}, \mathbf{s}_\perp)\quad \text{as } k\to\infty. \tag{17}$$

The asymptotic evaluation of $C_\nu^{(1)}(\mathbf{r}, \mathbf{s}_\perp)$ is carried out in the Appendix and one finds that as $k\to\infty$,

$$\left.\begin{aligned} C_\nu^{(1)}(\mathbf{r}, \mathbf{s}_\perp) &\sim S^{(0)}(\boldsymbol{\rho}_0, \nu)\exp(\mathrm{i}k\mathbf{s}\cdot\mathbf{r}), && \text{when } \boldsymbol{\rho}_0\in\sigma, \\ &= O\left(\frac{1}{k^{1/2}}\right), && \text{when } \boldsymbol{\rho}_0\notin\sigma, \end{aligned}\right\} \tag{18}$$

where

$$\boldsymbol{\rho}_0=\mathbf{r}_\perp-\frac{z}{s_z}\mathbf{s}_\perp \tag{19}$$

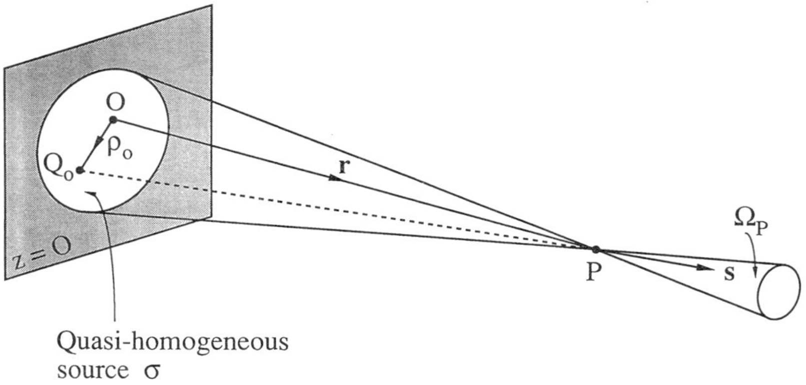

Figure 1. Illustrating the notation relating to the expression (21) for the spectral radiance of a field produced by a planar, secondary, quasi-homogeneous source σ. P is the point whose location is specified by the position vector $\mathbf{r}$. Q_0 is the point in the source plane whose position vector $\boldsymbol{\rho}_0$ is given by equation (19); it is the point of intersection with the source plane, of the line through the point P in the direction of the real unit vector $\mathbf{s}$.

represents a certain point Q_0 in the source plane; here $\mathbf{r}_\perp$ is the projection, considered as a two-dimensional vector, of the three-dimensional vector $\mathbf{r}$ on to the source plane. It follows from elementary geometry that the point Q_0 is precisely the point at which the line through the field point P, in the direction specified by the unit vector $\mathbf{s}$, intersects the source plane (see figure 1). It therefore follows from equations (10), (17) and (18) that as $k \to \infty$,

$$\mathscr{B}_\nu(\mathbf{r}, \mathbf{s}) \sim B_\nu(\mathbf{r}, \mathbf{s}), \tag{20}$$

where

$$B_\nu(\mathbf{r}, \mathbf{s}) = \begin{cases} k^2 s_z S^{(0)}(\boldsymbol{\rho}_0, \nu)\tilde{g}^{(0)}(k\mathbf{s}_\perp, \nu), & \text{when} \quad \mathbf{s} \in \Omega_{\mathbf{P}}, \\ 0, & \text{when} \quad \mathbf{s} \notin \Omega_{\mathbf{P}}, \end{cases} \tag{21}$$

here $\boldsymbol{\rho}_0$ is given by equation (19) and $\Omega_{\mathbf{P}}$ denotes the solid angle generated by the lines pointing from all the source points to the field point P. We will refer to $\Omega_{\mathbf{P}}$ as the geometrical optics cone subtended by the source at the point P.

Unlike the generalized radiance $\mathscr{B}_\nu(\mathbf{r}, \mathbf{s})$ the function $B_\nu(\mathbf{r}, \mathbf{s})$ given by expression (21) was shown in [2] to satisfy all the postulates of traditional radiometry in free space. In particular, it follows from equations (21) and (19) that $B_\nu(\mathbf{r}, \mathbf{s})$ is constant along straight line paths, i.e. that

$$B_\nu(\mathbf{r}, \mathbf{s}) = B_\nu^{(0)}[\mathbf{r}_\perp - (z/s_z)\mathbf{s}_\perp, \mathbf{s}], \tag{22}$$

where $B_\nu^{(0)}(\boldsymbol{\rho}, \mathbf{s})$ is the limiting value of the expression (21) as P approaches the plane $z=0$; that is

$$B_\nu^{(0)}(\boldsymbol{\rho}, \mathbf{s}) = k^2 s_z S^{(0)}(\boldsymbol{\rho}, z)\tilde{g}^{(0)}(k\mathbf{s}_\perp, \nu). \tag{23}$$

A rough derivation of equation (22) was given previously in [10].

Because the function $B_\nu(\mathbf{r}, \mathbf{s})$ given by equation (21) satisfies all the postulates of traditional radiometry in free space, we will refer to it as the (spectral) radiance of the field produced by the quasi-homogeneous source. It should be noted that expression (21) takes into account the spatial coherence properties of the source, which are ignored in the traditional theory.

3. Examples. Behaviour of the radiance function of fields produced by some model sources

In this section we will study, with the help of formulas (21) and (23), the behaviour of the radiance functions of fields produced by some commonly encountered planar, secondary sources, all of which will be assumed to be quasi-homogeneous.

3.1. *Lambertian source of uniform intensity*

Let us first consider a circular Lambertian source of radius $a \gg \lambda = c/\nu$, whose spectral density is the same at each source point, i.e.

$$S^{(0)}(\boldsymbol{\rho}, \nu)\begin{cases} \equiv S^{(0)}(\nu), & \text{when} \quad \rho \leqslant a, \\ =0 & \text{when} \quad \rho > a. \end{cases} \tag{24}$$

It is known that the degree of spectral coherence of a Lambertian source is given by [11]

$$g^{(0)}(\boldsymbol{\rho}_2 - \boldsymbol{\rho}_1, \nu) = \frac{\sin(k|\boldsymbol{\rho}_2 - \boldsymbol{\rho}_1|)}{k|\boldsymbol{\rho}_2 - \boldsymbol{\rho}_1|}. \tag{25}$$

The two-dimensional spatial Fourier transform of this expression, evaluated for the argument $k\mathbf{s}_\perp$, with $|\mathbf{s}_\perp|^2 < 1$, is known to be [12]

$$\tilde{g}^{(0)}(k\mathbf{s}_\perp, \nu) = \frac{1}{2\pi k^2 s_z}. \tag{26}$$

It follows from equations (23), (24) and (26) that the radiance function of such a source is given by the expression

$$B_\nu^{(0)}(\boldsymbol{\rho}, \mathbf{s}) = \begin{cases} S^{(0)}(\nu)/2\pi, & \text{when} \quad \rho \leqslant a, \\ 0, & \text{when} \quad \rho > a. \end{cases} \tag{27}$$

This formula shows that at each point in the source region ($\rho \leqslant a$), the radiance is independent of position ($\boldsymbol{\rho}$) and direction ($\mathbf{s}$).

To determine the radiance function at any point $\mathbf{r}$ in the half-space $z > 0$ into which the source radiates we only need to substitute from equations (24) and (26) into equation (21). We then find that

$$B_\nu^{(0)}(\boldsymbol{\rho}, \mathbf{s}) = \begin{cases} S^{(0)}(\nu)/2\pi, & \text{when} \quad \mathbf{s} \in \Omega_\mathrm{P}, \\ 0, & \text{when} \quad \mathbf{s} \notin \Omega_\mathrm{P}. \end{cases} \tag{28}$$

i.e. *the spectral radiance at a point* P *in the half-space $z > 0$ into which the source radiates has the same value for all* s*-directions which lie within the solid angle Ω_P generated by the lines pointing from the source to the point* P (*see figure 1*).

It will be useful to express this result in a somewhat different form by using spherical polar coordinates. We therefore set

$$\mathbf{r} \equiv (r \sin\theta \cos\phi,\ r \sin\theta \sin\phi,\ r \cos\theta), \tag{29 a}$$

$$\mathbf{s} \equiv (\sin\alpha \cos\beta,\ \sin\alpha \sin\beta,\ \cos\alpha), \tag{29 b}$$

where

$$0 \leqslant r < \infty, \qquad 0 \leqslant \theta \leqslant \pi/2, \qquad 0 \leqslant \phi < 2\pi, \tag{30 a}$$

$$0 \leqslant \alpha < \pi/2, \qquad 0 \leqslant \beta < 2\pi. \tag{30 b}$$

We will confine our attention to observation points in the x, z-plane (i.e. $\phi = 0$) and to $\mathbf{s}$-directions in this plane (i.e. $\beta = 0$). It follows from equation (28) that in this case

$$B_\nu^{(0)}(\mathbf{r}, \mathbf{s}) = \begin{cases} S^{(0)}(\nu)/2\pi, & \text{when} \quad \alpha_1 \leqslant \alpha \leqslant \alpha_2, \\ 0 & \text{otherwise}, \end{cases} \tag{31}$$

where

$$\alpha_1 = \arctan\left(\frac{x-a}{z}\right), \tag{32 a}$$

$$\alpha_2 = \arctan\left(\frac{x+a}{z}\right), \tag{32 b}$$

x and z being the Cartesian components of the vector $\mathbf{r}$.

In figure 2, polar diagrams of the spectral radiance, computed from equations (31) and (32), are shown at several different observation points, including the origin O. As was to be expected, the spectral radiance is constant at points located inside the

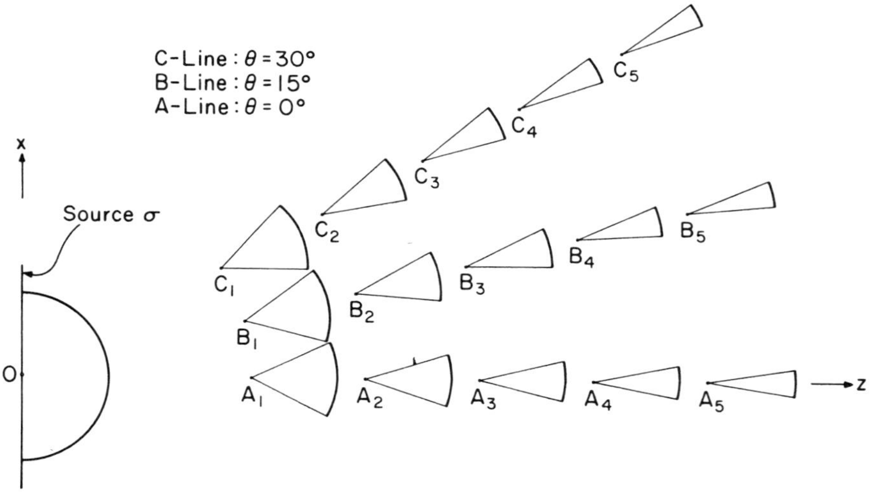

Figure 2. Polar diagrams of the spectral radiance $B_\nu(\mathbf{r}, \mathbf{s})$ at different points in the x, z-plane, produced by a uniform circular Lambertian source of radius $a = 2$ cm. The points with subscripts 1, 2, 3, 4 and 5 are at distances $r = 4$, 6, 8, 10 and 12 cm, respectively, from the centre O of the source.

geometrical optics cone Ω_P (denoted by thin lines) and is zero outside it. The narrowing of the geometrical optics cone with increasing r is evident from the figure and one also sees that as $r \to \infty$ in any particular direction θ, $B_\nu(\mathbf{r}, \mathbf{s})$ becomes non-zero only in the direction $\alpha = \theta$, i.e. for $\mathbf{s} \equiv (\sin\theta, 0, \cos\theta)$.

3.2. *Lambertian source with Gaussian intensity profile*

We again consider a Lambertian source or radius $a \gg \lambda$, but instead of assuming that its spectral density is the same at each source point we assume it is a truncated Gaussian function of position, i.e. that

$$S^{(0)}(\boldsymbol{\rho}, \nu) = S^{(0)}(\nu) \exp\left[-\boldsymbol{\rho}^2/2\sigma_s^2(\nu)\right] \operatorname{circ}(\boldsymbol{\rho}/a), \tag{33}$$

where

$$\operatorname{circ}(\mathbf{A}) = \begin{cases} 1, & \text{when } |\mathbf{A}| \leqslant 1, \\ 0, & \text{when } |\mathbf{A}| > 1, \end{cases} \tag{34}$$

and $\sigma_s(\nu) \gg \lambda$ is a function only of ν.

Since the source is Lambertian, the Fourier transform of its degree of spectral coherence is again given by equation (26), and hence equations (23) and (21) now give

$$B_\nu^{(0)}(\boldsymbol{\rho}, \mathbf{s}) = \frac{1}{2\pi} S^{(0)}(\nu) \exp\left[-\boldsymbol{\rho}^2/2\sigma_s^2(\nu)\right] \operatorname{circ}(\boldsymbol{\rho}/a) \tag{35}$$

and

$$B_\nu(\mathbf{r}, \mathbf{s}) = \begin{cases} \dfrac{1}{2\pi} S^{(0)}(\nu) \exp\left[-\left|\mathbf{r} - \dfrac{z}{s_z}\mathbf{s}_\perp\right|^2 \Big/ 2\sigma_s^2(\nu)\right], & \text{when } \mathbf{s} \in \Omega_P, \\ 0 & \text{when } \mathbf{s} \notin \Omega_P, \end{cases} \tag{36}$$

where we also made use of equation (19). We note that, according to equation (35), the spectral radiance in the source plane depends on position, but is again independent of direction.

To see more clearly the implications of equation (36) we proceed as in the previous case (section 3.1). We introduce spherical polar coordinates (equations (29) and (30)) and confine our attention to observation points in the x, z-plane ($\phi = 0$) and to $\mathbf{s}$-directions in this plane ($\beta = 0$). One finds from equations (29) and (36), after some algebra, that

$$B_\nu(\mathbf{r}, \mathbf{s}) = \begin{cases} \dfrac{1}{2\pi} S^{(0)}(\nu) \exp\left[-\left[\dfrac{r\sin(\theta - \alpha)}{\cos\alpha}\right]^2 \Big/ 2\sigma_s^2(\nu)\right], & \text{when } \alpha_1 \leqslant \alpha \leqslant \alpha_2, \\ 0 & \text{otherwise}, \end{cases} \tag{37}$$

where α_1 and α_2 are defined as in the previous example (see equations (32)).

In figure 3, polar diagrams of the spectral radiance, computed from equation (37), are shown at several different observation points. Two effects are evident from the figure. First, the spectral radiance inside the geometrical optics cone Ω_P (which is again indicated by thin lines) is no longer independent of the $\mathbf{s}$-direction (except at the centre O of the source); in fact, at every point except at O, the radiance has a maximum for the direction specified by $\alpha = \theta$, i.e. for $\mathbf{s} \equiv (\sin\theta, 0, \cos\theta)$. Also, the effect of the cut-off at the edge of the geometrical optics cone is not as drastic as in the previous example (section 3.1). This is because the spectral density at the boundary of the source is significantly smaller than it is at the centre of the source.

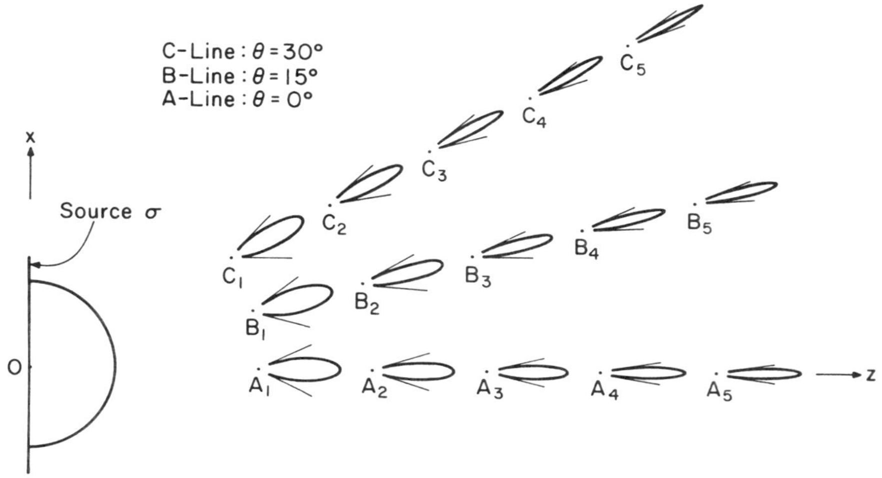

Figure 3. Polar diagrams of the spectral radiance $B_\nu(\mathbf{r}, \mathbf{s})$ at different points in the x, z-plane, produced by a circular Lambertian source of radius $a = 2$ cm, whose intensity profile is given by equation (33), with $\sigma_s(\nu) = 1$ cm. The location of the sixteen points $(\mathrm{O}, \mathrm{A}_1, \mathrm{B}_1, \ldots)$ are the same as those in figure 2.

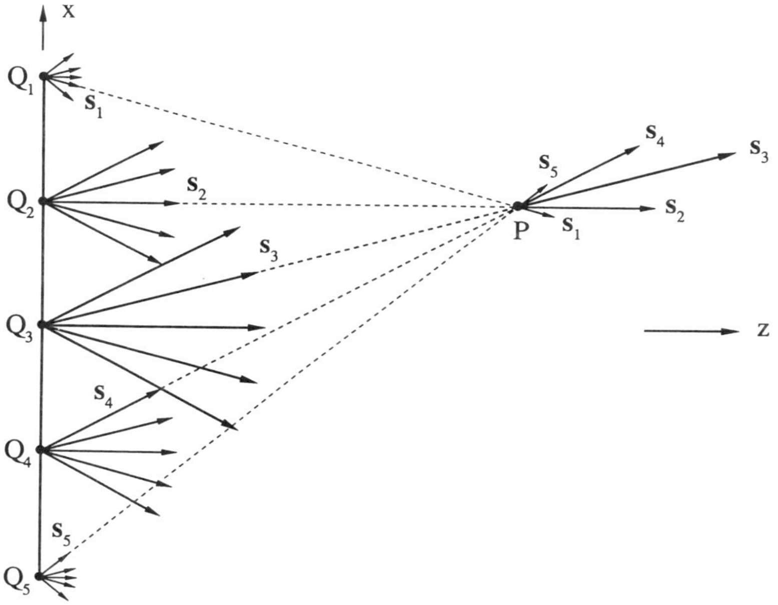

Figure 4. Diagram relating to a heuristic explanation, given in the text, of the propagation of radiance from a Lambertian source with Gaussian intensity profile.

A heuristic way of understanding the behaviour of the spectral radiance is illustrated in figure 4, where $\mathrm{Q}_1, \mathrm{Q}_2, \ldots, \mathrm{Q}_5$ are source points which lie on the x-axis and P is an observation point in the x, z-plane. At each of the source points five vectors are drawn in different directions; the magnitude of each vector is proportional to the local value of the spectral radiance for that particular direction. The vectors labelled by $\mathbf{s}_j$ $(j = 1, 2, \ldots, 5)$ represent the values $B_\nu^{(0)}(Q_j, \mathbf{s}_j)$ of the radiance along directions pointing from these source points to the field point P. The values of the radiance at P in the directions $\mathbf{s}_j$ are represented by the same vectors since, as is evident from equation (22),

$$B_\nu(P, \mathbf{s}_j) = B_\nu^{(0)}(Q_j, \mathbf{s}_j), \qquad j = 1, 2, 3, 4, 5. \tag{38}$$

It is clear from figure 4 that, considered as function of $\mathbf{s}$, $B_\nu(P, \mathbf{s})$ must be maximum in the $\mathbf{s}_3$-direction ($\alpha = \theta$) because the radiance delivered by P by the source point Q_3 is larger than that delivered by any other source point. It is also evident from the figure that, with P fixed, $B_\nu(P, \mathbf{s})$ decreases monotonically as $\mathbf{s}$ is rotated from the $\mathbf{s}_3$-direction and is zero when $\mathbf{s}$ lies outside the geometrical optics cone Ω_P.

3.3. *Gaussian-correlated source of uniform intensity*

As another example, we consider a Gaussian-correlated circular source of radius a, whose spectral density is same at each source point:

$$S^{(0)}(\boldsymbol{\rho}, \nu) = S^{(0)}(\nu)\, \mathrm{circ}\,(\boldsymbol{\rho}/a), \tag{39}$$

$$g^{(0)}(\boldsymbol{\rho}', \nu) = \exp\left[-\boldsymbol{\rho}'^2/2\sigma_g^2(\nu)\right]. \tag{40}$$

Here, $\sigma_g(\nu)$ is independent of $\boldsymbol{\rho}'$ and is assumed to be such that

$$\sigma_g(\nu) \ll a. \tag{41}$$

The two-dimensional spatial Fourier transform (defined by equation (11)) of the degree of spatial coherehence (40) is given by the expression

$$\tilde{g}^{(0)}(\mathbf{f}, \nu) = \frac{1}{2\pi}\sigma_g^2(\nu) \exp\left[-f^2\sigma_g^2(\nu)/2\right]. \tag{42}$$

Substituting from equations (39) and (42) into equations (23) and (21), we find that in the present case

$$B_\nu^{(0)}(\boldsymbol{\rho}, \mathbf{s}) = A(\nu) s_z \exp\left[-k^2\sigma_g^2(\nu)\mathbf{s}_\perp^2/2\right] \mathrm{circ}\,(\boldsymbol{\rho}/a) \tag{43}$$

and

$$B_\nu(\mathbf{r}, \mathbf{s}) = \begin{cases} A(\nu) s_z \exp\left[-k^2\sigma_g^2(\nu) s_\perp^2/2\right], & \text{when } \mathbf{s} \in \Omega_P, \\ 0 & \text{when } \mathbf{s} \notin \Omega_P, \end{cases} \tag{44}$$

where

$$A(\nu) = \frac{[k\sigma_g(\nu)]^2}{2\pi} S^{(0)}(\nu). \tag{45}$$

We see from equation (43) that, unlike in the two previous examples, the spectral radiance in the source plane now depends on the direction ($\mathbf{s}$); it is independent of position ($\boldsymbol{\rho}$) for points within the source area.

Let us again examine the spectral radiance by expressing the variables $\mathbf{r}$ and $\mathbf{s}$ in spherical polar coordinates (equations (29)). As before, we confine our attention to points in the x, z-plane ($\phi = 0$) and to $\mathbf{s}$-directions in that plane ($\beta = 0$). It readily follows from equations (44) and (29 *b*) that, in this case,

$$\left.\begin{aligned} B_\nu(\mathbf{r}, \mathbf{s}) &= A(\nu) \cos\alpha \exp\{-[k\sigma_g(\nu)\sin\alpha]^2/2\}, && \text{when } \alpha_1 \leqslant \alpha \leqslant \alpha_2, \\ &= 0, && \text{otherwise}, \end{aligned}\right\} \tag{46}$$

where α_1 and α_2 are again given by equations (32).

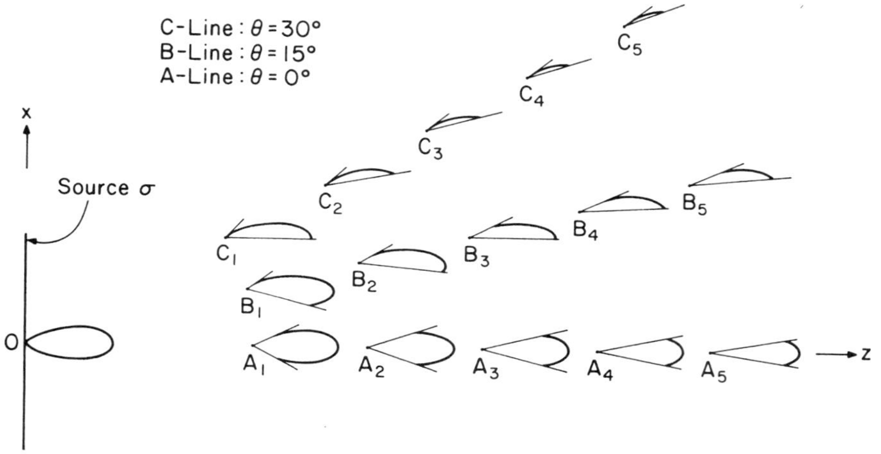

Figure 5. Polar diagrams of the spectral radiance $B_\nu(\mathbf{r}, \mathbf{s})$ at different points in the x, z-plane, produced by a circular Gaussian-correlated source of uniform intensity. The r.m.s. width of the Gaussian correlation function (equation (40)) was taken to be $\sigma_g(\nu)=0{\cdot}5\,\lambda$ and the radius of the source was assumed to be 2 cm. The locations of the sixteen points $(O, A_1, B_1, \ldots)$ are the same as those in figure 2.

Some polar diagrams calculated from equation (46) are presented in figure 5. Two features are obvious from the figure. First, for those field points P for which the associated cone Ω_P includes the forward directions, the radiance has its maximum in that direction ($\alpha=0°$). Secondly, the effect of the cut-off at the boundary of the geometrical optics cone Ω_P is again quite sharp, because the spectral density $S^{(0)}(\boldsymbol{\rho}, \nu)$ has (with ν fixed) a constant value right up to the boundary of the source (equation (39)).

A heuristic explanation of the behaviour of the spectral radiance is given in figure 6, where the radiance is represented in the same manner as in figure 4. In the present case, the radiance $B_\nu^{(0)}(Q_j, \mathbf{s})$ is sharply peaked at each source point in the forward direction ($\alpha=0°$) and decreases monotonically as $\mathbf{s}$ rotates away from this direction. The radiance 'diagrams' are the same at each source point (see equation (43)) and, because of the transport equation (38), the radiance at the point P must attain its maximum value in the s_z-direction ($\alpha=0°$), because the source point Q_2 is delivering the largest contribution to the point P. We also see from the figure that $B_\nu(P, \mathbf{s})$ will decrease monotonically as $\mathbf{s}$ rotates away from the forward direction and is zero if $\mathbf{s}$ lies outside the geometrical optics cone Ω_P.

3.4. *Gaussian-correlated source with Gaussian intensity profile*

As a final example, we again consider a Gaussian-correlated circular source of radius $a \gg \lambda$, but assume now that its spectral density is a truncated Gaussian function of position, i.e. that

$$S^{(0)}(\boldsymbol{\rho}, \nu)=S^{(0)}(\nu)\exp[-\rho^2/2\sigma_s^2(\nu)]\,\mathrm{circ}\,(\rho/a), \tag{47}$$

$$g^{(0)}(\boldsymbol{\rho}', \nu)=\exp[-\rho'^2/2\sigma_g^2(\nu)], \tag{48}$$

where $\sigma_s(\nu)$ and $\sigma_g(\nu)$ are independent of $\boldsymbol{\rho}$ and

$$\sigma_g(\nu) \ll \sigma_s(\nu), \qquad \sigma_g(\nu) \ll a. \tag{49}$$

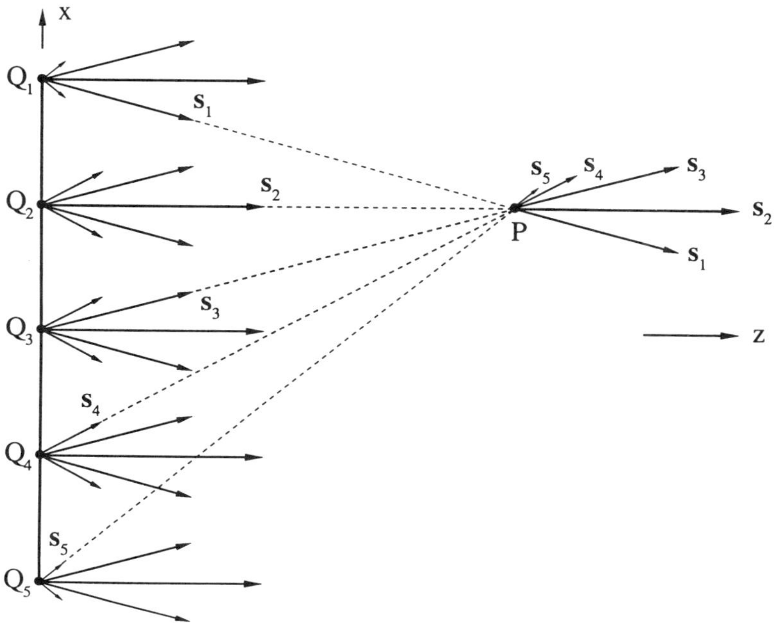

Figure 6. Diagram relating to a heuristic explanation, given in the text, of the propagation of a radiance function from a Gaussian-correlated source of uniform intensity.

The two-dimensional spatial Fourier transform of the expression (48) is again given by equation (42). Substituting from equations (42) and (47) into equations (23) and (21) we find that

$$B_\nu^{(0)}(\boldsymbol{\rho}, \mathbf{s}) = A(\nu) s_z \exp[-\rho^2/2\sigma_s^2(\nu)] \exp[-k^2\sigma_s^2(\nu)\mathbf{s}_\perp^2/2] \operatorname{circ}(\rho/a) \tag{50}$$

and

$$B_\nu(\mathbf{r}, \mathbf{s}) = \begin{cases} A(\nu) s_z \exp\left[-\left|\mathbf{r}_\perp - \dfrac{z}{s_z}\mathbf{s}_\perp\right|^2 \Big/ 2\sigma_s^2(\nu)\right] \exp[-k^2\sigma_g^2(\nu)/2], & \text{when } \mathbf{s} \in \Omega_P, \\ 0 & \text{otherwise,} \end{cases} \tag{51}$$

where $A(\nu)$is given by equation (45). We see from equation (50) that in the present case the spectral radiance in the source plane depends both on position ($\boldsymbol{\rho}$) and direction ($\mathbf{s}$).

As in the previous examples, we will express $\mathbf{r}$ and $\mathbf{s}$ in spherical polar coordinates (equations (29)) and confine our attention to points in the x, z-plane ($\phi = 0$) and $\mathbf{s}$-directions in that plane ($\beta = 0$). Using equation (29), equation (51) for the spectral radiance may be expressed in the form

$$B_\nu(\mathbf{r}, \mathbf{s}) = \begin{cases} A(\nu) \cos\alpha \exp\{-[r \sin(\theta - \alpha)/\cos\alpha]^2/2\sigma_s^2(\nu)\} & \\ \quad \times \exp\{-\tfrac{1}{2}[k\sigma_g(\nu)\sin\alpha]^2\}, & \text{when } \alpha_1 \leqslant \alpha \leqslant \alpha_2, \\ 0, & \text{otherwise,} \end{cases} \tag{52}$$

where α_1 and α_2 are again given by equations (32).

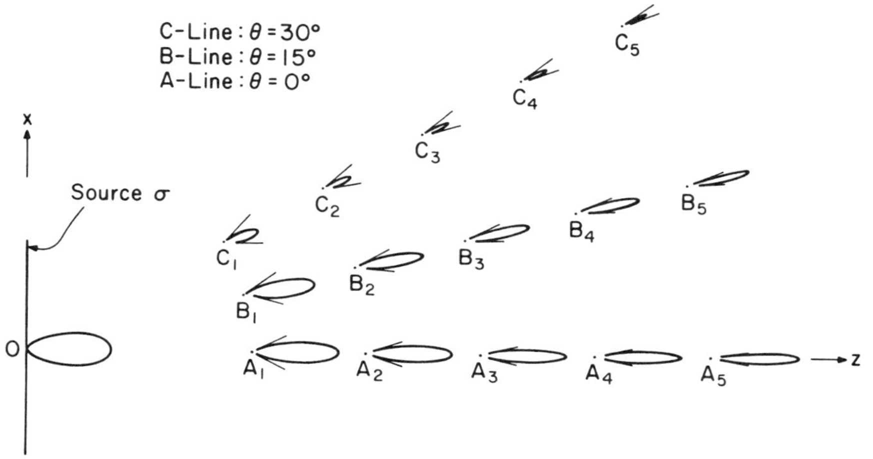

Figure 7. Polar diagrams of the spectral radiance $B_\nu(\mathbf{r}, \mathbf{s})$ at different points in the x, z-plane, produced by a Gaussian-correlated source with a Gaussian intensity profile. The r.m.s. width of the Gaussian correlation function (equation (48)) was taken to be $\sigma_g(\nu) = 0{\cdot}5\ \lambda$ and that of the spectral density of the source (equation (47)) as $\sigma_s(\nu) = 1$ cm. The radius a of the source was assumed to be 2 cm. The locations of the sixteen points (O, A_1, B_1, ...) are the same as those in figure 2.

Some polar diagrams calculated from equation (52) are presented in figure 7. As might be expected, these diagrams are a kind of compromise between those for the two previous examples. In the example discussed in section 3.2, the Gaussian spectral density of the source produced a radiance which peaked in the direction $\alpha = \theta$; in the example discussed in section 3.3, the Gaussian correlation produced a peak in the forward direction ($\alpha = 0$). In the present case, provided the field point P is not on the axis $\theta = 0$, the radiance is seen to peak at an intermediate angle α' $(0 < \alpha' < \theta)$.

4. Conclusions

Until fairly recently the influence of the state of coherence of sources on the radiometric properties of emitted fields have been ignored. In this paper, we derived explicit expressions for the radiance function of fields generated by some partially coherent planar sources. Using these expressions we obtained plots of the corresponding polar diagrams of the radiance functions of the fields at a number of observation points in the half-space into which the sources radiate. These plots reveal considerable differences between the polar diagrams of fields produced by uniform Lambertian sources and by sources of other kinds.

Acknowledgment

This research was supported by the Department of Energy under grant DE-FG02-90ER14119. The views expressed in this article do not constitute an endorsement by the Department of Energy.

Appendix. Asymptotic evaluation of $C_\nu^{(1)}(\mathbf{r}, \mathbf{s}_\perp)$ as $k\to\infty$ (equations (18) and (19))

We have, according to equation (14),

$$C_\nu^{(1)}(\mathbf{r}, \mathbf{s}_\perp)=\frac{kz}{2\pi \mathrm{i}} F_\nu(\mathbf{r}, \mathbf{s}_\perp), \tag{A 1}$$

where (if we write $\boldsymbol{\rho}'$ in place of $\boldsymbol{\rho}$)

$$F_\nu(\mathbf{r}, \mathbf{s}_\perp)=\int_\sigma S^{(0)}(\boldsymbol{\rho}', \nu)\frac{\exp[\mathrm{i}k\Phi(\mathbf{r}, \mathbf{s}_\perp, \boldsymbol{\rho}')]}{R^2}\,\mathrm{d}^2\boldsymbol{\rho}' \tag{A 2}$$

and

$$\Phi(\mathbf{r}, \mathbf{s}_\perp, \boldsymbol{\rho}')=R+\mathbf{s}_\perp\cdot\boldsymbol{\rho}', \tag{A 3}$$

with

$$R=|\mathbf{r}-\boldsymbol{\rho}'|. \tag{A 4}$$

Let

$$\mathbf{r}\equiv(x, y, z), \qquad \boldsymbol{\rho}'\equiv(x', y', 0). \tag{A 5}$$

Equation (A 2) becomes

$$F_\nu(\mathbf{r}, \mathbf{s}_\perp)=\int_\sigma f(x', y'; \mathbf{r}; \nu)\exp[\mathrm{i}kg(x', y'; \mathbf{r}; \mathbf{s}_\perp)]\,\mathrm{d}x'\,\mathrm{d}y', \tag{A 6}$$

where

$$f(x', y'; \mathbf{r}; \nu)=\frac{1}{R^2}S^{(0)}(x', y', \nu), \tag{A 7a}$$

$$g(x', y'; \mathbf{r}; \mathbf{s}_\perp)=s_x x'+s_y y'+R, \tag{A 7b}$$

with

$$R=[(x'-x)^2+(y'-y)^2+z^2]^{1/2}. \tag{A 8}$$

We are interested in the asymptotic behaviour of the integral (A 6) as $k\to\infty$. It is clear that, for large values of k, the exponential factor in the integrand on the right of equation (A 8) will vary more and more rapidly as the point (x', y') explores the domain of integration. On the other hand, the amplitude factor $f(x', y'; \mathbf{r}, \nu)$, being proportional to the source spectrum, may be expected, for large values of k, to be a well-behaved function of x', y'.† Under these circumstances the asymptotic behaviour of the integral may be determined by the application of the two-dimensional form of the principle of stationary phase ([13], Appendix III]). For this purpose, we must first determine the locations of the critical points of the first kind (if any) of the integrand. These are the points at which the functions $g(x', y'; \mathbf{r}; s_x, s_y)$ is stationary within the source region σ, with respect to the variation of (x', y'), i.e. where

$$g_{x'}=g_{y'}=0 \tag{A 9}$$

† Since the source is quasi-homogeneous, $S^{(0)}(x', y', \nu)$ will vary slowly over distances of the order of the spectral correlation length l_ν of the source. For realistic sources $l_\nu\gtrsim\lambda=c/\nu$; therefore, $S^{(0)}(x', y', \nu)$ will vary slowly over distances of the order of the wavelength λ.

within σ, with $g_{x'}$ and $g_{y'}$ denoting the first partial derivatives of g with respect to x' and y', respectively. Now , from equations (A 7 *b*) and (A 8) we find that

$$g_{x'}=s_x+\frac{x'-x}{R}, \qquad g_{y'}=s_y+\frac{y'-y}{R}. \tag{A 10}$$

From equations (A 10) and (A 8) it follows that if a point $x'=x'_0$, $y'=y'_0$ is a critical point of the first kind, it must satisfy the equations

$$x'_0-x=-s_x[(x'_0-x)^2+(y'_0-y)^2+z^2]^{1/2} \tag{A 11 a}$$

and

$$y'_0-y=-s_y[(x'_0-x)^2+(y'_0-y)^2+z^2]^{1/2}. \tag{A 11 b}$$

On squaring both sides of these equations, we obtain two simultaneous equations for the quantities $(x'_0-x)^2$ and $(y'_0-y)^2$. They can be readily solved and give

$$x'_0=x-\frac{s_x}{s_z}z, \qquad y'_0=y-\frac{s_y}{s_z}z, \tag{A 12 a}$$

or more explicitly, in vector form, with $\boldsymbol{\rho}'_0\equiv(x'_0, y'_0, 0)$,

$$\boldsymbol{\rho}'_0=\mathbf{r}_\perp-\frac{z}{s_z}\mathbf{s}_\perp. \tag{A 12 b}$$

The formulas (A 12) shows that the function $g(x', y'; \mathbf{r}; s_x, s_y)$ has one and only one stationary point in the source plane $z=0$. This point will be a critical point of the first kind of the integrand on the right-hand side of equations (A 2) only if it is located within the source region σ. The geometrical significance of this point is illustrated in figure 1 of the text, where it is denoted by Q_0.

According to a general formula [see [13], Appendix III, equations (20) and (21)] the asymptotic approximation to an integral of the form (A 6) is given by, if we ignore the dependence of the function $f(x', y', \nu)$ on ν, which is justified for reasons explained earlier,

$$F_\nu(\mathbf{r}, \mathbf{s}_\perp)\sim\frac{2\pi i\alpha}{k\sqrt{|\Delta|}}f(x'_0, y'_0; \mathbf{r}, \nu)\exp[ikg(x'_0, y'_0; \mathbf{r}, \mathbf{s}_\perp)], \qquad \text{as } k\to\infty, \tag{A 13}$$

with

$$\Delta=g_{x'x'}g_{y'y'}-g^2_{x'y'}\Big|_{x'_0y'_0} \tag{A 14}$$

and

$$\alpha=\begin{cases} +1, & \text{when } \Delta>0, \quad g_{x'x'}\Big|_{x'_0y'_0}>0, \\ -1, & \text{when } \Delta>0, \quad g_{x'x'}\Big|_{x'_0y'_0}<0, \\ -i, & \text{when } \Delta<0, \end{cases} \tag{A 15}$$

where it is assumed that $\Delta\neq 0$, and $g_{x'x'}$ denotes, of course, the second-order partial derivative of g with respect to x', etc. These derivatives are readily obtained on differentiating the expressions (A 10) and again using the expression (A 8) for R. One finds that

$$\left.\begin{aligned} g_{x'x'} &= \frac{1}{R}\left[1-\frac{(x'-x)^2}{R^2}\right], \\ g_{y'y'} &= \frac{1}{R}\left[1-\frac{(y'-y)^2}{R^2}\right], \\ g_{x'y'} &= -\frac{(x'-x)(y'-y)}{R^3}. \end{aligned}\right\} \tag{A 16}$$

The values of these quantities at the critical point $x'=x'_0$, $y'=y'_0$, which follow at once on substituting from equations (A 12 a) into equations (A 16), are

$$\left.\begin{aligned} g_{x'x'}\Big|_{x'_0y'_0} &= \frac{s_z}{z}(1-s_x^2), \\ g_{y'y'}\Big|_{x'_0y'_0} &= \frac{s_z}{z}(1-s_y^2), \\ g_{x'y'}\Big|_{x'_0y'_0} &= -\frac{s_xs_ys_z}{z}, \end{aligned}\right\} \tag{A 17}$$

where we used a fact which readily follows from equations (A 8) and (A 12 a), namely that

$$R\Big|_{x'_0y'_0} = \frac{z}{s_z}. \tag{A 18}$$

On substituting from equations (A 17) into the formulas (A 14) and (A 15), we find that

$$\Delta = \left(\frac{s_z^2}{z}\right)^2, \qquad \alpha = 1. \tag{A 19}$$

Further, on substituting from equations (A 12 a) into equations (A 7) and making use of equation (A 18), we obtain for $f(x'_0, y'_0; \mathbf{r}, \nu)$ and $g(x'_0, y'_0; \mathbf{r}, \mathbf{s}_\perp)$ the expressions

$$f(x'_0, y'_0; \mathbf{r}, \nu) = \left(\frac{s_z}{z}\right)^2 S^{(0)}(x'_0, y'_0, \nu), \tag{A 20}$$

$$g(x'_0, y'_0; \mathbf{r}, \mathbf{s}_\perp) = xs_x + ys_y + zs_z = \mathbf{r}\cdot\mathbf{s}. \tag{A 21}$$

Finally on substituting from equations (A 19), (A 20) and (A 21) into the general formula (A 13), we find that

$$F_\nu(\mathbf{r}, \mathbf{s}_\perp) \sim \frac{2\pi \mathrm{i}}{k}\frac{1}{z} S^{(0)}(\boldsymbol{\rho}'_0, \nu) \exp(\mathrm{i}k\mathbf{s}\cdot\mathbf{r}), \qquad \text{as} \quad k \to \infty, \tag{A 22}$$

provided that the point $\boldsymbol{\rho}'$, specified by equation (12 b), lies within the domain of integration (the source domain σ). If it does not, the contribution to the asymptotic approximation to $F_\nu(\mathbf{r}, \mathbf{s}_\perp)$ comes from critical points of the second kind, which are located on the boundary curve of σ (see [13], Appendix III) and, except perhaps for some special cases of symmetry, their contribution is

$$F_\nu(\mathbf{r}, \mathbf{s}_\perp) \sim O\left(\frac{1}{k^{3/2}}\right), \qquad \text{as} \quad k \to \infty. \tag{A 23}$$

Finally, on substituting from equations (A 22) and (A 23) into equation (A 1), we see that as $k \to \infty$,

$$C_\nu^{(1)}(\mathbf{r}, \mathbf{s}_\perp) \sim \begin{cases} S^{(0)}(\boldsymbol{\rho}'_0, \nu) \exp[ik\mathbf{s} \cdot \mathbf{r}), & \text{when} \quad \boldsymbol{\rho}'_0 \in \sigma, \\ O\left(\frac{1}{k^{1/2}}\right), & \text{when} \quad \boldsymbol{\rho}'_0 \notin \sigma, \end{cases} \tag{A 24}$$

where $\boldsymbol{\rho}'_0$ is given by equation (A 12 *b*).

Except for a trivial change in the notation ($\boldsymbol{\rho}'_0$ being written here in place of $\boldsymbol{\rho}_0$), equations (A 24) and (A 12 *b*) are equations (18) and (19) of the text.

References

[1] Wolf, E., 1978, *J. opt. Soc. Am.*, **68,** 6; Friberg, A. T., 1982, *Opt. Engng*, **21,** 927.
[2] Foley, J. T., and Wolf, E., 1985, *Optics Commun.*, **55,** 236.
[3] Kim, K., and Wolf, E., 1987, *J. opt. Soc. Am. A*, **4,** 1233; Agarwal, G. S., Foley, J. T., and Wolf, E., 1987, *Optics Commun.*, **62,** 67.
[4] Walther, A., 1968, *J. opt. Soc. Am.*, **58,** 1256.
[5] Walther, A., 1973, *J. opt. Soc. Am.*, **63,** 1622.
[6] Davenport, W. B., and Root, W. L., 1958, *An Introduction to the Theory of Random Signals and Noise* (New York: McGraw-Hill), p. 60.
[7] Goodman, J. W., 1968, *Introduction to Fourier Optics* (New York: McGraw-Hill), section 3.7.
[8] Carter, W. H., and Wolf, E., 1977, *J. opt. Soc. Am.*, **67,** 785.
[9] Mandel, L., and Wolf, E., 1976, *J. opt. Soc. Am.*, **66,** 529.
[10] Jannson, T., 1980, *J. opt. Soc. Am.*, **70,** 1544.
[11] Carter, W. H., and Wolf, E., 1975, *J. opt. Soc. Am.*, **65,** 1067.
[12] Wolf, E., and Carter, W. H., 1978, *Coherence and Quantum Optics IV*, edited by L. Mandel and E. Wolf (New York: Plenum), p. 415.
[13] Born, M., and Wolf, E., 1980, *Principles of Optics* (Oxford: Pergamon).

Coherence and radiometry*

Emil Wolf

Department of Physics and Astronomy, University of Rochester, Rochester, New York 14627

(Received 13 October 1977)

Recent researches have revealed that there exists an intimate connection between radiometry and the theory of partial coherence. In this paper a review is presented of some of these developments. After a brief discussion of various models for energy transport in optical fields and of some of the basic concepts of the classical theory of optical coherence, the following topics are discussed: the foundations of radiometry, the coherence properties of Lambertian sources, and the relationship between the state of coherence of a source and the directionality of the light that the source generates. Some very recent work is also described which reveals that certain sources that are spatially highly incoherent in a global sense will generate light that is just as directional as a laser beam.

I. INTRODUCTION

During a period of time that spans several centuries various models have evolved relating to the transport of energy by optical radiation. The oldest and conceptually the simplest one is at the heart of radiometry and of the so-called theory of radiative energy transfer. It is based on the simple notion of a light ray, i.e., of a geometrical trajectory along which radiant energy is assumed to be propagated. Vast areas of both theoretical and practical science concerned with quantitative aspects of optical radiation make use of the radiometric model in a fundamental way. It is employed to analyze problems ranging from very practical questions occurring in illumination engineering, relating, for example, to the distribution of radiant energy from various types of sources or to the performance of different kinds of radiation detectors, to sophisticated questions of astrophysics concerning the interior of stars or the nature of the stellar atmosphere. The strong intuitive appeal of the basic radiometric concepts has been largely responsible for a fact that is seldom noted: namely, that the relationship between radiometry and modern theories of radiation (i.e., Maxwell's electromagnetic theory and the quantum theory of radiation) has up to now not been clarified.

In the title of this talk I have coupled radiometry with a modern branch of physical optics, namely, coherence theory which, unlike radiometry, employs rather sophisticated concepts of probability theory and of the theory of random processes. You may well ask: "What could radiometry and coherence theory possibly have in common?" Actually, recent researches on the foundation of radiometry have revealed that there is a very intimate connection between these two fields. Not only have these researches answered some rather puzzling old questions about radiometry, but they have also lead to some interesting developments in coherence theory itself. Moreover, these researches have provided some insight into a question of a considerable practical importance at the present time: they clarify to what an extent radiometric concepts and radiometric laws, which have developed around conventional, rather incoherent thermal sources, also apply to radiation generated by highly coherent sources, such as lasers. Before presenting a review of some of this research let me recall how radiometry describes energy transport and let me also remind you of a few elementary results concerning energy in other theories of radiation.

**Editor's Note: This paper is the text of the 1977 Ives Medal Address, delivered at the Annual Meeting of the Optical Society of America in Toronto on October 13, 1977.*

II. MODELS FOR ENERGY TRANSPORT IN OPTICAL FIELDS

It is a basic assumption of radiometry that energy that is radiated from an element $d\sigma$ of a planar source[1] is distributed according to the elementary law

$$d\mathcal{E}_\nu = B_\nu(\mathbf{r},\mathbf{s}) \cos\theta \, d\sigma \, d\Omega \, dt. \tag{2.1}$$

Here $d\mathcal{E}_\nu$ represents the amount of energy, per unit frequency interval at frequency ν, that is propagated in a short time interval dt from a source element $d\sigma$ at a point Q, specified by position vector $\mathbf{r}$ into an element $d\Omega$ of solid angle around a direction specified by unit vector $\mathbf{s}$. Further, θ denotes the angle that the $\mathbf{s}$ direction makes with the unit normal $\mathbf{n}$ to the source plane (Fig. 1). The proportionality factor $B_\nu(\mathbf{r},\mathbf{s})$, which in general depends on position ($\mathbf{r}$), direction ($\mathbf{s}$), and frequency (ν), is known as the (spectral) *radiance* or *brightness*.

In the theory of radiative energy transfer this simple model for the distribution of radiant energy is somewhat generalized in the sense that the radiance function becomes a field quantity. The surface element $d\sigma$ is no longer restricted to coincide with a radiating surface element of a real source but rather is now an element of any (in general fictitious) surface in a region of space containing radiation, with $\mathbf{n}$ denoting the unit normal to $d\sigma$. Instead of radiance, one now speaks of the specific intensity of radiation, which is often denoted by $K_\nu(\mathbf{r},\mathbf{s})$ or $I_\nu(\mathbf{r},\mathbf{s})$. However, the elementary law (2.1) is assumed to retain its validity with this trivial formal change (radiance → specific intensity). By elementary considerations involving nothing more than simple intuitive quasi-geometrical arguments about energy conservation one is lead

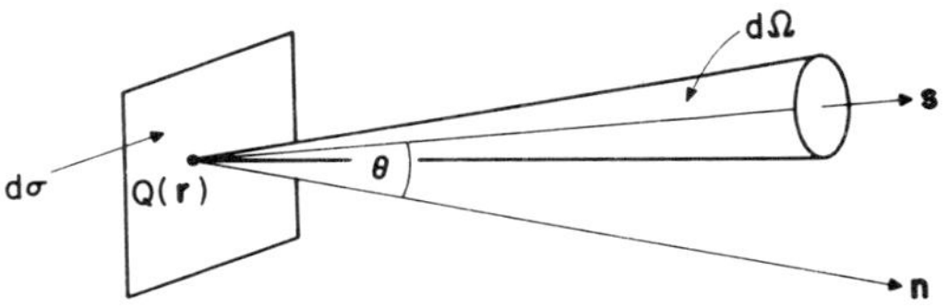

FIG. 1. Illustrating the definition of the radiance function $B(\mathbf{r},\mathbf{s})$.

 0030-3941/78/6801-0006$00.50

to the following integro-differential equation for the propagation of the specific intensity in any isotropic medium[2]:

$$\mathbf{s}\cdot\nabla_r K_\nu(\mathbf{r},\mathbf{s}) = -\alpha_\nu(\mathbf{r},\mathbf{s})K_\nu(\mathbf{r},\mathbf{s}) + \int\beta_\nu(\mathbf{r},\mathbf{s},\mathbf{s}')K_\nu(\mathbf{r},\mathbf{s}')\,d\Omega' + D_\nu(\mathbf{r},\mathbf{s}). \quad (2.2)$$

The left-hand side represents the rate of change of the specific intensity in the direction **s**. The first term on the right represents the rate of decrease of energy due to scattering and absorption, the second term on the right (with the integration extending over the surface of the unit sphere generated by the unit vector **s′**) represents the rate of increase in energy as a result of scattering from all other directions into the **s** direction, and $D_\nu(\mathbf{r},\mathbf{s})$, known as the source function, represents the rate at which energy is emitted due to spontaneous processes. The functions $\alpha_\nu(\mathbf{r},\mathbf{s})$ and $\beta_\nu(\mathbf{r},\mathbf{s},\mathbf{s}')$ are the so-called extinction coefficient and the differential scattering coefficient, respectively.

Equation (2.2) is known as the *equation of radiative transfer* and is one of the basic equations of astrophysics. How well it actually describes energy transport or, even more basically, how the various quantities $(K_\nu,\alpha_\nu,\beta_\nu,D_\nu)$ that appear in this equation are to be interpreted from the standpoint of modern theories of radiation is, to a large extent, an open question. Only in relatively recent times have attempts been made to clarify these questions; however, they have met with rather limited success.

The description of energy transport is quite different within the framework of Maxwell's electromagnetic theory, which is the basis of the whole classical physical optics. Here the quantity that is associated with energy transport is the Poynting vector $\mathbf{S} = (c/4\pi)(\mathbf{E}\times\mathbf{H})$, where **E** and **H** represent the electric and magnetic field vectors and c is the speed of light *in vacuo* (Gaussian system of units being used). It is generally assumed that the optical intensity may be identified with the magnitude of the Poynting vector. We note that there is a qualitative difference between the radiance function of the radiometric model and the optical intensity of physical optics. The radiance and the specific intensity are functions of position (**r**) and of direction (**s**), whereas the optical intensity is independent of direction. Actually, even the dependence of the Poynting vector and hence of the optical intensity on position is somewhat tenuous, because, according to Maxwell's theory, it is not the Poynting vector itself, but rather the integral of its normal component taken over a closed surface that has an unambiguous physical meaning. The integral represents the rate at which energy is transported across the surface. In fact, many paradoxes are known that arise from identifying the Poynting vector too closely with energy transport at a point in the field or because one assumes that the radiant energy "flows" along lines generated by the Poynting vector. Only under special circumstances (e.g., for a plane wave or in the far zone of a radiating source) can such a "hydrodynamic" model be strictly reconciled with electromagnetic theory.

The situation is different yet in the quantum theory of radiation. According to that theory, the radiation field consists of certain nonclassical particles, the photons, each carrying energy of amount $h\nu$, where h is Planck's constant. Although one can also speak of the optical intensity in the quantized field, by defining it via an appropriate operator, it seems more natural to associate energy transport with the photons themselves. However, it is well known that it is not possible to associate a position variable with a photon; in fact, the quantum theory of radiation does not even allow us to speak about the probability of finding a photon at a particular point in space.[3]

We observe that as we proceed from the older models to the more refined ones, less and less can be said about the detailed distribution of energy. According to the radiometric model, energy is localized both in space (**r** dependence of the radiance) and in direction (**s** dependence). Electromagnetic theory says nothing at all about the distribution of energy over directions and what it implies about its distribution throughout space has to be interpreted with caution because of our earlier remarks relating to the physical significance of the Poynting vector. Finally, the quantum theory of radiation tells us nothing at all about the localization of the elementary carriers of energy (the photons) throughout space.

The great physicist H. A. Lorentz in his classic book *The Theory of Electrons*, published in 1909—well before the formulation of the quantum theory of radiation—already warned against treating energy transport by light in a too mechanistic manner. This is what he said (on p. 25): ". . . in general it will not be possible to trace the paths of parts of elements of energy in the same sense in which we can follow in their course the ultimate particles of which matter is made up." And (on p. 26): "It might even be questioned whether in electromagnetic phenomena the transfer of energy really takes place in the way indicated by Poynting's law . . . ."

How is it then possible that the radiometric model, according to which energy is localized both in space and in directions, has been relatively successful—at least in connection with energy transport by light from conventional sources? The usual answer is that this is so because the wavelength of light is very short compared to the linear dimensions of the sources and of the bodies with which the light interacts. In spite of this widely held view no true justification for it has ever been given.

Recent researches have revealed that appreciably more is involved in the foundations of radiometry and of the theory of radiative energy transfer than a short wavelength limit. That this must be so can be seen from the following simple example: Suppose we compare the angular distribution of radiation from a conventional thermal source and from a well-stabilized laser. Under ordinary circumstances the thermal source will radiate in accordance with Lambert's law,

$$J(\theta) = J(0)\cos\theta, \quad (2.3)$$

where $J(\theta)$ denotes the radiant intensity in a direction making an angle θ with the normal **n** to the radiating surface [Fig. 2(a)]. This rather broad angular dependence is exhibited on a polar diagram in Fig. 2(c). On the other hand, the light generated by a laser [Fig. 2(b)] will be extremely directional: Practically all the radiated energy will be concentrated in a very narrow solid angle around the forward direction; typically for a laser with a cross-section diameter 2 mm say, 99% of the radiated energy will be concentrated within a cone of angular divergence of the order of 10^{-4} rad. Thus the polar diagram of the radiant intensity has a needle-like form [Fig. 2(d)]

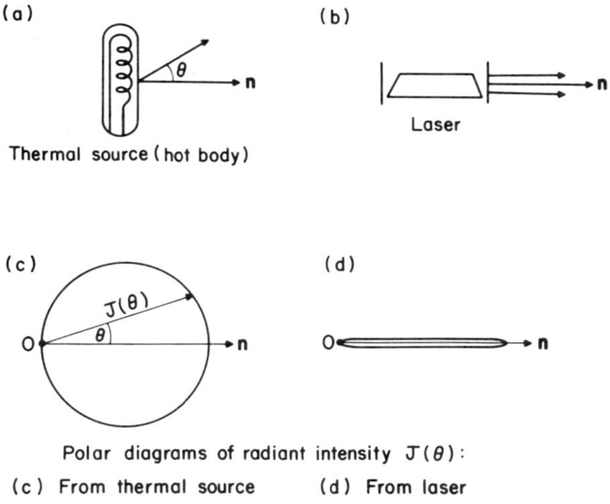

FIG. 2. Comparison of the angular distribution of radiant intensity from a thermal source and from a laser.

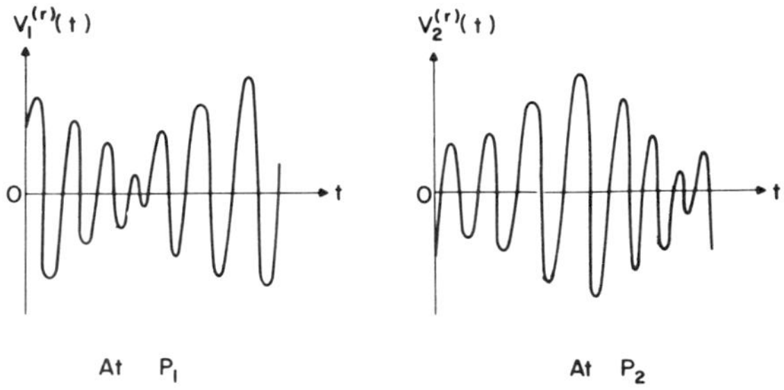

FIG. 3. Illustrating the temporal behavior of the field variable $V^{(r)}(t)$ at two points $P_1(\mathbf{r}_1)$ and $P_2(\mathbf{r}_2)$ in space.

which, roughly speaking, resembles the behavior of the Dirac delta function with its singularity at $\theta = 0$.

Now one of the chief differences between the two sources is that the thermal source is spatially almost completely incoherent, whereas the laser source is, of course, spatially highly coherent. Hence the comparison of the angular distribution of radiation from the two sources seems to indicate that there must be a close connection between the state of coherence of a source and the distribution of the radiant intensity of the light that the source generates. In turn, this result implies that there must be an intimate relationship between radiometry and the theory of partial coherence, a supposition that has been fully justified by recent researches. To appreciate this development it might be helpful to review at this point some of the basic concepts of optical coherence theory.

III. SOME BASIC CONCEPTS OF OPTICAL COHERENCE THEORY[4]

Every optical field has some fluctuations associated with it. The fluctuations are usually too rapid to be directly perceived by the eye or to be readily revealed by ordinary laboratory experiments. However, the existence of the fluctuations can be deduced indirectly, for example, from experiments involving interference effects. The concept of partial coherence is intimately related to a measure of statistical similarity—known more technically as correlation—between light fluctuations.

Imagine that we have two very good detectors that allow us to record the detailed time behavior of the optical field at two points P_1 and P_2, specified by position vectors $\mathbf{r}_1$ and $\mathbf{r}_2$, respectively. The result may be curves such as those indicated in Fig. 3, which show how the real field variable $V^{(r)}(\mathbf{r},t)$ varies at these points in the course of time. For simplicity we consider $V^{(r)}(\mathbf{r},t)$ to be a scalar, e.g., one of the Cartesian components of the electric field at the space-time point $(\mathbf{r},t)$. In the analysis of coherence effects it is convenient to associate with this real field variable a complex one, $V(\mathbf{r},t)$, known as the complex analytic signal. We will not discuss this step here; it is explained in many texts.[5]

The basic quantity used for the analysis of the simplest coherence effects in optical fields is the so-called *mutual coherence function*[6]

$$\Gamma_{12}(\tau) = \langle V_1(t+\tau)V_2^*(t)\rangle, \tag{3.1}$$

where the asterisk denotes the complex conjugate and the sharp brackets denote either a time average or an ensemble average. For the sake of simplicity we use abbreviated notations throughout this section: thus, for example, $V_1(t+\tau)$, is an abbreviation for $V(\mathbf{r}_1,t+\tau)$, etc. In terms of the mutual coherence function $\Gamma_{12}(\tau)$ one defines the so-called *complex degree of coherence* of the light vibrations at the two points P_1 and P_2 as

$$\gamma_{12}(\tau) = \Gamma_{12}(\tau)/\sqrt{I_1}\sqrt{I_2}, \tag{3.2}$$

where

$$I_j = \Gamma_{jj}(0) = \langle V_j(t)V_j^*(t)\rangle \qquad (j = 1,2) \tag{3.3}$$

is the (averaged) optical intensity at the two points, as customarily defined in scalar wave theory.[7] It may be shown that the absolute value of the complex degree of coherence obeys the inequality

$$0 \leqslant |\gamma_{12}(\tau)| \leqslant 1 \tag{3.4}$$

for all values of its arguments. The limiting cases $\gamma_{12}(\tau) = 0$ and $|\gamma_{12}(\tau)| = 1$ represent complete (second-order) incoherence and complete (second-order) coherence, respectively, of the vibrations at the two points P_1 and P_2, at times $t+\tau$ and t, respectively. When the two points coincide, the dependence of γ on τ reveals so-called *temporal coherence;* when τ is kept fixed (usually at zero value) the dependence of γ on the location of the two points P_1 and P_2 reveals what is known as *spatial coherence.*

Two well-known interference experiments, illustrated in Figs. 4 and 5, provide quantitative measures of temporal and spatial coherence. The sharpness and the location of the interference fringes in the plane $\mathcal{B}$ of the Michelson interferometer (Fig. 4) give information about the degree of temporal coherence of the light at the dividing mirror M_0 for the time delay τ associated with the relative path difference between the two beams that reach M_0 via the two other mirrors

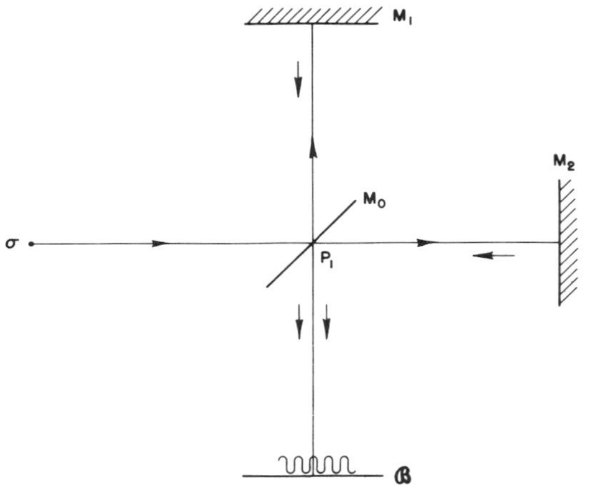

FIG. 4. Temporal coherence illustrated by means of a Michelson interferometer.

FIG. 6. The diffraction image of a star formed by a well-corrected telescope (courtesy of B. J. Thompson).

M_1 and M_2. The sharpness and the location of the fringes around the symmetrically situated point Q in the plane $\mathcal{B}$ of Young's interference experiment (Fig. 5) provide information about the degree of spatial coherence, for zero time delay ($\tau = 0$) of the light at the pinholes P_1 and P_2.

It seems that phenomena involving temporal coherence of light are generally reasonably well understood, largely, I believe, because they can usually be interpreted in an alternative way, in terms of the spectral properties of the light. Spatial coherence effects, on the other hand, do not appear to be so well understood, and because they play a basic role in some of the recent developments on the foundation of radiometry I will say a few words about them.

It is often asserted that thermal sources are spatially completely incoherent in the sense that no correlations exist between the light vibrations at two distinct points in the source. Actually, as we will soon see, this statement is not correct; in any radiating source the field is always correlated (i.e., spatially coherent), at least over distances of the order of the mean wavelength of the emitted light. However, the usual thermal sources may be regarded as *"globally" incoherent* in the sense that their linear dimensions are very large compared to the linear dimensions of the source region over which appreciable spatial coherence exists. Even though such sources are spatially incoherent in this sense, they may produce fields that will have a high degree of spatial coherence over arbitrarily large regions of space. Before explaining why this is so let me illustrate this fact by a simple example that I am sure is familiar to most of you, though you have probably not interpreted it before from the standpoint of coherence theory.

Suppose that we view a star on a good observing night through a well-corrected telescope. We will then observe in the focal plane of the telescope the diffraction image of the star formed by the telescope. It consists of a bright central disk surrounded by rings of rapidly diminishing intensity (Fig. 6). From the standpoint of coherence theory this image tells us something important about the spatial coherence of the starlight entering the telescope. According to elementary diffraction theory, the image may be considered as being formed by the combined effects of the so-called Huygens' secondary wavelets proceeding to the focal plane from the primary wave falling on the aperture of the telescope (Fig. 7). Now the fact that the diffraction image has pronounced intensity minima and maxima implies that the Huygens' wavelets are capable of strongly interfering with each other—in fact, their interference produces zero field along the dark rings of the diffraction image. This is only possible if the light entering the telescope is strongly spatially coherent across the whole aperture of the telescope. Yet, as we know, the light originated in very many atoms in the star, atoms that to a very high degree of approximation may be considered to have ra-

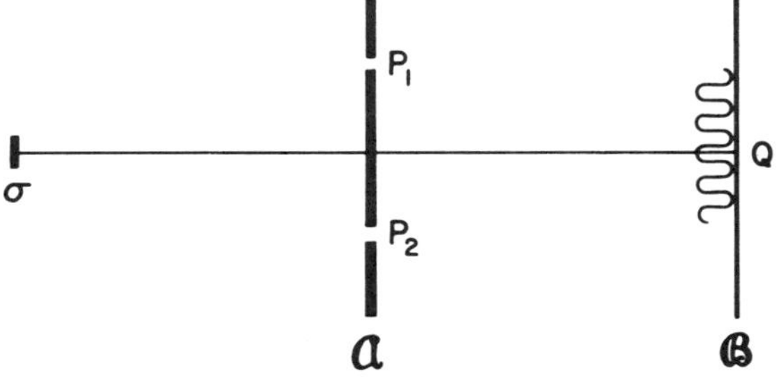

FIG. 5. Spatial coherence illustrated by means of Young's interference experiment.

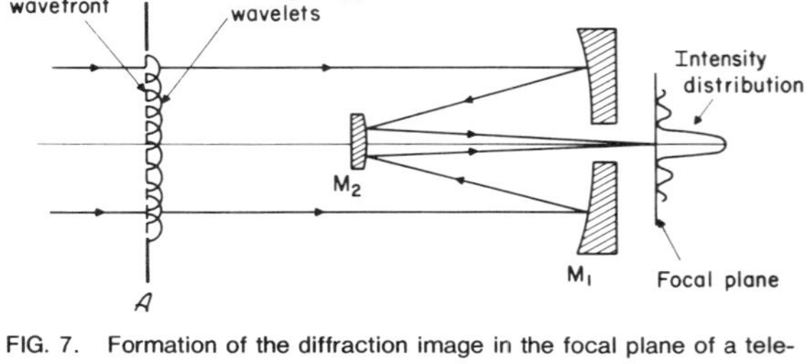

FIG. 7. Formation of the diffraction image in the focal plane of a telescope.

diated independently of each other. Thus the spatial coherence of the light entering the telescope is in this case not a manifestation of any coherence properties of the source itself, but rather—unlike in the case of laser light—it somehow must have been *generated in the process of propagation* of the light from the star to the telescope. It is not difficult to explain in mathematical language how this comes about. For our purposes it will be sufficient to give a rather simple, though necessarily less rigorous explanation, which, however, provides a good intuitive understanding of the underlying physics.

Consider two *independently* radiating point sources S_1 and S_2 and let us examine the behavior of the light at two points P_1 and P_2 in the field. We denote by A_1 and A_2 the wave trains arriving at P_1 and P_2 from the point source S_1 and by B_1 and B_2 the wave trains arriving at these points from the other point source, S_2 (Fig. 8). Since the point sources S_1 and S_2 are assumed to radiate independently, the light that they emit will be uncorrelated. We may express this fact symbolically by writing

$$\langle A_i B_j \rangle = 0 \qquad (i,j = 1,2). \tag{3.5}$$

Now for the sake of simplicity let us also assume that the distances S_1P_1 and S_1P_2 are approximately equal, as are the distances S_2P_1 and S_2P_2. Then we obviously have

$$A_2 \approx A_1, \tag{3.6a}$$

$$B_2 \approx B_1. \tag{3.6b}$$

Now the light vibrations at the points P_1 and P_2 are generated by superposition of the A and B wave trains, so that the fields V_1 and V_2 at the two points are

$$\text{at } P_1: \qquad V_1 = A_1 + B_1, \tag{3.7a}$$

$$\text{at } P_2: \qquad V_2 = A_2 + B_2. \tag{3.7b}$$

In view of the relations (3.6) we obviously have

$$V_1 \approx V_2. \tag{3.8}$$

This simple formula tells us that the light fluctuations at the points P_1 and P_2 will be very similar to each other, i.e., the field at these two points will have a high degree of spatial coherence, in spite of the fact that the two point sources S_1 and S_2 are completely uncorrelated. We see that this spatial coherence has been generated by superposition in the process of propagation.

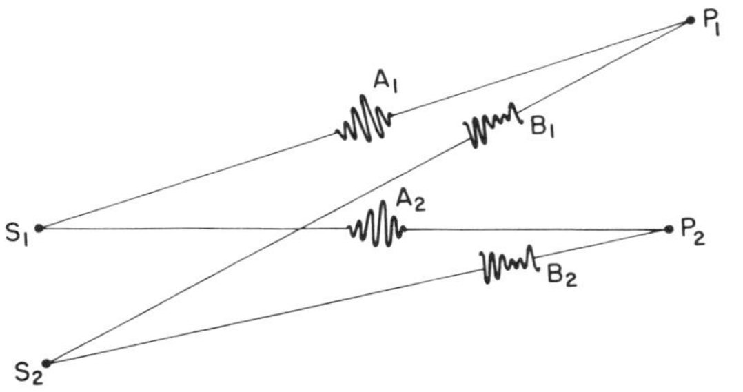

FIG. 8. Illustrating the generation of spatial coherence from two uncorrelated point sources.

One can easily extend this argument to radiation from any number of independent point sources and hence, in an appropriate limit, to radiation from an extended "globally" incoherent source, such as a star. The whole argument can be made mathematically quite rigorous and leads to the so-called van Cittert-Zernike theorem (Ref. 5, Sec. 10.4.2), which is one of the basic results of the classical theory of partial coherence. However, the physical origin of spatial coherence in a field from a highly uncorrelated source is quite clearly evident from the simple argument that I just gave.

There is another important conclusion that readily follows from the same type of argument. If we take the two field points P_1 and P_2 to be situated in the source region itself one finds that the light vibrations in that region must necessarily be correlated over distances of at least the order of magnitude of the mean wavelength of the light. This conclusion has a close bearing on some questions that arise with regard to the foundations of radiometry, to which we will now turn our attention.

IV. FOUNDATIONS OF RADIOMETRY

We saw earlier from our comparison of the angular distribution of radiation generated by a thermal source and by a laser, that there must be a close relationship between the radiometric properties of a source and its coherence properties. Considerable progress has been made in recent years to clarify this relationship, which turned out to be a rather subtle one, as we will see.

According to radiometry the total power radiated by a planar source per unit frequency interval at frequency ν is obtained at once from the elementary law (2.1) as

$$\mathcal{P}_\nu = \int_{(2\pi)} d\Omega \int_\sigma d\sigma B_\nu(\mathbf{r},\mathbf{s}) \cos\theta, \tag{4.1}$$

where the integrations extend over the source σ and over the 2π-solid angle formed by all the $\mathbf{s}$ directions that point into the half-space into which the source is radiating. Equation (4.1) may be rewritten in two equivalent forms

$$\mathcal{P}_\nu = \int_{(2\pi)} J_\nu(\mathbf{s})\, d\Omega = \int_\sigma E_\nu(\mathbf{r})\, d\sigma, \tag{4.2}$$

where

$$E_\nu(\mathbf{r}) = \int_{(2\pi)} B_\nu(\mathbf{r},\mathbf{s}) \cos\theta\, d\Omega, \tag{4.3}$$

and

$$J_\nu(\mathbf{s}) = \cos\theta \int_\sigma B_\nu(\mathbf{r},\mathbf{s})\, d\sigma. \tag{4.4}$$

The formulas (4.3) and (4.4) define two important radiometric quantities in terms of the radiance $B_\nu(\mathbf{r},\mathbf{s})$, namely the *radiant emittance* $E_\nu(\mathbf{r})$ and the *radiant intensity* $J_\nu(\mathbf{s})$.

Before proceeding any further we note that there is an implicit assumption built into Eq. (4.1), namely, that the total radiated power is obtained by simply adding the contributions from all elements of the source and from all directions. In other words, the various contributions are treated as if they were uncorrelated. From what we have learned earlier about coherence properties of sources and of optical fields, it is clear that this assumption is suspect.

The basic question then is: Can this quasi-geometrical radiometric model be reconciled with physical optics? And if so, what is the relation between the fundamental radiometric quantity, namely, the radiance, and the fundamental quantity of coherence theory, the mutual coherence function?

Since we are now considering the transport of energy at one particular frequency ν, it will be convenient to employ instead of the mutual coherence function its corresponding spectral component, known as the *cross-spectral density function,*

$$W_{12}(\nu) = \int_{-\infty}^{\infty} \Gamma_{12}(\tau)e^{2\pi i\nu\tau}d\tau. \tag{4.5}$$

It may be shown [Ref. 8, Eqs. (2.4) and (2.5a)] that this function is a measure of the correlation between the Fourier components $\hat{V}_1(\nu)$ and $\hat{V}_2(\nu)$ (interpreted in the sense of the theory of generalized functions) of the light vibrations at the two field points $P_1(\mathbf{r}_1)$ and $P_2(\mathbf{r}_2)$. In place of the complex degree of coherence $\gamma_{12}(\tau)$, defined by Eq. (3.2), we will now employ the spectral degree of spatial coherence[8]

$$\mu_{12}(\nu) = W_{12}(\nu)/[I_1(\nu)]^{1/2}[I_2(\nu)]^{1/2}, \tag{4.6}$$

where

$$I_j(\nu) = W_{jj}(\nu) \qquad (j = 1,2) \tag{4.7}$$

is a measure of the optical intensity at frequency ν at the point $P(\mathbf{r}_j)$. The absolute value of the spectral degree of spatial coherence may be shown to be bounded by zero and unity,

$$0 \leqslant |\mu_{12}(\nu)| \leqslant 1, \tag{4.8}$$

with the extreme limit zero representing complete spatial incoherence and the other extreme limit, unity, representing complete spatial coherence.

In an important paper[9] published in 1968, A. Walther proposed an expression for the radiance in terms of the cross-spectral density. In our notation this expression is

$$B(\mathbf{r},\mathbf{s}) = \left(\frac{k}{2\pi}\right)^2 \cos\theta \times \int W^{(0)}(\mathbf{r} + \mathbf{r}'/2, \mathbf{r} - \mathbf{r}'/2)e^{-ik\mathbf{s}_\perp \cdot \mathbf{r}'}\, d^2\mathbf{r}', \tag{4.9}$$

where

$$k = 2\pi\nu/c \tag{4.10}$$

is the free-space wave number associated with the frequency component ν. In (4.9) $W^{(0)}(\mathbf{r} + \mathbf{r}'/2, \mathbf{r} - \mathbf{r}'/2)$ is the cross-spectral density function of the field at points $\mathbf{r}_1 = \mathbf{r} + \mathbf{r}'/2$, $\mathbf{r}_2 = \mathbf{r} - \mathbf{r}'/2$ in the source plane at frequency ν (which we do not show explicitly from now on), and $\mathbf{s}_\perp$ is the two-dimensional vector obtained by projecting the three-dimensional unit vector $\mathbf{s}$ onto the source plane.

Walther's expression (4.9) for the radiance in terms of the cross-spectral density function of the field in the source plane is consistent with the radiometric formula (4.1) in the sense that when used in the integrand of Eq. (4.1) (but with the σ integration extending formally over the whole plane containing the planar source), it leads to the correct value for the total radiated power $\mathcal{P}$ when $\mathcal{P}$ is calculated on the basis of physical optics. Thus Walther's formula (4.9) appears at first sight to provide the missing link between radiometry and physical optics. However, a closer scrutiny shows that this cannot be the complete answer. For in the first place it is not difficult to find other (nonequivalent) expressions for a quantity $B(\mathbf{r},\mathbf{s})$ that lead to the correct value for the radiant power $\mathcal{P}$, (again calculated from physical optics), when substituted in the radiometric formula (4.1). In fact Walther himself introduced a few years later one such alternative expression.[10] Moreover, Marchand and I showed[11,12] that there are sources for which both the expressions for the radiance proposed by Walther will become negative for some values of their arguments, a result that is in contradiction with the physical significance attributed to the radiance. However, I wish to add that in spite of these conclusions, Walther's paper represents a major contribution to the foundation of radiometry and has become the nucleus of practically all subsequent research on this subject.[13]

Because one can find many different radiance functions that will lead to the correct radiant power $\mathcal{P}$, the question arises whether there is one amongst them that will be consistent with other (explicit or implicit) postulates of radiometry. Recently, one of my students (Friberg, Ref. 14) showed that this cannot be done. More specifically, if $\mathcal{L}$ denotes any linear transform, $J(\mathbf{s})$ denotes the radiant intensity as calculated on the basis of physical optics [Eq. (5.4) below], and $\mathbf{r}$ represents any point in the source plane, Friberg's theorem asserts that *there is no radiance function that satisfies the following four requirements for a planar source of any state of coherence:*

$$\text{(I)} \quad B(\mathbf{r},\mathbf{s}) = \mathcal{L}\{W^{(0)}(\mathbf{r}_1,\mathbf{r}_2)\},$$

$$\text{(II)} \quad B(\mathbf{r},\mathbf{s}) \geqslant 0,$$

$$\text{(III)} \quad B(\mathbf{r},\mathbf{s}) = 0 \text{ when } \mathbf{r} \notin \sigma,$$

$$\text{(IV)} \quad \cos\theta \int_\sigma B(\mathbf{r},\mathbf{s})\, d\sigma = [J(\mathbf{s})]_{\text{phys. opt.}}.$$

The first requirement expresses the fact that seems physically quite natural, namely, that the radiance should depend linearly on the cross-spectral density function of the light distribution across the source. The second and third requirements express two facts implied by the radiometric interpretation of the physical significance of the radiance; namely, that it should be non-negative and that it should have zero values at all points $\mathbf{r}$ in the source plane which lie outside the area occupied by the source. The fourth requirement expresses the condition that the radiometric expression (4.4) on the left-hand side should correctly represent the radiant intensity $[J(\mathbf{s})]_{\text{phys.opt.}}$ when the radiant intensity is calculated on the basis of physical optics[15] [Eq. (5.4) below].

It would thus seem that the radiometric model cannot be fully reconciled with physical optics. Although this result may perhaps appear rather surprising at first sight, it is consistent with the warning of Lorentz that I quoted earlier, about the impossibility of tracing, in general, the path of parts of energy through an optical field in too detailed a manner.

Friberg's theorem also contains a hint that the radiance is not a quantity that in the context of the theory of the quantized field would represent an observable. In fact there is a strong formal resemblance between Friberg's theorem and a theorem due to Wigner[16,17] relating to the so-called phase space representation of quantum mechanics.[18] Because this analogy seems to suggest the true significance of the radiance function, let me say a few words about Wigner's theorem.

For simplicity let us consider a quantum-mechanical system with only one degree of freedom. Let[19] $\hat{q}$ and $\hat{p}$ be the position and the momentum operators, $\hat{\rho}(\hat{q},\hat{p})$ the density operator that represents the state of the system, and $\hat{G}(\hat{q},\hat{p})$ an observable. The expectation value of the observable $\hat{G}$ is then given by

$$\langle \hat{G} \rangle = \mathrm{Tr}(\hat{\rho}\hat{G}), \tag{4.11}$$

where Tr denotes the trace. In the phase-space representation of quantum mechanics one maps all the operators onto c-number functions,

$$\begin{aligned} \hat{q} \to q, \qquad \hat{p} &\to p, \\ \hat{\rho}(\hat{q},\hat{p}) &\to \Phi(q,p), \\ \mathrm{G}(\hat{q},\hat{p}) &\to g(q,p), \end{aligned} \tag{4.12}$$

according to some prescribed rule that depends on the choice in which the noncommuting operators $\hat{q}$ and $\hat{p}$ are ordered in expressions involving their products.[20] The following question now arises: Can one find, among all such mappings (assumed to be linear) one that makes it possible to express the quantum-mechanical expectation value (4.11) of any observable $\hat{G}$ in the form of a classical c-number average with respect to the "phase-space" distribution function $\Phi(q,p)$, i.e., such that

$$\langle \hat{G} \rangle = \int\int g(q,p)\Phi(q,p)\, dq\, dp? \tag{4.13}$$

It is known that formally this can be done, in fact via many different mappings. However, *Wigner's theorem* asserts that *there is no phase space distribution function $\Phi(q,p)$ that will have all the properties of a true probability density.* In particular $\Phi(q,p)$ may become negative for some values of its arguments.

In spite of this fact, the phase-space representation of quantum mechanics has proved a very powerful technique for solving quantum-mechanical problems by methods of classical statistical mechanics. For example during the past fifteen years or so this technique has played an important role in various theories of the laser and in investigations concerning the statistical properties of light.

In its mathematical structure, radiometry has much in common with the phase-space representation of quantum mechanics. In particular both the phase-space distribution function $\Phi(q,p)$ and the radiance function $B(\mathbf{r},\mathbf{s})$ are functions of variables whose quantum-mechanical counterparts do not commute. Moreover, Walther's expression (4.9) for the radiance has a close formal similarity with the expression for the phase-space distribution function originally introduced by Wigner[16] and which in modern notation reads

$$\Phi(q,p) = \frac{1}{2\pi\hbar} \times \int \langle q - q'/2 | \hat{\rho} | q + q'/2 \rangle e^{ipq'/\hbar}\, dq', \tag{4.14}$$

where $\hbar$ is Planck's constant divided by 2π. It may well be that future work on the foundation of radiometry and the theory of radiative energy transfer will reveal that this correspondence is not a purely formal one, but that it is rooted in common physical principles. Moreover, Friberg's theorem and the analogy that I just spoke about suggest that although the radiance may not have the simple intuitive physical meaning that one traditionally attributes to it, it may nevertheless be used in calculating values of quantities that are truly measurable.

V. THE RADIANT INTENSITY FROM A SOURCE OF ANY STATE OF COHERENCE

In the preceding section we put forward evidence which indicates that the radiance $B(\mathbf{r},\mathbf{s})$ as customarily defined in radiometry does not represent a measurable physical quantity. One can show that the same is also true about the radiant emittance $E(\mathbf{r})$. However, the third basic radiometric quantity, the radiant intensity $J(\mathbf{s})$, acquires an unambiguous physical meaning as a measurable quantity if it is defined not via the radiometric formula (4.4) but more directly as representing the (averaged) rate at which energy is radiated by the source per unit solid angle around the $\mathbf{s}$ direction. The radiant intensity defined in this way, (denoted $[J(\mathbf{s})]_{\text{phys.opt.}}$ in Sec. IV), may readily be expressed in terms of the cross-spectral density function of the light in the source plane, irrespective of the state of coherence of the source. I will briefly indicate how this expression may be derived.

One can readily show that the radiant intensity as just defined may be expressed in terms of the energy flux vector $\mathbf{F}(\mathbf{r})$ of the field by the formula

$$J(\mathbf{s}) = \lim_{R\to\infty} [R^2 |\langle \mathbf{F}(R\mathbf{s})\rangle|]. \tag{5.1}$$

On expressing the flux vector in terms of the complex scalar field variable $V(\mathbf{r})$ (see Footnote 7), one finds that (with a suitable choice of units)

$$\begin{aligned} J(\mathbf{s}) &= \lim_{R\to\infty} [R^2 \langle V(R\mathbf{s}) V^*(R\mathbf{s})\rangle] \\ &= \lim_{R\to\infty} [R^2 W(R\mathbf{s},R\mathbf{s})], \end{aligned} \tag{5.2}$$

where $W(R\mathbf{s},R\mathbf{s})$ is, of course, a "diagonal" element of the cross-spectral density function $W(\mathbf{r}_1,\mathbf{r}_2)$ of the field. Now because the field obeys the wave equation one can show that in free space the cross-spectral density function obeys the two Helmholtz' equations[21]

$$\begin{aligned} \nabla_1^2 W(\mathbf{r}_1,\mathbf{r}_2) + k^2 W(\mathbf{r}_1,\mathbf{r}_2) &= 0, \\ \nabla_2^2 W(\mathbf{r}_1,\mathbf{r}_2) + k^2 W(\mathbf{r}_1,\mathbf{r}_2) &= 0, \end{aligned} \tag{5.3}$$

where ∇_1^2 is the Laplacian operator acting on the coordinates of $\mathbf{r}_1$, with similar interpretation of ∇_2^2. Using standard mathematical techniques for solving Helmholtz' equation one can readily express the cross-spectral density function at any pair of points in the far zone in terms of its values at all pairs of points in the source plane. If we make use of that relation on the right-hand-side of Eq. (5.2) we finally obtain the following expression for the radiant intensity:

$$J(\mathbf{s}) = (2\pi k)^2 \cos^2\theta\, \tilde{W}^{(0)}(k\mathbf{s}_\perp, -k\mathbf{s}_\perp). \tag{5.4}$$

In (5.4) $\tilde{W}^{(0)}(\mathbf{f}_1,\mathbf{f}_2)$ is the spatial Fourier transform of the

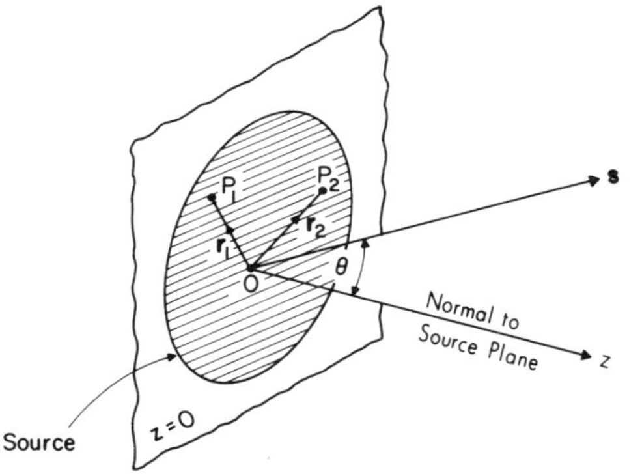

FIG. 9. Illustrating the notation relating to the formulas (5.4) and (5.5).

cross-spectral density function $W^{(0)}(\mathbf{r}_1,\mathbf{r}_2)$ of the field in the source plane $z = 0$ (see Fig. 9), viz.

$$\tilde{W}^{(0)}(\mathbf{f}_1,\mathbf{f}_2) = \frac{1}{(2\pi)^4} \times \int\int W^{(0)}(\mathbf{r}_1,\mathbf{r}_2)e^{-i(\mathbf{f}_1\cdot\mathbf{r}_1+\mathbf{f}_2\cdot\mathbf{r}_2)}\, d^2r_1\, d^2r_2\,, \quad (5.5)$$

the integration extending twice independently over the source and $\mathbf{s}_\perp$ denotes, as before, the two-dimensional vector obtained by projecting the three-dimensional unit vector $\mathbf{s}$ onto the source plane.

Equation (5.4) is an important formula. It expresses the radiant intensity in terms of the spatial Fourier transform of the cross-spectral density function of the light at the source and thus establishes a relation between a measurable radiometric quantity and a basic quantity of coherence theory. We note that according to Eq. (5.4) only those (two-dimensional) spatial frequency components $\mathbf{f}_1$ and $\mathbf{f}_2$ contribute to the radiant intensity for which

$$\mathbf{f}_1 = k\mathbf{s}_\perp, \qquad \mathbf{f}_2 = -k\mathbf{s}_\perp. \quad (5.6)$$

Thus we may say that only certain *"anti-diagonal" elements* of the spatial Fourier transform of the cross-spectral density function $W^{(0)}$ contribute to the radiant intensity. Moreover, since

$$|k\mathbf{s}_\perp| \leqslant |k\mathbf{s}| = k, \quad (5.7)$$

the spatial frequencies that contribute, all have magnitudes that do not exceed the wave number of the light. We call such frequencies *low frequencies.* Thus the formula (5.4) implies that *the radiant intensity is uniquely specified by the low-frequency anti-diagonal elements of the Fourier transform of the cross-spectral density function of the light at the source.*

With the help of the formula (5.4) one can obtain a great deal of insight into a variety of problems that involve radiation from planar sources of any state of coherence. I will conclude this talk by giving a few examples of this kind.

VI. COHERENCE AND THE DIRECTIONALITY OF LIGHT BEAMS

It is generally believed that in order to generate a highly directional light beam such as a laser beam it is necessary to have a source that is spatially completely coherent. Actually, as I remarked earlier, the relation between directionality of light and the state of coherence of the source[22] turns out to be a rather subtle one. I will now show that the formula (5.4) for the radiant intensity provides a clarification of this question and also that it leads to the rather surprising prediction that certain sources that are far from fully coherent will produce beams that are just as directional as laser beams.

Let

$$\mu^{(0)}(\mathbf{r}_1,\mathbf{r}_2) = W^{(0)}(\mathbf{r}_1,\mathbf{r}_2)/[I^{(0)}(\mathbf{r}_1)]^{1/2}[I^{(0)}(\mathbf{r}_2)]^{1/2} \quad (6.1)$$

be the spectral degree of spatial coherence of the light in the source plane.[23] For the sake of simplicity we will restrict our discussion to radiation from planar sources for which $\mu^{(0)}(\mathbf{r}_1,\mathbf{r}_2)$ depends on the position vectors $\mathbf{r}_1$ and $\mathbf{r}_2$ of points in the source plane only through the difference $\mathbf{r}_1 - \mathbf{r}_2$, i.e., $\mu^{(0)}(\mathbf{r}_1,\mathbf{r}_2)$ is of the form

$$\mu^{(0)}(\mathbf{r}_1,\mathbf{r}_2) = g^{(0)}(\mathbf{r}_1 - \mathbf{r}_2), \quad (6.2)$$

where $g^{(0)}(\mathbf{r}')$ is some function of a (two-dimensional) vector $\mathbf{r}'$. We will also assume that the (averaged) intensity $I^{(0)}(\mathbf{r})$ changes much more slowly with $\mathbf{r}$ than $g^{(0)}(\mathbf{r}')$ changes with $\mathbf{r}'$, remaining approximately constant over distances of the order of the correlation length Δ [the effective width Δ of $g^{(0)}$], as indicated for a one-dimensional source in Fig. 10. In addition we assume that the linear dimensions of the source are large compared both with the wavelength of the light and with the correlation length Δ. A source with these properties is said to be a *quasi-homogeneous source.*[24–26]

Let us now consider a quasi-homogeneous source with a Gaussian correlation function

$$g^{(0)}(\mathbf{r}_1 - \mathbf{r}_2) = e^{-|\mathbf{r}_1-\mathbf{r}_2|^2/2\sigma^2}. \quad (6.3)$$

According to Eqs. (6.1), (6.2), and (6.3) the cross-spectral density function of the light in the source plane is of the form

$$W^{(0)}(\mathbf{r}_1,\mathbf{r}_2) = [I^{(0)}(\mathbf{r}_1)]^{1/2}[I^{(0)}(\mathbf{r}_2)]^{1/2}e^{-|\mathbf{r}_1-\mathbf{r}_2|^2/2\sigma^2}. \quad (6.4)$$

On substituting from (6.4) into our general expression (5.4) for the radiant intensity one readily finds that the angular distribution of the radiant intensity is to a high degree of accuracy independent of the exact form of the optical intensity distribution $I^{(0)}(\mathbf{r})$ across the source, provided only that $I^{(0)}(\mathbf{r})$ varies sufficiently slowly with $\mathbf{r}$ as we assumed. The resulting expression for the radiant intensity is[27]

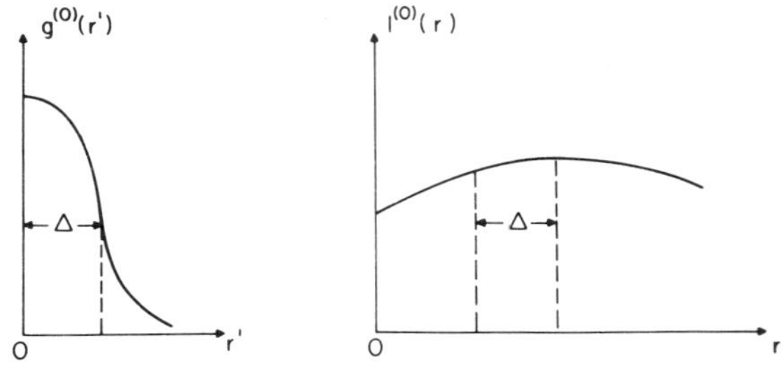

FIG. 10. Illustrating the concept of a quasi-homogeneous source. The degree of coherence $g^{(0)}(\mathbf{r}_1 - \mathbf{r}_2)$ of the light across the source changes much more rapidly with $r' = |\mathbf{r}_1 - \mathbf{r}_2|$ than the intensity $I^{(0)}(\mathbf{r})$ changes with $r = |\mathbf{r}|$.

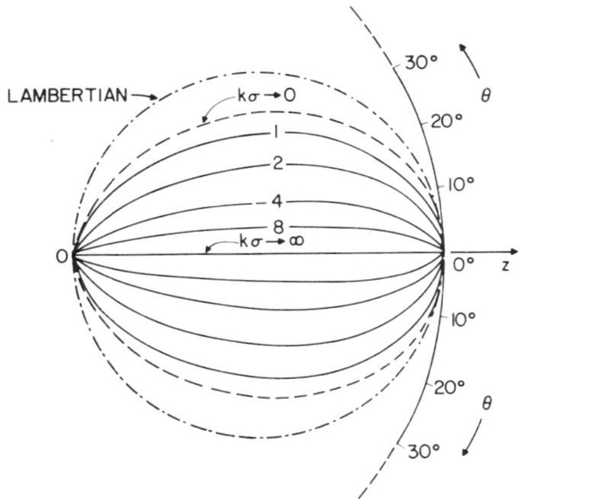

FIG. 11. Polar diagrams of the normalized radiant intensity $J(\mathbf{s})/J^{(0)}$ from a Gaussian-correlated quasi-homogeneous source [Eq. (6.5)] for different values of the rms width σ of the degree of coherence [Eq. (6.3)]. [After W. H. Carter and E. Wolf, J. Opt. Soc. Am. **67**, 790 (1977).]

$$\frac{J(\mathbf{s})}{J^{(0)}} \approx \cos^2\theta e^{-(1/2)(k\sigma)^2\sin^2\theta}, \tag{6.5}$$

where $J^{(0)}$ is the radiant intensity in the forward direction $\theta = 0$, which may readily be shown to be given by

$$J^{(0)} = \frac{(k\sigma)^2}{2\pi} \int I^{(0)}(\mathbf{r})\, d^2r. \tag{6.6}$$

In Fig. 11 polar diagrams are shown of the normalized radiant intensity from such Gaussian-correlated quasi-homogeneous sources, as determined from Eq. (6.5). The curves in Fig. 11 indicate that with increasing values of σ, i.e., with increasing correlation length of the light across the source, the radiant intensity becomes more and more peaked around the forward direction $\theta = 0$. Hence the light generated by a quasi-homogeneous Gaussian-correlated source becomes more and more directional as the area of coherence of the light in the source plane becomes larger and larger. More specifically the polar diagrams show that in order to generate a beam that is highly directional the linear dimensions of the regions of the source over which the light is highly correlated must be large compared with the wavelength—a fact that may be expressed by saying that the light must be *locally* coherent over distances of many wavelengths. However, because of our assumption of quasi-homogeneity, the linear dimensions of the source must be large compared to the correlation distance, i.e., such a source may be said to be spatially rather incoherent in the *global* sense. Thus we see that *complete spatial coherence of the source is not necessary to obtain a highly directional beam.*

A question that we may well ask at this point is the following one: is there a source of this class, i.e., one that is spatially highly incoherent in the global sense that would generate light whose radiant intensity distribution is identical with that produced by a laser? E. Collett and I have recently shown[28] that this is indeed so.

To indicate why this is possible and what the characteristics of such a source are let us return to the formula (5.4) for the radiant intensity from a source of any state of coherence, viz.

$$J(\mathbf{s}) = (2\pi k)^2 \cos^2\theta \tilde{W}^{(0)}(k\mathbf{s}_\perp, -k\mathbf{s}_\perp). \tag{6.7}$$

This formula implies that *any two sources with cross-spectral density functions whose four-dimensional spatial Fourier transforms have the same spatial low-frequency anti-diagonal elements will generate fields that have identical distributions of the radiant intensity.* (The two fields will have, in general, entirely different coherence properties, because these are determined by all the low-frequency elements,[29] not just by the anti-diagonal ones.) Our problem now is to try to determine a cross-spectral density function, say $W_Q^{(0)}(\mathbf{r}_1, \mathbf{r}_2)$, of a quasi-homogeneous source, which has a Fourier transform whose low-frequency anti-diagonal elements are identical with those of a laser source. It turns out that many such functions $W_Q^{(0)}$, representing different "equivalent" quasi-homogeneous sources, can be found.[30] I will only consider the one discussed in Ref. 28, which is particularly simple.

We assume that the laser generates an intensity distribution $I_L^{(0)}(\mathbf{r})$ across the output mirror (taken to be plane) that is Gaussian,

$$I_L^{(0)}(\mathbf{r}) = A_L e^{-r^2/2\sigma_L^2}, \tag{6.8}$$

where A_L and σ_L are positive constants and $r = |\mathbf{r}|$ denotes the radial distance from the axis of the laser. Since the laser light is spatially coherent and co-phasal over the output mirror, its degree of coherence is simply

$$g_L^{(0)}(\mathbf{r}_1 - \mathbf{r}_2) \equiv 1 \tag{6.9}$$

for all points $\mathbf{r}_1$ and $\mathbf{r}_2$ on the mirror. If, for the sake of simplicity, diffraction at the edge of the mirror is ignored, one finds that a quasi-homogeneous source with the following characteristics will generate light that has a distribution of radiant intensity which is identical with that of the laser beam:

$$g_Q^{(0)}(\mathbf{r}_1 - \mathbf{r}_2) = e^{-|\mathbf{r}_1 - \mathbf{r}_2|^2/8\sigma_L^2}, \tag{6.10}$$

$$I_Q^{(0)}(\mathbf{r}) = (\sigma_L/\sigma_Q)^2 A_L e^{-r^2/2\sigma_Q^2}, \tag{6.11}$$

where

$$\sigma_Q \gg 2\sigma_L. \tag{6.12}$$

We see from Eq. (6.10) that this quasi-homogeneous source is Gaussian correlated, with a root-mean-square (rms) width that is precisely twice the rms width σ_L of the Gaussian intensity distribution of the laser output. The intensity distribution of our quasi-homogeneous source is according to Eq. (6.11) also Gaussian, but we see from Eq. (6.12) that its rms width is very much larger than that of the laser output; hence our "equivalent" quasi-homogeneous source is much larger than the laser cross section. According to Eqs. (6.11), (6.12), and (6.8) the intensity distribution at the center of the quasi-homogeneous source is much smaller than at the center of the laser output mirror. These results are illustrated in Figs. 12 and 13. It seems remarkable that a source that is spatially so highly incoherent in the global sense should generate light that is just as directional as a laser beam.

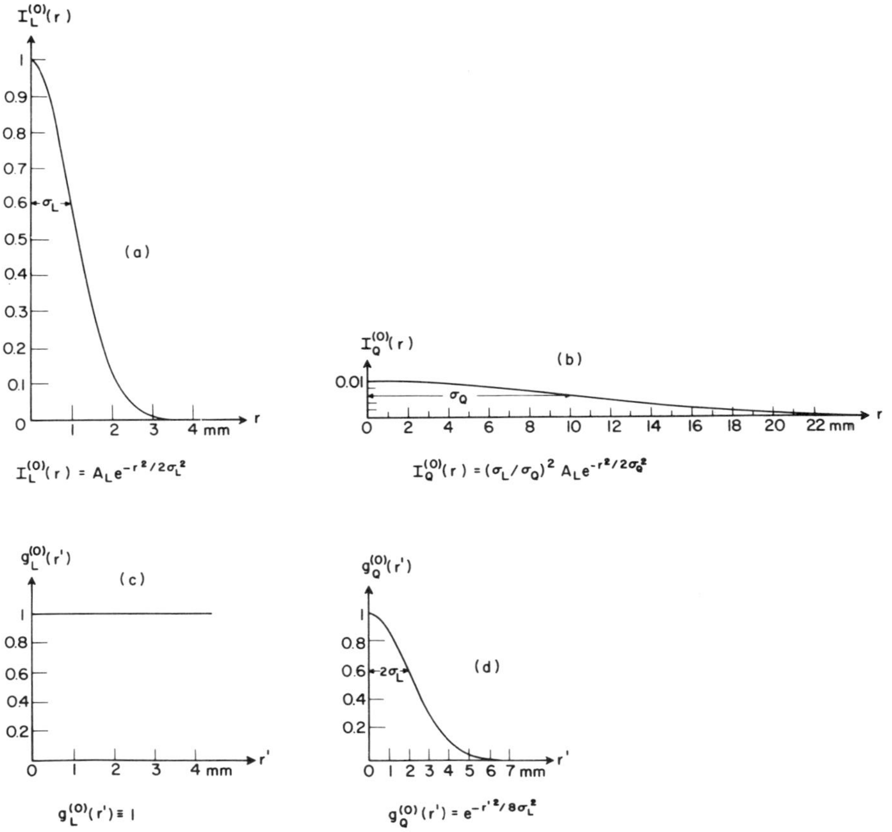

FIG. 12. Optical intensity distribution [(a)] and the spectral degree of spatial coherence [(c)] of a laser source and of a quasi-homogeneous source [(b) and (d)] which produce fields with identical distributions of the radiant intensity. The curves pertain to sources with $\sigma_L = 1$ mm, $\sigma_Q = 10$ mm. The optical intensity of the laser output is normalized to unity at the center of the output mirror. The vertical scale in (b) is ten times that of (a). [After E. Collett and E. Wolf, Opt. Lett. **2**, 27–29 (1978)].

One may well ask whether it is possible to produce a quasi-homogeneous source for testing these predictions. Up to the present time little work has been done on constructing sources with prescribed coherence properties, but some pioneering research in this direction has been carried out in the last few years by Bertolotti and his collaborators.[31] They constructed new types of secondary sources by scattering laser light on liquid crystals, under the application of a dc electric field. By varying the strength of the field, secondary sources are obtained whose coherence properties can to some extent be controlled by the strength of the applied fields. It seems plausable that some of our theoretical predictions could be tested with sources of this kind.

VII. COHERENCE PROPERTIES OF LAMBERTIAN SOURCES

Another situation that may readily be clarified with the help of the general formula (5.4) for the radiant intensity concerns Lambertian sources. Such sources will give rise to a radiant intensity that has the directional dependence

$$J(\mathbf{s}) = C \cos\theta, \tag{7.1}$$

(Lambert's law) where, as before, θ denotes the angle between the $\mathbf{s}$ direction and the normal to the source plane and C is a constant.

It is often asserted that sources of this type are spatially completely incoherent. This belief has presumably its origin in the fact that the usual Lambertian sources are thermal sources and that thermal sources at laboratory temperatures generate light chiefly by the highly random and uncorrelated process of spontaneous emission. However, we have seen earlier that even independent radiators produce field correlations in the source plane, and hence the question as to the kind of correlation that exists in Lambertian sources deserves attention.[32]

We will only consider planar Lambertian sources which are quasi-homogeneous, as is usually the case in practice. If we substitute from Eq. (7.1) into the left-hand side of Eq. (5.4) and make use of the relation

$$\cos\theta = (1 - s_x^2 - s_y^2)^{1/2} = (1 - \mathbf{s}_\perp^2)^{1/2}, \tag{7.2}$$

it follows after a straightforward though somewhat lengthy calculation that the cross-spectral density function of the light in the source plane must necessarily be of the form[33]

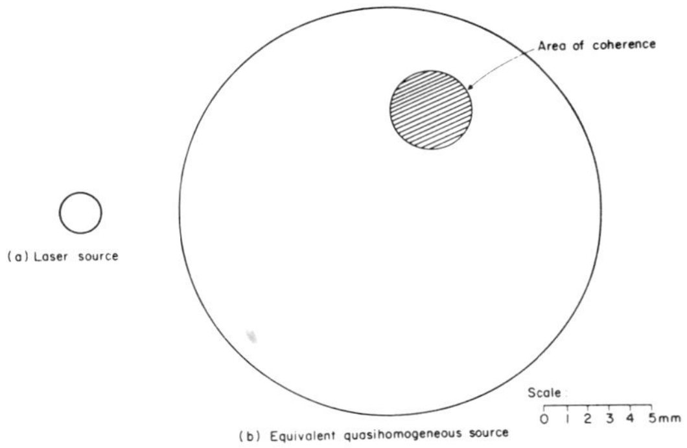

FIG. 13. Illustrating the effective sizes of the laser source (a) and of the "equivalent" quasi-homogeneous source (b) of Fig. 12. The effective area of coherence of the quasi-homogeneous source is shown shaded in Fig. (b).

$$W^{(0)}(\mathbf{r}_1,\mathbf{r}_2) = I^{(0)}[(\mathbf{r}_1+\mathbf{r}_2)/2] \times \frac{\sin k|\mathbf{r}_1-\mathbf{r}_2|}{k|\mathbf{r}_1-\mathbf{r}_2|} + [W^{(0)}(\mathbf{r}_1,\mathbf{r}_2)]^{HF}, \quad (7.3)$$

where $I^{(0)}(\mathbf{r})$ denotes the intensity distribution across the source and $[W^{(0)}(\mathbf{r}_1,\mathbf{r}_2)]^{HF}$ represents a high-frequency (nonradiating) part. If we ignore this nonradiating part we readily obtain from (7.3) the following expression for the spectral degree of spatial coherence [cf. Eqs. (6.2) and (6.1)]:

$$g^{(0)}(\mathbf{r}_1-\mathbf{r}_2) = \sin k|\mathbf{r}_1-\mathbf{r}_2|/k|\mathbf{r}_1-\mathbf{r}_2|. \quad (7.4)$$

[In deriving the expression (7.4) an obvious approximation was made that is implied by the quasi-homogeneity of the source.] Thus we see that *all quasi-homogeneous planar Lambertian sources have identical spatial coherence properties in the sense that their spectral degree of spatial coherence is given by Eq. (7.4)* (except possibly for a nonradiating high-frequency contribution that we ignored).

The behavior of the degree of spatial coherence given by Eq. (7.4) is shown in Fig. 14 as a function of the separation of the points $\mathbf{r}_1$ and $\mathbf{r}_2$ (in units of $1/k = \lambda/2\pi$, where λ is the wavelength of the light). Since its first zero occurs when $r' \equiv |\mathbf{r}'| = \pi/k = \lambda/2$ we see that light across a Lambertian source is not

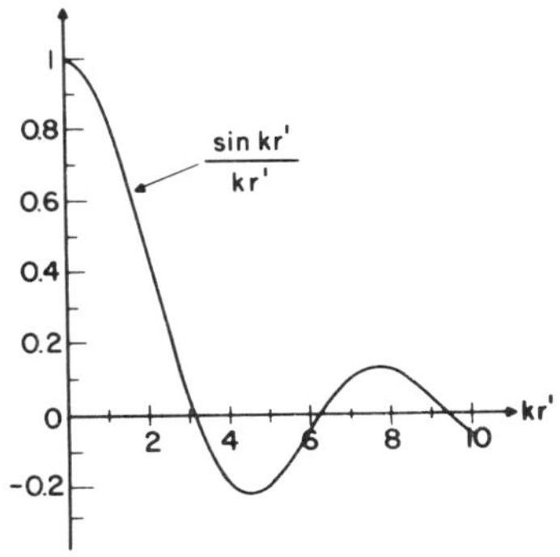

FIG. 14. The spectral degree of spatial coherence of a quasi-homogeneous Lambertian source. (The contributions from nonradiating spatial frequency components have been neglected.)

completely spatially incoherent but is correlated over distances of the order of the wavelength of the light. These results are found to be in agreement with known correlation properties of blackbody radiation.[34]

VIII. CONCLUDING REMARKS

It had originally been my intention to also review in this address recent researches on the foundations of the theory of radiative energy transfer which, as we noted earlier, may be regarded as a natural extension of radiometry. However, in preparing this talk it soon became apparent that it is not possible to do so in the available time. I will only mention, in this connection, that in spite of a good deal of research[35] much remains to be done in this area. Only in some special cases such as, for example, for transport in free space, has the relationship between the phenomenological theory of radiative energy transfer and the modern theories of radiation been clarified to a large extent.[36,37]

Returning to the foundations of radiometry I wish to stress a fact that should perhaps be obvious from our review: namely, that the connection between radiometry and physical optics is a rather subtle one and that no simple direct correspondence exists between all the quantities that describe energy transport in radiometry and in physical optics. In particular it is incorrect to identify irradiance with optical intensity, as has become rather customary in recent years in some publications.

I will conclude this talk by noting that although radiometry is one of the oldest branches of optics—predating Maxwell's electromagnetic theory and the quantum theory of radiation—there are many new and valuable results that have recently been obtained in this field, and undoubtedly more will follow. This development illustrates a recurring lesson that one can learn from the history of physics: namely, that subjects that often appear to be well understood and perhaps even a little old-fashioned have frequently some surprises in store for us.

[1]Throughout this talk we shall only be concerned with sources that generate radiation which is steady in the macroscopic sense. Such sources need not be, however, in thermal equilibrium with its surroundings.

[2]See, for example, E. Hopf, *Mathematical Problems of Radiative Equilibrium* (Cambridge U. P., Cambridge, 1934), Sec. 2.

[3]See, for example, A. I. Akhiezer and V. B. Berestetskii, *Quantum Electrodynamics* (Interscience, New York, 1965), Sec. 2.2, or W. Pauli, "Die Allgemeinen Principien der Wellenmechanik," in *Handbuch der Physik,* 2 Aufl., Band 24, 1 Teil, edited by H. Geiger and K. Scheel (Springer, Berlin, 1933), pp. 92 and 260.

[4]For a fuller account of optical coherence theory see, for example, L. Mandel and E. Wolf "Coherence properties of optical fields," Rev. Mod. Phys. **37,** 231–287 (1965) or Ref. 5 quoted below.

[5]See, for example, M. Born and E. Wolf, *Principles of Optics,* 5th ed. (Pergamon, New York, 1975), Sec. 10.2.

[6]The mutual coherence function is independent of t because of our earlier assumption (cf. Footnote 1) that the radiation is steady in the macroscopic sense. In the language of the theory of random processes this assumption is tantamount to the statement that the fluctuations can be described as a stationary random process.

[7]We mentioned earlier that within the framework of Maxwell's electromagnetic theory the optical intensity is usually identified with the magnitude of the energy flux vector (the Poynting vector). It would therefore seem more appropriate to identify the optical intensity in a complex scalar wavefield $V(\mathbf{r},t)$ with the magnitude of

the flux vector $\mathbf{F} = \alpha(\dot{V}\nabla V^* + \dot{V}^*\nabla V)$ associated with that field. (Here $\dot{V} = \partial V/\partial t$ and α is a constant, depending on the choice of units.) However, under experimental conditions frequently encountered in practice (e.g., when measurements are made in the far zone of a radiating system and the field is quasi-monochromatic), $\langle|\mathbf{F}|\rangle$ may be shown to be proportional to $\langle VV^*\rangle$ (at least to a high degree of accuracy).

[8]L. Mandel and E. Wolf, "Spectral coherence and the concept of cross-spectral purity," J. Opt. Soc. Am. **66,** 529–535 (1976).

[9]A. Walther, "Radiometry and coherence," J. Opt. Soc. Am. **58,** 1256–1259 (1968).

[10]A. Walther, "Radiometry and coherence," J. Opt. Soc. Am. **63,** 1622–1623 (1973).

[11]E. W. Marchand and E. Wolf, "Radiometry with sources of any state of coherence," J. Opt. Soc. Am. **64,** 1219–1226 (1974).

[12]E. W. Marchand and E. Wolf, "Walther's definition of generalized radiance," J. Opt. Soc. Am. **64,** 1273–1274 (1974); see also A. Walther, "Reply to Marchand and Wolf", J. Opt. Soc. Am. **64,** 1275 (1974).

[13]A somewhat different approach was described by A. S. Marathay, "Radiometry of partially coherent fields I," Opt. Acta **23,** 785–794 (1976); II, *ibid.* **23,** 795–798 (1976).

[14]A. T. Friberg, "On the existence of a radiance function for a partially coherent planar source," in *Proceedings of the Fourth Rochester Conference on Coherence and Quantum Optics,* edited by L. Mandel and E. Wolf (Plenum, New York, in press).

[15]The radiometric definition of the radiant intensity $J(\mathbf{s})$, via the formula (4.4), implies that $J(\mathbf{s})$ represents the average radiated power per unit solid angle around the $\mathbf{s}$ direction. One can calculate this radiated power directly from physical optics, without introducing any hypothetical radiance function $B(\mathbf{r},\mathbf{s})$, as will be discussed in Sec. 5. $[J(\mathbf{s})]_{\text{phys.opt.}}$ denotes here the radiant intensity when calculated in this more direct manner.

[16]E. Wigner, "On the quantum correction for thermodynamic equilibrium," Phys. Rev. **40,** 749–759 (1932).

[17]E. P. Wigner, "Quantum mechanical distribution functions revisited," in *Perspectives in Quantum Theory,* edited by W. Yourgrau and A. van der Merwe (M.I.T. Press, Cambridge, Mass., 1971), pp. 25–36.

[18]For a discussion of Wigner's theorem and of related researches, see M. D. Srinivas and E. Wolf, "Some nonclassical features of phase-space representations of quantum mechanics," Phys. Rev. D **11,** 1477–1485 (1975).

[19]Carets denote operators.

[20]A general theory of such mappings was formulated by G. S. Agarwal and E. Wolf, "Calculus of functions of noncommuting operators and general phase-space methods in quantum mechanics. I. Mapping theorems and ordering of functions of noncommuting operators," Phys. Rev. D **2,** 2161–2186 (1970); "II. Quantum mechanics in phase space," Phys. Rev. D **2,** 2187–2205 (1970); "III. A generalized Wick theorem and multitime mapping," Phys. Rev. D **2,** 2206–2225 (1970). These papers also contain an extensive bibliography of earlier publications on this subject.

[21]The two Helmholtz equations (5.3) for the cross-spectral density function may be obtained, for example, by taking the Fourier transform of the two wave equations that the mutual coherence function is known to satisfy (cf. Ref. 5, Sec. 10.7.1).

[22]This relationship appears to have been first considered, for the special case of radiation from large statistically homogeneous sources, by E. Wolf and W. H. Carter, "Angular distribution of radiant intensity from sources of different degrees of spatial coherence," Opt. Commun. **13,** 205–209 (1975).

[23]As in the previous sections we describe the coherence properties of a source in terms of correlation functions involving the field distribution in the source plane. Such a description may be employed whether the source is a primary or a secondary one. However, when the source is a primary one, one may characterize its coherence properties in an alternative way, by means of correlation functions involving the true source variable (e.g., the charge-current density distribution). For a primary scalar source this alternative approach is discussed in a forthcoming paper by W. H. Carter and E. Wolf, "Coherence and radiant intensity in scalar wavefields generated by fluctuating primary planar sources (submitted to J. Opt. Soc. Am.).

[24]W. H. Carter and E. Wolf "Coherence and radiometry with quasi-homogeneous planar sources," J. Opt. Soc. Am. **67,** 785–796 (1977).

[25]E. Wolf and W. H. Carter, "On the radiation efficiency of quasi-homogeneous sources of different degrees of spatial coherence," in *Proceedings of the Fourth Rochester Conference on Coherence and Quantum Optics,* edited by L. Mandel and E. Wolf (Plenum, New York, in press).

[26]Formally a slightly different but essentially equivalent class of sources has been considered by H. A. Ferwerda and M. G. van Heel "On the coherence properties of thermionic emission sources," Optik **47,** 357–362 (1977). See also H. A. Ferwerda and M. G. van Heel, "Determination of Coherence Length from Directionality," in *Proceedings of the Fourth Rochester Conference on Coherence and Quantum Optics,* edited by L. Mandel and E. Wolf (Plenum, New York, in press).

[27]The angular distribution of radiant intensity from some other types of model sources is discussed in the following papers: H. P. Baltes, B. Steinle, and G. Antes, "Spectral coherence and the radiant intensity from statistically homogeneous and isotropic planar sources," Opt. Commun. **18,** 242–246 (1976); B. Steinle and H. P. Baltes, "Radiant intensity and spatial coherence for finite planar sources," J. Opt. Soc. Am. **67,** 241–247 (1977); H. P. Baltes, B. Steinle, and G. Antes "Radiometric and correlation properties of bounded planar sources," in *Proceedings of the Fourth Rochester Conference on Coherence and Quantum Optics,* edited by L. Mandel and E. Wolf (Plenum, New York, in press); W. H. Carter and M. Bertolotti, "An analysis of the far-field coherence and radiant intensity of light scattered from liquid crystals" (J. Opt. Soc. Am., in press).

[28]E. Collett and E. Wolf, "Is complete spatial coherence necessary for the generation of highly directional light beams?," Opt. Lett. **2,** 27–29 (1978).

[29]E. W. Marchand and E. Wolf, "Angular correlation and the far-zone behavior of partially coherent fields," J. Opt. Soc. Am. **62,** 379–385 (1972), Eq. (34).

[30]Actually all of them have the same degree of spatial coherence $g_Q(\mathbf{r}_1 - \mathbf{r}_2)$. They can only differ by their intensity distributions $I_Q(\mathbf{r})$.

[31]See, for example, F. Scudieri, M. Bertolotti, and R. Bartolino, "Light scattered by a liquid crystal: A new quasi-thermal source," Appl. Opt. **13,** 181–185 (1974); M. Bertolotti, F. Scudieri, and S. Verginelli, "Spatial coherence of light scattered by media with large correlation length of refractive index fluctuations," Appl. Opt. **15,** 1842–1844 (1976).

[32]It has been shown that a spatially completely incoherent source would give rise to radiant intensity that falls off with θ in proportion to $\cos^2\theta$ rather than $\cos\theta$. [T. J. Skinner, Ph.D. Thesis (Boston University, 1965), p. 46; E. W. Marchand and E. Wolf, Ref. 11, Sec. V; W. H. Carter and E. Wolf, Ref. 24, Sec. III].

[33]The formula (7.3) is a generalization to quasi-homogeneous sources of an expression derived for large homogeneous sources by M. Beran and G. Parrent, "The mutual coherence function of incoherent radiation," Nuovo Cimento **27,** 1049–1063 (1963), Sec. 8; A. Walther, Ref. 9, Sec. 4; W. H. Carter and E. Wolf, Ref. 24, Sec. II.

[34]Ref. 24, Sec. II, especially Eqs. (31) and (32).

[35]See, for example, the following publications and the references quoted therein: (a) Yu. N. Barabanenkov, Yu. A. Kravtsov, S. M. Rytov, and V. I. Tatarskii, "Status of the theory of propagation of waves in randomly inhomogeneous medium," Sov. Phys.—Usp. **13,** 551–575 (1971); (b) V. I. Tatarskii, *The Effects of Turbulent Atmosphere on Wave Propagation* (U.S. Department of Commerce, National Technical Service, Springfield, Va., 1971), Sec. 63; (c) Yu. A. Kravtsov, C. M. Rytov, and V. I. Tatarskii, "Statistical problems in diffraction theory," Sov. Phys.—Usp. **18,** 118–130 (1975); (d) Yu. N. Barabanenkov, "Multiple scattering of waves by ensembles of particles and the theory of radiation transport," Sov. Phys.—Usp. **18,** 673–689 (1976); (e) A. Ishimaru, "Theory and application of wave propagation and scattering in random media," Proc. IEEE **65,** 1030–1061 (1977).

[36]E. Wolf, "New theory of radiative energy transfer in free electromagnetic fields," Phys. Rev. D **13,** 869–886 (1976).

[37]M. S. Zubairy and E. Wolf, "Exact equations for radiative transfer of energy and momentum in free electromagnetic fields," Opt. Commun. **20,** 321–324 (1977).

2024 OPTICS LETTERS / Vol. 19, No. 23 / December 1, 1994

Radiometric model for propagation of coherence

Emil Wolf

Department of Physics and Astronomy, University of Rochester, Rochester, New York 14627

Received July 25, 1994

A radiometric model for the propagation of coherence is formulated, which greatly simplifies the determination of the cross-spectral density and of the spectral degree of coherence of the field at an arbitrary distance from any planar, secondary, quasi-homogeneous source. The radiance function, which plays a central role in this model, satisfies all the postulates of traditional radiometry.

Calculations of the degree of coherence of fields at finite distances from a radiating source present formidable difficulties. If the source is planar, such calculations require the evaluation of four-folded integrals, whose integrands contain highly osculatory terms. The evaluation involves a prohibitive amount of work, even when modern computers are used. For these reasons, calculations of this kind have been limited to the domain of the paraxial approximation[1]; other investigations were restricted to statistically homogeneous fields,[2] sometimes without this being explicitly recognized (in this connection see Ref. 3). In some other publications models that appear to be generalizations of the traditional theory of radiative energy transfer were developed (see, for example, Ref. 4), but the radiance functions that are central to these theories do not obey all the postulates of traditional radiometry; they may acquire negative values, for example.

In this Letter a relatively simple approximate model for the propagation of coherence is presented that is free of such limitations and appreciably reduces the computational time needed to determine the cross-spectral density and the spectral degree of coherence of a field produced by a planar, quasi-homogeneous, secondary source at pairs of points that are situated at an arbitrary distance from the source. The model may be regarded as an extension of the traditional radiometric model relating to the propagation of energy. It involves only a relatively simple ray trace, followed by the evaluation of integrals of the type that are commonly encountered in diffraction theory with deterministic fields. Moreover, the radiance function that plays a central role in this model obeys all the postulates of traditional radiometry.

Let us consider a planar, secondary source σ of any state of coherence, occupying a finite portion of the plane $z = 0$ and radiating into the half-space $z > 0$ (Fig. 1). We assume that the source fluctuations can be represented by an ensemble that is statistically stationary, at least in the wide sense.[5] According to the coherence theory in the space-frequency domain, the cross-spectral density $W(\mathbf{r}_1, \mathbf{r}_2, \nu)$ at frequency ν at points P_1 and P_2 in the half-space $z > 0$, specified by position vectors $\mathbf{r}_1$ and $\mathbf{r}_2$, may be expressed in terms of an ensemble of monochromatic realizations $\{U(\mathbf{r}, \nu)\exp(-2\pi i\nu t)\}$ in the form[6]

$$W(\mathbf{r}_1, \mathbf{r}_2, \nu) = \langle [U(\mathbf{r}_1, \nu)]^* U(\mathbf{r}_2, \nu)\rangle, \tag{1}$$

where the angle brackets denote the ensemble average. Each realization $U(\mathbf{r}, \nu)$ may be represented in the form of an angular spectrum of plane waves, viz.,[7]

$$U(\mathbf{r}, \nu) = \int a(\mathbf{s}_\perp, \nu)\exp(ik\mathbf{s}\cdot\mathbf{r})\,\mathrm{d}^2 s_\perp, \tag{2}$$

where $\mathbf{r} \equiv (x, y, z > 0)$, $\mathbf{s} \equiv (s_x, s_y, s_z)$, $\mathbf{s}_\perp \equiv (s_x, s_y, 0)$,

$$s_z = +\sqrt{1 - \mathbf{s}_\perp^2} \quad \text{when } \mathbf{s}_\perp^2 \le 1, \tag{3a}$$

$$= +i\sqrt{\mathbf{s}_\perp^2 - 1} \quad \text{when } \mathbf{s}_\perp^2 > 1, \tag{3b}$$

and the integration in Eq. (2) is taken over the whole $\mathbf{s}_\perp$ plane. The contributions to integral (2) for which Eq. (3a) applies represent effects of homogeneous plane waves. Those for which Eq. (3b) applies represent effects of evanescent waves.

If we substitute from Eq. (2) into Eq. (1) and interchange the order of averaging and integration, we obtain the following expression for the cross-spectral density of the field, valid throughout the half-space $z \ge 0$:

$$W(\mathbf{r}_1, \mathbf{r}_2, \nu) = \int\int \mathcal{A}(\mathbf{s}_1, \mathbf{s}_2, \nu) \times \exp[ik(\mathbf{s}_2\cdot\mathbf{r}_2 - \mathbf{s}_1\cdot\mathbf{r}_1)]\,\mathrm{d}^2 s_{1\perp}\,\mathrm{d}^2 s_{2\perp}, \tag{4}$$

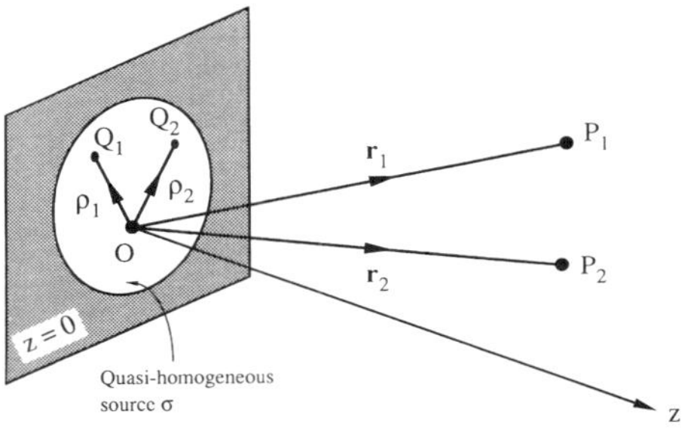

Fig. 1. Illustrating the notation.

0146-9592/94/232024-03$6.00/0

where

$$\mathcal{A}(\mathbf{s}_{1\perp}, \mathbf{s}_{2\perp}, \nu) = \langle a^*(\mathbf{s}_1, \nu) a(\mathbf{s}_2, \nu)\rangle \quad (5)$$

is the so-called angular correlation function.[8]

Representation (4) contains contributions from both homogeneous and evanescent waves. Because the contributions of the evanescent waves decay rapidly as the field propagates away from the plane $z = 0$, we can omit these contributions, as is customary in optics. Hence the integrations on the right-hand side of Eq. (4) extend over the domains $\mathbf{s}_{1\perp}^2 \le 1$, $\mathbf{s}_{2\perp}^2 \le 1$ only.

Let us now introduce new variables

$$\mathbf{s} = (\mathbf{s}_1 + \mathbf{s}_2)/2, \qquad \mathbf{s}' = \mathbf{s}_2 - \mathbf{s}_1. \quad (6)$$

Then $\mathbf{s}_1 = \mathbf{s} - \mathbf{s}'/2$, $\mathbf{s}_2 = \mathbf{s} + \mathbf{s}'/2$, and formula (4) becomes

$$W(\mathbf{r}_1, \mathbf{r}_2, \nu) = \int \mathrm{d}^2 s_\perp' \int \mathrm{d}^2 s_\perp \mathcal{A}\Big(\mathbf{s}_\perp - \mathbf{s}_\perp'/2, \mathbf{s}_\perp + \mathbf{s}_\perp'/2, \nu\Big) \exp\{ik[\mathbf{s}\cdot(\mathbf{r}_2 - \mathbf{r}_1) + \mathbf{s}'\cdot(\mathbf{r}_2 + \mathbf{r}_1)/2]\}, \quad (7)$$

where the subscript $\perp$ again indicates projections, considered as two-dimensional vectors, onto the z plane.

Let us now introduce a generalized radiance function $\mathcal{B}_\nu(\mathbf{r}, \mathbf{s})$ by the formula

$$\mathcal{B}_\nu(\mathbf{r}, \mathbf{s}) = s_z \int_{s_\perp^2 \le 4} \mathcal{A}(\mathbf{s}_\perp - \mathbf{s}_\perp'/2, \mathbf{s}_\perp + \mathbf{s}_\perp'/2) \times \exp(ik\mathbf{s}'\cdot\mathbf{r})\,\mathrm{d}^2 s_\perp'. \quad (8)$$

In terms of this quantity, formula (7) for the cross-spectral density of the field in the half-space $z > 0$ may be expressed in the form

$$W(\mathbf{r}_1, \mathbf{r}_2, \nu) = \int_{(2\pi)} \mathcal{B}_\nu\Big(\frac{\mathbf{r}_1 + \mathbf{r}_2}{2}, \mathbf{s}\Big) \times \exp[ik\mathbf{s}\cdot(\mathbf{r}_2 - \mathbf{r}_1)]\,\mathrm{d}\Omega, \quad (9)$$

where we have used the fact that $\mathrm{d}s_x \mathrm{d}s_y/s_z = \mathrm{d}\Omega$ is the element of solid angle formed by the $\mathbf{s}$ directions. The integration extends over the whole 2π solid angle subtended by all the unit vectors $\mathbf{s}$ pointing into the half-space $z > 0$.

In the special case when point $\mathbf{r}$ is located in the source plane $z = 0$, formula (8) reduces to a generalized radiance function introduced by Walther [Ref. 9, Eq. (22)]. Hence Eq. (8) may be regarded as its generalization for all pairs of points in the half-space $z > 0$.[10]

It is well known that the generalized radiance function does not satisfy all the postulates of traditional radiometry.[14] However, using results derived in Ref. 15, one can show that when the source is quasi-homogeneous[16] and the wavelength $\lambda = c/\nu$ is sufficiently short (more precisely, in the asymptotic limit as $k = 2\pi/\lambda \to \infty$), generalized radiance function (8) acquires all the properties attributed to it in traditional radiometry. We will denote this asymptotic limit of $\mathcal{B}_\nu(\mathbf{r}, \mathbf{s})$ by $B_\nu(\mathbf{r}, \mathbf{s})$. The value of $B_\nu(\mathbf{r}, \mathbf{s})$ at any point $\mathrm{P}(\mathbf{r})$ in the half-space $z \ge 0$ is then given by the expression

$$\left.\begin{aligned} B_\nu(\mathbf{r}, \mathbf{s}) &= k^2 s_z S^{(0)}(\boldsymbol{\rho}_0, \nu)\tilde{g}^{(0)}(k\mathbf{s}_\perp, \nu) && \text{when } \mathbf{s} \in \Omega_P \\ &= 0 && \text{when } \mathbf{s} \notin \Omega_P \end{aligned}\right\}, \quad (10)$$

where $\boldsymbol{\rho}_0 = \boldsymbol{\rho} - (\mathbf{s}_\perp z)/s_z$ is the position vector of the point Q_0 at which the line in the direction $\mathbf{s}$ through the field point $\mathrm{P}(\mathbf{r})$ intersects the source plane $z = 0$, $\mathbf{r} = (\rho, z)$, Ω_p denotes the solid angle generated by the lines that join all the source points to P (see Fig. 2), and

$$\tilde{g}^{(0)}(\mathbf{f}, \nu) = \frac{1}{(2\pi)^2} \int_\sigma g^{(0)}(\boldsymbol{\rho}', \nu) \exp(-if\cdot\boldsymbol{\rho}')\,\mathrm{d}^2\rho' \quad (11)$$

is the two-dimensional Fourier transform of $g^{(0)}(\boldsymbol{\rho}', \nu)$.

It is clear that one can make use of formula (9) for the cross-spectral density in the following manner: One evaluates $\mathcal{B}_\nu(\bar{\mathbf{r}}, \mathbf{s})$ at point $\bar{\mathrm{P}}$ with position vector

$$\bar{\mathbf{r}} = (\mathbf{r}_1 + \mathbf{r}_2)/2 \quad (12)$$

by use of the simple algebraic expression (10) (essentially using ray trace), for all $\mathbf{s}$ directions contained within the solid angle $\Omega_{\bar{P}}$ that the source subtends at point $\bar{\mathrm{P}}$. No other direction contributes to the integral because $\mathcal{B}_\nu(\bar{\mathbf{r}}, \mathbf{s})$ has the value zero for all $\mathbf{s}$ directions that are not contained within $\Omega_{\bar{P}}$.

When $\mathcal{B}_\nu$ in expression (9) may be replaced by expression (10), the integral over the solid angle may be transformed into an integral over the source. For this purpose we make use of the geometrical relations $\mathrm{d}\Omega = (\bar{z}/\bar{R})\,\mathrm{d}^2\rho_0/\bar{R}^2$, $s_z = \bar{z}/R$, where $\bar{z}$ is the distance of point $\bar{\mathrm{P}}$ from the source plane and $\bar{R}$ is the distance of $\bar{\mathrm{P}}$ from the corresponding source point $\bar{\mathrm{Q}}_0$. Expression (9) for the cross-spectral density then becomes

$$W(\mathbf{r}_1, \mathbf{r}_2, \nu) = k^2 \int_\sigma S^{(0)}(\bar{\rho}_0, \nu)\tilde{g}(k\mathbf{s}_\perp, \nu)\Big(\frac{\bar{z}}{\bar{R}}\Big)^2 \times \exp[ik\mathbf{s}\cdot(\mathbf{r}_2 - \mathbf{r}_1)]\frac{\mathrm{d}^2\bar{\rho}_0}{\bar{R}^2}, \quad (13)$$

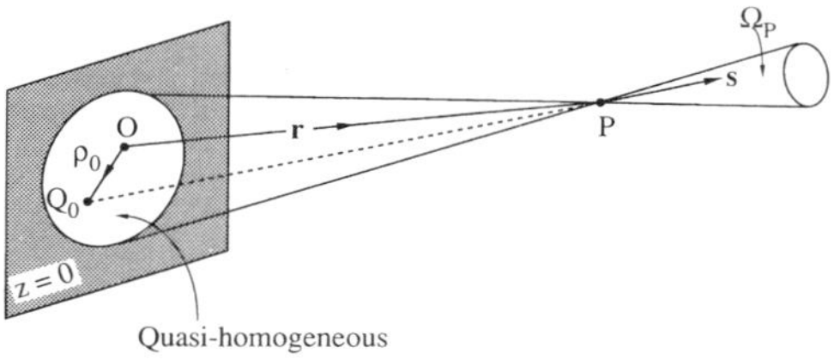

Fig. 2. Illustrating notation relating to formula (10).

2026 OPTICS LETTERS / Vol. 19, No. 23 / December 1, 1994

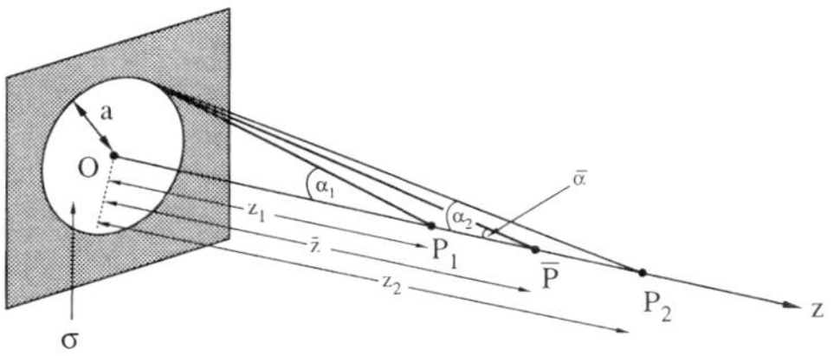

Fig. 3. Illustrating the meaning of the symbols appearing in Eqs. (19 and (20).

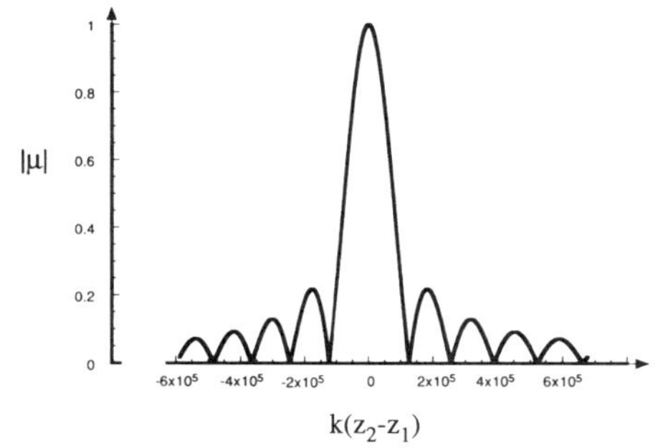

Fig. 4. Behavior of the modulus of the spectral degree of coherence at pairs of points P_1 and P_2 on the normal through the center of a uniform, circular, secondary, planar, quasi-homogeneous, Lambertian source σ, with $ka = 1.2 \times 10^5$, $kz_1 = 1.2 \times 10^7$.

where $\bar{\boldsymbol{\rho}}_0 = \bar{\mathbf{r}}_\perp - (\bar{z}/s_z)\mathbf{s}_\perp$ is the position vector of $\bar{Q}_0$.

As a simple example of the use of formula (13) let us determine the spectral degree of coherence of a field produced by a uniform, circular, secondary, quasi-homogeneous, planar, Lambertian source of radius a. For such a source one has, from Eq. (10) and from Eq. (4.17) of Ref. 17,

$$\left.\begin{aligned} B_\nu(\bar{\mathbf{r}}, \mathbf{s}) &= \frac{1}{2\pi} S^{(0)}(\nu) \quad \text{when } \mathbf{s} \in \Omega_{\bar{p}} \\ &= 0 \quad\quad\quad\quad\;\; \text{when } \mathbf{s} \notin \Omega_{\bar{p}} \end{aligned}\right\}. \tag{14}$$

Formula (9) now reduces to

$$W(\mathbf{r}_1, \mathbf{r}_2, \nu) = \frac{1}{2\pi} S^{(0)}(\nu) \int_{\Omega_{\bar{p}}} \exp[ik\mathbf{s} \cdot (\mathbf{r}_2 - \mathbf{r}_1)]\, \mathrm{d}\Omega. \tag{15}$$

In particular, when points P_1 and P_2 are located on the normal through the center of the source,

$$\mathbf{s} \cdot (\mathbf{r}_2 - \mathbf{r}_1) = (z_2 - z_1)\cos\alpha, \tag{16}$$

where z_1 and z_2 are the distances of the two points from the source plane and α is the angle that the **s** direction makes with the normal to the source plane. Since $\mathrm{d}\Omega = \sin\alpha \mathrm{d}\alpha \mathrm{d}\phi$, where ϕ is the azimuthal angle of the unit vector , formula (15) gives, if we now write $W(z_1, z_2, \nu)$ in place of $W(\mathbf{r}_1, \mathbf{r}_2, \nu)$,

$$\begin{aligned} W(z_1 z_2, \nu) = \frac{1}{2\pi} S^{(0)}(\nu) \int_0^{2\pi} \mathrm{d}\phi \int_0^{\bar{\alpha}} \sin\alpha \\ \times \exp[ik(z_2 - z_1)\cos\alpha]\mathrm{d}\alpha, \end{aligned} \tag{17}$$

where $\bar{\alpha}$ is the angular radius of the source as seen from midpoint $\bar{P}(\bar{\mathbf{r}})$ between the axial points P_1 and P_2 (Fig. 3). After a straightforward calculation one finds that in this case the spectral degree of coherence of the light at any two axial points is given by the expression

$$\begin{aligned} \mu(z_1 z_2, \nu) &\equiv \frac{W(z_1, z_2, \nu)}{[W(z_1, z_2, \nu)]^{1/2}[W(z_1, z_2, \nu)]^{1/2}} \\ &= \frac{\sin^2(\bar{\alpha}/2)}{\sin^2(\bar{\alpha}_1/2)\sin^2(\bar{\alpha}_2/2)} \left\{ \frac{\sin[k(z_2 - z_1)\sin^2(\bar{\alpha}/2)]}{[k(z_2 - z_1)\sin^2(\bar{\alpha}/2)]} \right\} \\ &\quad \times \exp[ik(z_2 - z_1)\cos^2(\bar{\alpha}/2)], \end{aligned} \tag{18}$$

where α_1 and α_2 are the angles that the radius of the source subtends at the axial points P_1 and P_2, respectively. In Fig. 4 $|\mu(z_1, z_2, \nu)|$ is plotted for some selected values of the parameters.

I am obliged to Weijian Wang for helpful comments regarding this manuscript and to Pramod Gupta for preparing Fig. 4. This research was supported by the U.S. Department of Energy under grant DE-FG02–90ER 14119. The views expressed in this Letter do not constitute an endorsement by the U.S. Department of Energy.

References

1. L. S. Dolin, Izv. Vyssh. Uchebn. Zaved. Radiofiz. **7,** 244 (1964).
2. V. I. Tatarskii, *The Effects of the Turbulant Atmosphere on Wave Propagation* (National Technical Information Service, Springfield, Va., 1971), Sec. 63.
3. E. Collett, J. T. Foley, and E. Wolf, J. Opt. Soc. Am. **67,** 465 (1977).
4. H. M. Pedersen, J. Opt. Soc. Am. A **8,** 176 (1991); **9,** 1626 (1992).
5. W. B. Davenport and W. L. Root, *An Introduction to the Theory of Random Signals and Noise* (McGraw-Hill, New York, 1958), p. 60.
6. E. Wolf, J. Opt. Soc. Am. **72,** 343 (1982); J. Opt. Soc. Am. A **3,** 76 (1986).
7. J. W. Goodman, *Introduction to Fourier Optics* (McGraw-Hill, New York, 1986), Sec. 3.7.
8. E. W. Marchand and E. Wolf, J. Opt. Soc. Am. **62,** 379 (1972).
9. A. Walther, J. Opt. Soc. Am. **58,** 1256 (1968).
10. Equivalent but formally different generalizations of Walther's definition, valid at points throughout the half-space $z > 0$, are considered in Refs. 11 and 12. The function $\mathcal{B}_\nu(\mathbf{r}, \mathbf{s})$ defined by Eq. (8) may be regarded as an analog of the Wigner distribution function well known in quantum mechanics.[13]
11. M. Bastians, J. Opt. Soc. Am. A **3,** 1227 (1986).
12. A. T. Friberg, Appl. Opt. **25,** 4547 (1986).
13. K. Imre, E. Özizmer, M. Rosenblum, and P. Zweifel, J. Math. Phys. **8,** 1097 (1967).
14. A. T. Friberg, J. Opt. Soc. Am. **69,** 192 (1979).
15. K. Kim and E. Wolf, J. Opt. Soc. Am. A **4,** 1233 (1987).
16. W. H. Carter and E. Wolf, J. Opt. Soc. Am. **67,** 785 (1977).
17. E. Wolf and W. H. Carter, in *Coherence and Quantum Optics IV,* L. Mandel and E. Wolf, eds. (Plenum, New York, 1978), p. 415.

1386 J. Opt. Soc. Am. B/Vol. 9, No. 8/August 1992 Friberg *et al.*

Statistical wave-theoretical derivation of the free-space transport equation of radiometry

Ari T. Friberg

Department of Technical Physics, Helsinki University of Technology, SF-02150 Espoo, Finland

Girish S. Agarwal

School of Physics, University of Hyderabad, Hyderabad, India

John T. Foley

Department of Physics and Astronomy, Mississippi State University, Mississippi State, Mississippi 39762

Emil Wolf

Department of Physics and Astronomy, University of Rochester, Rochester, New York 14627

Received October 22, 1991; revised manuscript received February 11, 1992

We are concerned with the derivation of the free-space form of the radiative transfer equation of traditional radiometry from statistical wave theory. It is shown that this equation governs the transport of all the generalized radiance functions of a wide class, for any field that is generated by a planar, secondary, quasi-homogeneous source, in the asymptotic limit as the wave number $k = 2\pi/\lambda \rightarrow \infty$.

1. INTRODUCTION

Since the publication of the first papers concerning the foundations of radiometry,[1-3] considerable progress has been made toward the clarification of the relationship between the radiometric model of energy transport and classical wave theory. These investigations have revealed a previously unsuspected fact that, in spite of the apparent simplicity of the radiometric model, some rather sophisticated concepts of statistical wave theory are needed to clarify that relationship.[4]

In spite of the progress that has been made so far regarding the foundations of radiometry, a number of important problems in this area remain unsolved. One of them concerns the following. A wide class of generalized radiance functions has been introduced, each of which has some of but not all the properties of the radiance that are postulated in traditional radiometry by heuristic arguments regarding energy transport. It was shown[5] that, when the source is planar, secondary, and quasi-homogeneous,[6] each of these generalized radiance functions acquires, at points in the source plane, all the properties of the radiance of traditional radiometry in the asymptotic limit as the wave number $k = 2\pi/\lambda \rightarrow \infty$, λ being the wavelength of the radiation. However, no corresponding result has been established so far for the radiance functions at points other than those located in the source plane.

In the present paper we show that in the asymptotic limit $k \rightarrow \infty$ all the generalized radiance functions of a wide class [characterized by so-called quadratic exponential filter functions defined by Eq. (4.8) below] of fields produced by any planar, secondary, quasi-homogeneous source become identical throughout the half-space into which the source radiates and, moreover, that they obey the transport equation of traditional radiometry. This result provides a rigorous basis for the foundations of radiometry of fields in free space, produced by planar, secondary, quasi-homogeneous sources. These are sources with which traditional radiometry is mainly concerned, although this fact was not explicitly recognized in the past.

2. GENERALIZED RADIANCE FUNCTIONS

Consider a steady-state (i.e. statistically stationary) planar, secondary source occupying a portion σ of the plane $z = 0$ and radiating into the half-space $z > 0$. The radiant intensity $J(\mathbf{s}, \nu)$ at frequency ν in the direction specified by the unit vector $\mathbf{s}$ is given by [Ref. 3, Eq. (42)]

$$J(\mathbf{s}, \nu) = (2\pi k s_z)^2 \tilde{W}^{(0)}(k\mathbf{s}_\perp, -k\mathbf{s}_\perp, \nu), \tag{2.1}$$

where s_z is the z component of the unit vector $\mathbf{s} = (s_x, s_y, s_z)$, $\mathbf{s}_\perp = (s_x, s_y, 0)$,

$$\tilde{W}^{(0)}(\mathbf{f}_1, \mathbf{f}_2, \nu) = \frac{1}{(2\pi)^4} \iint W^{(0)}(\boldsymbol{\rho}_1, \boldsymbol{\rho}_2, \nu) \times \exp[-i(\mathbf{f}_1 \cdot \boldsymbol{\rho}_1 + \mathbf{f}_2 \cdot \boldsymbol{\rho}_2)] \mathrm{d}^2\rho_1 \, \mathrm{d}^2\rho_2, \tag{2.2}$$

$$k = 2\pi\nu/c, \tag{2.3}$$

c being the speed of light *in vacuo*. Here $W^{(0)}(\boldsymbol{\rho}_1, \boldsymbol{\rho}_2, \nu)$ is the cross-spectral density of the field in the source plane $z = 0$. It may be expressed in terms of an ensemble $\{U^{(0)}(\boldsymbol{\rho}, \nu)\}$ of realizations of the field in that plane by[7]

$$W^{(0)}(\boldsymbol{\rho}_1, \boldsymbol{\rho}_2, \nu) = \langle U^{(0)}(\boldsymbol{\rho}_1, \nu) U^{(0)*}(\boldsymbol{\rho}_2, \nu) \rangle, \tag{2.4}$$

0740-3224/92/081386-07$05.00

where $\boldsymbol{\rho}_1$ and $\boldsymbol{\rho}_2$ are the position vectors of any two source points, Q_1 and Q_2, respectively.

In radiometry, the quantity that describes the energy transport from the source σ is the (spectral) radiance $B^{(0)}(\boldsymbol{\rho}, \mathbf{s}, \nu)$. It is defined as the rate at which energy at frequency ν is radiated from a source element of unit area at $\boldsymbol{\rho}$ into a unit solid angle around the $\mathbf{s}$ direction. From its definition, it follows that $B^{(0)}(\boldsymbol{\rho}, \mathbf{s}, \nu)$ has the following properties:

(i) $B^{(0)}(\boldsymbol{\rho}, \mathbf{s}, \nu) \geq 0$,
(ii) $B^{(0)}(\boldsymbol{\rho}, \mathbf{s}, \nu) = 0$ when $\boldsymbol{\rho} \notin \sigma$, and
(iii) The radiant intensity of physical optics is given by

$$J(\mathbf{s}, \nu) = s_z \int_\sigma B^{(0)}(\boldsymbol{\rho}, \mathbf{s}, \nu) \mathrm{d}^2\rho . \tag{2.5}$$

Furthermore, radiometry describes the propagation of energy by means of light rays rather than waves; in particular, the radiance $B(\mathbf{r}, \mathbf{s}, \nu)$ is assumed to be constant along straight-line paths in free space. Consequently its value at a point $\mathbf{r} = (\boldsymbol{\rho}, z)$ in the half-space $z > 0$ is related to the radiance in the source plane by the formula[8]

$$B(\mathbf{r}, \mathbf{s}, \nu) = B^{(0)}[\boldsymbol{\rho} - (z/s_z)\mathbf{s}_\perp, \mathbf{s}, \nu] . \tag{2.6}$$

In recent years efforts have been made to construct, from statistical wave theory, a function $\mathcal{B}^{(0)}(\boldsymbol{\rho}, \mathbf{s}, \nu)$, commonly referred to as a generalized radiance function (grf), that is linearly related to the cross-spectral density function of the source, has properties (i)–(iii), and obeys, in free space, the transport equation (2.6) on propagation. In particular, Walther proposed the following two different generalized radiance functions[9]:

$$\mathcal{B}_{\mathrm{W}}^{(0)}(\boldsymbol{\rho}, \mathbf{s}, \nu) = \left(\frac{k}{2\pi}\right)^2 s_z \int W^{(0)}\left(\boldsymbol{\rho} + \frac{1}{2}\boldsymbol{\rho}', \boldsymbol{\rho} - \frac{1}{2}\boldsymbol{\rho}'; \nu\right) \times \exp(-iks_\perp \cdot \boldsymbol{\rho}')\mathrm{d}^2\rho', \tag{2.7}$$

$$\mathcal{B}_{\mathrm{AS}}^{(0)}(\boldsymbol{\rho}, \mathbf{s}, \nu) = \left(\frac{k}{2\pi}\right)^2 s_z \int W^{(0)}(\boldsymbol{\rho}', \boldsymbol{\rho}; \nu) \times \exp[-ik\mathbf{s}_\perp \cdot (\boldsymbol{\rho}' - \boldsymbol{\rho})]\mathrm{d}^2\rho'. \tag{2.8}$$

Here the subscripts W and AS are used to stress a certain analogy with the so-called Weyl and antistandard correspondence, respectively, used in the theory of phase-space distributions in quantum-statistical mechanics.[10] Both of these functions have played a useful role in subsequent investigations relating to the foundations of radiometry.

It has been shown that there is no grf $\mathcal{B}^{(0)}(\boldsymbol{\rho}, \mathbf{s}, \nu)$ that is a linear mapping of the cross-spectral density of the source and has properties (i)–(iii) for fields of any state of coherence.[11] However, it has been demonstrated that, for a certain class of sources, the so-called quasi-homogeneous ones (discussed in Section 3), both of Walther's grf's have properties (i)–(iii) (see Ref. 6), and it has also been shown that for quasi-homogeneous sources they both satisfy the transfer equation (2.6) in the short-wavelength limit $\lambda \to 0$.[12,13] The question then arises whether there are other grf's that have the desired behavior.

Agarwal *et al.*[5] considered a wide class of grf's that include, as special cases, the ones proposed by Walther. They used the close connection between the theory of phase-space distributions used in statistical quantum mechanics and the theory of generalized radiance functions, a connection that had been pointed out previously by Wolf[4] (see also Ref. 10).

In quantum statistical mechanics, one can associate a class of phase-space distribution functions with a given density operator. In radiometry one can associate a class of grf's with a given cross-spectral density. Agarwal *et al.* carried this analogy further by introducing the Hilbert space position operator $\hat{\boldsymbol{\rho}} = (\hat{x}, \hat{y})$ and the "direction operator"

$$\hat{\mathbf{s}}_\perp = -i\lambda\!\!\!^{-}\nabla_\perp , \tag{2.9}$$

where

$$\lambda\!\!\!^{-} = \lambda/2\pi = 1/k ,$$

$$\nabla_\perp = \left(\frac{\partial}{\partial x}, \frac{\partial}{\partial y}\right). \tag{2.10}$$

The Cartesian components of $\hat{\boldsymbol{\rho}}$ and $\hat{\mathbf{s}}_\perp$ may be readily shown to obey the commutation relations

$$[\hat{x}, \hat{s}_x] = i\lambda\!\!\!^{-}, \qquad [\hat{y}, \hat{s}_y] = i\lambda\!\!\!^{-} , \tag{2.11}$$

which are analogous to the quantum-mechanical commutation relations for position and momentum operators. One can view the cross-spectral density function $W^{(0)}(\boldsymbol{\rho}_1, \boldsymbol{\rho}_2, \nu)$ as the coordinate space matrix element of a Hilbert space operator $\hat{G}$ (which is a function of the operators $\hat{\boldsymbol{\rho}}$ and $\hat{\mathbf{s}}_\perp$):

$$W^{(0)}(\boldsymbol{\rho}_1, \boldsymbol{\rho}_2, \nu) = \langle \boldsymbol{\rho}_1 | \hat{G} | \boldsymbol{\rho}_2 \rangle . \tag{2.12}$$

Agarwal *et al.* then introduced a wide class of grf's in a manner similar to that in which the phase-space distributions are introduced in quantum-statistical mechanics.[14] These generalized radiance functions are defined by

$$\mathcal{B}_{\Omega}^{(0)}(\boldsymbol{\rho}, \mathbf{s}, \nu) = \frac{s_z}{(2\pi)^4} \iiint \tilde{\Omega}(\mathbf{u}, \mathbf{v}) \times \exp\left[-i\left(\mathbf{u} \cdot \boldsymbol{\rho} + \mathbf{v} \cdot \mathbf{s}_\perp + \frac{1}{2}\lambda\!\!\!^{-}\mathbf{u} \cdot \mathbf{v}\right)\right] \times \exp(i\mathbf{u} \cdot \boldsymbol{\rho}_1) W^{(0)}(\boldsymbol{\rho}_1, \boldsymbol{\rho}_1 - \lambda\!\!\!^{-}\mathbf{v}, \nu)\mathrm{d}^2u\mathrm{d}^2v\mathrm{d}^2\rho_1 , \tag{2.13}$$

where the c-number function $\tilde{\Omega}(\mathbf{u}, \mathbf{v})$ characterizes the particular grf. The class of admissible $\tilde{\Omega}$ is defined by

$$\tilde{\Omega}(\mathbf{u}, \mathbf{v}) = [\Omega(-\mathbf{u}, -\mathbf{v})]^{-1}, \tag{2.14}$$

where the functions $\Omega(\mathbf{u}, \mathbf{v})$ are assumed to have the following properties:

(1) $\Omega(\mathbf{u}, \mathbf{v})$ is an entire analytic function of four complex variables u_x, u_y, v_x, v_y, which are the Cartesian components of the two-dimensional vectors $\mathbf{u}$ and $\mathbf{v}$,
(2) $\Omega(0, 0) = 1$,
(3) $\Omega(0, \mathbf{v}) = 1$ for all $\mathbf{v}$ [this requirement ensures that relation (2.5) between the radiant intensity and the radiance function is satisfied], and
(4) $\Omega(\mathbf{u}, \mathbf{v})$ has no zeros on the real u_x, u_y, v_x, v_y axes.

The choices

$$\Omega_{\mathrm{W}}(\mathbf{u}, \mathbf{v}) = 1, \tag{2.15}$$

$$\Omega_{\mathrm{AS}}(\mathbf{u}, \mathbf{v}) = \exp(-i\lambda\!\!\!^{-}\mathbf{u} \cdot \mathbf{v}/2) \tag{2.16}$$

1388 J. Opt. Soc. Am. B/Vol. 9, No. 8/August 1992

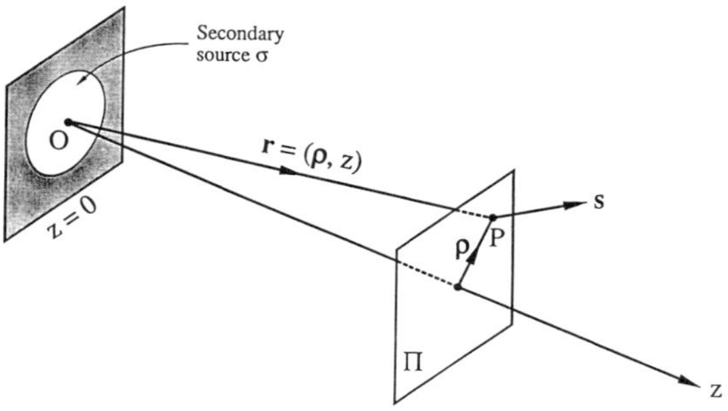

Fig. 1. Illustrating the notation.

yield, respectively, the grf's $\mathcal{B}_W^{(0)}(\boldsymbol{\rho}, \mathbf{s}, \nu)$ and $\mathcal{B}_{AS}^{(0)}(\boldsymbol{\rho}, \mathbf{s}, \nu)$ [Eqs. (2.7) and (2.8)]. Different functional forms for $\Omega(\mathbf{u}, \mathbf{v})$ are associated with different possible ways of ordering the noncommuting operators $\hat{x}$ and $\hat{s}_x$ and $\hat{y}$ and $\hat{s}_y$ to correspond to the relevant c-number functions. In the short-wavelength limit ($\lambda\!\!\!^{-} \to 0$) the operators all commute [see Eq. (2.11)], and $\Omega(\mathbf{u}, \mathbf{v}) \to 1$. Thus in the short-wavelength limit all the grf's reduce to the same function, namely, the short-wavelength limit of $\mathcal{B}_W^{(0)}(\boldsymbol{\rho}, \mathbf{s}, \nu)$. This function has the required properties of the radiance of radiometry, as shown in Ref. 13.

The above analysis pertains to points in the source plane, $z = 0$. One can define the generalized radiance function in any other plane $z > 0$ by replacing $W^{(0)}(\boldsymbol{\rho}_1, \boldsymbol{\rho}_2, \nu)$ under the integral sign in Eq. (2.13) by the cross-spectral density of the field in that plane, which can be expressed in a similar form as $W^{(0)}(\boldsymbol{\rho}_1, \boldsymbol{\rho}_2, \nu)$, viz. [cf. Eq. (2.4) and Ref. 15],

$$W(\mathbf{r}_1, \mathbf{r}_2, \nu) = \langle U(\mathbf{r}_1, \nu) U^*(\mathbf{r}_2, \nu) \rangle, \tag{2.17}$$

where $\mathbf{r}_j = (\boldsymbol{\rho}_j, z)$, $j = 1, 2$. Thus we define the grf at point $\mathbf{r} = (\boldsymbol{\rho}, z)$ as

$$\mathcal{B}_\Omega(\mathbf{r}, \mathbf{s}, \nu) = \frac{s_z}{(2\pi)^4} \iiint \tilde{\Omega}(\mathbf{u}, \mathbf{v}) \times \exp\left[-i\left(\mathbf{u} \cdot \boldsymbol{\rho} + \mathbf{v} \cdot \mathbf{s}_\perp + \frac{1}{2} \lambda\!\!\!^{-} \mathbf{u} \cdot \mathbf{v}\right)\right] \times \exp(i\mathbf{u} \cdot \boldsymbol{\rho}_1) W(\mathbf{r}_1, \mathbf{r}_1 - \lambda\!\!\!^{-} \mathbf{v}, \nu) \mathrm{d}^2u \mathrm{d}^2v \mathrm{d}^2\rho_1. \tag{2.18}$$

The purpose of the present investigation is to study the propagation of these grf's from the source plane $z = 0$ into the half-space $z > 0$. In particular, we are interested in determining under what conditions the grf's are constant along straight-line paths of propagation, i.e., under what condition they obey the transport equation (2.6).

In Appendix A we utilize the analogy between the short-wavelength limit ($\lambda\!\!\!^{-} \to 0$) of the grf's and the classical limit ($\hbar \to 0$) of quantum mechanics to provide a plausibility argument that strongly suggests that in the short-wavelength limit ($\lambda\!\!\!^{-} \to 0$) all the generalized radiances $\mathcal{B}_\Omega(\mathbf{r}, \mathbf{s}, \nu)$ given by Eq. (2.18) obey the transport equation (2.6). In Sections 3 and 4 we investigate this question by a more rigorous argument based on well-known technique of classical optics.

3. PROPAGATION OF THE GENERALIZED RADIANCE

Let us consider the situation illustrated in Fig. 1, where P represents a point in the plane z = constant > 0 (denoted by Π) parallel to the plane $z = 0$, which contains a secondary source σ. We denote by $\mathbf{r}$ the position vector of P, with longitudinal component z and transverse two-dimensional vector component $\boldsymbol{\rho}$, and $\mathbf{s}(s_z \geq 0)$ denotes a real unit vector specifying a direction along a particular line through P. Throughout the rest of this paper position vectors that refer to points in the plane $z = 0$ will be denoted by subscript zeros; e.g., the location of a point Q_1 in this plane will be specified by a position vector $\boldsymbol{\rho}_{01}$.

The grf $\mathcal{B}_\Omega(\mathbf{r}, \mathbf{s}, \nu)$ is defined by Eq. (2.18). For our purposes it will be preferable to express the grf in terms of a certain symmetrized form of the cross-spectral density in the plane Π. To obtain this expression we introduce the variables

$$\boldsymbol{\rho}' = \boldsymbol{\rho}_1 - (1/2)\lambda\!\!\!^{-}\mathbf{v}, \tag{3.1a}$$

$$\boldsymbol{\rho}'' = \lambda\!\!\!^{-}\mathbf{v}. \tag{3.1b}$$

In terms of these variables, Eq. (2.18) can be rewritten as

$$\mathcal{B}_\Omega(\mathbf{r}, \mathbf{s}, \nu) = \frac{k^2}{(2\pi)^4} s_z \iiint \tilde{\Omega}(\mathbf{u}, k\boldsymbol{\rho}'') \times \exp\{-i[\mathbf{u} \cdot (\boldsymbol{\rho} - \boldsymbol{\rho}') + k\mathbf{s}_\perp \cdot \boldsymbol{\rho}'']\} \times W\left(\mathbf{r}' + \frac{1}{2}\boldsymbol{\rho}'', \mathbf{r}' - \frac{1}{2}\boldsymbol{\rho}'', \nu\right) \mathrm{d}^2u \mathrm{d}^2\rho' \mathrm{d}^2\rho'', \tag{3.2}$$

where $\mathbf{r}' = (\boldsymbol{\rho}', z)$.

Inspection of Eq. (3.2) shows that the propagation of $\mathcal{B}_\Omega(\mathbf{r}, \mathbf{s}, \nu)$ is governed by the propagation of the symmetrized form $W[\mathbf{r}' + (1/2)\boldsymbol{\rho}'', \mathbf{r}' - (1/2)\boldsymbol{\rho}'', \nu]$ of the cross-spectral density. The propagation of that quantity can be described in the following way. The cross-spectral density at two points P_1 and P_2 in the plane Π, whose position vectors are $\mathbf{r}_1 = (\boldsymbol{\rho}_1, z)$ and $\mathbf{r}_2 = (\boldsymbol{\rho}_2, z)$, is given by Eq. (2.17), viz.,

$$W(\mathbf{r}_1, \mathbf{r}_2, \nu) = \langle U(\mathbf{r}_1, \nu) U^*(\mathbf{r}_2, \nu) \rangle, \tag{3.3}$$

where $U(\mathbf{r}_j, \nu)$ is the field at the point P_j ($j = 1, 2$). This field in the plane Π is related to the field in the plane $z = 0$ by a formula due to Rayleigh,[16] well known in diffraction theory:

$$U(\mathbf{r}, \nu) = \int G(\mathbf{r} - \boldsymbol{\rho}_0, \nu) U^{(0)}(\boldsymbol{\rho}_0, \nu) \mathrm{d}^2\rho_0. \tag{3.4}$$

Here $G(\mathbf{R}, \nu)$ is the free-space Green's function,

$$G(\mathbf{R}, \nu) = -\frac{1}{2\pi} \frac{\partial}{\partial z} \left[\frac{\exp(ikR)}{R}\right], \tag{3.5}$$

with $R = |\mathbf{R}|$. On using Eq. (3.4) in Eq. (3.3), we find that

$$W(\mathbf{r}_1, \mathbf{r}_2, \nu) = \iint G(\mathbf{r}_1 - \boldsymbol{\rho}_{01}, \nu) \times G^*(\mathbf{r}_2 - \boldsymbol{\rho}_{02}, \nu) W^{(0)}(\boldsymbol{\rho}_{01}, \boldsymbol{\rho}_{02}, \nu) \mathrm{d}^2\rho_{01} \mathrm{d}^2\rho_{02}. \tag{3.6}$$

In order to describe the propagation of the symmetrized form of the cross-spectral density, let us now make the change of variables involving the coordinates in the plane Π:

$$\boldsymbol{\rho}_1 = \boldsymbol{\rho}' + (1/2)\boldsymbol{\rho}'', \tag{3.7a}$$

$$\boldsymbol{\rho}_2 = \boldsymbol{\rho}' - (1/2)\boldsymbol{\rho}''; \tag{3.7b}$$

we also introduce the corresponding change of variables in the source plane:

$$\boldsymbol{\rho}_{01} = \boldsymbol{\rho}_0' + (1/2)\boldsymbol{\rho}_0'', \tag{3.8a}$$

$$\boldsymbol{\rho}_{02} = \boldsymbol{\rho}_0' - (1/2)\boldsymbol{\rho}_0''. \tag{3.8b}$$

In terms of these variables, Eq. (3.6) becomes

$$\begin{aligned} W[\mathbf{r}' + (1/2)\boldsymbol{\rho}'', \mathbf{r}' - (1/2)\boldsymbol{\rho}'', \nu] &= \iint G[\mathbf{r}' - \boldsymbol{\rho}_0' + (1/2)(\boldsymbol{\rho}'' - \boldsymbol{\rho}_0''), \nu] \\ &\times G^*[\mathbf{r}' - \boldsymbol{\rho}_0' - (1/2)(\boldsymbol{\rho}'' - \boldsymbol{\rho}_0''), \nu \\ &\times W^{(0)}(\boldsymbol{\rho}_0' + (1/2)\boldsymbol{\rho}_0'', \boldsymbol{\rho}_0' - (1/2)\boldsymbol{\rho}_0'', \nu)\mathrm{d}^2\rho_0'\mathrm{d}^2\rho_0'', \end{aligned} \tag{3.9}$$

where $\mathbf{r}' = (\boldsymbol{\rho}', z)$.

Let us now return, for a moment, to Eq. (3.2). If one substitutes expression (3.9) into Eq. (3.2), it can be shown (see Appendix B for details) that

$$\begin{aligned} \mathcal{B}_\Omega(\mathbf{r}, \mathbf{s}, \nu) = \frac{k^2}{(2\pi)^4} s_z \iiint & M(\mathbf{u}, \mathbf{r}', \boldsymbol{\rho}_0'', \nu) \\ &\times \exp\{-i[\mathbf{u} \cdot (\boldsymbol{\rho} - \boldsymbol{\rho}') + k\mathbf{s}_\perp \cdot \boldsymbol{\rho}_0'']\} \\ &\times \mathrm{d}^2u\mathrm{d}^2\rho'\,\mathrm{d}^2\rho'', \end{aligned} \tag{3.10}$$

where

$$\begin{aligned} M(\mathbf{u}, \mathbf{r}', \boldsymbol{\rho}_0'', \nu) = & \iint \tilde{\Omega}[\mathbf{u}, k(\boldsymbol{\rho}_b - \boldsymbol{\rho}_a + \boldsymbol{\rho}_0'')] \\ & \times \exp[-ik\mathbf{s}_\perp \cdot (\boldsymbol{\rho}_b - \boldsymbol{\rho}_a)]G(\mathbf{r}' - \boldsymbol{\rho}_a'', \nu)G^*(\mathbf{r}' - \boldsymbol{\rho}_b, \nu) \\ & \times W^{(0)}[½(\boldsymbol{\rho}_a + \boldsymbol{\rho}_b + \boldsymbol{\rho}_0''), (1/2)(\boldsymbol{\rho}_a + \boldsymbol{\rho}_b - \boldsymbol{\rho}_0''), \nu]\mathrm{d}^2\rho_a\mathrm{d}^2\rho_b. \end{aligned} \tag{3.11}$$

Let us now specialize the results expressed by Eqs. (3.10) and (3.11) to the case when the source is quasi-homogeneous. The cross-spectral density at any two points Q_1 and Q_2 in a quasi-homogeneous source may be written as[6]

$$W^{(0)}(\boldsymbol{\rho}_{01}, \boldsymbol{\rho}_{02}, \nu) = S^{(0)}[(1/2)(\boldsymbol{\rho}_{01} + \boldsymbol{\rho}_{02}), \nu]g^{(0)}(\boldsymbol{\rho}_{01} - \boldsymbol{\rho}_{02}, \nu), \tag{3.12}$$

where $S^{(0)}(\boldsymbol{\rho}_0, \nu)$ is the spectrum at a typical point Q_0 in the source plane and $g^{(0)}(\boldsymbol{\rho}_{01} - \boldsymbol{\rho}_{02}, \nu)$ is the degree of spectral coherence of the field at points Q_1 and Q_2. Moreover, for sources of this class $S^{(0)}(\boldsymbol{\rho}_0, \nu)$ changes so slowly with the position ($\boldsymbol{\rho}_0$) across the source that it is essentially constant over regions whose linear dimensions are of the order of the effective range of $|g^{(0)}(\boldsymbol{\rho}_{01} - \boldsymbol{\rho}_{02}, \nu)|$, i.e., of the order of the spectral correlation length, l_ν, say, of the light across the source. It is also assumed that the linear dimensions of the source are large compared with l_ν and with the wavelength $\lambda = c/\nu$.

It follows at once from Eq. (3.12) that, for a quasi-homogeneous source,

$$\begin{aligned} W^{(0)}[(1/2)(\boldsymbol{\rho}_a + \boldsymbol{\rho}_b + \boldsymbol{\rho}_0''), (1/2)(\boldsymbol{\rho}_a + \boldsymbol{\rho}_b - \boldsymbol{\rho}_0''), \nu] \\ = S^{(0)}[(1/2)(\boldsymbol{\rho}_a + \boldsymbol{\rho}_b), \nu]g^{(0)}(\boldsymbol{\rho}_0'', \nu). \end{aligned} \tag{3.13}$$

On substitution from Eq. (3.13) into Eq. (3.11), we find that for such sources

$$\begin{aligned} M(\mathbf{u}, \mathbf{r}', \boldsymbol{\rho}_0'', \nu) = g^{(0)}(\boldsymbol{\rho}_0'', \nu) & \iint \tilde{\Omega}[\mathbf{u}, k(\boldsymbol{\rho}_b - \boldsymbol{\rho}_a + \boldsymbol{\rho}_0'')] \\ & \times \exp[-ik\mathbf{s}_\perp \cdot (\boldsymbol{\rho}_b - \boldsymbol{\rho}_a)]G(\mathbf{r}' - \boldsymbol{\rho}_a, \nu) \\ & \times G^*(\mathbf{r}' - \boldsymbol{\rho}_b, \nu) \\ & \times S^{(0)}[(1/2)(\boldsymbol{\rho}_a + \boldsymbol{\rho}_b), \nu]\mathrm{d}^2\rho_a\mathrm{d}^2\rho_b. \end{aligned} \tag{3.14}$$

4. SHORT-WAVELENGTH LIMIT OF THE GENERALIZED RADIANCE OF A FIELD GENERATED BY A QUASI-HOMOGENEOUS SOURCE

Let us now consider the behavior of the generalized radiance function $\mathcal{B}_\Omega(\mathbf{r}, \mathbf{s}, \nu)$ for a field generated by a planar, secondary, quasi-homogeneous source in the short-wavelength limit, more precisely, in the asymptotic limit as $k = 2\pi/\lambda \to \infty$. For this purpose we first express the Green's function (3.5) in a more explicit form. On carrying out the differentiation one readily finds that

$$G(\mathbf{R}, \nu) = -\frac{1}{2\pi}\left[\left(ik - \frac{1}{R}\right)\frac{z}{R}\right]\frac{\exp(ikR)}{R}, \tag{4.1}$$

which, for sufficiently large values of kR, may be approximated by

$$G(\mathbf{R}, \nu) \sim -\frac{ikz}{2\pi R}\frac{\exp(ikR)}{R}. \tag{4.2}$$

On substituting this expression into Eq. (3.14), we find that

$$\begin{aligned} M(\mathbf{u}, \mathbf{r}', \boldsymbol{\rho}_0'', \nu) \sim \left(\frac{kz}{2\pi}\right)^2 g^{(0)}(\boldsymbol{\rho}_0'', \nu) & \iint \tilde{\Omega}[\mathbf{u}, k(\boldsymbol{\rho}_b - \boldsymbol{\rho}_a + \boldsymbol{\rho}_0'')] \\ & \times S^{(0)}\left[\frac{1}{2}(\boldsymbol{\rho}_a + \boldsymbol{\rho}_b), \nu\right] \\ & \times \frac{1}{|\mathbf{r}' - \boldsymbol{\rho}_a|^2}\exp[ik\phi(\mathbf{r}', \boldsymbol{\rho}_a)] \\ & \times \frac{1}{|\mathbf{r}' - \boldsymbol{\rho}_b|^2}\exp[-ik\phi(\mathbf{r}', \boldsymbol{\rho}_b)]\mathrm{d}^2\rho_a\mathrm{d}^2\rho_b, \end{aligned} \tag{4.3}$$

where

$$\phi(\mathbf{r}', \boldsymbol{\rho}_j) = |\mathbf{r}' - \boldsymbol{\rho}_j| + \mathbf{s}_\perp \cdot \boldsymbol{\rho}_j \qquad (j = a, b). \tag{4.4}$$

Since the integrand on the right-hand side of expression (4.3) contains factors such as $\exp[ik\phi(\mathbf{r}', \boldsymbol{\rho}_j)]$, it seems appropriate to apply the method of stationary phase to determine the asymptotic form of this integral as $k \to \infty$. However, this requires some care, since both $\tilde{\Omega}$ and $S^{(0)}$ depend on frequency.

First let us consider the behavior of $S^{(0)}(\boldsymbol{\rho}_0, \nu)$. The spatial correlation lengths, the l_ν, in realistic secondary sources are greater than or of the order of the wavelength $\lambda = c/\nu$. Therefore, since in a quasi-homogeneous source $S^{(0)}(\boldsymbol{\rho}_0, \nu)$ varies (at each frequency) slowly over distances of the order of l_ν, it must vary slowly over distances of the order of the wavelength. This implies that $S^{(0)}[(1/2)(\boldsymbol{\rho}_a + \boldsymbol{\rho}_b), \nu]$ will not contain terms such as $\exp(ik\phi)$ that would change the location of the critical point in the integral on the right-hand side of relation (4.3).

Next let us consider the behavior of the factor $\tilde{\Omega}[\mathbf{u}, k(\boldsymbol{\rho}_b - \boldsymbol{\rho}_a + \boldsymbol{\rho}_0'')]$. We know that

(a) $\mathbf{u}$ has the dimension of $(\text{length})^{-1}$ and $\mathbf{v}$ is dimensionless.

(b) $\Omega(\mathbf{u}, \mathbf{v})$ is dimensionless.

(c) Different Ω functions correspond to different orderings, and the only length that occurs in "reorderings" is $\lambda\!\!\!^-$.

It therefore appears that the variable $\mathbf{u}$ in $\Omega(\mathbf{u}, \mathbf{v})$ should enter only together with $\lambda\!\!\!^-$ (i.e., product $\lambda\!\!\!^-\mathbf{u}$) so that $\Omega(\mathbf{u}, \mathbf{v})$ has the functional form

$$\Omega(\mathbf{u}, \mathbf{v}) = f(\lambda\!\!\!^-\mathbf{u}, \mathbf{v}). \tag{4.5}$$

Expression (2.16) relating to antistandard ordering illustrates this fact. Since $\tilde{\Omega}(\mathbf{u}, \mathbf{v}) = [\Omega(-\mathbf{u}, -\mathbf{v})]^{-1}$, it follows that $\tilde{\Omega}(\mathbf{u}, \mathbf{v})$ will also depend on $\mathbf{u}$ only through the product $\lambda\!\!\!^-$; i.e., it will have the functional form

$$\tilde{\Omega}(\mathbf{u}, \mathbf{v}) = F(\lambda\!\!\!^-\mathbf{u}, \mathbf{v}). \tag{4.6}$$

Hence

$$\tilde{\Omega}(\mathbf{u}, k\boldsymbol{\rho}) = F(\lambda\!\!\!^-\mathbf{u}, k\boldsymbol{\rho}). \tag{4.7}$$

It follows from Eq. (4.7) that, in general, $\tilde{\Omega}[\mathbf{u}, k(\boldsymbol{\rho}_b - \boldsymbol{\rho}_a + \boldsymbol{\rho}_0'')]$ may contain terms of the form $\exp(ik\phi)$ that affect the location of the critical points of the integral in relation (4.3). For this reason we will restrict our analysis to the class of the so-called quadratic exponential filter function. For such functions

$$\tilde{\Omega}(\mathbf{u}, \mathbf{v}) = \exp(\alpha u^2 + \beta \mathbf{u} \cdot \mathbf{v} + \gamma v^2), \tag{4.8}$$

where α, β, and γ are, in general, complex. Such filter functions have played predominant roles both in previous investigations relating to the foundations of radiometry and in the phase-space representation of quantum mechanics.

It follows from condition (3) of Section 2 that in Eq. (4.8) $\gamma = 0$. It then follows from Eq. (4.6) that in this case

$$\tilde{\Omega}(\mathbf{u}, \mathbf{v}) = \exp(\alpha' \lambda\!\!\!^{-2} u^2 + \beta' \lambda\!\!\!^- \mathbf{u} \cdot \mathbf{v}), \tag{4.9}$$

where α' and β' are generally complex constants, which are independent of frequency. We see immediately from Eq. (4.9) that

$$\tilde{\Omega}(\mathbf{u}, k\boldsymbol{\rho}) = \exp(\alpha' \lambda\!\!\!^{-2} u^2 + \beta' \mathbf{u} \cdot \boldsymbol{\rho}), \tag{4.10}$$

and hence that such filter functions will not change the locations of the critical points of the integrals under consideration.

Let us now return to relation (4.3). It can be rewritten as

$$M(\mathbf{u}, \mathbf{r}', \boldsymbol{\rho}_0'', \nu) \sim \left(\frac{kz}{2\pi}\right)^2 g^{(0)}(\boldsymbol{\rho}_0'', \nu) \int N(\mathbf{u}, \mathbf{r}', \boldsymbol{\rho}_0'', \nu, \boldsymbol{\rho}_b) \times \frac{1}{|\mathbf{r}' - \boldsymbol{\rho}_b|^2} \exp[-ik\phi(\mathbf{r}', \boldsymbol{\rho}_b)] \mathrm{d}^2\rho_b, \tag{4.11}$$

where

$$N(\mathbf{u}, \mathbf{r}', \boldsymbol{\rho}_0'', \nu, \boldsymbol{\rho}_b) = \int \tilde{\Omega}[\mathbf{u}, k(\boldsymbol{\rho}_b - \boldsymbol{\rho}_a + \boldsymbol{\rho}_0'')] \times S^{(0)}\left[\frac{1}{2}(\boldsymbol{\rho}_a + \boldsymbol{\rho}_b), \nu\right] \times \frac{1}{|\mathbf{r}' - \boldsymbol{\rho}_a|^2} \exp[ik\phi(\mathbf{r}', \boldsymbol{\rho}_a)] \mathrm{d}^2\rho_a. \tag{4.12}$$

For the classes of filter functions and sources that we are considering, the locations of the critical points of the integral Eq. (4.12) are determined by the function $\phi(\mathbf{r}', \boldsymbol{\rho}_a)$. After a straightforward application of the method of stationary phase (see App. III of Ref. 17), we find that this integral has a critical point of the first kind when $\boldsymbol{\rho}_a = \boldsymbol{\rho}_c$, where

$$\boldsymbol{\rho}_c \equiv \boldsymbol{\rho}' - (z/s_z)\mathbf{s}_\perp, \tag{4.13}$$

and that

$$N(\mathbf{u}, \mathbf{r}', \boldsymbol{\rho}_0'', \nu, \boldsymbol{\rho}_b) \sim \frac{2\pi i}{kz} \exp(ik\mathbf{s} \cdot \mathbf{r}') \tilde{\Omega}[\mathbf{u}, k(\boldsymbol{\rho}_b - \boldsymbol{\rho}_c + \boldsymbol{\rho}_0'')] \times S^{(0)}\left[\frac{1}{2}(\boldsymbol{\rho}_c + \boldsymbol{\rho}_b), \nu\right] \quad \text{as } k \to \infty. \tag{4.14}$$

On substituting the result of Eq. (4.14) into Eq. (4.11), we find that

$$M(\mathbf{u}, \mathbf{r}', \boldsymbol{\rho}_0'', \nu) \sim \left(\frac{kz}{2\pi}\right)^2 g^{(0)}(\boldsymbol{\rho}_0'', \nu) \frac{2\pi i}{kz} \exp(ik\mathbf{s} \cdot \mathbf{r}') \times \int \tilde{\Omega}[\mathbf{u}, k(\boldsymbol{\rho}_b - \boldsymbol{\rho}_c + \boldsymbol{\rho}_0'')] \times S^{(0)}\left[\frac{1}{2}(\boldsymbol{\rho}_c + \boldsymbol{\rho}_b), \nu\right] \times \frac{1}{|\mathbf{r}' - \boldsymbol{\rho}_b|^2} \exp[-ik\phi(\mathbf{r}', \boldsymbol{\rho}_b)] \mathrm{d}^2\rho_b \quad \text{as } k \to \infty. \tag{4.15}$$

Further straightforward calculations utilizing the method of stationary phase lead to the result that the integral on the right-hand side of relation (4.15) has a critical point of the first kind when $\boldsymbol{\rho}_b = \boldsymbol{\rho}_c$ and that, as $k \to \infty$,

$$\begin{aligned} M(\mathbf{u}, \mathbf{r}', \boldsymbol{\rho}_0'', \nu) &\sim \left(\frac{kz}{2\pi}\right)^2 g^{(0)}(\boldsymbol{\rho}_0'', \nu) \frac{2\pi i}{kz} \times \exp(ik\mathbf{s} \cdot \mathbf{r}') \left[-\frac{2\pi i}{kz} \exp(-ik\mathbf{s} \cdot \mathbf{r}')\right] \\ &\quad \times \tilde{\Omega}[\mathbf{u}, k(\boldsymbol{\rho}_c - \boldsymbol{\rho}_c + \boldsymbol{\rho}_0'')] \times S^{(0)}\left[\frac{1}{2}(\boldsymbol{\rho}_c + \boldsymbol{\rho}_c), \nu\right] \\ &= \tilde{\Omega}(\mathbf{u}, k\boldsymbol{\rho}_0'') S^{(0)}(\boldsymbol{\rho}_c, \nu) g^{(0)}(\boldsymbol{\rho}_0'', \nu) \\ &= \tilde{\Omega}(\mathbf{u}, k\boldsymbol{\rho}_0'') W^{(0)}\left(\boldsymbol{\rho}_c + \frac{1}{2}\boldsymbol{\rho}_0'', \boldsymbol{\rho}_c - \frac{1}{2}\boldsymbol{\rho}_0'', \nu\right), \end{aligned} \tag{4.16}$$

where $\boldsymbol{\rho}_c$ is given by Eq. (4.13).

In order to find the asymptotic form of the generalized radiance function $\mathcal{B}_\Omega(\mathbf{r}, \mathbf{s}, \nu)$, let us use the result of relation (4.16) and the definition of $\boldsymbol{\rho}_c$ in Eq. (4.13). We find that, as $k \to \infty$,

$$\mathcal{B}_\Omega(\mathbf{r}, \mathbf{s}, \nu) \sim \frac{k^2}{(2\pi)^4} s_z \iiint \tilde{\Omega}(\mathbf{u}, k\boldsymbol{\rho}_0'') \times \exp\{-i[\mathbf{u}\cdot(\boldsymbol{\rho} - \boldsymbol{\rho}') + k\mathbf{s}_\perp \cdot \boldsymbol{\rho}_0'']\} \times W^{(0)}\left[\boldsymbol{\rho}' - (z/s_z)\mathbf{s}_\perp + \frac{1}{2}\boldsymbol{\rho}_0'', \boldsymbol{\rho}' - (z/s_z)\mathbf{s}_\perp - \frac{1}{2}\boldsymbol{\rho}_0'', \nu\right] \mathrm{d}^2u\,\mathrm{d}^2\rho'\,\mathrm{d}^2\rho_0''. \quad (4.17)$$

If we make the change of variables

$$\boldsymbol{\rho}_0' = \boldsymbol{\rho}' - (z/s_z)\mathbf{s}_\perp, \quad (4.18)$$

and note that

$$\boldsymbol{\rho} - \boldsymbol{\rho}' = \boldsymbol{\rho} - (z/s_z)\mathbf{s}_\perp - \boldsymbol{\rho}_0', \quad (4.19)$$

relation (4.17) can be rewritten as

$$\mathcal{B}_\Omega(\mathbf{r}, \mathbf{s}, \nu) \sim \frac{k^2}{(2\pi)^4} s_z \iiint \tilde{\Omega}(\mathbf{u}, k\boldsymbol{\rho}_0'') \times \exp(-i\{\mathbf{u}\cdot[\boldsymbol{\rho} - (z/s_z)\mathbf{s}_\perp - \boldsymbol{\rho}_0'] + k\mathbf{s}_\perp \cdot \boldsymbol{\rho}_0''\}) \times W^{(0)}\left(\boldsymbol{\rho}_0' + \frac{1}{2}\boldsymbol{\rho}_0'', \boldsymbol{\rho}_0' - \frac{1}{2}\boldsymbol{\rho}_0'', \nu\right) \times \mathrm{d}^2u\,\mathrm{d}^2\rho_0'\,\mathrm{d}^2\rho_0''. \quad (4.20)$$

In a straightforward manner one can show from Eq. (2.13) that the expression on the right-hand side of relation (4.20) is equal to $\mathcal{B}_\Omega^{(0)}[\boldsymbol{\rho} - (z/s_z)\mathbf{s}_\perp, \mathbf{s}, \nu]$. Hence we conclude that for quasi-homogeneous sources, and with quadratic exponential linear mappings [Eq. (4.8)],

$$\mathcal{B}_\Omega(\mathbf{r}, \mathbf{s}, \nu) \sim \mathcal{B}_\Omega^{(0)}[\boldsymbol{\rho} - (z/s_z)\mathbf{s}_\perp, \mathbf{s}, \nu], \quad \text{as } k \to \infty, \quad (4.21)$$

i.e., all the grf's of the class that we are considering obey the usual propagation law [Eq. (2.6)] of traditional radiometry in the asymptotic limit of large wave numbers (short wavelengths).

APPENDIX A: LIOUVILLE EQUATION FOR THE CROSS-SPECTRAL DENSITY OPERATOR AND ITS SHORT-WAVELENGTH LIMIT

In this appendix we present a heuristic argument that suggests the validity of the main result we just derived. For this purpose we first obtain an analog of the quantum-mechanical Liouville equation for the cross-spectral density operator $\hat{G}$ and then derive its short-wavelength limit.

We will make use of the angular spectrum representation [Eq. (A2) below] of a monochromatic field $U(\mathbf{r}, \nu)$, $[\mathbf{r} = (\boldsymbol{\rho}, z)]$ propagating from a finite secondary source located in the plane $z = 0$ into the half-space $z > 0$. Throughout that half-space the field will satisfy the Helmholtz equation

$$(\nabla^2 + k^2)U(\mathbf{r}, \nu) = 0, \quad (A1)$$

where $k = 2\pi\nu/c = 2\pi/\lambda$. Clearly the field will behave as an outgoing spherical wave at infinity in this half-space. As is well known, under fairly general conditions the field throughout the half-space $z > 0$ can then be represented in the form[18]

$$U(\mathbf{r}, \nu) = \int a(\mathbf{s}_\perp, \nu)\exp(ik\mathbf{s}\cdot\mathbf{r})\mathrm{d}^2s_\perp, \quad (A2)$$

where

$$\mathbf{s} = (s_x, s_y, s_z), \quad \mathbf{s}_\perp = (s_x, s_y, 0),$$
$$s_z = (1 - s_\perp^2)^{1/2} \quad \text{when } s_\perp \le 1$$
$$= i(s_\perp^2 - 1)^{1/2} \quad \text{when } s_\perp > 1, \quad (A3)$$

and the integration on the right-hand side of Eq. (A.2) extends over the whole $\mathbf{s}_\perp$ plane.

We note that, when $s_z = (1 - s_\perp^2)^{1/2}$,

$$\exp(ik\mathbf{s}\cdot\mathbf{r}) = \exp\{ikz[1 + (\nabla_\perp^2/k^2)]^{1/2}\}\exp(ik\mathbf{s}_\perp\cdot\boldsymbol{\rho}), \quad (A4)$$

where $\nabla_\perp = (\partial/\partial x, \partial/\partial y)$. Therefore, if we neglect evanescent waves, Eq. (A2) can be written as

$$U(\mathbf{r}, \nu) = \exp\{ikz[1 + (\nabla_\perp^2/k^2)]^{1/2}\}\int a(\mathbf{s}_\perp \cdot \nu) \times \exp(ik\mathbf{s}_\perp\cdot\boldsymbol{\rho})\mathrm{d}^2s_\perp = \exp\{ikz[1 + (\nabla_\perp^2/k^2)]^{1/2}\}U^{(0)}(\boldsymbol{\rho}, \nu). \quad (A5)$$

This is an operator form of the propagation equation of the field. Note that Eq. (A5) implies that

$$\left\{\frac{\partial}{\partial z} - ik[1 + (\nabla_\perp^2/k^2)]^{1/2}\right\}U(\mathbf{r}, \nu) = 0. \quad (A6)$$

We have thus reduced the Helmholtz equation to a partial differential equation, which is of first order in z. This is possible because we are considering propagation into the right half-space ($z > 0$) only.

We can now derive the evolution equation for the cross-spectral density $W(\mathbf{r}_1, \mathbf{r}_2, \nu)$, where $\mathbf{r}_j = (\boldsymbol{\rho}_j, z)$. For notational convenience, throughout the rest of this appendix we shall suppress the ν dependence of all functions except the grf's and denote the above-mentioned cross-spectral density as $W(\boldsymbol{\rho}_1, \boldsymbol{\rho}_2, z)$. It follows from Eq. (A6) that

$$W(\boldsymbol{\rho}_1, \boldsymbol{\rho}_2, z) \equiv \langle U(\boldsymbol{\rho}_1, z)U^*(\boldsymbol{\rho}_2, z)\rangle = \langle\exp\{ikz[1 + (\nabla_{1\perp}^2/k^2)]^{1/2}\}U(\boldsymbol{\rho}_1, 0)U^*(\boldsymbol{\rho}_2, 0) \times \exp\{-ikz[1 + (\nabla_{2\perp}^2/k^2)]^{1/2}\}\rangle, \quad (A7)$$

where the second operator on the right-hand side of Eq. (A7) operates on the function to its left. In view of Eq. (2.9), the operator version of Eq. (A7) is

$$\hat{G}(z) = \exp[ikz(1 - \hat{s}_\perp^2)^{1/2}]\hat{G}(0)\exp[-ikz(1 - \hat{s}_\perp^2)^{1/2}]. \quad (A8)$$

Equation (A8) is equivalent to

$$\frac{\partial\hat{G}(z)}{\partial z} = -\frac{i}{\lambda}[-(1 - \hat{s}_\perp^2)^{1/2}, \hat{G}(z)]. \quad (A9)$$

Evidently this equation may be identified with the Liouville equation for the cross-spectral density operator $\hat{G}$.

Starting from Eq. (A9) and using the usual method for transcribing operator equations into c-number equations,[14] we can derive the propagation equation for the grf $\mathscr{B}_\Omega(\mathbf{r}, \mathbf{s}, \nu)$. It should be noted that in deriving Eq. (A9) no paraxial approximation has been made.

We next consider the short-wavelength limit of Eq. (A9). This limit is analogous to taking the classical limit $\hbar \to 0$ of the quantum-mechanical equations of motion. For this purpose one can use Dirac's prescription,[19] according to which we replace commutators by Poisson brackets (PB):

$$-\frac{i}{\hbar}[\,,\,] \to \{\,\}_{\mathrm{PB}}. \tag{A10}$$

For the radiometric problem, this implies that we make the transition

$$-\frac{i}{\lambda}[\hat{A}, \hat{B}] \to A\left(\frac{\overleftarrow{\partial}}{\partial \rho}\frac{\overrightarrow{\partial}}{\partial \mathbf{s}_\perp} - \frac{\overleftarrow{\partial}}{\partial \mathbf{s}_\perp}\frac{\overrightarrow{\partial}}{\partial \rho}\right)B \tag{A11}$$

and replace the quantum density operator $\hat{G}$ by the classical phase-space distribution function. Thus we replace $\hat{G}(z)$ by $B(\mathbf{r}, \mathbf{s}, \nu)$, which is the short-wavelength limit of $\mathscr{B}_\Omega(\mathbf{r}, \mathbf{s}, \nu)$. Thus using Dirac's prescription reduces Eq. (A9) to

$$\begin{aligned}\frac{\partial B(\mathbf{r}, \mathbf{s}, \nu)}{\partial z} &= -(1 - \mathbf{s}_\perp{}^2)^{1/2}\left(\frac{\boldsymbol{\partial}}{\partial \mathbf{s}_\perp}\frac{\boldsymbol{\partial}}{\partial \boldsymbol{\rho}}\right)B(\mathbf{r}, \mathbf{s}, \nu)\\ &= -\frac{1}{(1 - \mathbf{s}_\perp{}^2)^{1/2}}\mathbf{s}_\perp \cdot \nabla_\perp B(\mathbf{r}, \mathbf{s}, \nu)\\ &= -\frac{1}{s_z}\mathbf{s}_\perp \cdot \nabla_\perp B(\mathbf{r}, \mathbf{s}, \nu),\end{aligned} \tag{A12}$$

i.e.,

$$\mathbf{s} \cdot \nabla B(\mathbf{r}, \mathbf{s}, \nu) = 0. \tag{A13}$$

Equation (A13) is precisely the free-space form of the radiative transfer equation, which implies the more explicit propagation law, Eq. (2.6).

APPENDIX B: DERIVATIONS OF EQS. (3.10) AND (3.11)

On substituting Eq. (3.9) into Eq. (3.2), we find that

$$\begin{aligned}\mathscr{B}_\Omega(\mathbf{r}, \mathbf{s}, \nu) = &\frac{k^2}{(2\pi)^4}s_z\int \mathrm{d}^2u\int \mathrm{d}^2\rho'\int \mathrm{d}^2\rho''\tilde{\Omega}(\mathbf{u}, k\boldsymbol{\rho}'')\\ &\times \exp\{-i[\mathbf{u}\cdot(\boldsymbol{\rho} - \boldsymbol{\rho}') + k\mathbf{s}_\perp \cdot \boldsymbol{\rho}'']\}\\ &\times \int \mathrm{d}^2\rho_0'\int \mathrm{d}^2\rho_0'' G[\mathbf{r}' - \boldsymbol{\rho}_0' + (1/2)(\boldsymbol{\rho}'' - \boldsymbol{\rho}_0''), \nu]\\ &\times G^*[\mathbf{r}' - \boldsymbol{\rho}_0' - (1/2)(\boldsymbol{\rho}'' - \boldsymbol{\rho}_0''), \nu]\\ &\times W^{(0)}[\boldsymbol{\rho}_0' + (1/2)\boldsymbol{\rho}_0'', \boldsymbol{\rho}_0' - (1/2)\boldsymbol{\rho}_0'', \nu].\end{aligned} \tag{B1}$$

In order to simplify this expression, it is convenient to make the change of variables

$$\boldsymbol{\delta} = \boldsymbol{\rho}'' - \boldsymbol{\rho}_0'', \tag{B2}$$

with $\boldsymbol{\rho}_0''$ fixed. Formula (B1) can then be rewritten in the form

$$\begin{aligned}\mathscr{B}_\Omega(\mathbf{r}, \mathbf{s}, \nu) = &\frac{k^2}{(2\pi)^4}s_z\iiint M(\mathbf{u}, \mathbf{r}', \boldsymbol{\rho}_0'', \nu)\\ &\times \exp\{-i[\mathbf{u}\cdot(\boldsymbol{\rho} - \boldsymbol{\rho}') + k\mathbf{s}_\perp \cdot \boldsymbol{\rho}_0'']\}\\ &\times \mathrm{d}^2u\,\mathrm{d}^2\rho'\,\mathrm{d}^2\rho_0'',\end{aligned} \tag{B3}$$

where

$$\begin{aligned}M(\mathbf{u}, \mathbf{r}', \boldsymbol{\rho}_0'', \nu) = &\iint \tilde{\Omega}[\mathbf{u}, k(\boldsymbol{\delta} + \boldsymbol{\rho}_0'')]\exp(-ik\mathbf{s}_\perp \cdot \boldsymbol{\delta})\\ &\times G[\mathbf{r}' - \boldsymbol{\rho}_0' + (1/2)\boldsymbol{\delta}, \nu]\\ &\times G^*[\mathbf{r}' - \boldsymbol{\rho}_0', - (1/2)\boldsymbol{\delta}, \nu]\\ &\times W^{(0)}[\boldsymbol{\rho}_0' + (1/2)\boldsymbol{\rho}_0'', \boldsymbol{\rho}_0' - (1/2)\boldsymbol{\rho}_0'', \nu]\mathrm{d}^2\rho_0'\mathrm{d}^2\delta.\end{aligned} \tag{B4}$$

Formula (B3) is Eq. (3.10) of the text.

For our purposes, it will be useful to express Eq. (B4) in a slightly different form. Let us make the change of variables

$$\boldsymbol{\rho}_a = \boldsymbol{\rho}_0' - (1/2)\delta, \tag{B5a}$$

$$\boldsymbol{\rho}_b = \boldsymbol{\rho}_0' + (1/2)\delta. \tag{B5b}$$

The inverse of this transformation is

$$\boldsymbol{\rho}_0' = (1/2)(\boldsymbol{\rho}_a + \boldsymbol{\rho}_b), \tag{B6a}$$

$$\boldsymbol{\delta} = \boldsymbol{\rho}_b - \boldsymbol{\rho}_a. \tag{B6b}$$

In terms of these new variables, Eq. (B4) can be rewritten as

$$\begin{aligned}M(\mathbf{u}, \mathbf{r}', \boldsymbol{\rho}_0'', \nu) = &\iint \tilde{\Omega}[\mathbf{u}, k(\boldsymbol{\rho}_b - \boldsymbol{\rho}_a + \boldsymbol{\rho}_0'')]\\ &\times \exp[-ik\mathbf{s}_\perp \cdot (\boldsymbol{\rho}_b - \boldsymbol{\rho}_a)]\\ &\times G(\mathbf{r}' - \boldsymbol{\rho}_a, \nu)G^*(\mathbf{r}' - \boldsymbol{\rho}_b, \nu)\\ &\times W^{(0)}[(1/2)(\boldsymbol{\rho}_a + \boldsymbol{\rho}_b + \boldsymbol{\rho}_0''), (1/2)(\boldsymbol{\rho}_a + \boldsymbol{\rho}_b - \boldsymbol{\rho}_0''), \nu]\mathrm{d}^2\rho_a\mathrm{d}^2\rho_b,\end{aligned} \tag{B7}$$

which is Eq. (3.11).

ACKNOWLEDGMENTS

Some of the research described in this paper was supported by the U.S. Department of Energy under grant DE-FG02-90ER14119.

We are obliged to WeiJian Wang for help in checking our calculations.

E. Wolf is also with The Institute of Optics, University of Rochester.

REFERENCES AND NOTES

1. A. Walther, J. Opt. Soc. Am. **58,** 1256 (1968).
2. A. Walther, J. Opt. Soc. Am. **63,** 1622 (1973).
3. E. W. Marchand and E. Wolf, J. Opt. Soc. Am. **64,** 1219 (1974).
4. E. Wolf, J. Opt. Soc. Am. **68,** 6 (1978).
5. G. S. Agarwal, J. T. Foley, and E. Wolf, Opt. Commun. **62,** 67 (1987).
6. W. H. Carter and E. Wolf, J. Opt. Soc. Am. **67,** 785 (1977).
7. E. Wolf, J. Opt. Soc. Am. **72,** 343 (1982).
8. Equation (2.5) is equivalent to requiring that $B(\mathbf{r}, \mathbf{s}, \nu)$ obey the free-space equation of radiative transfer: $\mathbf{s} \cdot \nabla B(\mathbf{r}, \mathbf{s}, \nu) = 0$.

9. $\mathcal{B}_W^{(0)}(\boldsymbol{\rho}, \mathbf{s}, \nu)$ was introduced in Ref. 1. $\mathcal{B}_{AS}^{(0)}(\boldsymbol{\rho}, \mathbf{s}, \nu)$ is the complex version of the generalized radiance function introduced in Ref. 2; in that paper the real part of $\mathcal{B}_{AS}^{(0)}(\boldsymbol{\rho}, \mathbf{s}, \nu)$ was used.
10. A. T. Friberg, "Phase-space methods for partially coherent wavefields," in *Optics in Four Dimensions—1980,* M. Machado and L. M. Narducci, eds., AIP Conf. Proc. **65,** 313 (1981).
11. A. T. Friberg, J. Opt. Soc. Am. **69,** 192 (1979).
12. J. T. Foley and E. Wolf, Opt. Commun. **55,** 236 (1985).
13. K. Kim and E. Wolf, J. Opt. Soc. Am. A **4,** 1233 (1987).
14. G. S. Agarwal and E. Wolf, Phys. Rev. D **2,** 2161, 2187, 2206 (1970).
15. E. Wolf, J. Opt. Soc. Am. A **3,** 76 (1986).
16. Lord Rayleigh, *The Theory of Sound* (reprinted by Dover, New York, 1945), Vol. II; Sec. 278 [with a modification appropriate to the time dependence $\exp(-2\pi i\nu t)$ used in the present paper].
17. M. Born and E. Wolf, *Principles of Optics,* 6th ed. (Pergamon, Oxford, 1980).
18. C. J. Bouwkamp, Rep. Prog. Phys. (London Phys. Soc.) **17,** 35 (1954).
19. P. A. M. Dirac, *The Principles of Quantum Mechanics,* 4th ed. (Clarendon, Oxford, 1958), Sec. 21.

Section 6 – Articles of Historical Interest

This short Section contains reprints of three papers based on lectures which can be loosely described as dealing with the history of physics. However, since I am not an historian of science, they should be regarded as having been written by an interested amateur.

The three papers are concerned with Einstein's researches on the nature of light (6.1), recollections of Max Born (6.2) and an account of the life and work of Christiaan Huygens (6.3).

Other accounts of some historical interest, based on previously unpublished lectures, are included in the next Section.

Section 6 – Articles of Historical Interest

6.1 "Einstein's Researches on the Nature of Light", *Optics News* **5**, 24–39 (1979). 536

6.2 "Recollections of Max Born", *Optics News* **9**, 10–16 (1983). 552

6.3 "The Life and Work of Christiaan Huygens", **Proc. Symposium Huygens' Principle 1690–1990: Theory and Applications**, H. Blok, H.A. Ferwerda and H.K. Kuiken, eds. (North-Holland, Amsterdam, 1992), 3–17. 559

Reprinted from

OPTICS NEWS

Volume 5, No. 1 Winter 1979, pages 24–39

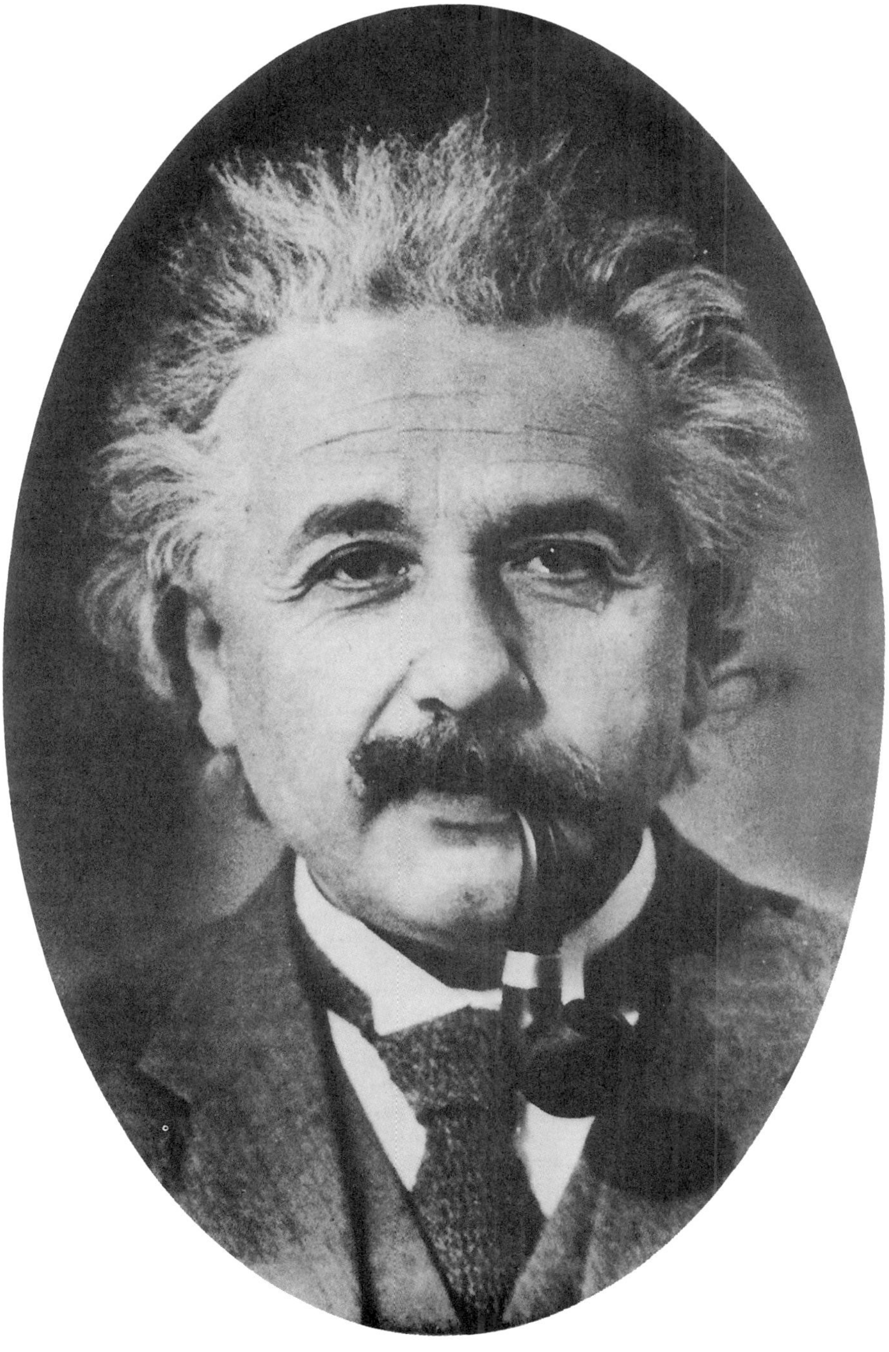

Einstein's
Researches on the Nature of Light*

EMIL WOLF

The fundamental contributions made by Albert Einstein toward our present-day understanding of the nature of light are reviewed. In order to place this work into proper perspective, the theories of Hooke, Huygens, Newton, Fresnel, and Maxwell are first summarized, and a brief account is given of the researches of Kirchhoff, Wien, Rayleigh, Jeans, and Planck on equilibrium radiation laws. The main publications of Einstein that led to the formulation of the quantum theory of radiation are then discussed. These include his investigations on the particle aspects of radiation, on the wave-particle duality, on the elementary processes of energy exchange between gas molecules and a radiation field, and on photon statistics.

"For the rest of my life I will reflect on what light is!"

A. Einstein, ca 1917

INTRODUCTION

In our time of ever-increasing specialization, there is a tendency to concern ourselves with relatively narrow scientific problems. The broad foundations of our present-day scientific knowledge and its historical development tend to be forgotten too often. This is an unfortunate trend, not only because our horizon becomes rather limited and our perspective somewhat distorted, but also because there are many valuable lessons to be learned in looking back over the years during which the basic concepts and the fundamental laws of a particular scientific discipline were first formulated.

To scientists and nonscientists alike, the name Albert Einstein is associated with a theory that has profoundly revolutionized man's ideas of space and time. His theory of relativity implied as basic a change in our conception of the universe as that which was brought about by Newton's theory of universal gravitation or Kepler's theory of the planetary system. For this work alone, Einstein will certainly always be remembered as one of the greatest geniuses of all times. What is not so well appreciated—even by many physicists—is that, quite apart from the theory of relativity, Einstein also made most basic contributions to our understanding of the nature of light and radiation in general. The present article is concerned with this aspect of Einstein's work. To place this work into a proper perspective, it seems appropriate first to recall certain milestones from the history of optics and radiation theory.

EARLY THEORIES CONCERNING THE NATURE OF RADIATION[1]

In the seventeenth century, two theories were put forward about the nature of light: the *wave* theory, whose chief proponents were Robert Hooke and Christian Huygens, and the *corpuscular* (or *emission*) theory, put forward by Isaac Newton.

According to the wave theory, light consists of rapid vibrations that are propagated in an elastic ether in a somewhat similar manner as a disturbance is propagated on the surface of water. According to the corpuscular theory, on the other hand, light is propagated from a luminous body by minute particles.

AIP Niels Bohr Library

Christian Huygens (1629–1695)

The wave theory, as then formulated, appeared to be incapable of explaining the phenomenon of polarization, discovered by Huygens himself in studying the refraction of light by crystals. Newton, on the other hand, was able to account for polarization on the basis of his corpuscular theory. It was largely for this reason that the wave theory was rejected for over a century in favor of the corpuscular theory.

In 1801 Thomas Young discovered the principle of interference of light. Seventeen years later Augustin Fresnel showed in a celebrated memoir that, by combining Young's principle of interference with a basic postulate of Huygens's theory, one is led to a wave theory of light that explains diffraction of light, a phenomenon that was not comprehensible on the basis of Newton's corpuscular theory. Within a few years after the publication of Fresnel's memoir and after experimental demonstrations of certain unsuspected predictions of his theory, Fresnel's wave theory became generally accepted and Newton's corpuscular theory fell into oblivion.

The formulation of the wave theory of light culminated in the work of

The author is with the Department of Physics and Astronomy, University of Rochester, Rochester, New York 14627

AIP Niels Bohr Library

Isaac Newton (1642–1727)

James Clerk Maxwell, who succeeded in 1865 in embodying all the laws of electricity and magnetism then known into a celebrated set of differential equations—now called *Maxwell's equations*. One of the consequences of these equations was a prediction that time-dependent electric and magnetic effects are transmitted from one region of space to another by means of waves—known now as electromagnetic waves. The speed of these waves in free space could be calculated from the results of purely electrical measurements, and it turned out to be of the order of magnitude of the speed of light, as then known from other experiments. This led Maxwell to conjecture that light waves are electromagnetic waves. In 1888 Heinrich Hertz demonstrated the existence of electromagnetic waves experimentally.

Augustin Fresnel (1788–1827)

AIP Niels Bohr Library

Thomas Young (1773–1829)

We may thus summarize this part of our brief historical introduction by saying that, toward the end of the nineteenth century, it appeared firmly established that light is an electromagnetic wave phenomenon.

Rather independently of the developments just mentioned, investigations were carried out concerning thermal or heat radiation, which eventually also turned out to be of fundamental importance for elucidating the nature of light.

In the period 1814–1817 Joseph Fraunhofer discovered dark lines in the solar spectrum, which have since been named after him. On the basis of experiments by Robert Bunsen and

AIP Niels Bohr Library

James Clerk Maxwell (1831–1879)

Heinrich Hertz (1857–1894)

Gustav Kirchhoff, they were interpreted around 1860 as absorption lines of certain gases in the solar atmosphere. In the course of his investigations of the solar spectrum, Kirchhoff derived from thermodynamics a number of fundamental results relating to radiation in thermal equilibrium with bodies at a fixed temperature, say T. Even under equilibrium conditions, when the system is thermally insulated from its surroundings, the bodies will emit and absorb radiation, or, as we say these days, there will be interaction between matter and the radiation field. The capacity of a body to emit and absorb radiation at some fixed frequency ν may be characterized by certain quantities known as the emission coefficient, ϵ_ν, and the absorption coefficient, α_ν. One of the laws, which Kirchhoff derived in 1859,

Table 1: Some of the Main Early Contributions Toward Elucidating the Nature of Light

Robert Hooke	1665	Wave theory
Christian Huygens	1678	
Isaac Newton	1671	Corpuscular theory
Thomas Young	1801	Interference of light
Augustin Fresnel	1818	Modernized scalar wave theory
Clerk Maxwell	1865	Electromagnetic wave theory
Heinrich Hertz	1888	Detection of electromagnetic waves

AIP Niels Bohr Library

Joseph Fraunhofer (1787–1826)

asserts that, under equilibrium conditions, the ratio of the emission and the absorption coefficients is independent of the nature of the bodies that interact with the radiation field, i.e., their ratio is some function $K_\nu(T)$ that depends only on the frequency and the temperature:

$$\epsilon_\nu/\alpha_\nu = K_\nu(T). \qquad (1)$$

This is clearly an important law, since it brings into evidence a universal function associated with radiation under equilibrium conditions.

In the special case in which the body is such that it absorbs all the radiation that falls on it, the absorption coefficient α_ν has the value unity. Such a body is said to be a *blackbody*, and the radiation emitted by it is said to be *blackbody radiation.*

It follows at once from Kirchhoff's law (1) that, for a blackbody,

$$(\epsilon_\nu)_{\text{blackbody}} = K_\nu(T), \qquad (2)$$

i.e., its emission coefficient, considered as a function of frequency and temperature, is precisely the universal function $K_\nu(T)$. One can show that $K_\nu(T)$ is related to the energy density $u_\nu(T)$ of blackbody radiation by the formula

$$u_\nu(T) = \frac{8\pi}{c} K_\nu(T), \qquad (3)$$

c being the speed of light in vacuum.

Because of the universality of the function $K_\nu(T)$, that is, because of its independence of the nature of the

AIP Niels Bohr Library

Gustav Kirchhoff (1824–1887)

bodies in equilibrium with the radiation field, a great deal of effort has gone into determining its form. The four decades that followed Kirchhoff's discovery of the law that we just discussed witnessed many unsuccessful attempts to derive the correct form of the function $K_\nu(T)$ or, what according to Eq. (3) amounts to the same thing, the spectrum of blackbody radiation as a function of frequency and temperature. Table 2 lists three of the most important formulas proposed for the function $K_\nu(T)$ in this period.

The first formula, derived by W. Wien in 1893 on the basis of thermodynamics, is known as *Wien's displacement law.* According to this law, the function $K_\nu(T)$ is equal to the product of the third power of the frequency and some function $F(\nu/T)$ of the ratio of the frequency ν to the temperature T. However, thermodynamics did not specify the form of this function.

Using a somewhat heuristic argument, Wien proposed three years later an explicit form for the function

Table 2 The Formulas of Wien and of Rayleigh and Jeans for the Specific Intensity $K_\nu(T) = (c/8\pi)u_\nu(T)$ of Blackbody Radiation

Wien's displacement law (1893)	$\nu^3 F(\nu/T)$
Wien's radiation law (1896)	$\alpha\nu^3 e^{-\beta(\nu/T)}$
Rayleigh–Jeans law (1900, 1905)	$(\nu/c)^2 kT$

Wilhelm Wien (1864–1928)

$F(\nu/T)$, leading to the so-called *Wien's radiation law.* It is listed in the second row in Table 2. The values of the two constants α and β that appear in this radiation law did not follow from Wien's argument and had to be determined empirically.

Subsequently, in 1900, Lord Rayleigh showed that the application of the equipartition theorem of statistical mechanics to electromagnetic vibrations in a cavity leads to quite a different formula. When later corrected by J. H. Jeans for a simple error, it has the form shown in the third row of Table 2 (where k is the Boltzmann constant and c is the speed of light *in vacuo*), known as the *Rayleigh–Jeans law.*

We note that both Wien's radiation law and the Rayleigh–Jeans law are of the form required by Wien's displacement law. However, careful measurements carried out around the turn of the century, notably by O. Lummer and E. Pringsheim and by H. Rubens and F. Kurlbaum, revealed that neither of them completely agrees with the experimentally determined energy distribution in the spectrum of blackbody radiation. It turned out that Wien's radiation law was a good approximation for sufficiently high frequencies and low-enough temperatures, whereas the Rayleigh–Jeans law was a good approximation in the other extreme case of low frequencies and high temperatures. The other law, Wien's displacement law, is, of course,

Lord Rayleigh (1842–1919)

Portrait by Sir George Reid. Published with permission of the Royal Society, London.

incomplete because it contains the unknown function $F(\nu/T)$.

This, then, was the situation in the closing years of the last century. Both thermodynamics and the theory of electricity and magnetism failed to predict the energy distribution in the spectrum of blackbody radiation.

The correct blackbody radiation law—one that is in complete agreement with experiment—was discovered by Max Planck in 1900, just a few months after the publication of Rayleigh's paper; and an assumption that went into its derivation marked the birth of one of the greatest revolutions that took place in science—the birth of the quantum theory.

Planck's radiation law may be written in the form

$$u_\nu(T) \equiv \frac{8\pi}{c} K_\nu(T) = \frac{8\pi h\nu^3}{c^3} \frac{1}{e^{h\nu/kT} - 1}, \quad (4)$$

where h and k are constants, the first of which is now known as *Planck's constant* and the other is Boltzmann's constant. An essential ingredient of Planck's original derivation of his radiation law was a fortunate guess relating to the entropy of a linear oscillator, which interacts with radiation. This guess was, in fact, equivalent to an interpolation between certain thermodynamic expressions appropriate to Wien's radiation law and to

James H. Jeans (1877–1946)

From Sir James Jeans by E.A. Milne (Cambridge University Press, 1962)

the Rayleigh–Jeans law. Having quickly realized that his new law was in an extremely good agreement with experiment, Planck set about to find a more satisfactory theoretical basis for it. Within a few weeks he succeeded, and he outlined the essential features of his new derivation at a meeting of the

Max Planck (1858–1947)

AIP Niels Bohr Library

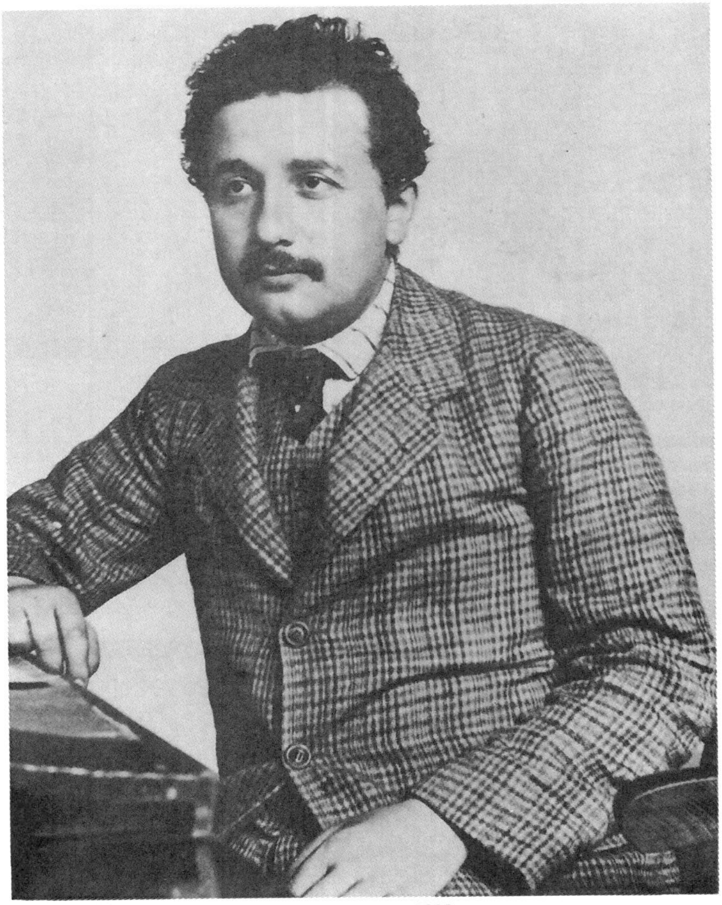

Albert Einstein in the Swiss Patent Office in Bern, ca 1905.

Einstein Archives

German Physical Society in Berlin on December 14, 1900.

According to Kirchhoff's law and formula (3), the equilibrium energy density $u_\nu(T)$ is independent of the exact nature of the bodies that interact with the radiation field. Planck chose one of the simplest possible models for these bodies, namely, harmonic oscillators. He then showed that the radiation law that he discovered a few weeks earlier may be derived in a systematic way from the laws of electromagnetic theory and thermodynamics if one makes the following assumption: An oscillator, vibrating with frequency ν, can take on only one of the energy values E_0, $2E_0, 3E_0, \ldots$, where $E_0 = h\nu$. Thus, in deriving the correct blackbody radiation law, Planck found it necessary to introduce the notion of a *quantum of energy* $E_0 = h\nu$, which represents the smallest amount of energy that an oscillator can emit or absorb. The need for introducing such a quantum of energy was in flat contradiction to Maxwell's electromagnetic theory, which places no restriction on the amount of energy that an oscillator can emit or absorb.

It may readily be shown that, when $\nu/T >> 1$ (i.e., in the limit of high frequencies and low temperatures), Planck's radiation law (4) reduces to

$$u_\nu(T) \equiv \frac{8\pi}{c} K_\nu(T) = \frac{8\pi}{c} \alpha \nu^3 e^{-\beta(\nu/T)}, \quad (5)$$

where

$$\alpha = \frac{h}{c^2}, \qquad \beta = \frac{h}{k}. \quad (6)$$

The formula (5) is just Wien's radiation law, but we now also obtain explicit expressions for the constants α and β. When $\nu/T << 1$ (i.e., in the limit of low frequencies and high temperatures), Eq. (4) reduces to

$$u_\nu(T) \equiv \frac{8\pi}{c} K_\nu(T) = \frac{8\pi}{c}\left(\frac{\nu}{c}\right)^2 kT, \quad (7)$$

which is nothing but the Rayleigh–Jeans law.

PARTICLE ASPECTS OF RADIATION

Even though Planck's introduction of the concept of an energy quantum led eventually to one of the greatest scientific revolutions of all times, his theory did not at first attract much attention. One of the first scientists who clearly recognized that Planck's discovery initiated a new era in physics was a young man, Albert Einstein, who around that time—in 1902—at the age of 23 was appointed to a post at the Swiss Patent Office. His appointment carried the title "Technical Expert, Third Class." A photograph of this young man from that particular period appears on this page.

In the spring of 1905, Einstein, then quite unknown to the scientific world, wrote to a former fellow student and friend, Conrad Habicht, asking for a copy of a thesis that Habicht had recently completed. Einstein's letter contains a passage that, in English translation, reads[2]:

"But why have you not sent me your thesis? Don't you know, you wretch, that I would be one of the few fellows who would read it with interest and pleasure? I can promise you in return four works, the first of which . . . is very revolutionary. . . ."

Three of these four papers that Einstein mentioned in this letter were published in one and the same volume, Volume 17 (4th series) of the journal *Annalen der Physik*, in 1905, having all been submitted within a period of only three and a half months. As Max Born noted,[3] this is one of the most remarkable volumes in the whole scientific literature; for each of the three papers of Einstein that appear in it is acknowledged as a masterpiece and the starting point of a new branch of physics. In our present-day terminology the subjects of these three papers, in the order in which they appear, are: (1) the particle nature of radiation, (2) the theory of Brownian motion, and (3) the special theory of relativity.

We will review only the first paper, since the other two deal with topics that do not come within the scope of this article. However, it seems appropriate to mention that the second paper probably did more than any other work to convince physicists of the reality of atoms and molecules and of the fundamental role that probability theory plays in the formulation and the interpretation of the basic laws of physics; and that in the third paper Einstein took the monumental steps by which the intuitive notions of absolute space and absolute time were abolished and replaced by the concept of a four-dimensional space-time continuum. This step has had the most profound consequences for our present-day understanding of the fundamental physical laws of the universe.

The first of these papers[4] has, in translation, the title: "On a heuristic point of view concerning the creation and conversion of light." It is this paper that Einstein himself referred to as "very revolutionary" in the letter to his friend Conrad Habicht. In modern textbooks it is usually referred to as "Einstein's paper on the photoelectric effect." Actually, this paper contains appreciably more. In fact, Einstein's whole discussion of the photoelectric effect covers less than four pages; but, as in most of his writing, Einstein was able to get to the root of a problem in

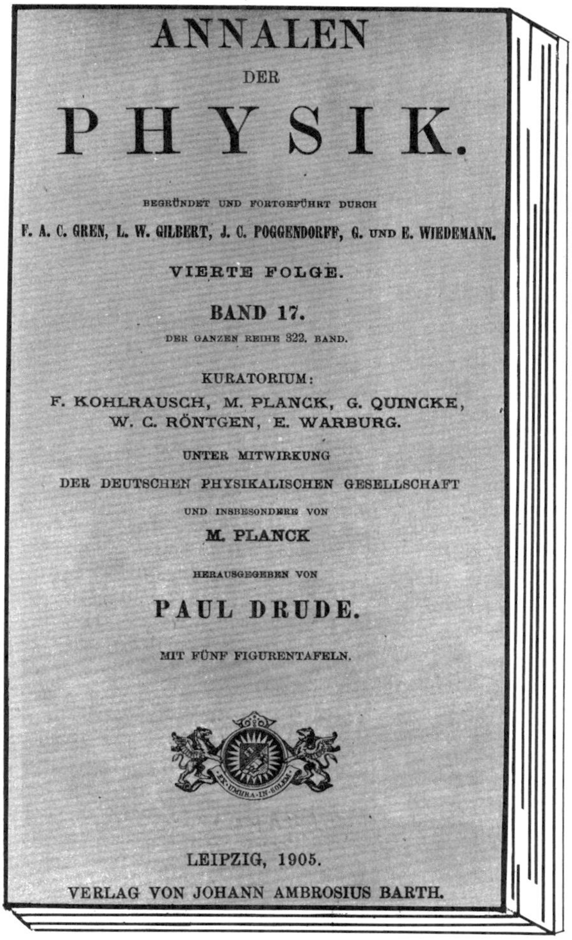

ANNALEN
DER
PHYSIK.

BEGRÜNDET UND FORTGEFÜHRT DURCH
F. A. C. GREN, L. W. GILBERT, J. C. POGGENDORFF, G. UND E. WIEDEMANN.

VIERTE FOLGE.

BAND 17.
DER GANZEN REIHE 322. BAND.

KURATORIUM:
F. KOHLRAUSCH, M. PLANCK, G. QUINCKE,
W. C. RÖNTGEN, E. WARBURG.

UNTER MITWIRKUNG
DER DEUTSCHEN PHYSIKALISCHEN GESELLSCHAFT
UND INSBESONDERE VON
M. PLANCK

HERAUSGEGEBEN VON
PAUL DRUDE.

MIT FÜNF FIGURENTAFELN.

LEIPZIG, 1905.
VERLAG VON JOHANN AMBROSIUS BARTH.

Title page of Vol. 17 (4th series) of Annalen der Physik, 1905, containing Einstein's three epoch-making papers—on the particle nature of radiation, the theory of Brownian motion, and the special theory of relativity.

a few lines, with simple language that was remarkably free of complicated mathematics.

Essentially what Einstein did in this paper was to put forward a great deal of evidence that not only do the processes of emission and absorption of radiation take place in discrete amounts of energy (as appears to have been established by Planck) but that radiation itself behaves under certain circumstances as if it consisted of a collection of particles, which in modern language are called *photons*. Thus in this paper Einstein reintroduced a corpuscular theory of light—first advocated by Newton in the 17th century. As we already noted, the corpuscular theory was completely discredited by Fresnel's wave theory, formulated almost 90 years before the publication of Einstein's paper; and the wave theory of light itself appeared to have been put on firm foundation by Maxwell about 40 years before the appearance of Einstein's paper.

In the introduction to his paper, Einstein discusses the success of the wave theory of light, which deals with continuous functions in space. Then Einstein goes on to say that neverthe-

Phillip E. A. Lenard (1862–1947)

Robert A. Millikan (1868–1953)

AIP Niels Bohr Library

less it is possible that this theory will lead to a contradiction with experience if it is applied to the phenomena of generation and conversion of light. He then continues as follows:

"In fact it seems to be that the observations on blackbody radiation, photoluminescence, the production of cathode rays by ultraviolet light, and other phenomena involving the emission or conversion of light, can be better understood on the assumption that the energy of light is distributed discontinuously in space. According to the assumption considered here, when a light ray starting from a point is propagated, the energy is not continuously distributed over an ever-increasing volume, but it consists of a finite number of energy quanta, localized in space, which move without being divided and which can be absorbed or emitted only as a whole."

We will now present the essence of some of the examples given by Einstein to support these views.

Suppose that we throw particles into a box of volume V. Let us select in the box some subregion of volume ΔV (not necessarily small). If we throw one particle into the box, the probability that it will land in the chosen subregion is clearly

$$p(1) = \frac{\Delta V}{V}. \qquad (8)$$

If we repeat this procedure n times, i.e., if we throw n particles into the box, one at a time, then, by an elementary rule of probability theory, the probability that all the n particles will end up in the subregion is

$$p(n) = \left(\frac{\Delta V}{V}\right)^n. \qquad (9)$$

Now instead of a system of particles, let us suppose that the box contains radiation under equilibrium conditions at temperature T. The energy density of the radiation field is given by Planck's law, but we know that for sufficiently high frequencies or sufficiently low temperatures it may be approximated by Wien's radiation law. Hence under these conditions the total energy of the radiation in the box is given by the formula

$$E = \frac{8\pi}{c}\alpha V \nu^3 e^{-\beta(\nu/T)}, \qquad (10)$$

where α and β are the constants given by Eq. (6).

We have already noted that, even under equilibrium conditions, there are fluctuations in the radiation field and hence in the course of time the energy is being redistributed throughout the box. Consequently, in the chosen subregion the energy will have sometimes a larger and sometimes a smaller value than its mean value. In fact there is some probability that at a given instant *all* the energy E will be concentrated in this subregion. By using Wien's radiation law [Eq. (10)] and only some general principles of thermodynamics, Einstein showed that this probability is given by

$$p(\text{all } E \in \Delta V) = \left(\frac{\Delta V}{V}\right)^{E/h\nu}. \qquad (11)$$

Comparison of this result with Eq. (9) shows that this probability is the same as if the radiation field consisted of n particles, where

$$n = \frac{E}{h\nu}, \qquad (12)$$

i.e., of n particles, each carrying energy $h\nu$. In Einstein's (translated) words:

"Monochromatic radiation of low density (within the range of validity of Wien's radiation law) behaves, in a thermodynamic sense, as if it consisted of mutually independent quanta of energy of magnitude[5] $h\nu$."

Another example that Einstein gave in this paper in support of his view regarding the corpuscular nature of radiation was, as already mentioned, the *photoelectric effect.* This is the phenomenon of ejection of electrons from a metal when electromagnetic radiation of short enough wavelength impinges on the metal surface. The effect was discovered in 1887 by Heinrich Hertz in the course of experiments referred to earlier, which played a decisive role in confirming the correctness of Maxwell's electromagnetic theory of light. In retrospect, there is some irony in this situation, since later, when the photoelectric effect was studied quantitatively, it was not possible to reconcile it with Maxwell's electromagnetic theory.

Some puzzling facts about the photoelectric effect were revealed largely by systematic experiments of P. Lenard, carried out in the period 1899–1902. Lenard found that the energy of an ejected electron is independent of the intensity of the light that illuminates the metal surface, but that it depends on the frequency of the light; and also that, when the light intensity is increased, there is an increase in the *number* of the ejected electrons, but not in their energies. The difficulties of trying to explain these observations by the wave theory of light become apparent when we recall that, according to the wave theory, the energy that is carried by a light wave is measured by the light intensity. Hence, as the intensity is increased, more energy becomes available to be imparted to the electrons, and hence their energy should also increase, contrary to Lenard's observations.

Einstein considered Lenard's observations as a clear demonstration of the corpuscular nature of light: If light consists of energy quanta, each of amount $h\nu$, the quanta penetrate into the surface layer of the metal and their energy is at least partly transformed into the kinetic energy of the electrons. The simplest situation is that in which a light quantum transfers all its energy to a single electron and the energy is sufficient to set the electron free. On leaving the metal, the electron loses part of this energy because some work (W, say) is required to remove it from the metal. Einstein was thus led to the following formula for the maximum kinetic energy of an ejected electron:

$$(E_{\mathrm{kin}})_{\mathrm{max}} = h\nu - W. \quad (13)$$

This formula, known now as *Einstein's photoelectric equation*, at once explains Lenard's observations.

We note that Einstein's photoelectric equation predicts that the maximum energy of the ejected photoelectrons is a *linear function* of frequency, whose slope is precisely Planck's constant. Hence the measurement of the dependence of the maximum electron energy on the frequency could be used to determine Planck's constant. In 1905, when Einstein put forward his photoelectric equation, quantitative studies of the photoelectric effect were in their infancy. It took nearly a decade of difficult experimentation before Einstein's equation could be fully tested. It was largely confirmed by the work of R. A. Millikan, who began by completely disbelieving Einstein's theory. In a paper[6] published in 1949 on the occasion of Einstein's seventieth birthday, this is what Millikan said: "I spent ten years of my life testing that 1905 equation of Einstein's and, contrary to all my expectations, I was compelled in 1915 to assert its unambiguous experimental verification in spite of its unreasonableness since it seemed to violate everything that we knew about the interference of light."

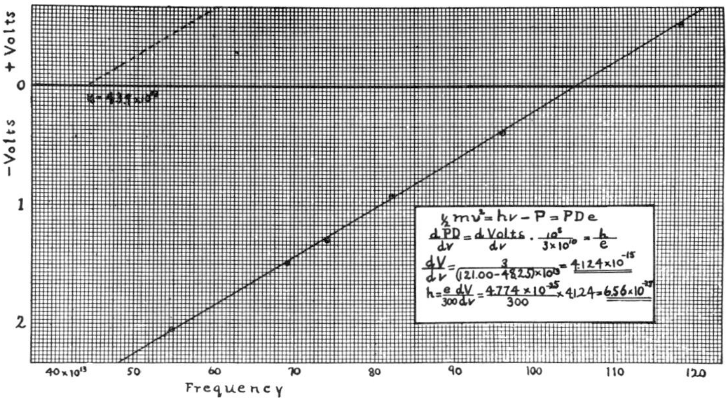

Fig. 1. The dependence of the photocurrent potential on the frequency of the incident light in experiments on the photoelectric effect [From R. A. Millikan, Phys. Rev. 7, 373 (1916)].

In Fig. 1 some of Millikan's experimental data are reproduced. They exhibit the linear relationship predicted by Einstein's photoelectric equation, the slope of the line being indeed the value of Planck's constant.

THE WAVE-PARTICLE DUALITY

The 1905 paper of Einstein, which we just discussed, clearly revealed the failure of classical physics to account for certain observed phenomena involving radiation. This paper showed a need for even more drastic changes than those brought about by Planck's assumption of quantized energy of the emitting and absorbing oscillators. Einstein's analysis indicated that not only do emission and absorption of energy take place in discrete energy quanta but that the radiation field itself behaves, in certain situations, as if it consisted of such corpuscles of energy. The situation was puzzling, for on one hand there were phenomena such as interference and diffraction of light, which seemed clearly to demonstrate the wave nature of radiation; but there were other phenomena such as the photoelectric effect, which reveal, as Einstein showed, that radiation has corpuscular nature.

In spite of the clarity and the simplicity of Einstein's arguments, his views about the particle structure of radiation were strongly opposed at that time—and for a long time afterward—by many eminent physicists, including Max Planck. Undeterred by the opposition, Einstein continued to explore the consequences of his corpuscular theory and was probing more deeply into the nature of radiation.

In 1909, four years after his "photoelectric paper," Einstein published a paper[7] with the title "On the present status of the problem of radiation," which became another milestone in the development of physics. In this publication Einstein showed, again by characteristically simple arguments typical of so much of his work, that Planck's radiation law itself implies that the radiation field exhibits not only wave features but also corpuscular features. This result was the first clear indication of the so-called *wave-particle duality* that many years later became an accepted feature of modern quantum physics.

The essence of Einstein's argument may be described as follows: Consider again blackbody radiation at temperature T in a thermally insulated cavity of volume V. Let us fix our attention on a subregion of volume ΔV inside the cavity. As we noted earlier, there will be energy fluctuations inside this subregion, caused by radiation moving in and out of it. Einstein inquired about the magnitude of these fluctuations.

Let E be the amount of energy in

AIP Niels Bohr Library

Back row: Einstein, Ehrenfest, de Sitter; front row: Eddington, Lorentz.

the subregion, at some fixed instant of time, in a frequency range $(\nu, \nu + \mathrm{d}\nu)$, say. The simplest measure of the energy fluctuations is the variance

$$\overline{(\Delta E)^2} = \overline{(E - \overline{E})^2}, \qquad (14)$$

where the bar denotes the statistical average.

Einstein first showed on the basis of thermodynamics that the variance may be expressed by a formula that can be written in the form

$$\overline{(\Delta E)^2} = kT^2 \frac{\partial \overline{E}}{\partial T}, \qquad (15)$$

where, as before, k is the Boltzmann constant. Now according to Planck's radiation law [Eq. (4)], the mean energy in the subregion of volume ΔV is given by the formula

$$\overline{E} = Z \frac{h\nu}{e^{h\nu/kT} - 1} \qquad (16)$$

where

$$Z = \frac{8\pi \Delta V}{c^3} \nu^2 \mathrm{d}\nu. \qquad (17)$$

On substituting from Eq. (16) into Eq. (15), Einstein obtained the following expression for the variance:

$$\overline{(\Delta E)^2} = h\nu \overline{E} + \frac{1}{Z} \overline{E}^2. \qquad (18)$$

Equation (18) is known today as *Einstein's fluctuation formula for blackbody radiation.*

Einstein drew some far-reaching consequences from this formula. He argued as follows: If the radiation field inside the cavity consisted of purely electromagnetic waves, as most scientists of that time believed, one could explain the energy fluctuations in the following way: The radiation field could be represented as a mixture of plane waves (normal modes) with different amplitudes, different phases, and different states of polarization, propagated in all possible directions. The fluctuations of the energy in the subregion ΔV could then be considered to be a manifestation of short-term interference effects between the different plane waves. Einstein indicated, by means of a simple dimensional argument that covers only a few lines, that the energy fluctuations produced in this manner would be given by just the second term on the right-hand side of his fluctuation formula, a result that may be expressed in the form

$$\overline{(\Delta E)^2_{\text{waves}}} = \frac{1}{Z} \overline{E}^2. \qquad (19)$$

Some years later H. A. Lorentz[8] verified this result by long explicit calculations (which cover more than six pages in print).

Einstein argued now as follows: Since his fluctuation formula, (18), derived from Planck's law, contains more than just the term proportional to $\overline{E}^2$ (which alone is comprehensible on the basis of the wave theory), something else must be going on apart from ordinary wave propagation. Suppose, Einstein argued, that the radiation consists of light quanta, each of energy $h\nu$, which behave like independent classical particles. If, at some given instant, there were n such quanta in the subregion ΔV at some fixed time, then elementary statistics (involving the Poisson distribution) gives the following expression for the variance $\overline{(\Delta n)^2} \equiv \overline{(n - \overline{n})^2}$ of their number n:

$$\overline{(\Delta n)^2} = \overline{n}. \qquad (20)$$

The total energy carried by the n quanta is

$$E = nh\nu, \qquad (21)$$

and the variance of this energy [which we denote by $\overline{(\Delta E)^2_{\text{particles}}}$] is therefore

$$\overline{(\Delta E)^2_{\text{particles}}} = (h\nu)^2 \, \overline{(\Delta n)^2}. \qquad (22)$$

Now, according to Eqs. (20) and (21), we have $\overline{(\Delta n)^2} = \overline{E}/h\nu$ and, on using this expression in Eq. (22), we obtain the formula

$$(\Delta E)^2_{\text{particles}} = h\nu \overline{E}, \qquad (23)$$

which is precisely the first term on the right-hand side of Eq. (18).

It follows that Einstein's fluctuation formula for blackbody radiation may be expressed in the form

$$\overline{(\Delta E)^2} = \overline{(\Delta E)^2_{\text{particles}}} + \overline{(\Delta E)^2_{\text{waves}}}. \qquad (24)$$

Thus Einstein showed that, in its statistical behavior, blackbody radiation exhibits both wave features and particle features.[9] This result is the first example of the famous *wave-particle duality* that much later (around 1926) followed in a consistent manner from modern quantum mechanics.

Radiation carries not only energy, but also momentum, which is manifested as pressure. In the same paper Einstein provided additional evidence for the wave-particle duality from an analysis of momentum fluctuations in blackbody radiation.

In a lecture that Einstein delivered in the same year (1909) at a scientific meeting in Salzburg, he clearly indicated, in the following words, the conclusions to which his investigations on the structure of radiation had led him so far[10]: "It cannot be denied that there is a broad group of facts concerning radiation, which show that light has certain fundamental properties that can be much more readily understood from the standpoint of the Newtonian emission theory than from the standpoint of the wave theory. It is, therefore, my opinion that the next stage of the development of theoretical physics will bring us a theory of light which can be regarded as a kind of fusion of the wave theory and the emission theory . . . a profound change in our views of the nature and constitution of light is indispensible."

ELEMENTARY PROCESSES OF INTERACTION BETWEEN RADIATION AND MATTER

During the next few years Einstein concentrated his efforts in different directions, and he developed his general theory of relativity. But in 1917 he returned to the radiation problem once again, and he published another fundamental paper in this field. By that time much progress had been made toward the understanding of the spectrum of atomic elements, chiefly as a result of some work by Niels Bohr. Before discussing the 1917 paper of Einstein, it may be useful to review briefly some of the background.

In 1911, Ernest Rutherford put forward a model of the atom according to which the atom consists of a small, heavy, charged central nucleus surrounded by a charge distribution of the opposite sign. However, the distribution of the charge was not understood. Bohr, in a well-known paper published in 1913, assumed that the atom can exist permanently only in one of a series of states—known as *stationary states*—characterized by discrete values of the energy. When the atom emits or absorbs radiation it undergoes a transition from one such stationary state to another. Bohr assumed that, if the emitted or absorbed radiation has frequency ν, the change in the energy of the atom is

$$E' - E'' = h\nu, \qquad (25)$$

where E' and E'' are the energies of the two stationary states and h is Planck's constant. Equation (25) is known as the *Bohr frequency condition.* This basic assumption of Bohr's theory is in direct contradiction with classical electrodynamics, which does not impose any such restriction on the emitted or absorbed energy. However, the Bohr theory gave the correct values of the wavelengths of the spectral lines of the hydrogen atom and of the Rydberg constant and also elucidated some features of the periodic table of atomic elements. But the theory also contained some rather mysterious features and gave no insight whatsoever into the laws that govern the *transitions* from one allowed state of the atom to another. It was Einstein's 1917 paper,[11] entitled "On the quantum theory of radiation," that provided the first real insight into this question.

In this paper Einstein considered a gas in equilibrium with radiation at temperature T. Bohr's theory suggests that the gas molecules can only exist in certain discrete states, with energies $E_1, E_2, E_3, \ldots$ Now even under equilibrium conditions there are fluctuations, caused by the exchange of energy between the molecules and the radiation field—but always in such a way that the equilibrium is maintained. The detailed mechanism of this process was at that time completely unknown. Einstein put forward a certain model for this energy exchange and showed that Planck's radiation law follows from it in a straightforward manner. This paper contains also a number of other remarkable results, as we will soon see.

Following Einstein, let us consider transitions between two possible states of the gas molecules, with energies E_n and E_m, where $E_m > E_n$. If a molecule makes a transition from the upper to the lower state, it will emit energy of the amount $E_m - E_n$. If it undergoes a transition from the lower to the upper state, it will absorb energy of this amount. The frequency of the emitted or absorbed radiation will again be denoted by ν. Einstein now made some assumptions about the mechanism responsible for such transitions. More specifically, he assumed that a transition between the allowed energy states occurs in one of three possible elementary processes of interaction:

AIP Niels Bohr Library

Niels Bohr (1885–1962), in early 1920's

(a) *Spontaneous Emission.* It had been known from the study of radioactivity since about the early years of this century that atoms may disintegrate into atoms of other kinds and that this process is completely unaffected by external circumstances—

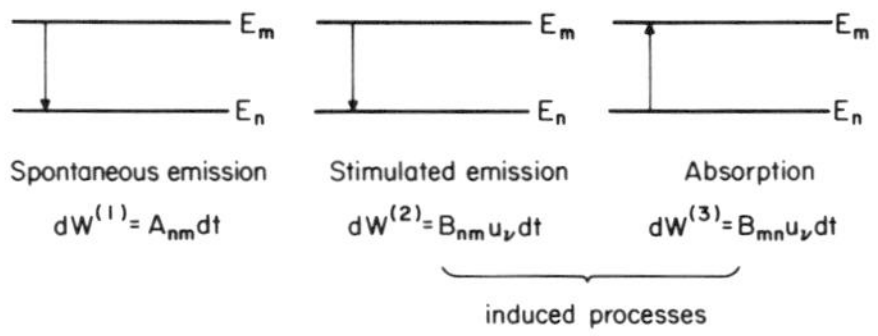

Fig. 2. The three elementary processes of interaction between gas molecules and radiation, which Einstein[11] assumed in his derivation of Planck's radiation law.

for example, the rate of disintegration is unaffected by temperature and pressure. From the analysis of experimental data on radioactive decay the following law was deduced: The probability, $dW^{(1)}$, that an atom disintegrates in a small time interval dt is simply proportional to dt, i.e.,

$$dW^{(1)} = A_{nm}\,dt, \qquad (26)$$

where A_{nm} is a constant, depending on the type of atom being considered. Einstein assumed that one of the elementary processes of energy exchange between the gas molecules and the radiation field is governed by such a law. In other words, a molecule may emit radiation into the field, and the emission is not affected by external circumstances; in particular, it is not affected by the nature of the radiation field that surrounds the molecule. This process is called *spontaneous emission.*

(b) *Induced Radiation Processes.* If an oscillator of frequency ν is placed in an electromagnetic field of the same frequency, then, according to classical electromagnetic theory, the energy of the oscillator may change as a result of transfer of energy between the oscillator and the field. Whether this change of energy is due to the emission or absorption of radiation by the oscillator depends on the relative phase difference between the oscillator and the field. These two processes are called *induced processes*, because the change in the energy of the oscillator can now be attributed to the presence of the field that surrounds it. Einstein assumed that such an energy exchange also takes place between the gas molecules and the equilibrium radiation field and, moreover, that such induced (or stimulated) emissions and absorptions are also governed by *statistical laws.* However, the elementary probabilities for the two induced processes, unlike those for spontaneous emission, can be expected to depend on the radiation field that surrounds the molecule. Specifically, Einstein postulated that the elementary probability for a molecule to undergo an *induced transition* in a small time interval of duration dt is

$$dW^{(2)} = B_{nm} u_\nu dt \qquad (27)$$

for stimulated emission and

$$dW^{(3)} = B_{mn} u_\nu dt \qquad (28)$$

for absorption, where B_{nm} and B_{mn} are constants (depending on the type of molecule being considered) and u_ν is the energy density of the radiation field at the frequency ν. These assumptions of Einstein are indicated schematically in Fig. 2.

Einstein now asked the following question: Assuming that the energy exchange takes place in this manner, what can one deduce about the energy density u_ν of the radiation field? He argued essentially as follows: Suppose that, in equilibrium, N_m of the molecules will, on the average, have energy E_m and N_n of them will have energy E_n. According to the principles of statistical mechanics,

$$\left.\begin{aligned} N_n &= C_n e^{-E_n/kT}, \\ N_m &= C_m e^{-E_m/kT}, \end{aligned}\right\} \qquad (29)$$

where k is again the Boltzmann constant and C_n and C_m are constants ("statistical weights") that are characteristic of the two quantum states of the molecules.

Since equilibrium is assumed to be maintained under the energy exchange, there must be on the average as many transitions per unit time from the upper to the lower state as from the lower to the upper state. This requirement is expressed mathematically by the equation

$$\underbrace{N_m(dW^{(1)} + dW^{(2)})}_{\text{emission}} = \underbrace{N_n\,dW^{(3)}}_{\text{absorption}}. \qquad (30)$$

On substituting into Eq. (30) for the three elementary probabilities from Eqs. (26)-(28) and for the average number of molecules from Eq. (29), one obtains the relation

$$C_m(A_{nm} + B_{nm}u_\nu)e^{-E_m/kT} = C_n B_{mn} u_\nu e^{-E_n/kT}. \qquad (31)$$

Assuming next, as Einstein did, that the energy density u_ν of the field will increase to infinity with T, one obtains from Eq. (31) the relation

$$C_m B_{nm} = C_n B_{mn}. \qquad (32)$$

If Eq. (31) is solved for u_ν and use is made of the relation (32), one obtains the following expression for the energy density of the equilibrium radiation field:

$$u_\nu = \frac{A_{nm}/B_{nm}}{e^{(E_m - E_n)/kT} - 1}. \qquad (33)$$

Now in order that this expression agree with Wien's displacement law, which we discussed earlier (See Table 2), the following relations must obviously hold:

$$\frac{A_{nm}}{B_{nm}} = \gamma\nu^3, \qquad (34)$$

$$E_m - E_n = h\nu, \qquad (35)$$

where γ and h are constants. On making use of these two relations in Eq. (33), one obtains for the energy density of the equilibrium radiation field the formula

$$u_\nu = \frac{\gamma\nu^3}{e^{h\nu/kT} - 1}. \qquad (36)$$

Equation (36) has the form of Planck's radiation law [Eq. (4)], except that the value of the constant γ has not been determined by this argument. It could be obtained, as Einstein also pointed out, by proceeding to the classical limit ($h\nu/kT << 1$); the expression (36) must then reduce to the Rayleigh-Jeans law (see Table 2). One then finds that γ has indeed the expected value,

$$\gamma = 8\pi h/c^3. \qquad (37)$$

This derivation by Einstein of the Planck radiation law is remarkable for a number of reasons:

In the first place, Einstein showed that Planck's law follows directly from his model of interaction between radiation and matter. The probabilities that Einstein assumed for each of the elementary processes taking part in the interaction are examples of what is known today as *transition probabilities* between states. In this connection we recall that Bohr's quantum theory of the hydrogen atom did not give any indication of the laws governing such transitions. The concept of transition probability originated in this paper of Einstein.

Einstein's derivation gave also as a by-product the relations (32) and (34) between the A- and B coefficients that specify the probabilities. These relations played an important role in the later development of the quantum theory of radiation.

Einstein's derivation also gave as a by-product the relation (35), which will be recognized as the Bohr frequency condition [Eq. (25)] that up to then was simply assumed.

These results alone would mark this paper of Einstein as a fundamental contribution to physics. However, much more was contained in this paper. Einstein also considered the question of momentum transfer between the gas molecules and the radiation field. He showed that, when a molecule absorbs or emits a quantum $h\nu$ under the influence of external radiation from a definite direction, momentum of magnitude $h\nu/c$ $(=h/\lambda$ where λ is the wavelength of the radiation associated with the frequency $\nu)$ is transferred to the molecule; and that the change in the momentum of the molecule is in the direction of the incident radiation if the energy is absorbed and is in the opposite direction if the energy is emitted. More surprisingly, Einstein showed that, if an energy quantum is emitted in the absence of any external influence (spontaneous emission), the momentum transfer is also a *directed process.* In Einstein's words: "There is no radiation in spherical waves. In a spontaneous emission process the molecule suffers a recoil of magnitude $h\nu/c$ in a direction that in the present state of the theory is determined only by 'chance'." He goes on to say that "These properties of the elementary processes . . . make it seem practically unavoidable that one must construct an essentially quantum mechanical theory of radiation."

University Archives, Washington University

Arthur H. Compton (1892–1962)

Satyendranath Bose (1894–1974)

Einstein's conclusions that a quantum of energy $h\nu$ carries a momentum whose magnitude is [12] $h\nu/c$ and has a definite direction were verified experimentally some years later. The well-known experiments of A. H. Compton[13] on scattering of x rays provided the first experimental confirmation of the correctness of some of these predictions.

THE BOSE-EINSTEIN STATISTICS AND MATTER WAVES

Successful as Einstein's notion of quanta of a radiation field was in elucidating various phenomena involving the interaction of light and matter, many puzzles surrounded it. All the derivations of Planck's radiation law, including Einstein's 1917 derivation, appealed at some point to classical electromagnetic theory. Yet the quantum features of the radiation field, which, as Einstein showed, are implicit in Planck's law, are in direct contradiction with the classical theory. Einstein himself was well aware of these difficulties, and he stressed over and over again the need to formulate a basically new theory that would fuse the wave features and the particle features of radiation. Such a theory, namely, modern quantum mechanics, was indeed formulated about eight years after the publication of the paper by Einstein that we have just discussed. Although today one mainly thinks of de Broglie, Schrödinger, Heisenberg, Born, Dirac, and Pauli as the founders of modern quantum mechanics, Einstein actually played a key role in preparing the ground for it, especially in connection with its wave-mechanical formulation. However, discussion of this development does not fall within the scope of the present article. We will, therefore, only briefly indicate a few important steps that Einstein took in this direction and that themselves have a bearing on the question of the nature of light, particularly with regard to its statistical properties.

AIP Niels Bohr Library

Louis de Broglie (born 1892)

Erwin Schrödinger (1887–1961)

In the summer of 1924 Einstein received a letter from an Indian physicist, S. N. Bose, which began as follows[14]:

"Respected Sir, I have ventured to send you the accompanying article for your perusal and opinion. I am anxious to know what you think of it. You will see that I have tried to deduce the coefficient $8\pi\nu^2/c^3$ in Planck's law independent of classical electrodynamics, only assuming that the ultimate elementary region in phase space has the content h^3. I do not know sufficient German to translate the paper. If you think the paper worth publication I shall be grateful if you arrange for its publication in *Zeitschrift für Physik*. Though a complete stranger to you, I do not feel any hesitation in making such a request. Because we are all your pupils though profiting only by your teaching through your writings. . . ."

In this manuscript Bose essentially treated the light quanta as particles of a gas, with the difference that those particles that belong to the same elementary cell of phase space, of volume h^3 (where h is Planck's constant), are intrinsically indistinguishable. This assumed property of the quanta led to a statistical procedure that differs from that of classical statistical mechanics and provided indeed a complete derivation of Planck's radiation law, without any appeal to classical electromagnetic theory.

Einstein translated Bose's manuscript and sent it to the Editor of *Zeitschrift für Physik* with a letter containing the following remark (which was printed in German at the end of the translated paper[15]): "In my opinion Bose's derivation of Planck's formula signifies an important advance. The method used also yields the quantum theory of the ideal gas, as I will show elsewhere."

On studying Bose's manuscript, Einstein seems to have realized that Bose's approach had far-reaching consequences. In a paper[16] that Einstein presented to the Prussian Academy only a few days after sending the translation of Bose's manuscript for publication, and in another paper[17] that Einstein published in the following year (1925), he applied Bose's method not to a gas of light quanta, but to a real gas, consisting of monatomic molecules. This work of Einstein is another of his masterpieces, and Martin J. Klein, the distinguished Yale historian of physics, said about the second of these two papers that it contains "as many ideas in its dozen pages as many an annual volume of the journals of physics."[18]

One of the many important results that Einstein derived in that paper is the following: He considered a gas in a vessel under equilibrium conditions. Let n be the number of molecules, with energies in the range $(E, E + dE)$, which are situated in some particular region of the vessel. This number will, of course, fluctuate, and Einstein showed that the variance of the fluctuations of n may be expressed in the same form as the variance of energy-fluctuations in blackbody radiation, which he derived in 1909 and which we discussed earlier. In other words, the variance is again expressible as the sum of two terms. The first term can be attributed to classical particles; this term may be understood on the basis of the Maxwell–Boltzmann statistics of noninteracting molecules. The second term, which is analogous to the contribution from wave interference in the radiation problem, cannot be understood from the classical (particle) theory. This is just the opposite situation to that encountered in the radiation case, where the interference term followed from classical (wave) theory, whereas the particle term was new. Einstein suggested that the second term in the expression for the variance of the fluctuations in an ideal gas may be interpreted in a similar way (as due to wave interference) if one associates a radiation field with the gas, and he goes on to say: "I shall discuss this interpretation in greater detail, because I believe that more than a mere analogy is involved here."

The paper by Bose and Einstein's two papers on the quantum theory of an ideal gas are the foundations of what today is known as the *Bose–Einstein statistics*. It applies to photons and to many other elementary particles, which were actually unknown in 1925. The formula for the variance, as the sum of contributions from classical particles and classical waves, is a fundamental consequence of the Bose–Einstein statistics.

Between the appearance of Einstein's first and second papers on the quantum theory of an ideal gas, a period spanning less than five months, another important development took place. In November 1924, Louis de Broglie submitted a thesis at the Sorbonne in which he put forward a theory according to which a wave—called a *matter wave*—was associated

Table 3 Einstein's Researches on the Nature of Radiation and the Wave–Particle Duality.

1905	PARTICLE NATURE OF RADIATION Particle features of blackbody radiation The photoelectric effect
1909	WAVE-PARTICLE DUALITY Energy fluctuations of blackbody radiation
1917	ELEMENTARY PROCESSES OF INTERACTION Spontaneous and induced processes, transition probabilities Momentum of energy quanta, directionality of interactions and their statistical aspects
1924, 1925	BOSE–EINSTEIN STATISTICS Quantum theory of an ideal gas; density fluctuations; matter waves

with the motion of every particle.[19] In this theory, the frequency ν and the wavelength λ of the matter wave were related to the energy E and to the magnitude p of the momentum of the particle by the formulas

$$E = h\nu, \qquad p = h/\lambda. \tag{38}$$

These two relations will be recognized as being identical with those that, according to Einstein, connect the particle features (E and p) of radiation with the frequency ν and the wavelength λ of the radiation field. Thus the association of matter waves with particles (de Broglie's theory) and the association of particles with a radiation field (Einstein's theory) are governed by the same basic relations, namely Eqs. (38).

Einstein received a copy of de Broglie's thesis from Paul Langevin[20] in December 1924 and at once realized that de Broglie's ideas were basically sound and that the appearance of the wave-interference term in his (Einstein's) new fluctuation formula for an ideal gas may be attributed to de Broglie's matter waves. He immediately wrote to Langevin, and his letter contains the prophetic remark that de Broglie "has lifted a corner of the great veil."[21] The very same month (December 1924) Einstein submitted for publication his second paper on the quantum theory of an ideal gas. He included in it a few remarks indicating the importance that he attached to de Broglie's work and outlining the connection that he believed to exist between de Broglie's hypothesis of matter waves and his own investigations on the quantum theory of the ideal gas. These remarks of Einstein are known to have stimulated Erwin Schrödinger only a few months later to develop one form of modern quantum mechanics—namely *wave mechanics*.[22]

Thus the structure that Einstein built with his 1905 "photoelectric paper," his 1909 paper containing the first clear evidence for the wave-particle duality, his 1917 paper on the elementary processes of interaction between molecules and radiation, and his 1924 and 1925 papers on the quantum theory of an ideal gas did not only elucidate the nature of light and

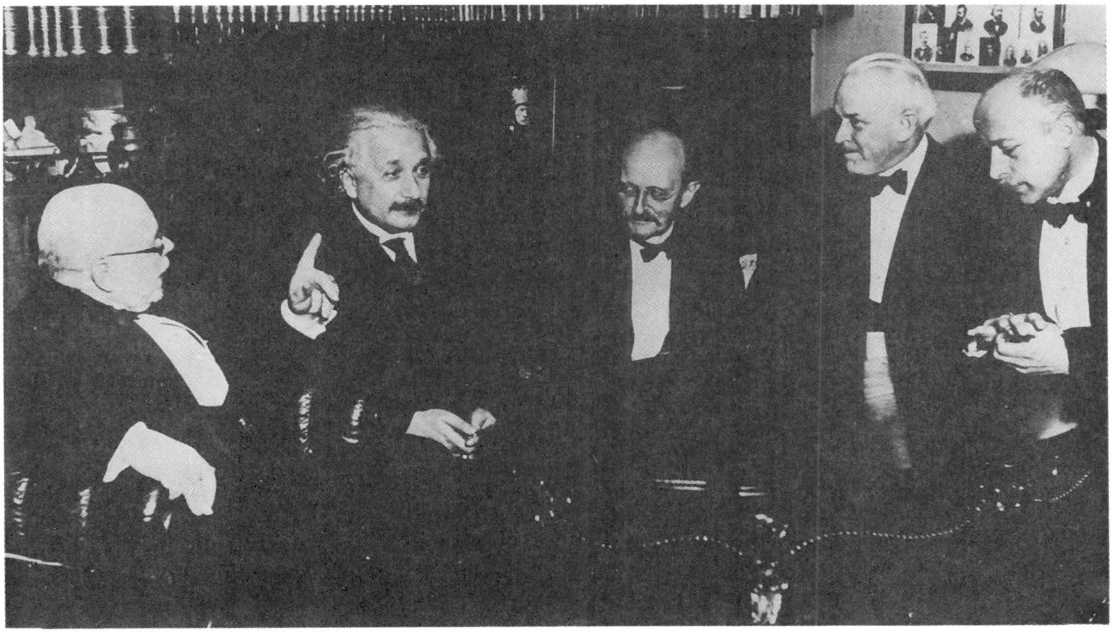

AIP Niels Bohr Library

Nernst, Einstein, Planck, Millikan, and von Laue (Berlin, ca 1928)

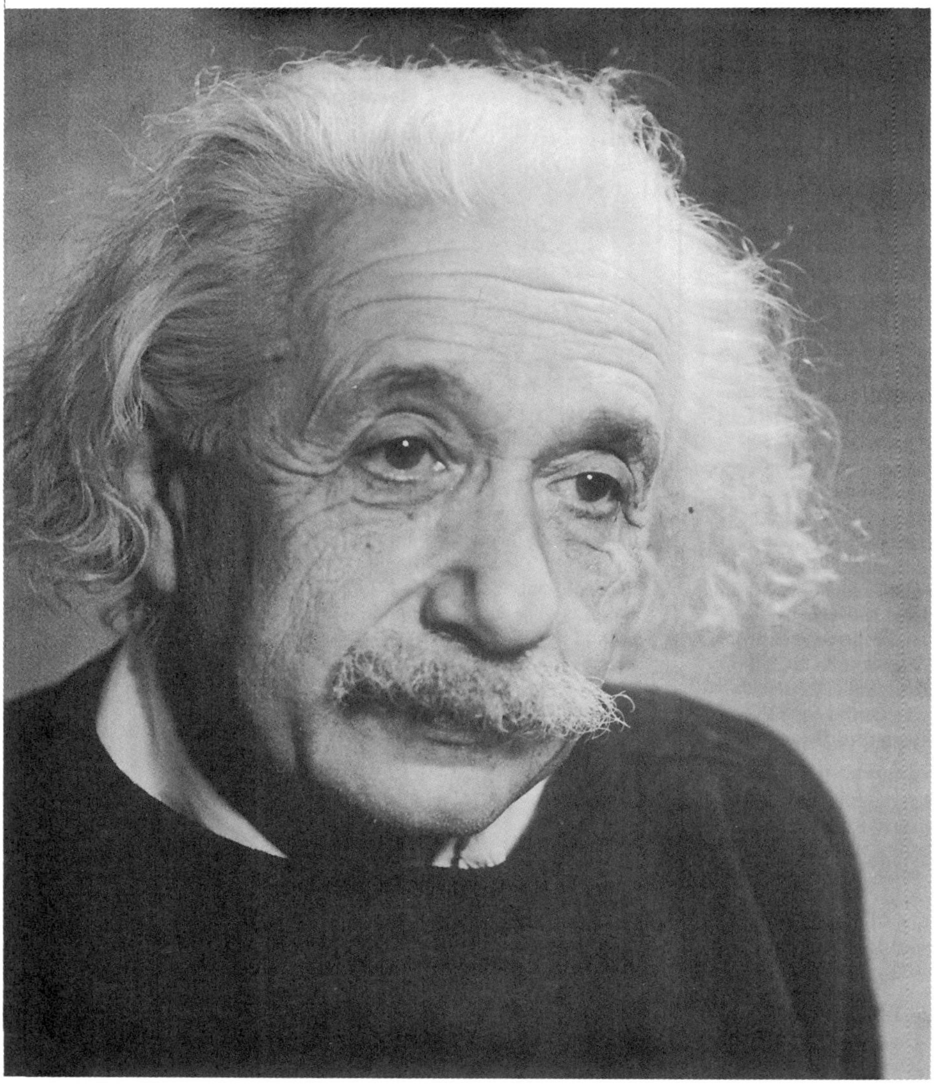

Albert Einstein
Ulm, March 14, 1879–Princeton, April 18, 1955

radiation in general, but also proved to be of fundamental importance to the formulation of wave mechanics.

It was a long journey and in some ways a rather lonely one, since Einstein encountered much opposition to his concept of quanta of radiation. We already noted the skepticism of Millikan. Another example is provided by a passage from a letter, which Max Planck wrote in 1913, in which Einstein was being proposed for membership in the Prussian Academy of Sciences.[23] After writing about Einstein in very glowing terms, this is what Planck said: "That he may sometimes have missed the target of his speculations, as for example in his hypothesis of light quanta, cannot really be held against him." How confident Einstein was to continue, in spite of such skepticism, to probe further and further along the path that he had entered upon as an unknown young scientist in the Swiss Patent Office around 1905!

Practically all Einstein's results relating to the nature of radiation have, in more recent times, been derived in a systematic manner from the modern theory of radiation, to the formulation of which Einstein contributed in such a fundamental way; but the heuristic models by means of which Einstein first obtained these results will forever remain wonderful examples of how truly great research in physics is done.

"All the fifty years of conscious brooding have brought me no closer to the answer to the question, 'What are light quanta?' Of course today every rascal thinks he knows the answer, but he is deluding himself."

A. Einstein, 1951

*This article is based on lectures given during the past two years to the Local Sections of the Optical Society of America. In addition to Einstein's original papers, a number of published accounts and discussions of Einstein's works influenced this presentation. I particularly wish to acknowledge my indebtedness to the following excellent sources:

F. Reiche, *The Quantum Theory* (Methuen, London, 1922).

M. Born, *Natural Philosophy of Cause and Chance* (Clarendon Press, Oxford, 1949; reprinted by Dover, New York, 1964).

M. J. Klein, "Einstein's first paper on quanta," in *The Natural Philosopher* (Blaisdell, New York), Vol. 2 (1963), pp. 57-86, and "Einstein and the wave-particle duality," ibid., Vol. 3 (1964), pp. 1-49.

M. Jammer, *The Conceptual Development of Quantum Mechanics* (McGraw Hill, New York, 1966).

REFERENCES

1. For references and fuller accounts of the early theories, see E. T. Whittaker, *A History of the Theories of Aether and Electricity*, Vol. I (*The Classical Theories*), revised and enlarged edition, 1951; Vol. II (*The Modern Theories 1900-1926*), 1953 (T. Nelson and Sons, London and Edinburgh). Excellent discussions of the radiation laws are given in the first chapter of Max Jammer, *The Conceptual Development of Quantum Mechanics* (McGraw Hill, New York, 1966) and in an article by Martin J. Klein, "Max Planck and the beginnings of the quantum theory," Archive for History of Exact Sciences **1**, 459 (1962).
2. Quoted in C. Seelig, *Albert Einstein, A Documentary Biography*, transl. M. Savill (Staples Press, London, 1956), p. 74.
3. Max Born, "Einstein's statistical theories," in *Albert Einstein: Philosopher-Scientist*, P. A. Schlipp, ed. (the Library of Living Philosophers, Evanston, Ill. 1949), pp. 161–177; republished by Harper & Brothers, New York, 1959.
4. A. Einstein, Ann. Physik, 4th Ser. **17**, 132–148 (1905); English translations: A. B. Arons and M. B. Peppard, Am. J. Phys. **33**, 367–374 (1965), and D. ter Haar, *The Old Quantum Theory* (Pergamon, Oxford and New York, 1967), pp. 91–107.
5. Actually Einstein expressed the energy of the quanta as $(R/N)\beta\nu$, where R is the universal gas constant, N is the Avogadro number, and β is the constant that appears in the exponential term of Wien's radiation law (second row of Table 2). The probable reason that Einstein did not make the formal identification of the proportionality factor $(R/N)\beta$ with Planck's constant h is discussed in Sec. 6 of M. J. Klein, "Einstein's first paper on quanta," in *The Natural Philosopher* (Blaisdell, New York) Vol. 2 (1963), pp. 57–86.
6. R. A. Millikan, Rev. Mod. Phys. **21**, 343–345 (1949).
7. A. Einstein, Phys. Z. **10**, 185–193 (1909).
8. H. A. Lorentz, *Les théories statistiques en thermodynamique* (Tuebner, Leipzig, 1916), Appendix IX, pp. 114–120.
9. Similar calculations based on the Rayleigh–Jeans law, (7), rather than on Planck's law, lead to the result

$$\overline{(\Delta E)^2} = \overline{(\Delta E)^2_{\text{waves}}},$$

whereas calculations based on the Wien's radiation law (5) give

$$\overline{(\Delta E)^2} = \overline{(\Delta E)^2_{\text{particles}}}.$$

If we recall that the Rayleigh–Jeans law and Wien's radiation law are good approximations when $h\nu/kT \ll 1$ and $h\nu/kT \gg 1$ respectively, we see that in the low-frequency–high-temperature limit wave features are predominant, whereas in the high-frequency–low-temperature limit the particle features predominate.
10. A. Einstein, Ver. Dsch. Phys. Ges. **11**, 482-500 (1909). The text of this lecture is also published in Phys. Z. **10**, 817-825 (1909), where an account is also given of the discussion that followed Einstein's paper (ibid, pp. 825–826).
11. A. Einstein, Phys. Z. **18**, 121–128 (1917). English translations in D. ter Haar, *The Old Quantum Theory* (Pergamon, Oxford and New York, 1967), pp. 167–183, and in B. L. van der Waerden, *Sources of Quantum Mechanics* (North-Holland, Amsterdam, 1967; reprinted by Dover, New York, 1968), pp. 63–77.
12. Suggestions that an energy quantum carries a momentum of magnitude $h\nu/c$ were actually made earlier on the basis of the special theory of relativity and electromagnetic theory (cf. M. Jammer, Ref. 1, loc. cit., p. 37).
13. A. H. Compton, Bull. Nat. Res. Council (U.S.) 4 (20), Part 2, pp. 1–56 (1922). Actually Einstein's ideas regarding the momentum of the energy quanta are not mentioned in Compton's paper. Their relevance for the interpretation of such experiments was, however, noted shortly afterward by P. Debye, Phys. Z. **24**, 161–166 (1923).
14. As quoted in M. Jammers, Ref. 1, loc. cit., p. 248.
15. (S. N.) Bose, Z. Phys. **26**, 178–181 (1924).
16. A. Einstein, Preuss. Akad. d. Wissenschaften, Berlin, Phys. Math. Kl., Sitzungsber., 1924, pp. 261–267.
17. A. Einstein, Preuss. Akad. d. Wissenschaften, Berlin, Phys. Math. Kl., Sitzungsber., 1925, pp. 3–14.
18. M. J. Klein, "Einstein and the wave particle duality," in *The Natural Philosopher* (Blaisdell, New York), Vol. 3 (1964), p. 33.
19. A brief preliminary account of the theory appeared in L. de Broglie, C. R. Acad. Sci. **177**, 507–510, 548–550, 630–632 (1923). An English summary of these three notes is given in Philos. Mag. **47**, 446–458 (1924).
20. A fuller account of Langevin's role in bringing de Broglie's theory to Einstein's attention is given in M. Jammer, Ref. 1, loc. cit., pp. 248-249.
21. Louis de Broglie, *New Perspectives in Physics*, transl. A.J. Pomerans (Basic Books, New York 1962), p. 139.
22. ". . . The whole thing would certainly not have originated yet, and perhaps never would have, (I mean not from me), if I had not had the importance of de Broglie's ideas really brought home to me by your second paper on gas degeneracy." [From a letter written by E. Schrödinger to A. Einstein on April 23, 1926, included in *Letters on Wave Mechanics*, K. Przibram, ed., translated and with an introduction by M. J. Klein (Philosophical Library, New York, 1967), p. 26.]
23. The letter was also signed by three other leading scientists from Berlin, W. Nernst, H. Rubens, and E. Warburg. [C. Kirsten and H.G. Körber, *Physiker über Physiker* (Akademie-Verlag, Berlin, 1975), pp. 201–208].

Cover photo: AIP Niels Bohr Library

Recollections of Max Born

By Emil Wolf

The invitation to address this commemorative meeting has given me the rare opportunity to set aside my customary activities and try to recall a period of my life several decades ago when I had the great fortune of being able to collaborate with Max Born. As the title of my talk suggests, this will be a rather personal account, but I will do my best to present a true image of a scientist who has contributed in a decisive way to modern physics in general and to optics in particular; it will also present glimpses of a man who, under a somewhat brusque exterior, was a very humane and kind person and who in the words of Bertrand Russell was brilliant, humble, and completely without fear in public utterances.

The early part of my story is closely interwoven with another great scientist, Dennis Gabor, through whose friendship I became acquainted with Born.

I completed my graduate studies in 1948 at Bristol University. My Ph.D. thesis supervisor was E. H. Linfoot, who at just about that time was appointed Assistant Director of the Cambridge University Observatory. He offered me, and I accepted, a position as his assistant in Cambridge. During the next two years while I worked in Cambridge I frequently traveled to London to attend the meetings of the Optical Group of the British Physical Society. They were usually held at Imperial College and were often attended by Gabor, whose office was in the same complex of buildings. From time to time I presented short papers at these meetings. At the end of some of the meetings Gabor would invite me to his office for a chat. He would comment on the talks, make suggestions regarding my work, and speak about his own researches. Gabor liked young people, and he always offered encouragement to them. He knew Born from Germany, and he had great admiration for him.

Through Gabor I learned in 1950 that Born was thinking of preparing a new book on optics, somewhat along the lines of his earlier German book *Optik*, published in 1933, but modernized to include accounts of the more important developments that had taken place in the nearly 20 years that had gone by since then. At that time Born was the Tait Professor of Natural Philosophy at the University of Edinburgh, a post he had held since 1936, and in 1950 he was 67 years old, close to his retirement. He wanted to find some scientists who specialized in modern optics and who would be willing to collaborate with him in this project. Born approached Gabor for advice, and at first it was planned that the book would be written jointly by him, Gabor, and H. H. Hopkins. The book was to include a few contributed sections on some specialized topics, and Gabor invited me to write a section on diffraction theory of aberrations, a topic I was particularly interested in at that time. Later it turned out that Hopkins felt he could not devote adequate time to the project, and in October of 1950, Gabor, with Born's agreement, wrote to Linfoot and me asking if either of us, or both, would be willing to take Hopkins' place. After some lengthy negotiations it was agreed that Born, Gabor, and I would co-author the book.

—**Emil Wolf** *is with the Department of Physics and Astronomy and the Institute of Optics of the University of Rochester.*

This article is essentially the text of lectures presented September 7, 1982 at the Max Born Centenary Conference held in Edinburgh, Scotland and October 21, 1982 at the Max Born Symposium held during the Annual Meeting of the Optical Society of America.

Dennis Gabor.

The start of collaboration

I was, of course, delighted with this opportunity, but there was the problem of my finding the necessary time to work on this project while holding a full-time appointment with Linfoot at Cambridge. I mentioned this to Gabor, and I told him that if there were any possibility of obtaining an appointment with Born, which would allow me to spend most of my time working on the book, I would gladly leave Cambridge and go to Edinburgh.

Gabor took up the matter with Born, who was interested. Toward the end of November 1950, Gabor wrote me that Born would be in London a few days later and that he (Gabor) was arranging for the three of us to meet the following weekend. It was agreed that I would come to Gabor's office at Imperial College on the following Saturday morning, December 2, 1950, and that we would then go to his home in South Kensington, within walking distance of Imperial College. Born was to come directly to Gabor's home from his London hotel, and the three of us and Mrs. Gabor would have lunch there.

I arrived at Gabor's office just before lunch, and I have a vivid recollection of that meeting. There was a long staircase leading to the entrance hall of the building. As we were walking down the staircase, Gabor suddenly became somewhat apprehensive. He knew that our luncheon meeting might lead to an appointment for me with Born, and he said to me, "Wolf, if you let me down, I will never forgive you. Do you know who Born's last assistant was? Heisenberg!" This statement was not accurate. Born had other assistants after Heisenberg, but the remark shows how nervous Gabor was on that particular occasion. Fortunately, all turned out well, and Gabor certainly seemed in later years well satisfied with the consequences of our luncheon with Born.

During that meeting Born asked me a few questions, mainly about my scientific interests, and before the lunch was over he invited me to become his assistant in Edinburgh, subject to the approval of Edinburgh University. It seemed to me remarkable that Born should have made up his mind so quickly, without asking for even a single letter of reference, especially since I had published only a few papers by that time and was quite unknown to the scientific community.

Later, when I got to know Born better, I realized that his quick decision was very much in line with one trait of his personality: he greatly trusted the judgment of his friends. Since Gabor recommended me, Born considered further inquiries about me to be superfluous. Unfortunately, as I also learned later, Born's implicit trust in people whom he considered to be his friends was occasionally misplaced and sometimes created problems for him.

A few days after our meeting I received a telegram from Born inviting me to a formal interview at Edinburgh University. The interview took place about two weeks later, and the next day Born wrote me saying that the committee which interviewed me recommended my appointment as his private assistant, beginning January 23, 1951. I resigned my post in Cambridge and took up the new appointment. Later I learned that committee approval was not really needed because my salary was to be paid from an industrial grant that was entirely at Born's disposal. However, on this occasion Born was careful, because some time earlier he had had on his staff Klaus Fuchs, who turned out to be a spy for the Russians, and Born got rather bad publicity from that.

Now, the name Fuchs means fox in German, and before inviting me to Edinburgh, Born apparently wrote to Sir Edward Appleton, the Principal of Edinburgh University at that time, saying that he felt the decision about this particular appointment should not be made by him alone; since he would like to appoint a Wolf after a Fox!

Arrival at Edinburgh

I arrived in Edinburgh toward the end of January 1951, eager to start on our project. Born's Department of Applied Mathematics was located in the basement of an old building on Drummond Street. I was surprised by the small size of the department. Physically it consisted of Born's office; an adjacent large room for all of his scientific collaborators, about five at that time; a small office for Mrs. Chester, his secretary; two rooms for the two permanent members of his academic staff, Robert Schlapp, a senior lecturer, and Andrew Nisbet, a lecturer; and one lecture room. The rest of the building was occupied by experimental physicists under the direction of Professor Norman Feather. In earlier days the building housed a hospital, in which Lord Lister, a famous surgeon known particularly for his work on antiseptics, also worked.

In spite of his advanced age Born was very active and, as throughout all his adult life, a prolific writer. He had a

The building on Drummond Street in Edinburgh that housed Max Born's Department of Applied Mathematics.

Max Born at his desk, ca 1950. (Credit: AIP Niels Bohr Library).

definite work routine. After coming to his office he would dictate to his secretary answers to the letters that arrived in large numbers almost daily. Afterward he would go to the adjacent room where all his collaborators were seated around a large U-shaped table. He would start at one end of it, stop opposite each person in turn, and ask the same question: "What have you done since yesterday?" After listening to the answer he would discuss the particular research activity and make suggestions. Not everyone, however, was happy with this procedure. I remember a physicist in this group who became visibly nervous each day as Born approached to ask his usual question, and one day he told me that he found the strain too much and that he would leave as soon as he could find another position. He indeed did so a few months later. At first I too was not entirely comfortable with Born's question, since obviously when one is doing research and writing there are sometimes periods of low productivity. One day when Born stood opposite me at the U-shaped table and asked, "Wolf, what have you done since yesterday?" I said simply, "Nothing!" Born seemed a bit startled, but he did not complain and just moved on to the next person, asking the same kind of question again.

Born was always direct in expressing his views and feelings, but he did not mind if others did the same, as this small incident indicates. There will be more examples of this later.

Work at Edinburgh

We started working on the optics book as soon as I came to Edinburgh. It was understood right from the beginning that Born's main contribution would consist of making material available from his German *Optik*, but he was to take part in the planning of the new book, make suggestions, and provide general advice. Most of the actual writing was to be done by Gabor and me and a few contributors. However, like Hopkins earlier on, Gabor soon found it difficult to devote the necessary time to the project, and it was mutually agreed that he would not be a co-author after all, but would just contribute a section on electron optics. So in the end it became my task to do most of the actual writing. Fortunately I was rather young then, and so I had the energy needed for what turned out to be a very large project. I was in fact 40 years younger than Born. This large age gap is undoubtedly responsible for a question I am sometimes asked – whether I am a son of the Emil Wolf who co-authored *Principles of Optics* with Max Born!

Although I did most of the writing, Born read the manuscript and made suggestions for improvements. We signed a contract with the publishers about a year after I came to Edinburgh, and we hoped to complete the manuscript by the time Born was to retire, one-and-a-half years later. However, we were much too optimistic. The writing of the book took about eight years altogether.

Throughout his life Born was a quick, prolific writer, publishing more than 300 scientific papers, about 31 books (not counting different editions and translations), apart from numerous articles on nonscientific topics.[1] In spite of my relative youth I could not compete with the speed with which Born wrote, even at his advanced age, and it soon became clear to me that he was not too pleased with my slow progress.

One day when I started writing an Appendix on Calculus of Variations, Born said that the best treatment of that subject he knew of was in his notes of lectures given by the great mathematician David Hilbert in Göttingen in the early years of this century. Born suggested that

David Hilbert, 1912.

he dictate the Appendix to me, following in the main Hilbert's presentation, and that we acknowledge this in the preface to our book. So we started in this way. After each dictating session I was to rewrite the notes and give them to Born the next day for his comments. But we did not get very far this way. After about two dictating sessions Born said he could prepare the whole Appendix himself much faster without my help, which he then did. It is essentially in this version, written by Born, that the Appendix on Calculus of Variations appears in our book.

Born's revered teacher

Incidentally, David Hilbert, whose presentation Born closely followed, was one of Born's great heroes. To physicists Hilbert is mainly known in connection with the concept of the Hilbert space and as co-author of the classic text *Methods of Mathematical Physics*, referred to generally as "Courant-Hilbert." But Hilbert contributed in a fundamental way to many branches of mathematics and was generally considered to have been the greatest mathematician of his time. Born became acquainted with Hilbert soon after coming to Göttingen in 1905, later becoming Hilbert's private assistant. In one of his later writings Born refers to Hilbert as his "revered teacher and friend," and in a biography of Hilbert by Constance Reid,[2] published in 1970, Born is quoted as saying that his job with Hilbert was to him "precious beyond description because it enabled [him] to see and talk to him every day."

Born had an encyclopedic knowledge of physics and whatever problem one brought to him, he was able to offer a useful insight or suggest a pertinent reference. He also knew personally all the leading physicists of his time and would often recall interesting stories about them.

Optics in those days—remember we are talking now about optics in pre-laser days—was not a subject that most physicists would consider exciting; in fact, relatively little advanced optics was taught at universities in those days. The fashion then was nuclear physics, particle physics, high energy physics, and solid state physics. Born was quite different in this respect from most of his colleagues. To him all physics was important, and rather than distinguish between "fashionable" and "unfashionable" physics he would only distinguish between good and bad physics research.

Born was equally broad-minded about the techniques used by physicists in their research. For example, when we were writing a section on certain mathematical methods needed to evaluate the performance of optical systems, we found that although the results given in a basic paper on this subject were correct, the derivation contained serious flaws. I was rather indignant about this, but Born just said something like, "In pioneering work everything is allowed, as long as one gets the right answer. Real justification can come later."

Max Born as *Privatdozent* in Göttingen. (Reproduced from *Hilbert* by Constance Reid, ref. 1).

Max Born in the 1920s.

One of the earliest occasions when many physics students encounter Born's name comes when they start studying quantum theory of scattering. Here they soon learn about the *Born approximation*. This term also occurs in many of the papers on potential scattering that have been published in the more than half a century that has gone by since Born wrote a basic paper on this subject. Yet Born was rather irritated when the Born approximation was mentioned. He once said to me, "I developed in that paper the whole perturbation expansion for the scattered field, valid to all orders, yet I am only given credit for the first term in that series!"

Resistance to new discoveries

It was not always easy for Born's collaborators to convince him quickly of new discoveries. Let me illustrate this by an example from my own experience. In the early 1950s I became very interested in problems of partial coherence. One day I found a result in this area of optics that seemed to me remarkable. I phoned Born from my home one morning, told him I had an exciting new result, and asked him for an appointment to discuss it. We arranged to have lunch together that day.

When I came to his office just before lunch, Born wanted to know straight away what the excitement was all about. I told him I had found that not only an optical field, but also its coherence properties, characterized by an appropriate correlation function (now known as the mutual coherence function), are propagated in the form of waves. Born looked at me rather skeptically, put his arm on my shoulder and said, "Wolf, you have always been such a sensible fellow, but now you have become completely crazy!" Actually after a few days he accepted my result, and I suspect he then no longer doubted my sanity.

This incident illustrates a fact well known to Born's collaborators—that Born had a certain resistance to accept new results obtained by others. Nonetheless, he continued thinking about them, and if they were correct he would eventually apologize for doubting them in the first place.

This trait of Born's personality is very well described by the Polish physicist Leopold Infeld, who collaborated with Born in Cambridge in the 1930s. I will quote shortly some very perceptive observations Infeld made about Born in his biography[3]; but before doing so I would like to mention a small incident relating to this book.

One day I browsed through a bookstore in Edinburgh and found a used copy of Infeld's book. I was astonished to note that the book had Born's signature on its first page. I purchased it and asked Born the next day whether he knew the book. He said, "Yes, I had a copy of it and there is a funny description of me in it; but I lent it to someone and it was never returned. I cannot remember whom I lent it to." The book I had purchased was obviously Born's missing copy, so I gave it to him, much to his delight.

In the book Infeld describes some of his experiences in Cambridge. He started working with Dirac but found him rather uncommunicative. Later Infeld attended some of Born's lectures. During one of them Born gave an account of some results that he had recently obtained. Infeld could not understand one of Born's arguments. He borrowed his notes so that he could study the argument more closely later. Let me now quote from Infeld's biography [ref. 2, p. 208 et seq.]:

On the evening of the day I received the paper the point suddenly became clear to me. I knew that the mass of the electron was wrongly evaluated in Born's paper and I knew how to find the right value. My whole argument seemed simple and convincing to me. I could hardly wait to tell it to Born, sure that he would see my point immediately. The next day I went to him after his lecture and said:

"I read your paper; the mass of the electron is wrong." Born's face looked even more tense than usual. He said: "This is very interesting. Show me why." Two of his audience were still present in the lecture room. I took a piece of chalk and wrote a relativistic formula for the mass density. Born interrupted me angrily:

"This problem has nothing to do with relativity theory. I don't like such a formal approach. I find nothing wrong with the way I introduced the mass." Then he turned toward the two students who were listening to our stormy discussion. "What do you think of my derivation?" They nodded their heads in full approval. I put down the piece of chalk and did not even try to defend my point. Born felt a little uneasy. Leaving the lecture room, he said: "I shall think it over."

Infeld then goes on to say:

I was annoyed at Born's behaviour as well as at my own and was, for one afternoon, disgusted with Cambridge. I thought: "Here I met two great physicists. One of them does not talk. I could as easily read his papers in Poland as here. The other talks, but he is rude." . . . . The next day I went again to Born's lecture. He stood at the door before the lecture room. When I passed him he said to me: "I am waiting for you. You were quite right. We will talk it over after the lecture. You must not mind my being rude. Everyone who has worked with me knows it. I have a resistance against accepting something from outside. I get angry and swear but always accept it after a time if it is right."

Our collaboration had begun with a quarrel, but a day later complete peace and understanding had been restored.

Mrs. Hedwig Born, 1961.

A little further on in his biography, Infeld speaks about Born again, and this is what he says:

I marveled at the way in which he managed his heavy correspondence, answering letters with incredible dispatch, at the same time looking through scientific papers. His tremendous collection of reprints was well ordered; even the reprints from cranks and lunatics were kept, under the heading "Idiots." Born functioned like an entire institution, combining vivid imagination with splendid organization. He worked quickly and in a restless mood. As in the case of nearly all scientists, not only the result was important but the fact that he had achieved it.

Infeld later continues:

There was something childish and attractive in Born's eagerness to go ahead quickly, in his restlessness and his moods, which changed suddenly from high enthusiasm to deep depression. Sometimes when I would come with a new idea he would say rudely, "I think it is rubbish," but he never minded if I applied the same phrase to some of his ideas. But the great, the celebrated Born was as happy and as pleased as a young student at words of praise and encouragement. In his enthusiastic attitude, in the vividness of his mind, the impulsiveness with which he grasped and rejected ideas, lay his great charm.

I regard these remarks of Infeld as a true and very perspective description of Born's mode of work and of Born's personality.

The house of Max and Hedwig Born in Edinburgh, at 84 Grange Loan.

Kind and compassionate

In spite of Born's occasional irritation and impatience, he was a person who cared deeply for the well-being of his fellow scientists and collaborators. His wife, Hedwig Born, was likewise a person with deep compassion for others. She too was a remarkable and gifted person. Mrs. Born published a number of books, especially poetry, and around 1938 became a Quaker, remaining active in the Quaker movement for the rest of her life.

I would like to give just one example from my own experience, which illustrates Born's concern for others. A few months after I began working with Born, I was getting married. In those days it was difficult to rent an apartment in Edinburgh. One day during the time when we were searching for a home I received a letter from Mrs. Born, who was then with Professor Born on a visit to Germany. She said that they had heard about our problem and were very concerned that we might have to postpone getting married if we did not find somewhere to live. She then offered to help us, suggesting that we share with them their small house in Edinburgh. In the end we found an apartment elsewhere; but this small episode is an indication of the warmth of their personalities and of their willingness to make a personal sacrifice to help, when help was needed.

I mentioned earlier that one of Born's great heroes was the mathematician David Hilbert. But there was another, even greater hero in Born's life: Albert Einstein, with whom he and also Mrs. Born maintained close personal friendships for almost half a century. Unfortunately, after Einstein left Europe for America in 1932 they did not see each other again, but they carried on extensive correspondence until Einstein's death in 1955. The letters they exchanged were published in 1971, together with Born's commentary, and the volume[4] is a

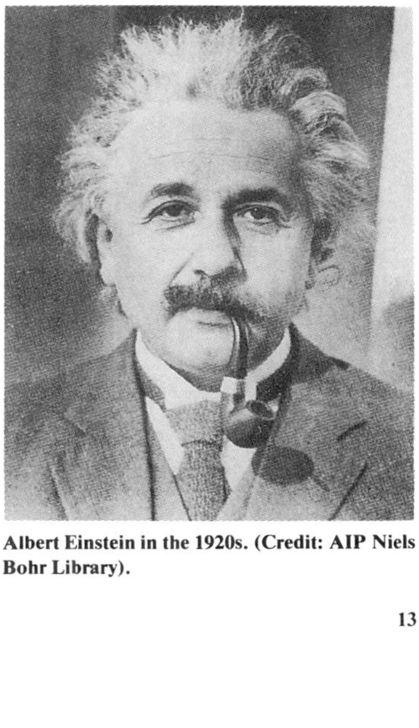

Albert Einstein in the 1920s. (Credit: AIP Niels Bohr Library).

precious contribution to the history of physics and of the times in which they lived.

There is an episode I would like to relate briefly in connection with Born's friendship with Einstein. In the early 1950s, when Sir Edmund Whittaker was preparing the second volume of his classic work *A History of the Theories of Aether and Electricity*, he sent Born the manuscript of a section dealing with the special theory of relativity. Whittaker's treatment placed a heavy emphasis on the work of Poincaré and Lorentz and dismissed Einstein's contribution as being of rather minor significance. Born, who himself wrote a book on the theory of relativity, was most unhappy with Whittaker's manuscript and sent him a long report in which he analyzed in detail the various contributions, indicating why he considered Einstein's contribution to be much more fundamental.

Sir Edmund Whittaker. (Reproduced by the courtesy of the University of Edinburgh).

However, Born did not succeed in changing Whittaker's opinion. In September of 1953, around the time Whittaker's book was published, Born wrote to Einstein about this. Let me quote from Born's letter:[6] "Many people may now think (even if you do not) that I played a rather ugly role in this business. After all it is common knowledge that you and I do not see eye to eye over the question of determinism."

Einstein was not concerned. This is what he said in his reply to Born:[7] "Don't lose any sleep over your friend's book . . . . If he manages to convince others, that is their own affair. I myself have certainly found satisfaction in my efforts . . . ." and then Einstein added, "After all, I do not need to read the thing."

> **"In an Age of mediocrity and moral pygmies, the lives of Albert Einstein and Max Born shine with an intense beauty. Something of this is reflected in their correspondence, and the world is the richer for its publication."**
>
> (From Bertrand Russell's Foreword to *The Born-Einstein Letters*, Ref. 4)

Born retired that year (in 1953). The accompanying photograph shows Born with the members of his department at the time of his retirement.

Life in retirement

Soon afterward the Borns left Edinburgh and settled in Bad Pyrmont, a spa in West Germany, not far from Göttingen, where they built a small house. When they left Edinburgh our book was far from finished. We corresponded about it, and I visited Born in his new home several times. Born was hoping that he and Mrs. Born would be able to lead a more quiet life in Bad Pyrmont, but he told me on one of my visits that this proved difficult to achieve. For example, soon after they settled in Bad Pyrmont, Born was invited to address a meeting of a West German physical society. He declined the invitation, saying that he was too old to travel. He received a reply stating that in view of this the meeting would be moved to Bad Pyrmont!

In 1954, the year after his retirement, Born was awarded the Nobel Prize. He was, of course, delighted, but I am quite sure he felt, as many others did, that this great recognition had come somewhat late. The Nobel Prize was awarded to him for contributions that he made almost 30 years earlier. However, as his son Gustav later noted in a postscript to Born's memoirs,[8] it came at the right time to add weight to his main retirement occupation, which was to educate thinking people in Germany and elsewhere in the social, economic, and political consequences of science and also of the dangers of nuclear weapons and re-armament.

In 1957 I was a Visiting Scientist at the Courant Institute of New York University, still working on our book. One day I received a letter from Born asking me why the book was not yet finished. I replied that practically the whole manu-

Members of Max Born's department at the time of his retirement (1953) from the Tait Chair of Natural Philosophy at the University of Edinburgh. Standing (from left to right): E. Wolf, D. J. Hooton, A. Nisbet. Sitting: Mrs. Chester (secretary), M. Born, R. Schlapp.

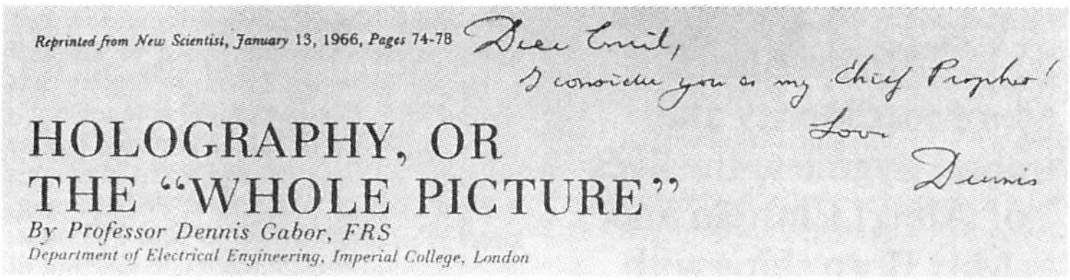
Reprinted from New Scientist, January 13, 1966, Pages 74-78

HOLOGRAPHY, OR THE "WHOLE PICTURE"

By Professor Dennis Gabor, FRS

Department of Electrical Engineering, Imperial College, London

A dedication from Dennis Gabor.

script was completed, except for a chapter on partial coherence on which I was still working. Born wrote back almost at once, saying something like, "Who apart from you is interested in partial coherence? Leave that chapter out and send the rest of the mansucript to the printers." Actually I completed that chapter shortly afterward and it was included in the book.

It so happened that within about two years after the publication of our book (in 1959) the laser was invented and optical physicists and engineers then became greatly interested in questions of coherence. Our book was the first that dealt in depth with this subject, and Born was then as pleased as I was that the chapter was included.

Our book was also one of the first textbooks containing an account of holography. Gabor was very happy about it. Later, when holography became popular and useful, he sent me a reprint of one of his papers with a charming dedication (see above).

As I approach the end of my reminiscences about Max Born, I would like to say that I hope my talk conveyed to you the warmth and the affection with which he remains in my memory—not only as a great scientist, but also as a kind and remarkable human being. My feelings about our collaboration are well described by exactly the same words that Born used when he spoke about his association with David Hilbert, quoted earlier, namely that my appointment with him was precious to me beyond description, because it enabled me to see and to talk to him every day.

Max Born in Bad Pyrmont feeding pigeons.

"Everybody who had contact with him remembers him not only as a brilliant scientist but also as a man of human warmth and greatness."

(From an introduction by Victor F. Weisskopf to an article by Max Born entitled "Man and the Atom," published by the Society for Social Responsibility of Science (Pamphlet #4) Southhampton, Pa., and the American Friends Service Committee, Philadelphia, Pa.

Max Born in front of his library at his home in Bad Pyrmont.

Olivia

Before ending I would like to show you a few pictures taken in Bad Pyrmont during Born's retirement and also to mention one more episode. One shows Professor and Mrs. Born with one of their daughters, Irene. Some years ago I learned that Irene is the mother of a lady who has achieved fame comparable to that of Max Born himself, but in an entirely different field. I am speaking of the pop singer Olivia Newton-John. Shortly after I learned that Olivia Newton-John was Max Born's granddaughter, I was on a sabbatical leave at the University of Toronto. Olivia was scheduled to give a concert in Toronto while I was there. I wrote to her, told her I had collaborated with her grandfather in the writing of a book, and asked her whether we could meet. I received a charming reply in which she invited me to meet her after the concert. We met then and talked mainly

Max Born with two of his grandsons, Max and Sebastian (children of his son Gustav) in Bad Pyrmont. (Credit: AIP Niels Bohr Library).

Hedwig Born and Max Born, with their daughter Irene Newton-John in Bad Pyrmont, 1957. (Credit: AIP Nields Bohr Library).

about her grandparents. Before I left Olivia gave me two autographed photos of herself. Let me add that to some of my students I am known not so much as the co-author of *Principles of Optics* but rather as the person who knows Olivia Newton-John and who has a picture of her hanging in his office signed "To Emil, Love, Olivia."

I cannot bring you the voice of Max Born, but I will end my presentation with one of the songs that made Olivia famous. (The lectures on which this article is based concluded with an excerpt from the song "If You Love Me Let Me Know.")

References

1. Bibliography of Born's scientific publications is given in "Max Born," by N. Kemmer and R. Schlapp in Biographical Memoirs of the Royal Society, *17*, London: the Royal Society, 1971, pp. 17-52. Born's autobiography cited as ref. 8 below was published posthumously, first in German in 1975 and is, therefore, not included in that bibliography.
2. C. Reid: *Hilbert* (Springer-Verlag, New York, Heidelberg, Berlin, 1970), p. 103.
3. L. Infeld: *Quest*, The Evolution of a Scientist (Doubleday, Doran and Co., New York, 1941).
4. *The Born-Einstein Letters*, with commentaries by Max Born (Walker and Company, New York, 1971).
5. Born's opinion on this question rather than Whittaker's is generally accepted. See, for example, D. Martin's biographical note about E. T. Whittaker in *Dictionary of Scientific Biography*, C. C. Gillespie, editor-in-chief (Charles Scribner's Sons, New York, 1976), Vol. XIV, p. 317; or A. Pais: *Subtle Is the Lord*, The Science and the Life of Albert Einstein (Clarendon Press, Oxford, and Oxford University Press, New York, 1982), p. 168.
6. Ref. 4, p. 197.
7. Ref. 4, p. 199.
8. Max Born: *My Life* (Taylor and Francis, London, and Charles Scribner's Sons, New York, 1978), p.296.

Acknowledgments: In preparing this article for publication I received assistance with obtaining some of the photographs, determining the approximate dates when they were taken and with checking some of the references. I am particularly obliged to G. V. R. Born (London University), R. M. Sillitto and S. D. Fletcher (University of Edinburgh), L. H. Caren (University of Rochester), and D. Dublin (American Institute of Physics) for their help.

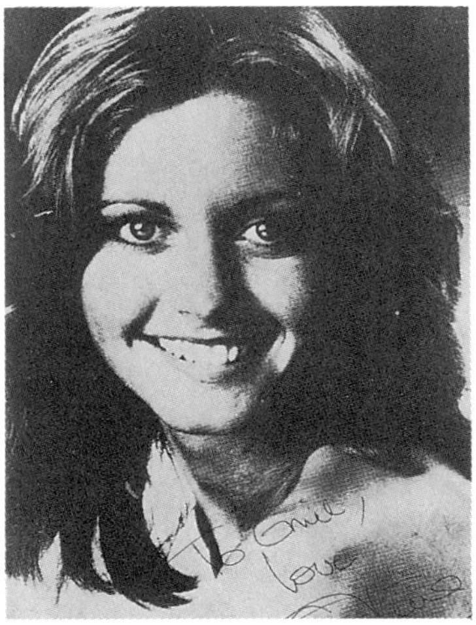

Olivia Newton-John,
granddaughter of Max Born.

Max Born.
(Credit: Lotte Meitner-Graf).

Huygens' Principle 1690–1990: Theory and Applications
H. Blok, H.A. Ferwerda, H.K. Kuiken (editors)

The life and work of Christiaan Huygens

Emil Wolf
Department of Physics and Astronomy
and
The Institute of Optics
University of Rochester
Rochester, NY 14627, USA

Modern science, as we know it today, had its beginning around the year 1600. In particular, the origin of modern physics can be traced to the work of Galileo Galilei, who was born in 1564. Galileo died in 1642 and the very same year another intellectual giant and one of the founders of modern physics, Isaac Newton, was born.

Christiaan Huygens, whose pioneering work in optics we are commemorating at this Symposium was born in 1629, thirteen years before Galileo's death and Newton's birth. He is, therefore, often said to be a transitional figure. But this really should be regarded only as a figure of speech, because many of Huygens' contributions were certainly not transitional. In fact, as we will soon see some of his contributions have continued to influence physics and mathematics to this day.

Christiaan Huygens' father, Constantin Huygens was a brilliant figure in the Dutch intellectual society, being talented in music, art and poetry. He was also a very good athlete. As a young man he travelled several times to England in various diplomatic capacities, studied at Oxford and in 1622, when only twenty-six years old, was knighted by the English King James I. He married his cousin Susanna van Baerle and they settled in the Hague. They had four sons and a daughter. Christiaan was their second son. When Christiaan was eight years old his mother died.

Among acquaintances of the family and visitors to the Huygens' home were some of the great contemporary figures, including Rembrandt, Hals, Spinoza and Descartes.

In 1645 Huygens entered Leiden University, as a student of law and mathematics. After two years at Leiden he joined one of his brothers at the

College of Breda. Whilst still a student he and his brother learned to grind lenses. Later they made telescopes, which were probably among the best in existence at that time. With one of them, Christiaan observed the planet Saturn and in 1655 he found a small star near Saturn that accompanied the planet. It was a satellite now known as Titan. Forty years earlier Galileo found something peculiar about Saturn. It seemed to consist of a central core with two small globes touching it at each side. However, in the following years Galileo discovered, much to his surprise that the globes diminished and then disappeared. In a letter to one of his acquaintances Galileo asked "Had Saturn devoured his children?". Figure 1 shows Galileo's sketches, indicating how Saturn appeared to him at different times.

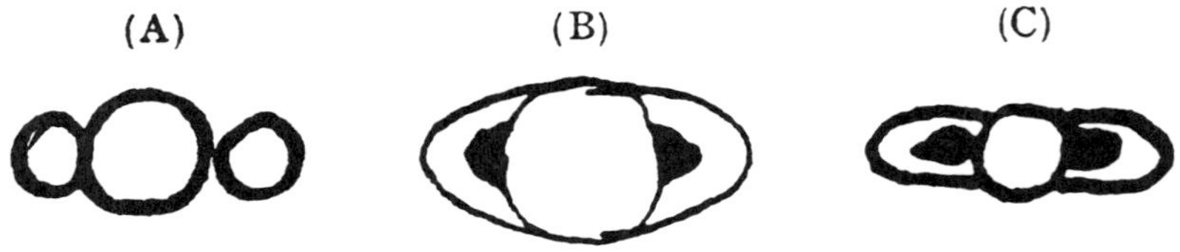

Figure 1. Galileo's sketches of Saturn [reproduced from J.J. Fahie, *Galileo* (John Murray, London, 1903)].

Using one of his telescopes Huygens solved the riddle and announced in 1656 that "*the core of the planet was griddled by a thin, flat ring, nowhere touching, inclined to the ecliptic*". Thus Huygens discovered the rings of the planet Saturn. Figure 2 shows some of Huygens' drawings of his perception of the rings. Today we know, of course a great deal more about Saturn, especially from photographs taken by the unmanned spacecraft Voyager ten years ago. Figure 3 shows a picture of Saturn and of its rings, taken in 1980, three-and-quarter century after Huygens discovered them. We see that his sketches come pretty close to Saturn's real likeness.

After completing his studies Huygens went to Paris, at the encouragement of his father. Because of his father's contacts and also because of his own work, Huygens was soon able to meet there leading French mathematicians and astronomers of that period. The visit stimulated his interest in astronomy and led him to become aware of the need for precise timekeeping, not only because this was required for astronomy, but also because accurate clocks were needed for determining longitude at sea. His desire to providea device for precise timekeeping led him to the study of pendulum motion

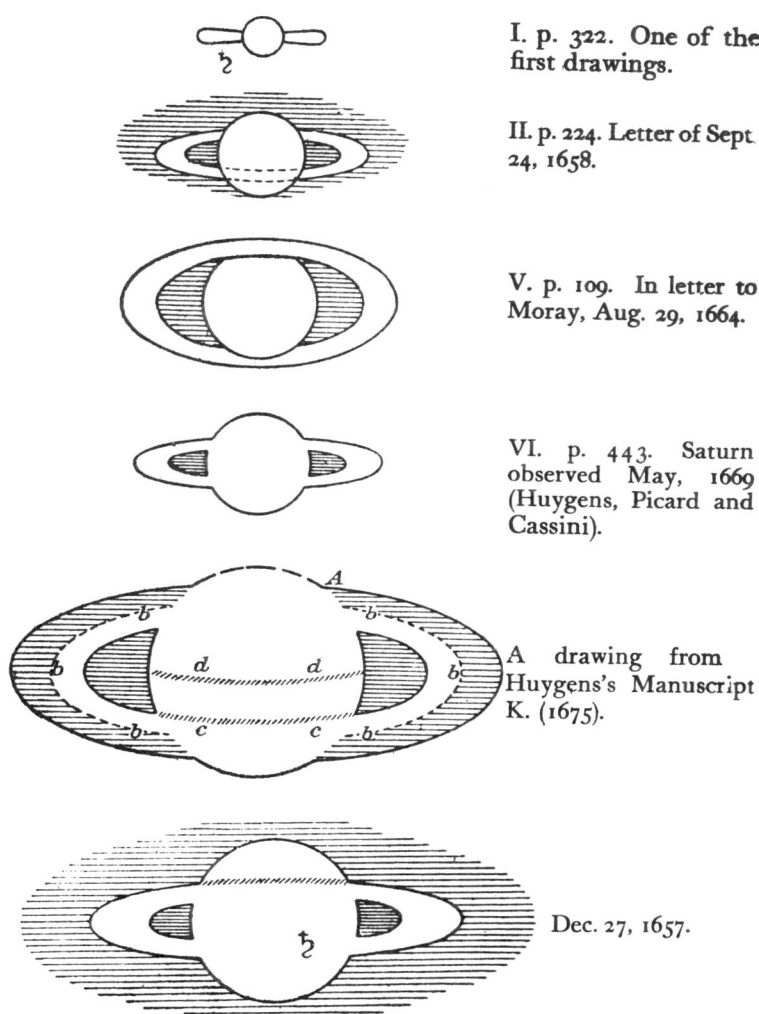

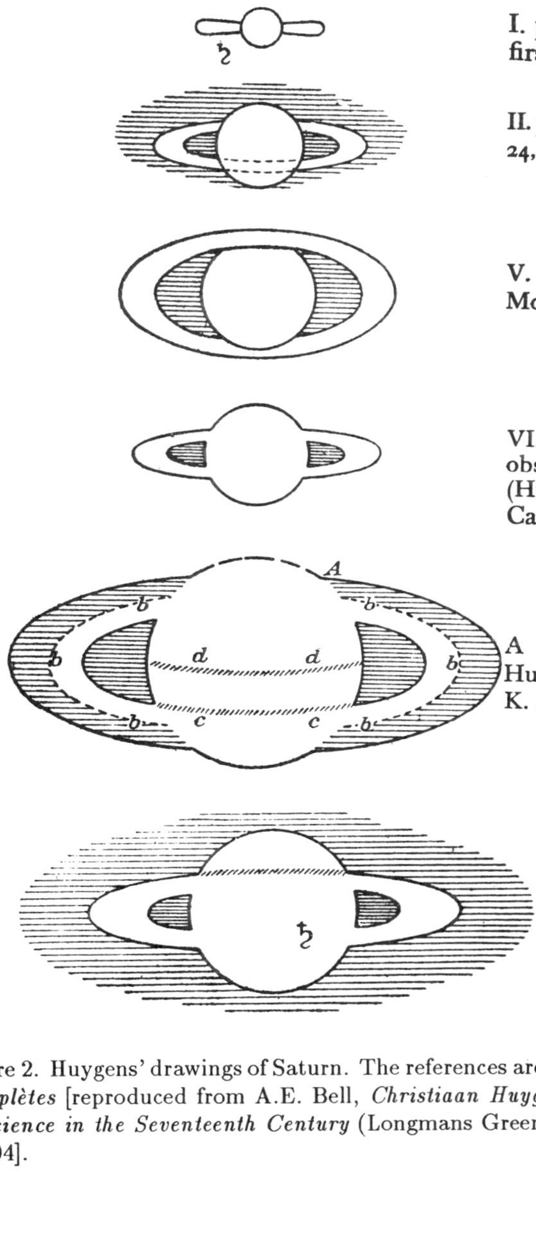

I. p. 322. One of the first drawings.

II. p. 224. Letter of Sept. 24, 1658.

V. p. 109. In letter to Moray, Aug. 29, 1664.

VI. p. 443. Saturn observed May, 1669 (Huygens, Picard and Cassini).

A drawing from Huygens's Manuscript K. (1675).

Dec. 27, 1657.

Figure 2. Huygens' drawings of Saturn. The references are to volumes of his *Oeuvres Complètes* [reproduced from A.E. Bell, *Christiaan Huygens and the Development of Science in the Seventeenth Century* (Longmans Green & Co., New York, 1947) p. 194].

Figure 3. Image of Saturn taken from Voyager spacecraft on October 18, 1980 from a distance of 34 million km [courtesy of NASA and JPL].

and to the invention of an accurate pendulum clock, to control clockwork movement. He completed his first pendulum clock in 1656 and he described it in a book entitled *Horologium* two years later. But he continued to study and to develop pendulum clocks. We will say more about this later.

I already mentioned some of Huygens' trip to Paris. He continued to travel frequently. He was in Paris again in 1660 and there he met many of the scientists who a few years later would constitute the Académie Royale des Sciences. In 1661 he visited London, where he conferred with some of the men who founded the Royal Society, which was chartered a year later. Huygens was elected a Fellow of the Society shortly afterwards. He also received an award from King Louis XIV for his development of the pendulum clock. In 1666 the Académie Royale des Sciences was founded and Huygens became one of its prestigious members. He was given an official position at the Academy and he resided in an apartment at the Bibliothèque Royale, where the Academy had its Headquarters.

During the next twenty-four years Huygens lived in Paris, but he also spent extended periods, sometimes lasting several years, in the Hague. In 1689 he revisited London where, among others, he met Newton. Huygens' return visits to the Hague were largely connected with his health. He enjoyed good health in his youth but from the age of about forty, he developed bouts of illness, accompanied by violent headaches and insomnia. On some of these occasions he would return to the Hague, hoping that the air in his native land would help to cure him.

Let us return to Huygens' scientific activities. I already spoke of his invention of the pendulum clock in 1656. He continued to work on this subject and on numerous topics in mechanics and they were described in his famous book *Horologium Oscillatorium*, published 17 years later, in 1673. It was dedicated to his patron in Paris, King Louis XIV. Let me say a few words about the background.

In 1581 Galileo noted the approximately isochronous nature of the pendulum swings and he tried to use it to develop a clock. But it turned out that an ordinary pendulum, whose bob moves on a portion of a circle does not provide sufficient accuracy. Huygens studied the relation between the amplitude of oscillation of a pendulum and the time. He found that to achieve isochronous oscillations, it was necessary for the bob of the pendulum to follow not a circle but a curve known to mathematicians as a cycloid. Huygens constructed a pendulum of this kind.

Actually the pendulum clock never proved practical for sea navigation, but Huygens' book *Horologium Oscillatorium* has made an outstanding contribution to the science of mechanics, containing, for example, the general idea of one of the most important principles of mechanics (and indeed the whole of physics), the law of conservation of energy. Newton who read *Horologium Oscillatorium* considered Huygens "the most elegant writer of his time".

In his studies of mechanics Huygens investigated the laws of collision and he introduced the important concepts of the moment of inertia and of centrifugal force; and his studies of pendulum motion made it possible for him to make the first accurate determination of the value of the acceleration of gravity and to show that it varied with latitude. There are other notable contributions which Huygens has made to mechanics, but as Scott E. Barr, a contemporary physics historian, states in an as yet unpublished book: *The People who made Physics:* "Most of Huygens' brilliant contributions to mechanics have been absorbed into textbooks with no mention that he originated them". However, in quite a different field, namely in optics, his contributions are clearly recognized to this day and his name is familiar to everyone who has ever taken any serious interest in the subject.

Within a few years of the publication of *Horologium Oscillatorium,* two important discoveries were made in optics, which undoubtedly rekindled Huygens' earlier interest in this subject, from the days when he and his brother built telescopes and studied properties of lenses and other optical devices, such as the microscope. The first of these discoveries was the phenomenon of double refraction made in 1669 by a Danish professor of mathematics and later also of medicine, Erasmus Bartholinus. Bartholinus obtained some beautiful crystals from a sailor who collected them in Iceland and when he viewed small objects through them, Bartholinus found that the objects appeared double. Let me quote what Bartholinus said about his discovery:

> *"Greatly prized by all man is the diamond and many are the joys which similar treasures bring, such as precious stones and pearls.*
>
> *... but he who, on the other hand, prefers the knowledge of unusual phenomena to these delights, he will, I hope, have no less joy in a new sort of body, namely, a transparent crystal, brought to us from Iceland, which perhaps is one of the greatest wonders nature has produced."*

Bartholinus discovered the origin of this phenomenon. If one sends a narrow beam of light – a light ray – into an ordinary transparent medium such as a piece of glass, it is refracted and then proceeds as a single beam. However, when it is refracted at the face of the Iceland spar, a second beam is generated and this is the reason for the appearance of the second image. Bartholinus suggested that one of the rays, which resembles the usual one in some ways be called the *ordinary* ray and the other one, which behaves in a somewhat unusual fashion be called the *extraordinary* ray, a terminology which has been used ever since.

Before discussing Huygens' contributions to the understanding of this phenomenon let me mention another important discovery of basic importance for optics which was made six years later, in 1675 by Olaf Römer, a Dane, who was at that time a Professor of Mathematics in France. Until then it was not known whether or not light propagates instantaneously. Römer discovered from the observations of the eclipses of the satellites of Jupiter that light propagates with finite speed.

Incidentally Römer studied astronomy and mathematics under Bartholinus, lived in his house and later married his daughter. He had a colorful career. He was at one time the Astronomer Royal of Denmark, the Mayor of Copenhagen and the Prefect of the police.

Around the time when Römer discovered that light propagates with finite speed, Huygens was ill again and he returned to the Hague from Paris in July 1676, staying in his country home for almost two years. It was during this period of recovery from his illness that Huygens formulated what is undoubtedly his major and lasting contribution to science, namely the wave theory of light; but it was not until twelve years later, in 1690, that he published an account of the theory in his famous book *Traité de la Lumière.* It is the publication of this book, which has become a landmark in the history of optics, that we are celebrating here today.

I may best explain the reasons for the long delay between Huygens' conception of these important ideas and his reporting them in published form by quoting his own words from the Preface of *Traité de la Lumière.*

> *"I wrote this Treatise during my sojourn in France twelve years ago, and I communicated it in the year 1678 to the learned persons who then composed the Royal Academy of Science, to the membership of which the King had done me the honour of calling me....*
>
> *...One may ask why I have so long delayed to bring this work to the light. The reason is that I wrote it rather carelessly in the Language in which it appears, with the intention of translating it into Latin, so doing in order to obtain greater attention to the thing.*
>
> *...But the pleasure of novelty being past, I have put off from time to time the execution of this design, .."*

Let me add here that anyone who is engaged in active research or in the supervision of graduate students will note from the last sentence quoted here how little has changed in this respect in the more than 300 years since Huygens wrote these words!

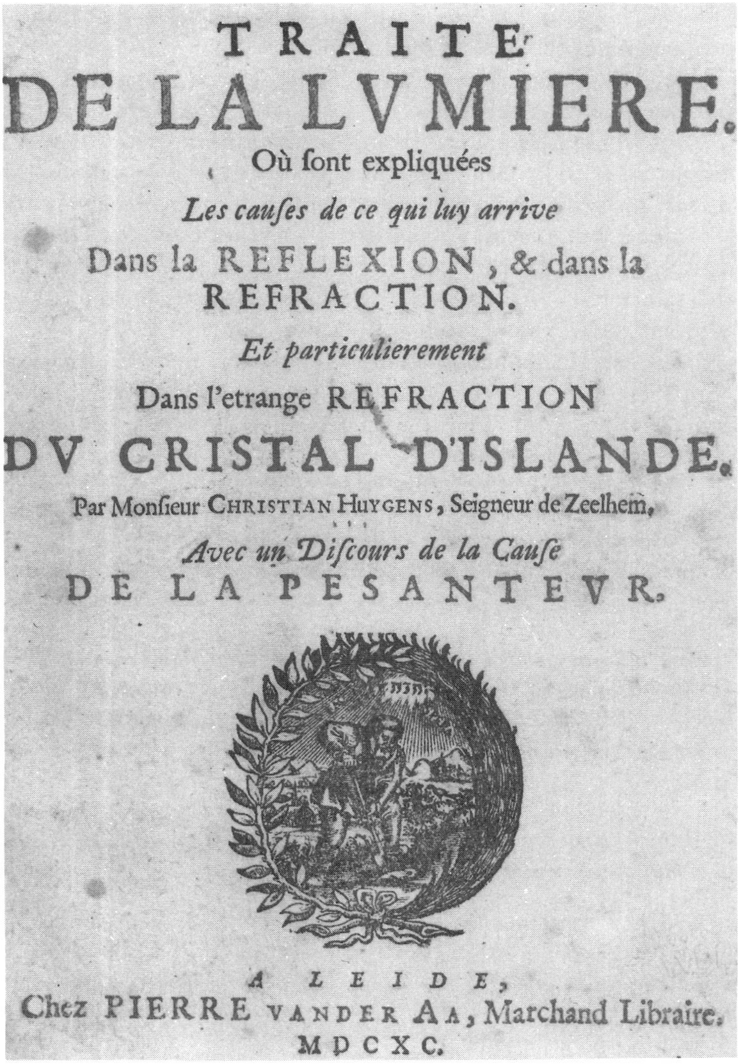

TRAITE

DE LA LVMIERE.

Où ſont expliquées

Les cauſes de ce qui luy arrive

Dans la REFLEXION, & dans la REFRACTION.

Et particulierement

Dans l'etrange REFRACTION

DV CRISTAL D'ISLANDE.

Par Monſieur CHRISTIAN HUYGENS, Seigneur de Zeelhem,

Avec un Diſcours de la Cauſe

DE LA PESANTEVR.

A LEIDE,

Chez PIERRE VANDER AA, Marchand Libraire.

MDCXC.

Figure 4. Title page from the original edition of Christiaan Huygens' *Traité de la Lumière.*

A very clear statement of why *Traité de la Lumière* has become an all-time classic of the optics literature was made by Ernst Mach, a philosopher of science who wrote in 1913 [1].

> *"In Huygens' "Traité de la Lumière", appears the first attempt to correlate and render consistent the different properties of light which had from time to time been discovered, and to explain some by others, so as to reduce the number of necessary fundamental properties and the number of fundamental concepts indispensable for a correct comprehension of the underlying processes. The principal fruits of Huygens' labours lay in his demonstration of the possibility of deriving all the essential features of rectilinear propagation, reflection, and simple and double refraction from the rate of propagation of light".*

The words *rate of propagation* in this quote from Mach's book is crucial here, because it stresses how basic in Huygens considerations was the finite speed of light, first demonstrated, as we saw, by Römer a few years earlier. It suggested that the spreading of light was either by means of some process resembling moving projectiles or by means of motion of some sort of matter (as Huygens put it), which would have to exist in the space between the luminous object and the human eyes.

In more modern terminology the "projectile model" is the basis of the socalled *corpuscular* or *emission theory* of light; the other is the nucleus of the *wave theory* of light. Both models have actually been suggested previously by others. Newton was prominently associated with the corpuscular theory, although he too interpreted some optical phenomena in terms of waves. Hooke proposed a primitive wave theory of light as early as 1664. However, neither Newton nor Hooke could provide satisfactory explanations of various optical phenomena then known by only one or the other of these models.

Of these two competing hypotheses, Huygens comes firmly on the side of the wave theory. The kinds of waves which were perhaps most familiar at that time were sound waves and Huygens believed that propagation of light resembles that of sound. But sound can only propagate in the presence of matter, which transmits the acoustical impulses via the elastic properties of the medium. On the other hand it was known from earlier experiments of Torricelli, that light propagates even in empty space. Huygens, therefore, conjectured that all space is filled with a "subtile" medium, which he called ethereal matter, consisting of tiny particles: and that light propagates through it by successive collisions of these particles, as a kind of a shock wave which is transmitted through a row of elastic particles. Utilizing the theory of impact which he formulated some years earlier in his studies of mechanics, Huygens showed that such a model would lead to finite speed of propagation of light.

In regards to the origin of these waves and the manner in which they spread let me describe this in Huygens own words and with reference to figures reproduced below from *Traité de la Lumière*. He says (see fig. 5a):

> *"... each little region of a luminous body, such as the sun, a candle, or a burning coal, generates its own waves of which that region is the centre. Thus in the flame of a candle, having distinguished the points A, B, C, concentric circles described about each of these points represent the waves which come from them. And one must imagine the same about every point of the surface and of the part within the flame".*

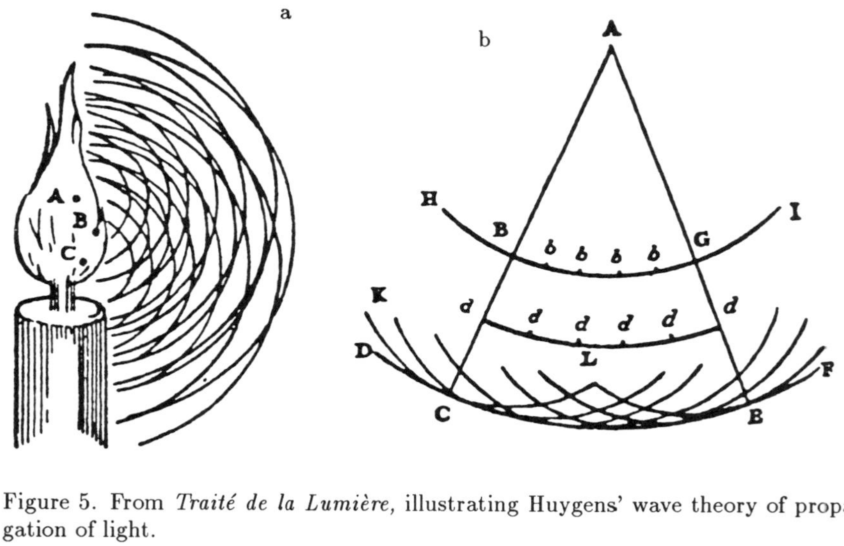

Figure 5. From *Traité de la Lumière,* illustrating Huygens' wave theory of propagation of light.

And a little further on, Huygens explained the manner in which light propagates by means of a new principle, which now bears his name and which appears in the title of this conference. It contains one of the most fruitful concepts of all of optics. It is illustrated in figure 5b and may be stated as follows:

In free space light travels from a point A of a luminous source in the form of an expanding spherical wave. Each point on the sphere which the light reaches at a particular instant of time – or as one now says the spherical wavefront – such as $HBGI$, becomes a source of a new elementary wave – called a partial wave or secondary wavelet. At any later instant of time, these secondary wavelets, such as KCL, produce a new wavefront, denoted by $DCBF$ say which, in mathematical language, is their envelope and which gives the correct position of the spherical wavefront at that time instant.

Further on Huygens formulated an extension of this principle from propagation in free space to propagation in media, the main modification arising from the assumption that the speed of propagation of light depends on the type of medium which is being considered. Using this principle he was able to explain the rectilinear propagation of light in free space and to deduce the laws governing reflection and refraction of light.

In the second part of *Traité de la Lumière*, Huygens investigated the phenomenon of double refraction, discovered by Bartholinus, which I already mentioned. We saw that Bartholinus associated one of the images observed through the crystal with ordinary rays and the other image with the extraordinary rays. Now Huygens showed in the first part of *Traité de la Lumière* that the laws governing the refraction of ordinary rays can be deduced from his Principle by assuming that the secondary wavelets are spherical. In the case of double refraction, the situation is obviously more complicated and Huygens solved the problem in a rather ingenious manner. He assumed that when light is incident on the Iceland spar, each element of it produces secondary wave surfaces which are no longer spherical but rather consist of two geometrical forms or, as one says, they consist of two sheets. One of the sheets is again spherical and is associated with the ordinary rays. The other sheet has the form of an ellipsoid and is associated with the extraordinary rays. Using this extended version of his Principle, Huygens explained all the new optical phenomena observed with the Iceland spar up to that time, except for one, which he himself discovered and described in the later part of *Traité de la Lumière.* I will return later to the phenomenon whose explanation eluded him.

As we have just seen Huygens' Principle in the form in which it was originally formulated in *Traité de la Lumière* enabled Huygens to explain both ordinary refraction and double refraction. This success, together with some subsequent developments, paved the way towards the general acceptance of the wave theory of light and the corpuscular theory fell into oblivion until the early years of this century.

But history of science teaches us that no scientific theory is ever quite complete and that as time goes on refinements and extensions take place, often as a consequence of new experimental observations. Huygens' Principle was no exception in this respect. In its original formulation Huygens' secondary wavelets behaved as if they were mutually independent and did not influence each other. More than a hundred years after the publication of *Traité de la Lumiére* a British physician, Thomas Young, discovered that when different wavelets reach the same points in space as, for example, happens when light emerges from two small openings, patterns of dark and bright bands appear known as interference fringes (see figure 6). The law governing the formation of such patterns is called the *Principle of Interference.*

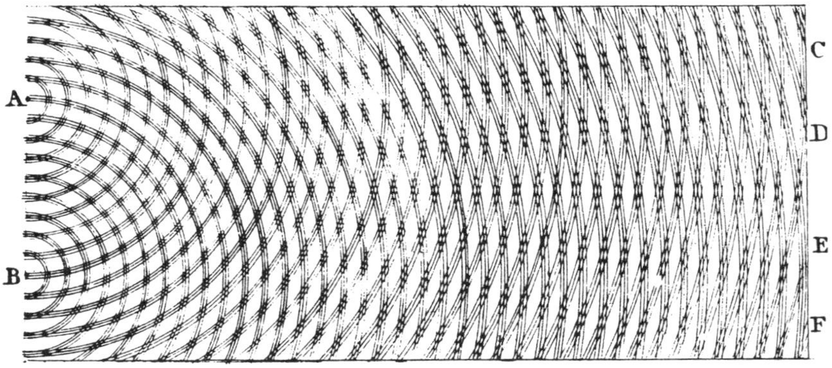

Figure 6. Young's illustration of his Principle of Interference, reproduced from the second volume of his treatise *A Course of Lectures on Natural Philosophy and the Mechanical Arts*, published in 1807.

About seventeen years later Augustin Fresnel, a French engineer and scientist, submitted to the French Academy of Sciences a celebrated memoir in which he provided a synthesis of Huygens' Principle and of Young's Principle of Interference. This synthesis enabled Fresnel to explain a rather puzzling phenomenon, known as diffraction, discovered by Grimaldi around 1660. It is manifested by the fact that the path of some light rays deviates from straight lines when they pass close to edges of apertures and boundaries of bodies or when they penetrate into shadow regions. Fresnel's explanation of these phenomena by means of Huygens' Principle, supplemented by Young's Principle of Interference, was one of the crowning triumphs of the wave theory of light.

I mentioned that there was one phenomenon which Huygens himself discovered and described in *Traité de la Lumière*, but which he has not succeeded in explaining. Because this discovery has played a major role in the subsequent development of the wave theory of light let me say a few words about it.

We saw that when light illuminates Iceland spar, two kinds of rays are formed, namely ordinary rays and extraordinary ones. If these rays are then allowed to fall on another piece of Iceland spar, one might expect that each would be again divided into two rays. However, on experimenting with such crystals Huygens found that this, in fact, did not happen. The property of light which later provided the explanation of this discovery is known as *polarization*. But it took another century before it was explained. The explanation was provided by Thomas Young who, as we saw, introduced the Principle of Interference. Young's explanation involved a major modification of the accepted view regarding the nature of light. Until then it was taken for granted that light waves resemble sound waves in that their oscillations are along their direction of propagation or, as one says, that they are *longitudinal* waves. Thomas Young suggested that the aetheral vibrations associated with the propagation of light are perpendicular to the direction of propagation, or as one says that they are *transverse* waves. The observations of Huygens which I just mentioned and which he was unable to explain were completely clarified by means of the concept of transverse waves. Although the concept of aetheral vibrations was eventually abandoned, the basic correctness of Young's view regarding the nature of polarization was established many years later, after a Scottish scientist James Clerk Maxwell discovered that light consists of transverse oscillations of electric and magnetic forces.

In the extended formulation which incorporates interference of the secondary wavelets, Huygens' Principle has continued to flourish and to play an important role in the development of physics. For example, in the 1920's physics has undergone a major revolution. One of its consequences was the discovery that elementary particles of matter, such as electrons, for example, may also exhibit wave-like behavior, and conversely that waves may under certain circumstances exhibit particle-like features. In a fundamental paper [2] on this subject known as quantum mechanics, published in 1926, Nobel laureate Erwin Schrödinger, one of the founders of quantum mechanics, invoked Huygens' Principle to elucidate the transition from the old to the new mechanics.

In another important paper [3] published in 1938, which dealt with quite a different subject, the Dutch Nobel laureate Frits Zernike used Huygens' Principle to elucidate how certain statistical features of light, known as its coherence properties, are transmitted on propagation.

In a 1948 publication [4] yet another Nobel laureate Richard Feynman made use of Huygens' Principle, in the so-called space-time formulation of quantum mechanics. Feynman notes in that paper that whilst Huygens' Principle is only approximately valid in optics, it is valid rigorously in quantum mechanics.

From the preceding remarks one might perhaps gain the impression that Huygens' Principle is only of interest to physicists. That this is not so is evident from the fact that even today, more than three centuries since Huygens annunciated it, mathematical texts are published which are devoted to its applications in the theory of differential equations [5].

Traité de la Lumière, whose 300th anniversary we are commemorating here today, was the culmination of Huygens' efforts to develop the science of optics in a clear and logical manner. In this he succeeded magnificently. After the publication of this treatise, Huygens continued with optical investigations, especially in connection with microscopes and telescopes. But in July 1694 he became ill again and his condition gradually worsened. He died in 1695 in this city.

It is clear that Huygens has left a rich legacy. Of course, like all theories, Huygens' views regarding the nature of light and, in particular, the celebrated Principle which bears his name were, as we saw, only partial truths. But as the sixteenth-century philosopher Giodano Bruno apparently said in a different context *"If it is not true, it was a very good idea"*.

Acknowledgement

I wish to express my thank to Mr. Kenn Harper, Head of the Physics/Astronomy/Optics Library at the University of Rochester for much assistance with obtaining some of the photographs and locating pertinent references needed in the preparation of this article. I am also grateful to Dr. E. McNevin of the Jet Propulsion Laboratory of the California Institute of Technology for help in securing the beautiful image of Saturn (fig. 3).

In addition to Huygens' own publications, a number of books and articles influenced considerably this presentation. I particularly wish to acknowledge the following sources:
E. Segrè, *From Falling Bodies to Radio Waves* (W.H. Freeman, New York, 1984).
H.A. Boorse and L. Motz, *The World of the Atom* (Basic Books Inc., New York, 1966), vol. I.
J.A. van Maanen, contribution in *Studies on Christiaan Huygens*, H.J.M. Bos, M.J.S. Rudewick, H.A.M. Snelders and R.P.W. Visser (Swets & Zeitlinger, Lisse, 1980).
A.E. Bell, *Christiaan Huygens and the Development of Science in the Seventeenth Century* (Longmans Green & Co., New York, 1947).

References

[1] E. Mach, *The Principles of Physical Optics* (Dutton, New York, 1925); reprinted by Dover, New York, 1953, chapt. XIII.

[2] E. Schrödinger, Ann. d. Phys. **79** (1926) 489 [§1].

[3] F. Zernike, Physica **5** (1938) 785 [§4].

[4] R. Feynman, Rev. Mod. Phys. **20** (1948) 367 [§7].

[5] See for example, P. Günther, *Huygens' Principle and Hyperbolic Equations* (Academic Press, New York, 1988).

Section 7 – Lectures

This Section contains accounts of five previously unpublished lectures, which I have given on various occasions to widely different audiences, consisting mainly of students, academic faculty and the staff of industrial laboratories.

Lecture 7.1 deals with the important subject of dispersion relations and describes its roots in the concepts of analyticity and causality. Lecture 7.2 is a somewhat informal presentation of the highlights from the life and work of twenty-one famous scientists who made major contributions to our understanding of the nature of light and whose work helped to lay the foundations of optics. Lecture 7.3 outlines developments which have lead to the understanding of optical coherence.

Lecture 7.4 is of a rather personal nature, concerned largely with my scientific career. The chapter concludes in a lighter vein, with lecture 7.5, which relates how I came to Rochester more than four decades ago and describes some of my observations and experiences at the University of Rochester over this long period of time.

Section 7 – Lectures

Lecture 7.1

Analyticity, Causality and Dispersion Relations *

Emil Wolf
University of Rochester

1 Introduction

There are many different kinds of theories in physics. Some of them are *ad hoc* theories which are formulated to solve a particular class of problems and are usually based on models specifically developed for that purpose. They generally involve various approximations, often based on intuition or on a good guess. But there are other kinds of theories that have the status of basic principles of physics. An obvious example is the theory of relativity. The theory of dispersion relations, the elements of which I will now review, also belongs to this second category. It is a beautiful theory which provides elegant and deep answers to some important questions, and it also reveals why analytic functions are the proper mathematical tools for the analysis of many problems encountered in physics and in engineering.

2 A brief historical review

Let me begin with a few historical remarks. Around 1925 there was a great deal of interest in problems involving the interaction between radiation and matter. This interest arose largely from attempts to understand the full significance of the various quantum rules which were formulated in the preceeding 25 years in a somewhat *ad hoc* manner, following Planck's introduction of the quantum of action. Very prominent among them were attempts by Kramers and some of his collaborators. One of them was Heisenberg. They were particularly interested in the dispersion of light by atoms. In the course of this work Kramers,[§] and around the same time Kronig,[†] independently discovered some relatively simple relations which connect the macroscopic parameters that characterize the

*This article is based on an evening lecture which I gave annually for many years in the 1980s and 1990s at the University of Rochester. The lecture supplemented a regular first-year graduate course on methods of mathematical physics which I taught. It was followed by a party at which drinks, cookies and cakes, prepared by my wife, Marlies, were served.

[§]H.A. Kramers, *Atti Congr. Int. Fis. Como* **2**, 545 (1927); reprinted in H.A. Kramers, *Collected Scientific Papers* (North Holland, Amsterdam, 1956), p. 333.

[†]R. de L. Kronig, *J. Opt. Soc. Am. and Rev. Sci. Instrum.* **12**, 547 (1926).

dispersion and the absorption properties of certain media. Let me state the main result obtained in these investigations.

Suppose that $\mathcal{E}$ is the electric field in the medium, $\mathcal{D}$ the electric displacement vector and $\mathcal{P}$ the induced polarization. These three vectors are related by the usual formula

$$\mathcal{D} = \mathcal{E} + 4\pi\mathcal{P}. \tag{1}$$

We assume that the three fields are monochromatic, i.e. that they are of the form

$$\mathcal{E}(\mathbf{r}, t) = \mathrm{Re}\left\{\mathbf{E}(\mathbf{r}, \omega)e^{-\mathrm{i}\omega t}\right\}, \tag{2}$$

etc., where Re denotes the real part and the frequency $\omega \geq 0$. For simplicity we also assume that the medium is homogeneous, isotropic and linear, with local response. Then at each point $\mathbf{r}$, the vectors $\mathbf{P}$ and $\mathbf{E}$ will be related by the formula

$$\mathbf{P}(\omega) = \eta(\omega)\mathbf{E}(\omega), \tag{3}$$

where, for simplicity, the spatial argument $\mathbf{r}$ is omitted. The quantity $\eta(\omega)$ is the dielectric susceptibility, which is a function of the frequency ω and is complex, in general, say

$$\eta(\omega) = \eta_1(\omega) + \mathrm{i}\eta_2(\omega), \tag{4}$$

where η_1 and η_2 are real. The real part, η_1, characterizes the dispersion properties of the medium (i.e. the spatial redistribution of the different frequency components of the light passing through the medium), whereas the imaginary part, η_2, characterizes the absorption properties (energy losses) on propagation through the medium. One can readily connect η_1 and η_2 with other well-known macroscopic parameters which characterize dispersion and absorption, e.g. to the real-valued refractive index, to the dielectric constant and to the absorption coefficient of the medium.

From a physical model for the processes of dispersion and absorption Kramers showed that $\eta_1(\omega)$ and $\eta_2(\omega)$ are not independent, but are coupled by the following two relations:

$$\eta_1(\omega) = \frac{2}{\pi}\mathrm{P}\int_0^\infty \frac{\omega'\eta_2(\omega')}{\omega'^2 - \omega^2}\mathrm{d}\omega', \tag{5a}$$

$$\eta_2(\omega) = -\frac{2}{\pi}\mathrm{P}\int_0^\infty \frac{\omega'\eta_1(\omega')}{\omega'^2 - \omega^2}\mathrm{d}\omega', \tag{5b}$$

where P denotes the Cauchy principal values of the integrals, taken at the singularity $\omega' = \omega$, of the integrands.

The relations (5) imply that if we knew the absorption properties of the medium characterized by η_1 for all frequencies, we could determine its dispersion properties characterized by η_2 and *vice versa*. These relations may be expressed in a more symmetric and mathematically more significant form if we define $\eta(\omega)$ formally also for negative frequencies, by the relation

$$\eta(-\omega) = \eta^*(\omega). \tag{6}$$

If we take the real and the imaginary parts of this relation we obtain definitions of $\eta_1(\omega)$ and $\eta_2(\omega)$ also for negative frequencies, namely

$$\eta_1(-\omega) = \eta_1(\omega), \tag{7a}$$

$$\eta_2(-\omega) = -\eta_2(\omega). \tag{7b}$$

Using the relations (7), the formulas (5) may readily be expressed in the following alternative and more symmetric form:

$$\eta_1(\omega) = \frac{1}{\pi}\mathrm{P}\int_{-\infty}^{\infty} \frac{\eta_2(\omega')}{\omega' - \omega}\mathrm{d}\omega', \tag{8a}$$

$$\eta_2(\omega) = -\frac{1}{\pi}\mathrm{P}\int_{-\infty}^{\infty} \frac{\eta_1(\omega')}{\omega' - \omega}\mathrm{d}\omega'. \tag{8b}$$

The relations (8) [or (5)] are known today as the *Kramers-Kronig relations* or *dispersion relations.* They couple the real and the imaginary parts of the dielectric susceptibility by a pair of linear transforms, namely Hilbert transforms. $\eta_1(\omega)$ is the Hilbert transform of $\eta_2(\omega)$ and $\eta_2(\omega)$ is the inverse Hilbert transform of $\eta_1(\omega)$.

As I already mentioned, Kramers and Kronig derived these relations from a physical model of the processes of dispersion and absorption. However, the coupling of $\eta_1(\omega)$ and $\eta_2(\omega)$ by the Hilbert transform relations has a much wider significance, as we will soon see. Later it was found that similar relations also hold in many other, and quite different, physical situations. For example they were found to apply to the complex reflectivities of various substances, to the complex amplitudes of waves scattered from various media, and also in connection with various problems in quantum physics, especially in high energy physics and in the theory of the scattering matrix. Because relations of this kind first appeared in connection with the dispersion of light, they are generally called *dispersion relations*, irrespective of the particular physical situation.

Since such relations are encountered in so many different branches of physics one might suspect that they have a common physical origin. A clue to the answer was provided already by Kramers in his paper in which these relations have first appeared. Kramers showed, on the basis of his model for the interaction of light with a linear medium, that *if an electromagnetic wave is incident on the medium, the induced polarization at any point in the medium remains zero until the wave reaches that point, i.e.*

$$\left.\begin{aligned} &\text{if} \quad \mathcal{E}(\mathbf{r}_0, t) = 0 \quad \text{for} \quad t < t_0, \\ &\text{then} \quad \mathcal{P}(\mathbf{r}_0, t) = 0 \quad \text{for} \quad t < t_0. \end{aligned}\right\} \tag{9}$$

This is a kind of causality requirement: "No output before input", and the dispersion relations were later found to be intimately related to this fact.

Subsequent research has shown that the true origin of the dispersion relations lies in

1. the linearity of the process,
2. the time translation invariance of the process,
3. causality,

and that mathematically these three properties together imply that the associated response functions are boundary values on the real axis of functions that are analytic and regular in one half of an associated complex plane. We will now show that this is so.

3 The response function of a linear, time-translation invariant causal system

It will be convenient to use the language of electrical engineering. Let us consider a system which transforms an input signal, $F(t)$, into an output signal $G(t)$. Symbolically,

$$G(t) = \mathcal{T}[F(t)]. \tag{10}$$

Here $F(t)$ is a generalization of the electric field $\mathcal{E}(\mathbf{r}_0, t)$, $G(t)$ is a generalization of the induced polarization $\mathcal{P}(\mathbf{r}_0, t)$ of the interaction problem treated by Kramers and Kronig and $\mathcal{T}$ is an operator which represents the transformation.

In mathematical language, the three assumptions that I just mentioned mean the following: If $F_1(t)$ and $F_2(t)$ are any two input signals, then

(1) Linearity

$$\text{(a)} \quad \mathcal{T}[F_1(t) + F_2(t)] = \mathcal{T}[F_1(t)] + \mathcal{T}[F_2(t)], \tag{11a}$$

$$\text{(b)} \quad \mathcal{T}[aF(t)] = a\mathcal{T}[F(t)], \quad (a = \text{any constant}). \tag{11b}$$

The two requirements (11a) and (11b) may be combined into one, but for our purposes it is more convenient to consider their implications separately.

(2) Time-translation invariance

$$\left.\begin{array}{ll} \text{If} & \mathcal{T}[F(t)] = G(t) \\ \text{then} & \mathcal{T}[F(t+t_0)] = G(t+t_0), \end{array}\right\} \tag{12}$$

where t_0 is any real number. This requirement implies that a temporal shift in the input as a whole will cause a corresponding temporal shift in the output. An example of this is the electromagnetic interaction with a medium whose macroscopic properties do not change in time, e.g. when the medium is a piece of glass, say, which is at rest.

(3) Causality (weak formulation)

$$\left.\begin{array}{llll} \text{If} & F(t) = 0 & \text{for} & t \le t_1 \\ \text{then} & G(t) = 0 & \text{for} & t \le t_1, \end{array}\right\} \tag{13}$$

i.e. *no output before input* (the effect cannot precede the cause).

These three assumptions can be shown to imply that the input and the output functions are necessarily related by a formula of the form

$$G(t) = \int_{-\infty}^{\infty} F(t')R(t-t')\mathrm{d}t', \tag{14}$$

where the function $R(t)$, which characterizes the system, satisfies the constraint

$$R(t) \equiv 0 \quad \text{for} \quad t < 0. \tag{15}$$

Because the relation (14), together with the constraint (15), are basic for the validity of the dispersion relations, let me sketch at least a rough, non-rigorous proof of it.

We may formally represent $F(t)$ in the form

$$F(t) = \int_{-\infty}^{\infty} F(t')\delta(t-t')\mathrm{d}t', \tag{16}$$

where δ is the Dirac delta function. Then, according to Eqs. (12) and (16),

$$\begin{aligned} G(t) = \mathcal{T}[F(t)] &= \mathcal{T}\left[\int_{-\infty}^{\infty} F(t')\delta(t-t')\mathrm{d}t'\right] \\ &= \int_{-\infty}^{\infty} \mathcal{T}[F(t')\delta(t-t')]\mathrm{d}t' \quad \text{(by linearity, Eq. (11a), with } \Sigma \to \textstyle\int) \\ &= \int_{-\infty}^{\infty} F(t')\mathcal{T}[\delta(t-t')]\mathrm{d}t'. \quad \text{(by linearity (11b))} \end{aligned} \tag{17}$$

Let us take for $F(t)$ the Dirac delta function, i.e. $F(t) = \delta(t)$. Equations (10) and (14) then imply that

$$\mathcal{T}[\delta(t)] \equiv R(t). \tag{18}$$

Because the system is assumed to be invariant with respect to time-translation [Eq. (12)],

$$\mathcal{T}[\delta(t-t')] = R(t-t'). \tag{19}$$

On substituting from Eq. (19) into (17) it follows at once that

$$G(t) = \int_{-\infty}^{\infty} F(t')R(t-t')\mathrm{d}t', \tag{20}$$

which is the relation (14). For obvious reasons [cf. Eq. (18)] $R(t)$ is called the *impulse response function* of the system.

So far we have used only the assumptions of linearity and of time-translation invariance. Let us now also impose the requirement of *causality*. For this purpose let us choose again

$$F(t) = \delta(t). \tag{21}$$

It then follows from Eq. (20) that

$$G(t) = R(t). \tag{22}$$

Since

$$\underbrace{\delta(t)}_{\substack{F(t) \\ \text{(input function)}}} = 0 \quad \text{for} \quad t < 0,$$

causality [Eq. (13)] then demands that the corresponding output function

$$\underbrace{R(t)}_{\substack{G(t)\\ \text{(output function)}}} = 0 \quad \text{for} \quad t < 0, \tag{23}$$

which is the requirement (15).

This completes the heuristic proof of Eqs. (14) and (15).

4 Mathematical consequences of Eqs. (14) and (15) and the dispersion relations

The relatively simple-looking results expressed by Eqs. (14) and (15) have some important mathematical consequences which we will now discuss. For simplicity we assume that the impulse response function $R(t)$ is square-integrable, i.e. that

$$\int_0^\infty |R(t)|^2 \mathrm{d}t < \infty. \tag{24}$$

$R(t)$ may then be expressed as a Fourier integral, i.e. in the form

$$\underbrace{R(t)}_{\text{impulse response function}} = \frac{1}{\sqrt{2\pi}} \int_{-\infty}^{\infty} r(\omega) e^{-\mathrm{i}\omega t} \mathrm{d}\omega, \tag{25}$$

where $r(\omega)$ is given by the inverse Fourier transform

$$\underbrace{r(\omega)}_{\text{frequency response function}} = \frac{1}{\sqrt{2\pi}} \int_0^\infty R(t) e^{\mathrm{i}\omega t} \mathrm{d}t, \tag{26}$$

with the lower limit being zero rather than $-\infty$ in view of Eq. (15); moreover, since $R(t)$ is square-integrable, $r(\omega)$, which is often called the frequency response function of the system, is also square-integrable, i.e.

$$\int_{-\infty}^{\infty} |r(\omega)|^2 \mathrm{d}\omega < \infty. \tag{27}$$

The appearance of the lower limit zero in Eqs. (26) and (24) is, of course, a consequence of our assumption of causality [Eq. (23)]. This innocent-looking fact has some far-reaching implications, as we will now show. For this purpose we formally replace in the integral (26) the real frequency ω by a complex frequency

$$\Omega = \omega + \mathrm{i}\zeta \quad (\omega,\ \zeta \quad \text{real}). \tag{28}$$

Then $e^{\mathrm{i}\omega t} \to e^{\mathrm{i}\Omega t}$ and since, for each t, the function $e^{\mathrm{i}\Omega t}$ is an entire analytic function of Ω, so is, therefore, the function $R(t)e^{\mathrm{i}\Omega t}$. Now a sum of regular analytic functions is also a regular analytic function and in an appropriate limit so will be the integral $\int_0^\infty R(t)e^{\mathrm{i}\Omega t}\mathrm{d}t$, provided $R(t)$ behaves sufficiently well and that the integral converges uniformly, which we will assume. Using Eq. (28), we have

$$\int_0^\infty R(t)e^{\mathrm{i}\Omega t}\mathrm{d}t = \int_0^\infty R(t)e^{\mathrm{i}\omega t}e^{-\zeta t}\mathrm{d}t \tag{29}$$

and we see that when $\zeta > 0$, i.e. when $\mathrm{Im}\,\Omega > 0$, the factor $e^{-\zeta t}$ increases the rapidity of convergence (when $\zeta < 0$ the integral diverges). This suggests that provided $R(t)$ is reasonably well-behaved, its Fourier transform $r(\omega)$, i.e. the frequency response function of the system given by the integral (26), will be the boundary value on the real ω-axis of a function that is analytic and regular in the upper half of the complex frequency plane; and from what we have learned in our class we can then expect that the real and the imaginary parts of $r(\omega)$ will be coupled by the Hilbert transform relations.

This entire discussion was rather heuristic. The exact mathematical formulation is given by a number of theorems that can be found in a well-known book by Titchmarsh on the Fourier integral and hence for this reason these theorems are collectively referred to by physicists as *Titchmarsh's theorem.* They may be stated as follows:[‡]

If $r(\omega)$ is a square-integrable function and fulfills any one of the following four conditions it fulfills all of them:

(I) The inverse Fourier transform $R(t)$ of $r(\omega)$ vanishes for all $t < 0$, i.e.,

$$R(t) = \frac{1}{\sqrt{2\pi}}\int_{-\infty}^\infty r(\omega)e^{-\mathrm{i}\omega t}\mathrm{d}\omega = 0 \quad \text{when} \quad t < 0.$$

(II) $r(\omega)$ is, for almost all ω (in the sense of the theory of sets), the limit as $\zeta \to +0$ of an analytic function $r(\omega + \mathrm{i}\zeta)$ which is regular in the upper half ($\zeta > 0$) of the complex frequency plane and is square-integrable over any line parallel to the real axis:

$$\int_{-\infty}^\infty |r(\omega + \mathrm{i}\zeta)|^2\mathrm{d}\omega < C, \qquad (\zeta > 0, \quad C = \text{constant})\,.$$

(III) Its real part $r_1(\omega)$ is the Hilbert transform of its imaginary part $r_2(\omega)$:

$$r_1(\omega) = \frac{1}{\pi}\mathrm{P}\int_{-\infty}^\infty \frac{r_2(\omega')}{\omega' - \omega}\mathrm{d}\omega'.$$

(IV) Its imaginary part $r_2(\omega)$ is the inverse Hilbert transform of its real part $r_1(\omega)$:

$$r_2(\omega) = -\frac{1}{\pi}\mathrm{P}\int_{-\infty}^\infty \frac{r_1(\omega')}{\omega' - \omega}\mathrm{d}\omega'.$$

We have seen earlier that the impulse response function $R(t)$ of a linear, time-translation invariant, causal system necessarily vanishes for all $t < 0$. Hence if we also assume that it is square-integrable, then its Fourier transform $r(\omega)$ (i.e. the frequency response function) is also square-integrable and according to the statement (II) of Titchmarsh's theorem, $r(\omega)$ is the boundary value

[‡]E.C. Titchmarsh, *Introduction to the Theory of Fourier Integrals*, 2nd ed. (Oxford Univ. Press, London and New York, 1948), pp. 125–129.

on the real frequency axis of a function that is analytic and regular in the upper half of the complex frequency plane. Moreover, accoring to the statements (III) and (IV) its real and imaginary parts are coupled by Hilbert transform relations.

It is quite remarkable that such a small number of assumptions about the system (linearity, time-translation invariance and causality), and the square integrability of its response function, which are satisfied in many physical situations, should have such very broad consequences. In particular these results explain to some extent why analytic functions are the natural mathematical tool for analyzing a great variety of problems that arise in physics. They also establish the validity of dispersion relations for many quantities of physical interest. These results imply as a by-product that the model which Kramers and Kronig used in their analysis of the interaction of light with matter was causal, as we already noted.

There are some generalizations of these results especially to response functions which are not square integrable. They are discussed, for example, in Section 1.7 of the book by H.M. Nussenzveig, cited in the list of references at the end of this article.

The main conclusions to which we have been led by our analysis are that the frequency response functions of all linear, time-translation invariant, causal systems necessarily satisfy dispersion relations of the Kramers-Kronig type. We have also seen that under these circumstances frequency response functions are boundary values on the real frequency axis of analytic functions which are regular in the upper half of the complex frequency plane.

We see, as a byproduct of our analysis, that there exists an intimate relation between a basic concept of physics – namely that of causality, and a wide class of functions – namely analytic functions of a complex variable. This is one of the reasons why analytic functions are the appropriate mathematical tool for the treatment of many problems that one commonly encounters in physics.

Some references on dispersion relations

1. J.S. Toll, "Causality and the Dispersion Relation: Logical Foundations", *Phys. Rev.* **104**, pp. 1760–1770 (1956).

2. H.M. Nussenzveig, *Causality and Dispersion Relations*, (Academic Press, New York, 1972), especially chapter 1.

3. H.M. Nussenzveig, "Causality and Analyticity in Optics", in *Optics in Four Dimensions – 1980* (ICO, Ensenada), M.A. Machado and L.M. Narducci, eds. (Conference Proceedings #65, American Institute of Physics, New York, 1981), pp. 9–30.

4. N.N. Bogoliubov and D.V. Shirkov, *Introduction to the Theory of Quantized Fields*, (Interscience, New York, 1959), Sec. 46.2.

5. R.L. Weaver and Y. Pao, "Dispersion Relations for Linear Wave Propagation in Homogeneous and Inhomogeneous Media", *J. Math. Phys.* **22**, pp. 1909–1918 (1981).

6. M. O' Donnell, E.T. Jaynes and J.G. Miller, "Kramers-Kronig Relationship between Ultrasonic Attenuation and Phase Velocity", *J. Acoust. Soc. Am.* **69**, pp. 696–701 (1981).

Lecture 7.2

Scientists Who Created the World of Optics*

Emil Wolf
University of Rochester

This lecture outlines the lives and the contributions of twenty-one great scientists, from Galileo Galilei to Paul Dirac, who have elucidated the nature of light and who have laid the foundations to the science of optics. The lecture is illustrated by portraits of these scientists and also by drawings and by photographs which help to place their work in proper perspective.

Let me begin by saying a little about the background which led me to prepare this lecture. Some years ago Dr. Duncan Moore, a former Director of the Institute of Optics at the University of Rochester suggested that I assemble a collection of photographs of famous optical scientists which could then be permanently displayed in the Institute. I was glad to take on this assignment, but little did I know how hard it is to get good pictures of so many scientists. Some of them lived long before photography was invented (which was in the first half of the 19th century) so I had also to search for paintings of some of them. The collection is now complete and the photographs and reproductions of the paintings are hanging in a hallway at the Institute, which leads to the Director's office.

In this talk I will be speaking about the 21 scientists whose pictures I have collected. But before doing so let me make some general remarks. Any selection of the sort that I was undertaking is obviously somewhat subjective, so I should mention the criteria which I have used. First, the wall space available in the Institute is obviously not unlimited, so I had to be somewhat strict as to whom to include. I have decided to select pictures of those scientists who were mainly responsible for having elucidated the nature of light or who have made other contributions to optics of a fundamental nature. Thus many scientists and engineers who have pioneered technological applications in fields such as lasers, opto-electronics or non-linear optics, for example, are not included.

Another criterion that I have used and which was actually very easy to implement in the selection process was that they should all be dead. This has undoubtedly avoided lots of potential arguments!

My talk is not meant to be a scholarly lecture about all the wonderful things which the 21 scientists who are included in our new picture gallery have discovered. Rather I intended this talk

*This article is based on a lecture entitled "The Institute of Optics New Gallery of Legendary Optical Scientists" given at the Institute of Optics of the University of Rochester on October 25, 1994 and later, under the present title, at several other Universities.

Figure 1: Galileo Galilei (1564–1642)

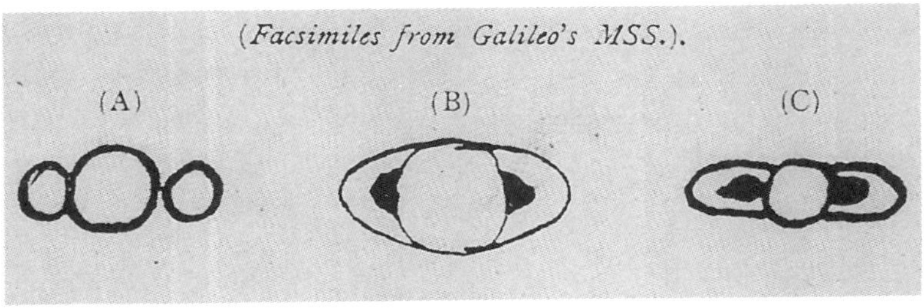

Figure 2: Galileo's drawings of the planet Saturn

to be on the lighter side, including a few stories about some of these remarkable people who have shaped our science.

Modern science as we know it today had its beginning around the year 1600. In particular the origin of modern physics can be traced to the work of Galileo Galilei (Fig. 1) who was born in Pisa in Italy in 1564 and whose picture is the first one in the collection. Galileo has made important contributions to many branches of physics, especially to mechanics. He invented the telescope which changed man's view of the universe. It also got him into a lot of trouble with the Church, as I am sure many of you know.

When Galileo observed with his telescope the planet Saturn, he found something peculiar about it. It seemed to consist of a central core with two small globes touching it at each side. However, in the following years Galileo discovered, much to his surprise, that the globes diminished and then disappeared altogether. In a letter to one of his acquaintances, Galileo asked: "Had Saturn devoured

Figure 3: Pierre Fermat (1601–1625)

his children?" The next slide (Fig. 2) shows Galileo's sketches, indicating how Saturn has appeared to him at different times. We will soon learn the reason for this strange behavior.

The next picture (Fig. 3) is that of Pierre Fermat, a French mathematician and physicist, who was born in 1601. He was actually not a professional mathematician, although he is generally regarded as having been the leading mathematician of the 17th century. He studied law and spent his working life as a magistrate in a small French town. In some of his leisure time Fermat was studying problems of maxima and minima of curves. In the course of these studies he discovered what is usually called the principle of least time. Its optical equivalent is known as the principle of stationary optical path which is at the root of all of geometrical optics.

A few years ago Fermat's name has been very much in the news, including, for example, in a front page article in the New York Times, in connection with the so-called Fermat's Last Theorem. This theorem, which may be regarded as a generalization of Pythagoras' theorem for right-angle triangles, asserts that there is no positive integer n greater than 2 for which a relation of the form $a^n + b^n = c^n$ can be satisfied by integers a, b and c. Fermat had the habit of recording his results in the margin of a certain book. At one time he stated the theorem in the margin and added that he had found a remarkable proof of it, but that the margin was too small to contain it. Well, for more than 350 years mathematicians have been trying to prove or to disprove his theorem. In 1908 a German industrialist left 100,000 German marks, then a large sum of money, to be awarded to the first person who gives a proof of Fermat's Last Theorem.

The newspaper articles which I have just mentioned reported last year that a British mathematician, Andrew Wiles, presented a proof of Fermat's Last Theorem in a lecture at Cambridge. His proof takes much more space than available in the margin of Fermat's book – the proof ran to more

Figure 4: Christiaan Huygens (1629–1695)

than 1,000 pages! But it was soon found that Wiles' proof was not complete. However, since then, Wiles with a collaborator succeeded in completing the proof. Unfortunately, Wiles will not receive much money from the prize that I mentioned. Inflation of the German mark after the First World War has reduced the prize to a fraction of a cent. Undoubtedly there will be other compensations for Wiles.

Our next luminary is Christiaan Huygens (Fig. 4) born in the Netherlands in 1629. He was the son of a distinguished Renaissance poet and his home in the Hague was frequented by some of the great contemporary figures, including the painter Rembrandt, the philosopher Spinoza and the natural scientist Descartes.

Whilst still a student, Huygens and one of his brothers learned how to grind lenses and later they made telescopes. With one of them Huygens observed the planet Saturn and in 1655 he found a small object near it. It was a satellite now known as Titan. He announced his discovery by the statement, "the core of the planet is griddled by thin flat rings, nowhere touching, inclined to the ecliptic." This was the discovery of the rings of the planet Saturn and the discovery had solved the riddle which puzzled Galileo. Obviously Huygens' telescope was superior to Galileo's. The next slide (Fig. 5) shows some of Huygens' drawings of his perception of the rings. Today we know, of course, a great deal more about them, especially from photographs taken by the spacecraft Voyager, fifteen years ago. Some pictures of Saturn and of its rings, taken in 1980, three-and-a quarter centuries after Huygens discovered them, are shown in Figs. 6 and 7.

The next slide (Fig. 8) shows Voyager's photograph of Saturn's moon Titan, which Huygens discovered.

Following his investigations regarding the planet Saturn, Huygens made another important

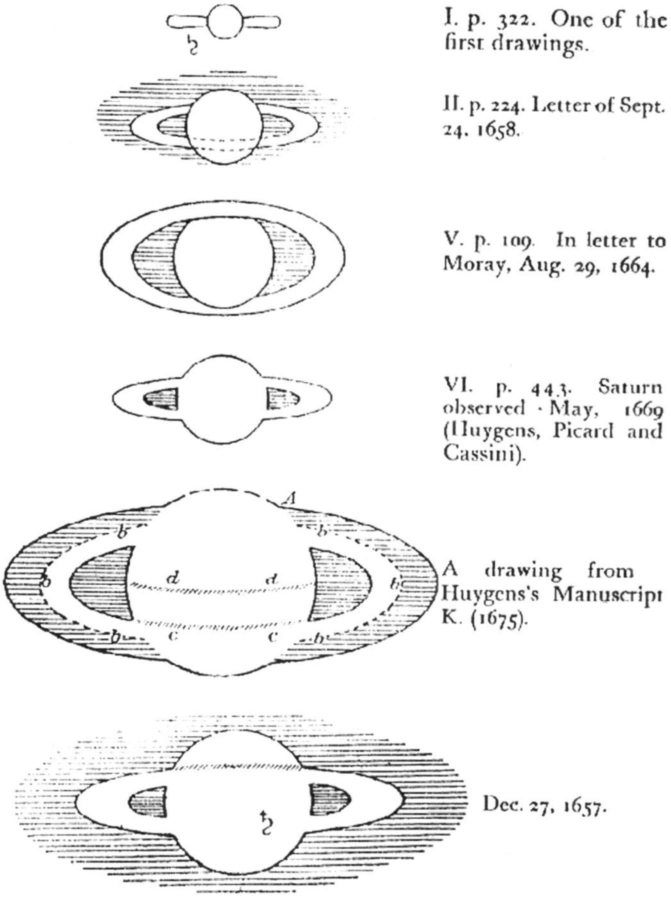

Figure 5: Huygens' drawings of Saturn's rings (from his *Œuvres Complètes*)

Figure 6: Saturn photographed by the spacecraft Voyager in 1980

Figure 7: Saturn rings photographed by the spacecraft Voyager in 1980

Figure 8: Saturn's moon Titan photographed by the spacecraft Voyager in 1980

scientific contribution, the invention of the pendulum clock. Because of limitations of time I will not discuss this topic here; I will only mention that his book on this subject has made an outstanding impact upon the science of mechanics. It contains, for example, the general idea of one of the most important principles of mechanics, the law of conservation of energy.

Huygens' greatest achievement was his development of the wave theory of light, described in his famous book *Traité de la Lumière* which was published in 1690. The title page of this book is reproduced in Fig. 9 and a translation of it is shown in Fig. 10.

Actually Huygens completed the writing of the book twelve years earlier. I can best explain the reason for its delay by quoting Huygens' own words from the preface of his book:

> *"I wrote this Treatise during my sojourn in France twelve years ago, and I communicated it in the year 1678 to the learned persons who then composed the Royal Academy of Science, to the membership of which the King had done me the honour of calling me...*
>
> *...One may ask why I have so long delayed to bring this work to the light. The reason is that I wrote it rather carelessly in the language in which it appears, with the intention of translating it into Latin, so doing in order to obtain greater attention to the thing.*
>
> *...But the pleasure of novelty being past, I have put off from time to time the execution of this design..."*

Let me add here that anyone who is engaged in active research or in the supervision of graduate students will notice from the last sentence which I just quoted here how little has changed in this respect in the more than 300 years since Huygens wrote these words!

Some years ago I was presented with a photograph of a page of the manuscript of Huygens' *Traité de la Lumière*, which is now in a library in Leiden in the Netherlands. The next slide (Fig. 11) shows that page, which contains a figure that illustrates the central idea of his wave theory:

TRAITE
DE LA LVMIERE.
Où font expliquées
Les caufes de ce, qui luy arrive
Dans la REFLEXION, & dans la
REFRACTION.
Et particulierement
Dans l'etrange REFRACTION
DV CRISTAL D'ISLANDE.
Par Monfieur CHRISTIAN HUYGENS, Seigneur de Zeelhem,
Avec un Difcours de la Caufe
DE LA PESANTEVR.

A LEIDE,
Chez PIERRE VANDER AA, Marchand Libraire.
MDCXC.

Figure 9: Title page of Huygens' *Traité de la Lumière*

TREATISE
ON LIGHT.
In which are explained
The causes of that which occurs
In REFLEXION, & in REFRACTION.
And particularly
In the strange REFRACTION
OF ICELAND CRYSTAL.
By CHRISTIAAN HUYGENS.
Rendered into English
By SILVANUS P. THOMPSON.

Figure 10: Translation of the title page of Huygens' *Traité de la Lumière*

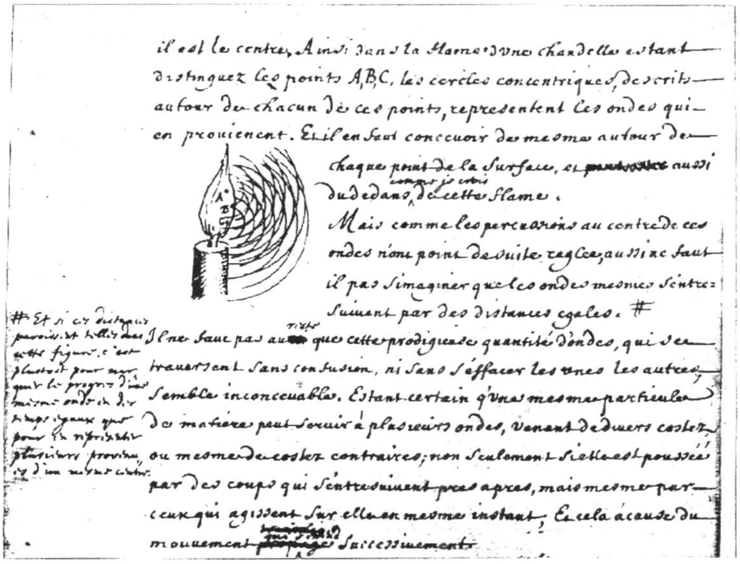

Figure 11: A page from Huygens' manuscript of *Traité de la Lumière*, with a famous drawing by which Huygens illustrated his basic ideas regarding the wave nature of light

light propagation in the form of spherical waves.

One could spend several lectures just discussing the numerous contributions to optics contained in Huygens' book – including the explanation of double refraction, for example, but again, because of limitations of time, I cannot go into this.

The next giant of science in our collection is Sir Isaac Newton (Fig. 12). He was born on Christmas day in 1642, the year Galileo died. He studied at Cambridge University but a few years later he left Cambridge because of the plague. He returned after the plague had died down and stayed there, altogether for almost 40 years. He became a Lucasian Professor at Cambridge University, later served as a member of the British parliament and then became Master of the British Mint.

Newton has made fundamental contributions to many fields, particularly to the theory of gravitation, mechanics, mathematics – having been a co-inventor of the calculus – and, of course, he has made basic contributions to optics. Actually his view regarding the nature of light has found little favor at his time and he frequently had to argue to explain and defend his ideas. In 1676 he seems to have had enough and he wrote to one of his contemporaries *"I see a man must either resolve to put out nothing new or become a slave to defend it"*. After he wrote that he did not publish anything on light for 28 years! But in 1704, after the 28 years, his classic book *Opticks* appeared whose title page is reproduced in Fig. 13.

A 1931 reprint of the fourth edition of *Opticks* contains several introductions by others, including one by Einstein. I will quote some passages from Einstein's foreword to the book because it beautifully summarizes what one of the greatest scientists of all times has said about the work of another of the greatest scientists:

Figure 12: Sir Isaac Newton (1642–1727)

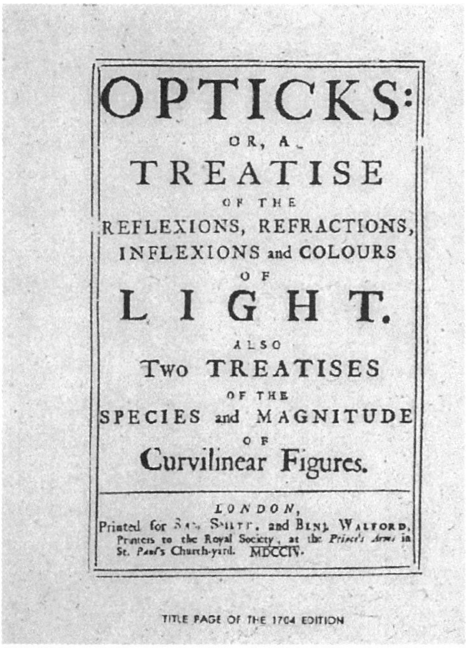

OPTICKS:
OR, A
TREATISE
OF THE
REFLEXIONS, REFRACTIONS,
INFLEXIONS and COLOURS
OF
LIGHT.
ALSO
Two TREATISES
OF THE
SPECIES and MAGNITUDE
OF
Curvilinear Figures.

LONDON,
Printed for [illegible], and Benj. Walford,
Printers to the Royal Society, at the *Prince's Arms* in
St. *Paul's* Church-yard. MDCCIV.

TITLE PAGE OF THE 1704 EDITION

Figure 13: Title page of Newton's *Opticks*

Figure 14: Thomas Young (1773–1829)

> *"In one person he combined the experimenter, the theorist, the mechanic and, not least, the artist in exposition. He stands before us strong, certain, and alone: his joy in creation and his minute precision are evident in every word and in every figure."*

And then Einstein goes on enumerating the contributions which Newton has presented in *Opticks*:

> *"Reflection, refraction, the formation of images by lenses, the mode of operation of the eye, the spectral decomposition and the recombination of the different kinds of light, the invention of the reflecting telescope, the first foundations of colour theory, the elementary theory of the rainbow pass by us in procession, and finally come his observations of the colours of thin films as the origin of the next great theoretical advance, which had to await over a hundred years, the coming of Thomas Young."*

Before leaving Newton let me mention that throughout his life he also had a deep interest in two other areas, namely in religion and alchemy. After his death two completed books and over 1,000 manuscript pages he wrote were discovered, devoted entirely to religious matters, and also books and extensive manuscripts on alchemy.

The next painting that we encounter in our gallery is that of Thomas Young (Fig. 14). Young was born in 1773 in England and was a child prodigy. He could read with considerable fluency at the age of 2 and by the time he was 13 years old he had a good knowledge of Latin, Greek, French and Italian. Soon afterward he learned Hebrew, Chaldean, Syriac, Samaritan, Arabic, Persian, Turkish and Ethiopic languages.

Young studied medicine and his early researches were concerned with the physiology of the

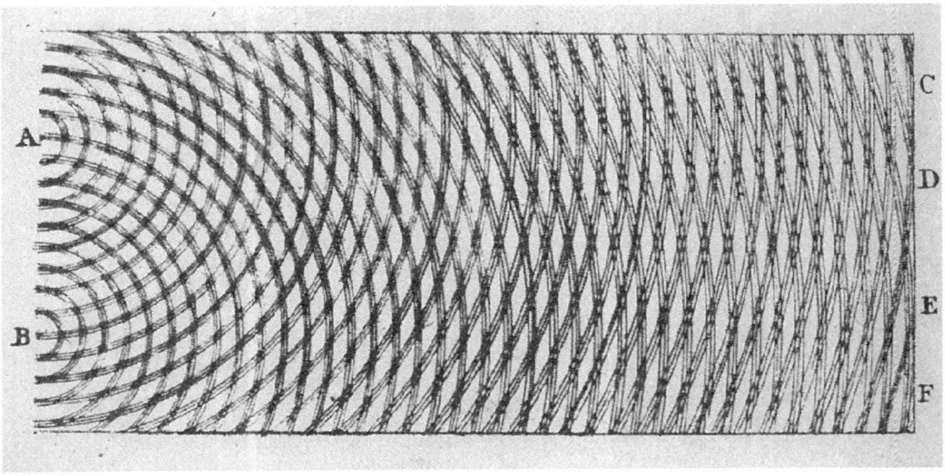

Figure 15: Young's interference fringes reproduced from his book, *'Course of Lectures on Natural Philosophy and the Mechanical Arts'*, published in 1807

eye and he gave the first description of the defect of astigmatism and of color sensation. He also contributed to mechanics – some of you may recall something called Young's modulus of elasticity, a concept due to him. But undoubtedly his most significant and lasting contribution to science arose from his discovery of the principle of interference of light, demonstrated by his classic two pinhole experiment. The next slide (Fig. 15) is a reproduction of a picture of his original drawing. The figure shows dark and bright bands formed on superposition of the light emerging from the pinholes – known today as interference fringes – and he correctly explained them as being due to constructive and destructive interference. This was indeed a fundamental contribution to optics, because at the time when he performed these experiments in 1801 and correctly interpreted the observed results, the wave theory of light was only one of several competing theories, and not yet generally established. Young's experiment, which brought into evidence the principle of interference of light, was one of the pillars which lead to the establishment and acceptance of the wave theory of light about 17 years later.

A few years after Young performed these experiments, he studied hieroglyphic writings and he helped to decipher the inscriptions on the Rosetta stone.* But he returned to optics again in about 1816 and made another fundamental contribution regarding the nature of light. It concerned some rather unexpected effects first observed more than a century earlier, when light was passed through a piece of Iceland spar. Huygens devoted much effort trying to understand this effect and he discussed it in his famous book *"Traité de la Lumière"*, which I mentioned earlier, but Huygens did not succeed in explaining it. Young found a solution to the puzzle and his explanation led immediately to a major modification of the wave theory of light. It was taken for granted until then that light waves resemble

*The Rosetta stone is a black basalt slab found in 1799 in Rosetta, a town near Alexandria in Egypt. It bore parallel inscriptions in Greek and in ancient Egyptian hieroglyphic and demotic characters. It provided a key to the deciphering of ancient Egyptian writings. It is now in the British Museum.

Figure 16: Josef Fraunhofer (1787–1826)

sound waves in that their oscillations are along their direction of propagation or, as one now says, they are *longitudinal waves*. Thomas Young explained the strange phenomena observed when light was passed through a piece of Iceland spar by assuming that the oscillations are perpendicular to the direction of propagation, i.e. that they are *transverse waves* – a completely novel concept, since such waves had not been encountered by that time. This assumption also turned out to be crucial to the formulation of the electromagnetic theory of light, about which I will speak later.

Our next luminary is Josef Fraunhofer (Fig. 16). He was born in 1787 in Germany, the youngest of 11 children. His father and several generations of his predecessors were involved with glass making. Both Fraunhofer's parents died when he was very young, leaving him an orphan when he was twelve years old. His guardians sent him to Munich to become an apprentice of a mirror-maker and grinder of ornamental glass.

In 1801, when Fraunhofer was 14 years old, the house where he worked collapsed and Fraunhofer was buried alive. Fortunately for him (and for all of us I should say), he was rescued. There was a huge public interest in the rescue operation (see Fig. 17) and, as a consequence of it, Fraunhofer was brought into contact with some local politicians and civil servants. One of them took Fraunhofer under his wings and played a major role in Fraunhofer's later career. Noticing Fraunhofer's enthusiasm for learning, he saw to it that Fraunhofer had access to textbooks on optics and on mathematics. From a gift that he had received, Fraunhofer bought a machine for grinding glass and he became an expert on making high-quality lenses. He later worked in a glass-manufacturing establishment in a former monastery near Munich. The establishment was owned by his benefactor; later, mainly due to Fraunhofer's activities there, it became a famous center for optics research. In 1814, whilst testing prisms to determine refractive indices of different types of glass, Fraunhofer observed that the Sun's spectrum was covered with fine dark lines, now called the Fraunhofer lines. He studied them

Figure 17: Rescue of Fraunhofer in 1701 from the ruins of a collapsed house

and he measured the position of 576 lines and named the main ones by letters A – Z. Today they are well-known lines in the spectrum of the Sun. The next slide (Fig. 18) is a picture of the solar spectrum drawn and colored by Fraunhofer himself, showing some of the lines that he discovered.

The curve above it, also sketched by Fraunhofer, represents the intensity distribution across the spectrum. I should add that some of the lines were discovered earlier (in 1802) by an English astronomer Wollaston, but Fraunhofer was the first who recognized that, in his own words, *"these lines and strips lie in the very nature of sunlight."* These discoveries paved the way towards a new branch of science – namely spectroscopy.

Between 1821 and 1823 Fraunhofer published two pioneering treatises on the diffraction of light.

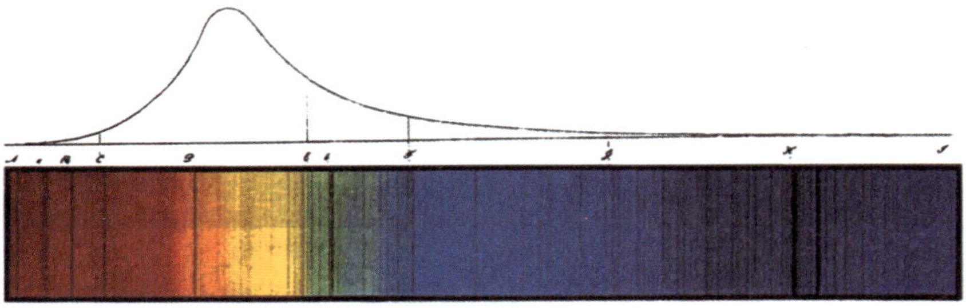

Figure 18: Solar spectrum sketched by Fraunhofer

Figure 19: Augustine Jean Fresnel (1788–1827)

The first was entitled, *"New theory describing how light is modified by the mutual interference and diffraction of light rays, and the laws governing such modifications."* He applied his theoretical results on diffraction to experiments aimed at measuring the wavelengths of different spectral components of light. For this purpose he employed diffraction gratings that he himself ruled with an astonishing precision.

The terms Fraunhofer lines in the solar spectrum and Fraunhofer diffraction have permanently entered the language of science.

The next optical scientist in our gallery of distinguished optical scientists is Augustine Fresnel (Fig. 19). Fresnel was a French engineer born in 1788. He grew up in the turbulent times of the French revolution. Fresnel supported the royalists during the years when Napoleon had been exiled by King Louis XVIII. When Napoleon returned from Elba, Fresnel, as a result of his royalistic leanings, lost his job.

In about 1814 Fresnel started studying optics and he has made numerous important contributions to it. Let me mention a few. He invented what is today known as the Fresnel biprism and the Fresnel lens. He gave the first indication of the origin of dispersion based on the concept of molecular structure of matter. He contributed significantly to our understanding of interference of light, unaware of the work of Thomas Young on this subject, which I mentioned earlier. Together with Arago, Fresnel investigated interference of two light rays polarized at right angles to each other and he found that they never interfere. This observation was later the key to the discovery by Thomas Young that light waves are transverse waves.

At the time when Fresnel started his researches there were two rival theories regarding the nature of light – the corpuscular theory (also called the emission theory) and the wave theory. The French Academy of Sciences used to award prizes for the elucidation of various scientific problems. The

Figure 20: Christian Doppler (1803–1853)

supporters of the corpuscular theory proposed diffraction as the topic for the competition in 1818, in the expectation that it would lead to a triumph of the corpuscular theory. (Diffraction, as most of you know is a phenomenon manifested by the fact that when light rays pass close to edges of apertures or boundaries of bodies, they deviate from rectilinear paths and penetrate into the shadow region).

The proponents of the corpuscular theory had a great surprise. In spite of strong opposition, the prize of the French Academy of Sciences was awarded to Fresnel, whose explanation of diffraction, as described in his essay, provided very strong support for the wave theory. The essence of Fresnel's treatment was the synthesis of Huygens' principle of light propagation and of Young's principle of interference. A particularly interesting prediction of Fresnel was that at the centre of the shadow of a small circular disk there should appear a bright spot. Many scientists who believed in the correctness of the corpuscular theory considered this prediction to be absurd, but experiments performed by Arago soon afterwards confirmed that Fresnel was right. The success of Fresnel's theory led, within a few years, to the universal acceptance of the wave theory of light – at least for another 100 years, until the concept of the photon and of the wave-particle duality made its appearance in the formulation of quantum theory of light.

Most of the scientists I spoke about so far made many contributions to optics. The one I am going to speak about now, Christian Doppler (Fig. 20) born in 1903 in Salzburg, Austria, is chiefly known for a single contribution, but it is a monumental one. Today this contribution is generally known as the Doppler effect or the Doppler Principle.

Doppler announced his discovery in a lecture entitled *"On the coloured light of the double stars and certain other heavenly bodies"* at a meeting of the Royal Bohemian Society of Sciences which was held in Prague more than 150 years ago. The next slide (Fig. 21) shows the title page of an

On the

Coloured Light of the Double Stars

and Certain Other

Stars of the Heavens

An Attempt at a General Theory which Incorporates Bradley's Theorem of Aberration as an Integral Part.

by

Christian Doppler

Professor of Mathematics and Practical Geometry at the Technical Institute and Associate Member of the Royal Bohemian Society of Sciences.

(Reprinted from the Proceedings of the Royal Bohemian Society of Sciences, Vth. Ed., Vol. 2)

Prague, 1842.

In commission with Borrosch & André.

Figure 21: English translation of the title page of Doppler's paper containing the announcement of his famous principle

English translation of the published version of his lecture, reprinted from the Proceedings of the Society.

The Doppler effect is the manifestation of the fact that when waves from a moving source are detected, the observed frequency depends on the velocity of the source relative to the observer. The effect can be most easily demonstrated with sound waves. When one hears, for example, a siren of a speeding ambulance or of a speeding police car that passes close by, the pitch of the siren goes up when the moving vehicle approaches us and it goes down when it passes us (Fig. 22).

Doppler was unable to verify his prediction with light but there is an amusing story about the experimental verification of the Doppler effect with sound waves. In 1845 a Dutch meteorologist, Christopher Buys-Ballot, persuaded a group of trumpeters to ride on the footplate of a moving

Figure 22: Illustrating Doppler's principle with a speeding ambulance

Figure 23: Sir William Rowan Hamilton (1805–1865)

locomotive while playing the same note with all their might. By listening carefully to the apparent change in the pitch as the train whizzed by musically-trained observers on each side of the track, Buys-Ballot convinced himself and others of the correctness of Doppler's prediction.

Today the Doppler effect is one of the pillars of modern astrophysics, because the change in the frequency of light due to the motion of astronomical objects relative to the Earth is manifested by shifts of spectral lines. From measurements of the magnitude of the shifts, astronomers can estimate how far the object is from us. But the Doppler effect has found numerous other uses, for example in medicine, where propagating ultrasonic waves with the help of the Doppler effect can provide information about how blood is flowing in some of the countless vessels in the human brain.

Let me conclude these brief remarks about Doppler by returning to the meeting of the Royal Bohemian Society of Sciences on the 25th May 1842 in Prague when Doppler presented his theory. You might perhaps think that the room was packed. Actually it was attended, apart from Doppler, by only five scientists and by the keeper of the minutes. I sometimes tell this to my students when they are preparing to give a talk at a scientific meeting, so that they do not get discouraged if only a few people should turn up to listen to them!

Let me turn to the next member of our gallery, Sir William Rowan Hamilton (Fig. 23), born in 1805 in Dublin, Ireland and generally regarded as the greatest scientist Ireland has ever produced. Like Thomas Young, he was a child prodigy. At the age of three he could read very well, at five he read and translated Greek and Hebrew, at eight he added Italian and French to his mastery of languages and before he was ten he had learned several oriental languages, including Arabic and Sanscrit.

Hamilton never attended any school before going to University. He entered Trinity College in Dublin when he was eighteen and he did extremely well. The ending of his undergraduate career was

Figure 24: Hermann von Helmholtz (1821–1894)

quite spectacular. A professorship of astronomy was being advertised at Trinity College at that time and there were some distinguished applicants for that post. But the Governing Board passed over all the applicants and elected Hamilton, then still an undergraduate, to the Professorship, though Hamilton had not even applied for it. (Obviously there were no tenure committees at that time!).

Hamilton started his Professorship brilliantly. Shortly after taking up his new appointment he published the first of a remarkable series of papers entitled "A theory of systems of rays". In these papers he introduced the concept of the so-called characteristic function, which provided a unified framework for the analysis of many problems of geometrical optics. One of the early successes of his theory was the prediction of conical refraction. This is a phenomenon associated with the propagation of light rays which proceed from the interior of some crystals in certain special directions. On emergence from the crystal the ray is divided into an infinite number of rays, lying on a conical surface. This unexpected consequence of Hamilton's theory was soon afterwards confirmed experimentally (by Humphrey Lloyd) and provided strong support to Fresnel's theory of light propagation in crystals.

Soon after Hamilton introduced his rather general analytic methods into optics he applied them also to dynamics. There is no time to say much about that subject here, except perhaps to recall some terms which many of you have undoubtedly heard in the context of dynamics – such as Hamilton's principle of least action or the Hamilton-Jacobi equations.

Almost exactly 100 years after Hamilton introduced the new analytical techniques into optics it was found that they are just the right mathematical tool needed to formulate the newly discovered laws of quantum physics, particularly those formulated by de Broglie and Schrödinger. I wonder how many quantum physicists realize that when they speak about the Hamiltonian of an atomic or molecular system they are paying tribute to an Irish optical scientist.

Figure 25: Gustav Kirchhoff (1824–1887)

The next luminary whom we encounter in our gallery is Hermann von Helmholtz (Fig. 24), born in 1821. He studied medicine in Berlin and had a rather unusual professional career. He was a military surgeon, later a teacher of Anatomy and then Professor of Physiology. Later he became the Director of a new Physico-Technical Institute in Berlin where, until the end of his life, he gave lectures as a Professor of Physics.

Helmholtz has made major contributions to two areas of science: physiology and physics. Particularly relevant for us are his inventions of the ophthalmoscope for inspecting the interior of the eye and of the opthalmometer for measuring the eye's curvature. He investigated accommodation of the lens of the eye, color vision and color blindness. In 1867 he published a book whose title in English translation is *Handbook of Physiological Optics*, which has become one of the classics of science. Helmholtz has also made major contributions to other fields, notably to the theory of sound, to the physiology of the ear, to thermodynamics – notably in connection with energy conservation – and to electrodynamics. There is a well-known equation used in optics named after him and one form of its general solution is the so-called Helmholtz-Kirchhoff integral theorem. Helmholtz discovered it in acoustics, Kirchhoff in optics.

Gustav Kirchhoff (Fig. 25), who was born in 1824, is actually the next distinguished scientist in our gallery. He was one of the foremost scientists of the 19th century. Whilst still a student he formulated the basic laws concerning the distribution of currents in networks, known today as Kirchhoff's laws.

I mentioned earlier Fraunhofer's discovery, made in 1814, of the lines in the spectrum of the Sun. However, the significance of this discovery was not clear to Fraunhofer nor to other physicists for many years. Its real significance only become apparent in 1860 from experiments of Bunsen and Kirchhoff. It then became clear that the Fraunhofer lines were characteristics of chemical elements;

Figure 26: James Clerk Maxwell (1831–1879)

more precisely, they were manifestations of absorption from the Sun's continuous spectrum of exactly those wavelengths which were emitted by cooler gases in the Sun's atmosphere. This discovery was the beginning of the science of spectroscopy. However, the relation between absorption and emission was not understood at that time, although a major step towards its elucidation was also made by Kirchhoff around that time, when he derived from thermodynamics what is today known as Kirchhoff's law of radiation. This law asserts that under equilibrium conditions, the ratio of certain coefficients which are measures of the rate of emission and of absorption is independent of the nature of the bodies which interact with the radiation field and that, moreover, it is a function, $\mathcal{K}_\nu(T)$ say, only of frequency ν and temperature T of the bodies. Because of the universality of this function, which turned out to be the energy distribution in the spectrum of blackbody radiation, many attempts were made to determine its explicit form, but it took another 40 years before this was done, and its determination lead to a major revolution in physics, as we will see later on.

The next great figure in our gallery is James Clerk Maxwell (Fig. 26), born in Scotland in 1831. Like Kirchhoff, Maxwell can be rightly considered to have been one of the greatest physicists of the 19th century. We are here mainly interested in his contributions to optics, but it would not be right to completely ignore his other great contributions. Let me just mention that he was one of the founders of the kinetic theory of gases, a theory which makes it possible to calculate the properties of gases from their molecular structure. One of the consequences of this theory is the well-known Maxwell's distribution law for molecular velocities in gases.

Maxwell's greatest work was undoubtedly in connection with electric and magnetic phenomena. In 1864 he succeeded in formulating a set of equations which covered all the laws of electricity and magnetism as then known. He noticed a certain asymmetry among the equations and he modified one of them in order to gain more symmetry. This was a guess, not based on any direct

Figure 27: Lord Rayleigh (J. W. Strutt) (1842–1919)

experimental evidence, but rather on belief in harmony; but it was one of the most fortunate guesses in the history of physics. His modified equations have solutions representing waves, which propagate with a certain velocity that turned out to be equal to the ratio of units measured by electrostatic experiments (using Coulomb's law) and electromagnetically (using Oersted's law). This ratio turned out to be equal to the speed of light, as then estimated from other experiments. Based on this result, Maxwell put forward the hypothesis that light is electromagnetic in nature, i.e. that light waves are electromagnetic waves. Many of you will know that the four Maxwell equations have become the basic equations of physics, and they encompass the whole of classical optics. Yet, during his lifetime Maxwell's great discoveries concerning electricity and magnetism were not generally appreciated and he died at the early age of 48, before the existence of electromagnetic waves was experimentally established. That was done by Heinrich Hertz in 1888, more than twenty years after Maxwell formulated his fundamental equations and nine years after his death.

The next great scientist in our gallery is Lord Rayleigh (Fig. 27), a British physicist born in 1842 as John William Strutt. He succeeded to his father's title at the age of 31. He graduated in mathematics from Cambridge University and he remained there after his graduation, but because of poor health he gave up academic life a few years later, to recuperate in a warmer climate. During his convalescence, which he spent on a houseboat on the Nile, Rayleigh wrote his two-volume *Theory of Sound* which remains to this day a classic in the literature on acoustics. On his return to England, Rayleigh built a laboratory next to his family home and carried out most of his work there, except for a period of five years, when he succeeded Maxwell as the Cavendish Professor of Experimental Physics at Cambridge University.

Rayleigh's researches covered many fields. The collected edition of his publications includes, apart from the two-volume *Theory of Sound*, altogether about 430 papers! In 1900 Rayleigh published a

Figure 28: Albert Abraham Michelson (1852–1931)

short paper, only 2 pages long, in which he derived an expression for the universal function $\mathcal{K}_\nu(T)$ whose existence was predicted by Kirchhoff, which represents the spectrum of blackbody radiation. Rayleigh derived his formula from the wave theory and from elementary thermodynamics. He made a small error in his derivation (involving a factor 8). This was noted by James Jeans who drew attention to the error in a short note, which he published a few years later. Because of this, the formula has been referred to as the Rayleigh-Jeans law ever since! It turned out that the Rayleigh-Jeans law did not correctly represent the blackbody spectrum at all wavelengths, but it played an important role in the development of radiation theory.

Apart from the Rayleigh-Jeans law, Rayleigh's name is associated with a law of scattering, according to which the intensity of sunlight scattered by particles in the atmosphere is proportional to the inverse fourth power of the wavelength – this being the cause as to why the sky appears blue. Rayleigh's name is also associated in optics with a theory of image formation in the microscope and with a well-known criterion for resolution.

In 1904 Rayleigh received the Nobel Prize for Physics for his investigations on the density of the more important gases and for his discovery of Argon. Rayleigh carried out some of this research in collaboration with William Ramsey, who received the Nobel prize for Chemistry the same year.

The next scientist we encounter in our gallery is Albert Abraham Michelson (Fig. 28). He was born in Poland in 1852 and was brought to the United States when he was four years old. He graduated from the United States Naval Academy, then studied in Europe and in 1882 he became the first Professor of Physics at the Case School of Applied Science in Cleveland, Ohio. In 1894 he became the head of the Physics Department at the University of Chicago where he remained for thirty-five years. In the last decade of his life he also carried out experiments at Mount Wilson and at the California Institute of Technology.

Whilst in Cleveland, Michelson collaborated with Edward Morley, who was Professor of Chemistry there, in the measurement of the speed of light, in an attempt to determine the absolute motion of the Earth with respect to the hypothetical aether. Their measurements failed to detect any such relative motion – a result which had profound impact on the later development of physics. The speed of light which Michelson and Morley obtained was the most accurate one available at that time and remained so for another 10 years, when Michelson himself made an even more accurate determination of it. In addition to his measurements of the speed of light, Michelson has made many other important contributions to optics, particularly in developing various high precision techniques for use in spectroscopy and in astronomy. He invented what today is known as Fourier spectroscopy and developed the first interferometric technique for measuring the diameters of stars.

In his private life, Michelson was somewhat temperamental. In a biography of her father, one of his daughters mentions that at one point during his first marriage, his wife wanted to have Michelson certified, i.e. to have him declared mentally ill, and the daughter adds *"after that, the marriage was never the same again"*!

Michelson was not entirely happy supervising graduate students. At one point he passed on all of them to his colleague, Robert Millikan, another famous physicist, with an explanation given in the following letter:

> *"If you can find some other way to handle it I don't want to bother any more with this thesis business. What these graduate students always do with my problems, if I turn them over to them, is either to spoil the problem for me because they haven't the capacity to handle it as I want it handled, and yet they make it impossible for me to discharge them and do the problem myself; or else, on the other hand, they get good results and at once begin to think the problem is theirs instead of mine, when in fact knowing of what kind of a problem it is worthwhile to attack is in general more important than the mere carrying out of the necessary steps. So I prefer not to bother with graduate students' theses any longer. I will hire my own assistant by the month, a man who will not think I owe him anything further than to see he gets his monthly check. You take care of the graduate students in any way you see fit and I'll be your debtor forever."*
>
> (A.A. Michelson to R.A. Millikan ca. 1905)

In 1907 Michelson becamse the first American Nobel Prize Winner in Physics.

We next encounter in our gallery Hendrik Antoon Lorentz (Fig. 29), a Dutch scientist born in 1853. He received his doctorate at the age of 22 and the subject of his thesis was the theory of reflection and refraction of light. It provided the solution to an important problem in the electromagnetic theory of light which Maxwell had left unsolved. Only two years later Lorentz was appointed Professor of Theoretical Physics at Leiden – the first chair in this field in the Netherlands and one of the first such chairs in Europe. Lorentz was largely responsible for shaping the development of theoretical physics in his lifetime.

One of Lorentz' major contributions was the formulation of a theory which is today generally referred to as the electron theory of matter. The theory explained many phenomena involving

Figure 29: Hendrik Antoon Lorentz (1853–1928)

the interaction of electromagnetic waves with matter. But probably his greatest contribution to physics and to optics was his demonstration that Maxwell's equations are not invariant under the usual transformation of coordinates based on certain classic notions of Galileo and Newton. Lorentz showed that Maxwell's equations are invariant under a different kind of transformation, known today as the *Lorentz transformation.* This transformation implied the strange notion that moving bodies contract in their direction of motion. It was only later that the full implication of Lorentz's transformation become apparent and its strange consequences were clarified. This was done by Einstein in 1905.

In 1902 Lorentz received the Nobel Prize in Physics.

Next we encounter in our gallery Max Planck (Fig. 30), born in 1858. Throughout much of his professional career Planck was a Professor of theoretical physics at the University in Berlin and his early research was on thermodynamics. Around 1896 he became interested in an important problem which I mentioned earlier when I spoke about Kirchhoff and Rayleigh. As I noted, Kirchhoff discovered the existence of a certain universal function, $\mathcal{K}_\nu(T)$, which expresses the ratio of the emission to the absorption coefficients under equilibrium conditions and which is equal to the energy distribution in blackbody radiation. In spite of many investigations on this subject the explicit form of this function remained unknown for a long time. Planck solved the problem in 1900, having found the correct formula

$$\begin{aligned} u_\nu(T) &= \frac{8\pi}{c}\mathcal{K}_\nu(T) \\ &= \frac{8\pi h\nu^3}{c^3}\frac{1}{e^{h\nu/K_BT}-1}. \end{aligned}$$

Figure 30: Max Planck (1858–1947)

Here K_B and h are constants, the first of which is Boltzman's constant and the second of which is now known as Planck's constant. In deriving this formula, Planck found it necessary to introduce the notion of a quantum of energy,

$$E = h\nu,$$

of the elementary oscillators, vibrating with frequency ν. It represents the smallest amount of energy that an oscillator can absorb or emit. The need for introducing such a quantum was in contradiction to Maxwell's theory and was the beginning of a major revolution in physics.

It was largely for his discovery of the quantum of energy that Planck received the 1918 Nobel Prize for Physics. He received many other honors, but his later years were nevertheless rather tragic. His eldest son died of wounds suffered in the first World War, his twin daughters died in child birth and his only surviving child was executed for alleged complicity in a 1944 plot to assassinate Hitler. His own home, containing an invaluable library, was destroyed in an air-raid on Berlin. He took shelter in the forest, until he was found by members of the US Army. He spent his last days in the home of his grand-niece in Göttingen. He died in 1947, when he was 89 years old.

When we come to the next member of our gallery of distinguished scientists, namely to Albert Einstein (Fig. 31), born in Ulm, Germany in 1879, I have run out of superlatives. It is probably no exaggeration to say that Einstein and Newton have been the greatest physicists of all times. Einstein's name is generally associated with the theory of relativity, which has profoundly revolutionized men's ideas of space and time. But as Max Born said in one of his articles, Einstein would probably have been one of the greatest scientists, even had he not written a single line on relativity.

Einstein made a spectacular appearance on the physics scene in 1905 when he published three papers in one and the same volume of the journal *Annalen der Physik*, the three papers having been

Figure 31: Albert Einstein (1879–1955)

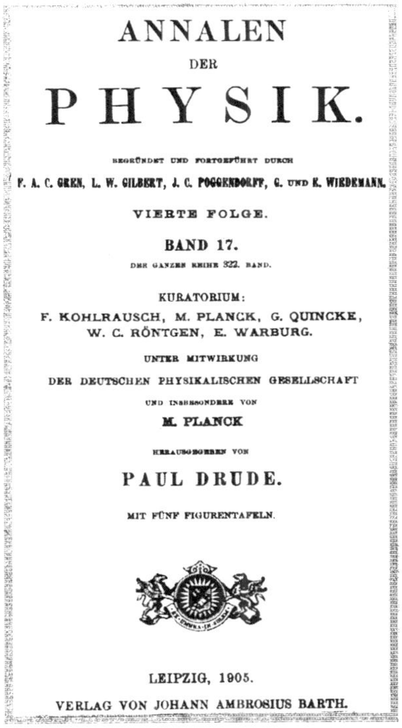

ANNALEN

DER

PHYSIK.

BEGRÜNDET UND FORTGEFÜHRT DURCH

F. A. C. GREN, L. W. GILBERT, J. C. POGGENDORFF, G. UND E. WIEDEMANN.

VIERTE FOLGE.

BAND 17.

DER GANZEN REIHE 322. BAND.

KURATORIUM:

F. KOHLRAUSCH, M. PLANCK, G. QUINCKE,
W. C. RÖNTGEN, E. WARBURG.

UNTER MITWIRKUNG

DER DEUTSCHEN PHYSIKALISCHEN GESELLSCHAFT

UND INSBESONDERE VON

M. PLANCK

HERAUSGEGEBEN VON

PAUL DRUDE.

MIT FÜNF FIGURENTAFELN.

LEIPZIG, 1905.

VERLAG VON JOHANN AMBROSIUS BARTH.

Figure 32: Title page of the 1905 volume of *Annalen der Physik*, containing three celebrated papers by Einstein.

1905	PARTICLE NATURE OF RADIATION • Photoelectric effect • Particle aspects of Wien's radiation law
1909	WAVE-PARTICLE DUALITY • Energy fluctuations in blackbody radiation • Momentum fluctuations in blackbody radiation
1910	STATISTICAL THEORY OF LIGHT SCATTERING • Critical opalescence
1917	ELEMENTARY PROCESSES OF INTERACTION • Spontaneous and induced processes • Transition probabilities
1924-1925	BOSE-EINSTEIN STATISTICS • Quantum theory of the ideal gas • Particle-number fluctuations • Matter waves

Table 1: The main topics in radiation theory to which Einstein made fundamental contributions

submitted within a period of only three and a half months. According to Max Born again, this is one of the most remarkable volumes in the whole scientific literature; for each of the three papers of Einstein which appear in this volume is considered to be a masterpiece, and the starting point of a new branch of physics. In our present day terminology the subject of these three papers are:

(*i*) the particle nature of radiation,

(*ii*) the theory of Brownian motion,

(*iii*) the special theory of relativity.

The next slide (Fig. 32) shows a photograph of the title page of that remarkable volume.

I will say a few words only about the first of these papers, because it has a particular bearing on optics. In translation its title is "On an heuristic point of view concerning the creation and conversion of light". In modern textbooks this paper is usually referred to as "Einstein's paper on the photoelectric effect". Actually, this paper contains appreciably more. In fact, Einstein's whole discussion of the photoelectric effect covers less than four pages; but, as in most of his writing, Einstein was able to get to the root of a problem in a few lines, with simple language that was remarkably free of complicated mathematics. Essentially what Einstein did in this paper was to put forward a great deal of evidence that not only the processes of emission and absorption of radiation take place in discrete amounts of energy, (as appears to have been established by Planck), but that also radiation itself behaves under certain circumstances as if it consisted of a collection of particles, which in modern language are called *photons*. Thus in this paper Einstein reintroduced a corpuscular

Figure 33: Max Born (1882–1970)

theory of light as first advocated by Newton in the 17th century. As I already noted, the corpuscular theory was completely discredited by Fresnel's wave theory, formulated almost 90 years before the publication of Einstein's paper, and the wave theory of light itself appeared to have been put on a firm foundation by Maxwell about 40 years before the publication of Einstein's paper. This paper, together with several others by Einstein, paved the way to the formulation of quantum mechanics and the quantum theory of radiation. Table 1 lists the main topics in radiation theory to which Einstein made fundamental contributions. In 1921 he was awarded the Nobel Prize in Physics.

The next distinguished scientist in our gallery, Max Born (Fig. 33), was a close friend of Einstein, although they never agreed on the true meaning of the laws of quantum mechanics, to which both have made major contributions. Max Born was born in 1882 in Breslau (then in Germany and now in Poland). He studied in Germany and in Switzerland and later became Professor of Physics at Góttingen University in Germany, where he founded a famous school of theoretical physics. Among his assistants were several scientists who later became leading physicists and Nobel laureates, including Heisenberg and Pauli. Soon after Hitler came to power Born emigrated to Great Britain and from 1936 until his retirement in 1953 he was Professor of Natural Philosophy at the University of Edinburgh in Scotland.

Born has made numerous important contributions to physics, having published about 360 scientific papers and about 30 books (not counting different editions and translations). One of his contributions which is particularly relevant to optics is his dynamical theory of crystal lattices which is important for the understanding of optical properties of solids and subjects such as dispersion and scattering of light, double refraction and optical rotation. In 1954 Born received the Nobel Prize in Physics for his fundamental research in quantum mechanics.

I have been very fortunate in having had the opportunity to collaborate with Max Born. I

Figure 34: Members of Max Born's Department at the time of his retirement in 1953 from the Tait Chair of Natural Philosophy at the University of Edinburgh. Standing (from left to right): E. Wolf, D.J. Hooton, A. Nisbet. Sitting: Mrs. Chester (secretary), M. Born, R. Schlapp.

Figure 35: Olivia Newton-John

Figure 36: Frits Zernike (1888–1966)

was actually his last assistant for about two years, before his retirement in 1953. He had a small Department at that time, but he was scientifically still very active. The next slide (Fig. 34) shows the members of his Department at the time of his retirement. (If you have a good imagination you might be able to recognize me in this picture).

Some of you may not know that Max Born had a granddaughter who achieved fame comparable to his own, but in quite a different field. It is the pop-singer Olivia Newton-John. Shortly after I learned that Olivia Newton-John was Max Born's granddaughter, I was on a sabbatical leave at the University of Toronto. Olivia was scheduled to give a concert in Toronto while I was there. I wrote to her, told her that I collaborated with her grandfather in the writing of a book, and I asked her whether we could meet. I received a charming reply in which she invited me to meet her after the concert. We met then and we talked mainly about her grandparents. Before I left Olivia gave me two autographed photos of herself. Here is a reproduction (Fig. 35) of one of them. Let me add that to some of my students I am known not so much as the co-author, with Max Born, of the book *Principles of Optics* but rather as the person who knows Olivia Newton-John and who has a picture of her hanging in his office, signed *"To Emil, Love, Olivia"*.

The next distinguished member of our gallery is Frits Zernike (Fig. 36), born in 1888 in the Netherlands. His doctoral thesis was concerned with critical opalescence, the phenomenon associated with the change in the physical properties of substances near their critical points – for example, when liquid changes into gas. It had been known for a long time that a completely transparent medium becomes milky in appearance near the critical point. The theory of scattering of light near a critical point has been the subject of considerable interest and many physicists, including Einstein, have investigated it. Zernike's contributions to this field, which began in his student days, are widely recognized and are still quoted today.

Figure 37: Dennis Gabor (1900–1979)

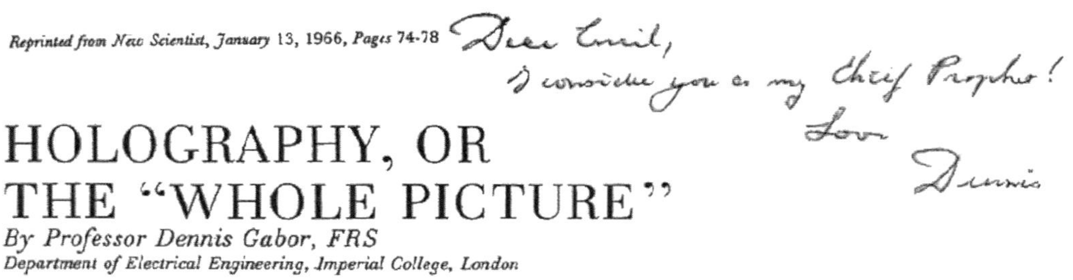

Figure 38: Dedication from Dennis Gabor

Zernike's interests throughout his life have centered on classical optics, particularly on diffraction. In 1935 he invented the phase-contrast microscope, which made it possible to directly observe phase changes introduced by transparent objects. For this invention Zernike was awarded in 1953 the Nobel Prize for Physics. Zernike has also made basic contributions to the theory of partial coherence.

The next scientist in our gallery is Dennis Gabor (Fig. 37). He was born in Hungary in 1900. Gabor studied mechanical and electrical engineering, first in Hungary and then in Germany, where he remained for the next six years, until Hitler came to power in 1933. He then came to England where, at first, he worked in industry again. In 1948 he obtained an academic appointment at Imperial College of London University and remained there for almost twenty years, until his retirement.

Much of Gabor's research efforts – but by no means all of them – were concerned with the improvement of the resolving power of the electron microscope. In this connection he invented holography, in about 1947. This invention was hardly noticed at that time or, for that matter, for more than 10 years afterwards. But after the development of the laser, which made it possible to

Figure 39: Paul A.M. Dirac (1902–1984)

greatly improve the technique, holography became a popular subject, now well-known even to the credit-card industry! By 1971, when Gabor received the Physics Nobel Prize for this invention, more than 2,000 scientific papers on that subject had been published.

I knew Gabor from the late 1940s until his death in 1979. It was actually Gabor who was responsible for my collaboration with Max Born. As I just mentioned, the invention of holography was hardly noticed at first, but I knew about it because of my acquaintance with Gabor and the book *Principles of Optics*, which Max Born and I co-authored, contained the first account of holography which appeared in a textbook. Gabor was very happy about it and when holography became popular and useful, he sent me a reprint of one of his papers with a charming dedication, shown on the next slide (Fig. 38).

The last scientist in our gallery is Paul Dirac (Fig. 39), born in England in 1902. He graduated in electrical engineering but did his Ph.D. work in mathematics. He is generally acknowledged as one of the finest and most creative theoretical physicists of this century. During the period 1925-1926 he formulated one version of quantum mechanics, now regarded as its definitive form. It well conforms to his philosophy of physics, according to which physical laws should have mathematical beauty.

About two years after he formulated the laws of quantum mechanics he made important contributions to the field of quantum electrodynamics, which is the theory of all electromagnetic interactions, in which the field is treated according to the laws of quantum mechanics. In particular he developed the technique of quantizing the electromagnetic field which elucidated the finer aspects of emission and absorption of radiation by an atomic system.

In 1928, Dirac set about reconciling quantum mechanics with Einstein's special theory of relativity. He formulated a relativistic theory of the electron, which led to new predictions, later verified experimentally, including that of the existence of anti-matter.

In 1933 Dirac shared, with Erwin Schrödinger, the Nobel Prize for Physics.

It seems to me rather fitting that Dirac's picture should be the last one in our gallery of famous optical scientists. For Dirac has brought the theory of radiation to its most refined and comprehensive form, capable of predicting even the most minute effects such as the so-called Lamb shift.

Of course a hundred years from now, Dirac's picture is unlikely to remain the last one in the gallery of scientists who have clarified the nature of light and its many properties. That I may be right in this prediction is suggested by the following citations which Einstein made on two occasions, 34 years apart.

> *"For the rest of my life I will reflect on what light is!"*
>
> Albert Einstein, ca. 1917

> *"All the fifty years of conscious brooding have brought me no closer to the answer to the question, 'What are light quanta?' Of course today every rascal thinks he knows the answer, but he is deluding himself."*
>
> Albert Einstein, ca. 1951

I would like to conclude with a few words of thanks. In searching for good pictures of the 21 scientists I spoke about I received much help from Mr. Kenn Harper, a former head of our physics-optics-astronomy library and also from Patricia Sulloff, the present head of our library.

I would like to express my appreciation to Dr. Weijian Wang, one of my graduate students, who has spent many hours climbing the stairs to the top of the Rush-Rhees library, searching for books which might contain the sort of pictures I needed.

I thank all of them for their help and I thank all of you for listening so patiently to this rather long lecture.

Some sources used for this lecture

1. J. Daintith, S. Mitchell, E. Tootill and D. Gjertsen, *Bibliographical Encyclopedia of Scientists* (in two volumes), (Institute of Physics Publishing, Bristol and Philadelphia, 2nd ed., 1994).
2. R. Weber (edited by J.M.A. Leniham), *Pioneers of Science* (The Institute of Physics, Bristol and London, 1980).
3. A. Eden, *The Search for Christian Doppler* (Springer-Verlag, Wien and New York, 1992).

∗ 4. E. Wolf, "Einstein's Researches on the Nature of Light", *Optics News* **5** (1979), 24–39.

∗ 5. E. Wolf, "Recollections of Max Born", *Optics News* **9** (1983), 10–16.

∗ 6. E. Wolf, "The Life and Work of Christiaan Huygens", *Proc. Symposium Huygens' Principle 1690-1990: Theory and Applications*, H. Blok, H.A. Ferwerda and H.K. Kuiken, eds. (North Holland, Amsterdam, 1992), 3–17.

*Reprinted in Section 6 of this volume.

In preparing this lecture for publication it seemed to me worthwhile to include the following list of Nobel Laureates (all in Physics) whose names are among those distinguished scientists whom I spoke about.

Hendrik Antoon Lorentz (with Pieter Zeeman), 1902 "in recognition of the extraordinary service they rendered by their researches into the influence of magnetism upon radiation phenomena."

Lord Rayleigh (John William Strutt), 1904 "for his investigations on the densities of the most important gases and for his discovery of Argon in connection with these studies."

Albert Abraham Michelson, 1907 "for his optical precision instruments and the spectroscopic and meteorological investigations carried out with their aid."

Max Planck, 1918 "in recognition of the services he rendered to the advancement of Physics by his discovery of energy quanta."

Albert Einstein, 1921 "for his services to Theoretical Physics, and especially for his discovery of the law of the photoelectric effect."

Paul Adrien Maurice Dirac (with Erwin Schrödinger), 1933 "for the discovery of new productive forms of atomic theory."

Frits Zernike, 1953 "for his demonstration of the phase-contrast method, especially for his invention of the phase-contrast microscope."

Max Born (with Walther Bothe), 1954 "for his fundamental research in quantum mechanics, especially for his statistical interpretation of the wave-function."

Dennis Gabor, 1971 "for his invention and development of the holographic method."

Lecture 7.3

The Development of Optical Coherence Theory*

Emil Wolf
University of Rochester

"... the image that will be formed in a photographic camera – i.e. the distribution of intensity on the sensitive layer – is present in an invisible, mysterious way in the aperture of the lens, where the intensity is equal at all points .."

[From remarks about coherence made by F. Zernike in a lecture published in Proc. Phys. Soc. (London) **61** (1948), 158.]

Optical coherence theory is probably one of the most poorly understood subjects in the whole of optics. There seems to be a widely held opinion that it is a playground for theoreticians, devoid of any useful applications. This view seems to have become even more prevalent after the invention of the laser in the 1960s, which provided scientists with intense highly coherent light. This development has lead many physicists to believe that now that such well-behaved light has become available it is not necessary to pay much attention to other kinds of light, especially because laser light has found and is continuing to find many useful and important applications in science, in technology, in medicine and in other fields. I am glad to have the opportunity to tell you in this lecture why I consider such views to be greatly misguided.

I will begin with a few remarks about *mechanics*, because the analogy between the development of mechanics and optics turns out to be rather instructive and helps to bring the development of the subject of optical coherence into proper perspective.

Classical mechanics is based on laws formulated by Sir Isaac Newton in his celebrated *Principia Mathematica* published in 1687. These laws had immediate impact on astronomy where they elucidated the origin of Keppler's law of planetary motion; and they permeate all the more recent formulations of mechanics, associated with the names of Hamilton, Lagrange and others.

Ordinary mechanics based on Newton's laws of motion is a strictly deterministic theory, concerned with particles and with forces that can be described by well-defined functions of position and time.

*Shortened version of a lecture first given at the Inaugural Conference of the School of Optics at the Center for Research and Education in Optics and Lasers, University of Central Florida (CREOL) on January 11, 1999 and later at several other Universities and conferences.

These laws are, however, not adequate to deal with systems consisting of many particles or with fluctuating media, such as liquids and gases. Such systems began to be systematically studied in the second half of the nineteenth century, when the atomistic nature of matter began to be understood. It culminated in the development of *classical statistical mechanics*, whose chief concepts and basic mathematical tools belong to the theory of probability and statistics. This chapter in the development of mechanics was completed by about the end of the 19th century, even though the atomistic and molecular structure of matter was at that time not yet generally accepted.

Because of the revolution which physics had undergone since Planck's discovery of the quantum of energy in 1900, it became later necessary to broaden the laws of statistical mechanics. This was done in the context of the density matrix formalism introduced by von Neuman in 1927, soon after the basic laws of quantum mechanics had been discovered. We note, in passing, that 240 years have gone by from the time that Newton formulated his laws of motion to von Neuman's introduction of the basic mathematical tool of quantum statistical mechanics.

Let us now turn from mechanics to optics. What Newton's equations of motion have done for mechanics, Maxwell's equations, formulated in 1864, have done for optics. Just like Newton's laws of motion, Maxwell's equations deal with deterministic situations, specifically with sources and with waves whose behavior can be described by deterministic functions of space and time. However, sources and fields which are found in nature and which are produced in laboratories, even the best behaved lasers, are never strictly deterministic, because the effect of spontaneous emission, for example, can never be completely eliminated. To deal with realistic situations Maxwell's equations must be generalized to take into account such randomness. One is then in the domain of *optical coherence theory.* It bears a similar relationship to Maxwell's theory as statistical mechanics bears to Newtonian mechanics. However, the transition from deterministic sources and deterministic fields to random ones has been made only very hesitantly and rather slowly in optics. In fact, until the early 1960s when the first lasers became available, very few scientists paid any attention to coherence,[1] but it is of interest to note that among those who did there were several Nobel prize winners, namely Max von Laue [3], Erwin Schrödinger [4] and Frits Zernike [5], and one could add to this group another Physics Nobel Prize winner, Albert Michelson, who, as Zernike said in a charming lecture [6], *"It may be said that he (Michelson) devoted much of his time during the last fifteen years of his life to measuring degrees of coherence – without knowing it."* This was so because the relevant statistical concepts and the appropriate mathematical techniques had not yet been developed.

Actually it was Zernike himself who, in the 1938 paper which I already mentioned, introduced the concept of the degree of coherence of light as a quantitative measure of the statistical correlation of field vibrations at two points P_1 and P_2 in an optical field. He first defined the degree of coherence operationally by means of Young's interference experiment. He identified the degree of coherence with the sharpness of the fringes in the interference pattern in the observation plane B when light reaches it from pinholes at the two points P_1 and P_2 in an opaque screen A (see Fig. 1), i.e.

[1]The first quantitative estimate relating to coherence of light appears to have been made by A. Verdet in 1865 (cf. ref. [1], pp. 555 and 576). Reference [1] contains also references to several other early publications concerning coherence of light, some of which are reprinted in [2].

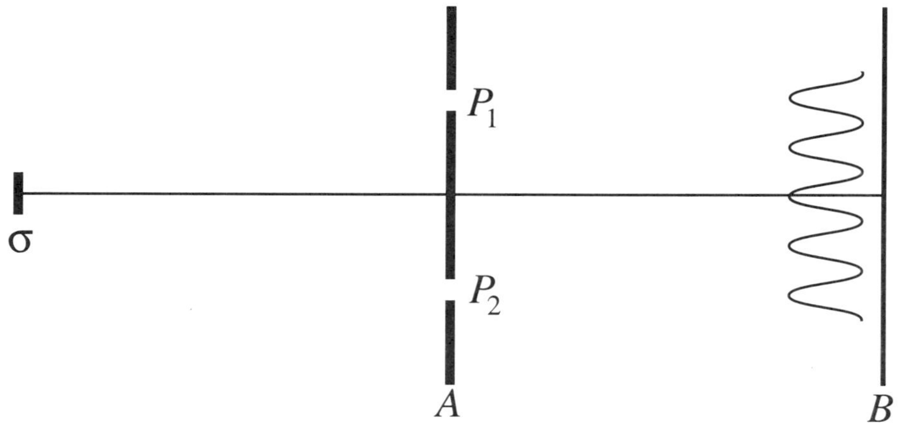

Figure 1: Young's interference experiment with partially coherent light.

$$\text{Degree of coherence } |j_{12}| \quad = \quad \text{visibility of fringes } \mathcal{V}. \tag{1}$$

The visibility $\mathcal{V}$ of the fringes, a concept due to Michelson, is defined by the expression

$$\mathcal{V} = \frac{I_{max} - I_{min}}{I_{max} + I_{min}}, \tag{2}$$

where I_{max} is the maximum and I_{min} the minimum of the intensity in the central portion of the interference pattern.

Next Zernike proceeded to define the degree of coherence statistically, having first introduced the concept of mutual intensity,

$$J_{12} = \langle V_1^*(t) V_2(t) \rangle \,. \tag{3}$$

Here $V_1(t)$ and $V_2(t)$ represent the light vibration at the pinholes P_1 and P_2, the asterisk denotes their complex conjugates[2] and the angular brackets denote the statistical (ensemble) average. In the theory of random processes, the mutual intensity is just a correlation function of the light vibrations $V_1(t)$ and $V_2(t)$ at the two points. Zernike next defined the (generally complex) degree of coherence of the light at P_1 and P_2 by the expression

$$j_{12} = \frac{J_{12}}{\sqrt{I_1}\sqrt{I_2}}, \tag{4}$$

where

$$I_j = \langle V_j^*(t) V_j(t) \rangle \,, \quad (j = 1, 2), \tag{5}$$

[2] In Zernike's original definition of the asterisk, which denotes complex conjugation, it was placed on $V_2(t)$, not on $V_1(t)$. The trivially modified definition we use in this lecture is in line with modern usage which is rooted in the rather dominant role which normal ordering of operators plays in quantum theory of coherence [see remarks following Eq. (20) below]. Also it is sometimes useful to write $V(P_j, t)$ or $V(\mathbf{r}_j, t)$ in place of $V_j(t)$ and $J(P_1, P_2)$ or $J(\mathbf{r}_1, \mathbf{r}_2)$ in place of J_{12}, where $\mathbf{r}_1$ and $\mathbf{r}_2$ are the position vectors of the points P_1 and P_2 respectively. We will freely use such more explicit notation when convenient.

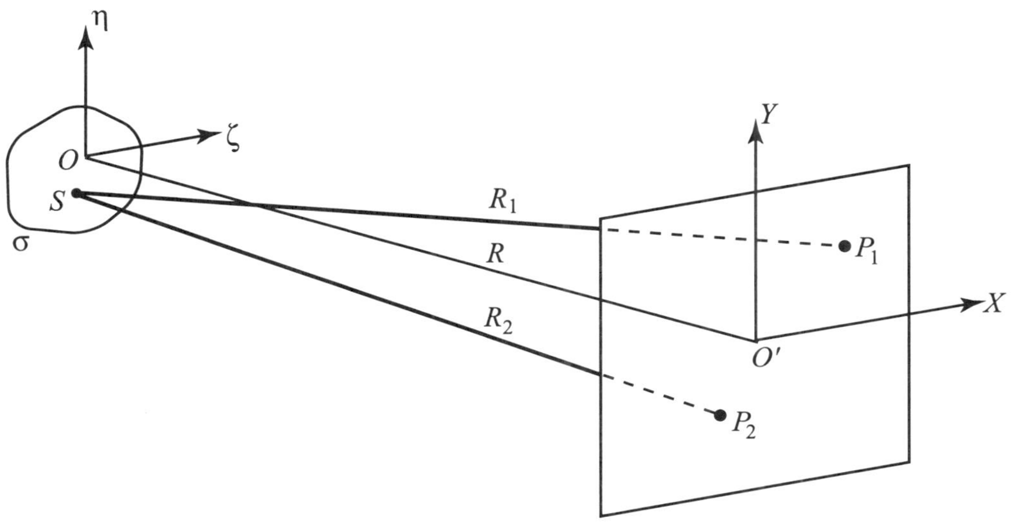

Figure 2: Notation relating to the van Cittert-Zernike theorem [Eq. (6)].

are the averaged intensities of the light at P_1 and P_2. He then showed that the degree of coherence, which he first introduced operationally by identifying it with fringe visibility, is just the absolute value of the quantity defined statistically by Eqs. (4) and (5). Later, the phase of the (complex) degree of coherence j_{12} was found also to have an operational significance, being related to the position of the intensity maxima in the fringe pattern.

In the same paper Zernike derived a number of important results concerning the mutual intensity and the complex degree of coherence. In particular he derived the following expression for the complex degree of coherence of light at two arbitrary points P_1 and P_2 in a field generated by a planar, incoherent source σ with intensity distribution $i(S)$:

$$j_{12} = \frac{1}{N}\int_\sigma i(S)\frac{e^{i\bar{k}(R_2-R_1)}}{R_1R_2}\mathrm{d}\sigma. \tag{6}$$

In this formula R_1 and R_2 are the distances to P_1 and P_2 from a typical source point S (See Fig. 2), $\bar{k}$ is the mean wave number of the light and N is a normalization constant. This result is remarkable for several reasons. First because it implies that even a completely incoherent source can generate a field which is highly coherent in some region of space. And secondly that the mutual intensity and also the complex degree of coherence are given by expressions which are analogous to well-known diffraction formulas based on Huygens' Principle for the field produced by diffraction of monochromatic light at an aperture in an opaque screen. The formula (6) is today known as the *van Cittert-Zernike theorem*, van Cittert's name being included because five years prior to the publication of Zernike's paper he obtained a similar result for a somewhat more restricted situation [7]. This theorem is a basic result of the elementary optical coherence theory and shows that coherence may

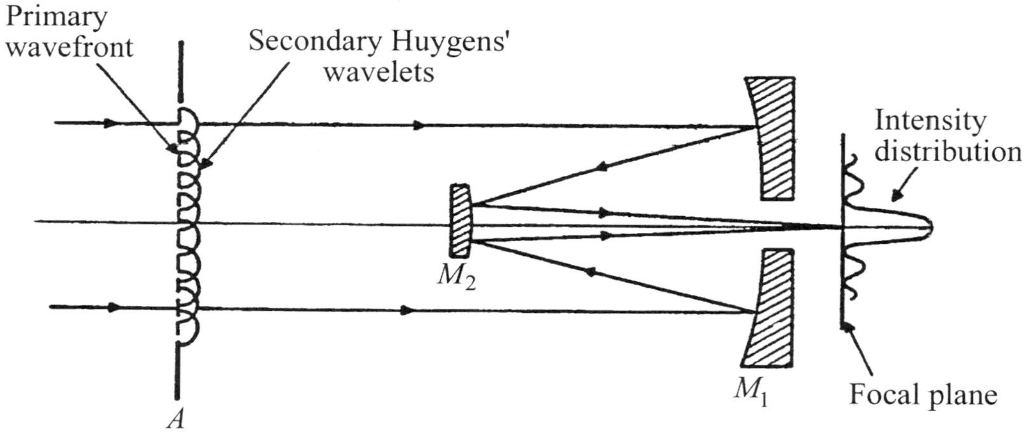

Figure 3: Illustrating the van Cittert-Zernike theorem by formation of the image by a telescope.

be generated in the process of propagation.[3] Although this is not generally appreciated, the fact that one does not need a laser to generate a field which is highly coherent over large regions of space is evident from the following simple example, illustrated in Fig. 3.

Consider the image of a star in the focal plane of a telescope. The light reaching the telescope originates in billions of atoms which radiate independently, largely by the process of spontaneous emission – the star is essentially an incoherent source. Yet the fact that, on a good observing night, one obtains a diffraction image of the star, with sharp minima in the focal plane of the telescope, shows that the light reaching the telescope must be highly coherent in order that the Huygens secondary wavelets proceeding from the aperture to the focal plane can cancel each other by destructive interference. This result can be quantitatively explained by the van Cittert-Zernike theorem.

Later in his paper, Zernike also obtained a theorem relating to the propagation of the mutual intensity. This result also proved to be of considerable importance for the later development of coherence theory.

I mentioned that the van Cittert-Zernike theorem resembles, in its mathematical structure, a formula well-known from elementary diffraction theory – namely an expression for the distribution of light in a pattern arising from the diffraction of a spherical wave at an aperture in an opaque screen. Zernike's law for propagation of the mutual intensity also has a close resemblance to a formula based on Huygens' Principle of elementary wave theory. I was puzzled for a long time by

[3]An experimental verification of the van Cittert-Zernike theorem was described in a paper by B.J. Thompson and E. Wolf, *J. Opt. Soc. Amer.* **47** (1957), 895. More recently the experiment was repeated with a high precision digital automatic system [D. Ambrosini, G. Schirripa Spagnolo, D. Paoletti and S. Vicalvi, *Pure Appl. Opt.* **7** (1998), 933]. Very good agreement with theory was found.

this similarity. Eventually I found the reason for it, which required a generalization of the correlation function introduced by Zernike in his basic 1938 paper.

The mutual intensity $J(\mathbf{r}_1, \mathbf{r}_2) = \langle V^*(P_1,t)V(P_2,t)\rangle$ is a measure of the correlation which exists between the field vibrations at points $P_1(\mathbf{r}_1)$ and $P_2(\mathbf{r}_2)$ at the *same* instants of time. In 1954 and 1955 I introduced more general correlation functions [8, 9] involving vibrations not only at two points in space, but also at two instants of time, t and $t+\tau$. For a scalar field such a function is known as the *mutual coherence function* and is defined by the formula

$$\Gamma(\mathbf{r}_1, \mathbf{r}_2, \tau) = \langle V^*(P_1,t)V(P_2,t+\tau)\rangle \,. \tag{7}$$

It is implicitly assumed here that the correlation function Γ is invariant with respect to the translation of the origin of time, as is usually the case. (In the terminology of the theory of random processes, this assumption means that the optical field is assumed to be statistically stationary in the wide sense). Further, unlike in Zernike's formulation, where the significance of the complex field variable $V(\mathbf{r},t)$ was somewhat vague in this more general formulation, it is defined as the complex analytic signal associated with the real scalar field ([1], Sec. 10.2, [10], Sec. 3.1) which may represent, for example, a Cartesian component of the electric field.

The generalization which leads from the mutual intensity to the mutual coherence function has important consequences. In particular, simple calculations show that in free space, the mutual coherence function obeys rigorously the two wave equations

$$\nabla_1^2\Gamma(\mathbf{r}_1, \mathbf{r}_2, \tau) = \frac{1}{c^2}\frac{\partial^2\Gamma(\mathbf{r}_1, \mathbf{r}_2, \tau)}{\partial\tau^2}, \tag{8}$$

$$\nabla_2^2\Gamma(\mathbf{r}_1, \mathbf{r}_2, \tau) = \frac{1}{c^2}\frac{\partial^2\Gamma(\mathbf{r}_1, \mathbf{r}_2, \tau)}{\partial\tau^2}, \tag{9}$$

where ∇_1^2 and ∇_2^2 are the Laplacian operators acting with respect to the coordinates of the two points $P_1(\mathbf{r}_1)$ and $P_2(\mathbf{r}_2)$, respectively, c being the speed of light in vacuum. These two equations imply that not only the propagation of the complex field amplitudes is governed by the wave equation, but also the propagation of the state of coherence of the field represented by the mutual coherence function $\Gamma(\mathbf{r}_1, \mathbf{r}_2, \tau)$. This fact provides the explanation of the analogy mentioned earlier, between Zernike's propagation law for the mutual intensity and the van Cittert-Zernike theorem on the one hand and some results of elementary diffraction theory on the other. For one can readily show that Zernike's propagation law for the mutual intensity and also the van Cittert-Zernike theorem are just approximate solutions of the wave equation for the mutual coherence function, just as the Huygens-Fresnel principle of classical diffraction theory is an approximate solution of the wave equation for the field.

Up to now we characterized the optical field by a random scalar function $V(\mathbf{r},t)$, i.e. we ignored the polarization properties of the field; they were also incorporated into the theory [9, 11, 12]. A simple way to understand how vector features can be included is to consider the effect of a compensating plate, C, and of a polarizer P on a fluctuating plane electromagnetic wave, as shown in Fig. 4. If one considers the intensity of the light which emerges from such a device [9] or the degree of polarization of the incident wave [11], one finds that these quantities depend on the averaged

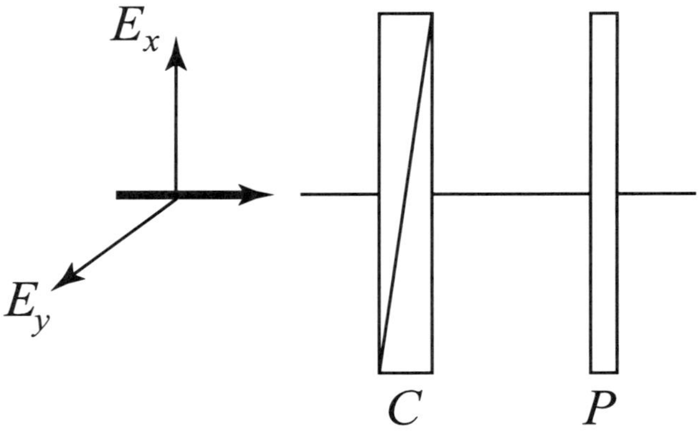

Figure 4: Illustrating the definition (11) of correlation in a fluctuating plane wave propagating along the z-direction. C = compensator, P = polarizer.

intensities

$$I_x = \langle E_x^*(t) E_x(t) \rangle \quad \text{and} \quad I_y = \left\langle E_y^*(t) E_y(t) \right\rangle \tag{10}$$

of two mutually orthogonal components $E_x(t)$, $E_y(t)$ of the incident field perpendicular to the direction of propagation and also on the correlation function

$$\mathcal{E}_{xy} = \langle E_x^*(t) E_y(t) \rangle \tag{11}$$

between them. This correlation function is the vector analogue of Zernike's mutual intensity function (3) of the scalar coherence theory.

The quantities defined by Eqs. (10) and (11) may be considered to be elements of the so-called coherence matrix or polarization matrix

$$\mathcal{E} = \begin{bmatrix} \langle E_x^*(t) E_x(t) \rangle & \langle E_x^*(t) E_y(t) \rangle \\ \left\langle E_y^*(t) E_x(t) \right\rangle & \left\langle E_y^*(t) E_y(t) \right\rangle \end{bmatrix}. \tag{12}$$

If one deals with more general fluctuating waves than a plane wave and one wishes to analyze more complicated situations than passage through simple devices such as a polarizer and a compensator, one has to employ more general correlation matrices (tensors), namely [9, 12]

$$\mathcal{E}_{jk}(\mathbf{r}_1, \mathbf{r}_2, \tau) = \left\langle E_j^*(\mathbf{r}_1, t) E_k(\mathbf{r}_2, t+\tau) \right\rangle, \quad (j, k = x, y, z), \tag{13}$$

$$\mathcal{H}_{jk}(\mathbf{r}_1, \mathbf{r}_2, \tau) = \left\langle H_j^*(\mathbf{r}_1, t) H_k(\mathbf{r}_2, t+\tau) \right\rangle, \quad (j, k = x, y, z). \tag{14}$$

These quantities represent correlations between the Cartesian components of the electric field $\mathbf{E}$ and of the magnetic field $\mathbf{H}$. For a more complete description one also has to introduce the "mixed" correlation matrices (tensors) with elements

$$\mathcal{M}_{jk}(\mathbf{r}_1, \mathbf{r}_2, \tau) = \left\langle E_j^*(\mathbf{r}_1, t) H_k(\mathbf{r}_2, t+\tau) \right\rangle, \tag{15}$$

$$\mathcal{N}_{jk}(\mathbf{r}_1, \mathbf{r}_2, \tau) = \left\langle H_j^*(\mathbf{r}_1, t) E_k(\mathbf{r}_2, t+\tau) \right\rangle. \tag{16}$$

These four coherence matrices (tensors) can be shown to be entirely adequate to characterize all the usual optical phenomena. As a consequence of Maxwell's equations these four tensors may readily be shown to be coupled by a set of partial differential equations. These equations are more general than Maxwell's equations because they take into account some of the statistical features of the field.

Unlike the electromagnetic field vectors, which in the optical range of the electromagnetic spectrum fluctuate too rapidly to be directly measured, the elements of these correlation matrices may generally be detected by suitable experiments. Unlike Maxwell's theory for deterministic fields, which deals with quantities in the optical regions of the electromagnetic spectrum that cannot be directly measured, coherence theory provides a basis for a description of optical phenomena in terms of measurable quantities.

The electromagnetic correlation tensors and the equations which they satisfy were introduced in 1954 and at first sight it seemed that this development completed the general formulation of optical coherence theory within the framework of classical optics. But almost as soon as this formulation was completed, an important new discovery was made, whose understanding required a further generalization of the theory. This was an experimental demonstration made by Hanbury Brown and Twiss in 1956 that under certain circumstances one can measure not only correlations between the field at different points and different times, but also correlations between the instantaneous intensities [13].

In their experiment, illustrated in Fig. 5, narrow-band light from a Mercury arc lamp was divided into two beams by a half-silvered mirror M and each of the beams illuminated a cathode of a photomultiplier C_1 and C_2. The current output of the two photomultipliers was fed into a correlator. Under appropriate circumstances the correlation between the two currents may be shown to be expressible in terms of the correlation function

$$\Omega_{12}(\tau) = \langle I(\mathbf{r}_1, t) I(\mathbf{r}_2, t+\tau) \rangle \tag{17}$$

of the instantaneous intensities of the light incident on the two photomultipliers. Conversely, from measurements of the correlation of the current outputs of the two photomultipliers, one can deduce information about correlations between the intensities of the incident light. Because the instantaneous intensity $I(\mathbf{r}, t)$ is equal to $V^*(\mathbf{r}, t)V(\mathbf{r}, t)$, Eq. (17) implies that

$$\Omega_{12}(\tau) = \langle V^*(\mathbf{r}_1, t)V(\mathbf{r}_1, t)V^*(\mathbf{r}_2, t+\tau)V(\mathbf{r}_2, t+\tau) \rangle \,. \tag{18}$$

Since this function contains the field variable $V(\mathbf{r}, t)$ in the fourth power it is an example of what is now known as a *fourth-order* correlation function of the field.

The Hanbury Brown-Twiss effect is rather subtle. In the early days which followed its discovery, there was a good deal of controversy about it, because photoelectric detectors do not measure fields – they detect photons. It was difficult, at first, to understand how such classical properties of the incident field as the intensity correlation function could be measured in this way. Eventually this came to be understood, mainly as a result of a very clear analysis of the effect by Purcell [14]. Important extension of Purcell's treatment was later made by Mandel [15].

At the time when experiments of this kind were first performed, the only light sources available produced *thermal light*, such as incandescent matter or gas discharge. The probabilities which govern

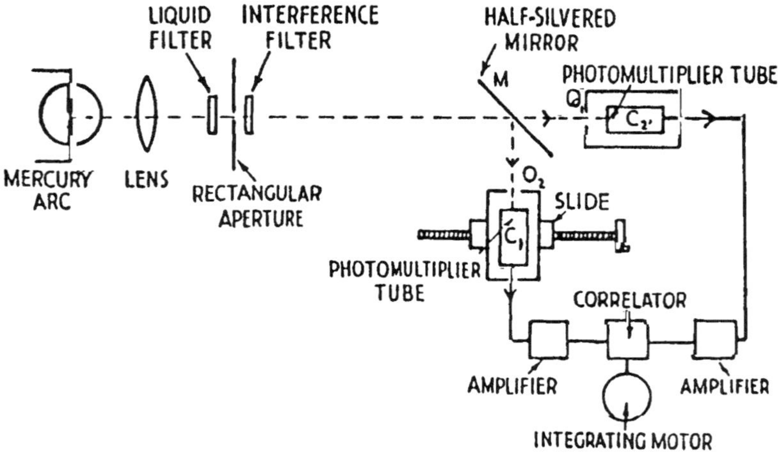

Figure 5: Apparatus for the demonstration of the Hanbury Brown-Twiss effect (reproduced from a paper by R. Hanbury Brown and R.Q. Twiss, *Nature* **177** (1956), 28).

the fluctuations of light of this kind may be shown to be all Gaussian functions and this fact makes it possible to determine the absolute value of the degree of coherence $\gamma_{12}(\tau)$ from measurements of the intensity correlation function $\Omega_{12}(\tau)$. As a consequence of this result, new experimental techniques were developed which have found useful applications, particularly in spectroscopy and in astronomy.

The Hanbury Brown-Twiss effect was the first example of a *higher-order* correlation effect that may exist in an optical field. Higher-order correlations have become of considerable interest since the development of the laser in the 1960s, because among other things they reveal some basic differences between the statistical properties of highly coherent light and thermal light.

Almost simultaneously, in 1963 and 1964, systematic descriptions of higher-order correlations were introduced both on the basis of the classical theory [16, 17] and on the basis of the theory of the quantized field [18]. Within the framework of classical wave theory a description of the higher correlations was introduced via natural generalizations of the correlation matrices ([12]). For example the coherence matrix (tensor) of order (m, n) of the electric field (not necessarily a field which is statistically stationary) is defined by the expectation value

$$\mathcal{E}^{(m,n)}_{j_1,j_2,\ldots j_{m+n}}(X_1, X_2, \ldots X_{m+n}) = \\ \left\langle E^*_{j_1}(X_1)E^*_{j_2}(X_2) \;\ldots\; E^*_{j_m}(X_m)E_{j_{m+1}}(X_{m+1})\ldots E_{j_{m+n}}(X_{m+n})\right\rangle . \tag{19}$$

In this formula the variables X_l, $(l = 1, 2, \ldots, m + n)$, represent the space-time points $(\mathbf{r}_l, t_l)$ and the subscripts $j_1, j_2, \ldots$ label Cartesian components. One can define, in a similar way, the coherence tensors of all orders of the magnetic field and also higher-order mixed coherence tensors involving both the electric and the magnetic fields. The whole hierarchy of such tensors of all orders completely characterizes the properties of a fluctuating electromagnetic field of any statistical properties. In the special but important practical case when the fluctuations of the electromagnetic field can be described as a multi-dimensional Gaussian random process (e.g. a thermal field), all the higher-order tensors may be expressed in terms of the lowest-order ones, i.e. in terms of those for which $m = n = 1$.

In the formulation based on quantum field theory, the central quantities which characterize coherence of arbitrary order are again certain tensors which have a striking formal resemblance to those of the classical theory. For example, the electric coherence tensor of order (m, n) is defined as

$$\mathcal{E}^{(m,n)}_{j_1,j_2,\ldots j_{m+n}}(X_1, X_2, \ldots X_{m+n}) = Tr\left\langle \hat{\rho}\hat{E}^{(-)}_{j_1}(X_1)\hat{E}^{(-)}_{j_2}(X_2)\ldots \times \hat{E}^{(-)}_{j_m}(X_m)\hat{E}^{(+)}_{j_{m+1}}(X_{m+1})\hat{E}^{(+)}_{j_{m+2}}(X_{m+2})\ldots\hat{E}^{(+)}_{j_{m+n}}(X_{m+n})\right\rangle . \tag{20}$$

Here $\hat{E}^{(+)}$ and $\hat{E}^{(-)}$ are the positive and negative frequency parts of the electric field operators $\hat{E}$ and $\hat{\rho}$ is the density operator which characterizes the state of the field. Unlike the formula (19) of the classical theory, the value of the quantum-mechanically defined electric coherence tensor depends on the order of the factors on the right-hand side of Eq. (20), because the positive and the negative parts of the electric field operator do not commute. The ordering on the right-hand side of Eq. (20) is the so-called normal ordering which appears naturally in the theory of photoelectric detection of light fluctuations.

It is clear from the preceding remarks that by 1964 a complete formulation of the coherence theory of optical fields was obtained, both on the basis of classical theory as well as on the basis of the theory of quantized fields. The theory provided the mathematical tools which were needed to formulate the laws of statistical optical wave fields. It is of interest to note that this happened almost exactly 100 years after Maxwell discovered his celebrated set of equations which govern the behavior of deterministic optical fields.

In the roughly thirty-five years which have gone by since then, numerous applications of the theory have been made and further new insights into the behavior of statistical wavefields have been obtained based on this formulation. Because of limitations of time, I will discuss now only one further development which has provided a new understanding of the behavior of fields of arbitrary state of coherence. It is a rather different formulation than the one initiated by Zernike and is known as coherence theory in the *space-frequency domain*, in contrast to the older formulation which may be called coherence theory in the *space-time domain.*

To explain the essence of this new formulation introduced in 1980 [19] we recall that we have implicitly assumed in the formulation of coherence theory in the space-time domain that we are dealing with steady-state fields which, in more technical terms, are said to be *statistically stationary.* This means that the probability distributions which characterize the fluctuations are invariant with respect to the translation of the origin of time. Since in stationary fields no particular instant of time, t, plays a preferential role, fields of this kind do not tend to zero as $t \to +\infty$ or $t \to -\infty$ and, therefore, do not possess a Fourier integral representation. Consequently the precise meaning of the spectrum, a basic attribute of an optical field, is somewhat obscure.

The problem of defining a spectrum of a stationary random process has been studied by mathematicians for many decades and several sophisticated mathematical techniques are now available for this purpose, such as the so-called generalized harmonic analysis, the use of the stochastic Stieltjes integral and the theory of generalized functions. Whilst these techniques lead to definitions of the spectrum of a steady-state (i.e. stationary) optical field, they do not provide an understanding of it which agrees with a physicists' intuition of the meaning of this concept.

The space-frequency formulation of coherence theory avoids the use of the rather sophisticated mathematical techniques previously developed for the spectral representation of stationary fields. The central role in the new theory in the space-frequency domain is played by the so-called cross-spectral density function $W(\mathbf{r}_1, \mathbf{r}_2, \omega)$, rather than its mutual coherence function $\Gamma(\mathbf{r}_1, \mathbf{r}_2, \tau)$, these two functions being just Fourier transforms of each other,[4]

$$W(\mathbf{r}_1, \mathbf{r}_2, \omega) = \frac{1}{2\pi} \int_{-\infty}^{\infty} \Gamma(\mathbf{r}_1, \mathbf{r}_2, \tau) e^{i\omega\tau} \mathrm{d}\tau. \tag{21}$$

It has been shown [19] that it is possible to introduce an *ensemble* (denoted by curly brackets)

$$\{U(\mathbf{r}, t)\} = \{U(\mathbf{r}, \omega) e^{-i\omega t}\} \tag{22}$$

of monochromatic realizations, all of the same frequency ω, such that the cross-spectral density $W(\mathbf{r}_1, \mathbf{r}_2, \omega)$ of the source and also of the field which the source generates are themselves correlation functions, namely

$$\langle U^*(\mathbf{r}_1, \omega) U(\mathbf{r}_2, \omega) \rangle_\omega = W(\mathbf{r}_1, \mathbf{r}_2, \omega). \tag{23}$$

The subscript ω on the angular brackets on the left-hand side of this equation indicates that the expectation value is taken over the ensemble of the space-dependent parts of the monochromatic realizations $U(\mathbf{r}, \omega)$, all of the same frequency ω. This ensemble must be clearly distinguished from the ensemble of the time-dependent realizations $V(\mathbf{r}, t)$ by means of which the mutual coherence function $\Gamma(\mathbf{r}_1, \mathbf{r}_2, \tau)$ is defined [Eq. (7)].

Members of the $\{U\}$-ensemble are introduced via the eigenfunctions $\phi_n(\mathbf{r}, \omega)$ and the eigenvalues $\lambda_n(\omega)$ of the cross-spectral density function $W(\mathbf{r}_1, \mathbf{r}_2, \omega)$. For example, for a two-dimensional, planar secondary source, occupying a finite portion σ of a plane,

$$U(\mathbf{r}, \omega) = \sum_n a_n(\omega) \phi_n(\mathbf{r}, \omega) \tag{24}$$

where

$$\int_\sigma W(\mathbf{r}_1, \mathbf{r}_2, \omega) \phi_n(\mathbf{r}_1, \omega)\, \mathrm{d}^2 r_1 = \lambda_n(\omega) \phi_n(\mathbf{r}_2, \omega), \tag{25}$$

and $a_n(\omega)$ are random coefficients such that

[4]Unlike the field variable $V(\mathbf{r}, t)$ of a stationary random field, the mutual coherence function $\Gamma(\mathbf{r}_1, \mathbf{r}_2, \tau)$ usually posesses a Fourier transform.

$$\langle a_n^*(\omega)a_m(\omega)\rangle = \lambda_n(\omega)\delta_{nm}, \tag{26}$$

δ_{nm} being the Kronecker delta symbol.

It follows at once from Eq. (23) that the spectral density $S(\mathbf{r},\omega)$ of the light at a typical source point is given by the expression

$$S(\mathbf{r},\omega) \equiv W(\mathbf{r},\mathbf{r},\omega) = \langle U^*(\mathbf{r},\omega)U(\mathbf{r},\omega)\rangle\,. \tag{27}$$

This expression agrees well with the physical intuition that the spectral density should be expressable in terms of an average of the squared modulus of monochromatic oscillations. However, as we have just seen, the monochromatic oscillations $U(\mathbf{r},\omega)e^{-\mathrm{i}\omega t}$ are not Fourier components (which do not exist) of a fluctuating steady-state field $V(\mathbf{r},t)$, but rather the expectation values of a suitably chosen statistical ensemble of monochromatic oscillations, defined by means of Eqs. (25) and (26).

One can define a *spectral degree of coherence*, usually denoted by $\mu(\mathbf{r}_1,\mathbf{r}_2,\omega)$, which is a measure of correlations in the space-frequency domain, by the formula [20]

$$\mu(\mathbf{r}_1,\mathbf{r}_2,\omega) = \frac{W(\mathbf{r}_1,\mathbf{r}_2,\omega)}{\sqrt{W(\mathbf{r}_1,\mathbf{r}_1,\omega)}\sqrt{W(\mathbf{r}_2,\mathbf{r}_2,\omega)}}. \tag{28}$$

It may be shown that $0 \leq |\mu(\mathbf{r}_1,\mathbf{r}_2,\omega)| \leq 1$ for all values of the arguments $\mathbf{r}_1$, $\mathbf{r}_2$, and ω, the upper bound, unity, representing complete coherence, the lower bound, zero, complete incoherence of the field at the two points and at the particular frequency ω. Just like the mutual coherence function $\Gamma(\mathbf{r}_1,\mathbf{r}_2,\tau)$ and the complex degree of coherence $\gamma(\mathbf{r}_1,\mathbf{r}_2,\tau)$, the cross-spectral density function $W(\mathbf{r}_1,\mathbf{r}_2,\omega)$ and the spectral degree of coherence $\mu(\mathbf{r}_1,\mathbf{r}_2,\omega)$ can be measured with the help of a Young's interference experiment, utilizing, in addition, narrow-band filters [21].

Coherence theory in the space-frequency domain, although of relatively recent origin, has already led to many interesting and useful developments. This is not the occasion for reviewing these recent developments. We only mention that the new formulation has clarified, for example, the coherence properties of transverse laser modes [22] and has led to the rather surprising prediction, not previously suspected but since then demonstrated by many experiments, that the spectrum of light may change as the light propagates, even in free space [23, 24]. This discovery has many potential ramifications.

It was the aim of this talk to outline the development of optical coherence theory. Because of limitations of time, it was only possible to describe the formulation and the basic concepts which clarified the notion of coherence, rather than the many interesting and useful applications of the theory. This is a circumstance I greatly regret. However, interested readers may find such topics discussed in several books which now exist [1, 10, 25] and also in those which will undoubtedly appear in the years to come.

References

[1] M. Born and E. Wolf, *Principles of Optics* (Cambridge, Cambridge University Press, 7th ed. , 1999).

[2] *Selected Papers on Coherence and Fluctuations of Light*, L. Mandel and E. Wolf, eds. (Dover, New York, 1970), Vol. I (1850–1960), vol II (1961–1966); reprinted by SPIE Optical Engineering Press Milestone Series, MS 19, Parts I and II (Bellingham, WA, 1990).

[3] M. von Laue, *Ann. d. Physik* **23** (1907), 1–43 and 795–797.

[4] E. Schrödinger, *Ann. d. Physik* **61** (1920), 69–86.

[5] F. Zernike, *Physica* **5** (1938), 785–795.

[6] F. Zernike, *Proc. Phys. Soc. (London)* **61** (1948), 158–164.

[7] P.H. van Cittert, *Physica* **1** (1934), 201–210.

[8] E. Wolf, *Proc. Roy. Soc.* A **230** (1955), 246–265.

[9] E. Wolf, *Nuovo Cimento* **12** (1954), 884–888.

[10] L. Mandel and E. Wolf, *Optical Coherence and Quantum Optics* (Cambridge, Cambridge University Press, 1995).

[11] E. Wolf, *Nuovo Cimento* **13** (1959), 1165–1181.

[12] E. Wolf, in *Proc. Symp. on Astronomical Optics*, ed. Z. Kopal (North-Holland, Amsterdam, 1956), 177–185.

[13] R. Hanbury Brown and R.Q. Twiss, *Nature* **177** (1956), 27–29.

[14] E.M. Purcell, *Nature* **178** (1956), 1449–1450.

[15] L. Mandel, in *Progress in Optics*, ed. E. Wolf (North-Holland, Amsterdam, v. II, 1963), 181–248.

[16] E. Wolf in *Proc. Symposium on Optical Masers*, ed. J. Fox (Brooklyn Polytechnic Press and Wiley, New York, 1963), 29–42.

[17] L. Mandel in *Quantum Electronics*, Proc. Third International Congress, ed. P. Grivet and N. Bloembergen (Dunod, Paris and Columbia University Press, New York, 1964), 101–108.

[18] R. Glauber, (a) *Phys. Rev.***130** (1963), 2529–2539; (b) *ibid.* **131** (1963), 2766–2788.

[19] E. Wolf, (a) in *Optics in Four Dimensions* (ICO Conference, Ensenada, Mexico, 1980), ed. M.A. Machado and L.M. Narducci (Conference Proceedings #65, American Institute of Physics, New York), 42–48; (b) *Optics Commun.* **38** (1981), 3–6; (c) *J. Opt. Soc. Amer.* **72** (1982), 343–351; (d) *ibid.* A **3** (1986), 76–85; (e) G.S. Agarwal and E. Wolf, *J. Mod. Opt.* **40** (1993), 1489–1496.

[20] E. Wolf and W.H. Carter, *Opt. Commun.* **13** (1975), 205–209; see also E. Wolf and W.H. Carter, *ibid.* **16** (1976), 297–302; L. Mandel and E. Wolf, *J. Opt. Soc. Amer.* **66** (1976), 529–535; M.J. Bastians, *Optica Acta* **24** (1977), 261–274.

[21] Methods for measuring the spectral degree of coherence are described, for example, by (a) E. Wolf, *Opt. Lett.* **8** (1983), 250–252; (b) D.F.V. James and E. Wolf, *Opt. Lett.***145** (1998), 1–4.

[22] E. Wolf and G.S. Agarwal, *J. Opt. Soc. Amer.* A **1** (1984), 541–546.

[23] E. Wolf, *Phys. Rev. Lett.* **56** (1986), 1370–1372.

[24] An account of the many investigations on this subject is given by E. Wolf and D.F.V. James, *Rep. Progr. Phys.* (IOP Publishing, Bristol and London) **59** (1996), 771–818.

[25] C. Brosseau, *Fundamentals of Polarized Light, A Statistical Optics Approach* (Wiley, New York, 1998).

Lecture 7.4
Recollections *

Emil Wolf
University of Rochester

Rector Jarab, Members of the Faculty of Palacký University, Ladies and Gentlemen:

The great honor which this old and famous University is bestowing upon me today is a particularly moving experience for me. For I left Czechoslovakia as a young school boy in 1939 for a world which was then quite unfamiliar to me, and I certainly would not have dreamt either at that time or later, for that matter, that more than half-a-century after leaving my home I would be returning to my beloved country under these auspicious circumstances.

Unfortunately having lived all these years in the West, my mother tongue, the Czech language, has become very rusty. I have had hardly any opportunity during all these years to speak Czech and my studies, my writings and my University lectures have all been in English. I must, therefore, ask you to forgive me for some clumsy expressions, mispronunciations and mistakes which I will undoubtedly make in the course of this talk and also for my having to read my speech.

I was asked to talk here about my scientific work. However, it is impossible for me to separate any success which I have had in this endeavor without mentioning the good fortune which I have had by being taught by excellent teachers and by having known and having collaborated with several outstanding scientists, some of whom have made a profound impact on the development of physics. I will, therefore, begin by first saying a few words about them and about the role which they have played in my career.

I completed my graduate studies in 1948 at Bristol University in England under the supervision of Dr. E.H. Linfoot, an outstanding mathematician and optical scientist. Later that year Linfoot became the Assistant Director of the Cambridge University Observatories. He offered me and I accepted a position as his assistant in Cambridge. During the next two years whilst I was in Cambridge I often travelled the relatively short distance to London, to attend meetings of the British Optical Society. Through these meetings I became acquainted with Dr. Dennis Gabor, who around that time became famous for the discovery of a new optical principle concerning the imaging of three-dimensional objects, known as holography. For this discovery Gabor later (in 1971) received the Nobel Prize in Physics.

*English translation of a speech (given in Czech) following the award of the degree of doctor of science *honoris causa* by Palacký University, Olomouc, Czechoslovakia on June 3, 1992.

Through Gabor I learned in 1950 that Max Born, one of the greatest physicists of this century and, at that time, the Professor of Natural Philosophy at the University of Edinburgh in Scotland, was looking for a collaborator to help him prepare a book on modern optics, a field in which I was working. Gabor encouraged me to apply for a position with Born to help him in the writing of such a book. I did this and I was very fortunate in becoming, at the beginning of 1951, Max Born's assistant and his collaborator on this project. Not only did this collaboration result in the publication, under our joint authorship, of a text called *Principles of Optics*, but it has given me the priceless opportunity of seeing and speaking to Born almost every day. In this way I was able to learn much more than I would have ever been able to do in any other way, about physics and how research is done. Incidentally, just like Gabor, Born received the Nobel Prize in Physics (in 1954), for his fundamental contributions to a field called quantum mechanics which, in about 1926, revolutionized physics.

Let me add that many years after the publication of our book I learned a little known fact, namely that Olivia Newton-John, a very famous, popular and beautiful singer, is Max Born's granddaughter. In this connection I should tell you that to some of my students I am known not so much as a co-author with Max Born of the book *Principles of Optics* but rather as the person who has a picture of Olivia Newton-John hanging in his office, signed *"To Emil, Love, Olivia."*

Much of the research which I did after I left Edinburgh and about which I will now speak, can be directly traced to my trying to clarify some questions which I felt were left unanswered in optics at the time when I collaborated in the writing of the book. And I am aware how much my style of research and my attitude to it were influenced by my close contact with Max Born.

Most of my research was in the field of optics – that is the branch of science which is concerned with light in all its aspects, for example in its generation, propagation and detection. In order to make the account of my work more easily understandable I will first say a little about some of the basic properties of light.

Light is produced by complicated processes involving atoms that constitute the light source, such as, for example, an electric bulb, and it then propagates into the space around it in the form of *light waves*. A pure light wave consists of periodic oscillations of electric and magnetic fields. The easiest way to think about waves is to consider the motion of a water surface disturbed, for example, by throwing a stone into a pond. As the disturbance propagates from the point where the stone hit the water surface, we observe certain repetitive patterns. The spatial periodicity of the waves, i.e. the separation between successive crests, is known as the *wavelength*. The temporal periodicity is characterized by the rate of these oscillations, that is to say by their frequency. Of course, with light we do not see the wave forms as directly as with water waves, but their existence can be deduced from appropriate experiments.

Light of different wavelengths gives rise to the psychological impressions of various colors. Actually light which we commonly encounter is not so pure but rather it consists of a mixture of different wavelengths. One can separate the various wavelength components by passing the light through a suitable optical device, such as a prism. The different wavelength components then become spatially separated by the prism and one observes on a screen placed in the path of the emerging light, a colored pattern, known as the spectrum of the light. I am sure that you are all familiar with spectra from observing the rainbow.

In the early years of my scientific career, around 1950, I investigated problems relating to the focusing of light, i.e. problems concerned with concentrating light into as small a region of space as possible, with the help of lenses, for example. The full understanding of the focal region is of great importance for technology. Some of our results on focusing, which my students and I have discovered, are even today, almost four decades later, frequently used for example in designing laser scanning systems or new types of microscopes, and in connection with the improvements of compact disks and other electronic devices.

In the 1960s I became interested in other problems, especially in the possiblity of determining the structures of semi-transparent objects such as tissues, various biological specimens, membranes, etc. from measurements of light and other types of radiation, for example sound waves, which passes through such objects or which are reflected, refracted or scattered by them. This work has contributed to a rapidly developing field known as diffraction tomography, now widely used in diagnostic medicine, for example in detecting the presence of brain lesions, cancerous breast tumors, and so on.

However, my main scientific interest throughout most of my career has been in a field known as optical coherence theory. The concept of optical coherence is associated with the order which exists in an optical source or in an optical field. In an ordinary light source such as an electric light bulb or the Sun, the atoms which produce the light radiate in a haphazard random manner, at different instants of time. Moreover the light waves which the atoms generate are not the pure waves about which I already spoke, but they consist of a range of wavelengths. Consequently the resulting wave-forms are complicated, varying rapidly and irregularly both in space and in time. We say that such light is incoherent.

In 1960, an entirely new light source, the laser, was invented and quickly developed; it has revolutionized modern technology. In such a source the atoms are forced to emit light in a highly ordered manner; one then says that the source is coherent. The light waves produced by a source of this type behave in a much more regular fashion than light which originates in traditional sources and the waves are then essentially the pure waves of definite wavelengths, which I already mentioned.

In a sense the traditional incoherent sources and the coherent laser sources are two extremes. Today there are many other types of sources which in their behavior are intermediate between the two – they are neither incoherent nor fully coherent. Such sources are said to be partially coherent. The subject of partial coherence is concerned with sources and fields of any degree of order, i.e. of any state of coherence.

Before proceeding further I would like to mention that the Department of Optics of this University, under the direction of Professor Jan Peřina, is well-known for the fine contributions which it has made and continues to make to this area of modern optics.

In the 1950s I was fortunate in having found a formulation of coherence theory which elucidates many phenomena involving light of different states of coherence. To avoid technicalities and also in order to keep this talk to a reasonable length I will discuss only one recent development based on this theory, although there are many others.

Earlier I spoke about the spectrum of light, namely its color components which are observed when the light is broken down into its constituent wavelengths by means of a suitable instrument. Actually each type of atom contained in the light source, for example the atoms of copper, has its own characteristic spectrum. It consists of lines, associated with definite wavelengths, known as spectral lines.

By analyzing the spectrum of light, i.e. by properly identifying the location of the different lines on the wavelength scale, one can learn a great deal about the nature of the light source and about the medium through which the light has passed. There is a whole branch of science, known as *spectroscopy*, which is concerned with this type of analysis. It has broad and important scientific and industrial uses.

To astronomers spectra provide basic information about the universe in which we live. By analyzing spectra of light that reaches us from stars, galaxies and other sources, astronomers can deduce, for example, what these objects are made of and what their temperatures are. Moreover, it is known that spectra can also provide information about how far away a particular stellar object is from us. This information is based on a phenomenon popularly known as the red shift of spectral lines. It is this topic with which our research during the last few years has been largely concerned.

The redshift of spectral lines is an effect that is analogous to something that many of you have undoubtedly encountered when you have heard a siren of an ambulance or of a police car that passes close by. The pitch of the siren goes up when the moving vehicle approaches you and it goes down again when it has passed you. This phenomenon is known as the Doppler effect, named after the nineteenth century physicist who first explained it as being due to the relative motion of a source with respect to an observer.

There is an amusing story about an experimental verification of the Doppler effect. In 1845 a Dutch meteorologist by the name of Christopher Buys-Ballot persuaded a group of trumpeters to ride on the footplate of a moving locomotive while playing the same note with all their might. By having musically trained observers stationed on each side of the track listen carefully to the apparent change in the pitch as the trumpeters whizzed by, Mr. Buys-Ballot convinced himself and others of the existence of this effect.

In the optical context the term red shift is used to refer to the shift of a spectral line towards a longer wavelength – i.e. towards the red end of the visible spectrum. If the shift is in the opposite direction, i.e. towards the shorter wavelengths (associated with the blue end of the visible spectrum) one speaks of a blue shift. Red shifts greatly predominate in the spectra of light which reaches the earth from astronomical objects.

The theory of the Doppler effect makes it possible to deduce the speed of a moving light source from the observed shifts of spectral lines. Moreover, according to a law postulated by an American astronomer, Edwin Hubble, in 1929, the speed of recession is proportional to the distance of the source from us. Thus according to this theory, the larger the observed red shift of a spectral line, the greater is the speed of recession and the further away is the source.

Astronomical observations interpreted on the basis of the Doppler effect and of Hubble's law provide strong support for the generally accepted cosmological theory according to which we live in a universe which originated in a Big Bang and which has been expanding ever since.

There is another physical mechanism for producing shifts of spectral lines of the same kind as that due to the Doppler effect, namely gravitation, predicted by Einstein many years ago. It arises when light passes close to a very massive body. However, this type of redshift is rather small and is of much less significance for astronomy than a redshift caused by the Doppler effect. Most astronomers do not believe that there is any other mechanism except the motion of the source relative to the observer or gravitation, that can give rise to redshifts of spectral lines.

In the 1950s astronomical sources were discovered which radiate radio waves rather than light waves. (Actually, radio waves of extraterrestrial origin were discovered in the early 1930s, but the discovery did not attract much attention at the time.) This was the origin of a new branch of astronomy, called radio astronomy. Some astronomical observations with radio waves were hard to understand on the basis of existing theories. In particular, new and rather mysterious astronomical sources were found, known today as quasars. The radiation from these sources is extremely large. If their redshifts were interpreted in terms of the Doppler effect then the Hubble law would indicate that they are extremely far away and they would produce a tremendous amount of radiation, typically thousands of times greater than is produced by an entire galaxy which consists of many billions of stars. Yet quasars are known not to be composite bodies like galaxies, but rather to be isolated objects. It is not easy to understand how such isolated objects can produce so much radiation. There are other puzzling features about quasars. For example, some of them appear to be connected to galaxies of much lower redshift. If they are connected, this would contradict Hubble's law, because according to that law sources of different redshifts have to be at different distances from us. Actually the subject of quasars is altogether a rather puzzling and controversial field.

In 1986 I predicted the existence of a new mechanism for producing redshifts of spectral lines. It arises from the coherence properties of sources and from somewhat similar properties of their atmospheres. What is perhaps most remarkable is that in some cases this mechanism may be shown to completely imitate the Doppler effect, even though the source is not moving away from us. This implies that if one detects a redshift of a spectral line from an astronomical source one cannot always be certain that the shift is due to recessional motion; it could be due to this new effect, which has nothing to do with motion.

The new phenomenon has attracted a good deal of attention and also some controversy. By now about 60 articles have been published on this subject. And the existence of this effect has been verified by experiments in several laboratories in the USA, in Italy, and in India. However, much more work needs to be done before one can be sure that this phenomenon is relevant to astronomy and that it can resolve some of the controversies concerning quasars about which I spoke.

I would like to conclude this talk by saying that I consider myself very fortunate in having known and having collaborated with distinguished scientists whom I have greatly admired. And also, for having chosen a career in a field which is alive and exciting. I hope that I was able to convey to you some of that excitement and the fun of having discovered something that others have not noticed before. But, of course, scientific discoveries are seldom made in isolation. One builds on what others have done before. As one of the greatest scientists of all time, Isaac Newton, said about a quarter of a millenium ago, towards the end of his life:

> *"I do not know what I may appear to the world, but to myself I seem to have been only like a boy playing on the seashore, and diverting myself in now and then finding a smoother pebble or a prettier shell than ordinary, while the great ocean of truth lay all undiscovered before me."*

In ending, I would like to express once again my deep appreciation to the authorities of the famous Palacký University, for the great honor they have done to me in having bestowed upon me a degree *honoris causa*. And I would also like to thank all of you for coming here to take part in this ceremony and to listen to my remarks.

Thank you.

Lecture 7.5
Commencement Remarks*

Emil Wolf
University of Rochester

President Jackson, Provost Phelps, new graduates, ladies and gentlemen,

I would like to begin with a brief story whose relevance will become clear shortly. There was a couple standing before a judge, the lady was suing for divorce. She was about 82 years old and her husband about 88. The judge asked the lady how long they have been married. She replied: "about 55 years". The judge said in astonishment: "I do not understand why, after 55 years, you are suing for a divorce?" She looked at him and said: *"enough is enough"*.

When I received a phone call from Provost Phelps about two months ago, he started by saying, "Emil, I have bad news and good news for you". Well, when I heard about the bad news I remembered this story and I thought, "Oh, no, I have been at the University of Rochester for 41 years and maybe now I am going to be told that *enough is enough* and that my appointment is being terminated." Fortunately, this was not so; the good news was, of course, that I will be receiving this prestigious award, the bad news was that I have to make a speech on this occasion.

Having told you that I have been in Rochester for 41 years, I would like to mention briefly how it came about that I am here at all and how I almost did not make it.

Before coming to Rochester I lived in England for 20 years, having managed to get there as a refugee from central Europe (from Czechoslovakia), just before the beginning of the Second World War. I received a doctorate from Bristol University, which entitles me to wear this very beautiful colored gown, and I then went on to various junior positions at Cambridge University, Edinburgh University, and the University of Manchester. After my appointment at Manchester ended, I was looking for more permanent employment. It was not easy in those days and I worried a great deal about it.

During the Easter vacation in 1958 I was away from Manchester, correcting proofs of a book I was working on and I asked a secretary to forward to me the proofs which were being sent to me by the printers. However, a batch of them did not arrive. When I returned to Manchester I asked the secretary whether she forwarded everything to me and she said, "of course". However I

*A speech given at the University of Rochester Commencement Ceremony on May 13, 2000, after having received the University Award for Excellence in Graduate Teaching.

was not convinced and when she went to lunch the next day I had a good look around her office. In one of the cupboards I found not only the missing proofs but also a letter from the University of Rochester addressed to me. I opened it and found that it was from Professor Robert Hopkins, then the Director of the Institute of Optics here at the University, asking me whether I would be interested to join the Rochester Faculty. He mentioned that he would be in England in about two weeks' time and if I was interested we could meet and talk about it. Those were days when faxes and e-mail did not exist, but I did manage to get a message to him in time. We met and I was happy to accept his offer. So you can see that it was only by a quirk of chance that I managed to get to Rochester and so be able to be present at this Ceremony which, of course, I would be most unhappy to have missed!

During the more than four decades that I have been at this University I have seen many changes. One of the obvious ones is the great increase in the cost of higher education. This is not a subject I feel qualified to comment on; I would just like to say that I hope all of you who are graduating today and, especially, your parents feel – or will eventually feel – that it was all worth it. Judging from the generosity that many alumni show to their *alma mater* this hope is probably justified.

Another change I have seen over the years is the increasing use of students' questionnaires to evaluate the performance of their Professors. I am sure that there must be some good use for such questionnaires, but I sometimes wonder whether they are perhaps administered rather hurridly and, consequently, may not necessarily reflect a careful judgment of the students – at least I hope so – when I recall the answer which one of my rather distinguished colleagues received in response to the question "What could one do to improve the quality of this course?" One of the answers was: "Shoot the instructor!"

Over the years I have also seen the increasing use of computers by students and faculty alike. This is, of course, in step with the way our Society is developing but it greatly favors students and younger Faculty over members of my own generation. I have to confess that, mainly because of great deal of help which I have received from my students, I have recently learned how to use e-mail, and if I have any problem with my computer I get help from my two young grandchildren, Ryan and Lauren, whom I am very pleased to see here in the audience. But, of course, I am not trying to deny the great opportunities in the educational system which come with the increasing use of computers. When it comes to e-mail, though, I often wonder what will happen to the art of writing and what sources future historians of science, for example, will rely on when letters become a rarity.

In the many years that I have been in Rochester I was telling my students from time to time that I would like them to follow the custom of a far away country where students provide economic support for their former Professors in their old age. (Unfortunately, I am not aware of any such country but it gives my students something potentially useful to think about.) Well, my suggestion seems to have backfired, because some of my former students are themselves now retired, whilst I still teach. I often wonder whether they now expect that the reverse will happen, namely that I will support them.

Let me conclude on a more serious note. Much as I feel honored by receiving this distinguished teaching award, I also feel that it cannot outdo the reward which comes from being fortunate in having interacted with many young students, and for having helped to launch them on their careers. And the careers of some have been truly spectacular.

Just to give an example before ending: one of my former students of about 30 years ago is now a world-renowned optical scientist and is the Director of an Indian National Physics Laboratory, with a staff of more than 1000 people. And he and some other of my former students stay in touch and occasionally return to U of R to collaborate with me and with my current graduate students on various scientific projects. That itself is a wonderful reward for me.

The teaching award which I have received today adds a silver lining to a teaching and research career which I have greatly enjoyed and, I should add, am still enjoying, after 41 happy years at the University of Rochester.

Thank you.

Section 8

Publications by Emil Wolf (up to Dec. 31, 2000)

(a) Papers

1. E. Wolf and W.S. Preddy, "On the Determination of Aspheric Profiles", *Proc. Phys. Soc.* **59**, 704–711 (1947).

2. E. Wolf, "On the Designing of Aspheric Surfaces", *Proc. Phys. Soc.* **61**, 494–503 (1948).

3. G.D. Wassermann and E. Wolf, "On the Theory of Aplanatic Aspheric Systems", *Proc. Phys. Soc. B* **62**, 2–8 (1949).

4. E.H. Linfoot and E. Wolf, "On the Corrector Plates of Schmidt Cameras", *J. Opt. Soc. Amer.* **39**, 752–756 (1949).

5. E. Wolf, "Diffraction Associated with Defocusing", *Nature* **164**, 924–925 (1949).

6. E. Wolf, "Light Distribution Near Focus in an Error-free Diffraction Image", *Proc. Roy. Soc.* A **204**, 533–548 (1951).

7. E. Wolf, "The Diffraction Theory of Aberrations", *Rep. Progr. Phys.* **14**, 95–120 (1951).

8. O. Theimer, G.D. Wassermann and E. Wolf, "On the Foundation of the Scalar Diffraction Theory of Optical Imaging", *Proc. Roy. Soc.* A **212**, 426–437 (1952).

9. E.H. Linfoot and E. Wolf, "On Telescopic Star Images", *Mon. Not. Roy. Astr. Soc.* **112**, 452–469 (1952).

10. E. Wolf, "On a New Aberration Function of Optical Instruments", *J. Opt. Soc. Amer.* **42**, 547–552 (1952).

11. A.B. Bhatia and E. Wolf, "The Zernike Circle Polynomials Occurring in Diffraction Theory", Letter in *Proc. Phys. Soc. B* **65**, 909–910 (1952).

12. E.H. Linfoot and E. Wolf, "Diffraction Images in Systems with an Annular Aperture", *Proc. Phys. Soc. B* **66**, 145–149 (1953).

13. E. Wolf, "The Xn and Yn Functions of Hopkins, Occurring in the Theory of Diffraction", Letter in *J. Opt. Soc. Amer.* **43**, 218 (1953).

14. E. Wolf, "A Macroscopic Theory of Interference and Diffraction of Light from Finite Sources", *Nature* **172**, 535–536 (1953).

15. E. Wolf, "Microwave Optics", *Nature* **172**, 615–618 (1953).

16. H.S. Green and E. Wolf, "A Scalar Representation of Electromagnetic Fields", *Proc. Phys. Soc.* A **66**, 1129–1137 (1953).

17. A.B. Bhatia and E. Wolf, "On the Circle Polynomials of Zernike and Related Orthogonal Sets", *Proc. Cambr. Phil. Soc.* **50**, 40–48 (1954).

18. E. Wolf, "Partially Coherent Optical Fields", in **Vistas in Astronomy**, Vol. 1, ed. A. Beer (Pergamon Press, London, 1955), 385–394.

19. A. Nisbet and E. Wolf, "On Linearly Polarized Electromagnetic Waves of Arbitrary Form", *Proc. Cambr. Phil. Soc.* **50**, 614–622 (1954).

20. E. Wolf, "A Macroscopic Theory of Interference and Diffraction of Light from Finite Sources – I. Fields with a Narrow Spectral Range", *Proc. Roy. Soc.* A **225**, 96–111 (1954).

21. E. Wolf, "Optics in Terms of Observable Quantities", *Nuovo Cimento* **12**, 884–888 (1954).

22. E. Wolf, "A Scalar Method for the Investigation of Electromagnetic Fields", in **Prepublication of Symposium on Microwave Optics**, Vol. I, Paper 17 (The Eaton Electronics Research Laboratory, McGill University, Montreal, Canada, 1953).

23. E. Wolf, "Fourier Integrals and a General Theory of Interference and Diffraction", in **Proc. of the McGill Symposium on Microwave Optics**, Part I, ed. B. S. Karasik, (Electronics Research Directorate, U. S. Air Force, Bedford, MA, 1959), 110–112.

24. E. Wolf, "A Macroscopic Theory of Interference and Diffraction of Light from Finite Sources – II. Fields with a Spectral Range of Arbitrary Width", *Proc. Roy. Soc.* A **230**, 246–265 (1955).

25. E. Wolf, "Recent Researches on the Foundations of Geometric Optics and Related Investigations in Electromagnetic Theory", *IRE Trans.* **AP-3**, 228–232 (1955).

26. E. Wolf, "The Coherence Properties of Optical Fields", in **Proc. Symposium on Astronomical Optics**, ed. Z. Kopal (North-Holland, Amsterdam, 1956), 177–185.

27. E. H. Linfoot and E. Wolf, "Phase Distribution Near Focus in an Aberration-free Diffraction Image", *Proc. Phys. Soc. B* **69**, 823–832 (1956).

28. B. Richards and E. Wolf, "The Airy Pattern in Systems of High Angular Aperture", Letter in *Proc. Phys. Soc. B* **69**, 854–856 (1956).

29. E.H. Linfoot and E. Wolf, "Phase Distribution in and Near the Bright Nucleus of an Aberration-free Diffraction Image", *Nature* **178**, 691–692 (1956).

30. E. Wolf, "Intensity Fluctuations in Stationary Optical Fields", *The Philosophical Magazine* **2**, 351–354 (1957).

31. B.J. Thompson and E. Wolf, "Two-Beam Interference with Partially Coherent Light", *J. Opt. Soc. Amer.* **47**, 895–902 (1957).

32. E. Wolf, "Reciprocity Inequalities, Coherence Time and Bandwidth in Signal Analysis and Optics", *Proc. Phys. Soc.* **71**, 257–269 (1958).

33. E. Wolf, "Some Aspects of Rigorous Scalar Treatment of Electromagnetic Wave Propagation", in **Proc. of International Symposium on Radio Wave Propagation** (Liege, 1958), (Academic Press Inc., London, 1960) 119–125.

34. E. Wolf, "A Scalar Representation of Electromagnetic Fields – Part II", *Proc. Phys. Soc.* **74**, 269–280 (1959).

35. E. Wolf, "Electromagnetic Diffraction in Optical Systems – I. An Integral Representation of the Image Field", *Proc. Roy. Soc.* A **253**, 349–357 (1959).

36. B. Richards and E. Wolf, "Electromagnetic Diffraction in Optical Systems – II. Structure of the Focal Region", *Proc. Roy. Soc.* A **253**, 358–379 (1959).

37. E. Wolf, "Coherence Properties of Partially Polarized Electromagnetic Radiation", *Nuovo Cimento* **13**, 1165–1181 (1959).

38. P. Roman and E. Wolf, "Correlation Theory of Stationary Electromagnetic Fields, Part I: The Basic Field Equations", *Nuovo Cimento* **17**, 462–476 (1960).

39. P. Roman and E. Wolf, "Correlation Theory of Stationary Electromagnetic Fields, Part II: Conservation Laws", *Nuovo Cimento* **17**, 477–490 (1960).

40. E. Wolf, "Correlation Between Photons in Partially Polarized Light Beams", *Proc. Phys. Soc.* **76**, 424–426 (1960).

41. K. Miyamoto and E. Wolf, "New Approach to Diffraction by Aperture", *J. Appl. Phys. (Japan)* **29**, 647–653 (1960) (in Japanese).

42. L. Mandel and E. Wolf, "Some Properties of Coherent Light", *J. Opt. Soc. Amer.* **51**, 815–819 (1961).

43. L. Mandel and E. Wolf, "Correlation in the Fluctuating Outputs from Two Square Law Detectors Illuminated by Light of Any State of Coherence and Polarization", *Phys. Rev.* **124**, 1696–1702 (1962).

44. K. Miyamoto and E. Wolf, "Generalization of the Maggi-Rubinowicz Theory of the Boundary Diffraction Wave, Part I", *J. Opt. Soc. Amer.* **52**, 615–625 (1962).

45. K. Miyamoto and E. Wolf, "Generalization of the Maggi-Rubinowicz Theory of the Boundary Diffraction Wave, Part II", *J. Opt. Soc. Amer.* **52**, 626–637 (1962).

46. E.W. Marchand and E. Wolf, "Boundary Diffraction Wave in the Domain of the Rayleigh-Kirchhoff Diffraction Theory", *J. Opt. Soc. Amer.* **52**, 761–767 (1962).

47. E. Wolf, "Is a Complete Determination of the Energy Spectrum Possible from Measurements of the Degree of Coherence?", *Proc. Phys. Soc.* **80**, 1269–1272 (1962).

48. Y. Kano and E. Wolf, "Temporal Coherence of Blackbody Radiation", *Proc. Phys. Soc.* **80**, 1273–1276 (1962).

49. L. Mandel and E. Wolf, "The Measures of Bandwidth and Coherence Time in Optics", *Proc. Phys. Soc.* **80**, 894–897 (1962).

50. E. Wolf, "Spatial Coherence of Resonant Modes in a Maser Interferometer", *Phys. Lett.* **3**, 166–168 (1963).

51. L. Mandel and E. Wolf, "Photon Correlations", *Phys. Rev. Lett.* **10**, 276–277 (1963).

52. E. Wolf, E.W. Marchand and K. Miyamoto, "A Boundary Wave Theory of Diffraction at an Aperture", in **Proc. of Symposium on Electromagnetic Theory and Antennas** (Copenhagen, 1962) (Pergamon Press, London, 1963), 109–111.

53. B. Karczewski and E. Wolf, "Polarization Properties of the Electromagnetic Field Diffracted from an Aperture", in **Proc. of Symposium on Electromagnetic Theory and Antennas** (Copenhagen, 1962)(Pergamon Press, London, 1963), 797–799.

54. E. Wolf, "Recent Researches on Coherence Properties of Light", in **Quantum Electronics III**, ed. N. Bloembergen and P. Grivet (Dunod, Paris; Columbia University Press, New York, 1964), 13–34.

55. E. Wolf, "Basic Concepts of Optical Coherence Theory", in **Proc. of Symposium on Optical Masers**, J. Fox ed. (Brooklyn Polytechnic Press and J. Wiley and Sons, Inc., New York, 1963), 29–42.

56. L. Mandel, E.C.G. Sudarshan and E. Wolf, "Theory of Photoelectric Detection of Light Fluctuations", *Proc. Phys. Soc.* **84**, 435–444 (1964).

57. L. Mandel and E. Wolf, "Detection of Laser Radiation", *J. Opt. Soc. Amer.* **53**, 1315 (1963).

58. B. Karczewski and E. Wolf, "Comparison of Three Theories of Electromagnetic Diffraction at an Aperture, Part I: Coherence Matrices", *J. Opt. Soc. Amer.* **56**, 1207–1214 (1966).

59. E. Karczewski and E. Wolf, "Comparison of Three Theories of Electromagnetic Diffraction at an Aperture, Part II: The Far Field", *J. Opt. Soc. Amer.* **56**, 1214–1219 (1966).

60. E. Wolf and E.W. Marchand, "Comparison of the Kirchhoff and the Rayleigh-Sommerfeld Theories of Diffraction at an Aperture", *J. Opt. Soc. Amer.* **54**, 587–594 (1964).

61. C.L. Mehta and E. Wolf, "Coherence Properties of Blackbody Radiation, Part I - Correlation Tensors of the Classical Field", *Phys. Rev.* **134**, A1143–A1149 (1964).

62. C.L. Mehta and E. Wolf, "Coherence Properties of Blackbody Radiation, Part II - Correlation Tensors of the Quantized Field", *Phys. Rev.* **134**, A1149–A1153 (1964).

63. A. Boivin and E. Wolf, "Electromagnetic Field in the Neighborhood of the Focus of a Coherent Beam", *Phys. Rev.* **138**, B1561–B1565 (1965).

64. E. Wolf and E. Marchand, "Multiply-diffracted Boundary Waves", *Acta Phys. Polonica* **27**, 147–152 (1965).

65. E. Wolf, "Light Fluctuations as a New Spectroscopic Tool", *Japan. J. Phys.* **4**, 1–14 (1965).

66. L. Mandel and E. Wolf, "Coherence Properties of Optical Fields", *Rev. Mod. Phys.* **37**, 231–287 (1965).

67. E. Wolf and C.L. Mehta, "Determination of the Statistical Properties of Light from Photoelectric Measurements", *Phys. Rev. Lett.* **13**, 705–707 (1964).

68. C.L. Mehta, E. Wolf and A. P. Balachandran, "Some Theorems on the Unimodular Complex Degree of Optical Coherence", *J. Math. Phys.* **7**, 133–138 (1966).

69. L. Mandel and E. Wolf, "Photon Statistics and Classical Fields", *Phys. Rev.* **149**, 1033–1037 (1966).

70. E. Wolf, "Some Recent Research on Coherence and Fluctuations of Light", *Optica Acta* **13**, 281–298 (1966).

71. E.W. Marchand and E. Wolf, "Consistent Formulation of Kirchhoff's Diffraction Theory", *J. Opt. Soc. Amer.* **56**, 1712–1722 (1966).

72. D. Dialetis and E. Wolf, "The Phase Retrieval Problem of Coherence Theory as a Stability Problem", *Nuovo Cimento* **47**, 113–116 (1967).

73. C.L. Mehta and E. Wolf, "Correlation Theory of Quantized Electromagnetic Fields, Part I: Dynamical Equations and Conservation Laws", *Phys. Rev.* **157**, 1183–1187 (1967).

74. C.L. Mehta and E. Wolf, "Correlation Theory of Quantized Electromagnetic Fields, Part II: Stationary Fields and Their Spectral Properties", *Phys. Rev.* **157**, 1188–1197 (1967).

75. A. Boivin, J. Dow and E. Wolf, "Energy Flow in the Neighborhood of the Focus of a Coherent Beam", *J. Opt. Soc. Amer.* **57**, 1171–1175 (1967).

76. E. Wolf, "Some Recent Research on Diffraction of Light", in **Proc. of Symposium on Modern Optics**, J. Fox ed. (Brooklyn Polytechnic Press, Brooklyn, NY and John Wiley and Sons, Inc., New York, 1967), 433–452.

77. C.L. Mehta and E. Wolf, "Coherence Properties of Blackbody Radiation, Part III: Cross Spectral Tensors", *Phys. Rev.* **161**, 1328–1334 (1967).

78. E. Wolf and J.R. Shewell, "The Inverse Wave Propagator", *Phys. Lett.* **25A**, 417–418 (1967: ibid. **26A**, 104–105 (1967).

79. G.S. Agarwal and E. Wolf, "Ordering Theorems and Generalized Phase-Space Distributions", *Phys. Lett.* **26A**, 485–486 (1968)

80. L. Mandel and E. Wolf, "Comments on a paper by E. Jakeman and E. R. Pike, 'The intensity-fluctuation Distribution of Gaussian Light' ", *Proc. Phys. Soc.* **1**, 625–627 (1968).

81. E. Wolf and G.S. Agarwal, "Ordering of Operators and Phase Space Descriptions in Quantum Optics", in **Polarisation, Matiere et Rayonnement**, (Livre de Jubile en l'honneur du Professeur A. Kastler) (Presses Universitaires de France, Paris, 1969), 541–556.

82. J.R. Shewell and E. Wolf, "Inverse Diffraction and a New Reciprocity Theorem", *J. Opt. Soc. Amer.* **58**, 1596–1603 (1968).

83. G.S. Agarwal and E. Wolf, "Quantum Dynamics in Phase Space", *Phys. Rev. Lett.* **21**, 180–183 (1968).

84. E.W. Marchand and E. Wolf, "Diffraction at Small Apertures in Black Screens", *J. Opt. Soc. Amer.* **59**, 79–90 (1969).

85. G.S. Agarwal and E. Wolf, "A Generalized Wick Theorem and Phase Space Representation of Operators", *Nuovo Cimento* **1**, 140–144 (1969).

86. L. Mandel and E. Wolf, "Terminology in Optics", *J. Opt. Soc. Amer.* **58**, 1678 (1968).

87. E. Wolf and J.R. Shewell, "Diffraction Theory of Holography", *J. Math. Phys.* **11**, 2254–2267 (1970).

88. E. Wolf, "Determination of the Amplitude and the Phase of Scattered Fields by Holography", *J. Opt. Soc. Amer.* **60**, 18–20 (1970).

89. E. Wolf, "Three-Dimensional Structure Determination of Semi-Transparent Objects from Holographic Data", *Opt. Commun.* **1**, 153–156 (1969).

90. G.S. Agarwal and E. Wolf, "Calculus for Functions of Non-commuting Operators and General Phase Space Methods in Quantum Mechanics. I: Mapping Theorems and Ordering of Functions of Non-commuting Operators", *Phys. Rev.* D **2**, 2161–2186 (1970).

91. G.S. Agarwal and E. Wolf, "Calculus for Functions of Non-commuting Operators and General Phase Space Methods in Quantum Mechanics. II: Quantum Mechanics in Phase Space", *Phys. Rev.* D **2**, 2187–2205 (1970).

92. G.S. Agarwal and E. Wolf, "Calculus for Functions of Non-commuting Operators and General Phase Space Methods in Quantum Mechanics. III: A Generalized Wick Theorem and Multitime Mapping", *Phys. Rev.* D **2**, 2206–2225 (1970).

93. R. Asby and E. Wolf, "Evanescent Waves and the Electromagnetic Field of a Moving Charged Particle", *J. Opt. Soc. Amer.* **61**, 52–59 (1971).

94. E.W. Marchand and E. Wolf, "Transmission Cross Section for Small Apertures in Black Screens", *J. Opt. Soc. Amer.* **60**, 1501–1510 (1970).

95. R. Asby and E. Wolf, "Theory of Cerenkovian Effects", *Phys. Teach.* **9**, 207–210 (1971).

96. E. Wolf, "Recent Work on the Ewald-Oseen Extinction Theorem", **Proceedings of the 1971 Rochester Symposium "Atomic and Molecular Optics"**, J. H. Eberly, ed. (University of Rochester, 1971).

97. É. Lalor and E. Wolf, "New Model for the Interaction Between a Moving Charged Particle and a Dielectric, and the Cerenkov Effect", *Phys. Rev. Lett.* **26**, 1274–1277 (1971).

98. G.S. Agarwal, D.N. Pattanayak and E. Wolf, "Structure of the Electromagnetic Field in a Spatially Dispersive Medium", *Phys. Rev. Lett.* **27**, 1022–1025 (1971).

99. E.W. Marchand and E. Wolf, "Angular Correlation and the Far-zone Behavior of Partially Coherent Fields", *J. Opt. Soc. Amer.* **62**, 379–385 (1972).

100. G.S. Agarwal, D.N. Pattanayak and E. Wolf, "Refraction and Reflection on a Spatially Dispersive Medium", *Opt. Commun.* **4**, 255–259 (1971).

101. G.S. Agarwal and E. Wolf, "Relation Between the Statistical Representations of Real and Associated Complex Fields in Optical Coherence Theory", *J. Math. Phys.* **13**, 1759–1764 (1972).

102. G.S. Agarwal, D.N. Pattanayak and E. Wolf, "Boundary Conditions on Exciton Polarization and Mode Coupling in a Spatially Dispersive Medium", *Opt. Commun.* **4**, 260–263 (1971).

103. É. Lalor and E. Wolf, "Exact Solution of the Equations of Molecular Optics for Refraction and Reflection of an Electromagnetic Wave on a Semi-infinite Dielectric", *J. Opt. Soc. Amer.* **62**, 1165–1174 (1972).

104. G.S. Agarwal, D.N. Pattanayak and E. Wolf, "A Generalized Extinction Theorem for Exciton Polarization in Spatially Dispersive Media", *Phys. Lett.* **40A**, 279–280 (1972).

105. D.N. Pattanayak and E. Wolf, "General Form and a New Interpretation of the Ewald-Oseen Extinction Theorem", *Opt. Commun.* **6**, 217–220 (1972).

106. E. Wolf, "A Generalized Extinction Theorem and its Role in Scattering Theory", in **Coherence and Quantum Optics**, eds. L. Mandel and E. Wolf (Plenum Press, New York, 1973), 339–357.

107. E.W. Marchand and E. Wolf, "Generalized Radiometry for Radiation from Partially Coherent Sources", *Opt. Commun.* **6**, 305–308 (1972).

108. A.J. Devaney and E. Wolf, "Multipole Expansions and Plane Wave Representations of the Electromagnetic Field", *J. Math. Phys.* **15**, 234–244 (1974).

109. A.J. Devaney and E. Wolf, "Radiating and Nonradiating Classical Current Distributions and the Fields They Generate", *Phys. Rev.* D **8**, 1044–1047 (1973).

110. L. Mandel and E. Wolf, "Constitutive Relations and the Electromagnetic Spectrum in a Fluctuating Medium", *Opt. Commun.* **8**, 95–99 (1973).

111. A.J. Devaney and E. Wolf, "New Representations for Multipole Moments and Angular Spectrum Amplitudes of Electromagnetic Fields", *Opt. Commun.* **9**, 327–330 (1973).

112. W.H. Carter and E. Wolf, "Degree of Polarization and Intensity Fluctuations in Thermal Light Beams", *J. Opt. Soc. Amer.* **63**, 1619–1620 (1973).

113. G.S. Agarwal, D.N. Pattanayak and E. Wolf, "Electromagnetic Fields in Spatially Dispersive Media", *Phys. Rev.* B **10**, 1447–1475 (1974).

114. E.W. Marchand and E. Wolf, "Radiometry with Sources of Any State of Coherence", *J. Opt. Soc. Amer.* **64**, 1219–1226 (1974).

115. E.W. Marchand and E. Wolf, "Walther's Definition of Generalized Radiance", *J. Opt. Soc. Amer.* **64**, 1273–1274 (1974).

116. E. Wolf, "Electromagnetic Scattering as a Non-local Boundary Value Problem", in **Symposia Mathematica**, Vol. XVIII (1976), (Academic Press, New York), 333–352.

117. G.S. Agarwal, D.N. Pattanayak and E. Wolf, "Structure of Electromagnetic Fields in Spatially Dispersive Media of Arbitrary Geometry", *Phys. Rev.* B **11**, 1342–1351 (1975).

118. M.D. Srinivas and E. Wolf, "Some Nonclassical Features of Phase-space Representations of Quantum Mechanics", *Phys. Rev.* D **11**, 1477–1485 (1975).

119. L. Mandel and E. Wolf, "Optimum Conditions for Heterodyne Detection of Light", *J. Opt. Soc. Amer.* **65**, 413–420 (1975).

120. E. Wolf and W.H. Carter, "Angular Distribution of Radiant Intensity from Sources of Different Degrees of Spatial Coherence", *Opt. Commun.* **13**, 205–209 (1975).

121. W.H. Carter and E. Wolf, "Coherence Properties of Lambertian and Non-Lambertian Sources", *J. Opt. Soc. Amer.* **65**, 1067–1071 (1975).

122. E. Wolf and W.H. Carter, "A Radiometric Generalization of the van Cittert-Zernike Theorem for Fields Generated by Sources of Arbitrary State of Coherence", *Opt. Commun.* **16**, 297–302 (1976).

123. D.N. Pattanayak and E. Wolf, "Scattering States and Bound States as Solutions of the Schrdinger Equation with Nonlocal Boundary Conditions", *Phys. Rev.* D **13**, 913–923 (1976).

124. E. Wolf, "New Theory of Radiative Energy Transfer in Free Electromagnetic Fields", *Phys. Rev.* D **13**, 869–886 (1976).

125. D.N. Pattanayak and E. Wolf, "Resonance States as Solutions of the Schrdinger Equation with a Nonlocal Boundary Condition", *Phys. Rev.* D **13**, 2287–2290 (1976).

126. L. Mandel and E. Wolf, "Spectral Coherence and the Concept of Cross-spectral Purity", *J. Opt. Soc. Amer.* **66**, 529–535 (1976).

127. E. Collett, J.T. Foley and E. Wolf, "On an Investigation of Tatarskii into the Relationship between Coherence Theory and the Theory of Radiative Transfer", *J. Opt. Soc. Amer.* **67**, 465–467 (1977).

128. W.H. Carter and E. Wolf, "Coherence and Radiometry with Quasihomogeneous Planar Sources", *J. Opt. Soc. Amer.* **67**, 785–796 (1977).

129. M.S. Zubairy and E. Wolf, "Exact Equations for Radiative Transfer of Energy and Momentum in Free Electromagnetic Fields", *Opt. Commun.* **20**, 321–324 (1977).

130. M.D. Srinivas and E. Wolf, "Stochastic Equations for Classical and Quantum Distribution Functions", in **Statistical Mechanics and Statistical Methods in Theory and Application**, ed. Uzi Landman (Plenum, New York, 1977), 219–251.

131. E. Wolf and W.H. Carter, "On the Radiation Efficiency of Quasi-homogeneous Sources of Different Degrees of Spatial Coherence", in **Coherence and Quantum Optics IV**, eds. L. Mandel and E. Wolf (Plenum, New York, 1978), 415–430.

132. E. Collett and E. Wolf, "Is Complete Spatial Coherence Necessary for the Generation of Highly Directional Light Beams?", *Opt. Lett.* **2**, 27–29 (1978).

133. E. Wolf, "Coherence and Radiometry", *J. Opt. Soc. Amer.* **68**, 6–17 (1978).

134. E. Wolf and W.H. Carter, "Coherence and Radiant Intensity in Scalar Wavefields Generated by Fluctuating Primary Planar Sources", *J. Opt. Soc. Amer.* **68**, 953–964 (1978).

135. E. Wolf and E. Collett, "Partially Coherent Sources which Produce the Same Far-field Intensity Distribution as a Laser", *Opt. Commun.* **25**, 293–296 (1978).

136. W.H. Carter and E. Wolf, "Some Relationships Between the Correlation Coefficients of Planar Sources and of Their Far Fields", *Opt. Commun.* **25**, 288–292 (1978).

137. E. Wolf, "The Radiant Intensity from Planar Sources of Any State of Coherence", *J. Opt. Soc. Amer.* **68**, 1597–1605 (1978).

138. J.T. Foley and E. Wolf, "A Note on the Far Field of a Gaussian Beam", *J. Opt. Soc. Amer.* **69**, 761–764 (1979).

139. E. Wolf and M.S. Zubairy, "Radiative Energy Transfer in Scalar Wave Fields", **Coherence and Quantum Optics IV**, eds. L. Mandel and E. Wolf (Plenum, New York, 1978), 457 (Abstract).

140. E. Collett and E. Wolf, "New Equivalence Theorems for Planar Sources which Generate the Same Distributions of Radiant Intensity", *J. Opt. Soc. Amer.* **69**, 942–950 (1979).

141. E. Wolf, "Einstein's Researches on the Nature of Light", *Optics News* **5**, 24–39 (1979).

142. W.H. Carter and E. Wolf, "Correlation Theory of Wavefields Generated by Fluctuating, Three-dimensional, Scalar Sources, Part I: General Theory", *Optica Acta* **28**, 227–244 (1981).

143. W.H. Carter and E. Wolf, "Correlation Theory of Wavefields Generated by Fluctuating, Three-dimensional, Scalar Sources, Part II: Radiation from Isotropic Model Sources", *Optica Acta* **28**, 245–259 (1981).

144. E. Collett and E. Wolf, "Beams Generated by Gaussian Quasi-homogeneous Sources", *Opt. Commun.* **32**, 27–31 (1980).

145. E. Collett and E. Wolf, "Symmetry Properties of Focused Fields", *Opt. Lett.* **5**, 264–266 (1980).

146. E. Wolf, "Phase Conjugacy and Symmetries in Spatially Bandlimited Wavefields Containing no Evanescent Components", *J. Opt. Soc. Amer.* **70**, 1311–1319 (1980).

147. A.J. Devaney and E. Wolf, "A New Perturbation Expansion for Inverse Scattering from Three-dimensional Finite-range Potentials", *Phys. Lett.* **89 A**, 269–272 (1982).

148. E. Wolf, "A New Description of Second-order Coherence Phenomena in the Space-frequency Domain", in **Optics in Four Dimensions** (I.C.O. Conference – Ensenada, Mexico, August 4–8, 1980), M. A. Machado and L. M. Narducci, eds. (Conference Proceedings #65, American Institute of Physics, New York, 1981), 42–48.

149. E. Wolf, A.J. Devaney and J.T. Foley, "On a Relationship Between Spatial Coherence and Temporal Coherence in Free Fields", in **Optics in Four Dimensions**, (I.C.O. Conference – Ensenada, Mexico, August 4–8, 1980), M. A. Machado and L. M. Narducci, eds. (Conference Proceedings #65, American Institute of Physics, New York, 1981), 123–130.

150. E. Wolf, "New Spectral Representation of Random Sources and of the Partially Coherent Fields that They Generate", *Opt. Commun.* **38**, 3–6 (1981).

151. L. Mandel and E. Wolf, "Complete Coherence in the Space-frequency Domain", *Opt. Commun.* **36**, 247–249 (1981).

152. E. Wolf and A.J. Devaney, "On a Relationship Between Spectral Properties and Spatial Coherence Properties of Light", *Opt. Lett.* **6**, 168–170 (1981).

153. E. Wolf and Y. Li, "Conditions for the Validity of the Debye Integral Representation of Focused Fields", *Opt. Commun.* **39**, 205–210 (1981).

154. E. Wolf and W.H. Carter, "Comments on the Theory of Phase-conjugated Waves", *Opt. Commun.* **40**, 397–400 (1982).

155. Y. Li and E. Wolf, "Focal Shifts in Diffracted Converging Spherical Waves", *Opt. Commun.* **39**, 211–215 (1981).

156. E. Wolf, "New Theory of Partial Coherence in the Space-frequency Domain: Part I: Spectra and Cross-spectra of Steady-state Sources", *J. Opt. Soc. Amer.* **72**, 343–351 (1982).

157. G.S. Agarwal and E. Wolf, "Theory of Phase Conjugation with Weak Scatterers", *J. Opt. Soc. Amer.* **72**, 321–326 (1982).

158. A. Starikov and E. Wolf, "Coherent Mode Representation of Gaussian Schell-model Sources and of Their Radiation Fields", *J. Opt. Soc. Amer.* **72**, 923–928 (1982).

159. Y. Li and E. Wolf, "Radiation from Anisotropic Gaussian Schell-model Sources", *Opt. Lett.* **7**, 256–258 (1982).

160. G.S. Agarwal, A.T. Friberg and E. Wolf, "The Effect of Backscattering in Phase conjugation with Weak Scatterers", *J. Opt. Soc. Amer.* **72**, 861–863 (1982).

161. Y. Li and E. Wolf, "Focal Shift in Focused Truncated Gaussian Beams", *Opt. Commun.* **42**, 151–156 (1982).

162. A.T. Friberg and E. Wolf, "Angular Spectrum Representation of Scattered Electromagnetic Fields", *J. Opt. Soc. Amer.* **73**, 26–32 (1983).

163. G.S. Agarwal, A.T. Friberg and E. Wolf, "Elimination of Distortions by Phase Conjugation Without Losses or Gains", *Opt. Commun.* **43**, 446–450 (1982).

164. G.S. Agarwal, A.T. Friberg and E. Wolf, "Scattering Theory of Distortion-correction by Phase Conjugation", *J. Opt. Soc. Amer.* **73**, 529–538 (1983).

165. E. Wolf, "Young's Interference Fringes with Narrow-band Light", *Opt. Lett.* **8**, 250–252 (1983).

166. E. Wolf, A.J. Devaney and F. Gori, "Relationship Between Spectral Properties and Spatial Coherence Properties in One-dimensional Free Fields", *Opt. Commun.* **46**, 4–8 (1983).

167. A.T. Friberg and E. Wolf, "Reciprocity Relations with Partially Coherent Sources", *Optica Acta* **30**, 1417–1435 (1983).

168. E. Wolf, "Recollections of Max Born", (a) *Optics News* **9**, 10–16 (1983); (b) reprinted in **Technology of Our Times**, ed. F. Su (SPIE Optical Engineering Press, Bellingham, WA, 1990), 15–29; (c) reprinted in *Astrophysics and Space Science* **227**, ed. A. Peratt (Kluwer Academic Publishers, Belgium, 1995), 277–297.

169. E. Wolf, G.S. Agarwal and A.T. Friberg, "Wavefront Correction and Scattering with Phase Conjugate Waves", in **Coherence and Quantum Optics V**, eds. L. Mandel and E. Wolf (Plenum Press, New York, 1984), 107–116.

170. A.J. Devaney and E. Wolf, "Non-radiating Stochastic Scalar Sources", in **Coherence and Quantum Optics V**, eds. L. Mandel and E. Wolf (Plenum Press, New York, 1984), 417–421.

171. E. Wolf and G.S. Agarwal, "Coherence Theory of Laser Resonator Modes", *J. Opt. Soc. Amer.* A **1**, 541–546 (1984).

172. Y. Li and E. Wolf, "Three-dimensional Intensity Distribution Near the Focus in Systems of Different Fresnel Numbers", *J. Opt. Soc. Amer.* A **1**, 801–808 (1984).

173. E. Wolf and W.H. Carter, "Fields Generated by Homogeneous and by Quasi-homogeneous Planar Secondary Sources", *Opt. Commun.* **50**, 131–136 (1984).

174. E. Wolf, "Completeness of Coherent-mode Eigenfunctions of Schell-model Sources", *Opt. Lett.* **9**, 387–389 (1984).

175. E. Wolf and M. Nieto-Vesperinas, "Analyticity of the Angular Spectrum Amplitude of Scattered Fields and Some of its Consequences", *J. Opt. Soc. Amer.* A **2**, 886–890 (1985).

176. W.H. Carter and E. Wolf, "An Inverse Problem with Quasi-homogeneous Sources", *J. Opt. Soc. Amer.* A **2**, 1994–2000 (1985).

177. M. Nieto-Vesperinas and E. Wolf, "Phase Conjugation and Symmetries with Wavefields in Free Space Containing Evanescent Components", *J. Opt. Soc. Amer.* A **2**, 1429–1434 (1985).

178. E. Wolf, "New Theory of Partial Coherence in the Space-frequency Domain. Part II: Steady State Fields and Higher-order Correlations", *J. Opt. Soc. Amer.* A **3**, 76–85 (1986).

179. J.T. Foley and E. Wolf, "Radiometry as a Short-wavelength Limit of Statistical Wave Theory with Globally Incoherent Sources", *Opt. Commun.* **55**, 236–241 (1985).

180. J.T. Foley, W.H. Carter and E. Wolf, "Field Correlations within a Completely Incoherent Primary Spherical Source", *J. Opt. Soc. Amer.* A **3**, 1090–1096 (1986).

181. M. Nieto-Vesperinas and E. Wolf, "Generalized Stokes Reciprocity Relations for Scattering from Dielectric Objects of Arbitrary Shape", *J. Opt. Soc. Amer.* A **3**, 2038–2046 (1986).

182. E. Wolf and R.P. Porter, "On the Physical Contents of Some Integral Equations for Inverse Scattering from Inhomogeneous Objects", *Radio Science* **21**, 627–634 (1986).

183. E. Wolf, "Invariance of the Spectrum of Light on Propagation", *Phys. Rev. Lett.* **56**, 1370–1372 (1986).

184. K. Kim, L. Mandel and E. Wolf, "Relationship Between Jones and Mueller Matrices for Random Media", *J. Opt. Soc. Amer.* A **4**, 433–437 (1987).

185. K. Kim and E. Wolf, "Non-radiating Monochromatic Sources and Their Fields", *Opt. Commun.* **59**, 1–6 (1986).

186. E. Wolf, "Coherent-mode Propagation in Spatially Band-limited Wave-fields", *J. Opt. Soc. Amer.* A **3**, 1920–1924 (1986).

187. R.W. Boyd, T.M. Habashy, A.A. Jacobs, L. Mandel, M. Nieto-Vesperinas, W. R. Tompkin and E. Wolf, "Nature of the Interference Pattern Produced on Reflection at a Phase-conjugate Mirror," *Opt. Lett.* **12**, 42–44 (1987).

188. E. Wolf, L. Mandel, R.W. Boyd, T.M. Habashy, M. Nieto-Vesperinas, "Interference Pattern Produced on Reflection at a Phase-conjugate Mirror. Part I: Theory", *J. Opt. Soc. Amer.* B **4**, 1260–1265 (1987).

189. A.A. Jacobs, W.R. Tompkin, R.W. Boyd and E. Wolf, "Interference Pattern Produced on Reflection at a Phase-conjugate Mirror. Part II: Experiment", *J. Opt. Soc. Amer.* B **4**, 1266–1268 (1987).

190. H.M. Nussenzveig, J.T. Foley, K. Kim and E. Wolf, "Field Correlations within a Fluctuating Homogeneous Medium", *Phys. Rev. Lett.* **58**, 218–221 (1987).

191. E. Wolf, "Non-cosmological Redshifts of Spectral Lines", *Nature* **326**, 363–365 (1987).

192. G.S. Agarwal, J.T. Foley and E. Wolf, "The Radiance and Phase-space Representations of the Cross-spectral Density Operator", *Opt. Commun.* **62**, 67–72 (1987).

193. K. Kim and E. Wolf, "Propagation Law For Walther's First Generalized Radiance Function and its Short Wavelength Limit with Quasi-homogeneous Sources", *J. Opt. Soc. Amer.* A **4**, 1233–1236 (1987).

194. F. Gori and E. Wolf, "Sources with Factorized Cross-spectral Densities", *Opt. Commun.* **61**, 369–373 (1987).

195. E. Wolf, "Redshifts and Blueshifts of Spectral Lines Caused by Source Correlations", *Opt. Commun.* **62**, 12–16 (1987).

196. B. Cairns and E. Wolf, "The Instantaneous Cross-spectral Density of Non-stationary Wave-fields", *Opt. Commun.* **62**, 215–218 (1987).

197. W.H. Carter and E. Wolf, "Far-zone Behavior of Electromagnetic Fields Generated by Fluctuating Current Distributions", *Phys. Rev.* A **36**, 1258–1269 (1987).

198. E. Wolf, "Red Shifts and Blue Shifts of Spectral Lines Emitted by Two Correlated Sources", *Phys. Rev. Lett.* **58**, 2646–2648 (1987).

199. Z. Dačić and E. Wolf, "Changes in the Spectrum of Partially Coherent Light Beam Propagating in Free Space," *J. Opt. Soc. Amer.* A **5**, 1118–1126 (1988).

200. A. Gamliel and E. Wolf, "Spectral Modulation by Control of Source Correlations", *Opt. Commun.* **65**, 91–96 (1988).

201. W.H. Carter and E. Wolf, "Scattering from Quasi-Homogeneous Media", *Opt. Commun.* **67**, 85–90 (1988).

202. J.T. Foley and E. Wolf, "Partially Coherent Sources which Generate the Same Far Field Spectra as Completely Incoherent Sources," *J. Opt. Soc. Amer.* A **5**, 1683–1687 (1988).

203. J. Jannson, T. Jannson and E. Wolf, "Spatial Coherence Discrimination in Scattering," *Opt. Lett.* **13**, 1060–1062 (1988).

204. E. Wolf, J.T. Foley and F. Gori, "Frequency Shifts of Spectral Lines Produced by Scattering from Spatially Random Media," *J. Opt. Soc. Amer.* A **6**, 1142 –1149 (1989).

205. A. Gamliel, K. Kim, A.I. Nachman and E. Wolf, "A New Method for Specifying Nonradiating, Monochromatic, Scalar Sources and their Fields," *J. Opt. Soc. Amer.* A **6**, 1388–1393 (1989).

206. E. Wolf and J.T. Foley, "Scattering of Electromagnetic Fields of Any State of Coherence from Space-time Fluctuations," *Phys. Rev.* A **40**, 579 –587 (1989).

207. J.T. Foley and E. Wolf, "Frequency Shifts of Spectral Lines Generated by Scattering from Space-time Fluctuations," *Phys. Rev.* A **40**, 588 – 598 (1989).

208. D.F.V. James and E. Wolf, "A Spectral Equivalence Theorem", *Opt. Commun.* **72**, 1 (1989).

209. E. Wolf, "Correlation-induced Doppler-like Frequency Shifts of Spectral Lines", *Phys. Rev. Lett.* **63**, 2220–2223 (1989).

210. B. Cairns and E. Wolf, "Comparison of the Born and the Rytov Approximations for Scattering on Quasi-homogeneous Media," *Opt. Commun.* **74**, 284–289 (1990).

211. W. Wang and E. Wolf, "Coherence Properties of Light Propagating in a One-dimensional Lorentz Medium," in **Coherence and Quantum Optics VI**, eds. J. Eberly, L. Mandel and E. Wolf (Plenum Press, New York, 1989), 1207–1212.

212. J.T. Foley and E. Wolf, "Scattering of Electromagnetic Fields of Any State of Coherence from Fluctuating Media," in **Coherence and Quantum Optics VI**, eds. J. Eberly, L. Mandel and E. Wolf (Plenum Press, New York, 1989), 309–314.

213. E. Wolf, "On the Possibility of Generating Doppler-like Frequency Shifts of Spectral Lines by Scattering from Space-time Fluctuations", in **Coherence and Quantum Optics VI**, eds. J. Eberly, L. Mandel and E. Wolf (Plenum Press, New York, 1989), 1235–1238.

214. B. Cairns and E. Wolf, "Comparison of Two Theories of Scattering from Quasi-homogeneous Media," in **Coherence and Quantum Optics VI**, eds. J. Eberly, L. Mandel and E. Wolf (Plenum Press, New York, 1989), 123–128.

215. D.F.V. James, M. Savedoff and E. Wolf, "Shifts of Spectral Lines Caused by Scattering From Fluctuating Random Media," *Astrophys. J.* **359**, 67–71 (1990).

216. E. Wolf and J.R. Fienup, "Changes in the Spectrum of Light Arising on Propagation Through a Linear Time-invariant System," *Opt. Commun.* **82**, 209–212 (1991).

217. E. Wolf, J. Jannson and T. Jannson, "Analogue of the van Cittert-Zernike Theorem for Statistically Homogeneous Wave Fields," *Opt. Lett.* **15**, 1032–1034 (1990).

218. D.F.V. James and E. Wolf, "Doppler-like Frequency Shifts Generated by Dynamic Scattering," *Phys. Letts.* A **146**, 167–171 (1990).

219. W. Wang and E. Wolf, "Effects of Spatial Coherence of Incident Fields on Reflection and Transmission," *Opt. Commun.* **79**, 131–138 (1990).

220. E. Wolf, "Influence of Source-correlations on Spectra of Radiated Fields", in **International Trends in Optics**, ed. J. W. Goodman (Academic Press, San Diego, California, 1991), 221–232.

221. D.F.V. James and E. Wolf, "Spectral changes produced in Young's interference experiment", *Opt. Commun.* **81**, 150–154 (1991).

222. D.F.V. James and E. Wolf, "Determination of Field Correlations from Spectral Measurements with Applications to Synthetic Aperture Imaging", *Radio Science* **26**, 1239–1243 (1991).

223. B. Cairns and E. Wolf, "Changes in the Spectrum of Light Scattered by a Moving Diffuser Plate", *J. Opt. Soc. Amer.* A **8**, 1922–1928 (1991).

224. W. Wang, R. Simon and E. Wolf, "Changes in the Coherence and Spectral Properties of Partially Coherent Light Reflected from a Dielectric Slab", *J. Opt. Soc. Amer.* A **9**, 287–297 (1992).

225. J.T. Foley and E. Wolf, "Radiance Functions of Partially Coherent Fields", *J. Mod. Opt.* **38**, 2053–2068 (1991).

226. E. Wolf, "Two Inverse Problems in Spectroscopy with Partially Coherent Sources and the Scaling Law", *J. Mod. Opt.* **39**, 9 – 20 (1992).

227. E. Wolf, "The Life and Work of Christiaan Huygens", **Proc. Symposium Huygens' Principle 1690–1990: Theory and Applications**, H. Blok, H. A. Ferwerda and H. K. Kuiken, eds. (North-Holland, Amsterdam, 1992), 3–17.

228. D.F.V. James and E. Wolf, "Some New Aspects of Young's Interference Experiment", *Phy. Letts.* A **157**, 6–10 (1991).

229. E. Wolf, "Some Recent Research on Optical Coherence", **Proc. Symp. Huygens' Principle 1690 – 1990: Theory and Applications**, H. Blok, H. A. Ferwerda and H. K. Kuiken, eds. (North-Holland, Amsterdam, 1992), 113–127.

230. E. Wolf and A. Gamliel, "Energy Conservation with Partially Coherent Sources which Induce Spectral Changes in Emitted Radiation", *J. Mod. Opt.* **39**, 927–940 (1992).

231. A.T. Friberg, G.S. Agarwal, J. T. Foley and E. Wolf, "Statistical Wave-theoretical Derivation of the Free-space Transport Equation of Radiometry", *J. Opt. Soc. Amer.* B **9**, 1386–1393 (1992).

232. E. Wolf, "Towards Spectroscopy of Partially Coherent Sources", in **Recent Developments in Quantum Optics**, R. Inguva ed. (Plenum Press, New York, 1993), 369–382.

233. M.W. Kowarz and E. Wolf, "Conservation Laws for Partially Coherent Free Fields", *J. Opt. Soc. Amer.* A **10**, 88–94 (1993).

234. W. Wang and E. Wolf, "Propagation of Gaussian Schell-model Beams in Dispersive and Absorbing Media", *J. Mod. Opt.* **39**, 2007–2021 (1992).

235. E. Wolf, "The Redshift Controversy and a New Mechanism for Generating Frequency Shifts of Spectral Lines", **Technical Bulletin of the National Physical Laboratory**, New Delhi, India, October 1991, pp. 1 – 15.

236. E. Wolf, "Coherence Effects in Radiometry and Spectroscopy", *Proceedings of the Tenth Symposium of Energy Sciences*, Argonne National Laboratory, Argonne, IL, Conf-9205147, 202–210 (1992).

237. R. Simon, A.T. Friberg and E. Wolf, "Transfer of Radiance by Twisted Gaussian Schell-model Beams in Paraxial System", *Pure Appl. Opt.* **5**, 331–343 (1996).

238. E. Wolf and T. Habashy, "Invisible Bodies and Uniqueness of the Inverse Scattering Problem", *J. Mod. Opt.* **40**, 785–792 (1993).

239. G.S. Agarwal and E. Wolf, "Higher-order Coherence Functions in the Space-frequency Domain", *J. Mod. Opt.* **40**, 1489–1496 (1993).

240. T. Habashy and E. Wolf, "Reconstruction of Scattering Potentials from Incomplete Data", *J. Mod. Opt.* **41**, 1679–1685 (1994).

241. D.G. Fischer and E. Wolf, "Inverse Problems with Quasi-homogeneous Random Media", *J. Opt. Soc. Amer.* A **11**, 1128–1135 (1994).

242. D.F.V. James and E. Wolf, "A Class of Scattering Media Which Generate Doppler-like Frequency Shifts of Spectral Lines", *Phys. Letts.* A **188**, 239–244 (1994).

243. W.H. Carter, A. Gamliel and E. Wolf, "Coherence Properties of the field Produced by an Infinitely Large, Uniform, Planar, Secondary Lambertian Source", *J. Mod. Opt.* **41**, 1973–1981 (1994).

244. H.C. Kandpal and E. Wolf, "Partially Coherent Sources which Generate Far Fields with the same Spatial Coherence Properties", *Opt. Commun.* **110**, 255–258 (1994).

245. E. Wolf, "Comments on the paper, 'Wolf Shifts and Their Physical Interpretation under Laboratory Conditions' by K. D. Mielenz," *NIST J. of Res.* **99**, 281–282 (1994).

246. E. Wolf, "Comments on 'Radiometric Measurements and Correlation-induced Spectral Changes', by K. A. Nugent and J. L. Gardner", *Metrologia* **31**, 311–313 (1994).

247. D.F.V. James, H.C. Kandpal and E. Wolf, "A New Method for Determining the Angular Separation of Double Stars", *Astrophys. J.* **445**, 406–410 (1995).

248. E. Wolf, "Radiometric Model for Propagation of Coherence", *Opt. Lett.* **19**, 2024–2026 (1994).

249. J.T. Foley and E. Wolf, "Radiometry with Quasihomogeneous Sources", *J. Mod. Opt.* **42**, 787–798 (1995).

250. W. Wang and E. Wolf, "Far-zone Behavior of Focused Fields in Systems with Different Fresnel Numbers", *Opt. Commun.* **119**, 453–459 (1995).

251. E. Wolf, "Spectral Invariance and Non-invariance of Light Generated by Partially Coherent Sources", *Appl. Physics* B **60**, 303–308 (1995).

252. A.T. Friberg and E. Wolf, "Relationships Between the Complex Degrees of Coherence in the Space-time and in the Space-frequency Domains", *Opt. Lett.* **20**, 623–625 (1995).

253. F. Gori and E. Wolf, "Sources which Generate Fields Whose Spectra are Invariant on Propagation", *Opt. Commun.* **119**, 447–452 (1995).

254. W. Wang, A.T. Friberg and E. Wolf, "Structure of Focused Fields in Systems with Large Fresnel Numbers", *J. Opt. Soc. Amer.* **12**, 1947–1953 (1995).

255. W. Wang and E. Wolf, "Invariance Properties of Random Pulses and of Other Random Fields in Dispersive Media", *Phys. Rev.* E **52**, 5532–5539 (1995).

256. E. Wolf, "Coherence Effects in Radiometry and Spectroscopy", (extended abstract), *Optical Review* **2**, 13 (1995).

257. E. Wolf, "Sudarshan's Optical Researches", (extended abstract), *Zt. f. Naturforschung* **52a**, 2 (1997).

258. W. Wang, A. Friberg and E. Wolf, "Focusing of Partially Coherent Light", from **Proceedings of the Seventh Rochester Conference on Quantum Optics**, J. Eberly, L. Mandel and E. Wolf, eds. (Plenum Press, New York, 1996), 695.

259. W. Wang, A. Friberg and E. Wolf, "Focusing of Partially Coherent Light in Systems of Large Fresnel Numbers", *J. Opt. Soc. Amer.* A **14**, 491–496 (1997).

260. E. Wolf, "Coherence, Interference and Spectra", from **Proceedings of the Seventh Rochester Conference on Quantum Optics**, J. Eberly, L. Mandel and E. Wolf, eds. (Plenum Press, New York, 1996), 259–264.

261. E. Wolf, "Principles and Development of Diffraction Tomography" in **Trends in Optics**, ed. A. Consortini (Academic Press, San Diego, CA, 1996), 83–110.

262. E. Wolf and D.F.V. James, "Correlation-induced Spectral Changes", *Rep. Prog. Phys.*, (IOP Publishing, Bristol and London), **59**, 771–818 (1996).

263. T. Habashy, C. Torres-Verdin, M. Oristaglio, A. de Hoop and E. Wolf, "An Overview of Recent Advances in the Inversion of Large-scale Electromagnetic Data", from **Proceedings of the International Symposium on Three-Dimensional Electromagnetics**, eds. M. Oristaglio and B. Spies, Schlumberger-Doll Research, Ridgefield, CN, October 4 – 6, 1995.

264. E. Wolf, "Nonimaging Optics and Quasihomogeneous Sources", (Letter to the Editor), *Optics & Photonics News*, May 1996.

265. G.S. Agarwal and E. Wolf, "Correlation-induced Spectral Changes and Energy Conservation", *Phys. Rev.* A **54**, 4424–4427 (1996).

266. T. Habashy, A.T. Friberg and E. Wolf, "Application of the Coherent-mode Representation to a Class of Inverse Source Problems", *Inverse Problems* **13**, 47–61 (1997).

267. D.G. Fischer and E. Wolf, "Theory of Diffraction Tomography for Quasi-homogeneous Random Objects", *Opt. Commun.* **133**, 17–21 (1997).

268. D.F.V. James and E. Wolf, "Cross-spectrally Pure Light and the Spectral Modulation Law", *Opt. Commun.* **138**, 257–261 (1997).

269. T. Visser and E. Wolf, "Scattering in the Presence of Field Discontinuities at Boundaries", *Phys. Lett.* A **234**, 1–4 (1997); erratum ibid. A **237**, 389.

270. E. Wolf, "Coherence Theory in Historical Perspective", in *Jemná Mechanika a Optika*, issue 11–12, 328–329 (1996) [*Fine Mechanics and Optics*, published in Czech by the Physics Institute of the Czech Academy of Sciences].

271. E. Wolf and D.F.V. James, "A New Technique for Remote Sensing", *Optics & Photonics News* **7**, 38 (1996).

272. E. Wolf, T. Shirai, H. Chen and W. Wang, "Coherence Filters and their Uses Part I: Basic Theory and Examples", *J. Mod. Opt.* **44**, 1345–1353 (1997).

273. E. Wolf, "Far-zone Spectral Isotropy in Weak Scattering on Spatially Random Media", *J. Opt. Soc. Amer.* A **14**, 2820–2823 (1997).

274. P.S. Carney, E. Wolf and G.S. Agarwal, "Statistical Generalizations of the Optical Cross-section Theorem with Application to Inverse Scattering", *J. Opt. Soc. Amer.* A **14**, 3366–3371 (1997).

275. A.J. Devaney, A.T. Friberg, A. Kumar and E. Wolf, "Decrease in Spatial Coherence of Light Propagating in Free Space", *Opt. Lett.* **22**, 1672–1673 (1997).

276. G. Gbur and E. Wolf, "Sources of Arbitrary State of Coherence that Generate Completely Coherent Fields Outside the Source", *Opt. Lett.* **22**, 943–945 (1997).

277. D.F.V. James and E. Wolf, "Determination of the Degree of Coherence of Light from Spectroscopic Measurements", *Opt. Commun.* **145**, 1–4 (1998).

278. E. Wolf, "The Redshift Controversy and Correlation-induced Spectral Changes", in **Waves, Information and Foundations of Physics**, R. Pratesi and L. Ronchi eds. Conference Proceedings vol. 60 (1998), 41–49 (Italian Physical Society, Bologna, Italy).

279. T. Shirai, E. Wolf, H. Chen and W. Wang, "Coherence Filters and Their Uses. II: One-dimensional Realizations", *J. Mod. Opt.* **45**, 799–816 (1998).

280. M. Berry, J.T. Foley, G. Gbur and E. Wolf, "Nonpropagating String Excitations", *Amer. J. Phys.* **66**, 121–123 (1998).

281. E. Wolf and J.T. Foley, "Do Evanescent Waves Contribute to the Far Field?", *Opt. Lett.* **23**, 16–18 (1998).

282. P.S. Carney and E. Wolf, "An Energy Theorem for Scattering of Partially Coherent Beams", *Opt. Commun.* **155**, 1–6 (1998).

283. A. Dogariu and E. Wolf, "Spectral Changes Produced by Static Scattering on a System of Particles", *Opt. Lett.* **23**, 1340–1342 (1998).

284. A. Dogariu and E. Wolf, "Spectral Changes Produced by Scattering of Polychromatic Light on Random Distribution of Particles", *Trends in Optics and Photonics*, vol. 22, **Biomedical Optical Spectroscopy and Diagnostics Therapeutic Laser Applications**, E.M. Sevick-Muraca, J.A. Izatt and M.N. Ediger, eds., pp. 26–29 (1998).

285. E. Wolf, "Correlation-induced Spectral Changes", in *Proceedings of the Sixteenth Symposium on Energy Engineering Sciences*, Argonne National Laboratory, Argonne, IL, CONF-980051, 8–15 (1998).

286. G. Gbur, D. James and E. Wolf, "Energy Conservation Law for Randomly Fluctuating Electromagnetic Fields", *Phys. Rev.* E **59**, 4594–4599 (1999).

287. T.D. Visser and E. Wolf, "Potential Scattering with Field Discontinuities at the Boundaries", *Phys. Rev.* E **59**, 2355–2360 (1999).

288. T.D. Visser, P. S. Carney and E. Wolf, "Remarks on Boundary Conditions for Scalar Scattering", *Phys. Lett.* A **249**, 243–247 (1998).

289. E. Wolf, T. Shirai, G. Agarwal and L. Mandel, "Storage and Retrieval of Correlation Functions of Partially Coherent Fields", *Opt. Lett.* **24**, 367–369 (1999).

290. D. James, P.W. Milonni, H. Fearn and E. Wolf, "Comments on 'The Ewald-Oseen Extinction Theorem' by M. Mansuripur", *Optics & Photonics News* **9**, (11), 4–5 (1998).

291. G. Gbur, J.T. Foley and E. Wolf, "Nonpropagating String Excitations– Finite Length and Damped Strings", *Wave Motion* **30**, 125–134 (1999).

292. G. Gbur and E. Wolf, "Phase Conjugation with Random Fields and with Deterministic and Random Scatterers", *Opt. Lett.* **24**, 10–12 (1999).

293. P.S. Carney, E. Wolf and G. S. Agarwal, "Diffraction Tomography Using Power Extinction Measurements", *J. Opt. Soc. Amer.* A **16**, 2643–2648 (1999).

294. G. Gbur and E. Wolf, "Determination of Density Correlation Functions from Scattering of Polychromatic Light", *Opt. Commun.* **168**, 39–45 (1999).

295. P.S. Carney, D.G. Fischer, J.T. Foley, A. T. Friberg, A. V. Shchegrow, T. D. Visser and E. Wolf, "Comment: 'Evanescent Waves Do Contribute to the Far Field'", *J. Mod. Opt.* **47**, 757–758 (2000).

296. S.A. Ponomarenko and E. Wolf, "Coherence Properties of Light in Young's Interference Pattern formed with Partially Coherent Light", *Opt. Commun.* **170**, 1–8 (1999).

297. T.M. Habashy, D.G. Dudley and E. Wolf, "Linear Inverse Problems in Wave Motion: Non-Symmetric First-Kind Integral Equations", *IEEE Transactions on Antennas and Propagation*, in press.

298. A.T. Friberg, T.D. Visser and E. Wolf, "A Reciprocity Inequality for Gaussian Schell-model Beams and Some of Its Consequences", *Opt. Lett.* **25**, 366–368 (2000).

299. A.V. Shchegrov and E. Wolf, "Partially Coherent Conical Beams", *Opt. Lett.* **25**, 141–143 (2000).

300. E. Wolf, "Coherence of Two Interfering Beams Modulated by a Uniformly Moving Diffuser", *J. Mod. Opt.* **47**, 1569–1573 (2000).

301. S. Ponomarenko and E. Wolf, "Light Beams with Minimum Phase-space Product", *Opt. Lett.* **25**, 663–665 (2000).

302. G.P. Agrawal and E. Wolf, "Propagation-induced Polarization Changes in Partially Coherent Optical Beams", *J. Opt. Soc. Am.* A **17**, 2019-2023 (2000).

303. C.M.J. Mecca, Y. Li and E. Wolf, "Interference of Converging Spherical Waves with Application to the Design of Compact Disks", *Opt. Commun.* **182**, 265–272 (2000).

304. T. Shirai and E. Wolf, "Transformation of Coherence and of the Spectrum of Light by a Moving Diffuser", *J. Mod. Opt.*, in press.

305. T. Visser, A.T. Friberg and E. Wolf, "Phase-space Inequality for Partially Coherent Beams", *Opt. Commun.*, in press.

306. S. Ponomarenko and E. Wolf, "Effective Spatial and Angular Correlations in Beams of any State of Coherence and an Associated Phase-space Product", *Opt. Lett.*, in press.

(b) Books (co-authored)

Max Born and Emil Wolf, **Principles of Optics**, Electromagnetic Theory of Propagation, Interference and Diffraction of Light (First edition, Pergamon Press, London and New York, 1959), xxviii + 808 pp.; (Seventh (expanded) edition, Cambridge University Press, Cambridge, 1999), xxiv + 952p.

Leonard Mandel and Emil Wolf, **Optical Coherence and Quantum Optics** (Cambridge University Press, Cambridge and New York, 1995), xxvi + 1166 pp.

(c) Books (edited)

Progress in Optics, (North-Holland and Elsevier Publishing, Amsterdam) Vol. I (1961); Vol II (1963); Vol. III (1964); Vol. IV (1965); Vol. V (1966); Vol. VI (1967); Vol. VII (1969); Vol. VIII (1970); Vol. IX (1971); Vol. X (1972); Vol. XI (1973); Vol. XII (1974); Vol. XIII (1976); Vol. XIV (1976); Vol. XV (1977); Vol. XVI (1978); Vol. XVII (1980); Vol. XVIII (1980); Vol. XIX (1981); Vol. XX (1983); Vol. XXI (1984); Vol. XXII (1985); Vol. XXIII (1986); Vol. XXIV (1987); Vol. XXV (1988); Vol. XXVI (1988); XXVII (1989); XXVIII (1990); XXIX (1991); XXX (1992), XXXI (1992), XXXII (1993), XXXIII (1994),XXXIV (1995),XXXV (1996), XXXVI (1996), XXXVII (1997), XXXVIII (1998),XXXIX (1999), XL (2000).

Selected papers on Coherence and Fluctuations of Light, L. Mandel and E. Wolf eds. (Dover Publications, Inc., New York, 1970), Vol. I (1850–1960); Vol. II (1961–1966); reprinted by SPIE Milestone Series, B.J. Thompson, ed. (SPIE Optical Engineering Press, Bellingham, WA, 1990).

Coherence and Quantum Optics, Proceedings of Third Rochester Conference on Coherence and Quantum Optics. L. Mandel and E. Wolf, eds. (Plenum Press, New York, 1973).

Coherence and Quantum Optics, Proceedings of Fourth Rochester Conference on Coherence and Quantum Optics, L. Mandel and E. Wolf, eds. (Plenum Press, New York, 1978).

Coherence and Quantum Optics, Proceedings of Fifth Rochester Conference on Coherence and Quantum Optics,, L. Mandel and E. Wolf, eds. (Plenum Press, New York, 1984).

Coherence and Quantum Optics, Proceedings of Sixth Rochester Conference on Coherence and Quantum Optics, J. Eberly, L. Mandel and E. Wolf, eds. (Plenum Press, New York, 1990).

Coherence and Quantum Optics, Proceedings of Seventh Rochester Conference on Coherence and Quantum Optics, J. Eberly, L. Mandel and E. Wolf, eds. (Plenum Press, New York, 1996).